Current Research in Acupuncture

Ying Xia • Guanghong Ding • Gen-Cheng Wu
Editors

Current Research in Acupuncture

Springer

Editors
Ying Xia
Department of Neurosurgery
The University of Texas Medical School at Houston
Houston, TX, USA

Gen-Cheng Wu
Department of Integrative Medicine and Neurobiology, Shanghai Medical College of Fudan University
Shanghai, China

Guanghong Ding
Shanghai Research Center for Acupuncture and Meridians, Fudan University
Shanghai, China

ISBN 978-1-4614-3356-9 ISBN 978-1-4614-3357-6 (eBook)
DOI 10.1007/978-1-4614-3357-6
Springer New York Heidelberg Dordrecht London

Library of Congress Control Number: 2012939116

Printed on acid-free paper

Springer is part of Springer Science+Business Media (www.springer.com)

Preface

Acupuncture has been practiced in China for thousands of years as part of the traditional Chinese medicine and has gradually spread across different regions of the world today. More and more patients worldwide are seeking therapeutic benefits from this ancient modality.

An exogenous stimulation, as with acupuncture, can trigger a signal that is transmitted through the body and elicits a biological effect. This forms the very basis of the effects of acupuncture in various disease states that it is employed in. A relevant analogy comes from a common clinical setting of deep brain stimulation (DBS), wherein brain stimulation by exogenous electrical currents has shown riveting therapeutic effects on Parkinson's disease and other neuropsychiatric disorders. This modality is broadly used in present clinical settings although the precise mechanisms underlying the therapy remain elusive. On the other hand, there is evidence showing that spinal cord stimulation instead of brain stimulation also induces similar benefits in the model of Parkinson's disease (refer to Science 323: 1578, 2009). Therefore, it is not entirely inconceivable to see a biological/medical effect being elicited by a mechanical and/or an electrical stimulation at various (acu)points over our body.

However, mechanisms of action of acupuncture, as of DBS, are not well understood. Elucidation of these mechanisms can help us improve this modality further and enable us to harness its full potential. Prior to 1997 when a Consensus Development Conference of acupuncture was held by NIH, acupuncture-related research was mainly performed in China and certain Asian countries (refer to our previous book entitled "Acupuncture Therapy for Neurological Diseases: A Neurobiological View"). But the past decade has witnessed a paradigm shift with many studies being published not only in the oriental world but also in the western medical community. To better understand the nature of acupuncture and improve its clinical application, a periodic review of the progress in research is essential to gain an integral perspective on the updated understanding and deficiencies of acupuncture. From this vantage point, I, together with Professors Ding and Wu, have contributed our combined effort to this book.

This book reviews recent advances in research in the field of acupuncture. It presents the modern scientific perspective through the contributions of over 60 scientists and clinicians from the USA, mainland China, Germany, Austria, Japan, Sweden, Portugal, and Hong Kong. The first five chapters discuss the basic mechanisms of acupuncture. Later chapters explore topics including acupuncture treatment and potential mechanisms for epilepsy, Parkinson's diseases, neurodegenerative disorders such as Alzheimer's disease, vascular cognitive impairment, aging, anxiety, polycystic ovary syndrome, pain, nerve root cervical spondylosis, stroke, inflammation, myocardial ischemia, and other cardiovascular diseases. Following the translational and clinical discussions, the next four chapters present prospective acupuncture speculations, theories, and applications. The final chapter discusses the pitfalls and problems of the previous studies with suggestions for revamping the design and analysis for improving the future of research on in-depth understanding of acupuncture, along with better application of acupuncture in modern medicine. This unique book provides a broad reference frame on the principles of acupuncture for neuroscientists and researchers. The laboratory and clinical investigations searching for ideal acupoints and optimal conditions provide an important resource for clinical acupuncturists to improve the efficacy of acupuncture. For a medical student, this book is a modern course in ancient Traditional Chinese Medicine, especially acupuncture.

I would like to apologize in advance to all those acupuncture research colleagues whose relevant work could not be included. This is partially because the subjects and contents presented in this book were chosen based on the authors' research interest and background. On the other hand, it is a pity that some excellent clinicians/scientists were unable to submit their chapters in time owing to a heavily loaded schedule conflicting with the submission deadlines. With their contributions, this book could have been all the more valuable a resource.

Although we are far from gathering the whole picture, this book however provides a useful guide to the scientific and medical community for promoting acupuncture research and directing it toward the ultimate goal. As I always say, and also state in the last chapter of this book, (1) acupuncture is effective in certain diseases but is not universal for all; (2) acupuncture is amazing but not mysterious; and (3) acupuncture mechanism is difficult to learn but can be explored by using modern scientific approaches.

I am particularly grateful to Professors Ding and Wu for their help in editing this book. I am indebted to all the authors for their significant contributions, considerable efforts, and invaluable knowledge that made this book happen. In addition, I highly appreciate the exhaustive efforts of our anonymous reviewers who closely reviewed our proposal over several rounds and gave helpful inputs and comments. Finally, I would like to thank Ms. Ann H Avouris, Ms. Simina Calin, and Ms. Samantha Lewis of Springer whose efforts helped in conjuring up this book to a reality from a mere proposal following a tireless series of e-mail/phone communications. Production of this book would not have been possible without the help of all the above-mentioned people.

Houston, TX, USA Ying Xia

Acknowledgments

This book would not have been possible without the support from

National Institutes of Health, USA
(AT-004422; HD-0034852)

National Basic Research Program, China
(09CB522901; 12CB518502)

Science and Technology Commission of Shanghai Municipality, China
(09DZ1976600; 10DZ1975800; 10DZ1975800; 11DZ1973300)

State Administration of Traditional Chinese Medicine, China

Contents

Contributors

Tetsuya Asakawa, M.D., Ph.D. Department of Neurosurgery, Hamamatsu University School of Medicine, Hamamatsu, Shizuoka, Japan

Florian Beissner, Dr. Phil. Nat. Pain and Autonomics–Integrative Research (PAIR), Clinic of Psychiatry and Psychotherapy, University Clinic Jena, Jena, Germany

Ronglin Cai, M.D. Research Center for Meridian-Viscera Correlationship, Anhui University of Traditional Chinese Medicine, Hefei, China

Anhui Institute of Acupuncture and Meridians, Anhui University of Traditional Chinese Medicine, Hefei, China

School of Acupuncture and Osteology, Anhui University of Traditional Chinese Medicine, Hefei, China

Hong Cao, M.D., Ph.D. Institute of Neurobiology, Institutes of Brain Research, Fudan University, Shanghai, China

State Key Laboratory of Medical Neurobiology, Fudan University, Shanghai, China

Boying Chen, M.D. Department of Integrative Medicine and Neurobiology, State Key Lab of Medical Neurobiology, Shanghai Medical College of Fudan University, Shanghai, China

Institute of Acupuncture Research, Institutes of Brain Science, Fudan University, Shanghai, China

Tao Chen, M.D., Ph.D. Department of Neurology, Hainan Provincial People's Hospital, Hainan, China

Dongman Chao, Ph.D. Department of Neurosurgery, The University of Texas Medical School at Houston, Houston, TX, USA

Yale University School of Medicine, CT, USA

Wei Yeung Chung, B.C.M., B.Sc., M.Phil. School of Chinese Medicine, The Chinese University of Hong Kong, Shatin, New Territory, Hong Kong

Guanghong Ding Department of Mechanics and Engineering Science, Fudan University, Shanghai, China

Shanghai Research Center for Acupuncture and Meridians, Fudan University, Shanghai, China

Li Fang, M.D., Ph.D. Department of Acupuncture and Moxibustion, The Third Affiliated Hospital of Zhejiang University of TCM, Hangzhou, Zhejiang, China

Yi Feng, Ph.D. Department of Integrative Medicine and Neurobiology, State Key Lab of Medical Neurobiology, Shanghai Medical College, Fudan University, Shanghai, China

Institute of Acupuncture Research (WHO Collaborating Center for Traditional Medicine), Institutes of Brain Science, Fudan University, Shanghai, China

Qiao Fu, M.D. Department of Rehabilitation Medicine, Hainan Provincial People's Hospital, Hainan, China

Linbao Ge, M.D. Shanghai Research Institute of Qigong, Shanghai, China

Shanghai Research Center of Acupuncture and Meridian, Shanghai, China

Henry Johannes Greten, M.D. Heidelberg School of Chinese Medicine, Heidelberg, Germany

Institute of Biomedical Sciences Abel Salazar (ICBAS), Porto University, Porto, Portugal

Quanbao Gu Laboratory of Cellular Electrophysiology, Shanghai Research Centre for Acupuncture and Meridians, Shanghai, China

Lu He, M.D. Research Center for Meridian-Viscera Correlationship, Anhui University of Traditional Chinese Medicine, Hefei, China

Anhui Institute of Acupuncture and Meridians, Anhui University of Traditional Chinese Medicine, Hefei, China

School of Acupuncture and Osteology, Anhui University of Traditional Chinese Medicine, Hefei, China

Ling Hu, M.D. Anhui Institute of Acupuncture and Meridian, Anhui University of Traditional Chinese Medicine, Hefei, China

School of Acupuncture and Osteology, Anhui University of Traditional Chinese Medicine, Hefei, China

Meng Huang, M.S. Department of Mechanics and Engineering Science, Fudan University, Shanghai, China

Shixiong Huang, M.D., Ph.D. Department of Neurology, Hainan Provincial People's Hospital, Hainan, China

Wenjing Huang, M.D. College of Acupuncture and Moxibustion, Chengdu University of TCM, Chengdu, Sichuan, China

Lei Lan, M.D. College of Acupuncture and Moxibustion, Chengdu University of TCM, Chengdu, Sichuan, China

Peng Li, M.D. Department of Medicine, Susan Samueli Center for Integrative Medicine, School of Medicine, University of California-Irvine, Irvine, CA, USA

Ying Li, Ph.D. College of Acupuncture and Moxibustion, Chengdu University of TCM, Chengdu, Sichuan, China

Fanrong Liang, Ph.D. College of Acupuncture and Moxibustion, Chengdu University of TCM, Chengdu, Sichuan, China

Jianfeng Liang, M.D., Ph.D. Department of Neurosurgery, The University of Texas Medical School at Houston, Houston, TX, USA

Yale University School of Medicine, CT, USA

Gerhard Litscher, M.Sc., Ph.D., M.D.Sc. TCM Research Center Graz and Research Unit of Biomedical Engineering in Anesthesia and Intensive Care Medicine, Medical University Graz, Graz, Austria

Mailan Liu, M.D. College of Acupuncture and Moxibustion, Chengdu University of TCM, Chengdu, Sichuan, China

Tao Liu, M.D. Department of Neurology, Hainan Provincial People's Hospital, Hainan, China

Zibing Liu, Ph.D. Research Center for Meridian-Viscera Correlationship, Anhui University of Traditional Chinese Medicine, Hefei, China

Anhui Institute of Acupuncture and Meridians, Anhui University of Traditional Chinese Medicine, Hefei, China

School of Acupuncture and Osteology, Anhui University of Traditional Chinese Medicine, Hefei, China

John C. Longhurst, M.D., Ph.D. Department of Medicine, Susan Samueli Center for Integrative Medicine, School of Medicine, University of California Irvine, Irvine, CA, USA

Tingting Ma, Ph.D. Department of Treatment for Undiseased and Sub-health, Teaching Hospital of Chengdu University of TCM, Chengdu, Sichuan, China

Yulan Ren, Ph.D. College of Acupuncture and Moxibustion, Chengdu University of TCM, Chengdu, Sichuan, China

Wolfgang Schwarz, Dr. Rer. Nat. Laboratory of Cellular Electrophysiology, Shanghai Research Centre for Acupuncture and Meridians, Shanghai, China

Institute for Biophysics, Organization Goethe University Frankfurt am Main, Frankfurt am Main, Germany

Xueyong Shen, M.D. College of Acupuncture and Moxibustion, Shanghai University of Traditional Chinese Medicine, Shanghai, China

Laboratory of Systemic Physiology for Acupuncture and Meridian, Shanghai Research Center for Acupuncture and Meridians, Shanghai, China

Elisabet Stener-Victorin, Ph.D. Department of Physiology/Endocrinology, Institute of Neuroscience and Physiology, The Sahlgrenska Academy at University of Gothenburg, Gothenburg, Sweden

Department of Obstetrics and Gynecology, The First Affiliated Hospital, Heilongjiang University of Chinese Medicine, Harbin, China

Zhiyan Sui Department of Neurology, Hainan Provincial People's Hospital, Hainan, China

Stephanie C. Tjen-A-Looi Department of Medicine, Susan Samueli Center for Integrative Medicine, School of Medicine, University of California Irvine, Irvine, CA, USA

Keming Wang, M.D. Research Center for Meridian-Viscera Correlationship, Anhui University of Traditional Chinese Medicine, Hefei, China

Lina Wang, Ph.D. College of Acupuncture and Moxibustion, Shanghai University of Traditional Chinese Medicine, Shanghai, China

Shu-Ming Wang, M.Sc., M.D. Department of Anesthesiology and Perioperative Care, University of California Irvine, Orange, CA, USA

Xianrui Wang, M.D., Ph.D. Department of Anesthesiology and Pain Management, Renji Hospital, Shanghai JiaoTong University School of Medicine, Shanghai, China

Guoqiang Wen, M.D. Department of Neurology, Hainan Provincial People's Hospital, Hainan, China

Chanji Wu, M.D. Department of Neurology, Hainan Provincial People's Hospital, Hainan, China

Huangan Wu, M.D. Shanghai Institute of Acupuncture Moxibustion and Meridian, Shanghai University of Traditional Chinese Medicine, Shanghai, China

Gen-Cheng Wu, M.D. Department of Integrative Medicine and Neurobiology, Shanghai Medical College of Fudan University, Shanghai, China

Xi Wu, Ph.D. College of Acupuncture and Moxibustion, Chengdu University of TCM, Chengdu, Sichuan, China

Zijian Wu, M.D. Research Center for Meridian-Viscera Correlationship, Anhui University of Traditional Chinese Medicine, Hefei, China

Anhui Institute of Acupuncture and Meridian, Anhui University of Traditional Chinese Medicine, Hefei, China

School of Acupuncture and Osteolog, Anhui University of Traditional Chinese Medicine, Hefei, China

Ying Xia, M.D., Ph.D. Department of Neurosurgery, The University of Texas Medical School at Houston, Houston, TX, USA

The University of Texas Graduate School of Biomedical Sciences at Houston, Houston, TX, USA

Yale University School of Medicine, New Haven, CT, USA

Mingshu Xu, Ph.D., M.D. State Laboratory of Cerebral Neurobiology, Shanghai Research Institute of Acupuncture and Meridian, Shanghai, China

Jie Yang, Ph.D. College of Acupuncture and Moxibustion, Chengdu University of TCM, Chengdu, Sichuan, China

Wei Yao, Ph.D. Department of Mechanics and Engineering Science, Fudan University, Shanghai, China

Yi Jie Zhan, M.Sc. School of Chinese Medicine, Hong Kong Baptist University, Kowloon Tong, Kowloon, Hong Kong

Di Zhang, Ph.D. Department of Mechanics and Engineering Science, Fudan University, Shanghai, China

Shanghai Research Center for Acupuncture and Meridians, Fudan University, Shanghai, China

Shi Ping Zhang, M.B., Ph.D. School of Chinese Medicine, Hong Kong Baptist University Kowloon Tong, Kowloon, Hong Kong

Yu-Qiu Zhang, M.D., Ph.D. Institute of Neurobiology, Institutes of Brain Research, Fudan University, Shanghai, China

State Key Laboratory of Medical Neurobiology, Shanghai Medical College of Fudan University, Shanghai, China

Dan Zhao, M.D. Shanghai Research Institute of Qigong, Shanghai, China

Ling Zhao, M.D., Ph.D. Shanghai University of Traditional Chinese Medicine, Shanghai, China

Shanghai Research Center for Acupuncture and Meridians, Shanghai, China

Zhi-Qi Zhao, Ph.D. Institute of Neurobiology, Institutes of Brain Research, Fudan University, Shanghai, China

State Key Laboratory of Medical Neurobiology, Shanghai Medical College of Fudan University, Shanghai, China

Hui Zheng, Ph.D. College of Acupuncture and Moxibustion, Chengdu University of TCM, Chengdu, Sichuan, China

Yongjun Zheng, M.D., Ph.D. Department of Anesthesiology and Pain Management, Renji Hospital, Shanghai JiaoTong University School of Medicine, Shanghai, China

Meiqi Zhou, Ph.D. Research Center for Meridian-Viscera Correlationship, Anhui University of Traditional Chinese Medicine, Hefei, China

School of Chinese Medicine, Anhui University of Traditional Chinese Medicine, Hefei, China

Yiping Zhou, M.D. Research Center for Meridian-Viscera Correlationship, Anhui University of Traditional Chinese Medicine, Hefei, China

Anhui Institute of Acupuncture and Meridian, Anhui University of Traditional Chinese Medicine, Hefei, China

Chapter 1
Acupuncture Modulation of Neural Transmitters/Modulators

Jianfeng Liang and Ying Xia

Abstract Acupuncture has historically been in clinical practice in China since a long time and today has spread to many other parts of the world as well. Accumulating evidence shows that manual acupuncture (MA) and electroacupuncture (EA) induce therapeutic effects on multiple disorders that have been demonstrated not only by the Oriental clinicians but also by Western medical community. A large body of literature shows that acupuncture affects the central nervous system (CNS) through regulation of neurotransmitters in terms of their synthesis, release, uptake, and degradation, suggesting an involvement of neurotransmitters in acupuncture therapy. In this chapter, we will update on these vantage points in view of the recent studies on acupuncture effects on neural transmitters/modulators. Besides technical development (e.g., fMRI and PET for measuring endogenous opioid concentrations in the brain), recent evidence suggests that (1) acupuncture can induce dual-directional regulation of endogenous opioids, monoamine substance P, and other transmitters/modulators, depending on acupuncture condition and body status (e.g., normal vs. disease); (2) acupuncture can increase the mRNA expression of brain-derived

J. Liang
Department of Neurosurgery, The University of Texas Medical School at Houston, Houston, TX 77030, USA

Yale University School of Medicine, New Haven, CT 06520, USA
e-mail: jianfeng.liang@uth.tmc.edu

Y. Xia (✉)
Department of Neurosurgery, The University of Texas Medical School at Houston, Houston, TX 77030, USA

The University of Texas Graduate School of Biomedical Sciences at Houston, Houston, TX 77030, USA

Yale University School of Medicine, New Haven, CT 06520, USA
e-mail: Ying.Xia@UTH.TMC.edu

Y. Xia et al. (eds.), *Current Research in Acupuncture*,
DOI 10.1007/978-1-4614-3357-6_1, © Springer Science+Business Media New York 2013

neurotrophic factor and enhance the expression of neurotrophic factor (NTF) receptors; (3) acupuncture downregulates the nitric oxide (NO) level in the CNS and elevates the NO content in plasma and certain organs; (4) EA can suppress the activation of acetylcholinesterase to attenuate acetylcholine degradation; (5) EA has a better effect on enhancing 5-HT mRNA expression than MA; (6) MA and EA have differential effects on different subtypes of glutamate receptors, leading to decreased expression of *N*-methyl-D-aspartate (NMDA) receptor and increased expression of AMPA receptor; and (7) EA increases the concentration of prostaglandins in plasma in some diseased states. These studies show that acupuncture induces comprehensive effects on transmitters/modulators in the CNS, and the outcome can be dual-directional depending on acupuncture approach and the body status.

Keywords Endogenous opioid peptides • 5-Hydroxytryptamine • Noradrenaline • Dopamine • Acetylcholine • Amino acids • Substance P • Prostaglandin • Cholecystokinin octapeptide-8 • Neurotrophic factor • Nitric oxide

1.1 Introduction

We have summarized major findings in past research on acupuncture-induced regulation of central neurotransmitters in our previous book (Xia et al. 2010). Over the recent years many studies focusing on the alterations in central neurotransmitters under various pathological conditions have generated new knowledge in this field. Furthermore, multidisciplinary new techniques, including transgenic and knockout techniques and functional magnetic resonance imaging (fMRI), offer diversified approaches for scientists and clinicians to establish new research models and conducting more informative studies for exploring the mechanisms of acupuncture mechanism. Almost all of these data further confirm that the mechanisms of acupuncture, either manual or electro-acupuncture (MA and EA), are indeed associated with an alteration(s) in the neurotransmitter activity in the brain and/or spinal cord, which, with other evidence, suggests that acupuncture induces its effects by modulating central neurotransmitters. In this chapter we will review the recent progress in acupuncture effects on central neurotransmitters/modulators, with a special focus on endogenous opioid peptide (EOP), 5-hydroxytryptamine, noradrenaline, dopamine, acetylcholine, amino acids, substance P, prostaglandin, cholecystokinin octapeptide-8 (CCK-8), neurotrophic factors (NTFs), and nitric oxide. Since the released central neurotransmitters and their metabolites may enter peripheral nervous system and blood plasma, we will, if relevant to acupuncture, briefly discuss this aspect and their potential effects on the body.

Since we have systemically reviewed the effects of acupuncture on central neurotransmitters and modulators in recent book chapters (Wen et al. 2010a, b), we will focus this chapter on major advancements in the research of these transmitters/modulators published in the past 3–5 years, instead of a comprehensive and systematic review, although we will reference to earlier literature in certain sections for an integrative and comparative discussion.

Because most of the literature we found was published in non-English journals, we have made 11 tables under all major sections to briefly reflect main findings in the published studies as well as acupuncture condition and experimental models for readers' convenience. Also for the purpose of informative coverage, we will cover most reports, including those with debatable data, though we may not necessarily agree with the conclusions drawn from these studies.

For the sake of convenience, we will use the term "acupuncture" as a general word that shall encompass MA and EA as well as other similar approaches unless clearly specified.

1.2 Acupuncture Effects on Endogenous Opioids

The endogenous opioid system is closely related to acupuncture effects and is one of the most studied neurotransmitter systems in acupuncture research. Earlier studies have recognized three classic kinds of opioid peptides in the body, namely, β-endorphin (β-EP), enkephalin (including methionine-enkephalin and leucine-enkephalin), and dynorphin. They cast a spectrum of effects in the pain management, emotions, internal secretions, urinary regulation, neuroprotection, cardiovascular activity, respiration, appetite modulation, digestion, sexual behavior, and many other functions (Feng et al. 2012). Orphanin FQ (also known as nociceptin) is the most recently introduced member, albeit discovered nearly 15 years ago, into the endogenous opioid family. Orphanin FQ influences many behavioral, psychological, and neurobiological processes including memory, anxiety, stress, and reward. Moreover, accumulating evidence shows that acupuncture can modulate the expression of opioid receptors. For example, electroacupuncture can upregulate the density of delta-opioid receptors in the ischemic brain (Tian et al. 2008).

1.2.1 Dual-Directional Regulation

The association between acupuncture and EOPs has become clear based on past studies, especially in acupuncture analgesia. Many earlier studies, including a group of investigators at former Shanghai Medical University (now Shanghai Medical College of Fudan University) (Xia et al. 2010), have revealed that MA or EA can promote the activity of EOP with an increase in EOP release. Han's group at former Beijing Medical University (now Peking University Health Science Center) further clarified the ambiguous relationship between EA frequency and the release of EOP subtypes. They demonstrated that EA given at 2 Hz induces an increase in the expression of β-EP, enkephalin, and endomorphin; EA at 15 Hz upregulates the release of the three kinds of EOP, and 100-Hz EA elevates the release of dynorphin only (Han 2003). These findings may need more verification and clarification

Table 1.1 Effect of acupuncture on endogenous opioids

Acupuncture	Acupoints	Frequency (Hz)	Disease states	Changes in disease states	Acupuncture effects	Subjects	References
EA	Baihui (DU-20) Dazhui (DU-14) Pishu (BL-20) Shenshu (BL-23)	150	Vascular dementia	Brain: β-EP↓ Plasma: β-EP—no change	Brain: β-EP↑ Plasma: β-EP—no change	Rat	Shao et al. (2008)
EA	Zusanli (ST-36) Yanglingquan (GB-34)	2/15	Obesity	Hypothalamus: β-EP↓ Plasma: β-EP↑	Hypothalamus: β-EP↑ Plasma: β-EP↓	Wistar rat	Chen et al. (2008c)
EA	Jiaji (Ex-B-2) Huantiao (GB-30) Weizhong (BL-40)	?	Lumbar disc herniation	Plasma: β-EP↓	Plasma: β-EP↑	Patient	Huang et al. (2008a)
MA	Zhongwan (CV-12) Zusanli (ST-36)		Visceral pain	Hypothalamus: β-EP↑ Plasma: PGE_2↑	Hypothalamus: β-EP↑ Plasma: PGE_2↓	Mouse	Yu et al. (2008)
MA	Neiguan (PC-6)		Myocardial ischemia and reperfusion injury	Plasma: β-EP↑	Plasma: β-EP↓	SD rat	Zhou et al. (2009)
EA	Neiguan (PC-6)	?	Myocardial ischemia and reperfusion injury	Cerebellomedullary: β-EP↑	Cerebellomedullary: β-EP↓	Rabbit	Zhang et al.(2007)
MA	Sanyinjiao (SP-6) Guanyuan (CV-4) Diji (SP-8)		Dysmenorrhea	Hypothalamus pituitary: β-EP↓ Ovary: β-EP↓	Hypothalamus pituitary: β-EP↑ Ovary: β-EP↑	Rat	Liu et al. (2008a)
MA	Sanyinjiao (SP-6) Guanyuan (CV-4) Diji (SP-8)		Dysmenorrhea	Plasma: β-EP↓	Plasma: β-EP↑	Rat	Liu et al. (2008b)
EA	Zusanli (ST-36) Sanyinjiao (SP-6)	2/100	Morphine dependence	Hypothalamus: β-EP↑ Plasma: β-EP ↑	Hypothalamus: β-EP↑ Plasma: β-EP—no change	Wistar rat	Zhu (2007)
EA	Shenmen (HT-7)	2/100	Morphine withdrawal	Locus coeruleus: M-ENK↓	Locus coeruleus: M-ENK↑	SD rat	Wang et al. (2008a)
EA	Shenshu (BL-23) Yaoyangguan (DU-3)	2/100	Transverse syndrome of the third lumbar	Vertebrae, spinal cord, and hypothalamus: β-EP↑, ENK↑	Vertebrae, spinal cord, and hypothalamus: β-EP↓, ENK↓	SD rat	Sun et al. (2008)

because of conflictive observations by other investigators. For example, several recent studies (Table 1.1) have detected an increased release of β-EP induced by high-frequency EA (e.g., 150 Hz). It is likely that EA differentially affects the release of EOP depending on different states of the body, i.e., "normal" vs. diseased.

Indeed, studies using various animal models for various human diseases have been conducted to better illustrate the relationship between acupuncture and EOP in the recent times (Table 1.1). These studies have generated lots of interesting data showing biphasic regulation of EOP release by acupuncture.

Shao et al. (2008) observed an increase in the cerebral β-EP levels following EA in the rat models for vascular dementia (a condition with reportedly low cerebral β-EP levels) and an associated improvement in the memory and learning ability. Previous studies have revealed that the plasma β-EP level is higher in obese patients (Givens et al. 1980; Ritter et al. 1992). Chen et al. (2008c) demonstrated that EA is an effective treatment for obesity and acts through the regulation of β-EP release. In their research on a rat model for obesity, they showed that EA at the Zusanli (ST-36) and Yanglingquan (GB-34) could raise the β-EP levels in the hypothalamus while simultaneously decreasing them in plasma. Therefore, they suggested that EA possibly reduces the release of β-EP in the hypothalamus in obese patients.

Several studies have consistently shown that MA or EA can increase the EOP concentrations in the plasma. Huang et al. (2008a) used bilateral EA at Jiaji (Ex-B2) and at Huantiao (GB-30) and Weizhong (BL-40) on the affected side for treatment of lumbar disc herniation and showed a good therapeutic effect. They also detected elevated levels of plasma β-EP. On the other hand, some cardiovascular diseases are associated with a high level of plasma β-EP, for example, increased plasma β-EP in the rat model for myocardial ischemia and reperfusion injury (Zhang et al. 2007; Zhou et al. 2009). There are reports showing that EA at Neiguan (PC-6) could decrease the cerebellomedullary and plasma β-EP, thereby protecting the heart muscles from myocardial ischemia and reperfusion injury (Zhang et al. 2007; Zhou et al. 2009). Furthermore, in the rat model of transverse section of the third lumbar vertebrae, EA could downregulate the levels of β-EP and ENK in the spinal cord and hypothalamus (Sun et al. 2008).

Due to its effects on EOP, EA is also effective in the treatment of drug addiction. Recent studies have shown that EA at 2/100 Hz can selectively enhance β-EP expression in the hypothalamus and increase M-ENK levels in the locus coeruleus while unaltering their plasma levels (Zhu 2007; Wang et al. 2008a).

In summary, we can infer that acupuncture increases the release of EOP in the central nervous system (CNS) with an elevated level of β-EP in cerebrospinal fluid and plasma. Under certain conditions, however, acupuncture may have different effects on the central and peripheral EOP. This is especially true with β-EP that assumes different roles in different pathological conditions. EA could elevate the plasma β-EP by positively regulating the release or reducing the plasma β-EP when it is baleful in certain diseases.

1.2.2 Role of Connexin in EA Modulation of EOP

Gap junctions are composed of two adjacent hemi-channels that directly connect the cytoplasmic compartments of adjacent cells and allow small molecules (<1 kDa) and ions to pass freely between the cells (Laird 2009). Considerable evidence suggests that "free" hemi-channels on the cell surface act as important communication channels between the cytosol and extracellular milieu (Laird 2009).

Several studies suggest that the effects of acupuncture are related to the distribution of gap junction and thus the intercellular communication. Acupuncture can significantly increase the expression of connexin 43 (Shang 1989; Huang et al. 2005). Some researchers investigated the potential connection between the acupuncture-induced release of EOP and the regulation of connexin 43 gene in acupoints. They observed low levels of β-EP in the rat model of dysmenorrhea and an improvement in symptoms following acupuncture via the enhancement of β-EP expression both in the CNS and plasma (Liu et al. 2008a, b). A local injection of interference plasmid pSilencer-connexin43-shRNA into the acupoints abrogated the acupuncture-induced β-EP expression. A similar outcome was noted in the research study on visceral pain. Yu et al. (2008) used the connexin 43 gene knockout (heterozygote) mice and documented that acupuncture could not alter the β-EP levels in these heterozygote mice. In a study on acupuncture analgesia, researchers also showed that acupuncture at Zusanli (ST-36) acupoint elevates the mRNA and protein levels of connexin 43, and the knockout of connexin 43 gene decreased acupuncture-induced analgesia (Huang et al. 2009). Recently, Luo et al. (2011) demonstrated that the expression of connexin 43 and glutamate transporter-1 (GLT-1, also named EAAT2, that is essential in regulating glutamate concentration in the synaptic cleft) was significantly enhanced after EA treatment in an ischemia model. Also noteworthy was the increase in postsynaptic density thickness in the EA treatment group as compared to the control. The above observations suggest that in acupuncture-induced regulation of EOP release connexin 43 plays a significant role in signal transduction from the site of needle puncture to the CNS.

However, additional experiments are needed to verify this phenomenon further. Even if it is true, the role of connexin may not be limited to the regulation of EOP because knocking out or knocking down connexin could alter tissue/cell structures around the acupoints and broad regions in the CNS, thus reducing general effects of acupuncture on neurotransmitters. It is important to specifically analyze the role of connexin 43 in the acupuncture action by more direct evidence. The gap junction protein connexin 43 is extensively expressed in the CNS and serves in signal transmission pathway between the glia and neurons (Sáez et al. 2005). There is no direct evidence as to how the acupuncture-induced changes in connexin 43 expression specifically affect EOP and/or other neurotransmitters/modulators.

1.2.3 Technical Development

Recent development of new techniques has opened a new window for investigating the effects of acupuncture on EOP. For instance, [^{11}C]diprenorphine is a radioactive substance that competes with innate EOP for the mu-, delta-, and kappa-opioid receptor binding site. The combined use of this radioactive substance with positron emission tomography (PET) delineates the brain regions showing changes in EOP receptor binding (Frost 2001). Lately, Dougherty et al. (2008) employed fMRI and PET to study the effects of acupuncture and reported an increase in the endogenous opioid concentration after EA, especially in the orbitofrontal cortex, medial prefrontal cortex, insula, thalamus, and the anterior cingulate cortex. In this study, authors applied a new integrative multimodal imaging approach (fMRI and [^{11}C]diprenorphine PET) which can promptly and directly report the level of EOP in the CNS. This provides a potential prospect for detecting EOP in vivo in acupuncture research.

1.3 5-Hydroxytryptamine

5-Hydroxytryptamine (serotonin) is a monoamine neurotransmitter. It plays a myriad of functions both under physiological conditions like the regulation of emotion, appetite, and sleep and under pathological conditions, including modulation of inflammation, analgesia, and cardiovascular disease.

Tsai et al. (1989) showed that EA increases the threshold of tail pressure pain in Sprague-Dawley rats, and this analgesic effect was abrogated by injection of *p*-chlorophenylalanine (a 5-HT synthesis inhibitor). In addition, many prior studies have shown that acupuncture can upregulate the level of 5-HT in the brain. Wang et al. (2005) studied the effectiveness of acupuncture on a rat model of the sciatic nerve compression. He treated the rats with EA once a day starting from the 3rd day after establishment of the model. Their results revealed that EA could increase the 5-HT levels in the brain of model rats, reaching the peak by the 14th day. In the control group, both EA and MA could increase the 5-HT levels in the brain of normal rats but reached the peak levels earlier in EA-treated group (14th day) compared to the MA-treated group (28th day). Although the 5-HIAA levels remained comparable in each group, a significant increase was observed in the group of the EA-treated model on the 28th day. Therefore, they concluded that electric stimulation is better than manual needling for modulating the brain 5-HT levels. Another study that was performed on a migraine rat model indicated that the EA-induced 5-HT expression spawns from an increased synthesis of 5-HT1F mRNA (Zhong and Li 2007).

On the other hand, strong corroborative evidence shows that the 5-HT levels in the brain increase in certain pathophysiological conditions and that acupuncture could downregulate the 5-HT levels (Wen et al. 2010b). Li et al. (2007a) documented that acupuncture at Baihui (DU-20) and Yintang (EX-HN-3) attenuated the

5-HT level in dorsal raphe nucleus in a rat model for neuroses. Also, Li and Li (2005) used the rat model for cerebral ischemia/reperfusion injury and demonstrated that 5-HT levels in the striatum increased after ischemic injury, and EA at 10 Hz could effectively suppress this increase.

In addition to CNS, acupuncture can regulate the 5-HT levels in the peripheral nervous system as well. Two groups have reported that EA can lower the 5-HT levels in the spinal cord. Wu and Yu (2005) investigated the effect of EA on the animal model of compressed lumbar nerve. Their data revealed that 5-HT level in the oppressed nerve root of EA-treated group was lower than the untreated group. Yi et al. (2009) demonstrated that EA could protect the heart from myocardial ischemia/reperfusion injury by suppressing the amounts of 5-HT in the spinal dorsal horn and root.

In the periphery, 5-HT acts as one of the mediators of inflammation. Many studies have revealed that the level of 5-HT is high in inflammatory conditions, such as gouty arthritis. Cui and Hou (2008) observed that EA decreased 5-HT levels in the synovium in a rabbit model for acute gouty arthritis. Moreover, acupuncture can also regulate 5-HT level in the plasma. A recent clinical study showed that "Jin's Three-Needling" has a significant beneficial effect on patients with generalized anxiety disorder, which probably involves regulation of platelet 5-HT and plasma adrenocorticotropic hormone concentrations (Luo et al. 2007).

The above-mentioned observations suggest that EA has a better efficacy than MA in terms of elevating 5-HT levels. In some cases, EA increases 5-HT level by enhancing the synthesis of 5-HT1F mRNA. However, the effects of acupuncture on 5-HT system are very complex as shown in Table 1.2. Therefore, acupuncture can up- or downregulate 5-HT system owing to influence of the concurrent physiological and/or pathological factors (Li and Li 2005; Li et al. 2007a; Zhong and Li 2007; Wu and Yu 2005). There is no clear evidence on the mechanism of acupuncture-induced downregulation at present and it needs further proof if acupuncture can downregulate 5-HT expression.

Taken together, acupuncture may differentially regulate the 5-HT system in the CNS and peripheral system and thus induce a beneficial effect to the body.

1.4 Noradrenaline

Noradrenaline (NA) is a vital monoamine neurotransmitter that exerts widespread effects across multiple areas of the brain and regulates arousal and stress responses. One of its functions in the cortical system is to balance the vigilant behavior with a focused attention on novel environmental stimuli and the state of arousal. The central NA system intrinsically interacts with the stress response system. Therefore, a dysregulation within the NA system is implicated in the pathogenesis of anxiety and depressive states.

Many previous studies have revealed that acupuncture can decrease the levels of noradrenaline in certain areas of the brain, such as nucleus coeruleus (Jin et al. 1979)

Table 1.2 Effect of acupuncture on 5-HT

Acupuncture	Acupoints	Frequency (Hz)	Disease states	Changes in disease states	Acupuncture effects	Subjects	References
EA	Baihui (DU-20) Yintang (EX-HN-3)	10	Neuroses	Dorsal raphe nucleus: 5-HT↑	Dorsal raphe nucleus: 5-HT↓	SD rat	Li et al. (2007a)
EA	Huantiao (GB-30) Yanglingquan (GB-34) Jiaji (Ex-B2)	2	Lumbar nerve and inflammatory medium	Oppressed nerve root: 5-HT↑, PGE_2↑	Oppressed nerve root: 5-HT↓, PGE_2↓	SD rat	Wu and Yu (2005)
MA	Neiguan (PC-6)		Myocardial ischemia/ reperfusion injury	Spinal dorsal horn and root: 5-HT↑	Spinal dorsal horn and root: 5-HT↓	SD rat	Yi et al. (2009)
EA	Brain or Quchi (LI-11) Hegu (LI-4) Zusanli (ST-36) Huantiao (GB-30)	10	Cerebral ischemia/ reperfusion injury	Striatum: 5-HT↓	Striatum: 5-HT↑	SD rat	Li and Li (2005)
EA	Zusanli (ST-36) Sanyinjiao (SP-6)	?	Acute gouty arthritis	Synovium: 5-HT↑, HA↑, NO↑	Synovium: 5-HT↓, HA↓, NO↓	Rabbit	Cui and Hou (2008)
EA	Taichong (LR-3) Yanglingquan (GB-34)	?	Migraine	Brain stem and trigeminal: 5-HT1F mRNA↓, iNOS mRNA↑	Brain stem and trigeminal: 5-HT1F mRNA↑, iNOS mRNA↓	SD rat	Zhong and Li (2007)
"Jin's Three-Needling"	Neiguan (PC-6) Shenmen (HT-7) Sanyinjiao (SP-6)		Generalized anxiety disorder	Hematoblast: 5-HT↓ Plasma ACTH↓	Hematoblast: 5-HT↑ Plasma ACTH↑	Patient	Luo et al. (2007)

and reticularis paragigantocellularis lateral (Wang et al. 1994a, b) in the "normal" brain. This is especially true with acupuncture-induced analgesia (Xia et al. 2010). However, recent studies investigating the effects of acupuncture on vascular dementia have shown low levels of NA in the disease, which improved after EA treatment (Table 1.3). For example, Wang et al. (2008d) demonstrated an upregulation of neural monoamines in the rat model of vascular dementia in response to EA. Liu et al. (2009) performed a clinical study to determine the effect of acupuncture therapy on somatization disorder and reported that acupuncture could stimulate the activation of monoamines in the plasma including NA, DA, and 5-HIAA. He et al. (2008) showed that EA could increase NA level in plasma on a rat model for acute myocardial ischemia.

Thus recent studies show that EA can upregulate NA activity in certain disease states, especially when NA activity is low such as in vascular dementia.

1.5 Dopamine

Dopamine is a catecholamine neurotransmitter that is produced in specific areas of the brain. It is commonly involved in the reward system of the brain that encompasses feelings of enjoyment and reinforcement to motivate a person to proactively perform certain activities. Dopamine excess is closely associated with addictive behaviors while an insufficiency can cause somatization disorder and Parkinson's disease.

Acupuncture, either MA or EA, can regulate both the brain and plasma levels of dopamine (Table 1.4). Liu et al. (2009) presented clinical evidence for the proposition that acupuncture could elevate the levels of dopamine in plasma and bring about an improvement in the neurological function on somatization disorder. High-frequency EA increases the dopamine level in the brain by reducing apoptosis in dopamine neurons, which is surmised as the probable mechanism underlying the EA treatment for Parkinson's disease (Wang et al. 2007b; Hao et al. 2009). In contrast, a study on psychological dependence on heroin showed that a low-frequency EA could successfully suppress the levels of dopamine in the hippocampus (Chen et al. 2008a).

These observations show that the effect of acupuncture on the dopamine could vary under different pathological conditions from upregulation in certain diseases like Parkinson's disease to downregulation in other conditions such as heroin addiction.

Table 1.3 Effect of acupuncture on noradrenaline

Acupuncture	Acupoints	Frequency (Hz)	Disease states	Changes in disease states	Acupuncture effects	Subjects	References
EA	Neiguan (PC-6) Shenmen (HT-7) Zhizheng (SI-7)	2	Acute myocardial ischemia	Plasma: NA↓, AChE↑	Plasma: NA↑, AChE↓	Rat	He et al. (2008)
EA	Ganshu (BL-18) Pishu (BL-20) Xinshu (BL-15) Zusanli (ST-36) Sanyinjiao (SP-6) Shenting (DU-24) Sishencong (EX-HN-1)	?	Somatization disorder	Plasma: DA↓, NA↓, 5-HIAA↓	Plasma: DA↑, NA↑, 5-HIAA↑	Patient	Liu et al. (2009)
EA	Shuigou (DU-26) Neiguan (PC-6) Fengchi (GB-20)	15	Vascular dementia	Brain: DA↓, NA↓, 5-HT↓	Brain: DA↑, NA↑, 5-HT↑	Wistar rat	Wang et al. (2008d)

Table 1.4 Effect of acupuncture on dopamine

Acupuncture	Acupoints	Frequency (Hz)	Disease states	Changes in disease states	Acupuncture effects	Subjects	References
MA	Ganshu (BL-18) Pishu (BL-20) Xinshu (BL-15) Zusanli (ST-36) Sanyinjiao (SP-6) Shenting (DU-24) Sishencong (EX-HN-1)		Somatization disorder	Plasma: DA↓, NE↓, 5-HIAA↓	Plasma: DA↑, NE↑, 5-HIAA↑	Patient	Liu et al. (2009)
EA	Shuigou (DU-26) Neiguan (PC-6) Fengchi (GB-20)	15	Parkinson's disease	Brain: DA↓, NE↓, 5-HT↓	Brain: DA↑, NE↑, 5-HT↑	Wistar rat	Wang et al. (2007b)
EA	Fengchi (GB-20) Hegu (LI-4) Taichong (LR-3)	120	Parkinson's disease	Chiasma opticum: DA↓	Chiasma opticum: DA↑	Wistar rat	Hao et al. (2009)
EA	Neiguan (PC-6) Sanyinjiao (SP6) Sishencong (EX-HN-1)	2	Heroin psychological dependence	Hippocampus: ACh↓, DA↑	Hippocampus: ACh↑, DA↓	Mouse	Chen et al. (2008a)

1.6 Acetylcholine

Acetylcholine is a neurotransmitter in the cholinergic synapses and is involved in the conduction of stimulatory impulses in the nervous system. It is synthesized by the enzyme choline acetyltransferase (ChAT) and eliminated from the synapse by acetylcholinesterase (AChE) (Fig. 1.1). Acetylcholine is the primary neurotransmitter associated with thought, learning, and memory functions of the brain (Table 1.5), and it is believed that neurodegenerative diseases could be associated with acetylcholine insufficiency.

Recent acupuncture research has borne witness to several studies focusing on elucidation of the mechanisms involved in the acupuncture treatment of Alzheimer's disease. Their results revealed that acupuncture causes an increase in acetylcholine and decrease in superoxide dismutase levels of the brain. The likely explanation is that acupuncture activates ChAT to reinforce the biosynthesis of acetylcholine and suppresses AChE activity to reduce hydrolyzation of acetylcholine (Ma et al. 2008a; Zeng et al. 2009; Wang et al. 2009b).

The dysfunction of cholinergic system is implicated as the major cause of vascular dementia. EA has been suggested to be therapeutic. Therefore, researchers raise the question if EA can regulate the cholinergic system in this disease. Indeed, some researchers demonstrated that low-frequency EA inhibits AChE function and slows the degradation of acetylcholine in the CNS (Wu et al. 2008b; Liu et al. 2008c). Several other groups of researchers have reported the similar results, i.e., EA decreases the concentration of AChE in the CNS (He et al. 2008; Zeng et al. 2009; Wang et al. 2009b). On the other hand, Yang et al. (2009b) reported an increase in both ChAT and AChE activities following EA in a rat model of vascular dementia. Their conflicting results however require further validation.

EA could also mediate its effect in the treatment of psychological dependence on heroin through the regulation of cholinergic system. Some studies showed that EA could regulate the activity of AChE in the hippocampus under heroin dependence. However, it is still a controversial issue owing to the conflict in reported results. Chen et al. (2008a) reported that EA at 2 Hz increased the AChE level, while opposite results were obtained by Xue (2009). Chen et al. (2008a) treated the mice model with EA at 2 Hz for 10 days and then sampled the hippocampus for AChE assay. In

Biosynthesis

Choline + Acetylcoenzyme A + ChAT → Ach + Coenzyme A

Degradation

Ach + Water + AchE → Choline + Acetic acid

Fig. 1.1 Biosynthesis and degradation of ACh

Table 1.5 Effect of acupuncture on acetylcholine

Acupuncture	Acupoints	Frequency (Hz)	Disease states	Changes in disease states	Acupuncture effects	Subjects	References
EA	Hegu (LI-4) Taichong (LR-3)	80/100	Alzheimer's disease	Brain: AChE↑, SOD↓	Brain: AChE↓, SOD↑	Wistar rat	Zeng et al. (2009)
EA	Baihui (DU-20) Zusanli (ST-36) Shenshu (BL-23)	2/4	Alzheimer's disease	Hippocampus and cortex: ACh↓, ChAT↓, AChE↑	Hippocampus and cortex: ACh↑, ChAT↑, AChE↓	SD rat	Wang et al. (2009b)
MA	Baihui (DU-20) Dazhui (DU-14)		Alzheimer's disease	Brain: AChE↑, SOD↓	Brain: AChE↓, SOD↑	Wistar rat	Ma et al. (2008a)
EA	Dazhui (DU-14) Baihui (DU-20) Shuigou (DU-26)	0.5	Vascular dementia	Brain: SOD↓, MDA↑, AChE↑	Brain: SOD↑, MDA↓, AChE↓	Wistar rat	Wu et al. (2008b)
EA	Baihui (DU-20) Sishencong (EX-HN-1) Shenting (DU-24) Benshen (GB-13)	2	Vascular dementia	Cortex: MDA↑, AChE↑	Cortex: MDA↓, AChE↓	SD rat	Liu et al. (2008c)
EA	Yingxiang (LI-20) Yintang (EX-HN-3)	80/100	Vascular dementia	Brain: ChAT↓, AChE↓	Brain: ChAT↑, AChE↑	Sprague-Dawley rat	Yang et al. (2009b)
EA	Neiguan (PC-6) Sanyinjiao (SP6) Sishencong (EX-HN-1)	2	Heroin psychological dependence	Hippocampus: AChE↓, DA↑	Hippocampus: AChE↑, DA↓	Mouse	Chen et al. (2008a)
EA	Zusanli (ST-36) Sanyinjiao (SP-6)	?	Learning and memory disorder of heroin dependence	Hippocampus: ACh↑, NOS↑	Hippocampus: ACh↓, NOS↓	SD rat	Xue (2009)
EA	Zusanli (ST-36)	?	Stress gastric ulcer	Mucosa: SOD↓, MDA↑	Mucosa: SOD↑, MDA↓	SD rat	Yu et al. (2007)
EA	Neiguan (PC-6) Shenmen (HT-7) Zhizheng (SI-7)	?	Acute myocardial ischemia	Plasma: NA↓, AChE↑	Plasma: NA↑, AChE↓	Rat	He et al. (2008)

contrast, Xue (2009) used the rat model and collected the hippocampal samples after 5 days of 2-/100-Hz EA treatment. The different experimental methods employed could probably account for the observed contrast in results. However, further studies remain a desideratum in order to elucidate this issue.

Yu et al. (2007) demonstrated that microinjection of acetylcholine into the nucleus tractus solitarius could enhance protection against stress-induced mucosal injury following EA treatment. Their results also indicate that the effect of EA treatment in this disease can be explained by its role in the cholinergic system.

In addition to the acupuncture effect on central acetylcholine, there is evidence suggesting that EA can regulate acetylcholine levels in the plasma. He et al. (2008) reported a decrease in the plasma level of acetylcholine in the rat model of acute myocardial ischemia following treatment with EA at 2 Hz. More studies are needed to further verify this interesting observation.

These recent studies prompt us to conclude that acupuncture differentially regulates the level of acetylcholine in the central and peripheral system. In acetylcholine insufficiency, acupuncture inhibits AChE in the brain thus suppressing the ACh degradation.

1.7 Amino Acids

Amino acids include both excitatory and inhibitory amino acids. Glutamate is the major excitatory neurotransmitter in the CNS (Komuro and Rakic 1996), and glutamate responses are mediated by ionotropic glutamate receptor (iGluR) and metabotropic glutamate receptor (mGluR) (D'Antoni et al. 2008). The iGluRs have three subtypes, namely, *N*-methyl-D-aspartate (NMDA) receptor, α-amino-3-hydroxy-5-methyl-4-isoxazolepropionic acid (AMPA) receptor, and kainate receptor.

Glutamate is essential for maintaining synaptic plasticity, learning, and development under physiological conditions (LoTurco et al. 1991). However, excessive glutamate exposure results in neuronal death, termed as excitotoxicity. Neural excitotoxicity might lead to neuronal damage under various neurological conditions including inflammation, ischemia, and neurodegenerative diseases. Glutamate-induced excitotoxicity is associated with the NMDA receptor. Excessive activation of NMDA receptors opens up the Ca^{2+} channels, resulting in a massive Ca^{2+} influx that activates many downstream signaling pathways, thereby damaging the neurons.

Many previous studies have demonstrated that EA can suppress the glutamate levels in CNS (Cao et al. 1993; Wang and Fan 1995). A recent study suggested that EA increases the astrocyte expression of GLT-1, which is essential for glutamate regulation, especially for glutamate uptake (Luo et al. 2011). One of the possibilities for the EA effect is that EA enhances astrocytic GLT-1 expression, and the increased GLT-1 in turn downregulates the accumulation of glutamate to protect the neurons from excitotoxicity. Besides suppression of the glutamate levels, recent studies also suggest an alteration in glutamate receptor after acupuncture. Meng et al. (2008) studied the effects of acupuncture on vascular dementia and revealed

Fig. 1.2 Acupuncture-induced brain protection via EOP-mediated reduction of excitotoxicity

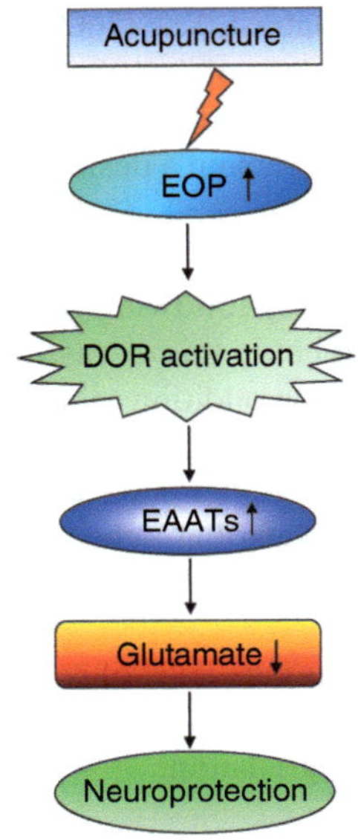

that EA pretreatment could mitigate the increased glutamate levels and suppress the mRNA expression of NMDA receptor 1. Peng et al. (2009) also reported that acupuncture downregulated NMDA receptor and thus relieve the gastric distension-induced pain. In addition, EA at 2 Hz could also downregulate glutamate levels in the lumbar segments (L4–L6) of spinal cord in a model of spared nerve injury (Yan et al. 2009). In contrast, GluR-2 (a subtype of AMPA receptors) activation can protect the neurons from excitotoxicity. Dong et al. (2007b) demonstrated that EA at Baihui (DU-20) and Shuigou (DU-26) could enhance the production of GluR-2 in the CA1 of ischemic rat brain.

In summary, recent reports suggest that acupuncture induces neuroprotection by increasing GLT-1 expression in astrocytes, inhibiting NMDA receptor expression and elevating AMPA receptor expression. In our recent work, we found that electroacupuncture can increase the expression of DOR (Tian et al. 2008), while DOR activation upregulates the expressions of GLT-1 (EAAT2) and glutamate aspartate transporter (GLAST, also named EAAT1) in the astrocytes and thus accelerates the reduction of extracellular glutamate concentration (Liang et al. 2010). This mechanism may, at least partially, play a role in acupuncture-induced brain protection, i.e., acupuncture upregulation of DOR signaling increases the expression of EAAT1 and EAAT2 and thus prevent the brain from excitotoxicity via an increase in the removal of excessive excitatory transmitters from extracellular space (Fig. 1.2).

In contrast to glutamate, GABA is the main inhibitory neurotransmitter in the CNS, and it has been shown to have neuroprotective effects in focal or global cerebral ischemia/hemorrhage in adults. Our previous review has showed that acupuncture could increase the levels of GABA in the brain (Wen et al. 2010a, b). Recently, Du et al. (2007) provided corroborative evidence by using a rat model for cerebral hemorrhage (Table 1.6).

Table 1.6 Effect of acupuncture on amino acids and their receptors

Acupuncture	Acupoints	Frequency (Hz)	Disease states	Changes in disease states	Acupuncture effects	Subjects	References
EA	Shuigou (DU-26) Neiguan (PC-6) Zusanli (ST-36)	2	Cerebral hemorrhage	Hippocampus: glutamate↑, GABA↓	Hippocampus: glutamate↓, GABA↑	Wistar rat	Du et al. (2007)
EA pretreatment	Baihui (DU-20) Shenshu (BL-23) Zusanli (ST-36)	1.67	Vascular dementia	Hippocampus: glutamate↑, NMDAR1↑	Hippocampus: glutamate↓, NMDAR1↓	SD rat	Meng et al. (2008)
EA	Baihui (DU-20) Shuigou (DU-26)	4/16	Ischemia	CA1: GluR-2↓	CA1: GluR-2↑	SD rat	Dong et al. (2007b)
MA	Zusanli (ST-36)		Gastric distension-induced pain	Spinal dorsal horn: SP↑, NMDAR↑	Spinal dorsal horn: SP↓, NMDAR↓	SD rat	Peng et al. (2009)
EA	Weizhong (BL-40) Huantiao (GB-30)	2	Spared nerve injury pain	Lumbar segment (L4–L6) of spinal cord: SP↓, glutamate↑	Lumbar segment (L4–L6) of spinal cord: SP↑, glutamate↓	SD rat	Yan et al. (2009)

1.8 Substance P

Substance P is a neuropeptide belonging to the tachykinin family of neuropeptides widely distributed in the brain. It is released from the terminals of specific sensory neurons and is found in the brain, spinal cord, and plasma. Substance P plays an important role under many physiological and pathological situations, such as cell growth, neurogenesis, vasodilation, pain, nociception, anxiety, inflammation, so on and so forth.

Many previous studies show that acupuncture can elevate substance P levels in the brain (Li et al. 1989; Cui et al. 1990). Recent studies also support this belief. EA can significantly elevate the expression of substance P in many nuclei such as the medial septal nucleus, lateral septal nucleus, ventromedial hypothalamic nucleus, and dorsomedial hypothalamic nucleus (Zhan et al. 2007). In a study on acute myocardial ischemia, Li et al. (2007b) also observed that EA at Neiguan (PC-6) elevates the level of substance P in the hypothalamus, medulla, and spinal lateral crus.

Substance P has been implicated for long in the mechanisms of pain causation and is found to be associated with EA analgesia. However, substance P has both antalgic and algogenic roles in pain control. N-terminal of substance P can alleviate pain, which can be abrogated by naloxone, signifying that the analgesic effect of substance P could be related to endogenous opioid mechanism. In contrast, the C-terminal of substance P induces nociception (Skilling et al. 1990; Li and Yi 2008) (Fig. 1.3). Most of the recent studies suggest that acupuncture inhibits the algogenic action of substance P.

Yan et al. (2009) showed that low-frequency (2 Hz) EA increases the pain threshold in the spared nerve injury model, which was associated with an upregulation of substance P level and downregulation of glutamate level in the lumbar spinal cord segments (L4–L6). However, it is unclear in this study if there is any causal connection between the increased level of substance P and decreased level of glutamate. Several studies have illustrated the algogenic effects of substance P and acupuncture-induced inhibition of substance P. Yang et al. (2009a) showed that acupuncture pretreatment suppresses the score of visceral traction pain, and the mechanism involved is through an increase in the amounts of leucine-enkephalin and reduction in the levels of substance P and AChE. Their data also suggest that MA has a better effect than EA in the treatment of visceral traction pain. In a study on chronic radicular pain in a rat model, Tu et al. (2008a, b) reported that low-frequency EA reduced the release of substance P in the dorsal root ganglion and induced an analgesic effect.

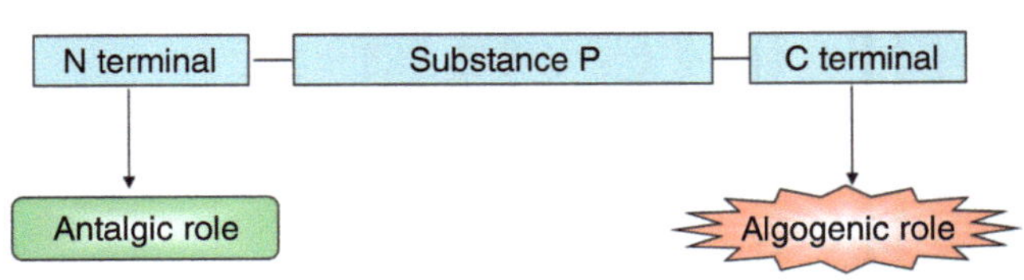

Fig. 1.3 Different roles of N- and C-terminals in substance P

In addition, they observed that the effect of EA stimulation at both Jiaji (EX-B2) and Yanglingquan (GB-34) is better than that of EA at either one of them.

Acupuncture at Zusanli (ST-36) could inhibit stress reaction induced by gastric distension and decrease pain scores, which is associated with the suppression of expression of substance P and NMDA receptor (Peng et al. 2009). In the chronic inflammatory pain model, EA at Yanglingquan (GB-34) and Zusanli (ST-36) acupoints inhibited the expression of substance P and calcitonin gene-related peptide (CGRP) in the L4–L6 lumbar segments of spinal cord, while intrathecal injection of buprenorphine enhances the effect of EA on inhibition of substance P and CGRP levels (Fan et al. 2007). Wu et al. (2008a) reported that warm needle moxibustion could suppress the pain indexes in the patients of lumbar intervertebral disc protrusion. They suggested the reduction in substance P levels in the plasma as the involved mechanism.

Acupuncture/EA has a good therapeutic effect on allergic rhinitis and effectively relieves the clinical symptoms. Recently, some investigators studied the mechanism of this treatment and showed that acupuncture/EA could downregulate the levels of substance P in the patient's plasma. They also reported that acupuncture directed at sphenopalatine ganglion had a better effect than the normal acupuncture treatment (Zheng et al. 2009; Li et al. 2007c).

There was a report suggesting that acupuncture therapy has a high efficacy in colonic irrigation in the slow transit constipation, as it can regulate the levels of gastrointestinal hormones—substance P and vasoactive intestinal polypeptide (Wang et al. 2009c). However, it is difficult to make solid causative conclusions associated with the effect of acupuncture on substance P and vasoactive intestinal polypeptide because of lacking strict control in this study.

As shown in Table 1.7 that lists the major results in recent research on the role of substance P in acupuncture, it seems that although substance P has both antalgic and algogenic roles in the pain control, acupuncture mainly suppresses its algogenic role mediated by the C-terminal of substance P in the nervous system. Additionally, acupuncture downregulates the level of substance P in plasma.

1.9 Prostaglandins

Prostaglandins (PGE) are found in a majority of human tissues and organs. They are synthesized in all nucleated cells except lymphocytes. They are autocrine and paracrine lipid mediators that act upon platelets, endothelium, uterine, and the mast cells. They mediate a wide range of physiological functions, such as control of blood pressure, contraction of smooth muscle, and modulation of inflammation.

The acute gastric mucosal injury results from the inhibition of the protective effect of prostaglandin E_2 (PGE_2). Xu et al. (2010) demonstrated an increase in the PGE_2 level of plasma and gastric mucosa following EA treatment in a mouse model for acute gastric mucosal injury. They also studied the alterations in these effects of EA treatment at different times of the day. Their data indicated that EA given

Table 1.7 Effect of acupuncture on substance P (SP)

Acupuncture	Acupoints	Frequency (Hz)	Disease states	Changes in disease states	Acupuncture effects	Subjects	References
EA	Neiguan (PC-6)	8/80	Acute myocardial ischemia	Paraventricular nucleus of hypothalamus, medullary bulb and vagi nerve complex, and spinal lateral crus: SP↓, NO↑	Paraventricular nucleus of hypothalamus, medullary bulb and vagi nerve complex, and spinal lateral crus: SP↑, NO↓	SD rat	Li et al. (2007b)
EA MA	Zusanli (ST-36)	2/15	Visceral traction pain	Ileum: SP↑, AChE↑, LEK—no change	Ileum: SP↓, AChE↓, LEK↑	SD rat	Yang et al. (2009a)
EA	Jiaji (Ex-B2) Yanglingquan (GB-34)	2	Chronic radicular pain	Dorsal root ganglion: SP↑	Dorsal root ganglion: SP↓	Wistar rat	Tu et al. (2008a)
EA	Jiaji (Ex-B2) Yanglingquan (GB-34)	2	Chronic radicular pain	Dorsal root ganglion: SP↑	Dorsal root ganglion: SP↓	Wistar rat	Tu et al. (2008b)
MA	Zusanli (ST-36)		Gastric distension-induced pain	Spinal dorsal horn: SP↑, NMDAR↑	Spinal dorsal horn: SP↓, NMDAR↓	SD rat	Peng et al. (2009)
Warm needle moxibustion	Jiaji (Ex-B2)		Lumbar intervertebral disc protrusion	Plasma SP?	Plasma SP↓	Patient	Wu et al.
EA	Yanglingquan (GB-34) Zusanli (ST-36)	2/100	Chronic inflammatory pain	Lumbar segment (L4–L6) of spinal cord: SP↑, CGRP↑	Lumbar segment (L4–L6) of spinal cord: SP↓, CGRP↓	SD rat	Fan et al. (2007)

EA	Weizhong (BL-40) Huantiao (GB-30)	2	Spared nerve injury pain	Lumbar segment (L4–L6) of spinal cord: SP↓, glutamate↑	Lumbar segment (L4–L6) of spinal cord: SP↑, glutamate↓	SD rat	Yan et al. (2009)
EA	Zusanli (ST-36)	50	Normal	Brain: SP—no change	Brain: SP ↑	SD rat	Zhan et al. (2007)
EA	Shenshu (BL-23) Yaoyangguan (DU-3)	2/100	Transverse process syndrome	Hypothalamus Spinal cord: SP↑, CCK-8↑	Hypothalamus Spinal cord: SP↓, CCK-8↓	SD rat	Hu et al. (2008)
MA	Yingxiang (LI-20) Shangxing (DU-23) Hegu (LI-4) Zusanli (ST-36) Sphenopalatine ganglion		Allergic rhinitis	Plasma: SP↑	Plasma: SP↓	Patient	Zheng et al. (2009)
EA	Yintang (EX-HN-3) Yingxiang (LI-20) Shangyingxiang (Ex-HN-8) Sphenopalatine ganglion	10–100	Allergic rhinitis	Plasma: SP↑	Plasma: SP↓	Patient	Li et al. (2007c)

between 7:00 and 9:00 pm had a better curative effect than EA given between 7:00 and 9:00 am. Since they carried out their project in a mice model and the biorhythm of mice is opposite to that of humans, Xu et al. (2010) postulated that EA at 7:00–9:00 am might have a better curative effect in a human patient (Table 1.8).

Qi (2007) reported that in a rat model for hyperlipidemia, EA upregulated the PGE level in plasma. Also, Guo et al. (2010) got the similar result and showed that EA significantly increased the level of 6-keto prostaglandin F1α (6-keto-PGF1α) in the patients with cerebral thrombosis. It seems that EA can increase 6-keto-PGF1α in plasma. In contrast, MA may have no such effect. Zheng et al. (2010) observed that MS did not significantly increase 6-keto-PGF1α in the plasma (Table 1.8), although inducing successful improvement of stroke symptoms. However, the combination of MA with mental imagery, an experience that resembles perceptual experience but occurs in the absence of the appropriate external stimuli for the relevant perception, could significantly elevate the level of 6-keto-PGF1α in the plasma and induce an even better outcome for the stroke patients (Zheng et al. 2010).

Therefore, EA and MA differentially regulate the level of 6-keto-PGF1α, which may be influenced by other psychological and biological factors.

1.10 Cholecystokinin Octapeptide-8

CCK-8 is synthesized in mucosal epithelium of the small intestine and secreted in the duodenum. Also, it is widely distributed throughout the brain and the spinal cord and plays multiple functions including antagonizing EA analgesia (Table 1.9).

It was reported that acupuncture at the Yingxiang (LI-20) and Yintang (EX-HN-3) can be used in treatment of the vascular dementia. Recent research suggests that the underlying mechanism is related to the elevation of CCK-8 in the hippocampus (Niu et al. 2009a). They also observed that a combination of EA and eugenol had a better curative effect than EA alone (Niu et al. 2009b).

He et al. (2007) reported that EA had positive effect on spike potential of gastrointestinal tract, which was associated with an upregulation of CCK-8 and motilin in plasma. Additionally, in a recent study on gastrointestinal dysfunctions, investigators created an animal model for gastrointestinal dysfunction model by injecting rabbits with atropine. Then they treated the disease model with EA at Sibai (ST-2), Yanglingquan (GB-34), or Zusanli (ST-36), located in the three-yin meridians of the foot. Their results implied that EA applied at three-yin meridians could improve the gastrointestinal function, and the mechanism is associated with the increased concentration of CCK-8 in the plasma and Oddi's sphincter (Zhang et al. 2009). There is also a recent study showing that EA successfully upregulated the CCK-8 concentrations in the hypothalamus and plasma of a rat model for obesity (Wang et al. 2008b).

These reports suggest that acupuncture enhances the level of CCK-8 in the CNS as well as the plasma in various conditions.

Table 1.8 Effect of acupuncture on prostaglandins

Acupuncture	Acupoints	Frequency (Hz)	Disease states	Changes in disease states	Acupuncture effects	Subjects	References
EA	Zusanli (ST-36)	?	Gastric mucosal injury	Plasma and gastric mucosa: PGE_2↓	Plasma and gastric mucosa: PGE_2↑	SD rat	Xu et al. (2010)
MA	Sutura parietotemporalis		Stroke	Plasma: 6-keto-PGF1α↓, TXB2↑	Plasma: 6-keto-PGF1α—no change, TXB2↓	Patient	Zheng et al. (2010)
EA	Zusanli (ST-36) Fenglong (ST-40) Sanyinjiao (SP-6) Taixi (KI-3) Quchi (LI-11) Neiguan (PC-6) Hegu (LI-4)	?	Cerebral thrombosis	Plasma: 6-keto-PGF1α?, TXB2?	Plasma: 6-keto-PGF1α↑, TXB2↓	Patient	Guo et al. (2010)
EA	Zusanli (ST-36) Neiguan (PC-6)	2	Hyperlipidemia	Plasma: 6-keto-PGF1α↓, TXB2↑	Plasma: 6-keto-PGF1α↑, TXB2↓	SD rat	Qi (2007)
EA	Baihui (DU-20) Zusanli (ST-36) Sishencong (EX-HN-1) Mingmen (DU-4) Dazhui (DU-14)	50	Vascular dementia	Plasma: 6-keto-PGF1α—no change, TXB2↑	Plasma: 6-keto-PGF1α↑, TXB2↓	Patient	Ji et al. (2007)
EA	Zusanli (ST-36) Fenglong (ST-40) Sanyinjiao (SP-6) Quchi (LI-11) Neiguan (PC-6) Hegu (LI-4)	?	Cerebral infarction	Plasma: 6-keto-PGF1α?, TXB2?	Plasma: 6-keto-PGF1α↑, TXB2↓	Patient	Chen et al. (2008d)
EA	Jianyu (LI-15)	2/100	Scapulohumeral periarthritis	Serum: PGE_2↑	Serum: PGE_2↓	Rabbit	Wang et al. (2007a)

Table 1.9 Effect of acupuncture on cholecystokinin octapeptide-8 (CCK-8)

Acupuncture	Acupoints	Frequency (Hz)	Disease states	Changes in disease states	Acupuncture effects	Subjects	References
XIU three needles	Yingxiang (LI-20) Yintang (EX-HN-3)	80/100	Vascular dementia	Hippocampus: CCK-8↓	Hippocampus: CCK-8↑	SD rat	Niu et al. (2009a)
EA	Yingxiang (LI-20) Yintang (EX-HN-3)	80/100	Vascular dementia	Cerebral marginal lobe: CCK-8↓	Cerebral marginal lobe: CCK-8↑	SD rat	Niu et al. (2009b)
EA	Sibai (ST-2) (ST-2), or Yanglingquan (GB-34) or Zusanli (ST-36)	4/50	Gastrointestinal dysfunction	Plasma and Oddi's sphincter tissues: CCK-8↓	Plasma and Oddi's sphincter tissues: CCK-8↑	New Zealand rabbit	Zhang et al. (2009)
EA	Zusanli (ST-36)	15	Normal	Plasma: CCK-8—no change, motilin—no change	Plasma: CCK-8↑, motilin↑	New Zealand rabbit	He et al. (2007)
EA	Shenshu (BL-23) Yaoyangguan (DU-3)	2/100	Transverse process syndrome	Hypothalamus and spinal cord: SP↑, CCK-8↑	Hypothalamus and spinal cord: SP↓, CCK-8↓	SD rat	Hu et al. (2008)
EA	Zusanli (ST-36) Sanyinjiao (SP-6) Guanyuan (CV-4) Zhongwan (CV-12)	100	Obesity	Hypothalamus and blood: CCK-8↓	Hypothalamus and blood: CCK-8↑	SD rat	Wang et al. (2008b)

1.11 Neurotrophic Factors

Neurotrophic factors (NTFs), for example, nerve growth factor (NGF), brain-derived neurotrophic factor (BDNF), glial cell line-derived neurotrophic factor (GDNF), neurotrophin-3 (NT-3), and neurotrophin-4 (NT-4), enhance the growth, function, and phenotypic development of neurons (Gall and Isackson 1989; Lindvall et al. 1994; Saavedra et al. 2008). They are also thought to play a crucial role in neuronal recovery from a pathological state (Connor and Dragunow 1998; Saavedra et al. 2008). Tropomyosin receptor kinase (Trk) and GDNF receptor are responsible for the NTF signaling. There are three subtypes in Trk family, i.e., TrkA, TrkB, and TrkC. Among them, TrkA has the highest affinity to bind NGF, and TrkB is activated by binding with BDNF and NT-4, while TrkC is mainly associated with NT-3 activity (Gall and Isackson 1989; Lindvall et al. 1994; Saavedra et al. 2008).

The past studies summarized in our previous review (Wen et al. 2010b) have demonstrated that acupuncture elevates the levels of NTFs in various conditions. Recent studies further endorse this phenomenon (Table 1.10). Tong et al. (2007) reported that NGF and its receptor along with basic fibroblast growth factor increased on the 3rd day after sciatic nerve injury, peaked on the 14th day, and then decreased. They also showed that EA and MA treatment could further delay the decrease that can prove to be beneficial for the sciatic nerve injury. Liu et al. (2010) reported that scalp acupuncture could elevate the level of BDNF in the dentate gyrus and proved that this was beneficial for the rehabilitation of neural functions in a rat model of middle cerebral artery occlusion. In a study on depression, Fu et al. (2009) reported that EA could bring about a significant improvement in the behavioral changes in a depression model. They suggested that EA could induce the expression of cAMP response element binding (CREB) and BDNF in the hippocampus, which could protect the neurons from degenerative death.

In addition, recent research suggests that acupuncture induces expression of NTFs via mRNA transcription. For example, EA on Zusanli (ST-36) and Shenshu can increase the mRNA and protein expression of NGF and can aid reparation of the sciatic nerve in diabetic peripheral neuropathy (Dong et al. 2007a). In the cerebral infarction model, the expressions of BDNF mRNA and Bcl-2 mRNA reached the peak levels when EA was administered in the 12–24 h time window after reperfusion, and it had maximal protective and rehabilitative effects on cerebral ischemia-reperfusion injury (Li et al. 2009). Ma et al. (2008b) showed that in the diabetic rat models, EA could increase the expression of NT-3 mRNA and positively expressing cells in the hippocampus and improve the learning ability.

EA could also regulate the expression of BDNF receptors. Wang et al. (2009a) reported that EA raised the excitability of the smooth muscles of the urinary bladder in a rat model for spinal cord injury-induced urinary retention. The investigators suggested the possible mechanism as EA-induced expression of BDNF and TrkB in the spinal cord. Also, the effects of EA at Guanyuan (RN-4) were found to be superior to those of EA at Shuidao (ST-28).

All the data suggest that MA or EA can upregulate NTFs and NTF receptors in various physiological and pathophysiological conditions.

Table 1.10 Effect of acupuncture on NTFs

Acupuncture	Acupoints	Frequency (Hz)	Disease states	Changes in disease states	Acupuncture effects	Subjects	References
EA	Baihui (DU-20) Dazhui (DU-14)	2/30	Cerebral infarction	Brain: BDNF mRNA↓, Bcl-2 mRNA↓	Brain: BDNF mRNA↑, Bcl-2 mRNA↑	SD rat	Li et al. (2009)
EA	Huantiao (GB-30) Zusanli (ST-36)	2	Sciatic nerve injury	Injured sciatic nerve: NTF?, NTFR?, bFGF?	Injured sciatic nerve: NTF↑, NTFR↑, bFGF↑	Wistar rat	Tong et al. (2007)
EA	Scalp acupuncture Anterior and posterior of oblique line of vertex-temporal	2/100	Cerebral ischemia-reperfusion	Dentate gyrus: BDNF↑	Dentate gyrus: BDNF↑	Wistar rat	Liu et al. (2010)
EA	Guanyuan (CV-4) Shuidao (ST28)	2/15	Urinary retention due to spinal cord injury	Spinal cord: BDNF↑, TrkB↑	Spinal cord: BDNF↑, TrkB↑	SD rat	Wang et al. (2009a)
EA	Zusanli (ST-36) Shenshu (BL-23)	2	Diabetic peripheral neuropathy	Sciatic nerve: NGF protein and mRNA↓	Sciatic nerve: NGF protein and mRNA ↑	Rat	Dong et al. (2007a)
EA	Hegu (LI-4) Taichong (LR-3)	2/20	Depression	Hippocampus: BDNF↓, CREB—no change	Hippocampus: BDNF↑, CREB↑	SD rat	Fu et al. (2009)
EA	Baihui (DU-20) Dazhui (DU-14)	0.23/0.43 (16/26 times/min)	Diabetic cognitive impairment	Hippocampus: NT-3 protein and mRNA↓	Hippocampus: NT-3 protein and mRNA↑	Wistar rat	Ma et al. (2008b)

1.12 Nitric Oxide

Nitric oxide (NO) is the first gas known to act as a transmitter in both the peripheral and central nervous systems. It is an important signal molecule in living organisms with diverse pathophysiological functions. Inducible NO synthase (iNOS) can be expressed in macrophage/microglia and is able to induce NO production.

As shown in Table 1.11, acupuncture can downregulate the NO levels in the CNS. Wang et al. (2008c) reported that after treatment with EA in rats with acute spinal cord injury, the level of NO decreased in the injured spinal cord. Their data also suggested that the EA given 2 h after the occurrence of spinal cord injury had a higher efficacy. In addition, moxibustion maintains the cellular form and ultramicrostructures of the dorsal root nerve ganglia and effectively inhibits the release of inflammatory factors iNOS and CGRP (Mi et al. 2009). Moreover, in the acute myocardial ischemia study, Li et al. (2007b) reported a decrease in the NO level in the paraventricular nucleus of hypothalamus, medullary bulb and vagi nerve complex, and the lateral spinal crus as a result of EA treatment.

Acupuncture also influences NO regulation in the plasma. Scalp acupuncture can increase the plasma NO level in hypertensive patients (Chen and Li 2009). In a study on acute gastric mucosal injury, Wang (2010) reported that EA at Zusanli (ST-36) and Zhongwan (RN-12) could increase NO and decrease gastrin levels both in the serum and gastric mucosa of the rat model. However, no significant increase was observed in the migraine patients (Chen et al. 2008b).

Acupuncture regulates the NO level in other tissues as well. For example, EA at Hegu (LI-4) and Taichong (LR3) could reduce iNOS and NO concentration and decrease the GSH-Px activity in the colonic tissues of the rat model for depression, which might be an illustration of its protective effect in the digestive tract against injury during depression (Huang et al. 2008b). EA at Neiguan (PC-6) could upregulate myocardial NO and NOS levels and downregulate myocardial intracellular Ca^{2+} concentration in the rat model of myocardial ischemic-reperfusion injury, which might partially explain its effect in relieving myocardial injury (Wang et al. 2010). In addition, EA increases the NO level in the prostatic tissue and curbs prostatic hyperplasia (Yuan et al. 2008). However, Lu et al. (2007) observed that although EA could improve the sexual behaviors in the rat model for depression, no significant change in the concentration of NO was seen in the penile tissues.

According to these recent studies, we conclude that acupuncture effectively reduces the NO level in the CNS and may enhance the NO content in plasma and some organs, but not in the penile tissues, suggesting differential effects of NO in various organs in response to acupuncture.

1.13 Concluding Remarks

Acupuncture has been extensively investigated over the past five decades, which has generated substantial data on its regulation of neurotransmitters/modulators. This chapter, together with our previous book chapters (Xia et al. 2010), prompts us

Table 1.11 Effect of acupuncture on nitric oxide (NO)

Acupuncture	Acupoints	Frequency (Hz)	Disease states	Changes in disease states	Acupuncture effects	Subjects	References
EA	Mingmen (DU-4) Dazhui (DU-14)	2	Acute spinal cord injury	Spinal cord: NO↑, SOD↓	Spinal cord: NO↓, SOD↑	SD rat	Wang et al. (2008c)
Scalp penetration acupuncture			Hypertension	Plasma: NO↓, CGRP↓	Plasma: NO↑, CGRP↑	Patient	Chen and Li (2009)
Scalp acupuncture			Migraine	Plasma: NO↓, CGRP↓	Plasma: NO—no change, CGRP↑	Patient	Chen et al. (2008b)
EA	Baihui (DU-20) Yintang (EX-HN-3)	2	Depression	Penile: NO—no change, cGMP↓	Penile: NO—no change, cGMP↑	SD rat	Lu et al. (2007)
MA/ moxibustion	Jiaji (Ex-B2)		Lumbar nerve root compression	Injured nerve root: NOS↑, CGRP↑	Injured nerve root: NOS↓, CGRP↓	SD rat	Mi et al. (2009)
EA	Neiguan (PC-6) Hegu (LI-4) Lieque (lU-7)	30/100	Myocardial ischemia-reperfusion injury	Myocardial tissue of the ischemic area: iNOS↓, NO↓, Ca2+↑	Myocardial tissue of the ischemic area: NOS↑, NO↑, Ca2+↓	Wistar rat	Wang et al. (2010)
EA	Hegu (LI-4) Taichong (LR3)	1.5/2	Depression	Colon: iNOS↑, NO↑, GSH-Px↓	Colon: iNOS↓, NO↓, GSH-Px↑	SD rat	Huang et al. (2008b)
EA	Zusanli (ST-36) Zhongwan (CV-12)	4/20	Acute gastric mucosal injury	Serum and gastric mucous: NO↓, GAS↑	Serum and gastric mucous: NO↑, GAS↓	SD rat	Wang (2010)
EA	Sanyinjiao (SP-6) Shenshu (BL-23) Guanyuan (CV-4) Zhibian (BL-54)	20	Benign prostatic hyperplasia	Prostate: NO↓	Prostate: NO↑	Wistar rat	Yuan et al. (2008)
EA	Neiguan (PC-6)	8/80	Acute myocardial ischemia	Paraventricular nucleus of hypothalamus, medullary bulb and vagi nerve complex, and spinal lateral crus: SP↓, NO↑	Paraventricular nucleus of hypothalamus, medullary bulb and vagi nerve complex, and spinal lateral crus: SP↑, NO↓	SD rat	Li et al. (2007b)

to believe without doubt that acupuncture, both MA and EA, can modulate sundry neurotransmitters/modulators in terms of their synthesis, release, and uptake as well as the activity of their receptors. In fact, this may not surprise those who have modern biomedical knowledge since acupuncture needles at acupoints can generate neural signals at local tissues that are transmitted to the CNS via nervous network in the body (Xia et al. 2010). It is equally interesting to note that acupuncture can affect various functions under both physiological and pathophysiological conditions since the acupuncture-affected transmitters/modulators are involved in broad functions of our body.

What is interesting is that acupuncture can induce dual-directional regulation of transmitters/modulators, which has been shown in many published studies reviewed in this chapter as well as in our previous reviews (Wen et al. 2010a, b). For example, the concentration of ACh is low in the hippocampus of Alzheimer's disease model while the application of acupuncture demonstrated an increase in ACh levels. In contrast, the ACh level is high in the hippocampus of the heroin-dependence model, which was shown to be suppressed after acupuncture treatment. Similar outcomes were also noted with endogenous opioids and dopamine. This dual-directional regulation may be the keystone in the basis of acupuncture treatment for multiple diseases. Indeed, the theory of Traditional Chinese Medicine has held this concept for a long time despite the lack of substantial scientific evidence. The content of this chapter may provide some clues for explaining this concept.

The dual regulation may mainly result from two aspects, i.e., acupuncture method and body status. The former includes acupoints, acupuncture approach (MA vs. EA or others), needle manipulation, stimulation parameters, etc., while the latter refers to physiological or pathophysiological conditions of the body during needle insertion and stimulation. Different acupuncture methods may lead to different outcomes of neural transmitters/modulators. On the other hand, the same acupuncture method may induce different effects on the body in different states.

It is worthy to clarify the difference between MA and EA. EA is a modified form of manual acupuncture. It is exhibited by an integrated effect of transcutaneous stimulation by electrical current. Most of the current studies use EA because it can be standardized by intensity, frequency, waveform, and length of the stimulation. Substantial data also show EA, even stimulation of peripheral nerves, can mimic traditional acupuncture with effective outcome. We made such observations in the 1980s (Kang and Xia 2010a, b; Wen et al. 2010a, b). Hsieh et al. (2000) reported that the electric stimulation of a peripheral nerve elicits an analgesic effect that is considered to be the characteristic effect of acupuncture. A major advantage of this approach is that different researchers can reproduce these results through a standardized stimulation protocol. However, some acupuncturists disagree with the substitution of acupuncture with EA. Many clinicians, especially those of the Oriental community, still prefer traditional MA instead of EA. Clearly, there are some differences between MA- and EA-induced effects on neural transmitters/modulators because of the difference in stimulation of acupoints with needles alone as opposed to needles + electricity. Yang et al. (2009a) studied acupuncture effect on visceral traction pain and observed an increased efficacy following a pretreatment with MA as compared to EA alone. Different effects of MA and EA have also been detected by several other studies (Yang et al. 2009a; Tong et al. 2007). It is sometimes difficult to distinguish the effect of needles at acupoints from that of needles at acupoints plus transcutaneous electricity (Lin and Chen 2008). More investigations are required to elucidate the MA- and EA-induced effect on neural transmitters/modulators.

Almost all the previous studies merely reported that MA/EA induced a change in neural transmitters/modulators but did not investigate the underlying mechanisms involved. The regulation of transmitters/modulators involves multiple steps at cellular and molecular levels, including biosynthesis, distribution, release, uptake, and degradation as well as changes in receptor expression and function. Acupuncture could induce one or several alterations in these steps. It will provide more insights into the mechanism of acupuncture by exploring this filed.

Since acupuncture condition is an important factor determining the responses of neural transmitters/modulators to acupuncture, it is essential to analyze the data based on this information. Unfortunately, some published articles did not describe the precise protocol involved, with respect to frequency, intensity, and wave width, whereas some other studies lacked a control group. Thus, it is difficult to compare the results of the grossly diverging studies by dissecting out methodological flaws. Additional pitfalls lay in the study designs and statistical analysis of some studies. All these drawbacks prevent us from making substantial conclusions, thereby forestall the progress of acupuncture research. We will discuss these issues in detail in the last chapter of this book (Chap. 23).

Nevertheless, currently available data have strongly demonstrated that both MA and EA can significantly modulate the activities of neural transmitters/modulators including their release, synthesis, and uptake, which certainly affect the function of the brain and the body in either physiological or pathophysiological conditions.

Acknowledgments This work was supported by the National Institutes of Health (AT-004422 and HD-034852) and the Vivian L. Smith Neurologic Foundation.

References

Cao W, Deng Y, Dong X, Wang Y, Lu Z (1993) Effects of electroacupuncture at different frequencies on the nociceptive response and central contents of GABA and glutamic acid in arthritic rats. Acupuncture Research 18: 48–52 (in Chinese with English abstract)

Chen H, Wu J, Li T et al (2008a) Effect of Low frequency electroacupuncture therapy on the density of hippocampus AchE and DA of addictive memory in Heroin psychological dependence of the mice. Journal of Chengdu University of Traditional Chinese Medicine 31(2): 29–33. (in Chinese with English abstract)

Chen J, Li J (2008b) Observations on the clinical therapeutic effect of Scalp Acupuncture on migraine and its regulating effect on CGRP and NO. Shanghai Journal of Acupuncture and Moxibustion. 27(12): 6–7 (in Chinese with English abstract)

Chen S, Gao Y, Meng F, Wang J et al (2008c) Effects of electroacupuncture with different parameters on body weight and contents of β- EP in plasma, hypothalamus and pituitary tissues in the rat. Liaoning Journal of Traditional Chinese Medicine 35(9): 1414–1416 (in Chinese with English abstract)

Chen S, Hu H, Bu Y (2008d) The effect of electroacupuncture on TXB2 and PGF1α in cerebral infarction patient. Chinese Journal of Traditional Medical Science and Technology 15(3): 166–167 (in Chinese)

Chen J, Li J (2009) Clinical study on therapeutic effect and adjustment to plasma CGRP and NO in the patient of essential hypertension with Scalp Penetration Acupuncture. Liaoning Journal of Traditional Chinese Medicine 36(9): 1570–1571 (in Chinese with English abstract)

Connor B, Dragunow M (1998) The role of neuronal growth factors in neurodegenerative disorders of the human brain. Brain Res Brain Res Rev 27:1–39

Cui L, Hou J (2008) Study of effect of electrical acupuncture on the animal model of acute gouty arthritis. Journal of Practical Traditional Chinese Medicine 24(6): 380–381 (in Chinese with English abstract)

Cui R, Ma C, Wang X, Tian Y (1990) The influence of P-chlorophenylalanine on the analgesia of electro-acupuncture and the level of SP in CNS of rats. Acupuncture Research 15(2): 109–111 (in Chinese with English abstract)

D'Antoni S, Berretta A, Bonaccorso C et al (2008) Metabotropic glutamate receptors in glial cells. Neurochem Res 33: 2436–2443

Dong H, Zhang Q, Yang Q et al (2007a) Experimental study of nerve growth factor of sciatic nerve of rats with diabetes peripheral neuropathy accommodated by electro-acupuncture. Chinese Journal of Rehabilitation Theory and Practice 13(8): 730–733 (in Chinese with English abstract)

Dong X, Luo T, Mao Q et al (2007b) Effect of electroacupuncture on GluR-2 expression of rat hippocampal neurons in brain ischemia. Chinese Journal of Histochemistry and Cytochemistry 16(6): 664–668 (in Chinese with English abstract)

Dougherty D, Kong J, Webb M et al (2008) A combined [11C]diprenorphine PET study and fMRI study of acupuncture analgesia. Behav Brain Res 193(1):63–8

Du Y, Tian Q, Sun G (2007) Effect of electrical needle by therapy of xiao-du tong-luo on contents of GLU and GABA in hippocampus area of cerebral hemorrhage rat Chinese Journal of the Practical Chinese with Modern Medicine 201(21): 1851–1852 (in Chinese with English abstract)

Fan L, Wang G, Yu Y (2007) Effects of electroacupuncture combined with intrathecal buprenorphine on expression of substance P and calcitonin gene-related peptide in spinal dorsal horn in rats with chronic inflammatory pain. Chin J Anesthesiol 27(4): 368–371 (in Chinese with English abstract)

Feng Y, He X, Yang Y et al (2012) Current research on opioid receptor function. Curr Drug Targets 13(2): 230–46

Frost JJ (2001) PET imaging of the opioid receptor: the early years. Nucl Med Biol 28(5): 509–513

Fu W, Liu J, Bai Y et al (2009) Effect of electroacupuncture on CREB-BDNF post receptor signal transduction pathway in hippocampus of depression rats. Chinese Journal of Gerontology 29: 3038–3043 (in Chinese with English abstract)

Gall C, Isackson P (1989) Limbic seizures increase neuronal production of messenger RNA for nerve growth factor. Science 245: 758–761

Givens J, Wiedemann E, Andersen R et al (1980) beta-Endorphin and beta-lipotropin plasma levels in hirsute women: correlation with body weight. J Clin Endocrinol Metab 50(5): 975–976

Guo S, Chen S, Li Y et al (2010) Effects of 9–11 Am and 1–3 Pm electroacupuncture on TXA2 and PGF1α and their quantitative relationships with the basic status in cerebral thrombosis patients. Shanghai J Acu-mox 29(8): 495–497 (in Chinese with English abstract)

Han J (2003) Acupuncture: neuropeptide release produced by electrical stimulation of different frequencies. Trends Neurosci 26(1): 17–22

Hao J, Li J, Wang M et al (2009) Effect of electroacupuncture on striatal neurotransmitter dopamine in a rat model of Parkinson's disease. Shanghai Journal of Acupuncture and Moxibustion 28(7): 418–420 (in Chinese with English abstract)

He G, Niu W, Liu H et al (2007) Effect of electrical acupuncture at "Zusanli" on the gastrointestinal myoelectric activity in rabbits and relationship between acupuncture effect and motilin and cholecystokinin. Chinese Journal Of Clinical Medicine 14(4): 504–506 (in Chinese with English abstract)

He L, Long D, Hu L et al (2008) Effects of electroacupuncture at different acupoints on plasma acetylcholine and noradrenalin in rabbits with acute myocardial ischemia. China Medical Herald 5(23): 23–25 (in Chinese with English abstract)

Hsieh C, Kuo C, Chen Y et al (2000) Analgesic effect of electric stimulation of peripheral nerves with different electric frequencies using the formalin test. Am J Chin Med 28: 291–299

Hu B, Liu L, Guo C et al (2008) Effect of acupotomy lysis on SP and CCK-8 contents in hypothalamus and spinal cord in rats with transverse process syndrome of the third lumbar vertebra. Acupuncture Research 33(1): 22–25 (in Chinese with English abstract)

Huang G, Zheng C, Yu W (2009) Involvement of connexin 43 in acupuncture analgesia. Chin Med J (Engl) 122(1): 54–60

Huang G, Zheng C, Zhang M (2005) Effect of acupuncture on expression of connexion 43 in "Zusanli" (ST-36) of the rat. Zhongguo Zhen Jiu 25(8): 565–568 (in Chinese with English abstract)

Huang G, Zhang H, Zhang T (2008a) Clinical study on treatment of Jiaji-balance-electroacupuncture in patients with lumbar disc herniation. China Journal of Rehabilitation Medicine 23(3): 235–237 (in Chinese with English abstract)

Huang X, Zhuo L, Ren L et al (2008b) Effect of electroacupuncture of "Hegu" (LI-4) "Taichong" (LR-3) on colonic nitric oxide synthase and glutathione peroxidase activity and nitric oxide content in depression rats. Acupuncture Research 33(3): 183–185 (in Chinese with English abstract)

Jin G, Han Y, Yu L et al (1979) The role of 5-HT and catecholamine in brain during acupuncture analgesia. Acta Physiol Sinica 31: 121–122 (in Chinese with English abstract)

Ji Y, Liu Y, Guo Z (2007) Clinical observation on electroacupuncture for treatment of vascular dementia. Chinese Archives of Traditional Chinese Medicine 25(5): 1070–1071 (in Chinese with English abstract)

Kang X, Xia Y (2010a) Effects of Acupuncture on Arrhythmia and Other Cardiac Diseases. In Xia et al (eds): Acupuncture Therapy for Neurological Diseases: A Neurobiological View, Springer-Tsinghua Press, Heidelberg, London, New York, Beijing, pp 263–288

Kang X, Xia Y (2010b) Acupuncture therapy for hypertension and hypotension. In Xia et al (eds): Acupuncture Therapy for Neurological Diseases: A Neurobiological View, Springer-Tsinghua Press, Heidelberg, London, New York, Beijing, pp 289–325

Komuro H and Rakic P 1996 Intracellular Ca2+ fluctuations modulate the rate of neuronal migration. Neuron 17: 275–85

Laird D (2009) The gap junction proteome and its relationship to disease. Trends Cell Biol 20(2): 92–101

Li D and Li M (2005) The effects of electrical acupuncture on 5-HT and regional cerebral blood flow in the rat model of ischemia reperfusion. Chinese Journal of Current Clinical Medicine 3(12): 1166–1167 (in Chinese)

Li H, Li X, Ruan H et al (1989) Relationship between the content of central substance P and the analgesic effect of electroacupuncture in rats. Acupuncture Research 14: 370–374 (in Chinese with English abstract)

Li H, Zhang H, Ma T et al (2007a) Effect of electro-acupuncture on 5-HT in dorsal raphe nucleus of mouse. Journal of Practical Traditional Chinese Internal Medicine 21(8): 23 (in Chinese)

Li M, Wang H, Cheng Z (2007b) Effect of electroacupuncture of Neiguan point on autonomic nerve, SP and NOS in AMI rats. China Journal of Traditional Chinese Medicine and Pharmacy 22(8): 543–546 (in Chinese with English abstract)

Li Y, Zhuang L, Lai X (2007c) Effects of electroacupuncture on plasma vasoactive intestinal peptide and substance P in perennial allergic rhinitis patients. Acupuncture Research 32(2): 136–138 (in Chinese with English abstract)

Li Z, Yi S (2008) Study progress in the relationship between brain gut peptide and acupuncture analgesia. Guiding Journal of TCM 14(5): 113–115 (in Chinese with English abstract)

Li M, Wang M, Mei Y et al (2009) Effect of electro-acupuncture in different time windows on the expression of Bcl-2 and brain derived neurotrophic factor in rats with cerebral infarction. Journal of Apoplexy and Nervous Diseases 26(5): 603–605 (in Chinese with English abstract)

Liang J, Chao D, Xia Y. The delta-opioid receptor agonist UFP-512 up-regulates excitatory amino acid transporters in astrocytes. Neuroscience Meeting Planner; Society for Neuroscience; San Diego, CA. 2010. Program No. 153.9/L1

Lin J, Chen W (2008) Acupuncture analgesia: a review of its mechanisms of actions. Am J Chin Med 36(4): 635–645
Lindvall O, Kokaia Z, Bengzon J (1994) Neurotrophins and brain insults. Trends Neurosci 17: 490–6
Liu F, Zheng C, Huang G et al (2008a) The effects of acupuncture on center and peripheral nervous system in rat model of menalgia. Zhejiang Journal of Traditional Chinese Medicine 43(8): 444–446 (in Chinese with English abstract)
Liu F, Xiong J, Huang G et al (2008b) Effects of acupuncture on the hypothalamus-pituitary-ovary axis in rats with dysmenorrhea. World J Acu-moxi 18(3): 26–35 (in Chinese with English abstract)
Liu Y, Li W, Miao J et al (2008c) Influences of electric acupuncture on malondialdehyde and acetylcholinesterase in cerebral tissues of rats with vascular demention. Tianjin Journal of Traditional Chin Medicine 25(6): 459–461 (in Chinese with English abstract)
Liu Y, Dong Y, Wang X et al (2009) The effects of acupuncture on plasma DA, NE, 5-HIAA in somatization disorder disease. Jiangsu Journal of Traditional Chinese Medicine 41(2): 2, 47–48 (in Chinese with English abstract)
Liu Y, Zhang H, Huang G et al (2010) Experimental study of scalp-acupuncture on BDNF expression in cerebral ischemia-reperfusion rats. Journal of Emergency in Traditional Chinese Medicine 19(5): 810–813 (in Chinese with English abstract)
LoTurco J, Blanton M, Kriegstein A (1991) Initial expression and endogenous activation of NMDA channels in early neocortical development. J Neurosci 11: 792–799
Lu J, Shi Y, Shi R et al (2007) Effect of electroacupuncture on nitric oxide and cyclic guanosine monophosphate levels in penile tissues of rats with depression. Journal of Beijing University of Traditional Chinese Medicine 30(7): 472–474 (in Chinese with English abstract)
Luo W, Liu H, Mei S et al (2007) Clinical study on "Jin's Three-needling" in treatment of generalized anxiety disorder. Chinese Journal of Integrated Traditional and Western Medicine 27(3): 201–203 (in Chinese with English abstract)
Luo Y, Xu N, Yi W et al (2011) Study on the correlation between synaptic reconstruction and astrocyte after ischemia and the influence of electroacupuncture on rats. Chin J Integr Med 2011 Jun 29. [Epub ahead of print]
Ma L, Zhao L, Cheng Y et al (2008a) The effect of Baihui point (DV-20) and Dazhui point (DV-14) on the SOD and the AChE contents in rats with aged dementia. Chinese Journal of Integrative Medicine on Cardio-/Cerebrovascular Disease 6(2): 166–167 (in Chinese with English abstract)
Ma P, Ji Z, Yu H et al (2008b) Effect of electroacupuncture on the learning and memory ability in diabetic rats with cognitive impairment and the expression of neurotrophin-3 in hippocampus. Chinese Journal of Comparative Medicine 18(8): 14–18 (in Chinese with English abstract)
Meng P, Sun G, Liu S et al (2008) Effect of electroacupuncture pretreatment on glutamate-NMDAR signal pathway in hippocampal neurons of vascular dementia rats. Acupuncture Research 33(2): 103–106 (in Chinese with English abstract)
Mi Y, Wu Y, Chen Y (2009) Effects of warm needle moxibustion on nerve root local inflammatory factors (NOS and CGRP) in the lumbar nerve root compress model rats. Chinese Acupuncture and Moxibustion 29(1): 48–52 (in Chinese with English abstract)
Niu W, Liu Z, Yang X et al (2009a) Effect of XIU three needles on learning - memory ability and CCK-8 content of hippocampus in vascular dementia rats. Journal of Sichuan of Traditional Chinese Medicine 127(16): 15–17 (in Chinese with English abstract)
Niu W, Liu Z, Yang X et al (2009b) Effect of combined acupuncture and medicine on learning-memory ability and free radical system of hippocampus in vascular dementia rats. Journal of Zhejiang University of Traditional Chinese Medicine 33(5): 425–427 (in Chinese with English abstract)
Peng Y, Yi S, Li Z (2009) Effect of acupuncture with varied intensity on NMDA and SP expression in spinal dorsal horn in rats with gastric distension-induced pain. World Chinese Journal of Digestology 17(13): 1339–1345 (in Chinese with English abstract)
Qi F (2007) The effects of electrical acupuncture on ET, TXA2, PGI2 and protecting the endotheliocytes in the blood plasma of SD rats with Hyperlipidemia. Journal of Clinical Acupuncture and Moxibustion 23(6): 44–46 (in Chinese with English abstract)
Ritter M, Richter W, Schwandt P (1992) Increase of beta-endorphin serum levels by human corticotropin-releasing factor does not affect beta-cell function in overweight women. Horm Metab Res 24(7): 349–50 (in Chinese with English abstract)

Saavedra A, Baltazar G, Duarte E (2008) Driving GDNF expression: the green and the red traffic lights. Prog Neurobiol 86(3): 186–215

Sáez J, Retamal M, Basilio D et al (2005) Connexin-based gap junction hemichannels: gating mechanisms. Biochim Biophys Acta 1711: 215–224

Shang C (1989) Singular point, organizing center and acupuncture point. Am J Chin Med 17(3–4): 119–27

Shao Y, Lai X, Gong Y et al (2008) Effects of electroacupuncture on plasma and cerebral somatostatin and β-EP contents and learning-memory ability in vascular dementia rats. Acupuncture Research 3(2): 98–102 (in Chinese with English abstract)

Skilling S, Smullin D, Larson A (1990) Differential effects of C- and N-terminal substance P metabolites on the release of amino acid neurotransmitters from the spinal cord: potential role in nociception. J Neurosci 10(4): 1309–1318

Sun H, Hu B, Li X et al (2008) Effect of acupotomy dissolution on pain threshold and ENK and β-EP in spinal cord and hypothalamus in rats with transverse syndrome of the third lumbar vertebrae. Journal of Beijing University of Traditional Chinese Medicine 3l(3): 176–179 (in Chinese with English abstract)

Tian XS, Zhou F, Yang R, Xia Y, Wu GC, Guo JC (2008) Electroacupuncture protects the brain against acute ischemic injury via up-regulation of delta-opioid receptor in rats. Zhong Xi Yi Jie He Xue Bao 6(6): 632–638 (in Chinese with English abstract)

Tsai H, Lin J, Inoki R (1989) Further evidence for possible analgesic mechanism of electroacupuncture: effects on neuropeptides and serotonergic neurons in rat spinal cord. Jpn J Pharmacol 49(2): 181–185

Tong S, Wang S, Sun Z, Liu J (2007) Changes in endogenous neurotrophic factor and its receptor after acupuncture. Journal of Clinical Rehabilitative Tissue Engineering Research 11(23): 4583–4586 (in Chinese with English abstract)

Tu W, Lou X, Jiang S et al (2008a) Effect of electroacupuncture of local plus distal acupoints in the same segments of spinal cord on spinal substance P expression in rats with chronic radicular pain. Acupuncture Research 33(1): 7–12 (in Chinese with English abstract)

Tu W, Jiang S, Lou X et al (2008b) Effect of low frequency transduction pulse electro-acupuncture on substance P in dorsal root ganglion in the rats with chronic radicular pain. Journal of Wenzhou Medical College 38(5): 429–431 (in Chinese with English abstract)

Wang C, Fang J, Shao X et al (2007a) Intervention of electroacupuncture on Jianyu acupoint to rabbit model with scapulohumeral periarthritis. Journal of Zhejiang University of Traditional Chinese Medicine 31(4): 478–481 (in Chinese with English abstract)

Wang Q, Tang C, Chen X et al (2007b) Effect of acupuncture on dopaminergic neuron apoptosis in rat model of Parkinson's Disease. Chinese Journal of Basic Medicine in Traditional Chinese Medicine 13(8): 621–622 (in Chinese with English abstract)

Wang C, Tian Y, Zhou D et al (2010) Influence of electroacupuncture on myocardial NO and NOS and intracellular Ca(2+) contents in myocardial ischemia-reperfusion injury rats. Acupuncture Research 35(2): 113–117 (in Chinese with English abstract)

Wang G, Wang M, Han Y (2008a) Effects of acupuncture on M-Enk in rat model of abstinence. Chinese Journal of Modern Drug Application 2(5): 38–39 (in Chinese with English abstract)

Wang S, Li W, Ge J et al (2008b) Effect of cholecystokinin on energy metabolism of obesity rats with acupuncture. Chinese Journal of Gerontology 12(28): 2314–2316 (in Chinese with English abstract)

Wang T, Wu J, Zhong Y et al (2008c) Influence of different electroacupuncture time on SOD and NO contents in rats with acute spinal cord injury. Journal of Beijing University of Traditional Chinese Medicine (Clinical Medicine) 15(6): 1–3 (in Chinese with English abstract)

Wang Z, Wu Z, Liu H et al (2008d) Effect of acupuncture on monoamines in rat model of vascular dementia. Chinese Journal of Practice Medicine 7(2): 116–117 (in Chinese with English abstract)

Wang H, Jiang J, Cao X (1994a) Changes of norepinephrine release in rat's nucleus reticularis paragigantocellularis lateralis in acupuncture analgesia. Acupuncture Research 19(1): 20–25 (in Chinese with English abstract)

Wang H, Wu G, Jiang J et al (1994b) Effect of norepinephrine in rat's nucleus reticularis paragigantocellularis lateralis during acupuncture analgesia. Journal of Shanghai Medical University 21: 409–413 (in Chinese with English abstract)

Wang J, Chen B, Yin J et al (2009a) Effect of electroacupuncture of different acupoints on the excitability of detrusor muscle and the expression of BDNF and TrkB in the spinal cord of rats with urinary retention due to spinal cord injury. Acupuncture Research 34(6): 387–392 (in Chinese with English abstract)

Wang L, Zhou L (2009b) Effects of electro-acupuncture treatment for AD model rats in the content of Ach, ChAT and AchE. Journal of Clinical Acupuncture and Moxibustion 25(6): 40–42 (in Chinese with English abstract)

Wang W, Liu L, Xu S (2009c) Study of correlation between plasma SP and VIP level in STC patients with acupuncture therapy and colonic irrigation. Chongqing Medical Journal 38(18): 2319–2320 (in Chinese with English abstract)

Wang L (2010) The effect of electroacupuncture of different point combinations of Zusanli and Zhongwan on NO, GAS, EGF in acute gastric mucosal injury in rats. Journal of Clinical Acupuncture and Moxibustion 26(7): 59–61 (in Chinese with English abstract)

Wang S, Sun Z, Tong S et al (2005) Effects of acupuncture on monoamine neurotransmitters in brain tissue of experimental rat models of sciatic nerve compression. Chinese Journal of Clinical Rehabilitation 9(33): 94–96 (in Chinese with English abstract)

Wang W, Fan H (1995) Effect of acupuncture on amino acid and peptide neurotransmitter. Journal of Jiangxi College of Traditional Chinese Medicine 7: 14–16 (in Chinese)

Wen G, He X, Lu Y, Xia Y (2010b) Effect of acupuncture on neurotransmitters/modulators. In Xia et al (eds): Acupuncture Therapy of Neurological Diseases: A Neurobiological View, Springer-Tsinghua Press, Heidelberg, London, New York, Beijing, pp 120–142

Wen G, Yang Y, Lu Y, Xia Y (2010a) Acupuncture-induced activation of endogenous opioid system. In Xia et al (eds): Acupuncture Therapy of Neurological Diseases: A Neurobiological View, Springer-Tsinghua Press, Heidelberg, London, New York, Beijing, pp 104–119

Wu C, Liu W, Liu J et al (2008a) Effect of warm needling moxibustion on pain index and substance P in patients with lumbar intervertebral disc protrusion. Chinese Journal of Traditional Medical Traumatology and Orthopedics 16(8): 35–36 (in Chinese with English abstract)

Wu C, Liu H, Xie X (2008b) Impact on SOD MDA and AChE of brain tissue of rats with vascular dementia by the treatment of "Praefect Yinao" acupuncture. Chinese Archives of Traditional Chinese Medicine 26(8): 1830–1832 (in Chinese with English abstract)

Wu Y and Yu Y (2005) Effect of acupuncture to the Ultra-micro structure of the injured rat lumbar nerve and inflammatory medium. Journal of Acupuncture and Tuina Science 3(12): 44–46

Xia Y, Cao X, Wu G et al (2010) Acupuncture Therapy for Neurological Diseases: A Neurobiological View. Springer-Tsinghua Press, Heidelberg, London, New York, Beijing

Xue M (2009) The relationship between the improvement of Han's acupuncture on the learning and memory disorder of heroin -dependent rats and hippocampus NOS and AchE. Acta Academiae Medicinae Xuzhou 29(2): 83–86 (in Chinese with English abstract)

Xu S, Peng C, Pan X et al (2010) Impacts on PGE_2, EGF and TGF-α in rats with acute gastric mucosal injury treated with acupuncture on Zusanli (ST36) at different time division. World Journal Of Integrated Traditional and Western Medicine 15(16): 495–497 (in Chinese with English abstract)

Yan L, Xu H, Ma C (2009) The role of electroacupuncture on the concentration of glutamate and substance P within related areas of the spinal cord and its effect on hyperalgia in rats with SNI. Acta Chinese Medicine and Pharmacology 137(14): 72–76 (in Chinese with English abstract)

Yang H, Guo T, Min Y et al (2009a) Effect of emulation acupuncture pretreatment on the behavior of rats with visceral traction pain and its mechanism research. Shanghai Journal of Acupuncture and Moxibustion 28(12): 732–737 (in Chinese with English abstract)

Yang X, Niu W, Liu Z et al (2009b) Effect of acupuncture and medicine on CHAT, ACHE in rat model of vascular dementia. Traditional Chinese of Shaanxi 30(10): 1413–1414 (in Chinese with English abstract)

Yi S, Zhou G, Lin Y et al (2009) The effect of acupuncture at Neiguan point on expression of 5-hydroxytryptamine in the dorsal spinal cord and dorsal root ganglion in myocardial ischemia/reperfusion injury in rats. Chin J TCM WM Crit Care 16(4): 195–199 (in Chinese with English abstract)

Yu W, Huang G, Zhang M et al (2008) Influence of connexin 43 gene knockout on the analgesic effect of acupuncture in visceral pain mice. Acupuncture Research 33(1): 3–6 (in Chinese with English abstract)

Yu Z, Wang P, Tan X et al (2007) The effects of microinjection of acetylcholine into nucleus tractus solitarius on electroacupuncture against stress mucosa injury in rats. Journal of Xianning University (Medical Sciences) 21(3): 211–214 (in Chinese with English abstract)

Yuan Q, Xiao X, Liang C et al (2008) Experimental study on therapeutic effect of electroacupuncture on benign prostate hyperplasia in rats. Chinese Journal of Rehabilitation Medicine 23(6): 532–534 (in Chinese with English abstract)

Zeng X, Zeng Y, Yang J et al (2009) Effect of electric-needling acupoint Siguan on AchE and SOD in brain of AD model rats. Medical Journal of the Chinese People's Armed Police Forces 20(4): 297–299 (in Chinese with English abstract)

Zhan S, Zhao Y, Guo X et al (2007) Influence of electroacupuncturing the "Zusanli" acupoint on expression of substance P in the brain of rat. Journal of Shandong University of Traditional Chinese Medicine 31(6): 492–495 (in Chinese with English abstract)

Zhang H, Yi S, Chang X et al (2009) Electroacupuncture effects on Oddi's sphincter movement and related brain-gut peptide in rabbits. Journal of Clinical Rehabilitative Tissue Engineering Research 13(7): 1329–1333 (in Chinese with English abstract)

Zhang H, Huang G, Zhou L (2007) The influence of electro - acupuncture (EA) on the content of β - endorphine in medullary bulb of myocardial ischemia reperfusion injury. Journal of Clinical Acupuncture and Moxibustion 23(7): 60–61 (in Chinese)

Zheng M, He F, Zai F et al (2009) Clinical observation of acupuncturing at Die'e point on allergic rhinitis and its effect on serum substance P. Journal of Fujian College of Traditional Chinese Medicine 19(4): 44–46 (in Chinese with English abstract)

Zheng S, Jiao J, Liu M et al (2010) Effect of acupuncture and mental imagery on upper limbs in stroke. Journal of Hubei College of Traditional Chinese Medicine 12(3): 48–50 (in Chinese with English abstract)

Zhong G and Li W (2007) Effects of acupuncture on 5-hydroxytryptamine$_{1F}$ and inducible nitric oxide synthase gene expression in the brain of migraine rats. Journal of Clinical Rehabilitative Tissue Engineering Research 11(29): 5761–5764 (in Chinese with English abstract)

Zhou G, Yi S, Lin Y et al (2009) Effects of different acupuncturing in Neiguan on reperfusion injury of MI and β-EP in rats. Journal of Traditional Chinese Medicine University of Hunan 29(4): 70–73 (in Chinese with English abstract)

Zhu Q (2007) On the effect of electroacupuncture on rats' morphine dependence. Journal of Anqing Teachers College (Natural Science Edition) 13(1): 98–100 (in Chinese with English abstract)

Chapter 2
Cellular Mechanisms in Acupuncture Points and Affected Sites

Wolfgang Schwarz and Quanbao Gu

Abstract The objective of our work is the elucidation of mechanisms underlying initiation, transmission and the final effects of acupuncture and moxibustion. In this chapter, we shall focus mainly on possible cellular events in tissue affected by Chinese medicine treatments using basically electrophysiological techniques combined with molecular biological and radioactive tracer techniques to measure activity of transmembrane transport.

In the first part we will review work of our laboratory suggesting that acupuncture-induced pain suppression involves interaction of δ-opioid receptor (DOR) with the neurotransmitter transporters for glutamate (EAAC1), γ-amino butyric acid (GAT1) and the sodium pump (Na^+, K^+-ATPase). Reduced activity of the transporters by co-expression of DOR resulted from intermolecular interaction with DOR. In addition, EAAC1 became stimulated in response to DOR activation, while GAT1 became inhibited. The Na^+, K^+-ATPase was not affected by DOR activation, but higher sensitivity to DOR agonist was found in response to sodium pump stimulation. Since endorphins are released in response to acupuncture, the effects described here may contribute to acupuncture-induced pain suppression.

W. Schwarz (✉)
Laboratory of Cellular Electrophysiology, Shanghai Research Center for Acupuncture and Meridians, Shanghai 201203, China

Institute for Biophysics, Goethe University Frankfurt am Main, Max-von-Laue-Str. 1, 60438 Frankfurt am Main, Germany
e-mail: wolfgang.schwarz@email.de

Q. Gu
Laboratory of Cellular Electrophysiology, Shanghai Research Center for Acupuncture and Meridians, Shanghai 201203, China
e-mail: guquanbao@126.com

Y. Xia et al. (eds.), *Current Research in Acupuncture*,
DOI 10.1007/978-1-4614-3357-6_2, © Springer Science+Business Media New York 2013

In the second part we will review work on effects of drugs on membrane transporters. In treatment of asthmatic rats by acupuncture cyclophilin A (CyPA) becomes over-expressed. Release of airway smooth muscle contraction by CyPA-induced inhibition of Na^+, Ca^{2+} exchanger may be the underlying mechanism. As an example for effects of Chinese herb extracts on the transporters described in Sect. 2.1, we will review the effect of *Acorus* extract on EAAC1 function showing that α-asarone effectively inhibits EAAC1-mediated current but stimulates glutamate transport. These effects may contribute to reduced excitatory activity supplementary to acupuncture.

Conclusion: The generally accepted effect of acupuncture is pain relief. We suggest that modulation of central nervous synaptic activity may be involved through indirect modulation of the activity of neurotransmitter transporters by endorphins. Membrane transporters may also be the target for drugs of Chinese herbs and for endogenous acupuncture-induced "drugs."

Keywords Neurotransmitter transporter • Opioid receptor • Na pump • Na^+, Ca^{2+} exchanger • Xenopus oocyte expression system • Mast cell • Electrophysiology

2.1 Introduction

Acupuncture/moxibustion but also treatments with Chinese herbal formulae are methods of Traditional Chinese Medicine. Stimulation of organ-specific acupuncture points (acupoints), located close to the body surface, leads to signals that seem to spread along a network of conduits beneath the so-called meridians to the affected sites (compare Fig. 2.1). Chinese medicine treatments may interfere with all of these structures. From the point of view of Western medicine, we are interested in the physiological and cellular counterparts of acupoints, conduits and affected sites. In our previous work, our laboratory has focused on exploration of cellular events occurring in acupoints (Zhang et al. 2008; Wang et al. 2010) as well as in affected sites (Xia et al. 2006; Deng et al. 2009).

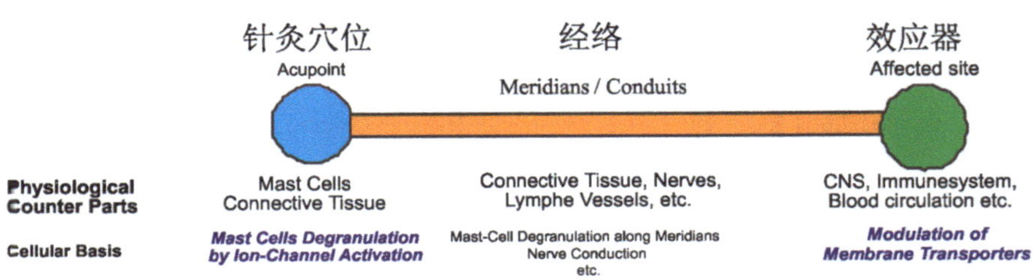

Fig. 2.1 Simplified scheme for underlying structures and mechanisms in acupuncture effects. Some examples of possible physiological and cellular counterparts are listed

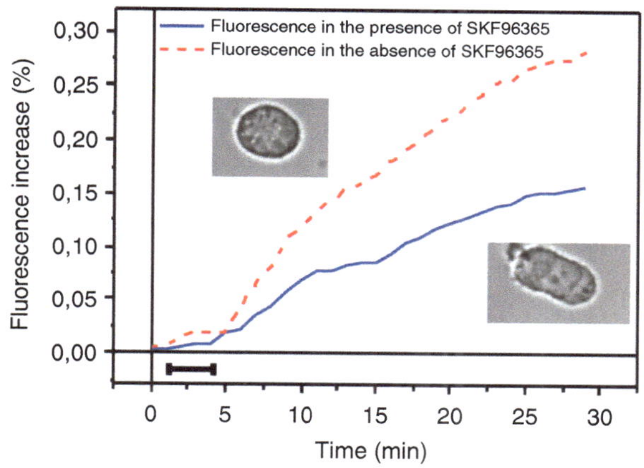

Fig. 2.2 Noxious heat (53°C) induced intracellular Ca^{2+} increase and associated mast-cell degranulation. Relative Calcium Green-1 AM (4 mM) fluorescence increase in HMC1 mast cells is shown after subtraction of background fluorescence. The *bar* indicates the time period of heat application. A selected cell before (*upper left inset*) and 20 min after the heat application (*lower right inset*) illustrates the heat-induced degranulation. [Zhang, D., Spielmann, A., Wang, L., Ding, G.H., Gu Q.B., Schwarz, W. Activation of mast-cell degranulation by different physical stimuli involves activation of transient-receptor-potential channel TRPV2 (unpublished)]

It seems quite established by now that degranulation of mast cells (Fig. 2.1) in response to mechanical stimulation of acupoints is a prerequisite for acupuncture-induced pain suppression in rat (Zhang et al. 2008). Also stimulation by noxious heat, as applied during moxibustion, or by red laser light, as applied in laser acupuncture, leads to mast-cell degranulation (Zhang et al. 2012).

In particular for the human mast-cell line HMC1, the transient potential receptor (valinoide-sensitive isoform 2), the TRPV2, had been demonstrated to be activated by these physical stimuli and to contribute to the mast-cell degranulation (Zhang et al. 2012). It could be shown that also activation of additional ion-selective channels that facilitate Ca^{2+} entry forms the basis for the mast-cell degranulation (Zhang et al. 2012; Wang et al. 2010).

Figure 2.2 shows the increase of intracellular Ca^{2+} in response to heat application using Calcium Green-1 AM as a Ca^{2+} indicator.[1] Part of the Ca^{2+} increase can be blocked by the TRPV2 channel blocker SKF96365. The two micrographs in Fig. 2.2 show the same cell before and 20 min after heat application. Similar results were obtained for a Cl^--permeable channel (Wang et al. 2010). Events occurring in acupoints are also dealt with in another chapter of this book by Ding et al. (Chap. 3).

[1] Zhang, D., Spielmann, A., Wang, L., Ding, G.H., Gu Q.B., Schwarz, W. Activation of mast-cell degranulation by different physical stimuli involves activation of transient-receptor-potential channel TRPV2 (unpublished)

Communication of mast cells with nerve endings is well known (Bauer and Razin 2000) and may form the next subsequent step within the acupoints for transmission along the conduits of the acupuncture stimuli to the affected organ (Fig. 2.1). This step is subject to current investigations in our laboratory. In this chapter we now like to review some of our speculations on acupuncture-induced effects in organ tissue. In Sect. 2.2 we will focus on molecular events that could contribute in the central nervous system (CNS) to acupuncture-induced pain suppression; this will be, in addition to neurotransmitter transporters, the Na^+, K^+-ATPase. In Sect. 2.3 we like to present some data on effects of selected drugs on membrane transporters. As an example of an endogenous "drug," we will present data showing a protein becoming over-expressed after acupuncture treatment of asthmatic rat inhibits Na^+, Ca^{2+} exchanger. In addition, we will show by another example that the transporters dealt with in Sect. 2.2 as target for acupuncture may also be a target for extracts of Chinese herbs.

2.2 Possible Role of Membrane Transporters in Pain Suppression by Acupuncture

Neurotransmitter transporters play a key role in the regulation of synaptic transmission (for illustration see Fig. 2.3). On the arrival of an action potential at a nerve terminus of the presynaptic neuron, neurotransmitter is released in response to Ca^{2+} entry. The transmitter will subsequently bind to and activate ionotropic and/or G-protein-coupled receptors at the postsynaptic membrane. This will result in depolarisation or hyperpolarisation of the membrane, depending on whether an excitatory or inhibitory transmitter, respectively, is released at the particular synapse. To terminate synaptic transmission, the neurotransmitter needs to be removed, and this is achieved by highly efficient, Na^+-gradient-driven neurotransmitter transporters in the presynaptic neuron and surrounding glia cells.

Glutamate and GABA are the dominating excitatory and inhibitory neurotransmitters in the mammalian brain, respectively. The predominate transporters controlling glutamate and GABA in the CNS are the excitatory neurotransmitter transporter EAAC1 (or also named EAAT3) and the GABA transporter GAT1.

It is generally accepted that pain sensation can be suppressed by acupuncture and that regulation of the glutamatergic and the GABAergic systems is involved in pain sensation. It could be demonstrated that inhibition of excitatory amino acid (EA) receptors resulted in pain suppression (Zhang et al. 2002). Reduction of EA-receptor activity may also be achieved by reduced glutamate concentration in the synaptic cleft, and reduction of glutamate concentration can be achieved by stimulating EAAC activity (compare Fig. 2.4).

In analogy to stimulation of EAAC1, we may expect for the GABAergic system that inhibition of the GABA transporter will result in elevation of GABA concentration in the synaptic cleft and hence in stimulation of GABA receptor activity (see Fig. 2.4); this could contribute to increased inhibitory synaptic transmission and also to reduced pain sensation. Indeed, experiments with transgenic mice with knockout or over-expressed GABA transporters GAT1 have demonstrated that the GAT1 is correspondingly involved in pain sensation (Hu et al. 2003).

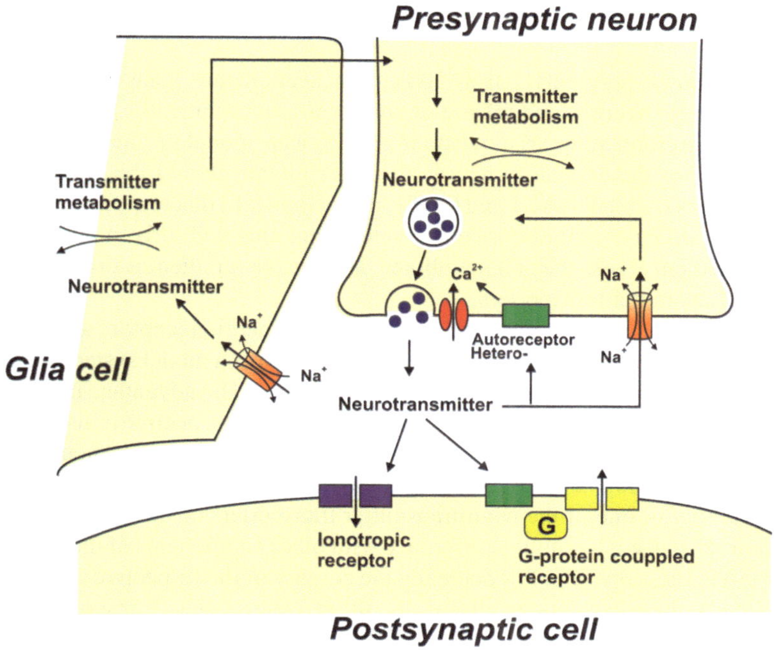

Fig. 2.3 Schematic illustration of processes at a synapse. The release of neurotransmitter from the presynaptic nerve terminus is initiated by the activation of Ca^{2+} channels. The neurotransmission is based on the subsequent activation of postsynaptic receptors. Neurotransmission is terminated by Na^{+}-driven neurotransmitter transporters

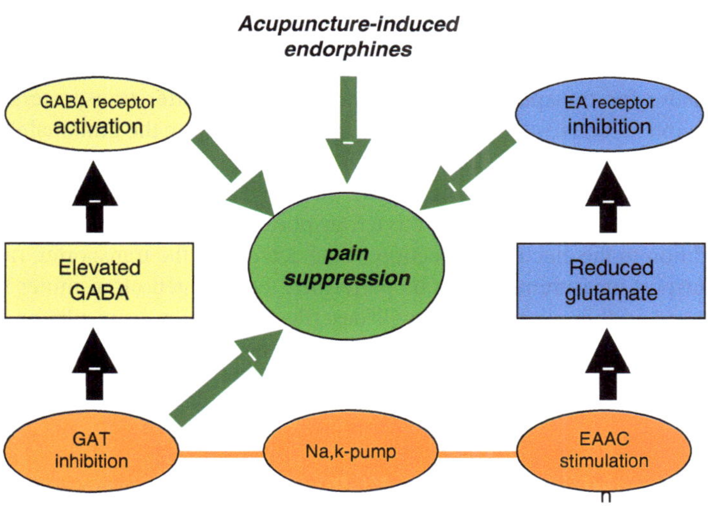

Fig. 2.4 Illustration of possible involvement of glutamatergic and GABAergic pathways in pain suppression. It was already demonstrated that inhibition of excitatory amino acid (EA) (Zhang et al. 2002) and reduced GAT1 activity (Hu et al. 2003) result in reduced pain sensation. Based on the knowledge that acupuncture-induced release of endorphins is involved in pain suppression, we may speculate that acupuncture interferes with these two neuronal systems via activation of opioid receptors

It was shown previously (Han 2004) that acupuncture leads to activation of enkephalinergic neurons and release of endogenous morphine, the endorphins. The question, therefore, may arise whether opioids can modulate the activity of the transporters for the excitatory neurotransmitter glutamate and the inhibitory neurotransmitter GABA, EAAC1 and GAT1, respectively. In this respect the activity of the Na^+, K^+-ATPase should not be neglected since this ATP-driven pump controls the Na^+ gradient across the cell membrane, and the Na^+ gradient is driving force for the neurotransmitter transporters.

To investigate possible interference of enkephalinergic receptors and the membrane transport proteins, *Xenopus* oocytes can be used as a model system to measure the activity of membrane transporters (Schwarz 2001). The advantage of this model system compared to the native cell is that function of the respective transport protein can be studied independent of the interference with other neuronal membrane proteins. With heterologously co-expressed opioid receptors, influence of the receptor on the activity of one of the neurotransmitter transporters or the sodium pump can be characterised, e.g. without interference with other components of the glutamatergic or GABAergic system. For detecting the activity of the respective transport protein, steady-state currents mediated by these electrogenic transporters can be recorded in voltage-clamp experiments (Schwarz 2001). The Na^+, K^+-ATPase pumps 3 Na^+ ions out of the cell and 2 K^+ ions into the cell [see, e.g. (Schwarz and Gu 1988)] per ATP molecule split and hence generates outward-directed net current. The neuronal glutamate transporter EAAC1 transports for 1 glutamate 3 Na^+ ions together with 1 H^+ into the cell and 1 K^+ out of the cell [see, e.g. (Zerangue and Kavanaugh 1996)], and hence generates inward-directed net current. The GABA transporter GAT1 transports for 1 GABA 2 Na^+ ions and 1 Cl^- into the cell [see, e.g. (Hilgemann and Lu 1999; Kavanaugh et al. 1992; Keynan and Kanner 1988; Krause and Schwarz 2005)], and hence also generates inward-directed net current. These currents can be measured as steady-state currents under voltage clamp and can serve as a measure for transporter activity. The neurotransmitter transporters can also operate in ion-selective channel modes without transporting the glutamate (Wadiche et al. 1995; Slotboom et al. 2001) or GABA [see, e.g. (Risso et al. 1996; Eckstein-Ludwig et al. 2000; Krause and Schwarz 2005)]. These currents contribute to the total current mediated by the respective transporter.

In addition to the electrical detection of the activity of the transporter, radioactive tracer transport measurements can be performed. For the neurotransmitter transporters discussed in this overview, we will describe in addition to results on electrical measurements also results on uptake measurements of ^3H-labelled glutamate [see, e.g. (Xia et al. 2006)] and GABA [see, e.g. (Krause and Schwarz 2005)], respectively.

2.2.1 Interaction of the Na^+, K^+ Pump and Co-expressed δ-Opioid Receptors (Deng et al. 2009)

Expression of δ-opioid receptor (DOR), but not other opioid receptors, resulted in reduced activity of the endogenous sodium pump in *Xenopus* oocytes [Fig. 2.5

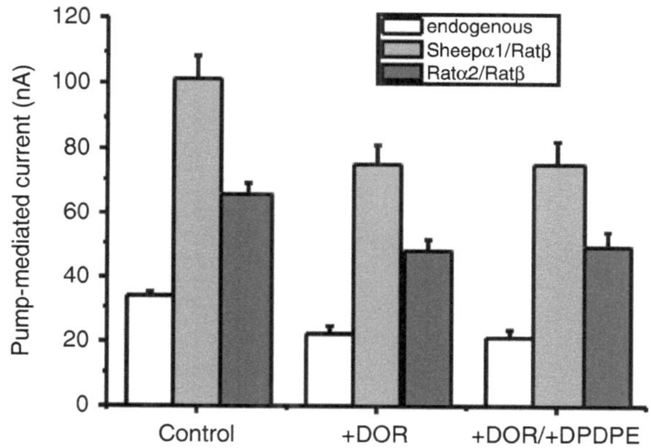

Fig. 2.5 Effect of co-expressed DOR on sodium pump current. Results are from three different isoforms and at −60 mV in *Xenopus* oocytes, the endogenous pump and heterologously expressed sheep and rat pumps. Data represent means ± SEM from 3 to 14 oocytes (based on data presented by Deng et al. 2009)

(open bars)]. Stimulation of DOR by 100 nM of the DOR agonist [D-Pen2,5]-encephalin (DPDPE) had no pronounced additional effect on pump activity. Qualitatively similar results were obtained for exogenously co-expressed sodium pumps (e.g. *Sheep* α1/*Rat* β1 or *Rat* α2/*Rat* β1, see Fig. 2.5).

The authors suggested that the reduction of pump activity in response to co-expression of DOR resulted from direct protein–protein interaction. Co-localisation of DOR and sodium pump could be demonstrated not only in the model system *Xenopus* oocytes but also in hippocampal neurons of rat by co-immunoprecipitation suggesting the physiological significance of the interaction of DOR with sodium pump. The reported reduction of sensitivity of DOR to the agonist DPDPE in response to pump activation (Deng et al. 2009) was also attributed to this interaction; for the *Sheep* α1/*Rat* β pump, e.g. the apparent affinity for DPDPE decreased from 5.9 ± 0.6 to 3.7 ± 0.5 μM^{-1} in response to activation of the sodium pump. To which extent the change in affinity has physiological significance needs further investigation.

2.2.2 *Interaction of the Glutamate Transporter EAAC1 and Co-expressed δ-Opioid Receptors (Xia et al. 2006)*

DOR co-expressed with the neuronal glutamate transporter EAAC1 in *Xenopus* oocytes, but not μ-opioid receptor, down-regulated EAAC1 function, and stimulation of DOR by 100 nM DPDPE could counteract the down-regulated function of EAAC1, as shown for glutamate uptake into *Xenopus* oocytes in Fig. 2.6.

As for the Na^+, K^+-ATPase and DOR, direct interaction of EAAC1 and DOR was suggested. Co-localisation of the two membrane proteins could be demonstrated by immunofluorescence microscopy and co-immunoprecipitation in the *Xenopus* oocytes. This could also be demonstrated in hippocampal neurons of rat supporting the physiological significance of the DOR–EAAC1 interaction. As for the reduced sodium pump activity, reduced EAAC1 function may be the result of the direct

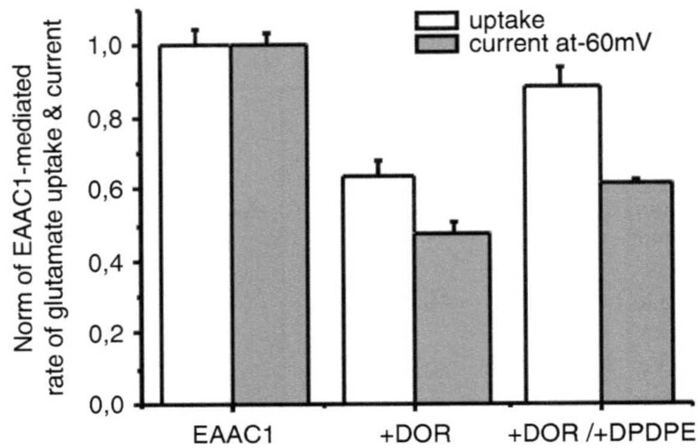

Fig. 2.6 Effect of co-expressed DOR on EAAC1-mediated uptake and current. DOR and EAAC1 were heterologously expressed in *Xenopus* oocytes. Data are normalised to uptake/current at −60 mV without co-expressed DOR and represent means ± SEM of at least 20 oocytes for uptake and 12 oocytes for current measurements (based on data presented by Xia et al. (2006)

interaction. The increased EAAC1 activity in response to activation of DOR may be the result of release of the inhibitory DOR–EAAC1 interaction.

2.2.3 *Interaction of the GABA Transporter GAT1 and Co-expressed δ-Opioid Receptors*

Co-expression of DOR with GAT1 led to a reduction in the number of fully functional transporters though the total number of expressed transporter was not changed as could be demonstrated by analysing transient charge movements by GAT1 and Western blot.[2]

In addition to this reduction in the number of functional transporters, rate of substrate translocation became dramatically reduced; also the current mediated by GAT1 became reduced by similar extent (Fig. 2.7). Activation of DOR by 100 nM DPDPE had no significant effect on rate of uptake, but current became slightly, but significantly, reduced.

In preliminary experiments,[3] inhibition of GAT1-mediated current in response to DOR activation could also be demonstrated in rat brain slices. The midbrain periaqueductal gray (PAG) is an important part of the CNS for controlling analgesia. Figure 2.8 shows the current mediated by GAT1 from a single neuron in a PAG slice of rat. Application of 100 nM DPDPE could completely block the current.

[2] Pu L., Xia, P., Fucke, T., Gu, Q.B., Pei, G., Schwarz, W. Regulation of GABA transporter by expression and activation of δ-opioid receptor (unpublished)

[3] Ren S.L., Gu, Q.B., Ding, G.H., Schwarz, W. Modulation of GAT1-mediated current in PAG neuron by DOR activation (unpublished)

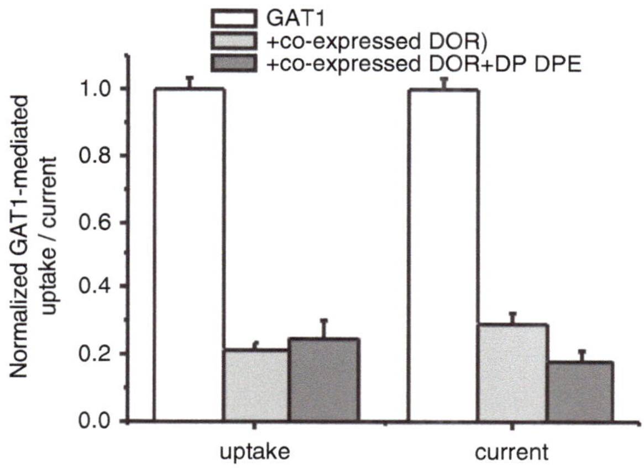

Fig. 2.7 Effect of co-expressed DOR on GAT1-mediated uptake and current. Data represent averages of 38–71 oocytes for uptake measurements and at least 6 oocytes for the current measurements at –60 mV. [Pu L., Xia, P., Fucke, T., Gu, Q.B., Pei, G., Schwarz, W.: Regulation of GABA transporter by expression and activation of δ-opioid receptor (unpublished)]

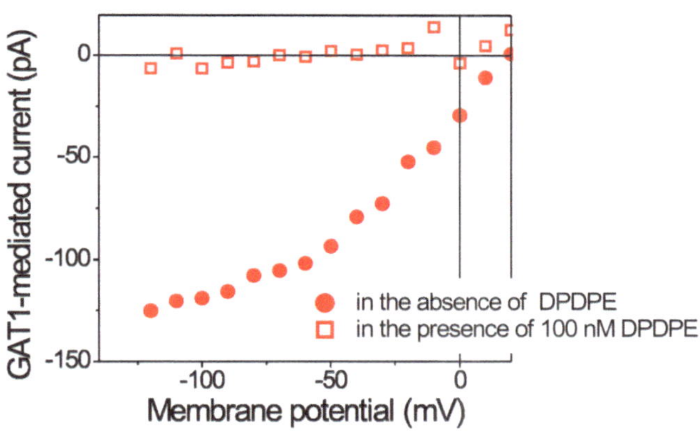

Fig. 2.8 Effect of DOR activation by DPDPE on GAT1-mediated current in rat brain slice from PAG area. Whole-cell patch-clamp recordings. The GAT1-mediated current was determined as Tiagabine-sensitive steady-state current. Presented data are from a single experiment, but similar results were obtained in four other PAG slices. [Ren S.L., Gu, Q.B., Ding, G.H., Schwarz, W. Modulation of GAT1-mediated current in PAG neuron by DOR activation (unpublished)]

In conclusion, we suggest that DOR can specifically and directly interact with a variety of membrane transport systems including the Na^+, K^+-ATPase, EAAC1 and GAT1. This interaction leads for all three transport systems to reduced transport. The sodium pump not only affects neuronal activity directly but also modulates through functional interaction with DOR the sensitivity of DOR to agonist, and hence sodium pump modulation may also modulate pain sensation. For GAT1, activation of DOR leads to inactivation of GAT1-mediated current; in PAG brain slices the current could even completely be blocked indicating complete inhibition of GAT1 activity in response to DOR activation. The stimulation of EAAC1 by DOR activation, via release of the inhibitory DOR–EAAC1 interaction, is opposite to the GAT1 inhibition. These effects of activation of DOR with GAT1 inhibition and EAAC1 stimulation would fit to the idea of acupuncture-induced pain suppression as illustrated in Fig. 2.4. In addition to the interaction of DOR with membrane transporters, the interaction between DOR and voltage-gated Na+ channels had been suggested and is also dealt with in Chap. 6.

2.3 Effects of Drugs on Membrane Transporters Modulated by Acupuncture

In the previous section we have described that membrane transport systems can be a target for acupuncture stimuli by interaction with opioid receptors that may become activated by acupuncture-induced elevation of endorphins. In this second section we will report that membrane transporters can also be targets for drugs, endogenous acupuncture-induced drugs that act on membrane transporter as well as components of Chinese medicine extracts.

2.3.1 *CyPA as an Acupuncture-Induced Inhibitor of Na^+, Ca^{2+} Exchanger*

In treatment of asthmatic rats by acupuncture, a protein could be identified that became over-expressed (Wang et al. 2009b). This protein turned out to be cyclophilin A (CyPA). If CyPA were involved in relief from asthma, we may ask for the target CyPA interacts with. Airway smooth muscle (ASM) cells have been an important target for asthma treatment in the past using drugs that interact with various plasma membrane receptors. Since smooth muscle contraction is governed by cytosolic Ca^{2+} (Ca_i^{2+}), control and modulation of Ca_i^{2+} are essential for the status of ASM cell contraction. Hence, all pathways involved in Ca_i^{2+} regulation may be likely candidate targets (for a review see Janssen 2009), in particular also for CyPA. An important protein involved in the regulation of Ca_i^{2+} is the Na^+, Ca^{2+} exchanger (NCX) that operates in a $3Na^+:1Ca^{2+}$ stoichiometry. In its normal mode, NCX uses the inward-directed gradient for Na^+ to remove Ca^{2+} from the cytoplasm. NCX can also operate in a reversed mode (Hirota et al. 2007) pumping Ca^{2+} into the cytoplasm. Muscle contraction is governed by release of Ca^{2+} from the sarcoplasmic reticulum (SR), and the refilling depends on the availability of Ca_i^{2+}. The reversed mode of the NCX of the plasma membrane has been demonstrated to contribute to refilling the Ca^{2+} stores in ASM cells (Hirota et al. 2007).

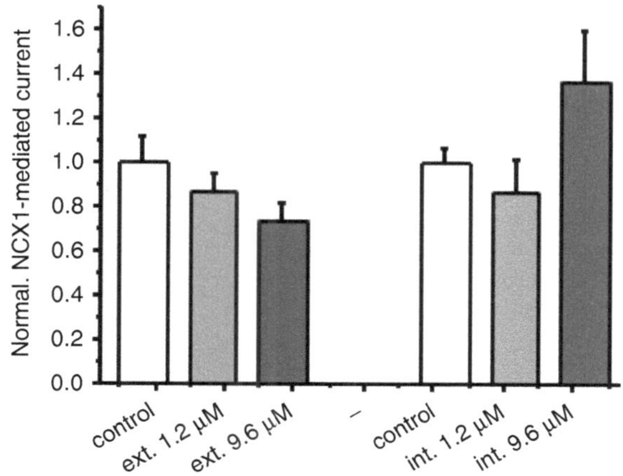

Fig. 2.9 Effect of externally applied and injected hCyPA on NCX1-mediated current. Data represent mean inward currents of 7–10 oocytes (±SEM) and were normalised to −60 mV. [Wang, Y., Gu, Q.B., Yang, Y.Q., Schwarz, W. (2009) Effects of human Cyclophilin A on Na^+/Ca^{2+} exchanger NXC1 in *Xenopus* oocytes (unpublished)]

To investigate modulation of NCX, *Xenopus* oocytes with heterologously expressed NCX1 were used as model system. NCX1-mediated current was determined as the current component sensitive to 2 mM Ni^{2+} in the external solution. Oocytes not expressing NCX1 did not exhibit Ni^{2+}-sensitive currents.

Figure 2.9 shows the effect of externally applied (overnight incubation) in 1.2–9.6 μM human CyPA (hCyPA) at −60 mV. The current became slightly, but significantly, reduced after the incubation with the hCyPA (Fig. 2.9).

If hCyPA were injected into *Xenopus* oocytes (final internal concentration up to 4.8 μM), NCX1-mediated current was hardly affected. Only at very high amounts (final internal concentration more than 9 μM), current became elevated (Fig. 2.9). In whole human blood sample hCyPA of about 200 nM has been reported, but in plasma only a very small fraction, if any, could be detected (Allain et al. 1995). On the other hand, in patients with inflammations cytoplasmic CyPA of several hundred nanometres has been found (Billich et al. 1997; Jin et al. 2004). Therefore, the inhibitory effect of CyPA on the activity of NCX1 may have physiological, therapeutic relevance. It has been reported (Hirota et al. 2007) that in ASM cells the reversed mode of NCX provides a source of Ca^{2+} for store refilling. Inhibition of NCX by acupuncture-induced CyPA elevation may, therefore, counteract the refilling and hence favour relaxation of the smooth muscle cells.

2.3.2 *Effects of α-Asarone on Membrane Transport*

In Chinese Medicine acupuncture is usually supplemented by treatment with Chinese herbal formula. The membrane proteins, which have been investigated in our laboratory as putative targets for acupuncture effects, may also be targets for such herbal extracts. In our laboratory, the *Xenopus* oocytes served also as an ideal model system for drug screening. As in the previous section, voltage-clamp and

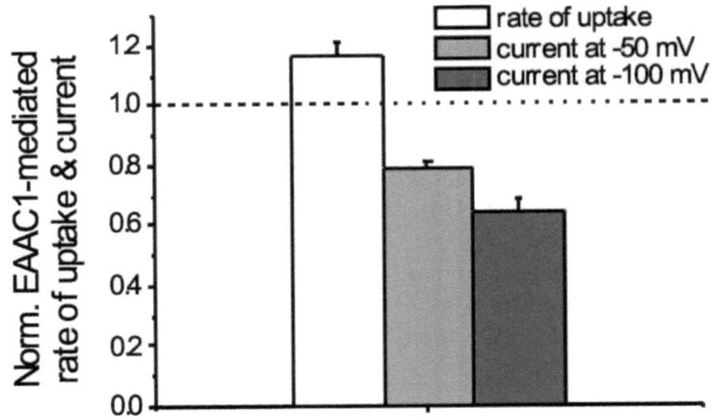

Fig. 2.10 Rate of glutamate uptake and glutamate-induced current in *Xenopus* oocytes expressing EAAC1. At 200 μM α-asarone, ^{3}H-glutamate uptake is stimulated, but glutamate-induced current is inhibited potential dependently (based on data presented by Gu et al. (2009)

tracer-flux experiments were performed to detect modulation of the Na^+, K^+-ATPase and the neurotransmitter transporters. As an example we will illustrate the effect of α-asarone on the glutamate transporter EAAC1 (Gu et al. 2009).

Different extracts of rhizomes of *Acorus tatarinowii* were tested with respect to their effects on the sodium pump, EAAC1 and GAT1. As the only effective component, α-asarone could be identified as modulator of EAAC1 activity; GAT1 and the Na^+, K^+-ATPase were not affected.

Figure 2.10 illustrates that glutamate uptake is stimulated by 200 μM α-asarone by about 15 %, while EAAC1-mediated current is inhibited with slight voltage dependence (stronger inhibition at more negative potentials). This voltage dependence reflects that α-asarone inhibits the Cl^--channel mode of EAAC1, which contributes to total EAAC1-mediated current with more negative potentials.

It has been reported that glutamatergic nerve terminals accumulate intracellularly Cl^- (Price and Trussell 2006). As a consequence, inhibition of Cl^- channels will hyperpolarise the membrane potential. Inhibition of the EAAC1-mediated Cl^--channel mode by α-asarone will, therefore, contribute to stabilising the membrane potential and, together with the stimulated rate of glutamate uptake, counteract excitatory synaptic transmission. This could contribute to the reported anticonvulsive (Wang et al. 1998; Yang et al. 2006) and neuroprotective (Cho et al. 2002) effect of α-asarone.

At similar concentrations, in the range of 100 μM, α-asarone inhibits also the purinergic receptor $P2X_7$ that is activated by ATP exceeding 100 μM concentrations (Spielmann et al. 2008). Activation of $P2X_7$ receptors has been shown to facilitate pro-inflammatory processes associated with arthritis (Dubyak and El-Motassim 1995; Carroll et al. 2009), and within the nervous system, pro-inflammatory processes may contribute to development and maintenance of chronic pain (Carroll et al. 2009; Hughes et al. 2007). In addition to the above described effect of α-asarone on EAAC1, the inhibition of $P2X_7$ could contribute to the previously reported observation that α-asarone can protect neuronal cells from excitotoxic cell death (Cho et al. 2002).

2.4 Concluding Remarks

The work of our laboratory suggests that the generally accepted effect of acupuncture in pain relief can at least partially be mediated by modulation of central nervous synaptic activity. This modulation may involve indirect modulation of GABA and glutamate transporter activity by acupuncture-induced release of endorphins. Since modulation of Na^+, K^+-ATPase activity can affect opioid sensitivity, this membrane protein should be considered as an interesting target in various kinds of treatment. Membrane transporters may also be a direct target for acupuncture-activated endogenous drugs as illustrated for the effect of hCypA on the Na^+, Ca^{2+} exchanger in treatment of asthma. The membrane transport systems that can be affected by acupuncture should also be considered as target for Chinese drugs.

In conclusion dysfunction of neurotransmitter transporters is involved in various neurological disorders, like the GABA transporter in epilepsy [see, e.g. (Fueta et al. 2003; Conti et al. 2004)] or the glutamate transporter in excitotoxicity [e.g. (Li et al. 1997; Maragakis and Rothstein 2004; Sheldon and Robinson 2007)]. Therefore, these transporters have become important target for Western medicines but may also be targets for curing neuronal diseases by acupuncture.

Acknowledgements We are very grateful to all the members of our research groups at the Max Planck Institute for Biophysics in Frankfurt and at the Shanghai Research Centre for Acupuncture and Meridians, in particular to H. Biehl, H. Fotis and H. M. Du for their excellent assistance; to Drs. T. Fucke, L. Pu, S. L. Ren, A. Spielmann, L. N. Wang, Y. Wang, P. Xia and D. Zhang for their unpublished contributions to this research field; and to Drs. G. H. Ding, G. Pei and Y. Q. Yang for their stimulating support.

References

Allain F, Boutillon C, Mariller C, Spik G (1995). Selective assay for CyPA and CyPB in human blood using highly specific anti-peptide antibodies. J Immunol Meth 178:113–120.

Bauer O, Razin E (2000). Mast Cell-Nerve Interactions. News Physiol Sc 15:213–218.

Billich A, Winkler G, Aschauer H, Roth A, Peichl P (1997). Presence of cyclophilin A in synovial fluids of patients with rheumatoid arthritis. J Exp Med 185:975–980.

Carroll WA, Donnelly-Roberts D, Jarvis MF (2009). Selective P2X7 receptor antagonists for chronic inflammation and pain. Purinergic Signal 5:63–73.

Cho J, Kim YH, Kong JY, ang CH, ark CG (2002). Protection of cultured rat cortical neurons from excitotoxicity by asarone, a major essential oil component in the rhizomes of Acorus gramineus. Life Sci 71:591–599.

Conti F, Minelli A, Melone M (2004). GABA transporters in the mammalian cerebral cortex: localization, development and pathological implications. Brain Res Rev 45:196–212.

Deng HP, Yang ZJ, Li YT, Bao GB, Friedrich T, Gu QB, Schwarz W (2009). Interaction of Na^+,K^+-ATPase and co-expression of δ-opioid receptor. Neurosc Res 65:222–227.

Dubyak GR, El-Motassim C (1995). Signal transduction via P2-purinergic receptors for extracellular ATP and other nucleotides. Am J Physiol 265:C577-C606.

Eckstein-Ludwig U, Fueta Y, Fei J, Schwarz W (2000). The neuronal GABA transporter GAT1 as a target for action of antiepileptic drugs. In: Suketa Y, et al., editors. Control and Diseases of Sodium Transport Proteins and Channels. Amsterdam:Elsevier Press. p. 373–376.

Fueta Y, Vasilets LA, Takeda K, Kawamura M, Schwarz W (2003). Down-regulation of GABA-transporter function by hippocampal translation products: Its possible role in epilepsy. Neuroscience 118:371–378.

Gu QB, Du HM, Ma CH, Fotis H, Wu B, Huang CG, Schwarz W (2009). Effects of α-asarone on the glutamate transporter EAAC1 in Xenopus oocytes. Planta Med 75:1–4.

Han J-S (2004). Acupuncture and endorphins. Neurosci Let 361:258–261.

Hilgemann DW, Lu C-C (1999). Gat1 ($GABA:Na^+:Cl^-$) Cotransport function - database reconstruction with an alternating access model. J Gen Physiol 114:459–475.

Hirota S, Pertens E, Janssen LJ (2007). The reverse mode of the Na^+/Ca^{2+} exchanger provides a source of Ca^{2+} for store refilling agonist-induced Ca^{2+} mobilization. Am J Physiol Lung Cell Mol Physiol 292:L438–L447.

Hu J-H, YangN., Ma Y-H, Zhou X-G, Jiang J, Duan S-H, Mei Z-T, Fei J, Guo L-H (2003). Hyperalgesic Effects of y-Aminobutyric Acid transporter I in Mice. Journal of Neuroscience Research 73:565–572.

Hughes JP, Hatcher JP, Chessell IP (2007). The role of P2X7 in pain and inflammation. Purinergic Signal 3:163–169.

Janssen LJ (2009). Asthma therapy: how far have we come, why did we fail and where should we go next? Eur Respir J 33:11–20.

Jin ZG, Lungu AO, Xie L, Wang M, Wong C, Berk BC (2004). Cyclophilin A is a proinflammatory cytokine that activates endothelial cells. Arterioscler Tromb Vasc Biol 24:1186–1191.

Kavanaugh MP, Arriza JL, North RA, Amara SG (1992). Electrogenic uptake of γ-aminobutyric acid by a cloned transporter expressed in Xenopus oocytes. J Biol Chem 267:22007–22009.

Keynan S, Kanner BI (1988). γ-Aminobutyric acid transport in reconstituted preparations from rat brain: Coupled sodium and chloride fluxes. Biochemistry 27:12–17.

Krause S, Schwarz W (2005). Identification and Selective Inhibition of the Channel Mode of the Neuronal GABA transporter 1. Molecular Pharmacology 68:1728–1735.

Li S, Mallory M, Alford M, Tanaka S, Masliah E (1997). Glutamate transporter alterations in Alzheimer disease are possibly associated with abnormal APP expression. J Neuropathol Exp Neurol901–911.

Maragakis NJ, Rothstein JD (2004). Glutamate transporters: animal models to neurologic disease. Neurobiol Disease 15:461–473.

Price GD, Trussell LO (2006). Estimate of the chloride concentration in a central glutamatergic terminal: A Gramicidin perforated-patch study on the Calyx of Held. J Neurosci 26:11432–11436.

Risso S, Defelice LJ, Blakely RD (1996). Sodium-dependent GABA-induced currents in GAT1-transfected HeLa cells. J Physiol 490:691–702.

Schwarz W (2001). Electrophysiology in Carrier Research. J Biol Membranes 19:48–58.

Schwarz W, Gu Q-B (1988). Characteristics of the Na^+/K^+-ATPase from Torpedo californica expressed in Xenopus oocytes: A combination of tracer flux measurements with electrophysiological measurements. Biochim Biophys Acta 945:167–174.

Sheldon AL, Robinson MB (2007). The role of glutamate transporters in neurodegenerative diseases and potential opportunities for intervention . Neurochem Int 51:333–355.

Slotboom DJ, Konings WN, Lolkema JS (2001). The structure of glutamate transporters shows channel-like features. FEBS Lett 492:183–186.

Spielmann A, Gu QB, Ma CH, Huang CG, Xuan LJ, Schwarz W (2008). Inhibition of P2X7 receptor by extracts of Chinese medicine. J Acupunct Tuina Sci 6:286–288.

Wadiche JI, Amara SG, Kavanaugh MP (1995). Ion fluxes associated with excitatory amino acid transport. Neuron 15:721–728.

Wang HZ, Chen YG, Fan CS (1998). Review of studies on chemical constituents and pharmacology of genus *Acorus* in China. Acta Bot Yunnan Suppl.X pp. 96–100.

Wang L, Ding GH, Gu QB, Schwarz W (2010). Single-channel properties of a stretch-sensitive chloride channel in the human mast cell line HMC1. Eur Biophys J 39:757–767.
Wang Y, Cui JM, Ma SL, Liu YY, Yin LM, Yang YQ (2009b). Proteomics analysis of component in serum with anti-asthma activity derived from rats treated by acupuncture. J Acupunct Tuina Sci 7:326–331.
Xia P, Pei G, Schwarz W (2006). Regulation of the glutamate transporter EAAC1 by expression and activation of δ-opioid receptor. Eur J Neurosci 24:87–93.
Yang LB, Li SL, Huang YZ, Wang YH, Liang JM, Zhang SQ (2006). Effect of Acorus gramineus and its active component α-asarone on apoptosis of hippocampal neurons in epileptic young rats. Chin Trad Herb Drugs 37:1196–1199.
Zerangue N, Kavanaugh MP (1996). Flux coupling in a neuronal glutamate transporter. Nature 383:634–637.
Zhang D, Ding G, Shen X, Yao W, Zhang Z, Zhang Y, Gu Q (2008). Role of mast cells in acupuncture effects: A pilot study. Explore 4:170–177.
Zhang D, Spielmann A, Wang L, Ding G, Huang F, Gu Q, Schwarz W (2012). Mast-cell degranulation induced by physical stimuli involves the activation of transient-receptor-potential channel TRPV2. Physiol Res 61:113–124.
Zhang Y-Q, Ji G-C, Wu G-C, Zhao Z-Q (2002). Excitatory amino acid receptor antagonists and electroacupuncture synergetically inhibit carrageenan-induced behavioral hyperalgesia and spinal fos expression in rats. Pain 99:525–535.

Chapter 3
Function of Collagen and Mast Cells in Acupuncture Points

Guanghong Ding, Di Zhang, Meng Huang, Lina Wang, and Wei Yao

Abstract The acupoint is the responsive point that is stimulated during acupuncture treatment, and its structural basis and action mechanism have been largely elucidated by researchers. Clinical medical imaging and extensive studies on anatomy have indicated that the structural and functional basis for the human body's meridians and acupoints is a complex system that primarily comprises connective tissues with numerous intertwining blood vessels, nerves, collagen fibers, and mast cells. Nevertheless, how the effective acupunctural signals transmit through acupoints to induce analgesia through a series of biochemical reactions along the meridians is a mystery. We have applied different acupuncture methods combined with morphology observations, mechanical signal detection, and animal pain threshold measurements to analyze the correlation between acupuncture analgesia and collagen fiber mechanical deformation as well as mast cell degranulation at acupoints. Furthermore, to explore the structural and functional basis of acupoints and the initiation of acupunctural signals, the patch clamp technique was used to investigate the dynamic

G. Ding (✉) • D. Zhang
Department of Mechanics and Engineering Science, Fudan University, 220 Handan Road, Shanghai 200433, China

Shanghai Research Center for Acupuncture and Meridians, Fudan University, 220 Handan Road, Shanghai 201203, China
e-mail: ghding@fudan.edu.cn; dizhang@fudan.edu.cn

M. Huang • W. Yao
Department of Mechanics and Engineering Science, Fudan University, 220 Handan Road, Shanghai 200433, China
e-mail: 071029004@fudan.edu.cn; weiyao@fudan.edu.cn

L. Wang
College of Acupuncture and Moxibustion, Shanghai University of Traditional Chinese Medicine, Shanghai, China
e-mail: linawang1103@163.com

Y. Xia et al. (eds.), *Current Research in Acupuncture*,
DOI 10.1007/978-1-4614-3357-6_3, © Springer Science+Business Media New York 2013

characteristics of mechanosensitive proteins in relation to acupuncture at the cellular level. The results indicated that acupuncture analgesia was induced by physical stimulation at acupoints, where collagen fibers were deformed by stress and the surrounding mast cells were activated. The acupoint response is based on the spatial conformations of the blood vessels–mast cells–nerves. Once activated, mast cells are degranulated and release several types of bioactive substances, which increase blood vessel permeability. In addition, they act on the surrounding peripheral nerve endings, which induce a local peripheral nerve impulse and then transfer the acupunctural signal to the central nervous system. During this process, a positive feedback signal is generated by neurotransmitters acting on the mast cells and distributed along the meridians. An analgesic effect is produced and transmitted through an indirect effect on the organs and tissues along the meridians.

Keywords Acupoint specificity • Mast cells • Connective tissue • Acupuncture analgesia • Signaling pathway • Mechanical stimuli

3.1 Introduction

With a history of more than 4,000 years, acupuncture is derived from Chinese ancestors' wisdom and experiences throughout the ages. Though its underlying scientific principles have not been entirely elucidated, there is no doubt that it constitutes a significant part of Chinese clinical medicine. Along with the development of acupuncture, Chinese medicine and health services have made incredible achievements despite a relatively low budget. For instance, Shanghai's urban GDP per capita was $10,529 in 2008 (in contrast with New York's $58,720); however, its average medical expenditure was only 1/30 of New York's, while the average life expectancy in Shanghai was 81 years, which is 2 years higher than in New York. Wide application of traditional Chinese medicine (TCM) clinical therapies, including acupuncture, has made crucial contributions to the above statistics.

Although the scientific theories on acupuncture are still largely unverified, the clinical effects of acupuncture have drawn extensive attention from followers of science and medicine. Many acupuncture clinical trials have been conducted using rigorous scientific methodology and have provided results that favor acupuncture over other controls. Individual trials showed a positive effect on the following conditions compared with sham controls: knee osteoarthritis (Berman et al. 2004; Manheimer et al. 2007; Witt et al. 2005), temporomandibular joint myofascial pain (Smith et al. 2007), chronic neck pain (Itoh et al. 2007), and chemotherapy-induced emesis (Shen et al. 2000). Meta-analyses also showed that acupuncture yielded statistically significant benefits in treating low back pain (Manheimer et al. 2005) and infertility (Manheimer et al. 2008).

Although scientists have reached a consensus in acknowledging the effects of acupuncture, an understanding of the logic behind the underlying mechanisms and mode of action remains elusive. The scope of acupuncture research has greatly expanded, and various hypotheses have emerged from the early studies on the neurophysiological mechanisms and chemical factors to the latest research at the molecular and genetic levels. For instance, in studying acupuncture analgesia, Han found that at specific points on the human body, known as acupoints (the human body is believed to have more than 360 acupoints), acupuncture stimulation induces the brain to release endogenous morphine, opioid peptide, and other substances that aid in the analgesic effect (Han 2004). Scrutinizing the mechanism of acupuncture at acupoints, Langevin showed that an acupuncture needle could cause the connective tissue to twist in the acupoints, thereby inducing signal transduction along the meridians at the cellular level (Langevin and Yandow 2002).

Acupuncture is the process of using special physical stimuli, which are either mechanical or electrical, at various acupoints in the human body that produce a therapeutic effect (e.g., analgesia). In this process, the acupoint converts mechanical or electrical stimuli into physiological signals that are disseminated throughout the human body. These signals are then transmitted to the central nervous system via a biological network-like neural system, and they may cause the release of various substances, such as morphine and opioid peptides, that produce effects such as analgesia. What role does the acupoint play in the overall process? This chapter will answer this question by systematically reviewing the research.

3.2 Scientific Study of Acupoints

As the basic elements in the meridian structure, acupoints serve as the initiating points for the TCM syndrome-differentiated treatments, especially in acupuncture. The patient experiences aching or tingling sensations when these points are stimulated. These points are linked by a network of meridians, which are distributed longitudinally throughout the body in both humans and other mammals. In TCM, energy, "Qi," is believed to flow through these meridians. If the meridian network is disturbed such that the "Qi" flow is blocked, illness results. Each meridian specifically relates to a particular organ. During acupoint stimulation, a feeling, including tingling, numbness, and heaviness, which is the so-called de-Qi, may be elicited. This phenomenon, which stimulates the local acupoints and induces a systemic and persistent acupuncture effect, has aroused the interests of researchers that are investigating its mechanism. However, the existence and function of acupoints are still a matter of controversy.

In the most recent half century, many scientific explanations about how acupuncture aids in the treatment process have been asserted by domestic and foreign scholars. For the aspect of acupoint structure, the "boundary between muscles" where meridians circulate, as described by the ancient recording *Miraculous Pivot*: *Meridian*, is consistent with the connective tissue structure. This establishes the connective tissue as the foundation for research on the morphology of meridians and acupoints. A research team led by Langevin suggested that local connective tissue deformation caused by lift-thrusting and rotation mechanical stimuli at the acupoints might result in signals at the cellular level along the connective tissue planes. Statistical analysis of ultrasound images showed that over 80 % of the acupoints in human arms coincide with the connective tissue plane boundaries (Langevin et al. 2001; Langevin and Yandow 2002). During the acupuncture process, the surrounding connective tissue could be effected by the needle body in two possible ways, contraction and binding. In vitro, isometric force transduction measurements showed that in less than a few minutes, thrombin fibroblast stimulation produced substantial tension by actin cytoskeleton integration, which caused the connective tissue to contract (Kolodney and Wysolmerski 1992). Close to the needle, the injured fascia, which is composed of collagen fibers, elastic fibers, fibroblasts, adipocytes, and mast cells, bound to the needles during manual acupuncture as shown by electro-microscopical analysis of the needle after removal (Kimura et al. 1992). Surrounding the acupoints, collagen is the most prevalent component in connective tissue. It can be folded into a structure with long-range three-dimensional order, which provides the collagen with a transport property similar to a liquid crystalline continuum, and its physical characteristics may become sensitive to small changes in certain environmental parameters (Ho and Knight 1998). Collagen is an important component of the extracellular matrix in addition to the integumentary system, and it is present in all types of systems (Aumailley and Gayraud 1998). Certain researchers suggest that the dense dermal connective tissues and subcutaneous loose connective tissues form a three-dimensional collagen fiber network, which connect the surface of the body to the internal organs (Yuan et al. 2005). It should not be overlooked that this network resembles the meridian network in both biochemical and structural properties, and it is capable of sensing environmental changes and directly connecting the periphery to the internal organs.

However, acupoints along the meridians are neither an isolated system, a biological tissue, nor a functional system composed of a single tissue. Instead, they comprise a multifunctional regulation system (Xu and Zheng 2003). This system not only carries ions and other materials but also transfers energy and information between the cells. Recent studies on acupoint morphology using various approaches, such as MRI, anatomy, and XCT, have shown that the structural basis of the meridians and acupoints is complex and composed of connective tissue with numerous intertwining blood vessels, nerves, mast cells, and lymphatic vessels (Ding et al. 2001; Fei et al. 1998; Zhang et al. 2004). The tissue surrounding acupoints has abundance of regularly distributed blood vessels. According to histological observations, 99.6 % of acupoints are located in tissues with an abundant distribution of small blood vessels and capillaries, and 45.5 % of acupoints are located close to large blood vessels (Mu 1990). For meridians, the same phenomena have also been observed. Through macro- and microdissection as well as systemic artery image pattern analysis using a radiographic technique, approximately five out of six total meridians were found in lower limb integument tissues in an area with a significantly higher blood vessel distribution density. Further, although the stomach meridian had no densely distributed blood vessels in the integument tissue, it was also located above the anterior tibial artery (Lou et al. 2006). Similar to blood vessels, most human acupoints in the lower limbs were along lymphatic vessels. Further, along most lymphatic vessels, two or three acupoints belong to a single meridian (Zhao

2000). At Hegu (L-14), the distribution of nerves and blood vessels is significantly denser in the areas with acupoints than without acupoints. In the acupoint-containing areas, substance P was observed in the nerve fibers that were closely related to the lymph-based neuron-immune system. This structure has been suggested to relate substance P to activation of the immune system, which could be related to the propagation sensation along the meridian after acupuncture (Yamada 1996). Neurons are the major sensory receptors for external stimuli, which are located throughout the body surface and internal organs. Through histological serial sections of acupoint-containing areas, most acupoints were observed near nerve fibers or nerve fiber joints (Xu et al. 1989). Through serial sections along the direction of tissue sample needling at the acupoint Shoushanli (LI-10) and using a normal insertion depth and the structural properties of adjacent tissue, it was observed that 93 % of the dense connective tissue was covered with nerve fibers derived from the radial nerve (Xu et al. 2004). The connective tissues surrounding acupoints may participate in transduction of the acupuncture signals to the neural system.

Needles inserted at acupoints may activate physiological responses through their unique structure to yield the therapeutic effects. But the following question remains: how are the signals of such local physiological responses transmitted to the inner and distant regions of the human body? Using a functional magnetic resonance imaging (fMRI) technology, Cho observed that needling a distal acupoint at the little toe, Zhiyin (UB-67) could produce brain activity in the visual cortex. However, they were not able to identify a signal passing from the needling site to the brain cortex (Cho et al. 2006).

One working hypothesis to answer this question was proposed, whereby mast cells are correlated with the transmission of the acupuncture effects. As the common cells in connective tissue, mast cells have attracted the interest of researchers since the 1980s (Song 1977). The following research findings were reported. A significantly large number of subcutaneous mast cells were found at the acupoints when compared with areas that do not contain acupoints (Deng et al. 1996; Zhong et al. 1994). A group led by Zhang found that there was a specialized junction structure between mast cells and neurons, which was different from simple contacts at the meridian in human skin (Zhang and Wang 1985). He suggested that, based on the axon reflex theory, this might be a structural foundation for the propagation sensation along the meridian. They found that in both humans and mice the neuromastocytic junction was on the meridian in the skin. Their hypothesis claimed that meridians were at an area of the skin with higher sensitivity to certain stimuli, and this hypothesis was supported by the observation that there are more mast cells distributed on the skin at the meridian compared with other areas of the skin. They also suggested in the same study that although acupuncture induced a propagation sensation that moved along the meridian instead of nerve afferents, it is a large-scale directional activity for which a directional structure was required in addition to neuron distribution. One possible candidate could be the synaptic-like junctions between nerve endings of adjacent dermatomes in skin, in which mast cells are also a major component (Chen and Zhang 1987). The distribution of mast cells was examined at the low-resistance area along the meridians and the area along the acupuncture-induced sensation propagation in both humans and rats dyed with toluidine blue. At different depths of a section along the entire range of the low-resistance area, mast cells were observed at significantly higher densities than in the control area (Zhu and Xu 1990). Mast cells are derived from bone marrow cells, and their precursors enter the blood circulation, drift to connective tissue at the periphery or in the mucosa of internal organs, go through a proliferation and differentiation process, and finally become mature mast cells (Levi-Schaffer and Shalit 1993). The heterogeneity of mast cell distribution, structure, and function hinders mast cell studies in vivo. During the last two decades, most of the studies focused on mast cells and acupoints used segregated cells (Metcalfe et al. 1997). Mast cells are easily identified though toluidine blue or neutral red staining because of their proteoglycan and protease-rich cytoplasmic granules. To distinguish between the separate types of mast cells, an immunohistochemical method can be employed to reveal the different serine proteases (Irani et al. 1987). The morphologies of the cells and granules can be observed using an electron microscope (Dvorak et al. 1993). Induced by certain specific stimuli from the microenvironment, mast cells can be activated to release granules and secrete their intracellular mediators. This entire process is referred to as degranulation. After degranulation, mast cells show either empty cavities or no granules in the cytoplasm (Pang 1992). However, this does not trigger the cell death process for all mast cells; instead, during the next few weeks, some of the cells are reloaded with new granules. After electroacupuncture stimuli at the acupoint Zusanli (ST-36) mast cells in the peritoneum had an increased ratio of degranulation but were decreased in number (Gao et al. 1984). However, using the twirling–reinforcing–reducing technique of manual acupuncture at Zusanli (ST-36), the total number of mast cells in the intestine increased significantly with an increased ratio of degranulation (Zou et al. 1993). It was suggested that acupuncture could induce the mast cell precursor in the mesentery to accelerate the proliferation process in the few days that followed. Those new, mature mast cells would replace cells that were dead after degranulation and finally outpopulate the control group. Compared with electroacupuncture, this twirling–reinforcing–reducing technique has an additional effect on target organ activation and regeneration of the balance of mast cells. These observations suggest that different application methods or techniques for acupoint stimulation yield different effects on the target organ or, at least, stimulate mast cells differently, which is reasonable considering that mast cells are activated differently according to the type and the strength of stimulation. Laser acupuncture at Zusanli (ST-36) also significantly increased the ratio of degranulated mast cells and produced an anesthesia effect similar to manual acupuncture (Cheng et al. 2009). An acupuncture effect was observed within the lower limb area of the left stomach meridian on vascular permeability at the stomach and an area along the meridian (Shen et al. 1998). The Shen group discovered that acupuncture induced an increase in vascular permeability not only near the stimulated section but also in a section further away along the same meridian. They suggested that this was a result of cooperation between neurons, mast cells, and blood vessels along the propagated acupuncture signal. The degranulation of mast cells was also observed after moxibustion at Zusanli (ST-36) along trunk section of the stomach meridian in mice. Upon stimulation, they found that the ratio of degranulated mast cells significantly increased along the ipsilateral meridian, and only part of the mast cells were degranulated along the contralateral meridian. For the control group, which was not treated with acupuncture, only a few mast cells were degranulated along both meridians (Hu et al. 2001; Liu et al. 2000). All of these findings suggested that mast cells and meridians are closely related in structure and function. The idea of a "mast cell–cytokine cascade reaction" was asserted, in which cytokines were released by mast cells and acted on the resident and recycling cells; the latter may further secrete cytokines, which leads to a positive feedback and an amplification effect (Galli 1993). After mast cell activation in the connective tissue, biologically active components are released in a complex intercellular signaling cascade (Metcalfe et al. 1997).

Further, using proton-induced X-ray emission (PIXE), one study reported that acupoints have a high level of calcium and suggested that acupoints might be special areas where metastable Ca was conserved for use in emergencies (Chen et al. 1998). Although calcium is predominantly a skeletal element, a small proportion is found inside the cells and in the extracellular fluid, and it acts as a chemical messenger between the cells, controlling motion and metabolism (Theobald 2005). All of the above studies and results have elucidated the basic structure of meridians and acupoints.

3.2.1 Specificity of Acupoints

Despite the considerable efforts in probing the anatomy of acupoints, structural characterization of acupoints remains elusive. Many methods have been employed in the research of acupuncture and its relationship with connective tissue structures, mast cells, capillaries, and nerves. To discern whether there is a difference between acupoint regions and the surrounding tissues, the morphological structure of acupoints was investigated. Using 15 lower limbs of human corpses fixed in formalin, several segments of blood vessels were found to enter the interosseous membrane (Zhang et al. 2009). The arterioles at classic acupoints appeared to be clustered together (Fig. 3.1a). Under the optical microscope, the capillaries in these areas of dense arterioles were arranged in a parallel pattern (Fig. 3.1b). Note that a small quantity of capillaries was arranged in a disorderly and unsystematic manner at non-acupoint regions (Fig. 3.1c). The electron microscopy image shows that the collagen fibers in these areas also form parallel patterns that extend in the same direction as the capillaries and are surrounded by an abundance of mast cells (Fig. 3.1d).

Rats were employed as animal models for further experiments. Sprague Dawley (SD) rats are widely used in experiments for studying pain mechanisms, including myalgia–skeletal pain (Wu et al. 2007), neuropathic pain (Staaf et al. 2009), immunological pain (Ndong et al. 2009), and the afferent pain mechanism (Hathway et al. 2009). The laboratory-born SD rats were provided by the Shanghai Experimental Animal Centre of the Chinese Academy of Science. They all weighed 180 ± 20 g

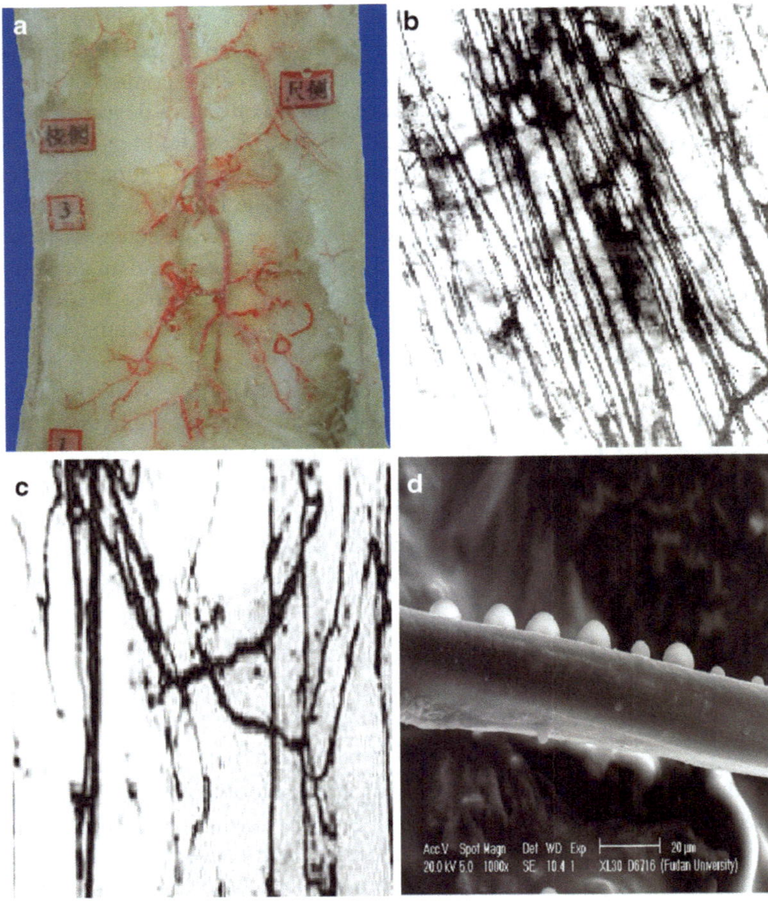

Fig. 3.1 Morphological observations of the anatomical structure of acupoints in human tissues. (**a**) Blood vessel distribution in front of the interosseous membrane in the upper limbs after perfusion with silica gel. (**b**) Parallel and closely arranged capillaries are located in the interosseous membrane at the acupoint. (**c**) A non-acupoint with a smaller quantity of capillaries arranged in a disorderly and unsystematic manner. The site 2 cm away from the acupoint was considered a non-acupoint. (**d**) The capillaries are surrounded by many mast cells (under the electron microscope) (Zhang et al. 2009)

and were housed in cages at a controlled temperature (20 ± 2 °C) with a 14/10 h light/dark cycle. All of the animals were handled with care to prevent infection and minimize stress.

The density of mast cells at the Zusanli (ST-36) acupoint and non-acupoints was significantly impaired in our experiment (Zhang et al. 2008). In the stained tissue slices, we observed the basophilic granules of the metachromatic mast cells (Fig. 3.2). The densities of the mast cells both in acupoint skin and muscle tissues

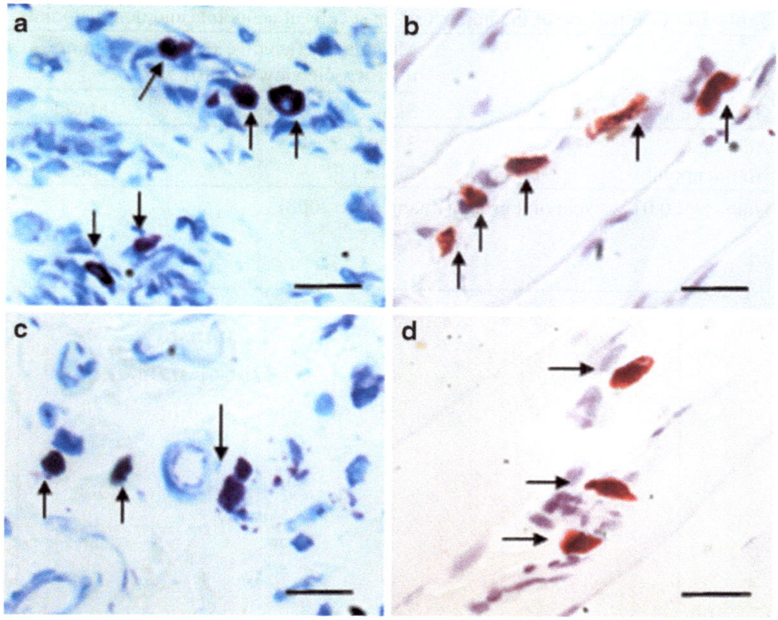

Fig. 3.2 Representative photomicrographs of stained mast cells (*arrows*) at Zusanli (ST-36) and non-acupoints in rats (*left panel* from skin stained with 0.5 % toluidine *blue*, *right panel* from muscle stained with 0.5 % neutral *red*). (**a**, **b**) acupoint Zusanli (ST-36): large rounded or elliptical mast cells were observed in skin and muscle with an intact cell membrane and were filled with intracellular basophilic granules. (**c**, **d**) non-acupoints: the mast cells have a similar appearance but lower density compared with those at the acupoints. The site 3 mm away from ST-36 (in the direction of the fibula) was considered a non-acupoint. The scale bar is 10 μm (Zhang et al. 2008)

were approximately 50 % higher than those at the sham point (Table 3.1). Our results confirm previous findings that mast cell densities are significantly higher at acupoints than non-acupoints, which was shown previously for several other acupoints, including Sanyinjiao (SP-6), Yanglingquan (GB-34), Hegu (LI-4), Quchi (LI-11), and Neiguan (PC-6) (Zhong et al. 1994).

Furthermore, synchrotron X-ray fluorescence (SXRF) analysis was used to probe the distribution of the four chemical elements in and around the acupoints in humans; two were located in the forearm and two in the lower leg (Yan et al. 2009). All four of the samples were provided by the Anatomy Teaching and Research Section of Second Military Medical University according to a protocol approved by the ethics committee of the Second Military Medical University for appropriate use of human tissues. Three of the four acupoints had significantly elevated concentrations of the elements Ca, Fe, Cu, and Zn compared with the levels in the surrounding tissue with similarly elevated Cu and Fe ratios (Fig. 3.3). The distribution of these elements implies that each acupoint is elliptical with the long axis along the meridian.

Table 3.1 Comparison of the number of mast cells at acupoints and non-acupoints

	MC number (average ± SEM) [0.16 mm²/microscopic view (×200)]	
Group name ($n = 20$)	Skin	Muscle
Acupoint	34.2 ± 2.5	12.4 ± 1.4
Non-acupoint	22.1 ± 1.7*	8.5 ± 1.0*

Note: *$P < 0.01$ vs. acupoint group (Zhang et al. 2008)

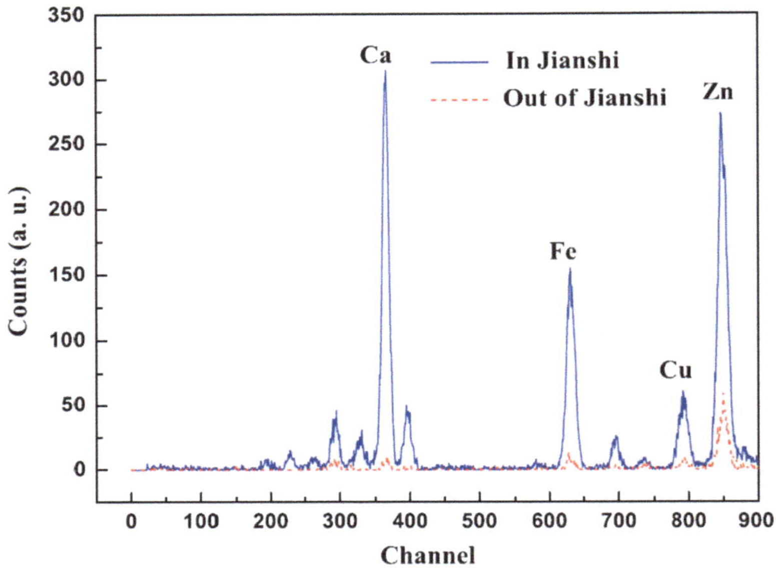

Fig. 3.3 SXRF spectra from the Jianshi (PC-5) acupoint (*solid line*) and the tissue surrounding Jianshi (PC-5) (*dotted line*). The site 2 cm away from the acupoint was considered a non-acupoint. The fluorescence intensity and Ca content were approximately 20 times higher in the Jianshi (PC-5) acupoint. Of the other three trace metals, Fe, Cu, and Zn, the integrated intensity was the highest for Zn followed by Fe and Cu. The spectra for the Tiaokou (ST-38) and Xiajuxu (ST-39) acupoints showed a similar pattern (Yan et al. 2009)

3.2.2 *The Function of Mast Cells in the Acupuncture Effects*

An arthritis model is a commonly used immunological hyperalgesia model. Rheumatoid arthritis is the most prevalent immune arthritis in humans (Gabriel et al. 1999). It is an autoimmune disease with lesions that are located in multiple peripheral joints, which are manifested by inflammation of the synovial membrane, increased accumulation of synovial fluid, and chronic as well as progressive bone erosion at the peripheral joints (Bergman 2009). Experimental animal arthritis models include both rats and mice, and they can be classified into the following two types according to the induction mechanism: the collagen-induced arthritis model (Brand et al. 2003) and adjuvant-induced arthritis model (Joe et al. 1999). In this study, the adjuvant-induced arthritis model was established by injecting a dose (0.05 ml) of Complete Freund's Adjuvant (CFA, Sigma) into a rat tibiotarsal joint on one side using intraperitoneal anesthesia with 10 % chloral hydrate (the proportion of chloral hydrate volume according to the weight of the rat is given in a dose of 0.04 ml/g). Twenty-four hours later, the rat's ankle was clearly swollen such that it would be difficult for the rat to walk, which was an index of successful modeling.

In our study, paw withdrawal latency (PWL) was induced by thermal stimulation and used to measure the analgesia effect. The paw withdrawal reflex is a typical pain reflex that involves the participation of a spinal nerve. The reflex was discovered by Woolfe and MacDonald in 1944 and designated as a specific measurement method by Hargreaves in 1988; it has been widely used in the development of a large number of pain-related drugs (Masters and Domb 1998) and pain mechanism studies (Ndong et al. 2009). During the measurement process, two primary factors can affect the measurement results (Dirig et al. 1997): an increase in light density lowers the pain threshold (PT) and both PT and measurement errors increase when the glass plate and room temperature are the same.

The Paw Stimulator Analgesia Meter (Model 336GT, IITC, USA) produces a noxious thermal stimulus. Unrestrained and conscious rats were placed in the plastic box within the instrument 20 min before the PWL evaluation to allow them to adjust to the testing apparatus. Radiant heat was then produced by a strong light beam applied to the animal's anklebone until the rat withdrew the leg. This interval was recorded as the PWL. The glass plate that the animals stood on was set at 30 ± 2 °C, while the environmental temperature was 25 ± 2 °C. The stimulator was programmed with a 15-s cutoff maximum to prevent tissue damage. Each animal was tested twice within an interval of 10 min, and the average was chosen as the baseline for the PT. In addition, to control for potential bias, the PWL test procedures were performed by an investigator blinded to the animal group assignments.

Manual acupuncture (MA) was applied during the major experiments. The following acupoints were involved in these experiments: Zusanli (ST-36), Kunlun (BL-60), Xiajuxu (ST-39), and Dubi (ST-35). Sterilized stainless steel acupuncture needles (0.25×25 mm, Suzhou Kangnian Medical Devices Co., Ltd., China) were used. For the twisting acupuncture, the needle was twisted at approximately two to three revolutions, and the revolution range was controlled within 90–180°. This procedure was performed for 20 min, and during each minute, a twist manipulation was applied for the first 30 s but not during following 30 s. For the lift-thrusting acupuncture, the needle was manipulated perpendicular to the skin surface within a 2 mm range and the insertion depth; the frequency and time period were the same as for the twisting manipulation.

A clinical monitoring system developed by our laboratory was used to quantify the manual acupuncture stimulation. The general construction of this system

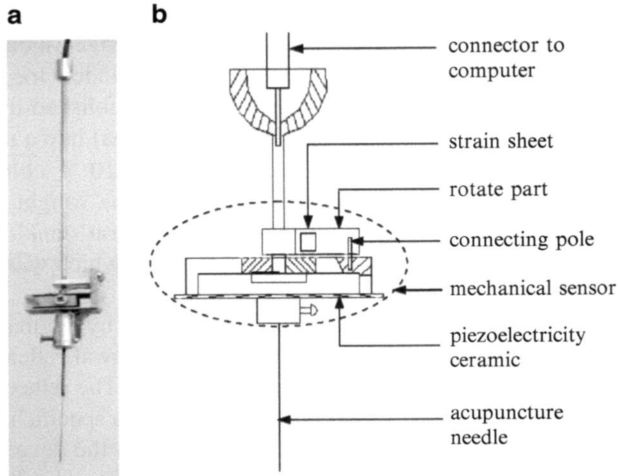

Fig. 3.4 The structure and principles of the real-time acupuncture needle force detection system. (**a**) A photograph of the detection needle. (**b**) A diagram of the mechanical sensor that was used to detect the force on the needle in real-time during acupuncture manipulation (Zhang et al. 2008)

involved an input interface that received the acupunctural signal collected by the detection needle and an output interface through which the processed data was forwarded to a personal computer (Fig. 3.4). During the application of acupuncture, we held the tail section of the detection needle firmly and applied either a twisting or lift-thrusting manipulation at the acupoints. Both types of forces were synchronously observed, and quantification indices of needling strength and frequency were recorded during the needling process.

To investigate the relationship between acupuncture analgesia and mast cell degranulation at acupoints, the acupoints were pretreated with the mast cell stabilizer disodium cromoglycate (DSCG) before acupuncture to induce analgesia. As we mentioned earlier, a change in the PT of the SD rats was adopted as an index of the analgesic effect. The Zusanli (ST-36) was selected as the site of the lift-thrusting and twisting manipulation for 30 min under the control of a real-time needle force monitor (Fig. 3.4). The degree of mast cell degranulation at acupoints before and after acupuncture was estimated by histological observations.

It was noticed that by keeping the stimulation intensity constant during acupuncture, the analgesia was better from stimulation at acupoints than at the nearby sham points (Table 3.2). The maximum PT ratio was reached after 20 min of acupuncture. After degranulation inhibition in acupoint mast cells using DSCG, the analgesic effect was significantly attenuated. The injections had a weak influence from which only a slight increase in the PT ratio was observed. No such reduction was observed when acupuncture was applied to the opposite leg from the injection side. Clearly,

Table 3.2 The influence of MC function during acupuncture before and after DSCG pretreatment

Group name ($n=6$)	Change in pain threshold (average ± SE) (%)
Control	−1.9±0.1
Acupuncture at ST-36	11.6±0.6*
Acupuncture at nearby ST-36	8.9±0.3*, ***
Injection of DSCG at ST-36	1.7±0.1*, **
Injection of NaCl at ST-36	1.4±0.1*, **
Acupuncture after DSCG at ST-36	6.2±0.4*, ***
Acupuncture after NaCl at ST-36	8.9±0.4*, †
Acupuncture opposite side after DSCG	9.3±0.1*, ‡

Note: *$P<0.05$ vs. control; **$P<0.01$, ***$P<0.05$ vs. acupuncture; †$P<0.01$, ‡$P<0.05$ vs. acupuncture after DSCG (Zhang et al. 2008)

local mast cell stabilization in acupoints caused a pronounced attenuation of the acupuncture-induced analgesia.

Specimens were prepared and histological evaluations were conducted for further investigation. All of the rats were decapitated under anesthesia immediately after the PT was measured. Tissue samples that were 5×5×5 mm from an ST-36-centered block were collected from the left hind limb. They were then immersion-fixed in 4 % formalin for 48 h before being processed into sequential paraffin slices (5 μm thick). The direction of the section was longitudinal to the skin and muscle tissue. Eight continuous slices that were taken from the region close to the needle site, excluding the crack caused by the needle insertion, were chosen for staining. Of these, four were stained with toluidine blue (TB) and the other four with Mallory. The morphologies of the collagen fibers and mast cells in the skin and muscle slices were observed at ×400 under a microscope. The mast cells with an incomplete cell membrane and surrounding granules were considered degranulated cells. For each animal, the skin from the four continuous slices was stained with TB, the degranulated cells were blindly counted, and the numbers were averaged. Group degranulation was calculated from the averaged data.

We found that in addition to the PT increase from acupuncture, facilitation of mast cell degranulation at the acupoints was also observed. Acupuncture increased the degree of mast cell degranulation in both the skin and muscular tissues at the acupoints (Fig. 3.5, Table 3.3). Under an electron microscope, we closely observed the degranulation in the tissue slices after acupuncture (Fig. 3.6) and noted that the cell membrane was thinned and extensive intragranular matrix material was released into the pericellular space with empty cavities in the cytoplasm. Following DSCG pretreatment, acupuncture was unsuccessful in significantly facilitating mast cell degranulation in the skin and muscular tissues of the acupoints. This suggests that mast cells at acupoints contribute to the acupuncture effect, and mast cell degranulation is an important link to the production of this effect.

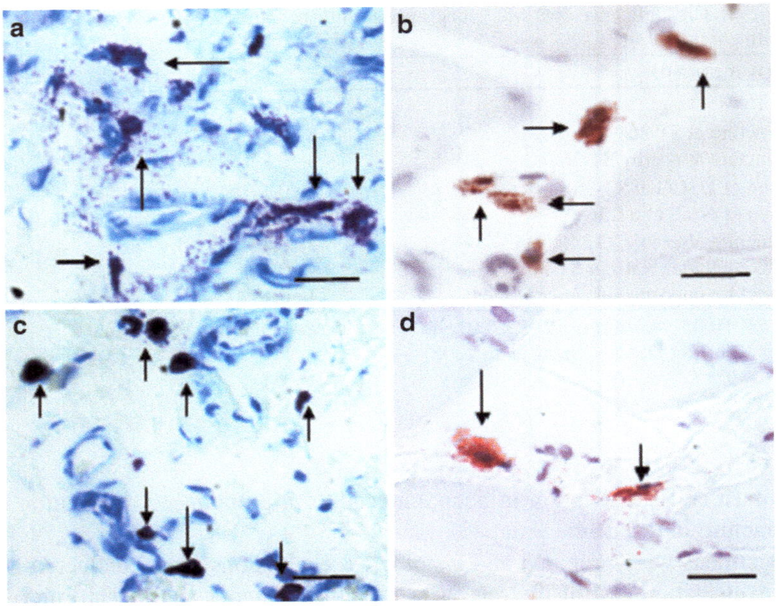

Fig. 3.5 Representative photomicrographs of stained mast cells (*arrows*) at Zusanli (ST-36) in rats (*left panel* from skin stained with TB, right panel from muscle stained with NR). (**a**, **b**) ST-36 after acupuncture; most of the mast cells show evidence of degranulation as judged by extensive changes in appearance and the release of intragranular matrix material into the pericellular space. (**c**, **d**) ST-36 after acupuncture with disodium chromoglycate (DSCG) pretreatment. The majority of mast cells remain non-degranulated after acupuncture, which shows that the mast cell stabilizer treatment was partially effective. Scale bar is 10 μm (Zhang et al. 2008)

Table 3.3 The ratio of degranulated mast cells at acupoints

	MC degranulation ratio (average ± SE) (%)	
Group name ($n=6$)	Skin	Muscle
Control	29.2 ± 1.2	26.3 ± 1.8
Acupuncture	48.9 ± 3.9*	32.2 ± 3.1**
Acupuncture after DSCG	33.7 ± 2.8***	28.5 ± 2.3
Acupuncture after NaCl	50.4 ± 2.5*	29.7 ± 2.3

Note: *$P<0.01$; **$P<0.05$ vs. control group; ***$P<0.01$ vs. acupuncture group (Zhang et al. 2008)

3.2.3 The Function of Collagen Fibers During Acupuncture Stimulation

Assuming that mast cells play an important role in the acupuncture analgesic effect, what is responsible for transmitting the mechanical signal to the mast cells, and what triggers their function? Through microstructure observations, the residues that were wound around the needle after being pulled out from the acupoints consisted

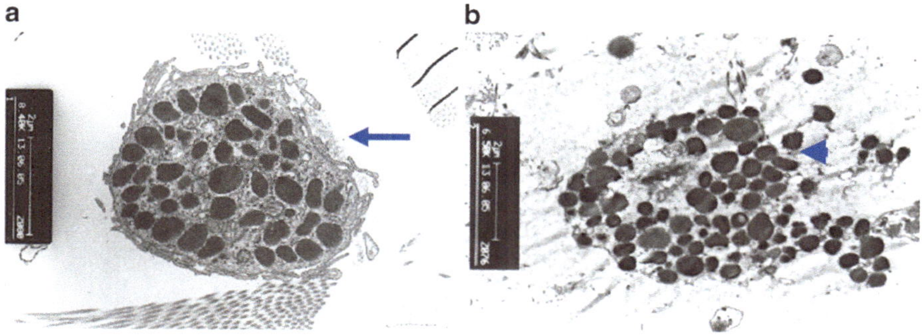

Fig. 3.6 Transmission electron microscopy photomicrographs of mast cells at Zusanli (ST-36) in control and acupuncture groups. (**a**) Skin mast cell without acupuncture, which is round in shape with an intact cytomembrane and densely packed intracellular granules (*arrow*). (**b**) Skin mast cell after acupuncture with a thin and broken cytomembrane and signs of degranulation, including granules outside of the cell (*arrowhead*) and an empty cavity in the cytoplasm (*asterisk*) (Zhang et al. 2008)

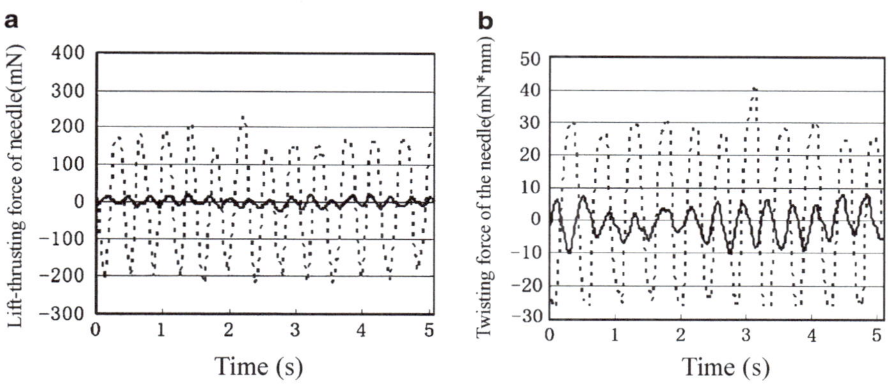

Fig. 3.7 Force changes in the acupuncture needle detected by the real-time monitoring system during two manipulations at Zusanli (ST-36) in rats. (**a**) Lift-thrusting force change. (**b**) Twisting force change. The *dotted* and *solid lines* indicate the force before and after the collagenase pretreatment, respectively. (Yu et al. 2009)

primarily of collagen fibers (Kimura et al. 1992). Hence, we investigated whether collagen fibers had a role in the transmission of signals to the acupoints.

The clinical monitoring system described above was used to record the real-time acupuncture needle force. The results showed the mean amplitudes of the lift-thrusting and twisting forces during acupuncture at Zusanli (ST-36) (dotted line in Fig. 3.7). Both of these forces notably decreased following pretreatment with type I collagenase at Zusanli (ST-36) (continuous line in Fig. 3.7). The data indicate that the collagen fibers wound and tightened around the needle during the lift-thrusting and rotation processes, thus inducing mechanical coupling between the needle and tissue as well as increasing the force of the needle grasp (Yu et al. 2009).

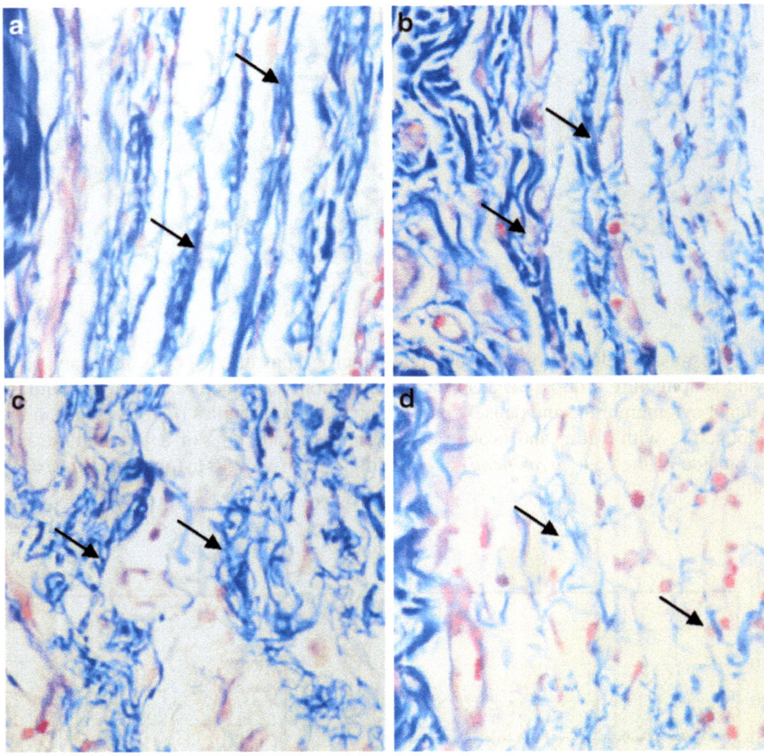

Fig. 3.8 Morphological observations of the stained collagen fibers in the skin at Zusanli (ST-36). (**a**) Control group. (**b**) Lift-thrusting acupuncture group. (**c**) Twisting acupuncture group. (**d**) Collagenase pretreatment group (Mallory stain; magnification ×400; collagens are *blue* and pointed with *arrow*; Muscle and red blood cells are *red*) (Yu et al. 2009)

Morphological observations revealed that the collagen fibers were curled and arranged in parallel bundles in normal skin tissue (Fig. 3.8a). They appeared dark blue after Mallory staining. After lift-thrusting manipulation, the surface of the collagen fibers became rough (Fig. 3.8b). After rotational manipulation, some of the collagen fibers were distorted and accompanied by contracture phenomena (Fig. 3.8d). However, after pretreatment with type I collagenase, the fibrous structures were blurred, the fibril arrangement was distorted, and the color of the stain had faded (Fig. 3.8d).

Considering that the lift-thrusting and the twisting manual acupuncture manipulations could differentially affect the collagen fibers, we analyzed them separately. The results showed that the PT was significantly increased after manual acupuncture irrespective whether the lift-thrusting or twisting mode was used (Table 3.4). There was no significant difference between the two manipulations. However, after pretreatment with type I collagenase, the destruction of collagen fibers at Zusanli (ST-36) weakened the analgesic effect of manual acupuncture considerably. In the

Table 3.4 The influence of collagenase pretreatment on the PT and MC degranulation during acupuncture

Group name ($n=8$)	Change in rate of PT (average ± SD)			Degranulation ratio of MCs in skin at ST-36 (average ± SD) (%)
	Before acu.	After acu.	Change rate (%)	
Control	9.19±0.65	–	–	10.6±2.6
Model	5.61±0.48	–	–	14.1±5.9
Lift-thrusting acupuncture (LA) at ST-36	6.13±0.82	8.05±1.56*	39.09	76.9±5.9‡
Twisting acupuncture (RA) at ST-36	5.44±0.37	7.52±0.65*	38.52	75.7±4.4‡
Injection of collagenase at ST-36	5.90±0.24	5.59±0.52	−5.06	11.9±4.7
LA after collagenase at ST-36	6.05±0.43	6.26±0.79	3.83**°	12.4±2.3§
RA after collagenase at ST-36	5.27±0.62	5.39±0.57	3.24***°	10.4±2.4¶
LA after collagenase at ST-35	5.46±0.42	7.91±0.89*	45.97†	60.5±11.6‡

Note: Change in rate of PT: *$P<0.05$ vs. before acupuncture process (t-test); **$P<0.05$ vs. change rate of LA group; ***$P<0.05$ vs. change rate of RA group; †$P<0.05$ vs. change rate of LA after collagenase group (one-way ANOVA); the change in the PT rate was significantly affected by the collagenase pretreatment (LA, RA, LA after collagenase, RA after collagenase, 2×2 factorial ANOVA, °$P<0.05$). Degranulation ratio: ‡$P<0.05$ vs. model group; §$P<0.05$ vs. LA group; ¶$P<0.05$ vs. RA group (Yu et al. 2009)

LA group after collagenase pretreatment at Dubi (ST-35), we chose a control point located nearest to Zusanli (ST-36) upstream along the stomach meridian. It was shown that the destruction of collagen fibers upstream along the meridian did not influence the local analgesic effect of acupuncture.

In addition, both manual manipulations increased the degranulation ratio of mast cells at local acupoints compared with the model group. Along with the attenuation of the analgesia effect, pretreatment with collagenase significantly counteracted the acupuncture-induced increase in degranulated mast cells (Table 3.4). We concluded that the collagen fibers twisted less due to the destruction caused by the collagenase pretreatment. The ratio of degranulated mast cells decreased with the decrease in the reception of acupunctural signals transmitted by the collagen fibers, which resulted in an attenuated acupuncture effect.

3.2.4 *Involvement of Mechanosensitive Channels in Mast Cell Activation*

All of the above results indicate that mast cells have a decisive role in the peripheral mechanism of acupuncture. Further, collagen fibers in the subcutaneous connective tissue act as a transduction media for mechanical signals between manual acupuncture stimulation and mast cells. What is the underlying mechanism that triggers mast cell degranulation after the mechanical signals are received? Research on mechanosensitive ion channels and the proteins that might play roles in the acupuncture effect was conducted at the cellular level using HMC-1 as the model cell.

Sensitivity to mechanical stimulation is an inherent property of many tissues and cells. Throughout living organisms' evolution, they have experienced a wide variety of mechanical stimuli. They need mechanisms which protect their delicate cell membranes from damaging mechanical stimuli; however, they also require mechanosensitivity (MS) to interact with their changing mechanical environment (Hamill and Martinac 2001). Since the first MS ion channel was reported in skeletal muscle in 1984 (Guharay and Sachs 1984), MS ion channels have been found in more than 30 cell types, including mammal cells, plant cells, and bacteria. MS ion channels are implicated in various physiological processes. For example, endothelial cells in blood vessels sense shear stress to control blood pressure (Lehoux and Tedgui 2004), and hair cells in the mammalian vestibular system sense gravity and speed to control hearing and balance (Colclasure and Holt 2003). MS channels are also involved in such processes as touch and pain sensations (Nakamura and Strittmatter 1996), salt and fluid balance (Bourque and Oliet 1997) as well as volume regulation (Pierce and Politis 1990). MS ion channels include K^+, Na^+, and Ca^{2+} cation channels. TREK and TRAAK (Maingret et al. 1999), ENAC (Sukharev and Corey 2004), and $TRPV_2$ (Muraki et al. 2003) as well as $TRPV_4$ (Jia et al. 2004) are typical proteins that mediate mechanosensitive cation movement across a lipid membrane. Cl^- channels, especially in the CLC family, are the primary mechanosensitive anion channels, which play an important role in volume regulation (Duan et al. 1997).

HMC-1 was the cell line successfully isolated and cultured from the blood of mast cell leukemia patients (Butterfield et al. 1988). As the only cell line similar to the human body's mast cells, it can be stained by metachromatic dyes, such as TB and alcian blue (Butterfield et al. 1988; Nilsson et al. 1994). Moreover, it expresses a series of known mast cell markers, such as tryptase, heparin, chondroitin sulfate (Butterfield et al. 1988), TNF-α, and IL-1βmRNA. HMC-1 expresses a tryptase+/chymase phenotype that has characteristics similar to pulmonary mast cells (Butterfield et al. 1988; Kitamura 1989). As to molecular expression at the membrane, HMC-1 resembles mastocytosis (Legler et al. 1996; Nilsson et al. 1994). In other aspects, such as the immunological phenotype, in combination with cytokines and their functions, HMC-1 also demonstrates a biological similarity to human mast cell (Butterfield et al. 1988; Harvima et al. 1994; Legler et al. 1996; Sillaber et al. 1993; Wedi et al. 1996). The HMC-1 used in this study was kindly provided by the laboratory of Dr. Butterfield (Mayo Clinic, Rochester, USA).

Human mast cells (HMC-1) were cultured in IMDM (Gibco, Invitrogen, Grand Island, NY, USA) supplemented with 2 mM L-glutamine, 25 mM HEPES, 10 % (v/v) fetal bovine serum (Gibco, Invitrogen, Australia), and 1 % penicillin and streptomycin (Gibco, Invitrogen, Grand Island, NY, USA) in a 95 % humidity-controlled incubator with 5 % CO_2 at 37 °C.

To observe the effects of osmotic stress on HMC-1 degranulation, inverted light microscopy (TE2000-U, Nikon, Japan) and a CCD video camera (ORCA-ER,

Hamamatsu, Japan) were used. Photos were taken every 5 min during the experiment at a magnification of ×400. For fluorescence experiments, HMC-1 cells were grown on glass cover slips coated with poly-L-lysine (Sigma Chemical Co.) for 10 min, loaded with 4 μM Calcium Green-1 AM in an IMDM loading buffer for 30 min and then washed with a standard bath solution thrice. All of the solutions used in the fluorescence experiment contained 0.1 mM probenecid. A 100 W super high-pressure mercury lamp (Nikon, C-SHG1) was used as a light source. The excitation light was passed through an interference filter (465–495 nm) and reflected by a dichroic mirror (cutoff wavelength, 505 nm) through a 40× plane objective. The filter and dichroic mirrors were from Nikon. Photos were taken every minute. The images were digitized, averaged (five frames), background-corrected, and analyzed by an image-processing system (Wasabi, Hamamatsu, Japan). The fluorescence intensities of individual cells in the field of view were determined by averaging the image intensities collected from regions of interest within each cell. All of the experiments were performed at room temperature.

Light microscopy analysis showed that a DIDS-sensitive pathway contributes to HMC-1 degranulation. After perfusion with a hypotonic solution (230 mOsm/kg H_2O) for up to 30 min, degranulation of HMC-1 cells was observed. The plasma membrane became rough, and the color of the cytoplasm appeared to be lighter due to a release of granules (Watanabe et al. 2002; Yang et al. 2005). The chloride channel blocker DIDS inhibited such osmotic stress-induced degranulation (Fig. 3.9). Using Calcium Green-1 AM as the Ca^{2+}-sensitive dye, we found that the fluorescence intensity of HMC-1 increased with exposure to a hypotonic bath solution (250 mOsm/kg H_2O) for 10 min, which was also diminished in the presence of DIDS (Fig. 3.10) (Wang et al. 2009).

A patch clamp technique was applied to investigate single-channel activation during mechanical stimulation. The patch clamp recording technique is described as follows. A seal of over a gigaohm (GΩ) between the tip of glass electrode and cell membrane was created by (1) the salt bridge formed by the negative charge of glass electrode tip and the cell membrane that contained a distribution of divalent cations, (2) the hydrogen bond formed by oxygen base on glass surface and oxygen or nitrogen base in membrane phospholipids, and (3) the van der Waals force. This seal separated the membrane patch inside of the electrode tip from the cell membrane outside of the electrode tip, which allowed for a record of the electric current through the ion channel at the picoampere (pA) level. Establishment of the patch clamp technique in 1976 has made the study of cell membrane proteins possible, and Neher and Sakmann, the founders of the patch clamp technique, won the Nobel Prize in Physiology and Medicine in 1991 (Schwarz and Rettinger 2003).

Outside-out patch voltage-clamp recording was performed (if not stated otherwise) with an EPC-10 amplifier using either Pulse (in the initial experiments) or PatchMaster software (HEKA Electronics, Lambrecht, Germany) at room temperature (25–26 °C). The outside-out conformation allowed for fast solution changes at the external membrane surface. Patch pipette electrodes had resistances of 3–6 MΩ. The Ag–AgCl reference electrode was connected to the bath via a 150 mM-KCl Agar Bridge. Access resistance was monitored continuously. Single-channel events

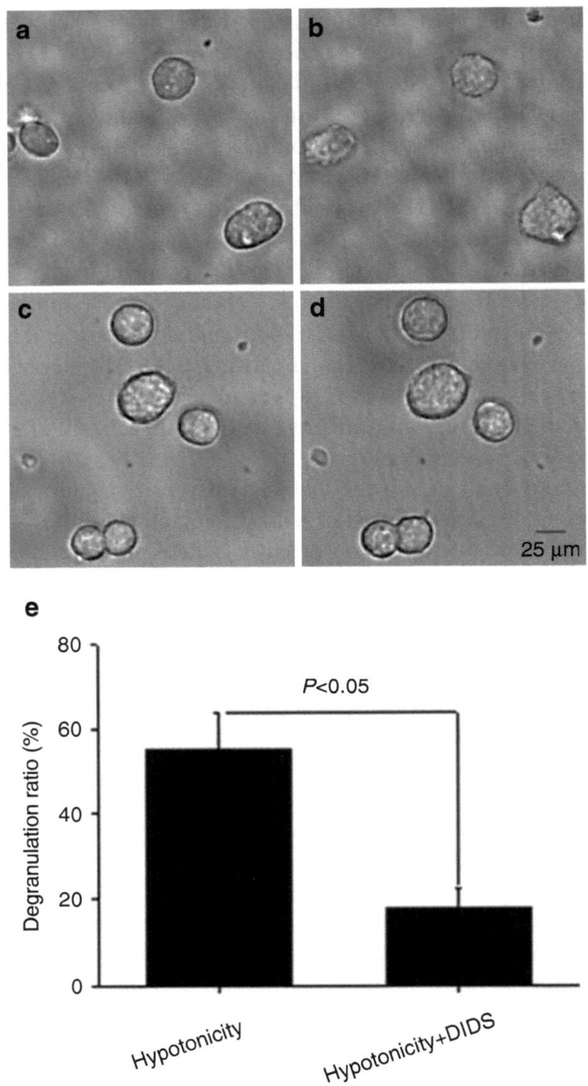

Fig. 3.9 DIDS inhibition of HMC-1 degranulation caused by osmotic stress. (**a**, **b**) HMC-1 cells under isotonic (310 mOsm/kg H_2O) and hypotonic conditions (230 mOsm/kg H_2O). (**c**, **d**) HMC-1 cells in an isotonic solution and a hypotonic solution containing 200 μM DIDS. Photos in *right panel* were taken when HMC-1 was incubated under hypotonic conditions for 30 min. (**e**) The ratio of degranulated HMC-1 cells in the hypotonic solution without ($n=4$) and with 200 μM DIDS ($n=4$) (Wang et al. 2009)

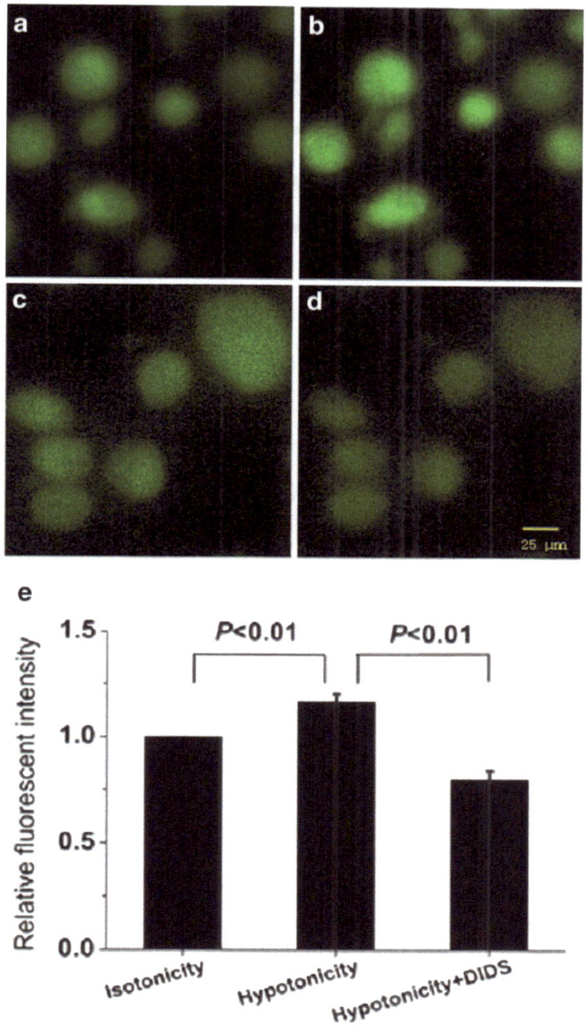

Fig. 3.10 Changes in the fluorescence intensity of the Ca-sensitive dye (Calcium Green-1 AM) after HMC-1 simulation by osmotic stress without and with 200 μM DIDS. (**a**, **b**) fluorescent image of HMC-1 in isotonic and hypotonic (250 mOsm/kg H_2O) bath solutions. (**c**, **d**) fluorescent image of HMC-1 under isotonic conditions and hypotonic conditions containing 200 μM DIDS. Photos in the *right panel* were taken when HMC-1 was incubated under hypotonic conditions for 10 min. (**e**) The relative fluorescence intensity of HMC-1 during exposure to hypotonic conditions without ($n=52$, in five independent experiments) and with ($n=29$, in three independent experiments) 200 μM DIDS (Wang et al. 2009)

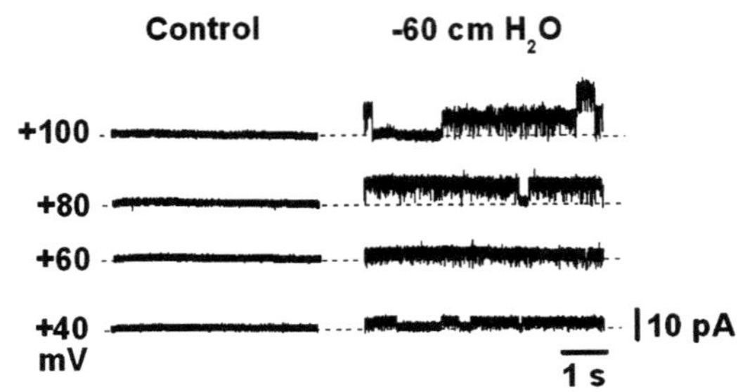

Fig. 3.11 Representative SA current traces recorded from HMC-1 cells in the outside-out mode. The SA channels were activated by −60 cm H_2O and outward currents. The *dashed lines* represent the closed state. The data were filtered at 1 kHz (Wang et al. 2009)

were recorded after the seal resistance approached 2–5 GΩ. Membrane potentials were clamped at 0 mV, and voltage pulses of 5 s duration were applied from −100 to +100 mV in 20 mV increments. The data were sampled at 10 kHz and filtered at either 1, 0.5, or 0.1 kHz. The membrane tension was altered by suction or blowing through the patch pipette using a micrometer-driven syringe; the final pressure was reached within 4 s.

In the electrophysiological experiments, standard extracellular bath solution for outside-out configuration contained (in mM) 150 NaCl, 5 KCl, 2 $CaCl_2$, 5 $MgCl_2$, 4 D-sorbitol, and 10 HEPES, pH 7.4 (adjusted with NaOH). The pipette solution was composed of (in mM) 140 CsCl, 1 $CaCl_2$, 2 $MgCl_2$, 5 EGTA, and 15 HEPES, pH 7.2 (adjusted with Tris). The osmolarity was 310 mOsm/kg H_2O. A stock solution of 4, 4′-diisothiocyanatostilbene-2, 2′-disulfonic acid (DIDS) (Sigma) (200 mM) was prepared in DMSO. The Calcium Green-1 AM (Invitrogen, USA) was dissolved in 20 % (w/v) Pluronic F-127 (Invitrogen, USA) for a 50 mM stock solution. The probenecid (Sigma) solution was prepared in 1 M NaOH for a 250 mM stock solution. All of the stock solutions were stored at −20 °C and diluted into a bath solution to the working concentrations for use.

The membrane tension was adjusted by reducing the NaCl concentrations of the bath solution to change the osmotic stress. The osmolarity of all solutions was regularly measured by a vapor pressure osmometer (Model 5520, Wescor, Logan, UT). During patch clamp recording, the membrane tension was altered by suction or blowing through the patch pipette using a micrometer-driven syringe; the final pressure was reached within 4 s. The pressure was monitored with a manometer (Model 8205, PCE Group oHG, Meschede, Germany). When the mechanical stretch of 30–60 cm H_2O was applied to excise the outside-out patches via the patch pipette, single-channel activity was observed in the excised patches. Figure 3.11 illustrates the representative current traces. The SA channel currents were voltage-dependent with a strong outward rectification. A chord conductance of approximately 55.1 ± 3.4 pS at +100 mV was observed as the dominant state. Further results showed that the SA currents were dependent on the extracellular Cl^- concentration and

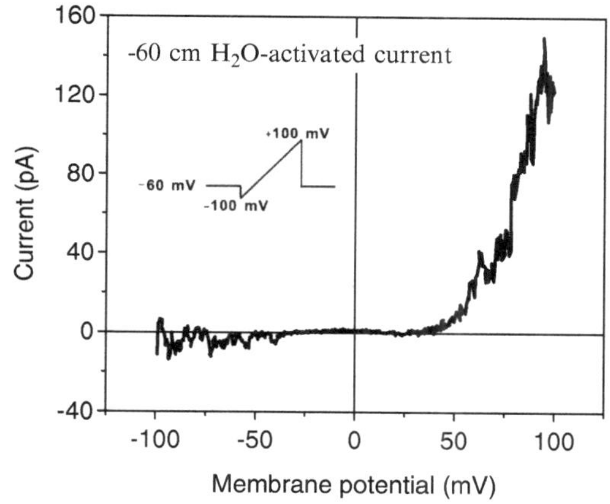

Fig. 3.12 Representative SA current traces recorded from mast cells in connective tissues from the ST-36 area rats using the whole-cell mode. SA current induced by –60 cm H_2O in response to a ramp voltage pulse (see insert) showed strong outward rectification (Wang et al. 2009)

DIDS sensitivity. Moreover, a stretch application in the whole-cell mode for HMC-1-induced TRPV2 currents (Zhang et al. 2012). Mast cells from the connective tissue of the ST-36 area in rats were also mechanosensitive (Fig. 3.12).

3.2.5 *The Effects of a Peripheral Nerve Block on Acupuncture Analgesia and MC Degranulation*

Taking into account the role of the central nervous system in acupuncture analgesia, we studied the pathway of acupuncture signals by blocking the nerves at acupoints along a meridian. Electric acupuncture (EA) and manual acupuncture (MA) were separately adopted to observe the analgesic effect from acupuncture at Zusanli (ST-36) in an AIA animal model. EA stimulation was delivered by a Han's Acupoint Nerve Stimulator (HANS) LH202H with the electrodes connected to two acupuncture needles. One needle was inserted at ST-36 with a 7 mm-insertion depth and the other at Kunlun (BL-60) (a point lying in the depression approximately halfway along a line between the tip of the lateral malleolus and the Achilles tendon) with a 2 mm-insertion depth. The EA treatment parameters were set at dense–disperse wave with a frequency of 2–100 Hz. Intensities of 0.5, 1.0, and 1.5 mA were applied for a duration of 10 min each. The acupoints were pretreated with a Lidocaine hydrochloride (2 %, 0.05 ml) injection to block nerve function (Huang et al. 2009).

When the nerves were blocked before acupuncture, EA and MA no longer increased PT, which demonstrated the indispensable role of the peripheral nerves in acupuncture analgesia (Table 3.5). Acupuncture analgesia at the peripheral level operates on the

Table 3.5 The influence of nerve blockage on the PT and MC degranulation

Groups ($n=8$)	Pain thresholds (average ± SE) (s)			Degranulation ratios (average ± SE) (%)	
	Before model	After model	After treatment	ST-35/ST-39	ST-36
Lido injection	9.72 ± 0.57	5.50 ± 0.83	6.18 ± 0.55	–	31.25 ± 3.66
EA	9.75 ± 0.62	5.44 ± 0.93	8.54 ± 0.93*	–	52.41 ± 3.50
Lido + EA	9.92 ± 0.52	6.03 ± 0.78	6.08 ± 0.54**	–	56.72 ± 6.1
ST-35 Lido + EA	9.64 ± 0.58	6.00 ± 0.68	6.25 ± 0.74**	36.00 ± 3.56	57.21 ± 5.82
ST-39 Lido + EA	9.44 ± 0.41	5.73 ± 0.62	8.40 ± 0.98*	35.46 ± 2.81	54.01 ± 3.34
MA	9.50 ± 0.40	6.08 ± 0.44	9.68 ± 0.82*	–	64.98 ± 6.18
Lido + MA	9.51 ± 0.29	5.92 ± 0.59	6.19 ± 0.51***	–	63.74 ± 3.43
ST-35 Lido + MA	9.44 ± 0.37	5.49 ± 0.56	6.30 ± 0.76***	38.50 ± 3.53	61.25 ± 9.01

*$P<0.01$ vs. Lido injection; **$P<0.01$ vs. EA group; ***$P<0.01$ vs. MA group (Huang et al. 2009)

premise that the nerve impulses produced by EA or MA and noxious stimulation are transferred to the acupoint center by the same afferent nerve. Thus, we blocked the nerves at ST-35 (on the proximal side along the same meridian as ST-36) and performed acupuncture at ST-36. As a result, the acupuncture analgesic effect was significantly attenuated. However, a similar effect was not seen when we blocked the nerves at ST-39 (on the distal side along the same meridian as ST-36). When EA was performed, the mast cell degranulation rate differences at the acupoint were not statistically significant either before or after Lidocaine injection, which shows that nerve blockage at the acupoint does not affect local EA-induced mast cell degranulation. A similar result was obtained using MA. All of the above results can be easily explained by Lidocaine-induced disruption of neural signal transfer. Therefore, we draw the conclusion that the production of acupuncture analgesia requires the integrity of the afferent nerves at the proximal end of the acupoint.

3.2.6 *The Effect of Histamine Release on MC Activation at the Acupoints*

Many regulatory factors affect mast cell degranulation in humans (Galli et al. 2005). Histamine is the most important substance among all of the by-products released by the mast cells. The effect of mast cell activation by histamine was observed through histamine injection at the acupoints. Compared with a saline injection, the histamine solution (100 μg/ml, 50 μl) increased the ratio of degranulated mast cells at the acupoint and produced an analgesic effect that resembled MA (Table 3.6). This further showed that mast cells participate in the production of the acupuncture analgesic effect. Substances released during degranulation, such as histamine, cytokines, and chemokines, generate biological stimuli at the acupoint, which produce acupunctural signals that are transmitted to the central nervous system through a peripheral nerve (Huang et al. 2010).

Table 3.6 The influence of histamine on PT and MC degranulation

Groups ($n=12$)	Pain thresholds (average ± SE) (s)			Degranulation ratios (average ± SE) (%)
	Before model	After model	After treatment	
Saline injection	9.20 ± 0.28	6.09 ± 0.30	6.61 ± 0.31	36.67 ± 2.36
Acupuncture	9.52 ± 0.18	6.58 ± 0.17	8.77 ± 0.26*	57.61 ± 1.42*
Histamine injection	9.21 ± 0.20	6.27 ± 0.22	8.50 ± 0.28*	57.03 ± 2.95*

*$P<0.01$ vs. saline injection group (Huang et al. 2010)

However, the precise mechanism of mast cell activation by histamine requires further exploration. The following are the probable methods that may lead to this activation. (A) The "Axon Reflex" mechanism is composed of reciprocal actions between the nervous and immunological systems. Histamine can lower the activation threshold of the peripheral nerve fiber to raise the activation rate of the sensory nerve (Coutaux et al. 2005). Thus, the ends of activated nerve fibers release several neuropeptide substances, including substance P, that can accelerate the release of histamine from mast cells (Metcalfe et al. 1997). Through these two events, a cross-reaction between the afferent fiber and mast cells is enabled, which may be responsible for the multi-signal transmission beyond the nerve structure in the periphery (Coutaux et al. 2005). (B) Histamine activates mast cells through a non-H3 receptor (Hegyesi et al. 2005; Xie and He 2004). (C) Histamine can also change the mast cell microenvironment to affect capillary permeability (e.g., the flow field for tissue fluid) (Ding et al. 2005). Thus, it can activate mast cells indirectly via a mechanical approach.

3.2.7 The Signaling Mechanism of Mast Cells in Response to Physical Stimuli in TCM

Our study revealed that local mast cells play an important role in the acupuncture analgesia effects because they have distinctly higher densities at acupoints compared with non-acupoints. Acupoint stimulation results in mast cell degranulation and release of biologically active mediators, which are proposed to act as body fluid system modulators. Based on the observation that the mast cell densities are significantly higher at acupoints than non-acupoints (Zhang et al. 2008), the effect of degranulation at acupoints is expected to be stronger than degranulation at non-acupoints. During the acupuncture process in the clinical setting (Fig. 3.13), lift-thrusting or twisting manipulation is applied at specific points that are known as analgesic acupoints. The needle may stimulate both the muscle membrane and the dense dermic connective tissue planes as well as cause collagen fiber deformation (Yu et al. 2009) and mechanosensitive mast cell degranulation at such sites (Wang et al. 2009). The substances in these granules activate the nerve receptors and generate upstream acupuncture analgesia signals (Huang et al. 2009). They may also flow along the meridian through the tissue fluid (Zhang et al. 2009) and induce

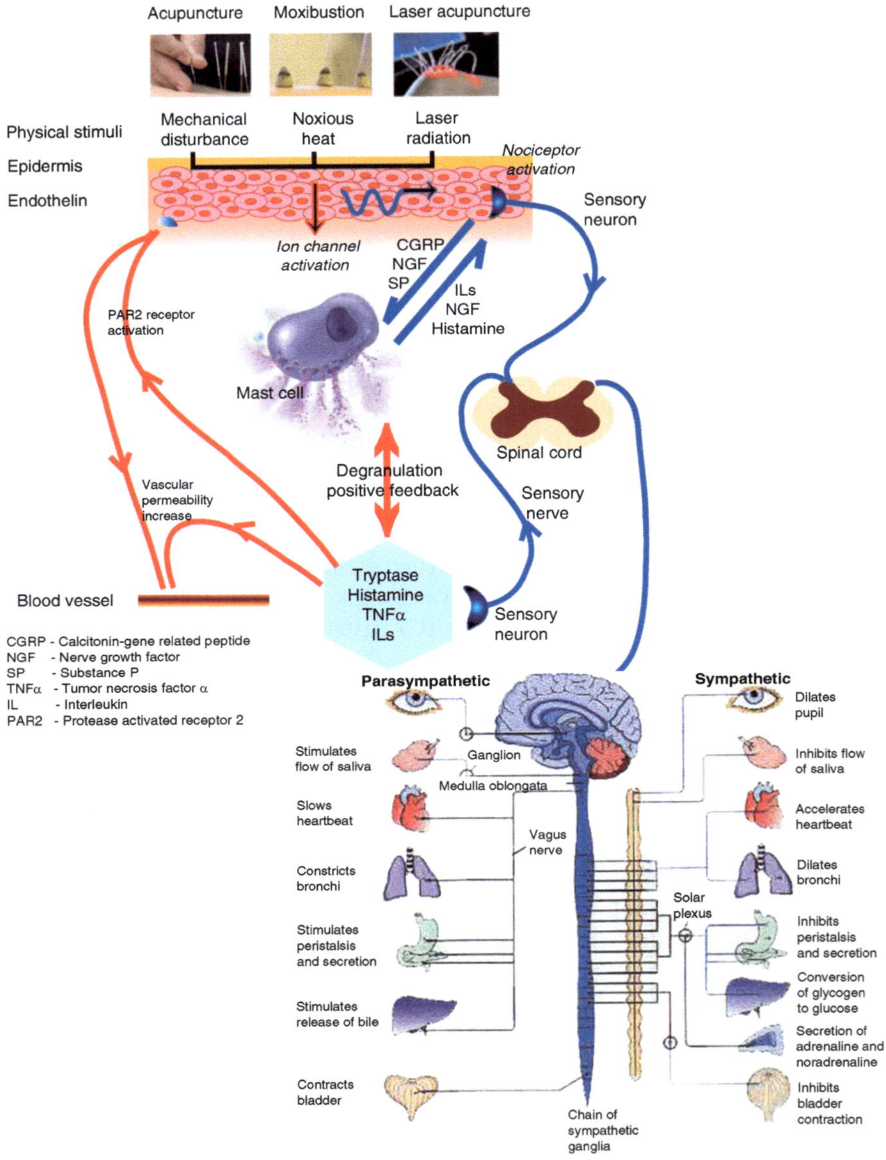

Fig. 3.13 The signaling mechanism of mast cells in response to physical stimuli in TCM. Mast cells are resident cells in tissues throughout the body and are found in association with structures such as blood vessels and nerves as well as in proximity to the skin interfacing with the external environment. Twisting and lift-thrusting acupunctures cause subcutaneous collagen fiber winding. Local mast cells are directly activated through the opening of a calcium-permeable channel, which leads to distinct signaling pathways and causes the extrusion of granules that have a positive feedback effect to further degranulation. All of the reactions from mechanical stimulation, mechanotransduction, and cellular responses combine and result in the clinical phenomena and curative effects

downstream mast cell degranulation, thereby generating the propagated sensation along the meridians. Once mast cell degranulation is inhibited or the collagen fibers are destroyed, mast cells at the acupoints cannot be further activated by manipulation. The transduction of signals at the cellular level and the acupuncture analgesia effect are blocked.

Miraculous Pivot • Jiuzhenshi'er Yuan: Nine Needles and Twelve Primary Acupoints reads, "What counts in acupuncture is the arrival of Qi and then it works." The basic principle behind acupuncture is to lead the "qi" to a lesion and let the "qi" cure the lesion. The Sensation starts at the acupoint after manual stimulation through a needle, propagates along a set meridian, travels down after reaching the lower portion of the meridian, and finally arrives at the target organ. After the acupoint stimulation, across-segment signal transduction is produced by the activation of nerve endings in adjacent neural segments, which is mediated by mast cells and various biologically active substances between the peripheral nerve axon and ending. Simultaneous excitation is then conducted between peripheral and spinal dorsal horn neurons (Zhao et al. 2002). In addition, the nerve growth factor has a significant enhancing effect on antigen-mediated histamine release, and when offered in vivo, it can clearly increase the number of mast cells in connective tissues and mucosa (Bienenstock et al. 1987). Experiments recorded the simultaneous action potential of the brain nerves upon activation by mast cells through the microiontophoresis approach combined with extracellular recording of a single cell (Koszegi et al. 2006). Further, it was observed that 43 % of hypothalamic neurons were excited, which indicates that mast cell degranulation had a direct impact on the excitation of neurons in a mammalian brain. Bidirectional signal transduction between the peripheral and central nervous system extends continuously and then forms an interactive process of orderly acupunctural signal transduction.

Numerous mast cells and nerves are distributed around blood vessels. The spatial conformations of these three components are significant, and they are mutually influenced functional units (Fig. 3.14). These three components participate in signal transduction at acupoints during the acupuncture process. Mast cells are activated by mechanical acupunctural signals through collagen fiber mediation and are consequently degranulated. Capillaries and smooth muscles generate a stress response to acupuncture stimulation at the acupoints. Different types of sensory nerve endings produce different responses. In addition, interactions among these three components form a complicated amplification system for acupunctural signals. The substances released by mast cell degranulation lead to telangiectasia, smooth muscle contraction, and the regulation of nerve ending activity (Metcalfe et al. 1997). Smooth muscle contraction and increased capillary permeability lead to an increase in osmotic pressure in extracellular spaces and a change in the distribution of connective tissue stress, both of which can stimulate nerve endings and mast cells (Shanahan et al. 1985; Wang et al. 2009). Several neuropeptides have direct effects on capillary permeability, and activated nerve fiber endings release several neuropeptide substances, including substance P, that can activate mast cells.

During activation of this spatial conformation by acupunctural signals, the three components also perform their respective functions. Mast cells have a specific

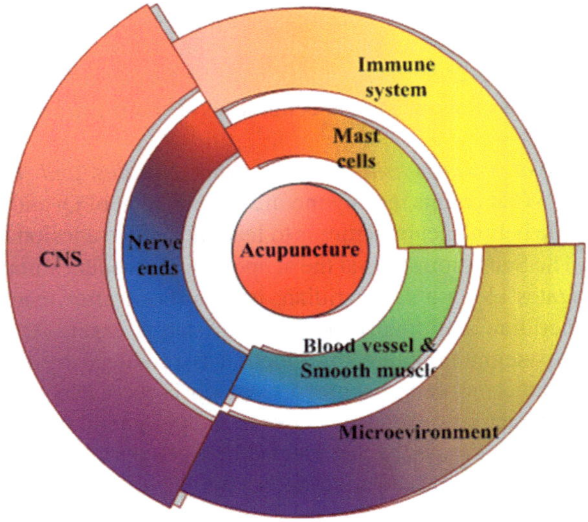

Fig. 3.14 A spatial conformation for the "blood vessel–mast cell–nerve" response materials at acupoints. The *inner circle* is the direct, primary response unit and includes mast cells, capillaries, smooth muscles, and various sensory nerve endings. The *outer circle* is the indirect secondary reaction. Mast cells dominate the regulation and initiation of the immune process. The local capillary and smooth muscle stress responses lead to regional muscle contraction, which results in a tight and sinking sense from the needle. The afferent nerve fiber transfers acupuncture-specific signals to the central nervous system. The color transition indicates the interactions between adjacent units

influence on the immune system; they are important response units for innate immunity and are key regulating units for the adaptive immune system (Galli et al. 2005). Local changes in the microenvironment at acupoints lead to regional muscle contraction, a tight and sinking sense from the needle through regional muscle contraction and visible tissue changes, such as a skin rash or bleeding (Sandberg et al. 2003). Nerve fibers transmit specific acupunctural signals continuously to the central nervous system (Huang et al. 2009).

3.3 Concluding Remarks

Acupuncture treatment is an important component of complementary alternative medicine and has attracted the attention of patients and the medical field worldwide because of its unique effectiveness. Contrary to the large benefits that modern medicine yields from the rapid development of basic research, the link between clinical and bench studies is far from beneficial for TCM. It lacks a self-contained and precisely described theory. No specific field of TCM research has formed. Researchers in every branch of TCM are anxiously looking for a breakthrough but to no avail.

Acupuncture as a science is the pioneer for globalization of TCM, which should be considered a part of life science and biology. Basic research on acupuncture can bridge TCM to modern science and lead the way for a better understanding of life for all nations.

Unlike modern medical treatments, acupuncture and moxibustion treatments focus on the acupoints where either needles or heat are applied to trigger biological changes and yield clinical benefits. Acupuncture effects are the biological responses to stimuli at the acupoints and meridians. An acupoint, as the initiating unit for all of these responses, has drawn the attention of researchers for decades. Based on the results thus far, we conclude that the responses of each meridian or acupoint are distinguishable and have different properties, which can be measured and quantified. The responses have a certain pattern, which could be affected by the structural characteristics of the acupoint treated and the manual acupuncture technique applied or the electroacupuncture parameter setting. Currently, our understanding of acupoint function in response to acupunctural signals is neither thorough nor complete. Acupoint structure and function require further in-depth exploration. Nevertheless, using a multidisciplinary approach, further research on acupoint mechanisms should include crucial issues, such as the development of a quantitative approach for each stage of acupuncture and new therapies for clinical practice. Extensive research has revealed that acupoint interactions rely on the transport and exchange of matter, energy, and information through the meridian and acupoint system. This system is a complex network that challenges attempts to methodically study it piece by piece. A multidisciplinary method with the latest technology is the best tool to fully understand this highly complex network. Further, international communication and cooperation between researchers in this field will benefit acupuncture research in the future. Acupoint mechanism studies are a stepping-stone for scientific development in acupuncture and moxibustion and have the potential to bring about a renaissance of TCM.

Possible directions for future research:

1. The acupoints are the locations that initiate the acupuncture-induced effects. In acupuncture research, it is rational to identify the influence of the acupuncture initiation process. This type of research will benefit the research that focuses on acupuncture effects and will aid in developing a scientific understanding of the meridian theory in TCM.
2. It has been more than half a century since the first wave of research that focused on acupuncture-induced effects. It has been shown that the tissues surrounding acupoints share certain common structural features, such as a high density of nerve endings, capillaries, and mast cells. However, based on the evidence thus far, no specific cell or tissue with structural or functional differences has been observed at the acupoints in humans or other animals. This indicates that all of the specific physical effects at acupoints might only be quantitatively different from non-acupoints and thus might be the same for other aspects of acupoint specificity. These observations should be considered in the future for both clinical and basic research on acupoint specificity.

3. The acupuncture process is a physical stimulus for the body and has both time and frequency characteristics. Such characteristics may be important in achieving an effective acupuncture response. In future studies, these characteristics should be considered instead of regarding acupuncture as a simple mechanical stimulus.
4. In clinical practice, acupuncture is performed according to the immediate response of the patient. Acupuncturists will modify the strength and the frequency based on the counterforce against the needle and the changes in the patient's heart rate, breathing rate, and pain, which create a feedback system between the patient and the acupuncturist. This feedback greatly improves the efficiency and effectiveness of the acupuncture treatment. Currently, in basic research, the effects of communication between the patient and acupuncturist have not been fully studied. These effects might be a crucial factor in producing reliable data in experiments after acupuncture treatment is standardized.
5. The effects of acupuncture can be classified into immediate effects and slow effects from a time point of view. Currently, most studies are focused on the immediate effects of acupuncture. Because in mammals the neural system is responsible for almost all immediate responses and given the studies on acupuncture's immediate effects, a fiction was produced that all acupuncture effects are achieved through the neural system. This statement might only be true for the immediate effects of acupuncture; however, these observations are logically incomplete given the slow effects of acupuncture, which are the primary motivation for the clinical practice of acupuncture. The discoveries thus far might only be the tip of the iceberg; the relationships between acupuncture and other functional structures in the body need to be studied.
6. Studies on the initiation acupuncture effects at acupoints were also based on the immediate effects of acupuncture. Mast cells release dozens of biochemical substances after activation during acupuncture. The effects of these substances released at acupoints are unclear. This modulation during acupuncture might be important for the slower effects of acupuncture and should be studied in the future.
7. Viewed from the entire body, the mechanism of acupuncture-induced signal transmission from the acupoints to the target organ is unclear. The acupuncture-specific input signals that are transmitted through afferents in the periphery and acupuncture-induced output signals transmitted through the efferents toward the target organ have been recorded after acupuncture treatment. However, the mechanism for how the central nerve system reacts to such input signals and produces an effective output signals remains elusive.
8. Clinically, acupuncture-induced effects were varied between different subjects, which indicates that acupuncture-induced effects are related not only to the acupoints and the acupuncture techniques but also to the condition of the treated subject. However, these factors have yet to be studied, which adds uncertainty to our understanding of acupuncture effects. The involvement of these factors in standardized research is a big challenge for further studies.

9. Acupuncture is founded on centuries of innovations and the practical experiences of ancient humans, which makes it scientifically remarkable. A broader understanding of acupuncture will not only facilitate clinical practice but also provide a more integrated view of health care. Such understanding might provide a revolutionary development in modern biology or uncover the biochemical veil of life and lead us directly to its true nature.

Acknowledgements This work was supported by the National Basic Research Program of China (973 Program, Grant No. 2012CB518502), the National Natural Science Foundation of China (Grant No. 81102630), the Shanghai Leading Academic Discipline Project (Grant Nos. S30304, B112), the Science Foundation of Shanghai Municipal Commission of Science and Technology (Grant Nos. 09DZ1976600, 09dZ1974303, 10DZ1975800), and the Fudan Science Foundation for Young (Grant No. 09FQ07).

References

Aumailley M, Gayraud B (1998) Structure and Biological Activity of the Extracellular Matrix. J Mol Med 76:(3–4): 253–265

Bergman MJ (2009) Assessing Adequate Treatment Response in Patients with Rheumatoid Arthritis. Clin Ther 31:(6): 1219–1231

Berman BM, Lao L, Langenberg P, Lee WL, Gilpin AM, Hochberg MC (2004) Effectiveness of Acupuncture as Adjunctive Therapy in Osteoarthritis of the Knee: A Randomized, Controlled Trial. Ann Intern Med 141:(12): 901–910

Bienenstock J, Tomioka M, Matsuda H, Stead RH, Quinonez G, Simon GT, Coughlin MD, Denburg JA (1987) The Role of Mast Cells in Inflammatory Processes: Evidence for Nerve/ Mast Cell Interactions. Int Arch Allergy Appl Immunol 82:(3–4): 238–243

Bourque CW, Oliet SH (1997) Osmoreceptors in the Central Nervous System. Annu Rev Physiol 59: 601–619

Brand DD, Kang AH, Rosloniec EF (2003) Immunopathogenesis of Collagen Arthritis. Springer Semin Immunopathol 25:(1): 3–18

Butterfield JH, Weiler D, Dewald G, Gleich GJ (1988) Establishment of an Immature Mast Cell Line from a Patient with Mast Cell Leukemia. Leuk Res 12:(4): 345–355

Chen EY, Shen XY, Dang RS, Chen HS, Cai DH, He WQ, Fei L (1998) A Relationship between Connecitve Tissue and Accumulation of Calcium with Points on Gb Channel Below Head. Shang Hai Zhen Jiu Za Zhi (Shanghai Journal of Acupuncture and Moxibustion) 17:(2): 36–37 (in Chinese with English abstract)

Chen LW, Zhang BZ (1987) Discovery of the Neuromastocytic Junction on the Modelled Meridian Line in Mouse Skin– a Study by Light and Electron Microscopical Immunohistochemistry. Sheng Jing Jie Pou Xue Za Zhi (Chinese Journal of Neuroanatomy):(2): 253–258 (in Chinese with English Abstract)

Cheng K, Shen XY, Ding GH, Wu F (2009) Relationship between Laser Acupuncture Analgesia and the Function of Mast Cells. Zhongguo Zhen Jiu (Chinese Acupuncture & Moxibustion) 29:(6): 478–483 (In Chinese with English abstract)

Cho ZH, Hwang SC, Wong EK, Son YD, Kang CK, Park TS, Bai SJ, Kim YB, Lee YB, Sung KK, Lee BH, Shepp LA, Min KT (2006) Neural Substrates, Experimental Evidences and Functional Hypothesis of Acupuncture Mechanisms. Acta Neurol Scand 113:(6): 370–377

Colclasure JC, Holt JR (2003) Transduction and Adaptation in Sensory Hair Cells of the Mammalian Vestibular System. Gravit Space Biol Bull 16:(2): 61–70

Coutaux A, Adam F, Willer JC, Le Bars D (2005) Hyperalgesia and Allodynia: Peripheral Mechanisms. Joint Bone Spine 72:(5): 359–371

Deng Y, Zeng T, Zhou Y, Guan X (1996) The Influence of Electroacupuncture on the Mast Cells in the Acupoints of the Stomach Meridian. Zhen Ci Yan Jiu (Acupuncuture Research) 21:(3): 68–70 (In Chinese with English abstract)

Ding GH, Shen XY, Yao W, Dang RS, Yang J, Chen EY (2005) The Dynamic Mechanism of Directional Flow of Plasma and the Phenomena of Human Meridian. Zi Ran Ke Xue Jin Zhan (Progress in Natural Science) 11:(1): 811–818 (in Chinese)

Ding GH, Yao W, Chu JH, Shen XY, Huang ZM, Liu H, Wang SZ, Yang J, Wei H, Fei L (2001) Spectral Characteristic of Infrared Radiations of Some Acupoint and Non-Acupoint Areas in Human Arm Surface. Ke Xue Tong Bao (Chinese Science Bulletin) 46:(23): 678–682 (in Chinese)

Dirig DM, Salami A, Rathbun ML, Ozaki GT, Yaksh TL (1997) Characterization of Variables Defining Hindpaw Withdrawal Latency Evoked by Radiant Thermal Stimuli. J Neurosci Methods 76:(2): 183–191

Duan D, Winter C, Cowley S, Hume JR, Horowitz B (1997) Molecular Identification of a Volume-Regulated Chloride Channel. Nature 390:(6658): 417–421

Dvorak AM, Furitsu T, Ishizaka T (1993) Ultrastructural Morphology of Human Mast Cell Progenitors in Sequential Cocultures of Cord Blood Cells and Fibroblasts. Int Arch Allergy Immunol 100:(3): 219–229

Fei L, Chen HS, Cai DH, Yang SX, Xu JR, Chen EY, Dang RS, Ding GH, Shen XY, Tang Y (1998) Experimental Exploration and Research Prospect of Physical Base and Functional Characteristics of Meridian. Ke Xue Tong Bao (Chinese Science Bulletin) 43:(6): 658–672 (in Chinese)

Gabriel SE, Crowson CS, O'Fallon WM (1999) The Epidemiology of Rheumatoid Arthritis in Rochester, Minnesota, 1955–1985. Arthritis Rheum 42:(3): 415–420

Galli SJ (1993) New Concepts About the Mast Cell. N Engl J Med 328:(4): 257–265

Galli SJ, Nakae S, Tsai M (2005) Mast Cells in the Development of Adaptive Immune Responses. Nat Immunol 6:(2): 135–142

Gao S, Ming C, Wu Y (1984) The Morphological Observation on Rat's Peritoneal Mast Cells under Electroacupuncture at Point "Zuan San Li". Zhen Ci Yan Jiu (Acupuncuture Research) 9:(1): 19–23 (in Chinese with English abstract)

Guharay F, Sachs F (1984) Stretch-Activated Single Ion Channel Currents in Tissue-Cultured Embryonic Chick Skeletal Muscle. J Physiol 352: 685–701

Hamill OP, Martinac B (2001) Molecular Basis of Mechanotransduction in Living Cells. Physiol Rev 81:(2): 685–740

Han JS (2004) Acupuncture and Endorphins. Neurosci Lett 361:(1–3): 258–261

Harvima IT, Horsmanheimo L, Naukkarinen A, Horsmanheimo M (1994) Mast Cell Proteinases and Cytokines in Skin Inflammation. Arch Dermatol Res 287:(1): 61–67

Hathway GJ, Vega-Avelaira D, Moss A, Ingram R, Fitzgerald M (2009) Brief, Low Frequency Stimulation of Rat Peripheral C-Fibres Evokes Prolonged Microglial-Induced Central Sensitization in Adults but Not in Neonates. Pain 144:(1–2): 110–118

Hegyesi H, Darvas Z, Thurmond R, Falus A (2005). Histamine Receptors and Signaling(ed) Biophysical Aspects of Transmembrane Signaling, Springer Berlin Heidelberg. Berlin. 265–291

Ho MW, Knight DP (1998) The Acupuncture System and the Liquid Crystalline Collagen Fibers of the Connective Tissues. Am J Chin Med 26:(3–4): 251–263

Hu JM, Ma WT, Tong H (2001). Acupuncture Researches in Last Decade, State Administration of Traditional Chinese Medicine of the People's Republic of China. Beijing (in Chinese)

Huang H, Zhan R, Yu XJ, Zhang D, Li WM, Ding GH (2009) Effects of Acupoint-Nerve Block on Mast Cell Activity, Manual Acupuncture- and Electroacupuncture-Induced Analgesia in Adjuvant Arthritis Rats. Zhen Ci Yan Jiu (Acupuncuture Research) 34:(1): 31–35, 56 (In Chinese with English abstract)

Huang M, Xie YY, Ding GH (2010) Acupoint-Injection of Histamine Induced Analgesic Effect in Acute Adjuvant-Induced-Arthritis Rats. Zhen Ci Yan Jiu (Acupuncuture Research) 35:(2): 99–103 (In Chinese with English abstract)

Irani AM, Craig SS, DeBlois G, Elson CO, Schechter NM, Schwartz LB (1987) Deficiency of the Tryptase-Positive, Chymase-Negative Mast Cell Type in Gastrointestinal Mucosa of Patients with Defective T Lymphocyte Function. J Immunol 138:(12): 4381–4386

Itoh K, Katsumi Y, Hirota S, Kitakoji H (2007) Randomised Trial of Trigger Point Acupuncture Compared with Other Acupuncture for Treatment of Chronic Neck Pain. Complement Ther Med 15:(3): 172–179

Jia Y, Wang X, Varty L, Rizzo CA, Yang R, Correll CC, Phelps PT, Egan RW, Hey JA (2004) Functional Trpv4 Channels Are Expressed in Human Airway Smooth Muscle Cells. Am J Physiol Lung Cell Mol Physiol 287:(2): L272–278

Joe B, Griffiths MM, Remmers EF, Wilder RL (1999) Animal Models of Rheumatoid Arthritis and Related Inflammation. Curr Rheumatol Rep 1:(2): 139–148

Kimura M, Tohya K, Kuroiwa K, Oda H, Gorawski EC, Hua ZX, Toda S, Ohnishi M, Noguchi E (1992) Electron Microscopical and Immunohistochemical Studies on the Induction Of "Qi" Employing Needling Manipulation. Am J Chin Med 20:(1): 25–35

Kitamura Y (1989) Heterogeneity of Mast Cells and Phenotypic Change between Subpopulations. Annu Rev Immunol 7: 59–76

Kolodney MS, Wysolmerski RB (1992) Isometric Contraction by Fibroblasts and Endothelial Cells in Tissue Culture: A Quantitative Study. J Cell Biol 117:(1): 73–82

Koszegi Z, Kovacs P, Wilhelm M, Atlasz T, Babai N, Kallai V, Hernadi I (2006) The Application of in Vivo Microiontophoresis for the Investigation of Mast Cell-Neuron Interactions in the Rat Brain. J Biochem Biophys Methods 69:(1–2): 227–231

Langevin HM, Churchill DL, Cipolla MJ (2001) Mechanical Signaling through Connective Tissue: A Mechanism for the Therapeutic Effect of Acupuncture. Faseb J 15:(12): 2275–2282

Langevin HM, Yandow JA (2002) Relationship of Acupuncture Points and Meridians to Connective Tissue Planes. Anat Rec 269:(6): 257–265

Legler DF, Loetscher M, Jones SA, Dahinden CA, Arock M, Moser B (1996) Expression of High- and Low-Affinity Receptors for C3a on the Human Mast Cell Line, Hmc-1. Eur J Immunol 26:(4): 753–758

Lehoux S, Tedgui A (2004) Shear and Signal Transduction in the Endothelial Cell. Med Sci (Paris) 20:(5): 551–556 (in French)

Levi-Schaffer F, Shalit M (1993) Proliferation and Functional Responses of Bone Marrow-Derived Mast Cells after Activation. Cell Immunol 148:(2): 435–443

Liu JL, Chen ZR, Hu XL, Rong PJ (2000) Progresses in the Study on Meridian Theory in 2000. Zhen Ci Yan Jiu (Acupuncuture Research) 26:(1): 72–76 (in Chinese with English abstract)

Lou XF, Mei J, Jiang SH, Shi Q, Zhang RF, Tang ML (2006) Study on Vascular Morphology of Integument Tissues of the Channel Area in the Leg. Zhongguo Zhen Jiu (Chinese Acupuncture & Moxibustion) 26:(9): 641–643 (In Chinese with English abstract)

Maingret F, Patel AJ, Lesage F, Lazdunski M, Honore E (1999) Mechano- or Acid Stimulation, Two Interactive Modes of Activation of the Trek-1 Potassium Channel. J Biol Chem 274:(38): 26691–26696

Manheimer E, Linde K, Lao L, Bouter LM, Berman BM (2007) Meta-Analysis: Acupuncture for Osteoarthritis of the Knee. Ann Intern Med 146:(12): 868–877

Manheimer E, White A, Berman B, Forys K, Ernst E (2005) Meta-Analysis: Acupuncture for Low Back Pain. Ann Intern Med 142:(8): 651–663

Manheimer E, Zhang G, Udoff L, Haramati A, Langenberg P, Berman BM, Bouter LM (2008) Effects of Acupuncture on Rates of Pregnancy and Live Birth among Women Undergoing in Vitro Fertilisation: Systematic Review and Meta-Analysis. Bmj 336:(7643): 545–549

Masters DB, Domb AJ (1998) Liposphere Local Anesthetic Timed-Release for Perineural Site Application. Pharm Res 15:(7): 1038–1045

Metcalfe DD, Baram D, Mekori YA (1997) Mast Cells. Physiol Rev 77:(4): 1033–1079

Mu X (1990) The Study of Meridian Substance -- Hyphothesis for Orderly Blood Capillary Network. Bei Jing Nong Xue Yuan Xue Bao (Journal of Beijing Agricultural College) 6:(1): 79–85 (in Chinese)

Muraki K, Iwata Y, Katanosaka Y, Ito T, Ohya S, Shigekawa M, Imaizumi Y (2003) Trpv2 Is a Component of Osmotically Sensitive Cation Channels in Murine Aortic Myocytes. Circ Res 93:(9): 829–838

Nakamura F, Strittmatter SM (1996) P2y1 Purinergic Receptors in Sensory Neurons: Contribution to Touch-Induced Impulse Generation. Proc Natl Acad Sci U S A 93:(19): 10465–10470

Ndong C, Pradhan A, Puma C, Morello JP, Hoffert C, Groblewski T, O'Donnell D, Laird JM (2009) Role of Rat Sensory Neuron-Specific Receptor (Rsnsr1) in Inflammatory Pain: Contribution of Trpv1 to Snsr Signaling in the Pain Pathway. Pain 143:(1–2): 130–137

Nilsson G, Blom T, Kusche-Gullberg M, Kjellen L, Butterfield JH, Sundstrom C, Nilsson K, Hellman L (1994) Phenotypic Characterization of the Human Mast-Cell Line Hmc-1. Scand J Immunol 39:(5): 489–498

Pang X (1992) The Environment Conditions and Morphological Changes of Mast Cell Degranulation. Nei Meng Gu Yi Xue Yuan Xue Bao (Acta Academiae Medicinae Neimongol) 14:(1): 68–70 (in Chinese)

Pierce SK, Politis AD (1990) Ca2(+)-Activated Cell Volume Recovery Mechanisms. Annu Rev Physiol 52: 27–42

Sandberg M, Lundeberg T, Lindberg LG, Gerdle B (2003) Effects of Acupuncture on Skin and Muscle Blood Flow in Healthy Subjects. Eur J Appl Physiol 90:(1–2): 114–119

Schwarz W, Rettinger, J., (2003). Foundations of Electrophysiology, Shaker Verl. Aachen, Germany

Shanahan F, Denburg JA, Fox J, Bienenstock J, Befus D (1985) Mast Cell Heterogeneity: Effects of Neuroenteric Peptides on Histamine Release. J Immunol 135:(2): 1331–1337

Shen DK, Chen XT, Xu GS (1998) Effects of Acupuncture at the Lower Limb Section of Foot-Yangming Meridian on Stomach Physiology and Vascular Permeability. Zhong Guo Zhong Yi Ji Chu Yi Xue Za Zhi (Chinese Journal of Basic Medicine in Traditional Chinese Medicine) 4:(1): 52–54 (in Chinese)

Shen J, Wenger N, Glaspy J, Hays RD, Albert PS, Choi C, Shekelle PG (2000) Electroacupuncture for Control of Myeloablative Chemotherapy-Induced Emesis: A Randomized Controlled Trial. Jama 284:(21): 2755–2761

Sillaber C, Bevec D, Butterfield JH, Heppner C, Valenta R, Scheiner O, Kraft D, Lechner K, Bettelheim P, Valent P (1993) Tumor Necrosis Factor Alpha and Interleukin-1 Beta Mrna Expression in Hmc-1 Cells: Differential Regulation of Gene Product Expression by Recombinant Interleukin-4. Exp Hematol 21:(9): 1271–1275

Smith P, Mosscrop D, Davies S, Sloan P, Al-Ani Z (2007) The Efficacy of Acupuncture in the Treatment of Temporomandibular Joint Myofascial Pain: A Randomised Controlled Trial. J Dent 35:(3): 259–267

Song JM (1977) Mast Cells and Meridian Phenomenon. Liao Ning Zhong Yi Za Zhi (Liaoning Journal of Traditional Chinese Medicine) 19:(2): 59–61 (in Chinese)

Staaf S, Oerther S, Lucas G, Mattsson JP, Ernfors P (2009) Differential Regulation of Trp Channels in a Rat Model of Neuropathic Pain. Pain 144:(1–2): 187–199

Sukharev S, Corey DP (2004) Mechanosensitive Channels: Multiplicity of Families and Gating Paradigms. Sci STKE 2004:(219): re4

Theobald HE (2005) Dietary Calcium and Health. Nutrition Bulletin 30:(3): 237–277

Wang LN, Ding GH, Gu QB, Schwarz W (2009) Single-Channel Properties of a Stretch-Sensitive Chloride Channel in the Human Mast Cell Line Hmc-1. Eur Biophys J 39:(5): 757–767

Watanabe Y, Todome Y, Ohkuni H, Sakurada S, Ishikawa T, Yutsudo T, Fischetti VA, Zabriskie JB (2002) Cysteine Protease Activity and Histamine Release from the Human Mast Cell Line Hmc-1 Stimulated by Recombinant Streptococcal Pyrogenic Exotoxin B/Streptococcal Cysteine Protease. Infect Immun 70:(7): 3944–3947

Wedi B, Elsner J, Czech W, Butterfield JH, Kapp A (1996) Modulation of Intercellular Adhesion Molecule 1 (Icam-1) Expression on the Human Mast-Cell Line (Hmc)-1 by Inflammatory Mediators. Allergy 51:(10): 676–684

Witt C, Brinkhaus B, Jena S, Linde K, Streng A, Wagenpfeil S, Hummelsberger J, Walther HU, Melchart D, Willich SN (2005) Acupuncture in Patients with Osteoarthritis of the Knee: A Randomised Trial. Lancet 366:(9480): 136–143

Wu HG, Yao W, Zhou EH, Dong GR, Zuo CT, Liu HR (2007) Remarks on the Research of Specificities of Acupoints. Zhong Guo Zhen Jiu (Chinese Acupuncture and Moxibustion) 27:(1): 59–62 (In Chinese with English abstract)

Xie H, He SH (2004) Research on Histamine Induced Mast Cell Activation. Zhong Hua Wei Sheng Wu He Mian Yi Xue Za Zhi (Chinese Journal of Microbiology and Immunology) 24:(3): 168 (in Chinese)

Xu MH, Qiao YM, Dai JD (1989) The Tissue Structure of Acupuncture Point and Its Functional Basis. Shang Hai Zhen Jiu Za Zhi (Shanghai Journal of Acupuncture and Moxibustion) 8:(3): 28–30 (in Chinese)

Xu XD, Cui HR, Lou XF, Jiang SH (2004) The Relation between Acupoint Shousanli and Radial Nerve & Vein. Zhen Jiu Lin Chuang Za Zhi (Journal of Clinical Acupuncture and Moxibustion) 20:(7): 47–48 (in Chinese)

Xu ZX, Zheng LY (2003) Present Status of the Relationship between Meridians and Connective Tissues. Shang Hai Zhen Jiu Za Zhi (Shanghai Journal of Acupuncture and Moxibustion) 22:(3): 45–47 (in Chinese with English Abstract)

Yamada K (1996) Relation between the Propagation Sensation Along Meridians and Lympatic Vessles. Guo Wai Yi Xue Zhong Yi Zhong Yao Fen Ce (Foreign Medical Sciences: volume of Traditional Chinese Medicine) 18:(2): 16–18 (in Chinese)

Yan XH, Zhang XY, Liu CL, Dang RS, Huang YY, He W, Ding GH (2009) Do Acupuncture Points Exist? Phys Med Biol 54:(9): N143–150

Yang PC, Zheng PY, Wang CS (2005) Lipopolysaccharide Activates Human Mast Cells to Induce Intestinal Epithelial Barrier Dysfunction Int J Gastroenterology 4:(1): 112–114

Yu XJ, Ding GH, Huang H, Lin J, Yao W, Zhan R (2009) Role of Collagen Fibers in Acupuncture Analgesia Therapy on Rats. Connect Tissue Res 50:(2): 110–120

Yuan L, Jiao PF, Tang L, Huang WH (2005) Anatomical Basic of Traditional Chinese Medical Channels. Zhong Guo Ji Chu Ke Xue (Chinese Basic Science) 7:(3): 5 (in Chinese)

Zhang BZ, Wang JM (1985) Discovery of the Neuro-Mastocytic Junction on the Meridian Line in Human Skin. I.Efferent Neuromastocytic Junction. Shen Jing Jie Pou Xue Za Zhi (Chinese Journal of Neurонanatomy) 1:(1): 47–52 (in Chinese with English abstract)

Zhang D, Ding GH, Shen XY, Yao W, Zhang Z, Zhang Y, Lin J, Gu QB (2008) Role of Mast Cells in Acupuncture Effect: A Pilot Study. Explore (NY) 4:(3): 170–177

Zhang D, Ding GH, Xue XY, Wang LN, Liu C (2004) Development of Researches on the Connective Tissues and Acupuncture. Zhen Ci Yan Jiu (Acupuncuture Research) 29: 77–81 (In Chinese with English abstract)

Zhang D, Spielmann A, Ding GH, Schwarz W (2012) Activation of Mast-Cell Degranulation by Different Physical Stimuli Involves Activation of the Transient-Receptor-Protential Channel Trpv2. Phys Res 61:(1): 113–124

Zhang D, Yao W, Ding GH, Yang J, Schwarz W, Fei Lun S, Liu F, Shen XY, Lao LY (2009) A Fluid Mechanics Model of Tissue Fluid Flow in Limb Connective Tissue--a Mechanism of Acupuncture Signal Transmission. Journal of Hydrodynamics, Ser. B 21:(5): 675–684

Zhao QC (2000) Observation of Superficial Lymph Vessels at 12 Acupoint in Lower Limp. Zhong Yi Yao Xue Bao (Acta Chinese Medicine and pharmacology) 28:(2): 56–57 (in Chinese)

Zhao Y, Shi WC, Sun QX, Zhang SH, Li XH, Jia J, Wang HS (2002) The Peripheral Neural Mechanism of the Sense of Propagation Along Meridian. Yi Xue Yan Jiu Tong Xun (Communication of Medical Research) 31:(12): 18–19 (in Chinese)

Zhong AM, Wu JL, Hu YL (1994) Study on Correlation between the Mast Cell and the Acupoint. World J Acupunc Mox 4:(4): 53–59

Zhu Z, Xu R (1990) Morphometric Observation on the Mast Cells under the Acupuncture Meridian Lines. Zhen Ci Yan Jiu (Acupuncture Research) 15:(2): 157–158 (In Chinese with English abstract)

Zou YH, He ZM, Tang JH, Zhong TJ, Zeng YL (1993) Effect of the Acupuncture of the Twirling Reinforcing- Reducing Method on Rat's Mast Cells in Transparent Fenestra of Mesentery. Guang Dong Jie Pou Xue Tong Bao (Guang Dong Bulletin of Anatomy) 15:(2): 107–111 (in Chinese with English abstract)

Chapter 4
Functional Magnetic Resonance Imaging in Basic Acupuncture Research

Florian Beissner

Abstract Functional magnetic resonance imaging (fMRI) has been used to study the underlying mechanisms of acupuncture for more than a decade. Until the present day, over 80 studies and a dozen review articles have been published on this topic. Emphasizing the methodological and statistical quality of the studies, this critical review evaluates the results obtained so far. After dividing the studies by their analysis strategies into hypothesis-driven and data-driven approaches, inclusion criteria were defined for a meta-analysis on the cortical activations reported by the hypothesis-driven studies. Due to the inhomogeneity of the methods the results from the data-driven studies were only evaluated qualitatively. Seventy-one out of 82 studies applied hypothesis-driven methods. However, only 14 of them met the inclusion criteria. The majority of these studies reported activations in cortical areas relevant to the processing of somatosensory, motor or pain signals, as well as areas related to the special senses. In contrast, data-driven approaches identified some cortical networks that were influenced strongly by acupuncture sometimes leading to sustained changes in cortical connectivity. So far it is impossible to say if cortical activations under acupuncture are part of an underlying mechanism or if they simply reflect the brain's processing of the acupuncture stimulus. Recent results from data-driven analyses, however, have started to shed light on the role that the brain may play in mediating acupuncture-related effects.

Keywords Acupuncture • Meta-analysis • Cerebral cortex • Brainstem • Functional magnetic resonance imaging • fMRI statistics

F. Beissner (✉)
Pain and Autonomics–Integrative Research (PAIR), Clinic of Psychiatry and Psychotherapy, University Clinic Jena, Philosophenweg 3, 07743 Jena, Germany
e-mail: florian.beissner@uni-jena.de

Y. Xia et al. (eds.), *Current Research in Acupuncture*, 89
DOI 10.1007/978-1-4614-3357-6_4, © Springer Science+Business Media New York 2013

4.1 Introduction

When taking a closer look at physiological systems of the human body that usually respond well to acupuncture treatments, it turns out that virtually all of them are neuronally regulated. Prominent examples include pain perception (Jones et al. 2003; Apkarian et al. 2005), the immune system (Nance and Sanders 2007; Tracey 2007), the control system of inner organ functions (Travagli et al. 2006; Samuels 2007) and the endocrine system (Wilder 1995). Disregarding the intricate details of neuronal pathways and connections the nervous system theoretically offers everything needed to explain the effects usually seen in acupuncture treatments: Putting a needle into the skin, a muscle or on a bone irritates local nerve fibres. This elicits an afferent signal that is transmitted to some relay station in the central nervous system (i.e. the spinal cord or the brain). Here the signal either gets modulated or directly transferred by efferent fibres to some other part of the body, where it can trigger a large variety of effects from inhibition of pain signals, through constriction or relaxation of smooth muscle cells, to excretion of hormones or other substances. Although there may exist other underlying mechanisms of acupuncture, it has been shown that cutting off the nerve supply to an acupuncture point by injection of a local anaesthetic can abolish the effect of acupuncture analgesia (Chiang et al. 1973), which underlines the importance of the nervous system at least for this most common acupuncture effect.

While scientist have tried to elucidate the underlying effects of acupuncture since the 1950s, research involving the human central nervous system has been limited by the lack of non-invasive methods to measure its activity. Although electrophysiological methods like electroencephalography (EEG) were available, the limited spatial resolution of this method has precluded the identification of structures involved in the more central parts of acupuncture pathways.

With the advent of functional magnetic resonance imaging (fMRI) in the early 1990s (Belliveau et al. 1990; Ogawa et al. 1990; Kwong et al. 1992), a new tool became available that allowed a non-invasive measurement of human brain activity with a comparatively high resolution of typically 3 mm. Since the late 1990s, this method has been successfully used to investigate the underlying mechanisms of acupuncture. While the idea had already been proposed in 1995 (Yoshida et al. 1995), first results were reported not until some years later. Kathleen Hui from Harvard Medical School and Zhang-Hee Cho from the University of California were the first to publish results of fMRI measurements obtained during acupuncture stimulation (Hui et al. 1997; Cho et al. 1998). Both authors focused on cortical areas of the brain and applied standard fMRI during manual acupuncture. Despite the similarity of their approaches, their results were markedly different: While Cho reported activations of the visual cortex under acupuncture of vision-related acupoints, Hui found activations in somatosensory areas when stimulating pain-related acupoints. Although such differences could in principle be explained by the different acupoints chosen, it will be shown later that they were in fact a warning sign to reconsider the research methodology. Despite receiving a lot of interest,

Cho's study was retracted by the authors in 2006 (Cho et al. 2006). Since these early experiments more than 80 studies and a dozen reviews have been published on the topic by research groups all over the world making acupuncture fMRI a vivid field of research.

This critical review will try to answer the question, what has been learned so far from using fMRI in acupuncture basic research? As the author has shown in a recent article (Beissner and Henke 2009), many studies have not adopted some of the methodological standards that are common practice in the field of fMRI research, which may have led to false-positive results that were unrelated to the applied acupuncture stimulation. Nevertheless, some studies were of excellent quality and have yielded important data on the role the brain plays in the underlying mechanisms of acupuncture.

4.2 The Basics of Functional Magnetic Resonance Imaging

4.2.1 Physical and Physiological Basics

In order to comprehend the basics of fMRI, three effects must be understood: The first effect is the so-called haemodynamic coupling, a physiological mechanism that adapts local cerebral perfusion to the varying needs during neuronal activity (Roy and Sherrington 1890; Attwell and Iadecola 2002). Thus, measuring these local blood flow changes allows an indirect measurement of neuronal activity.

The second prerequisite is the special magnetic behaviour of haemoglobin, the oxygen transporter of the human blood: This protein changes its magnetic properties from paramagnetic to diamagnetic when it binds oxygen molecules (Pauling and Coryell 1936). As brain tissue usually shows diamagnetic behaviour, a larger proportion of oxyhaemoglobin leads to a more homogeneous magnetic field.

The third important effect is the dependency of the so-called T_2* relaxation time of atomic nuclei on the homogeneity of the ambient magnetic field, which in turn is influenced by the amount of diamagnetic substances in the tissue, as shown above. T_2* is one of the physical quantities that is routinely measured in magnetic resonance imaging. As the neuronal need for oxygen is overcompensated by the haemodynamic coupling, an increase of neuronal activity thus leads to a rise of the oxyhaemoglobin concentration in the tissue, which in turn enhances the homogeneity of the magnetic field thus leading to a measurable increase of T_2*.

The combination of these three effects—one physiological and two physical—is the basis of the so-called BOLD contrast (blood oxygenation level dependency), which allows a non-invasive measurement of brain activity with the only prerequisite being a standard MRI scanner, as it is part of most clinical facilities today (Logothetis and Wandell 2004).

4.2.2 fMRI Study Design and Implications for Acupuncture Studies

In the analysis of fMRI data the researcher can choose between two fundamentally different approaches: hypothesis-driven and data-driven[1] ones. The two most frequent hypothesis-driven methods are the analysis based on the general linear model (GLM) (Friston et al. 1995) and the seed-voxel-based cross-correlation analysis (Biswal et al. 1995). Among the data-driven approaches, the independent component analysis (ICA) is the most popular method (McKeown et al. 1998).

4.2.2.1 Hypothesis-Driven Analyses

For hypothesis-driven approaches an a priori model of the time course of brain activity needs to be defined prior to analysis. This is usually achieved by convolving the time course of the external stimulus applied in the study with a so-called haemodynamic response function (HRF) that reflects the delay and temporal behaviour of the blood flow changes following an increase in neuronal activity (Glover 1999). In contrast to other functional imaging methods, like positron emission tomography (PET), the absolute value of the fMRI signal from any cortical area cannot be interpreted directly as it depends on technical circumstances (magnetic field strength, scanner model, coils, etc.), which differ from one measurement site to another. To circumvent this problem, special experimental designs are used in fMRI that follow a simple idea: The fMRI signal is measured for different levels of neuronal activity (e.g. stimulus "on" and stimulus "off"). Then the signals from the two levels are subtracted from each other. An example is the measurement of visual cortex activity during optical stimulation. Here, one performs a baseline measurement with no stimulation followed by a second measurement, where the subject is stimulated. The subtraction of the two signals then yields the fMRI signal related to the visual task. To achieve sufficient statistical power, the stimulation or task usually has to be repeated several times.

The need for stimulus repetition has led to three different groups of experimental designs in fMRI studies: (1) blocked designs, with comparatively long (~30 s) on-and-off periods of stimulation resulting in a boxcar-shaped stimulus curve; (2) event-related designs, where short (<1 s) single stimuli are applied that are separated by longer gaps; and (3) parametric designs, where the time course of the stimulation can be chosen freely as long as its variance is high enough to distinguish it from the background noise (Amaro and Barker 2006).

Although each design has been applied to acupuncture fMRI, the vast majority of studies have used blocked designs. However, as a repetition of the stimulus in an on-and-off design does not reflect the typical way how acupuncture is delivered,

[1] Also often called "model-free approaches".

this leads to problems not only in the interpretation of the results. As the exact time course of the acupuncture effect is unknown and cannot be measured so far, the implicit assumption of such on-and-off designs is that an acupuncture effect is only present as long as the needles are twisted, which is in contradiction to the standard mode of treatment, where needles are usually inserted and twisted followed by period of several minutes of rest with the needles in place. The use of blocked designs in acupuncture-fMRI studies has recently been criticized by several groups (Ho et al. 2008, 2009; Bai et al. 2009a, b, c; Yeo et al. 2010).

4.2.2.2 Data-Driven Analyses

The second approach to fMRI data analysis is data-driven (or "model-free") and relies on spatio-temporal analysis of the data for patterns that exhibit certain predefined characteristics. Methods like principal or independent component analysis (PCA, ICA) divide the variance into statistically maximally independent parts that are considered to be relevant signals in the data. Such an approach allows for a detection of brain activity patterns that are related to some stimulus, for which the exact time course is unknown (McKeown et al. 1998). However, the result of an ICA is strongly dependent on the estimated dimensionality of the signal space.[2] Therefore, over-fitting can easily occur, which results in signals being split up into more than one independent component. Such results can be very hard to interpret. In recent times, however, there has been considerable progress in both dimensionality estimation and fitting noise-free models to noisy data (Beckmann and Smith 2004).

As data-driven approaches do not require the specification of a predefined time course, they are useful alternatives when analysing fMRI data on acupuncture stimulation. If acupuncture activated a specific cortical network consistently over multiple subjects in an fMRI study cohort, an ICA-based analysis would probably be able to detect it, even if its time course was unspecified before. However, as the ICA results will contain several ICs, some a priori information will still be necessary to finally decide, which of them are related to the acupuncture stimulation and which are not.

4.2.2.3 The Problem of Multiple Comparisons

A problem that arises in both approaches, but mostly affects hypothesis-driven analyses, is that of multiple comparisons. As the usual spatial resolution of an fMRI measurement is about $3 \times 3 \times 3$ mm, an image of the whole brain comprises about 40,000 measurement points, which are called voxels. These are all evaluated independently when using a standard GLM model. In such an analysis, all voxels are tested, whether they exhibit a time course similar to that defined a priori. With a

[2] That is, the pre-chosen number of independent components.

conventional p-value of 0.05 we would find about 2,000 voxels to show a significant correlation (i.e. to be activated) by pure chance. In order to reduce these false-positive activations, special correction methods have been developed. These include voxel-wise control of the family wise error (FWE), voxel-wise control of the false discovery rate (FDR) and thresholding based on the size of the activation cluster. For preliminary results, p-values are sometimes set to arbitrary values like 0.001 (uncorr.) that, however, only allow for an unquantified control of false positives. To draw reliable conclusions, fMRI results should be reported at a corrected level using one of the methods mentioned above. Using uncorrected thresholds may only be appropriate, when there is a clear a priori hypothesis concerning the localization of the expected activation. One should keep in mind, however, that the use of statistical methods and thresholds has changed over the years and is still debated. For a recent summary on how to report the results of an fMRI study, see (Poldrack et al. 2008).

As the author has shown elsewhere, ignoring the multiple comparisons problem can easily lead to false-positive results, which are then often interpreted by the authors as being acupuncture-related (Beissner and Henke 2009). Hence, there are not many structures in the human brain that have not been linked to the underlying mechanisms of acupuncture by some fMRI study. As this leads to a large inhomogeneity of the results, it must be considered one of the major problems of acupuncture-fMRI studies so far. This also underlines once more the importance of a clear a priori hypothesis before designing or conducting any fMRI experiment.

4.3 Results from One Decade of Research

While all acupuncture-fMRI studies before 2006 have solely used GLM-based analyses (Qin et al. 2006), model-free methods have gained much popularity since then. Down to the present day,[3] there have been 82 acupuncture-fMRI studies published in peer-reviewed journals of English language (Hui et al. 1997, 2000, 2005, 2009; Cho et al. 1998, 2002, 2010; Wu et al. 1999, 2002, 2008; Alimi et al. 2002; Gareus et al. 2002; Kong et al. 2002, 2007a, b, 2009a, b, c; Siedentopf et al. 2002, 2005; Li et al. 2003a, b, 2004, 2005, 2006a, b, 2008, 2010; Zhang et al. 2003a, b, 2004, 2007, 2009a, b, c; Fang et al. 2004, 2008; Litscher et al. 2004; Liu et al. 2004, 2007, 2009a, b, 2010; Yoo et al. 2004, 2007; Jeun et al. 2005; Nakagoshi et al. 2005; Napadow et al. 2005a, b, 2007a, b, 2009a, b; Parrish et al. 2005; Ueda et al. 2005; Yan et al. 2005; Qin et al. 2006, 2008; Wang et al. 2006, 2007a, b; Xiao et al. 2006; Bai et al. 2007, 2009a, b, c, 2010; Chen et al. 2007; Lu et al. 2007; Schaechter et al. 2007; Beissner and Henke 2008; Deng et al. 2008; Dhond et al. 2008; Ho et al. 2008, 2009; MacPherson et al. 2008; Zhou and Jin 2008; Chae et al. 2009; Na et al. 2009; Park et al. 2009; Wesolowski et al. 2009; Asghar et al. 2010; Yeo et al. 2010). Seventy-one of these applied hypothesis-driven methods. Eleven studies either used ICA or other measures of functional connectivity (Qin et al. 2006, 2008; Bai et al.

[3] June 2010.

Table 4.1 Inclusion criteria for the meta-analysis of hypothesis-driven acupuncture-fMRI studies

Inclusion criteria for the meta-analysis
Minimum number of subjects (n): 12
Statistical analysis method: random-effects (or mixed effects) analysis
Statistical threshold (p-value): $p<0.001$ (uncorr.)

2007, 2009a, b; Dhond et al. 2008; Hui et al. 2009; Liu et al. 2009a, b, 2010; Zhang et al. 2009a, b).

Based on the brief introduction into fMRI statistics one can now motivate inclusion criteria for a meta-analysis of hypothesis-driven fMRI-acupuncture studies, which will help to assess the evidence gained during more than one decade of research. The three criteria that will be applied are summarized in Table 4.1. Except the demand for a random-effects analysis,[4] which had not yet become the gold standard, when some of the early studies were published, these criteria cannot be considered to be overly strict.

The minimum number of subjects in this context does not refer to the overall number of subjects included in the study but to those subjects whose data were actually used to calculate the statistical contrast that the authors' conclusions are based on. Many studies start with an adequate sample size of 25 or more subjects but divide it into several subgroups, e.g. to compare the effects of different acupoints. In each subgroup the number of subjects is then too low to generate activations that survive conservative thresholding. In fact, the uncorrected p-value of 0.001 chosen here is rather liberal and would not be considered for publication by many journals unless a clear a priori hypothesis has been stated before. The author is aware that statistical methods have undergone changes over the past years and that some early studies may be disfavoured by defining inclusion criteria based on today's knowledge. However, as in any emerging field of research, methodology has to be adapted to new knowledge.

4.3.1 *Meta-analysis of Acupuncture-Related Cortical Activations from Hypothesis-Driven fMRI Studies*

Although the inclusion criteria were chosen rather liberal, only 14 out of 71 studies fulfilled them (Li et al. 2003a, b, 2004, 2006a, b; Fang et al. 2004; Yan et al. 2005; Zhang et al. 2007, 2009a, b, c; Ho et al. 2008; MacPherson et al. 2008; Qin et al.

[4] When analysing fMRI data with the GLM approach, there are two basic statistical models: A fixed-effects model or a random-effects model. While a random-effects analysis (RFX) takes into account the between-subject variance of the neuronal activations found on the single subject level, in a fixed-effects analysis (FFX), only within-subject variance is considered. As a consequence FFX has a considerably higher susceptibility to outliers, i.e. single subjects with strong activations (Friston et al. 1999). Thus, for group analyses RFX is the method of choice, as the results from an FFX cannot be generalized from the sample to the population.

2008; Kong et al. 2009a, b, c; Na et al. 2009; Napadow et al. 2009a, b; Wesolowski et al. 2009). This shows that there is a paucity of high-quality studies in basic research on acupuncture as it has often been reported for clinical trials.

The studies that fulfilled the inclusion criteria were further analysed for their reported cortical activations under acupuncture stimulation. Areas that were reported by more than two studies were considered relevant (see Table 4.2). Activations and deactivations were treated separately. The area most often activated by acupuncture was the Secondary Somatosensory Cortex (S2), followed by the Insula, the Primary Somatosensory Cortex (S1), the Cerebellum, the Thalamus, the Primary Motor Cortex (M1), the Superior Temporal Gyrus (STG), the Visual Cortices, the Inferior Frontal Gyrus (IFG), the Pre-motor and Supplementary Motor Cortex [(pre-)SMA], the Basal Ganglia, the Medial Temporal Gyrus (MTG) and the Anterior Cingulate Cortex (ACC). For a summary see Fig. 4.1.

Although there is no one-to-one mapping of cortical areas and cognitive functions, a closer look at the reported activations (and deactivations) reveals that all of them can be subsumed under one of four basic systems: (1) the somatosensory system, (2) the motor system, (3) the sensory integration system, (4) the special senses (vision, hearing). For a summary see Fig. 4.2.

With this preliminary result, one can now hypothesize how these areas could explain possible underlying mechanisms of acupuncture on a cortical level. Many different models have been proposed so far. In the following, we will review some examples.

Li et al. (2006a, b), who conducted an fMRI study on stroke patients treated with acupuncture, conclude that the observed activations are the functional substrate of the therapeutical effect of acupuncture in stroke rehabilitation. Na et al. (2009) interpret the activation of the primary motor cortex, observed in their study, as a specific effect of acupuncture on motor function, while Napadow et al. (2009a, b) put forth the hypothesis of acupuncture as a somatosensory-guided mind–body therapy. Another popular way of interpreting a common finding in acupuncture-fMRI studies can be exemplified by the study of Zhang et al. (2007). Here the authors found activations in cortical areas belonging to the so-called pain matrix (Melzack 1990; May 2007). As the acupoints that were used in their study had an analgesic potential, they conclude that acupuncture at these points modulates the pain matrix and thereby pain perception (Zhang et al. 2007). Although the authors do not state what exactly is meant by "modulating the pain matrix," it is an important result that the majority of the cortical areas frequently activated under acupuncture stimulation also plays a role in the nociceptive (i.e. pain processing) system. In fact all areas of categories one to three (Fig. 4.2) belong to the pain matrix.

While it may be favourable from the viewpoint of acupuncture research to interpret the activation of nociceptive areas as an analgesic effect of acupuncture, a critical analysis must come to the conclusion that these activations could as well just be the response of the brain to the pain stimulus delivered by the acupuncture needle. Although the deqi sensation is usually considered not to be a simple pain stimulus,[5] it nevertheless can be described as a pain-like sensation. Even if the

[5]In fact some researchers clearly distinguish between deqi and "sharp pain" (Hui et al. 2007; Kong et al. 2007a, b).

Table 4.2 Results of a meta-analysis of GLM-based acupuncture-fMRI studies. Areas are shown, if they have been reported by more than two independent studies. One study (Wesolowski et al. 2009) did not find any activations. BA stands for Brodman area

	Li et al. (2003a, b)	Fang et al. (2004)	Li et al. (2004)	Yan et al. (2005)	Li et al. (2006a, b)	Zhang et al. (2007)	Ho et al. (2008)	MacPherson et al. (2008)	Qin et al.(2008)	Kong et al. (2009a, b)	Na et al. (2009)	Napadow et al. (2009a, b)	Zhang et al. (2009a, b, c)	Sum
Activations														
Secondary somatosensory cortex/parietal operculum/inferior parietal lobule/BA40		X			X	X	X		X			X	X	7
Insula		X				X	X		X			X	X	6
Primary sensory cortex/postcentral gyrus			X		X	X	X					X	X	6
Cerebellum				X	X	X		X			X		X	6
Thalamus		X				X	X		X		X		X	6
Primary motor cortex/precentral gyrus/BA4			X		X	X	X					X		5
Superior temporal gyrus				X		X						X	X	4
Visual cortex/cuneus/lingual gyrus/BA 17, BA18, BA19	X			X		X						X		4
Inferior frontal gyrus/frontal operculum/BA44, BA45		X				X						X		3
Premotor cortex/supplementary motor area/medial frontal cortex/BA6/BA8			X				X					X		3
Basal ganglia		X				X					X			3
Medial temporal gyrus				X		X						X		3
Anterior cingulate gyrus						X	X						X	3
Deactivations														
Visual cortex/cuneus/lingual gyrus/BA 17, BA18, BA19				X				X		X		X		4

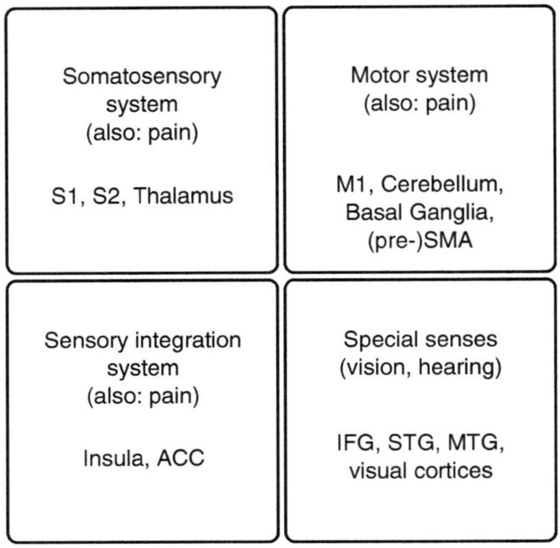

Fig. 4.1 Classification of areas frequently activated in acupuncture-fMRI studies

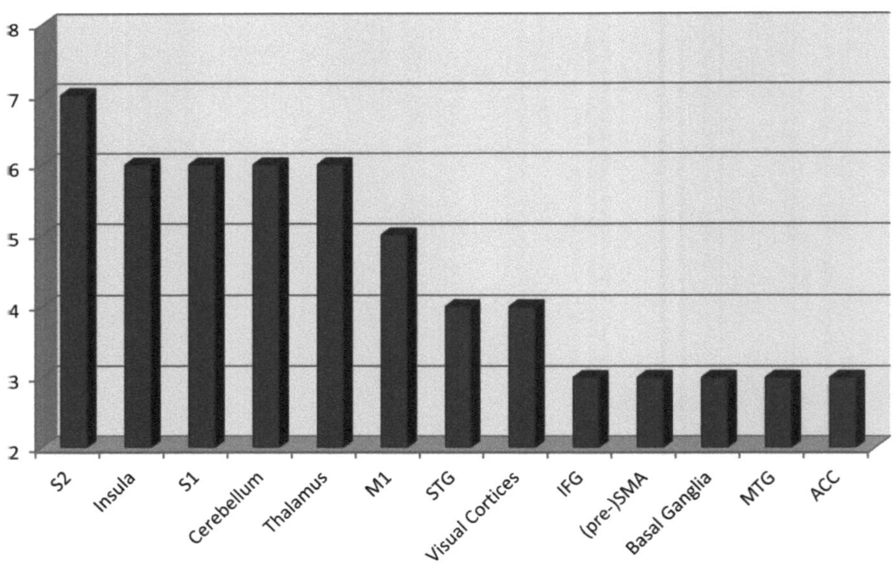

Fig. 4.2 The most frequently activated cortical areas under acupuncture stimulation. Data from 14 studies that fulfilled the inclusion criteria (see Table 4.1). Only areas that were activated more than twice are shown. For explanation of the abbreviations see the text above

insertion of the needle is not felt by the subject, its manipulation can often be unpleasant or even painful. Accordingly, activations in somatosensory and sensory integration areas in this context can be interpreted as the brain processing the sensory input of the needle manipulation, while motor activations can be explained by the fact that sensorimotor cortices are frequently activated in pain studies as well (May 2007).

Activations in areas of category four (special senses) have given reason to several papers with very similar results that were usually interpreted by the authors in the way that the observed activation, e.g. of the visual cortex under acupuncture stimulation offers an explanation for its effectiveness in eye diseases (Cho et al. 1998; Gareus et al. 2002; Li et al. 2003a, b; Litscher et al. 2004; Parrish et al. 2005; Kong et al. 2009a, b, c; Liu et al. 2009a, b). As the author has shown elsewhere, such an explanation is inconsistent, because eye-related diseases usually treated with acupuncture, like redness, pain and swelling of the eye (Deadman et al. 1998), have no pathophysiological connection to the visual cortex (Beissner and Henke 2009). Thus, a change in neuronal activity of this cortical area under acupuncture stimulation can hardly explain the therapeutic effect this stimulation may have on the eye. Furthermore, a point-specific effect is ruled out by the fact that both vision-related and non-vision-related acupoints activated the visual cortex (Parrish et al. 2005; Liu et al. 2009a, b). A more likely explanation for these activations, which has also been put forth by other authors, is that these activations and deactivations are not a direct result of acupuncture stimulation but rather of cross modal interactions, produced by needling-evoked somatosensory stimulation (Kong et al. 2009c). This is underlined by the fact that most of the areas of category four also play a role in sensory integration or attention.

Summing up, the evidence from the 14 hypothesis-driven acupuncture-fMRI studies that fulfilled the inclusion criteria does not allow a clear statement about the basic mechanism underlying acupuncture's effectiveness. In the best case, the cortical activations observed under acupuncture stimulation are the neuronal correlate of the underlying mechanism. In the worst case, however, these activations show nothing more than the brain processing the somatosensory or pain stimulus generated by the acupuncture needle being poked into the skin.

4.3.2 Additional Results from Functional Connectivity Studies

The limitations of hypothesis-driven studies have been mentioned above. The most fundamental problem is the need for a predefined acupuncture time course, which is usually not available. An exception is the online deqi rating method, recently introduced by Napadow et al. (2009a), who continuously recorded the subjects' intensity ratings of the deqi sensation on a visual analogue scale (Napadow et al. 2009a). Using the data from these ratings for the fMRI, his analysis revealed areas of the brain directly involved in processing of the acupuncture-specific sensation. How far this is related to the therapeutic effect of acupuncture still has to be elucidated.

In recent years new approaches like seed-voxel-based connectivity analyses and ICA have been increasingly applied to the study of acupuncture effects with fMRI. Although there is a large variability in methodology, the aim is always the same, namely to measure the functional connectivity between cortical regions, i.e. to identify networks of cortical areas that are involved in the same tasks. In a second step one can then try to answer the question, how these networks are influenced by acupuncture. Functional connectivity can either be analysed in a hypothesis-driven or in a data-driven way. In the first case an anatomical region of interest (ROI) is chosen as a seed region, whose mean time course is than compared to the rest of the cortex in order to find cortical areas that show a similar temporal behaviour. The ICA as the main example for a data-driven method to assess functional connectivity has already been discussed above (see Sect. 4.2.2.3). The rest of this section will summarize some results from the 11 studies that have applied these methods to measure cortical effects of acupuncture (Qin et al. 2006, 2008; Bai et al. 2007, 2009a, b; Dhond et al. 2008; Hui et al. 2009; Liu et al. 2009a, b, 2010; Zhang et al. 2009a, b).

Applying a seed-voxel-based approach using the amygdala as an ROI (Qin et al. 2006, 2008) found altered connectivity of the ventromedial prefrontal cortex, periaqueductal grey (PAG), ACC, hypothalamus and cerebellum after acupuncture stimulation. In some areas activity depended on the stimulation type (true or sham acupuncture): Insula, ACC and PAG showed higher connectivity after true acupuncture administration, while sham acupuncture increased the connectivity of the IFG, the secondary somatosensory cortex (S2), the superior parietal gyrus (SPG), the inferior temporal gyrus (ITG), the basal ganglia and the cerebellum. The same group also demonstrated differences between intrinsic connectivity networks before and after acupuncture administration using a network-theory-based method (Bai et al. 2007). The results were very similar to those of the earlier study. Later the authors found that the strength of the activation in the anterior insula showed a positive correlation with the intensity of the deqi sensation (Bai et al. 2009c). This is an interesting finding, as the anterior insula is an area that is known to differentiate C-fibre-mediated from Ad-mediated pain (Weiss et al. 2008). Activations of the anterior insula under acupuncture stimulation have also been emphasizes by other authors (Hui et al. 2009). Thus, the pivotal role of this cortical area may tell us something about the physiological basics of the deqi sensation.

Using a special algorithm to detect spontaneous cortical activity and combining it with a graph-theoretical analysis (Bai et al. 2009b) could show that acupuncture altered functional connectivity especially of the insula, the hippocampus, the putamen and caudate, the PAG, the hypothalamus and the amygdala. Dhond et al. (2008), who analysed the default mode network (DMN) and the sensory-motor network (SMN), found that activity in these networks was increased only by true and not by sham acupuncture. They further found that after acupuncture administration the DMN recruited additional cortical areas like the hippocampus. Furthermore, this area showed a positive correlation with the activity of the autonomic nervous system as measured by heart rate variability (HRV). Hui et al. (2009) came to the similar conclusion that acupuncture expanded the DMN. In their study the amygdala and the hypothalamus were deactivated in addition to the usual DMN areas. In a graph-theoretical approach applying partial correlation analysis (Liu et al. 2009b) found that different acupoints have different effects on the connectivity within the DMN. However, all acupoints had in common that they totally disrupted the connectivity between the PCC and the ACC. The same authors also investigated sustained effects of acupuncture (Liu et al. 2010), where they identified two distinct cortical networks of which only one was changing over time. In their results once more the anterior insula played an important role.

Summing up, there is a large variability among the results from functional connectivity studies of acupuncture effects. As some of the applied methods have not yet been extensively tested for their neuroscientific applicability, results should be interpreted with precaution. For example, the concept of anti-correlated networks as applied by some acupuncture-fMRI studies (Qin et al. 2006; Bai et al. 2009a, b, c; Hui et al. 2009, 2010) is currently under vigorous debate, as it may be an artifact of image preprocessing (Cole et al. 2010). In general, the exact meaning of intrinsic connectivity networks,[6] like DMN, SMN and others, is not yet well understood, although there is the hypothesis that they reflect the intrinsic energy demands of neuron populations that share a common functional purpose and have thus wired together through synaptic plasticity (Cole et al. 2010). As these networks have been observed to change, e.g. in learning tasks (Lewis et al. 2009), it may in fact be a valid approach to look for similar changes following acupuncture stimulation. However, one should never forget to ask the question, if such changes can really explain the effects of acupuncture seen in the clinical setting.

4.4 Concluding Remarks and Future Directions

The field of acupuncture-fMRI studies is rapidly developing. Following the general trend in neuroimaging, more and more studies nowadays apply data-driven and connectivity-based approaches. Although no clear mechanism has been identified so far, at least the involvement of cortical centres for somatosensory functions, motor functions and sensory integration as well as areas of the special senses has been clearly demonstrated. The importance of some of these areas, like the anterior insula, has also been corroborated by functional connectivity studies. If the involvement of these areas under acupuncture stimulation is the neuronal correlate of an underlying mechanism or shows nothing more than the brain's processing of the painful needling stimulus still has to be elucidated.

An article about fMRI would not be complete without mentioning the limitations of this technology. Besides the problem of possibly false-positive results, when ignoring the multiple comparisons problem, there is one other major drawback that

---

[6]Also called resting-state networks, because they are as well present, when the brain is "at rest."

must not be overlooked: fMRI is almost blind, when it comes to areas of the brain which are close to air-filled cavities[7] or in the vicinity of large arteries.[8] These include, e.g. the basal forebrain, the amygdala or the brainstem. Thus, activations in such areas will often be overlooked by standard fMRI methods. As the brainstem contains important centres for the maintenance of bodily homeostasis, it could be a crucial area to look for acupuncture-related changes. So far only three acupuncture-fMRI studies have specifically targeted this part of the brain, as it necessitates an adaptation of standard measurement methods (Napadow et al. 2005a, b, 2009a, b; Beissner and Henke 2008). The same holds true for the spinal cord (Li et al. 2005; Wang et al. 2006; Chen et al. 2007), although this area may be of even higher importance, as it has recently been shown to play a crucial role in placebo analgesia (Eippert et al. 2009).

Together with the upcoming advances in analysis methodology, extending fMRI to brainstem and spinal areas will hopefully lead to a more comprehensive understanding of neuronally mediated acupuncture effects.

Acknowledgements This work was supported by the Horst Görtz Foundation.

References

Alimi D, A G, Gardeur D (2002) Auricular acupuncture stimulation measured on functional magnetic resonance imaging. Medical Acupuncture 13(2):18–21

Amaro E, Barker GJ (2006) Study design in fMRI: Basic principles. Brain and Cognition 60:220–232

Apkarian AV, Bushnell MC, Treede R, Zubieta J (2005) Human brain mechanisms of pain perception and regulation in health and disease. Eur J Pain 9:463–484

Asghar AU, Green G, Lythgoe MF, Lewith G, MacPherson H (2010) Acupuncture needling sensation: the neural correlates of deqi using fMRI. Brain Res 1315:111–118

Attwell D, Iadecola C (2002) The neural basis of functional brain imaging signals. Trends Neurosci 25:621–625

Bai L, Qin W, Tian J, Dai J, Yang W (2009) Detection of dynamic brain networks modulated by acupuncture using a graph theory model. Prog Na. Sci 19:827–835

Bai L, Qin W, Tian J, Dong M, Pan X, Chen P, Dai J, Yang W, Liu Y (2009) Acupuncture modulates spontaneous activities in the anticorrelated resting brain networks. Brain Res 1279:37–49

Bai L, Qin W, Tian J, Liu P, Li L, Chen P, Dai J, Craggs JG, von Deneen KM, Liu Y (2009) Time-varied characteristics of acupuncture effects in fMRI studies. Hum Brain Mapp 30:3445–3460

Bai L, Tian J, Qin W, Pan X, Yang L, Chen P, Chen H, Dai J, Ai L, Zhao B (2007) Exploratory analysis of functional connectivity network in acupuncture study by a graph theory mode. Conf Proc IEEE Eng Med Biol Soc 2007:2023–2026

Bai L, Yan H, Li L, Qin W, Chen P, Liu P, Gong Q, Liu Y, Tian J (2010) Neural specificity of acupuncture stimulation at pericardium 6: evidence from an FMRI study. J Magn Reson Imaging 31:71–77

[7] Due to so-called susceptibility artifacts.

[8] Due to pulsation-induced motion artifacts.

Beckmann CF, Smith SM (2004) Probabilistic independent component analysis for functional magnetic resonance imaging. IEEE Trans Med Imaging 23:137–152
Beissner F, Henke C (2008) Investigation of Acupuncture Effects on Vegetative Brain Centres Using Functional Magnetic Resonance Imaging. J Acupunct Tuina Sci 6:264–265
Beissner F, Henke C (2009) Methodological problems in fMRI studies on acupuncture: A critical review with special emphasis on visual and auditory cortex activations. Evid Based Complement Alternat Med
Belliveau J, Rosen B, Kantor H, Rzedzian R, Kennedy D, McKinstry R, Vevea J, Cohen M, Pykett I, Brady T (1990) Functional cerebral imaging by susceptibility-contrast NMR. Magn Reson Med 14:538–546
Biswal B, Yetkin F, Haughton V, Hyde J (1995) Functional connectivity in the motor cortex of resting human brain using echo-planar MRI. Magn Reson Med 34:537–541
Chae Y, Lee H, Kim H, Sohn H, Park J, Park H (2009) The neural substrates of verum acupuncture compared to non-penetrating placebo needle: an fMRI study. Neurosci Lett 450:80–4
Chen YX, Kong KM, Wang WD, Xie CH, Wu RH (2007) Functional MR imaging of the spinal cord in cervical spinal cord injury patients by acupuncture at LI 4 (Hegu) and LI 11(Quchi). Conf Proc IEEE Eng Med Biol Soc 2007:3388–3391
Chiang CY, Chang CT, Chu HL, Yang LF (1973) Peripheral afferent pathway for acupuncture analgesia. Sci Sin 16:210–217
Cho S, Jahng G, Park S, Jung W, Moon S, Park J (2010) fMRI Study of Effect on Brain Activity According to Stimulation Method at LI11, ST36: Painful Pressure and Acupuncture Stimulation of Same Acupoints. J Altern Complement Med 16:489
Cho ZH, Chung SC, Jones JP, Park JB, Park HJ, Lee HJ, Wong EK, Min BI (1998) New findings of the correlation between acupoints and corresponding brain cortices using functional MRI. Proc Natl Acad Sci U S A 95:2670–3
Cho ZH, Chung SC, Lee HJ, Wong EK, Min BI (2006) Retraction. New findings of the correlation between acupoints and corresponding brain cortices using functional MRI. Proc Natl Acad Sci U S A 103:10527
Cho ZH, Son YD, Han JY, Wong EK, Kang CK, Kim KY, Kim HK, Lee BY, Yim YK, KH K (2002) fMRI Neurophysiological Evidence Of Acupuncture Mechanisms. Medical Acupuncture 14(1):16–22
Cole DM, Smith SM, Beckmann CF (2010) Advances and pitfalls in the analysis and interpretation of resting-state FMRI data. Frontiers in systems neuroscience 4:8
Deadman P, Baker K, Al-Khafaji M (1998) A Manual of Acupuncture. Journal of Chinese Medicine.
Deng G, Hou BL, Holodny AI, Cassileth BR (2008) Functional magnetic resonance imaging (fMRI) changes and saliva production associated with acupuncture at LI-2 acupuncture point: a randomized controlled study. BMC Complement Altern Med 8:37
Dhond RP, Yeh C, Park K, Kettner N, Napadow V (2008) Acupuncture modulates resting state connectivity in default and sensorimotor brain networks. Pain 136:407–418
Eippert F, Finsterbusch J, Bingel U, Buchel C (2009) Direct evidence for spinal cord involvement in placebo analgesia. Science 326:404
Fang J, Jin Z, Wang Y, Li K, Kong J, Nixon E, Zeng Y, Ren Y, Tong H, others (2008) The salient characteristics of the central effects of acupuncture needling: Limbic-paralimbic-neocortical network modulation. Hum Brain Mapp 6:1
Fang J, Krings T, Weidemann J, Meister I, Thron A (2004) Functional MRI in healthy subjects during acupuncture: different effects of needle rotation in real and false acupoints. Neuroradiology 46:359–362
Friston K, Holmes A, Worsley K, Poline J, CD (1995) Statistical parametric maps in functional imaging: a general linear approach. Hum Brain Mapp 2:189–210
Friston KJ, Holmes AP, Worsley KJ (1999) How Many Subjects Constitute a Study? Neuroimage 5:1–5
Gareus IK, Lacour M, Schulte A, Hennig J (2002) Is there a BOLD response of the visual cortex on stimulation of the vision-related acupoint GB 37? J Magn Reson Imaging 15:227–232

Glover GH (1999) Deconvolution of impulse response in event-related BOLD fMRI. Neuroimage 9:416–429

Ho T, Duann J, Chen C, Chen J, Shen W, Lu T, Liao J, Lin J (2008) Carryover effects alter FMRI statistical analysis in an acupuncture study. Am J Chin Med 36:55–70

Ho T, Duann J, Chen C, Chen J, Shen W, Lu T, Liao J, Lin Z, Shaw K, Lin J (2009) Temporally shifted hemodynamic response model helps to extract acupuncture-induced functional magnetic resonance imaging blood oxygenation-level dependent activities. Chin Med J 122:823–829

Hui K, Liu J, Makris N, Gollub R, Chen A, Moore C, Kennedy D, Rosen B, Kwong K (2000) Acupuncture modulates the limbic system and subcortical gray structures of the human brain: evidence from fMRI studies in normal subjects. Hum Brain Mapp 9:13–25

Hui K, Liu J, Marina O, Napadow V, Haselgrove C, Kwong K, Kennedy D, Makris N (2005) The integrated response of the human cerebro-cerebellar and limbic systems to acupuncture stimulation at ST 36 as evidenced by fMRI. Neuroimage 27:479–496

Hui K, Marina O, Claunch J, Nixon E, Fang J, Liu J, Li M, Napadow V, Vangel M, Makris N, others (2009) Acupuncture mobilizes the brain's default mode and its anti-correlated network in healthy subjects. Brain Res 1287:84–103

Hui KK, Liu J, Kwong KK (1997) Functional mapping of the human brain during acupuncture with magnetic resonance imaging somatosensory cortex activation. World J Acupunct Moxi 7(3):44–49

Hui KK, Marina O, Liu J, Rosen BR, Kwong KK (2010) Acupuncture, the limbic system, and the anticorrelated networks of the brain. Autonomic neuroscience : basic & clinical

Hui KK, Nixon EE, Vangel MG, Liu J, Marina O, Napadow V, Hodge SM, Rosen BR, Makris N, Kennedy DN (2007) Characterization of the "deqi" response in acupuncture. BMC Complement Altern Med 7:33

Jeun S, Kim J, Kim B, Park S, Lim E, Choi G, Choe B (2005) Acupuncture stimulation for motor cortex activities: a 3T fMRI study. Am J Chin Med 33:573–578

Jones AK, Kulkarni B, Derbyshire SW (2003) Pain mechanisms and their disorders. Br Med Bull 65:83–93

Kong J, Gollub R, Huang T, Polich G, Napadow V, Hui K, Vangel M, Rosen B, Kaptchuk TJ (2007) Acupuncture de qi, from qualitative history to quantitative measurement. J Altern Complement Med 13:1059–1070

Kong J, Gollub RL, Webb JM, Kong J, Vangel MG, Kwong K (2007) Test-retest study of fMRI signal change evoked by electroacupuncture stimulation. Neuroimage 34:1171–1181

Kong J, Kaptchuk TJ, Polich G, Kirsch I, Vangel M, Zyloney C, Rosen B, Gollub R (2009) Expectancy and treatment interactions: A dissociation between acupuncture analgesia and expectancy evoked placebo analgesia. Neuroimage 45(3):940–949

Kong J, Kaptchuk TJ, Polich G, Kirsch I, Vangel M, Zyloney C, Rosen B, Gollub RL (2009) An fMRI study on the interaction and dissociation between expectation of pain relief and acupuncture treatment. Neuroimage 47:1066–1076

Kong J, Kaptchuk TJ, Webb JM, Kong J, Sasaki Y, Polich GR, Vangel MG, Kwong K, Rosen B, Gollub RL (2009) Functional neuroanatomical investigation of vision-related acupuncture point specificity--a multisession fMRI study. Hum Brain Mapp 30:38–46

Kong J, Ma L, Gollub RL, Wei J, Yang X, Li D, Weng X, Jia F, Wang C, Li F, Li R, Zhuang D (2002) A pilot study of functional magnetic resonance imaging of the brain during manual and electroacupuncture stimulation of acupuncture point (LI-4 Hegu) in normal subjects reveals differential brain activation between methods. J Altern Complement Med 8:411–419

Kwong KK, Belliveau JW, Chesler Da, Goldberg IE, Weisskoff RM, Poncelet BP, Kennedy DN, Hoppel BE, Cohen MS, Turner R (1992) Dynamic magnetic resonance imaging of human brain activity during primary sensory stimulation. Proc Natl Acad Sci U S A 89:5675–9

Lewis CM, Baldassarre A, Committeri G, Romani GL, Corbetta M (2009) Learning sculpts the spontaneous activity of the resting human brain. Proc Natl Acad Sci U S A 106:17558–63

Li G, Cheung RT, Ma Q, Yang ES (2003) Visual cortical activations on fMRI upon stimulation of the vision-implicated acupoints. Neuroreport 14:669–673

Li G, Huang L, Cheung RT, Liu S, Ma Q, Yang ES (2004) Cortical activations upon stimulation of the sensorimotor-implicated acupoints. Magn Reson Imaging 22:639–644

Li G, Jack CR, Yang ES (2006) An fMRI study of somatosensory-implicated acupuncture points in stable somatosensory stroke patients. J Magn Reson Imaging 24:1018–1024

Li G, Liu H, Cheung RT, Hung Y, Wong KK, Shen GG, Ma Q, Yang ES (2003) An fMRI study comparing brain activation between word generation and electrical stimulation of language-implicated acupoints. Hum Brain Mapp 18:233–238

Li G, Ng MC, Wong KK, Luk KD, Yang ES (2005) Spinal effects of acupuncture stimulation assessed by proton density-weighted functional magnetic resonance imaging at 0.2 T. Magn Reson Imaging 23:995–999

Li K, Shan B, Xu J, Liu H, Wang W, Zhi L, Li K, Yan B, Tang X (2006) Changes in FMRI in the human brain related to different durations of manual acupuncture needling. J Altern Complement Med 12:615–623

Li L, Liu H, Li Y, Xu J, Shan B, Gong D, Li K, Tang X (2008) The human brain response to acupuncture on same-meridian acupoints: evidence from an fMRI study. J Altern Complement Med 14:673–678

Li L, Qin W, Bai L, Tian J (2010) Exploring vision-related acupuncture point specificity with multivoxel pattern analysis. Magn Reson Imaging 28:380–387

Litscher G, Rachbauer D, Ropele S, Wang L, Schikora D, Fazekas F, Ebner F (2004) Acupuncture using laser needles modulates brain function: first evidence from functional transcranial Doppler sonography and functional magnetic resonance imaging. Lasers Med Sci 19:6–11

Liu J, Qin W, Guo Q, Sun J, Yuan K, Liu P, Zhang Y, von Deneen KM, Liu Y, Tian J (2010) Distinct brain networks for time-varied characteristics of acupuncture. Neurosci Lett 468:353–358

Liu P, Qin W, Zhang Y, Tian J, Bai L, Zhou G, Liu J, Chen P, Dai J, von Deneen K, others (2009) Combining spatial and temporal information to explore function-guide action of acupuncture using fMRI. J Magn Reson Imaging 30:41–46

Liu P, Zhang Y, Zhou G, Yuan K, Qin W, Zhuo L, Liang J, Chen P, Dai J, Liu Y, Tian J (2009) Partial correlation investigation on the default mode network involved in acupuncture: an fMRI study. Neurosci Lett 462:183–187

Liu S, Zhou W, Ruan X, Li R, Lee T, Weng X, Hu J, Yang G (2007) Activation of the hypothalamus characterizes the response to acupuncture stimulation in heroin addicts. Neurosci Lett 421:203–208

Liu W, Feldman SC, Cook DB, Hung D, Xu T, Kalnin AJ, Komisaruk BR (2004) fMRI study of acupuncture-induced periaqueductal gray activity in humans. Neuroreport 15:1937–1940

Logothetis NK, Wandell BA (2004) Interpreting The BOLD Signal. Annu Rev Physiol 66:735–769

Lu N, Shan B, Xu J, Wang W, Li K (2007) Improved temporal clustering analysis method applied to whole-brain data in acupuncture fMRI study. Magn Reson Imaging 25:1190–1195

MacPherson H, Green G, Nevado A, Lythgoe MF, Lewith G, Devlin R, Haselfoot R, Asghar AU (2008) Brain imaging of acupuncture: comparing superficial with deep needling. Neurosci Lett 434:144–9

May A (2007) Neuroimaging: visualising the brain in pain. Neurol Sci 28 Suppl 2:S101–7

McKeown MJ, Makeig S, Brown GG, Jung TP, Kindermann SS, Bell AJ, Sejnowski TJ (1998) Analysis of fMRI data by blind separation into independent spatial components. Hum Brain Mapp 6:160–188

Melzack R (1990) Phantom limbs and the concept of a neuromatrix. Trends Neurosci 13:88–92

Na B, Jahng G, Park S, Jung W, Moon S, Park J, Bae H (2009) An fMRI study of neuronal specificity of an acupoint: electroacupuncture stimulation of Yanglingquan (GB34) and its sham point. Neurosci Lett 464:1–5

Nakagoshi A, Fukunaga M, Umeda M, Mori Y, Higuchi T, Tanaka C (2005) Somatotopic representation of acupoints in human primary somatosensory cortex: an FMRI study. Magn Reson Med Sci 4:187–9

Nance DM, Sanders VM (2007) Autonomic innervation and regulation of the immune system (1987–2007). Brain Behav Immun 21:736–745

Napadow V, Dhond R, Kim J, LaCount L, Vangel M, Harris R, Kettner N, Park K (2009) Brain encoding of acupuncture sensation–Coupling on-line rating with fMRI. NeuroImage 47:1055–1065
Napadow V, Dhond R, Park K, Kim J, Makris N, Kwong K, Harris R, Purdon P, Kettner N, Hui K (2009) Time-variant fMRI activity in the brainstem and higher structures in response to acupuncture. Neuroimage 47:289–301
Napadow V, Dhond R, Purdon P, Kettner N, Makris N, Kwong K, Hui K (2005) Correlating acupuncture FMRI in the human brainstem with heart rate variability. Conf Proc IEEE Eng Med Biol Soc 5:4496–4499
Napadow V, Kettner N, Liu J, Li M, Kwong K, Vangel M, Makris N, Audette J, Hui K (2007) Hypothalamus and amygdala response to acupuncture stimuli in carpal tunnel syndrome. Pain 130:254–266
Napadow V, Liu J, Li M, Kettner N, Ryan A, Kwong K, Hui K, Audette J (2007) Somatosensory cortical plasticity in carpal tunnel syndrome treated by acupuncture. Hum Brain Mapp 28:159–171
Napadow V, Makris N, Liu J, Kettner N, Kwong K, Hui K (2005) Effects of electroacupuncture versus manual acupuncture on the human brain as measured by fMRI. Hum Brain Mapp 24:193–205
Ogawa S, Lee T, Nayak A, Glynn P (1990) Oxygenation-sensitive contrast in magnetic resonance image of rodent brain at high magnetic fields. Magnetic Resonance in Medicine 14:68–78
Park S, Shin A, Jahng G, Moon S, Park J (2009) Effects of scalp acupuncture versus upper and lower limb acupuncture on signal activation of blood oxygen level dependent (BOLD) fMRI of the brain and somatosensory cortex. J Altern Complement Med 15:1193–1200
Parrish TB, Schaeffer A, Catanese M, Rogel MJ (2005) Functional magnetic resonance imaging of real and sham acupuncture. Noninvasively measuring cortical activation from acupuncture. IEEE Eng Med Biol Mag 24:35–40
Pauling L, Coryell CD (1936) The Magnetic Properties and Structure of Hemoglobin, Oxyhemoglobin and Carbonmonoxyhemoglobin. Proc Natl Acad Sci U S A 22:210–216
Poldrack R, Fletcher P, Henson R, Worsley K, Brett M, Nichols T (2008) Guidelines for reporting an fMRI study. Neuroimage 40:409–414
Qin W, Tian J, Bai L, Pan X, Yang L, Chen P, Dai J, Ai L, Zhao B, Gong Q, Wang W, von Deneen KM, Liu Y (2008) FMRI connectivity analysis of acupuncture effects on an amygdala-associated brain network. Mol Pain 4:55
Qin W, Tian J, Pan X, Yang L, Zhen Z (2006) The correlated network of acupuncture effect: a functional connectivity study. Conf Proc IEEE Eng Med Biol Soc 1:480–483
Roy C, Sherrington C (1890) On the regulation of the blood-supply of the brain. J Physiol 11:85
Samuels MA (2007) The brain-heart connection. Circulation 116:77–84
Schaechter JD, Connell BD, Stason WB, Kaptchuk TJ, Krebs DE, Macklin EA, Schnyer RN, Stein J, Scarborough DM, Parker SW, McGibbon CA, Wayne PM (2007) Correlated change in upper limb function and motor cortex activation after verum and sham acupuncture in patients with chronic stroke. J Altern Complement Med 13(5):527–532
Siedentopf CM, Golaszewski SM, Mottaghy FM, Ruff CC, Felber S, Schlager A (2002) Functional magnetic resonance imaging detects activation of the visual association cortex during laser acupuncture of the foot in humans. Neurosci Lett 327:53–56
Siedentopf CM, Koppelstaetter F, Haala IA, Haid V, Rhomberg P, Ischebeck A, Buchberger W, Felber S, Schlager A, Golaszewski SM (2005) Laser acupuncture induced specific cerebral cortical and subcortical activations in humans. Lasers Med Sci 20(2):68–73
Tracey KJ (2007) Physiology and immunology of the cholinergic antiinflammatory pathway. J Clin Invest 117(2):289–296
Travagli R, Hermann G, Browning K, Rogers R (2006) Brainstem circuits regulating gastric function. Annu Rev Physiol 68:279–305
Ueda Y, Hayashi K, Kuriowa K (2005) The application of fMRI to basic experiments in acupuncture. The effects of stimulus points and content on cerebral activities and responses. IEEE Eng Med Biol Mag 24:47–51

Wang S, Constable RT, Tokoglu FS, Weiss DA, Freyle D, Kain ZN (2007) Acupuncture-induced blood oxygenation level-dependent signals in awake and anesthetized volunteers: a pilot study. Anesth Analg 105:499–506

Wang W, Liu L, Zhi X, Huang J, Liu D, Wang H, Kong X, Xu H (2007) Study on the regulatory effect of electro-acupuncture on hegu point (LI4) in cerebral response with functional magnetic resonance imaging. Chin J Integr Med 13:10–16

Wang WD, Kong KM, Xiao YY, Wang XJ, Liang B, Qi WL, Wu RH (2006) Functional MR imaging of the cervical spinal cord by use of electrical stimulation at LI4 (Hegu). Conf Proc IEEE Eng Med Biol Soc 1:1029–31

Weiss T, Straube T, Boettcher J, Hecht H, Spohn D, Miltner WH (2008) Brain activation upon selective stimulation of cutaneous C- and Aδ-fibers. Neuroimage 41:1372–1381

Wesolowski T, Lotze M, Domin M, Langner S, Lehmann C, Wendt M, Usichenko TI (2009) Acupuncture reveals no specific effect on primary auditory cortex: a functional magnetic resonance imaging study. Neuroreport 20:116–120

Wilder R (1995) Neuroendocrine-immune system interactions and autoimmunity. Annu Rev Immunol 13:307–338

Wu M, Hsieh J, Xiong J, Yang C, Pan H, Chen Y, Tsai G, Rosen B, Kwong K (1999) Central Nervous Pathway for Acupuncture Stimulation: Localization of Processing with Functional MR Imaging of the Brain—Preliminary Experience. Radiology 212:133

Wu M, Sheen J, Chuang K, Yang P, Chin S, Tsai C, Chen C, Liao J, Lai P, Chu K, Pan H, Yang C (2002) Neuronal specificity of acupuncture response: a fMRI study with electroacupuncture. Neuroimage 16:1028–1037

Wu Y, Jin Z, Li K, Lu Z, Wong V, Han T, Zheng H, Caspi O, Liu G, Zeng Y, Zou L (2008) Effect of acupuncture on the brain in children with spastic cerebral palsy using functional neuroimaging (FMRI). J Child Neurol 23:1267–74

Xiao YY, Chen XK, Du L, Pei RQ, Chen FY, Liu GR, Wu RH (2006) The brain mapping on reinforcement acupuncture stimulation at ST36 (zusanli) evidenced by fMRI. Conf Proc IEEE Eng Med Biol Soc 1:1036–1039

Yan B, Li K, Xu J, Wang W, Li K, Liu H, Shan B, Tang X (2005) Acupoint-specific fMRI patterns in human brain. Neurosci Lett 383:236–240

Yeo S, Choe I, van den Noort M, Bosch P, Lim S (2010) Consecutive acupuncture stimulations lead to significantly decreased neural responses. J Altern Complement Med 16:481–7

Yoo S, Kerr CE, Park M, Im D, Blinder RA, Park H, Kaptchuk TJ (2007) Neural activities in human somatosensory cortical areas evoked by acupuncture stimulation. Complement Ther Med 15:247–54

Yoo S, Teh E, Blinder RA, Jolesz FA (2004) Modulation of cerebellar activities by acupuncture stimulation: evidence from fMRI study. Neuroimage 22:932–940

Yoshida T, Tanaka C, Umeda M, Higuchi T, Fukunaga M, Naruse S (1995) Non-invasive measurement of brain activity using functional MRI: toward the study of brain response to acupuncture stimulation. Am J Chin Med 23:319–325

Zhang J, Cao X, Lie J, Tang W, Liu H, Fenga X (2007) Neuronal specificity of needling acupoints at same meridian: a control functional magnetic resonance imaging study with electroacupuncture. Acupunct Electrother Res 32:179–193

Zhang J, Li J, Cao X, Feng X (2009) Can electroacupuncture affect the sympathetic activity, estimated by skin temperature measurement? A functional MRI study on the effect of needling at GB 34 and GB 39 on patients with pain in the lower extremity. Acupunct Electrother Res 34:151–164

Zhang W, Jin Z, Cui G, Zhang K, Zhang L, Zeng Y, Luo F, Chen AC, Han J (2003) Relations between brain network activation and analgesic effect induced by low vs. high frequency electrical acupoint stimulation in different subjects: a functional magnetic resonance imaging study. Brain Res 982:168–78

Zhang W, Jin Z, Huang J, Zhang L, Zeng Y, Luo F, Chen AC, Han J (2003) Modulation of cold pain in human brain by electric acupoint stimulation: evidence from fMRI. Neuroreport 14:1591–1596

Zhang W, Jin Z, Luo F, Zhang L, Zeng Y, Han J (2004) Evidence from brain imaging with fMRI supporting functional specificity of acupoints in humans. Neurosci Lett 354:50–53

Zhang Y, Liang J, Qin W, Liu P, von Deneen KM, Chen P, Bai L, Tian J, Liu Y (2009) Comparison of visual cortical activations induced by electro-acupuncture at vision and nonvision-related acupoints. Neurosci Lett 458:6–10

Zhang Y, Qin W, Liu P, Tian J, Liang J, von Deneen KM, Liu Y (2009) An fMRI study of acupuncture using independent component analysis. Neurosci Lett 449:6–9

Zhou Y, Jin J (2008) Effect of acupuncture given at the HT 7, ST 36, ST 40 and KI 3 acupoints on various parts of the brains of Alzheimer' s disease patients. Acupunct Electrother Res 33:9–17

Chapter 5
Chemical and Physical Characteristics of Moxibustion

Ling Zhao and Xueyong Shen

Abstract Moxibustion has been a part of acupuncture practice for 1,000 years, and the mechanisms of this traditional technique need scientific investigation and explanation. Up to the present, there has been no consensus of opinion on the mechanism of action which results in moxibustion's therapeutic effects in clinical settings. This chapter presents information on recent advances in the understanding of the mechanism of moxibustion's effect from the following three aspects: pharmacological action, thermal effect, and infrared physical properties. Previous studies have shown that there are many constituent elements in the smoke. In the literature on moxibustion's effects, there is significant support for the proposition that thermal effects play a key role in the therapeutic benefits of moxibustion. According to this theory, the moxibustion effect is mainly due to the warm–hot (thermal) stimulative effect of burning moxa. However, this thermal effect hypothesis has been challenged by researchers who question that this effect contributes to moxibustion's therapeutic benefits. Various studies indicate that the nonthermal action may affect the therapeutic effects significantly. Based on analysis of infrared radiation, the therapeutic effect produced by a traditional moxa stick may be the result of potent thermal action. The effects of indirect moxibustion with traditional media result both from thermal action and infrared radiation resonance between indirect moxibustion and acupuncture points. The apparatus with specific wavelength of infrared radiation

L. Zhao
Shanghai University of Traditional Chinese Medicine, Shanghai 201203, China

Shanghai Research Center for Acupuncture and Meridians, Shanghai 201203, China
e-mail: zhao099@hotmail.com

X. Shen (✉)
College of Acupuncture and Moxibustion, Shanghai University of Traditional Chinese Medicine, Shanghai 201203, China

Laboratory of Systemic Physiology for Acupuncture and Meridian, Shanghai Research Center for Acupuncture and Meridians, Shanghai 201203, China
e-mail: snowysh@hotmail.com

Y. Xia et al. (eds.), *Current Research in Acupuncture*,
DOI 10.1007/978-1-4614-3357-6_5,

similar to that of human acupoint was effective in rabbits with bradycardia and patients having osteoarthritis of knee. It should be further promoted in clinical application. Lastly, we will comment on several unsolved issues and future directions of the research concerning moxibustion treatment.

Keywords Moxibustion • Pharmacological action • Thermal effect • Infrared radiation • Spectrum

5.1 Introduction

Moxibustion entails burning the moxa herb, *Artemisia vulgaris*, on or above the skin at acupuncture points to warm the points in order to alleviate symptoms (Lao 1999). Moxibustion is an important part of the science of traditional Chinese medicine. The first recorded use of moxibustion was documented in the Han dynasty silk books the *Moxibustion Classic of 11 Foot-Hand Meridians* and the *Moxibustion Classic of 11 Yin Yang Meridians*. These books were discovered in 1973 in the No. 3 tomb of Mawangdui in Changsha, Hunan Province. There was no mention of acupuncture therapy in these writings. Therefore, we presume that the origin of moxibustion therapy predates that of acupuncture therapy. Doctors through the ages made considerable progress and published a great number of books on moxibustion. Moxibustion has been applied in treating a great range of diseases (Zhang 2004).

According to Traditional Chinese Medicine, moxa is warm in property and fragrant in flavor. It smoothes and facilitates circulation in the 12 meridians and regulates the qi and blood, expelling the coldness and dampness, warming the meridians, stopping bleeding, and preventing abortion. Moxibustion is very effective for treating yin syndrome, cold syndrome, and deficiency syndrome. It has significant advantages for the treatment of chronic diseases, hard-to-treat diseases, and the prevention of diseases. According to modern research, moxibustion therapy can promote metabolism, improve immune system functions (Ma and Chong 2008; Yuan and Tang 2007), and regulate the physiological function of internal organs. Recently reported randomized clinical trials and clinical observations have evaluated the effect of moxibustion on such conditions as fetus breech presentation (Joos et al. 2004; Neri et al. 2004; Ewies and Olah 2002; Cardini and Huang 1998) and acute lymphangitis (Zhou 2003). Moxibustion has also been used as an antiaging therapy and in the treatment of cancer (Song et al. 2006). Further, in recent years, studies have indicated moxibustion may produce therapeutic effects in the treatment of dementia (Xin et al. 2000) and stroke rehabilitation (Yu and Guo 2005; Lee et al. 2010). Preclinical basic science studies suggest that moxibustion can enhance immune function (Chen et al. 2003; Yamashita et al. 2001; Ma and Chong 2008) and improve such physiological functions as cerebral blood flow (Uchida et al. 2003) and arthritis (Yu et al. 2009) in animals.

There are various ways of practicing moxibustion. Commonly used techniques are burning a moxa stick directly over the skin at an acupuncture point (Cardini and

Huang 1998) and indirect moxibustion, that is, placing such insulating materials as monkshood cake or sliced ginger or garlic between the body and a burning moxa cone (Cheng 1999). However, the effects of these different techniques on the human body need scientific documentation.

5.2 Mechanism Study of Moxibustion

5.2.1 Pharmacological Action of Moxibustion

Recent reported Chinese and international research on the active ingredient of moxa concentrates mainly on the ingredients of the volatile oil and smoke of moxa. The volatile oil content of moxa varies greatly, depending upon the region where it was produced and the season in which it was grown. The volatile oil content in Qi-moxa grown in Qichun in Hubei Province reaches about 0.83 %, while the volatile oil content of the moxa from Ziyang in Sichuan Province and in Tangyin in Henan Province reaches only half that level. The trace element contents in Qi-moxa are also more than those of others (Mei et al. 1991).

The smoke from the combustion of moxa contains complex components. The volatile components are ammonia, alcohols (ethylene glycol, pentanol, and butanol), aliphatic hydrocarbons, aromatic hydrocarbons, and terpenoids and their oxides. These components may be derived from the incomplete combustion of moxa. The moxa oil and the products of its oxidation are also influencing the production of these components. (Li et al. 1993; Hong et al. 2007). Japanese scientists (Zhao et al. 1999) reported that the organic components of moxa are heptatriacontane alkanes and condensed tannin of catecholamine. Heptatriacontane alkanes are an important oxidant during combustion of moxa. The methanol extracts and the combustion products of moxa have the function of cleaning out free radicals and oxide lipids, with the combustion products of moxa appearing to have the stronger of the two effects (Daxijidai 1992). Through moxibustion treatment the compounds from moxa combustion can adhere to the skin and permeate the damaged skin through the moxibustion process (Xiguyuzi 1989).

Comparisons made between the Qi-moxa from Qichun in Hubei Province and those from Anguo in Hebei Province, Ziyang in Sichuan Province, and Tangyin in Henan Province (all of them are gathered during the 1991 Dragon Boat Festival) showed that the thermal discharge from the combustion of the Hubei-grown Qichun moxa was the highest while that from the moxa grown in Ziyang in Sichuan Province was the lowest, only 88.96 % of that of the level observed in Qi-moxa. The thermal discharge from the combustion of the moxa from Tangyin in Henan Province and that from Anguo in Hebei Province were 96.28 % and 96.03 % of that of Qi-moxa, respectively. (Mei and Wang 1994). Therefore, the thermal discharge from the combustion of the moxa rolls made from Qi-moxa is higher than those from other regions. Qi-moxa has a better heat penetrating power, so it may have a better curative effect in clinical practice.

Sliced ginger and garlic are commonly used between the body and a burning moxa cone during indirect moxibustion. In his research on the effects of intermediate substances used in moxibustion, Daxijidai (1989) burned moxa through sliced ginger or garlic by placing it on an evaporation pan; the results confirmed that the effective components, gingerol and garlicin, can permeate the human body through moxibustion and thus contribute to the therapeutic effect. However, there are still no studies about whether moxa smoke is harmful to the human body or whether the curative effect of the moxa is related with its smoke.

5.2.2 *Thermal Effect of Moxibustion*

The thermal effect of moxibustion is thought by some researchers to be the key to the therapeutic effect. For heat stimuli, people detect changes in temperature by at least three different types of sensory receptors, namely, cold receptors, warmth receptors, and pain receptors. Different stimuli activate relevant sensory receptors in the human body enabling the perception of gradations of thermal sensation (Lin and Wang 1999). Moxibustion can activate and modify the nervous system by stimulating receptors under the skin. Shang (2004), through her own clinical experience, concluded that the intensity of mild moxibustion is variable and based on the severity of disease, as well as the patient's constitution. Shang suggests it is best to account for these variabilities in treatment, rather than adhering to set formulas as prescribed in moxibustion texts.

At present, the surface temperature of burning moxa has still not been reported. Using Wien's displacement law, this temperature is estimated to be about 550 °C, which corresponds to the peak of the radiation spectrum (Wang and Sun 1994). The temperature could be up to 800 °C if we calculate by using the highest result reported by Hong et al. (2003). If such high temperatures were directly applied to the human body, the skin, the subcutaneous tissues, and even the muscle tissues would be severely burned in a very short time. Therefore, direct moxibustion is rarely used in clinical practice. Strict control over moxibustion intensity is required when using direct moxibustion to prevent burns. However, the high temperature of burning moxa wool is buffered by aconite, ginger, garlic, and other insulating materials in indirect moxibustion. Thus, the thermal effect is easier for patients to accept. During the burning process, the temperature can rise from room temperature to about 80 °C. In indirect moxibustion, the heat transfer characteristics may be affected by the texture, size, thickness and thinness, and/or the moisture content of the insulating material, which causes variations in temperature. Wei et al. (2007a, b, c) reported the maximum temperature during aconite cake-separated moxibustion (57.20 °C) was generally the same as that reported during ginger-separated moxibustion (57.03 °C). However, in aconite cake-separated moxibustion, the temperature rise was steeper than during ginger-separated moxibustion. As a consequence, patients feel pain more easily during ginger-separated moxibustion.

The fundamental physical research studies on moxibustion have mainly focused on investigations into the skin temperature changes at the acupoints after burning moxa (Dong et al. 1999a, b; Liu and Yang 1993). A study by Li et al. (2009a, b) has shown that skin temperature could rise to 55 °C while burning the moxa stick and up to 200 °C when burning the moxa cone. Okazaki et al. (1983; 1990) studied the effects on skin temperature of burning one or more zhuang of moxa cone (zhuang is the commonly used measurement unit in moxibustion; when one moxa cone is burnt, it is called one zhuang). They held that the temperature could rise with the bigger moxa cone (0.5–2 mg) when burning a single zhuang. The skin temperature rose to about 105 °C when burning a single zhuang (2 mg), while it could rise to about 125 °C when burning three zhuang (2 mg per zhuang). The temperature under the skin rose to about 56 °C when burning a single zhuang. Burning three zhuang had similar effects.

It can be concluded that the intensity of moxibustion in present clinical practice is higher than the pain threshold of thermal stimulation. (The threshold is about 45 °C.) Hecunguangding observed in his experiment on analgesia by moxibustion that indirect moxibustion rose to its highest temperature 1 min after being burned (56 ± 5 °C), while the analgesic effect is triggered at a temperature of 45 °C on the skin and 42.5 °C on the muscle (Li et al. 2009a, b).

A study by Huang (1989a, b) showed that changes in temperature were different in regional subcutaneous tissues and in muscle layers, demonstrating that the moxibustion stimuli could reach different depths in the human body. Dong et al. (1999a, b) found that many kinds of moxibustion significantly change the temperature from skin to muscle; different patterns and characteristic effects could be observed depending on the type of moxibustion therapy applied. Dong et al. (1996) thought that it might be the change of skin temperature which changed the activities of polymodal receptors inducing the effects of defervescence and heat dissipation. Dong et al. (1998, 2000) observed that the suspended moxibustion limited at 47–48 °C can induce the excitatory reaction of every unit observed of thermoreceptor, Aδ and C polymodal receptors (PRs); there was a close relationship between the therapeutic effect and the warm receptors (WRs) or polymodal receptors (PRs). Results showed that suspended moxibustion at 40 °C had no significant effect while at 47 °C it had a very striking antipyretic effect, thus indicating that the antipyretic effect of moxibustion is mediated primarily by PRs.

Xian et al. (2000) observed changes in rabbit's auricular temperature by using a thermometer (connected to a computer) before and after moxibustion stimulation. Results indicated that moxibustion caused an increase in skin temperature to 48 °C, which could inhibit the endotoxin and induce a lowering of auricular temperature. However, when moxibustion increased skin temperature to only 40 °C, no such effects were observed. It revealed that higher skin temperature moxibustion had a thermolytic effect on body fever, which had a close relation mainly with polymodal receptors rather than with temperature receptor. It also demonstrated that endotoxin-induced changes of auricular temperature could be used as an index for evaluating the thermolytic effect of moxibustion. Han et al. (2006) applied a computer moxibustion research test analysis system to a simultaneous test and comparative analysis of 12 factors, including mode of operation, quality of material, and properties of herbal cakes and moxa cones, involved in indirect moxibustion. Results showed different operation modes and processes, and material quality had the corresponding regularity of a change in temperature. The resulting research promises to uncover the optimum combination of correlative factors in clinical operation in order to guide and improve clinical procedure and the curative effect of moxibustion.

5.2.3 *The Infrared Physical Characteristics of Moxibustion and Demonstration of Its Effect*

5.2.3.1 The Infrared Physical Characteristics of Moxibustion

Some researchers believed that the therapeutic effect of moxibustion has little relationship to the pharmacological action (Bai and Lin 1991). The thermal effects produced by the different burning materials in moxibustion were the same (Zhang and Geng 1989). Since the effects that moxibustion exerts on the human body are thought to be produced by thermal radiation, we postulated that these effects are also related to the influence of infrared radiation on body caused by thermal radiation. Xu et al. (1988) found that neither the latent phase nor vocalization threshold had any significant change when the rat pain model was received thermal radiation between 40 °C and 43 °C. This indicated that not all thermal stimuli could bring a beneficial effect.

It is known that all objects in the universe, including human beings, produce infrared radiation. But the human body is not only a source of infrared radiation, it also absorbs it easily (Shen 1996). Infrared has strong penetrability and is easily absorbed by objects, including the human body. In the body the infrared radiation is transformed into internal energy. However, it is not known whether the therapeutic effect of traditional moxibustion is induced simply by the thermal irradiation of acupuncture points, nor is it known whether other materials can be substituted for those used in traditional indirect moxibustion.

Research conducted by Yang and Liu (1996a, b) found that the near-infrared radiation produced by the burning moxa wool could excite the hydrogen bonds of biomacromolecules inside the acupoints of the human body, producing an absorption effect of stimulated coherent resonance and transmitted energy to cells via the neurohormonal system. In addition to the thermal effect (Deng et al. 2004), it is posited that the nonthermal effect of light produced by the burning moxa also probably played an essential role in the treatment.

Results from the moxibustion radiation spectrum are varied, depending upon the detection methods and experimental conditions used by different researchers. Yuan and Zhuang (1995) used red color monochromatic measurements when burning moxa wool to find emissions of 0.8–5 μm infrared radiation with a peak round 1.5 μm. Zhang et al. (1999) applied a self-made detecting device, a visible light monochromator, and infrared monochromator to test three types of burning moxa sticks in three wave bands from visible light to 2.5 μm. They found that the peak of the infrared radiation was around 2.3 μm and the spectral characteristics of the burning moxa stick in the far-infrared radiation were unclear as the detecting range was within 2.5 μm. Hong et al. (2003) applied an opto-electrical chronometer to measure the temperature of the burning moxa stick indirectly. Calculating using Planck's formula, he found that the main part of the thermal radiation spectrum from the burning moxa stick was in the mid-infrared radiation range, which also included certain parts of the visible light spectrum, with a peak of around 2.8 μm. In a later study Hong et al. (2004) applied a grating spectrometer to a burning moxa stick in its natural state to directly measure the spectrums of thermal radiation in the wave bands of visible light and infrared light. He found that the radiating peak was round 3.5 μm, which indicated that the spectrum of the burning moxa stick was close to near-infrared but mainly in the far-infrared range.

Zhang et al. (1999) observed the continuous-wave electromagnetic spectrum of Chinese moxibustion by using a self-created and constructed testing compartment and a spectrometer to measure visible and infrared light. Experimental results showed that the heat which is given when Chinese moxibustion is burnt is from about 600 mill microns to far-infrared. Different kinds of Chinese moxibustion produce different electromagnetic spectrums. It was found that significant radiation also was present in the far-infrared radiation spectrum above 2,500 nm. Qian et al. (1991) also believed that since the emission spectrum of moxibustion covered visible light, near-infrared radiation, and far-infrared radiation and that it was very hard to detect the complete emission spectrum of moxibustion by one apparatus, as a result no such spectrum has ever been obtained.

Despite progress made in the study of the biophysical specificity of infrared radiation on the surface of the human body, the available methods were until recently only advanced to the level of infrared thermography (Zhang et al. 2002; Ovechkin et al. 2001; Hu et al. 2002; Klaus et al. 2005). The infrared thermography device has limited power and can only detect wavelengths between about between 3 and 6 μm at any given time; it cannot simultaneously cover the 1.5–16 μm infrared radiation spectrum of the human body to give a comprehensive analysis. Furthermore, that device is not sensitive enough to detect the physiological and pathological changes in the infrared radiation of the human body. Literatures about infrared radiation of indirect moxibustion are relatively less.

Shen and Ding (Shen et al. 2001, 2003; Ding et al. 2002) made a highly sensitive infrared radiation spectrum detection system by applying astronautic Hg1-xCdxTe infrared transducer and lock-phase amplifier techniques. They detected a 1.5–15 μm spectrum of a burning moxa stick and found that the peak of the spectrum was around 3.5 μm. The system was obtained from the Shanghai Institute of Technical Physics of the Chinese Academy of Sciences, Shanghai, China. The system consists of a narrow seam, a wave cutter, an infrared spectrometer, an Hg1-xCdxTe infrared transducer, a lock-phase amplifier, an oscilloscope, and a computer (Fig. 5.1).

This device features high sensitivity, low noise, a wide wavelength range, and stable performance, as previously reported (Deng et al. 2004; Liu and Shen 2002;

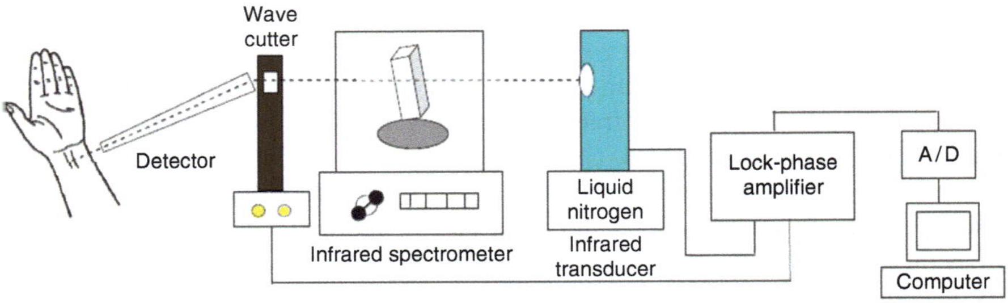

Fig. 5.1 Highly sensitive infrared radiation spectrum detecting system

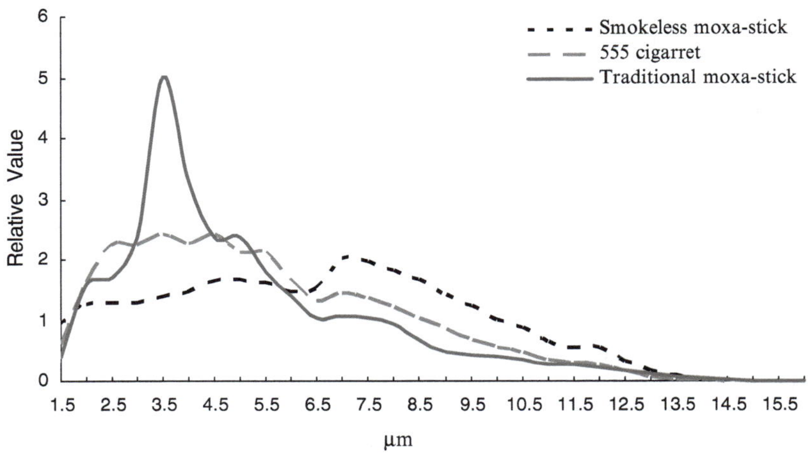

Fig. 5.2 Unified infrared radiation spectrums of a traditional moxa stick, a smokeless moxa stick, and a *555* cigarette

Wei et al. 2004). This highly sensitive infrared spectrum detection device was used to compare the spectrums of traditional moxibustion materials ($n=4$/group) to those of control materials ($n=4$/group) and to the spectrum at the surface area of an acupuncture point, Hegu (LI-4), on healthy volunteers ($n=7$). Results showed that Shen et al. (2006).

1. The infrared radiation intensity produced by a traditional moxa stick was 43,300.41 mV, with a peak on the infrared spectrum of 3.5 μm, while the respective radiation intensities of two controls, a smokeless moxa stick and the *555* cigarette, were 31.15 and 37.03 mV, with peaks of 7 and 3.5 μm (Fig. 5.2).
2. The infrared radiation intensities of the three traditional indirect moxibustion media—monkshood cake, ginger slices, and garlic slices—were 520.27, 594.79, and 681.87 mV, respectively, all with peaks around 7.5 μm and quite similar

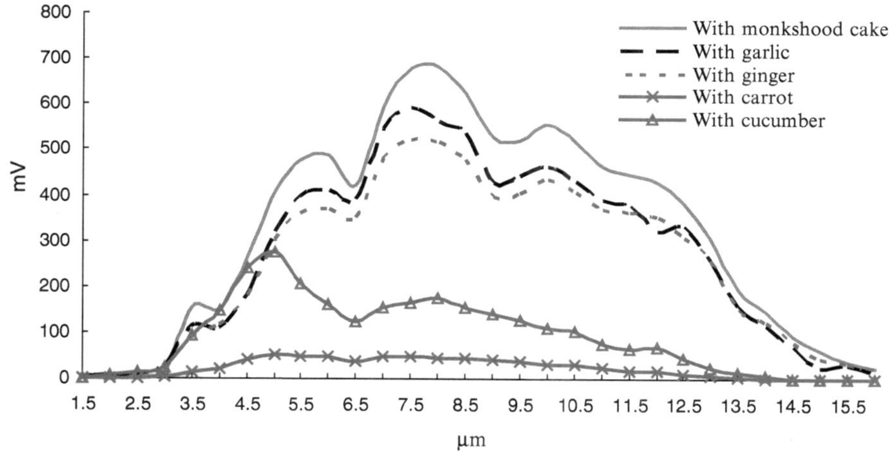

Fig. 5.3 Comparison of the infrared radiation spectrums of traditional and substitute indirect moxibustion media. The data were unified for this figure. Similar infrared radiation spectrum patterns can be seen among the three traditional media, while the patterns of the carrot and cucumber controls are clearly different

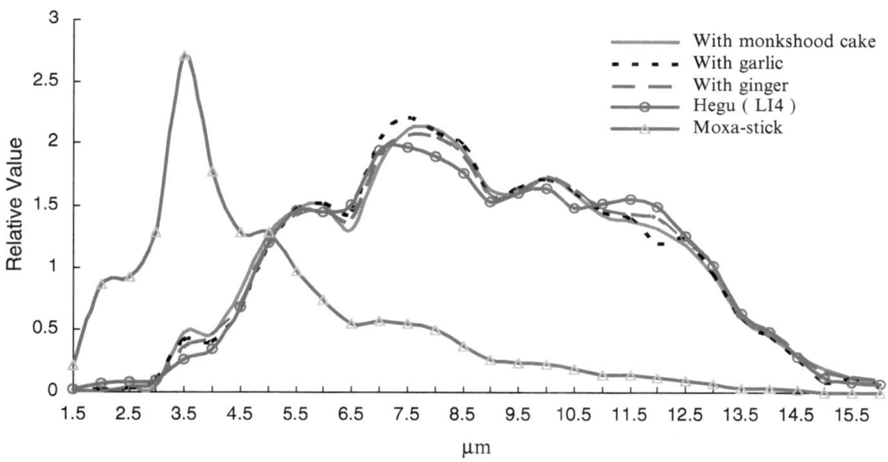

Fig. 5.4 Unified infrared radiation spectrums of an acupuncture point, direct moxibustion with a traditional moxibustion stick, and indirect moxibustion with three traditional media

spectrums. By contrast, the infrared radiation intensities of slices of cucumber and carrot, used as control media for indirect moxibustion, were 274.47 and 50.53 mV, respectively, quite different from that of the traditional media (Fig. 5.3).

3. Infrared radiation at Hegu (LI-4) was 20.40 mV, and it peaked on the infrared spectrum at about 7.5 μm (Fig. 5.4).

The experiment showed that the thermal action of the traditional moxa stick was more potent than that of indirect moxibustion. In addition, its radiation peak was different from that at the acupuncture point on the human body. By contrast, the thermal action of traditional indirect moxibustion was modest, and its radiation peak matched that at the acupuncture point.

The spontaneous infrared radiation data on Hegu (LI-4) of seven subjects (7 × 4) are presented in Table 5.1. Spectrum range was 1.5–16 μm, mainly distributed between 4.5 and 13.5 μm. The radiation peaked around 7.5 μm, and radiation intensity was 20.40 mV (Fig. 5.4).

The before-mentioned data from the highly sensitive newly developed infrared radiation spectrum detecting system show that compared to the controls the intensity of the traditional moxa stick is extremely strong, reaching 43,300.41 mV in comparison to the controls (only 31.15 mV from the smokeless moxa stick and 37.03 mV from the *555* cigarette).

The lesser thermal action produced by these substitutes may explain why these materials are not commonly used in clinical settings. On the other hand, the radiation peak in the traditional moxa stick spectrum is 3.5 μm, very different from the 7.5 μm of the acupuncture point (Fig. 5.4). (*Previous reports from Shen et al.* 2001 *have shown that the spectrum of a burning moxa stick and the acupuncture point was around 7.5 μm, but after modification, this wavelength should be around 10 μm.*) Hence the effect of the moxa stick may not due to sympathetic vibration; rather the therapeutic effect may due to its potent infrared radiation (Fig. 5.4).

The infrared radiation intensity produced by the three traditional indirect moxibustion media is significantly stronger than that of the controls (Table 5.1 and Fig. 5.2). The spectrum pattern of the three traditional media is very similar to that of acupoint LI-4 in that the radiation peaks at almost the same wavelength, around 7.5 μm (Table 5.1 and Fig. 5.4) (*After modification, this wavelength should be around 10 μm*). This similarity to the wavelength of the human acupuncture point may be the key to the therapeutic effect of traditional indirect moxibustion. Not only are the infrared radiation intensities produced by the controls low but their spectrum patterns are also completely different from that at the acupuncture point.

It has been posited that the moxibustion effect is associated both with thermal action and the pharmacological action of the materials used (Lin 1994). When the wavelength range of a source of infrared irradiation is similar to the infrared radiation emitted from the human body, body metabolism can be affected. This is because water molecules, and indeed all organic and biologic molecules of the body, can easily absorb infrared radiation, causing the molecules to vibrate and the energy level to change (Ma 1998). It has been hypothesized that moxibustion induces a sympathetic vibration in the molecules that corrects metabolic dysfunction and pathological conditions. The data show the effects of the thermal action and sympathetic vibration of moxibustion. It is possible, however, that all three actions, thermal action, sympathetic vibration, and pharmacological action, account for the therapeutic effects of moxibustion.

Table 5.1 Intensities and peak wavelength of the infrared radiation of traditional moxibustion, moxibustion with controls, and Hegu (LI-4)

	n	Intensity of radiation (mV)	Wavelength of the peak of radiation (μm)
Traditional moxa stick	4	43,300.41±425.15	3.5
Smokeless moxa stick	4	31.15±3.49*	7
555 cigarette	4	37.03±3.82*	3.5
Indirect moxibustion with monks-hood cake	4	681.87±47.52***,‡	8
Indirect moxibustion with ginger	4	520.27±68.22**,†	7.5
Indirect moxibustion with garlic	4	594.79±44.71***,‡	7.5
Indirect moxibustion with cucumber	4	274.47±19.61	5
Indirect moxibustion with carrot	4	50.53±4.68	5
Hegu (LI-4)	28	20.40±5.69	7.5

*Compared to the traditional moxa stick, $p=0.000$; **compared to indirect moxibustion with cucumber, $p=0.004$; ***compared to indirect moxibustion with cucumber, $p=0.000$; †compared to indirect moxibustion with carrot, $p=0.001$; ‡compared to indirect moxibustion with carrot, $p=0.000$

The infrared radiation spectrum of indirect moxibustion was unexpectedly similar to that of the acupoints of the human body, and both had their radiating peak round 10 μm. The resonance between the indirect moxibustion and the acupoint probably play a more important role in therapeutic effect of moxibustion than moxa's pharmacological and thermal radiation effects and the insulting materials separating the moxa.

As a natural infrared radiant source, the infrared radiation peak of the living human body is near 10 μm (Shen 1995), which is similar to the 10.6 μm wavelength of a CO_2 laser. The similarity to the wavelength of the human acupuncture point may be the key to the therapeutic effect of traditional indirect moxibustion. The CO_2 laser can be absorbed in the epidermis and cause a rapid and relative higher temperature rise in the skin. The CO_2 laser has an immediate warming effect as does moxibustion. The wavelength of the CO_2 laser is 10.6 μm, which can be absorbed by the epidermis to a depth of 0.2 mm and can greatly raise the temperature of the skin surface (Li and Zhang 2005; Zhao et al. 2006; Shen et al. 2009).

Low-level laser therapy (LLLT), characterized as noninvasive, painless, aseptic, safe, and easy to control, is now widely used in many countries. Most reported low-level lasers applying on acupoint in common clinical use today are *632.8* nm (Helium–Neon, gas), and other semiconductor lasers, in the visible light range and in the near-infrared region of the light spectrum (Whittaker 2004). These lasers can transverse the skin and make deep penetration into tissues, but their common problem is *nonthermal.* Thus the efficacy of "needling" is achieved while the efficacy of "moxibustion" could not be produced. The apparatus with specific wavelength of infrared radiation similar to that of human acupoint was made and applied to observe the effect of such radiation affecting rabbits with bradycardia and patients having osteoarthritis of knee.

5.2.3.2 The Efficacy of Laser Moxibustion Irradiation

The Efficacy of Moxibustion-Like Laser Irradiation on Bradycardia-Model Rabbits

Zhao et al. (2006) observed the effect of 10.6 μm CO_2 laser and 650 nm semiconductor laser irradiation on the acupoints of bradycardia-model rabbits. Ninety bradycardia-model rabbits were used in the study. The experiment applied 10.6 μm and 650 nm lasers separately and simultaneously. The data on heart rates of all rabbits at eight times are shown in Table 5.2. Means for heart rates are graphically displayed in Fig. 5.5.

The results showed that by 30 min after pituitrin injection, the heart rates of the laser Neiguan (PC-6) acupoint group are higher than that of the model control group, while the recorded heart rates in both lasers control and model control group are significantly lower than that of normal control group. It also indicated that the *rabbits' heart rates of* combined lasers Neiguan (PC-6) group recovered earlier at 30 min. Based on these results, it can be concluded that combined laser irradiation applied on the Neiguan (PC-6) of bradycardia rabbits at the described irradiation parameters caused an earlier recovery and produced a curative effect. At 30 min the heart rate of a single CO_2 laser Neiguan (PC-6) acupoint group is similar to that of normal group ($p>0.05$), which suggests that the CO_2 laser could accelerate the recovery of bradycardia. However, a single 650 nm laser irradiation on the Neiguan (PC-6) acupoint group does not produce such a therapeutic effect. It indicates that CO_2 laser may play a more important role in the effect of combined lasers. There may be a synergetic effect when the two kinds of lasers are applied together, so that both lasers are necessary to produce the therapeutic effect. Nevertheless, it seems from the research results that the CO_2 laser may play the more significant part in inducing the benefits.

Forty minutes after combined laser irradiation on the control point, the nonacupoint, a therapeutic effect was recorded, perhaps because the irradiation may influence the deep tissue of Neiguan (PC-6) in three-dimensional space. However the effect is reduced since the ulna acts as a barrier.

Studies done after moxibustion on the heat line along meridians showed that, although temperature on the body surface rose significantly, temperature on the corresponding deep tissue changed little due to the skin barrier (Zhang et al. 2003). Combined with the heat effect of the CO_2 laser, heat transfer along meridians should be more effective.

The Efficacy of Moxibustion-Like Laser Irradiation on Osteoarthritis

Shen et al. (2009) combined laser acupuncture with laser moxibustion on the knees of osteoarthritis (OA) patients. In the randomized, placebo-controlled, blinded pilot study, patients were irradiated with combined 650 nm–10.6 μm lasers. This procedure was compared to a placebo control group. The baseline characteristics of the patients are presented in Table 5.3.

Table 5.2 The change of heart rate after bradycardia produced by pituitrin ($\overline{X} \pm s$)

	Before pituitrin injection	After pituitrin injection						
		5 min	10 min	15 min	20 min	25 min	30 min	40 min
Normal ($n=15$)	252.40±29.06	251.33±32.01	243.93±35.28	246.67±31.93	241.33±34.81	241.93±34.17	239.13±33.57	244.33±32.79
Model ($n=15$)	273.47±47.60	147.00±27.56*	169.40±19.93*	177.33±21.52*	183.20±23.27*	190.67±25.33*	194.93±26.17*	204.53±30.50*
10.6 μm Laser Neiguan (PC-6) ($n=15$)	278.67±34.98	164.27±32.27*	176.07±23.32*	185.87±24.30*	195.13±21.82*	203.6±22.86*	215.93±20.54	226.6±21.88
650 nm Laser Neiguan (PC-6) ($n=15$)	269.4±47.50	150.07±30.70*	170.53±21.88*	182.27±22.63*	191.73±20.97*	198±26.44*	205.13±30.59*	216.53±29.77*
Lasers Neiguan (PC-6) ($n=15$)	271.53±41.30	155.20±29.19*	179.73±28.76*	192.20±33.48*	199.07±32.95*	209.47±35.56*	217.67±38.16**	225.27±37.23
Lasers control ($n=15$)	268.67±43.99	140.87±24.28*	163.47±20.18*	172.93±19.81*	183.53±20.27*	193.33±21.11*	203.20±22.51*	220.20±26.84

*Compare with normal control group at the same time, $p<0.05$; **compare with model control group at the same time, $p<0.05$

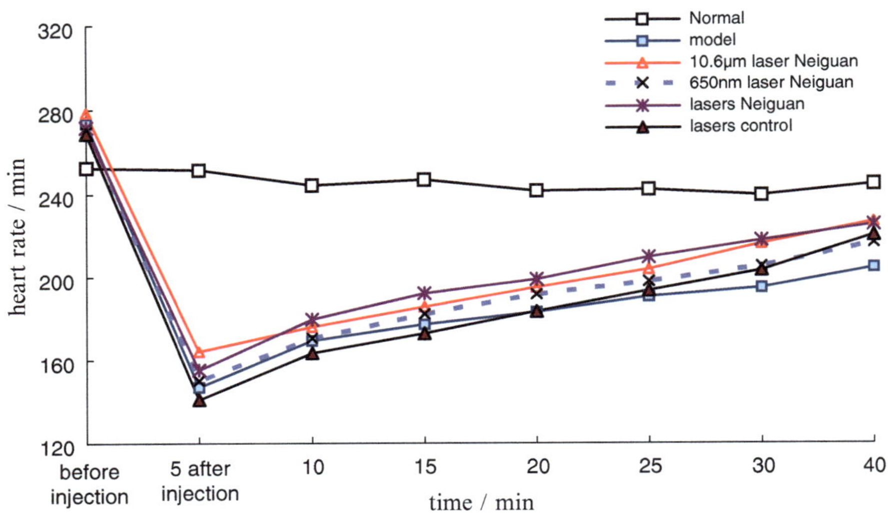

Fig. 5.5 The change of heart rate before and after bradycardia produced by pituitrin

Table 5.3 Baseline characteristics of patients

Group ($n=20$)	Age (year)	Disease duration (year)	Gender (female/male)	Affected knee (both/single)
Active laser	60.10±6.83	6.05±6.51	18/2	13/7
Placebo laser	56.40±7.41	4.24±6.65	18/2	10/10

Table 5.4 WOMAC index scores change at week 2 ($X \pm s$)*

Group	Pain difference (%)	Stiffness difference (%)	Function difference (%)
Active laser ($n=19$)	−49±34**	−3±126 ($n=14$)	−25±32
Placebo laser ($n=16$)	−13±62	56±218 ($n=12$)	4±65

*WOMAC index score reduction change (week 2–baseline)/baseline × 100 %

**Compared to the placebo laser (*t* test with Satterthwaite estimate of unequal variances): $p=0.047$

Forty patients with OA were randomly allocated to an active laser or placebo laser group (20 per group). They either received active or sham laser treatment at acupoint Dubi (ST-35) in a total of 12 sessions. Results showed that there was significant difference between the two groups in the WOMAC pain score, exhibiting a change from the baseline after 2 weeks of treatment. The pain reduction of the active laser treatment group was 49 %, whereas that of the placebo control group was only 13 % ($p=0.047$). However, due to the high patient dropout rate, the 4-week assessment cannot be analyzed (Table 5.4).

This pilot study tests a potentially useful laser device that combines thermal efficiency with laser acupuncture, thus solving two problems of clinical TCM. The thermal laser provides the warming effect of moxibustion without the problem of burning, smoke-producing moxa and the penetrating effect of acupuncture without the potential for pain or infection. The findings of this study could serve as the basis for future studies. Combined laser treatment seems beneficial to patients with knee OA. Based on this pilot study, a well-designed, multicenter clinical trial with a larger sample size is warranted to further investigate this new device.

5.3 Concluding Remarks

In conclusion, the mechanism of moxibustion effect is associated with the pharmacological action of the materials used, the thermal action, and the infrared radiation. The active ingredients of moxa include the ingredients of the volatile oil and smoke of moxa. The thermal action may be the key in the direct moxibustion effect. Moxibustion can differentially change the temperature of the local subcutaneous and muscular layers. The burning moxa emits 0.8–5 μm infrared radiation, and the high intensity of radiation covers the range from 1.5 to 3.5 μm. Using a highly sensitive infrared radiation spectrum detection system, the radiation peaks are found at 3.5 μm. The infrared radiation spectrums of indirect moxibustion and acupuncture points are unexpectedly similar. They share the same radiation peak at 10 μm. None of the substitute materials used as controls in the study possesses characteristics found by infrared analysis of the traditional materials. This suggests that selection of appropriate moxibustion material may be important for optimal therapeutic effect. Therefore, we presume that the infrared radiation resonance between indirect moxibustion and acupuncture points plays a more important role than the thermal effect in the therapeutic effects of traditional indirect moxibustion.

As the wavelength of the CO_2 laser is also about 10 μm, we hypothesized that it would share this sympathetic vibration with human body tissues. By applying an infrared laser therapeutic machine, with a specific wavelength similar to that of human acupuncture points, we verified the therapeutic effects of this infrared radiation spectrum on rabbits with bradycardia and human patients with knee osteoarthritis. It should be further promoted in clinical application.

Up to now, there are fewer studies about whether moxa smoke is harmful to the human body or whether the curative effect of the moxa is related with its smoke. In spite of the remarkable achievements in the thermal effect mechanism, the exact effect of the afferent information on thermal sensitive neurons and the involvement of the sympathetic nerve system remain unknown but are the important links of moxibustion treatment. It is thus worthy of further study. In the clinical literature, the limitations are its small sample size, its high dropout rate, and the fact that the blinding method needs improvement. Research about the factors that influence the effect of moxibustion therapy and systematic explanation of moxibustion regulation need multilevel, multidisciplinary collaboration to guide moxibustion clinical practice and improve moxibustion clinical effect.

Acknowledgements XS and LZ were partially supported by the National Basic Research Program of China (2009CB522901), the Key Program of the State Administration of Traditional Chinese Medicine of China and the Shanghai Municipal Science Foundation (11DZ1973300, 11ZR1436700).

References

Bai YH, Lin WZ (1991) Discussion on the relationship between moxibustion and thermal stimulation. Journal of acupuncture & moxibustion (4):10–11 (in Chinese)

Cardini F, Huang WX (1998) Moxibustion for correction of breech presentation: a randomized controlled trial. JAMA 280(18):1580–1584

Chen Y, Zhao C, Chen H, Qin H, Fang F (2003) Effects of "moxibustion serum" on proliferation and phenotypes of tumor infiltrating lymphocytes. Journal of Traditional Chinese Medicine 23(3):225–229

Cheng X (1999) Chinese Acupuncture and Moxibustion. Beijing, Foreign Languages Press. pp. 1590

Daxijidai. Translated by Liu PR (1989) Research about indirect moxibustion. Foreign Medical Sciences·Traditional Chinese Medicine Edition,14(3):60 (in Chinese)

Daxijidai. Translated by Zhao JP (1992) Anti-Free Radical Activity of moxa Burning Ashes. Foreign Medical Sciences·Traditional Chinese Medicine Edition, 14(3):60 (in Chinese)

Deng HP, Shen XY, Ding GH (2004) Characteristics Of infrared radiation of moxibustion and meridian- acupoints. Chinese acupuncture & moxibustion 24(2): 105–107 (in Chinese with English abstract)

Ding GH, Shen XY, Chu JH, Huang ZM, Yao W, Liu HP, Wang SZ, Fei L (2002) Chinese Journal of Biomedical Engineering 21(4):356–360 (in Chinese with English Abstract)

Dong QS (1996) Relationship between Moxibustion and Its Defervescence effect with Skin Receptors. Chinese Academic Periodical Abstracts 2(7): 115 (in Chinese)

Dong QS, Xian MQ, Dong XM, Zhang SH, Zheng XL (1998) Relation of moxibustion therapy to cutaneous receptors. World Journal of Acupuncture-Moxibustion 8(2): 32–39

Dong QS, Dong XM, Xian MQ, Zhang SH, Zheng XL (2000) Antipyretic action of moxibustion and its relation with acupoint receptors. World Journal of Acupuncture-Moxibustion 10(1):31–35

Dong XM, Dong QS, Xian MQ, Zhang SH (1999a) Effect of different moxibustion on the discharge of C polymodal receptors. World Journal of Acupuncture- Moxibustion 9(1):30- (in Chinese)

Dong XM, Dong QS, Xian MQ, Zheng XL, Zhang SH (1999b) Comparative Observation on the Temperature Changes of Acupoints by Different Moxibustions. Chinese Acupuncture & Moxibustio 19(1): 22–26 (in Chinese)

Ewies A, Olah K (2002) Moxibustion in breech version—a descriptive review. Acupuncture in Medicine 20(1):26–29

Han Z, Xu XQ, Xiao DB, Wu XD (2006) Test and analysis of the integrated effect of multiple factors in indirect moxibustion. Shanghai Journal of Acupuncture & Moxibustion 19(4):32–34 (in Chinese with English abstract)

Hong WX, Cai JH, Jing J (2004) A Research on Heat Radiation Spectrum Characteristics of Moxibustion. Therapy J. Applied Optics 25(4):1–3,44 (in Chinese with English Abstract)

Hong WX, Cai JH, Jing J, Li CW (2003) The Research on Heat Radiation Spectrum Characteristic of Moxibustion in Traditional. Chinese Medicine Journal of Biomedical Engineering Research 22(4): 27–30 (in Chinese with English Abstract)

Hong ZG, Nong YY, Yang ZT (2007) Qi-moxa Burning Smoke GC-MS Analysis of the Chemical Composition. Journal of South Central University for Nationalities (Nat. Sci. Edition) 26(1):10–13 (in Chinese)

Hu XL, Xu JS, Ye L Yang J, Wang PQ, Wu BH (2002) Elicitation of infrared radiant track along meridian courses over human body surface by local heating. Journal of Infrared And Millimeter Waves 21(1): 6–8 (in Chinese with English Abstract)

Joos S, Brinkhaus B, Maluche C, Maupai N, Kohnen R, Kraehmer N, Hahn EG, Schuppan D (2004) Acupuncture and moxibustion in the treatment of active Crohn's disease: a randomized controlled study. Digestion 69(3):131–139
Klaus PS, Walburg MO, Fritz AP (2005) Biophotonic in the Infrared Spectral Range Reveal Acupuncture Meridian Structure of the Body. The Journal of Alternative & Complement Medicine 11(1):171–173
Lao L (1999) Traditional Chinese Medicine, Part III, Chapter 12 Essentials of Complementary and Alternative Medicine. Wayne B. Jonas and Jeffery S. Levin, Eds. Philadelphia: Lippincott Williams & Wilkins, (b)
Lee MS, Shin BC, Kim JI, Han CH, Ernst E (2010) Moxibustion for Stroke Rehabilitation: Systematic Review. Stroke. 41(4):817–820
Li L, Zhu B, Rong PJ, Bi H, Gao XY, He W (2009a) Study on the Mechanisms of Thermal Moxibustion. Journal of Practical Training of Medicine 37(2): 70–71 (in Chinese)
Li Q, Jiang BC, Ma CH, Zhou JY, Wang KF, Pei TF (1993) Experimental Analysis of chemical composition of Moxibustion smog. Heilongjiang Journal of Traditional Chinese Medicine (4): 44–46 (in Chinese)
Li X, Hu J, Wang XY, Zhang HR, Liund JP (2009b) Moxibustion and other acupuncture point stimulation methods to treat breech presentation: a systematic review of clinical trials Chin Med 4: 4.
Li Z, Zhang Z (2005) The Mechanism on Laser Acupuncture and Moxibustion. Acta Laser Biology Sinica 1: 8–11
Lin WZ (1994) Experimental Acupuncture & Moxibustion. Shanghai, Publishing house of Shanghai Science & Technology. pp 58–63 (in Chinese)
Lin WZ, Wang P (1999) Experimental Acupuncture. Shanghai Scientific and Technology Press (in Chinese)
Liu HP, Shen XY (2002) Application of infrared thermography in the research of acupuncture & moxibustion. Journal of Modern Traditional Chinese Medicine (5): 63–66 (in Chinese)
Liu NY, Yang ZM (1993) Effect of Moxibustion on Skin Temperature. Jiangsu Journal of Traditional Chinese Medicine 14(11): 31–32(in Chinese)
Ma BX, Chong GQ (2008) Summary of investigated development on immune regulation of moxibustion. Jiangsu Journal of Traditional Chinese Medicine, 40(1):86–87 (in Chinese)
Ma JS (1998) Effect of Middle-Far Infrared Electric-Magnetic Irradiation on Biochemical Reaction. Chinese Journal of Nature 21(2): 85–88 (in Chinese)
Mei QX, Dong PR, Wang J, Tan XC, Liu YW (1991) Comparison of volatile oil and microelement content in Moxa leaves from Different producing area. China Journal of Chinese Materia Medica 16(12):718 (in Chinese)
Mei QX, Wang J (1994) Comparison of thermal discharge of combustion of the moxa from Different producing area. Chinese medicinal materials 17(9):46–47 (in Chinese)
Neri I, Airola G, Contu G, Allais G, Facchinetti F, Benedetto C (2004) Acupuncture plus moxibustion to resolve breech presentation: a randomized controlled study. Journal of Maternal-Fetal & Neonatal Medicine 15(4):247–252
Okazaki M, Furuya E, Kasahara T, Sakamoto K (1983) Effects of single moxibustion on phagocytic activity in mice. Am J Chin Med 11(1–4):112–122
Okazaki M, Sakamoto H, Suzuki M, Oguchi K (1990) Effects of single and multiple moxibustions on activity of platelet function, blood coagulation and fibrinolysis in mice. Am J Chin Med 18(1–2):77–85
Ovechkin A, Lee SM, Kim KS (2001) Thermovisual evaluation of acupuncture points. Acupuncture Electrotherapy Research 26(1–2):11–23
Qian CZ, Qian J, Bai YH (1991) A study of near infrared simulator moxibustion & and its clinical curative effect. Infrared Technology 13(6): 27–32 (in Chinese)
Shang XK (2004) Clinical Cases on Gastrointestinal Diseases Treating by Long-time Mild Moxibustion. Shanghai Journal of Acupuncture and Moxibustion 23(3): 22–23 (in Chinese)
Shen X (1995) Physical problems in the life sciences. Physics 3, 166–71
Shen X (1996) The physical hypostasis of super-weak irradiance of biologic system. Physics of life sciences. Beijing, Publishing house of Beijing University: pp 49–69 (in Chinese)

Shen XY, Ding GH, Chu JH, Huang ZM, Yao W, Zhou Y, Wei JZ, Chu LX, Zhang HM, Fei L (2003) Comparison of infrared radiation spectrum of traditional moxibustion, substitute moxibustion and acupoints of human body. Journal of Infrared and Millimeter Waves 22(2): 123–126 (in Chinese with English Abstract)

Shen XY, Ding GH, Chu JH, Yao W, Huang ZM, Yang WY, Fei L (2001) Human acupoints and moxibustion infrared radiation spectrum with infrared transmission in acupoints. Acta Universitatis Traditionismedicalis Sinensis Pharmacologiaeque Shanghai 15(4): 33–35 (in Chinese with English abstract)

Shen XY, Ding GH, Wei JZ, Zhao L, Zhou Y, Deng HP, Lao LX (2006) An infrared radiation study of the biophysical characteristics of traditional moxibustion. Complementary Therapies in Medicine 14(3):213–219

Shen XY, Zhao L, Ding GH, Tan M, Gao JP, Wang LZ, Lao LX (2009) Effect of combined laser acupuncture on knee osteoarthritis: a pilot study. Lasers in Medical Science 24(2): 129–136

Song JL, Fan FJ, Han ZP, Hong WX (2006) Advances of studies on traditional moxibustion therapy for treatment of cancer, Chinese Acupuncture & Moxibustion 26(3):227–229 (in Chinese with English abstract)

Translated by Huang QF (1989a) Burning Temperature of Moxa and the changes of Temperature in the Body (I)—the Burning Region's Temperature of Subcutaneous Base Foreign Medicine-Section of Traditional Chinese Medicine11(5): 48 (in Chinese)

Translated by Huang QF (1989b) The organic components of AI. Foreign Medical Sciences·Traditional Chinese Medicine Edition 11(5): 47 (in Chinese)

Uchida S. Suzuki A. Kagitani F. Nakajima K. Aikawa Y (2003) Effect of moxibustion stimulation of various skin areas on cortical cerebral blood flow in anesthetized rats. American Journal of Chinese Medicine 31(4):611–621

Wang DS, Sun ML (1994) Preliminary Observation on Temperature Measurement about Two Different Warm Needlings. Information Compilation 95(14): 54–58 (in Chinese)

Wei JZ, Shen XY, Ding GH, Zhao L (2007a) Investigation of pathway and mechanism of heat stimulating action of sandwiched moxibustion. Chinese Acupuncture & Moxibustion 27(5): 391–393 (in Chinese with English abstract)

Wei JZ, Shen XY, Ding GH, Zhao L, Cheng K (2007b) Analysis of the Temperature-time Curve of Indirect Moxibustion. Shanghai Journal of Acupuncture & Moxibustion 26(12):34–35

Wei JZ, Shen XY, Ding GH, Zhao L, Cheng K (2007c) Analysis of the Temperature-time Curve of Indirect Moxibustion. Shanghai Journal of Acupuncture and Moxibustion. 26(12): 34–35 (in Chinese)

Wei JZ, Shen XY, Ding GH, Zhao L, Ying J, Zhou Y (2004) Spontaneous infrared radiation spectrum of Acupoint Taiyuan. Acta Universitatis Traditionismedicalis Sinensis Pharmacologiaeque Shanghai 18(4): 33–35 (in Chinese with English abstract)

Whittaker P. (2004) Laser acupuncture: past, present, and future. Lasers in Medical Science 19:69–80

Xian MQ, Dong QS, Zhang SH, Dong SM (2000) Thermolytic effect of moxibustion and its relation to acupoint receptors. World Journal of Acupuncture- Moxibustion 10(2):24–31

Xiguyuzi. Translated by Huang QF (1989) Antioxidation products of AI Burning Ashes. Foreign Medical Sciences·Traditional Chinese Medicine Edition 11(5):47–48 (in Chinese)

Xin BY, Hong LF, Shi XM, Li XW (2000) Research Progress on Anti aging by Moxibustion Shanghai Journal of Acupuncture& Moxibustion 19(5): 44–46 (in Chinese with English abstract)

Xu W, Chen ZQ, Yan YS (1988) The effect on pain reaction of big mouse produced by treating its Changqiang (GV1) with heat or moxa-oil. Research of acupuncture (3) Supplement: 190 (in Chinese)

Yamashita H, Ichiman Y, Tanno Y (2001) Changes in peripheral lymphocyte subpopulations after direct moxibustion. American Journal of Chinese Medicine 29(2):227–235

Yang HY, Liu TY (1996) Preliminary study on biophysics mechanisms of moxibustion therapy. Chinese Academic Periodical Abstracts 16(10): 17–18 (in Chinese)

Yang HY, Liu TY (1996) Preliminary study on biophysics mechanisms of moxibustion therapy. Chinese acupuncture & moxibustion 16(10):17–18 (in Chinese)

Yu BX, Guo XJ (2005) Moxibustion treatment for Stroke in recent 10 years. Nei Mongol Journal of Traditional Chinese Medicine, 24(3):35–36 (in Chinese)
Yu S, Song XG, Hu L (2009) General situation of the studies of the effect of moxibustion on experimental arthritis. Shanghai Journal of Acupuncture & Moxibustion 28(4):187–190 (in Chinese)
Yuan J, Tang ZL (2007) Research Progress of the immune effect of moxibustion Regulation. Journal of Anhui Traditional Chinese Medicine College 26(2): 60–62 (in Chinese)
Yuan ZQ, Zhuang YL (1995) Traditional Chinese Moxibustion therapy. Medical Equipment (1):24–25 (in Chinese)
Zhang D, Zhu YG, Wang SY (2002) Infrared thermoimages display of body surface temperature reaction in experimental cholecystitis. World J Gastroenterol 8(2):323–328 (in Chinese)
Zhang D, Wang SY, Wang YC (2003) Measurement on the corresponding deep tissue under the heat line along meridians. Chinese Journal of Basic Medicine in Traditional Chinese Medicine 10: 62–4
Zhang FR, Geng JY (1989) Chinese moxibustion therapy, the first edition, Beijing: People's Medical Publishing, pp. 25 (in Chinese)
Zhang H, Cheng S, Liu Y(1999) Analysis of Chinese moxibustion in visible and infrared spectra in process of clinical application. Spectroscopy and Spectral Analysis 19 (3):344–346 (in Chinese with English abstract)
Zhang R (2004) History and current state of moxibustion. Journal of Chinese Integrative Medicine 2(6): 466–473 (in Chinese with English abstract)
Zhao JX, Jia CS, Tian YX, Li XH (1999) Recent 10-years Progress of Moxibustion Experimental Research in Japan. Chinese Acupuncture & Moxibustion 8:507–511(in Chinese with English abstract)
Zhao L, Shen XY, Gao JP, Ding GH, Wei JZ, Deng HP, Wang LZ, Zhao XY (2006) Effect of different LLLT (Low Level Laser Therapy) on pituitrin-induced bradycardia in the rabbit. Lasers in Medical Science 21(2):61–66
Zhou W (2003) Acute lymphangitis treated by moxibustion with garlic in 118 cases. Journal of Traditional Chinese Medicine 23(3):198

Chapter 6
Acupuncture Treatment of Epilepsy

Dongman Chao and Ying Xia

Abstract Epilepsy is one of the commonest brain disorders characterized by recurrent unprovoked seizures that result from the abnormal, excessive, and synchronous activity of clusters of nerve cells in the brain. Epilepsy is very prevalent and has a high incidence. Intensive efforts have been made in the treatment for epilepsy, including medical, surgical, and other treatment strategies such as ketogenic diet and vagus nerve stimulation. However, these treatment options have their own limitations and complications. About one-third of patients show no response to pharmacological treatment. Concern regarding the adverse effects of these treatments has drawn people toward other therapies for treatment of epilepsy. Acupuncture, one of the popular therapeutic techniques in traditional Chinese medicine (TCM), has proven to be safe, efficient, and convenient with minimal side effects. Thus, it has gained popularity in treatment of various diseases, including epilepsy, in the West. However, due to the limited availability of high-quality research papers documented in English and a difference in the understanding of antiepileptic effects of acupuncture between the scientists of East and West, the application of acupuncture in the treatment of epilepsy is limited in Western community. In this chapter,

D. Chao
Department of Neurosurgery,
The University of Texas Medical School at Houston,
6431 Fannin Street, Houston, TX 77030, USA

Yale University School of Medicine, New Haven, CT 06520, USA
e-mail: dongman.chao@uth.tmc.edu

Y. Xia (✉)
Department of Neurosurgery,
The University of Texas Medical School at Houston,
6431 Fannin Street, Houston, TX 77030, USA

The University of Texas Graduate School of Biomedical Sciences at Houston,
6431 Fannin Street, Houston, TX 77030, USA

Yale University School of Medicine, New Haven, CT 06520, USA
e-mail: ying.xia@uth.tmc.edu

Y. Xia et al. (eds.), *Current Research in Acupuncture*,
DOI 10.1007/978-1-4614-3357-6_6,

we attempt to summarize the clinical observations on acupuncture in the treatment of epilepsy. Based on the underlying mechanisms of epilepsy and epileptogenesis, we have outlined the probable modes of action of acupuncture antiepilepsy in this chapter. Epileptic seizures and epileptogenesis are complex processes that result from changes in the excitability of neurons and involve synchronous discharges of a population of hyperexcitable neurons propagated through normal or pathological pathways. The increased excitability is ascribed to the changes that occur in both neurons and glial cells. In neurons, it is mainly due to (1) the innate ability of intrinsically bursting neurons in the cortex and hippocampus, (2) alterations in the membrane properties, (3) imbalance between excitatory and inhibitory transmission, and (4) alterations in existing synaptic contacts/circuits and/or establishment of new synaptic contacts/circuits. In glial cells, the most striking change is their impaired ability to maintain glutamate and K^+ homeostasis. The most probable mechanism that underlies the antiepileptic effect of acupuncture is normalization of this disrupted neuronal excitability through multiple effects at different levels (from cells to molecules) with different time course (short term vs. long term) and during different stages of epilepsy (early/acute vs. later/chronic). These strategies include the modification of neuronal electrical activity and the activity of therapeutic network, the regulation of the activity of biomediators (e.g., amino acids, neuropeptides, monoamines, NO, Ca^{2+} and Ca^{2+}-dependent activity, hormones, and humoral factors), the prevention from seizure-induced neuronal injury/loss and the subsequent aberrant neurogenesis and synaptic reorganization/dendritic remodeling in seizure-susceptible areas, and the regulation of ion channel activity.

Keywords Acupuncture/electroacupuncture • Epilepsy • Seizures • Epileptogenesis • Hyperexcitability • Homeostasis • Functional normalization

6.1 Introduction

Epilepsy is one of the commonest brain disorders characterized by recurrent unprovoked seizures that result from the abnormal, excessive, and synchronous activity of clusters of nerve cells in the brain (Hirtz et al. 2007). There are many kinds of epilepsy depending on different classification criteria (e.g., causes, observable manifestations, locations in the brain, events that trigger seizures). Based on clinical and EEG observations (rather than the underlying pathophysiology or anatomy), the International League Against Epilepsy (ILAE) proposed a classification scheme for epilepsy and epileptic syndromes, categorizing over 40 different types of epilepsy, and is still being updated (Epilepsia 30:389–399; 1989). Practically, based on how the abnormal brain activity begins, epilepsy is generally divided into two major categories—focal seizures and generalized seizures, in which many different subtypes of seizures are included. Due to a large variety of epilepsy types, the symptoms also vary depending on the types of seizures. In epilepsy, the normal pattern of neuronal activity is disturbed, causing different patterns of symptoms, including temporary confusion, a staring spell, strange sensations and emotions, uncontrollable jerking movements of limbs, and loss of consciousness. Epilepsy is also a disorder with varied etiology. An epileptic seizure could be genetic or acquired. In fact, epilepsy has no identifiable causes (idiopathic epilepsy) in about half of the patients. In the other half, epilepsy may be traced to various factors, including genetic factors, head injuries that occur during birth or from accidents during youth and adulthood (e.g., car accident, lead poisoning), developmental disorders (e.g., autism, Down syndrome), sporadic central nervous system (CNS) infection (e.g., meningitis, AIDS, and viral encephalitis), stroke or any other type of damage to the brain (e.g., dementia, tumor, surgery), alcohol, drugs, and many others (Sander and Shorvon 1996; Blümcke 2009; Lu and Wang 2009).

Epilepsy is a highly prevalent disorder. Reported prevalence rates vary in the range of 1.5–57/1,000 for active epilepsy. The annual incidence rates of epilepsy are usually between 40 and 70/100,000 in developed countries and between 122 and 190/100,000 in developing countries. Lifetime prevalence rates of epilepsy are much higher than the incidence or prevalence of active epilepsy, which up to 5% of any population experiences nonfebrile seizures at some point in life (Sander and Shorvon 1996; Sander 2003; Hauser and Beghi 2008). The incidence rates are higher in children than in adults and decrease in children with their growth but were seen to increase in the elderly during the 1970s to 2003 in the developed countries. Also, there exists a difference in the incidence rates based on sex and race with a slightly higher rate in males than in females and higher rates in black African populations (Sander and Shorvon 1996; Sander 2003; Hauser and Beghi 2008). People with epilepsy often suffer from a poor quality of life (e.g., in danger for driving, cognitive, behavioral, and emotion problems, transient blindness, or even paralysis). Epilepsy patients are also at special risk for four life-threatening conditions, i.e., status epilepticus, suicide associated with depression, trauma from seizures, and sudden unexpected death in epilepsy (SUDEP) (Walczak et al. 2001; Hitiris et al. 2007; Lathers 2009).

As with its heterogeneous classification, the biologic bases of epilepsy are variable and complex to understand. However, cerebral imbalance of the major regulatory antagonistic systems, i.e., decreased activity of GABAergic inhibitory amino acid system associated with increased excitatory glutamatergic pathways, has been found to play a key role in initiating seizures (Cordero et al. 1994; Liu and Cheng 1995; Liu et al. 1997; Fukao et al. 1998; Brailowsky and Garcia 1999; Yu et al. 2006; Butuzova and Kitchigina 2008; Szyndler et al. 2008; Campbell and Hablitz 2008; Garrido-Sanabria et al. 2008; Knopp et al. 2008; Sandow et al. 2009; Tyzio et al. 2009; Lopantsev et al. 2009). This imbalance functionally results in the de-inhibition of postsynaptic neurons, leading to the hyperexcitability and highly synchronous discharges of a cluster of neurons in the epileptic focus that propagate these discharges to the neighborhood, thus leading to an extensive excitation of the innocent brain. During epileptic seizures, the electrolyte homeostasis gets disrupted between the intra- and extracellular space (e.g., Na^+ Ca^{2+}, K^+, and Cl^-) and the activities of ion channels are altered. For example, there is an increased mRNA expression of Na^+ channels, increased amplitude of voltage-dependent I_{Na}, a shortened phase of inactivation, and shift of inactivation currents toward depolarization in neurons from epileptic seizures (Sashihara et al. 1994; Tian et al. 1995;

Lombardo et al. 1996; Vreugdenhil et al. 1998; Xia et al. 2003; Guo et al. 2008a; Blumenfeld et al. 2009; Hargus et al. 2011). Besides, many other mechanisms, like the neurotransmitter imbalance (such as enkephalin and dynorphin, release of nitric oxide) (Kovács et al. 2009) and disrupted calcium homeostasis (Delorenzo et al. 2005), are also involved in the epileptic seizures and epileptogenesis.

Intensive efforts have been made for the treatment for epilepsy. These include medication treatment, surgical treatment, and other treatment strategies such as ketogenic diet, vagus nerve stimulation, and complementary and alternative medicine (CAM). Recently, some new types of therapies, including stem cell transplantation, use of a device that could predict seizures up to 3 min before they begin, are under investigation to help patients control seizures (Boison 2009; Maisano et al. 2009; Waldau et al. 2010; Massachusetts Institute of Technology 2006; Thomas Jefferson University 2007; Park et al. 2011). Among all the treatments, medical treatment is the principal one of epilepsy. Based on the biological basis of epilepsy, a number of drugs, including many conventional antiepileptic drugs (AEDs) and approximately 20 recently licensed novel third-generation AEDs have been developed for the treatment of epilepsy (Malawska 2005; Bialer 2006; Luszczki 2009), and some potential AEDs are in the development pipeline (Rogawski 2006). The majority of AEDs are designed to modify the neural excitability by enhancing the inhibitory GABAergic system, decreasing excitatory glutamatergic system, or targeting the ion channels, particularly Na^+ channels (Reckziegel et al. 1999; Vreugdenhil and Wadman 1999; Czapiński et al. 2005; Errington et al. 2005; Malawska 2005; Bialer 2006; Luszczki 2009). Each drug potentially can have more than one mechanism of action (Czapiński et al. 2005). The long-established AEDs can control seizures in 50% of patients who develop partial seizures and in 60–70% of those with generalized seizures (Malawska 2005). Though AEDs have great efficacy with a successful control of seizures, up to 30% of patients remain resistant to the available AEDs (Engel 1996; Malawska 2005; Beleza 2009; Luszczki 2009). In addition, all AEDs have some side effects (Perucca and Meador 2005). The mild side effects include fatigue, dizziness, weight gain, loss of bone density, skin rashes, loss of coordination, and speech problems. There are more severe side effects in some patients, especially those with prolonged use of AEDs, such as cognitive impairment, depression, suicidal thoughts and behaviors, severe rash, and inflammation of certain organs.

Surgery is an option for patients whose seizures remain resistant to the treatment with AEDs (Engel 1996; Blümcke 2009). This treatment requires accurate identification and localization of epileptogenic regions. The removal of epileptic focus (a small area of the brain where seizures originate), or a discrete lesion of intracortical connections which spread seizures in generalized epilepsy with seizures originate in part of the brain that cannot be removed, can significantly reduce or even halt seizures for some patients with intractable epilepsy. The most common form of intractable epilepsy in adults is temporal lobe epilepsy (TLE) with hippocampal sclerosis, and the most common type of surgical treatment for it is the anterior temporal lobectomy to remove the front portion of the temporal lobe including the amygdala and hippocampus. However, even after a successful surgery, some medicines are always needed although maybe with fewer drugs or reduced dose. Surgery for epilepsy may also carry some amount of risks as any other kind of surgery does. It may cause some potential postoperative neurological deficits or complications such as permanently altered cognitive abilities or personality changes (Engel 1996). The direct/indirect expenditure of medical care is also higher for surgical treatment.

Ketogenic diet, a high-fat and low-carbohydrate diet developed in the 1920s, is mainly used in the treatment of children with severe, medically intractable epilepsy. Studies have shown that in some cases, children may experience fewer seizures if they strictly maintain a ketogenic diet (Neal et al. 2008). However, maintenance of a ketogenic diet is a challenge for children. It may have some side effects like dehydration, constipation, retarded growth due to nutritional deficiency, and buildup of uric acid, which can lead to kidney stones.

Vagus nerve stimulation is another potential therapy for patients with intractable epilepsy with a reported reduction in seizure frequency of >50% (Milby et al. 2008). This therapy uses a stimulating electrode wrapped around the vagus nerve in the neck and a programmable pulse generator and battery pack surgically implanted underneath the skin of chest like a pacemaker. The battery-powered device delivers short bursts of electrical energy to the brain through the vagus nerve (Milby et al. 2008). The precise mechanism of this therapy is unknown. Most patients may need additional antiepileptic medication. Despite its efficacy in the treatment of refractory epilepsy, vagus nerve stimulation has some adverse effects, which include hoarseness, throat pain, coughing, shortness of breath, tingling, muscle pain, nausea, and vomiting (Milby et al. 2008).

Obviously, various treatment options, including AEDs, surgery, ketogenic diet, and vagus nerve stimulation, have their own limitations and complications. Concern regarding the adverse effects of these treatments has attracted people to explore other therapies for epilepsy. CAM is one of the popular options for epilepsy treatment (Cohen 2003; Schachter 2008; McElroy-Cox 2009). The practice of CAM in the US society has substantially increased in the past years. For example, a survey showed that in 1997, 42% of US adults had used CAM in the past 12 months with costs of $36 billion to $47 billion on CAM therapies (Eisenberg et al. 1998), while the percentage of US adults using CAM in the past year had increased to 62% in 2002 (Barnes et al. 2004). Acupuncture is the most commonly used form of CAM.

Acupuncture is one of the most popular therapeutic techniques in TCM. It involves insertion of fine needles into specific body areas (acupoints or meridian points) and swift manual manipulation (e.g., rotating, lifting, thrusting, retaining) until the appearance of the *Qi* phenomenon or *De Qi* (acquisition of energy, i.e., a mixed sensation of soreness, numbness, swelling, sinking, and hot sensation appears in acupoints) that is an important component of acupuncture for therapeutic effect and clinical efficacy (Kong et al. 2007a; Mao et al. 2007; Lin 2008). Acupuncture has been practiced for epilepsy treatment for thousands of years in China. The first known description of epilepsy and acupuncture therapy appeared in *Huang Di Nei Jing* (*The Yellow Emperor's Classic of Internal Medicine*), an ancient Chinese medicine book written by a group of Chinese physicians around 770–221 BC. This book consists of 2 volumes: *Shu Wen* and *Ling Shu*. In *Ling Shu*, there was a record of treatment of convulsion by acupuncture at the acupoint Tianzhu (BL-10). In fact, acupuncture treatment for epilepsy has been employed throughout all the dynasties in China (Yang and Cheng 2010). In modern Chinese medicine, acupuncture has been widely practiced to treat epilepsy at a number of hospitals in China. Acupuncture is also widely practiced in Japan, Korea, and other Asian countries and immigrant communities of oriental origin in the West. Over the past two decades, acupuncture popularity has grown in the United States and Europe. In contrast to the conventional Western medicine and surgical treatment, the most significant advantages of acupuncture treatment are safety, efficiency, convenience, and minimal side effects (NIH Consensus Development panel on Acupuncture 1998; Jindal et al. 2008; Wu et al. 2008).

In this chapter, we will summarize the clinical application of acupuncture for epilepsy and then discuss the possible underlying mechanisms. Because multiple methods and terms of acupuncture were developed and variably used by the investigators/acupuncture physicians in their research/clinical treatment, for the purposes of easy expression, "acupuncture" is generally used as a reflection of the modality with various acupuncture treatments including classical acupuncture and electroacupuncture.

6.2 Clinical Application of Acupuncture for Epilepsy

6.2.1 Traditional Chinese Medicine Theory of Epilepsy and Acupuncture Therapy

The philosophy of acupuncture therapy is based on the TCM theory of epilepsy etiology. Based on the concepts of TCM, epilepsy is known as *Dian Xian* in Chinese that means "falling sickness" (*Dian*) and convulsion (*Xian*). It is described as a disorder of, in most cases, disharmony of emotion and *Qi* (energy flow/vital energy), which leads to an imbalance between *Yin* and *Yang*, and dysfunction of the activities of *Zang-fu* (organ systems), for example, phlegm retention in the liver, spleen, and stomach, and internal "wind" invasion in *Zang-fu*. The emotional irritation will be converted into phlegm and lead to liver phlegm retention and *Qi* stagnation, which make liver, spleen, and kidney unharmonious, upstirring the wind-*yang*, and ultimately make the clear orifice misted, the mind disturbed, and the channels invaded; thus seizures occur (Yang and Cheng 2010). Here, phlegm in TCM refers to congealed moisture that is a kind of pathological products of the body's activities acting upon *Qi* and moisture produced during the course of disease due to certain pathogenic factors. Phlegm flows throughout the whole body following the ascent and descent of *Qi*. Phlegm is divided into two types: the visible and the invisible one. The visible phlegm usually refers to the sputum that often occurs with coughs, while the invisible phlegm refers to the thicker retained water existing in *Zang-fu*, channels, or skin and muscles. According to this pathogenesis of epilepsy, it is believed that acupuncture, by stimulating various acupoints of the governor and conception channels (GV, CV) and internal organ (*Zang-fu*, i.e., liver, gallbladder, stomach, etc.) channels, can correct the disrupted flow of *Qi*, restore the imbalance between *Yin* and *Yang* and natural internal homeostasis to refresh the brain, eliminate "wind," clear phlegm retention, and restore consciousness, thus relieving the symptoms and curing epilepsy.

6.2.2 Commonly Used Methods of Acupuncture and Acupoints for Treatment

Currently, there are many methods available to perform acupuncture. Besides the traditional fine needle acupuncture, various other forms of acupuncture have been developed, including catgut implantation at the acupoints, acupuncture plus Chinese herbs or AEDs, electroacupuncture, laser acupunctures, acupressure, gold bead/wire implants at the acupoints, magnetic chips implantation at acupoints, injection of substance to acupoints, capsicum patches at acupoints, moxibustion at acupoints, cupping at acupoints, etc. (Yang and Cheng 2010).

There are varieties of acupoints that can be chosen for the treatment of epilepsy. Majority of these acupoints are distributed around the skull and trunk, in particular along the governor vessel, such as Dazhui (GV-14), Baihui (GV-20), Renzhong/Shuigou (GV-26), Shenting (GV-24), Jinsuo (GV-8), Changqiang (GV-1), etc. Acupoints such as Jiuwei (CV-15), Shaoshang (LU-11), Shenmen (HT-7), Xinshu (BL-15), Houxi (SI-3), Juque (CV-14), Yinbai (SP-1), Shenmai (BL-62), Yongquan (KI-1), Zhongwan (CV-12), Jianshi (PC-5), Laogong (PC-8), Jinmen (BL-63), Zhaohai (KI-6), Tianjing (SJ-10), etc., are also chosen for epilepsy treatment. Practically, it is unusual to apply acupuncture at a single point; therefore, several points are chosen and simultaneously stimulated, i.e., several acupoints are used together as the primary acupoints, while others as subsidiary ones. For example, Baihui (GV-20), Dazhui (GV-14), Shenting (GV-24), Yaoqi (ExB-9), and Taichong (LR-3) were used as primary acupoints. Taiyang (Ex-HN-5), Fengchi (GB-20), Anmian (Ex-HN-22), Shenmen (HT-7), Neiguan (PC-6), Fenglong (ST-40), Taodao (GV-13), Jinsuo (GV-8), and Changqiang (GV-1) were used as subsidiary acupoints. Practically, the acupoints of Renzhong (GV-26), Yongquan (KI-1), Shenting (GV-24), Baihui (GV-20) are chosen for restoration of consciousness, restriction of abnormal excitement of the brain, elimination of the "wind", and opening of the orifices. Dazhui, the crossing point of governor and conception channels, is needled to dredge the jammed communication of *Qi* between governor vessels and conception vessels in order to maintain a balance between *Yin* and *Yang*. Needling at Jiuwei (CV-15), the connecting acupoint of CV channels, can arrest the epileptic episode, relieve mental stress, harmonize the interaction of *Qi* and phlegm in stomach, and turn over turbidity. Taichong (LR-3) soothes the liver phlegm retention and eliminates the liver "wind". Fenglong (ST-40) makes stomach harmonized turbid *Qi* downward, resolves phlegm, and reduces heat.

6.2.3 Clinical Observations on Acupuncture Therapy

Despite substantial evidence provided through animal research in support of the antiepileptic effects of acupuncture (Chen and Huang 1984; Klide et al. 1987; van Niekerk and Eckersley 1988; He et al. 1990a; Panzer and Chrisman 1994; Huang et al. 1999; Chao et al. 2001; Jin et al. 2005a; Shu et al. 2005; Guo et al. 2008b, c; Kang and Xia 2008; Kim et al. 2008b; Zhang et al. 2008b, c; Goiz-Marquez et al. 2009), numerous case reports have also shown that patients have benefitted from acupuncture as it was successful in controlling their seizures, especially in cases of refractory epilepsy (Chen et al. 1983; Shi et al. 1987; Yang 1990; Lin and Hong 1999; Leng et al. 2000; Wen 2000; Deng et al. 2001; Lin et al. 2001; Wang 2001; Ma et al. 2001; Mao and Guo 2005; Zhang et al. 2006; Zhuang et al. 2006a; Ren 2006; Xu 2006; Song et al. 2006; Han et al. 2008), though there are some reports that show negative results. For example, a study from a Norway group fails to corroborate the above findings (Kloster et al. 1999).

The number of cases reported in these clinical observations varies from several to over 100 (Chen et al. 1983; Shi et al. 1987; Yang 1990; Leng et al. 2000; Wen 2000; Deng et al. 2001; Lin et al. 2001; Wang 2001; Ma et al. 2001; Mao and Guo 2005; Zhang et al. 2006; Zhuang et al. 2006a; Ren 2006; Xu 2006; Song et al. 2006; Han et al. 2008). The patients that were treated also differed in their age groups, varying from infants to adults and the elderly. Types of epilepsy treated in these patients also varied from absent seizures, febrile convulsions, and clonic–tonic seizures to generalized and even status epilepticus. Many patients treated were reported to be AEDs-resistant. Overall, in these cases reported, the effects of acupuncture principally manifested as reduction in symptoms (e.g., a reduction of seizure frequency, shortness of episodes), functional recovery (e.g., smoothed breath from short and quick one and occasionally breath stop, recovery from unconsciousness), improved EEG (e.g., reduction of spike wave, desychronization), and/or decreased epilepsy scores. The efficacy of acupuncture therapy was reported to be as high as 98%. Various acupoints were chosen, including Fengfu (GV-16), Jinsuo (GV-8), Renzhong (GV-26), Baihui (GV-20), Dazhui (GV-14), Jizhong (GV-6), etc., and various methods for therapy were used in these case reports, including traditional manual needle acupuncture, electroacupuncture, and catgut implantation at the acupoints, acupuncture plus Chinese herbs or AEDs, etc. Of all these methods, catgut implantation at acupoints is one of the most effectively used therapies believed to be with a high long-term efficacy (Li and Zhuang 2007).

6.2.4 Comments

Despite the large number of case reports that are in favor of acupuncture treatment for epilepsy, few clinical observations oppose the curative effects of acupuncture on epilepsy. For example, a Norway group compared acupuncture treatment ($n=15$) for 8 weeks with 12 weeks follow-up with sham acupuncture ($n=14$) in 29 patients with chronic intractable epilepsy. They found a reduction in the seizure frequency in both groups with no statistical significance and a significant increase in the number of seizure-free weeks in the sham group (Kloster et al. 1999). They were unable to prove a beneficial effect of acupuncture in chronic intractable epilepsy (Kloster et al. 1999). This group also reported that there was no significant effect of acupuncture on the quality of life of patients with intractable epilepsy in the same trials (Stavem et al. 2000). A case of 25-year-old healthy male with convulsive syncope associated with acupuncture was reported in which an episode of convulsive syncope characterized by irregular clonic–tonic movements with unconsciousness for about 1 min immediately after the insertion of acupuncture needles into the bilateral Zusanli (ST-36) acupoints was observed (Cole et al. 2002). Indeed, there are still some concerns on the efficacy of acupuncture treatment for epilepsy in these reports. In a recently updated review, Cheuk and Wong (2008) selected 11 clinical study papers and analyzed the efficacy of acupuncture treatment for epilepsy based on their own criteria, and concluded that acupuncture has not yet been proven to effectively and safely treat people with epilepsy.

Considering the above discussion, one of the biggest disputes on acupuncture treatment that has come forth is the methodology of the clinical assessment. Furthermore, some of the case reports have a poor quality in terms of their experimental designs. For example, there is a lack of control or inadequate control in some observations; the sample size is small; the clinical background (disease history, types, severity, etiology, therapy taken before, AEDs kinds and doses taken, etc.) of the patients before acupuncture treatment has sometimes been neglected or not properly assessed in some reports. These may jeopardize the allocation of subjects under the principle of randomization, and thus bias the results and interpretations. In addition, non-standardized criteria for outcome measurement and effectiveness overestimate the efficacy with a low confidence. Besides, several additional issues need consideration as well. Firstly, there is a lack of adequate follow-up in most of the case reports. The follow-up in many clinical observations is relatively short, ranging from a few weeks to few months. It is very important to determine the long-term effects of acupuncture because unlike the AEDs that directly target the ion channels or enhance inhibitory transmission and attenuate excitatory one to lower the excitability of nerve cells in the brain, effects of acupuncture are not immediate. According to the TCM philosophy, acupuncture is a modality of recuperation that harmonizes the functions of various parts of the body and can take time for the patients to appreciate its effects to the full potential, just like a fitness regime for improvement of cardiorespiratory functions that requires persistence and perseverance. People who expect a quick fix will undoubtedly be disappointed, and those who are compliant with the treatment have a better chance of success (Yang and Cheng 2010). Secondly, the question on acupoint specificity remains unanswered. Kloster et al. (1999) found that needling at acupoints and at sites away from the classical locations of acupoints had similar outcomes in patients with intractable epilepsy, and they did not support the concept of the acupoint specificity. However, acupoint specificity is tacitly accepted, but without validation, in most clinical practices (Liu et al. 2009a, b; Ren et al. 2010), although animal studies have clearly demonstrated acupoint specificity (Fang et al. 2004; Kang and Xia 2008; Guo et al. 2008b, c). Thirdly, what are the optimal parameters for treatment? Due to the heterogeneity in methods used, it is hard to ascertain the optimal parameters in practice. In addition, acupuncture is a modality requiring very strong practices of the manipulation (e.g., the rate, intensity, duration of swift rotation, lifting, thrusting, retaining) and is largely dependent on the personal experience of acupuncturists. This makes the evaluation for optimal parameters further difficult.

Despite some issues on acupuncture treatment for epilepsy, we believe that acupuncture, a safe, convenient and relatively inexpensive modality, is effective for epilepsy treatment, at least for certain types of epilepsy though it requires experienced hands for appropriate manipulation. Firstly, ancient acupuncture physicians had documented their successful cases as per clinical improvement based on a large number of clinical trials. They focused on controlling seizures and how to control seizures better by comparing different acupuncture methods with each other and acupuncture alone with acupuncture plus other therapies (Yang and Cheng 2010). Therefore, there is no doubt that ancient acupuncture physicians had validated the effects of acupuncture in treatment of epilepsy. Secondly, we must acknowledge that treatment of all types of epilepsy, especially drug-resistant and intractable one, would be difficult to achieve through any kind of therapy, including acupuncture. Determining whether acupuncture treatment can reach a satisfactory curative effect is dependent on multiple factors, including the correct diagnosis of epilepsy, the experience and operating technique of the acupuncturist, the correct choice of acupoints based on the diagnosis, the level of expertise of the acupuncturists in needling and manipulation, and even the belief and cooperation of the patient. Thirdly, the studies contradicting efficacy of acupuncture in epilepsy treatment are not without flaws as well, for example, small sample size can affect the determination of efficacy and jeopardize the statistical power to detect changes in parameters given the heterogeneity with regard to seizure types and frequency of occurrence in patients. In this respect, though the Norway group did not show any beneficial effect of acupuncture in intractable epilepsy in their study, they acknowledged that the small sample size could be responsible in part for this observation (Kloster et al. 1999). Even the phenomenon of convulsive syncope associated with acupuncture that was observed in only one healthy male immediately after the insertion of acupuncture needles into acupoints could have a plausible alternative explanation, for example, needle phobia and anxiety (Cole et al. 2002). Fourthly, despite some issues with the clinical observations/studies on acupuncture treatment, accumulated evidence from animal experiments at multiple levels from behavior (Klide et al. 1987; Shu et al. 2005; Kang and Xia 2008), morphology (He et al. 1990b; Wen and Huang 2008; Kim et al. 2008b; Guo et al. 2008b, c), electrophysiology (Chen and Huang 1984; He et al. 1990a; Huang et al. 1999; Chao et al. 2001; Zhang et al. 2008b, c; Goiz-Marquez et al. 2009), neurochemistry (Chao et al. 2001; Jin et al. 2005a; Shu et al. 2005), to molecular biology (Yang et al. 1996; Guo et al. 2008b, c; Kim et al. 2008b) have demonstrated the beneficial effects of acupuncture treatment in experimental epilepsy, and more and more scientific aspects of acupuncture treatment are being unveiled. Fifthly, in clinical/veterinary practice, simultaneous application of acupuncture can reduce the dose of AEDs used, thus reducing side effects of the drugs (Klide et al. 1987).

Finally, there is strong need for a comprehensive review of all published literature on acupuncture for epilepsy treatment. In a recent review paper, the authors disregarded acupuncture as a treatment for epilepsy through an assessment of 11 trials, chosen based on their own criteria (Cheuk and Wong 2008). Though some of their viewpoints (e.g., poor methodological quality; small sample size, short follow-up, etc., in these trials) are agreeable, there are some points regarding their conclusion that appear unreasonable. Firstly, the reviewers considered the effects of acupuncture as beneficial or effective only if all the parameters (seizure score, seizure frequency and duration, percentage of efficacy) and all the endpoints of treatment were statistically significant. However, in the trials they analyzed with their own criteria, some of the parameters were actually significant. For example, the treatment group had more patients with complete seizure control (Yu et al. 1999; Leng et al. 2000; Deng et al. 2001; Shi and Gao 2001; Peng 2003; Zhuang et al. 2006a), and/or who achieved ≥75% (Zhou 2000; Ma et al. 2001; Peng 2003; Xiong et al. 2003; Zhang et al. 2006) or ≥50% (Yu et al. 1999; Zhou 2000; Deng et al. 2001; Zhang et al. 2006) reduction in seizure frequency, and/or ≥50% reduction in seizure duration (Zhou 2000; Ma et al. 2001), or an improved EEG (Zhou 2000; Ma et al. 2001; Deng et al. 2001; Shi and Gao 2001), even though the total efficacy was not significantly higher in the treatment group than the control group in some trials. Therefore, it cannot be determined that acupuncture is not effective in treating people with epilepsy. Secondly, in their review, Cheuk and Wong (2008) pooled four trials (Kloster et al. 1999; Ma et al. 2001; Xiong et al. 2003; Zhang et al. 2006) together to determine the effects of acupuncture on seizures and concluded that no statistical difference was found between the treatment and control groups, despite the fact that three trials reported positive outcomes and one trial reported a negative result. However, it is hard to agree to this pooled analysis because these trials (Kloster et al. 1999; Ma et al. 2001; Xiong et al. 2003; Zhang et al. 2006) were heterogeneous as the authors have also acknowledged, with respect to the treatment, comparison groups used, the age of patients, underlying epilepsy types, the outcomes chosen, etc.

Overall, it is believable that acupuncture is an effective option that is safe, convenient, and cost effective for the treatment of epilepsy. However, improvements in the experimental design and assessment of clinical studies are required in the future.

6.3 Mechanisms of Acupuncture on Epilepsy

6.3.1 Pathological Genesis of Epilepsy

About 40% of epilepsies have genetic factors, and other 60% of epilepsies are acquired epilepsies caused by events including serious prenatal events, major perinatal trauma, serious head injury, stroke, encephalitis, bacterial meningitis, and tumors (Armijo et al. 2005). Despite its inherited or acquired type, epileptogenesis is a complex process and is due to the changes in the excitability of neurons in the brain, i.e., a population of neurons are hyperexcitable and in highly synchronous activities that are propagated through normal or pathological pathways. The increased excitability is ascribed to the changes occurring in both neurons and glial cells. For neurons, it is mainly due to (1) the innate ability of intrinsically bursting neurons in the cortex and hippocampus, (2) alterations in membrane properties, (3) imbalance between excitatory and inhibitory transmission, and (4) alterations in existing synaptic contacts/circuits and/or establishment of new synaptic contacts/circuits. In glial cells, the most striking change is the impaired ability to maintain glutamate and K^+ homeostasis (Wetherington et al. 2008; David et al. 2009).

6.3.1.1 Role of Intrinsically Bursting Neurons in Epileptogenesis

Epileptic discharges do not begin at random but rather origin in particular subsets of the brain (Prince and Connors 1984). The intrinsic burst-generating neurons are observed in pyramidal neurons in the hippocampal CA3 area and layers IV–V in the cerebral cortex and have unique properties in generating paroxysmal discharges (Prince and Connors 1984; Avanzini and Franceschetti 2003; Reid et al. 2008). For example, these neurons have very high-gain intrinsic membrane properties (low-threshold Na^+ spikes and more persistent conductance for Na^+ and Ca^{2+} activated in the subthreshold), and both somatic and dendritic membranes of these neurons have the capability to generate intrinsic burst discharges. These cells have direct excitatory synaptic connections or mutually excitatory circuits among them. As a result of widespread expansion of their axonal arborization, intrinsic burst-generating neurons establish synaptic contacts with a large number of neighboring neurons and sustain highly synchronized activities with other groups of cells through their output, therefore serving as pacemakers for spontaneous epileptiform discharges (Prince and Connors 1984; Avanzini and Franceschetti 2003; Reid et al. 2008). Therefore, in patients resistant to AEDs, these intrinsic burst-generating neurons are surgically resected to control epilepsy.

6.3.1.2 Alterations in Membrane Properties

Neuronal excitability is at least determined by two major factors. The first determinant is the balance of synaptic input from neighboring cells within the network through chemical transmission via ligand-gated ion channels and G-protein-coupled metabotropic receptors and through rapid electrical transmission via gap junctions (see Sect. 6.3.1.3). The second and also the most important determinant is the electrogenic property intrinsic to each neuron, which is based on the activities of ion channels such as Na^+, K^+, and Ca^{2+} channels in the membrane (Errington et al. 2005). Any abnormal alternations in the activities (opening, closing, and inactivation), expression, or even the structure of ion channels may be responsible for the hyperexcitability of neurons and the consequent epilepsy.

Voltage-gated Na^+ channels are responsible for the initiation and propagation of action potential and are critical determinants of intrinsic neuronal excitability. For Na^+ channels, the rapid activation and timely inactivation are essential for the determination of neuronal firing frequency. Fast transit cycle of these two states facilitates high-frequency neuronal firing and contributes to the generation of synchronous bursting that characterizes epilepsy (Errington et al. 2005). For example, heat in fever increases the gating rate of voltage-gated Nav1.2 channels that contributes to the evolution of febrile seizures (Thomas et al. 2009). In addition, increased amplitude of voltage-dependent I_{Na}, shortened phase of inactivation, shift toward depolarization of inactivation currents, and more negative activation have been associated with the development of some types of epilepsy (Tian et al. 1995; Vreugdenhil et al. 1998; Avanzini and Franceschetti 2003; Armijo et al. 2005; Errington et al. 2005; Thomas et al. 2009). Even few Na^+ channels that fail to inactivate carry a small yet enough persistent fraction of I_{Na} to drive the membrane toward the firing threshold and increase the excitability of neurons associated with epilepsy (Avanzini and Franceschetti 2003; Armijo et al. 2005; Stafstrom 2007; Fekete et al. 2009; Hargus et al. 2011). Increased expression of Na^+ channels in neurons is also associated with hyperexcitability and epileptic phenotype (Sashihara et al. 1994; Lombardo et al. 1996; Xia et al. 2003). Mutations of the genes encoding Na^+ channel α-subunits (e.g., *Scn1a, Scn2a, Scn3a*) as well as *Scnb1*-encoding β1-subunit can also lead to the development of epilepsies such as benign familial neonatal/infantile convulsions, generalized epilepsy with febrile seizures, and severe myoclonic epilepsy of infancy (Catterall et al. 2010; Escayg and Goldin 2010; Meisler et al. 2010). Implantation of tetrodotoxin (TTX), a potent Na^+ channel blocker, in rat neocortex before inducing cortical trauma has been shown to prevent injury-induced epileptiform discharges in the cortex (Graber and Prince 1999). Therefore, neuronal voltage-gated Na^+ channels represent one of the most important targets for AEDs action, and most of the AEDs exert their antiepileptic role through inhibition of the activities of Na^+ channels (Armijo et al. 2005; Errington et al. 2005).

K^+ channels are a structurally and functionally diverse family of ion channels that are particularly important for many key cellular processes, including the maintenance of the resting membrane potential, the termination of the action potential and repolarization of the neuron, the modulation of the frequency and pattern of repetitive firing, and the regulation of presynaptic neurotransmitter release. Inhibition of K^+ current, especially M current (that can be inhibited by activation of muscarinic acetylcholine receptors), leads to massive depolarization, which favors burst discharges (Avanzini and Franceschetti 2003; Armijo et al. 2005; Errington et al. 2005). Perfusion of K^+ channel blockers or higher extracellular K^+ concentration in the brain slices can induce epileptiform discharges (Avanzini and Franceschetti 2003; Armijo et al. 2005; Errington et al. 2005). In addition, Kv1.1 potassium channel knockout mice have a seizure-sensitive predisposition at a very young age (P10), before either spontaneous seizure activity or changes in c-fos mRNA expression can be demonstrated. Approximately half of the Kv1.1 potassium channel knockout mice develop recurrent spontaneous seizures within the first postnatal month with many of the characteristics of limbic system seizures and increased early mortality and neuropathological changes, and therefore represent a clinically relevant model of early-onset epilepsies (Rho et al. 1999). Hippocampal granule cells become intrinsically hyperexcitable in TLE via changes in ion channels. However, granule cells in the sclerotic focus of TLE have reduced hyperexcitability. This is because neurons in the sclerotic focus appear to acquire anticonvulsive properties by increased resting conductance due to upregulation of the activity and expression of the inward rectifier K^+ (Kir2, Twik) channels (Young et al. 2009). Mutations in *KCNQ2* and *KCNQ3* genes (which encode delayed rectifier K^+ channel subunits) are associated with the phenotype of benign familial neonatal convulsions and mutations in *KCNA1* (which encodes a slow rectifier voltage-gated K^+ channel) to episodic ataxia and an increased risk of epilepsy in these patients (Armijo et al. 2005; Errington et al. 2005). Functional studies have shown that the mutations of the K^+ channels result in significantly less K^+ passages, leading to an impairment of membrane repolarization and prolonged depolarization, thereby increasing neuronal hyperexcitability (Armijo et al. 2005). Therefore, potentiation of K^+ currents could unveil an important angle to lower neuronal excitability. Thus, K^+ channels represent a largely unexploited, yet potential target for the future development of AEDs (Armijo et al. 2005; Errington et al. 2005).

Voltage-gated calcium channels are pivotal for some neuronal functions. Ca^{2+} influx through presynaptic membrane Ca^{2+} channels plays a critical role in translating transient electrical signals into biochemical events (neurotransmitter release). In postsynaptic neurons, Ca^{2+} currents increase intracellular Ca^{2+} that has been associated with several factors underlying epileptogenesis, including sustained depolarization, long-term potentiation, kindling, and excitotoxicity (Armijo et al. 2005). Intrinsic burst-generating neurons in hippocampal CA3 give rise to repetitive spiking and epileptiform discharges, which are mainly mediated by Ca^{2+} influx through high voltage-activated Ca^{2+} channels (Prince and Connors 1984; Armijo et al. 2005). Overactivity of the low-voltage-gated Ca^{2+} channels in thalamocortical relay neurons is responsible for the aberrant synchronized depolarization, which underlies the epileptiform activity and absence epilepsy (Avanzini and Franceschetti 2003; Armijo et al. 2005). In hippocampal epileptogenesis, there was an upregulation of Ca^{2+} currents and an increased density of Ca^{2+} channels (Jones 2002; Armijo et al. 2005). Mutations of Ca^{2+} channel genes have been associated with certain types of epilepsy. For example, mutation of *CACNB4* (encodes the β4-subunit of voltage-gated Ca^{2+} channel) is associated with juvenile myoclonic epilepsy; *CACNA1A* (which encodes the α1-subunit of P/Q-type Ca^{2+} channel) with idiopathic generalized epilepsy, childhood absence epilepsy, and cerebellar ataxia in child; and *CACNA1H* (which encodes α1-subunit of T-type Ca^{2+} channel) with childhood absence epilepsy (Jones 2002; Armijo et al. 2005; Errington et al. 2005). However, there are only a few available AEDs that target the voltage-gated Ca^{2+} channels to exert their anticonvulsant effects owing to the difficulties in assessment of Ca^{2+} channel involvement in epileptogenesis (Avanzini and Franceschetti 2003; Errington et al. 2005).

6.3.1.3 Imbalanced Excitatory and Inhibitory Transmission

Cerebral imbalance between the inhibitory and the excitatory system plays a key role in initiating seizures. GABA and glutamate are the principal inhibitory and excitatory neurotransmitters of the CNS, respectively. Under physiological conditions, there exists a perfect balance between these two antagonistic major regulatory systems by the regulation of their metabolisms such as synthesis, release, and uptake. However, under pathological conditions, this balance is disrupted and believed to be responsible for the initiation of seizures and the development of chronic epilepsy (Bradford 1995; Olsen and Avoli 1997; Yu et al. 2006; Sierra-Paredes and Sierra-Marcuño 2007; Maciejak et al. 2010). Changes associated with epilepsy in these two major regulatory antagonistic systems include the exaggerated glutamatergic excitatory activities, decreased GABAergic inhibitory activities, or a combination of these two activities.

Excessive glutamatergic activity plays a central role in both the induction and the expression of hyperexcitability and epilepsy at three levels—the chronic subconvulsive hyperactivity in the epileptic focus, amplified excitatory activity leading to initiation of the seizures and the recruitment of neighboring excitatory neurons, and the spread and generalization of the hyperactivity to precipitate a full seizure (Bradford 1995; Aroniadou-Anderjaska et al. 2008). Excessive glutamatergic activity could hypothetically result from changes in enzymatic activity in the glutamate synthesis, release and uptake, receptor population and sensitivity, receptor channel properties, and increased excitatory synapse numbers and synaptic transmission efficacy. In experimental model, status epilepticus can be induced by injection of kainic acid, which strongly activates the kainate subtype of glutamate receptors. In fact, activation of both ionotropic and metabotropic glutamate receptors has been proven to be involved in the hyperexcitation and seizures (Bradford 1995; Camón et al. 1998; Semyanov and Godukhin 2001; Alexander and Godwin 2006; Reid et al. 2008). In cultured cortical neurons, overactivation of NMDA receptors with glutamate results in a decreased expression of glutamic acid decarboxylase (GAD), a key enzyme for GABA production, leading to reduced cortical GABA synthesis and a reinforced hyperexcitability in the cortex (Asada et al. 1997; Monnerie and Le Roux 2008). Blockade of ionotropic glutamatergic receptors or their subunits can prevent seizures and epileptogenesis (Kolarova et al. 2003; Koh et al. 2004; Chen et al. 2007a). Glutamine, an immediate precursor for neurotransmitter glutamate, significantly increases in the extracellular spaces during kindled seizures. This could represent an enhanced glutamate synthesis by neurons in order to replenish the transmitter pool following an increased release of glutamate during the seizure episode (Bradford 1995). Remarkable increases in extracellular glutamate in epileptic foci in both patients and animal models have been reported (Maciejak et al. 2010). Inhibition of presynaptic glutamate release has been indicated to prevent EEG hyperactivity and motor seizures (Bradford 1995). Undoubtedly, extracellular glutamate, whether synaptically released or from other local cellular sources, is likely to be causatively responsible for the epileptic hyperactivity (Young et al. 1992; Bradford 1995; Aroniadou-Anderjaska et al. 2008). However, it is possible that the local neuronal depolarization in the epileptic focus following initiation by the glutamate action could further lead to a subsequent glutamate release (Bradford 1995). The significantly increased glutamate levels in the extracellular spaces result from an excessive release from the nerve cells (Bradford 1995; Liu and Cheng 1995; Liu et al. 1997; Szyndler et al. 2008) as well as an impaired uptake by neurons and glial cells. For the latter, the dysfunction of astrocyte glutamate transporters plays a key role (Tanaka et al. 1997; Wetherington et al. 2008; Campbell and Hablitz 2008; Sarac et al. 2009) (see Sect. 6.3.1.5). Decreased glutamate transporter expression increases neuronal excitability (Campbell and Hablitz 2008), while upregulated expression of glutamate transporters in neurons of brain regions involved in the circuit of the seizures (e.g., hippocampal CA1, CA2, CA3 areas, DG granule cells, cerebral cortex and septum) could be a protective mechanism against seizures and the potentially deleterious consequences of seizure-enhanced glutamatergic neurotransmission (Voutsinos-Porche et al. 2006). In addition, the altered population size and/or sensitivity of excitatory receptors and their subunits (e.g., postsynaptic NMDA and AMPA/KA ionotropic glutamate receptors, presynaptic metabotropic glutamate receptors) can also be responsible for the neuronal hyperexcitability and epileptogenesis. Increase in ionotropic glutamate receptors and the density of their subunits is reported in the epileptogenic hippocampus and cortex of humans and animal seizure models (Bradford 1995; Brines et al. 1997; de Lanerolle et al. 1998; Babb et al. 2000; Lynd-Balta et al. 2004). Presynaptic metabotropic glutamate receptors (glutamate auto-receptors) are very crucial for the modulation of glutamate release via feedback inhibition. A long-lasting impairment of mGlu2/3 expression may contribute to abnormal presynaptic plasticity, exaggerated glutamate release, and hyperexcitability in TLE, and agonistic activation of this kind of glutamate receptors can prevent the exaggerated glutamate release, hyperexcitability, and hence the seizures (Bradford 1995; Folbergrová et al. 2005; Garrido-Sanabria et al. 2008). Postsynaptic metabotropic glutamate receptor activation seems to produce severe and delayed seizures by increasing intracellular Ca^{2+} via G-protein-second message pathways (Camón et al. 1998). In chemically or traumatically injured brain, epileptogenesis always follows the lowered seizure threshold and overexcitation of neurons in the hippocampus and cortex (Golarai et al. 2001; Topolnik et al. 2003; Chen et al. 2006c; Nita et al. 2006; Griesemer and Mautes 2007; Ding et al. 2008; Reid et al. 2008; Statler et al. 2008). This kind of epilepsy secondary to the brain injury mainly depends on the NMDA receptor-mediated facilitation of glutamatergic synaptic transmission, enhancement of synaptic strength, increase in excitatory synapse circuitry, and supersensitivity of glutamate receptors due to the deafferentation and disruption of homeostatic synaptic plasticity in the injured brain areas (Bradford 1995; Shetty et al. 2005; Cao et al. 2006; Chen et al. 2006c; Cross and Cavazos 2007; Williams and Dudek 2007; Avramescu and Timofeev 2008; Morgan and Soltesz 2008). Moreover, gene mutations of AMPA and KA receptors may allow receptor-linked ion channels possessing more Ca^{2+} permeability and Ca^{2+} influx, thus playing a role in epileptogenesis (Armijo et al. 2005).

Alterations in GABAergic activity also contribute to epileptogenesis in both animals and humans (Bradford 1995; Olsen and Avoli 1997; Sierra-Paredes and Sierra-Marcuño 2007; Takechi et al. 2009). In general, GABA plays an inhibitory role in the brain by hyperpolarizing neurons via $GABA_A$ receptor-mediated increase in Cl^- conductance (rapid inhibition) or via G-protein-coupled $GABA_B$ receptor-mediated increase in K^+ conductance (slow inhibition). Widespread existence of GABAergic neurons or terminals in the CNS, particularly in the forebrain structures, indicates an important role of GABA in stabilizing brain excitability. Altered GAD regulation and reduced GABA synthesis are often associated with epilepsy (Obenaus et al. 1993; Neder et al. 2002; Knopp et al. 2008). The severe damage and/or loss of GABAergic neurons or nerve terminals as well as markedly altered expression of receptor subunits have been demonstrated in the cortex, amygdala, and hippocampus of animal epilepsy models and patients with TLE (Bradford 1995; Sierra-Paredes and Sierra-Marcuño 2007; Aroniadou-Anderjaska et al. 2008). Inhibitory interneurons are more vulnerable to insults than other neurons and loss of inhibitory interneurons or their function can cause a reduction of tonic inhibition in seizure-sensitive areas such as the cortex, hippocampus, and amygdala (Prince and Connors 1984; Yu et al. 2006; Dudek and Suluta 2007; Ogiwara et al. 2007; Knopp et al. 2008; Fritsch et al. 2009; Chen et al. 2010; Zhang et al. 2010). This could lead to the epileptogenesis triggered by acute brain insults such as traumatic head/brain injury, stroke, or an episode of status epilepticus (Knopp et al. 2008; Fritsch et al. 2009; Chen et al. 2010). Genetic mutations of GABA receptors or application of GABA receptor antagonists induce epileptic seizures, while enhancement of GABAergic inhibition prevents or retards epileptogenesis, suggesting that GABAergic system can control the induction of epileptogenesis (Fisher 2004; Armijo et al. 2005; Sierra-Paredes and Sierra-Marcuño 2007; Aroniadou-Anderjaska et al. 2008). Besides, a reduction in the number of GABA transporters has been reported in both human and rat epileptic tissues, which could result in a decrease in glutamate uptake-triggered GABA release through reversal of GABA transporters (Bradford 1995; Héja et al. 2009). Actually, extracellular GABA levels in the epileptic brain tissues are significantly decreased or slightly increased with an associated significant increase in the glutamate/GABA ratio that shifts the neurons to the dominance of overexcitation (Liu and Cheng 1995; Bradford 1995; Maciejak et al. 2010). In addition, aberrant GABAergic reorganization in the hippocampal dentate gyrus and CA3 area may contribute to TLE (Sierra-Paredes and Sierra-Marcuño 2007). Since GABA generally plays an inhibitory role in the brain and a decrease in GABAergic inhibitory activities can induce epileptic seizures, drugs that enhance GABAergic activity are a part of anti-epileptic therapy. For example, novel AEDs such as tiagabine and vigabatrin exert their anticonvulsant activity mainly through enhancing GABAergic action by raising extracellular GABA levels or increasing receptor functions (Bradford 1995; Duffy et al. 2004; Czapiński et al. 2005; Rogawski 2006; Madsen et al. 2010). However, it is important to point out that GABAergic enhancement does not exert antiepileptic action at all times. In some cases, GABAergic activity may be excitatory and play a synchronizing role in the neuronal networks involved in seizure generation (Olsen and Avoli 1997). The disinhibition in reticular neurons of thalamocortical circuit is a good illustration of this fact. In this circuit, reticular neurons that are depressed by the neighboring reticular neurons through $GABA_A$ receptors inhibit thalamic relay neurons through $GABA_B$ receptors (Fig. 6.1). The disinhibition of these interneurons may cause burst activation of the cortex and absence seizures (Chang and Lowenstein

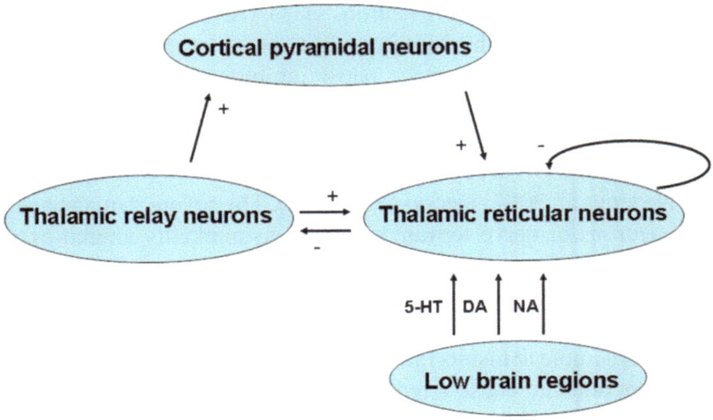

Fig. 6.1 Thalamocortical circuit. This circuit involves mainly three neuronal populations: thalamic relay neurons, thalamic reticular neurons, and cortical pyramidal neurons. The thalamic relay neurons can activate the cortical pyramidal neurons in either tonic or burst mode depending on the input of the thalamic reticular neurons. The thalamic reticular neurons receive the projects from both thalamic relay neurons and cortical pyramidal neurons to form a circuit. The reticular neurons also can be inhibited through the recurrent collaterals from neighboring reticular neurons. The disinhibition of these interneurons may cause burst activation of the cortex and absence seizure

2003). GABAergic receptor desensitization, by repeated GABA application to hippocampal pyramidal cells or during repetitive electrical activation of normal cortex, can result in depression of inhibitory synaptic process and subsequent increase in excitability (Prince and Connors 1984). This may be one of the reasons for the genesis of refractory epilepsy. Infants and children are more prone to seizures than adults. One possible reason is that GABAergic activity in the cortex is excitatory in infants and children but inhibitory in adults due to a difference in the intracellular Cl^- levels and $GABA_A$ receptor-mediated Cl^- conductance between immature neurons and mature ones; namely, intracellular Cl^- in cortical neurons is very high in immature neurons while low in mature neurons, and activation of $GABA_A$ receptors leads to outward Cl^- current and depolarization in immature neurons but inward Cl^- current and hyperpolarization in mature neurons in the cortex (Glykys et al. 2009). Therefore, in certain conditions, GABAergic activity can also contribute to epileptogenesis, depending on multiple factors such as neuronal development, neuronal conditions, and local circuitry.

6.3.1.4 Alterations in Existing Synaptic Contacts/Circuits and/or Establishment of New Synaptic Contacts/Circuits

The "chicken and egg" relationship between synaptic reorganization and epilepsy makes it hard to reason as to which is the cause and which is the effect. Actually, recurrent seizures may lead to neuronal loss (Chen et al. 2010; Zhang et al. 2010) and

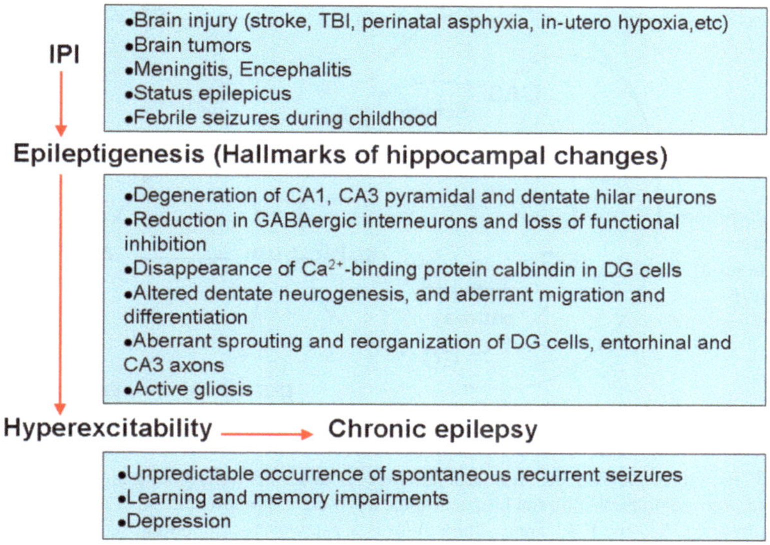

Fig. 6.2 Initial precipitating injury (IPI)-induced hippocampal changes and temporal lobe epilepsy with hippocampal sclerosis

subsequent sprouting to establish new synaptic contacts in compensation for the disrupted activity due to the neuronal loss. In addition, new formation of sprouting and abnormal connections is responsible for the development of brain injury-induced chronic epilepsy in acquired epilepsies caused by events that make severe brain injury such as serious prenatal events (e.g., hypoxia in utero, poisoning), major perinatal trauma, serious head injury, stroke, (even a single) status epilepticus, etc. Therefore, synaptic reorganization and epilepsy form a vicious circle in which both interchangeably act as the cause and the consequence. Figure 6.2 shows a typical relationship between the brain injury and epilepsy where the initial precipitating injury (IPI) induces hippocampal changes and TLE with hippocampal sclerosis (Fig. 6.2).

Axon sprouting following neuronal injury and loss, as shown by Timm's staining and other anatomical techniques (e.g., dynorphin immunocytochemistry), is a ubiquitous feature of an injured brain that is not usually present in a normal animal or human brain (Houser et al. 1990; Cavazos and Cross 2006; Frotscher et al. 2006). During the "silent period," between the initial brain injury and the occurrence of spontaneous seizures, neurons in the epileptic foci undergo intensive axon sprouting and synaptic reorganization (Fig. 6.2). For example, the presence of mossy fiber sprouting in the hippocampus is a marker of formation of new recurrent synapses in other areas of the brain and of the potential development of epilepsy. Normally, hippocampal dentate gyrus (DG) granule cells accept input from entorhinal cortex (EC) via perforant pathway that arises from layer 2 of the EC and relay their activity to CA3 pyramidal cells via the mossy fibers. The CA3 pyramidal cells modify the activities of CA1 pyramidal cells via Schaffer collaterals, and

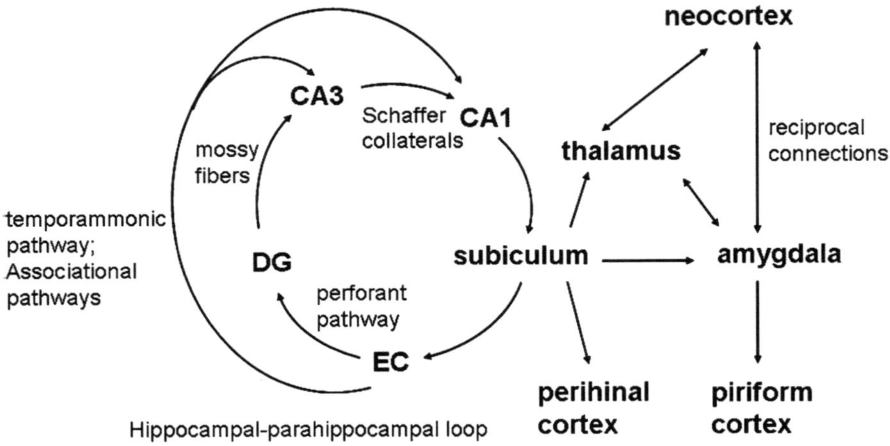

Fig. 6.3 Paradigm showing the structural and functional connections among the epileptic seizure-sensitive brain regions. Within the hippocampus, a typical trisynaptic circuit is formed in between dentate gyrus, CA3, and CA1 areas. Subiculum is a gateway for the output of the hippocampus, which has widespread connections with many other brain regions

the axons of CA1 pyramidal cells extend into the subiculum and deeper layers of the EC (Fig. 6.3), thus forming a trisynaptic circuit within the hippocampus. Glutamate is the principal synaptic neurotransmitter of the neurons in the hippocampus. In human TLE and in both kainate and pilocarpine animal models for status epilepticus, or stroke and trauma model of epilepsy, extensive cell loss was found in the hippocampus (Sutula et al. 1994; Shetty et al. 2005; Otani et al. 2006; Swartz et al. 2006; Cavazos and Cross 2006; Frotscher et al. 2006; Deshpande et al. 2007; Liang et al. 2007; de Oliveira et al. 2008; Mouri et al. 2008; Zhang et al. 2009b). In response to deafferentation of mossy cells' target cells, which include granule cells and interneurons, mossy fibers sprout new axon collaterals that direct back into the inner molecular layer of the dentate gyrus where the dendrites of granule cells are normally situated (McNamara 1994; Larner 1995; Chang and Lowenstein 2003: Lynd-Balta et al. 2004; Williams and Dudek 2007). The majority of newly formed synaptic contacts are excitatory though inhibitory synapses can be formed as well (Liu et al. 2003; Zhang et al. 2009b). These aberrant mossy fibers instigate a recurrent excitatory circuit by forming synapses between the dendrites of neighboring granule cells. A small cluster of pathologically interconnected neurons in this aberrant recurrent network is capable of generating powerful hypersynchronous bursts of discharges to initiate epileptogenesis via a kindling effect, and allows the development of discharges that spread throughout the limbic system, finally leading to TLE (Bragin et al. 2000; Williams et al. 2004; Morgan and Soltesz 2008). Similar aberrant sprouting and robust synaptic reorganization were also found in the dentate gyrus and other hippocampal areas like CA1, CA3, etc.

(Siddiqui and Joseph 2005; Ma et al. 2006; Cavazos and Cross 2006; Cross and Cavazos 2007), and the malformed or deafferented cortex (Luhmann and Raabe 1996; Chevassus-au-Louis et al. 1999; Babb et al. 2000; Houweling et al. 2005). This further contributes to the progressive increase in the excitability and brain injury-associated epileptogenesis. Since the aberrant sprouting and robust network modeling are due to the loss of target cells, transplantation of target cells (CA3 neurons) has demonstrated a vigorous increase in the formation of reciprocal circuitry between the dentate granule cells (DGCs) and the grafted CA3 pyramidal neurons, and efficient prevention of the aberrant mossy fiber sprouting in kainate-lesioned CA3 region of the adult rat hippocampus, which is important to prevent the development of TLE following hippocampal injury (Shetty et al. 2005). Rapid ablation of mossy cells with a fast, cell-specific ablation technique before axon sprouting is proven to cause hypoexcitability rather than hyperexcitability in the granule cell populations located at sites both near and distant to those of mossy cell lesions (Ratzliff et al. 2004). Therefore, axon sprouting and the establishment of new synaptic contacts/circuits may cause hyperexcitability and contribute to the epileptogenesis associated with brain injury.

Besides, alterations in the existing synaptic contacts/circuits can also initiate epileptogenesis. A good example of this is kindling. Kindling is a process in which repeated focal applications of initially subconvulsive electrical/chemical stimulations eventually lead to progressive, highly reliable, permanent, and intensive partial or generalized convulsive seizures that are similar to human TLE (Sutula et al. 1994; Morimoto et al. 2004). One of the most striking characteristics of kindling is kindling-induced potentiation that resembles long-term potentiation (LTP) and can persist for months, and increase in the strength of synaptic connections (Sutula et al. 1994; Morimoto et al. 2004). This process does not necessarily require establishment of new synapses or circuitry, but rather functional alterations in pre- and postsynaptic glutamate receptors/transporters and changes in the spatiotemporal pattern of activation of the existing synapses and circuitry (Morimoto et al. 2004). In high K^+ model of epilepsy, it was found that synaptic transmission between the areas CA3 and CA1 greatly changed shortly after three ictal discharges induced by transient elevation of extracellular K^+ in mouse hippocampal slices (Lopantsev et al. 2009). This change manifested as the discharges initiated in the CA3 region and propagated via the Schaffer collaterals into CA1 where they were associated with sustained membrane depolarization and bursts of action potentials in CA1 pyramidal cells and enhanced LTP of Schaffer collateral-evoked field excitatory postsynaptic potentials (EPSPs) in the CA1 area (Lopantsev et al. 2009). Thus, during LTP, epileptogenesis in response to repetitive stimulation is enhanced (Lopantsev et al. 2009). The lasting augmentation of the synaptic excitability of the CA1 network involves reduced efficacy of GABAergic inhibition and increased excitatory transmission at Schaffer collateral synapses and may contribute to early phases of progressive epileptogenesis (Lopantsev et al. 2009).

6.3.1.5 Astrocytic Dysfunction in Maintenance of K^+ and Glutamate Homeostasis

Astrocytes are the predominant type of glial cells with remarkably heterogeneous populations in the CNS. It was not until the term "tripartite synapse" was coined, which comprised of the pre- and postsynaptic elements along with astrocytic process forming a structural and functional unit (Araque et al. 1999), that newer roles of astrocytes beyond the traditionally considered "brain glue" (i.e., a structural support and passive elements because of their electrical silence, lacking transmitter receptors and expressing a limited set of ion channels) came into light (Araque et al. 1999; Volterra and Meldolesi 2005; Seifert et al. 2006; Perea et al. 2009). In fact, astrocytes are very active participants and direct communication partners of neurons for normal brain functions. Astrocytes provide metabolic support to neurons via intercellular networks through activity-dependent intracellular trafficking of lactate necessary to sustain glutamatergic synaptic activity (Rouach et al. 2008). In addition, they are also vital for maintaining the homeostasis of neuronal microenvironments, including maintenance of low interstitial glutamate concentrations, control of the brain interstitial volume, and regulation of extracellular K^+ concentration and local blood flow (Kofuji and Newman 2004; Volterra and Meldolesi 2005; Abbott et al. 2006; Koechler et al. 2009). Strikingly, astrocytes can sense neuronal activity via G-protein-coupled receptors (e.g., type I metabotropic glutamate and ATP purinergic $P_{2\gamma1}$ receptors) in their membrane and send feedback responses to neurons to modulate the CNS signaling via the gliotransmitters (e.g., glutamate, D-serine, and ATP) (Fellin et al. 2004; Volterra and Meldolesi 2005; Jourdain et al. 2007; Perea and Araque 2007; Barres 2008; Benarroch 2009; Perea et al. 2009; Allen et al. 2012; Min and Nevian 2012). The network organization from extensive intercellular communication between astrocytes through gap junction channels formed by connexins (particularly Cx43) provides a robust tool for astrocytes as a functional syncytium to rapidly disperse small molecules such as K^+ or glutamate and therefore prevent their extracellular accumulation during neuronal firing (Wallraff et al. 2006; Orthmann-Murphy 2008). Because of abundance of astrocytes in the brain along with their critical roles in regulation of neuronal microenvironments and modulation of the CNS signaling, dysfunctions of astrocytes result in neurological disorders (Seifert et al. 2006; Barres 2008; Ricci et al. 2009).

Throughout the history of epilepsy research, astrocyte dysfunction has repeatedly been incriminated as the primary cause of epilepsy, but it has never been much highlighted due to the long-standing view of astrocytes as passive elements in the brain (D'Ambrosio 2004). In the past decade, neuroscientists have made some exciting findings regarding the functions of astrocytes. For example, they found that astrocytes act as important communication elements in the CNS and play a critical role in the modulation of synaptic strength and neuronal function (Fig. 6.4). These recent developments have forced the investigators to rethink the role of astrocytes in the pathology of epilepsy. Dysfunction of astrocytes is now thought to be the central cause of the epileptic seizures and epileptogenesis (Jabs et al. 2008; Wetherington et al. 2008; Benarroch 2009; Seifert et al. 2010). Astrocytic contributions to epilepsy include excessive astrocyte–neuron signaling-induced synchronization of neuronal

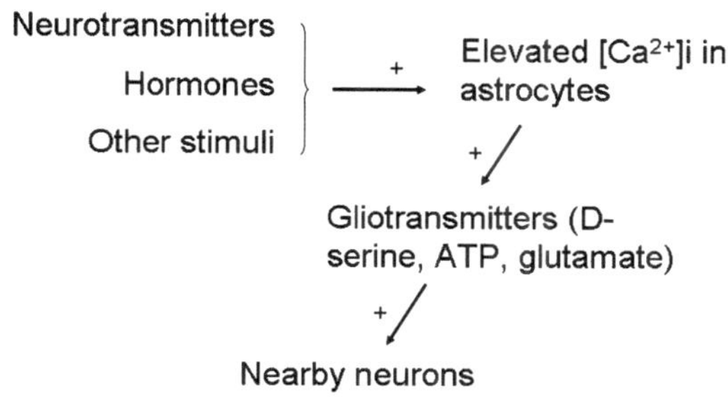

Fig. 6.4 Neuron–astrocyte interactions. Pre- and postsynaptic elements of neurons with astrocytic process together form a structural and functional unit termed as "tripartite synapse," which confers astrocytes more roles beyond the traditionally considered as "brain glue." And astrocytes are now believed to be important communication elements in the CNS and play a critical role in modulating synaptic strength and neuronal function via gliotransmitters

activity that leads to increased neuronal excitability and an impaired ability to maintain the K^+ and glutamate homeostasis.

Neuronal activity-dependent elevation of intracellular Ca^{2+} in astrocytes stimulates the release of a number of gliotransmitters (Perea and Araque 2007; Gómez-Gonzalo et al. 2010; Hamilton and Attwell 2010; Min and Nevian 2012). Compelling evidence has demonstrated that astrocytic glutamate release can activate glutamatergic receptors in neurons and modulate the strength of synaptic transmission at nearby synapses (Jourdain et al. 2007; Perea and Araque 2007; Min and Nevian 2012) (Fig. 6.4). In the hippocampus, astrocyte-released glutamate induces slow inward currents (SICs) that are similar to "paroxysmal depolarizing shift" in pyramidal neurons (Tian et al. 2005; Wetherington et al. 2008). Though the role of SICs is epilepsy is uncertain, it may induce local, microsynchronous firing among populations of two to four adjacent neurons. Because the fine terminal processes of a single astrocyte reach tens of thousands of synapses simultaneously, astrocyte-released glutamate can increase the excitability of neuron–astrocyte network, favor neuronal synchronization, and ultimately predispose neurons to seizures (Jabs et al. 2008; Wetherington et al. 2008; Benarroch 2009; Seifert et al. 2010; Gómez-Gonzalo et al. 2010).

During neuronal spiking, extracellular glutamate increases. High extracellular glutamate can lower the seizure threshold and induce the development of epilepsy. Thus, timely clearance is important to assure continuation of normal neuronal activity. As previously mentioned, there is a remarkable increase in the extracellular glutamate in epileptic foci in both patients with epilepsy and many animal models (see Sect. 6.3.1.3). This results mainly from the impaired uptake of glutamate. Astrocyte-specific glutamate transporters (EAAT1/GLAST, EAAT2/GLT1) are largely responsible for glutamate clearance from the extracellular space (Wetherington et al. 2008). Genetic deletion of astrocytic glutamate transporter EAAT2/GLT1 results in the elevation of residual glutamate in the mice brain and the development of epilepsy as well as exacerbated brain injury (Tanaka et al. 1997). In contrast, upregulation of astrocyte glutamate transporter expression before the onset of epilepsy decreases extracellular glutamate levels, neuronal death, and seizure frequency (Zeng et al. 2010). The alteration in the activity of astrocytic glutamate transporters is a common feature seen in TLE, though the reported results are still controversial (Wetherington et al. 2008; Seifert et al. 2010). In addition, damaged glutamate–glutamine cycle between neurons and astrocytes also contributes to the elevated extracellular glutamate levels and the development of epilepsy (Eid et al. 2008). Normally, extracellular glutamate is largely taken up via astrocytic glutamate transporters to astrocytes in which glutamate is converted to glutamine via the astrocyte-specific enzyme, glutamine synthetase. Glutamine is then released into the extracellular space and transferred to neurons via glutamine transporters as the direct precursor of glutamate synthesis in neurons and as the source of glutamate for GABA synthesis in the inhibitory interneurons (Eid et al. 2008). Patients with mesial temporal lobe epilepsy are severely deficient in glutamine synthetase in the hippocampus with an elevated extracellular glutamate levels resulting from the slow conversion of glutamate to glutamine and the subsequent accumulation of glutamate in extracellular space and astrocytes (Eid et al. 2008). Inhibition of glutamine synthetase in astrocytes and/or glutamine transporters in neurons has been demonstrated to reduce amplitudes of evoked IPSCs and GABA releases from neurons (Seifert et al. 2010; Ortinski et al. 2010), which also increases the neuronal excitability (Ortinski et al. 2010). Therefore, impairment of the glutamate–glutamine cycle plays an important role in the development of epilepsy and offers a fertile ground for the development of new therapeutic modalities in future (Eid et al. 2008; Seifert et al. 2010).

Active neurons release K^+ into the extracellular space and induce an elevation of extracellular K^+ ($[K^+]o$) from baseline (~3 mM) up to the "ceiling level" (approximately 10–12 mM) (Chao and Xia 2010). High $[K^+]o$, if not cleared timely, can compromise the normal neuronal firing, increase neuronal excitability, and induce epileptiform activity resulting in the onset of seizures (Somjen and Müller 2000; Semyanov and Godukhin 2001; Gabriel et al. 2004; Ivens et al. 2007; Florence et al. 2009; Lopantsev et al. 2009). In fact, excessive extracellular K^+ accumulation is the single best-characterized proconvulsant agent (D'Ambrosio 2004). "High K^+ model" of epileptiform activity has been widely accepted ever since its initial proposal (Green 1964). Glial cells, especially astrocytes, play a key role in the maintenance of K^+ homeostasis by buffering excessive extracellular K^+, which includes either K^+ uptake or K^+ spatial buffering (Kofuji and Newman 2004). In astrocytes, the K^+ uptake is mediated predominantly by the inward rectifying K^+ channels Kir4.1 that are expressed in abundance (Olsen and Southeimer 2008). The classical property of Kir channels is that K^+ flows inward when the resting membrane potential is negative compared to the equilibrium potential for K^+, but outward currents are inhibited at potentials that are more positive. This provides the driving force for glial uptake of K^+ that is released during neuronal activity, a key function of astrocytes (Butt and Kalsi 2006). K^+ spatial buffering involves the gap junctions between adjacent astrocytes, which make astrocytes to act as a functional syncytium allowing rapid intercellular K^+ diffusion and thus preventing extracellular K^+ accumulation during neuronal firing (Kofuji and Newman 2004). Multiple gap junction proteins (connexins) are expressed by the astrocytes. Among them, Cx43 is the principal gap junction protein that constitutes astrocytic gap junctions (Wallraff et al. 2006; Orthmann-Murphy 2008). Due to their key role in K^+ homeostasis, the properties of astrocytic K^+ buffering have been keenly investigated in both experimental and human epilepsy. Impaired K^+ buffering due to the dysfunction of Kir channels and/or connexins has been demonstrated to induce epilepsy (see reviews by Binder and Steinhäuser 2006; Wetherington et al. 2008; Seifert et al. 2010). Either genetic deletion of astrocyte gap junctions or genetic knockout of Kir4.1 or Kir4.1 null lead to impaired K^+ buffering and the development of spontaneous epileptiform activity and severe seizure behavioral phenotype prior to animal death (Wallraff et al. 2006; Djukic et al. 2007). In TLE (Bordey and Spencer 2004; Heuser et al. 2010) and other epilepsy syndromes such as tuberous sclerosis complex (Jansen et al. 2005; Xu et al. 2009), tumor-associated epilepsy (Olsen and Southeimer 2004), posttraumatic epilepsy (D'Ambrosio et al. 1999; Garga and Lowenstein 2006; Tomkins et al. 2008), etc., impaired astrocytic Kir channel function or gap junction coupling was observed and was always accompanied by the impaired K^+ homeostasis and excessively increased neuronal excitability. Perturbations in neurovascular integrity and breakdown of the blood–brain barrier lead to neuronal hypersynchronization and epileptiform activity (Ivens et al. 2007; Tomkins et al. 2008). This phenomenon mainly results from the direct brain exposure to serum albumin and transforming growth factor β receptor-mediated astrocyte uptake of serum albumin. Astrocytic uptake of serum albumin leads to the downregulation of Kir 4.1 channels in astrocytes and results in the reduced buffering of extracellular K^+ and activity-dependent accumulation of extracellular K^+, thus subsequently facilitating NMDA receptor-mediated neuronal hyperexcitability and, eventually, epileptiform activity (Ivens et al. 2007; Seifert et al. 2010). Besides the decrease in K^+ buffering in astrocytic dysfunction, genetic downregulation of Kir4.1 or reduced Kir activity greatly decreases the ability of astrocytes to clear extracellular glutamate (Djukic et al. 2007; Kucheryavykh et al. 2007; Inyushin et al. 2010), which synergically potentiates neuronal excitability and the development of epilepsy with high $[K^+]o$ due to impaired K^+ buffering of astrocytes.

6.3.1.6 Other Potential Factors

Besides the above-mentioned major factors that determine the neuronal excitability, epileptogenesis, and seizure activities, other potential factors also contribute to the development of epilepsy. Some neurotransmitters/neuromodulators (other than excitatory glutamate and inhibitory GABA) and signaling molecules, such as neuropeptide systems (e.g., endogenous opioids, somatostatin, cholecystokinin (CCK), neuropeptide Y), monoamine neurotransmitters (serotonin, dopamine, norepinephrine), amino acids (traurine, glycine, aspartate), acetylcholine, NO, Ca^{2+} and Ca^{2+}-dependent activity, hormone and humoral factors, and neurotrophic factors and cytokines, also play various roles in epileptogenesis (see the next Sect. 6.3.2.2). These factors together with the above-discussed ones compose a very complicated system of determining the process of epilepsy.

6.3.2 Potential Mechanisms of Acupuncture Treatment for Epilepsy

To understand the mechanisms of acupuncture treatment, it is necessary to understand the maintenance of homeostasis in the body. As we know, homeostasis of internal environment of the body is a very important concept of normal physiological function. Disruption of homeostasis leads to abnormal functions and even diseases if it lasts long. Epilepsy, like many other brain diseases, is also associated with disrupted neuronal activities and is characterized by neuronal overexcitation resulting from a variety of neuropathological alterations in the brain. In TCM terminology, epilepsy is a disorder caused, in most cases, by the disharmony in emotions and *Qi* (energy flow/vital energy), which leads to an imbalance between *Yin* and *Yang* and dysfunction of the activities of *Zang-fu* and ultimately makes the clear orifice misted, the mind disturbed, and the channels invaded, thus ending up in seizures (Yang and Cheng 2010). Accordingly, in modern Western medicine terms, acupuncture treatment is to help keep the disrupted homeostasis of internal environment normalized or recovered. That is, acupuncture may play the antiepileptic roles in a way to *normalize* the disrupted neuronal excitability (by lowering the overexcited neuronal activity through multiple strategies like enhancing the inhibitory system and attenuating the excitatory one in the brain). In TCM terms, acupuncture helps correct the disrupted flow of *Qi* (vital energy) and restore the imbalance between *Yin* and *Yang* and the natural internal homeostasis to refresh the brain and restore consciousness. Acupuncture does not only relieve the symptoms but also treats the cause of illness. Acupuncture treats the patient as a whole and helps the body to heal itself and restore the balance between the physical, emotional, and spiritual aspects. Here, the concept of "functional normalization" by acupuncture needs special emphasis because the effects of acupuncture on disrupted neuronal activities may be bidirectional depending on the alterations in the brain function and activities (Hui et al. 2010). As a modality and interventional procedure, and also as a complex somatosensory stimulation, acupuncture can influence neuronal activity and the intrinsic connections between the so-called default mode network and the anti-correlated task-positive network (also called "sensorimotor network" or "acupuncture-related network") in the brain (Dhond et al. 2008; Bai et al. 2007, 2009, 2010; Liu et al. 2009a, b; Hui et al. 2010) at different levels (from cells to molecules), with different time course (short term vs. long term) and at different stages of epilepsy (early/acute vs. later/chronic). The acute/short-lived effects of acupuncture on neurons may be related to the modification of neuronal electrical activity and the release of neurotransmitters/modulators, while the long-lasting/sustained effects of acupuncture may involve the modification of neurotransmitter/modulator synthesis, the expression of receptors/transporters/ion channels of neurons, the mobilization of therapeutic neuronal networks, and the regulation of neuronal injury-related events such as neuronal death/loss, altered neurogenesis, abnormal axonal reorganization and dendritic remodeling, active gliosis, etc. The following section will outline the possible mechanisms underlying the antiepileptic effects of acupuncture treatment.

6.3.2.1 Acupuncture Modification of Neuronal Electrical Activity and the Activity of Therapeutic Networks

A recently proposed network interaction hypothesis of electrical stimulation therapy for CNS disorder (Faingold 2008) lends us some ideas for explaining the mechanism of acupuncture treatment for epilepsy. According to this hypothesis, CNS neuronal networks not only control normal physiological functions (e.g., locomotion, respiration, learning, and memory) but also mediate the pathophysiology of many CNS disorders. Stimulation therapies including acupuncture can activate "therapeutic" networks that were dormant before the stimulatory treatment and make them compete with the pathophysiological neuronal networks to diminish or prevent the disease symptoms (Faingold 2008). In this hypothesis, conditional multireceptive (CMR) neurons in CMR brain regions (including brainstem reticular formation, amygdala, hippocampus, cerebellum, and association areas of cerebral cortex, etc.) are the key elements because these neurons have extensive connections to numerous brain areas with extensively changeable responsiveness to any given modality depending on stimulus parameter type, strength, and repetition rate, vigilance state, and pharmacological treatment conditions, which allow these regions to potentially participate in many networks (Faingold 2008). The same CMR regions either get recruited into disease networks in response to the input from neuronal networks that initiate CNS disorders, or become a functional part of a previously dormant therapeutic network in response to the external stimulation of "therapeutic" sites. The activation of previously dormant therapeutic networks may produce both short-term plasticity (e.g., changes in neuronal firing rates and patterns and synaptic efficacy) and long-term plasticity (e.g., network expansion) to compete the activities of disease networks to terminate and prevent disease symptoms mediated by disease networks (Faingold 2008).

There are extensive studies with EEG-functional magnetic resonance imaging (EEG-fMRI) showing the existence of neuronal networks of epilepsy in patients with epilepsy (Aghakhani et al. 2004; Berman et al. 2010; Moeller et al. 2010; Szaflarski et al. 2010; Zhang et al. 2010). Thalamocortical circuit-mediated absence epilepsy in both human beings and experimental and genetically mutant rodent models is a good example (Freestone et al. 2009; also see reviews by Chang and Lowenstein 2003; Avanzini and Franceschetti 2003). This circuit involves mainly three neuronal populations: thalamic relay neurons, thalamic reticular neurons, and cortical pyramidal neurons (Fig. 6.1). The thalamic relay neurons can activate the cortical pyramidal neurons in either tonic or burst mode, depending upon the inputs from the thalamic reticular neurons. The thalamic reticular neurons receive the projections from both thalamic relay neurons and cortical pyramidal neurons to form a circuit. The recurrent collaterals from neighboring reticular neurons can inhibit these thalamic reticular neurons (Fig. 6.1). Dysfunction of thalamocortical network, probably due to the Ca^{2+} channel dysfunction in the relay neurons, can lead to abnormal paroxysmal episodes of rhythmic cortical activation, i.e., absence seizures (Chang and Lowenstein 2003; Avanzini and Franceschetti 2003). Another good example is the kindling phenomenon. Kindling epilepsy has been extensively investigated as a chronic model of TLE in many previous studies. Daily repetitions of subthreshold electrical stimulation of the seizure-susceptible regions (e.g., hippocampus, amygdala) or injection of sub-dose convulsants induce progressive and dramatic hyperexcitability within the limbic system and cortex and eventually trigger generalized and spontaneous seizures. During this procedure, repeated electrical/chemical stimulation gradually facilitates the activity of networks between hippocampus–cortex–amygdala (Fig. 6.3), leading to kindled and even spontaneous seizures that are widely propagated all over the network, most dramatically in the temporal lobe structures such as hippocampus, amygdala, piriform, entorhinal, and prefrontal cortex (Lothman 1994; Sutula et al. 1994; Sloviter 1999; Aroniadou-Anderjaska et al. 2008; Szyndler et al. 2010). This process also results in the impaired cognition and emotional response due to the alterations in networks.

The use of fMRI in the brain research provides powerful insight into acupuncture effect on the brain networks (Hui et al. 2010). Recent studies using fMRI revealed that acupuncture stimulation specifically evokes the deactivation of certain networks (e.g., limbic–paralimbic–neocortical network, amygdala, and hypothalamus) and the activation of sensorimotor networks in healthy subjects that are closely matched with the so-called default mode network or anti-correlated brain network (Fang et al. 2004; Campbell 2006; Dhond et al. 2008; Bai et al. 2007, 2009, 2010; Liu et al. 2009a, b; Hui et al. 2010; Ren et al. 2010). The competition between these networks integrates the effects of acupuncture in the brain at multiple levels and in a bidirectional way to provide the potential therapeutic effects (Hui et al. 2010).

As with acupuncture analgesia and the manifestations of its integrative processes at multiple levels in the CNS between afferent impulses from pain regions and impulses from acupoints in pain control (Dhond et al. 2008; Zhao 2008; Hui et al. 2010), acupuncture can also play the antiepileptic roles by integrating the neuronal electrical activities in between seizure-related networks and acupuncture stimulation-activated therapeutic networks at multiple levels of brain regions. Substantial evidence has indicated that acupuncture can regulate neuronal electrical activity and the functional states of the CNS (Iasnetsoy and Prardivtsev 1982; Hamba and Toda 1985; Kukushkin and Reshetniak 1986; Ishibashi 1986; Gao et al. 1990; Fang et al. 1993; Li et al. 1993; Zhu et al. 1993; Chen et al. 2006a; Xing et al. 2007; Napadow et al. 2007; Bai et al. 2009; Chae et al. 2010; Shen et al. 2010). For example, acupuncture stimulation to the sacral segment induced EEG changes from light anesthesia-like state to deep anesthesia state, and this involves GABA-mediated suppression of the firing rate in noradrenergic locus coeruleus neurons (Wang et al. 2007). Functional MRI analysis revealed that acupuncture stimulation at real acupoints (Liv-3 and GV-40) (not the sham one) activates several brain regions including secondary somatosensory cortical areas, frontal areas, the right side of the thalamus, and the left side of the cerebellum in healthy subjects (Fang et al. 2004). In the animal epilepsy model, acupuncture effectively inhibits the cortical epileptiform discharges and restores abnormal EEG. For example, in both electroconvulsive shock- and penicillin-induced convulsion models, application of electroacupuncture at Fengfu (GV-16) and Jinsuo (GV-8) acupoints inhibited the epileptiform discharges and presented in EEG as the decrease in amplitude and frequency, shortening of duration, and the restoration of increased total power of EEG in the rat models of seizures with electroacupuncture stimulation (He et al. 1990a; Huang et al. 1999; Chao et al. 2001). These effects are probably associated with the enhancement of inhibitory process in the brain with electroacupuncture stimulation (He et al. 1990a; Huang et al. 1999; Chao et al. 2001). Similar observations are also reported by other investigators in either animal models (Lou 1989; Zhang et al. 1992; Liu and Cheng 1995; Zhuang et al. 2006b; Guo et al. 2008c) or epileptic patients (Chen et al. 1983). In the hippocampus, intracerebroventricular injection of coriaria lactone-induced epileptiform unit discharges can be classified into three types according to the phase correlation between the hippocampal unit discharge frequency and the occurrence of hippocampal EEG epileptiform, i.e., (1) positive-correlated unit discharges that are characterized by the coappearance of high-frequency spike bursts and disappearance in between spike bursts; (2) negative-correlated unit discharges that just appear in interictal but disappear in spike bursts; and (3) irrelevant unit discharges that have no corresponding relationships in between unit discharges and spike phase and frequency (Gao et al. 1991). Electroacupuncture at Dazhui (GV-14) or contralateral Taichong Kunlun (LR-3-BL-60) selectively inhibits most of the positive-correlated unit discharges but excites most of the negative-correlated unit discharges with no effects on most of the irrelevant unit discharges in the hippocampus (Gao et al. 1991). This is a typical example of bidirectional regulation and functional "normalization" of neuronal electrical activities by acupuncture stimulation. The authors thus suggested that acupuncture plays its antiepileptic role by inhibiting the discharges of intrinsic burst-generating neurons and activating the CNS inhibitory system, and thereby promoting the changes of brain EEG from synchronized epileptiform-like discharges to desynchronous discharges (i.e., functional normalization of brain by acupuncture) (Gao et al. 1991). Wu and his coworkers (Wu et al. 1986; Ma and Wu 1986; Wu 1992) reported that acupuncture stimulation at some governor acupoints (Dazhui, Yaoyu) and limb acupoints (Zusanli) attenuated the cortical epileptiform discharges induced by topical application of penicillin on rat sensorimotor cortex. This attenuation manifested as a reduction in frequency, amplitude, and duration of the discharges in the cortex. They also observed that penicillin application on the surface of cerebral cortex diminishes the potential in sensorimotor cortex evoked by antidromic stimulation of pyramidal tract. This potential has been proven to be the postsynaptic potential produced by recurrent inhibition of pyramidal tract. This diminution in potentials is followed by the epileptiform discharges in penicillin-applicated cerebral cortex and can be restored by electroacupuncture. Therefore, the authors concluded that the antiepileptic effect of electroacupuncture can be explained by the restoration of recurrent inhibition of pyramidal tract (Wu et al. 1986; Ma and Wu 1986; Wu 1992). These investigators also demonstrated that the antiepileptic effect of electroacupuncture involved a wide range of brain structures, including cortex, hippocampus, hypothalamic arcuate nucleus, striatum, substantia nigra, the brain stem raphe nucleus, periaqueductal gray, and nucleus accumbens, since selective blockade/lesion of these nuclei attenuated or abolished the effects of electroacupuncture on epileptic seizures while activation or direct stimulation of these nuclei mimicked the effects of electroacupuncture (Wu et al. 1986; Ma and Wu 1986; Wu 1992). The seizure suppressing effect of electroacupuncture at Zusanli (ST-36) may involve the pathway of hypothalamic arcuate nucleus to the brain stem raphe nucleus and finally to the hippocampus, because the suppressing effect of electroacupuncture at Zusanli (ST-36) on hippocampal epileptic bursts can be attenuated by lesion, and mimicked by stimulation, of arcuate nucleus and raphe nucleus, and the seizure suppressing effect during stimulation of hypothalamic arcuate nucleus became weak after the lesion of raphe nucleus (Wu 1992). Here, the pathway of hypothalamic arcuate nucleus to the brain stem raphe nucleus and finally to the hippocampus seems to form a therapeutic network as suggested in the hypothesis of electrical stimulation therapy for CNS disorders (Faingold 2008).

Like acupunctural stimulation at acupoints, direct brain stimulation to modify neuronal activity is also antiepileptic in some models (Hamani et al. 2008; Tass et al. 2009; Fisher et al. 2010). Multisite electrical stimulation in a rat hippocampal slice with desynchronizing coordinated reset stimulation caused a long-lasting desynchronization between hippocampal neuronal populations together with a widespread decrease in the amplitude of the epileptiform activity induced by Mg^{2+} omission in ACSF (Tass et al. 2009). Electrical stimulation can also inhibit hyperexcitability and epileptic discharges in cortical neurons resulting from deafferentation (Houweling et al. 2005; Madhavan et al. 2006). These changes can probably be attributed to the stimulation-induced reshaping of the connectivity pattern of neural or oscillator networks with synaptic plasticity and network shift from one attractor to another (network competition) as a result of altered neuronal firing within networks (Madhavan et al. 2006; Tass et al. 2009).

Thalamus has abundant connections with other brain areas prone to epilepsy, such as amygdala, hippocampus, and cortex (Fig. 6.3), and is one of the epilepsy sensitive areas (Blume 2007). Thalamus is believed to be involved in human absence seizures and TLE (Chang and Lowenstein 2003; Avanzini and Franceschetti 2003; Blume 2007; Mueller et al. 2010) as well as other brain diseases (Eusebio et al. 2009). Acupuncture stimulation can influence the neuronal firing rate in the thalamus (Ishibashi 1986; Fang et al. 2004) and stimulate the interaction between neighboring thalamic regions, especially with a form of somatosensory competition (de Carvalho 1994). Direct modification of the neuronal activity by electrical stimulation of the anterior nucleus of the thalamus before the onset of status epilepticus significantly increases the latency for seizures in pilocarpine-induced seizures (Hamani et al. 2008) and also has beneficial effects on controlling seizures in patients with medically refractory partial and secondary generalized seizures (Fisher et al. 2010). In addition, a recent report indicated that thalamic neurons show epileptiform activities after injection of pentylenetetrazole via the jugular vein in rats. Electroacupuncture at Dazhui (GV-14) has mixed effects on pentylenetetrazole-induced epileptic discharges but has a predominant inhibition of these epileptiform discharges in the thalamic neurons, and this effect was instantaneous and short-lived (Zhang et al. 2008b). The same authors reported these observations while studying the cortical neurons of pentylenetetrazole-induced epilepsy rats (Zhang et al. 2008c). Therefore, the authors suggested that electroacupuncture exerts effects on a wide range of brain structures that may directly or indirectly affect thalamic neuronal response (here, thalamic neurons serve as conditional multireceptive neurons in the therapeutic networks) and interrupt the genesis and spread of epileptiform activities across this key subcortical gate to the cerebral cortex (Zhang et al. 2008b, c). The acute effects of electroacupuncture seen in this model are due to the direct mobilization of neuronal circuits and are most likely attributed to the activation of the inhibitory neuronal circuits rather than changes in factors such as neurotransmission systems (expression of transmitters/modulators, cytokines, and their receptors) and even morphological changes that would generally take time (Zhang et al. 2008b, c).

Most interestingly, Zhang et al. found that electroacupuncture at Dazhui acupoint is comparable to vagus nerve stimulation in inhibiting epileptiform activities in the thalamus and cortex in the animal models (Zhang et al. 2008b, c). Vagus nerve stimulus (VNS) treatment is a well-known potential alternative for patients with intractable epilepsy (Cakmak 2006; Milby et al. 2008). Nucleus tractus solitarii (NTS) plays a key role in mediating the effects of VNS in epilepsy. NTS is not only the site of vagal afferent projections, but it also receives the endings from glossopharyngeal, facial and trigeminal nerves, and even the stimulation signals from extremities (Cakmak 2006). NTS also has anatomically or functionally reciprocal connections with cerebral cortex, hippocampus, amygdala, and thalamus; thus it acts as an integrating site for many physiological signals (Cakmak 2006). Therefore, NTS serves as a "perfect" conditional multireceptive brain site involving the competition between stimulation-activated therapeutic networks and pathophysiological networks, as Faingold suggested (Faingold 2008). VNS may play its antiepileptic role by altering the synaptic transmission in NTS that regulates seizure susceptibility, impeding the development of convulsive evolution and secondary generalization, and disrupting the epileptiform activity by deactivating the epilepsy sensitive regions (e.g., thalamus, parahippocampal gyrus, hippocampus, and amygdala) and desynchronizing interconnected cortical regions implicated in seizures (Cakmak 2006; Milby et al. 2008). Intriguingly, auricular electroacupuncture has been shown to stimulate vagal activity in a randomized trial in healthy men (La Marca et al. 2010) and to be antiepileptic in animals (Panzer and Chrisman 1994; Hu et al. 1998a; Liu and Huang 1998; Shu et al. 2004, 2005). Therefore, a stimulation of vagal activity by acupuncture can end with similar outcomes as with VNS in the treatment of epilepsy. In addition, electroacupuncture stimulation also alters sleep patterns and improves sleep disturbances which always occur in epilepsy patients (Bazil 2003; de Weerd et al. 2004). In this process, NTS is involved (Yi et al. 2004). Indeed, acupuncture, whether directly through head/facial acupoints (e.g., earpoints), or indirectly via extremity acupoints (e.g., Zusanli, ST-36, see Noguchi and Hayashi 1996), stimulates NTS through the auricular branch of vagus, trigeminal, glossopharyngeal or cervical spinal nerves, as well as somatic nerves (Cakmak 2006). In this way, it promotes the activation of previously dormant therapeutic networks by modifying the activities of neuronal electrical activity in the circuitry to compete with epilepsy networks existing in the limbic system or cortex to reduce the severity of seizure, as seen in Zhang et al.'s study (Zhang et al. 2008b, c). Therefore, Zhang et al.'s results (Zhang et al. 2008b, c) are most interesting but not surprising. Thus, Zhang et al. suggests that acupuncture could be a good alternative therapy to VNS in the management of epilepsy considering the potential side effects, comfort, and the financial costs of VNS treatment (see Milby et al. 2008) (Zhang et al. 2008c). In addition, activation of somatic afferents in the stimulated tissues is a common feature of acupuncture treatment. Various kinds of afferent fibers have been found to transmit the bioinformation (e.g., intense, rate, and duration of acupuncture stimulation, sensation of *Qi*) of acupuncture from the acupoints to the brain (Lu 1983).

In summary, various conditional multireceptive neurons exist (e.g., NTS, thalamic neurons) in the brain that modify their neuronal activity, in response to the impulses from acupoints that they receive through different kinds of afferent fibers, to recruit/activate inhibitory system in the brain (as a therapeutic network). Acupuncture, by modulating the activities of these conditional multireceptive neurons, functionally enhances inhibitory networks and thereby suppresses the hyperexcitability and epileptic firing by competing with seizure networks in the seizure vulnerable brain regions, thus alleviating the symptoms of epilepsy as a result. Therefore, it seems that acupuncture plays its antiepileptic effects by modifying the electrical activities of neurons and the therapeutic networks between the involved brain regions, and this effect of acupuncture on neuronal electrical activities is instantaneous and short-lived and disappears before the permanent changes in network plasticity occur.

6.3.2.2 Acupuncture Regulation of the Activity of Biomediators

Many studies have shown that acupuncture contributes to the biochemical balance in the CNS and the maintenance or recovery of homeostasis. Many neurotransmitters and signaling molecules are involved in the regulation of neuronal excitability and seizure activities. These biomediators also mediate the antiepileptic effect of acupuncture. In the following section, we will first pay our attention to the roles of these biochemicals in the epilepsy and then move on to discuss their possible involvement and potential roles in the antiepileptic effect of acupuncture.

Excitatory and Inhibitory Amino Acids

As we have discussed previously, cerebral imbalance between the inhibitory and the excitatory system plays a key role in initiating seizures. This imbalance manifests as alterations in the activities of the two major amino acids, i.e., the excitatory glutamate and the inhibitory GABA, at multiple levels including their biosynthesis and contents in the brain, their release and clearance, their receptor number, sensitivity and expression, the receptor-coupled channel properties, and excitatory/inhibitory synapses/circuitry (see Sect. 6.3.1.3). Besides these two principal antagonistic amino acids, other amino acids, such as aspartate, glycine, and taurine, also play important roles in epilepsy. It is generally accepted that taurine is an inhibitory neurotransmitter and has anticonvulsant properties (Baran 2006; El Idrissi and L'Amoreaux 2008; Junyent et al. 2009, 2011), while aspartate is an excitatory neurotransmitter and is proepileptic (Ma et al. 2008; Losi et al. 2010). Glycine can act as either an inhibitory neurotransmitter or as a co-agonist of NMDA-dependent excitatory neurotransmission and has either pro- or anticonvulsant properties in different rodent models of epilepsy (Meldrum 1984; Dannhardt and Kohl 1998; Löscher 1998; Böhme and Lüddens 2001; Kalinichev et al. 2010). During seizures, the concentration of aspartate in the brain is significantly higher. In contrast, as an antagonistic counterpart, traurine is also significantly increased; the levels of glycine are either increased or not changed (Kaura et al. 1995; Sejima et al. 1997; Szyndler et al. 2008; Maciejak et al. 2009, 2010; Junyent et al. 2011). Together with the changes in concentrations of glutamate and GABA (see Sect. 6.3.1.3), there is a shift in the balance between these neurotransmitters toward increased production of excitatory amino acids and overexcitation of neurons in epilepsy.

Acupuncture can restore the disrupted balance between excitatory and inhibitory amino acids in a variety of ways. Firstly, acupuncture can modulate the concentrations of these amino acids in the brain. Several lines of evidence have indicated that acupuncture decreases the levels of excitatory amino acids and/or increases the inhibitory amino acids in the brain during seizures in various rodent models of epilepsy. For example, in penicillin-induced seizure models, the levels of both the excitatory amino acids (glutamate, aspartate) and inhibitory amino acids (GABA, taurine) were found to be significantly increased in the hippocampus during seizures; acupuncture stimulation at specific acupoints not only significantly diminished the glutamate and aspartate levels but also enhanced the increase in taurine, GABA, and glycine (Wang and Cheng 1994c; Shu et al. 2004; Deng et al. 2006). In kainic acid-induced epilepsy model, taurine was significantly increased and GABA and glycine tended to increase, while glutamate and aspartate tended to decrease in the hippocampus after electroacupuncture treatment (Liu and Cheng 1995; Yang et al. 2006). In pentylenetetrazole-kindled epilepsy rats, after electroacupuncture treatment, GABA, taurine, and glycine levels significantly increased, while glutamate contents greatly decreased, in the homogenate of the whole brain and in several brain regions such as sensorimotor cortex, visual cortex, reticular nucleus of the thalamus, and reticular part of substantia nigra (Yang et al. 2002a; Zhang et al. 2002). In hyperbaric oxygen-induced convulsion, GABA levels in the brain homogenate were significantly lower, with the severity of convulsions paralleling the degree of decrease in GABA levels. This decrease in GABA levels as well as the convulsions was reversed after electroacupuncture stimulation at Renzhong and Chengjiang (Wu et al. 1992). Most importantly, acupuncture does not only alter the levels of these amino acids but also increases the GABA/glutamate ratio in epilepsy (Yan et al. 1999). In addition, the restoration of the imbalance between the excitatory and inhibitory amino acids has been shown not only in epilepsy but also in cerebral ischemia (Zhao and Cheng 1997; Pang et al. 2003; Gan et al. 2005; Guo et al. 2006; Zhang et al. 2007; Meng et al. 2008; Lee et al. 2010b). This could partially be responsible for the antiepileptic effect of acupuncture in epilepsy secondary to cerebral infarction (Mao and Guo 2005). Obviously, acupuncture inhibits seizures at least by downregulating excitatory amino acids and upregulating inhibitory ones to restore the disrupted balance between them.

Secondly, the antiepileptic effect of acupuncture can be enhanced or attenuated by chemical manipulation of amino acid neurotransmission. For example, exogenous taurine itself is not only antiepileptic but also enhances the anticonvulsive effect of electroacupuncture stimulation (Li et al. 2005). In the penicillin-induced seizure model, selective blockade of either NMDA receptors or non-NMDA receptors by hippocampal microinjection of either AP5 (a NMDA receptor antagonist) or DNQX (a non-NMDA receptor antagonist) greatly enhanced, while blockade of $GABA_A$ receptors with bicuculline largely attenuated, the antiepileptic effect of electroacupuncture (Wang and Cheng 1994c; Liu and Cheng 1997), suggesting the involvement of ionotropic glutamate receptors and $GABA_A$ receptors in the antiepileptic role of acupuncture. In pentylenetetrazole-induced epilepsy rats, it seems that different subtypes of GABA receptors play antagonistic roles in mediating the antiepileptic effect of electroacupuncture. For example, $GABA_A$ receptors enhance, while $GABA_B$ receptors attenuate, the effect of electroacupuncture because peripheral or central administration of $GABA_A$ receptor agonist, muscimol, shortened the duration of epileptic discharge and further potentiated the effect of electroacupuncture, whereas the application of $GABA_B$ agonist, baclofen, abolished the effect of acupuncture (Wan et al. 1997, 1999). Furthermore, intraperitoneal or intracerebroventricular or intracerebral (into the GABA-enriched brain regions like caudate nucleus, globus pallidus, and substantia nigra) injection of 3-mercaptopropionic acid, an inhibitor of GAD (the rate-limiting enzyme for GABA synthesis), markedly attenuated the antiepileptic effect of electroacupuncture (Wu 1992; Zhang et al. 1992). All these findings suggest that the antiepileptic effect of acupuncture is considerably dependent on amino acid neurotransmission.

Thirdly, acupuncture can upregulate the expression of biosynthesis enzymes and transporters of inhibitory amino acids and downregulate the receptors of excitatory amino acid. GAD is the rate-limiting enzyme for GABA synthesis and is a marker of GABAergic neurons. GAD67 is the enzymatically active form of GAD (Yogeeswari et al. 2005). It has been shown that GAD67 expression is either decreased or increased in various models of epilepsy, and the increase of GAD67 expression has been suggested as a self-protective, anticonvulsive mechanism in limbic epilepsy (Schwarzer and Sperk 1995; Esclapez and Houser 1999; Freichel et al. 2006; Li et al. 2007; Gill et al. 2010). Electroacupuncture can increase GAD67 expression in hippocampal dentate gyrus of pilocarpine-induced epileptic rats (Guo et al. 2008c) and hippocampal CA1 and CA3 areas of kainic acid-induced epilepsy in mice (Kim et al. 2008b). These findings suggest that electroacupuncture inhibits epileptic seizures and hippocampal GABAergic neuronal death by increasing GAD expression and GABA synthesis. Taurine transporter is crucial for the maintenance of normal taurine levels in tissues. Taurine transporter knockout mice had a 70–90% decrease in taurine levels in various tissues including the brain, and these mice presented with many pathological features (Warskulat et al. 2007). Taurine transporters have been found to be present in hippocampal astrocytes 3–7 days after kainic acid treatment in mice, which is responsible for the increased taurine levels in the hippocampus induced by kainic acid treatment (Junyent et al. 2011). It has been shown that electroacupuncture could increase taurine transporter levels in the hippocampus of rats with kainic acid- or penicillin-induced seizures, and this change in taurine transporter levels has been suggested to be related to the antiepileptic effect of electroacupuncture (Li et al. 2005; Jin et al. 2005a). In addition, electroacupuncture is known to inhibit ischemia-induced increase in glutamate levels and down-regulate NMDA receptor 1 expression in the hippocampus (Meng et al. 2008). This could partially contribute to the antiepileptic effect of acupuncture in the epilepsy secondary to cerebral infarction (Mao and Guo 2005).

In summary, the antiepileptic effect of acupuncture involves the modulation of excitatory and inhibitory amino acid systems, which restores the balance between the excitatory and inhibitory neurotransmission.

Neuropeptides (Opioid, Somatostatin, CCK, NPY)

Neuropeptides include a large family of small protein-like molecules with a length of 3–40 amino acids and function as neurotransmitters, such as opioids, somatostatin, CCK, and neuropeptide Y (NPY). Neuropeptides and their corresponding receptors are widespread in the CNS and the peripheral nervous system and play various roles in regulating physiological and pathophysiological activities. A large number of studies have indicated the involvement of opioids, somatostatin, and CCK systems in regulating seizure activities and mediating the inhibitory effect of acupuncture on seizures (see the following sections). NPYergic system is also a very important peptidergic regulator of epilepsy. Particularly, in the hippocampus, NPY is mainly produced and released by GABAergic interneurons to inhibit glutamatergic neurotransmission in the excitatory trisynaptic circuit and therefore acts as an endogenous anticonvulsant (see reviews by Xapelli et al. 2006; Dubé 2007; Nadler et al. 2007; Sperk et al. 2007). Though NPY is also reported to mediate the effect of acupuncture treatment in various conditions (Tian et al. 2006; Chae et al. 2008; Huang et al. 2009), and acupuncture can increase the expression of NPY in brain regions including the hippocampus (Bucinskaite et al. 1994; Lim et al. 2003; Kim et al. 2002), no information is available regarding the role of NPY system in the antiepileptic effect of acupuncture.

Opioids

Classic opioid systems include δ-, μ- and κ-opioid and their corresponding receptors, DOR, MOR and KOR. A fourth opioid system, i.e., nociceptin/ophanin FQ and its receptors (NOR), has been added recently to the opioid family. Endogenous opioid peptides enkephalin, β-endorphin, dynorphin, and nociceptin in the brain preferentially bind to DOR, MOR, KOR, and NOR, respectively, in physiological concentrations and have multiple functions in the brain.

The roles of opioid systems in epileptic seizures and epileptogenesis have been suggested with very complex and mixed effects. The accumulated evidence indicates that seizure activity exerts profound changes on the metabolism of opioid peptides in the brain, as manifested by the markedly increased release of enkephalin and dynorphin during epileptic seizures and long-term increase in the biosynthesis of enkephalin and increase or decrease in dynorphin synthesis after epileptic insults in different seizure models (Tortella and Long 1985; Kanamatsu et al. 1986; Hong et al. 1988; Bartenstein et al. 1993; Bing et al. 1997; Koepp et al. 1998). Opioid receptors are upregulated in several brain regions such as the hippocampus and parahippocampal cortex in both animal models of TLE and patients with TLE following epileptic seizures (Madar et al. 1997; Skyers et al. 2003; Hammers et al. 2007; Rocha et al. 2007a). However, it is not well clear if the increased release of opioid neuropeptides and upregulated receptor binding represent a compensatory mechanism against epilepsy because multiple factors affect the roles of opioids in epilepsy in the brain. For δ-opioid system, both anti- (Tortella et al. 1985, 1988; Koide et al. 1992; Madar et al. 1997) and proconvulsant effects (Grecksch et al. 1999; Yajima et al. 2000; Danielsson et al. 2006; Jutkiewicz et al. 2006), even no effect (Lee et al. 1989; Lasoń et al. 1994; Negus et al. 1998; Saboory et al. 2007) with DOR activation have been proposed in various types of epilepsy. The reasons for these complex and mixed effects of δ-opioid system on seizures are not well understood but could partially be related to the factors such as animal species, seizure types, the methods for drug administration, dose used, and target neurons. For example, in pilocarpine-induced seizures in rats, pretreatment with a non-peptide DOR agonist, SCN80, was found to be anticonvulsant at a dose of 30 mg/kg and proconvulsant at a dose of 60 mg/kg (Bausch et al. 2005). Others found that SCN80, either in low or high dose, produced convulsion in rats but had little effects on rhesus monkeys (Negus et al. 1998; Jutkiewicz et al. 2006; Danielsson et al. 2006). The convulsive effect of SCN80 in rats could be minimized by slowing the rate of SNC80 administration (Jutkiewicz et al. 2005). Also, the action of δ-opioids depends on the site where the opioids act. In the hippocampus, inhibitory GABAergic interneurons have DOR expression, and DOR activation in these neurons leads to facilitation, rather than suppression of seizures via postsynaptic de-inhibition (Simmons and Chavkin 1996; Drake et al. 2007). Mu- (Honar et al. 2004; Liu et al. 2004a; Saboory et al. 2007; Roshanpour et al. 2009; Zhang and Ko 2009) and nociceptin (Feng et al. 2004; Carmona-Aparicio et al. 2007) opioid systems were also found to have either anti- or proconvulsant action similar to that observed in δ-opioid system. For κ-opioid system, the endogenous KOR agonist, dynorphin, is abundant in the hippocampus (Houser et al. 1990; Simonato and Romualdi 1996; Simmons and Chavkin 1996; Drake et al. 2007), and KOR is also abundantly distributed in hippocampal granule cells and perforant path (Simmons and Chavkin 1996; Drake et al. 2007). Dynorphin, by acting at presynaptic KOR, may inhibit glutamate release from perforant path and mossy fiber terminals and thus inhibits granule cell-mediated excitatory neurotransmission in the hippocampus (Simmons and Chavkin 1996; Terman et al. 2000; Drake et al. 2007). Although the majority of studies support the anticonvulsant effect of κ-opioid system, especially for complex partial seizures (Simmons and Chavkin 1996; Drake et al. 2007), there are few reports that suggest a proconvulsant effect (Houser et al. 1990; McDermott and Schrader 2011).

Several lines of evidence have demonstrated that opioid systems are involved in the anticonvulsant effect of acupuncture. However, the role of different opioid systems in the acupuncture suppression of epilepsy appears to be varied and complex. Wu and his coworkers demonstrated, using a penicillin-induced seizure model, that the inhibitory effect of electroacupuncture on cortical epileptiform discharges could be reversed by an injection of a broad spectral opioid receptor antagonist, naloxone, given via various routes (e.g., intraperitoneal, intracerebroventricular, into periaqueductal gray matter of midbrain, nucleus accumbens, or even directly into the cortical site where penicillin was applied), while an injection of naloxone into the reticular formation of midbrain and the reticular nucleus of thalamus had little effect on acupuncture antiepilepsy. An injection of naloxone into the above-mentioned sites without application of penicillin on the cortex also did not induce epileptic burst (Wu et al. 1986; Wu 1992), suggesting a specific involvement of opioid systems in certain brain regions in acupuncture antiepilepsy. Therefore, the authors concluded that endogenously released opioid peptides mediate the suppressive effect of electroacupuncture on seizures though the specific types of the released opioid peptides are not clear (Wu et al. 1986; Wu 1992). The reports on enhancement of biosynthesis of central enkephalins by electroacupuncture in normal rats (Yuan and Han 1985), increased dynorphin concentration in hippocampal perfusate, and increased dynorphin immunoreactivity in hippocampal mossy fiber and hilus by electroacupuncture in rats with penicillin-induced seizures or electrical shock-induced convulsion (Wang and Cheng 1992, 1994a, 1995) all support the above viewpoint. In contrast, seizure-induced increase in leu-enkephalin and β-endorphin levels in hippocampal perfusate and greatly enhanced expression of preproenkephalin mRNA in the brain regions such as entorhinal cortex, subiculum, hippocampal CA1 area, amygdaloidal nucleus, and piriform cortex are greatly suppressed by electroacupuncture in experimental seizure rats (He et al. 1989; He and Cao 1989; Wang et al. 1994; Wang and Cheng 1994a). Activation of KOR with U50488H in hippocampus enhances, while the blockade of KOR with KOR antagonist MR2266 or anti-dynorphin 1–13 serum abolishes the antiepileptic effect of electroacupuncture in electroconvulsive rats (Gao and Cheng 1998). These findings suggest that electroacupuncture may play antiepileptic role by increasing dynorphin synthesis/release and KOR activity and decreasing the synthesis/release of enkephalin and/or β-endorphin in the hippocampus. In addition, the alleged sigma opioids are also involved in electroacupuncture antiepilepsy since microinjection of 1,3,di(2-tolyl) guanidine, an agonist of sigma receptors, into the hippocampus reduced the epileptiform activities and enhanced the electroacupuncture anticonvulsion significantly, whereas dextrorphan, an antagonist of the sigma receptors, provoked epileptiform activities and resulted in disappearance of inhibitory effects of electroacupuncture on seizures (Liu et al. 2007; Gao and Wang 2010).

Radioreceptor-binding assay combined with autoradiography revealed that repeated electroconvulsive shock resulted in epileptiform EEG and behavioral seizures that were accompanied by the increase in opioid receptor densities in caudate nucleus, hippocampus, habenula nucleus, and amygdala in rats, while electroacupuncture at Fengfu (GV-16) and Jingsuo (GV-8) suppressed the seizure activities and also significantly decreased the increase in opioid receptor densities in above brain regions, except caudate nucleus in which a tendency to decrease in opioid receptor densities was observed (He et al. 1990b). Most interestingly, direct subacute high-frequency electrical stimulation of parahippocampal cortex in patients with intractable mesial temporal lobe epilepsy also produces antiepileptic effect, and the high effectiveness of the antiepileptic effects induced by this high-frequency electrical stimulation is associated with reduced opioid peptide binding including ^{3}H-DAMGO, ^{3}H-DPDPE, and ^{3}H-nociceptin in the same brain regions (Rocha et al. 2007a). These findings suggest the involvement of opioid receptors in mediation of the antiepileptic effects of acupuncture.

Altogether, it is clear that acupuncture can regulate the activities of endogenous opioid neuropeptides and their receptors in both animal and human brain with epilepsy. Therefore, it can be concluded that acupuncture exerts its antiepileptic effect by regulating the activities of opioid systems.

Somatostatin

Somatostatin, also known as growth hormone-inhibiting hormone (GHIH) or somatotropin release-inhibiting factor (SRIF), can act either as a peptide hormone that regulates the endocrine system or as a neuropeptide transmitter to affect neurotransmission and cell proliferation via interaction with G-protein-coupled somatostatin receptors. Somatostatin is highly expressed in the brain regions involved in epilepsy. In the hippocampus, somatostatin is always co-localized in GABAergic interneurons in the hilus of dentate gyrus that are sensitive to seizure-induced death (Zhang et al. 2009b; Halabisky et al. 2010). Seizures significantly increase the biosynthesis and release of somatostatin in the hippocampus, particularly in dentate gyrus (Binaschi et al. 2003; Tallent and Qiu 2008). The expression of somatostatin receptors in the hippocampus and other brain regions is markedly altered in both epileptic animals and patients with TLE (Hu et al. 1998b; Kang et al. 2003; Csaba et al. 2005; Kwak et al. 2008). Studies on somatostatin ligands and somatostatin receptor knockout mice indicate that somatostatin has antiepileptic properties (Moneta et al. 2002; Cammalleri et al. 2004; also see reviews by Binaschi et al. 2003; Tallent and Qiu 2008) while few reports show a proconvulsant effect of this neuropeptide (Zhao and Huang 2000). In the hippocampus, somatostatin neurons, in the dentate gyrus, are part of a negative feedback circuit modulating the firing rate of granule cells in CA1, and their feedback inhibition reduces the excitatory drive on pyramidal neurons (Binaschi et al. 2003; Tallent and Qiu 2008). Somatostatin released in the hippocampal CA3 area interacts with both presynaptic receptors located on associational/commissural terminals and postsynaptic receptors located on pyramidal cell dendrites, thereby reducing excitability of pyramidal neurons (Binaschi et al. 2003; Tallent and Qiu 2008). Therefore, somatostatin system plays a role in the inhibitory modulation of hippocampal circuitries under normal conditions. Loss of somatostatin function in the dentate gyrus could contribute to epileptogenesis and seizure susceptibility. For example, hippocampal somatostatin system undergoes damage and neuron loss in the dentate hilus following status epilepticus and chronic TLE, which compromises inhibitory mechanisms in the dentate gyrus (Zhang et al. 2009b; Halabisky et al. 2010). However, surviving somatostatin neurons showed significant alteration of morphological and electrophysiological properties. These alterations include the significant increases in somatic area and dendritic length, axon collateral sprouting in the molecular layer, forming new synapses with granule cells, a small but significant decrease in resting membrane potential, increases in time constant and whole cell capacitance, increased frequency of slowly rising spontaneous EPSCs, increased evoked EPSC amplitude, and increased spontaneous action potential generation (Zhang et al. 2009b; Halabisky et al. 2010). All these changes in both morphological and electrophysiological properties of somatostatin interneurons in the epileptic dentate gyrus suggest the mechanisms for surviving hilar interneurons with increased excitatory innervation on their distal dendrites and increased output to inhibit more granule cells, therefore compensate partially for the loss of vulnerable interneurons in the epileptogenic hippocampus.

Somatostatin system is implicated in the antiepileptic effects of acupuncture. For example, in a penicillin-induced seizure model, somatostatin concentration in the hippocampus increased significantly during seizures, and the expression of somatostatin mRNA was strongly upregulated at 24 h after seizures in several brain regions that are prone to seizures, including piriform cortex, frontal cortex, cingulated gyrus, lateral septal nucleus, basal amygdaloid nucleus, pyramidal layer of CA1-4 areas of hippocampus and granule cell layer, and polymorphical layer in dentate gyrus (Liu and Huang 1998; Shu et al. 2004). Ear-point electrical stimulation could decrease the seizure-induced elevation of somatostatin levels in the hippocampus and the enhanced expression of somatostatin mRNA in the frontal cortex, basal amygdaloid nucleus, hippocampus dentate gyrus, and frontal piriform cortex (Liu and Huang 1998; Shu et al. 2004). In electroconvulsive rats, a significant increase in the expression of somatostatin mRNA and a decrease in somatostatin immunoreactivity were identified in the hippocampus, particularly in CA4 area and subiculum following electroconvulsive stimulation, suggesting the increased biosynthesis and release of somatostatin and enhanced expression of somatostatin gene. These changes were reversed with electrical stimulation of the great auricular nerve along with a significant decrease in amplitude and frequency of epileptiform spikes and effectively inhibited convulsive behaviors (Hu et al. 1998a). These studies suggest that the inhibitory effect of electroacupuncture involves the regulation of the expression of somatostatin gene in the hippocampus (Liu and Huang 1998; Hu et al. 1998a; Shu et al. 2004). In pilocarpine-induced chronic epilepsy, newly generated neurons in the hippocampus expressed somatostatin and electroacupuncture could reduce/inhibit this expression (Wen and Huang 2008). The authors thus concluded that somatostatin is involved in epileptogenesis and electroacupuncture inhibits it by decreasing somatostatin localized in newborn neurons in the hippocampus (Wen and Huang 2008). This conclusion seems arguable, given that the increased somatostatin interneuron innervations on their distal dendrites and the increased output of the surviving neurons that inhibit more granule cells represent a compensatory mechanism for the loss of vulnerable interneurons in the epileptogenic hippocampus (Zhang et al. 2009b; Halabisky et al. 2010).

Electroacupuncture-induced inhibition of the expression of somatostatin in newly generated neurons in the chronic epilepsy rat model, as reported by Wen and Huang (2008), can also be explained as newly generated neurons do not need to offer more compensatory mechanisms in such a condition because epileptic seizures has already been suppressed by electroacupuncture treatment. Nevertheless, these studies strongly suggest the involvement of somatostatin system in the antiepileptic role of acupuncture.

Cholecystokinin

CCK is widely distributed in the CNS and acts both as a transmitter and a modulator. In some seizure-prone brain regions (e.g., hippocampus and cerebral cortex), CCK is always co-localized in inhibitory interneurons (e.g., GABAergic basket cells) with some other neuropeptides such as somatostatin and neuropeptide Y (Kawaguchi and Kondo 2002; Sun et al. 2007; Wyeth et al. 2010). Studies have demonstrated that CCK is involved in the regulation of seizure activity. For example, in animal model of TLE, status epilepticus-induced neuronal loss has been found in CCK-immunoreactivity-positive neurons in the piriform cortex (Chen et al. 2007b) and hippocampal CA1 area where the perisomatic inhibition of pyramidal cells is selectively reduced due to the loss of CCK-expressing basket cells in the mice with spontaneous seizures (Wyeth et al. 2010). Seizure activities also induce a marked increase in the CCK concentration and the expression of CCK mRNA in many brain regions including the hippocampus and cortex. This increase is believed to represent a compensatory anticonvulsant adaptation and the self-defense mechanisms for subsequent epileptic episodes (Gruber et al. 1993; Zhang et al. 1996). In the brain of seizure-prone rodents, CCK immunoreactivity and mRNA expression were either increased or downregulated depending on animal models and brain regions (Zhang et al. 1993; Fetissov et al. 2003). Gene transfer of CCK in the brain has been shown to correct the increased susceptibility to audiogenic seizures in rats (Zhang et al. 1997). Pretreatment with CCK significantly suppressed the seizure activities in various seizure models and potentiated the anticonvulsant effect of other compounds such as AEDs, and these effects could be blocked by CCK receptor antagonists (Legido et al. 1995; Ferraro and Sardo 2009). Therefore, CCK is a potent anticonvulsant neuropeptide in certain seizure models.

CCK mediates various effects of acupuncture through central mechanisms (Zhao 2008; Kim et al. 2010). Electrical stimulation at Zusanli (ST-36) can induce CCK release in the thalamus of conscious rats (Xu et al. 1990). However, little information is available on the involvement of CCK in acupunctural antiepilepsy. In penicillin-induced seizure rats, markedly increased CCK mRNA levels were found in the dentate gyrus and CA3 area in both diencephalic and mesencephalic sections of the hippocampus along with the subiculum in mesencephalic sections of the hippocampus, 4 h after seizures (Yang et al. 1996). Electroacupuncture stimulation at Fengfu (GV-16) and Jinsuo (GV-8) attenuated the increased CCK mRNA levels in dentate gyrus and CA3 area in diencephalic hippocampus but enhanced the increased CCK mRNA concentrations in the subiculum, dentate gyrus, and CA3 area in mesencephalic hippocampus (Yang et al. 1996). In addition, the EEG epileptiform discharges and seizure behaviors were significantly reduced (Yang et al. 1996). Therefore, the authors concluded that the inhibitory effect of electroacupuncture on seizure activity could be related to the alteration of CCK gene expression in the different areas of hippocampus (Yang et al. 1996).

Monoaminergic Systems (5-HT, DA, NE)

Monoamines are neurotransmitters and neuromodulators, including serotonin (5-HT), dopamine (DA), and norepinephrine (NE), that contain one amino group connected to an aromatic ring by a two-carbon chain ($–CH_2–CH_2–$). They are abundant in the CNS and modulate neuronal excitability via G-protein-coupled receptor signaling. A large number of studies have suggested the involvement of monoaminergic systems in epilepsy. Generally, 5-HT, DA, and NE are regarded to act as endogenous anticonvulsants (see reviews by Weinshenker and Szot 2002; Theodore 2003; Bagdy et al. 2007). Substantial evidence indicates that an increase in extracellular 5-HT, DA, and NE levels in the brain (e.g., blocking their reuptake by inhibiting or genetically deleting their transporters or inhibiting their degradation by monoamine oxidase inhibitors) inhibits both limbic and generalized seizures (Mishra et al. 1993; Löscher et al. 1999; Clinckers et al. 2004; Kaminski et al. 2005; Merrill et al. 2007; Smolders et al. 2008; Szyndler et al. 2010), whereas genetic or chemical depletion of brain monoamines markedly lowers the threshold for audiogenically, chemically or electrically evoked convulsions (Szot et al. 1996, 1999; Bagdy et al. 2007; Trindade-Filho et al. 2008). Activation of certain subtypes of 5-HT, DA, or NE receptors exerts antiepileptic effects in a variety of seizure models while chemically blocking or genetically downregulating them enhances seizures (Weinshenker et al. 2001; Clinckers et al. 2004; Witkin et al. 2007; Krishnakumar et al. 2009; Gholipour et al. 2010). In addition, the activities of these monoamine neurotransmission systems, particularly serotoninergic system (e.g., rate-limiting synthesizing enzymes, transporters, receptors), were also found to be altered in both epileptic patients (Broderick et al. 2000; Giovacchini et al. 2005; Rocha et al. 2007b; Hasler et al. 2007; Manna et al. 2007; Kauffman et al. 2009; Bateman et al. 2010; Stefulj et al. 2010) and genetically epilepsy-prone rats (Szot et al. 1996; Merrill et al. 2007; Shin et al. 2009). Therefore, an impaired monoaminergic transmission in the brain may be involved in the initiation and propagation of seizure activity.

Monoaminergic systems may mediate acupuncture effects in various conditions, including exercise (Chang et al. 1996; Lee et al. 2002), pain (Zhao 2008), hypertension (Guo et al. 2008d), depression (Han 1986; Xu et al. 2007c), obesity (Liu et al. 2000; Wei and Liu 2003), ethanol/drug abuse and withdrawal (Yoon et al. 2004; Kim et al. 2005; Zhao et al. 2006), spinal cord injury (Yan et al. 2011), Parkinson's disease (Huang et al. 2006), and cognitive impairment (Chang et al. 1999; dos Santos et al. 2005). However, there is lack of information about the involvement of dopaminergic or epinephrinergic systems in the antiepileptic effects of acupuncture. Available information shows that only serotoninergic system is involved in mediating these effects of acupuncture (Ma and Wu 1986; Wu 1992; dos Santos et al. 2005). Firstly, acupuncture can activate serotoninergic neurons and increase 5-HT levels in the brain of normal animals (Wenhe and Yucun 1981; Fukuda et al. 2005; Yoshimoto et al. 2006; Guo et al. 2008d). Secondly, acupuncture alterations of EEG in humans are reported to be associated with increased 5-HT turnover in the brain (Danielczyk 1976). Thirdly, in penicillin-induced seizures, blocking the enzymatic synthesis of 5-HT with *p*-chlorophenylalanine attenuates the electroacupunctural restoration of seizure-induced decline of evoked recurrent inhibitory surface wave in cortical sensorimotor area and the inhibitory effect of electroacupuncture on epileptiform discharges (Ma and Wu 1986; Wu 1992). Electrolytic lesion of the dorsal raphe nucleus that enrich in serotoninergic neurons leads to marked attenuation of electroacupuncture effect, while direct electrical stimulation of this nucleus results in the distinct inhibition of the epileptiform discharges (Ma and Wu 1986; Wu 1992). Finally, in pilocarpine-epileptic rats, blocking 5-HT synthesis with *p*-chlorophenylalanine is able to block the preventive effect of electroacupuncture on epilepsy-induced tissue atrophy of some limbic structures (e.g., the basolateral nucleus of the amygdale, substantia nigra, and perirhinal cortex) and seizure-associated cognitive deficits (dos Santos et al. 2005). All these findings suggest that serotoninergic system participates in the antiepileptic effect of acupuncture.

Nitric Oxide (NO)

NO is an important gaseous messenger molecule involved in many physiological and pathological processes both beneficial and detrimental within the mammalian body, and exerts a strong functional modulation of several neurotransmitter systems (Hou et al. 1999; Ferraro and Sardo 2004). NO is also proposed to have both pro- and anticonvulsive effects (see review by Ferraro and Sardo 2004). NO as well as its synthase, namely, nitric oxide synthase (NOS), is also reported to be involved in mediating the pro- or anticonvulsive effect of other chemicals (Khavandgar et al. 2002; Yahyavi-Firouz-Abadi et al. 2007; Akula et al. 2008; Aslan et al. 2010; Ghasemi et al. 2010). Therefore, the roles of NOS/NO system in epilepsy appear to be complex and variable depending on the models of seizures and different experimental seizure paradigms.

Substantial evidence demonstrates the involvement of NOS/NO in mediating the acupuncture effects under many physiological and neuropathological situations (Ma et al. 2005; Almeida and Duarte 2008; Kim et al. 2008a; Han et al. 2009). In penicillin-induced seizure model, in addition to the epileptic EEG, a continuous increase in NO concentrations in the hippocampus lasting up to 4 h was also observed (Huang et al. 1999). Treatment with either neuronal NOS inhibitor 7-nitroindazole (i.c.v.) or electroacupuncture at Fengfu (GV-16) and Jinsuo (GV-8) significantly attenuated the increase in NO concentrations and epileptic discharges in EEG (Huang et al. 1999). RT-PCR analysis further indicated that the NO increase in the hippocampus of penicillin-induced seizure rats involved the increase of neuronal NOS expression and the production of inducible NOS (vs. no changes in epithelial NOS expression), and electroacupuncture at the same acupoints could markedly decrease the expression of these two forms of NOS (Yang et al. 2000). Similar observations were reported in both the brain and liver of rats with epilepsy (Jin et al. 2005b). In pentylenetetrazole-kindled rats, pretreatment with either electroacupuncture or nonselective NOS inhibitor *N*-nitro-L-arginine methyl ester 30 min before the last intraperitoneal administration of pentylenetetrazole abolished the increase of cGMP in the hippocampus during epilepsy, while L-arginine (a NO precursor) reversed the effects of electroacupuncture on cGMP (Yang et al. 2008). All these findings suggest that acupuncture executes its antiepileptic effect by modulating the activities of NOS/NO system, i.e., mainly by inhibiting the expression of NOS and NO production.

Ca^{2+} and Ca^{2+}-Dependent Activity

Brain injury-induced alterations in intracellular Ca^{2+} concentrations ([Ca^{2+}]i) and Ca^{2+} homeostatic mechanisms play a role in the development and maintenance of acquired epilepsy (see reviews by Delorenzo et al. 2005; Nagarkatti et al. 2009). A decrease in Ca^{2+}-binding proteins, such as calbindin D-28k and calretinin, and Ca^{2+}/calmodulin-dependent protein kinase II activity has been shown to be related to the alterations in [Ca^{2+}]i and Ca^{2+} homeostasis and has been causally linked to the development of acquired epilepsy characterized by spontaneous, recurrent seizures in human as well as experimental animals (Butler et al. 1995; Montpied et al. 1995; Semyanov and Godukhin 2001; Sadowski et al. 2002; Carter et al. 2006, 2008; Tóth et al. 2010).

Strong evidence has shown that Ca^{2+} is involved in the antiepileptic effects of acupuncture. In a pentylenetetrazole-kindled rat model of epilepsy, both electroacupuncture at Baihui (GV-20) and Dazhui (GV-14) for 7 days and treatment with nimodipine (a dihydropyridine calcium channel blocker) markedly prolonged the latency and decreased the frequency of epileptic EEG seizures. Moreover, intracellular Ca^{2+} level in the hippocampus was greatly lowered with both electroacupuncture and nimodipine treatment (Yang et al. 2009b). The authors therefore concluded that electroacupuncture has an obvious antiepileptic effect, which is closely related to its effect on downregulation of hippocampal Ca^{2+} level in pentylenetetrazole-kindled epilepsy rats (Yang et al. 2009b). Similar observation was made in a stroke model with vascular dementia (Zhang et al. 2008a). Obviously, these studies suggested that increased [Ca^{2+}]i mainly resulted from the influx of extracellular Ca^{2+} via Ca^{2+} channels, while electroacupuncture plays its role by decreasing Ca^{2+} entry through Ca^{2+} channels directly or indirectly. In fact, it has been shown in a cross-perfusion experiment that electroacupuncture can attenuate the increase in hippocampal [Ca^{2+}]i by humoral factors (Zhang et al. 2009a) (see next Subsect. Hormones and Humoral Factors in Sect. 6.3.2.2). Therefore, acupuncture exerts its antiepileptic role by regulating the levels of [Ca^{2+}]i in seizure-sensitive regions in the brain. Besides, calmodulin also seems to participate in the effect of acupuncture on epilepsy. In rats with epilepsy induced by pentylenetetrazole, hippocampal expression of calmodulin was greatly increased. This alteration is markedly inhibited by electroacupuncture or nimodipine treatment, which results in significant prolongation of the latency and a decreased frequency of epileptic seizures (Yang et al. 2009a). The authors therefore suggested that electroacupuncture plays its antiepileptic effect by regulating the expression of neuronal calmodulin in the hippocampus in rats with epilepsy (Yang et al. 2009a). In addition, it was reported that in an ischemic stroke model, electroacupuncture effectively alleviated ischemia-induced neuronal apoptosis in the hippocampus CA3 area and upregulated the expression of decreased calbindin D-28k (Ma et al. 2010). Therefore, electroacupuncture in this model could have contributed to the maintenance of Ca^{2+} homeostasis and might be related to the antiepileptic effect of acupuncture on epilepsy secondary to cerebral infarction (Mao and Guo 2005).

Hormones and Humoral Factors

It is widely accepted that seizure activity influences endogenous hormonal activities and this altered hormonal activity, in turn, affects seizure activity (Frye 2008; Pack 2010). Cross-perfusion experiments showed that cultured hippocampal neuronal cells displayed an increase in $[Ca^{2+}]i$ after adding the serum of the normal rats, while $[Ca^{2+}]i$ in cultured hippocampal neurons decreased when adding the serum of the rats given electroacupuncture at "Baihui" (GV-20), "Zusanli" (ST-36), "Quchi" (LI-11), and "Sanyinjiao" (SP-6) for 2 weeks (Zhang et al. 2009a). Obviously, there are some active substances in the serum of electroacupuncture-treated animals that obviously decrease intracellular Ca^{2+} levels, providing a direct evidence for the role of humoral factors in acupuncture treatment. Similarly, it is reasonable to assume that acupuncture can modulate the hormone activities to exert its antiepileptic action. Indeed, this idea is supported by the results of some studies done in animals and patients with epilepsy (Chao et al. 2001; Chen et al. 2008).

Melatonin, a hormone synthesized and secreted by pineal gland, whether endogenously produced or exogenously administrated, prevents neurons against insults and has antiepileptic effect though few reports have shown a contrasting proconvulsant effect (Chao and Cheng 2001a, b). Melatonin levels are altered in the plasma and the brain with a disruption of its secretion pattern in patients with epilepsy, which could influence seizure activities (seizure threshold, seizure types, circadian rhythm, etc.) (see reviews by Chao and Cheng 2001a; Baxendale 2009). Melatonin might be one of the hormonal factors that mediate the antiepileptic effects of acupuncture (Chao et al. 2001). Actually, melatonin is also involved in additional effects of acupunctures in various other conditions (Zhou et al. 2001; Spence et al. 2004; Li et al. 2008; Nordio and Romanelli 2008; Cakmak 2009). In the rat model of seizures induced by intrahippocampal injection of penicillin, we observed that melatonin levels were elevated in pineal gland and hippocampus, but no change was seen in the serum during the early phases of seizures, and then a significant elevation was found with seizure crisis lasting. The increased level of melatonin was further enhanced with 30 min electroacupuncture stimulation at Fengfu (GV-16) and Jinsuo (GV-8), which was accompanied by a significant improvement of epileptic EEG and seizure behaviors (Chao et al. 2001). The upregulation of melatonin levels in the brain and blood by acupuncture was also observed in cerebral ischemia and neuropsychiatric disorder, in which neuronal damage and bad neurological performance were improved by acupuncture treatment (Spence et al. 2004; Li et al. 2008). Since melatonin is considered as an antistressor and a natural downregulator of epileptiform activity (see review by Chao and Cheng 2001a), we postulate that the elevation in melatonin levels during seizures is likely to be an endogenous protective mechanism that counteracts convulsions and seizure-induced stress, and further elevation of melatonin levels with electroacupuncture treatment suggests that melatonin might be one of the possible mediators of electroacupuncture antiepilepsy (Chao et al. 2001).

Besides, pituitary-adrenal axis hyperactivity is observed in patients with clinically active epilepsy (Marek et al. 2010), and the serum concentrations of ACTH and cortisol during sleep seizures are linked with pre-ictal and ictal EEG changes in epilepsy patients (Zhang and Liu 2008). Corticotropin-releasing factor (CRF) can influence neuronal connectivity and function of immature hippocampus and increase the neuronal excitability, and thus plays a role in human developmental epilepsies (Hollrigel et al. 1998; Chen et al. 2004). It has been reported that the disrupted plasma levels of cortisol and other hormones (estradiol, progesterone, and prolactin) can be improved by acupuncture treatment in patients with epilepsy (Chen et al. 2008). In addition, electroacupuncture has been shown to effectively downregulate increased serum cortisol levels, hypothalamic CRF mRNA, and pituitary ACTH mRNA expression and upregulate decreased hypothalamic glucocorticoid receptor mRNA expression in rats with cerebral ischemia-perfusion injury (Cai et al. 2009). These effects of acupuncture could contribute to the alleviated brain injury and poststroke seizures and epilepsy (Mao and Guo 2005; Cai et al. 2009).

Therefore, some hormones and humoral factors are also involved in the antiepileptic effects of acupuncture.

Neurotrophic Factors and Cytokines

Neurotrophic factors, also called neurotrophins, are a class of growth factors secreted by nerve cells, including nerve growth factor, brain-derived neurotrophic factor, and neurotrophin-3, neurotrophin-4, etc. Substantial evidence has indicated that neurotrophins have dual roles in affecting seizure susceptibility (Binder et al. 2001; Koyama and Ikegaya 2005; Simonato and Zucchini 2010; Terauchi et al. 2010).

About 60% of epilepsy is acquired epilepsy that is largely caused by severe brain damage such as stroke, traumatic brain injury, and status epilepticus (Delorenzo et al. 2005). Reactive glial activity and neuroinflammatory reaction are the most common features of the injured brain. During neuroinflammatory reaction, reactive glial cells produce and release a large amount of proinflammatory cytokines (such as IL-1, IL-6, TNF-alpha, etc.) and upregulate the expression of the receptors of the cytokines, leading to hyperexcitability in the injured brain regions and consequently spontaneous recurrent epilepsy (see reviews by Choi and Koh 2008; Vezzani et al. 2008; Rao et al. 2009).

So far, there is no direct evidence available on the role of neurotrophins and cytokines in acupuncture antiepilepsy. However, some indirect evidence suggests the involvement of neurotrophins and cytokines in the acupunctural protection against neuronal injury and brain damage-induced seizures and epileptogenesis. Growing evidence has indicated that neurotrophins are the main mediators of acupuncture in neural activity, plasticity, and repair following neurodegeneration and/or traumatic injury (see reviews by Du 2008; Manni et al. 2010). Acupuncture can prevent neuronal death induced by cerebral ischemia and other injuries, and increase neuronal survival by regulating the expression of neurotrophins (Ou et al. 2001; Liang et al. 2002; Yi et al. 2006; Gao et al. 2006). On the other hand, acupuncture can decrease inflammatory reaction and neuroinflammation-induced neuronal injury and death and improve the functional recovery by downregulating the proinflammatory cytokines in the brain suffered from injuries like stroke (Xu et al. 2002; Liu et al. 2004b; Kong et al. 2007b; Li and Lai 2007; Zhou et al. 2009). Thus, these findings put forth the possibility that acupuncture regulation of neurotrophins and cytokines could indirectly contribute to its antiepileptic effects on the epilepsy secondary to cerebral ischemic damage (Mao and Guo 2005).

Other Possible Biomediators

Besides the above-discussed biomediators, some other factors such as second messenger systems (cAMP, cGMP), immediate early genes (IEG), as well as acetylcholine (Ach) system are also involved in the antiepileptic effects of acupuncture.

Cyclic AMP and cGMP are the second messengers that are important in many biological processes. These two kinds of second messengers as well as their downstream signaling molecules (e.g., protein kinases) are the major modulators of synaptic transmission and are involved in molecular and cellular events leading to epileptogenesis (Ferrendelli et al. 1980; Folbergrová 1980; Yokoyama et al. 1989; Boulton et al. 1993; Higashima et al. 2002; Vázquez-López et al. 2005; Ure and Altrup 2006). For example, increased cAMP and cGMP levels in the brain are consistently observed in all epilepsy animals, and a stimulated increase in cAMP can, in turn, induce epileptic discharges and seizures. Cyclic GMP is always involved in the dual roles of NO in epilepsy. Some AEDs such as carbamazepine and valproic acid have been found to inhibit the activity of adenylyl cyclase (an enzyme catalyzing the conversion of ATP to cAMP), decrease cAMP levels, and inhibit the downstream effects (e.g., PKA-dependent long-term potentiation of mossy fiber to CA3 pyramidal cell transmission in the hippocampus) (Chen et al. 1996; Chang et al. 2010). Acupuncture has been shown to play its antiepileptic role by modulating the levels of cAMP and cGMP in the brain. In pentylenetetrazole-kindled rats, the levels of cAMP and cGMP in the hippocampus increased greatly during epileptic seizures, and a pretreatment with electroacupuncture at Baihui (GV-20), Fengchi (GB-20), and Fengfu (GV-16) 30 min before the last i.p. injection of pentylenetetrazole significantly attenuated the increase in cAMP and cGMP, and the most prominent decrease in cAMP levels was found in 30 min after electroacupuncture treatment and that of cGMP occurred in 30–60 min after electroacupuncture treatment (Yang et al. 2002b; 2008). Intracerebroventrical injection of cAMP elicited high-amplitude and high-frequency epileptic discharges in the cortex, and these discharges were inhibited by electroacupuncture stimulation at bilateral Zusanli (ST-36) (Lou 1989). These findings suggest that cAMP and cGMP are actively involved in the execution of the antiepileptic effects of acupuncture.

Immediate early genes (IEGs) are genes activated transiently and rapidly in response to a wide variety of cellular stimuli. They represent a standing response mechanism that gets activated at the transcription level in the first round of response to stimuli before any new proteins are synthesized and are therefore called the "gateway to the genomic response." The earliest known and best-characterized IEGs include c-fos and c-jun genes. The expression of c-fos and c-jun has been shown to be greatly increased during seizures and acts as molecular switches turning on a plasticity (kindling) program in neurons in a fashion similar to their induction of developmental programs in the dividing cells, thereby contributing to the development of spontaneous seizures (see reviews by Dragunow et al. 1989; Kiessling and Gass 1993). Whether the increased expression of c-fos and c-jun is neuroprotective or detrimental to the brain is not very clear. However, a relationship between the increased expression of c-fos and c-jun and hippocampal neuronal death induced by kainic acid is proposed (Smeyne et al. 1993; Lee et al. 2001; Kim et al. 2008b). Some AEDs change seizure-induced c-fos and c-jun mRNA expression. For example, valproic acid suppresses, while lamotrigine enhances, the seizure-induced c-fos and c-jun mRNA expression in rats despite a similar anticonvulsant effect of these two AEDs (Szot et al. 2005). Acupuncture can regulate IEGs in various conditions including cerebral ischemia and epilepsy (Medeiros et al. 2003; Jiang et al. 2004; Lee et al. 2004, 2010a). Acupuncture can modulate seizure-induced expression of c-fos and c-jun and prevent hippocampal neuronal death along with an alleviation of seizure behavior (Wang and Cheng 1994b; Kim et al. 2008b). In penicillin-induced seizures, the expression of c-fos protein was evident in several seizure-sensitive regions including hippocampus (CA1, CA3, and dentate gyrus), piriform cortex, dorsal part of entorhinal cortex, and amygdala, and the electroacupuncture treatment further enhanced the expression of c-fos protein in above-mentioned brain regions, except hippocampal CA1 subfield where c-fos expression was suppressed by electroacupuncture (Wang and Cheng 1994b). In contrast, the acupuncture treatment in kainic acid-induced epileptic rats completely abated the increased expression of c-fos and c-jun in hippocampal CA1 and CA3 areas. Meanwhile, seizure behavior and seizure-induced hippocampal cell death were also suppressed by acupuncture (Kim et al. 2008b). The reasons for this difference in effects of acupuncture on c-fos expression at the same brain regions (CA1 and CA3 area) between these two models are not known but could reasonably be ascribed to the differences in acupunctural methods (manual vs. electroacupuncture), choice of acupoints (HT-8 vs. GV-16 and GV-8), acupuncture parameters used, etc. Despite this, acupuncture implements antiepileptic effect through regulation of the IEGs expression.

Both increased acetylcholine (ACh) transmission and mutations in nicotinic acetylcholine receptor subunits are causally linked to certain kinds of seizures (see reviews by King and LaMotte 1989; Savolainen and Hirvonen 1992; Okada 2010). Acupuncture can decrease the ACh level, inhibit neural pathway of cholinergic transmission, and enhance the activity of cholinesterase (an enzyme that catalyze the hydrolysis of acetylcholine into choline and acetic acid), resulting in a reduction in epileptiform discharges in an epileptic brain (Yang and Cheng 2010). Therefore, cholinergic system also contributes to the antiepileptic effect of acupuncture.

6.3.2.3 Acupuncture Prevention of Seizure-Induced Brain Injury

Despite the differences in the incidence and degree of seizure-induced brain injury in various types of epilepsy, substantial evidence in animal models (Cavazos et al. 1994; Sloviter et al. 1996; Bengzon et al. 1997; Fujikawa et al. 1999; Kondratyev and Gale 2004; Chen et al. 2010; Zhang et al. 2010) and humans (Mathern et al. 2002; Scott et al. 2003; Swartz et al. 2006; Choi et al. 2009) indicates that seizures induce neuronal damage via a diverse cellular and molecular mechanisms. The most overt form of seizure-induced brain damage is neuronal death, including necrosis and apoptosis (Kondratyev and Gale 2004). Human mesial temporal lobe epilepsy (MTLE) is always accompanied by hippocampal sclerosis, which is caused by neuronal loss and active gliosis in the hilus of the dentate gyrus, CA1, and CA3 hippocampal subfields (Mathern et al. 2002; Chandra et al. 2010; Marucci et al. 2010; Alonso-Nanclares et al. 2011). Animal models also provide convincing evidence that seizures can lead to neuronal death and loss (Cavazos et al. 1994; Sloviter et al. 1996; Bengzon et al. 1997; Fujikawa et al. 1999; Kondratyev and Gale 2004; Chen et al. 2010; Zhang et al. 2010). It is seizures per se that induce neuronal injury/death and loss. For instance, in an intra-amygdala injection of kainic acid-induced seizure model, neuronal damage was not only limited to the injected amygdala but also extended to CA1 and CA3 areas of the ipsilateral hippocampus and a small number of ipsilateral cortical neurons (Mouri et al. 2008). Obviously, the damage in the hippocampus and cortex in this model is seizure-related rather than a result of diffusion of kainic acid from the injection site. In a febrile seizure model, neuronal injury in the central nucleus of the amygdala and hippocampal CA3 and CA1 pyramidal cell layers was very prominent in rats with hyperthermic seizures but was completely prevented in rats in which the hyperthermic seizures were prevented by the pretreatment with pentobarbital before hyperthermia induction (Toth et al. 1998). It is widely accepted that prolonged seizures (e.g., status epilepticus) can produce irreversible neuronal death (Sloviter 1999; Fujikawa 2005; Chen et al. 2010; Zhang et al. 2010), but recurrent brief seizures or even single brief events can also induce neuronal death and loss (Cavazos et al. 1994; Sutula et al. 1994; Bengzon et al. 1997; Norwood et al. 2010). Inhibitory interneurons (e.g., GABAergic basket cells, somatostatinergic neurons in the hippocampus) are more vulnerable to injury, and seizures can lead to the loss of the inhibitory interneurons, causing reduced tonic inhibition in the seizure-sensitive areas such as the cortex, hippocampus, and amygdala (Prince and Connors 1984; Dudek and Suluta 2007; Knopp et al. 2008; Fritsch et al. 2009; Chen et al. 2010; Zhang et al. 2010). Seizure-induced neuronal loss is more profound in mature brain compared to immature brain because the latter is more resistant to certain mechanisms of brain damage due to a number of unique metabolic features of immature brain (Wasterlain and Shirasaka 1994).

Besides neuronal death and loss, seizures can result in some other neuropathological alterations in the brain, for example, dendritic abnormalities (Swann et al. 2000; Wong 2005), altered neurogenesis (Bengzon et al. 1997; Parent 2002; Smith et al. 2005; Hattiangady and Shetty 2008), aberrant axonal sprouting of DGCs (Dudek and Sutula 2007), and reactive gliosis (Benkovic et al. 2004; Mouri et al. 2008). In human epilepsy and animal models, epileptic foci in the neocortex and hippocampus show a variety of dendritic abnormalities, most commonly a loss of dendritic spines and sometimes changes in the dendritic length, shape, and branching patterns (Swann et al. 2000; Wong 2005, 2008). These changes in dendritic spines, particularly in the hippocampus, are closely linked to learning and memory dysfunction in the epilepsy patients (Wong 2005). Normal neurogenesis (the birth of new neurons) in the hippocampal dentate gyrus (DG) and forebrain subventricular zone (SVZ) occurs throughout life in all species, including humans, and exhibits a close relationship with the function of learning and memory (Hattiangady and Shetty 2008). However, altered neurogenesis is a hallmark in hippocampal subgranular zone of DG and in SVZ in a variety of seizure models of adult animals (chemoconvulsant-induced status epilepticus, electrical stimulation-induced kindling model) and human TLE after an IPI such as brain injury (due to stroke, brain trauma, perinatal asphyxia, in utero hypoxia, etc.), brain tumors, meningitis, encephalitis, status epilepticus, and febrile seizures during childhood (Parent 2002; Hattiangady and Shetty 2008) (Fig. 6.2). The altered neurogenesis exhibits a dramatic increase in the production of new neurons in DG during the early phase after IPI. This was followed by a substantial decrease in new neuron generation (up to 64–81% reduction below the baseline level) weeks or months after IPI when chronic TLE developed (Parent 2002; Hattiangady and Shetty 2008) (Fig. 6.5). Enhanced neurogenesis is always accompanied with increased neuronal death. However, the early enhancement of dentate gyrus neurogenesis is demonstrated to be evoked by brief seizure activity alone in the absence of neuronal death or by even a single afterdischarge induced by hippocampal stimulation in adult rodents (Bengzon et al. 1997; Smith et al. 2005). Seizure activities stimulate neurogenesis in two possible ways. One is the direct stimulation of mitotic activity by electrical activation and the other is indirect enhancement of neurogenesis through injury-associated mechanisms (Parent 2002). The altered neurogenesis leads to structural abnormality in the epileptic hippocampal formation. In the adult rodent pilocarpine or KA model of TLE, a substantial fraction of newly generated cells in subgranular zone ectopically migrate to either dentate granule cell (DGC) layers or dispersed in the hilus and inner molecular layers that are normally absent of granule-like neurons (the phenomenon of DGC dispersion in epilepsy) (Parent 2002; Hattiangady and Shetty 2008). Most of these ectopically migrated newborn DG cells and almost all of the newly generated cells that ectopically migrated from forebrain SVZ to hippocampal CA1 and CA3 subfield differentiated into glial cells (vs. neurons in normal age-matched control) (i.e., the phenomenon of active gliosis). This is

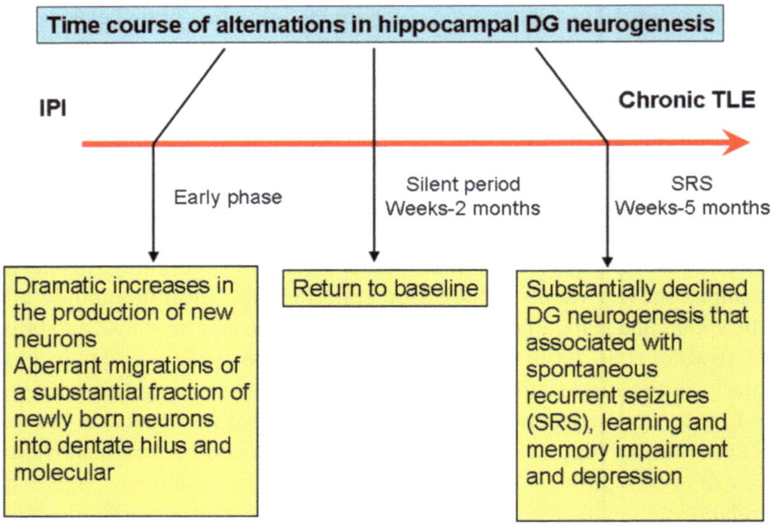

Fig. 6.5 Altered neurogenesis in hippocampal dentate gyrus during epileptogenesis after initial precipitating brain injury

responsible, to a considerable degree, for the hippocampal sclerosis in chronic TLE patients (Parent 2002; Hattiangady and Shetty 2008) (Fig. 6.2). New granular-like neurons together with mature granule cells in granule cell layer are largely responsible for the aberrant outgrowth or sprouting of axons to atypical locations (axonal reorganization) (Frotscher et al. 2006; Dudek and Sutula 2007) that contributes to the injury-induced hyperexcitability in the hippocampal CA3 pyramidal cells and subsequent chronic TLE (McNamara 1994; Dudek and Suluta 2007) (Figs. 6.2 and 6.5). DGC dispersion due to the aberrant migration of DGC precursors is another structural abnormality in epileptic hippocampal formation. Unlike the normal mature granule cells in DGC layer with apical dendrites directly projecting into CA3 pyramidal neurons, ectopic newborn hilar granule cells exhibit persistent basal dendrites (a feature believed to contribute toward the establishment of recurrent excitatory circuitry), and the remarkably increased number of excitatory synapses on dendrites of hilar ectopic granule-like neurons in epileptic rats (dendritic remodeling) (Ribak et al. 2000) displays hyperexcitability in the form of abnormal burst firing occurring synchronously with CA3 pyramidal cells. This also contributes to spontaneous recurrent seizures in chronic TLE (Parent 2002; Hattiangady and Shetty 2008). Substantially decreased neurogenesis in chronic epilepsy could contribute to the persistence of spontaneous seizures, cognitive deficits, and depression as seen in TLE patients (Hattiangady and Shetty 2008). It has been reported that only a small number of newly generated cells (14%) in DG differentiate into inhibitory GABAergic basket neurons in the hippocampal neuronal network (Liu et al. 2003). A decrease in the production of this kind of neurons, as seen in patients and animal models of TLE, undoubtedly leads to hyperexcitability in neuronal network and seizures (McNamara 1994; Dudek and Suluta 2007). Substantial evidence has indicated a direct relationship between the maintenance of hippocampus-dependent learning and long-term memory and continuous integration of newly generated DG cells into DGC networks in the adults (Shors et al. 2001; van Praag et al. 2002; Aimone et al. 2006; Kee et al. 2007). Thus, seizure-associated alteration in DGC neurogenesis and subsequent dysfunction of circuits necessary for learning and memory (acquisition and retrieval) may cause cognition deficits found in chronic TLE patients (Parent 2002; Hattiangady and Shetty 2008). In general, seizures may result in brain injury and abnormalities in various forms, including neuronal death/loss, altered neurogenesis, abnormal axonal reorganization and dendritic remodeling, dispersion of granule cell layer, and active gliosis.

Acupuncture has been demonstrated to decrease seizure-induced neuronal death, increase the survival of neurons in seizure-sensitive brain regions, reduce supragranular mossy fiber sprouting, modulate neurogenesis in hippocampal dentate gyrus, prevent atrophy of some brain structures that are prone to seizures, and improve the cognitive deficits in kainate- or pilocarpine-induced rodent seizures with the features similar to human TLE. For example, Kim et al. (2008b) showed in kainate-induced seizures that acupuncture at Shaofu (HT-8) greatly decreased kainate-induced expression of c-fos and c-jun, and the neuronal death in the CA1 and CA3 area of hippocampus in mice. Histological analysis with Nissl and TUNEL staining in pilocarpine-induced rat model revealed that electroacupuncture at acupoints of Baihui (GV-20), Dazhui (GV-14), Yaoshu (GV-2), and Yin Tang (M-HN-3) as well as Zusanli (ST-36) eliminated seizure-induced neuronal apoptosis in the cerebral cortex and hippocampal CA1 area and also abolished the tissue shrinkage in dorsal hippocampus, basolateral nucleus of the amygdala, substantia nigra, and perirhinal cortex that is accompanied with the improvement of cognitive deficits manifested as good performance in both learning and memory (acquisition and retention) (dos Santos et al. 2005; Wang et al. 2006b; Tang et al. 2008). In rats with status epilepticus, therapy with catgut implantation at acupoints was reported to significantly inhibit seizure-induced apoptosis of hippocampal neurons (Chen et al. 2006d). Taurine-deficient rats were more susceptible to kainic acid-induced epilepsy and cell death in the CA3 and dentate gyrus area of the hippocampus, and electroacupuncture stimulation at Renzhong (GV-26) and Yongquan (KI-1) greatly suppressed epileptic seizures and neuronal death in the hippocampus (Jin et al. 2005a). Mossy fiber sprouting in the dentate gyrus supragranular layer of the hippocampus (as manifested by Timm's staining) was seen to be significantly reduced by electroacupuncture stimulation at bilateral Zusanli (ST-36) in rats with pilocarpine-induced epilepsy (Guo et al. 2008c). Along with these changes secondary to acupuncture stimulation, the seizure behaviors and/or epileptic EEG were also significantly decreased (Wang et al. 2006b; Kim et al. 2008b; Guo et al. 2008c). In addition to the alleviation of seizures, acupuncture also seems to modify the neurogenesis in hippocampal dentate gyrus. By using bromodeoxyuridine (BrdU) labeling and immunohistochemistry method, electroacupuncture stimulation at Dazhui (GV-14) and Baihui (GV-20) was reported to significantly reduce BrdU-positive cells (newborn cells) in hippocampal CA1, CA3, and dentate gyrus subfields in rats with spontaneous recurrent seizures after 2 weeks of seizures (Wen et al. 2006), and some of the newborn cells in the hippocampus were somatostatin immunopositive (Wen and Huang 2008). Therefore, the authors concluded that electroacupuncture inhibits epileptogenesis by decreasing the number of newborn neurons that express somatostatin (Wen et al. 2006; Wen and Huang 2008). Strangely, another report from the same group showed in the same model that electroacupuncture efficiently increased the proliferation of neural precursor cells in the dentate gyrus 7 and 14 days after epileptic seizures (Wang et al. 2006a). This controversy in part could be attributed to the differences over the choice of acupoints (Dazhui and Baihui by Wen et al. vs. Dazhui only by Wang et al.), electrical parameters applied (50 Hz and 20 mA for 20 min, 20 min/time with 2 times/daily from day 2 to day 5 after i.p. injection of pilocarpine by Wang et al. vs. disperse-dense waves with 80 Hz in frequency and 5 mA in intensity, for 20 min/day at 6, 13, 27 days after i.p. pilocarpine injection, or alternatively 6.25 Hz and 1.8–4.9 mA with the duration of 2.08 s for dense wave and 3.85 Hz and 1.4–2.0 mA with the duration of 1.28 s for disperse wave, 20 min daily from 3 to 28 day after injection of pilocarpine by Wen et al.), and animals used (180–200 g by Wen et al. vs. 200–250 g by Wang et al.). However, except this, the results reported by Wen et al. (2006) and Wang et al. (2006b) are still not concrete enough because of the obscurity in description of methods and results in their papers. Nevertheless, the modification of cell proliferation and differentiation by acupuncture stimulation is substantially manifested in both healthy (Hwang et al. 2010) and injured brain (Kim et al. 2001; Yang et al. 2005; Cheng et al. 2009) as well as injured spinal cord (Yan et al. 2011). However, it should be pointed out that the modification of neurogenesis with acupuncture stimulation is not the endpoint for the beneficial effects of acupuncture on seizure-induced neuronal injury. The key process in acupuncture treatment is the prevention of the aberrant proliferation and differentiation of the newborn cells, lowering the subsequent aberrant axonal reorganization and dendritic remodeling, and decreasing the aberrant synaptic plasticity of these newborn neurons. In this respect, it is reported that electroacupuncture can suppress the mossy fiber sprouting in the dentate gyrus supragranular layer and probably improve the synaptic plasticity in pilocarpine-epileptic rats (dos Santos et al. 2005; Guo et al. 2008c), and accelerate the differentiation of newborn cells into neurons (but not glias) and their functional maturation in subgranular zone of dentate gyrus in healthy animals (Hwang et al. 2010) as well as that of transplanted stem cells in injured spinal cord (Yan et al. 2011). Stabilizing brain structure has been suggested to be a novel therapeutic approach for epilepsy (Wong 2008); therefore, these changes are essentially relevant to the decreased seizures and brain injury-related secondary epileptogenesis for chronic epilepsy and the improvement in cognitive deficits conferred by acupuncture stimulation (dos Santos et al. 2005; Wang et al. 2006b; Cheng et al. 2008; Guo et al. 2008c; Hwang et al. 2010).

Ischemic brain injury largely accounts for the poststroke seizures and epilepsy with the frequency of early postischemic seizures (which resulted from cellular biochemical dysfunction, e.g., increased extracellular glutamate level) ranging from 2% to 33% and that of late postischemic seizures (which are caused by gliosis and the development of a meningocerebral cicatrix, e.g., changes in membrane properties, deafferentation, selective neuronal loss, and collateral sprouting) ranging from 3% to 67%, with an overall rate of postischemic stroke epilepsy of 2–4% (Camilo and Goldstein 2004). Substantial evidence has indicated that acupuncture is neuroprotective against ischemic brain injury and the resultant brain dysfunction (Kim et al. 2001; Ou et al. 2001; Wu et al. 2001; Jang et al. 2003; Yang et al. 2005; Yi et al. 2006; Chen et al. 2009; Chuang et al. 2007; Cheng et al. 2009; Liu et al. 2009c; Luo et al. 2009; Shen et al. 2009; Wang et al. 2009; Zhou et al. 2009; Kang et al. 2010). Thus, the neuroprotection of acupuncture against ischemic brain injury may also contribute to the antiepileptic effect of acupuncture on poststroke seizures and epilepsy. The neuroprotective effects of acupuncture on ischemic brain injury include reduced ischemic neuronal death (Jang et al. 2003; Pang et al. 2003; Wang et al. 2009; Kang et al. 2010), enhanced neuronal proliferation, differentiation and survival in ischemic brain regions (Kim et al. 2001; Yang et al. 2005; Cheng et al. 2009), increase in the expression of signal molecules beneficial to neuronal survival and decrease in those promoting cell death (Jang et al. 2003; Ou et al. 2001; Chen et al. 2009), inhibition of the inflammatory reaction and active gliosis (Liu et al. 2009c; Luo et al. 2009; Zhou et al. 2009), and the maintenance of normal brain structures and functions (Wu et al. 2001; Yi et al. 2006; Chuang et al. 2007; Peng et al. 2007; Shen et al. 2009). It should be pointed out here that the neuroprotective effect of acupuncture on ischemia-induced brain inflammatory reaction and active gliosis, as well as brain structures/functions, may be of high relevance to poststroke seizures and epilepsy. It has been demonstrated that acupuncture stimulation markedly improves the blood supply to cerebral tissues (Shi et al. 1998; Pang et al. 2003; Chen et al. 2006b; Hsieh et al. 2006; Zhou et al. 2011), which can alleviate hypoxia-/ischemia-associated neuronal injury and inhibit ischemia-induced swelling of astrocytes and the increase in the number of astrocytes as well as glia-related inflammatory reactions in the brain (Liu et al. 2009c; Luo et al. 2009; Zhou et al. 2009). In addition, acupuncture stimulation prevents ischemic brain atrophy (Chuang et al. 2007), protects synaptic ultrastructure in ischemic cerebral cortex, maintains the integrity of blood–brain barrier (BBB), prevents extravasation in ischemic brain (Wu et al. 2001; Yi et al. 2006; Chuang et al. 2007; Peng et al. 2007; Shen et al. 2009), and improves memory impairment in cerebral multi-infarction (Wang et al. 2009). As we have discussed in the preceding section, astrocytic dysfunction-related disruption of glutamate/K^+ homeostasis as well as inflammatory reaction is one of the most important factors of epileptogenesis (also see reviews by Wetherington et al. 2008; Seifert et al. 2010). Disruption of BBB also plays an important role in the pathogenesis of epilepsy (Marchi et al. 2007; van Vliet et al. 2007; Tomkins et al. 2008). Therefore, the effects of acupuncture on the structural/functional changes in ischemic brain are likely to contribute to the antiepileptic effects of acupuncture on poststroke seizures and epilepsy. The clinical observation that acupuncture combined with medicine significantly attenuates the epilepsy secondary to cerebral infarction and improves the side effects of medical treatment (Mao and Guo 2005) supports the above-stated fact.

The clinical and experimental benefits of acupuncture against neuronal injury have been demonstrated not only in epilepsy and cerebral ischemia but also in a variety of other neurological diseases, such as Parkinson's disease (Liang et al. 2002; Park et al. 2003; Liu et al. 2004b; Jia et al. 2009), Alzheimer's disease (Zhou and Jin 2008; Xue et al. 2009; Shen et al. 2010), spinal cord injury (Wong et al. 2003; Zhao et al. 2008; Choi et al. 2010; Yan et al. 2011), and amyotrophic lateral sclerosis (Wasner et al. 2001; Yang et al. 2010). Though the mechanisms underlying the brain injuries in different types of neurological diseases are not consistently the same, acupuncture treatment seems to share some common neuroprotective mechanisms against brain injury in these diseases. No matter what kind of brain injury it is, acupuncture offers neuroprotection to the brain, and this common mechanism could be the underlying antiepileptic effect of acupuncture treatment relevant to brain injury-induced seizures and epileptogenesis.

6.3.2.4 Acupuncture Regulation of Ion Channel Activity

As we have discussed before, abnormal alterations in the activities (opening, closing, and inactivation), expression, and even the structures of ion channels may be responsible for the hyperexcitability of neurons and the consequent epilepsy. Therefore, the conclusion that the antiepileptic effect of acupuncture could involve the regulation of ion channel activities seems reasonable. This hypothesis is based on indirect evidence.

Sodium channel mutation has a known association with human epilepsy (Mantegazza et al. 2005; Yamakawa 2006; Xu et al. 2007a, b; Catterall et al. 2010; Escayg and Goldin 2010; Meisler et al. 2010). Upregulation of Na^+ channels has been demonstrated to result in epileptic hyperexcitability and seizures (Sashihara et al. 1992; Agrawal et al. 2003; Xia et al. 2003). Interestingly, we observed that in the mutant brain exhibiting spontaneous epilepsy, Na^+ channel is upregulated (Xia et al. 2003), while δ-opioid receptor (DOR) is downregulated (Zhao et al. 2005), suggesting a potential role of DOR impairment in the pathophysiology of epilepsy associated with genetic abnormality. Indeed, there is strong evidence suggesting that DOR activation is anticonvulsant in both animals and humans (Lee et al. 1984; Tortella and Long 1985; Koide et al. 1992; Madar et al. 1997) though there are some controversial observations also being reported (Danielsson et al. 2006; Jutkiewicz et al. 2006). In some experimental models of epilepsy, endogenous enkephalin levels are markedly increased in the brain with epileptic seizures (Tortella and Long 1985; Madar et al. 1997), and opioid receptors are upregulated following spontaneous epileptic seizures (Hammers et al. 2007), which may represent a compensatory mechanism against epilepsy. In addition, our studies with co-expression of Na^+ channels and DOR in *Xenopus* oocytes directly demonstrated that activation of DOR inhibits Na^+ channel function by decreasing the amplitude of sodium currents and increasing its threshold for activation (Kang et al. 2009). Our work also indicated that DOR activation stabilizes hypoxic/ischemic disruption of Na^+–K^+ homeostasis chiefly mediated by Na^+ channels in the cortex (Chao et al. 2007a, b, 2008, 2009, 2012; Kang et al. 2009; also see review by Chao and Xia 2010). It is also well known that disrupted homeostasis of electrolytes inside and outside the cells (e.g.,

Na^+, Ca^{2+}, K^+, and Cl^-) is one of the most common phenomena during seizures, and a lasting high extracellular K^+ could compromise normal neuronal firing, increase neuronal excitability, and induce epileptiform activity and the onset of seizures (Somjen and Müller 2000; Semyanov and Godukhin 2001; Gabriel et al. 2004; Ivens et al. 2007; Florence et al. 2009; Lopantsev et al. 2009). Therefore, these findings imply a positive role of DOR in controlling seizures by interacting with Na^+ channels. More interestingly, we along with others observed that electroacupuncture protects the brain from ischemic injury via DOR system (Zhao et al. 2002; Xiong et al. 2007; Tian et al. 2008a, b). Since ischemia can induce stroke-related epilepsy (Camilo and Goldstein 2004), and Na^+ channels play a major role in neuronal excitability including epileptic hyperexcitability and most of AEDs are actually inhibitors of Na^+ channels (Remy et al. 2003; Armijo et al. 2005), and since acupuncture can activate opioid system in the brain (Han 2003; Wang et al. 2008) either by accelerating the production of endogenous enkephalins (Yuan and Han 1985) or by upregulating DOR expression (Tian et al. 2008a), and DOR activation, in turn, can inhibit Na^+ channels and stabilize Na^+–K^+ homeostasis, we hence believe that these findings provide a novel acuity to the acupuncture therapy for epilepsy.

6.4 Concluding Remarks

Acupuncture is widely practiced in the oriental countries (e.g., China, Japan, and Korea) and immigrant communities of oriental origin in the West. Over the past two decades, acupuncture has grown in popularity in the United States and Europe. In the Western medicine, the focus of acupuncture is in pain control (pain relief), particularly postoperative pain and cancer pain. Therefore, a huge gap exists between the oriental acupuncturists and the Western physicians in practice of acupuncture, owing to lingering recognition and acceptance of the theory of acupuncture treatment, the unavailability of high-quality papers written in English, etc. The theories on acupunctural treatment of epilepsy in TCM have a more philosophical rather than biological approach. Some of the theories and concepts (e.g., *Yin-Yang* theory, five elements theory, meridian and collateral theory, internal wind invasion) aim at an integrative diagnosis rather than specific one based on the pathology of diseased organs. Therefore, acupuncture treatment in TCM aims at harmonizing a disturbed energy balance. Though the TCM theories and the treatment philosophy of acupuncture are unique, they are not easily understood by Western scientists and physicians who pursue specific diagnoses and treatment. In addition, most of the existing good case reports and useful experiences are written and communicated in Chinese, thus making it inaccessible to the Western peers. All these hinder the acceptance and spread of acupuncture treatment for epilepsy in the Western medicine. Therefore, it is of urgent need to review the existing literature in a better way to communicate the experiences between oriental and Western community of acupuncturists and physicians.

In this chapter, we have discussed, in modern scientific language, pathophysiological genesis of epilepsy, the clinical application of acupuncture, and the potential mechanisms underlying acupunctural treatment of epilepsy. Epileptic seizures and epileptogenesis are complex processes that involve the altered excitability of neurons in the brain. As a modality and interventional procedure and a complex somatosensory stimulation, acupuncture influences the neuronal activity and intrinsic connectivity via multiple mechanisms and at multiple levels in order to normalize the disrupted neuronal activities in a bidirectional way and thus lower brain hyperexcitability and alleviate/control epilepsy. Both animal studies and clinical observations support that acupuncture is a safe, efficient, and convenient modality with minimal side effects for the treatment of epilepsy.

However, some issues still prevail. One of the biggest disputes in acupuncture treatment, as we have discussed before (see Sect. 6.2.4 in Sect. 6.2), is on the methodology for the clinical assessment. In this respect, more reliable, detailed, and comprehensive assessment standards are absolutely necessary. This requires extensive efforts on part of acupuncturists and physicians. Future research and clinical treatments with acupuncture also need to reasonably set up control without bias (e.g., selection criteria, sham needle device, placebo control), restrictively define patient population (e.g., criteria for investigation, the disease history, pretreatment used), and follow up patients in the long term. Because acupuncture is a practice requiring strong personal experiences on behalf of acupuncturists/physicians and the patients, the outcomes of acupuncture treatment largely depend on the acupuncture techniques (e.g., diagnoses accuracy, acupoint selection, stimulation method, treatment sessions and durations, proficiency, clinical experiences) adopted by the acupuncturists/physicians and belief and cooperation of patients. These factors increase the difficulties of understanding the acupuncture treatment. Therefore, more efforts are also required to share the experiences by both the acupuncturists/physicians and the patients.

Acknowledgements This work was supported by the National Institutes of Health (HD-034852 and AT-004422) and the Vivian L Smith Neurologic Foundation.

References

Abbott NJ, Rönnbäck L, Hansson E (2006) Astrocyte–endothelial interactions at the blood–brain barrier. Nat Rev Neurosci 7:41–53.

Aghakhani Y, Bagshaw AP, Bénar CG, Hawco C, Andermann F, Dubeau F, Gotman J (2004) fMRI activation during spike and wave discharges in idiopathic generalized epilepsy. Brain 127:1127–1144.

Agrawal N, Alonso A, Ragsdale DS (2003) Increased persistent sodium currents in rat entorhinal cortex layer V neurons in a post-status epilepticus model of temporal lobe epilepsy. Epilepsia 44:1601–1604.

Aimone JB, WilesJ, Gage FH (2006) Potential role for the adult neurogenesis in the encoding of time in new memories. Nat Neurosci 9:723–727.

Akula KK, Dhir A, Kulkarni SK (2008) Nitric oxide signaling pathway in the anti-convulsant effect of adenosine against pentylenetetrazol-induced seizure threshold in mice. Eur J Pharmacol 587:129–134.

Alexander GM, Godwin DW (2006) Metabotropic glutamate receptors as a strategic target for the treatment of epilepsy. Epilepsy Res 71:1–22.

Allen NJ, Bennett ML, Foo LC, Wang GX, Chakraborty C, Smith SJ, Barres BA (2012) Astrocyte glypicans 4 and 6 promote formation of excitatory synapses via GluA1 AMPA receptors. Nature 486:410–414.

Almeida RT, Duarte ID (2008) Nitric oxide/cGMP pathway mediates orofacial antinociception induced by electroacupuncture at the St36 acupoint. Brain Res 1188:54–60.

Alonso-Nanclares L, Kastanauskaite A, Rodriguez JR, Gonzalez-Soriano J, Defelipe J (2011) A stereological study of synapse number in the epileptic human hippocampus. Front Neuroanat 5:8.

Araque A, Parpura V, Sanzgiri RP, Haydon PG (1999) Tripartite synapses: glia, the unacknowledged partner. Trends Neurosci 22:208–215.

Armijo JA, Shushtarian M, Valdizan EM, Cuadrado A, de las Cuevas I, Adin J (2005) Ion channels and epilepsy. Curr Pharm Design 11:1975–2003.

Aroniadou-Anderjaska V, Brita Fritsch B, Qashu F, Braga MFM (2008) Pathology and pathophysiology of the amygdale in epileptogenesis and epilepsy. Epilepsy Res 78:102–116.

Asada H, Kawamura Y, Maruyama K, Kume H, Ding R, Kanbara N, Kuzume H, Sanbo M, Yagi T, Obata K (1997) Cleft palate and decreased brain γ-aminobutyric acid in mice lacking the 67-kDa isoform of glutamic acid decarboxylase. Proc Natl Acad Sci USA 94:6496–6499.

Aslan A, Yildirim M, Ayyildiz M, Güven A, Agar E (2010) Interaction of leptin and nitric oxide pathway on penicillin-induced epileptiform activity in rats. Brain Res 1321:117–124.

Avanzini G, Franceschetti S (2003) Cellular biology of epileptogenesis. Lancet Neurol 2:33–42.

Avramescu S, Timofeev I (2008) Synaptic strength modulation after cortical trauma: A role in epileptogenesis. J Neurosci 28:6760–6772.

Babb TL, Ying Z, Mikuni N, Nishiyama K, Drazba J, Bingaman S, Wyllie E, Wylie CJ, Yacubova K (2000) Brain plasticity and cellular mechanisms of epileptogenesis in human and experimental cortical dysplasia. Epilepsia 41(Suppl. 6):S76-S81.

Bagdy G, Kecskemeti V, Riba P, Jakus R (2007) Serotonin and epilepsy. J Neurochem 100:857–873.

Bai L, Qin W, Tian J, Dong M, Pan X, Chen P, Dai J, Yang W, Liu Y (2009) Acupuncture modulates spontaneous activities in the anticorrelated resting brain networks. Brain Res 1279:37–49.

Bai L, Tian J, Qin W, Pan X, Yang L, Chen P, Chen H, Dai J, Ai L, Zhao B (2007) Exploratory analysis of functional connectivity network in acupuncture study by a graph theory mode. Conf Proc IEEE Eng Med Biol Soc 2007:2023–2026.

Bai L, Yan H, Li L, Qin W, Chen P, Liu P, Gong Q, Liu Y, Tian J (2010) Neural specificity of acupuncture stimulation at pericardium 6: evidence from an FMRI study. J Magn Reson Imaging 31:71–77.

Baran H (2006) Alterations of taurine in the brain of chronic kainic acid epilepsy model. Amino Acids 31:303–3077.

Barnes PM, Powell-Griner E, McFann K, Nahin RL (2004) Complementary and alternative medicine use among adults: United States, 2002. Advance Data from Vital and Health Statistics No.343. Hyattsville, Maryland: National Center for Health Statistics, May 27, 2004.

Barres BA (2008) The mystery and magic of glia: A perspective on their roles in health and disease. Neuron 60:430–440.

Bartenstein PA, Duncan JS, Prevett MC, Cunningham VJ, Fish DR, Jones AK, Luthra SK, Sawle GV, Brooks DJ (1993) Investigation of the opioid system in absence seizures with positron emission tomography. J Neurol Neurosurg Psychiatry 56:1295–1302.

Bateman LM, Li CS, Lin TC, Seyal M (2010) Serotonin reuptake inhibitors are associated with reduced severity of ictal hypoxemia in medically refractory partial epilepsy. Epilepsia 51:2211–2214.

Bausch SB, Garland JP, Yamada J (2005) The delta opioid receptor agonist, SNC80, has complex, dose-dependent effects on pilocarpine-induced seizures in Sprague-Dawley rats. Brain Res 1045:38–44.

Baxendale S (2009) Seeing the light? Seizures and sunlight. Epilepsy Res 84:72–76.

Bazil CW (2003) Epilepsy and sleep disturbance. Epilepsy Bahav 41(suppl 2):S39–S45.

Beleza P (2009) Refractory epilepsy: A clinically oriented review. Eur Neurol 62:65–71.

Benarroch EE (2009) Astrocyte-neuron interactions: Implications for epilepsy. Neurology 73:1323–1327.

Bengzon A, Kokaia Z, Elmer E, Nanobashvili A, Kokaia M, Lindvall O (1997) Apoptosis and proliferation of dentate gyrus neurons after single and intermittent limbic seizures. Proc Natl Acad Sci USA 94:10432–10437.

Benkovic SA, O'Callaghan JP, Miller DB (2004) Sensitive indicators of injury reveal hippocampal damage in C57BL/6J mice treated with kainic acid in the absence of tonic-clonic seizures. Brain Res 1024:59–76.

Berman R, Negishi M, Vestal M, Spann M, Chung MH, Bai X, Purcaro M, Motelow JE, Danielson N, Dix-Cooper L, Enev M, Novotny EJ, Constable RT, Blumenfeld H (2010) Simultaneous EEG, fMRI, and behavior in typical childhood absence seizures. Epilepsia 51:2011–2022.

Bialer M (2006) New antiepileptic drugs that are second generation to existing antiepileptic drugs. Exp Opin Investig Drugs 15:637–647.

Binaschi A, Bregola G, Simonato M (2003) On the role of somatostatin in seizure control: clues from the hippocampus. Rev Neurosci 14:285–301.

Binder DK and Steinhäuser C (2006) Functional changes in astroglial cells in epilepsy. Glia 54:358–368.

Binder DK, Croll SD, Gall CM, Scharfman HE (2001) BDNF and epilepsy: too much of a good thing? Trends Neurosci 24:47–53.

Bing G, Wilson B, Hudson P, Jin L, Feng Z, Zhang W, Bing R, Hong JS (1997) A single dose of kainic acid elevates the levels of enkephalins and activator protein-1 transcription factors in the hippocampus for up to 1 year. Proc Natl Acad Sci USA 94:9422–9427.

Blümcke I (2009) Neuropathology of focal epilepsies: A critical review. Epilepsy Behav 15:34–39.

Blume WT (2007) Thalamus: the "inner chamber" reveals its secrets. Epilepsy Curr 8:123–124.

Blumenfeld H, Lampert A, Klein JP, Mission J, Chen MC, Riera M, Dib-Hajj S, Brennan AR, Hains BC, Waxman SG (2009) Role of hippocampal sodium channel Nav1.6 in kindling epileptogenesis. Epilepsia 50:44–55.

Böhme I, Lüddens H (2001) The inhibitory neural circuitry as target of antiepileptic drugs. Curr Med Chem 8:1257–1274.

Boison D (2009) Engineered adenosine-releasing cells for epilepsy therapy: Human mesenchymal stem cells and human embryonic stem cells. Neurotherapeutics 6:278–283.

Bordey A, Spencer DD (2004) Distinct electrophysiological alterations in dentate gyrus versus CA1 glial cells from epileptic humans with temporal lobe sclerosis. Epilepsy Res 59:107–122.

Boulton CL, McCrohan CR, O'Shaughnessy CT (1993) Cyclic AMP analogues increase excitability and enhance epileptiform activity in rat neocortex in vitro. Eur J Pharmacol 236:131–136.

Bradford (1995) Glutamate, GABA and epilepsy. Prog Neurobiol 47:477–511.

Bragin A, Wilson CL, Engel J Jr (2000) Chronic epileptogenesis requires development of a network of pathologically interconnected neuron clusters: A hypothesis. Epilepsia 42(suppl 6):S144–S152.

Brailowsky S, Garcia O. (1999) Ethanol, GABA and epilepsy. Arch Med Res 30:3–9.

Brines ML, Sundaresan S, Spencer DD, de Lanerolle NC (1997) Quantitative autoradiographic analysis of ionotropic glutamate receptor subtypes in human temporal lobe epilepsy: up-regulation in reorganized epileptogenic hippocampus. Eur J Neurosci 9:2035–2044.

Broderick PA, Pacia SV, Doyle WK, Devinsky O (2000) Monoamine neurotransmitters in resected hippocampal subparcellations from neocortical and mesial temporal lobe epilepsy patients: in situ microvoltammetric studies. Brain Res 878:48–63.

Bucinskaite V, Lundeberg T, Stenfors C, Ekblom A, Dahlin L, Theodorsson E (1994) Effects of electro-acupuncture and physical exercise on regional concentrations of neuropeptides in rat brain. Brain Res 666:128–132.

Butler LS, Silva AJ, Abeliovich A, Watanabe Y, Tonegawa S, and McNamara JO (1995) Limbic epilepsy in transgenic mice carrying a Ca^{2+}/calmodulin-dependent kinase II alpha-subunit mutation. Proc Natl Acad Sci USA 92:6852–6855.
Butt AM, Kalsi A (2006) Inwardly rectifying potassium channels (Kir) in central nervous system glia: a special role for Kir4.1 in glial functions. J Cell Mol Med 10:33–44.
Butuzova MV, Kitchigina VF (2008) Repeated blockade of GABAA receptors in the medial septal region induces epileptiform activity in the hippocampus. Neurosci Lett 434:133–138.
Cai YY, Liu ZS, Wang S, Wang QF, Cui XM, Qu L (2009) Influence of electroacupuncture of meridian acupoints on the related hormones of the hypothalamus-pituitary-adrenal axis in rats with cerebral ischemia reperfusion injury. Zhen Ci Yan Jiu 34:297–303. (In Chinese with English abstract)
Cakmak YO (2006) Epilepsy, electroacupuncture and the nucleus of the solitary tract. Acupunct Med 24:164–168.
Cakmak YO (2009) How acupuncture may relieve infantile colic symptoms--melatonin, serotonin and circadian rhythmicity. Acupunct Med 27:134.
Camilo O, Goldstein LB (2004) Seizures and epilepsy after ischemic stroke. Stroke 35:1769–1775.
Cammalleri M, Cervia D, Langenegger D, Liu Y, Dal Monte M, Hoyer D, Bagnoli P (2004) Somatostatin receptors differentially affect spontaneous epileptiform activity in mouse hippocampal slices. Eur J Neurosci 20:2711–2721.
Camón L, Vives P, de Vera N, Martínez E (1998) Seizures and neuronal damage induced in the rat by activation of group I metabotropic glutamate receptors with their selective agonist 3,5-dihydroxyphenylglycine. J Neurosci Res 51:339–348.
Campbell A (2006) Point specificity of acupuncture in the light of recent clinical and imaging studies. Acupunct Med 24:118–122.
Campbell SL, Hablitz JJ (2008) Decreased glutamate transport enhances excitability in a rat model of cortical dysplasia. Neurobiol Dis 32:254–261.
Cao R, Hasuo H, OoBa S, Akasu T, Zhang X (2006) Facilitation of glutamatergic synaptic transmission in hippocampal CA1 area of rats with traumatic brain injury. Neurosci Lett 401:136–141.
Carmona-Aparicio L, Peña F, Borsodi A, Rocha L (2007) Effects of nociceptin on the spread and seizure activity in the rat amygdala kindling model: their correlations with ^{3}H-leucyl-nociceptin binding. Epilepsy Res 77:75–84.
Carter DS, Haider SN, Blair RE, Deshpande LS, Sombati S, DeLorenzo RJ (2006) Altered calcium/calmodulin kinase II activity changes calcium homeostasis that underlies epileptiform activity in hippocampal neurons in culture. J Pharmacol Exp Ther 319:1021–1031.
Carter DS, Harrison AJ, Falenski KW, Blair RE, DeLorenzo RJ (2008) Long-term decrease in calbindin-D28K expression in the hippocampus of epileptic rats following pilocarpine-induced status epilepticus. Epilepsy Res 79:213–223.
Catterall WA, Kalume F, Oakley JC (2010) Nav1.1 channels and epilepsy. J Physiol 588: 1849–1859.
Cavazos JE, Cross DJ (2006) The role of synaptic reorganization in mesial temporal lobe epilepsy. Epilepsy Behav 8:483–493.
Cavazos JE, Das I, Sutula TP (1994) Neuronal loss induced in limbic pathways by kindling: evidence for induction of hippocampal sclerosis by repeated brief seizures. J Neurosci 14:3106–3121.
Chae Y, Park HJ, Habm DH, Lee BH, Park HK, Lee H (2010) Spatiotemporal patterns of neural activity in response to electroacupuncture stimulation in the rodent primary somatosensory cortex. Neurol Res 32(suppl 1):64–68.
Chae Y, Yeom M, Han JH, Park HJ, Hahm DH, Shim I, Lee HS, Lee H (2008) Effect of acupuncture on anxiety-like behavior during nicotine withdrawal and relevant mechanisms. Neurosci Lett 430:98–102.
Chandra PS, Bal C, Garg A, Gaikwad S, Prasad K, Sharma BS, Sarkar C, Singh MB, Padma VM, Tripathi M (2010) Surgery for medically intractable epilepsy due to postinfectious etiologies. Epilepsia 51:1097–1100.
Chang BS, Lowenstein DH (2003) Epilepsy. N Engl J Med 349:1257–1266.

Chang P, Chandler KE, Williams RS, Walker MC (2010) Inhibition of long-term potentiation by valproic acid through modulation of cyclic AMP. Epilepsia 51:1533–1542.

Chang YH, Hiseh MT, Wu CR (1999) Effects of acupuncture at pai-hui on the deficit of memory storage in rats. Am J Chin Med 27:289–298.

Chang YH, Hsieh MT, Cheng JT (1996) Increase of locomotor activity by acupuncture on Bai-Hui point in rats. Neurosci Lett 211:121–124.

Chao D, Balboni G, Lazarus LH, Salvadori S, Xia Y (2009) Na^{+} mechanism of DOR protection against anoxic K^{+} derangement in the cortex. Cell Mol Life Sci 66:1105–1115.

Chao D, Bazzy-Asaad A, Balboni G, Salvadori S, Xia Y (2008) Activation of DOR attenuates anoxic K^{+} derangement via inhibition of Na^{+} entry in mouse cortex. Cereb Cortex 18:2217–2227.

Chao D, Bazzy-Asaad A, Balboni G, Xia Y (2007a) δ-, but not μ-, opioid receptor stabilizes K^{+} homeostasis by reducing Ca^{2+} influx in the cortex during acute hypoxia. J Cell Physiol 212:60–67.

Chao D, Donnelly DF, Feng Y, Bazzy-Asaad A, Xia Y (2007b) Cortical δ-opioid receptors potentiate K^{+} homeostasis during anoxia and oxygen-glucose deprivation. J Cereb Blood Flow Metab 27:356–368.

Chao D, He X, Yang Y, Bazzy-Asaad A, Lazarus LH, Balboni G, Kim DH, Xia Y (2012) DOR activation inhibits anoxic/ischemic Na^{+} influx through Na^{+} channels via PKC mechanisms in the cortex. Exp Neurol 236:228–239.

Chao D, Xia Y (2010) Ionic storm in hypoxic/ischemic stress: Can opioid receptors subside it? Prog Neurobiol 90:439–470.

Chao DM, Chen G, Cheng JS (2001) Melatonin might be one possible medium of electroacupuncture anti-seizures. Acupunct Electro-Ther Res 26:39–48.

Chao DM, Cheng JS (2001a) Melatonin and epilepsy. Chin J Clin Neurosci 9:413–416. (In Chinese)

Chao DM, Cheng JS (2001b) Preventive effect of melatonin against benzylpenicillin- induced seizures in rats. Chin J Neurosci 17:67–71.

Chen AC, Liu FJ, Wang L, Arendt-Nielsen L (2006a) Mode and site of acupuncture modulation in the human brain: 3D (124-ch) EEG power spectrum mapping and source imaging. Neuroimage 29:1080–1091.

Chen F, Yan ZK, Yang B (2009) Effects of electroacupuncture on cerebral Bcl-2 and caspase-3 expression after cerebral ischemia reperfusion in rats. Zhen Ci Yan Jiu 34:363–367. (In Chinese with English abstract)

Chen G, Pan B, Hawver DB, Wright CB, Potter WZ, Manji HK (1996) Attenuation of cyclic AMP production by carbamazepine. J Neurochem 67:2079–2086.

Chen K, Chen G, Feng X (1983) Observation of immediate effect of acupuncture on electroencephalograms in epileptic patients. J Tradit Chin Med 3:121–124.

Chen Q, He S, Hu XL, Yu J, Zhou Y, Zheng J, Zhang S, Zhang C, Duan W, Xiong ZQ (2007a) Differential roles of NR2A- and NR2B-Containing NMDA receptors in activity-dependent brain-derived neurotrophic factor gene regulation and limbic epileptogenesis. J Neurosci 27:542–552.

Chen RC, Huang YH (1984) Acupuncture on experimental epilepsies. Proc Natl Sci Counc Repub China B 8:72–77.

Chen S, Fujita S, Koshikawa N, Kobayashi M (2010) Pilocarpine-induced status epilepticus causes acute interneuron loss and hyper-excitatory propagation in rat insular cortex. Neuroscience 166:341–353.

Chen S, Kobayashi M, Honda Y, Kakuta S, Sato F, Kishi K (2007b) Preferential neuron loss in the rat piriform cortex following pilocarpine-induced status epilepticus. Epilepsy Res 74:1–18.

Chen W, Gu HW, Ma WP, Li QS, Yu Q, Liu XQ, Liu SH, Li WH, Liu HL, Dai MT (2006b) Multicentral randomized controlled study on effects of acupuncture at Zusanli (ST 36) and Xuanzhong (GB 39) on cerebrovascular function in the patients of ischemic stroke. Zhongguo Zhen Jiu 26:851–853. (In Chinese with English abstract)

Chen WF, Chang H, Huang LT, Lai MC, Yang CH, Wan TH, Yang SN (2006c) Alterations in long-term seizure susceptibility and the complex of PSD-95 with NMDA receptor from animals previously exposed to perinatal hypoxia. Epilepsia 47:288–296.

Chen WH, Zhuang LX, Ding XH (2006d) Effect of acupoint embedding therapy on apoptosis of hippocampal neurons in rats with status epilepsy. Chin J Clin Rehabilit 10(36):96–98.1. (In Chinese with English abstract)

Chen XH, Zhang W, Yang HT (2008) Effects on endocrinal function treated with acupuncture based on Qijie theory in epilepsy patients. World J Acu Moxi 18:1–8.

Chen Y, Bender RA, Brunson KL, Pomper JK, Grigoriadis DE, Wurst W, Baram TZ (2004) Modulation of dendritic differentiation by corticotropin-releasing factor in the developing hippocampus. Proc Natl Acad Sci USA 101:15782–15787.

Cheng H, Yu J, Jian Z, Zhang X, Liu C, Peng Y, Chen F, Qu Y, Jia Y, Tian Q, Xiao C, Chu Q, Nie K, Kan B, Hu X, Han J (2008) Acupuncture improves cognitive deficits and regulate the brain cell proliferation of SAMP8 mice. Neurosci Lett 432:111–116.

Cheng S, Ma M, Ma Y, Wang Z, Xu G, Liu X (2009) Combination therapy with intranasal NGF and electroacupuncture enhanced cell proliferation and survival in rats after stroke. Neurol Res 31:753–758.

Cheuk D, Wong V (2008) Acupuncture for epilepsy. Cochrane Database Syst Rev Issue 4. Art. No.: CD005062. DOI: 10.1002/14651858.CD005062.pub3.

Chevassus-au-Louis N, Jorquera I, Ben-Ari Y, Represa A (1999) Abnormal connections in the malformed cortex of rats with prenatal treatment with methylazoxymethanol may support hyperexcitability. Dev Neurosci 21:385–392.

Choi DC, Lee JY, Moon YJ, Kim SW, Oh TH, Yune TY (2010) Acupuncture-mediated inhibition of inflammation facilitates significant functional recovery after spinal cord injury. Neurobiol Dis 39:272–282.

Choi J, Koh S (2008) Role of brain inflammation in epileptogenesis. Yonsei Med J 49:1–18.

Choi J, Nordli DR Jr, Alden TD, DiPatri A Jr, Laux L, Kelley K, Rosenow J, Schuele SU, Rajaram V, Koh S (2009) Cellular injury and neuroinflammation in children with chronic intractable epilepsy. J Neuroinflammation 6:38.

Chuang CM, Hsieh CL, Li TC, Lin JG (2007) Acupuncture stimulation at Baihui acupoint reduced cerebral infarct and increased dopamine levels in chronic cerebral hypoperfusion and ischemia-reperfusion injured Sprague-Dawley rats. Am J Chin Med 35:779–791.

Clinckers R, Smolders I, Meurs A, Ebinger G, Michotte Y (2004) Anticonvulsant action of hippocampal dopamine and serotonin is independently mediated by D and 5-HT receptors. J Neurochem 89:834–843.

Cohen MH (2003) Regulation, religious experience, and epilepsy: a lens on complementary therapies. Epilepsy Behav 4:602–606.

Cole M, Shen J, Hommer D (2002) Convulsive syncope associated with acupuncture. Am J Med Sci 324:288–289.

Cordero ML, Ortiz JG, Santiago G, Negron A, Moreira JA (1994) Altered GABAergic and glutamatergic transmission in audiogenic seizure-susceptible mice. Mol Neurobiol 9:253–258.

Cross DJ, Cavazos JE (2007) Synaptic reorganization in subiculum and CA3 after early-life status epilepticus in the kainic acid rat model. Epilepsy Res 73:156–165.

Csaba Z, Pirker S, Lelouvier B, Simon A, Videau C, Epelbaum J, Czech T, Baumgartner C, Sperk G, Dournaud P (2005) Somatostatin receptor type 2 undergoes plastic changes in the human epileptic dentate gyrus. J Neuropathol Exp Neurol 64:956–969.

Czapiński P, Blaszczyk B, Czuczwar SJ (2005) Mechanisms of action of antiepileptic drugs. Cur Top Med Chem 5:3–14.

D'Ambrosio R (2004) The role of glial membrane ion channels in seizures and epileptogenesis. Pharmacol Ther 103:95–108.

D'Ambrosio R, Maris DO, Grady MS, Winn HR, Janigro D (1999) Impaired K^+ homeostasis and altered electrophysiological properties of post-traumatic hippocampal glia. J Neurosci 19:8152–8162.

Danielczyk W (1976) EEG, 5-HTP-metabolism and acupuncture. J Neural Transm 38:303–311.

Danielsson I, Gasior M, Stevenson GW, Folk JE, Rice KC, Negus SS (2006) Electroencephalographic and convulsant effects of the delta opioid agonist SNC80 in rhesus monkeys. Pharmacol Biochem Behav 85:428–434.

Dannhardt G, Kohl BK (1998) The glycine site on the NMDA receptor: structure-activity relationships and possible therapeutic applications. Curr Med Chem 5:253–263.

David Y, Cacheaux LP, Ivens S, Lapilover E, Heinemann U, Kaufer D, Friedman A (2009) Astrocytic dysfunction in epileptogenesis: Consequence of altered potassium and glutamate homeostasis. J Neurosci 29:10588–10599.

de Carvalho LA (1994) Modeling the thalamocortical loop. Int J Biomed Comput 35:267–296.

de Lanerolle NC, Eid E, von Campe G, Kovacs I, Spencer DD, Brines M (1998) Glutamate receptor subunits GluR1 and GluR2/3 distribution shows reorganization in the human epileptogenic hippocampus. Eur J Neurosci 10:1687–1703.

de Oliveira DL, Fischer A, Jorge RS, da Silva MC, Leite M, Goncalves CA, Quillfeldt JA, Souza DO, Souza TM, Wofchuk S (2008) Effects of early-life LiCl-pilocarpine-induced status epilepticus on memory and anxiety in adult rats are associated with mossy fiber sprouting and elevated CSF S100B protein. Epilepsia 9:842–852.

de Weerd A, de Haas S, Otte A, Trenité DK, van Erp G, Cohen A, de Kam M, van Gerven J (2004) Subjective sleep disturbance in patients with partial epilepsy: a questionnaire-based study on prevalence and impact on quality of life. Epilepsia 45:1397–1404.

Delorenzo RJ, Sun DA, Deshpande LS (2005) Cellular mechanisms underlying acquired epilepsy: The calcium hypothesis of induction and maintenance of epilepsy. Pharmacol Therap 105: 229–266.

Deng Y, Liu W, Liang W, Zhou P (2006) Influence of acupuncture on power of brain electrical activity mapping and hippocampal amino acid transmitters in acute epilepsy rats. J Guangzhou Univ Tradit Chin Med 23:509–512. (In Chinese with English abstract)

Deng YJ, Wang JJ, Lin YP, Liu WY, Wang LH (2001) Clinical observation on treatment of epilepsy general tonic-clonic attack with catgut implantation at acupoint plus antiepileptic Western Medicine of small dose. Zhongguo Zhen jiu 21:271–273. (In Chinese with English abstract)

Deshpande LS, Lou JK, Mian A, Blair RE, Sombati S, DeLorenzo RJ (2007) In vitro status epilepticus but not spontaneous recurrent seizures cause cell death in cultured hippocampal neurons. Epilepsy Res 75:171–179.

Dhond RP, Yeh C, Park K, Kettner N, Napadow V (2008) Acupuncture modulates resting state connectivity in default and sensorimotor brain networks. Pain 136:407–418.

Ding CM, Lo EH, Stanley GB (2008) Sustained focal cortical compression reduces electrically-induced seizure threshold. Neuroscience 154:551–555.

Djukic B, Casper KB, Philpot BD, Chin LS, McCarthy KD (2007) Conditional knock-out of Kir4.1 leads to glial membrane depolarization, inhibition of potassium and glutamate uptake, and enhanced short-term synaptic potentiation. J Neurosci 27:11354–11365.

Dos Santos JG Jr, Tabosa A, do Monte FH, Blanco MM, de Oliveira Freire A, Mello LE (2005) Electroacupuncture prevents cognitive deficits in pilocarpine-epileptic rats. Neurosci Lett 384:234–238.

Dragunow M, Currie RW, Faull RL, Robertson HA, Jansen K (1989) Immediate-early genes, kindling and long-term potentiation. Neurosci Biobehav Rev 13:301–313.

Drake CT, Chavkin C, Milner TA (2007) Opioid systems in the dentate gyrus. Prog Brain Res 163:245–263.

Du J (2008) The messengers from peripheral nervous system to central nervous system: involvement of neurotrophins and cytokins in the mechanisms of acupuncture. Zhen Ci Yan Jiu 33:37–40. (In Chinese with English Abstract)

Dubé C (2007) Neuropeptide Y: potential role in recurrent developmental seizures. Peptides 28:441–446.

Dudek FE, Suluta TP (2007) Epileptogenesis in the dentate gyrus: a critical perspective. Prog Brain Res 163:755–773.

Duffy S, Nguyen PV, Baker GB (2004) Phenylethylidenehydrazine, a novel GABA-transaminase inhibitor, reduces epileptiform activity in rat hippocampal slices. Neuroscience. 126:423–432.

Eid T, Williamson A, Lee TSW, Petroff OA, de Lanerolle NC (2008) Glutamate and astrocytes–key players in human mesial temporal lobe epilepsy? Epilepsia 49(Suppl. 2):42–52.

Eisenberg DM, Davis RB, Ettner SL, Appel S, Wilkey S, van Rompay M, Kessler RC (1998) Trends in alternative medicine use in the United States, 1990–1997. JAMA 280:1569–1575.

El Idrissi A, L'Amoreaux WJ (2008) Selective resistance of taurine-fed mice to isoniazide-potentiated seizures: in vivo functional test for the activity of glutamic acid decarboxylase. Neuroscience 156:693–699.

Engel J Jr (1996) Surgery for seizures. N Eng J Med 334:647–652.

Errington AC, Stöhr T, Lees G (2005) Voltage gated ion channels: Targets for anticonvulsant drugs. Cur Top Med Chem 5:15–30.

Escayg A, Goldin AL (2010) Sodium channel *SCN1A* and epilepsy: Mutations and mechanisms. Epilepsia 51:1650–1658.

Esclapez M, Houser CR (1999) Up-regulation of GAD65 and GAD67 in remaining hippocampal GABA neurons in a model of temporal lobe epilepsy. J Comp Neurol. 412:488–505.

Eusebio A, Pogosyan A, Wang S, Averbeck B, Gaynor LD, Cantiniaux S, Witjas T, Limousin P, Azulay J, Brown P (2009) Resonance in subthalamo-cortical circuits in Parkinson's disease. Brain 132:2139–2150.

Faingold CL (2008) Electrical stimulation therapies for CNS disorders and pain are mediated by competition between different neuronal networks in the brain. Med Hypotheses 71:668–681.

Fang JL, Krings T, Weidemann J, Meister IG, Thron A (2004) Functional MRI in healthy subjects during acupuncture: different effects of needle rotation in real and false acupoints. Neuroradiology 46:359–362.

Fang Z, Yu Q, Li Y (1993) The correlation between postsynaptic inhibition and GABA, opioid peptides, SP in electroacupuncture. Zhen Ci Yan Jiu 18:64–67. (In Chinese with English abstract)

Fekete A, Franklin L, Ikemoto T, Rozsa B, Lendrai B, Vizi ES, Zelles T (2009) Mechanism of the persistent sodium current activator veratridine-evoked Ca^{2+} elevation: Implication for epilepsy. J Neurochem 111:745–756.

Fellin T, Pascual O, Gobbo S, Pozzan T, Haydon PG, Carmignoto G (2004) Neuronal synchrony mediated by astrocytic glutamate through activation of extrasynaptic NMDA receptors. Neuron 43:729–743.

Feng Y, Chao DM, Li WM, Cao YX, Wang YQ, Wu GC (2004) Inhibition of nociceptin/orphanin FQ on penicillin-induced seizures in rats. Brain Res 1020:214–219.

Ferraro G, Sardo P (2004) Nitric oxide and brain hyperexcitability. In Vivo 18:357–366.

Ferraro G, Sardo P (2009) Cholecystokinin-8 sulfate modulates the anticonvulsant efficacy of vigabatrin in an experimental model of partial complex epilepsy in the rat. Epilepsia 50:721–730.

Ferrendelli JA, Blank AC, Gross RA (1980) Relationships between seizure activity and cyclic nucleotide levels in brain. Brain Res 200:93–103.

Fetissov SO, Jacoby AS, Brumovsky PR, Shine J, Iismaa TP, Hökfelt T (2003) Altered hippocampal expression of neuropeptides in seizure-prone GALR1 knockout mice. Epilepsia 44:1022–1033.

Fisher JL (2004) A mutation in the GABAA receptor alpha 1 subunit linked to human epilepsy affects channel gating properties. Neuropharmacology 46:629–637.

Fisher R, Salanova V, Witt T, Worth R, Henry T, Gross R, Oommen K, Osorio I, Nazzaro J, Labar D, Kaplitt M, Sperling M, Sandok E, Neal J, Handforth A, Stern J, DeSalles A, Chung S, Shetter A, Bergen D, Bakay R, Henderson J, French J, Baltuch G, Rosenfeld W, Youkilis A, Marks W, Garcia P, Barbaro N, Fountain N, Bazil C, Goodman R, McKhann G, Babu Krishnamurthy K, Papavassiliou S, Epstein C, Pollard J, Tonder L, Grebin J, Coffey R, Graves N; SANTE Study Group (2010) Electrical stimulation of the anterior nucleus of thalamus for treatment of refractory epilepsy. Epilepsia 51:899–908.

Florence G, Dahlem MA, Almeida AG, Bassani JWM (2009) The role of extracellular potassium dynamics in the different stages of ictal bursting and spreading depression: a computational study. J Theoret Biol 258:219–228.

Folbergrová J (1980) Cyclic GMP and cyclic AMP in the cerebral cortex of mice during seizures induced by 3-mercaptopropionic acid: effects of anticonvulsant agents. Neurosci Lett 16:291–296.

Folbergrová J, Druga R, Otáhal J, Haugvicová R, Mareš P, Kubova H (2005) Seizures induced in immature rats by homocysteic acid and the associated brain damage are prevented by group II metabotropic glutamate receptor agonist (2R,4R)-4-aminopyrrolidine-2,4-dicarboxylate. Exp Neurol 192:420–436.

Freestone DR, Grayden DB, Lai A, Nelson TS, Halliday A, Burkitt AN, Cook MJ (2009) The thalamocortical circuit and the generation of epileptic spikes in rat models of focal epilepsy. Conf Proc IEEE Eng Med Biol Soc 2009:1533–1536.

Freichel C, Potschka H, Ebert U, Brandt C, Löscher W (2006) Acute changes in the neuronal expression of GABA and glutamate decarboxylase isoforms in the rat piriform cortex following status epilepticus. Neuroscience 141:2177–2194.

Fritsch B, Qashu F, Figueiredo TH, Aroniadou-Anderjaska V, Rogawski MA, Braga MFM (2009) Pathological alterations in gabaergic interneurons and reduced tonic inhibition in the basolateral amygdale during epileptogenesis. Neuroscience 163:415–429.

Frotscher M, Jonas P, Sloviter RS (2006) Synapses formed by normal and abnormal hippocampal mossy fibers. Cell Tissue Res 326:361–367.

Frye CA (2008) Hormonal influences on seizures: basic neurobiology. Int Rev Neurobiol 83:27–77.

Fujikawa DG (2005) Prolonged seizures and cellular injury: understanding the connection. Epilepsy Behav 7:S3–S11.

Fujikawa DG, Shinmei SS, Cai B (1999) Lithium-pilocarpine-induced status epilepticus produces necrotic neurons with internucleosomal DNA fragmentation in adult rats. Eur J Neurosci 11:1605–1614.

Fukao K, Momiyama T, Ishihara K, Ujihara H, Fujita Y, Taniyama K, Serikawa T, Sasa M (1998) Inhibition by γ-aminobutyric acid system activation of epileptic seizures in spontaneously epileptic rats. JPN J Pharmacol, 76:387–396.

Fukuda F, Shinbara H, Yoshimoto K, Yano T, Kuriyama K (2005) Effect of moxibustion on dopaminergic and serotonergic systems of rat nucleus accumbens. Neurochem Res 30: 1607–1613.

Gabriel S, Njunting M, Pomper JK, Merschhemke M, Sanabria ERG, Eilers A, Kivi A, Zeller M, Meencke H, Cavalheiro EA, Heinemann U, Lehmann T (2004) Stimulus and potassium-induced epileptiform activity in the human dentate gyrus from patients with and without hippocampal sclerosis. J Neurosci 24:10416–10430.

Gan P, Cheng JS, Ng YK, Ling EA (2005) Role of GABA in electro-acupuncture therapy on cerebral ischemia induced by occlusion of the middle cerebral artery in rats. Neurosci Lett 383: 317–321.

Gao C, Liao W, Yang J (1991) Effects of electroacupuncture on epileptiform hippocampal unit activity by intracerebroventricular injection of coriaria lactone. Zhen Ci Yan Jiu 16:54–60. (In Chinese with English abstract)

Gao H, Cheng J (1998) Role of intrahippocampal kappa opioid receptors in inhibiting the seizure by electroacupuncture (EA) in rats. Zhen Ci Yan Jiu 23:122–125. (In Chinese with English abstract)

Gao H, Guo J, Zhao P, Cheng J (2006) Influences of electroacupuncture on the expression of insulin-like growth factor-1 following, focal cerebral ischemia in monkeys. Acupunct Electro-Ther Res 31:259–272.

Gao H, Wang C (2010) Role of Intrahippocampal Sigma Receptor in Inhibiting Seizure by Electroacupuncture in Rats. Med Acupunct 22:19–24.

Gao M, Li Q, Zhang J, Liu Y (1990) Effect of changing the functional state of frontal cortex by GABA on the acupuncture effect of PAG neurons. Zhen Ci Yan Jiu 15:264–268. (In Chinese with English abstract)

Garga N, Lowenstein DH (2006) Posttraumatic epilepsy: a major problem in desperate need of major advances. Epilepsy Curr 6:1–5.

Garrido-Sanabria ER, Otalora LFP, Arshadmansab MF, Herrera B, Francisco S, Ermolinsky BS (2008) Impaired expression and function of group II metabotropic glutamate receptors in pilocarpine-treated chronically epileptic rats Brain Res 1240:165–176.

Ghasemi M, Shafaroodi H, Nazarbeiki S, Meskar H, Ghasemi A, Bahremand A, Ziai P, Dehpour AR (2010) Inhibition of NMDA receptor/NO signaling blocked tolerance to the anticonvulsant effect of morphine on pentylenetetrazole-induced seizures in mice. Epilepsy Res 91:39–48.
Gholipour T, Ghasemi M, Riazi K, Ghaffarpour M, Dehpour AR (2010) Seizure susceptibility alteration through 5-HT(3) receptor: modulation by nitric oxide. Seizure 19:17–22.
Gill DA, Ramsay SL, Tasker RA (2010) Selective reductions in subpopulations of GABAergic neurons in a developmental rat model of epilepsy. Brain Res 1331:114–123.
Giovacchini G, Toczek MT, Bonwetsch R, Bagic A, Lang L, Fraser C, Reeves-Tyer P, Herscovitch P, Eckelman WC, Carson RE, Theodore WH (2005) 5-HT 1A receptors are reduced in temporal lobe epilepsy after partial-volume correction. J Nucl Med 46:1128–1135.
Glykys J, Dzhala VI, Kuchibhotla KV, Feng F, Kuner T, Augustine G, Bacskai BJ, Staley KJ (2009) Differences in cortical versus subcortical GABAergic signaling: A candidate mechanism of electroclinical uncoupling of neonatal seizures. Neuron 63:657–672.
Goiz-Marquez G, Caballero S, Solis H, Rodriguez C, Sumano H (2009) Electroencephalographic evaluation of gold wire implants inserted in acupuncture points in dogs with epileptic seizures. Res Vet Sci 86:152–161.
Golarai G, Greenwood AC, Feeney DM, Connor JA (2001) physiological and structural evidence for hippocampal involvement in persistent seizure susceptibility after traumatic brain injury. J Neurosci 21:8523–8537.
Gómez-Gonzalo M, Losi G, Chiavegato A, Zonta M, Cammarota M, Brondi M, Vetri F, Uva L, Pozzan T, de Curtis M, Ratto GM, Carmignoto G (2010) An excitatory loop with astrocytes contributes to drive neurons to seizure threshold. PLoS Biol. 8:e1000352. doi:10.1371/journal.pbio.1000352
Graber KD, Prince DA (1999) Tetrodotoxin prevents posttraumatic epileptogenesis in rats. Ann Neurol 46:234–242.
Grecksch G, Becker A, Schroeder H, Höllt V (1999) Involvement of delta-opioid receptors in pentylenetetrazol kindling development and kindling-related processes in rats. Naunyn Schmiedebergs Arch Pharmacol 360:151–156.
Green JD (1964) The hippocampus. Physiol Rev 44:561–608.
Griesemer D, Mautes AM (2007) Closed head injury causes hyperexcitability in rat hippocampal CA1 but not CA3 pyramidal cells. J Neurotrauma 24:1823–1832.
Gruber B, Greber S, Sperk G (1993) Kainic acid seizures cause enhanced expression of cholecystokinin-octapeptide in the cortex and hippocampus of the rat. Synapse 15:221–228.
Guo F, Yu N, Cai JQ, Quinn T, Zong ZH, Zeng YJ, Hao LY (2008a) Voltage-gated sodium channel Nav1.1, Nav1.3 and β1 subunit were up-regulated in the hippocampus of spontaneously epileptic rat. Brain Res Bull 75:179–187.
Guo J, Liu J, Fu W, Ma W, Xu Z, Yuan M, Song J, Hu J (2008b) The effect of electroacupuncture on spontaneous recurrent seizure and expression of GAD67 mRNA in dentate gyrus in a rat model of epilepsy. Brain Res 1188:165–172.
Guo J, Liu J, Fu W, Ma W, Xu Z, Yuan M, Zhou X, Hu J (2008c) Effect of electroacupuncture stimulation of hindlimb on seizure incidence and supragranular mossy fiber sprouting in a rat model of epilepsy. J Physiol Sci 58:309–315.
Guo J, Zhao P, Xia Y, Zhou F, Yang R, Cheng J (2006) Involvement of taurine in cerebral ischemia and electroacupuncture anti-ischemia. Adv Exp Med Biol 583:395–400.
Guo ZL, Moazzami AR, Tjen-A-Looi S, Longhurst JC (2008d) Responses of opioid and serotonin containing medullary raphe neurons to electroacupuncture. Brain Res 1229:125–136.
Halabisky B, Parada I, Buckmaster PS, Prince DA (2010) Excitatory input onto hilar somatostatin interneurons is increased in a chronic model of epilepsy. J Neurophysiol 104:2214–2223.
Hamani C, Hodaie M, Chiang J, del Campo M, Andrade DM, Sherman D, Mirski M, Mello LE, Lozano AM (2008) Deep brain stimulation of the anterior nucleus of the thalamus: effects of electrical stimulation on pilocarpine-induced seizures and status epilepticus. Epilepsy Res 78:117–123.
Hamba M, Toda K (1985) Effects of electroacupuncture on the neuronal activity of the arcuate nucleus of the rat hypothalamus. Exp Neurol 87:118–128.

Hamilton NB, Attwell D (2010) Do astrocytes really exocytose neurotransmitters? Nat Rev Neurosci 11:227–238.

Hammers A, Asselin M, Hinz R, Kitchen I, Brooks DJ, Duncan JS, Koepp MJ (2007) Upregulation of opioid receptor binding following spontaneous epileptic seizures. Brain 130:1009–1016.

Han DX, Sun M, Zhuang LX (2008) Observation on therapeutic effect of 85 cases of generalized epilepsy treated with acupoint catgut embedding therapy. J Clin Acu-Mox 24:35–36. (In Chinese)

Han JS (1986) Electroacupuncture: an alternative to antidepressants for treating affective diseases? Int J Neurosci 29:79–92.

Han JS (2003) Acupuncture: neuropeptide release produced by electrical stimulation of different frequencies. Trends Neurosci 26:17–22.

Han YJ, Li WX, Jia BH, Shi YJ, Tu Y (2009) Effect of electroacupuncture on hippocampal NO-cGMP signaling pathway in depression rats. Zhen Ci Yan Jiu 34:236–241. (In Chinese with English abstract)

Hargus NJ, Merrick EC, Nigam A, Kalmar CL, Baheti AR, Bertram III EH, Patel MK (2011) Temporal lobe epilepsy induces intrinsic alterations in Na channel gating in layer II medial entorhinal cortex neurons. Neurobiol Dis 41:361–376.

Hasler G, Bonwetsch R, Giovacchini G, Toczek MT, Bagic A, Luckenbaugh DA, Drevets WC, Theodore WH (2007) 5-HT1A receptor binding in temporal lobe epilepsy patients with and without major depression. Biol Psychiatry 62:1258–1264.

Hattiangady B, Shetty AK (2008) Implications of decreased hippocampal neurogenesis in chronic temporal lobe epilepsy. Epilepsia 49(Suppl 5):26–41.

Hauser WA, Beghi E (2008) First seizure definitions and worldwide incidence and mortality. Epilepsia 49 (Suppl 1): 8–12.

He XP, Cao XD (1989) Effects of intrahippocampal delta-receptors on inhibition of electroconvulsive shock by electro-acupuncture. Zhongguo Yao Li Xue Bao 10:197–201. (In Chinese with English abstract)

He XP, Cao XD, Shen LL, Jiang YM (1990a) Effect of electroacupuncture on experimental epilepsy-power spectrum analysis on EEG. Sheng Li Xue Bao 42:141–148. (In Chinese with English abstract)

He XP, Chen BY, Zhu JM, Cao XD (1989) Change of Leu-enkephalin- and B-endorphin-like immunoreactivity in the hippocampus after electroconvulsive shock and electroacupuncture. Acupunct Electro-Ther Res 14:131–139.

He XP, Zhu JM, Huang DK, Li KY, Cao XD (1990b) Effect of electroacupuncture on electroconvulsive shock--an autoradiographic study for opioid receptors. Sheng Li Xue Bao 42:149–154. (In Chinese with English abstract)

Héja L, Barabás P, Nyitrai G, Kékesi KA, Lasztóczi B, Töke O, Tárkány G, Madsen K, Schousboe A, Dobolyi A, Palkovits M, Kardos J (2009) Glutamate uptake triggers transporter-mediated GABA release from astrocytes. PLoS ONE 4: e7153. doi:10.1371/journal.pone.0007153.

Heuser K, Nagelhus EA, Taubøll E, Indahl U, Berg PR, Lien S, Nakken S, Gjerstad L, Ottersen OP (2010) Variants of the genes encoding AQP4 and Kir4.1 are associated with subgroups of patients with temporal lobe epilepsy. Epilepsy Res 88:55–64.

Higashima M, Ohno K, Koshino Y (2002) Cyclic AMP-mediated modulation of epileptiform afterdischarge generation in rat hippocampal slices. Brain Res 949:157–161.

Hirtz D, Thurman DJ, Gwinn-Hardy K, Mohamed M, Chaudhuri AR, Zalutsky R (2007). How common are the 'common' neurologic disorders? Neurology 68: 326–337.

Hitiris N, Mohanraj R, Norrie J, Brodie M J (2007) Mortality in epilepsy. Epilepsy Behavior 10: 363–376.

Hollrigel GS, Chen K, Baram TZ, Soltesz I (1998) The pro-convulsant actions of corticotropin-releasing hormone in the hippocampus of infant rats. Neuroscience 84:71–79.

Honar H, Riazi K, Homayoun H, Sadeghipour H, Rashidi N, Ebrahimkhani MR, Mirazi N, Dehpour AR (2004) Ultra-low dose naltrexone potentiates the anticonvulsant effect of low dose morphine on clonic seizures. Neuroscience 129:733–742.

Hong JS, McGinty JF, Grimes L, Kanamatsu T, Obie J, Mitchell CL (1988) Seizure-induced alterations in the metabolism of hippocampal opioid peptides suggest opioid modulation of seizure-related behaviors. NIDA Res Monogr 82:48–66.

Hou YC, Janczuk A, Wang PG (1999) Current trends in the development of nitric oxide donors. Curr Pharm Des 5:417–441.

Houser CR, Miyashiro JE, Swartz BE, Walsh GO, Rich JR, Delgado-Escueta AV (1990) Altered patterns of dynorphin immunoreactivity suggest mossy fiber reorganization in human hippocampal epilepsy. J Neurosci 10:267–282.

Houweling AR, Bazhenov M, Timofeev I, Steriade M, Sejnowski TJ (2005) Homeostasis synaptic plasticity can explain post-traumatic epileptogenesis in chronically isolated neocortex. Cereb Cortex 15:834–845.

Hsieh CL, Chang QY, Lin IH, Lin JG, Liu CH, Tang NY, Lane HY (2006) The study of electroacupuncture on cerebral blood flow in rats with and without cerebral ischemia. Am J Chin Med 34:351–361.

Hu J, Li K, Huang X (1998a) Changes of somatostatin in the hippocampus following electroconvulsion and the anticonvulsion effect caused by great auricular nerve electrostimulation. Chin J Histochem Cytochem 7:125–129. (In Chinese with English abstract)

Hu JY, Li YG, Li KY, Huang XF (1998b) Changes of somatostatin receptor in the rat brain following electroconvulsion. Chin Pharmacol Bull 14:408–409. (In Chinese)

Huang W, Wu XP, Ding Y, Wang YW, Wang SJ, Sun GJ (2009) Influence of electroacupuncture of "Shuigou" (GV 26) on hippocampal CGRP and NPY contents in cerebral ischemia rats. Zhen Ci Yan Jiu 34:13–15, 42. (In Chinese with English abstract)

Huang Y, Jiang XM, Li DJ (2006) Effects on electro-scalp acupuncture on cerebral dopamine transporter in patients with Parkinson's disease. Zhongguo Zhong Xi Yi Jie He Za Zhi 26:303–307. (In Chinese with English abstract)

Huang ZN, Yang R, Chen G, Cheng JS (1999) Effect of electroacupuncture and 7-NI on penicillin-induced epilepsy and their relation with intrahippocampal NO changes. Sheng Li Xue Bao 51:508–514. (In Chinese with English abstract)

Hui KK, Marina O, Liu J, Rosen BR, Kwong KK (2010) Acupuncture, the limbic system, and the anticorrelated networks of the brain. Auton Neurosci. 157:81–90.

Hwang IK, Chung JY, Yoo DY, Yi SS, Youn HY, Seong JK, Yoon YS (2010) Comparing the effects of acupuncture and electroacupuncture at Zusanli and Baihui on cell proliferation and neuroblast differentiation in the rat hippocampus. J Vet Med Sci 72:279–284.

Iasnetsoy VV, Prardivtsev VA (1982) Effect of opioid peptides, morphine and electroacupuncture on the neuronal activity of the sensorimotor cortex and brain stem reticular formation. Biull Eksp Biol Med 94:53–56. (In Russian with English abstract)

Inyushin M, Kucheryavykh LY, Kucheryavykh YV, Nichols CG, Buono RJ, Ferraro TN, Skatchkov SN, Eaton MJ (2010) Potassium channel activity and glutamate uptake are impaired in astrocytes of seizure-susceptible DBA/2 mice. Epilepsia 51:1707–1713.

Ishibashi S (1986) The effect of auricular electroacupuncture on the neuronal activity of the thalamic and hypothalamic neurons of the rat. Acupunct Electro-Ther Res 11:15–23.

Ivens S, Kaufer D, Flores LP, Bechmann I, Zumsteg D, Tomkins O, Seiffert E, Heinemann U, Friedman A (2007) TGF-β receptor-mediated albumin uptake into astrocytes is involved in neocortical epileptogenesis. Brain 130:535–547.

Jabs R, Seifert G, Steinhäuser C (2008) Astrocytic function and its alteration in the epileptic brain. Epilepsia 49(Suppl. 2):3–12.

Jang MH, Shin MC, Lee TH, Lim BV, Shin MS, Min BI, Kim H, Cho S, Kim EH, Kim CJ (2003) Acupuncture suppresses ischemia-induced increase in c-Fos expression and apoptosis in the hippocampal CA1 region in gerbils. Neurosci Lett 347:5–8.

Jansen LA, Uhlmann EJ, Crino PB, Gutmann DH, Wong M (2005) Epileptogenesis and reduced inward rectifier potassium current in tuberous sclerosis complex-1-deficient astrocytes. Epilepsia 46:1871–1880.

Jia J, Sun Z, Li B, Pan Y, Wang H, Wang X, Yu F, Liu L, Zhang L, Wang X (2009) Electroacupuncture stimulation improves motor disorders in Parkinsonian rats. Behav Brain Res 205:214–218.

Jiang K, Zhao Z, Shui Q, Xia Z (2004) Electro-acupuncture preconditioning abrogates the elevation of c-Fos and c-Jun expression in neonatal hypoxic-ischemic rat brains induced by glibenclamide, an ATP-sensitive potassium channel blocker. Brain Res 998:13–19.

Jin HB, Li B, Gu J, Cheng JS, Yang R (2005a) Electro-acupuncture improves epileptic seizures induced by kainic acid in taurine-depletion rats. Acupunct Electro-Ther Res 30:207–217.

Jin YZ, Li ZR, Ma C, Shen MH, Niu WM (2005b) Effects of electroacupuncture on nitric oxide and nitric oxide synthase in brain and liver of rats with epilepsy. Chin J Clin Rehabilit 9:114–115. (In Chinese with English abstract)

Jindal V, Ge A, Mansky PJ (2008) Safety and efficacy of acupuncture in children: A review of the evidence. J Pediatr Hematol Oncol 30:431–442.

Jones OT (2002) Ca^{2+} channels and epilepsy. Eur J Pharmacol 447:211–225.

Jourdain P, Bergersen LH, Bhaukaurally K, Bezzi P, Santello M, Domercq M, Matute C, Tonello F, Gundersen V, Volterra A (2007) Glutamate exocytosis from astrocytes controls synaptic strength. Nat Neurosci 10:331–339.

Junyent F, De Lemos L, Utrera J, Paco S, Aguado F, Camins A, Pallàs M, Romero R, Auladell C (2011) Content and traffic of taurine in hippocampal reactive astrocytes. Hippocampus 21:185–197.

Junyent F, Utrera J, Romero R, Pallàs M, Camins A, Duque D, Auladell C (2009) Prevention of epilepsy by taurine treatments in mice experimental model. J Neurosci Res 87:1500–1508.

Jutkiewicz EM, Baladi MG, Folk JE, Rice KC, Woods JH (2006) The convulsive and electroencephalographic changes produced by nonpeptidic delta-opioid agonists in rats: comparison with pentylenetetrazol. J Pharmacol Exp Ther 317:1337–1348.

Jutkiewicz EM, Traynor JR, Woods JH (2005) Separation of the convulsions and antidepressant-like effects produced by the delta-opioid agonist SNC80 in rats. Psychopharmacology (Berl) 182: 588–596.

Kalinichev M, Starr KR, Teague S, Bradford AM, Porter RA, Herdon HJ (2010) Glycine transporter 1 (GlyT1) inhibitors exhibit anticonvulsant properties in the rat maximal electroshock threshold (MEST) test. Brain Res 1331:105–113.

Kaminski RM, Shippenberg TS, Witkin JM, Rocha BA (2005) Genetic deletion of the norepinephrine transporter decreases vulnerability to seizures. Neurosci Lett 382:51–55.

Kanamatsu T, Obie J, Grimes L, McGinty JF, Yoshikawa K, Sabol S, Hong JS (1986) Kainic acid alters the metabolism of Met5-enkephalin and the level of dynorphin A in the rat hippocampus. J Neurosci 6:3094–3102.

Kang KA, Shin ES, Hur J, Hasan MR, Lee H, Park HJ, Park HK, Kim YJ (2010) Acupuncture attenuates neuronal cell death in middle cerebral artery occlusion model of focal ischemia. Neurol Res 32 (Suppl 1):84–87.

Kang TC, An SJ, Park SK, Hwang IK, Seo MO, Kim HS, Kang JH, Kwon OS, Won MH (2003) The somatostatin receptors in the normal and epileptic hippocampus of the gerbil: subtype-specific localization and its alteration. Brain Res 986:91–102.

Kang X, Chao D, Gu Q, Ding G, Wang Y, Balboni G, Lazarus LH, Xia Y (2009) δ-opioid receptors protect from anoxic disruption of Na^+ homeostasis via Na^+ channel regulation. Cell Mol Life Sci 66:3505–3516.

Kang XZ, Xia Y (2008) Effect of electroacupuncture on experimental epilepsy: Roles of different acupoints and stimulation parameters. J Acupunct Tuina Sci 6:279–280.

Kauffman MA, Consalvo D, Gonzalez-Morón D, Aguirre F, D'Alessio L, Kochen S (2009) Serotonin transporter gene variation and refractory mesial temporal epilepsy with hippocampal sclerosis. Epilepsy Res 85:231–234.

Kaura S, Bradford HF, Young AM, Croucher MJ, Hughes PD (1995) Effect of amygdaloid kindling on the content and release of amino acids from the amygdaloid complex: in vivo and in vitro studies. J Neurochem 65:1240–1249.

Kawaguchi Y, Kondo S (2002) Parvalbumin, somatostatin and cholecystokinin as chemical markers for specific GABAergic interneuron types in the rat frontal cortex. J Neurocytol 31: 277–287.

Kee N, Teixeeira CM, Wang AH, Frankland PW (2007) Preferential incorporation of adult-generated granule cells into spatial memory networks in the dentate gyrus. Nat Neurosci 10:355–362.

Khavandgar S, Homayoun H, Dehpour AR (2002) The role of nitric oxide in the proconvulsant effect of delta-opioid agonist SNC80 in mice. Neurosci Lett 329:237–239.

Kiessling M, Gass P (1993) Immediate early gene expression in experimental epilepsy. Brain Pathol 3:381–393.
Kim EH, Jang MH, Shin MC, Lim BV, Kim HB, Kim YJ, Chung JH, Kim CJ (2002) Acupuncture increases cell proliferation and neuropeptide Y expression in dentate gyrus of streptozotocin-induced diabetic rats. Neurosci Lett 327:33–36.
Kim EH, Kim YJ, Lee HJ, Huh Y, Chung JH, Seo JC, Kang JE, Lee HJ, Yim SV, Kim CJ (2001) Acupuncture increases cell proliferation in dentate gyrus after transient global ischemia in gerbils. Neurosci Lett 297:21–24.
Kim JI, Kim YS, Kang SK, Kim C, Park C, Lee MS, Huh Y (2008a) Electroacupuncture decreases nitric oxide synthesis in the hypothalamus of spontaneously hypertensive rats. Neurosci Lett 446:78–82.
Kim MR, Kim SJ, Lyu YS, Kim SH, Lee Y, Kim TH, Shim I, Zhao R, Golden GT, Yang CH (2005) Effect of acupuncture on behavioral hyperactivity and dopamine release in the nucleus accumbens in rats sensitized to morphine. Neurosci Lett 387:17–21.
Kim SK, Kim J, Woo HS, Jeong H, Lee H, Min BI, Nam S, Bae H (2010) Electroacupuncture induces Fos expression in the nucleus tractus solitarius via cholecystokinin A receptor signaling in rats. Neurol Res 32(Suppl 1):116–119.
Kim ST, Jeon S, Park HJ, Hong MS, Jeong WB, Kim JH, Kim Y, Lee HJ, Park HJ, Chung JH (2008b) Acupuncture inhibits kainic acid-induce hippocampal cell death in mice. J Physiol Sci 58:31–38.
King JT Jr, LaMotte CC (1989) El mouse as a model of focal epilepsy: a review. Epilepsia 30:257–265.
Klide AM, Farnbach GC, Gallagher SM (1987) Acupuncture therapy for the treatment of intractable, idiopathic epilepsy in five dogs. Acupunct Electro-Ther Res 12:71–74.
Kloster R, Larsson PG, Lossius R, Nakken KO, Dahl R, Xue XL, Zhou WX, Kinge E, Røssberg E (1999) The effect of acupuncture in chronic intractable epilepsy. Seizure 8:170–174.
Knopp A, Frahm C, Fidzinski P, Witte OW, Behr J (2008) Loss of GABAergic neurons in the subiculum and its functional implications in temporal lobe epilepsy. Brain 131:1516–1527.
Koechler RC, Roman RJ, Harder DR (2009) Astrocytes and the regulation of cerebral blood flow. Trends Neurosci 32:160–169.
Koepp MJ, Richardson MP, Brooks DJ, Duncan JS (1998) Focal cortical release of endogenous opioids during reading-induced seizures. Lancet 352:952–955.
Kofuji P, Newman EA (2004) Potassium buffering in the central nervous system. Neuroscience 129:1045–1056.
Koh S, Tibayan FD, Simpson JN, Jensen FE (2004) NBQX or topiramate treatment after perinatal hypoxia-induced seizures prevents later increases in seizure-induced neuronal injury. Epilepsia 45:569–575.
Koide S, Onishi H, Yamagami S, Kawakita Y (1992) Effects of morphine and D-Ala2-Leu5-enkephalin in the seizure-sensitive E1 mouse. Neurochem Res 17:779–783.
Kolarova A, Ringer R, Täuber MG, Leib SL (2003) Blockade of NMDA receptor subtype NR2B prevents seizures but not apoptosis of dentate gyrus neurons in bacterial meningitis in infant rats. BMC Neuroscience 4:21.
Kondratyev A, Gale K (2004) Latency to onset of status epilepticus determines molecular mechanisms of seizure-induced cell death. Mol Brain Res 121:86–94.
Kong J, Gollub R, Huang T, Polich G, Napadow V, Hui K, Vangel M, Rosen B, Kaptchuk TJ. (2007a) Acupuncture de qi, from qualitative history to quantitative measurement. J Altern Complement Med 13:1059–1070.
Kong LH, Sun GJ, Mao JJ, Su FR, Gao S (2007b) Effects of acupoint-embedding of medicinal thread and electroacupuncture on contents of TNF-alpha and IL-6 of hippocampus tissue in rats with focal cerebral ischemia-reperfusion injury. Zhen Ci Yan Jiu 32:301–305. (In Chinese with English abstract)
Kovács R, Rabanus A, Otáhal J, Patzak A, Julianna Kardos J,3 Klaus Albus K, Heinemann U, Kann O (2009) Endogenous nitric oxide is a key promoting factor for initiation of seizure-like events in hippocampal and entorhinal cortex slices. J Neurosci 29:8565–8577.

Koyama R, Ikegaya Y (2005) To BDNF or not to BDNF: that is the epileptic hippocampus. Neuroscientist 11:282–287.

Krishnakumar A, Abraham PM, Paul J, Paulose CS (2009) Down-regulation of cerebellar 5-HT(2C) receptors in pilocarpine-induced epilepsy in rats: therapeutic role of Bacopa monnieri extract. J Neurol Sci 284:124–128.

Kucheryavykh YV, Kucheryavykh LY, Nichols CG, Maldonado HM, Baksi K, Reichenbach A, Skatchkov SN, Eaton MJ (2007) Downregulation of Kir4.1 inward rectifying potassium channel subunits by RNAi impairs potassium transfer and glutamate uptake by cultured cortical astrocytes. Glia 55:274–281.

Kukushkin ML, Reshetniak VK (1986) Effect of electroacupuncture on the nature of changes in the activity of neurons in the 2d somatosensory region of the cerebral cortex. Biull Eksp Biol Med 102:517–519. (In Russian with English abstract)

Kwak SE, Kim JE, Choi HC, Song HK, Kim YI, Jo SM, Kang TC (2008) The expression of somatostatin receptors in the hippocampus of pilocarpine-induced rat epilepsy model. Neuropeptides 42:569–583.

La Marca R, Nedeljkovic M, Yuan L, Maercker A, Ehlert U (2010) Effects of auricular electrical stimulation on vagal activity: evidence from three-armed randomized trial. Clin Sci (Lond.) 118:537–546.

Larner AJ (1995) Axonal sprouting and synaptogenesis in temporal lobe epilepsy: possible pathogenetic and therapeutic roles of neurite growth inhibitory factors. Seizure 4:249–258.

Lasoń W, Przewłocka B, Coenen A, Przewłocki R, Van Luijtelaar G (1994) Effects of mu and delta opioid receptor agonists and antagonists on absence epilepsy in WAG/Rij rats. Neuropharmacology 33:161–166.

Lathers GM (2009) Epilepsy and sudden death: Personal reflections and call for global action. Epilepsy Behav 15:269–277.

Lee B, Shim I, Lee H, Yin CS, Park HK, Yang JS, Hahm DH (2010a) Morphine-induced locomotor response and Fos expression in rats are inhibited by acupuncture. Neurol Res 32(Suppl 1):107–110.

Lee GJ, Yin CS, Choi SK, Choi S, Yang JS, Lee H, Park HK (2010b) Acupuncture attenuates extracellular glutamate level in global ischemia model of rat. Neurol Res 32(Suppl 1):79–83.

Lee HJ, Lee B, Choi SH, Hahm DH, Kim MR, Roh PU, Pyun KH, Golden G, Yang CH, Shim I (2004) Electroacupuncture reduces stress-induced expression of c-fos in the brain of the rat. Am J Chin Med 32:795–806.

Lee MC, Rho JL, Kim MK, Woo YJ, Kim JH, Nam SC, Suh JJ, Chung WK, Moon JD, Kim HI (2001) c-JUN expression and apoptotic cell death in kainate-induced temporal lobe epilepsy. J Korean Med Sci 16:649–656.

Lee PH, Obie J, Hong JS (1989) Opioids induce convulsions and wet dog shakes in rats: mediation by hippocampal mu, but not delta or kappa opioid receptors. J Neurosci 9:692–697.

Lee RJ, Bajorek JG, Lomax P (1984) Similar anticonvulsant, but unique, behavioural effects of opioid agonists in the seizure-sensitive Mongolian gerbil. Neuropharmacology 23:517–524.

Lee SH, Chung SH, Lee JS, Kim SS, Shin HD, Lim BV, Jang MH, Kim H, Kim EH, Kim CJ (2002) Effects of acupuncture on the 5-hydroxytryptamine synthesis and tryptophan hydroxylase expression in the dorsal raphe of exercised rats. Neurosci Lett 332:17–20.

Legido A, Adler MW, Karkanias C, Geller EB, Bradley E, Greenstein JI, Grover WD (1995) Cholecystokinin potentiates morphine anticonvulsant action through both CCK-A and CCK-B receptors. Neuropeptides 28:107–113.

Leng YL, Pan QR, Chen XN, Yang TF (2000) Epilepsy treated by point thread embedding. Report of 86 cases. J Ext Ther Tradit Chin Med 9:16. (In Chinese with English abstract)

Li C, Zhu L, Li W, Ji C (1993) Relationship between the presynaptic depolarization effect of acupuncture and γ-aminobutyric acid, opioid peptide and substance P. Zhen Ci Yan Jiu 18:178–182. (In Chinese with English abstract)

Li Q, Guo JC, Jin HB, Cheng JS, Yang R (2005) Involvement of taurine in penicillin-induced epilepsy and anti-convulsion of acupuncture: a preliminary report. Acupunct Electro-Ther Res 30:1–14.

Li Q, Guo M, Xu X, Xiao X, Xu W, Sun X, Tao H, Li R (2007) Rapid decrease of GAD 67 content before the convulsion induced by hyperbaric oxygen exposure. Neurochem Res 33:185–193.

Li W, Lai XS (2007) Changes of interleukin-1beta and TNF-alpha contents in the hippocampus and the interventional effect of electroacupuncture in vascular dementia rats. Zhen Ci Yan Jiu 32:34–37. (In Chinese with English abstract)

Li YZ, Zhuang LX (2007) Current clinical research progress of study on the epilepsy treated with acupuncture. Zhen Jiu Lin Chuang Za Zhi 23:47–48. (In Chinese)

Li ZR, Shen MH, Niu WM (2008) Involvement of melatonin in the adjusting effect of electroacupuncture in resisting oxygen stress in cerebral ischemia-reperfusion injury rats. Zhen Ci Yan Jiu 33:164–168. (In Chinese with English abstract)

Liang LP, Beaudoin ME, Fritz MJ, Fulton R, Patel M (2007) Kainate-induced seizures, oxidative stress and neuronal loss in aging rats. Neuroscience 147:1114–1118.

Liang XB, Liu XY, Li FQ, Luo Y, Lu J, Zhang WM, Wang XM, Han JS (2002) Long-term high-frequency electro-acupuncture stimulation prevents neuronal degeneration and up-regulates BDNF mRNA in the substantia nigra and ventral tegmental area following medial forebrain bundle axotomy. Brain Res Mol Brain Res 108:51–59.

Lim S, Ryu YH, Kim ST, Hong MS, Park HJ (2003) Acupuncture increases neuropeptide Y expression in hippocampus of maternally-separated rats. Neurosci Lett 343:49–52.

Lin J, Deng QP, Zhang JW (2001) Observation on therapeutic effect of 160 cases of epilepsy treated with acupoint catgut embedding therapy. Zhongguo Zhen jiu 21(11):14–15. (In Chinese with English abstract)

Lin YC, Hong GG (1999) Acupuncture treatment of pediatric status epilepticus. Med Acupunct 11:poster presentation. (also see http://www.medicalacupuncture.org/aama_marf/journal/vol11_2/poster.htmlposter presentation).

Lin ZN (2008) What is the primary entity of Meridian? J Accord Interg Med 4:243–247.

Liu H, Gao HM, Zhang WQ, Tang YY, Song HS (2004a) Effects of chronic administration of PL017 and beta-funaltrexamine hydrochloride on susceptibility of kainic acid-induced seizures in rats. Sheng Li Xue Bao 56:101–106. (In Chinese with English abstract)

Liu J, Cheng JS (1995) Changes of amino acids release in rat's hippocampus during kainic acid induced epilepsy and acupuncture. Zhen Ci Yan Jiu 20:50–54. (In Chinese with English abstract)

Liu J, Cheng JS (1997) Hippocampal non-NMDA and GABA-A receptors in benzylpenicillin-induced epilepsy and electroacupuncture antiepilepsy. Zhongguo Yao Li Xue Bao 18:189–191. (In Chinese with English abstract)

Liu P, Qin W, Zhang Y, Tian J, Bai L, Zhou G, Liu J, Chen P, Dai J, von Deneen KM, Liu Y (2009a) Combining spatial and temporal information to explore function-guide action of acupuncture using fMRI. J Magn Reson Imaging 30:41–46.

Liu P, Zhang Y, Zhou G, Yuan K, Qin W, Zhuo L, Liang J, Chen P, Dai J, Liu Y, Tian J (2009b) Partial correlation investigation on the default mode network involved in acupuncture: an fMRI study. Neurosci Lett 462:183–187.

Liu RY, Huang XF (1998) Changes of somatostatin mRNA in rat frontal brain after penicillin induced epilepsy and ear acupuncture. Chin J Histochem Cytochem 7:271–275. (In Chinese with English abstract)

Liu S, Wang J, Zhu D, Fu Y, Lukowiak K, Lu YM (2003) Generation of functional inhibitory neurons in adult rat hippocampus. J Neurosci 23:732–736.

Liu XY, Zhou HF, Pan YL, Liang XB, Niu DB, Xue B, Li FQ, He QH, Wang XH, Wang XM (2004b) Electro-acupuncture stimulation protects dopaminergic neurons from inflammation-mediated damage in medial forebrain bundle-transected rats. Exp Neurol 189:189–196.

Liu YQ, Gao HM, Wang SP (2007) Effects of intrahippocampal sigma receptor on inhibition of electroconvulsive seizure by electroacupuncture (EA) in rats. China Med Herald 4:12–14. (In Chinese with English abstract)

Liu Z, Fang JQ, Zeng C, Fang F (2009c) Development of studies on the underlying mechanism of acupuncture intervention in reducing post-ischemic inflammatory reaction and the related new research though. Zhen Ci Yan Jiu 34:61–66. (In Chinese with English abstract)

Liu Z, Stafstrom CE, Sarkisian MR, Yang Y, Hori A, Tandon P, Holmes GL (1997) Seizure-induced glutamate release in mature and immature animals: an in vivo microdialysis study. NeuroReport 8:2019–2023.

Liu Z, Sun F, Han Y (2000) Effect of acupuncture on level of monoamines and activity of adenosine triphosphatase in lateral hypothalamic area of obese rats. Zhongguo Zhong Xi Yi Jie He Za Zhi 20:521–523. (In Chinese with English abstract)

Lombardo AJ, Kuzniecky R, Powers RE, Brown GB (1996) Altered brain sodium channel transcript levels in human epilepsy. Mol Brain Res 35:84–90.

Lopantsev V, Both M, Draguhn A (2009) Rapid plasticity at inhibitory and excitatory synapses in the hippocampus induced by ictal epileptiform discharges. Eur J Neurosci 29:1153–1164.

Löscher W (1998) Pharmacology of glutamate receptor antagonists in the kindling model of epilepsy. Prog Neurobiol 54:721–741.

Löscher W, Lehmann H, Teschendorf HJ, Traut M, Gross G (1999) Inhibition of monoamine oxidase type A, but not type B, is an effective means of inducing anticonvulsant activity in the kindling model of epilepsy. J Pharmacol Exp Ther 288:984–992.

Losi G, Cammarota M, Chiavegato A, Gomez-Gonzalo M, Carmignoto G (2010) A new experimental model of focal seizures in the entorhinal cortex. Epilepsia 51:1493–1502.

Lothman EW (1994) Seizure circuits in the hippocampus and associated structures. Hippocampus 4:286–290.

Lou Z (1989) Inhibitory effect of electroacupuncture on cAMP induced ECOG epileptiform waves. Zheng Ci Yan Jiu 14:319–322, 314. (In Chinese with English abstract)

Lu GW (1983) Characteristics of afferent fiber innervation on acupuncture points zusanli. Am J Physiol 245:R606-R612.

Lu Y, Wang X (2009) Genes associated with idiopathic epilepsies: a current overview. Neurol Res 31:135–143.

Luhmann HJ, Raabe K (1996) Characterization of neuronal migration disorders in neocortical structures: I. Expression of epileptiform activity in an animal model. Epilepsy Res 26:67–74.

Luo Y, Xu NG, Yi W, Du YX (2009) Effect of electroacupuncture on astrocytes in the marginal zone of cerebral ischemia locus in rats. Zhen Ci Yan Jiu 34:101–105. (In Chinese with English abstract)

Luszczki JJ (2009) Third-generation antiepileptic drugs: mechanisms of action, pharmacokinetics and interactions. Pharmacol Rep 61:197–216.

Lynd-Balta E, Pilcher WH, Joseph SA (2004) AMPA receptor alternations precede mossy fiber sprouting in young children with temporal lobe epilepsy. Neuroscience 126:105–114.

Ma DL, Tang YC, Chen PM, Chia SC, Jiang FL, Burgunder J, Lee WL, Tang FR (2006) Reorganization of CA3 area of the mouse hippocampus after pilocarpine induced temporal lobe epilepsy with special reference to the CA3-septum pathway. J Neurosci Res 83:318–331.

Ma HF, Ren XJ, Tu Y, Zhou L (2010) Effect of electroacupuncture on the expression of hippocampal calbindin-D 28 K in hyperlipemia rats with concurrent cerebral ischemia. Zhen Ci Yan Jiu 35:22–26. (In Chinese with English abstract)

Ma JY, Wu DZ (1986) The role of 5-HT in the electroneedling suppression of penicillin-induced epileptiform discharges. Sheng Li Xue Bao 38:377–382. (In Chinese with English abstract)

Ma R, Zhang X, Liu Y, Li X, Yang C, Xiong J (2001) Clinical observation on treatment of tonoclonic attack of infantile epilepsy with acupuncture plus Xi Feng capsule. J Tradit Chin Med 42:276–278. (In Chinese with English abstract)

Ma SX, Ma J, Moise G, Li XY (2005) Responses of neuronal nitric oxide synthase expression in the brainstem to electroacupuncture Zusanli (ST 36) in rats. Brain Res 1037:70–77.

Ma Y, Yun SR, Nam SY, Kim YB, Hong JT, Kim Y, Choi H, Lee K, Oh KW (2008) Protective effects of sanjoinine A against N-methyl-D-aspartate-induced seizure. Biol Pharm Bull 31: 1749–1754.

Maciejak P, Szyndler J, Turzyńska D, Sobolewska A, Bidziński A, Kołosowska K, Płaźnik A (2010) Time course of changes in the concentrations of amino acids in the brain structures of pentylenetetrazole-kindled rats. Brain Res 1342:150–159.

Maciejak P, Szyndler J, Turzyńska D, Sobolewska A, Taracha E, Skórzewska A, Lehner M, Bidziński A, Hamed A, Wisłowska-Stanek A, Płaźnik A (2009) The effects of group III mGluR ligands on pentylenetetrazol-induced kindling of seizures and hippocampal amino acids concentration. Brain Res 1282:20–27.
Madar I, Lesser RP, Krauss G, Zubieta JK, Lever JR, Kinter CM, Ravert HT, Musachoi JL, Mathews WB, Dannals RF, Frost JJ (1997) Imaging of delta- and mu-opioid receptors in temporal lobe epilepsy by positron emission tomography. Ann Neurol 41:358–367.
Madhavan R, Chao ZC, Wagenaar DA, Bakkum DJ, Potter SM (2006) Multi-site stimulation quiets network-wide spontaneous bursts and enhances functional plasticity in cultured cortical networks. Conf Proc IEEE Eng Med Biol Soc 1:1593–1596.
Madsen KK, White HS, Schousboe A (2010) Neuronal and non-neuronal GABA transporters as targets for antiepileptic drugs. Pharmacol Ther 125:394–401.
Maisano X, Carpentino J, Becker S, Lanza R, Aaron G, Grabel L, Naegele JR (2009) Embryonic stem cell-derived neural precursor grafts for treatment of temporal lobe epilepsy. Neurotherapeutics 6:263–277.
Malawska B (2005) New anticonvulsant agents. Cur Top Med Chem 5:69–85.
Manna I, Labate A, Gambardella A, Forabosco P, La Russa A, Le Piane E, Aguglia U, Quattrone A (2007) Serotonin transporter gene (5-Htt): association analysis with temporal lobe epilepsy. Neurosci Lett 421:52–56.
Manni L, Albanesi M, Guaragna M, Barbaro Paparo S, Aloe L (2010) Neurotrophins and acupuncture. Auton Neurosci 157:9–17.
Mantegazza M, Gambardella A, Rusconi R, Schiavon E, Annesi F, Cassulini RR, Labate A, Carrideo S, Chifari R, Canevini MP, Canger R, Franceschetti S, Annesi G, Wanke E, Quattrone A (2005) Identification of an Nav1.1 sodium channel (SCN1A) loss-of-function mutation associated with familial simple febrile seizures. Proc Natl Acad Sci USA 102:18177–18182.
Mao JJ, Farrar JT, Armstrong K, Donahue A, Ngo J, Bowman MA (2007) De qi: Chinese acupuncture patients' experiences and beliefs regarding acupuncture needling sensation--an exploratory survey. Acupunct Med 25:158–165
Mao KY, Guo Z (2005) Observations on the efficacy of combined acupuncture and medicine for treating epilepsy secondary to cerebral infarction. Shanghai Zhen jiu Za Zhi 24:17–18. (In Chinese with English abstract)
Marchi N, Angelov L, Masaryk T, Fazio V, Granata T, Hernandez N, Hallene K, Diglaw T, Franic L, Najm I, Janigro D (2007) Seizure-promoting effect of blood-brain barrier disruption. Epilepsia 48:732–742.
Marek B, Kajdaniuk D, Kos-Kudła B, Kapustecki J, Swietochowska E, Ostrowska Z, Siemińska L, Nowak M, Głogowska-Szelag J, Borgiel-Marek H, Ciesielska-Kopacz N, Foltyn W, Pierzchała K, Krysiak R, Bienek R (2010) Mean daily plasma concentrations of beta-endorphin, leu-enkephalin, ACTH, cortisol, and DHEAS in epileptic patients with complex partial seizures evolving to generalized tonic-clonic seizures. Endokrynol Pol 61:103–110.
Marucci G, Rubboli G, Giulioni M (2010) Role of dentate gyrus alterations in mesial temporal sclerosis. Clin Neuropathol 29:32–35.
Massachusetts Institute of Technology (2006, September 12). Epilepsy Breakthrough On Horizon: MIT Developing Device To Detect Oncoming Seizures. *ScienceDaily*. Retrieved June 29, 2012, from http://www.sciencedaily.com /releases/2006/09/060912215213.htm
Mathern GW, Adelson PD, Cahau LD, Leite JP (2002) Hippocampal neuron damage in human epilepsy: Meyer's hypothesis revisited. Prog Brain Res 135:237–251.
McDermott CM, Schrader LA (2011) Activation of opioid receptors increases intrinsic excitability of dentate gyrus granule cells. J Physiol 589:3517–3532.
McElroy-Cox C (2009) Alternative approaches to epilepsy treatment. Curr Neurol Neurosci Rep 9:313–318.
McNamara JO (1994) Cellular and molecular basis of epilepsy. J Neurosci 14:3413–3425.
Medeiros MA, Canteras NS, Suchecki D, Mello LE (2003) c-Fos expression induced by electroacupuncture at the Zusanli point in rats submitted to repeated immobilization. Braz J Med Biol Res 36:1673–1684.

Meisler MH, O'Brien JE, Sharkey LM (2010) Sodium channel gene family: epilepsy mutations, gene interactions and modifier effects. J Physiol 588:1841–1848.
Meldrum B (1984) Amino acid neurotransmitters and new approaches to anticonvulsant drug action. Epilepsia 25(Suppl 2):S140–9.
Meng PY, Sun GJ, Liu SH, Yan HM (2008) Effect of electroacupuncture pretreatment on glutamate-NMDAR signal pathway in hippocampal neurons of vascular dementia rats. Zhen Ci Yan Jiu 33:103–106. (In Chinese with English abstract)
Merrill MA, Clough RW, Dailey JW, Jobe PC, Browning RA (2007) Localization of the serotonergic terminal fields modulating seizures in the genetically epilepsy-prone rat. Epilepsy Res 76:93–102.
Milby AH, Halpern CH, Baltuch GH (2008) Vagus nerve stimulation for epilepsy and depression. Neurotherapeutics 5:75–85.
Min R, Nevian T (2012) Astrocyte signaling controls spike timing-dependent depression at neocortical synapses. Nat Neurosci 15:746–753.
Mishra PK, Kahle EH, Bettendorf AF, Dailey JW, Jobe PC (1993) Anticonvulsant effects of intracerebroventricularly administered norepinephrine are potentiated in the presence of monoamine oxidase inhibition in severe seizure genetically epilepsy-prone rats (GEPR-9s). Life Sci 52:1435–1441.
Moeller F, Muhle H, Wiegand G, Wolff S, Stephani U, Siniatchkin M (2010) EEG-fMRI study of generalized spike and wave discharges without transitory cognitive impairment. Epilepsy Behav 18:313–316.
Moneta D, Richichi C, Aliprandi M, Dournaud P, Dutar P, Billard JM, Carlo AS, Viollet C, Hannon JP, Fehlmann D, Nunn C, Hoyer D, Epelbaum J, Vezzani A (2002) Somatostatin receptor subtypes 2 and 4 affect seizure susceptibility and hippocampal excitatory neurotransmission in mice. Eur J Neurosci 16:843–849.
Monnerie H, Le Roux PD (2008) Glutamate alteration of glutamic acid decarboxylase (GAD) in GABAergic neurons: The role of cysteine proteases. Exp Neurol 213:145–153.
Montpied P, Winsky L, Dailey JW, Jobe PC Jacobowitz DM (1995) Alteration in levels of expression of brain calbindin D-28k and calretinin mRNA in genetically epilepsy-prone rats. Epilepsia 36:911–921.
Morgan RJ, Soltesz I (2008) Nonrandom connectivity of the epileptic dentate gyrus predicts a major role for neuronal hubs in seizures. Proc Natl Acad Sci USA 105:6179–6184.
Morimoto K, Fahnestock M, Racine RJ (2004) Kindling and status epilepticus models of epilepsy: rewiring the brain. Prog Neurobiol 73:1–60.
Mouri G, Jimenez-Mateos E, Engel T, Dunleavy M, Hatazaki S, Paucard A,, Matsushima S, Taki W, Henshall DC (2008) Unilateral hippocampal CA3-predominant damage and short latency epileptogenesis after intra-amygdala microinjection of kainate acid in mice. Brain Res 1213:140–151.
Mueller SG, Laxer KD, Barakos J, Cheong I, Finlay D, Garcia P, Cardenas-Nicolson V, Weiner MW (2010) Involvement of the thalamocortical network in TLE with and without mesiotemporal sclerosis. Epilepsia 51:1436–1445.
Nadler JV, Tu B, Timofeeva O, Jiao Y, Herzog H (2007) Neuropeptide Y in the recurrent mossy fiber pathway. Peptides 28:357–364.
Nagarkatti N, Deshpande LS, DeLorenzo RJ (2009) Development of the calcium plateau following status epilepticus: role of calcium in epileptogenesis. Expert Rev Neurother 9:813–824.
Napadow V, Liu J, Li M, Kettner N, Ryan A, Kwong KK, Hui KK, Audette JF (2007) Somatosensory cortical plasticity in carpal tunnel syndrome treated by acupuncture. Hum Brain Mapp 28:159–171.
Neal EG, Chaffe H, Schwartz RH, Lawson MS, Edwards N, Fitzsimmons G, Whitney A, Cross JH (2008) The ketogenic diet for the treatment of childhood epilepsy: a randomized controlled trial. Lancet Neurol 7:500–506.
Neder L, Valente V, Carlotti CG (2002) Glutamate NMDA receptor subunit R1 and GAD mRNA expression inhuman temporal lobe epilepsy. Cell Mol Neurobiol 22:689–698.

Negus SS, Gatch MB, Mello NK, Zhang X, Rice K (1998) Behavioral effects of the delta-selective opioid agonist SNC80 and related compounds in rhesus monkeys. J Pharmacol Exp Ther 286:362–375.

NIH Consensus Development Panel on Acupuncture (1998) Acupuncture. JAMA 280: 1518–1524.

Nita DA, Cissé Y, Timofeev I, Steriade M (2006) Increased propensity to seizures after chronic cortical deafferentation in vivo. J Neurophysiol 95: 902–913.

Noguchi E, Hayashi H (1996) Increases in gastric acidity in response to electroacupuncture stimulation of the hindlimb of anesthetized rats. JPN J Physiol 46:53–58.

Nordio M, Romanelli F (2008) Efficacy of wrists overnight compression (HT 7 point) on insomniacs: possible role of melatonin? Minerva Med 99:539–547.

Norwood BA, Bumanglag AV, Osculati F, Sbarbati A, Marzola P, Nicolato E, Fabene PF, Sloviter RS (2010) Classic hippocampal sclerosis and hippocampal-onset epilepsy produced by a single "cryptic" episode of focal hippocampal excitation in awake rats. J Comp Neurol 518: 3381–3407.

Obenaus A, Esclapez M, Hauser CR (1993) Loss of glutamate decarboxylase mRNA containing neurons in the rat dentate gyrus following pilocarpine-induced seizures. J Neurosci 13: 4470–4485.

Ogiwara I, Miyamoto H, Morita N, Atapour N, Mazaki E, Inoue I, Takeuchi T, Itohara S, Yanagawa Y, Obata K, Furuichi T, Hensch TK, Yamakawa K (2007) Nav1.1 localizes to axons of parvalbumin-positive inhibitory interneurons: a circuit basis for epileptic seizures in mice carrying a Scn1a gene mutation. J Neurosci 27:5903–5914.

Okada M (2010) Generation of epilepsy animal model bearing a genetic abnormality identified in autosomal dominant nocturnal frontal lobe epilepsy (ADNFLE) of humans. Nihon Shinkei Seishin Yakurigaku Zasshi 30:9–14. (In Japanese with English abstract)

Olsen ML, Southeimer H (2004) Mislocalization of Kir channels in malignant glia. Glia 46: 63–73.

Olsen ML, Southeimer H (2008) Functional implications for Kir4.1 channels in glial biology: from K^+ buffering to cell differentiation. J Neurochem 107:589–601.

Olsen RW, Avoli M (1997) GABA and epileptogenesis. Epilepsia 38:399–407.

Orthmann-Murphy JL (2008) Gap junctions couple astrocytes and oligodendrocytes. J Mol Neurosci 35:101–116.

Ortinski PI, Dong J, Mungenast A, Yue C, Takano H, Watson DJ, Haydon PG, Coulter DA (2010) Selective induction of astrocytic gliosis generates deficits in neuronal inhibition. Nat Neurosci 13:584–591.

Otani N, Nawashiro H, Fukui S, Ooigawa H, Ohsumi A, Toyooka T, Shima K, Gomi H, Brenner M (2006) Enhanced hippocampal neurodegeneration after traumatic or kainate excitotoxicity in GFAP-null mice. J Clin Neurosci 13:934–938.

Ou YW, Han L, Da CD, Huang YL, Cheng JS (2001) Influence of acupuncture upon expressing levels of basic fibroblast growth factor in rat brain following focal cerebral ischemia--evaluated by time-resolved fluorescence immunoassay. Neurol Res 23:47–50.

Pack AM (2010) Implications of hormonal and neuroendocrine changes associated with seizures and antiepileptic drugs: a clinical perspective. Epilepsia 51(Suppl 3):150–153.

Pang J, Itano T, Sumitani K, Negi T, Miyamoto O (2003) Electroacupuncture attenuates both glutamate release and hyperemia after transient ischemia in gerbils. Am J Chin Med 31: 295–303.

Panzer RB, Chrisman CL (1994) An auricular acupuncture treatment for idiopathic canine epilepsy: a preliminary report. Am J Chin Med 22:11–17.

Parent JM (2002) The role of seizure-induced neurogenesis in epileptogenesis and brain repair. Epilepsy Res 50:179–189.

Park HJ, Lim S, Joo WS, Yin CS, Lee HS, Lee HJ, Seo JC, Lee MK, Son YS, Kim YJ, Kim CJ, Kim YS, Chung JH (2003) Acupuncture prevents 6-hydroxydopamine-induced neuronal death in the nigrostriatal dopaminergic system in the rat Parkinson's disease model. Exp Neurol 180:93–98.

Park Y, Luo L, Parhi KK, Netoff T (2011) Seizure prediction with spectral power of EEG using cost-sensitive support vector machines. Epilepsia 52:1761–1770.
Peng YJ, Zhou F, Gu J, Yang R, Yang YQ, Cheng JS, Guo JC (2007) Regulative effect of electroacupuncture on aquaporin-4 in rats with focal cerebral ischemia/reperfusion. Zhen Ci Yan Jiu 32:83–87. (In Chinese with English abstract)
Peng YS (2003) Clinical study on treatment of infantile primary epilepsy with acupoint catgut-implantation therapy. World J Acu-Mox 13:38–41.
Perea G, Araque A (2007) Astrocytes potentiate transmitter release at single hippocampal synapses. Science 317:1083–1086.
Perea G, Navarrete M, Araque A (2009) Tripartite synapses: astrocytes process and control synaptic information. Trends Neurosci 32:421–431.
Perucca E, Meador KJ (2005) Adverse effects of antiepileptic drugs. Acta Neurol Scand 112 (Suppl. 181): 30–35.
Prince DA, Connors BW (1984) Mechanisms of epileptogenesis in cortical structures. Ann Neurol 16(suppl): s59-s64.
Proposal for revised classification of epilepsies and epileptic syndromes. Commission on Classification and Terminology of the International League Against Epilepsy. Epilepsia 30:389–399. 1989.
Rao RS, Prakash A, Medhi B (2009) Role of different cytokines and seizure susceptibility: a new dimension towards epilepsy research. Indian J Exp Biol 47:625–634.
Ratzliff ADH, Howard AL, Santhakumar V, Osapay I, Soltesz I (2004) Rapid deletion of mossy cells does not result in a hyperexcitable dentate gyrus: Implications for epileptogenesis. J Neurosci 24:2259–2269.
Reckziegel G, Beck H, Schramm J, Urban BW, Elger CG (1999) Carbamazepine effects on Na^+ currents in human dentate granule cells from epileptogenic tissue. Epilepsia 40:401–407.
Reid CA, Adams BEL, Myers D, O'Brien TG, Williams DA (2008) Sub region-specific modulation of synchronous neuronal burst firing after a kainic acid insult in organotypic hippocampal cultures. BMC Neuroscience 9:59.
Remy S, Urban BW, Elger CE, Beck H (2003) Anticonvulsant pharmacology of voltage-gated Na^+ channels in hippocampal neurons of control and chronically epileptic rats. Eur J Neurosci 17:2648–2658.
Ren Y (2006) Acupuncture treatment of Jacksonian epilepsy-A report of 98 cases. J Tradit Chin Med 26:177–178.
Ren Y, Bai L, Feng Y, Tian J, Li K (2010) Investigation of acupoint specificity by functional connectivity analysis based on graph theory. Neurosci Lett 482:95–100.
Rho JM, Szot P, Tempel BL, Schwartzkroin PA (1999) Developmental seizure susceptibility of Kv1.1 potassium channel knockout mice. Dev Neurosci 21:320–327.
Ribak CE, Tran PH, Spigelman I, Okazaki MM, Nadler JV (2000) Status epilepticus-induced hilar basal dendrites on rodent granule cells contribute to recurrent excitatory circuitry. J Comp Neurol 428:240–253.
Ricci G, Volpi L, Pasquali L, Petrozzi L, Siciliano G (2009) Astrocyte-neuron interactions in neurological disorders. J Biol Phys 35:317–336.
Rocha L, Cuellar-Herrera M, Velasco M, Velasco F, Velasco AL, Jiménez F, Orozco-Suarez S, Borsodi A (2007a) Opioid receptor binding in parahippocampus of patients with temporal lobe epilepsy: its association with the antiepileptic effects of subacute electrical stimulation. Seizure 16:645–652.
Rocha L, Lorigados-Pedre L, Orozco-Suárez S, Morales-Chacón L, Alonso-Vanegas M, García-Maeso I, Villeda-Hernández J, Osorio-Rico L, Estupiñán B, Quintana C (2007b) Autoradiography reveals selective changes in serotonin binding in neocortex of patients with temporal lobe epilepsy. Prog Neuropsychopharmacol Biol Psychiatry 31:1208–1218.
Rogawski MA (2006) Diverse mechanisms of antiepileptic drugs in the development pipeline. Epilepsy Res 69:273–294.

Roshanpour M, Ghasemi M, Riazi K, Rafiei-Tabatabaei N, Ghahremani MH, Dehpour AR (2009) Tolerance to the anticonvulsant effect of morphine in mice: blockage by ultra-low dose naltrexone. Epilepsy Res 83:261–264.

Rouach N, Koulakoff A, Abudara V, Willecke K, Giaume C (2008) Astroglial metabolic networks sustain hippocampal synaptic transmission. Science 322:1551–1555.

Saboory E, Derchansky M, Ismaili M, Jahromi SS, Brull R, Carlen PL, El Beheiry H (2007) Mechanisms of morphine enhancement of spontaneous seizure activity. Anesth Analg 105:1729–1735.

Sadowski M, Lazarewicz JW, Jakubowska-Sadowska K, Wisniewski HM, Mossakowski MJ, Brown WT (2002) Long-term changes in calbindin D28K immunoreactivity in the rat hippocampus after cardiac arrest. Neurosci Lett 321:90–94.

Sander JW (2003) The epidemiology of epilepsy revisited. Curr Opin Neurol 16:165–170.

Sander JW, Shorvon SD (1996) Epidemiology of the epilepsies. J Neurol Neurosurg Psychiatry 61:433–43.

Sandow N, Zahn RK, Gabriel S, Heinemann U, Lehmann TN (2009) Glutamine induces epileptiform discharges in superficial layers of the medial entorhinal cortex from pilocarpine-treated chronic epileptic rats in vitro. Epilepsia 50:849–858.

Sarac S, Afzal S, Broholm H, Madsen FF, Ploug T, Laursen H (2009) Excitatory amino acid transporters EAAT-1 and EAAT-2 in temporal lobe and hippocampus in intractable temporal lobe epilepsy. APMIS 117:291–301.

Sashihara S, Yanagihara N, Izumi F, Murai Y, Mita T (1994) Differential upregulation of voltage-dependent Na^+ channels induced by phenytoin in brains of genetically seizure-susceptible (E1) and control (ddy) mice. Neuroscience 62:803–811.

Sashihara S, Yanagihara N, Kobayashi H, Izumi F, Tsuji S, Murai Y, Mita T (1992) Overproduction of voltage-dependent Na^+ channels in the developing brain of genetically seizure-susceptible E1 mice. Neuroscience 48:285–291.

Savolainen KM, Hirvonen MR (1992) Second messengers in cholinergic-induced convulsions and neuronal injury. Toxicol Lett 64–65:437–445.

Schachter SC (2008) Complementary and alternative medical therapies. Curr Opin Neurol 21: 184–189.

Schwarzer C, Sperk G (1995) Hippocampal granule cells express glutamic acid decarboxylase-67 after limbic seizures in the rat. Neuroscience 69:705–709.

Scott CR, King MD, Gadian DG, Neville BGR, Connelly A (2003) Hippocampal abnormalities after prolonged febrile convulsion: A longitudinal MRI study. Brain 126:2551–2557.

Seifert G, Carmignoto G, Steinhäuser C (2010) Astrocyte dysfunction in epilepsy. Brain Res Rev 63:21–221..

Seifert G, Schilling K, Steinhauser C (2006) Astrocyte dysfunction in neurological disorders: a molecular perspective. Nat Rev Neurosci 7:194–206.

Sejima H, Ito M, Kishi K, Tsuda H, Shiraishi H (1997) Regional excitatory and inhibitory amino acid concentrations in pentylenetetrazol kindling and kindled rat brain. Brain Dev 19:171–175.

Semyanov A, Godukhin O (2001) Epileptiform activity and EPSP-spike potentiation induced in rat hippocampal CA1 slices by repeated high-K^+: involvement of ionotropic glutamate receptors and Ca^{2+}/calmodulin-dependent protein kinase II. Neuropharmacology 40:203–211.

Shen MH, Li ZR, Xiang XR, Niu WM (2009) Effect of electroacupuncture on cerebral cortex ultrastructure in rats with cerebral ischemia-reperfusion injury. Zhen Ci Yan Jiu 34:167–170. (In Chinese with English abstract)

Shen MH, Tang QQ, Li ZR, Ma C (2010) Effect of electroacupuncture on hippocampal LTP in Alzheimer' s disease rats induced by Abeta(25–35). Zhen Ci Yan Jiu 35:3–7. (In Chinese with English abstract)

Shetty AK, Zaman V, Hattiangady B (2005) Repair of the injected adult hippocampus through graft-mediated modulation of the plasticity of the dentate gyrus in a rat model of temporal lobe epilepsy. J Neurosci 25:8391–8401.

Shi JF, Gao M (2001) Absence seizures of epilepsy treated by scalp acupuncture combined with hydro-acupuncture at Renying point. Shaanxi J Tradit Chin Med 22:43–44. (In Chinese)

Shi R, Ji G, Zhao L, Wang S, Dongjun (1998) Effects of electroacupuncture and twirling reinforcing-reducing manipulations on volume of microcirculatory blood flow in cerebral pia mater. J Tradit Chin Med 18:220–224.

Shi Z, Gong B, Jia Y, Huo Z (1987) The efficacy of electro-acupuncture on 98 cases of epilepsy. J Tradit Chin Med 7:21–22.

Shin EJ, Jeong JH, Chung YH, Kim TW, Shin CY, Kim WK, Ko KH, Kim HC (2009) Decrease in the kainate-induced wet dog shake behavior in genetically epilepsy-prone rats: possible involvement of an impaired synaptic transmission to the 5-HT(2A) receptor. J Pharmacol Sci 110:401–404.

Shors TJ, Miesegaes G, Beylin A, Zhao M, Rydel T, Could E (2001) Neurogenesis in the adult is involved in the formation of trace memories. Nature 410:372–376.

Shu J, Liu RY, Huang XF (2004) The effects of ear-point stimulation on the contents of somatostatin and amino acid neurotransmitters in brain of rat with experimental seizure. Acupunct Electro-Ther Res 29:43–51.

Shu J, Liu RY, Huang XF (2005) Efficacy of ear-point stimulation on experimentally induced seizures. Acupunct Electro-Ther Res 30:43–52.

Siddiqui AH, Joseph SA (2005) CA3 axonal sprouting in kainate-induced chronic epilepsy. Brain Res 1066:129–146.

Sierra-Paredes G, Sierra-Marcuño G (2007) Extrasynaptic GABA and glutamate receptors in epilepsy. CNS Neurol Dis-Drug Targets 6:288–300.

Simmons ML, Chavkin C (1996) Endogenous opioid regulation of hippocampal function. Int Rev Neurobiol 39:145–196.

Simonato M, Romualdi P (1996) Dynorphin and epilepsy. Prog Neurobiol 50:557–583.

Simonato M, Zucchini S (2010) Are the neurotrophic factors a suitable therapeutic target for the prevention of epileptogenesis? Epilepsia 51(Suppl 3):48–51.

Skyers PS, Einheber S, Pierce JP, Milner TA (2003) Increased mu-opioid receptor labeling is found on inner molecular layer terminals of the dentate gyrus following seizures. Exp Neurol 179: 200–209.

Sloviter RS (1999) Status epilepticus-induced neuronal injury and network reorganization. Epilepsia 40(Suppl. 1):S34–39.

Sloviter RS, Dean E, Sollas AL, Goodman TH (1996) Apoptosis and necrosis induced in different hippocampal neuron populations by repetitive perforant path stimulation in the rat. J Comp Neurol 366:516–533.

Smeyne RJ, Vendrell M, Hayward M, Baker SJ, Miao GG, Schilling K, Robertson LM, Curran T, Morgan JI (1993) Continuous c-fos expression precedes programmed cell death in vivo. Nature 363:166–169.

Smith PD, McLean KJ, Murphy MA, Turnley AM, Cook MJ (2005) Seizures, not hippocampal neuronal death, provoke neurogenesis in a mouse rapid electrical amygdale kindling model of seizures. Neuroscience 136:405–415.

Smolders I, Clinckers R, Meurs A, De Bundel D, Portelli J, Ebinger G, Michotte Y (2008) Direct enhancement of hippocampal dopamine or serotonin levels as a pharmacodynamic measure of combined antidepressant-anticonvulsant action. Neuropharmacology 54:1017–1028.

Somjen GG, Müller M (2000) Potassium-induced enhancement of persistent inward current in hippocampal neurons in isolation and in tissue slices. Brain Res 885:102–110.

Song YP, Yang W, Guo HM, Han YY (2006) Clinical observation on acupuncture combined with medicine for treatment of infantile febrile convulsion. Zhongguo Zhen jiu 26:561–562. (In Chinese with English abstract)

Spence DW, Kayumov L, Chen A, Lowe A, Jain U, Katzman MA, Shen J, Perelman B, Shapiro CM (2004) Acupuncture increases nocturnal melatonin secretion and reduces insomnia and anxiety: a preliminary report. J Neuropsychiatry Clin Neurosci 16:19–28.

Sperk G, Hamilton T, Colmers WF (2007) Neuropeptide Y in the dentate gyrus. Prog Brain Res163:285–297.

Stafstrom CE (2007) Persistent sodium current and its role in epilepsy. Epilepsy Curr 7:15–22.

Statler KD, Swank S, Abildskov V, Bigler ED, White HS (2008) Traumatic brain injury during development reduces minimal clonic seizure thresholds at maturity. Epilepsy Res 80: 163–170.

Stavem K, Kloster R, Røssberg E, Larsson PG, Dahl R, Kinge E, Lossius R, Nakken KO (2000) Acupuncture in intractable epilepsy: lack of effect on health-related quality of life. Seizure 9:422–426.

Stefulj J, Bordukalo-Niksic T, Hecimovic H, Demarin V, Jernej B (2010) Epilepsy and serotonin (5HT): variations of 5HT-related genes in temporal lobe epilepsy. Neurosci Lett 478:29–31.

Sun C, Mtchedlishvili Z, Bertram EH, Erisir A, Kapur J (2007) Selective loss of dentate hilar interneurons contributes to reduced synaptic inhibition of granule cells in an electrical stimulation-based animal model of temporal lobe epilepsy. J Comp Neurol 500:876–893.

Sutula TP, Cavazos JE, Woodart AR (1994) Long-term structural and functional alternations induced in the hippocampus by kindling: Implications for memory dysfunction and the development of epilepsy. Hippocampus 4:254–258.

Swann J, Al-Noori S, Jiang M, Lee CL (2000) Spine loss and other dendritic abnormalities in epilepsy. Hippocampus 10:617–625.

Swartz BE, Houser CR, Tomiyasu U, Walsh GO, DeSalles A, Rich JR, Delgado-Escueta A (2006) Hippocampal cell loss in posttraumatic human epilepsy. Epilepsia 47:1373–1382.

Szaflarski JP, Difrancesco M, Hirschauer T, Banks C, Privitera MD, Gotman J, Holland SK (2010) Cortical and subcortical contributions to absence seizure onset examined with EEG/fMRI. Epilepsy Behav 18:404–413.

Szot P, Reigel CE, White SS, Veith RC (1996) Alterations in mRNA expression of systems that regulate neurotransmitter synaptic content in seizure-naive genetically epilepsy-prone rat (GEPR): transporter proteins and rate-limiting synthesizing enzymes for norepinephrine, dopamine and serotonin. Brain Res Mol Brain Res 43:233–245.

Szot P, Weinshenker D, White SS, Robbins CA, Rust NC, Schwartzkroin PA, Palmiter RD (1999) Norepinephrine-deficient mice have increased susceptibility to seizure-inducing stimuli. J Neurosci 19:10985–10992.

Szot P, White SS, Shen DD, Anderson GD (2005) Valproic acid, but not lamotrigine, suppresses seizure-induced c-fos and c-Jun mRNA expression. Brain Res Mol Brain Res 135:285–289.

Szyndler J, Maciejak P, Turzyńska D, Sobolewska A, Bidziński A, Płaźnik A (2010) Time course of changes in the concentrations of monoamines in the brain structures of pentylenetetrazole-kindled rats. J Neural Transm 117:707–718.

Szyndler J, Maciejak P, Turzyńska D, Soholewska A, Lehner M, Taracha E, Walkowiak J, Skórzewska A, Wislowska-Stanek A, Hamed A, Bidziński A, Plaźnik A (2008) Changes in the concentration of amino acids in the hippocampus of pentylenetetrazole-kindled rats. Neurosci Lett 439:245–249.

Takechi K, Fujiwara A, Watanabe Y, Kamei Y (2009) Participation of GABA-ergic system in epileptogenic activity induced by teicoplanin in mice. Epilepsy Res 84:127–134.

Tallent MK, Qiu C (2008) Somatostatin: an endogenous antiepileptic. Mol Cell Endocrinol 286: 96–103.

Tanaka K, Watase K, Manabe T, Yamada K, Watanabe M, Takahashi K, Iwama H, Nishikawa T, Ichihara N, Kikuchi T, Okuyama S, Kawashima N, Hori S, Takimoto M, Wada K (1997) Epilepsy and exacerbation of brain injury in mice lacking the glutamate transporter GLT-1. Science 276: 1699–1702.

Tang W, Wang W, Zou SF, Bai Y, Wang ZR (2008) Experimental study on protective effect of pre-acupuncture on neuron injury after epilepsy. Chin J Tradit Med Sci Technol 15:406–407. (In Chinese with English abstract)

Tass PA, Silchenko AN, Hauptmann C, Barnikol UB, Speckmann EJ (2009) Long-lasting desynchronization in rat hippocampal slice induced by coordinated reset stimulation. Phys Rev E Stat Nonlin Soft Matter Phys. 80(1 Pt 1):011902. DOI:10.1103/PhysRevE.80.011902

Terauchi A, Johnson-Venkatesh EM, Toth AB, Javed D, Sutton MA, Umemori H (2010) Distinct FGFs promote differentiation of excitatory and inhibitory synapses. Nature 465:783–787.

Terman GW, Drake CT, Simmons ML, Milner TA, Chavkin C (2000) Opioid modulation of recurrent excitation in the hippocampal dentate gyrus. J Neurosci 20:4379–4388.

Theodore WH (2003) Does serotonin play a role in epilepsy? Epilepsy Curr 3:173–177.

Thomas EA, Hawkins RJ, Richards KL, Xu R, Gazina EV, Petrou S (2009) Heat opens axon initial segment sodium channels : A febrile seizure mechanism? Ann Neurol 66:219–226.

Thomas Jefferson University (2007). Brain Implant Could Predict And Stop Epilepsy Seizures. *ScienceDaily*. Retrieved June 29, 2012, from http://www.sciencedaily.com/releases/2007/07/070726142003.htm

Tian GF, Azmi H, Takano T, Xu Q, Peng W, Lin J, Oberheim N, Lou N, Wang X, Zielke HR, Kang J, Nedergaard M (2005) An astrocytic basis of epilepsy. Nat Med 11:973–981.

Tian LM, Otoom S, Alkadhi KA (1995) Endogenous bursting due to altered sodium channel formation in rat hippocampal CA1 neurons. Brain Res 680:164–172.

Tian N, Wang F, Tian DR, Zou Y, Wang SW, Guan LL, Shi YS, Chang JK, Yang J, Han JS (2006) Electroacupuncture suppresses expression of gastric ghrelin and hypothalamic NPY in chronic food restricted rats. Peptides 27:2313–2320.

Tian XS, Zhou F, Yang R, Xia Y, Wu GC, Guo JC (2008a) Electroacupuncture protects the brain against acute ischemic injury via up-regulation of delta-opioid receptor in rats. Zhong Xi Yi Jie He Xue Bao 6:632–638. (In Chinese with English abstract)

Tian XS, Zhou F, Yang R, Xia Y, Wu GC, Guo JC (2008b) Role of intracerebral δ-opium receptors in protecting ischemic injury in rats by cumulative electro-acupuncture. Shanghai J Tradit Chin Med 42:71–74. (In Chinese with English abstract)

Tomkins O, Shelef I, Kaizerman I, Eliushin A, Afawi Z, Misk A, Gidon M, Cohen A, Zumsteg D, Friedman A (2008) Blood brain barrier disruption in post-traumatic epilepsy. J Neurol Neurosurg Psychiatry 79:774–777.

Topolnik L, Steriade M, Timofeev I (2003) Hyperexcitability of intact neurons underlies acute development of trauma-related electrographic seizures in cats in vivo. Eur J Neurosci 18:486–496.

Tortella FC, Echevarria E, Robles L, Mosberg HI, Holaday JW (1988) Anticonvulsant effects of mu (DAGO) and delta (DPDPE) enkephalins in rats. Peptides 9:1177–1181.

Tortella FC, Long JB (1985) Endogenous anticonvulsant substance in rat cerebrospinal fluid after a generalized seizure. Science 228:1106–1108.

Tortella FC, Robles L, Holaday JW (1985) The anticonvulsant effects of DADLE are primarily mediated by activation of delta opioid receptors: interactions between delta and mu receptor antagonists. Life Sci 37:497–503.

Tóth K, Erőss L, Vajda J, Halász P, Freund TF Maglóczky Z (2010) Loss and reorganization of calretinin-containing interneurons in the epileptic human hippocampus. Brain 133:2763–2777.

Toth Z, Yan XX, Haftoglou S, Ribak CE, Baram TZ (1998) Seizure-induced neuronal injury: vulnerability to febrile seizures in an immature rat model. J Neurosci 18:4285–4294.

Trindade-Filho EM, de Castro-Neto EF, de A Carvalho R, Lima E, Scorza FA, Amado D, Naffah-Mazzacoratti Mda G, Cavalheiro EA (2008) Serotonin depletion effects on the pilocarpine model of epilepsy. Epilepsy Res 82:194–199.

Tyzio R, Khalilov I, Represa A, Crepel V, Zilberter Y, Rheims S, Aniksztejn L, Cossart R, Nardon R, Mukhtarov M, Minlebaev M, Epsztein J, Milh M, Becq H, Jorquera I, Bulteau C, Fohlen M, Olivier V, Dulac O, Dorfmuller G, Delalande O, Ben-Ari Y, Khazipor R (2009) Inhibitory actions of the gamma-aminobutyric acid in pediatric Sturge-Weber syndrome. Ann Neurol 66:209–218.

Ure A, Altrup U (2006) Block of spontaneous termination of paroxysmal depolarizations by forskolin (buccal ganglia, Helix pomatia). Neurosci Lett 392:10–15.

van Niekerk J, Eckersley N (1988) The use of acupuncture in canine epilepsy. J S Afr Vet Assoc 59:5.

van Praag H, Schinder AF, Christie BR, Toni N, Palmer TD, Gage FH (2002) Functional neurogenesis in adult hippocampus. Nature 415:1030–1034.

van Vliet EA, da Costa Araújo S, Redeker S, van Schaik R, Aronica E, Gorter JA (2007) Blood-brain barrier leakage may lead to progression of temporal lobe epilepsy. Brain 130:521–534.
Vázquez-López A, Sierra-Paredes G, Sierra-Marcuño G (2005) Role of cAMP-dependent protein kinase on acute picrotoxin-induced seizures. Neurochem Res 30:613–618.
Vezzani A, Balosso S, Ravizza T (2008) The role of cytokines in the pathophysiology of epilepsy. Brain Behav Immun 22:797–803.
Volterra A, Meldolesi J (2005) Astrocytes, from brain glue to communication elements: The revolution continues. Nat Rev Neurosci 6:626–640.
Voutsinos-Porche B, Koning E, Clément Y, Kaplan H, Ferrandon A, Motte J, Nehlig A (2006) EAAC1 glutamate transporter expression in the rat lithium-pilocarpine model of temporal lobe epilepsy. J Cereb Blood Flow Metab 26:1419–1430.
Vreugdenhil M, Faas GC, Wadman WJ (1998) Sodium currents in isolated CA1 neurons after kindling epileptogenesis. Neuroscience 86:99–107.
Vreugdenhil M, Wadman WJ (1999) Modulation of sodium currents in rat CA1 neurons by carbamazepine and valproate after kindling epileptogenesis. Epilepsia, 40:1512–1522.
Walczak TS, Leppik IE, D'Amelio M, Rarick J, So E, Ahman P, Ruggles K, Cascino GD, Annegers JF, Hauser WA (2001). Incidence and risk factors in sudden unexpected death in epilepsy: a prospective cohort study. Neurology 56:519–525.
Waldau B, Hattiangady B, Kuruba R, Shetty AK (2010) Medial ganglionic eminence-derived neural stem cell grafts ease spontaneous seizures and restore GDNF expression in a rat model of chronic temporal lobe epilepsy. Stem Cells 28:1153–1164.
Wallraff A, Kohling R, Neinemann U, Theis M, Willecke K, Steinhäuser C (2006) The impact of astrocytic gap junctional coupling on potassium buffering in the hippocampus. J Neurosci 26:5438–5447.
Wan P, Sun XD, Wu DZ, Zhang ZX (1997) The role of substantia nigra, thalamus and GABAA receptor in acupuncture inhibition during petit mal epilepsy. Liaoning J Tradit Chin Med 24:470–471. (In Chinese)
Wan P, Sun XD, Wu DZ, Zhang ZX (1999) Roles of reticular thalamic nucleus and GABA B receptor in absence epilepsy. Chin J Appl Physiol 15:59–61. (In Chinese with English abstract)
Wang BE, Cheng JS (1992) The relationship between dynorphon and acupuncture anticonvulsion in hippocampus. Chin Sci Bull 37:1321–1323. (In Chinese with English abstract)
Wang BE, Cheng JS (1994a) Alteration of dynorphin1–8 and leu-enkephalin in rat hippocampus during seizure and electroacupuncture. Zhongguo Yao Li Xue Bao 15:155–157. (In Chinese with English abstract)
Wang BE, Cheng JS (1994b) c-fos expression in rat brain during seizure and electroacupuncture. Zhongguo Yao Li Xue Bao 15:73–75. (In Chinese with English abstract)
Wang BE, Cheng JS (1994c) Changes of hippocampal amino acids levels and NMDA receptor during epilepsy and electroacupuncture. Acta Acad Med Shanghai 21:10–13. (In Chinese)
Wang BE, Cheng JS (1995) Release of hippocampal dynorphin during electro-stimulated seizure and acupuncture anti-convulsion. Shanghai Zhen Jiu Za Zhi 14:33–34. (In Chinese)
Wang BE, Yang R, Cheng JS (1994) Effect of electroacupuncture on the level of preproenkephalin mRNA in rat during penicillin-induced epilepsy. Acupunct Electro-Ther Res 19:129–140.
Wang H, Tanaka Y, Seki H, Jodo E, Kayama Y, Kawauchi A, Miki T, Otsuki M (2007) Acupuncture stimulation to the sacral segment affects state of vigilance in rats. Neurosci Res 57:531–537.
Wang JC (2001) One hundred and twenty cases of epilepsy treated by three acupoints on back. Shanghai Zhen Jiu Za Zhi 20:20. (In Chinese)
Wang JC, Huang YG, Wen XN, Wang WD, Xia F, Wang XM, Yang F, Wu SD (2006a) Effect of electroacupuncture applied at acupoint on neurogenesis in hippocampal dentate gyrus and behavior of rats with lithium-pilocarpine induced status epilepticus. J Fourth Mil Med Univ 27:441–444. (In Chinese with English abstract)
Wang JC, Wang WD, Liu XD, Xia F, Wen XN, Huang YG (2006b) Effect of electroacupuncture applied at acupoint on cognitive function in lithium-pilocarpine induced spontaneous recurrent epileptic rat. J Xi'an Jiaotong Univ (Med Sci) 27:380–382, 386. (In Chinese with English abstract)

Wang SM, Kain ZN, White P (2008) Acupuncture analgesia: I. The scientific basis. Anesth Analg 106:602–610.
Wang T, Liu CZ, Yu JC, Jiang W, Han JX (2009) Acupuncture protected cerebral multi-infarction rats from memory impairment by regulating the expression of apoptosis related genes Bcl-2 and Bax in hippocampus. Physiol Behav 96:155–161.
Warskulat U, Heller-Stilb B, Oermann E, Zilles K, Haas H, Lang F, Häussinger D (2007) Phenotype of the taurine transporter knockout mouse. Methods Enzymol 428:439–458.
Wasner M, Klier H, Borasio GD (2001) The use of alternative medicine by patients with amyotrophic lateral sclerosis. J Neurosci Sci 191:151–154.
Wasterlain CG and Shirasaka Y (1994) Seizures, brain damage and brain development. Brain Dev 16:279–295.
Wei Q, Liu Z (2003) Effects of acupuncture on monoamine neurotransmitters in raphe nuclei in obese rats. J Tradit Chin Med 23:147–150.
Weinshenker D, Szot P (2002) The role of catecholamines in seizure susceptibility: new results using genetically engineered mice. Pharmacol Ther 94:213–233.
Weinshenker D, Szot P, Miller NS, Palmiter RD (2001) Alpha(1) and beta(2) adrenoreceptor agonists inhibit pentylenetetrazole-induced seizures in mice lacking norepinephrine. J Pharmacol Exp Ther 298:1042–1048.
Wen T (2000) Treatment of 78 cases of epilepsy in narcotic abstaining. Zhongguo Zhen jiu 20(4):34. (In Chinese)
Wen XN, Huang YG (2008) Influence of electroacupuncture on somatostatin expression of newborn neurons in the hippocampus of chronic epilepsy rats. J Xi'an Jiaotong Univ (Med Sci) 29:370–373. (In Chinese with English abstract)
Wen XN, Huang YG, Wang JC, Yang F (2006) Effect of electroacupuncture on hippocampal neurogenesis of rats with experimental epilepsy. Chin J Neuromed 5:240–243. (In Chinese with English abstract)
Wenhe Z, Yucun S (1981) Change in levels of monoamine neurotransmitters and their main metabolites of rat brain after electric acupuncture treatment. Int J Neurosci 15:147–149.
Wetherington J, Serrano G, Dingledine R (2008) Astrocytes in the epileptic brain. Neuron 58:168–178.
Williams P, Dou P, Dudek FE (2004) Epilepsy and synaptic reorganization in a perinatal rat model of hypoxia-ischemia. Epilepsia 45:1210–1218.
Williams PA, Dudek FE (2007) A chronic histopathological and electrophysiological analysis of a rodent hypoxic-ischemic brain injury model and its use as a model of epilepsy. Neuroscience 149:943–961.
Witkin JM, Baez M, Yu J, Barton ME, Shannon HE (2007) Constitutive deletion of the serotonin-7 (5-HT(7)) receptor decreases electrical and chemical seizure thresholds. Epilepsy Res 75:39–45.
Wong AMK, Leong CP, Su TY, Yu SW, Tsai WC, Chen CPC (2003) Clinical trial of acupuncture for patients with spinal cord injuries. Am J Phys Med Rehabil 82:21–27.
Wong M (2005) Modulation of dendritic spines in epilepsy: cellular mechanisms and functional implications. Epilepsy Behav 7:569–577.
Wong M (2008) Stabilizing dendritic structure as a novel therapeutic approach for epilepsy. Expert Rev Neurother 8:907–915.
Wu D (1992) Mechanism of acupuncture in suppressing epileptic seizures. J Tradit Chin Med 12:187–192.
Wu DZ, Ma JY, Li WJ (1986) Inhibitory effect of electro-acupuncture on penicillin-induced cortical epileptiform discharges. Sheng Li Xue Bao 38:325–331. (In Chinese with English abstract)
Wu XD, Du LN, Wu GC, Cao XD (2001) Effects of electroacupuncture on blood-brain barrier after cerebral ischemia-reperfusion in rat. Acupunct Electro-Ther Res 26:1–9.
Wu Y, Shen Q, Zhang Q (1992) The effect of acupuncture on OHP-induced convulsion and its relationship to the brain GABA concentration in mice. Zhen Ci Yan Jiu 17:104–109. (In Chinese with English abstract)

Wu Y, Zou LP, Han TL, Zheng H, Caspi O, Wong V, Su Y, Shen KL (2008) Randomized controlled trial of traditional Chinese medicine (acupuncture and *tuina*) in cerebral palsy: Part 1-Any increase in seizure in integrated acupuncture and rehabilitation group versus rehabilitation group? J Altern Complement Med 14:1005–1009.

Wyeth MS, Zhang N, Mody I, Houser CR (2010) Selective reduction of cholecystokinin-positive basket cell innervation in a model of temporal lobe epilepsy. J Neurosci 30:8993–9006.

Xapelli S, Agasse F, Ferreira R, Silva AP, Malva JO (2006) Neuropeptide Y as an endogenous antiepileptic, neuroprotective and pro-neurogenic peptide. Recent Pat CNS Drug Discov 1: 315–324.

Xia Y, Zhao P, Xue J, Gu XQ, Sun X, Yao H, Haddad GG (2003) Na^+ channel expression and neuronal function in the Na^+/H^+ exchanger 1 null mutant mouse. J Neurophysiol 89:229–236.

Xing GG, Liu FY, Qu XX, Han JS, Wan Y (2007) Long-term synaptic plasticity in the spinal dorsal horn and its modulation by electroacupuncture in rats with neuropathic pain. Exp Neurol 208:323–332.

Xiong LZ, Yang J, Wang Q, Lu ZH (2007) Involvement of delta- and mu-opioid receptors in the delayed cerebral ischemic tolerance induced by repeated electroacupuncture preconditioning in rats. Chin Med J 120:394–399.

Xiong X, Zhang GZ, Huang WY, Sun JQ (2003) Clinical observation of acupuncture and Chinese medicine for treatment of epilepsy in children. Chin J Info Tradit Chin Med 10:62–63. (In Chinese)

Xu L, Zeng L, Wong M (2009) Impaired astrocytic gap junction coupling and potassium buffering in a mouse model of tuberous sclerosis complex. Neurobiol Dis 34:291–299.

Xu M, Aiuchi T, Nakaya K, Arakawa H, Maeda M, Tsuji A, Kato T, Takeshige C, Nakamura Y (1990) Effect of low-frequency electric stimulation on in vivo release of cholecystokinin-like immunoreactivity in medial thalamus of conscious rat. Neurosci Lett 118:205–207.

Xu R, Thomas EA, Gazina EV, Richards KL, Quick M, Wallace RH, Harkin LA, Heron SE, Berkovic SF, Scheffer IE, Mulley JC, Petrou S (2007a) Generalized epilepsy with febrile seizures plus-associated sodium channel β1 subunit mutations severely reduce beta subunit-mediated modulation of sodium channel function. Neuroscience 148:164–174.

Xu R, Thomas EA, Jenkins M, Gazina EV, Chiu C, Heron SE, Mulley JC, Scheffer IE, Berkovic SF, Petrou S (2007b) A childhood epilepsy mutation reveals a role for developmentally regulated splicing of a sodium channel. Mol Cell Neurosci 35:292–301.

Xu SF, Zhuang LX, Tang CZ, Yang JJ (2007c) Effects of acupuncture and embedding thread on central monoamine neurotransmitters in the depression model rat. Zhongguo Zhen Jiu 27:435–437. (In Chinese with English abstract)

Xu ZF (2006) Clinical observation on treatment of epileptic seizure by combined catgut embedding and herbal medicine. Shanghai Zhen Jiu Za Zhi 25:13–14. (In Chinese with English abstract)

Xu ZF, Wu GC, Cao XD (2002) Effect of electroacupuncture on the expression of interlukin-1beta mRNA after transient focal cerebral ischemia. Acupunct Electro-Ther Res 27:29–35.

Xue WG, Zhang Z, Bai LM, Xu H, Wu HX (2009) Effect of electroacupuncture on the behavior and the expression of amyloid beta-protein, amyloid precursor protein and ChAT in APP 695 V 717 I transgenic mice. Zhen Ci Yan Jiu 34:152–158. (In Chinese with English abstract)

Yahyavi-Firouz-Abadi N, Tahsili-Fahadan P, Riazi K, Ghahremani MH, Dehpour AR (2007) Melatonin enhances the anticonvulsant and proconvulsant effects of morphine in mice: role for nitric oxide signaling pathway. Epilepsy Res 75:138–144.

Yajima Y, Narita M, Takahashi-Nakano Y, Misawa M, Nagase H, Mizoguchi H, Tseng LF, Suzuki T (2000) Effects of differential modulation of mu-, delta- and kappa-opioid systems on bicuculline-induced convulsions in the mouse. Brain Res 862:120–126.

Yamakawa K (2006) Na channel gene mutations in epilepsy--the functional consequences. Epilepsy Res 70:S218–222.

Yan L, Li J, Guo C, Yin K (1999) Effects of acupuncture on contents of γ-aminobutyric acid and glutamic acid in the brain stem of rats of acute experimental epilepsy. Zhongguo Zhen jiu 19(4):43–45,4. (In Chinese with English abstract)

Yan Q, Ruan JW, Ding Y, Li WJ, Li Y, Zeng YS (2011) Electro-acupuncture promotes differentiation of mesenchymal stem cells, regeneration of nerve fibers and partial functional recovery after spinal cord injury. Exp Toxicol Pathol 63:151–156.

Yang EJ, Jiang JH, Lee SM, Hwang HS, Lee MS, Choi SM (2010) Electroacupuncture reduces neuroinflammatory responses in symptomatic amyotrophic lateral sclerosis model. J Neuroimmunol 223:84–91.

Yang F, Wang B, Shen DK, Xu GN, Ang WP, Zhang FQ, Liu XG (2008) Effect of electroacupuncture on ON-cGMP pathway in hippocampus of rats with epilepsy induced by pentylenetetrazol. Chin J Tradit Med Technol 15:325–326. (In Chinese with English abstract)

Yang F, Xu G, Wang P, Shi J, Zhang F, Ang W, Shen D, Liu X (2002a) Effect of electroacupuncture on the contents of amino acids in the brain in PTZ kindled epilepsy rats. Zhen Ci Yan Jiu 27:174–176, 227. (In Chinese with English abstract)

Yang F, Xu GL, Feng PZ, Shen DK, Hu JP, Yang YQ (2009a) Effects of electroacupuncture on behaviors and neuronal expression of calmodulin at hippocampus in rats with epilepsy induced by pentylenetetrazole. J Emerg Tradit Chin Med 18(1):77–79. (In Chinese with English abstract)

Yang F, Xu GL, Wang P, Zhang FQ, Shi J, Ang WP, Sheng DK, Liu XG (2002b) Effect of electroacupuncture on contents of cAMP and cGMP in hippocampus of rat with epilepsy. Chin J Tradit Med Sci Tech 9:199–200. (In Chinese with English abstract)

Yang F, Xu GL, Yang YQ, Shen DK, Feng PZ, Wang P, Liu XG (2009b) Effect of electroacupuncture on epileptic EEG and intracellular Ca^{2+} content in the hippocampus in epilepsy rats. Zhen Ci Yan Jiu 34:163–166. (In Chinese with English abstract)

Yang J (1990) Treatment of status epilepticus with acupuncture. J Tradit Chin Med 10:101–102.

Yang R, Cheng J (2010) Effect of acupuncture on epilepsy. In Xia Y, Cao X, Wu G, Cheng J Eds: Acupuncture therapy for neurological diseases. pp.326–364. Springer & Tsinghua University Press.

Yang R, Huang ZN, Cheng JS (2000) Anticonvulsion effect of acupuncture might be related to the decrease of neuronal and inducible nitric oxide synthases. Acupunct Electro-Ther Res 25: 137–143.

Yang R, Li Q, Guo JC, Jin HB, Liu J, Wang BE, Cheng JS (2006)Taurine participates in the anticonvulsive effect of electroacupuncture. Adv Exp Med Biol 583:389–394.

Yang R, Wang B, Cheng J (1996) Effect of electroacupuncture on cholecystokinin gene expression in rat hippocampus during penicillin-induced epileptic seizure. Zhen Ci Yan Jiu 21:62–66. (In Chinese with English abstract)

Yang ZJ, Shen DH, Guo X, Sun FY (2005) Electroacupuncture enhances striatal neurogenesis in adult rat brains after a transient cerebral middle artery occlusion. Acupunct Electro-Ther Res 30:185–199.

Yi PL, Tsai CH, Lin JG, Liu HJ, Chang FC (2004) Effects of electroacupuncture at "Anmian (extra)" acupoints on sleep activities in rats: the implication of the caudal nucleus tractus solitarius. J Biomed Sci 11:579–590.

Yi W, Xu NG, Wang GB (2006) Experimental study on effects of electroacupuncture in improving synaptic plasticity in focal cerebral ischemia rats. Zhongguo Zhong Xi Yi Jie He Za Zhi 26:710–714. (In Chinese with English abstract)

Yogeeswari P, Sriram D, Vaigundaragavendran J (2005) The GABA shunt: an attractive and potential therapeutic target in the treatment of epileptic disorders. Curr Drug Metab 6:127–139.

Yokoyama N, Mori N, Kumashiro H (1989) Chemical kindling induced by cAMP and transfer to electrical kindling. Brain Res 492:158–162.

Yoon SS, Kwon YK, Kim MR, Shim I, Kim KJ, Lee MH, Lee YS, Golden GT, Yang CH (2004) Acupuncture-mediated inhibition of ethanol-induced dopamine release in the rat nucleus accumbens through the GABAB receptor. Neurosci Lett 369:234–238.

Yoshimoto K, Fukuda F, Hori M, Kato B, Kato H, Hattori H, Tokuda N, Kuriyama K, Yano T, Yasuhara M (2006) Acupuncture stimulates the release of serotonin, but not dopamine, in the rat nucleus accumbens. Tohoku J Exp Med 208:321–326.

Young CC, Stegen M, Bernard R, Műller M, Bischofberger J, Veh RW, Haas CA, Wolfart J (2009) Upregulation of inward rectifier K^+ (Kir2) channels in dentate gyrus granule cells in temporal lobe epilepsy. J Physiol 587:4213–4233.

Young RSK, During, MJ, Aquila WJ, Tendler D, Ley E (1992) Hypoxia Increases Extracellular Concentrations of excitatory and inhibitory neurotransmitters in subsequently induced seizure: In vivo microdialysis study in the rabbit. Exp Neurol 117:204–209.

Yu DS, Ping XM, Zhao XL (1999) Epilepsy treated by acupuncture with growing fluid and extinguishing wind. Report of 30 cases. Jiangsu J Tradit Chin Med 20:37. (In Chinese)

Yu FH, Mantegazza M, Westenbroek RE, Robbins CA, Kalume F, Burton KA, Spain WJ, McKnight GS, Scheuer T, Catterall WA (2006) Reduced sodium current in GABAergic interneurons in a mouse model of severe myoclonic epilepsy in infancy. Nat Neurosci 9:1142–1149.

Yuan H, Han JS (1985) Electroacupuncture accelerates the biogenesis of central enkephalins in the rat. Sheng Li Xue Bao 37:265–273. (In Chinese with English abstract)

Zeng LH, Bero AW, Zhang B, Holtzman DM, Wong M (2010) Modulation of astrocyte glutamate transporters decreases seizures in a mouse model of Tuberous Sclerosis Complex. Neurobiol Dis 37:764–771.

Zhang HN, Ko MC (2009) Seizure activity involved in the up-regulation of BDNF mRNA expression by activation of central mu opioid receptors. Neuroscience 161:301–310.

Zhang HZ, Zhang LX, She YF, Li WL, Zhang XJ, Zhang RJ (2008a) Effect of electroacupuncture on apoptosis of hippocampus tissue in mice with vascular dementia. Zhen Ci Yan Jiu 33:377–381. (In Chinese with English abstract)

Zhang J, Li YZ, Zhuang LX (2006) Observation on therapeutic affect of 90 tonic-clonic epilepsy patients treated by catgut implantation therapy. J Clin Acu-Mox 22:8–10. (In Chinese)

Zhang JL, Guo Y, Li RW, Li CH, Luo MF (2009a) Protecting action of acupuncture serum on calcium over-loaded neurons of the hippocampus in rats of ischemia-reperfusion. Zhongguo Zhen Jiu 29:45–47. (In Chinese with English abstract)

Zhang JL, Zhang SP, Zhang HQ (2008b) Antiepileptic effect of electroacupuncture vs. vagus nerve stimulation in the rat thalamus. Neurosci Lett 441:183–187.

Zhang JL, Zhang SP, Zhang HQ (2008c) Antiepileptic effect of electroacupuncture vs. vagus nerve stimulation on cortical epileptiform activities. J Neurol Sci 270:114–121.

Zhang LX, Li XL, Smith MA, Post RM, Han JS (1997) Lipofectin-facilitated transfer of cholecystokinin gene corrects behavioral abnormalities of rats with audiogenic seizures. Neuroscience 77:15–22.

Zhang LX, Smith MA, Kim SY, Rosen JB, Weiss SR, Post RM (1996) Changes in cholecystokinin mRNA expression after amygdala kindled seizures: an in situ hybridization study. Brain Res Mol Brain Res 35:278–284.

Zhang LX, Zhou Y, Du Y, Han JS (1993) Effect of CCK-8 on audiogenic epileptic seizure in P77PMC rats. Neuropeptides 25:73–76.

Zhang SW, Liu YX (2008) Changes of serum adrenocorticotropic hormone and cortisol levels during sleep seizures. Neurosci Bull 24:84–88.

Zhang TS, Yang L, Hu R, Qiao XL, Yang X, Liu XG (2007) Effect of electroacupuncture on the contents of excitatory amino acids in cerebral tissue at different time courses in rats with cerebral ischemia and reperfusion injury. Zhen Ci Yan Jiu 32:234–236. (In Chinese with English abstract)

Zhang W, Yamawaki R, Wen X, Uhl J, Diaz J, Prince DA, Buckmaster PS (2009b) Surviving hilar somatostatin interneurons enlarge, sprout axons, and form new synapses with granule cells in a mouse model of temporal lobe epilepsy. J Neurosci 29:14247–14256.

Zhang Y, Wu D, Wan P (2002) Effect of electroacupuncture on brain amino acid contents during epileptic petit mal in the rat. Zhen Ci Yan Jiu 27:177–179. (In Chinese with English abstract)

Zhang Z, Lu G, Zhong Y, Tan Q, Chen H, Liao W, Tian L, Li Z, Shi J, Liu Y (2010) fMRI study of mesial temporal lobe epilepsy using amplitude of low-frequency fluctuation analysis. Hum Brain Mapp 31:1851–1861.

Zhang Z, Yu Z, Zhang H (1992) Inhibitory effect of electro-acupuncture on penicillin-induced amygdala epileptiform discharges. Zheng Ci Yan Jiu 17:96–98. (In Chinese with English abstract)

Zhao P, Cheng J (1997) Effects of electroacupuncture on extracellular contents of amino acid neurotransmitters in rat striatum following transient focal cerebral ischemia. Acupunct Electro-Ther Res 22:119–126.

Zhao P, Guo JC, Hong SS, Bazzy-Asaad A, Cheng JS, Xia Y (2002) Electro-acupuncture and brain protection from cerebral ischemia: The role of delta-opioid receptor. Program No. 490.13, SfN Abstract.

Zhao P, Ma MC, Qian H, Xia Y (2005) Down-regulation of delta-opioid receptors in Na^+/H^+ exchanger 1 null mutant mouse brain with epilepsy. Neurosci Res 53:442–446.

Zhao RJ, Yoon SS, Lee BH, Kwon YK, Kim KJ, Shim I, Choi KH, Kim MR, Golden GT, Yang CH (2006) Acupuncture normalizes the release of accumbal dopamine during the withdrawal period and after the ethanol challenge in chronic ethanol-treated rats. Neurosci Lett 395:28–32.

Zhao W, Zhao Q, Liu J, Xu XY, Sun WW, Zhou X, Liu S, Wang TH (2008) Electro-acupuncture reduces neuronal apoptosis linked to Bax and Bcl-2 expression in the spinal cords of cats subjected to partial dorsal root ganglionectomy. Neurochem Res 33:2214–2221.

Zhao WJ, Huang XF (2000) The role of somatostatin in area CA1 of rat hippocampus in experimental epilepsy. Sheng Li Xue Bao 52:131–136. (In Chinese with English abstract)

Zhao ZQ (2008) Neural mechanism underlying acupuncture analgesia. Prog Neurobiol 85:355–375.

Zhou F, Guo J, Cheng J, Wu G, Xia Y (2011) Electroacupuncture increased cerebral blood flow and reduced ischemic brain injury: dependence on stimulation intensity and frequency. J Appl Physiol 111:1877–1887.

Zhou L, Zhang HX, Wang Q, Liu LG, Yang X, Yang M, Liu YN, Li X (2009) Effect of scalp acupuncture on the expression of NF-KappaB mRNA, COX-2 mRNA and their proteins in rats with acute cerebral ischemia-reperfusion injury. Zhen Ci Yan Jiu 34:304–308. (In Chinese with English abstract)

Zhou MM, Yu CX, Wang MZ, Cao XD, Wu GC (2001) Alteration of orphanin FQ immunoreactivity and ppOFQ mRNA by combination of melatonin with electroacupuncture. Acupunct Electro-Ther Res 26:49–58.

Zhou Y, Jin J (2008) Effect of acupuncture given at the HT 7, ST 36, ST 40 and KI 3 acupoints on various parts of the brains of Alzheimer' s disease patients. Acupunct Electro-Ther Res 33:9–17.

Zhou YL (2000) Epilepsy treated by acupuncture with complementation with Yin and Yang. Report of 60 cases. J Tradit Chin Med 41:441.

Zhu LX, Li CY, Ji CF, Yang B, Li WM (1993) The role of substance P and somatostatin in acupuncture and moxibustion-induced postsynaptic inhibition. Zhen Ci Yan Jiu 18:290–295. (In Chinese with English abstract)

Zhuang LX, Zhang J, Li YZ (2006a) Clinical observation on catgut implantation at acupoint for treatment of general paroxysmal epilepsy. Zhongguo Zhen jiu 26:611–613. (In Chinese with English abstract)

Zhuang M, Bai Y, Ding S, Huang Z, Guo M, Qing N (2006b) Effect of electroacupuncture stimulation in Jinsuo on EEG in animal model of grand mal and status epilepticus. Chin J Phys Med Rehabilit 28:673–675. (In Chinese with English abstract)

Chapter 7
Acupuncture Treatment for Parkinson's Disease

Tetsuya Asakawa and Ying Xia

Abstract Acupuncture is widely accepted as an alternative therapy for Parkinson's disease (PD) in East Asia. The aim of this chapter is to review the current progress of the clinical and experimental acupuncture studies. A number of clinical studies in China and Korea proved that the efficacy of acupuncture lies in improving the motor impairments and complications of PD and reducing the long-term medical side effects; however, certain problems in the study design possibly exaggerated the efficacy of acupuncture and crippled the power of these evidences. More strictly designed clinical studies following the principles of randomization, control, double blindedness, and placebo are needed, to furnish powerful evidence in order to verify the efficacy of acupuncture. Several bench studies have tried to clarify the proposed mechanisms of acupuncture therapy like delaying the dopamine (DA) neuron depletion, improving the dopaminergic (DA-ergic) system in the PD state, neuroprotective effects, amelioration of the motor control network in the basal ganglia, and to attenuate the oxidative stress in PD state. The limited present-day bench studies still have not been able to provide sufficient evidence for revealing the secrets of acupuncture. At the end of this chapter, a new experimental strategy has been introduced. On combining the basis of the traditional acupuncture theory with the observation that the stimulation of the somatosensory pathways affects the cerebral DA level, a modified electroacupuncture method—the

T. Asakawa (✉)
Department of Neurosurgery, Hamamatsu University School of Medicine,
Handayama, 1-20-1, Higashi-ku, Hamamatsu-city, Shizuoka 431-3192, Japan
e-mail: asakawat1971@gmail.com

Y. Xia (✉)
Department of Neurosurgery, The University of Texas Medical School at Houston,
Houston, TX 77030, USA

The University of Texas Graduate School of Biomedical Sciences at Houston,
Houston, TX 77030, USA

Yale University School of Medicine, New Haven, CT 06520, USA
e-mail: Ying.Xia@uth.tmc.edu

Y. Xia et al. (eds.), *Current Research in Acupuncture*,
DOI 10.1007/978-1-4614-3357-6_7,

so-called superficial electrical stimulation (SES), is proposed. We hope that general PD researchers would accept the acupuncture therapy to a greater extent in the future.

Keywords Parkinson's disease • Acupuncture • Acupoint • Dopamine neuron • Degeneration • Neuroprotection • Deep brain stimulation • Motor control network • Basal ganglia

7.1 Introduction

Parkinson's disease (PD), first reported by James Parkinson in 1817, is a neurodegenerative disorder that results from a progressive degeneration of striatum-projecting dopamine (DA) neurons in the substantia nigra pars compacta (SNpc) manifesting as static tremor, rigidity, bradykinesia, gait dysfunction, and postural instability (Hirsch et al. 1988; Przedborski 2005; Parkinson 2002). PD is classified as the second most common neurodegenerative disease following the Alzheimer's disease (AD), with a lifetime risk of 2 % (Schapira 2009). The pathogenesis involved in PD onset is still unclear. However, it is generally accepted by most of the authors that the probable mechanisms lie in the genetic factors and environmental toxins (George et al. 2009; Schapira 2009) (Fig. 7.1.)

At present, there is no satisfactory therapy for PD. The available traditional treatment options include medical and surgical therapies. The medical therapies at hand

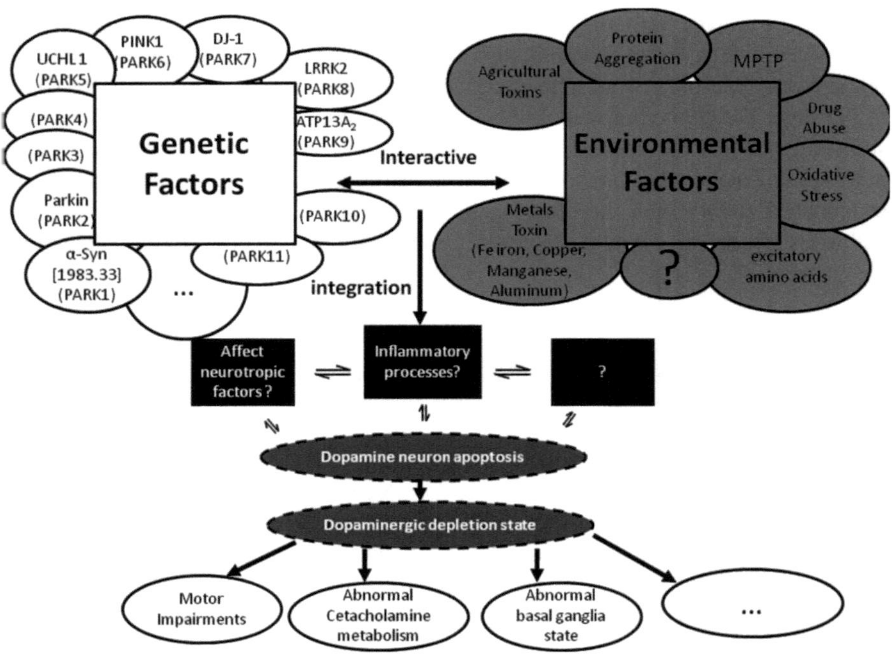

Fig. 7.1 The possible mechanisms involved in the PD pathogenesis

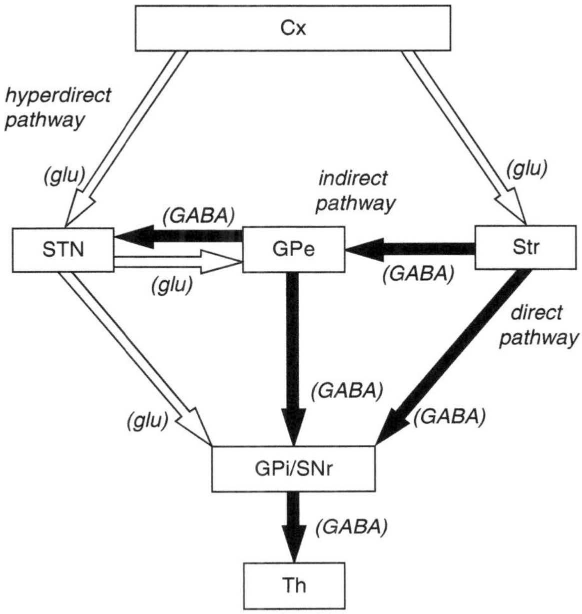

Fig. 7.2 The motor control network in basal ganglia (Nambu et al. 2002)

include: (1) Administration of levodopa (L-dopa), which is the most useful drug for idiopathic PD; (2) Direct agonists of DA-ergic receptors administration, such as selective D_2 receptor agonists like bromocriptine; D_1- and D_2-receptor agonists like pergolide; (3) Administration of monoamine oxidase-B (MAO-B) inhibitors such as rasagiline and selegiline, which were reported to possess additional neuroprotective and anti-apoptotic effects (Lam et al. 2008; Palhagen et al. 2006); (4) Other medications such as amantadine, anticholinergic medications (artane), etc., are known to efficacious as well. The main mechanism of the drug therapy is correction of the abnormal catecholamine metabolism and enhancing the DA-ergic effects on the pre/postsynaptic structures. The most commonly used neurosurgical treatments include primitive surgical ablation (pallidotomy or thalamotomy) and high-frequency deep brain stimulation (DBS) of certain structures such as subthalamic nucleus (STN). The exact mechanisms of surgical ablation and DBS are still unclear. STN plays a critical role in the basal ganglia and thalamocortical circuitry (Fig. 7.2.). Anatomically, it acts as an input and output station for basal ganglia. The STN receives γ-aminobutyric acid (GABA)-ergic input from the external segment of the globus pallidum (GPe) and sends the excitatory glutamatergic output to GPe, substantia nigra pars reticulate (SNpr), and the internal segment of the globus pallidum (GPi). In PD state, the overactive STN inhibits the thalamus through the hyperdirect and indirect pathways that are regarded as the key elements of motor control network in the basal ganglia. Then, the outputs from the thalamus become inhibitory, thus causing motor impairments. It is hypothesized that unilateral surgical ablations break the circuits, while DBS corrects the overactive state of STN in PD patients. In early studies, high-frequency stimulations of the STN reportedly caused inhibition of the STN itself or of the GPi and SNr. The inhibition caused by high-frequency stimulations seemed to be in reasonable accordance with Alexander–DeLong diagram. Recently, Hashimoto et al. and Nambu et al. have shown that high-frequency STN stimulation does not simply inhibit the GPi, but rather excites or causes an excitation–inhibition–excitation pattern in the GPi (Alexander et al. 1986, 1990; Fang et al. 2006a; Hashimoto et al. 2003; Nambu et al. 2002).

Recently, newer therapies such as stem cell transplantation and gene therapy have been developed. Stem cell transplantation is also considered as a promising new therapy for PD. The research on stem cell transplantation as a prospective PD therapy has boomed with the introduction of induced pleuripotent stem (iPS) cells, and has also successfully resolved the issue regarding the supply of stem cells. Some groups have reportedly succeeded in inducing the embryonic stem cells (ES cells) into DA neurons in vitro with a high success rate (Kim et al. 2002). However, the main difficulties lie with increasing the life of the induced neurons and reducing the cancerous trend in vivo. Stem cell transplantation still has a long way to go to mature into an established therapy that could be effectively used for the cure of patients. Gene therapy is another upcoming strategy, albeit only a few genes including α-synuclein (PARK1) and Parkin (PARK2) have been deeply investigated in the animal PD models (George et al. 2009). With the increasing knowledge of the vital genes involved in the neurodegenerative process, gene therapy has also developed considerably. An important aspect of gene therapy is increasing the gene expression of certain neurotrophic factors. It has been reported that a reduction in certain neurotrophic factors can possibly cause DA neuron degeneration. Some authors found that the administration of brain-derived neurotrophic factor (BDNF) was beneficial for the survival of the DA neurons and protected them from injury by neurotoxins, both, in vivo and in vitro (Frim et al. 1994; Shults et al. 1995; Spina et al. 1992; Liang et al. 2002). The Glial cell line-derived neurotrophic factor (GDNF) is another important neuroprotective factor reported to increase the survival rate of a DA neuron in a neurotoxic environment (Burke et al. 1998; Clarkson et al. 1999). However, different clinical trials using the same protocol of intra-putamen infusion of GDNF got inconsistent results, which indicates the need for verification of the efficacy of GDNF therapy (Lang et al. 2006; Gill et al. 2003; Slevin et al. 2005). Another notable observation of gene therapy is the increase in GABA-ergic effects in STN with an increased expression of GABA-ergic genes. Multiple evidences have indicated that increased GABA-ergic effects in STN can bring about an improvement in the symptoms of Parkinson's disease (Kaplitt et al. 2007; Levy et al. 2001; Windels et al. 2005; Feigin et al. 2007). It was observed that with the transfer of glutamic acid decarboxylase (GAD) gene by adeno-associated virus (AAV) into the STN of patients with Parkinson's disease, the symptoms of these patients improved with an associated upregulation of GABA expression in STN (Kaplitt et al. 2007). Based on the excitatory/inhibitory mechanisms involved in the basal ganglia (Nambu et al. 2002), microinjection of a drug or drug combination at the stations of the motor control network is potentially a new therapy, e.g., using GABA agonist or/and glutamate inhibitor at STN, GPi, or striatum; or using GABA inhibitor or/and glutamate agonist at the thalamus and GPe. There are several other therapies under development including anti-oxidant treatment, neuroprotective treatment, and metal dysregulation treatment such as improvement of the metal-associated protein (Ferritin), metal chelation therapies, and metal protein attenuating compounds (MPAC) (George et al. 2009). The efficacy and safety of these therapies is arguable and therefore needs to be verified by more clinical and bench investigations.

However, there are several limitations to the above strategies that make PD an intractable disease. (1) *The limitation of the symptomatic therapy*: Except some uncertain neuroprotective roles of the neurotrophic factors and the stem cell transplantation, both mature medical and surgical strategies, nowadays, are chiefly symptomatic therapies and have a little efficacy on stopping the neurodegenerative process (Hilker et al. 2003, 2005). (2) *The long-term efficacy is unfavorable*: The most effective therapy till now, i.e., L-dopa administration, is well known for its "wearing off" phenomenon, which means the efficacy of L-dopa weakens with the progression of PD and the dose of L-dopa has to be increased in order to maintain the same efficacy, however, along with the side effects. The long-term efficacy of DBS is also controversial. Though many reports suggest that DBS is a beneficial long-term treatment (Kleiner-Fisman et al. 2003; Krack et al. 2003), yet the battery life of the stimulator and its side-effects have to be taken into account. (3) *Cost and risk* associated with the surgical therapies are the limitations that significantly affect a patient's selection of surgical treatment as the preferred modality, especially, in a developing country which lacks a good health insurance system. (4) In advanced-stage PD patients, as the disease symptoms progress along with the accumulation of motor side-effects of L-dopa, the motor impairments get further complicated, intractable and sometimes, incurable.

In view of the above limitations, to maintain a good motor function and long-term medical efficacy, most of the patients undergo comprehensive treatments (e.g., L-dopa plus MAO-B inhibitor, L-dopa plus DBS, etc.), which are analogous to the drug cocktail therapy of AIDS. In this context, some alternative therapies are being adopted. It was reported that 40 % of the PD patients in the USA and 38.7 % in the UK underwent complementary and alternative medical therapies, of which 7–10 % patients had selected acupuncture (Rajendran et al. 2001; Ferry et al. 2002; Lam et al. 2008). Indeed, acupuncture is widely used as an adjuvant in the treatment of PD in East Asia including China, Japan, Korea, and Singapore. It is thought to benefit by improving the symptoms and complications of PD.

7.2 Clinical Application of Acupuncture in PD

Substantial evidence suggests that acupuncture can stimulate peripheral nerves (Zhou et al. 2010) and thus generate neural signals to regulate the activities of neurotransmitter systems, including the synthesis and release of dopamine system, in the central nervous system (Wen et al. 2010a, b). Therefore, acupuncture is broadly used in the treatment of neurological diseases (Xia et al. 2010) such as stroke (Guo et al. 2010) and other cardiovascular disorders (Kang and Xia 2010a, b) in clinical settings, especially in oriental countries.

PD had been reported in an ancient classic of traditional Chinese medicine (TCM)—The Yellow Emperor's Inner Classic (Huangdi Neijing), nearly 2,000 years ago. It was believed that the tremor was derived from the "liver dysfunction" according to the TCM theory, and the stiffness was caused by the over strong element, "wind." In the current TCM theory, the pathogenesis of PD lies in the weakness of "qi and blood," insufficiency of the "liver" and "kidney" (in the TCM theory) and inner movements of wind and phlegm. Acupuncture use in the treatment of PD has a long history in the East Asian countries of China, Japan, Korea, and Singapore. Liu, in 1974, first reported the application of acupuncture in central nervous system disorders, including PD (Liu and Sadove 1974).

The various styles of acupuncture involved in treating PD include scalp acupuncture, body acupuncture, electroacupuncture (EA), alternative acupuncture, and combination therapy. A typical acupuncture operator always learns the techniques from a predecessor teacher or from the classics of TCM and then gathers experience by treating patients. The traditional pattern and practice of the acupuncture have lead to a lack of objective standard in diagnosis and therapy. The selection of therapy styles usually depends upon the operator's experience (sometimes arbitrary preference) rather than an objective standard, which is considered as a drawback of all traditional medicines.

Compared to the diseases such as stroke (and its sequelae) (Guo et al. 2010), spinal injury, and Bell's palsy for which acupuncture is more widely used, acupuncture is always selected as a complementary therapy in most of the PD cases to optimize the efficacy/dose of DA-ergic treatments, to reduce the side-effects, or to improve the motor deficits. Currently, there is no report available that suggests adoption of acupuncture as an independent main therapy for PD in the clinical practice. Although most authors of the traditional medicine from China and Korea believe in the efficacy of acupuncture, some authors consider the role of acupuncture in treatment of movement disorders, especially PD, as quite a limited one (Shulman et al. 2002; Lam et al. 2008; Lee et al. 2008; Cristian et al. 2005). In the following sections, we try to provide an objective overview on these studies, rather than drawing a conclusion for disputation.

7.2.1 *Scalp Acupuncture*

With regard to the acupuncture treatment for PD, the scalp acupuncture is undividedly accepted to hold the highest status. It is a common belief of many Chinese investigators that scalp acupuncture can achieve much better results more swiftly when compared to other routes. There are two methods to stimulate the scalp acupoints. One is to insert and operate the needles manually and another is to use electroacupuncture (EA). As manual acupuncture relies more on the therapists' experience, therefore, EA stimulation has become much more popular in the recent years (EA is discussed in Sect. 7.2.3).

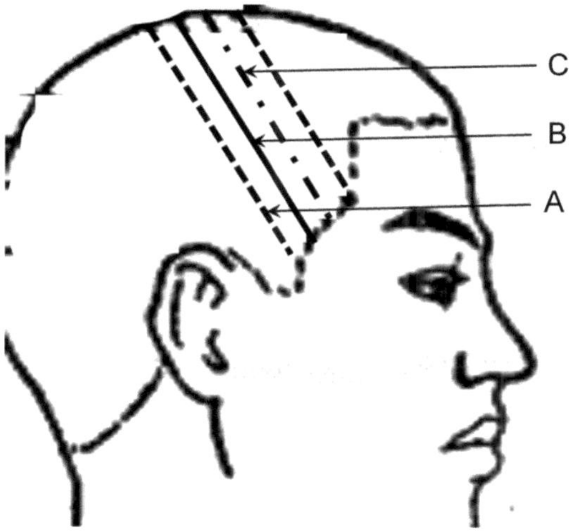

Fig. 7.3 The Chorea-Tremor controlling area defined by Dr Jiao Sunfa. *A line* depicts the sensory controlling area (SCA), which is at 2 cm posterior to the middle line of the skull. *B line* depicts the motor controlling area (MCA), which is at 0.5 cm posterior to the middle line of the skull and is considered as an extension line of the precentral gyrus. *C line* depicts the location of the Chorea-Tremor controlling area (CTCA), which is at 1.5 cm anterior to the *B line* (The figure was referred and modified from http://yibian.hopto.org/shu/?sid=8162 with permission)

The acupoints are selected based on the theory of differential symptoms (Bian-Zheng-Shi-Zhi) of TCM. Stimulation of the scalp acupoints has an overall effect of "calming the liver wind" that is believed to benefit the symptoms of tremor and stiffness. The so-called Chorea-Tremor controlling area (CTCA)—first described by Dr. Sunfa Jiao in 1975, is regarded as the most important scalp acupoint for treating the extra pyramidal disorders including PD (Wang 2000). Jiao defined several scalp acupuncture controlling areas according to the cortex area extension line and clinical experience. The most important areas are the sensory controlling area (SCA), motor controlling area (MCA), and the CTCA (Fig. 7.3). It has been widely reported that stimulation of the CTCA improves the tremor and involuntary movements of the contralateral limbs (Shi et al. 2002; Wang 2000; Yu 1994). Acupuncture at the CTCA along with other acupoints can achieve a better efficacy. Shi reported that stimulating the CTCA together with SCA would benefit the sensory impairments, whereas stimulating the CTCA with MCA would improve the motor impairments (Shi et al. 2002).

Other commonly used acupoints are shown in the Fig. 7.4. Xi et al. (1996), reported an improvement in the tremor myoelectric potential by electrical stimulation at the acupoint Baihui (DU-20) in Parkinson's patients. Wang et al. reported that EA at the acupoints Xuanli (GB-6), Xuanlu (GB-5), Fengfu (DU-16), Tianzhu (BL-10), and Fengchi (GB-20) significantly improved the electromyography (EMG) (Wang et al. 2003), superoxide dismutase (SOD), and lipid peroxides (LPO) in the PD patients

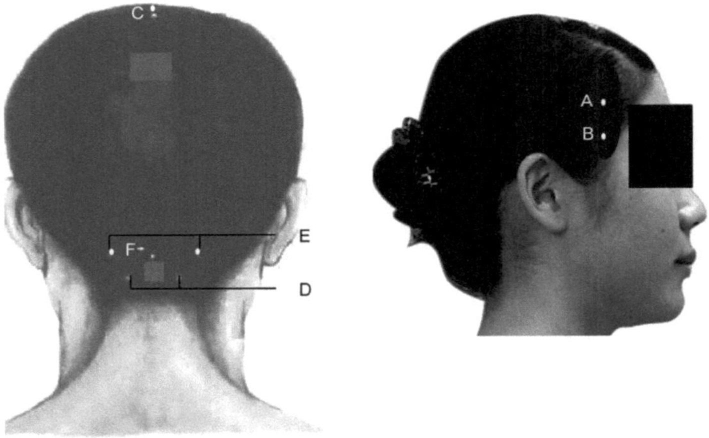

Fig. 7.4 The usually used acupoints in scalp acupuncture. *A*=Xuanlu (GB-5), *B*=Xuanli (GB-6), *C*=Baihui (DU-20), *D*=Tianzhu (BL-10), *E*=Fengchi (GB-20), *F*=Fengfu (DU-16) (The figures were referred and modified from http://www.21nx.com/c/am/pic/fnaoh.html with permission)

(Wang et al. 2006b). Jiang observed an improvement of the L-dopa side effects such as hidrosis, constipation, etc., and postponement of the "on-off" sign with the stimulation of areas MS-6, MS-4, MS-8, MS-9, and MS-14 (Jiang et al. 2006b).

All of the above studies show that the scalp acupuncture occupies a very important position in the acupuncture therapy for PD. In the actual clinical practice, scalp acupuncture is usually performed together with the body acupuncture.

7.2.2 *Body Acupuncture*

Body acupuncture is usually adopted in association with the scalp acupuncture to increase its efficacy in the treatment of PD. Occasionally, few special acupoints are employed for treating certain specific symptoms or for controlling local body symptoms. Tremor usually occurs in the muscular groups of limbs and trunk; therefore, the acupoints in the meridians of Shaoyang, Hand Jueyin, Foot Yangming, and Foot Taiyang are selected to treat the tremor. The acupoints in bilateral back regions of Foot Taiyang Meridian are used to relax the rigidity (Ren 2008). Xu et al. (2002) reported that Yanglingquan (GB-34) and Taichong (LR-3) were useful in improving the tremor and rigidity. Zhu (2001) reported that the stimulation of the Zusanli (ST-36), Hegu (LI-4), and Sanyinjiao (SP-6) acupoints increased the efficacy of L-dopa and reduced its side effects. However, the selection of appropriate acupoints needs expertise and a deeper understanding of the TCM theory. The commonly used acupoints for body acupuncture in PD are listed in Table 7.1.

Table 7.1 Commonly used acupoints for body acupuncture in the treatment of PD overviewed and summarized from the works of Ren (2008), Lam et al. (2008) and Shulman et al. (2002).

Meridians and collaterals	Acupoints
Lung meridian of hand Taiyin (LU)	Chize (LU-5)
Large intestine meridian of hand Yangming (LI)	Hegu (LI-4), Yangxi (LI-5), Quchi (LI-11), Jianyu (LI-15)
Stomach meridian of foot Yangming (ST)	Zusanli (ST-36), Fenglong (ST-40), Jiexi (ST-41)
Spleen meridian of foot Taiyin (SP)	Sanyinjiao (SP-6), Yinlingquan (SP-9),
Heart meridian of hand Shaoyin (HT)	Shaohai (HT-3),
Small intestine meridian of hand Taiyang (SI)	Yanglao (SI-6), Jianzhen (SI-9),
Bladder meridian of foot Taiyang (BL)	Yuzhen (BL-9), Chengfu (BL-36), Weizhong (BL-40), Kunlun (BL60)
Kidney meridian of foot Shaoyin (KI)	Taixi (KI-3), Dazhong (KI-4), Yingu (KI-10)
Pericardium meridian of hand Jueyin (PC)	Quze (PC-3), Neiguan (PC-6), Daling (PC-7)
Triple energizer meridian of hand Shaoyang (TE)	Yangchi (TE-4), Jianliao (TE-14),
Gallbladder meridian of foot Shaoyang (GB)	Huantiao (GB-30), Yanglingquan (GB-34)
Liver meridian of foot Jueyin (LR)	Taichong (LR-3), Zhongfeng (LR-4),
Meridians of hand Shaoyang (SJ)	Waiguan (SJ-5)
Du channel (DU)	Yaoyangguan (DU-3), Xuanshu (DU-5), Jizhong (DU-6), Lingtai (DU-10), (DU-17),
Extra points	Jiaji (EX-B-2)

7.2.3 *Electroacupuncture*

Recently, EA has become the most extensively used method in the treatment of PD via acupuncture and has a strong potential to replace the traditional manual acupuncture. EA stimulation is easier to manipulate than traditional manual acupuncture. The parameter setting is relatively more objective and relies less on experience. Another advantage of EA, as accepted by many Chinese authors, is its safety. It is believed that EA causes less injury and side effects, thereby, increasing the level of its acceptability among the patients.

In contrast to the treatments with objective parameters setting such as DBS, there are great variations in the parameters selected by different therapists. The parameter setting relies on both, the experience of the therapist and the patient's subjective experience. Nowadays, these have become more uniform as most of the investigators have come to adopt the stimulation conditions listed in Table 7.2 (Jiang et al. 2006b; Wang et al. 2003, 2006b; Xi et al. 1996).

7.2.4 *Other Alternative Therapies Related to Acupuncture*

In addition to typical acupuncture, some Chinese investigators selected other alternative therapies related to the classical acupuncture. Xu et al. (2002) injected *Mailuoning*, a mixture of several Chinese medicines, at Zusanli (ZT-36) and Yanglingquan (GB-34) acupoints in 23 PD patients. They got a total effectivity rate

Table 7.2 Commonly adopted EA stimulation conditions by the Chinese therapeutists

Parameters	Data
Depth from the skin	30–40 mm
Frequency	2–100 Hz
Current intensity	Top strength endured by the patient
Wave Pattern	Condensation wave or disperse-dense wave
Time of one therapy	Stimulation for 30 min, retaining the needle for 60 min, one therapy per day
Time of the whole course	30 days (consecutive or consecutive 5 days with a 2-day interval on the weekends)
Diameter and material of the needle	0.3–0.4 mm, stainless steel

of 65 % from the injection of these acupoints. They also found that the acupoints' injection brought about an alteration in the monoamine neurotransmitters in the cerebrospinal fluid (CSF); however, these variations were not significantly correlated to the efficacy. Zhang et al. (2005) performed herbs-partitioned moxibustion at the Shenque (CV-8) in 54 PD patients who were undergoing the classical medicine treatment, and compared them to 36 PD control patients who received only classical medicine. They found that there were significant improvements in PD rating scales in the patients of treatment group than the controls (those who only received classical medicine treatment). Therefore, they concluded that herbs-partitioned moxibustion was an effective therapy for PD. Huang (2000a, b) performed eye acupuncture, which is a special method of acupuncture that stimulates certain areas around the eyes, on 45 patients (37 primary PD cases and 8 Parkinsonism cases). She compared the scores of the Chinese PD rating scales before and after eye acupuncture, and found significant improvements in 40 cases. She also found a better efficacy in the early-stage PD patients (Huang 2000a). There are several other alternative methods to stimulate the acupoints such as, bloodletting therapy (Wang 2000); stimulation of the special areas in ear—presumably the same as those used for brain mapping (Su 2002); acupuncture based on the theory of strengthening the body by corresponding the Taiji structure in man and the universe (Qi and Wang 2007; Wu 2002); and stimulation of the acupoints by Dianxue, which is a special method to hit specific acupoints (Wu et al. 1996; Dai 2004).

The methods described above are the "alternative" therapies in the acupuncture, which in itself is considered as an "alternative" therapy for PD. Such treatments rely to a great extent on experience and lack sufficient objectivity that could verify their efficacy. If an investigator wants to evaluate a new therapy, he must provide enough believable evidence to prove the efficacy and safety. This issue is discussed in the Sect. 7.2.6.

7.2.5 Combination Therapy

The L-dopa administration is still regarded as the most effective therapy for PD. It is for this reason that all the other therapies are based on their effects on L-dopa whether they prolong the effective period of L-dopa, reduce its dose, increase its efficacy, abate the side effects, or ameliorate the complications and rehabilitation in the advanced-stage PD cases. Except a few studies, which adopted acupuncture as a temporary monotherapy for a limited experimental period in order to clarify the acupunctural mechanism (Wang et al. 2000), we did not find a single report in the available English and Chinese literature that adopted acupuncture as a unique main treatment even in the early-stage mild PD patients in the actual clinical practice. This shows that acupuncture is chiefly used as an alternative treatment for PD. The meaning of "combination therapy" in this section refers to the combination of different styles of therapies offered in TCM.

Most of the acupuncture therapists adopt the combination therapy in real-time clinical practice. The commonest combination is the scalp acupuncture plus body acupuncture. First, the area for scalp acupuncture is selected, which is almost always CTCA (the most important area) although the MCA and SCA are also considered as per the symptoms. Next, the acupoints for the body acupuncture are selected. Since each individual PD patient is different from another, the acupoint selection also presents a great individuality. All the acupoints must be selected based on the theory of differentiate symptoms in the TCM. Therefore, the step of acupoint selection is of vital importance and needs an in-depth understanding of the TCM theory along with an excellent experience in acupuncture. The most commonly used acupoints for body acupuncture are listed in the Table 7.1.

Another widely used combination is the acupuncture and TCM. Liu and Li (2001) compared the efficacy of the acupuncture therapy with the surgical therapy via ventral pallidotomy. They performed scalp acupuncture at CTCA and MS6, and administrated a mixture of several Chinese medicines including Tianma (rhizome-gastrodiae), Gouteng (uncaria), Zhenzhumu (mother of pearl), and Jiangcan (cicada) in the TCM group. They found that both the therapies significantly improved the PD symptoms. The effects of surgical operation presented sooner but were accompanied with severe side effects as well. In contrast, the onset of the effects in the TCM group was late but associated with fewer side effects. Zhu (2001) used scalp acupuncture at CTCA combined with several Chinese herbal medicines for PD patients who were receiving normal L-dopa therapy. They found that in comparison with the patients receiving regular treatment with L-dopa, the patients receiving a lower dose of L-dopa plus acupuncture and Chinese herbal medicines achieved the same efficacy. Li and Yang (2005) treated PD with comprehensive therapies. Fifty-five patients were divided into two groups at random. The control group underwent only L-dopa administration and the treatment group additionally underwent the comprehensive treatments, including both body and scalp acupuncture, and the Chinese medicine (i.e., pills of six ingredients with rehmannia plus Zuogui pills). He did not find any significant difference in the total scores on the PD rating scale between the two groups after treatment. The treatment group got the same efficacy as the control group; however, the dose of L-dopa in the treatment group was significantly lower than that of control group.

The reports above indicate that the combination therapy is a successful strategy for acupunctural treatment of PD. More and more western clinicians have now accepted acupuncture as one of the plausible options in treating the intractable, advanced-stage PD patients, as it is indeed a prickly endeavor to choose an effective method that can bring further improvements. For some terminal stage PD patients, on whom all the classical therapies have been exhausted, acupuncture might be the last ray of hope, at least in bringing psychological comfort to the patient and his family.

7.2.6 Comments

7.2.6.1 Factors that Influence the Acupuncture Efficacy

Most of the current studies focus on the validation of the efficacy of acupuncture as an alternative therapy for PD. There is no enough information yet on the efficacy of acupuncture in different conditions (e.g., during the different EA parameters, selection of the patients, acupoints selection, and so on), which is necessary to delineate the factors that might influence the efficacy of acupuncture. Here, we try to summarize the influence factors based on data currently available in the literature.

(a) Stage of the patients: Several studies have reported that acupuncture has a better efficacy in early-stage PD than the advanced stage (Ma 1999; Wang 2000; Jiang et al. 2006b; Liu and Ren 1993). This tendency shows a trend comparable to other therapies such as DBS and Medical.
(b) Clinical subtype of PD: A good efficacy was observed in the younger patients with chief symptoms of tremor or rigidity and who had lesser complications. Poor efficacy was noted in those who were old, suffered from more complications, and those who had Parkinsonism (Liu and Ren 1993); however, as per the results obtained by Wang, there was no significant difference among the patients based on their presenting symptoms (Wang 2000). The conflictive results make it difficult to draw any conclusions. Further studies are therefore required to shed more light on this issue.
(c) Protocol of acupuncture: Since PD is a progressive disease with a long course, protocol selection is crucial for improving the efficacy. Xi and Cai (1995) proposed a long protocol which involved acupuncture every 2 days for a minimum of 60 days. Zhang and Han (1996) observed an improvement in symptoms after a continuous acupuncture q.d. for 40 days. Wang believed that the protocol should be prolonged to 2–3 years (Wang 2000). Though the discrepancy on the length of course of acupuncture among different authors is large, all the authors agree on one fact that the onset of the effects of acupuncture is extremely slow and that it should be continued for a long time to keep the efficacy up.
(d) Choosing the acupoints: As we have discussed several times in the previous sections, the selection of acupoints is extremely difficult and equally crucial. In some English studies performed by non-Chinese investigators, they made strict trial designs but still got negative results on the acupuncture efficacy which were largely different from the results obtained by the Chinese investigators (Cristian et al. 2005; Shulman et al. 2002). Besides the study design (see next section),

we also found that the acupoints selected in these studies were quite different from those used in Chinese/Korean reports. Thus, the selection of acupoints should be considered quite seriously when designing an acupuncture study.

7.2.6.2 Problems with the Experimental Design of the Studies Evaluating the Efficacy/Safety of Acupuncture

Evaluation of a new therapy is practiced through evidence-based medicine. Strong evidence should be gathered based on the following principles: (1) randomization; (2) control; (3) double blinding; and (4) placebo. For making a reasonable statistical deduction the following additional conditions must be achieved: (5) enough sample size; (6) appropriate statistical method; (7) repeatability, and (8) full introduction about the adverse events and the patients who cannot finish the trial.

Notwithstanding the legends of acupuncture being mysterious and charming, it must be subjected to the same objective evaluation standards as all other medical treatments. If not, any conclusion made on the acupuncture efficacy would be dubious.

As with other alternative therapies based on the TCM, the efficacy of acupuncture was sometimes overestimated or exaggerated by some improperly conducted studies, which raised a question not only on the acupuncture therapy but also on the validity of TCM. What is worse is that the data reported by some Chinese/Korean investigators are possibly losing the trust of many investigators on grounds of "high publication bias" (Moher et al. 2003; Vickers et al. 1998).

Lee et al. (2008) searched the keywords "acupuncture and Parkinson disease" through main English/Chinese/Korean electronic database in the USA, UK, Korea, and China within 1 year. They found that only 11 of the 103 studies were randomized controlled trials (RCTs) with subjective outcome measure. Only 1 of these 11 studies described a double-blinded method (Cristian et al. 2005), and only 2 of these described the patient blinding (Lee et al. 2008). Interestingly, the double-blinded study by Cristian and another rigorous study by Shulman failed to prove the efficacy of acupuncture, albeit the negative results maybe a consequence of erroneous selection of the acupoints (see Sect. 7.2.6.1) (Cristian et al. 2005; Shulman et al. 2002). Lee et al. (2009) searched the same keywords through ten Korean electronic databases; only one study from South Korea proved that the constitutional acupuncture is more effective than standard acupuncture based on Unified PD rating scale (UPDRS) and freezing gate; however, no further evidence was found to prove the efficacy of acupuncture. Lam et al. (2008) searched the cross-referenced data on acupuncture therapy in PD both as MeSH and as text words in English/Chinese through several electronic databases. Only 10 of the 784 studies could be defined as rigorous RCTs studies. Although, nine of the ten studies suggested a "significant" efficacy of acupuncture in treating PD, Lam pointed out there was insufficient evidence to support this conclusion due to certain methodological flaws in the studies. It is true that, so far, we have not been able to find a well-designed study that provides formidable positive results on acupuncture efficacy in treating PD.

A recently performed systematic review on various Chinese studies evaluating the potency of acupuncture treatment in PD, reached at an analogous conclusion. Yang et al. (2010) analyzed 832 cases from 13 randomized clinical studies. Although many authors found that the efficacy of medication plus acupuncture was far better than that of medication alone, however, the shortcomings in the designs of several studies weakened this conclusion.

The main problems in the experimental design of the acupuncture studies in treating PD are listed as below:

1. Absence of or incorrect usage of blinding methods is considered to be the most commonly recurring problem with the experimental design. Most of the authors use some subjective rating scales, such as UPDRS and Webster rating scale to evaluate the PD status. If there is no "blinding" applied to the observers, there is an increased possibility of occurrence of observational bias. The patients tend to report a positive variation under the influence of psychological suggestion in the absence of blinding known as the "placebo effect." Some studies tried to demonstrate the acupuncture efficacy by comparing acupuncture plus drug therapy with drugs only (i.e., C+A vs. C, C=classical therapy; A=acupuncture). If double blinding is not employed, it is quite natural to get the results in the form of C+A>C, which is possibly attributable to the placebo effect and/or observational bias. The efficacy of A is thereby overestimated. In this regard, to exclude the so-called placebo effect, the study design of C+A vs. C should be ameliorated by C+A vs. C+P (P=placebo acupuncture, which might be a false acupuncture or an ineffective acupuncture, such as using a false needle, or inserting the needle to a non-acupoint area). However, achieving full double blinding in physical therapies is deemed to be difficult (Lam et al. 2008). Only under one circumstance is the single-blinded-patients-design acceptable, that is, if the evaluating index is objective and is not affected by the inklings of the observer. Several methods were adopted to perform the single-blinded trials, for instance, by use of gauze to prevent the patients from seeing the acupuncture procedure; sham device, placebo needles, and inserting the needles into the blank areas with respect to the acupoints were also used for patient blinding (Cristian et al. 2005; Hammerschlag 1998; McManus et al. 2007; Park et al. 2002).
2. Inadequate randomization and too small sample sizes are other big issues associated with the trial design. It is a tendency of the investigators to select milder cases for the treatment group, whereas severer cases for the control group. Inadequate randomization will cause such a selection bias, which can confound the results and overestimate the efficacy. However, to perform complete randomization one needs enough sample size to maintain a baseline balance between the treatment and the control groups. Too small a sample size will lead to a difference at the baseline before even giving the treatments, which leaves the data after treatments incomparable. In terms of statistical analysis, too small size will lead type II error.
3. Insufficient account of the adverse events and the dropouts or withdrawal reduce the reliability of the whole study. Though acupuncture seems to be safe (Jiang

et al. 2006b; Wang et al. 2006b), it is not a totally harmless therapy. Minor adverse effects such as stabbing pain, bleeding, and hematoma have been reported, along with some potentially serious side effects like infection and pneumothorax (Lam et al. 2008; Yamashita et al. 1998; Odsberg et al. 2001; White et al. 2001; White and Ernst 2001).

4. Other methodological weaknesses, such as inappropriate statistical analysis, inadequate control group setting, etc., might occur and can reduce the worthiness of the study.

For obtaining powerful evidence on the efficacy and safety of acupuncture therapy in PD, well-designed studies with less methodological flaws are of great desideratum. According to the guidelines for clinical trials of acupuncture (Birch 2004; MacPherson et al. 2002a, b), a perfect study for evaluating the efficacy and safety of acupuncture in treating PD should be a large, double-blinded (or single-blinded to patients with an objective evaluating criterion), rigorously randomized, and placebo-controlled study. Detailed recording of the adverse events and correct selection of statistical methods should be seriously taken into account. Long-term, multicentric, and multinational large clinical trials should be designed and performed to obtain formidable evidence for verifying the efficacy and safety of acupuncture in treating PD.

7.3 Potential Mechanisms of Acupuncture Therapy for PD

The mechanisms of acupuncture treatment for PD are still unclear at present. Although there are numerous clinical reports from China and Korea that have claimed its efficacy and safety, the studies aiming to uncover its mechanism are relatively few. We found several reports regarding the bench studies for the mechanisms of acupuncture on PD. It is difficult, however, to make solid conclusions from these studies because of limited size of samples, lack of repeatability, or problematic experimental design (see Sect. 7.2.6.2). Owing to these drawbacks, more in-depth studies are needed to validate the potential mechanisms summarized below based on the current studies.

7.3.1 Delaying the Progress of DA Neuron Degeneration by Neuroprotective Effects

The most important causative factor of PD is the progressive degeneration of the DA neurons in SNpc, which "triggers" the successive pathophysiology. "To arrest" the DA neuron degeneration is, hence, the best strategy for treating PD. Unfortunately, even the most effective therapies including L-dopa administration and STN-DBS cannot contribute much to this effect (see Sect. 7.1).

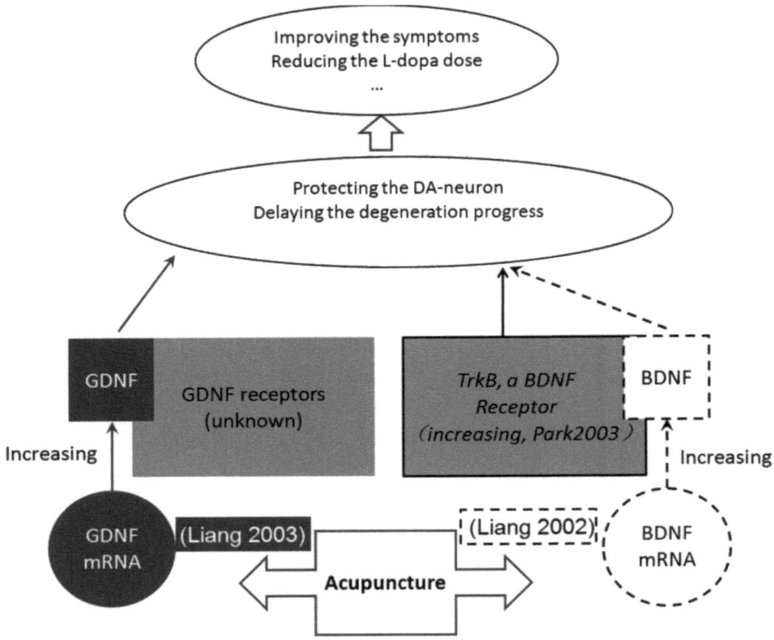

Fig. 7.5 Mechanisms of delayed DA neuronal degeneration secondary to acupuncture through neurotrophic factor enhancement

Since the depletion of neurotrophic factors plays an important role in the DA neuron degeneration, addition of neurotrophic factors such as GDNF or BDNF might be a key to abate the DA neuron degeneration (see Sect. 7.1). It was hypothesized that acupuncture was beneficial in PD by affecting these neurotrophic factors. There are three studies that focus on this hypothesis (see Fig. 7.5).

Liang et al. (2002) performed EA at Dazhui (DU-14) and Baihui (DU-21) in the Medial Forebrain Bundle (MFB) lesioned rats. High (100 Hz) and low (2 Hz) frequency stimulation was adopted to stimulate the rats for a total of 24 sessions (28 days after lesioning), during which the current varied from 1 to 3 mA. The DA neurons in the substantia nigra compacta (SNc) and the ventral tegmental area (VTA) were counted after tyrosine hydroxylase (TH) staining. The gene expression of BDNF mRNA in VTA, SNc and substantia nigra reticulate (SNr) was localized by in situ hybridization. The results showed that the number of DA neurons in the 100 Hz EA stimulation group (60.7 ± 5.1 %.) was significantly ($p < 0.05$) higher than that in 0 Hz stimulation group (40.46 ± 4.9 %) and 2 Hz stimulation group (52.76 ± 4.7 %) (Fig. 7.6a). Similar results were obtained by Jia et al. (2009) in the same laboratory with Liang (Fig. 7.6b). On comparing the values of integrated optic density (IOD) on the lesioned side in 100 Hz EA group and control group ($p < 0.01$), a significant difference in BDNF mRNA expression was observed. It was then concluded that the high-frequency EA increased the expression of BDNF mRNA, which contributed in delaying the DA neuron degeneration. Liang et al. (2003) used an analogous protocol to investigate mRNA expression of another neurotrophic

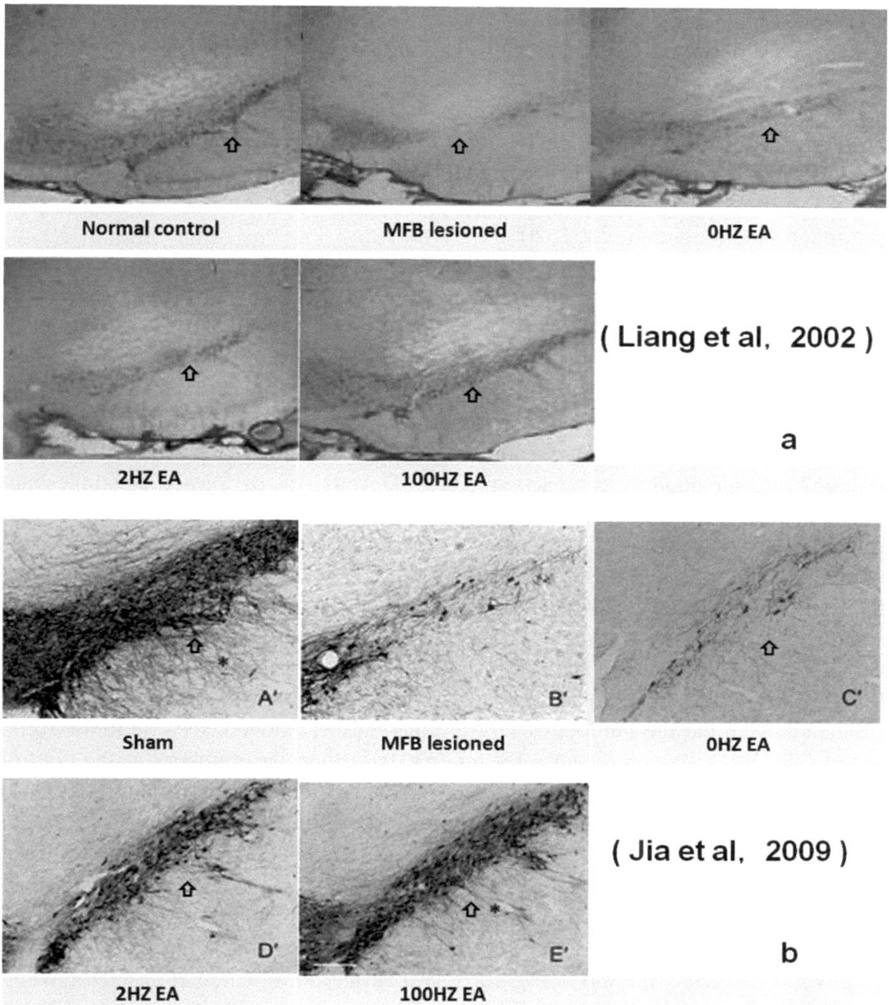

Fig. 7.6 EA increased the DA neurons in SN by delaying DA neuron degeneration. (lesioned side). (**a**) Shows a significant increase in DA neurons in SN by 100 Hz EA (Liang et al. 2002); (**b**) Shows that both 2 and 100 Hz EA significantly improved the DA depletion in SN (Jia et al. 2009). Such favorable results suggest EA to be superior to the traditional treatments such as L-dopa administration and STN-DBS. However, this conclusion needs to be verified by analogous studies from different laboratories since the repeatability of such results is still under question

factor, GDNF. Amphetamine-induced rotational movements confirmed the efficacy of EA. At an early stage (14 days after lesioning), both 2 and 100 Hz stimulation improved the rotational movements significantly ($p<0.05$ of 2 Hz; $p<0.01$ of 100 Hz), while at 28 days after lesioning only 100 Hz stimulation produced a significant improvement ($p<0.05$). They measured the striatal DA concentration by HPLC but could not find any variation between 2 and 100 Hz EA stimulations.

The expression of GDNF mRNA was evaluated by in situ hybridization and calculation of the GDNF-positive cells. Interestingly, 2 Hz EA improved GDNF mRNA expression on the unlesioned side in SN and VTA, while 100 Hz improved that on both lesioned and unlesioned sides. Both of Liang's studies suggested that EA manifests a neuroprotective effect against the degeneration of DA neurons by activating endogenous neurotrophic factors such as BDNF and GDNF. The increased expression of BDNF mRNA and GDNG mRNA after EA may be a plausible mechanism of action of acupuncture (Liang et al. 2002, 2003).

Park et al. (2003) performed acupunctures at two groups of acupoints in the striatal-lesioned rats. Acu1, which included Yanglingquan (GB-34) and Taichong (LR-3), was set as an effective acupuncture for PD; Acu2, which included Hegu (LI-4) and Quchi (LI-11)—the acupoints for gastrointestinal functions, was set as an ineffective acupuncture. Five groups of rats were investigated at random. (1) Acu1 on 6-OHDA lesioned models was set as the effective treatment group; (2) Acu2 on 6-OHDA lesioned models was set as noneffective treatment group; (3) control group was the 6-OHDA lesioned rats that underwent no acupuncture; (4) Acu1 was performed on some sham operated rats, which was set as SAcu1 group. Sham group underwent only sham operation and did not receive any acupuncture. Acupunctures were performed for 13 consecutive days with each treatment lasting for 60 s. Apomorphine-induced rotational movements were used to evaluate the motor status. The DA neurons in the SN were calculated after the TH staining 2 weeks after lesioning. In addition, trkB (regarded as a high-affinity receptor for BDNF) was evaluated by counting the trkB-positive cells in SN after immunohistochemistry specific for the tyrosine kinase (trkB) receptor. The results showed that the number of DA neurons on the lesioned SN side in the Acu1 group (78.6 %) was significantly higher ($p<0.01$) than the numbers in the control (45.7 %) and Acu2 (58.8 %) groups; the trkB-positive cells in Acu1 group (56.8±4.4) were significantly higher than the contralateral part of the same group ($p<0.01$) and ipsilateral part of the control group ($p<0.05$). The data indicated that acupuncture at Yanglingquan and Taichong significantly reduced the DA neurons degeneration, which was possibly caused by the activation of the BDNF through increasing BDNF receptor (trkB) expression (Park et al. 2003).

Several Chinese (Ma and Wang 2001; Ma et al. 2003b; Chen et al. 2010; Wang et al. 2010; Zhao et al. 2010) and Korean authors (Joh et al. 2010; Hong et al. 2010; Jeon et al. 2008) reported similar results. Kim et al. (2005) performed acupuncture at Zusanli (ST-36) in striatal PD rat models. They found that the Zusanli acupuncture could improve the behavioral test and TH staining in SN. They, therefore, concluded that acupuncture at the ST-36 is a useful strategy for treating PD. Jeon et al. (2008) investigated whether acupuncture alters protein expression in the SN to favor attenuation of neuronal degeneration in C57BL/6 mice treated with MPTP (30 mg/kg/day). They performed EA at Yanglingquan (GB-34), observed an improvement in the behavioral impairment and reduction in tyrosine TH-positive DA neurodegeneration. Moreover, the levels of cyclophilin A (CypA), which is a neuroprotective agent, were unchanged by MPTP treatment but were increased in MPTP-EA mice. All these studies showed positive results in demonstrating the neuroprotective effects of acupuncture. These studies followed the randomized-control design and the evaluating criteria were also objective. However, several existing methodological flaws decreased the reliability of these studies. These were: (1) In the study of Liang et al. (2002), they did not perform the behavioral tests to confirm the efficacy of EA. Only three rats from each group were selected "randomly" to perform the pathological assay, the sample size here appears to be insufficient. The same problem was found in Kim's study (five rats form each group). The study of Liang et al. (2003) confirmed the EA efficacy by amphetamine-induced rotational movements and measuring the striatal DA levels. However, there were no observations on the number of DA neurons in SN, which is a vital aspect for confirming the successful establishment of a PD model. It is therefore, impossible to establish a relationship between the DA neurons in SN and the behavioral performance, which is important to evaluate the EA efficacy. Without behavioral evaluation, such positive results (significantly increased DA neurons in 100 Hz EA group) could also be derived from an "unsuccessful" PD model, in which surgical lesioning did not effectively destroy DA neurons. (2) Park et al. (2003) and Kim et al. (2005) used the striatal model system. They performed all the experiments 2 weeks after lesioning. However, we found that a time gap of 2 weeks after lesioning was far too less for using the striatal models in the experimentation. It takes at least 3 weeks for a 6-OHDA striatal injection to reach its peak toxic effect at the soma of DA neurons and hence to make a stable experimental rat model (Fang et al. 2006b). (3) In the studies of Liang et al. (2002) and Jia et al. (2009), acupuncture enhanced the 30–40 % number of DA neurons in SN. However, the rotational behavioral test showed a significant improvement in Jia's results (Jia et al. 2009). In comparison to the previous studies using the rotational behavioral test, Jia's results are out of our expectation. Fang et al. (2006a, b) and Sun et al. (2010) performed behavioral tests in 6-OHDA lesioned rats. They made one-lesion striatal models (24 % DA neurons depletion in SN) and four-lesion striatal models (89 % DA neuron depletion in SN), but did not find any difference in the rotational behavioral tests between these two groups. It is believed that the rotational behavioral test is not sensitive enough to evaluate the DA neuron depletion. Furthermore, if acupuncture actually has the capacity to increase DA neurons in SN by 30 % (or above), it could prove to be a great advancement in the treatment of PD and would be superior to currently available therapies except the potential stem cell transplantation. Thus, the results of Liang and Jia need further validation before being accepted widely. There is a compelling need for conducting well-designed, large sample size studies with appropriate PD models to further study the neuroprotective effects of acupuncture.

7.3.2 Improvement of the Dopaminergic System in PD Patients and Animal Models

Accumulating evidence shows that acupuncture profoundly affects the release, synthesis, reuptake, and degradation of the central neurotransmitters, including dopamine and modifies functions of DA receptors (Wen et al. 2010a).

Wang et al. (2009) found that the expression of TH mRNA and dopamine transport (DAT) mRNA in the SN of 6-OHDA-treated rats was upregulated upon treatment with electroacupuncture at "Baihui" (GV-20)-through-"Taiyang" (EX-NH-5), when given in two courses (one course = once per day for 6 days). This finding suggests that EA attenutes neuronal degeneration, thus increasing DA neurons in SN.

Some authors believe that acupuncture improves the DA-ergic system by reducing the neuronal degeneration in PD state, as DA neuron degeneration is the pathognomonic dysfunction of the DA-ergic system seen in PD. If acupuncture contributes to the reduction of DA neuronal depletion (see Sect. 7.3.1), it is reasonable to conclude that the DA-ergic system can consequently be improved. Furthermore, other researchers have reported that peripheral electrical stimulation of certain parts of the animal body surface directly affected the cerebral DA levels (see Sect. 7.4).

7.3.2.1 Acupuncture Increases DA in the Brain

Sundry reports by Chinese investigators have indicated that acupuncture application at certain acupoints increases the cerebral DA level. Other investigators unrelated to TCM and acupuncture have also accounted that peripheral electrical stimulation influenced the cerebral DA levels (see Sect. 7.4). Thus, it is accepted that stimulation of certain parts of the body will affect the DA levels in the brain, albeit the exact mechanisms and processes are still unclear. It also remains unfathomed as to whether these observations are attained by the direct effects of acupuncture, or due to secondary variations through an increase in the number of DA neurons. In this section, we list only some of these Chinese reports; we will discuss the peripheral electrical stimulation in the next section.

Xu et al. (2002) performed acupuncture at Yanglingquan (GB-34), Neiguan (PC-6), Taichong (LR-3), and Baihui (DU-21) in PD patients and measured the concentrations of DA, noradrenalin (NA), and 5-HT in the patients' CSF by HPLC-EC. They reported that the levels of DA and NA in the effective acupuncture patients were significantly increased ($p<0.001$), but 5-HT levels showed no change. Zhu et al. (1996) performed acupuncture at Baihui (DU-21) in 1-methyl 4-phenyl 1,2,3,6-tetrahydropyridine (MPTP) treated mice. They measured DA and its metabolite, dihydroxyphenylacetic acid (DOPAC) in the caudate nucleus, and mesencephalon by HPLC-EC. They found that acupuncture significantly increased DA in the caudate nucleus and mesencephalon and normalized the increased DOPAC/DA ratio in the PD state. Wang et al. (2006a) performed acupuncture at the Baihui (DU-21), Fengfu (DU-16) of the lesioned side, and bilateral Yanglingquan (LR-3) in the striatal PD rat models. Apomorphine-induced rotational movements were used to confirm the successful establishment of the PD models and the acupuncture efficacy. The striatal DA and its metabolites DOPAC and homovanillic acid (HVA) were measured by HPLC-EC. They found that acupuncture at the above acupoints significantly increased the behavioral performance as well as the striatal DA, DOPAC, and HAV levels.

Yin et al. (2009) found that the application of EA (frequency from 2 to 100 Hz, voltage from 2 to 4 V) at Hegu (LI-4) and Taichong (LV-3) increased the expression of protein kinase A (PKA), and cAMP response element binding protein (CREB1) in the SNc of MPTP-treated mice. Since PKA and CREB1 play important roles in maintaining the DA balance, it is possible that they are also involved in the DA changes seen after acupuncture treatment in the PD animals.

It was also found that DA neuron depletion results in insufficient DA in the auditory conduction pathway, which leads to abnormal auditory brainstem responses (ABRs). Prolongation of latencies (wave III or V) and interpeak latencies (IPLs; I–III, I–V, and III–V IPL) are regarded as the characteristic abnormalities in PD. Such abnormalities can be improved by the administration of L-dopa; thus, ABRs can be employed as indirect, yet objective evidence to assess the cerebral DA levels (O'Donnell et al. 1987; Kodama et al. 1999). Wang et al. (2000) studied the ABRs in the PD patients and observed the abnormalities including prolonged latencies of wave V and interpeak latencies of I–V, III–V IPL. All these patients were then subjected to the experimental single acupuncture treatment. The results showed that the efficacy of acupuncture was remarkable and there was a significant positive correlation between the ABRs and Webster rating scores. It was therefore concluded that acupuncture improves the cerebral DA and activates the DA neurons.

7.3.2.2 Acupuncture Activates the Dopamine Transport

Zhang et al. (2006) investigated the expression and activity of DAT affected by EA (frequency from 2 to 100 Hz, voltage from 2 to 4 V) in MPTP-treated mice by immunohistochemistry staining. They found that after stimulating the Hegu (LI-4) and Taichong (LR-3) acupoints for three courses (7 days, 20 min/day), the number of cells with DAT-positive expression significantly increased (24.67 ± 13.01) in the treatment group (vs. 2.33 ± 3.22 in the control group) along with the integral optical density (26088.88 ± 4498.05 in the treatment group vs. 15248.27 ± 4195.38 in the control group) in SNc. This finding demonstrated that the activation of DAT could be an important mechanism of action underlying the acupuncture treatment for PD.

Jiang et al. (2006a) studied the changes in DAT in the striatum by conducting single photon emission computer tomography (SPECT) in five PD patients. All the patients underwent electro-scalp acupuncture (ESA) for 6 weeks at MS6, MS4, MS8, MS9, and MS14 (hemi-PD patients received hemi-ESA; bilateral PD patients received bilateral ESA). The ESA parameters were the same as described in Table 7.2. SPECT scans were performed before and after ESA treatment in each case. They used ^{99m}Tc-TRODAT as the tracer agent for DAT (125 mCi, i.v). Regions of interest (ROI) were set as bilateral striatum (ST) and unilateral occipital lobe (OC). The ratio of radioactivity in the ST to OC was regarded as an index of DAT density or functional state. The results showed that bilateral ST/OC improved after ESA, but not significantly ($p > 0.05$). It was concluded that the ESA was instrumental in activating bilateral striatal DAT or decreasing the loss of DAT (Fig. 7.7). Later works from this lab using the same experimental system showed that patients who received L-dopa and complementary acupuncture had increased regional cerebral blood flow (rCBF) in the following areas on the affected side after ESA: frontal lobe,

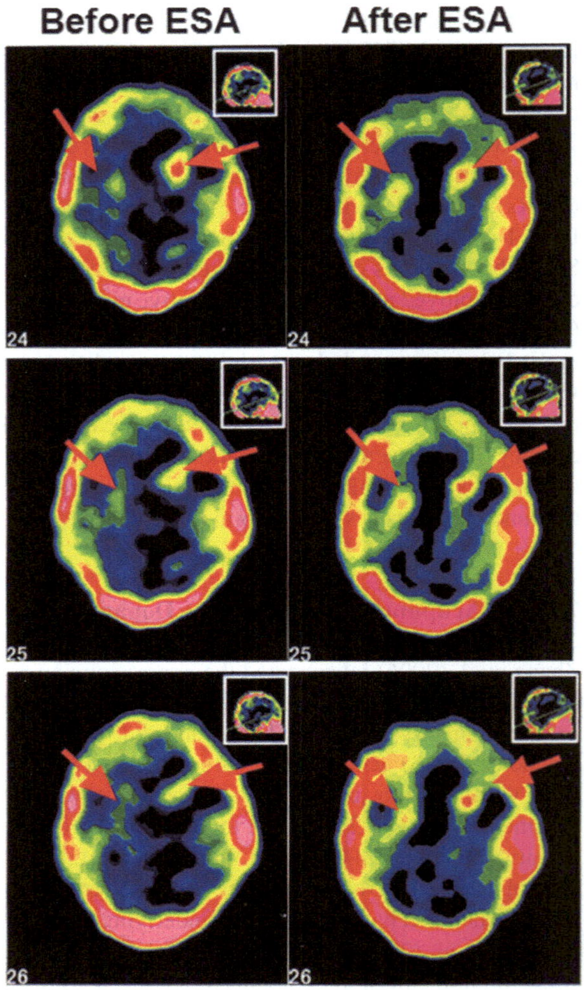

Fig. 7.7 SPECT images of striatal DAT before and after ESA in a hemisection of the brain of a PD patient. *Left arrow* indicates contralateral side to the abnormal limbs. *Right arrow* indicates the ipsilateral side to the abnormal limbs in each image. The SPECT imaging showed ESA increased the bilateral striatal DAT though the increase was not significant ($p>0.05$) (Jiang et al. 2006a)

occipital lobe, basal ganglion, and cerebellum when compared to the baseline, but there were no observed changes in the basal ganglia DAT levels. Treatment with L-dopa alone did not change rCBF but it did increase the basal ganglia DAT activity (Huang et al. 2010).

In this section, we have summarized several Chinese studies, which reported that the acupuncture improved the DA-ergic system including the DA concentrations in CSF, striatum, auditory conduction pathway, and the DAT. No study was found to be successful in analyzing the relationship between acupuncture and the changes in DA receptors, which is another important element of the DA-ergic system.

7.3.3 Acupuncture Improvement of the Motor Control Network in the Basal Ganglia

As a consequence of DA neuron depletion, the abnormal state of the motor controlling network (Fig. 7.2.) is regarded as an important pathological element of PD (Hashimoto et al. 2003; Nambu et al. 2002). Some studies reported that acupuncture has a direct contribution to correct the basal ganglia circuit.

Liang et al. (2003) reported that EA balances the activity of each nucleus in the basal ganglia circuit. Jia et al. (2009) performed the same EA procedure as Liang (Liang et al. 2002) and confirmed the successful establishment of PD rat models and the efficacy of EA by rotational behavioral test. The ventral midbrain glutamate decarboxylase-67 (GAD 67) mRNA was evaluated by real-time quantitative RT-PCR; substance P (SP) and enkephalin (ENK) of ventral midbrain and globus pallidus (GP) was measured by a radioimmunoassay (RIA) test. The results showed that 100 Hz EA normalized the SP, which is decreased in PD ($p<0.05$) and decreased the GAD 67 mRNA level in the midbrain ($p<0.01$), however it did not affect the ENK in GP. These data suggest that high-frequency EA stimulation results in neuropeptide normalization in the basal ganglia circuit. EA appears to normalize the MFB lesion-induced variations in GAD 67 mRNA expression in the ventral midbrain, thereby, affecting the inhibitory GABA transmission and counteracting the overactive output structures of the basal ganglia in PD. In a recent study by the same group, Jia investigated the GABA content of midbrain in the MFB transected rat models during effective EA (100 Hz) and ineffective EA (2 Hz) at Dazhui (DU-14) and Baihui (DU-21) by HPLC. He found that the effective EA improved the GABA content in the midbrain but had no effect on the GABA content of the globus pallidus, which indicated that the effective EA improved the motor controlling network by increasing the GABA-ergic inhibition in the output structure of basal ganglia (Jia et al. 2010). These processes might share the same mechanisms as the surgical therapies including STN-DBS and microinjection.

7.3.4 Acupuncture Attenuation of Oxidative Stress in PD State

Ever since Przedborski first reported in 1992 that oxy-radicals play a role in the MPTP-induced neurotoxicity, the oxidative stress has been largely accepted as a pathogenic factor in PD. It is believed that increased SOD and decreased LPO activities act as protective factors that confer resistance to oxidative stress and oxidative damage in the PD state (Przedborski et al. 1992; Verma and Nehru 2009; Jenner and Olanow 1996; Reynolds et al. 2007).

Zhang et al. (1997) performed scalp acupuncture at CTCA and Baihui (DU-21) in MPTP-treated C57 mouse models. The activities of superoxide dismutase (T-SOD), glutathione peroxidase (GSH-PX), and catalase (CAT), and the levels of malondialdehyde (MDA) were measured. The results suggested that the scalp acupuncture increased the antioxidase activities in the caudate nucleus and mesencephalon and normalized the overactive lipid peroxidation. Yuan et al. (2010) evaluated the expression of SOD, GSH-PX, and MDA in 6-OHDA lesioned rats. They got similar results as obtained in the mice. Acupuncture significantly raised the effect of SOD and GSH-PX, and attenuated the expression of MDA.

Dong et al. (2001) investigated the blood LPO and SOD levels in the 57 PD patients who underwent acupuncture treatment. The acupoints included MS6, Waiguan (TE-5), Hegu (LI-4), Yanglingquan (GB-34), Taichong (LR-3), and Zusanli (ST-36). Three courses of acupuncture were given with each course comprising of acupuncture given once daily for ten sessions. The plasma LPO and SOD of red blood cells were then measured. The results showed that the acupuncture significantly improved SOD levels in the senior (over 60 years, $p<0.05$) and female cases ($p<0.01$). There was an improving trend of SOD seen in the early-stage PD patients (onset within 3 years, $p\approx0.05$). Wang et al. (2006b) performed ESA at the Xuanli (GB-5), Xuanlu (GB-6), Fengfu (DU-16), Tianzhu (BL-10), and Fengchi (GB-20) in the 37 PD patients who received normal L-dopa administration. The other 39 control patients received L-dopa treatment only. The SOD and LPO levels of all the cases were investigated before and after ESA. The results showed that acupuncture significantly improved the scores on the rating scale ($p<0.01$). SOD in the treatment group was significantly increased while the LPO was decreased ($p<0.01$).

Above two bench studies and the two clinical studies showed that acupuncture increased the SOD while decreased the LPO, which is suggestive of a possible mechanism of acupuncture action underlying the DA neuroprotection from oxidative stress and oxidative damage.

7.3.5 *Other Possible Mechanisms Involved in the Acupuncture Treatment of PD*

We introduced the most widely investigated mechanisms of acupuncture involved in treating PD in the previous sections. As with other neurological diseases, many interactive factors and processes might be involved thereby making these mechanisms complicated. In this section, we list other possible mechanisms that need to undergo further investigation.

7.3.5.1 Regulation of Endogenous Opioids and Their Receptors

Endogenous opioid peptides (EOP) produce a myriad of physiological functions in the central nervous system by binding with different opioid receptors, including μ, δ, and κ-opioid receptors (MOR, DOR, and KOR) (Wen et al. 2010b). The DOR and its endogenous ligands (e.g., enkephalins) are widely expressed in the basal ganglia, STN, striatum, GP, and SNr (Abou-Khalil et al. 1984; Aubert et al. 2007;

Mansour et al. 1993). DOR has been well recognized for its neuroprotection (Chao et al. 2008, 2009; Ma et al. 2005; Yang et al. 2009; Zhang et al. 2000, 2002; Feng et al. 2009). It is for this reason that DOR has been proposed as a therapeutical target in PD (Mabrouk et al. 2008). Hallett and Brotchie (2007) reported that enhanced DOR transmission activates an endogenous compensatory mechanism, which inhibits the overactivated indirect pathway in PD. Mabrouk found that a DOR agonist is beneficial in ameliorating the motor dysfunction in PD rat models (Mabrouk et al. 2008). Sun et al. (2011) observed that opioid treatment protects against DA-ergic dysfunction. All of these studies emphasize the important role of EOP in the pathogenesis of PD.

Currently no study is being conducted to investigate the DOR changes in PD patients undergoing acupuncture treatment. However, several lines of indirect evidence indicates that the activation of EOP system, especially DOR, is a prominent underlying aspect of acupuncture treatment for PD. Chao et al. (2008) showed that the activation of DOR induces a neuroprotective effect by attenuating the disruption of K^+ homeostasis seen in pathologic states. Tian et al. found that EA increases DOR density and reduces ischemic injury in the brain (Tian et al. 2008). Since acupuncture/EA activates EOP system and in turn affects neural functions (Wen et al. 2010b), it is very likely that acupuncture contributes to the improvement of PD via DOR-mediated neuroprotection against neurodegenerative damage.

7.3.5.2 The Neurotoxic Effects of Excitatory Amino Acids

The role of the excitatory amino acids (EAA) in PD has been intensively studied ever since Turski et al. (1991) first demonstrated the role of EAA in PD pathogenesis. It was reported that the excitatory Glu-ergic system is overactivated in PD state; hence the correction of the overactive Glu-ergic system might prove to be useful in treating PD (Zipp et al. 1998). One Chinese study reported that acupuncture influences the glutamine (Glu) in striatum. Ma et al. (2003a) performed low frequency EA (2 Hz) at Taichong (LR-3) and Fengfu (DY-16) in SN-lesioned rat models. After 7 days of EA treatment, the striatal Glu concentration was measured. The results showed that EA reduced the striatal glutamate significantly. Whether acupuncture affects other elements in the Glu-ergic system is still unknown.

7.3.5.3 Inflammatory Processes

MPTP activates the microglia in the SNpc, which is thought to play a vital role in many neurodegenerative diseases including PD. This activation is associated with the expression of cyclooxygenase-2 (COX-2) and inducible nitric oxide synthase (iNOS) (Dehmer et al. 2000). The upregulation of iNOS expression in the glial cells of the SNpc is strongly implicated in the pathogenesis of PD (Hunot et al. 1996; Dehmer et al. 2000). Inflammatory processes associated with increased expression of COX-2 and elevated levels of prostaglandin E2 (PGE_2) also contribute to neurodegeneration (Teismann et al. 2003a, b).

Kang et al. (2007) performed acupuncture at Yanglingquan (GB-34) and Taichong (LR-3) in MPTP-treated C57BL/6 mice. In accordance with the previous studies, acupuncture generated a neuroprotective effect in the SN and ST at 1, 3, and 7 days after MPTP injection. The increasing expression of COX2 and iNOS on the days 1, 3, and 7 was down-regulated. The remaining striatal DA in the acupuncture group was measured and found to be 78 % at the seventh day compared to that of 46 % in the control group. Therefore, it was concluded that acupuncture could be used as a neuroprotective intervention for inhibiting microglial activation and inflammatory processes in PD.

7.3.5.4 Acupuncture Induces Peripheral Improvements in PD

Besides affecting the central nervous system, acupuncture also contributes to the amelioration of the peripheral abnormalities caused by PD.

Several Chinese clinical studies have reported that ESA stimulation improves the abnormal tremor myoelectric potential. Xi et al. (1996), Huang (2000b), and Wang et al. (2003) compared the electromyography (EMG) before and after the effective ESA stimulation. The acupoints selected were as follows: Xi used Baihui (DU-21); Huang used bilateral CTCA and MCA; Wang selected Fengchi (GB-20), Xuanli (GB-6), Qianding (GV-21), and Xuanlu (GB-5). In all these studies, the selected patients suffered from a serious tremor. It was observed that after ESA stimulation, the tremor symptoms in each of the three studies significantly improved ($p<0.05$). After stimulation, in 39 % (Xi), 25.8 % (Huang), and 79 % (Wang) patients' abnormal tremor myoelectric potential disappeared; in 33 % (Xi), 51.6 % (Huang), and 21 % (Wang) improvement was obtained, which was defined as a 50 % reduction in the frequency and swing range; while there were 28 % (Xi), 22.6 % (Huang), and 0 % (Wang) who did not improve. The improvements in EMG coincided with the variations of the tremor symptoms. It was concluded that ESA was beneficial in correcting the abnormal myoelectric potential in PD state. EMG can be employed as an effective objective index for evaluating the tremor in PD.

7.3.6 *Comments*

In this section, we summarized the potential mechanisms of the acupuncture treatment in PD. The most important mechanism is the reduction of the DA neuron degeneration by increasing some neurotrophic factors, which is quite superior to the classic treatments including L-dopa administration and DBS. Though many studies from China and Korea published rather excellent results showing significant improvements of the DA neurons in SN the weaknesses in the experimental design and the lack of peer confirmation of these results in other laboratories reduced the reliability of such reports.

We believe that the mechanics of acupuncture involve many complex factors and processes rather than a single mechanism. It is reasonable to believe that acupuncture can potentially affect different aspects of the pathogenesis of PD. Uncovering the secret of these mechanisms still has a long way to go.

7.4 Prospective Strategy: DBS vs. Modified Electroacupuncture

DBS has been widely accepted as a mainstream surgical therapy since Benabid initially reported it 11 years ago (Benabid et al. 2000). However, DBS is an invasive procedure with a high surgical risk. The limitation of the battery life of the stimulator embedded under the skin forces the patients to make a tough decision either to undergo another surgical operation to change the battery or adopt a palliative pattern by reducing the stimulation current to save the battery. In some developing countries without a perfect health insurance system, DBS is an extremely expensive operation. In addition to the cost, due to the above flaws of DBS many PD clinicians today hope to find a new effective procedure that is less invasive and can be performed by electrically stimulating certain easily accessible superficial structures (Xia et al. 2010).

Acupuncture proves to be an exactly such procedure ideal for PD, conducted by stimulating the acupoints on the body surface. However, EA is not so popular other than the Asian countries and is effectively practiced by those who can really understand the theory of TCM. The difficult theories of TCM hinder the understanding and acceptance of acupuncture. In this section, we would like to suggest a new electrical stimulation strategy, which is a modified EA mainly based on the conventional medical theory but closely related to the TCM theory, which we will refer to as the superficial electrical stimulation (SES). We hope that this would come across as an appropriate and acceptable alternative to the general PD investigators.

Several reports, other than those studies based on the acupoints stimulation, have demonstrated that peripheral electrical stimulation can influence the cerebral DA release. There is enough evidence on the direct interaction between the somatosensory system and motor system both in the PD patients and animal models. Back in 1977, Nieoullon first demonstrated that electrical stimulation of the forepaw caused an increased DA release in ipsilateral basal ganglia in the animals (Nieoullon et al. 1977). Rothblat and Schneider (1993) later pointed out that sensory stimulus affected the spontaneous activity of caudate nucleus units, which reflects the overall motor condition of the animal. Schultz found that peripheral noxious stimulation influenced the mesencephalic DA neurons by depressing or activating them (Schultz and Romo 1987). Inoue et al. (2004) performed sensory stimulation at the forepaw in six normal cats. [^{11}C]-raclopride positron emission tomography (PET) was used to detect the functional status of D_2-like receptor, which is commonly employed as an indirect index for evaluating DA release during the stimulation. The results showed that the sensory stimulation obviously influenced the endogenous DA release in the contralateral nucleus accumbens and striatum (Fig. 7.8). Fuentes et al. (2009) lately proposed a new spinal cord electrical stimulation method for the treatment of PD. Dorsal column stimulation (DCS) was performed on acute MPTP-treated mice and chronic 6-OHDA treated rats. The motor impairments in the mice and rats were seen to improve significantly, especially, the initiation of the locomotion.

All of the findings above suggest that stimulating certain peripheral "stations" in specific somatosensory pathways will affect the cerebral DA release, which can

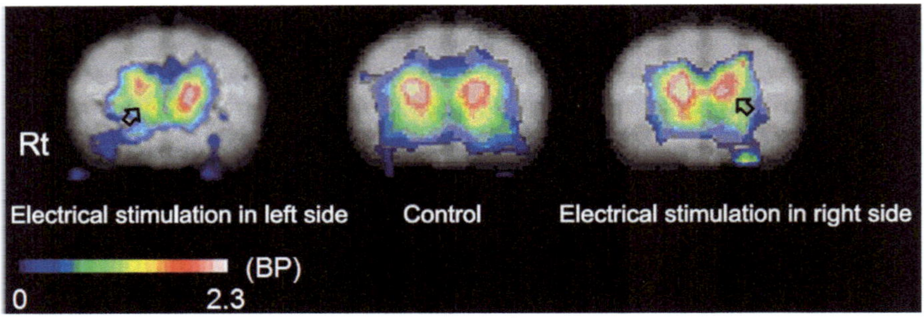

Fig. 7.8 The striatal D2-like receptor activity affected by contralateral forepaw electrical stimulation. Parametric images of binding potential (BP) in coregistered MRI; *Left image*: The electrical stimulation in the left paw decreased the BP value in the right basal ganglia. *Right image*: The electrical stimulation in the right paw decreased the BP value in the left basal ganglia. *Center image*: A parametric image of BP without stimulation. The BP value is shown in the *color bar*. Rt: The right side of the cat brain (Inoue et al. 2004).

possibly be beneficial in improving the PD symptoms. The anatomical connections of somatosensory pathways, including the distribution on the body surface, are not fully understood. Inoue's data suggested the possible anatomical pathways exist between the mesencephalic DA-ergic nuclei and the sensory system causing the observed modulation of DA release in the basal ganglia. One plausible connection is from the sensory areas in the neocortex contralateral to the stimulation side projects back into the ipsilateral striatum. One such pathway is the sensory area of the neocortex that projects back into the ipsilateral striatum. This connection activates the ipsilateral DA release from mesencephalic DA-ergic nuclei. Another possible anatomical connection could be the projecting fibers between the nuclei intralaminares thalami and the SNpc through the striatum on contralateral side. In addition, there is another potential pathway between the VTA and the mesencephalic central gray, which is innervated collaterally by the spinothalamic tract. (Inoue et al. 2004). The dorsal column in the spinal cord is believed to act as a connecting "station" in a certain somatosensory pathway. Stimulation of the dorsal column is thought to recruit brainstem arousal systems, leading to sufficient cortical and striatal desynchronization required for the voluntary initiation of movements (Fanselow et al. 2000; Fuentes et al. 2009). These findings also provide enough clues for establishing the SES approach. In our future studies, we will try to find such "special" somatosensory pathways and the related peripheral "stations" which may prove to be advantageous in improving locomotion in the PD animal models and patients by SES.

7.4.1 Animal Model Selection

Rodent animals (rat and mouse), cats, and non-human primate animals (rhesus monkey or cynomolgus monkey) are most commonly recruited for setting up of PD models. Because of a difference in DA metabolic enzyme systems, the selective DA neuron neurotoxins are different among different animals. Mouse, cat, and non-human primate animal PD models are usually established by MPTP administration, while the rat and pig PD models are set up by 6-OHDA injection, or surgical lesioning (such as MFB transection). Ideal animal models for SES studies should not only closely imitate the symptoms but also the natural process of PD in a human being. Hence, chronic–low dose–toxin administration in an aged animal seems to be a reasonable approach for theoretically establishing the PD model, yet it is rarely adopted considering its experimental efficiency and cost.

The following factors have to be contemplated upon while selecting an appropriate PD model for SES studies:

1. MPTP administration is quite dangerous for the experimenters and requires expertise along with a good ventilation system. For safety reasons, rats might be a better choice for a larger sample required in the early stage of the study; while MPTP-treated monkey models are good for the terminal animal experiments before applying certain new therapies or new medicines to human beings.
2. A monkey model can imitate all the symptoms observed in humans, whereas it is difficult to produce certain symptoms such as tremor and gesture failure in the non-primates animals.
3. For functional imaging studies such as PET and SPECT, the brain size is of vital importance in differentiating the variations in tiny structures. Too small brain will cause a large bias in the data acquisition, absorption correction, image reconstruction, and statistical analysis. Mouse model is not good for such imaging studies for SES.
4. The compliance to the behavioral tests. The behavioral evaluation techniques in rodent animals and monkeys are mature and reliable, and they are thought to have good compliance with the task training.

We therefore selected rat and monkey models after minutely considering the above factors. Rat models were used during the early stage of the study, to detect the appropriate stimulating points. Monkey models were selected for the advanced stage, when the experimental system has been established and the efficacy has been confirmed. Monkey models also proved to be a good choice for carrying out the functional imaging studies before planning a human study.

Among the rat models, we had MFB-lesioned models (complete kill model), SN-lesioned models, and striatal models. In our previous studies, we found that the D_2-like receptor in MFB-lesioned rats was upregulated during early stages after lesioning (4 weeks after lesioning), then down-regulated in the later stages (6 months after lesioning), which is close to the dynamic changes of D_2-like receptor seen in the PD patients (Sun et al. 2010). We, hence, believe that MFB-lesioned models are better than striatal and SN-lesioned for the initiation of SES study.

7.4.2 *Behavioral Assessments*

Behavioral tests have come to play increasingly important roles in designing experimental PD studies. Behavioral tasks are always used for verifying the success of establishing an animal model, or assessing the efficacy of a new therapy. An ideal behavioral test should have the following features: (1) objectivity and repeatability; (2) quantitative; (3) representativeness; (4) operability; and (5) sensitivity.

For SES study, we have to detect the appropriate stimulation targets first; thus, symptoms that are easily observed should be taken into account. The initiation time of the stepping test in rats and tremor termination in the monkeys are appropriate for target selection.

When optimizing the targets to be stimulation, more behavioral tests should be tried, including the stepping test (Nikkhah et al. 1993; Fang et al. 2006b), Rotarod test (Rozas et al. 1997; Fang et al. 2006b), and amphetamine-induced rotational movement in rats and the food-reaching test (FRT) in the monkeys (unpublished data). We also considered developing a new behavioral test appropriate for SES efficacy assessment in the study.

7.4.3 Exploring the Appropriate Superficial Targets for SES

This is the most important procedure of the whole study. Just like it has been described in the former sections, SES is different yet in a way related, to the traditional EA. The target identification for SES can be performed by either of the two approaches: First is by following the potential somatosensory pathways: find and stimulate the important peripheral "stations" through the pathway; and second is by following the theory of TCM—stimulating the various acupoints. Conversely, we can track all the acupoints with good efficacy, comprehensively analyze them, and find new peripheral distributions of their associated somatosensory pathways.

7.4.4 Adjusting the Parameters for Stimulation

The selection of stimulation parameters is another critical issue in SES. First, while detecting the stimulating targets, we must decide on a "potentially" effective stimulating condition, which is decided based on the experiences of the previous studies, such as high stimulating frequency (over 100 Hz) and strong current intensity (1–2 mA). After the stimulating targets are set, we should compare the different conditions of stimulation: (1) low frequency vs. high-frequency stimulation; (2) the appropriate current intensity; and (3) stimulations using different wave patterns, such as rectangular wave in DBS or disperse-dense wave in EA.

7.4.5 Efficacy Evaluation Using Molecular Imaging Approaches

In the traditional animal experiments, animals sometimes had to be killed to perform a pathological assay. We cannot compare the index such as the DA neurons in SN

Table 7.3 Currently available PET tracers for PD

Radionuclide	Tracers	Function
^{11}C	^{11}C-dopa	Presynaptic DA synthesis
	PE2I	DAT
	^{11}C-CFT	DAT
	^{11}C-raclopride	DA D_2 like receptor
	^{11}C-SCH23390	DA D_1 like receptor
^{18}F	FDG	Glucose metabolism
	FLT	Cell proliferation
^{15}O	H_2O	Regional cerebral blood flow

before and after treatment using one rat. An alternative method is to set several parallel control/Sham groups. However, if the sample size is not large enough, the conclusions can be distorted by individual differences (selection bias). Molecular imaging methods can absolve these limitations.

With the development of new tracers, PET scan is beginning to be used in the evaluation of certain new therapies. PET is a noninvasive method and exempts the need to sacrifice the animals; the biggest advantage of PET, however, lies in the fact that it can quantitatively or semiquantitatively detect the dynamic real-time variations in certain substances/indices in a living animal, or during certain treatments such as DBS or acupuncture. Variations in different indices induced by a treatment can be dynamically evaluated by using multiple tracers in the same animal. Such features are appropriate for evaluating SES, since the dynamic changes in the DA-ergic system induced by SES in vivo are critical for evaluation of the SES efficacy.

The available tracers commonly used for current PET studies on PD are listed in Table 7.3.

7.5 Concluding Remarks

In this chapter, we have summarized the clinical applications of acupuncture in PD treatment by reviewing the current studies from all over the world and have also tried to find clues on the potential mechanisms from the limited information available. A number of studies from China and Korea have reported positive efficacy and fewer side effects of acupuncture in PD treatment. Several studies focusing on the different aspects of the potential mechanisms involved in the acupuncture are being conducted. If such effects of acupuncture are rigorously validated, acupuncture can be introduced to the mainstream academia for a commonly used alternative therapy for PD.

Like with other alternative therapies, the efficacy and safety of acupuncture have not been strictly verified. Due to methodological weaknesses, the observations and results thus obtained have overestimated the treatment efficacy and concealed the possible adverse events, which reduces the reliability of these studies. This is a major reason for the unacceptability of some of the alternative therapies. What is worse is that not only is the therapy itself under doubt, but it has also prejudiced the viewpoint of the involved researchers against the whole TCM theory system, including those from its motherland. An illustration of this fact is that despite so many Chinese/Korean studies on PD, which have proved the efficacy of acupuncture (essentially, a kind of peripheral stimulation), it has still not stepped into the view of mainstream PD academia, as the PD clinicians are still pondering to find a noninvasive new electrical stimulation method for PD, and have not yet accepted acupuncture as an option. It is for this reason that TCM today faces an awkward stand even in China. Hence, we wish to raise this interest and bring forth the need for well-designed, rigorous clinical and experimental studies to provide formidable scientific evidence validating the efficacy and safety of acupuncture in the treatment of PD.

At fag end of this chapter, a modified EA method conceived from the combination of the theories of western medicine and TCM, called SES, has been suggested. In future, we will concentrate on developing this new therapy, which we believe would be accepted by most PD researchers including the general medical investigators.

Acknowledgments TA was supported by Japan Society for the Promotion of Science (Grant-in-Aid for Young Scientists, Type B, #20791025 and Grant-in-Aid for Scientific Research General, #24592157). YX was supported by the National Institutes of Health (AT-004422 and HD-034852) and the Vivian L. Smith Neurologic Foundation.

References

Abou-Khalil B, Young AB, Penney JB **(1984)** Evidence for the presynaptic localization of opiate binding sites on striatal efferent fibers. Brain Res 323 (1):21–29

Alexander GE, Crutcher MD, DeLong MR **(1990)** Basal ganglia-thalamocortical circuits: parallel substrates for motor, oculomotor, "prefrontal" and "limbic" functions. Prog Brain Res 85:119–146

Alexander GE, DeLong MR, Strick PL **(1986)** Parallel organization of functionally segregated circuits linking basal ganglia and cortex. Annu Rev Neurosci 9:357–381

Aubert I, Guigoni C, Li Q, Dovero S, Bioulac BH, Gross CE, Crossman AR, Bloch B, Bezard E **(2007)** Enhanced preproenkephalin-B-derived opioid transmission in striatum and subthalamic nucleus converges upon globus pallidus internalis in L-3,4-dihydroxyphenylalanine-induced dyskinesia. Biol Psychiatry 61 (7):836–844

Benabid AL, Koudsie A, Benazzouz A, Fraix V, Ashraf A, Le Bas JF, Chabardes S, Pollak P **(2000)** Subthalamic stimulation for Parkinson's disease. Arch Med Res 31 (3):282–289

Birch S **(2004)** Clinical research on acupuncture. Part 2. Controlled clinical trials, an overview of their methods. J Altern Complement Med 10 (3):481–498

Burke RE, Antonelli M, Sulzer D **(1998)** Glial cell line-derived neurotrophic growth factor inhibits apoptotic death of postnatal substantia nigra dopamine neurons in primary culture. J Neurochem 71 (2):517–525

Chao D, Balboni G, Lazarus LH, Salvadori S, Xia Y **(2009)** Na+ mechanism of delta-opioid receptor induced protection from anoxic K+ leakage in the cortex. Cell Mol Life Sci 66 (6):1105–1115

Chao D, Bazzy-Asaad A, Balboni G, Salvadori S, Xia Y **(2008)** Activation of DOR attenuates anoxic K+ derangement via inhibition of Na+ entry in mouse cortex. Cereb Cortex 18 (9):2217–2227

Chen L, LI J, Li X, Chen B, Mo L **(2010)** Tendons acupuncture points Yanglingquan substantia nigra in Parkinson's mouse model of TH expression. Guiding Journal of Traditional Chinese Medicine and Pharmacology 16 (11):67–68 (in Chinese)

Clarkson ED, Edwards-Prasad J, Freed CR, Prasad KN **(1999)** Immortalized dopamine neurons: A model to study neurotoxicity and neuroprotection. Proc Soc Exp Biol Med 222 (2):157–163

Cristian A, Katz M, Cutrone E, Walker RH **(2005)** Evaluation of acupuncture in the treatment of Parkinson's disease: a double-blind pilot study. Mov Disord 20 (9):1185–1188

Dai LJ **(2004)** The progress of treating Parkinson's disease by acupuncture. Yunnan Journal of Traditional Chinese Medicine and Materia Medica 25 (5):48–49 (in Chinese)

Dehmer T, Lindenau J, Haid S, Dichgans J, Schulz JB **(2000)** Deficiency of inducible nitric oxide synthase protects against MPTP toxicity in vivo. J Neurochem 74 (5):2213–2216

Dong XT, Wang SK, Ren XQ, Yu MJ, Yang JH, Liu AH, Liu JY, Chen GP, Wang YJ **(2001)** Effects of Acupuncture Treatment on Blood LPO and SOD Contents in Parkinsonism Patients. Acupuncture Research 26 (1):28–30 (in Chinese with English abstract)

Fang X, Sugiyama K, Akamine S, Namba H **(2006a)** Improvements in motor behavioral tests during deep brain stimulation of the subthalamic nucleus in rats with different degrees of unilateral parkinsonism. Brain Res 1120 (1):202–210

Fang X, Sugiyama K, Akamine S, Namba H **(2006b)** The stepping test and its learning process in different degrees of unilateral striatal lesions by 6-hydroxydopamine in rats. Neurosci Res 55 (4):403–409

Fanselow EE, Reid AP, Nicolelis MA **(2000)** Reduction of pentylenetetrazole-induced seizure activity in awake rats by seizure-triggered trigeminal nerve stimulation. J Neurosci 20 (21):8160–8168

Feigin A, Kaplitt MG, Tang C, Lin T, Mattis P, Dhawan V, During MJ, Eidelberg D **(2007)** Modulation of metabolic brain networks after subthalamic gene therapy for Parkinson's disease. Proc Natl Acad Sci U S A 104 (49):19559–19564

Feng Y, Chao D, He X, Yang Y, Kang X, Lazarus LH, Xia Y **(2009)** A novel insight into neuroprotection against hypoxic/ischemic stress. Sheng Li Xue Bao 61 (6):585–592

Ferry P, Johnson M, Wallis P **(2002)** Use of complementary therapies and non-prescribed medication in patients with Parkinson's disease. Postgrad Med J 78 (924):612–614

Frim DM, Uhler TA, Galpern WR, Beal MF, Breakefield XO, Isacson O **(1994)** Implanted fibroblasts genetically engineered to produce brain-derived neurotrophic factor prevent 1-methyl-4-phenylpyridinium toxicity to dopaminergic neurons in the rat. Proc Natl Acad Sci U S A 91 (11):5104–5108

Fuentes R, Petersson P, Siesser WB, Caron MG, Nicolelis MA **(2009)** Spinal cord stimulation restores locomotion in animal models of Parkinson's disease. Science 323 (5921):1578–1582

George JL, Mok S, Moses D, Wilkins S, Bush AI, Cherny RA, Finkelstein DI **(2009)** Targeting the progression of Parkinson's disease. Curr Neuropharmacol 7 (1):9–36

Gill SS, Patel NK, Hotton GR, O'Sullivan K, McCarter R, Bunnage M, Brooks DJ, Svendsen CN, Heywood P **(2003)** Direct brain infusion of glial cell line-derived neurotrophic factor in Parkinson disease. Nat Med 9 (5):589–595

Guo JC, Cheng JS, Xia Y (2010) Acupuncture therapy for stroke. In: Acupuncture Therapy for Neurological Diseases: A Neurobiological View. Heidelberg, London, New York, Beijing, pp 226–262

Hallett PJ, Brotchie JM **(2007)** Striatal delta opioid receptor binding in experimental models of Parkinson's disease and dyskinesia. Mov Disord 22 (1):28–40

Hammerschlag R **(1998)** Methodological and ethical issues in clinical trials of acupuncture. Journal of Alternative and Complementary Medicine 4 (2):159–171

Hashimoto T, Elder CM, Okun MS, Patrick SK, Vitek JL **(2003)** Stimulation of the subthalamic nucleus changes the firing pattern of pallidal neurons. J Neurosci 23 (5):1916–1923

Hilker R, Portman AT, Voges J, Staal MJ, Burghaus L, van Laar T, Koulousakis A, Maguire RP, Pruim J, de Jong BM, Herholz K, Sturm V, Heiss WD, Leenders KL **(2005)** Disease progression continues in patients with advanced Parkinson's disease and effective subthalamic nucleus stimulation. J Neurol Neurosurg Psychiatry 76 (9):1217–1221

Hilker R, Voges J, Ghaemi M, Lehrke R, Rudolf J, Koulousakis A, Herholz K, Wienhard K, Sturm V, Heiss WD **(2003)** Deep brain stimulation of the subthalamic nucleus does not increase the striatal dopamine concentration in parkinsonian humans. Mov Disord 18 (1):41–48

Hirsch E, Graybiel AM, Agid YA **(1988)** Melanized dopaminergic neurons are differentially susceptible to degeneration in Parkinson's disease. Nature 334 (6180):345–348

Hong MS, Park HK, Yang JS, Park HJ, Kim ST, Kim SN, Park JY, Song JY, Jo DJ, Park SW, HwanYun D, Ban JY, Chung JH (**2010**) Gene expression profile of acupuncture treatment in 1-methyl-4-phenyl-1,2,3,6-tetrahydropyridine-induced Parkinson's disease model. Neurol Res 32 Suppl 1:74–78

Huang WY (**2000a**) Eye acupuncture in 45 Parkinsonian Patients, a clinical report. Shanghai Journal of Acupuncture and Moxibustion 19 (4):20 (in Chinese)

Huang WY (**2000b**) Influence of electric stimulation of scalp acupuncture on electromyogram in tremor pattern of Parkinson's disease. Acta Universitatis Traditionis Medicalis Sinensis Pharmacologiaeque Shanghai 14 (3):38–39 (in Chinese)

Huang Y, Jiang X, Zhuo Y, Wik G (**2010**) Complementary acupuncture in Parkinson's disease: a spect study. Int J Neurosci 120 (2):150–154

Hunot S, Boissiere F, Faucheux B, Brugg B, Mouatt-Prigent A, Agid Y, Hirsch EC (**1996**) Nitric oxide synthase and neuronal vulnerability in Parkinson's disease. Neuroscience 72 (2):355–363

Inoue M, Katsumi Y, Hayashi T, Mukai T, Ishizu K, Hashikawa K, Saji H, Fukuyama H (**2004**) Sensory stimulation accelerates dopamine release in the basal ganglia. Brain Res 1026 (2):179–184

Jenner P, Olanow CW (**1996**) Oxidative stress and the pathogenesis of Parkinson's disease. Neurology 47 (6 Suppl 3):S161–170

Jeon S, Kim YJ, Kim ST, Moon W, Chae Y, Kang M, Chung MY, Lee H, Hong MS, Chung JH, Joh TH, Park HJ (**2008**) Proteomic analysis of the neuroprotective mechanisms of acupuncture treatment in a Parkinson's disease mouse model. Proteomics 8 (22):4822–4832

Jia J, Li B, Sun ZL, Yu F, Wang X, Wang XM (**2010**) Electro-acupuncture stimulation acts on the basal ganglia output pathway to ameliorate motor impairment in Parkinsonian model rats. Behav Neurosci 124 (2):305–310

Jia J, Sun Z, Li B, Pan Y, Wang H, Wang X, Yu F, Liu L, Zhang L (**2009**) Electro-acupuncture stimulation improves motor disorders in Parkinsonian rats. Behav Brain Res 205 (1):214–218

Jiang XM, Huang Y, Li DJ, Tang AW, Wang SX, Zhuo Y, Li QS, Chen J, Gao YP (**2006a**) [Effect of electro-scalp acupuncture on cerebral dopamine transporter in the striatum area of the patient of Parkinson's disease by means of single photon emission computer tomography]. Zhongguo Zhen Jiu 26 (6):427–430 (in Chinese with English abstract)

Jiang XM, Huang Y, Zhuo Y, Gao YP (**2006b**) [Therapeutic effect of scalp electroacupuncture on Parkinson disease]. Nan Fang Yi Ke Da Xue Xue Bao 26 (1):114–116 (in Chinese with English abstract)

Joh TH, Park HJ, Kim SN, Lee H (**2010**) Recent development of acupuncture on Parkinson's disease. Neurol Res 32 Suppl 1:5–9

Kang JM, Park HJ, Choi YG, Choe IH, Park JH, Kim YS, Lim S (**2007**) Acupuncture inhibits microglial activation and inflammatory events in the MPTP-induced mouse model. Brain Res 1131 (1):211–219

Kang XZ, Xia Y (**2010a**) Acupuncture therapy for hypertension and hypotension. In: Acupuncture Therapy for Neurological Diseases: A Neurobiological View. Springer-Tsinghua Press, Heidelberg, London, New York, Beijing, pp 289–325

Kang XZ, Xia Y (**2010b**) Effects of Acupuncture on Arrhythmia and Other Cardiac Diseases. In: Acupuncture Therapy for Neurological Diseases: A Neurobiological View. Springer-Tsinghua Press, Heidelberg, London, New York, Beijing, pp 263–288

Kaplitt MG, Feigin A, Tang C, Fitzsimons HL, Mattis P, Lawlor PA, Bland RJ, Young D, Strybing K, Eidelberg D, During MJ (**2007**) Safety and tolerability of gene therapy with an adeno-associated virus (AAV) borne GAD gene for Parkinson's disease: an open label, phase I trial. Lancet 369 (9579):2097–2105

Kim JH, Auerbach JM, Rodriguez-Gomez JA, Velasco I, Gavin D, Lumelsky N, Lee SH, Nguyen J, Sanchez-Pernaute R, Bankiewicz K, McKay R (**2002**) Dopamine neurons derived from embryonic stem cells function in an animal model of Parkinson's disease. Nature 418 (6893):50–56

Kim YK, Lim HH, Song YK, Lee HH, Lim S, Han SM, Kim CJ (**2005**) Effect of acupuncture on 6-hydroxydopamine-induced nigrostratal dopaminergic neuronal cell death in rats. Neurosci Lett 384 (1–2):133–138

Kleiner-Fisman G, Fisman DN, Sime E, Saint-Cyr JA, Lozano AM, Lang AE (**2003**) Long-term follow up of bilateral deep brain stimulation of the subthalamic nucleus in patients with advanced Parkinson disease. J Neurosurg 99 (3):489–495

Kodama Y, Ieda T, Hirayama M, Koike Y, Ito H, Sobue G (**1999**) Auditory brainstem responses in patients with autonomic failure with Parkinson's disease and multiple system atrophy. J Auton Nerv Syst 77 (2–3):184–189

Krack P, Batir A, Van Blercom N, Chabardes S, Fraix V, Ardouin C, Koudsie A, Limousin PD, Benazzouz A, LeBas JF, Benabid AL, Pollak P (**2003**) Five-year follow-up of bilateral stimulation of the subthalamic nucleus in advanced Parkinson's disease. N Engl J Med 349 (20):1925–1934

Lam YC, Kum WF, Durairajan SS, Lu JH, Man SC, Xu M, Zhang XF, Huang XZ, Li M (**2008**) Efficacy and safety of acupuncture for idiopathic Parkinson's disease: a systematic review. J Altern Complement Med 14 (6):663–671

Lang AE, Gill S, Patel NK, Lozano A, Nutt JG, Penn R, Brooks DJ, Hotton G, Moro E, Heywood P, Brodsky MA, Burchiel K, Kelly P, Dalvi A, Scott B, Stacy M, Turner D, Wooten VG, Elias WJ, Laws ER, Dhawan V, Stoessl AJ, Matcham J, Coffey RJ, Traub M (**2006**) Randomized controlled trial of intraputamenal glial cell line-derived neurotrophic factor infusion in Parkinson disease. Ann Neurol 59 (3):459–466

Lee MS, Shin BC, Choi SM, Kim JY (**2009**) Randomized clinical trials of constitutional acupuncture: a systematic review. Evid Based Complement Alternat Med 6 Suppl 1:59–64

Lee MS, Shin BC, Kong JC, Ernst E (**2008**) Effectiveness of acupuncture for Parkinson's disease: a systematic review. Mov Disord 23 (11):1505–1515

Levy R, Lang AE, Dostrovsky JO, Pahapill P, Romas J, Saint-Cyr J, Hutchison WD, Lozano AM (**2001**) Lidocaine and muscimol microinjections in subthalamic nucleus reverse Parkinsonian symptoms. Brain 124 (Pt 10):2105–2118

Li ZT, Yang WB (**2005**) Treatment of Parkinson's disease mainly by comprehensive therapy: A Clinical Observation of 30 Cases. Journal of New Chinese Medicine 37 (10):52–54 (in Chinese)

Liang XB, Liu XY, Li FQ, Luo Y, Lu J, Zhang WM, Wang XM, Han JS (**2002**) Long-term high-frequency electro-acupuncture stimulation prevents neuronal degeneration and up-regulates BDNF mRNA in the substantia nigra and ventral tegmental area following medial forebrain bundle axotomy. Brain Res Mol Brain Res 108 (1–2):51–59

Liang XB, Luo Y, Liu XY, Lu J, Li FQ, Wang Q, Wang XM, Han JS (**2003**) Electro-acupuncture improves behavior and upregulates GDNF mRNA in MFB transected rats. Neuroreport 14 (8): 1177–1181

Liu JY, Ren XQ (**1993**) The observation of the efficiency in the 159 Parkinson's disease cases treated by acupuncture. Journal of Clinical Acupuncture and Moxibustion 9 (5):10–11 (in Chinese)

Liu TH, Sadove MS (**1974**) Scalp needle therapy--acupuncture treatment for central nervous system disorders. Am J Chin Med (Gard City N Y) 2 (3):261–269

Liu YTS, Z.D., Li SJ (**2001**) Contrast study of microelectrode guided post - ventrual pallidotomy and Chinese medicine for treatment in Parkinson disease. Chinese Journal of Stereotactic and Functional Neurosurgery 14 (3):731 (in Chinese)

Ma J (**1999**) The theories of the Parkinson's disease in traditional Chinese medicine. Journal of Hubei College of Traditional Chinese Medicine 1 (1):62–63 (in Chinese)

Ma J, Tian Q, Wang YC, Wang H (**2003a**) Influence of electroacupuncture on the content of glutamine in striatum of Parkinson's disease rats. Shanghai Journal of Acupuncture and Moxibustion 22 (1):36 (in Chinese)

Ma J, Wang H (**2001**) Effect of electroacupuncture on TH content in mesencephalic substantia nigra of PD rats. Acupuncture Research 26 (4):288–291 (in Chinese with English abstract)

Ma J, Wang YC, Gan SY (**2003b**) Influence of electroacupuncture at Fengfu point and Taichong point on apoptosis of black substance neuron of PD rats. Hubei Journal of Traditional Chinese Medicine 25 (11):5–6 (in Chinese)

Ma MC, Qian H, Ghassemi F, Zhao P, Xia Y (**2005**) Oxygen-sensitive {delta}-opioid receptor-regulated survival and death signals: novel insights into neuronal preconditioning and protection. J Biol Chem 280 (16):16208–16218

Mabrouk OS, Volta M, Marti M, Morari M **(2008)** Stimulation of delta opioid receptors located in substantia nigra reticulata but not globus pallidus or striatum restores motor activity in 6-hydroxydopamine lesioned rats: new insights into the role of delta receptors in parkinsonism. J Neurochem 107 (6):1647–1659

MacPherson H, White A, Cummings M, Jobst K, Rose K, Niemtzow R **(2002a)** Standards for reporting interventions in controlled trials of acupuncture: The STRICTA recommendations. STandards for Reporting Interventions in Controlled Trails of Acupuncture. Acupunct Med 20 (1):22–25

MacPherson H, White A, Cummings M, Jobst KA, Rose K, Niemtzow RC **(2002b)** Standards for Reporting Interventions in Controlled Trials of Acupuncture: the STRICTA recommendations. J Altern Complement Med 8 (1):85–89

Mansour A, Thompson RC, Akil H, Watson SJ **(1993)** Delta opioid receptor mRNA distribution in the brain: comparison to delta receptor binding and proenkephalin mRNA. J Chem Neuroanat 6 (6):351–362

McManus CA, Kaptchuk TJ, Schnyer RN, Goldman R, Kerr CE, Nguyen LT, Stason WB **(2007)** Experiences of acupuncturists in a placebo-controlled, randomized clinical trial. J Altern Complement Med 13 (5):533–538

Moher D, Pham B, Lawson ML, Klassen TP **(2003)** The inclusion of reports of randomised trials published in languages other than English in systematic reviews. Health Technol Assess 7 (41):1–90

Nambu A, Tokuno H, Takada M **(2002)** Functional significance of the cortico-subthalamo-pallidal 'hyperdirect' pathway. Neurosci Res 43 (2):111–117

Nieoullon A, Cheramy A, Glowinski J **(1977)** Nigral and striatal dopamine release under sensory stimuli. Nature 269 (5626):340–342

Nikkhah G, Duan WM, Knappe U, Jodicke A, Bjorklund A **(1993)** Restoration of complex sensorimotor behavior and skilled forelimb use by a modified nigral cell suspension transplantation approach in the rat Parkinson model. Neuroscience 56 (1):33–43

O'Donnell BF, Squires NK, Martz MJ, Chen JR, Phay AJ **(1987)** Evoked potential changes and neuropsychological performance in Parkinson's disease. Biol Psychol 24 (1):23–37

Odsberg A, Schill U, Haker E **(2001)** Acupuncture treatment: side effects and complications reported by Swedish physiotherapists. Complement Ther Med 9 (1):17–20

Palhagen S, Heinonen E, Hagglund J, Kaugesaar T, Maki-Ikola O, Palm R **(2006)** Selegiline slows the progression of the symptoms of Parkinson disease. Neurology 66 (8):1200–1206

Park HJ, Lim S, Joo WS, Yin CS, Lee HS, Lee HJ, Seo JC, Leem K, Son YS, Kim YJ, Kim CJ, Kim YS, Chung JH **(2003)** Acupuncture prevents 6-hydroxydopamine-induced neuronal death in the nigrostriatal dopaminergic system in the rat Parkinson's disease model. Exp Neurol 180 (1):93–98

Park J, White A, Stevinson C, Ernst E, James M **(2002)** Validating a new non-penetrating sham acupuncture device: two randomised controlled trials. Acupunct Med 20 (4):168–174

Parkinson J **(2002)** An essay on the shaking palsy. 1817. J Neuropsychiatry Clin Neurosci 14 (2):223–236; discussion 222

Przedborski S **(2005)** Pathogenesis of nigral cell death in Parkinson's disease. Parkinsonism Relat Disord 11 Suppl 1:S3–7

Przedborski S, Kostic V, Jackson-Lewis V, Naini AB, Simonetti S, Fahn S, Carlson E, Epstein CJ, Cadet JL **(1992)** Transgenic mice with increased Cu/Zn-superoxide dismutase activity are resistant to N-methyl-4-phenyl-1,2,3,6-tetrahydropyridine-induced neurotoxicity. J Neurosci 12 (5):1658–1667

Qi XJ, Wang S **(2007)** The progress in the clinical researches of the acupuncture therapy on Parkinson's disease. Journal of Clinical Acupuncture and Moxibustion 23 (7):68–69 (in Chinese)

Rajendran PR, Thompson RE, Reich SG **(2001)** The use of alternative therapies by patients with Parkinson's disease. Neurology 57 (5):790–794

Ren XM **(2008)** Fifty cases of Parkinson's disease treated by acupuncture combined with madopar. J Tradit Chin Med 28 (4):255–257

Reynolds A, Laurie C, Mosley RL, Gendelman HE **(2007)** Oxidative stress and the pathogenesis of neurodegenerative disorders. Int Rev Neurobiol 82:297–325

Rothblat DS, Schneider JS (**1993**) Response of caudate neurons to stimulation of intrinsic and peripheral afferents in normal, symptomatic, and recovered MPTP-treated cats. J Neurosci 13 (10):4372–4378

Rozas G, Guerra MJ, Labandeira-Garcia JL (**1997**) An automated rotarod method for quantitative drug-free evaluation of overall motor deficits in rat models of parkinsonism. Brain Res Brain Res Protoc 2 (1):75–84

Schapira AH (**2009**) Neurobiology and treatment of Parkinson's disease. Trends Pharmacol Sci 30 (1):41–47

Schultz W, Romo R (**1987**) Responses of nigrostriatal dopamine neurons to high-intensity somatosensory stimulation in the anesthetized monkey. J Neurophysiol 57 (1):201–217

Shi Y, Liu HR, Wu HG (**2002**) The clinical study of the mechanism of acupuncture therapy for Parkinson's disease. Chinese Journal of Clinical Rehabilitation 6 (December):3598–3599 (in Chinese with English abstract)

Shulman LM, Wen X, Weiner WJ, Bateman D, Minagar A, Duncan R, Konefal J (**2002**) Acupuncture therapy for the symptoms of Parkinson's disease. Mov Disord 17 (4):799–802

Shults CW, Kimber T, Altar CA (**1995**) BDNF attenuates the effects of intrastriatal injection of 6-hydroxydopamine. Neuroreport 6 (8):1109–1112

Slevin JT, Gerhardt GA, Smith CD, Gash DM, Kryscio R, Young B (**2005**) Improvement of bilateral motor functions in patients with Parkinson disease through the unilateral intraputaminal infusion of glial cell line-derived neurotrophic factor. J Neurosurg 102 (2):216–222

Spina MB, Squinto SP, Miller J, Lindsay RM, Hyman C (**1992**) Brain-derived neurotrophic factor protects dopamine neurons against 6-hydroxydopamine and N-methyl-4-phenylpyridinium ion toxicity: involvement of the glutathione system. J Neurochem 59 (1):99–106

Su EL (**2002**) The clinical application and experience of scalp acupuncture at the specia areas of head imitating the viscus mapping in the tongue. Liaoning Journal of Traditional Chinese Medicine 29 (5):289–290 (in Chinese with English abstract)

Sun W, Sugiyama K, Fang X, Yamaguchi H, Akamine S, Magata Y, Namba H (**2010**) Different striatal D2-like receptor function in an early stage after unilateral striatal lesion and medial forebrain bundle lesion in rats. Brain Res 1317:227–235

Sun YM, Hoang T, Neubauer JA, Walters AS (**2011**) Opioids protect against substantia nigra cell degeneration under conditions of iron deprivation: a mechanism of possible relevance to the Restless Legs Syndrome (RLS) and Parkinson's disease. J Neurol Sci 304 (1–2):93–101

Teismann P, Tieu K, Choi DK, Wu DC, Naini A, Hunot S, Vila M, Jackson-Lewis V, Przedborski S (**2003a**) Cyclooxygenase-2 is instrumental in Parkinson's disease neurodegeneration. Proc Natl Acad Sci U S A 100 (9):5473–5478

Teismann P, Vila M, Choi DK, Tieu K, Wu DC, Jackson-Lewis V, Przedborski S (**2003b**) COX-2 and neurodegeneration in Parkinson's disease. Ann N Y Acad Sci 991:272–277

Tian XS, Zhou F, Yang R, Xia Y, Wu GC, Guo JC (**2008**) Electro-acupuncture protects the brain against ischemic injury via up-regulation of delta-opioid receptors. J Chin Integr Med 6:632–638 (in Chinese with English abstract)

Turski L, Bressler K, Rettig KJ, Loschmann PA, Wachtel H (**1991**) Protection of substantia nigra from MPP+ neurotoxicity by N-methyl-D-aspartate antagonists. Nature 349 (6308):414–418

Verma R, Nehru B (**2009**) Effect of centrophenoxine against rotenone-induced oxidative stress in an animal model of Parkinson's disease. Neurochem Int 55 (6):369–375

Vickers A, Goyal N, Harland R, Rees R (**1998**) Do certain countries produce only positive results? A systematic review of controlled trials. Control Clin Trials 19 (2):159–166

Wang LL (**2000**) Acupuncture therapy of Parkinson's disease. Modern Rehabilitation 4 (March):346–348 (in Chinese)

Wang LL, Chong HE, Liu YG, Zhu LL (**2000**) Effects of acupuncture on brain stem evoked potential with Parkinson's disease. J Nanjing Tradit Chin Med Univ 16 (4):229–231 (in Chinese with English abstract)

Wang Q, Chen XH, Tang CZ, Yang JJ, Xian ZH (**2006a**) The Effect of Acupuncture on the Content of Dopamine, HVA and DOPAC in Striate Body of Rats with Parkinson's Disease. Chinese Journal of Integrative Medicine on Cardio/Cerebrovascular Disease 4 (6):92–93 (in Chinese)

Wang S, Cai YY, Shang YJ, Jin-rong L (**2006b**) [Effects of head point-through-point electroacupuncture on SOD and LPO in the patient of Parkinson's disease]. Zhongguo Zhen Jiu 26 (4):240–242 (in Chinese with English abstract)

Wang S, Qi XJ, Han D (**2009**) [Effect of electroacupuncture scalp point-through-point therapy on the expression of tyrosine hydroxylase and dopamine transporter mRNAs in substantia nigra of Parkinson's disease model rats]. Zhongguo Zhen Jiu 29 (5):391–394 (in Chinese with English abstract)

Wang S, Zhou ZK, Hu BC (**2003**) Clinical study on head point-through-point electroacupuncture for treatment of Parkinson's disease. Zhongguo Zhen Jiu 23 (3):129–131 (in Chinese with English abstract)

Wang YC, Cheng YH, Ma J, Gan SY, Wang SJ, Zhou H, Du YJ, Yang M, Shen F (**2010**) [Effects of electroacupuncture on the expression of GDNF and Ret in Parkinson's disease model rats]. Zhongguo Zhen Jiu 30 (9):739–743 (in Chinese with English abstract)

Wen GQ, He XZ, Lu Y, Xia Y (**2010a**) Effect of acupuncture on neurotransmitters/modulators. In: Acupuncture Therapy for Neurological Diseases: A Neurobiological View. Springer-Tsinghua Press, Heidelberg, London, New York, Beijing,, pp 120–142

Wen GQ, Yang YL, Xia Y (**2010b**) Acupuncture-induced activation of endogenous opioid system. In: Acupuncture Therapy for Neurological Diseases: A Neurobiological View. Springer-Tsinghua Press, Heidelberg, London, New York, Beijing,, pp 104–119

White A, Ernst E (**2001**) Adverse events associated with acupuncture reported in 2000. Acupunct Med 19 (2):136–137

White A, Hayhoe S, Hart A, Ernst E (**2001**) Survey of adverse events following acupuncture (SAFA): a prospective study of 32,000 consultations. Acupunct Med 19 (2):84–92

Windels F, Carcenac C, Poupard A, Savasta M (**2005**) Pallidal origin of GABA release within the substantia nigra pars reticulata during high-frequency stimulation of the subthalamic nucleus. J Neurosci 25 (20):5079–5086

Wu Q (**2002**) Refractory senile diseases and the needling method of strengthening the body resnoscope and correspondance of Taiji structure in man and universe. Tianjin Journal of Traditional Chinese Medicine 19 (2):63–66 (in Chinese)

Wu WG, Li CB, Sun P (**1996**) Treating the Parkinson's disease by hitting the acupoints. Journal of Clinical Acupuncture and Moxibustion 12 (4):39–40 (in Chinese)

Xi GF, Cai DH (**1995**) The analysis of the clinical efficacy in the Parkinson's disease cases treated by acupuncture and traditional Chinese medicine. Shanghai Journal of Acupuncture and Moxibustion 14 (6):243 (in Chinese)

Xi GF, Cai DH, Chen GM, Yu AZ (**1996**) The influence of the tremor myoelectricpotential by electrical stimulation at the scalp acupoints in Parkinson's Patients. Shanghai Journal of Acupuncture and Moxibustion 15 (3):5–6 (in Chinese)

Xia Y, Cao XD, Wu GC, Cheng JS (**2010**) Acupuncture Therapy for Neurological Diseases: A Neurobiological View (Research Monograph). Springer-Tsinghua Press, Heidelberg, London, New York, Beijing,

Xu B, Ma C, Chen GZ (**2002**) Clinical Observation on Regulative Action of Acupuncture on Monoamine Neurotransmitters in Cerebrospinal Fluid in the Patient of Parkinson Disease. Zhongguo Zhen Jiu 22 (3):183–185 (in Chinese with English abstract)

Yamashita H, Tsukayama H, Tanno Y, Nishijo K (**1998**) Adverse events related to acupuncture. JAMA 280 (18):1563–1564

Yang L, Du Y, Xiong J, Liu J, Wang Y, Li Y, Li L (**2010**) Acupuncture treatment for Parkinson disease: A systematic review. Chinese Journal of Evidence-Based Medicine 10 (6):711–716 (in Chinese with English abstract)

Yang Y, Xia X, Zhang Y, Wang Q, Li L, Luo G, Xia Y (**2009**) delta-Opioid receptor activation attenuates oxidative injury in the ischemic rat brain. BMC Biol 7:55

Yin H, Tang Y, Chen J, Yu S (**2009**) Increased expression of PKA and CREB1 in substantia nigra of Parkinson's disease mice by electroacupuncture therapy. Chinese Journal of Gerontology 29 (3):263–264 (in Chinese with English abstract)

Yu ZQ (**1994**) Using the scalp plus Siguan acupunture to treat Parkinson's disease. Journal of Zhejiang University of Traditional Chinese Medicine 031 (6):45 (in Chinese)

Yuan H, Liu P, Zheng C, Wang X, Zhang Z, Li C, Jiao S, Bai L, Liang L (**2010**) The possible mechanism of oxidative stress on acupuncture treatment in the rat of Parkinson's disease. Chinese Journal of Health Care and Medcine 12 (1):45–47 (in Chinese with English abstract)

Zhang J, Gibney GT, Zhao P, Xia Y (**2002**) Neuroprotective role of delta-opioid receptors in cortical neurons. Am J Physiol Cell Physiol 282 (6):C1225–1234

Zhang J, Haddad GG, Xia Y (**2000**) delta-, but not mu- and kappa-, opioid receptor activation protects neocortical neurons from glutamate-induced excitotoxic injury. Brain Res 885 (2):143–153

Zhang JF, Sun GS, Zhao GH (**2005**) [Observation on therapeutic effect of herbs-partitioned moxibustion on Parkinson disease of 54 cases]. Zhongguo Zhen Jiu 25 (9):610–612 (in Chinese with English abstract)

Zhang L, Tang Y, Yu S (**2006**) Increased expression of dopamine transporter in Parkinson's Disease mice substantia nigra by electroacupuncture therapy. Journal of Chengdu University of Traditional Chinese Medicine 29 (3):30–33 (in Chinese with English abstract)

Zhang L, Xi GF, He BM (**1997**) Effects of the cerebral antioxidase by scalp-acupuncture and traditional Chinese medicine in Parkinsonian mouse models. Shanghai Journal of Acupuncture and Moxibustion 16 (6):32–33 (in Chinese)

Zhang NZ, Han XH (**1996**) The observation of the clinical efficiency in the 35 cases of Parkinson's disease treated by scalp and body acupuncture. Zhongguo Zhen Jiu 15 (2):5 (in Chinese with English abstract)

Zhao Y, Sun Z, Huang L, Li H (**2010**) Impact of acupuncture and moxibustion on behaviors and apoptosis in rats with Parkinson's disease. World Journal of Integrated Tradititional and Western Medicine 5 (3):209–211 (in Chinese with English abstract)

Zhou F, Huang DK, Xia Y (**2010**) Neuroanatomic basis of acupuncture points. In: Acupuncture Therapy for Neurological Diseases: A Neurobiological View. Springer-Tsinghua Press, Heidelberg, London, New York, Beijing,, pp 32–80

Zhu WX, Xi GF, Ju JH (**1996**) The effect of the cerebral dopamine in the Parkinson's disease mouse models treated by acupuncture combined with traditional Chinese medicines. Acupuncture Research 21 (4):46–48 (in Chinese)

Zhu Z (**2001**) Observation of the rehabilitation efficiency by therapy of acupuncture with Chinese medicine in Parkinson's disease. Modern Rehabilitation 5 (11):125 (in Chinese)

Zipp F, Demisch L, Derouiche A, Fischer PA (**1998**) Glutamine synthetase activity in patients with Parkinson's disease. Acta Neurol Scand 97 (5):300–302

Chapter 8
Can Acupuncture Treat Alzheimer's Disease and Other Neurodegenerative Disorders?

Tetsuya Asakawa and Ying Xia

Abstract The high morbidity, prevalence, and disability rates of neurodegenerative diseases (NDDs), especially Alzheimer's disease (AD)—the most common of all NDDs, with limited effective therapies, make it necessary to pay attention to alternative therapies, such as acupuncture. In fact, in the East Asian countries such as China and Korea, the practice of acupuncture for treating NDDs dates long back and continues to be popular in modern day clinical practice. There has been some progress in clinical and experimental studies on acupuncture therapy for NDDs. Although a number of Chinese studies claim that acupuncture is an effective therapy for AD, the flaws in their experimental design reduce the worthiness of these studies. Mechanistic studies have also generated several hypotheses with regard to acupuncture for AD, e.g., cholinergic hypothesis, amyloid mechanism, tau protein theory, etc. However, rigorous and well-designed studies are required to investigate the underlying mechanism and testify the clinical efficacy. In this chapter, we provide a concise review on the progress of clinical and bench studies in this field and the future prospects of acupuncture in the treatment of AD. Also, we have briefly presented the clinical and experimental studies on acupuncture treatment for other NDDs, including motor neuron disease, multiple system atrophy, hereditary ataxia, and syringomyelia.

T. Asakawa (✉)
Department of Neurosurgery, Hamamatsu University School of Medicine, Handayama, 1-20-1, Higashi-ku, Hamamatsu, Shizuoka 431-3192, Japan
e-mail: asakawat1971@gmail.com

Y. Xia (✉)
Department of Neurosurgery, The University of Texas Medical School at Houston, Houston, TX 77030, USA

The University of Texas Graduate School of Biomedical Sciences at Houston, Houston, TX 77030, USA

Yale University School of Medicine, New Haven, CT 06520, USA
e-mail: Ying.Xia@uth.tmc.edu

Y. Xia et al. (eds.), *Current Research in Acupuncture*,
DOI 10.1007/978-1-4614-3357-6_8,

Keywords Neurodegenerative diseases • Alzheimer's disease • Motor neuron disease • Amyotrophic lateral sclerosis • Multiple system atrophy • Hereditary ataxia • Syringomyelia

8.1 Introduction

Neurodegenerative diseases (NDDs) cover a wide spectrum of disorders with various symptoms and pathological features. The word "degeneration" in these disorders reflects their characteristic presentation of chronic, progressive, irreversible, and systemic neuronal loss in the central nervous system (CNS) including brain and spinal cord. Over time, with progressive neuronal depletion certain dysfunctions and disabilities emerge. Based on their symptomatology, NDDs are generally divided into two groups. One group includes the motor dysfunction diseases such as Parkinson's disease (PD), Huntington's disease (HD), motor neuron diseases (MNDs), etc., and the other, the cognitive impairment group with resultant dementia, such as Alzheimer's disease (AD), Pick's Disease, and Creutzfeldt–Jakob disease (CJD).

The etiology and pathogenesis of the NDDs are complex and still largely unclear. Recently, with the developments in molecular biology, immunopathology, and genetic technology, the links between AD, PD, and the other NDDs have been widely investigated. It has been largely accepted that these diseases do share a common pathophysiologic mechanism which results in a derangement of processing, functioning, and/or folding of specific proteins (Uversky 2009). Genetic and environmental factors like age, toxins, oxidative stress, etc., also possibly contribute to the mechanism of progressive neuronal death and protein aggregation, and thus cause multifarious clinical phenotypic effects.

The high morbidity and disability rates of NDDs have made it a global burden. Reports of WHO predict that NDDs will become the second leading cause of death by the middle of this century due to an increase in the aged population (Menken et al. 2000). However, due to our limited acumen in the pathogenesis of NDDs, we cannot "eliminate" the precise "toxic factors" that lead to the neuronal death. Therefore, the process of cell death and protein aggregation remains irreversible after the onset of NDDs, and we cannot stop the progress of the disease and aggravation of the symptoms. Till now, no effective cure for NDDs has been found. In this regard, besides exploring the mechanisms of pathogenesis, two lines of work have become obviously very important. One is to develop specific and sensitive diagnostic biomarkers for NDDs, since early diagnosis can open a window of opportunity for application of neuroprotective/neurorestorative treatments at an early stage and thereby prevent the progression of the disease due to neuronal loss (Grunblatt 2008), and another is to explore new therapies against NDDs, including some alternative therapies, such as acupuncture.

Acupuncture, a mode of treatment based on the theory of Traditional Chinese Medicine (TCM), has been widely used for NDDs in many East Asian countries including China, Korea, Japan, and Vietnam along with some areas in America and Europe. Analogous to the application of acupuncture in PD that we presented in the previous chapter, in this chapter we will focus on the acupuncture use in AD and other NDDs.

8.2 Acupuncture for Alzheimer's Disease

AD is the most common disorder of all NDDs. However, it lacks specific and promising therapy. Acupuncture has long since been proposed as an important alternative therapy for AD. We therefore performed a systemic review of the progress in acupuncture therapy for AD and have presented our observations in this section.

8.2.1 An Overview of AD

Alzheimer's disease (AD), first reported in 1901, is a disorder characterized by cognitive impairment (especially dementia and/or dysmnesia) and mainly affects the aging population. AD is the most prevalent of all NDDs. More than 5 million patients in the USA are suffering from AD and these numbers are expected to rise up to 16 million by 2050 (Weiner et al. 2010). Globally, it was estimated that 24.3 million people were afflicted with AD in 2005 and were expected to increase to 42.3 million in 2020 and 81.1 million in 2040 (Ferri et al. 2005). AD has become a momentous public health problem that requires development of effective therapies to slow the progression of AD pathology or ameliorate the dementia symptoms.

Although the pathogenesis of AD is still unclear, aging is considered as a main risk factor for AD. AD incidence rises from 2.8 per 1,000 person-years in the people aged 65–69 to a 56.1 per 1,000 person-years in the people older than 90 (Kukull et al. 2002). Some genetic variants are believed to play roles in the mechanisms of AD onset. The early-onset familial forms of AD were reported to have an autosomal dominant inheritance linked to three genes: APP, PSEN1, and PSEN2. APOEε4 is most highly associated with AD cases with a family history of dementia (Bekris et al. 2010). In the newest study by the Alzheimer Disease Genetics Consortium (ADGC), five new genes, namely, MS4A4, MS4A6E, CD2AP, CD33, and EPHA1 have been found to be associated with late-onset AD (Naj et al. 2011). In contrast with PD, except for education level, no other specific environmental risk factor was considered to be consistently associated with AD (Grunblatt 2008). Albeit the roles of these genes in AD pathogenesis have been studied deeply, the specific pathology involved in the progression of AD is still unclear. AD is therefore still regarded as a genetically and environmentally complex disease (Bekris et al. 2010) (Fig. 8.1).

Histologically, the first hallmark of AD is the formation of amyloid plaques, which results from the extracellular amyloid-β (Aβ) depositions. Aβ is cleaved from the large amyloid-precursor protein (APP) by secretase. It produces a 42-amino acid peptide [Aβ (1–42)] along with a 40-residue peptide [Aβ (1–40)] that accumulates in the brain through the amyloidogenic pathways (Humpel 2011). Numerous studies have demonstrated the critical role of Aβ in AD pathogenesis. Aggregated forms of the Aβ peptide amyloid-like structure are neurotoxic to the cortical cells

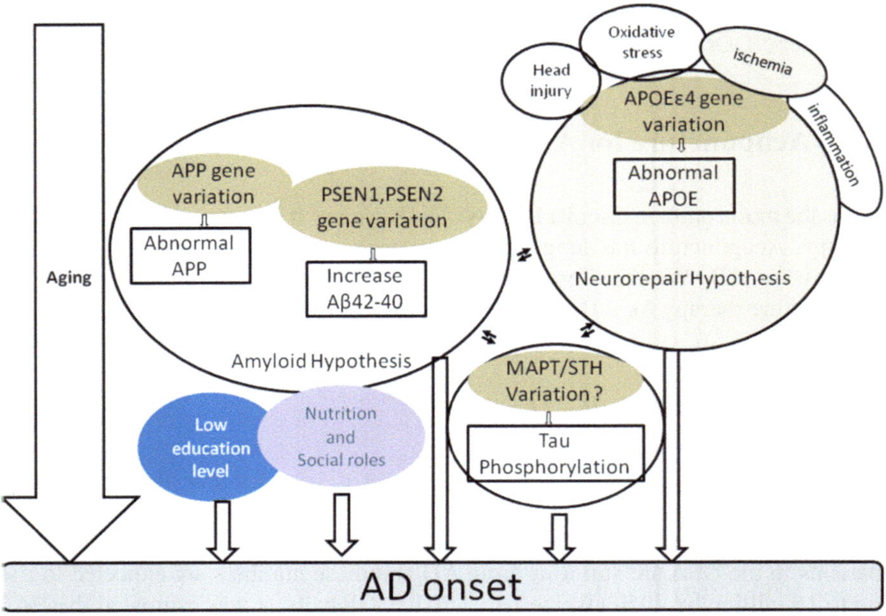

Fig. 8.1 The possible mechanisms of the pathogenesis of AD

(Hartley et al. 1999; Simmons et al. 1994; Uversky 2009; Walsh et al. 2002). Aβ (1–42) is more pathogenic due to its higher concentrations in the amyloid plaques of sporadic AD cases, especially early-onset AD (Gravina et al. 1995; Hardy 1997; Uversky 2009). Some in vitro studies suggested that Aβ (1–42) has a greater tendency of aggregating and precipitating as amyloid (Barrow and Zagorski 1991; Harper and Lansbury 1997). There is strong evidence that shows that Aβ-peptides alone are sufficient enough to induce AD-like phenotypes (Wentzell and Kretzschmar 2010). Crowther et al. (2005) pointed out that learning defects do not necessarily need a visible plaque formation. They observed that before the appearance of large extracellular deposits the cognitive impairments do correlate with the intracellular accumulation of Aβ (1–42). Some recent studies have put forth and provided evidence that intracellular accumulation of Aβ causes synaptic dysfunction, which is pathognomonic of AD (Crowther et al. 2005; Nimmrich and Ebert 2009). All the evidence emphasizes on the toxicity of amyloid. However, the neurotoxic mechanisms of amyloid are complicated and remain poorly understood. One example is that there is a poor correlation between cerebral amyloid level and cognitive impairment. The removal of amyloid cannot stop the decline of the cognitive function (Fletcher et al. 2007). Other possible mechanisms of cognitive impairment caused by the amyloid aggregation are space occupation and inhibition of neuronal outgrowth and post developmental arborization by cell–cell interactions (Leyssen et al. 2005; Li et al. 2004). Amyloidosis is also thought of as trigger for other pathophysiologic processes that finally lead to neuronal loss.

The second hallmark of AD is the formation of neurofibrillary tangles (NFTs) caused by accumulation of hyperphosphorylated tau protein. Tau is a microtubule-associated protein, which is abundant in the axons and promotes assembly and stability of the microtubules (Dupiereux et al. 2009). In AD patients, tau protein is hyperphosphorylated and in turn leads to a disruption of microtubular structure and axonal transport function. Some recent studies reported that hyperphosphorylation of tau has a toxic effect that not only interferes with its interaction with tubulin but also increases its ability to form NFTs, which further results in tau aggregates (Alonso et al. 2008). The mechanism of tau neurotoxicity is quite complex. Several kinases and phosphatases, such as glycogen synthase-3 (GSK-3β), cdk5, protein kinase A (PKA), and microtubule-affinity regulating kinase (MARK) are implicated in the regulation of tau hyperphosphorylative processes (Lee et al. 2001). Some studies, utilizing fly models, suggested the role of gene variation in tau pathology. Nishimura et al. (2004) found a positive correlation between the expression of PAR-1 and the toxic effects of tau phosphorylation. It was reported that tau hyperphosphorylation in flies is mediated, at least in part, by GSK-3β which is encoded by the *shaggy* (sgg) gene in *Drosophila* (Wentzell and Kretzschmar 2010). Pan-neuronal expression of wild-type human tau in the flies induced neuronal degeneration but did not form NFTs (Wittmann et al. 2001). When tau was coexpressed with sgg, phosphorylation of tau increased along with degeneration and NFTs formation in the eyes of the fly (Jackson et al. 2002). Chau et al. (2006) reported coexpression of sgg and cdk-5 enhances tau phosphorylation and its degenerative effects. Compared with amyloid plaques, Tau-induced NFTs seem to correlate more with the progression and severity of AD (Arriagada et al. 1992). Hyperphosphorylated tau protein-induced NFTs are found not only in AD but also in HD, Pick's Diseases, which suggests that taupathies play an important role in the pathophysiology of the NDDs.

Apolipoprotein (apo) E4 is another important causative factor of AD, which plays a significant role in the Aβ peptide and tau phosphorylation (Mahley and Huang 2006; Mahley et al. 2006). The interactions between Aβ peptide and apoE4 can stimulate the deposition of Aβ in the plaques and inhibit the clearance of deposits, which are critical mechanisms underlying the development of AD (Huang et al. 2004). Ye et al. (2005) found that apoE4 increases the Aβ production. Furthermore, Mahley and Huang (2006) proposed that the structure of apoE4 determines the pathological outcome of neurons, including apoE4 potentiation of amyloid beta-induced lysosomal leakage and apoptosis, and the proteolytic cleavage of apoE synthesized by neurons. On the other hand, ApoE4 is believed to be involved in the tau phosphorylation. Increased tau phosphorylation was observed in the transgenic mice expressing human apoE in their neurons along with intracellular NFT-like inclusion formation stimulated by apoE (Tesseur et al. 2000a, b; Harris et al. 2003). In addition to these specific mechanisms involved in AD, other apoE peptides like apoE4 (1–272) and apoE4 (127–272) are also neurotoxic (Mahley et al. 2006).

Besides the three major hypotheses on the pathogenesis of AD, there are other important factors that may be related to AD. The earliest known is the "cholinergic hypothesis" which implies that AD develops as a result of reduced acetylcholine (Ach) synthesis that leads to loss of neurons in the basal ganglia, along with a dramatic decrease in the cerebral cholinergic receptors. However, the poor efficacy of cholinergic therapy in ameliorating the symptoms in AD patients has resulted in a rising disfavor against this hypothesis (Maltsev et al. 2011). Oxidative stress is known to play a vital role in the pathogenesis of NDDs including AD and PD. Oxidative injury increases with age in response to the age-related changes in the generation and/or elimination of reactive oxygen species (ROS). It has been reported that ROS increases with aging in the organs including brain (Driver et al. 2000). In addition to the direct toxic effect to the cell membranes, proteins, and DNA, a secondary damage introduced by errors in protein synthesis or dysregulation of gene expression also plays an important role in the AD pathogenesis (Shi and Gibson 2007; Grunblatt 2008). The gene expression in AD is found to be abnormal in two aspects: (1) Genes that respond to oxidative stress were abnormally induced including antioxidant defense, detoxification, inflammation, etc.; and (2) genes that are associated with brain energy metabolism are mostly down-regulated. Such abnormal gene expression will damage the brain function (Shi and Gibson 2007). Inflammation is also implicated in the pathogenesis of AD. Several inflammatory processes including complement activation, glial cell activation, acute-phase protein synthesis, major histocompatibility antigen complex expression, etc., increase in the AD and are believed to contribute to its pathogenesis (Bonifati and Kishore 2007; Wojtera et al. 2005; Grunblatt 2008).

As to the diagnosis of AD, two things are extremely important. First, to develop more sensitive diagnostic methods to enable detection in the preclinical phase or early stage of AD in order to increase the efficacy and benefits of neuroprotective and neurorestorative therapies; and secondly, to find specific biomarkers to distinguish AD from dementia of other pathophysiological processes, such as vascular dementia (e.g., Binswanger's Syndrome), Pick's disease, Huntington's disease, etc. Thus, the diagnosis of AD should be a systemic and complex task (Fig 8.2). We have listed some of the useful biomarkers in the Table 8.1. Recently, molecular imaging methods have gained importance in making early diagnosis of AD. Positron emission tomography (PET) and functional magnetic resonance imaging (fMRI) are often used in the AD studies. It has been well documented that hypometabolism is the dominant change in the preclinical and early-stage AD, which can be observed with classical ^{18}F-fluorodeoxyglucose (FDG) PET. The cross-sectional and longitudinal studies demonstrate that the reduction in FDG uptake is closely related to cognitive impairment (Alexander et al. 2002; Hampel et al. 2010; Hirono et al. 2004; Mega et al. 2005; Stefanova et al. 2006). On the other hand, the PET studies are always performed in AD patients for investigating the cholinergic state (Namba et al. 2002). The most popular tracers are *N*-^{11}C-methylpiperidine-4-ylpropionate (^{11}C-PMP) and *N*-^{11}C-methylpiperidyl-4-yl acetate (^{11}C-MP4A). It is reported that the acetylcholinesterase (AchE) activity in AD cases is lower than that of normal individuals (Herholz et al. 2007). Recently, with the development of tracers, the PET studies for investigating the amyloidosis have become more and more popular. 2-(1-{6-[(2-^{18}F-fluoroethyl)(methyl)amino]-2-naphthyl}ethylidene) malononitrile (^{18}F-FDDNP) was the first tracer used for cerebral amyloid plaque, and its binding was noted to be increased in the temporal, parietal, and frontal regions of the AD brain (Shoghi-Jadid et al. 2002). Pittsburgh compound B (^{11}C-PIB), a derivative of

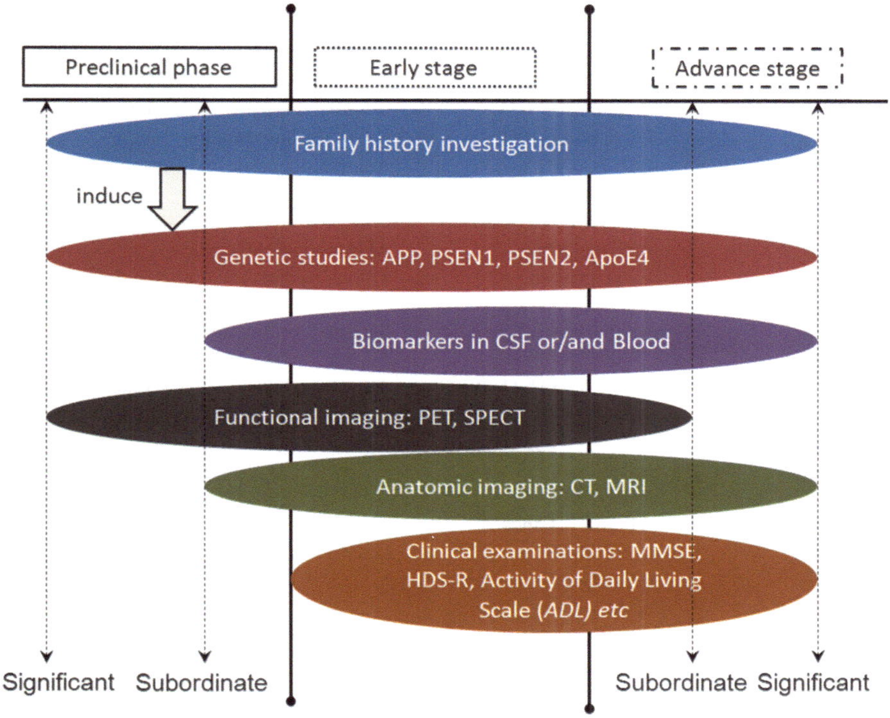

Fig. 8.2 The diagnosis of AD at different stages

thioflavin T amyloid dye, binds with high affinity and specificity to neuritic Aβ plaques (Mathis et al. 2002). New tracers such as ^{18}F-BAY-94-9172, ^{11}C-SV13, ^{1}C-2-[2-(2-dimethylaminothiazol-5-yl)ethenyl]-6-[2-(fluoro)ethoxy]benzoxazole (^{11}C-BF-227), (^{11}C-AZD-2184), ^{18}F-flutemetamol, and (*E*)-4-(2-(6-(2-(2-(2-^{18}F-fluoroethoxy)ethoxy)ethoxy)pyridin-3-yl)vinyl)-*N*-methylbenzeneamine, which are amyloid PET ligands, are being tested in AD patients (Kadir and Nordberg 2010). Besides PET, fMRI is also used to determine the distribution and content of amyloid in the brain with AD (Amatsubo et al. 2010). Such systemic methods including genetic studies, CSF/blood assay and imaging studies should be employed not only for the diagnosis/differential diagnosis of AD but also for evaluation of the efficacy of AD treatment.

There is no ideal therapy for AD. Until now, none of the treatments have been able to stop the progression of neuronal loss and amyloidosis. Though some treatments are likely to improve the cognitive function temporarily, the dementia gradually becomes severe and will eventually require assisted living arrangements. The classical therapy nowadays includes an AchE inhibitor and a *N*-Methyl-D-aspartate (NMDA) receptor antagonist that target the neurotransmitters. As we described above, "cholinergic hypothesis" is the oldest hypothesis for AD. The drugs such as

Table 8.1 Useful biomarkers for AD in different spheres

Spheres	Level	Biomarkers	Introduction	Variations in AD
CSF	Specific (core)	Aβ(1-42)	Amyloid plaque	Marked reduction
		Total tau	Neuronal injury	Marked increase
		Phospho-tau-181	Neurofibrillary tangles	
	Nonspecific	CSF BACE1	Amyloidogenesis	Increase
		Truncated amyloid-β isoforms		
		APP isoforms	APP products	
		F2-isoprostane	Mitochondrial dysfunction	
		Biomarkers of synaptic degeneration	Synaptic dysfunction	
Blood	Potential	Aβ (1-42), Aβ (1-40) or ratio	Amyloid plaque	Controversial
Genetic	Autosomal dominant	APP	Early onset of AD	
		PSEN1		
		PSEN2		
	Risk factor	APOE4 (ε4 andε2)		
		MAPT/STH	Later onset of AD	
		CYP2D6		
		A2M		
		NOS1		
		BDNF rs6265		
		GSTM1		
		TF rs1049296		
		Others (GAB2, CHRNB2, CH25H, and CALHM1)		

Cited/modified from works of Grunblatt (2008) and Wattamwar and Mathuranath (2010)

tacrine, donepezil (Aricept®), rivastigmine (Exelon®), and galantamine are employed in the AD treatment and work by improving the cholinergic activity through AchE inhibition. Such AchE inhibitors bring about a modest, however efficient, amelioration of behavioral or cognitive symptoms in a minority of patients (Bird 1993). Memantine, a NMDA receptor antagonist, has shown some effectiveness in the treatment of moderate to severe AD (Reisberg et al. 2003; Tariot et al. 2004). Lopez et al. (2009) reported that cholinesterase inhibitor therapies caused a delay in placement of patients in nursing homes, and this effect was amplified by concurrent use of cholinesterase inhibitors and memantine. Such therapies show limited efficacy in altering the progression of the disease. The next generation of medications that are underway are expected to slow down the disease course and/or improve clinical efficacy (Grill and Cummings 2010). With the postulation of the amyloid, neurorepair, and tau hypotheses, new therapeutic targets are coming into picture (Table 8.2). Some of them, like antidepressant medication anti-inflammatory agents (NSAIDs), estrogens, nerve growth factors, ginkgo biloba, statins, BACE inhibitors, and antioxidants, are still under investigation (Bird 1993; Overshott and Burns 2005; Klafki et al. 2006; Masters and Beyreuther 2006). Recently, some surgical therapies have

Table 8.2 The potential therapeutic targets for AD

Amyloid hypothesis	Reduce the Aβ production	α-secretase enhancement
		β-secretase inhibition
		ϒ-secretase inhibition
		ϒ-secretase inhibition
	Inhibit the Aβ degradation	Neprilysin activation
		Insulin-degrading enzyme activation
	Promote the Aβ removal	Vaccination
		Passive immunization
		Receptor-mediated removal from CNS
		Prevent entry from periphery
		Prevent aggregation by binding Aβ
	Preventing Aβ toxicity	Prevent oligomerization through metal protein attenuation
Tau hypothesis		Prevent tau aggregation
		Prevent tau hyperphosphorylation
		Facilitate tau phosphatase
		Microtubule stabilization
Neurorepair hypothesis		Inhibit the preferential binding of apoE4
		Convert apoE4 to apoE3-like molecule
		Small molecules to stabilize transthyretin
		Homotetramers and reduce their susceptibility
		Prevent the apoE4 potentiation of Aβ-induced
		Lysosomal leakage and apoptosis
Other (neuroprotection)		Growth factor administration and/or activation
		Anti-apoptotic agents
		Block inflammatory disease processes
		Antioxidative stress

Cited/modified from works of Grill and Cummings (2010) and Mahley et al. (2006)

also been developed. Hamani et al. (2008) performed bilateral hypothalamic deep brain stimulation and observed a surprising improvement in the autobiographical memories in AD patients. Electroencephalographic source localization indicated that hypothalamic deep brain stimulation activates the mesial temporal lobe structures. They suggested that in these patients the hypothalamic stimulation modulates the limbic activity and improves certain memory functions. More recently, this group performed a phase I clinical trial of deep brain stimulation of memory circuits in Alzheimer's disease and found that DBS activated the memory circuit, including the entorhinal and hippocampal areas and activated the brain's default mode network. PET scans showed an early and striking reversal of the impaired glucose utilization in the temporal and parietal lobes that was maintained after 12 months of continuous stimulation. Evaluation on the Alzheimer's Disease Assessment Scale-Cognitive Subscale and the Mini Mental State Examination suggested that some patients treated with 6–12 month DBS therapy showed an improvement and/or a protracted rate of decline in cognitive faculties (Laxton et al. 2010). Some "new generation" therapies such as stem cells or induced pluripotent stem cells (iPS cells) transplantation for treating AD are under development and could potentially change the face of future clinical AD therapy. However, without the in-depth understanding of the

Different Levels		Abnormal state in AD	Research Methods
Gene		Genetic Variation	Biochemical technology
Protein	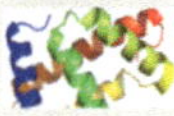	Abnormal Expression	Biochemical technology Molecular imaging Biomakers in CSF or blood
Cell		Cell Death Neuron Loss (necrosis, apoptosis and amyloidosis) Synapse Loss	Classical pathological stain Immunohistochemistry
Brain	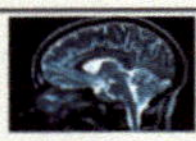	Atrophy Amyloid Plaques Tau Tangles	Anatomic imaging (CT, MRI, etc) Functional imaging (PET)
Clinical		Dementia and/or Dysmnesia	Cognitive evaluating Scale (MMSE, *Hasegawa* Dementia Scale (HDS-R))
Epidemiology		The Commonest NDDs, Great Public Health Problem.	Retrospective *research* Prospective research Clinical trial

Fig. 8.3 The comprehensive understanding of AD at different levels

pathogenesis of AD, it is hard to comprehend the biological behavior of such induced cells after in vivo transplantation. Moreover, alternative therapies are the best available options for refractory AD and include TCM and acupuncture. We have introduced the acupuncture for AD in the following sections.

As discussed above, AD mechanisms are complex processes (Fig. 8.1). We have only limited understanding of the interactions between the risk factors and the pathophysiologic sequence of the changes in AD. To uncover the mysteries of AD, it is vital to systemically and comprehensively improve our understanding of it from microcosm (molecular level) to macrocosm (epidemiological level), i.e., the sequence of pathophysiologic variations that lead to AD and dementia (Fig. 8.3).

8.2.2 *TCM View of AD Pathogenesis*

We will briefly review the concept of AD in TCM theory in this section. It is needed to note that the organs and tissues mentioned in this section, including marrow, blood, phlegm, brain, heart, lungs, liver, spleen, and kidneys, are concepts and terms used in TCM theory and their interpretation is usually different from modern physiology and medicine.

In the ancient TCM times, there was no separate concept of "AD," "vascular dementia," or neurodegeneration. Diseases with the characters of "progressive cognitive impairments with aging" were considered as one disease. Nearly

2,000 years ago, in two classical TCM books—*The Yellow Emperor's Inner Classic* (Huangdi Neijing) and *The Classic of Miraculous Pivot* (Ling Jiu), the concept of humans being governed by "head, brain, and sea of marrow (Suihai)" was introduced. It was believed that an insufficiency and/or weakness of the sea of marrow could cause various symptoms that eventually lead to dementia. Another great TCM classic—*The Classic of Plain Questions* (Su Wen), reported that the mental and cognitive functions are affected by the functions of other viscera (Theory of Zang and Fu). It was believed that "Heart controls the spirit; lungs store the corporal soul; liver stores ethereal soul; spleen controls the thinking; and the kidneys store the mind" and the weakness in the visceral function at an older age is responsible for the development of dementia. The Tian Nian chapter in The Classic of Miraculous Pivot describes, "The Liver function (liver-qi) begins to enervate in the age of 50s in humans; then heart function (heart-qi) begins to weaken in 60s; spleen function (spleen-qi) weakens in the 70s; and lung function (lung-qi) weakens in the 80s, which causes dementia and the dysfunction of verbal function." Later during the Qing dynasty (1636–1911), Dr. Shiduo Chen (1687) contemplated that human intelligence was a result of an interaction between the heart and the kidneys, instead of the organs themselves. The dysfunction of this interaction will result in a loss of intelligence. Dr Qingren Wang (1830), in an attempt to correct the errors in medical practice, pointed out in his book that cognition is the function of brain and refuted the prevalent belief of it being a function of the heart (Yi Lin Gai Cuo). He believed that the memory loss was due to the gradual exhaustion of the brain marrow with aging. Dr. Xichun Zhang (1909) believed that the weakness of "spirit and qi" together with the insufficiency of "blood and qi" in older individuals causes the "yang producing wind" then "qi and blood retrogradation upward" which leads to the "Phlegm and Blood Stagnation" rush to the brain and cause injury, thereby resulting in cognitive dysfunction.

The modern TCM theories propose that dementia is essentially a dysfunction of the brain along with other organs including heart, liver, spleen, and kidneys. The factors involved in its pathogenesis are wind, fire, phlegm, stagnation, and weakness. Weakness of the brain and the five main organs (Wu Zang) namely, the heart, liver, spleen, lungs, and kidneys, are the primary pathogenesis factors while the phlegm, stagnation, and stagnant fire are the secondary factors (Dong 1999; Zhang 2009). (1) *Deficiency of heart and kidneys* causes the phlegm stagnation and stagnant fire. The weakness of heart and kidneys leads to a poor nutritional supply to the brain and results in dementia. This theory is supported by Xia and Deng (Xia 1991; Deng and Yuan 1991). They believed that dementia is caused by three factors, i.e., (a) malnutrition of the brain from a weakened marrow caused by deficient kidney in older patients; (b) dysfunction of the organs leads to insufficient "qi and blood" that exhausts the nutrition to the brain and marrow; and (c) phlegm stagnation and stasis cause the dysfunction of a clear mind. (2) *Weakness of liver, spleen, and kidneys* leads to phlegm stagnation and then causes cognitive impairment. (3) *Weakness of qi and insufficiency of spirit* caused the wind, fire, phlegm, and blood stasis. As these pathogenic factors accumulate, they block the brain collaterals, which results in the loss of qi and blood supply to the brain thereby ending in development of dementia (Deng and Yuan 1991). (4) *Stagnation and/or weakness of qi* leads to blood stasis and could contribute to AD evolution (Zhang 2009; Yan and Lv 1991; Yan et al. 2002). Zhu and Hu (2007) summarized that dementia could present in both deficiency and excess patterns. The deficiency type is characterized mainly by impaired liver-yin or kidney-yin, deficiency of the heart and spleen-qi, and yang deficiency. The excess type is characterized mainly by the obstruction of the intelligence opening with phlegm and blood stasis, combined with the pathogenic fire and depression.

In accordance with the pathogenetic mechanisms described above and based on the theory of differential symptoms of TCM (Bian Zheng Shi Zhi), the treatment of AD is based upon two principles: dredging and tonifying (Zhang 2009). The principle of dredging involves removal of obstruction in the collaterals and washing up the phlegm; clearing the blood stasis; eliminating fire and reducing the grime; and adjusting qi and eliminating phlegm. The principle of tonifying involves tonification of the kidney and heart; complementing the spirit, blood and qi; and besteading the marrow (Zhang 2009).

8.2.3 *Clinical Application of Acupuncture in AD*

According to the principles of AD treatment we described in the Sect. 8.2.2, acupuncture has an important therapeutic role in TCM and is considered as an important alternative therapy for neurological diseases (Xia et al. 2010b), including AD. The usually selected acupoints are Shenmen (HT-7), Baihui (DU-20), Xinshu (BL-15), Jianshi (PC-5), Shangxing (DU-23), etc. The channels and collaterals that are usually involved are the Du Channel (DU) and Bladder Meridian of Foot Taiyang (BL) (Dong 1999; Lu 2006). Most of these reports show application of body acupuncture or electroacupuncture (EA) (Zhou and Hang 2006).

8.2.3.1 Manual Acupuncture

Manual acupuncture has been the most popular form of acupuncture for AD before EA came into light. Most of manual acupunctures were applied to the acupoints through DU Meridian, Heart Meridian of Hand Shaoyin (HT), and Kidney Meridian of Foot Shaoyin (KI). But some of them were applied to the acupoints involved several collaterals. The acupoints are different for the deficiency and excess types. However, the treatment principle for both is essentially the same. "*Restoring the consciousness and opening the intelligence*" *are the most important principles followed in treating AD.* Based on this principle, Wang (1997) performed acupuncture in 31 cases of senile dementia patients and found that 12 cases were "cured," 9 cases showed profound improvement, 8 cases were improved, and 2 cases remained unaffected. Jiang et al. (2004) also selected the acupoints resting on this principle. They selected Baihui (DU-20) or Sishenchon (EX-HN1), Fengchi (GB-20), Dazhui (GV-14), and Shenshu (BL-23) as the main acupoints, which were combined with the

Table 8.3 Commonly used EA parameters for AD in Chinese reports

Parameters	Data
Depth from the skin	5–15 mm
Frequency	1–200 HZ
Current intensity	Top strength can be endured by patients (1–40 mA)
Wave pattern	Condensation wave or disperse-dense wave or continuous wave, or alternation
Time of one therapy	Stimulation 1–40 min, reservation the needle 25–40 min, one therapy per day
Time of the whole course	25–28 days (consecutive or consecutive 5 days with 2 days intervals in weekend)
Diameter and material of the needle	NA

other acupoints according to the differential symptoms (Bian Zheng) of patients. They performed acupuncture on 24 patients for 4 weeks, 5 times/week q.d. and found that the MMSE, CDR, and ADL scores significantly improved in the patients who underwent acupuncture. Since weakness in a certain organ is a contributing factor of the pathogenesis of AD. *Tonifying and complementing is another important principle.* Ouyang et al. (1999) selected the acupoints based on the principle of tonifying kidney and complementing marrow. They used Baihui (DU-20) or Sishencong (EX-HN1) and Shenshu (BL-23) as the main acupoints, while Taichong (LR-3), Guanyuan (RN-4), Sanyinjiao (SP-6), and Zusanli (ST-36) as the secondary acupoints. They performed acupuncture for 8 weeks in 30 AD patients and found that the improvements in ADL, HDS, and Wechsler Memory Scale (WMS) scores were not different when compared to the control group that composed of patients who underwent Nimodipine administration. They concluded that acupuncture has the same efficacy as Nimodipine administration. Methods of needle insertion in the body acupuncture are very complex and require expertise (Sect. 8.2.3.6). Therefore, EA has nearly overtaken classical body acupuncture and is being widely practiced by most therapeutists lately.

8.2.3.2 Electroacupuncture

As discussed in Sect. 8.2.3.1, the difficult comprehension of methods of needle insertion acts as the major limitation of classical acupuncture. EA, on the other hand, has more objective parameters and relies less on the subjective experience, and is thus being widely preferred due to better operability. Since EA has the same treatment principle as the classical body acupuncture, we consider it as a special type of body acupuncture. Just as in PD (see Sect. 7.2.3), here we introduce the commonly used acupuncture parameters for AD based on some Chinese reports (Table 8.3).

Dong et al. (1997) followed the principles of "restoring the consciousness and opening the intelligence" together with tonifying and complementing, and selected Baihui (DU-20) or Sishencong (EX-HN1), bilateral Fengchi (GB-20) Dazhui (GV-14), and Shenshu (BL-23) as the main acupoints, and the secondary acupoints according to the differential symptoms in each case. They performed EA with condensation wave or disperse-dense wave for 15 min, and then changed to continuous wave for 40 min (180 Hz, maximum current intensity can be endured) in 32 senile dementia patients. After 3 courses of treatment (1 course = q.d. × 5 days/week × 4 weeks), the scores on MMSE significantly ameliorated (Dong and Bai 2002). Fu et al. (2005b), Luo et al. (2006), and Zhou and Jin (2008) used an analogous EA protocol and found an improvement in the fMRI, MRI, and blood concentration of Aβ.

8.2.3.3 Scalp Acupuncture

Compared with PD, scalp acupuncture is seldom adopted for treatment of AD. Some researchers believe that scalp acupuncture could be beneficial in ameliorating the cerebral blood circulation and metabolism (Zhou and Hang 2006). Lv and Xu (1997) performed scalp acupuncture in 60 cases of senile dementia. They selected anterior oblique line of vertex-temporal (MS-6), posterior oblique line of vertex-temporal (MS-7), lateral line 1 of vertex (MS-8) and lateral line 2 of vertex (MS-9) as the main areas to receive acupuncture, and the speech, vertigo, and auditory areas along with Baihui (DU-20), Fengchi (GB-20), and Sishencong (EX-HN1) as the secondary acupoints. After acupuncture for 10 days (1 time/day), they reported that 81.7 % patients improved. Xiao (2004) performed scalp acupuncture at the vasomotor and the foot motor sensory areas with the same secondary acupoints. The efficacy of scalp acupuncture was observed to be better in the acupuncture group when compared to the control group (effective rate 78.8 % vs. 45.0 %, $p<0.01$).

8.2.3.4 Acupuncture Combined with Other Therapeutics

For a better efficacy, acupuncture is often performed hand in hand with other therapeutic modalities, such as normal medicine, TCM, and other alternative therapies.

Acupuncture Along with the Medications

Ouyang et al. (2001) performed EA in combination with the administration of perphenazine in 16 AD cases with cognitive impairment. They followed the principle of "restoring the consciousness and opening the intelligence" and selected Baihui (DU-20), Yingtang (EX-HN3), and Dazhui (GV-14) as the main acupoints. They reported that 81.25 % patients improved with EA, and a significantly reduced dose of perphenazine.

Acupuncture is based on the theory of TCM, and selects the appropriate acupoints following the principles of differential symptoms (Bian Zheng Shi Zhi). It is not out of ordinary to perform acupuncture together with Chinese medicines, which are also selected and administrated based on the principles of differential symptoms. Xu and Xu (1995) performed acupuncture at Shisangui points that are a combination of Shuigou (DU-26), Shaoshang (LU-11), Yinbai (SP-1), Daling (PC-7), Shenmai (BL-62), Fengfu (DU-16), Xiache (ST-6), Chengjiang (RN-24), Laogong (PC-8), Shangxing (DU-23), Huiyin (RN-1), Quchi (LI-11), Haiquan (EX-HN11), Jianshi (PC-5), and Houxi (SI3), employed in the treatment of mental symptoms in one senile dementia case. They performed two courses of acupuncture (15 times. q.d. × 15 days) together with Chinese medicines selected by following the principles of opening the intelligence and tonifying the marrow. They achieved a good efficacy at the end. Mu and Li (2001) and Li et al. (2002) performed acupuncture on 24 AD cases. They compared the efficacy in the group that underwent EA plus Chinese medicine *Danggui Shaoyao San* with the group that underwent EA only. They found that the improvement of activities of daily living (ADL) in the combination group was 61.9 %, while that of EA only group was 52.4 %. It was thus concluded that for mild and medium cases of AD the combined treatment was better than EA alone. Ma (2003) selected acupoints and medicines on the basis of principles of "restoring the consciousness and opening the intelligence." He performed acupuncture at Shenting (DU-24), Baihui (DU-20), Fengchi (GB-20), Shenmen (HT-7), Dazhui (GV-14), Dazhong (KI-4), Taixi (KI-3), and Xuanzhong (GB-39) together with *Anshenxingnao soup* in 56 senile dementia patients. The control group was treated with the administration of *Anshenxingnao soup* only. He observed that 29 cases achieved perfect efficacy, 19 cases reached some efficacy, and 8 cases failed. It was observed that the group that underwent combined treatments attained better efficacy than the control group. Recently, Peng and Dong (2009) performed acupuncture at Baihui (DU-20), Sishencong (EX-HN 1), Dazhui (GV-14), and Guanyuan (CV-4) in addition to the administration of *Yizhi Jiannao Granules* in 28 AD cases. They found that the MMSE and ADL scores improved with the combined treatment. In the combined treatment group, 85.7 % patients improved, while only 71.4 % improved in the single medication group. The combination treatment reported a significant enhancement in efficacy ($p<0.05$).

Acupuncture Along with Moxibustion

Another popular combination is acupuncture plus moxibustion. Yang (1996a) selected the acupoints based on the principle of "restoring the consciousness and opening the intelligence." He selected Shuigou (GV-26), Baihui (DU-20), Dazhui (GV-14) Fengchi (GB-20), and Neiguan (PC-6) as the main acupoints and Waiguan (SJ-5), Taixi (KI-3), and Xuanzhong (GB-39) as the secondary acupoints. The needles were kept for 20 min; then moxibustion was given at Baihui (GV-20) and Dazhui (GV-14) q.d. × 10 days. He got a 92.3 % total effective rate in 26 senile dementia patients. Shen (1996) selected Shenting (DU-24), Hegu (LI-4), Jianshi (PC-5), Shenmen (HT-7), Zusanli (ST-36), Taichong (LR-3), and Sanyinjiao (SP-6) as acupoints, combined with moxibustion at Baihui (DU-20), and attained an 85 % efficacy rate. Mei et al. (2003) performed acupuncture in 76 senile dementia cases at Baihui (Du-20), Sishencong (EX-HN1), Dazhui (GV-14), and Guanyuan (RN-4), then reserved the needle at Baihui (DU-20) and Sishencong (EX-HN1) for 5 h/day and 6 days/course. The results showed that 7 cases were cured, 33 cases showed a significant improvement, 26 cases showed some efficacy, and 10 cases were unaffected. Ji et al. (2007) conducted a clinical study on 91 senile dementia patients. They performed acupuncture together with moxibustion at Baihui (DU-20) and bilateral Neiguan (PC-6) and found that acupuncture significantly improved the scores of HDS-R with 89.3 % of the patients being improved as compared to 63.6 % in the control group.

There are other combinations such as acupuncture plus injection medicine in acupoints (Dong et al. 1997; Yang et al. 2008), acupuncture with music therapy, etc., that are available; but such therapies are not yet popular in treating AD (Liu and Yuan 2005).

8.2.3.5 Other Acupuncture-Related Alternative Therapies

Besides acupuncture, there are other alternative therapies that are related to acupuncture and acupoints. Yang (1995) injected genseng and Chinese angelica solution into the acupoints of bilateral Shenshu (BL-23), Zusanli (ST-36), Meichong (BL-13), and Sanyinjiao (SP-6) in 26 senile dementia patients and observed a good efficacy. Yang (1996b) used so-called moving cupping (a kind of Chinese medicine cupping jar) to stimulate the areas of DU from the seventh cervical vertebra (C7) to sacroiliac, and then used so-called bodybuilding cupping (another cupping jar) to simulate Dazhui (GV-14) and bilateral Shenshu (BL-23) in 18 senile dementia patients. He found that 6 cases showed success, 10 cases had some efficacy, and 2 cases failed to improve. Yu (1996) used naprapathy and Dianxue, a special method to hit acupoints, on one senile dementia patient. A good efficacy was observed after 4 courses of treatment. Because these treatments are unconventional therapies, they are not widely adopted even in China.

8.2.3.6 Factors that Influence the Acupuncture Efficacy

Several factors affect the acupuncture efficacy in AD treatment. These factors should be taken into account during the clinical practice. In this section, we have compiled together some of these factors that are reported to be important for achieving high efficacy and reducing the side effects.

Stage of the AD

Several studies have reported that the efficacy of acupuncture varies with different stages of AD: greatest with in the milder cases and the least in advanced cases. Those who have shorter duration of onset get better results than those with longer intervals since onset (Mei et al. 2003; Matsukawa et al. 2005; Guo et al. 2002). This is comparable to the observations in PD (see the Sect. 7.2.6.1).

Selection of the Acupoints

The selection of the acupoints is crucial to the success of acupuncture. The selection criteria vary among the therapeutists in China and in the other countries. Acupoint selection in Japan has always been based on some objective indices, such as the acupoints achieving improvement in the cerebral Ach (Sawada et al. 2001), or causing activation at certain brain areas (Guo et al. 2000). Thus, only a limited number of acupoints are frequently used in Japan. In China, the selection of the acupoints relies more on the therapeutists' experience and understanding of the principles of differential symptoms. The lack of an objective standard makes the acupoint selection so complex and varied in AD treatment.

Methods of Needle Insertion in Acupuncture

The method of needle insertion significantly affects the efficacy of acupuncture (Zhou and Hang 2006). This requires a high level of expertise and cooperation between the therapist and patient, although many a times it is difficult to achieve. Different acupoints need different inserting methods. For example, strong stimulation is suggested for Shuigou (DU-26) and Neiguan (PC-6); an action involving powerful lifting and thrusting is used for Waiguan (TE-5); a complex action of twirling and rotating is used for Xuanzhong (GB-39) and Dazhui (GV-14); alternate reinforcing and weakening actions are applied at Fengchi (GB-20) and Baihui (DU-20) (Yang 1996a; Zhu and Hu 2007). The complexity of insertion methods has posed as a hurdle to the widespread application of acupuncture. Lately, EA application has emerged and expanded in countries practicing it with relatively objective parameters.

8.2.3.7 Comments

In this section, we reviewed the clinical applications of acupuncture AD treatment. We abandoned the studies using terms like "Vascular Dementia (VD)," although there is no documented difference between AD and VD as far as TCM is concerned, as it uses a more vague term "senile dementia," which technically covers both. Thus, all the studies involving "senile dementia" are likely to include the non-AD dementia, which is an important limitation of this section.

It is well known that AD is a refractory NDD. None of the reports in the mainstream medicine claim "to have cured AD," nevertheless, some of the Chinese reports we have previously cited claim to have "cured" or "significantly improved" AD, which is believed to be an exaggeration of efficacy or a publication bias. We have discussed these problems of the experimental design in a greater detail in the previous chapter on PD (see Sect. 7.2.6.2). Unfortunately, acupuncture studies on AD also fall as victim to the same problem which was found to be the case in other neurological diseases (Xia et al. 2010b). Liu et al. (2009a) searched the China

Biology Medicine (CBM) disc with the keywords "acupuncture" and "senile dementia" and found 20 papers. Only four of them introduced randomization of which only one used the double-blind design and reported the withdrawn cases. All of them were low-quality studies having Jadad scores of only 1–2 (5 being the best). The flaws in the experimental design reduce the worthiness of these studies. Lee et al. (2009) searched 17 databases with the terms of "acupuncture" and "Alzheimer" and found 40 papers. Thereinto 24 out of these 40 were excluded after screening the abstract; 13 of the remaining 16 were abandoned due to various methodological flaws, leaving only 3 papers that were randomized clinical trials (RCTs). Even these three RCTs had poor methodological quality. None of them applied randomization, report the withdrawn cases or blinding methods. The limited number of RCTs and too small sample size make it impossible to draw firm conclusions. These flaws in the experimental design result in higher susceptibility to bias and provides little useful information to evaluate the efficacy of acupuncture in AD. These flaws in the experimental design are likely to be the problems common to most of the studies involved in the alternative therapies that end up exaggerating the efficacy and dodge the potential adverse events. It reduces the worthiness of these studies and questions the credibility of alternative therapies. Again, as we pointed in the PD chapter, long-term, multicentric, multinational, large, and well-designed clinical trials are required to provide more formidable evidence to evaluate potential of acupuncture in AD treatment in the future.

8.2.4 *Potential Mechanisms of Acupuncture Therapy for AD*

With the progress in the knowledge of the mechanisms involved in AD, the potential mechanisms of acupuncture therapy for AD developed as well; however, it is still a long way to go to fully uncover the intricacies of acupuncture use in AD. In this section, we will try to discuss the potential mechanisms based on the limited bench studies that we have at hand. Most of the available studies are based on the various "hypotheses" on the pathogenesis of AD, which we discussed in the Sect. 8.2.1. We will discuss the probable mechanisms of action of acupuncture on the basis of these hypotheses, in this section.

8.2.4.1 Amelioration of Impaired Cholinergic Function by Acupuncture

As we discussed earlier, the classical "cholinergic hypothesis" has gradually faded out of sight, but it is the one that first motivated clinical application, thus most of the currently available bench studies were carried out leaning on this hypothesis. In some countries, the cholinergic indices are employed as an objective standard to select acupoints (Matsukawa 2005). It is known that acupuncture influences the cholinergic activity by complex mechanisms. The activities of AchE, Ach, and cholinergic receptors varied with acupuncture proceeding (Wen et al. 2010a).

The detailed mechanisms in AD are still poorly understood. One possibility is that acupuncture affects cholinergic function through endogenous opioid system. It is widely accepted that opioids have a direct impact on the release of Ach, substance P, and a number of other neurotransmitters that have been implicated in the pathogenesis of AD (Barg et al. 1993). Albeit heretofore no study focused on the direct correlation between cholinergic function, endogenous opioid system, and acupuncture in AD objects; it has been known that acupuncture activates the endogenous opioid system (Wen et al. 2010b), which was verified in the ischemic brain (Tian et al. 2008). We still do not know if it is true that acupuncture activates the endogenous opioid system such as delta opioid receptor (DOR) primarily in AD, and then affects the cholinergic function secondarily. More studies should be employed to investigate the roles of the endogenous opioid system that play after acupuncture in AD objects. A number of Chinese studies reported improvements in the cholinergic indices with acupuncture in animal studies. Yang et al. (1999) found an elevated activity of AchE in the hippocampus of a senile dementia mouse model. After acupuncture at Renzhong (DU-26) and Neiguan (PC-6), the cognitive functions improved, the striatal and hippocampal Ach increased, and normalized the abnormally activated AchE. Dong and Bai (2002) performed EA (1 Hz, continuous wave, 30 min/day × 30 days, current intensity at the lowest current to cause limb dithering) at Dazhui (GV-14) and Baihui (DU-20) in AD rat models. They found that the hippocampal AchE activity improved with EA. Zhang et al. (2003) performed EA (3 Hz, continuous wave, 10 min/day × 15 days, current intensity at the lowest current to cause needle dithering) at Baihui (DU-20) and Shenshu (BL-23) in AD rat models. They found that EA improved the cognitive performance, enhanced the content of cerebral Ach, and decreased the ratio of cerebral AchE/Ach in rat AD models. Liu et al. (2004) made AD rat models by transecting the fimbria/fornix that damaged the cholinergic system and caused a dysfunction of memory and learning. They found that EA at Baihui (DU-20) and Yongquan (KL-1) improved the cognitive impairments caused by the cholinergic damage, which provided indirect evidence that EA could potentially ameliorate the dysfunction of cholinergic system in AD. Zhao et al. (2008) performed acupuncture at Baihui (DU-20) and Dazhui (GV-14) in AD rat models. They found significantly decreased cerebral AchE level after EA ($p < 0.01$). Besides the studies regarding Ach and AchE, Tang et al. (2006) reported that acupuncture has a neuroprotective effect on cholinergic neurons. They performed EA (20 Hz, continuous wave, 30 min/day × 20 days, current intensity 2 mA) at Baihui (DU-20), Taixi (KI-3), Yongquan (KI-1), and Xuehai (SP-10) in aged AD rats. They also found that the acetyltransferase activity, nerve growth factor expression, and c-fos proteins in the EA group were significantly higher than those in the control group. Thus they concluded that EA exerts neuroprotective effects on hippocampal cholinergic neurons in aged AD rats. Promoting c-fos protein synthesis and increasing the nerve growth factor expression could also contribute to this effect. More recently, Liu et al. (2009b) performed acupuncture at "Xiusanzhen" including bilateral Yingxiang (LI-20) and Yintang (EX-HN 3) in AD rat models. They noticed that this acupuncture improved the cognitive function in addition to increasing the hippocampal ChAT and AchE activity.

There are a limited number of studies conducted on human patients. Tang et al. (2001) treated eight senile dementia patients with acupuncture at Sishencong (EX-HN1), Shenmen (HT-7) and Taixi (KI-3) for 3 courses (1 course=30 min/day×7 days). The efficacy of acupuncture was verified by MMSE. They investigated the plasma activity of AchE. Although they reported that AchE activity increased, however, there was no significant difference. But previous studies from Japan reported different results. Inaniwa (1993) used low-power laser (LPL) to stimulate forehead in the midline, a Jiaosun (SJ-20) and Yamen (DU-15), and found that the hippocampal Ach increased in response along with increased regional cerebral blood flow. Sawada et al. (2001) reported that the simulation at Quchi (LI-11), Shousanli (LI-10), and Hegu (LI-4) contributes to the increase in the cerebral Ach and cortical blood flow (Sawada et al. 2001; Matsukawa 2005).

8.2.4.2 Acupuncture Relieving the Amyloid Neurotoxicity

With the postulation of the "amyloid hypothesis," the role of acupuncture in the amyloidogenic processes has also gained an increasing interest. Luo et al. (2003) found that EA can significantly improve the plasma levels of Aβ and transforming growth factor-α (TGF-α) in AD rat models. In their later clinical study, they found that EA reduced the plasma Aβ level in senile dementia patients (Luo et al. 2006). Feng (2007) performed acupuncture (q.d.×28 days) at Baihui (DU-20), Dazhui (GV-14), Shenshu (BL-23), and Zusanli (ST-36) in AD rat models. The plasma Aβ level, cerebral amyloid expression, and neuron apoptosis were examined. She found that EA significantly improved the cognitive impairments in AD rats. Furthermore, EA significantly reduced the plasma Aβ level and cerebral amyloid accumulation as well as inhibited the neuronal apoptosis. Cai and Xiao (2008) selected Sishencong (EX-HN1), Baihui (DU-20), and Shenting (DU-24) as the main acupoints and bilateral Shenmen (HT-7), Neiguan (PC-6), Hegu (LI-4), Sanyinjiao (SP-6), Zusanli (ST-36), and Taichong (LR-3) as the secondary acupoints. They performed acupuncture on 50 cases of senile dementia (q.d. 5 days/week×4 weeks). They observed that the plasma concentrations of Aβ and homocysteine (HCY) significantly increased following EA ($p<0.01$). More recently, Xue et al. (2009) performed EA (2/100 Hz, 3 mA, 15 min/day×90 days) at Baihui (DU-20) and Yongquan (KI-1) in APP 695 V 717 I transgenic mice. They reported that the APP expression levels in various areas (bilateral cortex and hippocampal CA 1 area) and Aβ, the hippocampal CA1 area, were significantly reduced ($P<0.05$), whereas the ChAT activity significantly increased ($P<0.05$ in cerebral cortex and $P<0.01$ in hippocampal CA 1 area) after EA. They then concluded that EA could improve the cognitive functions in AD mice by increasing the Ach activity and lowering the APP and Aβ levels in the brain.

The above studies used an index of plasma Aβ which was later believed to be an inappropriate index for the diagnosis and evaluation of AD (Humpel 2011). In future, better indices, such as CSF Aβ, should be taken into account.

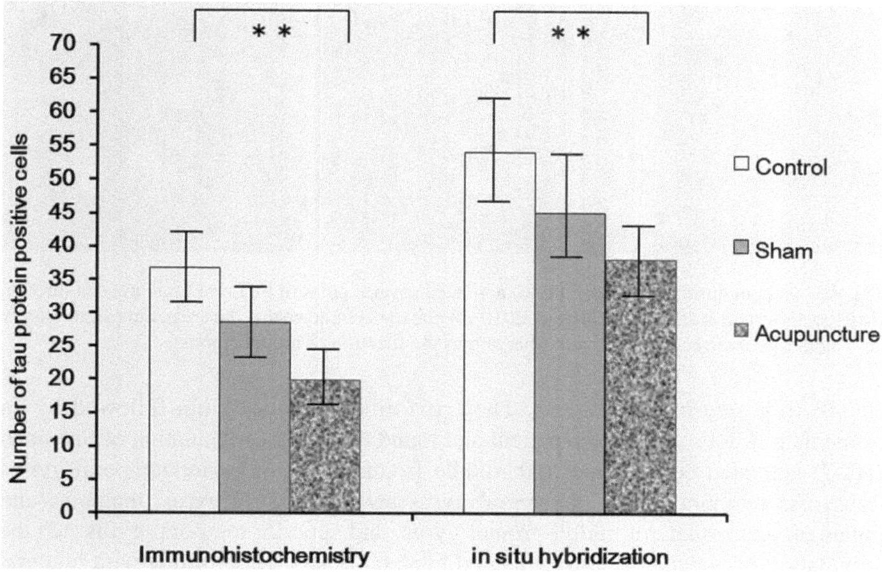

Fig. 8.4 Acupuncture reduces the tau protein expression in the CA1 area of hippocampus. Note that a significant decrease in the acupuncture group was detected by both immunohistochemistry and in situ hybridization [This figure is reproduced from those of others' studies (Jiang et al. 2008; Liu et al. 2010)]. *Double asterisk* means $p<0.01$. Note that there is a typo in the results section of the original text

8.2.4.3 Acupuncture and Hyperphosphorylated Tau Protein

Jiang et al. (2008) selected Baihui (DU-20), Dazhui (GV-14), Shenshu (BL-23), Taixi (KI-3), and Zusanli (ST-36) to perform EA (30 Hz, continuous wave, 30 min/day × 30 days, and current intensity at the lowest current to cause limb dithering) in AD rat models. Hippocampal Tau protein expression was measured by both immunohistochemistry staining technique and in situ hybridization. They found a reduced number of cells with positive expression of tau protein in the CA1 area of hippocampus after EA. Therefore, they concluded that the inhibition of tau expression in hippocampus plays a role in the mechanism of action of acupuncture in AD treatment (Fig. 8.4).

8.2.4.4 Activation of Cerebral Memory and Learning Areas by Acupuncture

Recently, some researchers used fMRI and other functional imaging techniques and found that effective acupuncture could activate several areas in the brain controlling the memory and learning functions, although the mechanism and involved pathways of this activation are poorly understood.

Guo et al. (2000) effectively performed EA on 20 senile dementia cases and reported that the stimulation of Jingming (BL-1) activated hippocampus and hypothalamus. Fu et al. (2005b) performed EA (1 Hz, continuous wave, 1 min × 3 times with 1 min interval, current 10–30 mA) at Shenmen (HT-7) (20 cases) and Neiguan

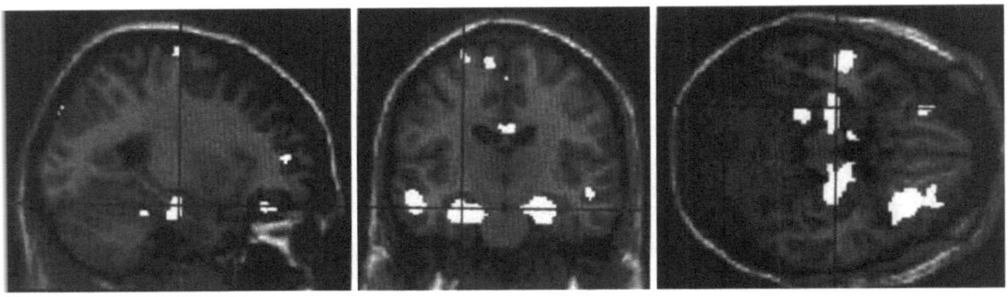

Fig. 8.5 Acupuncture at Neiguan (PC-6) activated several areas of bilateral brain as seen through fMRI imaging (Fu et al. 2005a; Liu et al. 2010). *White areas* indicate that the areas were significantly activated. The intersection of the crossbar shows the location of hippocampus

(PC-6) (6 cases) in AD patients. They performed EA for 1 min followed by an immediate fMRI scan on each patient and found that (1) the stimulation of Shenmen (HT-7) activated contralateral left middle frontal gyrus, superior temporal gyrus, transverse temporal gyrus, precentral gyrus and postcentral gyrus, thalamus and putamen, and ipsilateral middle frontal gyrus, and superior temporal gyrus; (2) the stimulation of Neiguan (PC-6) activated bilateral hippocampus along with bilateral temporal and frontal lobes (Fig. 8.5) (Fu et al. 2005a). Zhou and Jin (2008) performed EA (2.7 Hz discontinuous wave, 30 s × 6 with 30 s interval, current 1 mA) at Shenmen (HT-7), Zusanli (ST-36), Fenglong (ST-40), and Taixi (KI-3) in 26 AD cases. They noticed activation following EA, of the contralateral temporal lobe including hippocampus, insular lobus and part of parietal lobe along with parts of ipsilateral temporal lobe, parietal lobe and cerebellum, which were related to the cognitive functions.

These studies indicate that acupuncture can activate certain brain regions involved in cognitive function. This is a potential mechanism of acupuncture on AD. The outcome data of the EA treatment have two features: (1) The efficacy can be evaluated immediately, during stimulation that will last for some time even after stopping. (2) The observed effects on the brain are bilateral, however, not symmetrical. Unfortunately, these are only the initial reports on use of functional imaging technique for acupuncture research. We expect more and better reports in the future.

8.2.4.5 Acupuncture Reducing Oxidative Stress

Increase in oxidative stress due to aging is regarded as one of the factors in the pathogenesis of AD. Several Chinese studies believe that abating the free radical generation might be one of the mechanisms of acupuncture therapy for AD.

Gao et al. (2001) performed acupuncture at Zusanli (ST-36) and Guanyan (RN-4) in aged normal rats and found significantly increased amounts of superoxide dismutase (SOD) and decreased malondialdehyde (MDA). Ouyang et al. (2002) performed acupuncture on dementia rat models and found that the cerebral SOD increased, whereas lipid peroxides (LPO) decreased. Liu et al. (2002) performed acupuncture at Shuigou (DU-26), Neiguan (PC-6), and Taichong (LR-3) in acute brain atrophy mouse models. They analyzed the changes in cerebral MDA, SOD

Table 8.4 Acupuncture induced antioxidative effects in AD animal models

Author	Acupoints	Animal	Results
Sun (2002)	Yongquan (KL-1)	Aging normal rats	SOD↑, LPO↓
Gao (2003)	Baihui (DU-20) Dazhui (GV-14) Mingmen (GV-4)	Subacute aging mouse	Inhibiting MDA↑, NO↑ and SOD↓
Wang (2005)	Baihui (DU-20) Dazhui (GV-14) Shenshu (BL-23) Xinshu (BL-15)	AD rat models	MDA↓ and SOD↑ GSH-PX↑
Zhao (2005)	Baihui (DU-20) Sishencong (EX-HN1)	AD rat models	SOD↑
Tang (2005)	Dazhui (GV-14) Shenshu (BL-23) Taixi (KI-3) Zusanli (ST-36)	AD rat models	SOD↑ and GSH-PX↑, MDA↓
Liu (2010)	Stimulating olfaction with volatile oil of Acorus Gramineus	AD rat models	SOD↑ and GSH-PX↑, MDA↓

Cited and summarized from later Chinese reports (Liu et al. 2010; Sun et al. 2002; Gao et al. 2003; Wang et al. 2005; Zhao et al. 2005; Tang et al. 2005b).

glutathione peroxidase (GSH-PX), and catalase (CAT) before and after acupuncture and found that acupuncture significantly increased the cerebral SOD, activated CAT and GSH-PX, and reduced the MDA levels. Thus, they concluded that the antioxidative stress effect of acupuncture involves a complex mechanism instead of only one or two antioxidant enzymes. Several later reports using different acupuncture protocols got the analogous results (Table 8.4).

8.2.4.6 Acupuncture Ameliorating the Disordered Neurotransmitters in AD State

Some Chinese studies have reported that acupuncture could correct several neurotransmitter abnormalities that occur in AD. It could be a potential mechanism involved in the acupuncture therapy, albeit we do not know whether such changes in neurotransmitters are the causes of AD, or are caused by AD.

Shi et al. (1996) followed the principle of "restoring the consciousness and opening the intelligence," and performed acupuncture in the mouse models of acute dementia. They noticed that acupuncture was beneficial in increasing the cerebral DA and 5-hydroxytryptamine (5-HT) that contributed in improving cognition in AD. Wang et al. (2003) performed acupuncture in two groups of rats. The acupoints of one group were selected based on the principle of tonifying the "kidney," while the other group was selected based on strengthening the "spleen." They analyzed the cerebral catecholamine (CA) neurotransmitter levels before and after acupuncture and found that acupuncture based on the principle of tonifying the kidney had more potential to improve the dementia symptoms and elevate the CA levels than that of the acupuncture performed on the basis of the principle of strengthening the spleen. They, thereby, concluded that the principle followed and acupoint selection are the key aspects of acupuncture practice. Bao and Lv (2003) performed acupuncture at Fengfu (DU-16) in senile dementia mouse models. They found acupuncture at Fengfu (DU-16) contributes to increase the cerebral levels of 5-HT, DA, and norepinephrine (NE). Indeed, there appears to be sufficient evidence demonstrating that acupuncture can potentially increase the central 5-HT activity in most cases (Wen et al. 2010a).

8.2.4.7 Other Potential Mechanisms of Acupuncture for AD

Besides the mechanisms suggested by the studies described above, there are some studies that have focused on finding other potential mechanisms. It is still uncertain what their exact roles are. In this section we will briefly introduce these studies.

Lun and Yang (2000) selected scalp areas that bore lesions as projected by the CT scans of senile patients as the primary acupoints. They also selected Shenmen (HT-7) and Zusanli (ST-36) as the secondary acupoints and found that the rCBF increased after effective EA. Shang and Ding (2003) performed acupuncture at Baihui (DU-20), Dazhui (GV-14), Danzhong (RN-17), and Guanyuan (RN-4). The hemodynamic indices were measured with transcranial Doppler (TCD). They reported that after an effective acupuncture, the cerebral blood supply improved. Therefore, one of the potential mechanisms of acupuncture is to improve the blood supply to the AD brain.

Liu (2003) performed acupuncture at Baihui (DU-20), Yongquan (KL-1), Taixi (KI-3), and Xuehai (SP-10) in AD rat models. They observed that effective acupuncture could regulate the hippocampal signal conduction transmitters. In AD rat models, the blood choline acetylase (ChAT) and the hippocampal protein kinase C (PKC), protein tyrosine kinase (PTK), 3′-5′-cyclic guanosine monophosphate (cGMP), and neuronal nitric oxide synthetase (nNOS) were significantly reduced. An effective EA can improve the hippocampal nNOS activity as well as the blood ChAT. Furthermore, it contributes to activate the PKC and PTK in the membrane of hippocampus. In a later study performed in the same laboratory, they investigated the hippocampal expression of nNOS and cGMP using the same animal models and acupuncture protocol. They found after an effective acupuncture, the nNOS-positive cells in the hippocampus significantly increased ($p<0.01$), whereas the hippocampal cGMP showed increasing trend ($p>0.05$) (Yu et al. 2004).

Tang et al. (2005a) performed acupuncture at Dazhui (GV-14), Shenshu (BL-23), Taixi (KI-3), and Zusanli (ST-36) in AD rat models. They observed that acupuncture decreased the levels of several inflammatory factors (IL-1 and IL-6) in blood. Zhu and Sun (2009) performed acupuncture at Baihui (DU-20), Dazhui (KI-3), and Zusanli (ST-36) in AD rat models; they found that an effective acupuncture reduced the number of the activated glial cells in the hippocampus. These results indicate that acupuncture successfully contributes to the relief of inflammatory processes in AD state.

Jiang (2004) performed acupuncture at Baihui (DU-20), Dazhui (GV-14), and Zusanli (ST-36). He compared the hippocampal expression of NMDA receptor mRNA with in situ hybridization technique before and after an effective acupuncture. He reported that acupuncture affects the long-term potentiation (LTP) of the hippocampal synapses by modulating the expression of NMDA receptors and protein kinase. This serves as an evidence for the effect of acupuncture on the morphosis of the hippocampal synapse and could be one of its functions in AD treatment.

Jia et al. (2005) selected Shenshu (BL-23), Xuanzhong (GB-39), Baihui (DU-20), Fengfu (DU-16), and Yongquan (KL-1) to perform acupuncture. They found that the cerebral somatostatin and nerve growth factor (NGF) increased with effective acupuncture. Zhang et al. (2007) performed acupuncture at Baihui (DU-20) and Fengfu (DU-16) in senile dementia rat models. They also found that the cerebral SS levels improved after acupuncture. They elevated the levels of SS and NGF normalized the disturbances in neuronal signal transduction in AD, which is one of the effects of acupuncture.

Zeng et al. (2006) performed EA at Baihui (DU-20) and Quanyong (KL-1) in senile dementia rat models. They found an effective acupuncture-activated mitochondrial succinate dehydrogenase (SDH), Na^+–K^+-ATPase, and Ca^{2+}–Mg^{2+}-ATPase in the hippocampus, which can improve the cerebral energy metabolism.

Zhou and Wang (2008) performed EA at Baihui (DU-20), Zusanli (ST-36), and Shenshu (BL-23). They reported that an effective EA enhanced the Nestin expression of the hippocampal neural stem cells (NSCs) in AD rat models. In another study of the same laboratory, they reported an upregulation of basic fibroblast growth factor (bFGF) positive cell expression in the hippocampus after acupuncture. Thus, they concluded that an effective EA enhances the bFGF expression of the NSCs in the hippocampus (Wang and Zhou 2008). The mechanisms and exact interpretations of these findings are still unclear.

Cheng et al. (2008) performed a modified "Yi-Qi-Tiao-Xue and Fu-Ben-Pei-Yuan" acupuncture (a special acupuncture with the aim of besteading qi alone with adjusting the blood state and supporting the Vital Essence), which includes the acupoints, namely, Shanzhong (CV-17), Zhongwan (CV-12), Qihai (CV-6), Zusanli (ST-36), and Xuehai (SP-10) (Ding et al. 2006), in senescence-accelerated mouse prone 8 (SAMP8) senile dementia mouse models. They observed that an effective acupuncture enhanced the cell proliferation in dentate gyrus (DG) (Fig. 8.6a). Moreover, the newly proliferated cells were seen to be distributed along the dorsum of alveus hippocampi (Alv) in the form of "a stream-like distribution," which was considered as a potential explanation of the mode of action of acupuncture in treating AD (Fig. 8.6b) (Cheng et al. 2008).

8.3 Acupuncture Therapy for Other Neurodegenerative Diseases

We have discussed the role of acupuncture in treating the most common NDDs including AD and PD in the above sections. Certainly, acupuncture is also adopted for other NDDs such as MNDs, olivopontocerebellar atrophy (OPCA), etc.

These diseases occur rarely when compared to AD and PD. Our current understanding of these studies is controversial and still under progress. Therefore, we have only limited information on evaluating the acupuncture therapy for treating these diseases. In this section, we have presented a brief review based on the small number of papers available.

8.3.1 Acupuncture for Treatment of Motor Neuron Diseases

MNDs are a large group of progressive neurodegenerative disorders, which selectively destroy motor neurons that control voluntary muscle activities such as walking, gripping, speaking, swallowing, and other general body movements. The most common MND is amyotrophic lateral sclerosis (ALS). In addition, MNDs include several other uncommon types, i.e., primary lateral sclerosis (PLS), progressive muscular atrophy (PMA) and progressive bulbar palsy (PBP), etc. The clinical hallmark common to all MNDs is the progressive loss of the motor activities such as walking, moving,

a

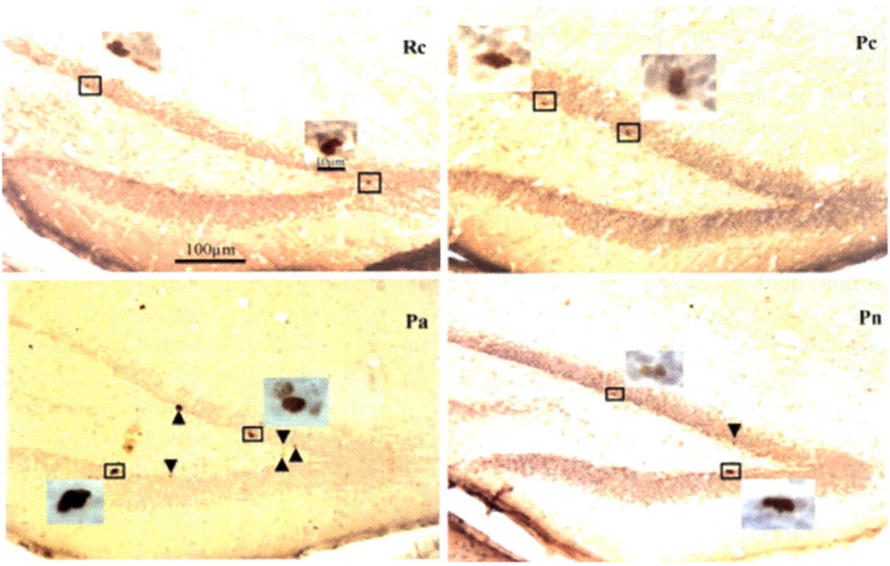

Fig. 8.6 (**a**) Acupuncture improves the cell proliferation in dentate gyrus (DG) (Cheng et al. 2008). The cell density of the Pa group was significantly increased by acupuncture (vs. Pn group). SAMR1 normal control group (Rc), SAMP8 control group (Pc), SAMP8 acupuncture group (Pa), and SAMP8 non-acupoint control group (Pn). (**b**) A stream-like distribution observed along the dorsum of alveus hippocampi (Alv) in the acupuncture group (Pa) (Cheng et al. 2008). The *brown* cells with *arrowheads* and *rectangles* show the newly proliferated cells. SAMR1 normal control group (Rc), SAMP8 control group (Pc), SAMP8 acupuncture group (Pa), and SAMP8 non-acupoint control group (Pn)

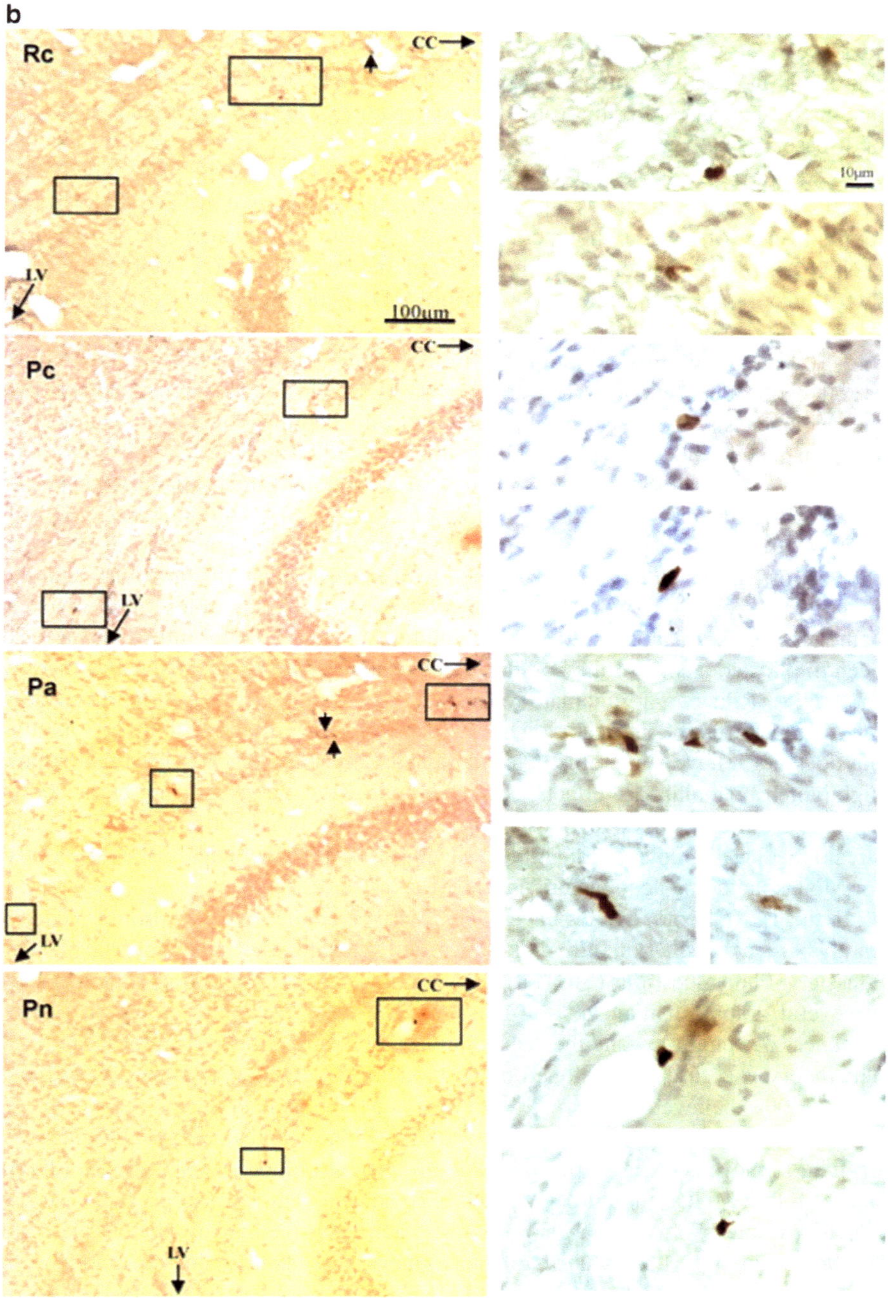

Fig. 8.6 (continued)

even swallowing, and breathing at the advanced stage. The muscular weakness and atrophy are progressive and irreversible. Cruel as it may seem, in most of cases the sensations and cognitive functions are affected *only* in the terminal stage, which means that the patients are suffering from afflictions with a clear mind during a long course of disease. Recently, it was reported that 50% cases died within 30 months, while 20% cases survived 5–10 years from the symptom onset (Kiernan et al. 2011). The pathogenesis and causes of MNDs are complex and under controversy. The regional and familial onset of several subtypes indicates a genetic involvement in the onset of MNDs. As to ALS, 13 genes are implicated in the etiology of ALS (Maruyama et al. 2010). It is reported that the mutations in SOD1 gene contribute to 20 % familiar ALS and 5 % sporadic disease (Rosen 1993); Mutations in TARDBP gene contribute to 5–10 % of familial ALS, mutations in FUS gene contributes 5 %, and mutations in ANG contributes about 1 % (Kiernan et al. 2011). Most of researchers believe that the pathogenesis of MNDs is multifactorial with complex interlaced interactions between genetic and several molecular pathways (Vucic and Kiernan 2009; Pasinelli and Brown 2006). Inflammatory processes [activation of microglia, disturbance of inflammatory factors such as tumor necrosis factor-α (TNFα) and interleukin (IL)-6], neurotoxic effects of excitatory amino acids, oxidative stress (generation of free radicals, abnormal SOD1 enzymes combined with mitochondrial dysfunction), and disruption of axonal transport processes are believed to play roles in the MNDs pathogenesis (Kiernan et al. 2011). Autoimmune reaction to viral infections has also been implicated as one of the possible mechanisms. There is no satisfactory therapy for MNDs. Riluzole, a glutamate release inhibitor, is the only FDA-approved drug proven to be effective in treating ALS; however, it only prolongs life by 2–3 months but does not relieve symptoms (Cleveland and Rothstein 2001). Other reports have highlighted potential side effects associated with Riluzole (Weishaupt et al. 2006). Thus, exploring an effective and safe therapy is of key desideratum. Since the mainstream medical therapies are so limited, several alternative therapies are being adopted for MNDs including acupuncture and TCM.

In the TCM theory, MNDs have been described as the "Wei Syndrome" (flaccid paralysis)—a dysfunction of liver, kidneys, and spleen. In the ancient classic of "The Classic of Plain Questions (Su Wen)," it was referred to as "Yangming often conceals at several organs. It manages the muscle function"; "Weakness of organs causes weakness of Yangming, and then causes muscular dysfunction and motor disability." The theory was further developed in the modern TCM (Zhou and Ma 2009). Jiang et al. (2000a) believed there were four types of MND, namely: "weakness of spleen and deficiency of qi" type, "deficiency of spleen, kidneys and yang" type, "deficiency of liver, kidneys and yin" type and "phlegm-heat and blood stasis" type. Li (2005) put forth that MNDs should belong to Wei Syndrome deficiency type. In TCM theory, the "kidneys" are in charge of bone and generates the marrow that is related to the brain function. Disturbances of the kidneys will result in loss of nutrition to bones. Spleen controls the muscles of limbs. Weakness of spleen will lead to loss of nutrition to these muscles thereby causing Wei syndrome. Thus the principles of treating MNDs should include tonifying kidneys and liver together with activating spleen and stomach. Wang (2005) thought that MNDs are caused by the damage of vital essence of the kidneys. The organs of TCM theory that are affected are kidneys, spleen, liver, and correlative channels and collaterals. The dysfunctions of the kidneys, spleen, and liver will cause the weakness of tendon, muscle, and bones. Chen et al. (2005) believed that MNDs are caused by damage to the Eight Extra Channels, that results in loss of supply of "Xue and qi, yin and yang" in all of the 12 channels. The treatment principle of "supplying the archaeus and removing atrophy" and "tonifying to generate the muscles" should thus be followed. For treating MND, scalp acupuncture is the most popular form, though body acupuncture is also adopted; some investigators use combined therapy such as acupuncture plus Chinese medicine, and acupoint injection therapy has also been reported (Sun and Cui 2005).

As ALS is the most common MND, many authors selected ALS as their target to test acupuncture efficacy in MND treatment. Wu et al. (2000) used acupuncture with a special method (*Shaoshanhuo method*) to treat 15 cases of ALS and PMA. They selected four groups of acupoints for acupuncture. (1) Baihui (DU-20), Shenshu (BL-23), Pishu (BL-20), Mingmen (GV-4), and Taixi (KI-3); (2) Shanzhong (CV-17), Guanyuan (CV-4), Qihai (CV-6), and Sanyinjiao (SP-6); (3) Dazhui (GV-14), Quchi (LI-11), Waiguan (SJ-5), Hegu (LI-4), and Zusanli (ST-36); (4) Fengchi (GB-20), Shousanli (LI-10), Neiguan (PC-6), Yanglingquan (GB-34), and Taichong (LR-3). Twenty days were set as 1 course. A single course lasted for 20 days. Initial improvement was recorded in all the patients after 1–2 courses. Other investigators adopted combined treatments for ALS. Li (1998) used acupuncture together with naprapathy to treat ten ALS cases. He found that one patient was "cured," three cases showed profound efficacy, three cases showed some efficacy, and three cases were unaffected. Zhen (2004) performed acupuncture in 20 ALS patients. He selected Dazhui (GV-14), Shousanli (LI-10), Quchi (LI-11), Hegu (LI-4), Zusanli (ST-36), Yanglingquan (GB-34), and Xuanzhong (GB-39) as the main acupoints; and Mingmen (GV-4), Qihai (CV-6), Xuehai (SP-10) Taichong (LR-3), Pishu (BL-20), Guanyuan (CV-4), Neiguan (PC-6), and Yuji (LU-10) as the secondary acupoints. He injected Vitamin B_{12} and Chinese angelica, and then embedded threads at these acupoints. He noted that four cases were "cured," ten cases "profoundly improved," four cases improved and a failure of therapy was seen in two cases. Furthermore, Wang (2005) used analogous acupuncture, acupoint injection with moxibustion, and Jiang et al. (2000b) used acupuncture together with Chinese medicines. Both of them reported good efficacy with these methods. In addition, there are other types of acupuncture that have been used in ALS treatment. Xie et al. (1995) reported that embedding threads at Guanyuan (CV-4), Qihai (CV-6), Pishu (BL-20), and Ganshu (BL-18) brought beneficial results for ALS. Liang (1999) reported using several Chinese medicines to inject Dazhui (GV-14), Neiguan (PC-6), and Zusanli (ST-36) in ALS patients and found that two cases were cured, nine cases had profound efficacy, one case showed some efficacy, and two cases were unaffected. Ma et al. (2001) used analogous acupoints and got the 43.3 % efficacy.

The reports on other types of MND are limited. Zhao (1997) used scalp acupuncture for treating 30 cases of Progressive Bulbar Palsy (PBP). He selected Neiguan (PC-6), Shuigou (DU-16), Shangxing (DU-23), Baihui (DU-20), Yingtang (EX-HN3), Fengchi (GB-20), Wangu (GB-12), Tianzhu (BL-10), Yifen (SJ-17), Lianquan (RN-23), and the scalp motor area as the main acupoints. He performed acupunctures for 3 courses (1 course = twice/day × 30 days) and reported that 3 cases were cured, 9 cases achieved profound efficacy, 15 cases got some efficacy, and 5 cases failed. The reported efficacy was 83 %. Jie (2002) performed acupuncture adopting the same protocol as Zhao (1997). Yuan et al. (2005a) performed acupuncture with intravenous administration of Jiweiling injection; Peng et al. (2005) performed acupuncture at Lianquan (RN-23), Fengchi (G-20), Hegu (LI-4), and Zusanli (ST-36); Pei (2006) performed acupuncture at Fengchi (GB-20). All of these studies claimed a good rate of efficacy.

The bench studies on studying the involved mechanisms of acupuncture for MNDs are being conducted/are under progress. The efficacy of Riluzole indicates that neuroinflammation underlies the pathophysiological process of MNDs. As acupuncture at Zusanli (ST-36) has been accepted to be effective in regulating the immune system by inhibiting the inflammatory reactions and exerts neuroprotective effects in PD (Jeon et al. 2008; Moon et al. 2007; Wilms et al. 2007), Yang et al. (2010) performed EA (2 Hz, rectangular wave for duration time of 1 ms, current intensity 1 mA) at Zusanli (ST-36) in hSOD1 transgenic (hG93ASOD1) mouse models of ALS. They found that EA significantly improved the motor activity. In the mice which underwent effective EA, the microglial activity and TNF-α expression in the brain stem (Fig. 8.7) and spinal cord (Fig. 8.8) significantly decreased, along with an improvement in neuronal loss. Therefore, they concluded that a reduction of the inflammatory processes are probable mechanisms of EA in ALS treatment (Fig. 8.9).

8.3.2 *Acupuncture for Treatment of Multiple System Atrophy*

Multiple system atrophy (MSA) is a generic term for a group of NDDs that are characterized by a combination of following symptoms: ataxia, Parkinsonism, and impairments of autonomic functions. The generalized definition of MSA includes Shy-Drager syndrome (an old term to define a group of patients whose dominative symptoms are autonomic dysfunctions); OPCA, and striatonigral degeneration. The pathogenesis of MSA is unknown. The patients always present with a group of symptoms that result from the degeneration of neurons in specific cerebral areas. The common symptoms include motor impairments (bradykinesia, stiffness, etc., PD-like symptoms); cerebellar dysfunction (ataxia) and autonomic dysfunction such as dysfunction of blood pressure regulation, constipation, abnormal sweating, and sleep disorders. There is no effective therapy for MSA and all present treatments are symptomatic only.

In TCM, two types of MSA are described. First is the deficiency type. It is believed that MSA results due to the deficiency of five main organs including kidneys, liver, heart, spleen, and lungs. Song (2008) reported that the weakness of kidneys is the commonest (80.26 %), followed by liver (34.21 %), and weakness of the other organs occurs less often. This type always accompanies with a deficiency of Qi, blood, Yin, and Yang. The different combinations of these deficiencies give rise to a diverse array of symptoms. The second type is sthenia type. However,

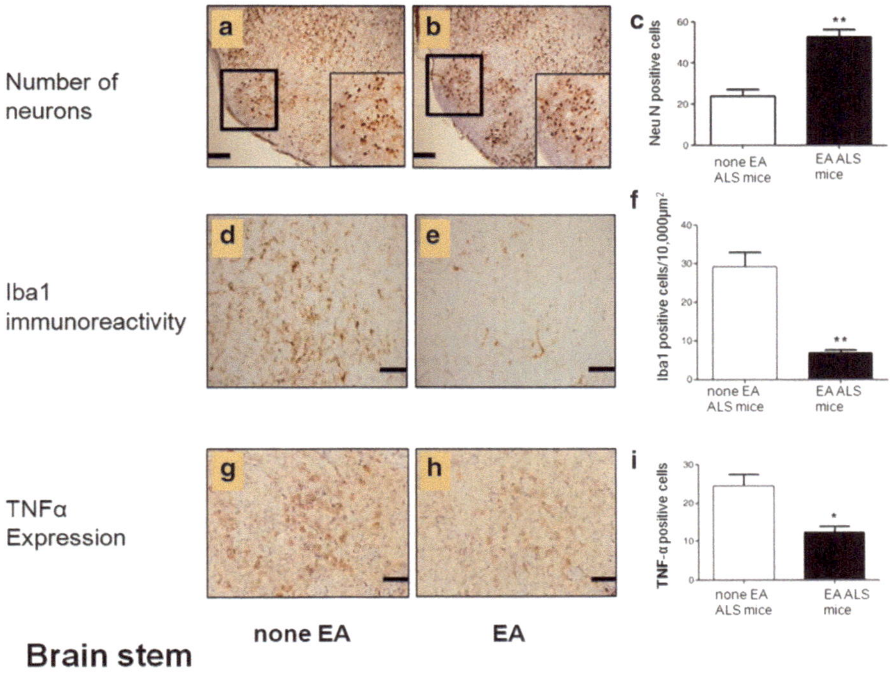

Fig. 8.7 Cytological improvement in the brain stem induced by EA (Yang et al. 2010). (**a–c**) Effects of EA on the neurons in brain stem (**a** shows the cells in the none EA group, **b** shows the cells increased by EA in the EA group, **c** is the result of cell counting, $**p<0.005$, Scale bars = 100 μm). (**d–f**) EA decreases the Iba1 immunoreactivity in brain stem (**d** shows the Iba1 expression positive cells in the none EA group, **e** shows the Iba1 expression positive cells decreased by EA in the EA group, **f** is the result of cell counting, $**p<0.001$, Scale bars = 100 μm). (**g–i**) EA decreases TNF-α expression in brain stem (**g** shows the TNF-α expression positive cells in the none EA group, **h** shows the TNF-α expression positive cells decreased by EA in the EA group, **i** is the result of cell counting, $*p<0.05$, Scale bars = 100 μm). Cited and modified from Yang et al. (2010)

such "sthenia" is not because of the vigorous organ function but rather a false appearance caused by the weakness of organs (Wang 2010). Song also pointed out that in the sthenia type, phlegm dampness type is the commonest (42.11 %); followed by the phlegm dampness and accumulated heat type (39.47 %) with other types presenting less commonly (Song et al. 2008; Song 2008). Wang (2010) summarized previous TCM theories on MSA, and regarded weakness of liver, spleen, and kidneys as the main causes of MSA with phlegm playing a secondary role in the pathogenesis of MSA.

Several clinical reports have claimed varying success in the use of acupuncture for treating MSA. Ma et al. (1998) proposed that acupuncture at Zusanli (ST-36), Sanyinjiao (SP-6), and Yanglingquan (GB-34) could be beneficial for MSA. Sakakibara et al. (2007) performed moxibustion at Zhongji (RN-3), bilateral Qichong (ST-30), bilateral Sanyinjiao (SP-6), and bilateral Shuiquan (KI-5) in a

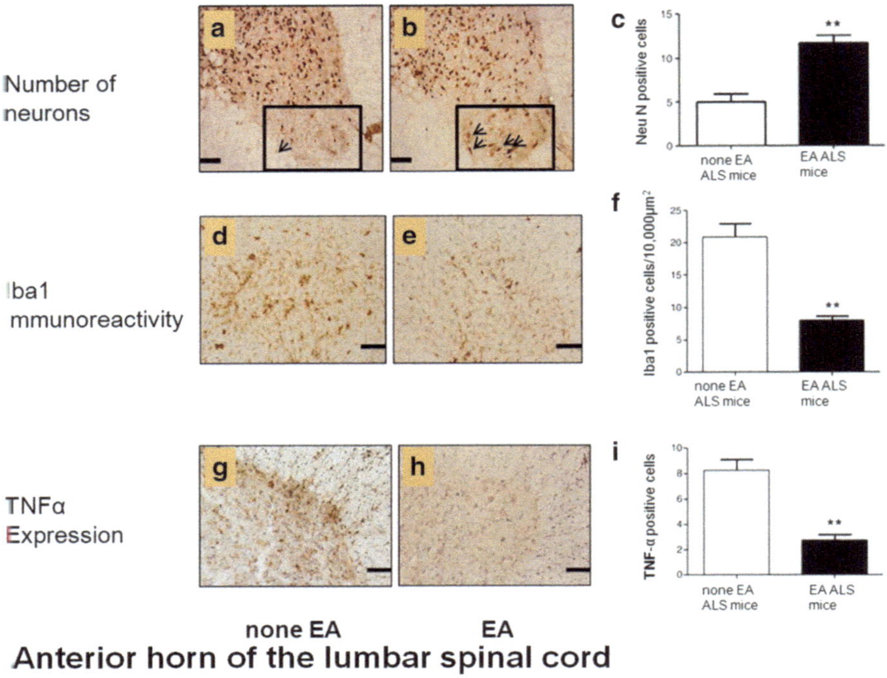

Fig. 8.8 Cytological improvement in the anterior horn of the lumbar spinal cord induced by EA (Yang et al. 2010). (**a–c**) EA enhances the numbers of neurons in anterior horn of the lumbar spinal cord (**a** shows the cells in the none EA group, **b** shows the cells increased by EA in the EA group, **c** is the result of cell counting, $^{**}p<0.005$, Scale bars = 200 μm). (**d–f**) EA decreases the Iba1 immunoreactivity in anterior horn of the lumbar spinal cord (**d** shows the Iba1 expression positive cells in the none EA group, **e** shows the Iba1 expression positive cells decreased by EA in the EA group, **f** is the result of cell counting, $^{**}p<0.001$, Scale bars = 200 μm). (**g–i**) EA decreases TNF-α expression in anterior horn of the lumbar spinal cord (**g** shows the TNF-α expression positive cells in the none EA group, **h** shows the TNF-α expression positive cells decreased by EA in the EA group, **i** is the result of cell counting, $^{*}p<0.001$, Scale bars = 100 μm). Cited and modified from Yang et al. (2010)

65-year-old MSA patient who lost the circadian arginin vasopressin rhythm (C-AVP-R) and nocturnal polyuria. After being treated for 6 months (3 times/day. q.d.), the daily rhythm reverted to normal and the nocturnal polyuria was significantly improved. Cao and Liu (2007) followed the principle of "nourishing the sea of marrow" and selected Yingtang (EX-HN3), fengfu (DU-16), Baihui (Du-20), Taiyang (EX-HN5), Shenting (DU-24), and Touwei (ST-8) to perform acupuncture for 1 month (5 times/week, q.d.). They observed improvement in symptoms and ADL in all patients. Recently, Cheng and Bian (2010) performed acupuncture following the principle of "restoring the consciousness and opening the intelligence" and reported good levels of efficacy. Other investigators who used acupuncture along with Chinese medicine also reported good efficacy. Niu and Duan (2005) selected Renyin (ST-9), Guanyuan (RN-4), and Zusanli (ST-36) as the main acupoints.

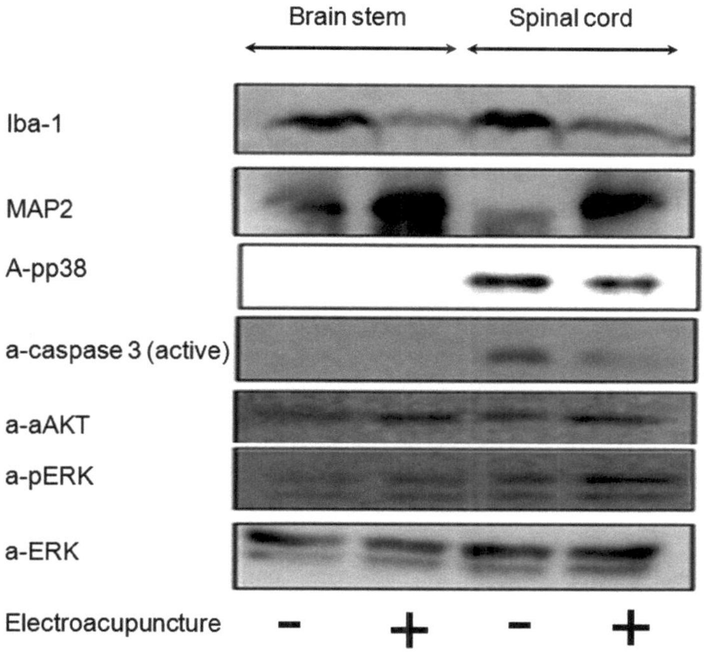

Fig. 8.9 Results of western blotting ameliorated by EA (Yang et al. 2010). The changes in brain stem and spinal cord showed a similar trend. Note that EA increased activities of MAP2, AKT, and ERK, indicating a neuroprotective effect. The activities of Iba-1, p38 mitogen-activated protein kinase, and active caspase-3 decreased by effective EA, which showed EA may release the inflammatory effects. Cited and modified from Yang et al. (2010)

Acupuncture was performed together with oral administration Chinese medicine. The selection of Chinese medicine followed the principles of differential symptoms of TCM (Bian Zheng Shi Zhi). They found that combined treatment of acupuncture with Chinese medicine attained better results. Finally, there are a few reports on treatment of the less common types of MSA with acupuncture. Zhou and Han (2003) selected Chinese medicine based on the principles of "Supplying Yang and nourishing Yin, tonifying Primary Qi and replenishing Vital Essence" and formulated a mixture of medicines called the "Shengmaiwenyatang" (a medicine soup to generate the sphygmus and stabilize the blood pressure, which is composed of *Radix Panacis Quinquefolii*, *Fructus Schinsan-drae*, *Radix Ophiopogonis, Radix Aconiti Lateralis Preparata, Rhi-zorna Polygonati*, and *Radix Glycyrrhizae Preparata*). They administrated the medicine mixture along with acupuncture at Zusanli (ST-36), Sanyinjiao (SP-6) and Pishu (BL-20) in 18 cases of patients with Shy-Drager syndrome for 1 course (1 month). They found that the combined treatment improved the postural hypotension and other symptoms in 94.44 % patients. Yang et al. (2005) performed scalp acupuncture at scalp motor area, balance area, and Baihui (DU-20); body acupuncture at Hegu (LI-4), Zusanli (ST-36), and Yinlingquan (SP-9) together with oral administration of *Wen Dan Tang* (A Chinese medicine soup to warm up the gallbladder) in 32 cases of cerebellar atrophy. After 1 course (30 min/time q.d. × 30 days) of treatment, they found that 10 cases reported profound improvement, 18 cases reached some improvement, and 4 cases failed. They reported that totally, 87.5 % patients improved, thereinto 31.25 % achieved notable improvement. Recently, Li and Zhou (2010) reported use of acupuncture to treat a case of OPCA. They used a combination therapy of body acupuncture, scalp acupuncture, and Chinese medicine. They found that after 29 days of treatment the motor function and language expression of this patient showed notable improvement. Thus, they concluded that acupuncture is efficacious in treating the motor impairment in OPCA.

In this section, we have introduced most of the studies that reported the use of acupuncture for MSA treatment. No bench study has been conducted in this regard. In the future, more studies should be carried out to clarify the efficacy, side effects and mechanisms of acupuncture for MSA treatment.

8.3.3 Acupuncture in Treating the Other NDDs

In our extensive review of literature, the studies conducted to test acupuncture application in treatment of other NDDs were hard to come by; for instance, we could not find even a single paper on acupuncture use for Huntington's disease (HD) and progressive supranuclear palsy (PSP). The probable reason for this could be the difficult diagnosis and/or rare onset of such diseases. Here, we will only introduce the topic in essence, based on the reports available.

Albeit there are several studies that mention treatment of "ataxia" by acupuncture, however, most of the etiology of such ataxia is secondary to other diseases such as stroke. Hence we tried to find studies on ataxia caused by neurodegeneration such as spinocerebellar ataxia type 3 (SCA3, Machado–Joseph Disease) or Friedreich's ataxia, although such studies were scarcely available. Dr Sheng was the first to report "cure" of hereditary ataxia (HA) in one Friedreich's ataxia case (Sheng 1964), and several follow-up studies reconfirmed the efficacy of acupuncture in HA. It is believed that the scalp acupuncture has better efficacy than body acupuncture, but currently, most of the researchers adopted a combined treatment (Zhao et al. 2010). Li and Ma (1992) performed scalp acupuncture on 8 HA cases. The main acupoints were scalp balance area, sensory area, and the first, second, and third speech areas. The secondary acupoints were frontoparietal zone, vertex-temporal zone, and parietooccipital zone. Acupuncture was performed for 15 min/time q.d. × 10 days. They reported that eight cases were "cured," three cases showed profound improvement, and five cases reached some improvement, only one case failed (Li and Ma 1992; Li et al. 2008). Luo and Guo (1996) performed body acupuncture in 30 cases of HA patients. They selected two groups of acupoints. The first group included Baihui (DU-20), Houding (DU-19), Fengfu (DU-16), Jugu (LI-16), Quchi (LI-11), Xiaohai (SI-8), Waiguan (SJ-5), Yuji (LU-10), Dubi (ST-35), Yanglingquan (GB-34), Weizhong (BL-40), Zusanli (ST-36), and Sanyinjiao (SP-6); the second group included Baihui (DU-20), Naohu(DU-17), Dazhui (GV-14), Fengchi (GB-20),

Jianyu (LI-15), Jianzhen (SI-9), Jianliao (SJ-14), Quze (PC-3), Zusanli (ST-36), Weiyang (BL-39), Taixi (KI-3), Kunlun (BL-60), and Taichong (LR-3). After being treated over 5 courses (1 course = 30 min/time, q.d. × 10 days), 9 cases reached profound efficacy, 19 cases showed some efficacy, and 2 cases were unaffected. They reported that totally 93 % cases improved. There are several studies adopting other combined treatment protocol to treat HA. Fu and Sang (1991) treated 50 cases of HA patients using scalp acupuncture combined with injection of acupoints with *Acelglutamide Injection* and *Ligustrazine Injection*; Sun (1994) treated ten cases of HA using acupuncture combined with moxibustion. Both of them reported success with 26 "cured" cases; 20 cases with profound improvement; 8 cases with certain improvement, and 6 cases with failed results. Finally 90 % patients gained improvement. Recently, Xia et al. (2010a) reported using Frenkel training in combination with acupuncture and naprapathy to treat 12 cases of cerebellar ataxia. They found that such combined therapies could improve the symptoms and enhance the ADL. There is, however, no bench study involved in studying the mechanisms of acupuncture for ataxia (till August 2010).

Syringomyelia (including syringobubia) is a generic term for a group of diseases with a hallmark of cyst or cavity formation in the spinal cord. The cyst or cavity, for which we used the term "syrinx," can expand or extend through the spinal cord with the progression of the disease and in the process cause symptoms such as pain and abnormal sensations. The pathogenesis of syringomyelia is unclear. Some of the cases are caused by Chiari malformation. There is a controversy over the inclusion of syringomyelia into the classic NDDs. From the TCM point of view, syringomyelia also belongs to the "Wei Syndrome" (flaccid paralysis) or/and "Bi Syndrome" (palsy) spectrum. Li et al. (1983) first adopted acupuncture for the treatment of syringomyelia and found that the sensation of "De Qi" in the acupuncture is essential to attain good results. Wu and Guo (1986) used combined treatments of acupuncture with Chinese medicine for syringomyelia. Wang and his team adopted acupuncture combined with Chinese medicine to treat large numbers of syringomyelia cases (445 cases from 2000 to 2003 and 98 cases from 2003 to 2004). They used a Chinese medicine called "YiSuiLing capsule" (a mixture to benefit marrows composed of *ginseng, deer horn, dodder, tuber fleeceflower root, medlar, spinal cord of goat, scorpion and suberect spatholobus sem*) and selected an area in the middle of the spinal column (which is believed as the pathway of DU channel) to perform moxibustion. Acupuncture was also performed at *HuaTuoJiaJi* acupoints (EX-B2) together with Shenshu (BL-23), Pishu (BL-20), and Ganshu (BL-18). They reported good results with these combination therapies. In the 2005 report, 32 cases achieved notable improvement, 57 cases got improvement, and 9 cases failed to report any benefit. They summarized that 90.82 % of the cases improved (Wang et al. 2004; Yuan et al. 2005b; Ping et al. 2005). Based on these studies, Wang and Chen (2006) summarized that syringomyelia was caused by deficiency of the "kidney" essence and weakness of the DU channel substantially; however, the patients often presented with the symptoms of blood stasis along with weakness of the collaterals. Therefore, the therapeutic principles should be (1) supplying the vital essence and tonifying the marrow, thus nourishing the DU channel, which is the curative therapy; (2) relieving spasm and removing obstruction in collaterals, which is a symptomatic therapy. Later studies after postulation of Dr Wang's theory also showed successful results (Wang 2007). Besides Wang's team, there are several available reports from other teams. Huang's report performed acupuncture at DU channel in a syringomyelia case (Chiari malformation type) based on Wang's theory (Huang and Zhu 2010); Wang et al. (2008) selected Dazhui (GV-14), Lingtai (DU-10), Jinsuo (GV-8), Jizhong (DU-6), Mingmen (GV-4), Yaoyangguan (DU-3), Yaoshu (DU-2), and Jiaji (EX-B2) to perform acupuncture in combination with Chinese medicine *SanCaiFengSuiDan* (a mixture of *asparagus fern, prepared rhizome of rehmannia, ginseng, amboynas, fructus amomi and liquorice*) in 40 syringomyelia cases. They found the outcomes of combined treatment group were better than those of only acupuncture and only medicine groups (97.5 % improved vs. 75 % and 77 % improved, respectively). They also noted that the efficacy in early stage syringomyelia was better compared to advanced stage cases. The principles of acupoints and medicine selection were in sync with Wang's theory. There is no bench study involved in the acupuncture for syringomyelia.

There are several studies involved in using acupuncture to treat dystonia and tics-coprolalia syndrome. Due to multifactorial causation of these symptoms, they are conventionally excluded from NDDs; therefore, we have not discussed them in this chapter.

8.4 Concluding Remarks

In this chapter, we reviewed the progress made in the clinical and bench studies on acupuncture for AD and the other NDDs. Most of the studies focus on AD and the studies on the other NDDs are extremely limited. Unfortunately, most of these studies were of poor quality and minimal useful information could be extracted from them.

In the clinical studies, the flaws in the experimental design and methodology declined the credibility of the findings that potentially exaggerated the efficacy and understated the side effects caused by acupuncture. This problem has also been described in the PD chapter and presents as a common drawback of the studies for evaluation of alternative therapies. On the contrary, such flaws always air skepticism against the alternative therapies, leading to their exclusion from the sight of mainstream medical academia. According to the studies I introduced in this section, we cannot make an affirmative conclusion if acupuncture is actually beneficial for AD and other NDDs. We thus suggest large-sample, multicentric, long-term, and well-designed clinical trials for evaluating the clinical worthiness of acupuncture treatment for these diseases.

On the other hand, the recent bench studies involved in acupuncture for AD and the other NDDs are illuminating yet preliminary. It is hypothesized that multiple factors might govern the functioning of acupuncture since the mechanisms of NDDs are quite complex. Although we tried to introduce the potential mechanisms in the round, yet none of the studies at hand provided formidable and explicit evidence for clarifying the mechanisms. The bench studies must catch up on the "hot topics" of the mainstream research on NDDs, and use the new technologies of the biochemistry and/or molecular imaging in order to provide a solid evidence for the mechanisms involved in the acupuncture for AD and other NDDs. Objective methods of acupoints selection, objective standard in evaluating the efficacy and objective acupuncture parameters should be established to attract more scientists to step into this field. Objectivizing all the items in the experimental design (evaluation standard, experimental methods, parameters, etc.) might be a challenge for the acupuncture research nowadays. Fortunately, more and more TCM researchers have understood the situation and have begun improvising to find appropriate solution to this problem. Rigorous, well-designed studies on acupuncture for NDDs can be expected in the near future.

Acknowledgments TA was supported by Japan Society for the Promotion of Science (Grant-in-Aid for Young Scientists, Type B, #20791025 and Grant-in-Aid for Scientific Research General, #24592157). YX was supported by the National Institutes of Health (AT-004422 and HD-034582) and the Vivian L. Smith Neurologic Foundation.

References

Alexander GE, Chen K, Pietrini P, Rapoport SI, Reiman EM (**2002**) Longitudinal PET Evaluation of Cerebral Metabolic Decline in Dementia: A Potential Outcome Measure in Alzheimer's Disease Treatment Studies. Am J Psychiatry 159 (5):738–745

Alonso AC, Li B, Grundke-Iqbal I, Iqbal K (**2008**) Mechanism of tau-induced neurodegeneration in Alzheimer disease and related tauopathies. Curr Alzheimer Res 5 (4):375–384

Amatsubo T, Yanagisawa D, Morikawa S, Taguchi H, Tooyama I (**2010**) Amyloid imaging using high-field magnetic resonance. Magn Reson Med Sci 9 (3):95–99

Arriagada PV, Growdon JH, Hedley-Whyte ET, Hyman BT (**1992**) Neurofibrillary tangles but not senile plaques parallel duration and severity of Alzheimer's disease. Neurology 42 (3 Pt 1):631–639

Bao YX, Lv GH (**2003**) Effect of acupuncture on dysmnesia and monoamine neurotransmitter of dementia mice. Shanghai Journal of Acupuncture and Moxibustion 22 (7):23–25 (in Chinese with English abstract)

Barg J, Belcheva M, Rowinski J, Ho A, Burke WJ, Chung HD, Schmidt CA, Coscia CJ (**1993**) Opioid receptor density changes in Alzheimer amygdala and putamen. Brain Res 632 (1–2):209–215

Barrow CJ, Zagorski MG (**1991**) Solution structures of beta peptide and its constituent fragments: relation to amyloid deposition. Science 253 (5016):179–182

Bekris LM, Yu CE, Bird TD, Tsuang DW (**2010**) Genetics of Alzheimer disease. J Geriatr Psychiatry Neurol 23 (4):213–227

Bird TD (**1993**) Alzheimer Disease Overview. http://www.ncbi.nlm.nih.gov/entrez/query.fcgi?cmd=Retrieve&db=PubMed&dopt=Citation&list_uids=20301340.

Bonifati DM, Kishore U (**2007**) Role of complement in neurodegeneration and neuroinflammation. Mol Immunol 44 (5):999–1010

Cai T, Xiao DL (**2008**) Clinical Observations on the Intervening Effect of Acupuncture on Hcy and Aβ in Senile Dementia Patients. Shanghai Journal of Acupuncture and Moxibustion 27 (12):3–5 (in Chinese with English abstract)

Cao Y, Liu ZS (**2007**) Treating multiple system atrophy by electroacupuncture followed the principle of nourishing the sea of marrow. Liaoning Journal of Traditional Chinese Medicine 34 (5):651 (in Chinese)

Chau KW, Chan WY, Shaw PC, Chan HY **(2006)** Biochemical investigation of Tau protein phosphorylation status and its solubility properties in Drosophila. Biochem Biophys Res Commun 346 (1):150–159

Chen JL, Wang DH, Zhou SL **(2005)** A clinical observation using Jiweiling injection to treat motor neuron disease. Chinese Journal of Integrated Traditional and Western Medicine 25 (1):66–67 (in Chinese)

Cheng H, Yu J, Jiang Z, Zhang X, Liu C, Peng Y, Chen F, Qu Y, Jia Y, Tian Q, Xiao C, Chu Q, Nie K, Kan B, Hu X, Han J **(2008)** Acupuncture improves cognitive deficits and regulates the brain cell proliferation of SAMP8 mice. Neurosci Lett 432 (2):111–116

Cheng Y, Bian LN **(2010)** Using acupuncture to treat C-type multiple system atrophy. Liaoning Journal of Traditional Chinese Medicine 37 (6):1113–1114 (in Chinese)

Cleveland DW, Rothstein JD **(2001)** From Charcot to Lou Gehrig: deciphering selective motor neuron death in ALS. Nat Rev Neurosci 2 (11):806–819

Crowther DC, Kinghorn KJ, Miranda E, Page R, Curry JA, Duthie FA, Gubb DC, Lomas DA **(2005)** Intraneuronal Abeta, non-amyloid aggregates and neurodegeneration in a Drosophila model of Alzheimer's disease. Neuroscience 132 (1):123–135

Deng ZM, Yuan YJ **(1991)** Storke and dementia. China Journal of Traditional Chinese Medicine and Pharmacy 6 (3):13–15 (in Chinese)

Ding X, Yu J, Yu T, Fu Y, Han J **(2006)** Acupuncture regulates the aging-related changes in gene profile expression of the hippocampus in senescence-accelerated mouse (SAMP10). Neurosci Lett 399 (1–2):11–16

Dong HT **(1999)** Treatment of dementia in senior person by Traditional Chinese Medicine, a systemic review. Journal of Traditional Chinese Medical Literature 023 (4):39–41 (in Chinese)

Dong HT, Bai Y **(2002)** Influence of Acupunture Therapy to Alzheimer's Disease in the Score of MMSE Measuring Scale. Journal of External Therapy of Traditional Chinese Medicine 11 (3):6–7 (in Chinese)

Dong YX, Fu Q, Li XC, Hang CX **(1997)** Acupuncture combined with injection of cerebrolysin at acupoints in treating senile dementia. Journal of Apoplexy and Nervous Diseases 14 (4):242–242 (in Chinese)

Driver AS, Kodavanti PR, Mundy WR **(2000)** Age-related changes in reactive oxygen species production in rat brain homogenates. Neurotoxicol Teratol 22 (2):175–181

Dupiereux I, Zorzi W, Quadrio I, Perret-Liaudet A, Kovacs GG, Heinen E, Elmoualij B **(2009)** Creutzfeldt-jakob, Parkinson, lewy body dementia and Alzheimer diseases: from diagnosis to therapy. Cent Nerv Syst Agents Med Chem 9 (1):2–11

Feng LF **(2007)** Effects of electroacupuncture to the cognitive impairment and amyloid β in senile dementia rat models. Hubei University of Chinese Medicine, Wuhan (in Chinese with English abstract)

Ferri CP, Prince M, Brayne C, Brodaty H, Fratiglioni L, Ganguli M, Hall K, Hasegawa K, Hendrie H, Huang Y, Jorm A, Mathers C, Menezes PR, Rimmer E, Scazufca M **(2005)** Global prevalence of dementia: a Delphi consensus study. Lancet 366 (9503):2112–2117

Fletcher PT, Powell S, Foster NL, Joshi SC **(2007)** Quantifying metabolic asymmetry modulo structure in Alzheimer's disease. Inf Process Med Imaging 20:446–457

Fu JZ, Sang BL **(1991)** Scalp acupuncture combined with acupoint injection to treat cerebellar ataxia, a 50-case clinical report. Chinese Acupuncture & Moxibustion 11 (1):23–24 (in Chinese)

Fu P, Jia JP, Min BQ **(2005a)** Acupuncture at Neiguan acupoint for brain functional MRI of patients with Alzheimer disease. Chinese Journal of Neurology 38 (2):118–119 (in Chinese)

Fu P, Jia JP, Wang M **(2005b)** Acupuncture at shenmen acupoint for brain functional MRI of patients with Alzheimer disease. Chinese Journal of Clinical Rehabilitation 9 (1):120–121 (in Chinese with English abstract)

Gao HQ, Wang Y, Han YS **(2001)** An experimental study to investigate the effects of acupuncture at Zusanli and Guanyuan to the cerebral SOD, MDA and NO in senior rats. Journal of Clinical Acupuncture and Moxibustion 17 (11):47–48 (in Chinese with English abstract)

Gao XY, Jia ZL, Ai BW **(2003)** The anti-aging effects of acupuncture in subacute aging mouse model. Jiangsu Journal of Traditional Chinese Medicine 24 (2):53–55 (in Chinese)

Gravina SA, Ho L, Eckman CB, Long KE, Otvos L, Jr., Younkin LH, Suzuki N, Younkin SG **(1995)** Amyloid beta protein (A beta) in Alzheimer's disease brain. Biochemical and immunocytochemical analysis with antibodies specific for forms ending at A beta 40 or A beta 42(43). J Biol Chem 270 (13):7013–7016

Grill JD, Cummings JL **(2010)** Current therapeutic targets for the treatment of Alzheimer's disease. Expert Rev Neurother 10 (5):711–728

Grunblatt E **(2008)** Commonalities in the genetics of Alzheimer's disease and Parkinson's disease. Expert Rev Neurother 8 (12):1865–1877

Guo Y, Shi X, Uchiyama H, Hasegawa A, Nakagawa Y, Tanaka M, Fukumoto I **(2002)** A study on the rehabilitation of cognitive function and short-term memory in patients with Alzheimer's disease using transcutaneous electrical nerve stimulation. Front Med Biol Eng 11 (4):237–247

Guo Y, Shi XM, Utiyama N, Hasegawa A, M. T, Fukumoto Y **(2000)** Study of a New Rehabilitation Method of Dementia with Electrostimulation on the JingMing-Point. Journal of the Japan Eastern Medical Association 16 (4):17–30 (in Japanese with English abstract)

Hamani C, McAndrews MP, Cohn M, Oh M, Zumsteg D, Shapiro CM, Wennberg RA, Lozano AM **(2008)** Memory enhancement induced by hypothalamic/fornix deep brain stimulation. Ann Neurol 63 (1):119–123

Hampel H, Frank R, Broich K, Teipel SJ, Katz RG, Hardy J, Herholz K, Bokde AL, Jessen F, Hoessler YC, Sanhai WR, Zetterberg H, Woodcock J, Blennow K **(2010)** Biomarkers for Alzheimer's disease: academic, industry and regulatory perspectives. Nat Rev Drug Discov 9 (7):560–574

Hardy J **(1997)** Amyloid, the presenilins and Alzheimer's disease. Trends Neurosci 20 (4):154–159

Harper JD, Lansbury PT, Jr. **(1997)** Models of amyloid seeding in Alzheimer's disease and scrapie: mechanistic truths and physiological consequences of the time-dependent solubility of amyloid proteins. Annu Rev Biochem 66:385–407

Harris FM, Brecht WJ, Xu Q, Tesseur I, Kekonius L, Wyss-Coray T, Fish JD, Masliah E, Hopkins PC, Scearce-Levie K, Weisgraber KH, Mucke L, Mahley RW, Huang Y **(2003)** Carboxyl-terminal-truncated apolipoprotein E4 causes Alzheimer's disease-like neurodegeneration and behavioral deficits in transgenic mice. Proc Natl Acad Sci U S A 100 (19):10966–10971

Hartley DM, Walsh DM, Ye CP, Diehl T, Vasquez S, Vassilev PM, Teplow DB, Selkoe DJ **(1999)** Protofibrillar intermediates of amyloid beta-protein induce acute electrophysiological changes and progressive neurotoxicity in cortical neurons. J Neurosci 19 (20):8876–8884

Herholz K, Carter SF, Jones M **(2007)** Positron emission tomography imaging in dementia. Br J Radiol 80 Spec No 2:S160–167

Hirono N, Hashimoto M, Ishii K, Kazui H, Mori E **(2004)** One-year change in cerebral glucose metabolism in patients with Alzheimer's disease. J Neuropsychiatry Clin Neurosci 16 (4):488–492

Huang SY, Zhu GQ **(2010)** Features and Analysis in Treatment of Bronchial Asthma Treated by Acupoint Catgut Embedding Therapy. Journal of Guiyang College of Traditional Chinese Medicine 32 (3):45–46 (in Chinese)

Huang Y, Weisgraber KH, Mucke L, Mahley RW **(2004)** Apolipoprotein E: diversity of cellular origins, structural and biophysical properties, and effects in Alzheimer's disease. J Mol Neurosci 23 (3):189–204

Humpel C **(2011)** Identifying and validating biomarkers for Alzheimer's disease. Trends Biotechnol 29 (1):26–32

Inaniwa C **(1993)** Using low power laser in treating dementia a SPECT study and 3-case report. Paper presented at the The 14th Annual Meeting of Japan Society for Laser Surgery and Medicine, Osaka (in Japanese)

Jackson GR, Wiedau-Pazos M, Sang TK, Wagle N, Brown CA, Massachi S, Geschwind DH **(2002)** Human wild-type tau interacts with wingless pathway components and produces neurofibrillary pathology in Drosophila. Neuron 34 (4):509–519

Jeon S, Kim YJ, Kim ST, Moon W, Chae Y, Kang M, Chung MY, Lee H, Hong MS, Chung JH, Joh TH, Park HJ **(2008)** Proteomic analysis of the neuroprotective mechanisms of acupuncture treatment in a Parkinson's disease mouse model. Proteomics 8 (22):4822–4832

Ji FY, Gao HT, Wang F, Xiong FL (**2007**) Treating senile dementia by acupuncture at Baihui and Neiguan. Gansu Journal of Traditional Chinese Medicine 20 (7):14–15 (in Chinese)

Jia CW, Zhang M, Zhao WF (**2005**) The acutherapy of "tonifying the kidney to benefit marrow": Its effects on CNS signal transduction of model of dementia rats. Journal of Shaanxi College of Traditional Chinese Medicine 28 (6):38–41 (in Chinese with English abstract)

Jiang GH (**2004**) The study of the effect of electro-acupuncture on the learning and memory signal transduction mechanism and brain biochemistry in laboratory rats with senile dementia. Heilongjiang University of Chinese Medicine, Harbin (in Chinese with English abstract)

Jiang GH, Chen ZZ, Pan WY, Yang WH (**2000a**) Combination of Chinese Medicine and acupuncture to treat motor neuron disease, a 45-case clinical observation. Journal of New Chinese Medicine 32 (8):27 (in Chinese)

Jiang GH, Pan WY, Chen ZZ, Yang WH (**2000b**) Using acupuncture combined with traditional Chinese Medicines to treat motor neuron disease Journal of New Chinese Medicine 32 (8): 27–28 (in Chinese)

Jiang GH, Xu Q, Zhang Y (**2004**) Clinical study on the neuroethology of senile dementia treated by acupuncture. Journal of Clinical Acupuncture and Moxibustion 20 (4):1–3 (in Chinese)

Jiang XC, Jiang GH, Yu Y (**2008**) The effect of the acupuncture to the cerebral tau protein expression in the senile dementia rat models. Journal of Clinical Acupuncture and Moxibustion 24 (11):38–39 (in Chinese)

Jie ZH (**2002**) Using acupuncture in treating progressive bulbar palsy, a 37 cases clinical observation. Journal of Clinical Acupuncture and Moxibustion 18 (8):19 (in Chinese)

Kadir A, Nordberg A (**2010**) Target-specific PET probes for neurodegenerative disorders related to dementia. J Nucl Med 51 (9):1418–1430

Kiernan MC, Vucic S, Cheah BC, Turner MR, Eisen A, Hardiman O, Burrell JR, Zoing MC (**2011**) Amyotrophic lateral sclerosis. Lancet 377 (9769):942–955

Klafki HW, Staufenbiel M, Kornhuber J, Wiltfang J (**2006**) Therapeutic approaches to Alzheimer's disease. Brain 129 (Pt 11):2840–2855

Kukull WA, Higdon R, Bowen JD, McCormick WC, Teri L, Schellenberg GD, van Belle G, Jolley L, Larson EB (**2002**) Dementia and Alzheimer disease incidence: a prospective cohort study. Arch Neurol 59 (11):1737–1746

Laxton AW, Tang-Wai DF, McAndrews MP, Zumsteg D, Wennberg R, Keren R, Wherrett J, Naglie G, Hamani C, Smith GS, Lozano AM (**2010**) A phase I trial of deep brain stimulation of memory circuits in Alzheimer's disease. Ann Neurol 68 (4):521–534

Lee MS, Shin BC, Ernst E (**2009**) Acupuncture for Alzheimer's disease: a systematic review. Int J Clin Pract 63 (6):874–879

Lee VM, Goedert M, Trojanowski JQ (**2001**) Neurodegenerative tauopathies. Annu Rev Neurosci 24:1121–1159

Leyssen M, Ayaz D, Hebert SS, Reeve S, De Strooper B, Hassan BA (**2005**) Amyloid precursor protein promotes post-developmental neurite arborization in the Drosophila brain. EMBO J 24 (16):2944–2955

Li QJ, Ma FM (**1992**) Using scalp acupuncture for cerebellar ataxia, an 8 case report. Shanxi Journal of Traditional Chinese Medicine 8 (6):37 (in Chinese)

Li SC, Jiang CC, Chen GB (**1983**) The relationship between needling sensation and acupuncture effects, with special reference to their ascending pathway in the spinal cord. Acupunct Electrother Res 8 (2):105–110

Li SJ (**1998**) Using acupuncture and naprapathy to treat ALS. Journal of New Chinese Medicine 30 (8):24–25 (in Chinese)

Li SQ, Zhou HF (**2010**) Clinical observation of treating olivopontocerebellar atrophy (OPCA) mainly by acupuncture in one OPCA case. Journal of Liaoning University of Traditional Chinese Medicine 12 (1):89–90 (in Chinese)

Li Y, Liu T, Peng Y, Yuan C, Guo A (**2004**) Specific functions of Drosophila amyloid precursor-like protein in the development of nervous system and nonneural tissues. J Neurobiol 61 (3):343–358

Li YJ, Fan XS, Wang S, Shi X, M. (**2008**) To summarize acupuncture treating ataxia. Journal of Clinical Acupuncture and Moxibustion 24 (7):50–51 (in Chinese)

Li ZR, Mu YY, Ouyang Q **(2002)** A case-control study on the treatment of Alzheimer's disease by the Combination of Acupuncture and DangguiShaoyaoSan. Chinese Journal of Clinical Rehabilitation 6 (19):2848–2849 (in Chinese with English abstract)

Li ZT **(2005)** Treating motor neuron diseases following the principles of tonifying kidney and liver together with invigorating spleen and stomach in 27 cases. Journal of Sichuan Traditional Chinese Medicine 23 (7):37 (in Chinese)

Liang M **(1999)** Using Traditional Chinese Medicine combined with acupoints injection to treat ALS. Henan Traditional Chinese Medicine 19 (6):48 (in Chinese)

Liu G, Yuan LX **(2005)** Clinical observation on acupuncture combined with music for treatment of Alzheimer disease. Chinese Acupuncture & Moxibustion 25 (6):390–392 (in Chinese with English abstract)

Liu JL, Du YH, Li B, Xiong J, Hu YC **(2009a)** Evaluation of the studies on the acupuncture in treating senile dementia with principles of evidence-based medicine. Jilin Journal of Traditional Chinese Medicine 29 (10):855–857 (in Chinese)

Liu YF, Shi X, M., Han JX, Wang S, Li P, Liu QZ, Zhao LR, Yang ZR, Zhao JH, Li XK **(2002)** Effect of Acupuncture on Activity of Antioxydase in the of the Brain Senescence Accelerated En- cephalatrophy Model Mouse. Chinese Acupuncture & Moxibustion 22 (5):327 (in Chinese with English abstract)

Liu YX **(2003)** Studies on the mechanism of hippocampal signal conduction transmitters adjusted by electroacupuncture in senile dementia rat models. Chengdu University of Traditional Chinese Medicine, Chengdu (in Chinese with English abstract)

Liu YX, Liang FR, Yu SG, Luo S, Han T **(2004)** An experimental study of acupuncture effects in dementia senile model rats. Chinese Journal of Rehabilitation Medicine 19 (3):188 (in Chinese with English abstract)

Liu ZB, Niu WM, Yang XH, Niu XM, Wang Y **(2009b)** [Effect of "Xiusanzhen" on learning-memory ability and hippocampal ChAT and AChE activity in Alzheimer disease rats]. Zhen Ci Yan Jiu 34 (1):48–51 (in Chinese with English abstract)

Liu ZB, Niu WM, Yang XH, Wang Y, Wang WG **(2010)** Study on perfume stimulating olfaction with volatile oil of Acorus gramineus for treatment of the Alzheimer's disease rat. J Tradit Chin Med 30 (4):283–287

Lopez OL, Becker JT, Wahed AS, Saxton J, Sweet RA, Wolk DA, Klunk W, Dekosky ST **(2009)** Long-term effects of the concomitant use of memantine with cholinesterase inhibition in Alzheimer disease. J Neurol Neurosurg Psychiatry 80 (6):600–607

Lu F **(2006)** Role of Du Channel in treating senile dementia. J Tradit Chin Med 26 (2):104–105

Lun X, Yang WH **(2000)** A clinical observation of treating dementia with acupuncture located by CT lesions. Shanghai Journal of Acupuncture and Moxibustion 19 (1):20 (in Chinese)

Luo DH, Lai XS, Tang CZ **(2003)** Effects of Electropuncture Therapy on Levels of β-AP and TGF-α in Rats of AD Model. Chinese Journal of Information on Traditional Chinese Medicine 10 (10):12–14 (in Chinese with English abstract)

Luo DH, Zhou T, Huang QY **(2006)** A clinical observation for the changes of the amyloid β affected by the electroacupuncture in senile dementia patients (type of kidney weakness). Medicine world (9):64–66 (in Chinese)

Luo JH, Guo JM **(1996)** Using scalp acupuncture for cerebellar ataxia, a 30-case clinical report. Shanghai Journal of Acupuncture and Moxibustion 15 (4):25–26 (in Chinese)

Lv JX, Xu DD **(1997)** Using scalp acupuncture in treating senile dementia, a 60-case report. Xinjiang Journal of Traditional Chinese Medicine 15 (1):23–24 (in Chinese)

Ma L, Song WG, Wang CG **(1998)** Multiple system atrophy in senior people. Chinese Acupuncture & Moxibustion 18 (9):554 (in Chinese)

Ma MQ **(2003)** Treating senile demetia by acupuncture combined with Anshenxingnao soup, a 56-case report. Shaanxi Journal of Traditional Chinese Medicine 23 (8):701–702 (in Chinese)

Ma SF, Wu XE, Ding CL **(2001)** Clinical observation on treating neuronal diseases with acupoint injection. Information on Traditional Chinese Medicine 18 (2):37 (in Chinese)

Mahley RW, Huang Y **(2006)** Apolipoprotein (apo) E4 and Alzheimer's disease: unique conformational and biophysical properties of apoE4 can modulate neuropathology. Acta Neurol Scand Suppl 185:8–14

Mahley RW, Weisgraber KH, Huang Y **(2006)** Apolipoprotein E4: a causative factor and therapeutic target in neuropathology, including Alzheimer's disease. Proc Natl Acad Sci U S A 103 (15):5644–5651

Maltsev AV, Bystryak S, Galzitskaya OV **(2011)** The role of beta-amyloid peptide in neurodegenerative diseases. Ageing Res Rev

Maruyama H, Morino H, Ito H, Izumi Y, Kato H, Watanabe Y, Kinoshita Y, Kamada M, Nodera H, Suzuki H, Komure O, Matsuura S, Kobatake K, Morimoto N, Abe K, Suzuki N, Aoki M, Kawata A, Hirai T, Kato T, Ogasawara K, Hirano A, Takumi T, Kusaka H, Hagiwara K, Kaji R, Kawakami H **(2010)** Mutations of optineurin in amyotrophic lateral sclerosis. Nature 465 (7295):223–226

Masters CL, Beyreuther K **(2006)** Alzheimer's centennial legacy: prospects for rational therapeutic intervention targeting the Abeta amyloid pathway. Brain 129 (Pt 11):2823–2839

Mathis CA, Bacskai BJ, Kajdasz ST, McLellan ME, Frosch MP, Hyman BT, Holt DP, Wang Y, Huang GF, Debnath ML, Klunk WE **(2002)** A lipophilic thioflavin-T derivative for positron emission tomography (PET) imaging of amyloid in brain. Bioorg Med Chem Lett 12 (3):295–298

Matsukawa H **(2005)** Study on symptom progression in senile dementia of the Alzheimer type with TCM. Tianjin University of Traditional Chinese Medicine, Tianjin of P.R China (in Chinese with English abstract)

Matsukawa H, Liu GW, Yang YX, Yuan ZM **(2005)** Discussion on Course of Alzemeimer's Dementia with Traditional Chinese Medicine. Tianjin Journal of Traditional Chinese Medicine 22 (6):514–516 (in Chinese with English abstract)

Mega MS, Dinov ID, Porter V, Chow G, Reback E, Davoodi P, O'Connor SM, Carter MF, Amezcua H, Cummings JL **(2005)** Metabolic patterns associated with the clinical response to galantamine therapy: a fludeoxyglucose f 18 positron emission tomographic study. Arch Neurol 62 (5):721–728

Mei J, Liu CG, Sun L **(2003)** Seventy-six Patients with Senile Dementia Treated by Acupuncture. China's Naturopathy 11 (12):13–14 (in Chinese)

Menken M, Munsat TL, Toole JF **(2000)** The global burden of disease study: implications for neurology. Arch Neurol 57 (3):418–420

Moon PD, Jeong HJ, Kim SJ, An HJ, Lee HJ, Yang WM, Park SK, Hong SH, Kim HM, Um JY **(2007)** Use of electroacupuncture at ST36 to inhibit anaphylactic and inflammatory reaction in mice. Neuroimmunomodulation 14 (1):24–31

Mu YY, Li ZR **(2001)** The treatment of Alzheimer's disease by the combination of acupuncture and medicine. Shanghai Journal of Acupuncture and Moxibustion 20 (3):3–5 (in Chinese with English abstract)

Naj AC, Jun G, Beecham GW, Wang LS, Vardarajan BN, Buros J, Gallins PJ, Buxbaum JD, Jarvik GP, Crane PK, Larson EB, Bird TD, Boeve BF, Graff-Radford NR, De Jager PL, Evans D, Schneider JA, Carrasquillo MM, Ertekin-Taner N, Younkin SG, Cruchaga C, Kauwe JS, Nowotny P, Kramer P, Hardy J, Huentelman MJ, Myers AJ, Barmada MM, Demirci FY, Baldwin CT, Green RC, Rogaeva E, George-Hyslop PS, Arnold SE, Barber R, Beach T, Bigio EH, Bowen JD, Boxer A, Burke JR, Cairns NJ, Carlson CS, Carney RM, Carroll SL, Chui HC, Clark DG, Corneveaux J, Cotman CW, Cummings JL, Decarli C, Dekosky ST, Diaz-Arrastia R, Dick M, Dickson DW, Ellis WG, Faber KM, Fallon KB, Farlow MR, Ferris S, Frosch MP, Galasko DR, Ganguli M, Gearing M, Geschwind DH, Ghetti B, Gilbert JR, Gilman S, Giordani B, Glass JD, Growdon JH, Hamilton RL, Harrell LE, Head E, Honig LS, Hulette CM, Hyman BT, Jicha GA, Jin LW, Johnson N, Karlawish J, Karydas A, Kaye JA, Kim R, Koo EH, Kowall NW, Lah JJ, Levey AI, Lieberman AP, Lopez OL, Mack WJ, Marson DC, Martiniuk F, Mash DC, Masliah E, McCormick WC, McCurry SM, McDavid AN, McKee AC, Mesulam M, Miller BL, Miller CA, Miller JW, Parisi JE, Perl DP, Peskind E, Petersen RC, Poon WW, Quinn JF, Rajbhandary RA, Raskind M, Reisberg B, Ringman JM, Roberson ED, Rosenberg RN, Sano M, Schneider LS, Seeley W, Shelanski ML, Slifer MA, Smith CD, Sonnen JA, Spina S, Stern RA, Tanzi RE, Trojanowski JQ, Troncoso JC, Van Deerlin VM, Vinters HV, Vonsattel JP, Weintraub S, Welsh-Bohmer KA, Williamson J, Woltjer RL, Cantwell LB, Dombroski BA, Beekly D, Lunetta KL, Martin ER,

Kamboh MI, Saykin AJ, Reiman EM, Bennett DA, Morris JC, Montine TJ, Goate AM, Blacker D, Tsuang DW, Hakonarson H, Kukull WA, Foroud TM, Haines JL, Mayeux R, Pericak-Vance MA, Farrer LA, Schellenberg GD **(2011)** Common variants at MS4A4/MS4A6E, CD2AP, CD33 and EPHA1 are associated with late-onset Alzheimer's disease. Nat Genet

Namba H, Fukushi K, Nagatsuka S, Iyo M, Shinotoh H, Tanada S, Irie T **(2002)** Positron emission tomography: quantitative measurement of brain acetylcholinesterase activity using radiolabeled substrates. Methods 27 (3):242–250

Nimmrich V, Ebert U **(2009)** Is Alzheimer's disease a result of presynaptic failure? Synaptic dysfunctions induced by oligomeric beta-amyloid. Rev Neurosci 20 (1):1–12

Nishimura I, Yang Y, Lu B **(2004)** PAR-1 kinase plays an initiator role in a temporally ordered phosphorylation process that confers tau toxicity in Drosophila. Cell 116 (5):671–682

Niu HY, Duan HT **(2005)** Treating multiple system atrophy by acupuncture combined with Chinese Medicine. Chinese Acupuncture & Moxibustion 25 (6):399–400 (in Chinese)

Ouyang Q, Li ZR, Mu YY, Long HW, Zhao D, Wu Q **(2001)** Clinical observation of electric acupuncture combined with fentazin in treating mental symptom of Alzheimer's disease. Journal of Clinical Acupuncture and Moxibustion 17 (1):25–26 (in Chinese)

Ouyang Q, Li ZR, Mu YY, Ma P, Shen MH **(2002)** Relationship Between Promoting Effect of Acupuncture on Learning Memory of Rats and Its Anti-oxidation Effect. Journal of Nanjing University of Traditional Chinese Medicine 18 (2):110–112 (in Chinese with English abstract)

Ouyang Q, Li ZR, Mu YY, Zhao D, Long HW, Wu Q **(1999)** A case-control study for the clinical efficacy of acupuncture for Alzheimer's disease. Chinese Acupuncture & Moxibustion 19 (7):399–401 (in Chinese)

Overshott R, Burns A **(2005)** Treatment of dementia. J Neurol Neurosurg Psychiatry 76 Suppl 5:v53–59

Pasinelli P, Brown RH **(2006)** Molecular biology of amyotrophic lateral sclerosis: insights from genetics. Nat Rev Neurosci 7 (9):710–723

Pei L **(2006)** Acupuncture at Fengchi (GB-20) to treat dysphagia in a case of progressive bulbar palsy. Shanghai Journal of Acupuncture and Moxibustion 25 (7):36 (in Chinese)

Peng GP, He GX, Wang ER **(2005)** Using acupuncture in treating progressive bulbar palsy, a 30 case clinical study. Shaanxi Journal of Traditional Chinese Medicine 26 (6):566–567 (in Chinese)

Peng XW, Dong KL **(2009)** [Clinical observation on acupuncture combined with Yizhi Jiannao granules for treatment of Alzheimer's disease]. Zhongguo Zhen Jiu 29 (4):269–271 (in Chinese with English abstract)

Ping Y, Wang DH, Yuan XS **(2005)** Treating 98 cases of syringomyelia with Yisuiling capsule plus acupuncture. Chinese Journal of the Practical Chinese With Modern Medicine 18 (16):725–726 (in Chinese)

Reisberg B, Doody R, Stoffler A, Schmitt F, Ferris S, Mobius HJ **(2003)** Memantine in moderate-to-severe Alzheimer's disease. N Engl J Med 348 (14):1333–1341

Rosen DR **(1993)** Mutations in Cu/Zn superoxide dismutase gene are associated with familial amyotrophic lateral sclerosis. Nature 364 (6435):362

Sakakibara R, Murakami E, Katagiri A, Hayakawa S, Uchiyama T, Yamamoto T, Hattori T **(2007)** Moxibustion, an alternative therapy, ameliorated disturbed circadian rhythm of plasma arginine vasopressin and urine output in multiple system atrophy. Intern Med 46 (13):1015–1018

Sawada T, Sawada C, Fukuda F, Yano T, Tanzawa S **(2001)** Effect of TEAS (transcutaneous Electrical Acupuncture Point Stimulation) on Intellect and Daily Activity of the Aged. Journal of the Japan Society of Acupuncture and Moxibustion 51 (1):561–564 (in Japanese with English abstract)

Shang Y, Ding N **(2003)** A 24-case clinical observation by TCD to evaluate acupuncture for senile dementia. Chinese Journal of Traditional Medical Science and Technology 10 (4):229 (in Chinese)

Shen WD **(1996)** A preliminary study on treating senile dementia by acupuncture. Shanghai Journal of Acupuncture and Moxibustion 15 (5):5–6 (in Chinese)

Sheng CR **(1964)** Using acupuncture to treat Friedreich's ataxia a one case clinical observation. Guangdong Medical Journal 4 (4):7 (in Chinese)

Shi Q, Gibson GE **(2007)** Oxidative stress and transcriptional regulation in Alzheimer disease. Alzheimer Dis Assoc Disord 21 (4):276–291

Shi XM, Han JX, Li P, Li L, Wang S, Zhao LR, Liu QZ, Zhao JH, Miao Y **(1996)** The effects of Xingnaokaiqiao acupuncture in acute dementia mouse models. Chinese Journal of Integrated Traditional and Western Medicine (S1):61–64 (in Chinese)

Shoghi-Jadid K, Small GW, Agdeppa ED, Kepe V, Ercoli LM, Siddarth P, Read S, Satyamurthy N, Petric A, Huang SC, Barrio JR **(2002)** Localization of neurofibrillary tangles and beta-amyloid plaques in the brains of living patients with Alzheimer disease. Am J Geriatr Psychiatry 10 (1):24–35

Simmons LK, May PC, Tomaselli KJ, Rydel RE, Fuson KS, Brigham EF, Wright S, Lieberburg I, Becker GW, Brems DN, et al. **(1994)** Secondary structure of amyloid beta peptide correlates with neurotoxic activity in vitro. Mol Pharmacol 45 (3):373–379

Song CX **(2008)** A preliminary study for multiple system atrophy in the viewpoint of TCM. Beijing University of Chinese Medicine, Beijing (in Chinese with English abstract)

Song CX, Guo GH, Wang GX, Gu WH, Zou YH **(2008)** Thoughts about Syndrome Study of Muhiple System Sclerosis. Chinese Journal of Basic Medicine in Traditional Chinese Medicine 14 (4):278–279 (in Chinese)

Stefanova E, Wall A, Almkvist O, Nilsson A, Forsberg A, Langstrom B, Nordberg A **(2006)** Longitudinal PET evaluation of cerebral glucose metabolism in rivastigmine treated patients with mild Alzheimer's disease. J Neural Transm 113 (2):205–218

Sun H, Cui LY **(2005)** Temporary state of acupuncture treatment of motor neuron disease. Chinese Journal of Clinical Rehabilitation 9 (21):174–176 (in Chinese)

Sun SM, Yu SG, Zeng DB, Song KY **(2002)** A Study on metabolism of free radicals in senile rats by management with electrical acupuncture on Yongquan point. Chinese Journal of Natural Medicine 4 (2):80–82 (in Chinese with English abstract)

Sun Y **(1994)** Using scalp acupuncture combined with moxibustion to treat Friedreich's ataxia, a 10-case clinical report. Journal of Guiyang College of Traditional Chinese Medicine 16 (3):60 (in Chinese)

Tang CZ, Lai XS, Lin ZH, Yang JJ, Lin JQ, Xian ZH, Li HY **(2005a)** Effects of electro-acupuncture on IL-1, IL-6 in the brain tissue of Rat with Alzheimer's disease. Chinese Journal of Basic Medicine in Traditional Chinese Medicine 11 (7):532–533 (in Chinese)

Tang CZ, Lai XS, Yang JJ, Lin JQ, Lai XL, Wang L, Lin ZH **(2005b)** Exprimental study on the effects of acupuncture to cerebral SOD,MDA and GSH-Px in senile dementia rat models. Shaanxi Journal of Traditional Chinese Medicine 26 (2):180–181 (in Chinese)

Tang Y, Yu SG, Chen J, Zhang H **(2001)** The Effect of Acupuncture on Gnosia and Acetylcholinesterase in Senile Dementia. Shanghai Journal of Acupuncture and Moxibustion 30 (3):6–7 (in Chinese with English abstract)

Tang Y, Yu SG, Luo S, Han T, Yin HY **(2006)** [Neuroprotective effect of electroacupuncture therapy on damaged cholinergic neurons in rats with Alzheimer's disease]. Zhong Xi Yi Jie He Xue Bao 4 (4):374–377 (in Chinese with English abstract)

Tariot PN, Farlow MR, Grossberg GT, Graham SM, McDonald S, Gergel I **(2004)** Memantine treatment in patients with moderate to severe Alzheimer disease already receiving donepezil: a randomized controlled trial. JAMA 291 (3):317–324

Tesseur I, Van Dorpe J, Bruynseels K, Bronfman F, Sciot R, Van Lommel A, Van Leuven F **(2000a)** Prominent axonopathy and disruption of axonal transport in transgenic mice expressing human apolipoprotein E4 in neurons of brain and spinal cord. Am J Pathol 157 (5):1495–1510

Tesseur I, Van Dorpe J, Spittaels K, Van den Haute C, Moechars D, Van Leuven F **(2000b)** Expression of human apolipoprotein E4 in neurons causes hyperphosphorylation of protein tau in the brains of transgenic mice. Am J Pathol 156 (3):951–964

Tian XS, Zhou F, Yang R, Xia Y, Wu GC, Guo JC **(2008)** Electro-acupuncture protects the brain against ischemic injury via up-regulation of delta-opioid receptors. J Chin Integr Med 6: 632–638 (in Chinese with English abstract)

Uversky VN **(2009)** Intrinsic disorder in proteins associated with neurodegenerative diseases. Front Biosci 14:5188–5238

Vucic S, Kiernan MC **(2009)** Pathophysiology of neurodegeneration in familial amyotrophic lateral sclerosis. Curr Mol Med 9 (3):255–272

Walsh DM, Klyubin I, Fadeeva JV, Rowan MJ, Selkoe DJ **(2002)** Amyloid-beta oligomers: their production, toxicity and therapeutic inhibition. Biochem Soc Trans 30 (4):552–557

Wang DH **(2007)** Treatment of 60 cases of syringomyelia with Guiluyisui capsule and acupuncture. Journal of Traditional Chinese Medicine 48 (6):530–531 (in Chinese)

Wang DH, Chen JL **(2006)** The mechanisms of treating syringomyelia with Traditional Chinese Medicine. Liaoning Journal of Traditional Chinese Medicine 33 (1):25–26 (in Chinese)

Wang DH, Chen JL, Huang T **(2004)** Treatment of 445 cases of syringomyelia with yisuiling capsule and acupuncture. New Journal of Traditional Chinese Medicine 36 (10):58–59 (in Chinese)

Wang HY **(2010)** A overview of treating multiple system atrophy by TCM. Medical Information 23 (9):165–166 (in Chinese)

Wang MK, Hao JD, Wang YP **(2008)** Treating 40 cases of syringomyelia with Sancaifengsuidan combined with acupuncture. Medical Journal of the Chinese People's Armed Police Forces 19 (5):476–477 (in Chinese)

Wang SJ, Kang SB, Li AY, Cao G **(2005)** The influence of acupuncture on intracerebral free radical oxidation injury in simulate AD rats. Shanghai Journal of Acupuncture and Moxibustion 24 (4):35–37 (in Chinese with English abstract)

Wang XH, Dong CC, Han JX **(2003)** Influence of Acupuncture on Catecholamine Neurotransmitter in Brain Tissues of SAMP8. Shanghai Journal of Acupuncture and Moxibustion 22 (1):21–23 (in Chinese with English abstract)

Wang XY, Zhou LS **(2008)** Effects of electroacupuncture on bFGF of AD model rat's hippocampal neural stem cells and study memory ability. Journal of Hubei University of Chinese Medicine 10 (1):12–14 (in Chinese with English abstract)

Wang ZL **(1997)** Following the principle of restoring the consciousness and opening the intelligence combined with body acupuncture in treating senile dementia. Journal of Clinical Acupuncture and Moxibustion 13 (9):16–17 (in Chinese)

Wang ZX **(2005)** Treating 12 motor neuron diseases cases using combined Chinese medicine and acupuncture. Chinese Acupuncture & Moxibustion 25 (8):553 (in Chinese)

Wattamwar PR, Mathuranath PS **(2010)** An overview of biomarkers in Alzheimer's disease. Ann Indian Acad Neurol 13 (Suppl 2):S116–123

Weiner MW, Aisen PS, Jack CR, Jr., Jagust WJ, Trojanowski JQ, Shaw L, Saykin AJ, Morris JC, Cairns N, Beckett LA, Toga A, Green R, Walter S, Soares H, Snyder P, Siemers E, Potter W, Cole PE, Schmidt M **(2010)** The Alzheimer's disease neuroimaging initiative: progress report and future plans. Alzheimers Dement 6 (3):202–211 e207

Weishaupt JH, Bartels C, Polking E, Dietrich J, Rohde G, Poeggeler B, Mertens N, Sperling S, Bohn M, Huther G, Schneider A, Bach A, Siren AL, Hardeland R, Bahr M, Nave KA, Ehrenreich H **(2006)** Reduced oxidative damage in ALS by high-dose enteral melatonin treatment. J Pineal Res 41 (4):313–323

Wen GQ, He XZ, Lu Y, Xia Y **(2010a)** Effect of acupuncture on neurotransmitters/modulators. In: Acupuncture Therapy for Neurological Diseases: A Neurobiological View. Springer-Tsinghua Press, Heidelberg, London, New York, Beijing, pp 120–142

Wen GQ, Yang YL, Xia Y **(2010b)** Acupuncture-induced activation of endogenous opioid system. In: Acupuncture Therapy for Neurological Diseases: A Neurobiological View. Springer-Tsinghua Press, Heidelberg, London, New York, Beijing, pp 104–119

Wentzell J, Kretzschmar D **(2010)** Alzheimer's Disease and tauopathy studies in flies and worms. Neurobiol Dis 40 (1):21–28

Wilms H, Zecca L, Rosenstiel P, Sievers J, Deuschl G, Lucius R **(2007)** Inflammation in Parkinson's diseases and other neurodegenerative diseases: cause and therapeutic implications. Curr Pharm Des 13 (18):1925–1928

Wittmann CW, Wszolek MF, Shulman JM, Salvaterra PM, Lewis J, Hutton M, Feany MB **(2001)** Tauopathy in Drosophila: neurodegeneration without neurofibrillary tangles. Science 293 (5530):711–714

Wojtera M, Sikorska B, Sobow T, Liberski PP **(2005)** Microglial cells in neurodegenerative disorders. Folia Neuropathol 43 (4):311–321

Wu GF, Ren F, Zhang S **(2000)** Using acupuncture to treat ALS and PMA 15 cases. Chinese Acupuncture & Moxibustion 20 (7):400 (in Chinese)

Wu YX, Guo RY **(1986)** Treatment of syringomyelia by acupuncture and Chinese medicinal herbs. J Tradit Chin Med 6 (1):13–16

Xia CY **(1991)** The clinical experience of Prof Yao Peifa in treating senile dementia. Journal of New Chinese Medicine 23 (1):55–56 (in Chinese)

Xia QG, Zhang Y, Ma GZ **(2010a)** Using Frenkel training combined with acupuncture and naprapathy to treat cerebellar ataxia, a 12-case clinical report. Acta Chinese Medicine and Pharmacology 38 (1):89–90 (in Chinese)

Xia Y, Cao XD, Wu GC, Cheng JS **(2010b)** Acupuncture Therapy for Neurological Diseases: A Neurobiological View (Research Monograph). Springer-Tsinghua Press, Heidelberg, London, New York, Beijing,

Xiao JF **(2004)** Observation on the therapeutic effect of acupuncture on encephalatrophy. Chinese Acupuncture & Moxibustion 8 (24):541–543 (in Chinese with English abstract)

Xie YY, Zhang JR, Wang JH, Zhang FL, Tang XJ **(1995)** The diagnosis and treatment of motor neuron disease by TCM. Chinese Journal of Information on Traditional Chinese Medicine 2 (9):25 (in Chinese)

Xu J, Xu SW **(1995)** Using acupuncture combined with Chinese medicine to treat senile dementia in one case. Jiangxi Journal of Traditional Chinese Medicine 2 (3):51 (in Chinese)

Xue WG, Zhang Z, Bai LM, Xu H, Wu HX **(2009)** [Effect of electroacupuncture on the behavior and the expression of amyloid beta-protein, amyloid precursor protein and ChAT in APP 695V 717 I transgenic mice]. Zhen Ci Yan Jiu 34 (3):152–158 (in Chinese with English abstract)

Yan DX, Lv LY **(1991)** The relationship of senile dementia and blood stasis. Liaoning Journal of Traditional Chinese Medicine (8):37 (in Chinese)

Yan SY, Yang SY, Huang JQ **(2002)** The experience of Dr Qiu Zuping in treating senile dementia. Liaoning Journal of Traditional Chinese Medicine (3):131 (in Chinese)

Yang DR, Peng L, Mu JP, Li SZ, Liu MJ, Hu XY **(2008)** Effect of Acupuncture and Acupoint Injection on Dementia in Elder. Chinese Journal of Rehabilitation Theory and Practice 14 (1):81 (in Chinese with English abstract)

Yang EJ, Jiang JH, Lee SM, Hwang HS, Lee MS, Choi SM **(2010)** Electroacupuncture reduces neuroinflammatory responses in symptomatic amyotrophic lateral sclerosis model. J Neuroimmunol 223 (1–2):84–91

Yang GW, Yan BC, Wu SZ **(2005)** Using acupuncture combined with Chinese Medicine to treat cerebellar atrophy, a 32-case clinical observation. Shaanxi Journal of Traditional Chinese Medicine 26 (4):359–360 (in Chinese)

Yang SC **(1995)** Treating denile dementia by a water acupuncture a 26-case report. Zhejiang Journal of Traditional Chinese Medicine 30 (8):357–357 (in Chinese)

Yang XT **(1996a)** Acupuncture by the principle of restoring the consciousness and opening the intelligence in treating senile dementia. Chinese Acupuncture & Moxibustion 16 (11):3 (in Chinese)

Yang ZR, Lai LP, Li P, Wang S, Liu QZ, Zhao LR **(1999)** The effects of acupuncture on the cerebral Ach concentration and AchE activity in acute aging dementia mouse model. Tianjin Journal of Traditional Chinese Medicine 16 (3):20–22 (in Chinese)

Yang ZZ **(1996b)** Using moving cupping for back-shu points in treating senile dementia, a 18-case report. New Journal of Traditional Chinese Medicine 28 (12):31 (in Chinese)

Ye S, Huang Y, Mullendorff K, Dong L, Giedt G, Meng EC, Cohen FE, Kuntz ID, Weisgraber KH, Mahley RW **(2005)** Apolipoprotein (apo) E4 enhances amyloid beta peptide production in cultured neuronal cells: apoE structure as a potential therapeutic target. Proc Natl Acad Sci U S A 102 (51):18700–18705

Yi NK **(1996)** Treating senile demetia by naprapathy. Fujian Journal of Traditional Chinese Medicine 27 (5):32–32 (in Chinese)

Yu SG, Liu YX, tang Y, Liang FR, Zhao JL, Zhou QZ, Han T, Luo S **(2004)** Research on the NO-cGMP signal transduction pathway mechanism of improving learning & memory ability by electro-acupuncture in senile dementia rats. Chinese Journal of Gerontology 24 (7):626–628 (in Chinese with English abstract)

Yuan XS, Chen JL, Wang DH, Li HX **(2005a)** Treating progressive bulbar palsy by acupuncture combined with Jiweiling solution intravenous injection, a 30-case clinical report. Chinese Acupuncture & Moxibustion 25 (9):622 (in Chinese)

Yuan XS, Wang DH, Chen JL **(2005b)** Treatment of 111 cases of syringomyelia with yisuiling capsule and acupuncture. Shandong Journal of Traditional Chinese Medicine 24 (5):283–284 (in Chinese)

Zeng F, Zhao JL, Zhou QZ, Tang Y, Yin HY, Yu SG **(2006)** Influence of electro-acupuncture therapy on the mitochondrial enzymatic activities of hippocampal neuron of animal model of Alzheimers' disease. Chinese Journal of Gerontology 26 (1):68–71 (in Chinese with English abstract)

Zhang CB, Li ZR, Shen MH **(2003)** Mechanisms of treatment combined acupuncture and Chinese medicine improves the memory and learning abilities in dementia rat models. Jiangsu Journal of Traditional Chinese Medicine 24 (2):51–52 (in Chinese)

Zhang D, Zhang JY, Liu D **(2007)** The cognitive function and cerebral somatostatin level affected by performing electroacupuncture at Baihui (DU-20) and Fengfu (DU-16) in senile dementia rat models. Information on Traditional Chinese Medicine 24 (1):52–53 (in Chinese)

Zhang HF **(2009)** A review of the pathogenesis mechanisms and differential symptoms treatments by Traditional Chinese Medicine in the senior dementia patients. Chinese Medicine Modern Distance Education of China 7 (4):6–7 (in Chinese)

Zhao JH, Wang X, Hu JY, Cheng N **(2010)** Progress of using TCM to treat hereditary ataxia. Clinical Journal of Traditional Chinese Medicine 22 (8):70–73 (in Chinese)

Zhao LG, Ma L, Cheng WP, Wu Q, Zhen ZY, Mei CJ **(2008)** The effect of acupuncture on Baihui and Dazhui on cognition behavior and AchE level in brain of the aged dementia rats. Chinese Journal of Gerontology 28 (6):549–550 (in Chinese)

Zhao LG, Ma L, Zheng ZY, Zhang J **(2005)** Effect of acupuncture of "Baihui"(GV 20) and "Sishencong"(EX-HN 1) on memory and cerebral SOD activity in Alzheimer's disease rats. Acupuncture Research 30 (1):26–28 (in Chinese with English abstract)

Zhao XM **(1997)** Using acupuncture to treat bulbar palsy, a 30 case clinical observation. Shanghai Journal of Acupuncture and Moxibustion 16 (1):14 (in Chinese)

Zheng L **(2004)** Using acupuncture to treat motor neuron diseases, a 20 case report. Shanghai Journal of Acupuncture and Moxibustion 23 (11):28 (in Chinese)

Zhou HH, Han YZ **(2003)** Treatment of Shy - Drager syndrome by Shengmai Wenya Tang combined with warm needling: a clinical observation of 18 cases. New Journal of Traditional Chinese Medicine 35 (3):19–20 (in Chinese)

Zhou LS, Wang XY **(2008)** Effects of electroacupuncture on AD model rat hippocampus neural stem cells nestin and impact on learning and memory capacity. Journal of Jianghan University 36 (2):61–64 (in Chinese with English abstract)

Zhou P, Ma XM **(2009)** A review of progress of using acupuncture to treat motor neuron diseases. Journal of Yunnan College of Traditional Chinese Medicine 32 (2):68–70 (in Chinese)

Zhou Y, Jin J **(2008)** Effect of acupuncture given at the HT 7, ST 36, ST 40 and KI 3 acupoints on various parts of the brains of Alzheimer' s disease patients. Acupunct Electrother Res 33 (1–2):9–17

Zhou YL, Hang HY **(2006)** The progress of clinical research of the acupucture in treating Alzheimer's disease. Journal of Clinical Acupuncture and Moxibustion 22 (11):50–53 (in Chinese)

Zhu SX, Sun GJ **(2009)** [Effects of electroacupuncture on learning and memory ability and glial cells of the hippocampus in the rat of Alzheimer disease]. Zhongguo Zhen Jiu 29 (2):133–136 (in Chinese with English abstract)

Zhu WM, Hu HY **(2007)** A survey of TCM treatment for Alzhermer's disease. Journal of Traditional Chinese Medicine 27 (3):226–232

Chapter 9
Acupuncture Therapy for Vascular Cognitive Impairment

Guoqiang Wen, Tao Liu, Shixiong Huang, Qiao Fu, Tao Chen, Chanji Wu, and Zhiyan Sui

Abstract Acupuncture has been widely practiced in treating vascular cognitive impairment (VCI) in China. A great body of evidence has proved that acupuncture is a promising therapy to coping with VCI: it is effective and safe, and it is simple to operate, thus it is cost-effective. The available experiments have explored several aspects on treating VCI with acupuncture from morphology and molecular biology to electrophysiology in the specific brain areas and structures. The data indicate a series of mechanism for acupuncture to exert its effect, such as promoting the local blood flow and metabolism, alleviating the injury caused by free radicals, and modulating the neurotransmitter activities. On the other side, even though the therapeutic phenomenon has been observed and being applied clinically, the therapy method still needs further validation and verification through large-sample, multicenter, double-blind, and random controlled study. Further molecular and integrated neurobiological research is also warranted to better understand the mechanism. This chapter provides an introduction on the clinical manifestation and characteristics of VCI, the evolution of the definition and diagnosis for VCI, and a systematic literature review of the basic research. Based on it, the application and operation of acupuncture for the clinical treatment of VCI, as well as the possible neurobiological mechanisms, are discussed in details.

Keywords Vascular cognitive impairment • Acupuncture • Stroke

---

G. Wen (✉) • T. Liu • S. Huang • T. Chen • C. Wu • Z. Sui
Department of Neurology, Hainan Provincial
People's Hospital, Hainan 570311, China
e-mail: wenguoqiang901@hotmail.com; ltao829@163.com;
neurology0898@163.com; ctxwyc@163.com; meteor911@126.com;
yanyandedong@hotmail.com

Q. Fu
Department of Rehabilitation Medicine, Hainan Provincial
People's Hospital, Hainan 570311, China
e-mail: hkfuqiao@hotmail.com

Y. Xia et al. (eds.), *Current Research in Acupuncture*,
DOI 10.1007/978-1-4614-3357-6_9,

9.1 Introduction

Cognitive impairment is one of the major diseases that affect the health and quality of life for elderly people and has aroused mounting research concern due to its progressively increasing morbidity. Vascular dementia (VD) is a dementia syndrome resulting from all kinds of cerebrovascular diseases, but it is generally acknowledged that inconspicuous cerebrovascular diseases such as subcortical white matter ischemia, cerebral small-vessel disease (SVD), and lacunars infarct, as well as risk factors for cerebrovascular diseases like hypertension and diabetes, may all induce cognitive impairment and dementia. However, the syndrome often has missed the crucial period for early intervention when it finally meets the diagnosis. In recent years, vascular cognitive impairment (VCI), a new terminology, has been put forward instead of VD. It is believed that intensive study of VCI can help people better understand cognitive impairment and dementia which originated from vascular factors and diseases, and it would be conducive for early diagnosis and treatment.

VCI is categorized in dementia just like VD in terms of the clinical features, and there has not yet been an equivalent expression in the literature of the Traditional Chinese Medicine for it (Ma and Zhou 2003). In 2000, the Gerontology Committee, the Society of Traditional Chinese Medication, developed the standard for VD syndromes as seven criteria including kidney-essence deficiency, phlegm turbidity block, static blood blocking collaterals, upper hyperactivity of liver yang, heat pathogen and internal disturbance, *Fuzhi Zhuoliu* (fu organs stagnation and wetness-evil stasis), and blood deficiency that usually show overlapping occurrence in clinical situation. To the issue of treatment, in complementary with the western medicine, other alternative therapies of the Traditional Chinese Medicine such as acupuncture have been more popularly accepted. Acupuncture therapy for VCI works on benefiting Qi for activating blood circulation, rejuvenating essence and source that are critical for health. The main points for acupuncture vary from scalp to eyes over the body. Though such therapy for dementia has boasted for a long history in practice with positive effects yet little adverse reaction, its clinical operation and study have been far from standardization.

9.2 Vascular Cognitive Impairment

9.2.1 VCI

For quite a long time, the concept and definition of VCI have been uncertain and debated. It evolved through a series of concepts including cerebral arteriosclerosis, chronic cerebral ischemia, multi-infarct dementia (MID), VD, and finally VCI. For deeper research and accurate description on VCI, the US National Institute for Neurological Disorders and Stroke (NINDS) and the Canadian Stroke Network (CSN) organized a collaboration study among the physicians, researchers, and specialists in the fields of clinical diagnosis, epidemiology, neuropsychology, neuroimaging, animal experiment model, biological markers, genetics, and clinical trial methodology. The study reached a consensus on the definition of VCI: it was a kind of induced cognitive impairment by vascular or vascular-related factors; it might occur independently or affiliated to Alzheimer's disease (AD); the severity of the cognitive impairment varied on different levels and types such as vascular cognitive impairment with no dementia (VCIND), VD, and mixed dementia (MD) and so on (Hachinski et al. 2006).

9.2.2 *Epidemiologic Studies of VCI*

A survey conducted by the Canadian study of Health and Aging (CSAH) with random sampling of 10,263 community residents and inpatients has showed a nearly 5 % morbidity of VCI among elderly people at the age of 65 or above. Meanwhile a positive correlation was found between age and morbidity: the highest ratio of VCIND occurred between ages 65 and 84, and the morbidity of mixed AD and VD increased among those who were above the age of 85. The 5-year follow-up found out that 46 % of VCI patients might develop dementia and 52 % died. A higher hospitalization rate was detected in VCI patients than that in noncognitive impairment ones but lower than that in AD patients, while VCI with impaired cognition had a higher fatality rate than that in noncognitive impairment patients but equal to it in AD patients. No significant differences were reported in the subtypes among VCIND and mixed AD and VD patients in terms of hospitalization rate and fatality rate after 5 years. The follow-up research also showed that some VCIND might alleviate automatically, hinting that VCIND did not always result in VD (Rockwood 2002; Wentzel et al. 2001). Longitudinal studies by Finnish researchers on ischemic stroke patients indicated a 16.7 % drop in cognitive level 3 months after stroke (Pohjasvaara et al. 1997) and that risk factors for cognitive impairment included age, education background, diabetes, movement disorder, and subcortical multiple infarction (Szatmari et al. 1999). Tatemichi et al. (1994) reported a 35.2 % cognitive impairment incidence after stroke yet only 3.8 % in non-stroke control group. Roman (2002) pointed out that most VCI cases were caused by VD and ischemic heart disease (IHD), and 26 % of VCI patients would suffer from cognitive impairment of varying degrees that resulted from IHD-induced congestive heart failure (CHF). Besides, 80–90 % of the patients who had gone through coronary artery bypass grafting (CABG) were reported of a decrease in cognitive level as well as a 42 % repeated incidence after 5 years, which might be attributed to cerebral perfusion.

An epidemiologic study of random sampling in Singapore showed a 2–4 % overall incidence of dementia, 40–50 % of which was vascular dementia (Kua and Ko 1995). In 2002, a follow-up study of 252 transient ischemic attack (TIA) and nondisabled ischemic stroke patients in Singapore reported an intial ratio of 56 % with no cognitive impairment, 40 % VCIND, and 4 % dementia, whereas 31 % of VCIND patients regained cognitive impairment symtoms, 10 % with no cognitive impairment patients developed into VCIND, and 1 % of VCIND patients grew into dementia after 1 year (Tham et al. 2002). The study also disclosed a close correlation between cerebrovascular disease and cognitive impairment with a lower incidence of early dementia than VCIND, although not all VCIND cases would develop into VD.

9.2.3 Clinical Manifestations of VCI

VCI is a syndrome with various forms of manifestation featuring from mild cognitive impairment to dementia. Mild VCI refers to impaired attention and executive function but relatively normal memory. It is often accompanied possibly by action and information processing retardation or depression, emotional instability, apathy, etc. Moderate VCI is characterized by one or two impaired functions such as language, memory, vision, spatial skill, emotion, personality, and cognition (summarization, calculation, judgment, etc.), but the dementia symptoms have not yet reached the prevailing diagnostic criteria, and the focal signs of nervous system may be minimal or instable. Severe VCI patients display obvious focus and focal signs of nervous system (Jia 2005). Sachdev et al. (2004) pointed out that the intellectual impetus speed and frontal lobe executive function of VD patients were the most affected; Rockwood (2002) found that the most frequent VCI behavorial symptoms were lowered motivation (61 %) and depression (3 %). Only evident symptoms could be noticed by family members.

9.2.4 Neuroimaging Manifestations of VCI

Neuroimaging consists of structural imaging and functional imaging. Structural imaging technology can display clearly the relevance between cerebral structural change and cognitive impairment. VCI patients are characterized by cerebral atrophy with comparatively mild cortical atrophy, expansion in cerebral ventricle asymmetry, and single or multiple ischemic lesions of inequal sizes which are most likely located in parietal cortex, angular gyrus, occipital lobe, and basal ganglia area. MRI detects long T2 signals of basal ganglia, thalamus, and brain stem white matter demyelination; meanwhile, low-density shadows of white matter can often be found associated with focal infarction. Our previous researches reported that magnetic resonance DTI could be used as a basis for early diagnosis and treatment of cognitive impairment patients with acute cerebral infarction, that is, a correlation could be found between such impairment and the focal location. Cerebral hemisphere injuries in left side, forepart, and cortex were more likely to lead to cognitive impairment, whereas joint application of ADC volume and FA volume could contribute to implementing more precise clinical stages for cerebral infarction, estimating its occurrence time, and deciding on its prognosis (Liao et al. 2007; Wen et al. 2006, 2007). Functional imaging examination works better on early pathological observation of dementia than structural imaging because it can detect any functional abnormality before structural changes. Frequently used functional imaging examination methods include single-photon emission computed tomography (SPECT) and positron emission tomography (PET). The former is mainly used for evaluating cerebral blood perfusion, while the latter is for observing oxygen/glucose metabolism of brain tissue and cerebral blood perfusion. Moreover these two methods can distinguish the ischemic zone from infarction zone aside from positioning. Therefore, functional imaging can help analyze the relevance between ischemic focus and cognitive impairment. VCI focus locations are normally unfixed; the forms are varied. This shows a close correlation between the changes of functional imaging and the pathological basis for cerebrovascular diseases. PET examination results often suggest the drop in single or multiple focal metabolism and decreased blood perfusion and show asymmetry mostly on left side as well as encephalopathy in the whole brain (Meyer et al. 2005).

9.2.5 Pathogenesis of VCI

VCI is prone to occur through multiple channels on the ground that VCI varies in blood vessels and tissues; therefore the neuromechanisms diversify.

9.2.5.1 Vascular Injury

Ischemic and hemorrhagic cerebrovascular diseases can all give rise to VCI. In terms of the hypothetical neuromechanism, VCI comprises three ischemic subtypes (Leblanc et al. 2006). These are the following: (1) relevant single or multiple infarctions or limbic cortex infarctions, (2) minor subcortex infarctions damaging cortex–subcortex circuit loop, and (3) white matter impairment. Although these changes may vary in manifestations, they are often coccurrent as the vascular risk factors are congenerous. The macroangiopathy is very likely to result in dementia due to direct damages. Cerebral hemorrhage generally refers to multiple hemorrhage caused by extensive brain injuries and consequent dementia. Aneurysm rupture of anterior communicating artery may lead to chronic brain syndrome characterized by frontal lobe syndrome. Studies on SVD found relevance between SVD and lacunars infarct, white matter impairment, and cortical cerebral atrophy and the critical role that the volume, location, and quantity of vascular lesion play in the pathogenic process of SVD (Bowler 2004). Major SVDs include arteriolosclerosis (hypertrophy and failure of small arterial wall, also referring to fiber or fat hyaline degeneration), cerebral amyloid angiopathy (vessel wall impairment caused by β-amyloid protein deposition), and CADASIL (cerebral autosomal dominant arteriopathy with subcortical infarction and leukoencephalopathy). Vessel wall impairment may result in vascular occlusion and minor lacunar infarct, while vessel rupture may lead to cerebral hemorrhage and diffuse white matter changes. Craniocerebral MRI of elderly people often displays SVD-related white matter hyperintensity (WMH) and lacunar infarct. Van der Flier et al. (2005) studied on the effects of WMH and lacunar infarct on the overall aged cognitive function and determined WMH and lacunar infarct as independent related factors to the decline in the overall aged cognitive function.

9.2.5.2 CBF and Metabolism Disorder

Early application of PET in cerebral blood flow (CBF) studies failed to reflect long-term cerebral ischemia. A mild fall in CBF was confirmed in VD, but the patients' oxygen uptake rate proved normal. CBF should be matched with reduced metabolism needs under such circumstances. The use of xenon-enhanced CT in detection for deep white matter area of AD and VD patients demonstrated a drop in both CBFs, with virtually equal degrees, revealing a mechanism of long-term cerebral ischemia-induced white matter ischemia. However, PET results failed to match this viewpoint. The fact that local CBF changes did not intensify as in normal circumstances after the use of vasodilator acetazolamide indicated a comparatively low vascular reserve capacity. Meanwhile, for those carotid artery occlusion and ipsilateral TIA patients, their local nervous function is not damaged or is recovering from impairment, but the cognitive function deficiency is permanent (Bakker et al. 2000). Besides, thromboembolism and chronic cerebral hypoperfusion were believed to have played a significant role in VCI (Klijn et al. 1997). It was also believed that cerebral metabolism disorder was related to vascular cognitive function. Bakker et al. (2003) used ^{1}H-MR spectral analysis to detect the ipsilateral semiovale center of carotid artery occlusion patients and discovered a decrease in both the mean value of CO_2 reactivity and the average specific value of *N*-acetylaspartic acid and creatine. Moreover, the result that one-third of the patients were detected of lactate in non-infarction zones confirmed the correlation between lactate and cognitive impairment. In addition, PET tests in cerebral quiescent condition found glycometabolism in local areas as well as drop in CBF, which explained the fundamental pathophysiological mechanism for cognitive function decline. Functional magnetic resonance imaging (fMRI) can be used to detect the changes in regional blood flow of brain tissue, provide an indirect assessment for neuronic function, and search for potential causes.

9.2.5.3 Biochemical Factors

Biochemical studies of VCI and its pathogenesis have long been a major concern for researchers. Rise of Tau protein in cerebrospinal fluid was once considered a biochemical indicator for AD and neuraxial degeneration or damage. A recent research showed a noticeable rise of Tau protein in cerebrospinal fluid for subcortex vascular dementia patients (Schonknecht et al. 2003). Results of another recent research suggested a lower serum Aβ density of both vascular dementia patients and stroke patients in acute phase than that of normal humans and presumed there was an Aβ metabolism disorder in both groups (Cheng et al. 2004). Significant rise of cerebroside in cerebrospinal fluid for subcortex vascular dementia patients might hint progressive white matter demyelinating diseases (Blennow et al. 1994). Chui et al. (2000) discovered an increase of neural mercerizing protein in cerebrospinal fluid for both vascular dementia and AD patients, with marked rise in the former in particular. As suggested by still another research, lipid metabolism and hemorheology abnormalities were often observed on VCI intermediate and advanced patients, such as marked rise in apolipoprotein (APO) B, coagulation disorder, whole-blood viscosity, etc. Meanwhile it was noticed that blood plasma cysteine of VD patients was higher than that of the age-matching cerebral infarction group, and hyperhomocysteinemia was a risk factor leading to obstructive large-vessel diseases (Bottiglieri et al. 2001). However, all these above-mentioned abnormal indices are not typical of and characterized by vascular dementia; their diagnostic value for early VCI is yet to be evaluated.

9.2.5.4 Neural Network Impairment

Research on neural networks and the synergistic effects on impairment have shed light on the pathogenesis of VCI. In whole neural networks, impairment in a connection along a loop circuit may mean the start of other loop circuits and only minor clinical symptoms. Although potential ischemic impairment, especially in the white matter, is not likely to give rise to apparent neurological hypofunction, impairment in other locations may result in marked clinical symptoms. Such impairment may damage not only the normal loop circuits but also the newly built circuits (Bowler 2005; van der Flier et al. 2005). Based on neural network theory, in cases with multiple foci, the earliest cognitive impairment would impact functions most dependent on neural networks (Bowler 2005).

Diaschisis and its important role in cognitive impairment were not generally acknowledged until a century ago. With the development of neuroimaging technologies, fMRI can aid in quantitative analysis of cortical diaschisis originating from afferent pathway destruction.

9.2.5.5 Impairment in Cognitive Structure Domain

Damage to certain psychological functions related to cerebral structures can result in cognitive impairment. (1) Injuries to major structures along the Papez loop circuit, especially in the anterior thalamic nuclei, result in memory disorders (Saito et al. 2003). (2) The prefrontal cortex (PFC) and corpus striatum loop circuit: dopaminergic dysfunction maybe reduce working-memory capacity, so does dopamine (DA)-dependent loop circuit affect cognitive processes (Gabrieli 1996). Neuropathological technologies can help reduce acetylcholine (ACh) activity in brain regions such as the cortex, hippocampus, corpus striatum, cerebrospinal fluid, etc., for 40 % of vascular dementia (VD) patients, as ischemia can destroy cholinergic structures (Court et al. 2002). (3) White matter abnormalities caused by periventricular and diffusive white matter demyelination leading to cognitive function deficiency can impair instruction processing speed as well as executive function and both immediate and delayed memory (Gold et al. 2005), while overall intelligence remains unaffected. Critical location WMH can destroy cholinergic fiber connections and cause cognitive function decline. Damage to white matter integrity may affect interactions between the PFC and other structures, such as the hippocampus and corpus striatum. (4) Some studies have demonstrated the importance of the hippocampus and medial temporal lobe (MTL) structures in that they are integral to memory function, and damage to these areas can lead to severe memory disorders. (5) Severe cerebellar diseases may result in cerebellar cognitive and affective disorder syndrome (Schmahmann and Sherman 1998) characterized by disorders in executive function, spatial cognition, language and emotional adjustment, and control behaviors. (6) Autopsy data have verified that gray matter lacunar infarction is an independent predictor of aged VCI in cases in which other types of dementia such as Alzheimer's disease (AD) can be excluded (Gold et al. 2005, 2007).

9.2.5.6 Genetic Factors and Epigenetic Mechanisms

The brain, consisting of more than 20 billion neurons, is the main organ involved in cognitive function and self-awareness. About 25 % of human genes are expressed in the brain. VCI genetic research aims to identify a quantitative phenotype to provide objective and reliable diagnoses for large-sample cohorts.

There are two different types of VCI susceptibility genes: (1) susceptibility genes that are likely to cause cerebrovascular diseases such as hypertension and stroke and (2) genes that determine the reactivity of tissues to cerebrovascular diseases, reflect the tolerance and susceptibility of tissues to ischemia, and are involved in the ability of tissues to recover from ischemic impairment. CADASIL is a typical genetic disease that often induces cognitive impairment, resulting from a Notch3 stereotypic mutation when Notch3 proteins are mainly expressed on vascular smooth muscle cells (Leblanc et al. 2006). Vascular lesions are characterized by progressive degeneration of vascular smooth muscle cells, granular osmiophilic material (GOM) accumulation on basilar membranes, and Notch3 protein accumulation on cell membranes (Dubroca et al. 2005), all of which may lead to additional vascular hindrance. Such arteriolar changes cause decreases in CBF, especially in the white matter (Chabriat et al. 2000; Bruening et al. 2001; Tuominen et al. 2004). The inevitable result is periventricular white matter degeneration, deep white matter changes, and basal ganglia lacunar infarction (Kalimo et al. 2002). Previous studies have shown that 3–15 months after a stroke, most cognitive function decline is related to apolipoprotein E4 (APO E4) gene expression, although the effect of APO E4 has been disputed.

Previous neurobiological studies have suggested that epigenetic mechanisms may play an important role in synaptic plasticity and the memory formation process

(Levenson and Sweatt 2005). Epigenetics is a generic term for chromatin structure and DNA methylation and is a product of the interaction between the environment and the genome. Chromatin is a dynamic structure that can integrate signals transmitted via the cell surface and influence relative transcription reactions. Both chromatin and DNA themselves constantly accrue epigenetic marks, which is also an important part of information integration. These processes are effective at the genomic level. Changes in the epigenetic conditions of the chromatin will have long-term effects on individual behaviors. In addition, there is a correlation between cognitive impairment and epigenetic mark function disorder. Therefore, sufficient knowledge of the epigenetic regulatory mechanism for neuron function is crucial for a thorough understanding of human memory formation and cognitive processes. The role of this mechanism in VCI diseases has yet to be confirmed. In summary, VCI may be a result of interactions between the organism, vascular risk factors, cerebral vascular lesions, and genetic factors, and it is related to multiple fields, including neurosurgery, neuropathophysiology, neuromorphology, neurobiology, and genetics. Further in-depth studies are required to determine the detailed pathogenesis of VCI.

9.3 Clinical Research of Acupuncture Therapy for VCI

Acupuncture therapy for stroke has been used in Asia for nearly 3,000 years. Systematic review and meta-analysis of the application of acupuncture in 38 random samples of post-stroke recovery by Wu et al. (2010) before 2009 indicated that acupuncture is effective in post-stroke recovery, especially with regard to motor function (OR = 4.33, 95 % CI: 3.09–6.08). In terms of point location and acupuncture methods, clinical acupuncture therapies include body acupuncture, scalp acupuncture, electroacupuncture, and composite therapy. Chen et al. (2000a) reported that the most frequently prescribed master meridian in acupuncture therapy for vascular dementia is the *Du* meridian (Governor Meridian), followed by the gallbladder meridian, pericardium meridian, and extra-point meridians. However, the most commonly prescribed master meridians are *Baihui* (DU-20), followed by *Fengchi* (GB-20), *Shenmen* (HT-7), *Renzhong* (DU-26), and *Dazhui* (DU-14). The most frequently prescribed master meridians in body acupuncture therapy are the bladder meridian, followed by the stomach meridian, liver meridian, and gallbladder meridian. However, the most commonly prescribed master meridian was the bladder meridian, followed by the gallbladder meridian, stomach meridian, *Fenglong* (ST-40), *Taichong* (LR-3), *Hegu* (LI-4), and *Sanyinjiao* (SP-6). Acupuncture therapy targeted toward these meridian points for senile dementia was also shown to be effective, but there have been few systematic and objective reports on the curative effects of acupuncture on cognitive impairment in stroke patients. Therefore, this report presents the most common forms of clinical acupuncture therapy for cognitive impairment associated with stroke along with a discussion of related clinical research.

9.3.1 *Body Acupuncture Therapy*

9.3.1.1 Point Selection Differentiated by Syndrome

Zhao et al. (2000) treated 36 cases based on the acupoint selection principles of *Yishen Tiansui* (tonifying kidney and filling up marrow), *Xingnao Qizhi* (clearing the mind and inspiring intelligence), and *Ditan Huayu* (clearing phlegm and removing blood stasis), with the acupuncture group showing a much higher total effective rate compared to the Dihydroergotoxine (DHET) control group. Yu et al. (2005) adopted the *Tiaoshen Yizhi* (regulating mind and brain-tonifying) method and randomly divided 42 patients into a pure medication group and an acupuncture plus medication group. Both groups showed notable improvement in intelligence and curative effects after treatment ($P<0.01$). However, the acupuncture plus medication group showed a better curative effect compared to the medication-only group ($P<0.01$). He (2001) applied *Bushen Tianjing* (invigorating kidney for consolidating semen) and *Jiannao Yizhi* (brain-tonifying) therapies to acupuncture points, such as *Zusanli* (ST-36) and *Sanyinjiao* (SP-6), with a 95 % total effective rate. Liu et al. (1997) randomly divided 100 VD patients into a *Huazhuo Yizhi* (eliminating wetness-evil and brain-tonifying) group and a traditional acupuncture group. Hasegawa Dementia Score (HDS) and Hachinski Ischemic Integral Scale were tested before and after the therapy. The *Huazhuo Yizhi* therapy included acupuncture points such as *Zhongwan* (RN-12), *Fenglong* (ST-40), *Neiguan* (PC-6), *Yongquan* (KI-1), *Renying* (ST-9), and *Fengchi* (GB-20), and traditional therapy included acupuncture points such as *Baihui* (DU-20), *Fengchi* (GB-20), *Dazhong* (KI-4), and *Shenmen* (HT-7). The results indicated that *Huazhuo Yizhi* therapy showed a marked improvement in total clinical effective rate, HDS, and responses to a questionnaire concerning social function activities ($P<0.05$). Better performance has also been reported in extending the Rheoencephalogram (REG) and lowering whole-blood viscosity, hematocrit, and platelet maximum agglutination in subjects receiving *Huazhuo Yizhi* therapy. Cao et al. (2007) examined the effects of *Xingnao Huatan* (inducing resuscitation and removing phlegm) therapy on cognitive memory function and activities of daily living (ADL) in AD patients, and the results showed a 66.7 % effective rate for the observation group and 51.6 % for the controls ($P<0.01$). Scores on both the Mini-Mental State Exam (MMSE) and ADL before and after therapy were also shown to be improved in the observation group relative to the controls ($P<0.01$). Shi examined the effects of *Xingnao Kaiqiao* (inducing resuscitation) therapy in a total of 9,005 stroke patients and reported that after 3–5 courses of treatment, 5,337 patients showed full recovery (59.27 %), 2,085 showed a beneficial effect (23.15 %), 1,453 showed improvement (16.14 %), 40 had no response (0.44 %), and 90 died (1.0 %). The total effective rate of acupuncture treatment was 98.56 %. Clinical laboratory tests for relevant indices indicated remarkable differences after compared with before therapy (Shi 2005).

9.3.1.2 Point Combination for Special Experience Acupuncture Points

Zhao et al. (2004) investigated a combination of modalities including on *Taixi* (KI-3), *Feiyang* (BL-58), and other matching acupoints in 34 cases and reported a total effective rate of 85.3 %. Lai et al. (2000b) conducted a control study between *Qijingbamai* (Eight Extraordinary Meridians) acupuncture therapy, which combined *Qijingbamai* acupoints with *Lingguibafa* (Eight Methods of Intelligent Turtle) and Jin's three-needling treatment, and found no significant total effective rate on HDS, ADL, or MMSE rating scales in either group. Zhao et al. (2000) compared acupuncture therapy on master meridians including *Shenshu* (BL-43), *Geshu* (BL-17), and *Baihui* (DU-20) with oral administration of DHET. Both groups showed a marked rise in MMSE score, but the acupuncture group showed a better effect than the medication group ($P<0.05$). Ma (2001) adopted *Sannao Liutou Jiushen acupoints* (GB-19, DU-17 double, GB-15 double, ST-8 double, GB-11 double, GB-13 double, EX-HN1 four, HT-7 double, DU-24 double) therapy in 83 cases of VD, and the acupoint locations included *Naokong* (GB-19). *Naohu* (DU-17, double), *Toulinqi* (GB-15, double), *Touwei* (ST-8, double), *Touqiaoyin* (GB-11, double), *Benshen* (GB-13, double), *Sishencong* (EX-HN-1), *Shenmen* (HT-7, double), and *Shenting* (DU-24, double). Twenty-five patients showed an effect as measured by the HDS rating scale, 39 were valid, 11 improved, and 8 were unaffected, with a total effective rate of 90.36 %. Jin's three needling consisted of different therapies. *Yizhi Qingshen* (brain-tonifying and clearing head) therapy featuring *Zhisanzhen* (DU-24, GB-13 double), *Sishenzhen* (EX-HN1 four), *Naosanzhen* (DU-17, GB-19 double), and *Niesanzhen* (Temporal three needles) showed favorable curative effects for cognitive impairments such as pediatric mental retardation, senile dementia, VD, and amnesia (Wang 2004; Li 2005). Ruan et al. (1999) divided 60 stable early stroke patients randomly into acupuncture and control groups and selected acupoints including *Shesanzhen* (Tongue Three Needles), *Sishencong* (EX-HN-1), and *Benshen* (GB-13) for the acupuncture therapy group. Cognitive function markedly improved after 3–4 months compared to the control group ($P<0.01$), indicating that body acupuncture therapy on certain acupoints can enhance the cognitive function of early stroke patients. Huang et al. (2008) divided 80 post-cerebral infarction VCI patients randomly into acupuncture and control groups and selected the *Baihui* (DU-20), *Renzhong* (DU-26), and *Shenmen* (HT-7) acupoints for the acupuncture therapy group. MMSE, Clock Drawing Task (CDT), and Block Test results indicated that body acupuncture therapy had a beneficial effect on VCI.

9.3.2 Scalp Acupuncture Therapy

According to the Theory of Meridians and Collaterals, the brain is the center of the meridian system and is the confluence of all *yang channels*. Scalp acupuncture can stimulate not only the head meridians but also the areas of projection of cerebral cortex function on the scalp. In addition, the head is the location of onset for stroke.

Therefore, scalp acupuncture therapy for stroke has long been practiced in clinical medicine. Liu and Liao (1994) treated 15 cases on acupoints such as Middle Line of the Vertex on the uninjured side and Anterior and Posterior Oblique Lines of Vertex-Temporal and reported a total effective rate of 80 %. Xu et al. (2002) employed both scalp and body acupuncture for treatment of 22 patients with hemiplegic cognitive impairment in combination with physical therapy (PT), occupational therapy (OT), speech therapy (ST), and cognitive training (CT) for 4 weeks, while the control group received only PT, OT, ST, and CT, respectively. The results showed greater increases in scores on the Cognitive Capacity Screening Examination (CCSE) and the Loewenstein Occupational Therapy Cognitive Assessment Battery (LOTCA) in the combination therapy group than in the controls. As the periphery of the projection areas for CT focuses on the ipsilateral scalp as the acupuncture point, Lun et al. (2006) applied horizontal insertion of needles to the center of the projection areas, combined with the *Shenmen* (HT-7) and *Zusanli* (ST-36) acupoints, and rotated the needles to gauge patients' responses. The patients were then connected to the G6805-I electroacupuncture apparatus, with a wave of condensation and rarefaction and current strength based on each patient's tolerance. The traditional acupuncture group received the *Sishencong* (EX-HN-1) and *Fengchi* (GB-20) as master points, in combination with *Shenmen* (HT-7) and *Zusanli* (GB-20). Total effective rates of 96.49 % and 75.00 % were observed for the CT-localization scalp acupuncture group and the traditional acupuncture group, respectively (both $P<0.05$).

9.3.3 Electroacupuncture Therapy

Jin et al. (2002) suggested that electroacupuncture therapy for VD would enhance cerebral cortex excitability, reduce blood viscosity, and improve microcirculation to enhance cerebral blood circulation and facilitate energy metabolism of degenerated neurons. He and Jia (1999) suggested that international standard scalp electroacupuncture and acupoint therapy may be a useful new approach for treatment of VD. Based on clinical observations, Jiang et al. (1998) reported that electroacupuncture could enhance clinical effects, increase superoxide dismutase activity (SOD) activity, and reduce lipoperoxidation (LPO) and nitric oxide (NO) levels. Mo et al. (2000) studied the effects of electroacupuncture on the cognitive structure of VD patients and found that this type of therapy could boost the recovery of cognitive and social activity function and enhance patients' quality of life. Moreover, electroacupuncture therapy showed better recent curative effects than medication with DHET. Peng et al. (2009) conducted a randomized controlled trial (RCT) of scalp electroacupuncture therapy in 90 VD patients and selected four acupoints: *Sishencong* (EX-HN-1), *Baihui* (DU-20), *Shenting* (DU-24), and *Fengchi* (GB-20). After 4 weeks, a marked improvement was observed in MMSE scale in the electroacupuncture group, which was much greater than that of the control group ($P<0.05$). Peng et al. (2005) used electroacupuncture as an opsonization for *Suihai* (KI-14) and selected acupoints such as *Baihui* (DU-20), *Fengfu* (DU-16), *Fengchi* (GB-20), *Taiyang* (EX-HN-5),

Hegu (LI-4), *Taichong* (LR-3), and *Shangyintang* (EX-HN-3) in 12 severe VD patients. The total effective rates were 91.6 % and 50 % for MMSE and ADL scales, respectively, which represented significant improvement in the cognitive function and ADL self-care ability of the VD patients. Chen et al. (2000b) divided 46 VD patients randomly and evenly into an electroacupuncture group and the nimodipine medication group. Master points included *Sishencong* (EX-HN-1), *Benshen* (GB-13; double), *Baihui* (DU-20), and *Fengchi* (GB-20; double), selected in accordance with the symptoms, with HDS as the evaluation criterion. The results showed that electroacupuncture therapy had actual curative effects on VD patients, and outcome was better than that seen with medication. Wang (2007) used nimodipine and DHET as control medications and examined the effectiveness of electroacupuncture therapy for patients with VD. Both ADL and MMSE rating scales were tested and assessed. Marked elevation of ADL and cognition were observed after treatment ($P<0.01$). In addition, the experimental group showed a more favorable outcome than the control group, with total effective rates of 89 % and 74 %, respectively. Zhao et al. (2007) adopted an RCT protocol to examine the effects of scalp electroacupuncture on VD patients. In terms of improvement in cognition, the total effective rates were 66.7 %, 73.9 %, and 76.9 % for pure medication, electroacupuncture, and mixed therapy groups, respectively.

9.3.4 Combined Therapies

Zheng et al. (2000) employed acupuncture therapy in combination with self-made restorative tonic in a study population of 30 VD patients. Acupoint selection for one group included Renzhong (DU-26), *Sishencong* (EX-HN-1), Vestibule, *Zusanli* (ST-36), etc., and *Baihui* (DU-20), *Dazhui* (DU-14), and *Ganshu* (BL-18) as alternatives for one another. The control group took nimodipine orally, and HDS and blood plasma NO changes were assessed before and after treatment. The combined therapy group showed a more favorable outcome than the pure medication group, and it also helped to reduce the NO content in the blood plasma of VD patients. Yuan et al. (2002) divided 90 patients randomly into a therapy group (electroacupuncture combined with tonic) and a control group (oral piracetam tablets), based on their specific conditions. Master acupuncture points included *Sishencong* (EX-HN-1), *Benshen* (GB-13), *Shenting* (DU-24), *Baihui* (DU-20), and *Fengchi* (GB-20), and the adjuncts consisted of *Zusanli* (ST-36), *Taixi* (KI-3), *Shenshu* (BL-23), *Xuanzhong* (GB-39), *Shuigou* (DU-26), *Shenting* (DU-24), and *Mingmen* (DU-4). The patients were then given a continuous wave stimulus upon needling responses, spontaneously supplied with corresponding tonic therapy according to traditional Chinese syndrome differentiation analysis. The therapy group showed a more favorable result than the control group after 2 months with reference to *Criteria for Diagnosis, Syndrome Differentiation, and Curative Effect Evaluation of Aged Dementia*. Gao et al. (1999) divided 63 aged VD patients randomly into an acupuncture therapy group and a medication-only control group (oral piracetam tablets) and observed changes in HDS grades, P300 REG, electroencephalogram (EEG), red blood cell SOD, and blood plasma LPO. Acupoint selection for the acupuncture therapy group included *Sishencong* (EX-HN-1), *Niesanzhen*, *Renzhong* (DU-26), *Neiguan* (PC-6), *Sanyinjiao* (SP-6), and *Fenglong* (ST-40), which was conducted by professional physicians. Mild-warm moxibustion therapy was administered on *Baihui* (SU-20), *Shenmen* (HT-7), *Shenque* (RN-8), and *Zusanli* (ST-36) for 15 min each once a day. The control group took piracetam tablets orally (0.8 g) three times a day. Therapies lasted 8 weeks for both groups. The total effective rate for the acupuncture group was 80.6 %, which was much higher than the value of 25 % for the control group, and indicated a better recent curative effect by acupuncture therapy, particularly for recent intelligence rehabilitation of VD patients. Wang (2001) divided 84 VD patients randomly and evenly into a treatment group (acupuncture combined with piracetam tablets) and control group (piracetam only) and used HDS as a measure to assess the effectiveness of treatment. The treatment group showed more favorable results than the controls. Yang and Zhang (1997) applied citicoline injections on acupoints such as *Shenting* (DU-24) and *Toulinqi* (GB-15) for 25 patients, resulting in a total effective rate of 92.0 %. Zhao et al. (1995) applied citicoline injection on acupoints including *Baihui* (DU-20) and *Fengchi* (GB-20) in 234 patients and reported a total effective rate of 85 %. Wang et al. (2000) injected nerve growth factor into bioholographic acupuncture points such as the heart, kidney, and scalp as master points in addition to body acupuncture therapy. They reported a total effective rate of 91.4 %, which was significantly higher than the value of 51.4 % for the control group ($P<0.01$). Wan (2004) examined the effects of combined therapy consisting of scalp acupuncture point block with Cerebrolysin and traditional Chinese medicine in VD patients and reported a total effective rate of 96.3 % in the treatment group, which was markedly higher than that in the medication-only control group.

In conclusion, there has been a great deal of progress in clinical research on acupuncture therapy for VCI, and there is clinical evidence that acupuncture therapy is capable of alleviating post-stroke cognitive impairment as well as enhancing patients' ADL. However, there have been very few multicenter or large-sample RCTs. Yuan and Zhang (2010) evaluated the quality of all published clinical RCTs on acupuncture therapy for post-stroke cognitive impairment, and only 9 studies in a total of 620 patients fulfilled the quality criteria. In addition, the quality of research methodologies used to date has been far from satisfactory. Due to the adoption of different neurological function defect scales (NFDS), only four publications adopted the MMSE rating scale for meta-analysis. These research findings suggested that MMSE scores are improved in the acupuncture therapy group compared to the control group (OR = 8.59, 95 % CI: 4.76–15.51). The analysis indicated that some clinical studies showed that acupuncture therapy alleviated post-stroke cognitive impairment as well as enhancing patients' ADL. Nevertheless, it is still too early to come to definitive conclusions as there have been too few clinical trials and insufficient numbers of cases, and high-quality large-sample RCTs are required to further explore the curative effects of acupuncture therapy.

9.4 Mechanisms of Acupuncture Therapy for VCI

Medical researchers have found that acupuncture can ameliorate cognitive abilities of both experimental animals and clinical patients to a greater degree than the improvement seen in the cerebral functions.

9.4.1 Effects of Acupuncture on Cerebral Blood Flow and Hemorheology

Previous studies have indicated that acupuncture therapy can increase the blood flow in ischemic areas and that it shows protective effects upon post-cerebral ischemia neuronal damage (Shen et al. 1998). Acupuncture therapy can also help to increase the microcirculatory blood flow of infarction patients, enhance blood flow, increase tissue perfusion, ameliorate the supply of blood and oxygen for cells around foci of infection (these cells are in a partial ischemic state during the early pathological period), and activate their remaining functions to compensate for functions of ischemia-induced necrotic cerebral cells, such as language competence, memory, orientation force, and ability to perform complex tasks (Lin et al. 2003).

Du et al. (1998) applied laser Doppler flowmetry in rats with acute focal cerebral ischemia at the middle cerebral artery (MCA) for dynamic detection of self-discipline movement amplitude and frequency of blood capillaries in the ischemic areas on the cerebral surface, local CBF, velocity, and kinetic blood corpuscle number in unit volumes. The results showed that acupuncture on the *Neiguan* (PC-6) and the *Renzhong* (DU-26) at the prophase of ischemia (within 3 h) could effectively relieve microvascular spasms and create conditions for peripheral collateral compensatory blood to enter the ischemic areas. Meanwhile, the resumed blood flow provided blood and oxygen to the ischemic tissues and alleviated the ischemic injury of blood capillaries and prevented the appearance of "speedy but ineffective oscillation" and "anesthesia hypodynamic movement." Therefore, a beneficial cycle emerged between blood capillaries and compensatory blood flow. Shi and Ji (1997) reported that acupuncture at the *Renzhong* (DU-26), *Neiguan* (PC-6), *Zusanli* (ST-36), and *Quchi* could significantly improve the microcirculatory blood flow of ligated cerebral arteries and demonstrated relative acupoint idiosyncrasy, with most effects on the *Renzhong* (DU-26) and *Neiguan* (PC-6). Ji et al. (1996) reported that acupuncture at the *Zusanli* (ST-36) and *Quchi* (LI-11) markedly accelerated the CBF in dogs.

Arteriosclerosis is pathological for most cerebral vessel diseases, but dyslipidemia is the major risk factor for arteriosclerosis and can lead to hemorheological abnormalities. Studies have shown that acupuncture has beneficial effects on blood lipids, which is one of the mechanisms of acupuncture therapy for stroke. In a study performed in a population of 400 ischemic cerebral stroke patients before and after therapy, Xu (1992) found that acupuncture significantly reduced the levels of blood lipids, including total cholesterol (TC), triglycerides (TG), and β-lipoprotein. Peng et al.

(1998) reported that acupuncture reduced the quantity of TC in the blood lipids of cerebral thrombosis (CT) patients but increased the high-density lipoprotein (HDL) level. Liu et al. (2002a) showed that acupuncture could improve the whole-blood viscosity and hematocrit levels in rabbits with experimentally induced cerebral ischemia. Wang et al. (1995) applied Xingnao Kaiqiao acupuncture to treat TIAs, and the results indicated that with the exception of the K-equation of erythrocyte sedimentation rate, all hemorheological indices showed statistically significant improvements.

9.4.2 *Effects of Acupuncture Therapy on Cerebral Metabolism*

The functions of the hippocampus are linked to learning and memory, especially to the storage and processing of spatial information (Gillani et al. 2010). As an area sensitive to ischemia and hypoxia, the hippocampus shows a high level of energy metabolism and damage to this area leads to memory impairment, particularly affecting short-term and spatial memory (Nikonenko et al. 2009).

Active cerebral metabolism requires an energy supply. Glucose is virtually the only energy source for cerebral tissue. Due to a lack of endogenous energy stores, the maintenance of cerebral functions depends on a sustainable blood supply. Any abnormalities in the cerebral blood vessels or blood flow may cause a reduction of the glucose supply or inefficiency of its utilization resulting in damage to neural function. Glut-1 and Glut-3 are the major transport proteins in the brain responsible for glucose delivery. As a specific protein of the blood–brain barrier (BBB), Glut-1 is usually expressed in BBB capillary endothelial cells. Glut-1 also regulates the blood sugar stability of cerebral tissues at the molecular level and determines intracellular glucose transport and metabolism in cerebral tissue (Tanaka et al. 2010). Glut-3 is usually expressed in neurons (Zovein et al. 2004). Glut-1 and Glut-3 expression are regulated not only by the concentration of blood sugar to maintain stable cerebral energy metabolism but also by hypoxia and ischemia (Tanaka et al. 2010; Zovein et al. 2004; Vemula et al. 2009) (Fig. 9.1). Under normal conditions, oxidative phosphorylation is the rate-limiting step through which glucose is translated into energy, as long as there are adequate oxygen and glucose supplies. Under conditions of cerebral hypoxia and ischemia, glucose is transported across the BBB as the rate-limiting step. Zhang et al. (2010) reported increased Glut-1 and Glut-3 expression in MID rat hippocampi, indicating that thrombosis led to obstruction of the blood supply, and tissues showed accelerated glucose transportation to compensate for the insufficient energy supply caused by hypoxia and ischemia. Therefore, compensatory expression levels of Glut-1 and Glut-3 increased. Moreover, acupuncture therapy was shown to ameliorate the hypoxic and ischemic status of tissues, promote Glut-1 and Glut-3 synthesis, and accelerate not only the transportation of glucose from blood into cerebral tissue but also improve glucose absorption in neurons. As a result, cerebral energy exhaustion could be alleviated, supply of substances related to cerebral energy could be enhanced, and death of neurons after ischemia could be relieved.

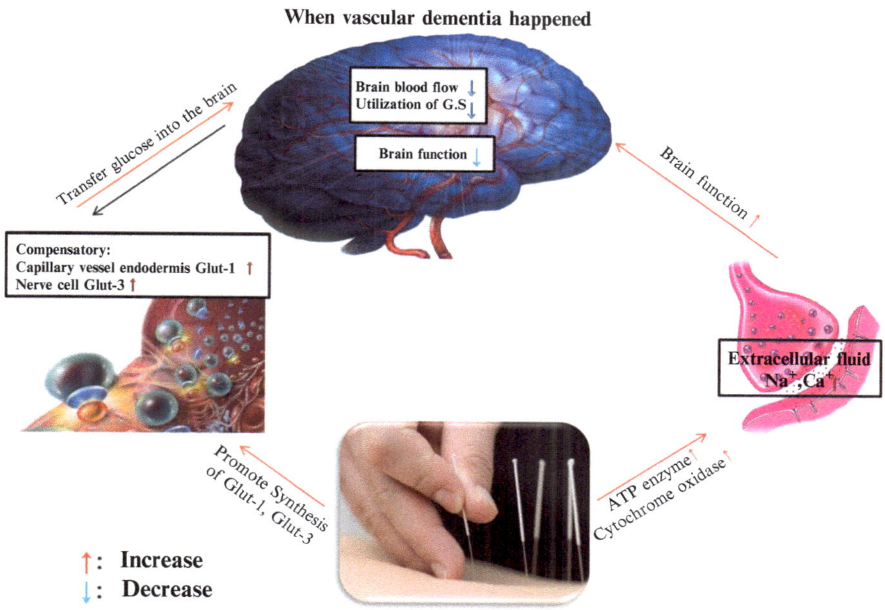

Fig. 9.1 Effects of acupuncture therapy on cerebral metabolism. Active cerebral metabolism requires an energy supply. Glucose is virtually the only energy source for cerebral tissue. Glut-1 and Glut-3 are the major transport proteins in the brain responsible for glucose delivery. Acupuncture therapy was shown to ameliorate the hypoxic and ischemic status of tissues, promote Glut-1 and Glut-3 synthesis, and accelerate not only the transportation of glucose from blood into cerebral tissue but also improve glucose absorption in neurons

Acupuncture can also maintain the stability of internal and external ions in cerebral cells. Ma et al. (1997) reported that acupuncture at 12 *Jing* acupoints may reduce the elevation of K^+ and H^+ concentrations as well as reduce the concentrations of Ca^{2+} and Na^+ in the extracellular fluid in ischemic areas of the cerebral cortex in MCA-closed rats. These effects could alleviate cytotoxic edema due to the imbalance between internal and external ions. Li et al. (1996) used flame atomic absorption spectrophotometry to measure the ion changes associated with whole-brain ischemic reperfusion injury in rats and reported that electroacupuncture may significantly reduce the content of Ca^{2+} in the cerebral tissues, and tended to reduce Na^+ content, and may ultimately ameliorate hydrocephalus.

Du and Shi (1999) reported that Nissl bodies in the neurons of cerebral ischemic areas were greatly reduced, and even lost completely in some cases. The loss of Nissl bodies was markedly alleviated after restoring consciousness acupuncture therapy, and their functions gradually recovered, suggesting that acupuncture can protect Nissl bodies and provide the necessary energetic substances for the functional activity of neurons to alleviate ischemia-induced neuronal injury. They also observed that ATPase and cytochrome oxidase (CCO) activities in the cerebral tissues were markedly reduced after ischemia, and the cerebral cells would show impairment of membrane ion transport and respiratory function. Although such impairment may recover to some extent after 3 or 6 h, these enzyme activities remained very low, indicating that the organism's own compensatory mechanisms were insufficient to avoid injury. The increases in activities of these two enzymes after acupuncture therapy suggested that acupuncture could improve the enzymatic metabolism of ischemic cerebral tissues and have a protective effect on the transmembrane transport of Na^+ and K^+, as well as cellular respiratory function.

9.4.3 Regulation of Cerebral Neuroelectrical Physiological Activities by Acupuncture

The electrical activities of the brain are among the best measures reflecting the cerebral function status. Changes in the functions of nerve cells are associated with changes in their electrical activities. Therefore, dynamic observation of the electrical activities in the brain can be applied to evaluate the extent of brain damage. Mou et al. (1994) examined the brain stem auditory evoked potentials (BAEP) of TIA patients before and after acupuncture therapy and reported that the latent period of BAEP was earlier, and there was a marked difference before and after I-wave acupuncture therapy. Their findings indicated that scalp acupuncture stimulated the conduction of acoustic nerves under pathological conditions and exerted a greater impact on peripheral than on central nerves, indicating that scalp acupuncture regulated the physiological abnormalities. Dong et al. (1994) performed a neuroelectrical physiological study to examine the immediate effects of clinical acupuncture therapy on cerebral hemorrhage. They concluded that acupuncture may stimulate those shocked or dormant cerebral neurons, the excitability of which was restrained by hemorrhagic stimulation and hematoma oppression and that their excitability could be recovered rapidly. Sun et al. (1994) examined the influence of acupuncture for scalp acupoint motor areas upon motor evoked potentials (MEPs) and found that direct stimulation of the central motor nervous system by scalp acupuncture through certain transmissions was the main mechanism of treating brain-derived nervous system diseases; the closer the location to the anterior central gyrus, the easier it was to induce MEPs.

The somatosensory evoked potential (SEP) is a neuroelectrical activity obtained at the sensorimotor areas corresponding to the cerebral cortex by stimulating sensory nerves. SEP reflects the integrity of nerve bodies' sensory channel structure and functions and can be used to evaluate the changes in cerebral function impairment. Nie (1999) reported significant changes in SEP in cerebral hemorrhage patients receiving electro-"di"-acupuncture therapy. In addition, the latent period was delayed and shortened, while evanescent waves reappeared and low repeatability amplitudes rose. Tang et al. (1996) reported that electro-"di"-acupuncture at scalp acupoints *Baihui* (DU-20) to *Qubin* (GB-7) reduced the SEP amplitudes in acute cerebral infarction patients, reestablished the lost waves, shortened the delayed latent period, recovered the central conduction time, and rehabilitated the abnormal

SEP to some extent. Therefore, electro-"di"-acupuncture could exert significant regulatory effects on electrophysiological activities of cerebral nervous cells, and such therapy could help to restore the damaged cerebral functions after cerebral ischemia. In their MCA occlusion and reperfusion experiments in rats, Si et al. (1998) reported that SEP, cerebral damage volume, and degree of thalamencephalon atrophy in the acupuncture group were significantly reduced compared to the ischemia group after 85 days of cerebral ischemia, suggesting that electroacupuncture may have a protective effect against anaphase trauma in cases of cerebral ischemia. Jin and Cheng (1998) reported that electroacupuncture could promote the recovery of SEP and reduce the cerebral infarct volume after cerebral ischemia, and its effects on SEP at the acupoints along the *Du* meridian were better than those of body acupuncture. The clinical research of Wang et al. (1993) also showed that scalp acupuncture could gradually normalize the electroencephalographic activities of ischemic cerebral stroke patients, remarkably shorten the centrum conduction duration of patients, regulate the N2- and P3-related nervous conduction passage and cortical functions of SEP, and improve the conduction passage from the thalamencephalon to the cortex.

9.4.4 Molecular Biological Effects of Acupuncture on the Brain

9.4.4.1 Change of Proteins

Shi et al. (1999) performed sodium dodecyl sulfate polyacrylamide gel electrophoresis (SDS-PAGE) to examine the changes in electrophoretic patterns of brain cellular nucleoproteins and non-histone chromatin proteins (NHCP) in senescence-accelerated mice with encephalatrophy after acupuncture at *Shuigou* (DU-26) and *Neiguan* (PC-6) and measured the ratios of protein zones to their apices by scanning densitometry of the gels. The results showed that acupuncture could reduce the contents of membrane proteins (the zone of cellular nucleoprotein with a molecular weight of 62 kDa) and increase the content of NHCP (the zone of chromatin protein with molecular weight of 24 kDa). These findings indicated that the mechanism of action of acupuncture at *Shuigou* (DU-26) and *Neiguan* (PC-6) to defer senility and improve cerebral functions may be due to promotion of cerebral chromatinic NHCP synthesis, which affected gene expression, strengthened the releasing effects of histone repression, and increased the transcriptional activity of DNA templates. Jin et al. (2009) reported that acupuncture at *Baihui* (Hundred Convergences) could reduce the overexpression of S100β protein in the brains of rats with mild cognitive impairment and reduce neuritic macula formation. Luo et al. (2003) investigated the effects of electroacupuncture therapy on the levels of serum β-amyloid and growth factors in a rat model of AD, and the results indicated that electroacupuncture therapy could effectively reduce serum β-amyloid levels, which may be one of the molecular biological mechanisms by which this type of treatment can improve the learning and memory capabilities of AD rats.

Li et al. (2006) showed that both restoring consciousness acupuncture and routine acupuncture could significantly reduce the thickness of C-reactive protein (CRP), suggesting that both of these approaches had beneficial anti-inflammatory effects. However, when the former approach was applied, the extent of reduction in each index was greater than with the latter approach, and the difference for CRP was significant. These observations indicated that the anti-inflammatory effects of the restoring consciousness acupuncture approach were stronger than those of routine acupuncture. Therefore, they suggested that restoring consciousness acupuncture could better prevent and control the development of ischemic cerebrovascular diseases, and such therapy could play an important role in reducing the occurrence of cerebrovascular disease and improving patient prognosis.

Neuropeptide Y (NPY), which consists of 36 amino acid residues, acts as a vasoconstrictor in the central nervous system (CNS). Cerebral ischemia may increase the release of NPY, which in turn aggravates ischemia. Liu and Yang (2003) modeled a 4-vessel occlusion (4-VO) method and found that the density of NPY in the rat brain was significantly reduced after electroacupuncture at the *Du*, *Taixi* (KI-3), and *Geshu* (BL-17) scalp acupoints, and cerebral vessel contraction was also ameliorated. As a result, the learning and memory competence of VD rats showed marked improvement.

Cerebral arginine vasopressin (AVP) is a type of NPY that is closely related to learning and memory. In a series of animal experiments, Mo et al. (2002) found that acupuncture could elevate the content of AVP and induce its redistribution, thus ameliorating the learning and memory deficits of hypertensive VD rats. Wang (2002) examined the effects of acupuncture on learning and memory behaviors along with the cerebral heme oxygenase (HO) activity of VD rats and showed that acupuncture could regulate the HO activity of the cortex and hippocampus to improve learning and memory. Using molecular hybridization technology, Ma et al. (2001) examined the level of heat shock protein 70 (HSP70) mRNA transcription in ischemic cerebral tissues in experimental cerebral infarction rats and the confounding effects of restoring consciousness acupuncture on HSP70 mRNA expression in ischemic cerebral tissues. The results showed that restoring consciousness acupuncture might increase the expression of cortical, striatum corpora, and hippocampal HSP70 gene expression in the infarction areas, and they suggested that this may be related to the mechanism underlying the cerebral cellular protection effect of restoring consciousness acupuncture therapy.

Through bilateral carotid ligation and reperfusion, Lu et al. (2003) applied acupuncture at *Baihui* (DU-20), *Zhongwan* (RN-12), and *Zusanli* (ST-36) to VD rats. After Chinese herbal Yang-invigorating soup lavage, they found that the serum calcitonin gene-related peptide (CGRP) level in VD rats showed a marked increase and endothelin (ET) level was markedly decreased, while the number of errors in the step down test was significantly reduced, indicating that acupuncture protected nerves by altering the contents of centrum vasoexcitor materials (VEM) and by changing the blood supply to cerebral tissues.

By 4-VO and global brain ischemia (GBI) reperfusion, Zhang et al. (Zhang et al. 2001b, c; Zhang and Xiao 2000) prepared VD rats and showed that after auriculoacupuncture therapy, the loss of neurons at CA1 of the rat hippocampus was reduced, the expression of bc1-2 rose sharply, the expression of c-Fos protein increased, and the expression of b-nitric oxide synthase (NOS) protein decreased.

9.4.4.2 Changes in Free Radicals

Since it was first proposed by Harmer, the free radical damage theory has attracted a great deal of research attention, and a number of studies have addressed the correlation between LPO and diseases such as thrombus disease, stroke, arteriosclerosis, dementia, and senescence (McCord 1985). Lai et al. (2000a) examined the effects of electroacupuncture on learning and memory, SOD activity, and malondialdehyde (MDA) levels in rats with experimentally induced VD. Their findings indicated that LPO was due to the effects of hypoxia and ischemia on cellular and organelle membranes. As a result, large amounts of free radicals were released, large amounts of SOD were consumed, and large quantities of MDA were generated. However, free radical production was decreased, SOD activity was increased, and MDA content was reduced after electroacupuncture in this model. Tian et al. (2002) examined the SOD contents in the cerebral tissues of cerebral hemorrhage rats and found that acupuncture markedly increased the SOD level. It was suggested that acupuncture at *Baihui* (DU-20) and *Dazhui* (DU-14) may reduce the activity of NOS in cerebral tissues, increase SOD levels, reduce NO levels, and decrease the amounts of free radicals in cerebral tissues (Wang et al. 2005; Zheng et al. 2005).

Zhang et al. (1999) showed experimentally that acupuncture could increase the activity of SOD and decrease the contents of MDA and LPO, indirectly indicating that restoring consciousness acupuncture could alleviate free radical damage caused by cerebral ischemia and reperfusion. Liu et al. (2002b) reported that restoring consciousness acupuncture affected antioxidant activity, and this type of therapy could not only strengthen the activity of Cu–Zn SOD in the cerebral tissues of experimental animals but could also increase those of catalase (CAT) and glutathione peroxidase (GSH-Px). Consequently, the whole system involved in preventing oxidative damage was upregulated, which might be one of the reasons why acupuncture alleviates damage related to free radicals.

9.4.4.3 Regulative Effects of Acupuncture on Neurotransmitters

A great deal of research has shown that electroacupuncture activates or strengthens the functions of some central neurotransmitters and neuromodulators and weakens or blocks those of other neurotransmitters by regulating the release of neurotransmitters and neuromodulators and degrading or altering their contents in the CNS. Electroacupuncture usually strengthens the functions of endogenous opioid peptides (EOPs), serotonin (5-HT), DA, ACh, and inhibitory amino acids (GABA, aminoacetic acid, aminoethylsulfonic acid, and alanine), while weakening the functions of methyl epinephrine and excitatory amino acids (glutamic acid and aspartic acid).

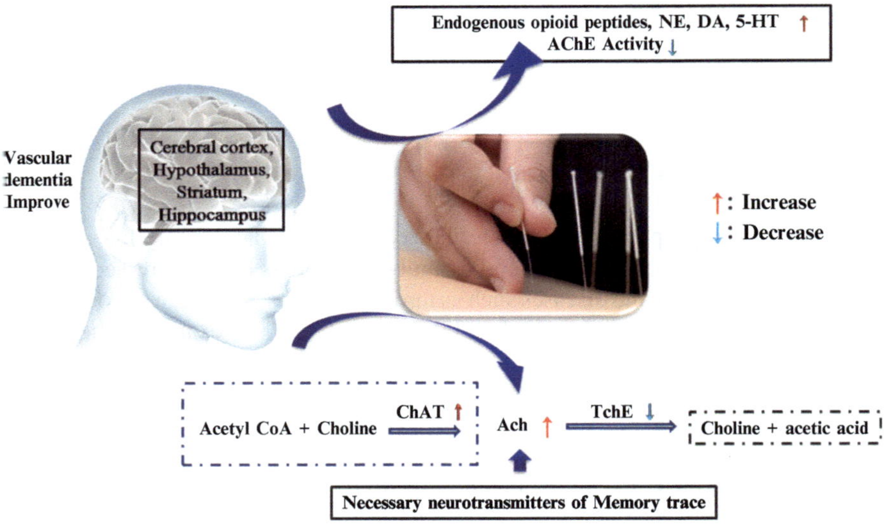

Fig. 9.2 Regulative effects of acupuncture on neurotransmitters. Electroacupuncture usually strengthens the functions of endogenous opioid peptides, serotonin (5-HT), DA, acetylcholine (ACh), and inhibitory amino acids (GABA, aminoacetic acid, aminoethylsulfonic acid, and alanine). Choline acetyltransferase (ChAT), a specific marker of cholinergic neurons, also serves as the rate-limiting enzyme in catalyzing the synthesis of ACh. ChAT is usually used to study the cholinergic neuron contents in the CNS or as an indirect index of ACh content. ACh is synthesized by the enzymatic reaction between acetyl coenzyme A and choline catalyzed by ChAT and is decomposed into choline and acetic acid by the action of AChE

Long-term acupuncture can also induce excessive cholecystokinin 8 (CCK-8) expression, exhaust the substance produced by prestimulation, and promote electroacupuncture tolerance. In addition, electroacupuncture may also regulate the expression and functions of receptors (Xia and Li 1984; Xia et al. 1985; Chen et al. 1984; Zhang et al. 1997).

With regard to the pathogenesis of VD, alterations in neurotransmitter levels are known to play very important roles in the development and prognosis of disease. Changes have been demonstrated in many types of neurotransmitter in the brains of VD patients; the nervous activity is significantly reduced in the cerebral cortex, hypothalamus, striatum corpora, and hippocampus, and reductions in ACh and monoamine neurotransmitter levels in the cerebral cortex are closely related to the decline in cognition associated with VD (Mo et al. 2001; Wang et al. 2003). Tang et al. (2007) reconfirmed that acupuncture can increase the concentrations of neurotransmitters such as ACh, norepinephrine (NE), DA, and 5-HT in the cerebral tissues of VD rats and reduce the activity of acetylcholine esterase (AChE) in the rat brain, which may be the biochemical mechanism by which acupuncture improves the cognitive impairment seen in VD rats.

Choline acetyltransferase (ChAT), a specific marker of cholinergic neurons, also serves as the rate-limiting enzyme in catalyzing the synthesis of Ach (Fig. 9.2). ChAT is usually used to study the cholinergic neuron contents in the CNS or as an indirect index of ACh content. ACh is synthesized by the enzymatic reaction between acetyl coenzyme A and choline catalyzed by ChAT and is decomposed into choline and acetic acid by the action of AChE. ACh has been shown to be the neurotransmitter required for forming mnemonic traces and is the physiological basis of long-term memory. Changes in ACh contents can be reflected in the activity of its hydratase TChE. TChE participates directly in vegetative neurofunctional regulation, cerebral cogitation, and memory as a key enzyme of the cholinergic system, the function of which is directly reflected by the TChE activity level. When the level of TChE activity changes, the functions of the above-mentioned tissues and organs will also change. The contents of TChE and its activity are equivalent to those released each time the nerve fires. The reduction of ACh content can lead to a reduction of synthesis or release, leading to a corresponding drop in TChE activity (Yu et al. 2002). The levels of cholinergic neurotransmitters in the brains of VD patients and the loss of ACh are related to the number of age pigments and intelligence damage (Wang et al. 2007). Therefore, attempts have been made to increase the contents of ACh in the CNS by restricting the activity of TChE to improve the clinical symptoms of VD patients. The experimental results of Zheng et al. (2009) showed that acupuncture therapy has a significant regulatory effect on the cholinergic system and could markedly improve cognitive function in the rat VD model. Acupuncture could not only reduce the activity of TChE but also increase the activity of ChAT and the contents of ACh in the brains of these model animals. Acupuncture may eventually be able to alleviate VD by compensating for the loss of cholinergic neurons and the reduction of ACh content in the brain during the development of VD.

9.4.5 Effects of Acupuncture on Cerebral Structure

Liu and Li (2000) reported that nerve cells of rabbits with experimentally induced ischemia showed blur sketch while the rough endoplasmic reticulum of the cells expanded, in slit shapes or intensive stripes. Other changes were also observed as follows: the superior nucleus proteasomes became desquamated; the Golgi apparatus became swollen; cells and vesicae disappeared; mitochondria showed obvious swelling; ground substances were reduced; cristae became transformed, distorted, or even disappeared altogether presenting a vacuole-like appearance; the nuclear membrane was not continuous and showed pitting; chromatin became condensed or dissolved, with a loosening or disappearance of the nucleolus; lysosome numbers increased markedly, canaliculi became distorted, and microfilaments dissolved; part of the neurofibril myelin sheath was disordered, showing demyelination; the number of synaptic vesicles in the presynaptic membrane increased, cystica or substantive mitochondria became transformed or broken, postsynaptic membranes showed edema, and the degeneration of nerve terminals occured, manifesting as broken synaptic vesicles and blurred structures; osmiophily of the whole terminals enhanced and electron density increased; and the number of visible dense bodies in synaptic clefts increased. All of these changes were alleviated to some extent after restoring consciousness acupuncture. Hu et al. (1996) also reported similar findings.

Liu et al. (2005) reported that acupuncture could improve these pathological indices of cerebral tissues in rats with ischemic cerebral injury, as well as promote the recovery of synapse number and structure. Luo et al. (1996) and Xu et al. (2001) also confirmed that acupuncture could alleviate the disorder of nerve cells in the cerebral ischemic zones, lessen the degree of swelling of neurocellular mitochondria, and restore their number. That is, acupuncture may have significant effects on the ultrastructure of neurocellular mitochondria in the cerebral tissues of patients with ischemic cerebral stroke. Using in situ end-labeling, Zhang et al. (2001a) observed apoptotic changes in the ischemic areas of MCA occlusion rats and reported that apoptosis occurred in the ischemic zone after 1 h reaching a peak at approximately 24 h. The apoptotic cells were mainly confined to the "penumbral region" around the central necrotic zone, and the apoptosis in the "penumbral region" may have been related to expansion of the infarct focus. Restoring consciousness acupuncture markedly reduced apoptosis and decreased the infarct area and thus had a protective effect on the brain.

Taken together, the findings discussed above indicate that the changes occurring in response to acupuncture are multilayered, multi-targeted, and multisided. Acupuncture may reinforce the functions of multiple anti-trauma factors in the brain crucial and thus show efficacy in VCI. However, additional fundamental research is required to explore the signal transduction processes and mechanisms underlying the effects of acupuncture. Such studies will clarify the mechanisms by which acupuncture shows its effects, encourage adoption of electroacupuncture therapy, and facilitate the enhancement of its effects.

9.5 Concluding Remarks

In this chapter, the authors introduce the concept and clinical manifestations of VCI and explore the pathogenesis of VCI from the perspectives of Neuroanatomy, Neural pathophysiology, Neuromorphology, Neurobiology and Genetics. The chapter focuses on the clinical applications and experimental research of acupuncture therapy for VCI. Although the experimental research of acupuncture therapy for VCI started very late, many valuable accomplishments have been achieved. The research involves the effects of acupuncture therapy on VCI in the ways of cerebral morphology and molecular biology and regulation on neuroelectrical physiological activities. It points out that the multilayered, multi-targeted and multisided mechanisms can be effectively used to cure VCI. These mechanisms include that acupuncture can promote the synthesis increase of Glut-1 and Glut-3 and improve the cerebral metabolism by improving the CBF, reduce the level of serum β amyloid and neuroactive peptide NPY, promote the expression of bcl-2 and c-fos, reduce the damage of free radicals, regulate the functions of EOP, 5-HT, dopamine (DA) and ACh, and inhibitory amino acid.

Since the pathological mechanisms of VCI diversify, and Western medical treatments at present can only be directed to one specific aspect of these pathological mechanisms, there has been no affirmative therapy reported so far. On the other hand, Chinese acupuncture therapy is convenient and safe and boasts a long history and high-cost performance, and numerous clinical research has confirmed that acupuncture therapy works effectively on VCI. However, it must also be noted that there are a great many things to be done for its clinical practice and research such as poor research design, lack of concurrent control, insufficient sample size, low data validity, and lack of long-term follow-up data. Furthermore, the acupoints of acupuncture therapy are very complicated and jumbled, and there have been no definite criteria for mastering and matching. In addition, the needle manipulation has not yet been standardized. Henceforth, the modern scientific research apparatus and methods should be used to strengthen the fundamental research of acupuncture therapy, unify the acupuncture therapy standards, and promote high-quality, large-sample, polycentric double-blind RCTs so as to carry forward Chinese acupuncture therapy.

Acknowledgments This work was partially supported by the grant of Health Department of Hainan Province (2008-35).

References

Bakker FC, Klijn CJ, Jennekens-Schinkel A, Kappelle LJ (2000) Cognitive disorders in patients with occlusive disease of the carotid artery: a systematic review of the literature. J Neurol 247: 669–676

Bakker FC, Klijn CJ, Jennekens-Schinkel A, van der Tweel I, van der Grond J, van Huffelen AC, Tulleken CA, Kappelle LJ (2003) Cognitive impairment is related to cerebral lactate in patients with carotid artery occlusion and ipsilateral transient ischemic attacks. Stroke 34: 1419–1424

Blennow K, Wallin A and Ekman R (1994) Neuron specific enolase in cerebrospinal fluid: a biochemical marker for neuronal degeneration in dementia disorders? J Neural Transm Park Dis Dement Sect 8: 183–191

Bottiglieri T, Parnetti L, Arning E, Ortiz T, Amici S, Lanari A, Gallai V (2001) Plasma total homocysteine levels and the C677T mutation in the methylenetetrahydrofolate reductase (MTHFR) gene: a study in an Italian population with dementia. Mech Ageing Dev 122: 2013–2023

Bowler JV (2004) Vascular cognitive impairment. Stroke 35: 386–388

Bowler JV (2005) Vascular cognitive impairment. J Neurol Neurosurg Psychiatry 76: v35-44

Bruening R, Dichgans M, Berchtenbreiter C, Yousry T, Seelos KC, Wu RH, Mayer M, Brix G, Reiser M (2001) Cerebral autosomal dominant arteriopathy with subcortical infarcts and leukoencephalopathy: decrease in regional cerebral blood volume in hyperintense subcortical lesions inversely correlates with disability and cognitive performance. AJNR Am J Neuroradiol 22: 1268–1274

Cao JM, Bi QL and Zhao JW (2007) Clinical observations on vascular dementia treated by Xingnao Huatan acupuncture therapy. Henan Traditional Chinese Medicine 27: 29–30

Chabriat H, Pappata S, Ostergaard L, Clark CA, Pachot-Clouard M, Vahedi K, Jobert A, Le Bihan D, Bousser MG (2000) Cerebral hemodynamics in CADASIL before and after acetazolamide challenge assessed with MRI bolus tracking. Stroke 31: 1904–1912

Chen BY, Pan XP, Jiang CC, Fan GC (1984) Correlation of pain threshold and level of β-endorphin-like immunoreactive substance in human csf during electro-acupuncture analgesia. Acta Physiologica Sinica 36: 183–187

Chen JF, Li TY, Chen LH (2000) Acupuncture therapy patterns for senile dementia. Journal of Clinical Acupuncture and Moxibustion 16: 2–7

Chen ZH, Jiang GH, Lai XS (2000) Clinical effect observation on electroacupuncture therapy for vascular dementia . Journal of Clinical Acupuncture and Moxibustion 16: 18–20

Cheng OM, Zhong SJ, Yan Y, Chen WR, Li GQ (2004) Measurement of β-amyloid in stroke patient's plasma . Chongqing Medical Journal 33: 555–556

Chui HC, Mack W, Jackson JE, Mungas D, Reed BR, Tinklenberg J, Chang FL, Skinner K, Tasaki C, Jagust WJ (2000) Clinical criteria for the diagnosis of vascular dementia: a multicenter study of comparability and interrater reliability. Arch Neurol 57: 191–196

Court JA, Perry EK, Kalaria RN (2002) Neurotransmitter control of the cerebral vasculature and abnormalities in vascular dementia. In: Erkinjuntti T. Gauthier S. eds. Vascular Cognitive Impairment. London, UK: Martin Duniz Ltd: 167–185.

Dong GR, Wang Z, Wu BZ, Zhang X, Zhang CY, Yang Y (1994) Discussion on the mechanisms of immediate effect of head needling on acute encephalorrhagia. Chinese Acupuncture & Moxibustion 14: 26–29

Dubroca C, Lacombe P, Domenga V, Maciazek J, Levy B, Tournier-Lasserve E, Joutel A, Henrion D (2005) Impaired vascular mechanotransduction in a transgenic mouse model of CADASIL arteriopathy. Stroke 36: 113–117

Du YH, Zhai N, Shi XM (1998) Study of acute cerebral ischemia treated with acupuncture (I)—influence of acupuncture on vasomotion of MCA model rats. Acupuncture Research 23: 275–278

Du YH, Shi XM (1999) The influence of acupuncture on cerebral ATP enzyme and cytochrome oxidase in rats with acute cerebral ischemia. Shanghai Journal of Acupuncture and Moxibustion 18: 38–39

Gabrieli JD (1996). Memory systems analyses of mnemonic disorders in aging and age-related diseases. Proc Natl Acad Sci U S A 93: 13534–13540

Gao HY, Yan LF, Liu BB, Wang Y, Wei XL, Sun LY, Cui HQ (1999) Clinical research on acupuncture treatment for senile vascular dementia. Journal of Traditional Chinese Medicine 40: 471–473

Gillani RL, Tsai SY, Wallace DG, O'Brien TE, Arhebamen E, Tole M, Schwab ME, Kartje GL (2010) Cognitive recovery in the aged rat after stroke and anti-Nogo-A immunotherapy. Behav Brain Res 208: 415–424

Gold G, Kovari E, Herrmann FR, Canuto A, Hof PR, Michel JP, Bouras C, Giannakopoulos P (2005) Cognitive consequences of thalamic, basal ganglia, and deep white matter lacunes in brain aging and dementia. Stroke 36: 1184–1188

Gold G, Kovari E, Hof PR, Bouras C, Giannakopoulos P (2007) Sorting out the clinical consequences of ischemic lesions in brain aging: a clinicopathological approach. J Neurol Sci 257: 17–22

Hachinski V, Iadecola C, Petersen RC, Breteler MM, Nyenhuis DL, Black SE, Powers WJ, DeCarli C, Merino JG, Kalaria RN, Vinters HV, Holtzman DM, Rosenberg GA, Wallin A, Dichgans M, Marler JR, Leblanc GG (2006) National Institute of Neurological Disorders and Stroke-Canadian Stroke Network vascular cognitive impairment harmonization standards. Stroke 37: 2220–2241

He XJ (2001) Clinical observations of 20 cases on treatment of vascular dementia with combined Use of acupuncture and medicine. Study Journal of Traditional Chinese Medicine 19: 267–267

He J, Jia SJ (1999) International standard scalp electroacupuncture and acupoint treatment for vascular dementia . Sichuan Journal of Traditional Chinese Medicine 17: 51–52

Hu GQ, Tian F, Li P, Zhao LR (1996) Effect of xingnao kaiqiao acupuncture method on free radicals and ultrastructure of cerebral tissue of rabbit with cerebral ischemia and reperfusion. Chinese Critical Care Medicine 8: 5–7

Huang F, Liu Y, Zhou FX, Yao GX, He QT (2008) Effects of acupuncture therapy on post-cerebral infarction vascular cognitive impairment. Guangdong Medical Journal 29: 1918–1920

Ji GC, Zhao LM, Shi RH, Liu YF, Wang SJ, Wu FD (1996) Effects of electrical acupuncture on the cerebral blood flow and the pial microcirculatory blood flow in dogs. Acupuncture Research 21: 43–46

Jia JP (2005) Early diagnosis and management of vascular cognitive impairment . Chinese Journal of Neurology 38: 4–6

Jiang GH, Lai XS, Mo FZ (1998) Effect of electroacupuncture on vascular dementia . Journal of Guangzhou University of Traditional Chinese Medicine 15: 110–113

Jin HS, Huo L, Guo XX (2009) Effects of electro-acupuncture Baihui(DU 20) on S100β protein in the rat model brain of Mild Cognitive Impairment. Liaoning Journal of Traditional Chinese Medicine 36: 648–650

Jin XQ, Song SL, Yu MH, Lu BF (2002) Clinical observations on the treatment of vascular dementia by scalp electroacupuncture plus acupoint injection. Shanghai Journal of Acupuncture and Moxibustion 21: 7–9

Jin ZQ, Cheng JS (1998) The effects of needling different acupoints on cortical SEP and cerebral infarction volume in rats with transient cerebral ischemia. Shanghai Journal of Acupuncture and Moxibustion 17: 37–39

Kalimo H, Ruchoux MM, Viitanen M, Kalaria RN (2002) CADASIL: a common form of hereditary arteriopathy causing brain infarcts and dementia. Brain Pathol 12: 371–384

Klijn CJ, Kappelle LJ, Tulleken CA, van Gijn J (1997) Symptomatic carotid artery occlusion. A reappraisal of hemodynamic factors. Stroke 28: 2084–2093

Kua EH, Ko SM (1995) Prevalence of dementia among elderly Chinese and Malay residents of Singapore. Int Psychogeriatr 7: 439–446

Lai XS, Wang L, Jiang XH, Chen ZH, Zeng Q (2000) Effects of electroacupuncture on the learning memory, SOD and MDA in experimental vascular dementia rats. Chinese Acupuncture & Moxibustion 20: 497–500

Lai XS, Yu J, Zhuang LX, Li JQ, Jiang GH, Zheng ZC, Zhang JW (2000) Short-term clinical effect of acupuncture therapy for vascular dementia. Journal of Guangzhou University of Traditional Chinese Medicine 17: 301–305

Leblanc GG, Meschia JF, Stuss DT, Hachinski V (2006) Genetics of vascular cognitive impairment: the opportunity and the challenges. Stroke 37(1): 248–255

Levenson JM, Sweatt JD (2005) Epigenetic mechanisms in memory formation. Nat Rev Neurosci 6: 108–118

Li SZ, Han M, Pi M, Cai T (2006) Clinical observations on effects of Xingnao Kaiqiao needling on C-reactive protein of acute cerebral infarction patients. Chinese Journal of Management in Chinese Medicine 14: 59–61

Li W, Fan JM, Jia SQ, Chen GH, Zhou HX (1996) Protective effects of electroacupuncture on the injury induced by cerebral ischemia and reperfusion in the rat. Chinese Acupuncture & Moxibustion 16: 21–22

Li YC (2005) Clinical practice of Zhi's needling treatment. Journal of Clinical Acupuncture and Moxibustion 21: 31–31

Liao XP, Wen GQ, Chen T, You ZG, Ou YF, Li TL, Wang BY (2007) Study of the related factors between location of cerebral infarction and cognitive impairment . Journal of Apoplexy and Nervous Diseases 24: 71–72

Lin HY, Xu HQ, Liu GS (2003) Cognitive function and motor function of cerebral infarction patients in recovery period. Zhejiang Journal of Preventive Medicine 15: 72–72,74

Liu B, Tang Q, Li J (2005) Effect of electroacupuncture on the cerebral ischemia in rats . Chinese Journal of Rehabilitation Theory & Practice 11: 514–515

Liu HA, Hou DF, Diao ZY, Wang Y (1997) Observation on the clinical curative effects of turbit-clearing and intelligence-improving acupuncture therapy on vascular dementia and the study on its mechanisms . Chinese Acupuncture & Moxibustion 9: 521–525

Liu J, Liao SL (1994) Preliminary observation of acupuncture therapy for 15 cases of cerebral vascular dementia. Chengdu Medical Journal 20: 45–47

Liu LY, Yang J (2003) The Influences on study and memory of synthetic vascular dementia rats under electro-acupuncturing governor meridian mostly. Journal of Chengde Medical College 20: 17–20

Liu XN, Li YZ (2000) The effect of acupuncturing on ultrastructure of ICVD in experimental rabbit. Journal of Mudanjiang Medical College 21: 1–5

Liu XN, Li YZ, Sun C (2002) The effect of acupuncturing on ultrastructure of BBB and rheology of ICVD in experimental rabbits. Journal of Mudanjiang Medical College 23: 1–4

Liu YF, Shi XM, Han JX, Wang S, Li P, Liu QZ, Zhao LR, Yang ZR, Zhao JH, Li XK (2002) Effect of acupuncture on activity of antioxydase in the brain of senescence accelerated encephalatrophy model mouse. Chinese Acupuncture & Moxibustion 22: 327–330

Lun X, Rong L, Fu B (2006) Effect of encircled skull acupuncture under CT location on the concomitant symptoms and hemorheological changes in stroke patients during restoration period. Chinese Journal of Clinical Rehabilitation 10: 140–142

Luo DH, Lai XS, Tang CZ (2003) Effects of electropuncture therapy on levels of β-AP and TGF-α in rats of AD model. Chinese Journal of Information on TCM 10: 22–23

Luo Y, Huang ZX, Xu C (1996) Mitochondria rian of experimental cortical infarction in rats. Shanghai Journal of Acupuncture and Moxibustion 15: 28–30

Lu HY, Xuan Y, Liu HL, Zhou DA, Zhang L (2003) Effects of the combination of acupuncture and drugs on serum calcitonin gene-related peptide and endothelin in the rats with vascular dementia. Journal of Beijing University of Traditional Chinese Medicine 26: 86–87

Ma T (2001) 83-case study of therapeutic effect of acupuncture on vascular dementia. Clinical Journal of Traditional Chinese Medicine 13: 410

Ma YF, Wang S, Lu B, Zhang CH, Han JX, Shi XM (2001) Study on HSP_(70)mRNA expression intervened by Xingnao Kaiqiao needling method in rats of experimental cerebral infarction. Chinese Acupuncture & Moxibustion 21: 107–112

Ma YF, Guo Y, Zhang YJ, Xu TP, Bao JZ (1997) Guantitative pathomorphologic study on inhibitory effects of acupuncture on experimental hyperplasia of mammary glands in mice. Chinese Acupuncture & Moxibustion 17: 562–564

Ma YZ, Zhou XQ (2003) Review of the Chinese and Western medical researches on vascular dementia . Journal of Henan College of Traditional Chinese Medicine 18: 74–77

McCord JM (1985) Oxygen-derived free radicals in postischemic tissue injury. N Engl J Med 312: 159–163

Meyer JS, Quach M, Thornby J, Chowdhury M, Huang J (2005) MRI identifies MCI subtypes: vascular versus neurodegenerative. J Neurol Sci 229-230: 121–129

Mo FZ, Li JQ, Chen CH, Lai XS, Zhang JW (2002) Effect of electroacupuncture on vascular dementia and AVP in brain of rats. Chinese Journal of Rehabilitation Theory & Practice 8: 129–131

Mo FZ, Li JQ, Lei LP, Chen CH, Lai XS, Zhang JW (2001) Effect of electroacupuncture and hydergine on acetylcholinesterase in encephalon of experimental vascular dementia in rats. Chinese Journal of Gerontology 21: 119–121

Mo FZ, Li JQ, Zhu L, Lei LP, Lai XS, Liu SH (2000) Effect of the electroacupuncture treatment of vascular dementia and its P_(300). Chinese Journal of Rehabilitation Theory & Practice 6: 65–69

Mou SL, Sun XL, Song CL (1994) Research on effects of head acupoint needling on brainstem evoked potential—observations on curative effect of 21 vertebrobasilar artery blood supply deficiency patients. Liaoning Journal of Traditional Chinese Medicine 21: 185–186

Nie H (1999) Dynamic observation on electroacupuncture needling treatment for apoplectic hemiplegia patients SEP. Chinese Acupuncture & Moxibustion 19: 369–370

Nikonenko AG, Radenovic L, Andjus PR, Skibo GG (2009) Structural features of ischemic damage in the hippocampus. Anat Rec (Hoboken) 292: 1914–1921

Peng WN, Ma XJ, Liu ZS (2005) 12 Patiens with severe vascular dementia treated with electroacupuncutre regulating Suihai. Journal of Clinical Acupuncture and Moxibustion 21: 34–35

Peng XH, Zhao Y, Liu Y, Liu J, Li QL (2009) Research on scalp electric acupuncture in improving the cognition function of vascular dementia patients. Journal of Sichuan of Traditional Chinese Medicine 27: 113–115

Peng XM, Jin R, Zhang JW, Lai XS (1998) Effects of acupuncture on blood lipid, hemorheology, and superoxide dismutase in cerebral thrombosis patients. Chinese Acupuncture & Moxibustion 18: 12–14

Pohjasvaara T, Erkinjuntti T, Vataja R, Kaste M (1997) Dementia three months after stroke. Baseline frequency and effect of different definitions of dementia in the Helsinki Stroke Aging Memory Study (SAM) cohort. Stroke 28: 785–792

Rockwood K (2002) Vascular cognitive impairment and vascular dementia. J Neurol Sci 203–204: 23–27

Roman GC (2002) Vascular dementia may be the most common form of dementia in the elderly. J Neurol Sci 203-204: 7–10
Ruan JW, Zheng PY, Wen M, Rao ZD (1999) Clinical observations on the effects of acupuncture on the cognition of the early stroke patients. Acupuncture Research 24: 223–226
Sachdev PS, Brodaty H, Valenzuela MJ, Lorentz L, Looi JC, Wen W, Zagami AS (2004) The neuropsychological profile of vascular cognitive impairment in stroke and TIA patients. Neurology 62: 912–919
Saito K, Kimura K, Minematsu K, Shiraishi A, Nakajima M (2003) Transient global amnesia associated with an acute infarction in the retrosplenium of the corpus callosum. J Neurol Sci 210: 95-97
Schmahmann JD, Sherman JC (1998) The cerebellar cognitive affective syndrome. Brain 121: 561–579
Schonknecht P, Pantel J, Hartmann T, Werle E, Volkmann M, Essig M, Amann M, Zanabili N, Bardenheuer H, Hunt A, Schroder J (2003) Cerebrospinal fluid tau levels in Alzheimer's disease are elevated when compared with vascular dementia but do not correlate with measures of cerebral atrophy. Psychiatry Res 120: 231–238
Shen DK, Hou SW, Xu NG (1998) Pathological and histological study on the protective effect of electroacupuncture therapy on neuronic injury of rat's local cerebral ischemia. Chinese Journal of Traditional Medical Science and Technology 5: 269–270
Shi RH, Ji GC (1997) Effects of electroacupuncture on plasma monoamine neurotransmitters in rabbits of acute myocardial ischemia. Chinese Acupuncture & Moxibustion 17: 606–607
Shi XM, Han JX, Zhao HJ, Yang ZR, Li P, Wang S, Liu QZ, Zhao LR, Zheng HY (1999) Effects of acupuncture on nucleoproteins and chromatinic nonhistone of brain cells in mice of encephalatrophy. Chinese Acupuncture & Moxibustion 19: 43–46
Shi XM (2005) Clinical research on the treatment of 9,005 cases of apoplexy with the acupuncture method of sharpening mind and inducing consciousness. Guiding Journal of Traditional Chinese Medicine and Pharmacy 11: 3–5
Si QM, Wu GC, Cao XD (1998) Protective effect of electroacupuncture on cerebral ischemia. Shanghai Journal of Acupuncture and Moxibustion 17: 40–41
Sun ZR, Huo LG, Liu LL, Li SL (1994) Study on MEP produced by acupuncture scalp. Chinese Journal of Traditional Medical Science and Technology 1: 14–18
Szatmari S, Fekete I, Csiba L, Kollar J, Sikula J, Bereczki D (1999) Screening of vascular cognitive impairment on a Hungarian cohort. Psychiatry Clin Neurosci 53: 39–43
Tanaka T, Wakamatsu T, Daijo H, Oda S, Kai S, Adachi T, Kizaka-Kondoh S, Fukuda K, Hirota K (2010) Persisting mild hypothermia suppresses hypoxia-inducible factor-1alpha protein synthesis and hypoxia-inducible factor-1-mediated gene expression. Am J Physiol Regul Integr Comp Physiol 298: R661–671
Tang M, Yang XL, Wang L, Li XY (2007) The effect of acupuncture on behavior and neurotransmitter in rats with vascular dementia. World Journal of Integrated Traditional and Western Medicine 2: 338–340
Tang Q, Feng J, Zhang CY, Yang Y, Meng QG, Cheng WP (1996) Study of 60 acute cerebral infarction SEP cases on different rotating time for head acupoint penetration needling . Chinese Acupuncture & Moxibustion 16: 1–4
Tatemichi TK, Desmond DW, Stern Y, Paik M, Sano M, Bagiella E (1994) Cognitive impairment after stroke: frequency, patterns, and relationship to functional abilities. J Neurol Neurosurg Psychiatry 57: 202–207
Tham W, Auchus AP, Thong M, Goh ML, Chang HM, Wong MC, Chen CP (2002) Progression of cognitive impairment after stroke: one year results from a longitudinal study of Singaporean stroke patients. J Neurol Sci 203-204: 49–52
Tian Q, Ma J, Liu YX, Sun GJ (2002) Effects of electroacupuncture on the cerebral water content and SOD level of experimental cerebral ischemia rats during acute period. Hubei Journal of Traditional Chinese Medicine 24: 3–4
Tuominen S, Miao Q, Kurki T, Tuisku S, Poyhonen M, Kalimo H, Viitanen M, Sipila HT, Bergman J, Rinne JO (2004) Positron emission tomography examination of cerebral blood flow and glucose metabolism in young CADASIL patients. Stroke 35: 1063–1067

van der Flier WM, van Straaten EC, Barkhof F, Verdelho A, Madureira S, Pantoni L, Inzitari D, Erkinjuntti T, Crisby M, Waldemar G, Schmidt R, Fazekas F, Scheltens P (2005) Small vessel disease and general cognitive function in nondisabled elderly: the LADIS study. Stroke 36: 2116–2120

Vemula S, Roder KE, Yang T, Bhat GJ, Thekkumkara TJ, Abbruscato TJ (2009) A functional role for sodium-dependent glucose transport across the blood-brain barrier during oxygen glucose deprivation. J Pharmacol Exp Ther 328: 487–495

Wan J (2004) Clinical observations of 108 cases of scalp acupoint block plus traditional Chinese medicine for senile vascular dementia. Chinese Journal of Clinical Rehabilitation 8: 5324

Wang GX, Liu F, Zhao FL, Zhang S (1993) Clinical effects of scalp acupuncture on ischemic stroke. Chinese Acupuncture & Moxibustion 13: 4

Wang H, Dai ZP, Zhu MF, Yang DD, Wang H (2007) Relationship between AchE activity in hippocampus and effect of ketamine on spatial learning and memory of early developing rats. Chinese Journal of Clinical Pharmacology and Therapeutics 12: 672–675

Wang HM (2007) Clinical effect observation on electroacupuncture treatment for vascular dementia . Tianjin Journal of Traditional Chinese Medicine 24: 218–220

Wang L (2002) Effects of electroacupuncture on learning and cerebral heme oxygenase activity in rats with vascular dementia. Journal of Guangzhou University of Traditional Chinese Medicine 19: 286–288,291

Wang LC, Liu HY, Wang ZY, Li BD, Ge JY, Ma L (2005) Effect of acupuncture on cerebral NO content and NOS activity and hemorheology in vascular dementia rats. Acupuncture Research 30: 195-198

Wang QY (2004) Academic features of Jin's three needling treatment for cerebral palsy . Shanghai Journal of Acupuncture and Moxibustion 23: 3–4

Wang RJ (2001) Clinical analysis of acupuncture therapy for 42 vascular dementia patients. Journal of Clinical Acupuncture and Moxibustion 17: 8–9

Wang S, Shi XM, Zhang CS (1995) Hemorheological observations on the treatment of apoplexy aura with the acupuncture method of sharpening mind and inducing consciousness. Journal of Clinical Acupuncture and Moxibustio 11: 25

Wang SF, Ma ZJ, Yang YG, Yang JH, Zhao SX (2000) Biological holographic acupoint injection of neural live element for 70 early vascular dementia patients. Chinese Journal of Information on TCM 7: 44–45

Wang XH, Dong CC, Han JX (2003) Influence of acupuncture on catecholamine neurotransmitter in brain tissues of SAMP8. Shanghai Journal of Acupuncture and Moxibustion 22: 21–23

Wen GQ, Liao XP, Chen T, Wang BY, Ou YF, Long ZG, Li JJ, Li CZ, Deng YD, He XY (2006) Diagnostic value of diffusion tensor magnetic resonance imaging in patients with cerebral infarction . Chinese Journal of Cerebrovascular Diseases 3: 450–453

Wen GQ, Liao XP, Chen T, Wang BY, Ou YF, Li JJ, Li CZ, Huang SX, Huang YL (2007) Clinical application of diffusion tensor imaging in cognitive impairment of patients with acute cerebral infarction . Chinese Journal of Neurology 40: 437–440

Wentzel C, Rockwood K, MacKnight C, Hachinski V, Hogan DB, Feldman H, Ostbye T, Wolfson C, Gauthier S, Verreault R, McDowell I (2001) Progression of impairment in patients with vascular cognitive impairment without dementia. Neurology 57: 714–716

Wu P, Mills E, Moher D, Seely D (2010) Acupuncture in poststroke rehabilitation: a systematic review and meta-analysis of randomized trials. Stroke 41: e171–179

Xia Y, Guo XQ, Zhang AZ, Cao XD, Li P (1985) Inhibitory effect of analogous electro-acupuncture on experimental arrhythmia. Acupunct Electrother Res 10: 13–34

Xia Y, Li P (1984) Effects of microinjection of 5-HT, morphine and other drugs into the medial medulla on ventricular extrasystoles induced by hypothalamic stimulation in the rabbits. Acta Physiol Sinica 36: 538–545

Xu JP, Xiao SJ, Guo HZ (2002) Clinical observations of encircled acupuncture on head and body acupuncture for hemiplegia cognitive impairment . Journal of Clinical Acupuncture and Moxibustion 18: 4–5

Xu NG, Shen DK, Zhou YP, Wang YL (2001) Effects of electroacupuncture on cortical somatosensory evoked potential and cellular ultrastructure in rats of local cerebral ischemia . Journal of Traditional Chinese Medicine 42: 342–343

Xu YM (1992) Effects of acupuncture on blood lipid changes of ishemic stroke patients. Shanghai Journal of Acupuncture and Moxibustion 12: 13

Yang YH, Zhang JF (1997) Scalp acupoint injection treatment for 25 vascular dementia patients. Hubei Journal of Traditional Chinese Medicine 19: 43

Yu SH, Wang CM, Liu ZF, Zang ZL (2005) Clinical observations of 21 cases on the treatment of vascular dementia by the acupuncture therapy of regulating spirit and benefiting intelligence . Journal of Clinical Acupuncture and Moxibustion 21: 25–26

Yu SG, Zhou QZ, Zhang XR, Zhao JL, Wei JL, Tang Y (2002) Experimental study on the effect of acupuncture on AchE activity of senile dementia rats. Journal of Clinical Acupuncture and Moxibustion 18: 43–45

Yuan SH, Xu J, Zhang Y (2002) Curative effect observation on electroacupuncture plus decoction treatment for vascular dementia . Journal of Clinical Acupuncture and Moxibustion 18: 40-41

Yuan SS, Zhang SY (2010) Meta analysis of acupuncture therapy for post stroke cognitive impairment. Chinese Journal of Ethnomedicine and Ethnopharmacy 19: 47–48

Zhang CH, Wang S, Zhao HJ, Liu QZ, Zheng HY, Yang ZR (1999) Effects of acupuncture on brain tissue SOD and MDA of experimental cerebral ischemia rats. Journal of Tianjin University of Traditional Chinese Medicine 18: 39–41

Zhang CH, Wang S, Zheng HY, Li T, Zhang L, Shi XM (2001) Effect of acupuncture on cerebral apotosis in rats with local cerebral ischemia . Acupuncture Research 26: 102–105

Zhang LX, Li XL, Wang L, Han JS (1997) Rats with decreased brain cholecystokinin levels show increased responsiveness to peripheral electrical stimulation-induced analgesia. Brain Res 745: 158–164

Zhang XZ, Xiao ML (2000) Correlation between auricular acupuncture and dysmnesia and c-fos expression in vascular dementia rats. Journal of Clinical Acupuncture and Moxibustion 16: 34–36

Zhang XZ, Xiao ML, Sun GJ (2001) Correlation between auricular acupuncture induced improvement of memory and bcl-2 expression in vascular dementia rats. Acupuncture Research 26: 106–110

Zhang XZ, Xiao ML, Sun GJ, He ZG, Lv MZ (2001) The Improvement of memory of vascular dementia rats by ear-acupuncture and its relationship with the expression of nNOS. Shanghai Journal of Acupuncture and Moxibustion 20: 39–41

Zhang XZ, Yang HB, Zhao L, Wu BQ, Yu JC, Cheng HY, Han JX (2010) Effect of acupuncture on cognition and glucose transport in brain of multiple-infarct dementia rat. Acta Chinese Medicine and Pharmacology 38: 79–82

Zhao BY, Yue XL, Fu BZ (1995) Acupoint injection therapy for 234 cerebral vascular dementia patients. Shanghai Journal of Acupuncture and Moxibustion 14: 202

Zhao H, Sun ZR, Sun YZ, Xing YL, Wang DY (2004) Observation on therapeutic effect of acupuncture on vascular dementia. Chinese Acupuncture & Moxibustion 24: 525–527

Zhao JX, Tian YX, Cheng YW, Ma WX, Wang BY, Tong JJ (2000) Clinical observations on treating VD with kidney-supplementing & mind-clearing acupuncture method . Hebei Journal of Traditional Chinese Medicine 22: 844–846

Zhao L, Zhang H, Zheng Z, Huang J (2007) The clinical research on scalp electric acupuncture improving the cognition function of 90 vascular dementia patients . Journal of Sichuan of Traditional Chinese Medicine 25: 105–108

Zheng J, Zhou XM, Guo XX, Jin HS (2009) Effect of needling Baihui and Taixi on cognitive function of rat model with vascular dementia. Chinese Archives of Traditional Chinese Medicine 27: 1037–1039

Zheng L, Zhuang LX, Li YH (2000) Observations on the curative effect of acupuncture and herbs on vascular dementia. Shanghai Journal of Acupuncture and Moxibustion 19: 8–10

Zheng ZY, Zhao LG, Mei CJ, Wang LC (2005) Experimental study on effects of acupuncture on brain tissue SOD activity and NO of vascular dementia rats. New Journal of Traditional Chinese Medicine 37: 90–92

Zovein A, Flowers-Ziegler J, Thamotharan S, Shin D, Sankar R, Nguyen K, Gambhir S, Devaskar SU (2004) Postnatal hypoxic-ischemic brain injury alters mechanisms mediating neuronal glucose transport. Am J Physiol Regul Integr Comp Physiol 286: R273–282

Chapter 10
Acupuncture and Moxibustion for Antiaging

Huangan Wu

Abstract Mechanisms under aging and its related diseases have increasingly been a research focus for the potential of leading to novel approaches to antiaging, which is of great significance of human health. In recent years, much attention has been given to complementary and alternative therapies, among which are traditional Chinese medicine (TCM) treatments such as acupuncture and moxibustion. While substantial research evidence is available supporting their antiaging effects, acupuncture and moxibustion are characterized by easy administration and cost-effectiveness. According to biomedical theories, moxibustion and acupuncture are supposed to influence aging processes through various mechanisms, such as elimination of free radical damage and regulation of immunity, neuroendocrine, lipid metabolism, rheological characteristics of blood flow, trace elements, telomerase, and carbonyl poisoning. These findings reveal the potential value of acupuncture and moxibustion for antiaging in experimental research as well as practical application.

Keywords Acupuncture • Moxibustion • Aging • Review

10.1 Introduction

Aging or senescence is a slow, continuous process, characterized by systemic and progressive degeneration in tissues and cells with increasing years after an organism's maturation (Zhang and Yu 2000). During the processing of aging, an organism

HG. Wu (✉)
Shanghai Institute of Acupuncture Moxibustion and Meridian, Shanghai University of Traditional Chinese Medicine, Shanghai 201203, China
e-mail: wuhuangan@126.com

Y. Xia et al. (eds.), *Current Research in Acupuncture*,
DOI 10.1007/978-1-4614-3357-6_10, © Springer Science+Business Media New York 2013

tends to lose functions in various organs, suffer from diseases and finally come to death. Aging-related diseases represent a heavy burden on families and society. So it is of great importance to reveal mechanisms underlying aging and its related diseases, thereby developing effective, safe, and reliable methods which can slow down, or even reverse, the aging process.

In exploring the mechanism of aging, research attention has been shifted from studying the whole organ to cell biology and molecular biology. Various putative theoretical models have been put forward for understanding the aging process, including, for example, genetics theory of aging, cross-linking theory of aging, immune theory of aging, neuroendocrine theory of aging, free radical theory of aging, nonenzymatic glycation (NEG)-induced aging theory, telomerase–telomere theory, pineal gland theory, carbonyl poisoning theory, etc.

As for measures for aging prevention, modern biomedicine emphasizes lifestyle factors such as maintaining a good mood, having scientific diet, paying attention to rational nutrition, keeping body building exercising, and scientific living pattern. Moreover, it is now clear that free radical scavengers, antioxidants, and some agents that can lower serum cholesterol can also to some extent slow down aging processes. However, these medications do not always elicit reliable antiaging effects, and related side effects provoke great controversy.

Taken together, there have been great advances in basic-science studies of aging mechanisms. However, antiaging methods and medications are developed rather slowly. Acupuncture and moxibustion have evidence-based effects of anti-inflammation and immunoregulation, which not only forms the basis for acupuncture and moxibustion to be clinically used for preventing and treating various types of immune disorders and acute or chronic inflammatory diseases but also demonstrates the potential value of acupuncture and moxibustion in deferring aging and preventing disease.

10.2 Clinical Application of Acupuncture and Moxibustion for Antiaging and Clinical and Experimental Research in This Area

10.2.1 Commonly Used Acupoints

1. Acupoints in DU meridian and bladder meridian of foot-taiyang.

   According to traditional Chinese medicine (TCM) theories, DU meridian (Fig. 10.1) governs all yang meridians in the body. Human Zang-fu organs are believed to be governed by the qi of DU meridian via back-located points in bladder meridian of foot-taiyang (BL) (Fig. 10.2). Acupoints, such as Shenshu (BL-23), Mingmen (DU-4), and Dazhui (DU-14), of the two meridians are thought to have effects of tonifying kidney (or qi of kidney) and consolidating source qi. Most yang qi deficiency, typically with old patients, is suitable for moxibustion. Moxa is acrid and hot (pertaining to yang) in property, which can

Fig. 10.1 DU meridian (DU)

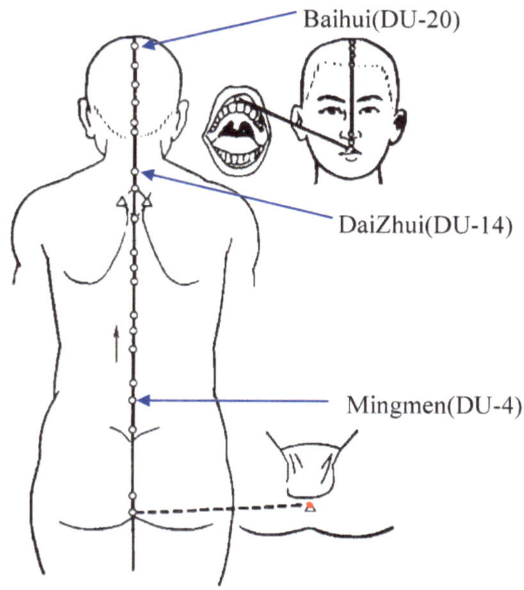

tonify and strengthen yang in the body and expel deep-rooted cold evil, thus effectively preventing and curing diseases. Gao et al. studied the effect of moxibustion on immune function in aging mice, finding that moxibustion applied on acupoints of DU meridian, such as Mingmen (DU-4), Dazhui (DU-14), and Baihui (DU-20), could delay the atrophy of immune organs and boost immune function of humans (Gao and Wang 2004).

2. Acupoints on RN meridian.

   According to TCM, RN meridian (Fig. 10.3) has the function of governing all yin meridians of the body. Shimen or Dantian is known as "the source of vital qi." Other points, such as Zhongji (RN-3), Guanyuan (RN-4), Qihai (RN-6), Shenque (RN-8), located in lower abdomen can as well tonify qi, strengthen the body, and prolong life. Zhongwan (RN-12) is the alarm acupoint of stomach meridian of foot-yangming (ST), moxibustion on which can warm and activate spleen yang, tonifying and replenishing middle qi. Many aged people suffer from cold, with the symptom of spontaneous sweat due to deficiency and lassitude, resulting from deficiency in spleen, kidney, and essential qi. When moxibustion is applied on acupoints such as Qihai (RN-6), Guanyuan (RN-4), and Zhongwan (RN-12), it can tonify source qi, harmonize the nutrient and defense, and rehabilitate the function of Zang-fu organs. Chen et al. applied ginger-partitioned moxibustion on Qihai (RN-6), Guanyuan (RN-4) and mild moxibustion on Dazhui (DU-14), and Zusanli (ST-36) in elder patients, showing that the moxibustion group was better than the control group in terms of alleviating aging-related symptoms. Moxibustion could increase the serum SOD content and decrease LPO content, and improve the ratio of estradiol/testosterone and

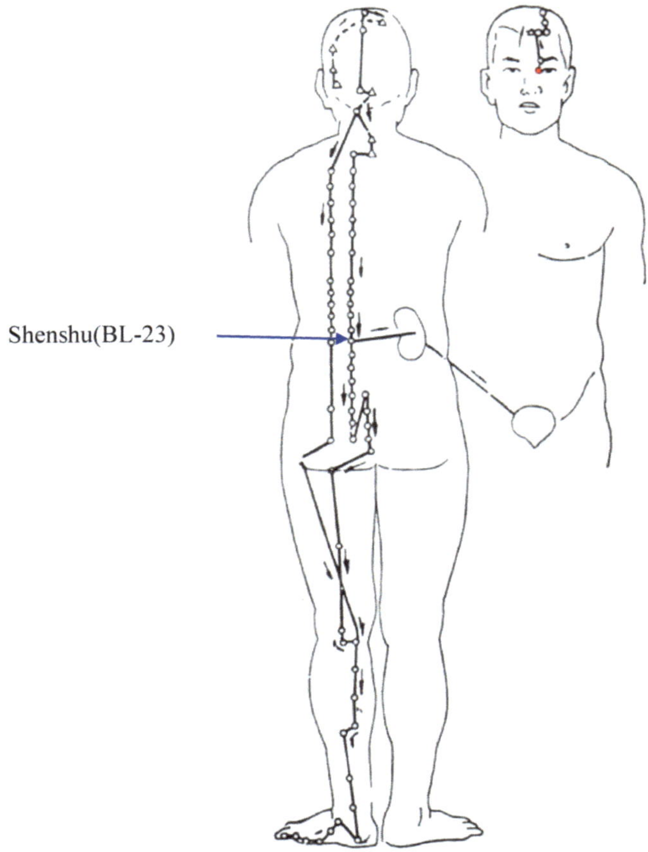

Fig. 10.2 Bladder meridian of foot-taiyang (BL)

zinc/copper (Chen and Tang 1994). Wang et al. applied moxibustion on Shenque (RN-8) in middle-aged and aged patients, reporting that moxibustion could increase the number of CD2+ and CD4+ of T cell subsets in the serum, decrease the number of CD8+, increase ratio of CD4+ and CD8+, and regulate serum IgG and serum IgA and humoral immunity (Wang et al. 1996).

3. Acupoints on stomach meridian of foot-yangming.

According to TCM, stomach meridian of foot-yangming (ST) (Fig. 10.4) is abundant in qi and blood and is closely related with spleen meridian. Stomach, representing exterior (yang), and spleen, representing interior (yin), are thought as the acquired foundation and origin of qi and blood, so most acupoints of stomach meridian can be used to regulate and tonify spleen and stomach—that is, tonify qi and replenish blood. Moxibustion can effectively strengthen the function of spleen and stomach. Applying moxibustion on Zhongwan (RN-12) and

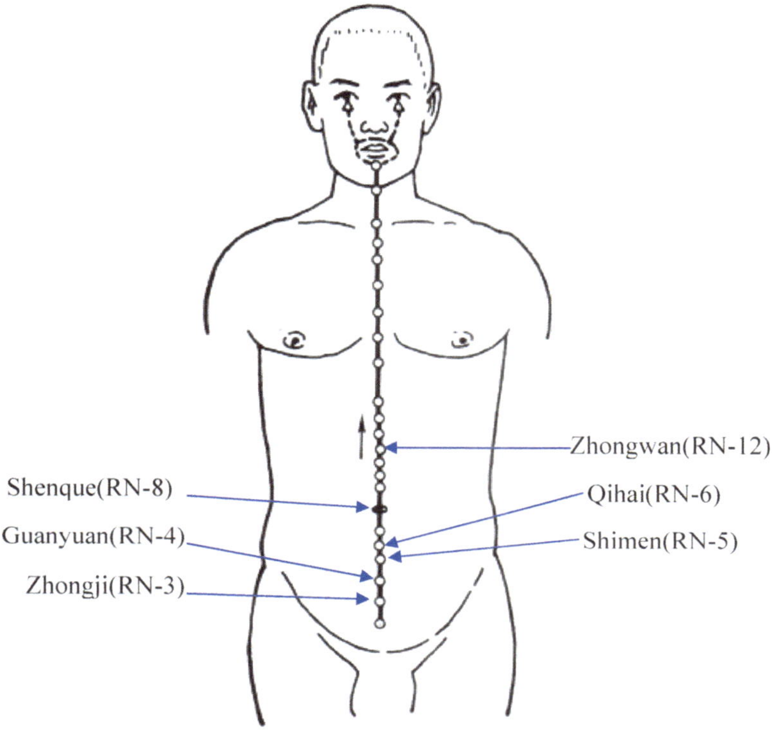

Fig. 10.3 Ren meridian (RN)

Zusanli (ST-36) can produce therapeutic effects in elder patients suffering from poor appetite and lassitude due to spleen deficiency. Because Zusanli (ST-36) is the lower sea point of stomach meridian of foot-yangming, moxibustion on it can improve transportation and transformation of spleen, invigorate spleen, warm middle abdomen, and tonify acquired foundation.

10.2.2 Effective Methods

As unique meridian-based treatments in TCM, acupuncture and moxibustion have been used for health maintenance and life prolonging for more than 2,000 years, especially moxibustion spreads far and wide (Zhu et al. 2006). Moxibustion, which is considered to be able to reinforce healthy qi and eliminate pathogenic factors through heating power and drug action (Wu 2006), is one of the most commonly used therapeutic methods for treating diseases. Moxibustion is a kind of external treatment by burning moxa and roasting on superficial areas, achieving the purpose

Fig. 10.4 Stomach meridian of foot-yangming (ST)

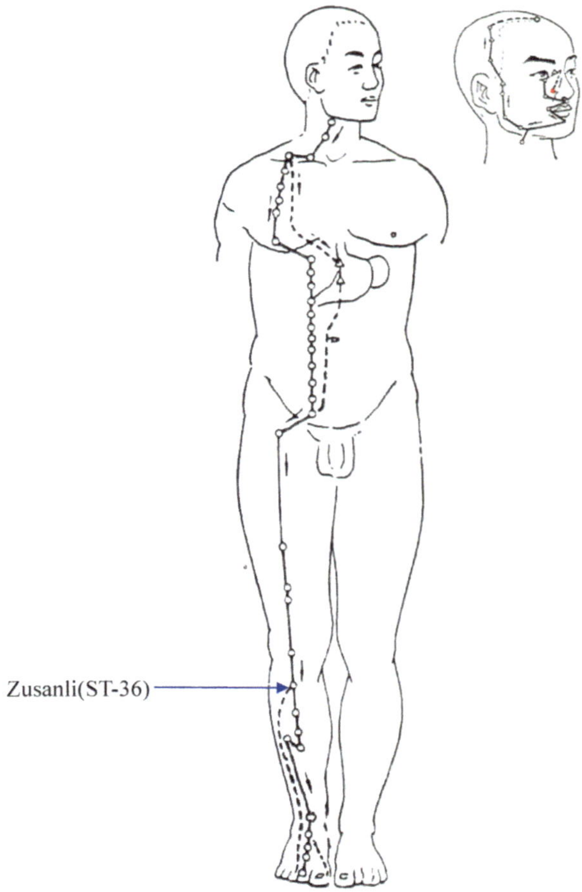

of treating diseases and maintaining health by warming and activating qi and blood. Moxa is warm in nature and neutral and bitter in flavor, acting on liver, spleen, and kidney meridians. Moxa, aromatic in smell and easily ignited, has mild fire power when being ignited. Its fire power can penetrate into the skin and muscle with aromatic smell, so it can regulate qi and blood, expel cold and dampness, and dredge meridians. Clinical studies found that moxibustion with moxa could make elder patients biologically younger than chronological age, significantly alleviating symptoms relating to old age and improving physiological functions. Interestingly, patients obtained better treatment outcomes if they started moxibustion for health care in the early stage of gerontism (Liao et al. 2004). Studies exploring the mechanism underlying antiaging effects of moxibustion have shown that moxibustion may take effect through different manners, including enhancing immune function, regulating the free radical and lipid metabolisms (Gao et al. 2003), influencing the blood flow and trace element, and regulating apoptosis (Du and Sun 2005; Du et al. 2006a, b).

Fig. 10.5 Mild moxibustion

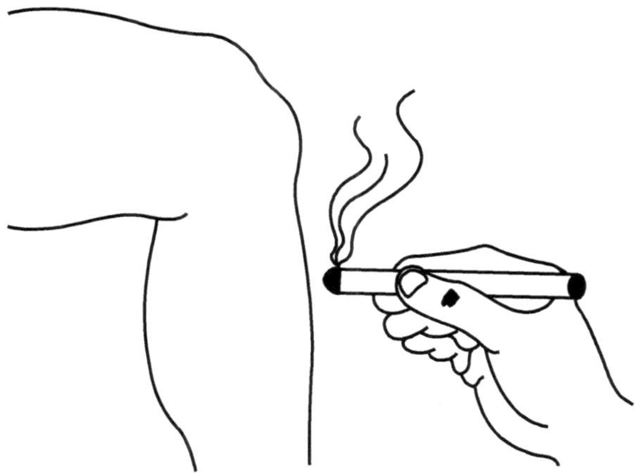

In clinical practice, moxibustion is administered in different forms, including suspended moxibustion, direct moxibustion, postulating moxibustion, indirect moxibustion, and natural moxibustion.

1. Mild moxibustion (Fig. 10.5)
   Acupoints: Guanyuan (RN-4), Shenque (RN-8), Zusanli (ST-36).
   Methods: Ignite moxa roll and hold 2–3 cm over an acupoint. It is suitable for patients to feel warm without pain. Apply moxibustion on each acupoint for 10–15 min each time until the skin turns flushed. The treatment is provided once every day and seven times at the very beginning of each month.
2. Indirect moxibustion (Fig. 10.6)
   Acupoints: Guanyuan (RN-4), Qihai (RN-6), Zusanli (ST-36), Shenque (RN-8).
   - Ginger-partitioned moxibustion
     Methods: Cut a slice of fresh ginger about 0.2–0.3 cm thick, punch several holes in it and place it on the points selected. On top of this, a moxa cone is placed and ignited. When the patient feels scorching, raise it up. Apply five cones on one acupoint until the skin turns flushed. The treatment is done for 10 days, then, after an interval of 3 days, repeat another course.
   - Aconite cake-partitioned moxibustion
     Methods: Grind aconite into powder, and then mix 2.5 g powders with appropriate amount of yellow wine into thick paste. After that, press the thick paste to make a cake of 0.5 cm in thickness and 2.5 cm in diameter by mold. Punch several holes in the center of the cake with the needle, and put a moxa cone upon the cake for moxibustion. Apply 1–2 cones on each acupoint for each time. The treatment is done once every 2 days and 10 times make up of one treatment course. There is a 3–5-day interval between the two courses.
   - Drug-cake-partitioned moxibustion
     Methods: Grind the herbs for invigorating spleen and tonifying kidney and activating blood to dredge collaterals, such as Bu Gu Zhi, Dang Gui, Sheng Di, Xian Mao, Xian Ling Pi, and Rou Gui into powders. Mix them with 80 % alcohol,

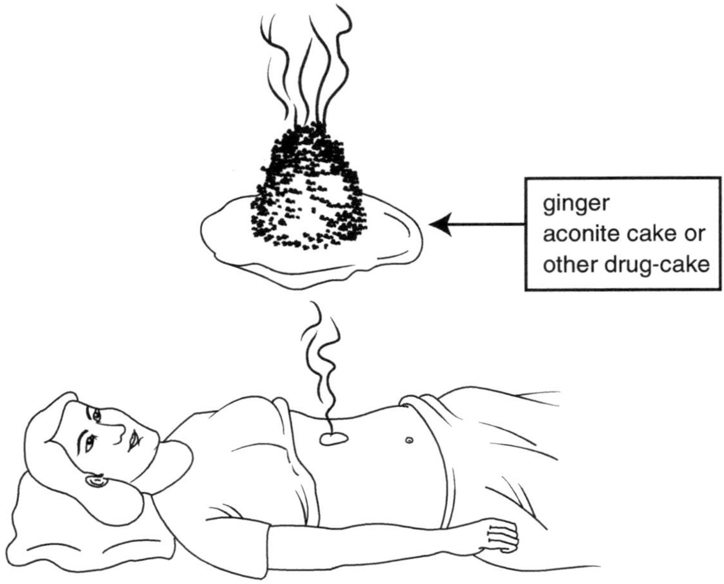

Fig. 10.6 Indirect moxibustion

and then make cakes of 2.5 cm in diameter and 0.5 cm in thickness by mold. Put a moxa cone upon the cake for moxibustion. Apply 1–2 cones on each acupoint for each time. The treatment is done once every 2 days. The number of cones should be more than 480 on each acupoint for each person every year.

3. Festering moxibustion (Fig. 10.7)
   Acupoint: Zusanli (ST-36), Shousanli (LI-10), etc.
   Methods: 7–9 moxa cones of 0.7 cm high are burnt directly on the acupoint. After one cone burns out, clean the local area of the acupoint by normal saline, and then apply the next one. Gently clap on the nearby area of the acupoint to relieve pain when applying postulating moxibustion. It will cause aseptic suppuration on the acupoint; after moxibustion, put a small medicinal paste on it till the scar heals. When the scar heals, the treatment may be applied again. Infection must be avoided, and it is necessary to strengthen nutrition and keep light diet.
4. Warming needle moxibustion (Fig. 10.8)
   Acupoint: Mingmen (DU-4), Shenshu (BL-23), Zusanli (ST-36), etc.
   Methods: After sterilizing local skin of acupoint with alcohol, insert the needle of 1.5 cun into skin to a depth of 0.8–1.2 cun. After the arrival of qi and with the needle retained in the point, get a small section of a moxa stick (about 2 cm long) and put it on the handle of the needle, and then ignite the moxa stick from its bottom and let it burn out. After that, remove the scrap and pull out the needle. The treatment is done once every 2 days and 10 times make up of one treatment course. There is a 1-week interval between two courses.
5. Taiyi miraculous moxa stick (thunder–fire miraculous moxa stick) (Fig. 10.9)
   Acupoint: Qihai (RN-6), Zhongwan (RN-12), Zusanli (ST-36).

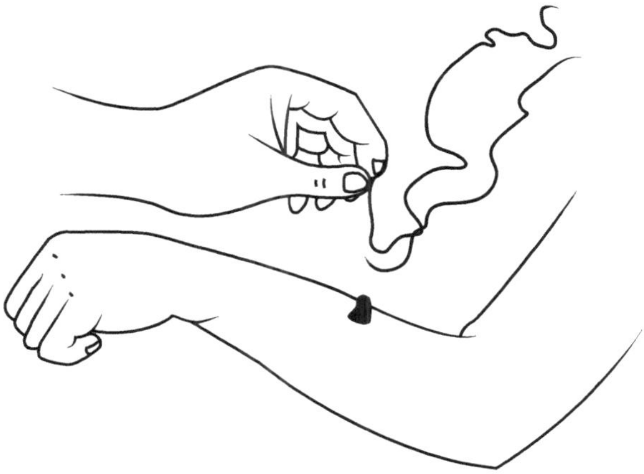

Fig. 10.7 Festering moxibustion

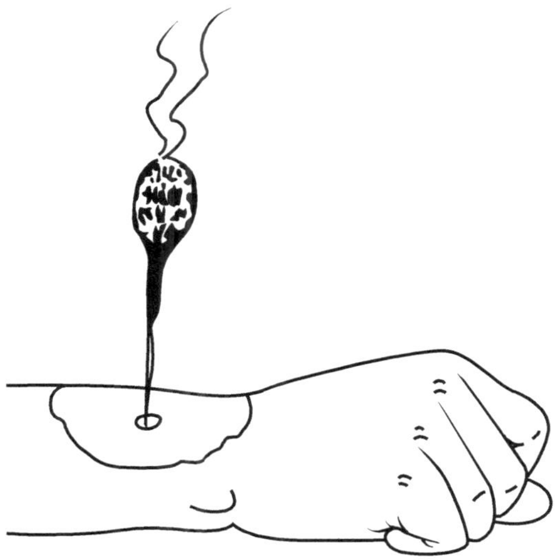

Fig. 10.8 Warming needle moxibustion

Methods: Spread the 5–7 layer cotton clothes on the local area for moxibustion. Ignite one end of the Taiyi miraculous moxa stick, Thunder–fire miraculous moxa stick, and aim at the acupoint and press it tightly onto the cotton, transmitting the warmth and pharmaceutic substance into the deep areas. If the patient feels burning pain in the local place, the practitioner may lift up the needle a little, and go on after the heat abates. If the condition allows, two sticks may be ignited and be used by turns, improving therapeutic efficacy by transmitting the continuous pharmaceutic effects into the skin with the heat power. Apply 5–7 cones on each acupoint for each time. The treatment is done once every day and one treatment course consists of ten times; continue moxibustion in 3–5-day interval between the two courses.

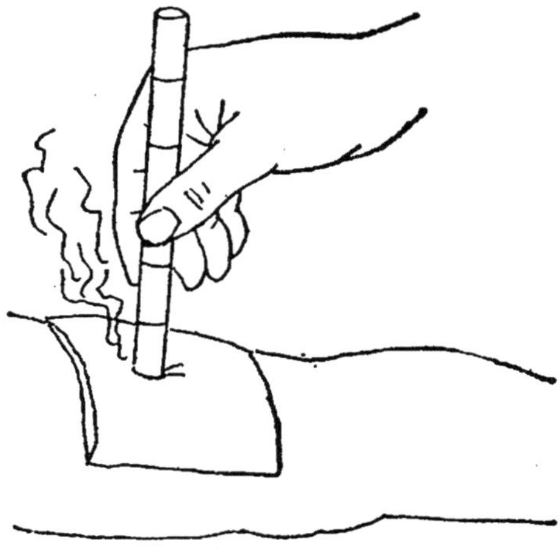

Fig. 10.9 Taiyi moxibustion (Thunder–fire moxibustion)

10.3 Mechanisms Underlying Antiaging Effects Induced by Acupuncture and Moxibustion

10.3.1 Immune-Regulatory Effects of Acupuncture and Moxibustion on Aging

Numerous studies have demonstrated that acupuncture and moxibustion can alter specific as well as nonspecific immunity and exert significant influences on both immune cells and immune molecules. Experimental studies conducted in aging animals showed that acupuncture/moxibustion has holistic and bidirectional regulatory effects on immune function.

Roy Wallord, the first to put forward the immune theory of aging 40 years ago, held that aging of normal humans and animals was associated with the decline of immune function and the presence of autoimmune response (Van den Berg et al. 2000). With the increasing age and under the influence of complex factors, the immune system tends to take some self-tissues as antigen, resulting in autoimmune responses. These responses exert harmful influences on cells, tissues, and organs of an organism and thereby lead to autoimmune diseases, accelerating the organism's aging and death (Noga and Michal 2009). With the increasing age, the function of the immune system declines—for example, lowered function of T lymphocyte results in the decline of an organism's resistance to disease—and so does the reliability of the immune system (DiCarlo et al. 2009). Moreover, in the process of aging, the thymus gland begins to degenerate and secretion of thymosin decreases, and then antibodies produced by autoimmunity indiscriminately destroy self and nonself cells. These are reasons responsible for aging (Yang et al. 2005). The functional decrease of the immune system is not only the result of aging but also a factor contributing to the acceleration of aging. Considering these findings, an in-depth understanding of antiaging effects of acupuncture and moxibustion, from the immunological viewpoint, has became an important part of research efforts made to find new ways for preventing aging. Increased research attention is being given to this area which will prove to be a promising field of the integration of acupuncture into modern biomedicine.

10.3.1.1 Influences of Acupuncture and Moxibustion on Immune-Regulatory Molecules

The immune molecules include immunoglobulin, complement system, cytokines, major histocompatible antigen, and other molecules. They play different roles in humans and animals.

IgG, the immunoglobulin, can sterilize bacteria and eliminate virus. IgM plays a critical role in early defense, bacteriolysis and hemolysis. Yim et al. found that electroacupuncture decreased serum IgG and serum IgM in rats with arthritis mediated by type II collagen and simultaneously decreased serum IL-6, serum TNF-α, serum INF-γ, antibody to type II collagen, and the amount of CD69 +/CD3e + and CD11a +/CD19 + in the lymph nodes of rats (Yim et al. 2007a, b). They also observed therapeutic effects of acupoint injection of Zisu on asthma in rats, possibly due to that point injection effectively decreased the number of CCR3+ cells in the bronchoalveolar fluids and inhibited mRNA expression of the inflammatory cytokines such as IL-4 (Yim et al. 2007a, b). Cytokines, which is synthesized and secreted by activated immune cell (mononuclear macrophage, T cell) or nonimmune cell (vascular endothelial cell, epithelial cell), can mediate interactions between immune cells, regulate physiological functions of various cells, and mediate inflammatory responses that participate in immune responses and tissue repair. Chae et al. found changes in cytokines level after acupuncture and moxibustion in rats with palm edema induced by carrageenan. These findings showed that after moxibustion the expression of IL-6, β-NGF, and TIMP-1 was significantly lowered, but no significant reductions in other 16 cytokines (Chae et al. 2007).

10.3.1.2 Influences of Acupuncture and Moxibustion on Immune Cells

Immune cells generally refer to all the cells and their precursor cells which participate in, or associated with, immune responses, mainly consisting of lymphocytes, mononuclear macrophages, granulocyte, erythrocyte, mastocyte, and hematopoietic stem cells. They all play important roles in immune responses except that hematopoietic stem cells, the ancestral cell of other immune cells, do not participate in immune responses.

1. Influences of acupuncture and moxibustion on regulating T lymphocytes.

   T lymphocytes are the important immune cells of organisms, and thus changes in their amount and functions are an important indicator when evaluating the cellular immune function. Numerous studies have confirmed that one possible mechanism mediating therapeutic effects of acupuncture and moxibustion is that they can regulate functions of T lymphocytes. Acupuncture and moxibustion can not only increase the rate of T lymphocytes and their subgroup in peripheral blood but also enhance their activities. Feng et al. observed the influence of acupuncture and moxibustion on the amount of T lymphocyte subgroups in peripheral venous blood in rats with spleen deficiency indicated by the reduction in the amount of CD3, CD4, and CD8. After the application of acupuncture and moxibustion in Tianshu (ST-25), the amount of CD3 and CD4 and the ratio of CD4/CD8 were significantly increased (Jiang et al. 2005). Gao et al. studied the influence of moxibustion on the immune function of subacute aging mice caused by D-galactose. After 6 weeks treatment (once a day, 5 times per week, and 3 min per time) in acupoints of DU meridian, such as Dazhui (DU-14), Mingmen (DU-4), and Baihui (DU-20), they reported that moxibustion could delay the influence of modeling on the index of thymus and spleen, in terms of reversing the disorder of T-cell subgroups, decreasing the rate of CD8 and increasing the rate of CD3 and CD4 and the ratio of CD4+/CD8+. There are no significant differences between moxibustion and normal groups but significant differences between moxibustion and model groups (Gao and Wang 2004).

2. Influences of acupuncture on the erythrocyte immunity.

   Zhan et al. found that when the mouse received acupuncture on Zusanli (ST-36), with the rate of positive coagulation measured by adopting yeast blood coagulation induced by complements at coagulation titer of 1:16, there were significant differences between control and acupuncture groups (Zhan and Hua 2001). Gao et al. stimulated Zusanli (ST-36) of rats by electroacupuncture and detected the immune function of erythrocytes by the methods of erythrocyte C3b receptor-yeast rosette and RBC-IC rosette test, reporting that the percent of CD4+ cells, rate of C3b receptor rosette, and rate of RBC-IC rosette in Zusanli (ST-36) group were significantly higher than those of the control group (Gao et al. 2000).

3. Influences of acupuncture and moxibustion on phagocytosis of macrophage.

   Zhu et al. investigated the function of liver macrophages after needling bilateral Zusanli (ST-36) and Guanyuan (RN-4) of elder rats. They found that acupuncture increased liver macrophages not only in amount but also in volume, and the liver macrophage was in the activated state (Zhu et al. 2003). Zhu et al. divided 130 mice into a control group, a control group treated by moxibustion, an immnuodeficiency group, and an immnuodeficiency group treated by moxibustion, evaluating the phagocytosis of macrophages by the method of smear test under microscope and flow cytometry. They found that moxibustion on Dazhui (DU-14) did not influence the macrophage phagocytosis of healthy mice, but it significantly strengthened macrophage phagocytosis of mice with immnuodeficiency-deficiency (Zhu and Liu 2005).

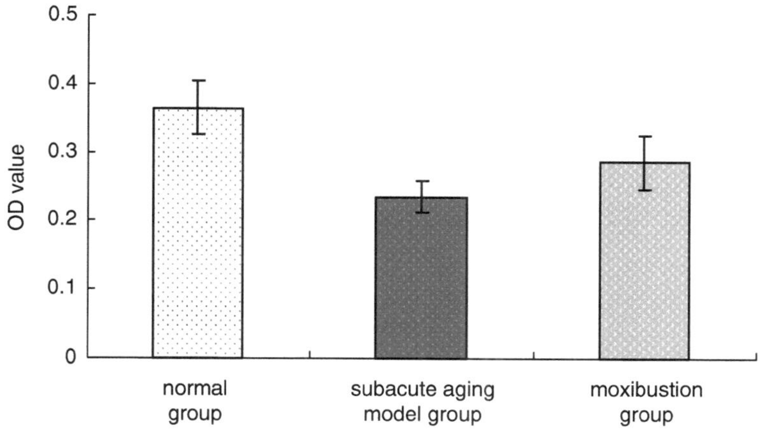

Fig. 10.10 The telomerase activity of rat tissues in subacute aging model was significantly lower than that of normal group. The telomerase activity in moxibustion group was significantly higher than that of subacute aging model, and there were significant differences between them

10.3.2 Regulatory Effects of Acupuncture and Moxibustion on Telomerase Activity in the Tissue of Aging Rats and Their Underlying Mechanisms

Many studies have shown that telomerase activity closely relates to cellular growth, development, and aging (Bayne and Liu 2005; Kajstura et al. 2006; Oh et al. 2004; Haendeler et al. 2004; Young et al. 2003; Djojosubroto et al. 2003; Hwang 2002; Dahse et al. 1997). Telomerase is synthesized by reverse transcription and protects telomeres, playing an important role in regulating life span and cellular proliferation.

Telomerase, a rare ribonucleoprotein compound of micromolecular RNA and protein, is a reverse transcriptase. It is now clear that human telomerase consists of three components, including human telomerase RNA (hTR), human telomerase protein (hTP), and human telomerase reverse transcriptase (hTRT/hTERT). Telomerase can maintain repeated sequence of telomeres at the end of chromosome, thereby counteracting consumption of telomeric DNA induced by cell division. Furthermore, it can also repair the broken end of chromosome (preventing incision of excision enzyme on chromosomal DNA) to maintain stability of genome inheritance.

Wu reported that telomerase activity of rat tissues in subacute aging models was significantly lower than that in normal rats. Telomerase activity in the moxibustion group was significantly higher than that in the subacute aging group, and there were significant differences between them (Fig. 10.10). Subacute aging model, an artificial model, is based on trauma theory which is caused by carbohydrate metabolism disorder. However, natural aging models should be much better than artificial models in characterizing physiological processes of human aging. Does the same result can be obtained in natural aging animal models? To address this question, we conduced

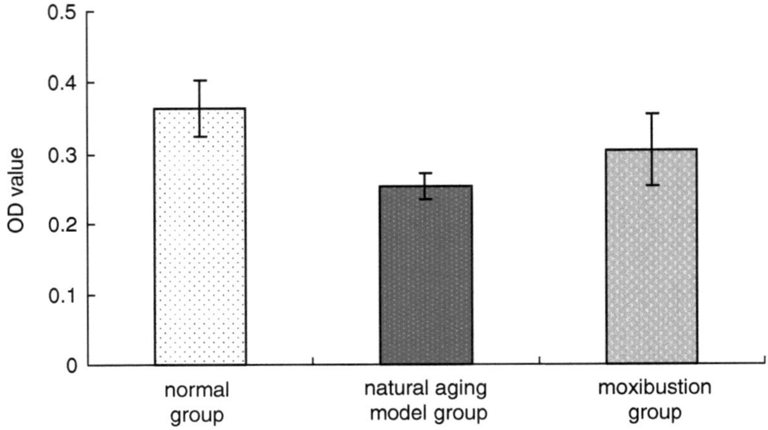

Fig. 10.11 The telomerase activity in natural aging model was significantly lower than that of normal group, and the telomerase activity in the moxibustion group was higher than that of the group of natural aging, and there are significant differences between them

another study in natural aging rats, showing that the telomerase activity in natural aging models was significantly lower than that in normal rats, and the telomerase activity in the moxibustion group was higher than that in the natural aging group, and there were significant differences between them (Fig. 10.11). These findings indicate that the decline in telomerase activity may be one of the key links responsible for tissue aging in subacute and natural aging rats, and it may represent an important pathway that regulates telomerase activity of tissues in subacute and natural aging rats.

TERT expression is associated with telomerase activity (Shkreli et al. 2007), and therefore is an indispensable component for the inhibition of telomerase activity (Weinrich et al. 1997). Given the correlation of telomerase activity with TERT transcription (Kyo and Inoue 2002), some scholars use changes in TERT as an indicator of telomerase activity. On the basis of this consideration, we explored the role of changes in telomerase activity in senescence and in anti-senescence by detecting TERT in situ hybridization. The result indicated that the area of positive expression of telomerase TERT and optical density of positive expression of telomerase TERT in rat tissues in subacute aging models were lower than those in the normal group. Areas of positive expression in the moxibustion group were greater than those in the subacute aging group, and there were significant between-group differences (Table 10.1). Do natural aging models generate the same results? Another study conducted in natural aging rats indicated that the area of positive expression in the moxibustion group was greater than that in the subacute aging group, and there were significant between-group differences (Table 10.2). These findings strongly suggest that changes in gene expression of telomerase TERT are one of possible mechanisms responsible for the decline in telomerase activity in tissues of subacute aging rats. It is likely to be a pathway by which warm moxibustion can regulate telomerase activity through influencing the expression of telomerase TERT gene.

Table 10.1 Comparison of the expression of telomerase TERT gene in subacute aging rats of each group ($\bar{X} \pm S$)

Group	*N*	Positive areas (μm²)	Positive optical density
Normal group	8	54,872.04 ± 5,207.9521	16.67 ± 1.2724
Subacute aging model group	8	33,459.04 ± 5,178.2378**	11.25 ± 2.0759**
Moxibustion group	8	47,403.58 ± 5,887.1047*, ***	12.71 ± 0.4521**

TERT gene: versus normal group, *$P<0.05$, **$P<0.01$; versus subacute aging model group, ***$P<0.01$

Table 10.2 Comparison of the expression of telomerase TERT gene in natural aging rats of each group ($\bar{X} \pm S$)

Group	*N*	Positive areas (μm²)	Positive optical density
Normal group	8	56,239.96 ± 14,527.4489	10.54 ± 0.6886
Natural aging model group	8	20,295.58 ± 5,595.5182**	7.58 ± 1.3063**
Moxibustion group	8	44,342.83 ± 8,017.0207*, ***	8.88 ± 1.4901**

TERT gene: versus normal group, *$P<0.05$, **$P<0.01$; versus natural aging model group, ***$P<0.01$

It is noteworthy that telomerase activity is subject to the influence of numerous factors, including expression of telomerase gene, protein interaction after transcription and protein phosphorylation. Some proto-oncogenes and stress-induced tumor-inhibitory factors directly or indirectly regulate telomerase activity, including C-Myc, B cell lymphoma gene-2 (Bcl-2), P21wafl, Rb, P53, protein kinase C (PKC), Akt/PKB, and protein phosphatase (Liu 1999; Cong et al. 2002).

Some studies reported that inactivation of P53/Rb pathway might lead to the activation of telomerase (Lu and Zhang 2002; Lantuejoul et al. 2005). P53, through hindering repression of hTERT core promoter mediated by E2F-Rb, can induce expression of hTERT in cells whose telomerase is inactivated (Beitzinger et al. 2006). In turn, the overexpression of Rb and P53 can downregulate telomerase activity, thus inhibiting telomerase activity and aging. An animal study conducted by our institute on in situ hybridization and immunohistochemistry of Rb and P53 demonstrated that the expression of Rb, P53 mRNA, and protein in rat tissues in the natural aging group is higher than that in the normal group. By contrast, the expression of Rb, P53 mRNA, and protein in moxibustion group is lower than that in the natural aging group, with significant between-group differences (Tables 10.3, 10.4, 10.5, and 10.6). A combination of these findings indicates that warm moxibustion may regulate telomerase activity and thereby defer senility through inhibiting the expression of P53 and Rb.

Bcl-2 is an important anti-apoptosis gene. Because Bcl-2, called longevity gene, can inhibit apoptosis induced by various stimuli, it is often considered as an indicator of aging (Konigsberg et al. 2004). Numerous studies have shown that the expression of Bcl-2 declines in the aging cell (Zhang et al. 2007; Mak et al. 2006; Crescenzi et al. 2003), whereas overexpression of Bcl-2 can prolong the life span of hematopoietic cells after removal of growth factor (Chen et al. 2002).

Table 10.3 The expression of Rb mRNA in natural aging rats from each group ($\overline{X} \pm S$)

Group	N	Positive areas (μm²)	Positive optical density
Normal group	8	9,368.79 ± 1,728.7085	6.58 ± 0.4273
Natural aging model group	8	28,050.33 ± 5,958.2020**	10.21 ± 1.3794**
Moxibustion group	8	13,885.29 ± 3,875.2379***, *	7.50 ± 0.7766***

Rb mRNA: versus normal group, *$P<0.05$, **$P<0.01$; versus natural aging model group, ***$P<0.01$

Table 10.4 The expression of P53 mRNA in natural aging rats from each group ($\overline{X} \pm S$)

Group	N	Positive areas (μm²)	Positive optical density
Normal group	8	15,205.13 ± 9303.9797	6.58 ± 0.6108
Natural aging model group	8	28,102.71 ± 3826.3526*	9.54 ± 0.8898**
Moxibustion group	8	20,115.33 ± 4999.0013***, *	8.04 ± 1.3503†, **

P53: versus normal group, *$P<0.05$, **$P<0.01$; versus natural aging model group, ***$P<0.05$, †$P<0.01$

Table 10.5 The expression of Rb protein in natural aging rats from each group ($\overline{X} \pm S$)

Group	N	Positive areas (μm²)	Positive optical density
Normal group	8	11,871.83 ± 1,064.0501	8.20 ± 0.3959
Natural aging model group	8	30,265.08 ± 1,654.9417*	12.04 ± 0.5756*
Moxibustion group	8	20,824.92 ± 1,903.9519**, *	9.38 ± 0.9332**, *

Rb protein: versus normal group, *$P<0.01$; versus natural aging model group, **$P<0.01$

Table 10.6 The expression of P53 protein in natural aging rats from each group ($\overline{X} \pm S$)

Group	N	Positive areas (μm²)	Positive optical density
Normal group	8	12,412.96 ± 838.9290	6.04 ± 0.4520
Natural aging model group	8	25,972.29 ± 499.8160*	9.67 ± 0.5039*
Moxibustion group	8	22,089.79 ± 584.7327**, *	8.21 ± 0.6651**, *

P53: versus normal group, *$P<0.01$; versus natural aging model group, **$P<0.01$

In addition, PKC, a calcium–phospholipid-dependent protein kinase, participates in the regulation of cellular activity (Battaini and Pascale 2005). The activated PKC, an important molecule of signal pathway, can phosphorylate residue of serine and threonine. Several experiments demonstrated that the function of PKC in aging cell (Van der Zee et al. 2004; Centurione et al. 2003) determined occurrence and development of aging cell (Di Giulio et al. 2007). PKC, widely distributed in organs, tissues, and cells of mammals, influences the intracellular transmission of biological information and plays a key role in regulating cellular metabolism, differentiation, proliferation, and even death (Li et al. 2001; Serova et al. 2006). Thus, PKC, which can activate intracellular phosphorylation of polypeptide in many cells, is one of the

Table 10.7 The expression of Bcl-2 mRNA in natural aging rats from each group ($\overline{X} \pm S$)

Group	*N*	Positive areas (μm²)	Positive optical density
Normal group	8	87,832.54±6090.5696	10.13±0.8153
Natural aging model group	8	53,477.96±12536.5392**	7.17±1.1819**
Moxibustion group	8	74,106.50±9973.8226***, *	8.96±1.4302***

Bcl-2: versus normal group, *$P<0.05$, **$P<0.01$; versus natural aging model group, ***$P<0.01$

Table 10.8 The expression of PKC mRNA in natural aging rats from each group ($\overline{X} \pm S$)

Group	*N*	Positive areas (μm²)	Positive optical density
Normal group	8	52,107.17±3,521.2429	8.42±0.3450
Natural aging model group	8	18,870.96±5,322.1633**	6.04±0.9332**
Moxibustion group	8	36,848.13±4,067.8285***, **	7.96±0.5176***, *

PKC: versus normal group, *$P<0.05$, **$P<0.01$; versus natural aging model group, ***$P<0.01$

Table 10.9 The expression of Bcl-2 protein in natural aging rats from each group ($\overline{X} \pm S$)

Group	*N*	Positive areas (μm²)	Positive optical density
Normal group	8	51,153.08±1,426.6478	9.71±0.8439
Natural aging model group	8	31,335.88±714.7557*	6.13±0.8345*
Moxibustion group	8	46,653.21±1,230.4301**, *	7.92±0.7506**, *

Bcl-2: versus normal group, *$P<0.01$; versus natural aging model group, **$P<0.01$

Table 10.10 The expression of PKC protein in natural aging rats from each group ($\overline{X} \pm S$)

Group	*N*	Positive areas (μm²)	Positive optical density
Normal group	8	78,015.46±4,665.8159	11.79±0.7333
Natural aging model group	8	41,051.92±14,731.0526**	8.58±1.6785**
Moxibustion group	8	63,966.29±17,034.9533†, *	10.46±0.6409***, **

PKC: versus normal group, *$P<0.05$, **$P<0.01$; versus natural aging model group, ***$P<0.05$, †$P<0.01$

critical molecules in intracellular signal transmission. Consistent with these findings, several animal studies in our institute on in situ hybridization and immunohistochemistry of Rb and P53 showed that the expression of Bcl-2, PKC mRNA, and protein in rat tissues in the natural aging group was lower than that in the normal group. However, the expression of Bcl-2, PKC mRNA, and protein in rat tissues in the moxibustion group was lower than that in the normal group (Tables 10.7, 10.8, 10.9, and 10.10).

In all, warm moxibustion can inhibit the expression of Rb and P53 and enhance the expression of Bcl-2 and PKC, suggesting that warm moxibustion may influence the gene expression of telomerase TERT, regulate telomerase activity, and thereby defer senility through regulating Rb, P53, Bcl-2, and PKC.

10.4 Effects of Acupuncture and Moxibustion on Carbonyl Poison Reaction and Their Signal Regulation Mechanisms in Aging Rats

The theory of carbonyl-poisoning aging is proposed on the basis of the theory of free radical aging and the theory of NEG aging. After organism protein is injured, it will be repaired, eliminated by protein degradation system, or accumulated. However, carbonylation modification of protein is an exception. As an irreversible modification, injured protein can be either degraded or accumulated (Nystrom 2005). Therefore, the carbonylation modification, if not eliminated in time, will cause the accumulation and polymerization of injured protein, which in turn damages the functions of cells and tissues and induces pathophysiological changes, resulting in the acceleration of aging processes. Because carbonylated protein can be more easily degraded by protein degradation system than unoxidized protein (Grune et al. 2004), the rapid carbonylation of abnormal protein might ensure that these abnormal peptides are directly degraded by proteolysis system.

Numerous studies have shown that the damage of micromolecules such as protein caused by carbonyl poisoning plays an important role in aging (Feng et al. 2008). A series of studies performed by Levine demonstrated that protein carbonyl content was significantly increased in aged animals, up to 40%–50% under extreme conditions. Carbonylation of proteolytic enzyme reduces the organism's ability to eliminate oxidation protein, resulting in accumulation of oxidation protein which accelerates aging of cells and organisms. Consistent with this idea, domestic studies on aging models induced by D-galactose have shown that protein carbonyl content in model rats was significantly higher than that in normal rats in the process of modeling with D-galactose, indicating that the damage of carbonyl poisoning to an organism caused the loss of functions of rat proteins and resulted in a series of aging changes (Wang et al. 2008). Although mechanisms underlying aging are yet to be explicitly explained, carbonyl poisoning is likely to play a pivotal role in aging (Li and Yin 2008).

Studies carried out by our research team are in agreement with those from home and abroad (Levine 2002; Das et al. 2001; Liu et al. 2001, 2002), confirming that the protein carbonyl content in modeled rats was significantly higher than that in the normal rats in the process of modeling with D-galactose. Carbonyl poisoning leads to the loss of functions of rat proteins, resulting in a series of aging changes. Warm moxibustion and vitamin can reduce protein carbonyl content to some degree and thereby prevent the damage of carbonyl poisoning to an organism, thus deferring senility.

It is well established that P19ARF/P53/P21Cip1 is one of the pathways that mediate signal transduction of cell aging. If key regulatory factors of this pathway are mutated, presenility of cell will occur or the cell will bypass aging process and continue to proliferate. The latter may further result in formation or production of tumor (Massari and Murre 2000; Lundberg et al. 2000). At the core of this aging-induced pathway are products of some antioncogenes, including P53, P19ARF, P21CiP1, and Rb tumor suppressor protein.

In the pathway of P19ARF/P53/P21Cip1 signal transduction, P53 is a tumor suppressor gene. It can inhibit cell proliferation and thus is associated with organic differentiation and aging, due to that it is an important regulatory gene in cell cycle and apoptosis and is a negative growth factor of cell growth. While there is evidence showing that P53 is inactivated in more than half of human tumor, its activity is increased in aging cell (Itahana et al. 2001, 2004). If it is activated, the P53 will bind special sequence of many downstream target genes and enhance expression of them. Many products of expression of target genes participate in the regulation of cell growth, including cell cycle arrest, cellular apoptosis, and inhibition of vascular formation (Papazoglu and Mills 2007). Therefore, P53 plays a critical role in cell aging, the activation of which involves complex modifying processes after transcription, including phosphorylation and acetylation. The level of P53 is mainly determined by ubiquitination and degradation rate of subsequent proteinase. MDM2, aimed at ubiquitination ligase of E3-type P53, can contribute to proteinase degradation mediated by ubiquitination. MDM2, an oncogene which is often overexpressed in sarcoma and some other tumors, can directly bind transcriptional activation domain (TAD) of P53 and downregulate transcriptional activity. On the other hand, P53 can activate expression of MDM2 gene. MDM2 gene consists of DNA binding site of P53 and an element of gene responding, so the expression of MDM2 is regulated by the level of wild P53 protein. MDM2 can be combined with P53 and form a compound which can decrease the action of positive transcription factor of MDM2. P53 regulates transcription of MDM2 at transcriptional level, whereas MDM2 adjusts the function of P53 at protein level. If the function of MDM2 is impaired, growth arrest and apoptosis mediated by P53 can in turn be activated (Momand et al. 2000).

Activated P53 can influence the expression of many genes. P21, a target gene of P53, plays an important role in regulation of cell cycle. P21CiP1 protein, a broad-spectrum cyclin, is a cyclin-dependent kinase (CDK) inhibitor. It can decrease the level of Rb phosphorylation and reduce the release of transcription factor E2F through inhibiting phosphorylation activity of cyclin–CDK compound, which finally inhibits DNA synthesis and results in cell division stagnate in G1 stage, failing to enter S stage. So P21CiP1 protein can regulate the process of cell cycle and participate in cellular growth, differentiation, aging, and death.

P53 can upregulate P21CiP1, and the expression of P21 in aging fibroblast is raised by 10–20 times. P21CiP1 is a downstream effecter of P53, and its level of protein expression is found to accrue in many aging cells (Ohtani et al. 2007; Sun et al. 2006). It has been established that P19ARF, a tumor-suppressing gene combined with oncogenic protein (negative growth factor of P53), can block up the P53 degradation induced by MDM2 (i.e., enhance P53 stability), which can increase intracellular level of P53 and induce cell cycle arrest and apoptosis and inhibit the growth of tumor (Jackson et al. 2001).

In line with these findings, studies by our research team demonstrated that the expression of P19ARF, P53, P21CiP1, Rb mRNA, and protein in rat liver tissues from subacute aging model was significantly higher than that in normal rats; the expression of MDM2 mRNA and protein was significantly lower than that in normal

Table 10.11 The expression of P19ARF mRNA from each group [M(Q_{25}, Q_{75})]

Group	*N*	P19ARFmRNA (%)
Normal group	12	0.701(0.425,1.018)
Aging model group	12	2.703(1.191,8.565)*
Moxibustion group	12	0.795(0.727,1.105)***
Vitamin E group	12	1.148(0.533,1.977)**

P19ARF mRNA: versus normal group, *$P<0.01$; versus aging model group, **$P<0.05$, ***$P<0.01$

Table 10.12 The expression of P53 mRNA from each group [M(Q_{25}, Q_{75})]

Group	*N*	P53mRNA (%)
Normal group	12	0.567(0.396,0.801)
Aging model group	12	3.694(3.271,4.809)**
Moxibustion group	12	1.698(1.123,2.121)*,†
Vitamin E group	12	1.705(0.655,3.172)***

P53 mRNA: versus normal group, *$P<0.05$, **$P<0.01$; versus aging model group, ***$P<0.05$, †$P<0.01$

Table 10.13 The expression of P21CiP1 mRNA from each group [M(Q_{25}, Q_{75})]

Group	*N*	P21CiP1mRNA (%)
Normal group	12	1.64±1.03
Aging model group	12	3.95±1.78*
Moxibustion group	12	2.33±1.66**
Vitamin E group	12	2.80±1.57

P21CiP1 mRNA: versus normal group, *$P<0.01$; versus aging model group, **$P<0.05$

Table 10.14 The expression of MDM2mRNA from each group [M(Q_{25}, Q_{75})]

Group	*N*	MDM2mRNA (%)
Normal group	12	1.135(0.978,2.920)
Aging model group	12	0.845(0.675,1.118)*
Moxibustion group	12	1.575(1.065,1.893)**
Vitamin E group	12	1.640(0.943,2.378)**

MDM2 mRNA: versus normal group, *$P<0.05$; versus aging model group, **$P<0.05$

rats, and moxibustion and vitamin E can significantly decrease expression of P19ARF, P53, P21CiP1, Rb mRNA, and protein in liver tissues in subacute aging rats and increase expression of MDM2 mRNA and protein (Tables 10.11, 10.12, 10.13, and 10.14). These findings suggest that warm moxibustion can influence pathways of signal regulation and defer aging through regulating expression of P19ARF, P53, P21CiP1, Rb, and MDM2 mRNA.

10.5 Concluding Remarks

Recent research shows that acupuncture and moxibustion can have not only positive effects on objective indicators of aging but also holistic regulatory effects on the entire body. Many studies, based on various theories of senescence, have observed objective antiaging effects of acupuncture and moxibustion and explored their underlying mechanisms with different outcome measures. However, the ultimate goal of studies of acupuncture and moxibustion for antiaging should be set as to improve life quality of the middle aged as well as elderly and increase longevity. In terms of acupoint selection, acupuncture and moxibustion for antiaging should be focused on Zusanli (ST-36), Guanyuan (RN-4), Qihai (RN-6), Dazhui (DU-14), Shenque (RN-8), Zhongwan (RN-12), Shuigou (DU-26), Neiguan (PC-6), Yongquan (KI-1), etc. As for treatment methods, acupuncture and moxibustion for antiaging should adopt needling, direct moxibustion, and indirect moxibustion, combined with herbal, electric heat, warm needling, ear acupoint, etc. In clinical practice, according to acupoint features, how to achieve the best effect of antiaging by the integration of different treatment methods, syndrome differentiation, and disease differentiation needs further investigation. In summary, acupuncture and moxibustion for antiaging have great advantages, although many critical questions remain to be answered. Acupuncture and moxibustion will prove to be major treatment options for antiaging in the future if much more research attention is given to them.

References

Battaini F, Pascale A (2005) Protein kinase C signal transduction regulation in physiological and pathological aging. Ann N Y Acad Sci 1057: 177–192

Bayne S, Liu JP (2005) Hormones and growth factors regulate telomerase activity in ageing and cancer. Mol Cell Endocrinol 240: (1–2): 11–22

Beitzinger M, Oswald C, Beinoraviciute KR, Stiewe T (2006) Regulation of telomerase activity by the p53 family member p73. Oncogene 25: (6): 813–826

Centurione L, Di Giulio C, Cacchio M, Rapino M, Bosco D, Grifone G, Sabatini N, Bianchi G, Castorina S, Antonucci A, Cataldi A (2003) Correlations between protein kinase C zeta signaling and morphological modifications during rat heart development and aging. Mech Ageing Dev 124: (8–9): 957–966

Chae Y, Hong MS, Kim GH, Hahm DH, Park HJ, Ha E, Kim MJ, Park HJ, Yang J, Lee H (2007) Protein array analysis of cytokine levels on the action of acupuncture in carrageenan-induced inflammation. Neurol Res 29: (1): S55–58.

Chen YQ, Tang LM (1994) Clinical observation on tian kui moxibustion in retarding the process of aging. Practical Geriatrics 8: (3): 115–116 (in Chinese with English abstract).

Chen ZS, Wu ZK, Cai HG, Xu J, Chen YY, Lv XX (2002) Effect of the drugs for reinforcing the kidney and promoting generation of blood on expression of Bcl-2mRNA in aging mice. Journal of Traditional Chinese Medicine 43: (5): 382–384 (in Chinese with English abstract)

Cong YS, Wright WE, Shay JW (2002) Human telomerase and its regulation. Microbiol Mol Bio Rev 66: (3): 407–425

Crescenzi E, Palumbo G, Brady HJ (2003) Bcl-2 activates a programme of premature senescence in human carcinoma cells. Biochem J 375: (Pt 2): 263–274

Dahse R, Fiedler W, Ernst G (1997) Telomeres and telomerase. Pathologe 18: (6): 425–429.

Das N, Levine RL, Orr WC, Sohal RS (2001) Selectivity of protein oxidative damage during aging in Drosophila melanogaster. Biochem J 360: (pt 1): 209–216

Di Giulio C, Rapino M, Zingariello M, Antonucci A, Cataldi A (2007) PKC alpha-mediated CREB activation is oxygen and age-dependent in rat myocardial tissue. Histochem Cell Biol 127: (3): 327–333

DiCarlo AL, Fuldner R, Kaminski J, Hodes R (2009) Aging in the context of immunological architecture function and disease outcomes. Trends in Immunology 30: (7): 293–294

Djojosubroto MW, Choi YS, Lee HW, Rudolph KL (2003) Telomeres and telomerase in aging, regeneration and cancer. Mol Cells 15: (2): 164–175

Du YJ, Song GJ, Kang YP (2006) Studies on TLC of Huangzhi Oral Liquids. Chinese Archives of Traditional Chinese Medicine 24: (9): 1679–1680 (in Chinese with English abstract)

Du YJ, Sun GJ (2005) Effect of moxibustion on mitochondrial membrane potentials and neuronal apoptosis in aged rats. Acupuncture Research 30: (4): 212–214 (in Chinese with English abstract)

Du YJ, Song J, Sun GJ, Kang YP (2006) Effect of moxibustion on CasPase-3 in apoptosis signal transduction iter in aged rats. Hubei Journal of Traditional Chinese Medicine 28: (7): 3–4 (in Chinese).

Feng J, Navratil M, Thompson LV, Arriaga EA (2008) Principal component analysis reveals age related and muscle type related differences in protein carbonyl profiles of muscle mitochondria. J Gerontol A Biol Sci Med Sci 63: (12): 1277–1288

Gao W, Huang YX, Zhao NX, Wang QL, Zhang HX (2000) Influences of electroacupuncture (EA) on T lymphocyte subgroups of peripheral blood and erythrocyte immune function in rats. Journal of the Fourth Military Medical University 21: (4): 414–416 (in Chinese with English abstract)

Gao XY, Feng LH, Li WH, Liu ZW (2003) The effect of free radical metabolism and immune organ of subacute senile mouse by moxibustion on the du meridian points. Journal of Henan College of Traditional Chinese Medicine 18: (5): 22–24 (in Chinese with English abstract)

Gao XY, Wang Y (2004) Effects of moxibustion at points of the Governor Vessel on immune functions in the D-galactose-induced subacute aging mouse. Chinese Acupuncture & Moxibustion 24: (7): 488–490 (in Chinese with English abstract)

Grune T, Jung T, Merker K, Davies KJ (2004) Decreased proteolysis caused by protein aggregates, inclusion bodies, plaques, lipofuscin, ceroid, and "aggresomes" during oxidative stress, aging, and disease. Int J Biochem Cell Bio 36: (12): 2519–2530

Haendeler J, Hoffmann J, Diehl JF, Vasa M, Spyridopoulos I, Zeiher AM, Dimmeler S (2004) Antioxidants inhibit nuclear export of telomerase reverse transcriptase and delay replicative senescence of endothelial cells. Circ Res 94: (6): 768–775

Hwang ES (2002) Replicative senescence and senescence-like state induced in cancer-derived cells. Mech Ageing Dev 123: (12): 1681–1694

Itahana K, Campisi J, Dimri GP (2004) Mechanisms of cellular senescence in human and mouse cells. Biogerontology 5: (1): 1–10

Itahana K, Dimri G, Campisi J (2001) Regulation of cellular senescence by p53. Eur J Biochem 268: (10): 2784–2791

Jackson MW, Lindstrom MS, Berberich SJ (2001) MdmX binding to ARF affects Mdm2 protein stability and p53 transactivation. J Biol Chem 276: (27): 25336–25341

Jiang ZQ, Wang FC, Yan XK (2005) The experimental research of the influence of immunological function by acupuncture Tianshu points to the Rat of splenasthenic diarrhea. Jiangsu Journal of Traditional Chinese Medicine 26: (4): 27

Kajstura J, Rota M, Urbanek K, Hosoda T, Bearzi C, Anversa P, Bolli R, Leri A (2006) The telomere-telomerase axis and the heart. Antioxid Redox Signal 8: (11–12): 2125–2141

Konigsberg M, Lopez-Diazguerrero NE, Aguilar MC, Ventura JL, Gutiérrez-Ruiz MC, Zentella A (2004) Senescent phenotype achieved in vitro is indistinguishable, with the exception of Bcl-2 content, from that attained during the in vivo aging process. Cell Biol Int 28: (8–9): 641–651

Kyo S, Inoue M (2002) Complex regulatory mechanisms of telomerase activity in normal and cancer cells: how Can we apply them for cancer therapy? Oncogene 21: (4): 688–697

Lantuejoul S, Soria JC, Morat L, Lorimier P, Moro-Sibilot D, Sabatier L, Brambilla C, Brambilla E (2005) Telomere shortening and telomerase reverse transcriptase expression in preinvasive bronchial lesions. Clin Cancer Res 11: (5): 2074–2082

Levine RL (2002) Carbonyl modified proteins in cellular regulation, aging, and disease. Free Radic Biol Med 32: (9): 790–796

Li GL, Yin DZ (2008) Protein carbonylation and aging. Chinese Journal of Gerontology (28): 2070–2073 (in Chinese)

Li Y, Liu JH, Zhao Q, Ma ZX, Liu PY, Yu YM, Fan LQ (2001) Changes of cAMP and PKC in As203 induced apoptosis in gastric cells. Journal of Hebei Medical University 22: (4): 206–209 (in Chinese)

Liao FZ, Zhong L, Ai SC (2004) Effects of electric heat medicine-separated moxibustion on clinical aging symptoms in the aged. Chinese Acupuncture & Moxibustion 24: (3): 161–164 (in Chinese with English abstract)

Liu JP (1999) Studies of the molecular mechanisms in the regulation of telomerase activity. FASEB J 13: (15): 2091–2104

Liu XQ, Li WD, Tang HQ, Lian ZC (2001) The free radical metabolism of the rat aging models induced by D-galactose. Chinese Journal of Gerontology 21: (6): 456–458 (in Chinese with English abstract)

Liu XQ, Li WD, Tang HQ, Lian ZC (2002) Carbonyl Toxification of the Aging Model Induced by Injection of D-Galactose in Rats. Chinese Journal of Laboratory Animal Science 12: (3): 141–143 (in Chinese with English abstract)

Lu DD, Zhang XR (2002) The study of abnormal-telomerase activation, p53 transgenation, rb gene and PCNA in PHC. Journal of Modern Laboratory Medicine 17: (3): 27–28 (in Chinese)

Lundberg AS, Hahn WC, Gupta P, Weinberg RA (2000) Genes involved in senescence and immortalization. Curr Opin Cell Biol 12: (6): 705–709

Mak YT, Chan WY, Lam WP, Yew DT (2006) Immunohistological evidences of Ginkgo biloba extract altering Bax to Bcl-2 expression ratio in the hippocampus and motor cortex of senescence accelerated mice. Microsc Res Tech 69: (8): 601–605

Massari ME, Murre C (2000) Helix-loop-helix proteins: regulators of transcription in eucaryotic organisms. Mol Cell Biol 20: (2): 429–440

Momand J, Wu HH, Dasgupta G (2000) MDM2-master regulator of the p53 tumor suppressor protein. Gene 242: (1–2): 15–29

Noga Ron-Hare N, Schwartz M (2009) Immune senescence and brain aging: can rejuvenation of immunity reverse memory loss? Trends in Neurosciences 32: (7): 367–375

Nystrom T (2005) Role of oxidative carbonylation in protein quality control and senescence. EMBO J 24: (7): 1311–1317

Oh BK, Lee CH, Park C, Park YN (2004) Telomerase regulation and progressive telomere shortening of rat hepatic stem-like epithelial cells during in vitro aging. Exp Cell Res 298: (2): 445–454

Ohtani N, Imamura Y, Yamakoshi K, Hirota F, Nakayama R, Kubo Y, Ishimaru N, Takahashi A, Hirao A, Shimizu T, Mann DJ, Saya H, Hayashi Y, Arase S, Matsumoto M, Kazuki N, Hara E (2007) Visualizing the dynamics of p21(Waf1/Cip1) cyclin-dependent kinase inhibitor expression in living animals. Proc Natl Acad Sci U S A 104: (38): 15034–15039

Papazoglu C, Mills AA (2007) p53: at the crossroad between cancer and ageing. J Pathol 211: (2): 124–133

Serova M, Ghoul A, Benhadji KA, Cvitkovic E, Faivre S, Calvo F, Lokiec F, Raymond E (2006) Preclinical and clinical development of novel agents that target the protein kinase C family. Semin Oncol 33: (4): 466–478

Shkreli M, Dambrine G, Soubieux D, Kut E, Rasschaert D (2007) Involvement of the oncoprotein c-Myc in viral telomerase RNA gene regulation during Marek's disease virus-induced lymphomagenesis. J Virol 81: (9): 4848–4857

Sun MM, Liu J, Chen EQ, Jiang JG (2006) Expression of P21 ~ (WAF-1) of neuron in second-acute brain aging models. Laboratory Animal and Comparative Medicine 26: (4): 223–226 (in Chinese with English abstract)
Van den Berg EJ, Russel FG, Bos RP, Smits P, Kramers C (2000) Lavage almost never indicated after an autointoxication. Ned Tijdschr Geneeskd 144: (20): 916–918
Van der Zee EA, Palm IF, O'Connor M, Maizels ET, Hunzicker-Dunn M, Disterhoft JF (2004) Aging-related alterations in the distribution of Ca^{2+}-dependent PKC isoforms in rabbit hippocampus. Hippocampus 14: (7): 849–860
Wang FL, Li H, Wei ZX, Wang XH, Wang QM, Li J, Xia XQ, Zhao R, Tian YY (1996) Effects of immunologic function and general body state by moxibustion on Shenque acupoint in aged people. Chinese Acupuncture & Moxibustion (7): 39–40 (in Chinese)
Wang X, Liao Y, Li G, Yin D, Sheng S (2008) A comparative study of artificial ceroid/lipofuscin from different tissue materials of rats. Exp Aging Res 34: (3): 282–295 (in Chinese with English)
Weinrich SL, Pruzan R, Ma L, Ouellette M, Tesmer VM, Holt SE, Bodnar AG, Lichtsteiner S, Kim NW, Trager JB, Taylor RD, Carlos R, Andrews WH, Wright WE, Shay JW, Harley CB, Morin GB (1997) Reconstitution of human telomerase with the template RNA component hTR and the catalytic protein subunit h TRT . Nat Genet 17: (4): 498–502
Wu HG (2006) Chinese moxibustion. Shanghai technology publishing company (in Chinese)
YangT, Zhang C, Chen QX (2005) Progress in Studies of Aging Mechanism. Progress in Biotechnology 25: (3): 6–11 (in Chinese with English abstract)
Yim YK, Lee H, Hong KE, Kim YI, Ko SK, Kim JE, Lee SY, Park KS (2007) Anti-inflammatory and immune-regulatory effects of subcutaneous perillae fructus extract injections on OVA - induced asthma in mice. eCAM 7: (1): 79–86
Yim YK, Lee H, Hong KE, Kim YI, Lee BR, Son CG, Kim JE (2007) Electro-acupuncture at acupoint ST36 reduces inflammation and regulates immune activity in Collagen-Induced Arthritic Mice. eCAM 4(1):51–57
Young AT, Lakey JR, Murray AG, Mullen JC, Moore RB (2003) In vitro senescence occurring in normal human endothelial cells can be rescued by ectopic telomerase activity. Transplant Proc 35: (7): 2483–2485
Zhan XJ, Hua XN (2001) Effect of acupuncture at Zusanli acupoint on erythrocyte immunity in mice. Medical Journal of the Chinese People's Armed Police Forces 12: (5): 267–268 (in Chinese with English abstract)
Zhang HQ, Yu WX (2000) Chinese anti-aging medicine. Beijing science publishing company: 3 (in Chinese)
Zhang ZB, Cai CY, Tian SP, Li M, Zhuang RH (2007) Lipid peroxidation affects serum T and Bcl-2 expression in the testis of aging male rats. National Journal of Andrology 13: (1): 46–49
Zhu HJ, Jiang WB, Yang L (2006) Review on puncture of delaying and preventing body's aging. journal of traditional chinese medicine 21: (12): 72–73 (in Chinese with English abstract)
Zhu M, Gao HQ, Liu RF, Jin H, Ma JX, Li YD, Zhou ZG (2003) The experimental research on the kupffer cell function in liver by acupuncturing "Zusanli" and "Guanyuan" of old rats. Journal of Clinical Acupuncture and Moxibustion 19: (6): 52–53 (in Chinese with English abstract)
Zhu WL, Liu RQ (2005) Effects of moxa-cone moxibustion at Dazhui acupoint on the phagocytic function of macrophages in hypoimmune mice. Journal of Beijing University of Traditional Chinese Medicine 28: (1): 89–90 (in Chinese with English abstract)

Chapter 11
Development and Progression of Auricular Acupuncture as a Treatment for Anxiety and Pain

Shu-Ming Wang

Abstract Auricular acupuncture is commonly used for various clinical conditions. It can be used as a primary mode of treatment or in conjunction with body acupuncture or other forms of interventions. The external auricle is believed to organize as the entire body as an inverted fetus. Every part of the body has a reflect point located on the external body. In other words, every point on the ear represents a specific part or organ of the body. This chapter includes history and development of auricular acupuncture, the potential mechanism of auricular acupuncture, and review of the efficacy of clinical trials.

Keywords Auricular acupuncture • Ear somatotopic map

11.1 Introduction

Auricular acupuncture is a practice of applying physical stimuli onto various points on the external auricular to relieve various medical conditions. The common physical stimuli used for auriculotherapy are similar to body acupuncture, including finger pressure, pressure bead, needles, and electrical stimulations. Auricular therapy was introduced in the late 1950s by a French physician Dr. Paul Nogier. The external auricle is used to diagnose illnesses of the body parts or organs. The flakiness of the skin cover of the external auricle, presence of engorged blood vessel, or formation of nodule frequently represents the severity and duration of the illnesses. Auricular acupuncture is frequently used as the primary treatment or as an adjunctive treatment for body acupuncture. When there is a body pathology, the so-called reflect point commonly exhibits pressure allodynia.

S.-M. Wang (✉)
Department of Anesthesiology and Perioperative Care, University of California-Irvine, Orange, CA 92868, USA
e-mail: shuminw1@uci.edu

Y. Xia et al. (eds.), *Current Research in Acupuncture*,

DOI 10.1007/978-1-4614-3357-6_11,

11.2 The History and Development of Auricular Acupuncture (Lu 2005; Frank and Soliman 2006; Gori and Firenzuoli 2007)

The ear is an important structure in traditional Chinese acupuncture. According to the traditional textbook, six yang meridian skirt around the ear and the six yin meridians are indirectly connect trough ear (Fig. 11.1). However, the practice of auricular acupuncture and related interventions has never been the main focus of traditional Chinese acupuncture.

The practice of auricular cauterization as a treatment for illnesses has been reported among Egypt and certain tribes in Arabia. The Egyptologist Alexandre Varille has documented that women in ancient Egypt cauterized their external ear as a form of contraception. Mediterranean sailors wore gold earrings not just used as decorations but were used to improve vision (Soliman and Frank 1999). Hippocrates treated male impotence and leg pain by making a small opening in the vein situated behind the ear. In the medical literature of the Egyptians, Greeks and Romans clearly documented the healing value of the scarification of the outer ear, e.g., to treat sciatica pain and sexually related disease by cauterization of the external ear. Similar auricular intervention was also described in literature during the Renaissance period. In 1637,

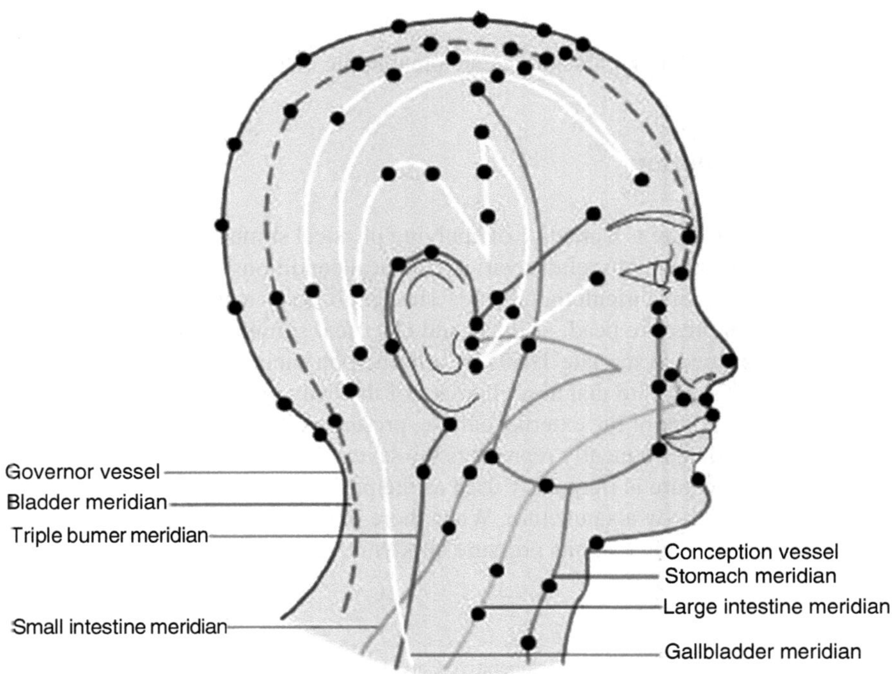

Fig. 11.1 The ear and the Chinese meridians (copy from http://www.faqs.org/health/Healthy-Living-V2/Alternative-Medicine-Acupuncture.html)

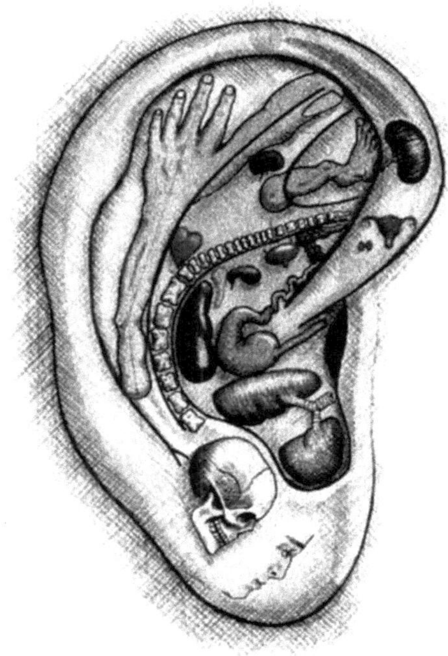

Fig. 11.2 The Nogier's ear map (copy from http://omhs.biz/addictiontherapy/history.html)

Portuguese physician Zacatus Lusitanus described the treatment of sciatica pain by cauterization of the ear. In 1717, *Aura Humanus Tractatus* described scarification of antitragus as the treatment of toothache. In 1810, Professor Colla of Parma reported the observation of a man stung by a bee in antihelix, which resulted in dramatic relief of pain in the legs. In the same year, an Italian physician, Dr. Cecconi, performed cauterization to help treat sciatica pain. In 1850, the French *Journal Des Connaissances Medico-Chirurgicales* reported 13 different cases of sciatic pain that had been treated by cauterization with a hot iron applied to the ear. Only 1 of the 13 patients did not improve completely.

The current auriculotherapy was not developed until 1957 in modern France (Nogier 1983). Dr. Paul Nogier observed the scars in the ears of patients who were successfully treated for sciatica pain by French lay practitioners. Dr. Nogier developed the somatotopic map of the ear based on the concept of inverted fetus orientation (Fig. 11.2) (Oleson 1996).

Later, his work was translated into Chinese. Once Nogier's chart arrived in China, the Nanking Army Ear Acupuncture Research Team conducted a massive study to verify the clinical accuracy of Nogier's auricular map (Nogier 1983). As a result, there are inconsistencies between the French and Chinese Auricular maps. In the early 1980s, a study conducted at UCLA indicated that there were high correlations between the tender areas of the ear and bodily pain or dysfunction (75 %) (Oleson et al. 1980). Since then, auricular acupuncture has been used to treat alcohol, narcotic, and nicotine withdrawal symptoms, subsequently leading to the formation of the National Acupuncture Detoxification Association.

Current auriculotherapy consists of utilizing ear diagnosis as a guide for treatment and applying various different stimulation techniques such as finger pressure, seeds, needle acupuncture, heat, electrical stimulation, laser stimulation, magnets, and other intervention techniques to the external auricular point.

11.3 The Theories of Auriculotherapy (Oleson 1996)

The following theories have been proposed as possible mechanisms of auriculotherapy.

11.3.1 Embryological Theory

Dr. Nogier has noted that the ear is composed of tissue from each of the three primary embryological tissues (Nogier 1983). He proposed that every type of embryological tissue on the ear accounts for the different somatotopic functions related to that auricular area. For example, the endodermal tissue is found in the concha of the ear, and this embryological tissue also forms the intestinal digestive tract, lungs, tonsils, and internal organs such as liver, pancreas, urinary bladder, urethra, thyroid gland, parathyroid gland, and the thymus gland. The mesodermal tissue is found on the antihelix, scaphoid, and triangular fossa. This embryological tissue also forms the skeletal muscles, cardiac muscles, smooth muscles, connective tissue, joints, blood cells from bone marrow, blood vessels, lymphatic tissue, adrenal cortex, and internal organs such as kidneys, ureter, and gonads. Lastly, the ectodermal tissue is found on the earlobe and helix tail, and the tissue also forms skin, hair, glands, cornea, eye lens, nose epithelium, teeth, peripheral nerves, spinal cord, subcortical brain, cerebral cortex, pituitary gland, pineal gland, and adrenal medulla. As a result, by applying stimulation at the somatotopic region of the external auricle, one can relieve the pathology of the affected organ.

11.3.2 Neurophysiological Theory

Tsun-Nin Lee proposed the Thalamic Neuron Theory that thalamus is the reflex connections between acupuncture points and brain. The hypothesis suggested that pathological changes in peripheral tissue would eventually lead to dysfunctional neural firing patterns (Lee 1994). Others suggested that there are three principal nerves that innervate the external auricle. The trigeminal nerve innervates the antihelix, helix root, superior helix, and the mesodermal region of the auricle. The auricular branch of vagus nerve innervates the concha, the endodermal region of the auricle.

The cerebral cervical plexus nerves innervate the lobe, superior helix, helix tail, and ectodermal region of the auricle. The midbrain reticular formation has been suggested as a site of integration with the central nervous system, where input from the nerves of the auricle may interact with collateral fibers that project to other areas of the body. Some investigators imply that the release of endogenous opiates may also play an important role in auricular analgesia.

11.3.3 Microsystem Theory

Ralph Alan Dale, Ph.D., indicated that not only the ear but also every part of the gross anatomy, e.g., iris, scalp, hand, foot, face, or nose, could function as an energy system for diagnosis and therapy. Subsequently, Dale called this system the microsystem of acupuncture, differentiating it from the traditional Chinese system of energy channels (Dale 1976). Dale suggested that every microsystem has a holographic distribution of these acupoints, and the topology of these acupoints reiterates the anatomy of the body. The microsystem reflex map of the body replicates the anatomical arrangement of the whole body (an invert homunculus). In every microsystem of acupuncture, the acupoints exhibit reflex responses useful for diagnosis and effective for treatment. These systems may be engaged by finger pressure, needle acupuncture, heat, electrical stimulation, laser stimulation, magnets, or other interventional techniques utilized by traditional acupuncture. It is not uncommon for a patient to complain of tenderness when a simple finger pressure is applied to an "active" reflex point.

11.4 The Scientific Evidence of Auriculotherapy

Early animal studies indicated that an electrical stimulation to the auricular lobe, corresponding to the jaw and teeth of rabbits, produced a significant decrease in behavioral reflexes and in the cortical response to tooth pulp stimulation. For most animals, the suppression of the behavioral and neurophysiological nociceptive responses was abolished by injection of the opiate antagonist naloxone, suggesting endorphinergic mechanisms (Kalyuzhnyi 1990). Fedoseeva et al. (1990) also found auricular electroacupuncture led to a reduction in the amplitude of cortical somatosensory potentials evoked by tooth pulp stimulation. The intravenous administration of the opiate antagonist naloxone was found to diminish the analgesic effect of auricular electroacupuncture at 15 Hz stimulation frequencies, but not at 100 Hz stimulation. Conversely, an injection of saralasin, an antagonist of angiotensin II, blocked the analgesic effect of 100 Hz auricular acupuncture, but not 15 Hz stimulations. This experiment illustrated that the mechanism of auriculotherapy for pain treatment may be similar to body acupuncture with neurophysiological basis. Using continuous, noninvasive measurements of infrared spectroscopy and multidirectional transcranial Doppler sonography, scientists were able to continually monitor changes in cerebral oxygen metabolism following different frequencies of laser stimulations to specific ear acupuncture points (Szeles and Litscher 2004). Szeles and Litscher suggested that the changes in cerebral function related to auricular stimulation might be the underlying mechanism for pain relief (Szeles and Litscher 2004). Although above observations have been mentioned in the literature, the mechanism of auriculotherapy for pain and anxiety remains unclear. Thus, future studies should be conducted in this area.

11.5 The Clinical Evidence of Auricular Acupuncture

11.5.1 The Use of Auricular Acupuncture as a Treatment for Anxiety

Several studies have been conducted to determine the anxiolytic effect of auricular acupuncture and related techniques. Anxiety is defined as a "persistent feeling of dread, apprehension and impending disaster or tension and uneasiness" (Ohman 2000). Anxiety disorders have been used to describe a number of conditions including panic disorder, phobias, obsessive-compulsive disorder, general anxiety, traumatic stress disorder, and anxiety disorder due to general medical conditions. Extreme anxiety can affect an individual's ability to carry out the activities of normal life. The persistent and ongoing symptoms can affect both the physical and mental status of the individual, causing a "snowball" effect on the well-being of the patient (Ohman 2000). A survey indicated that stress or anxiety constitutes an effectiveness gap or an area of clinical practice in which available treatments are not fully effective (Eisenberg et al. 1998). In fact, anxiety, depression, and insomnia are among the most common reasons for individuals to seek treatment with complementary therapies. A study indicates that 65.9 % of those seen by a conventional practitioner had also used CAM therapies to treat their anxiety, and the perceived helpfulness of CAM therapies was similar to that of conventional therapies. The use of acupuncture as a treatment for anxiety was relatively low at 0.7 %, compared with relaxation techniques, herbal and nutritional supplements, massage, and spiritual healing (Pilkington et al. 2007). There is a paucity of data to support or deny the use of acupuncture as a treatment or an adjunct treatment for anxiety disorders. However, several randomized control studies indicate that auricular acupuncture and related techniques can be used as treatment for situation anxiety.

Situational anxiety is defined as a state of apprehension, discomfort, and anxiety, precipitated by the experience of new or changed situation or event, which is different from an anxiety disorder. Although situational anxiety may not be abnormal or require treatment, at times the symptoms associated with situational anxiety can easily have a negative influence, e.g., parental anxiety may affect their child's anxiety in response to medical intervention, aggravation of pain, or nausea and vomiting.

Several randomized control studies have been conducted to explore the use of auricular acupuncture as a treatment for anxiety related to surgery. A total of eight randomized controlled trials (RCTs) are identified.

11.5.1.1 Auricular Acupuncture for Situational Anxiety

Wang and Kain (2001) applied bilateral auricular acupuncture at a relaxation point, a 'sham' point (tip of the concha) and at the Shen men point in a small group of operating room staff (Fig. 11.3).

Press needles were inserted and retained for 48 h. Acupuncture at a relaxation point was found to be more effective at 30 min and at 24 h, based on State-Trait Anxiety Inventory (STAI) and physiological measures, than the other two interventions and better than the Shenmen intervention at 48 h ($p=0.042$). However, the sham intervention was likely to elicit a response since the point used was a secondary choice for treating anxiety. This study demonstrates that auricular acupuncture is a simple and safe potential effective intervention for situation anxiety. In a second study, Wang and colleagues (2001) employed a similar technique but in this case used three TCM points, three relaxation points or three points unrelated to anxiety. The results of this larger and adequately powered study ($n=91$) suggested that auricular acupuncture at relaxation points was more effective based on STAI. No physiological measures were taken in this study, and the effect sizes appear relatively small. Another study involved mothers of children about to undergo an operation (Wang et al. 2004). Sixty-seven mothers and their children were enrolled in the study which was well designed with adequate randomization, allocation concealment, and blinding of assessors. Auricular acupuncture at anxiety-relieving points ($n=34$) was compared with auricular acupuncture at sham points ($n=33$) at least 30 min before induction of anesthesia. Mothers were assessed using the STAI, while children were assessed using the Modified Yale Preoperative Anxiety Scale. Maternal anxiety in the acupuncture group was significantly lower, and children

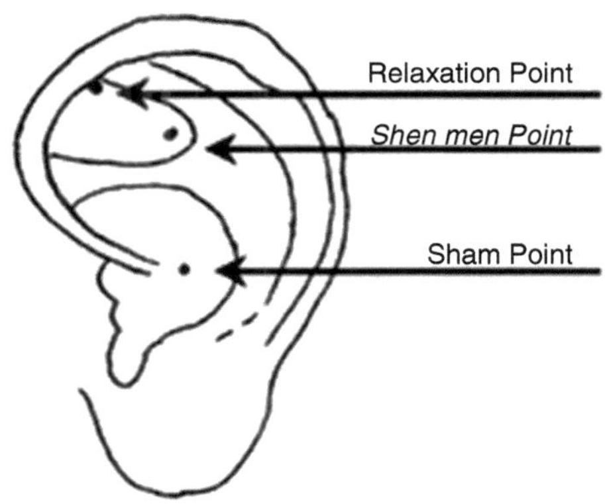

Fig. 11.3 The relaxation point (copyright Anesth & Analg 2001; 92:548–553)

were also significantly less anxious. In the most recent study, auricular acupuncture was compared with intranasal midazolam for dental anxiety. Karst and colleagues (2007) randomized 67 patients who were scheduled to receive a dental procedure. Auricular acupuncture at anxiety-relieving points ($n=19$) was compared with auricular acupuncture at the sham points ($n=19$), intranasal midazolam ($n=19$), and no treatment ($n=19$). The researchers were able to demonstrate that in the treatment for dental anxiety, auricular acupuncture and intranasal midazolam were superior to no treatment. Again, the anxiety-reducing effect started as early as 30 min after the insertion of needles or application of midazolam. The duration of sedation was less prolonged in the patients who received auricular acupuncture treatment than those receiving intranasal midazolam.

21.5.1.2 Auricular Acupressure for Situational Anxiety During Prehospital Transport

The first study was published in 1987 (Lewis 1987). This RCT included 90 patients undergoing various surgical procedures. Auricular acupressure was compared with diazepam and with progressive relaxation for preoperative anxiety. No difference was demonstrated between the three interventions based on subjective assessment of anxiety by the patient and an observer. However, it is unclear whether the trial had sufficient power to detect a difference in this outcome. There was also no difference in doses of anesthetic drug, systolic blood pressure, or pulse rate, but there was a highly significant difference in palmar sweating ($P<0.005$), and the authors suggest that this indicates better control of anxiety with auricular acupressure. Kober et al. (2003) focused on anxiety related to hospital transfer. In a well-designed trial, a single treatment of bilateral auricular acupressure at a "relaxation" point was found to be more effective than bilateral acupressure using a sham point in patients suffering from gastrointestinal illnesses but with no pain. This appears a potentially simple and safe intervention, but outcomes were measured using a visual analogue scale only and the trial was small ($n=36$). A subsequent study also indicated that auricular acupressure was effective in decreasing anxiety and pain for elderly patient suffering from hip fracture during the ambulance transport to the hospital (Barker et al. 2006). The investigators enrolled 38 patients aging between 80 and 95 years with hip fracture. The investigator found that auricular acupressure could reduce the anxiety ($F=4.3, p=0.018$) and pain ($F=28, p=0.0001$) and lower heart rate ($F=18$, $p=0.0001$). Thus the investigators encouraged physicians, health-care providers, and emergency rescuers to learn this easy and inexpensive technique. Mora et al. (2007) enrolled 100 patients with renal calculi who were transported to the local hospital by ambulance. All these patients were randomized into relaxation and sham treatment groups. They found that patients who received auricular acupressure with a 1 mm plaster ball at the relaxation point had significant reduction of anxiety and more optimistic about the outcome of treatment.

Auriculotherapy seems to show a promising effect in reducing the situational anxiety and the onset of anxiolytic effect is about 30 min. Additional research should focus on whether the similar interventions found to be effective in situational anxiety can be utilized as a treatment or serve as an adjuvant treatment for patients with various types of anxiety disorders.

11.5.2 The Use of Auricular Acupuncture as a Treatment for Pain

Pain represents a significant, complex problem and a major obstacle to successful rehabilitation. Multiple pharmacologic interventions are prescribed for pain, but the fear of side effects, including tolerance and addiction, has influenced practitioners as well as patients to limit use of pain medications. Thus, nonpharmacologic analgesic techniques would be a valuable addition to the current repertoire. Acupuncture has become increasingly popular in western pain clinics. Auricular acupuncture and related interventions are also found to be effective as a treatment for pain in various clinical entities. Auriculotherapy has been applied as treatment for musculoskeletal pain despite the fact that there is no clear mechanism of auricular analgesia. At present, several high-quality RCTs have been conducted to explore the efficacy of auriculotherapy as treatment for various pain conditions.

11.5.2.1 Auriculotherapy for Acute Postoperative Pain

Usichenko and colleagues (2005) enrolled 61 patients undergoing elective total hip arthroplasty under standardized general anesthesia. These patients were randomized into either active auricular acupuncture (auricular acupuncture) or sham auricular acupuncture (control) groups (Fig. 11.4).

The patients in the auricular acupuncture group received retained auricular needles at the Shenmen, thalamus, lung, and hip points. All auricular interventions were performed before surgery and retained in situ until the following morning. The amount of the rescue analgesic over the first 3 days after surgery was the outcome measurement. The results indicated that the patients in the control group require more piritramide than those of the auricular acupuncture group on day 1 and over a 3-day period. The same investigators (Usichenko et al. 2007) also conducted another study on a total of 120 patients undergoing ambulatory arthroscopic knee surgery with standardized general anesthesia. These patients were randomized into two groups: active auricular acupuncture (auricular acupuncture) and invasive needle intervention (control). The patients in the auricular acupuncture group received retained auricular needles at the Shenmen, lung, and knee points. All auricular interventions were performed before surgery and retained in situ until the following

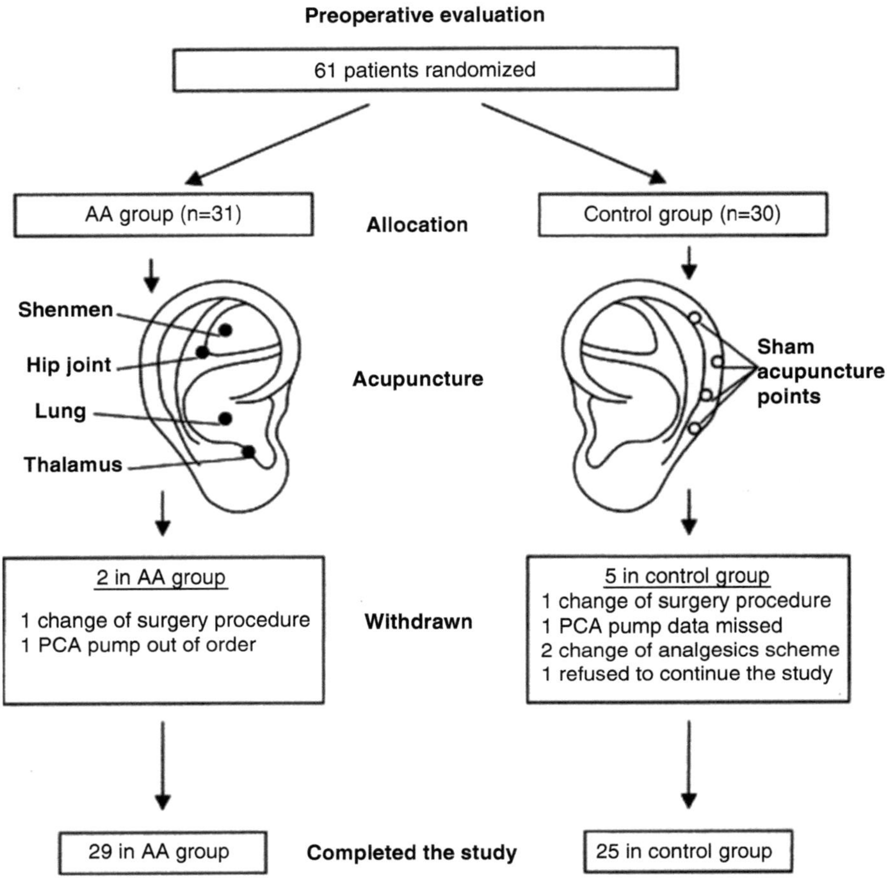

Fig. 11.4 The protocol and locations of auricular acupoints (copyright Pain 2005; 114: 320–327)

morning. The investigators also found that the patients in the control group required more ibuprofen than those of the auricular acupuncture group. Michalek-Sauberer and colleagues (2007) conducted an RCT on 149 patients undergoing third molar extraction either with auricular acupuncture or auricular acupuncture with electrical stimulation. They did not find any differences between the two interventions in regard to the severity of pain or analgesia consumption. In contrast, Likar and colleagues (2007) conducted an RCT on 41 patients undergoing laparoscopic nephrectomy to determine the effect of electrical auricular acupuncture stimulations. The investigator found that patients who received electrical auricular acupuncture stimulation required a less amount of morphine than those who received a sham treatment.

11.5.2.2 Auriculotherapy for Procedural Pain

Lewis and colleagues (1990) conducted a crossover study on a total of 11 burn patients undergoing dressing changes. The patients either received bilateral electrical auricular acupuncture first or a placebo pill. The investigator discovered that the intensity of pain was consistently lower after patients received electrical auricular acupuncture. Sator-Katzenschlager and colleagues (2006) conducted an RCT in a group of 94 women undergoing oocyte aspiration. All the women were randomized into one of three intervention groups: (1) electrical auricular acupuncture stimulation, (2) retained auricular acupuncture, or (3) a placebo group. They found that the pain relief was more pronounced in the women who received electrical auricular acupuncture intervention than those women in the other two groups. The women in the placebo group had the highest consumption of analgesia.

11.5.2.3 Auriculotherapy as a Treatment for Low Back Pain

Sator-Katzenschlager and colleagues (2004) conducted a 6-week study comparing the differences in the analgesic effect of weekly auricular acupuncture with and without electrical stimulations on a group of patients suffering from chronic low back pain (Fig. 11.5). They found that patients who received 6 weeks, at 48 h per

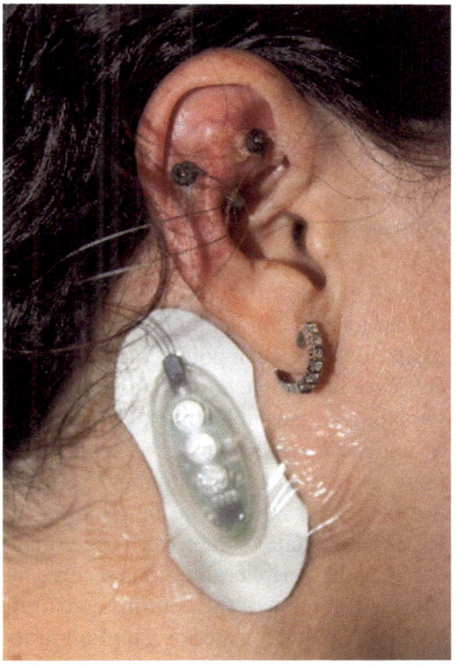

Fig. 11.5 The electrical stimulator for auricular acupoints

week, of auricular acupuncture with electrical stimulation had a low intensity of pain and disability during the study period and at a 3-month follow-up than those who received 48 h auricular acupuncture without electrical stimulation. Suen and colleagues conducted a clinical RCT on a group of 60 elderly patients suffering from low back pain (Suen et al. 2007). These patients were randomized into one of two treatment groups: control with semen vaccariae or magnetic pellets to seven auricular points. The researcher found that over a period of 4 weeks, patients who received magnetic pellets had a significant improvement in their overall disability level, pain/sensation, and physical and functional abilities. The investigator concluded that there were longitudinal changes in the improvement of disability levels among low back pain patients after receiving auriculotherapy. Wang and colleagues (2009) conducted a clinical RCT to determine whether auricular acupuncture could be effective as a treatment for posterior pelvic and low back pain in pregnant women. All 159 patients were randomized into one of three treatment groups. The investigator found that pregnant women who received 1 week of auricular acupuncture treatment had less pain and disability level than those who received auricular acupuncture at the sham point or no intervention. Although the sample size of these clinical RCTs was relatively small, these studies showed promising results that auriculotherapy could reduce the pain in both chronic and acute pain conditions as well as minimize adverse effects. However, many of these clinical trials did not address the issue of potential bias. Frequently only one acupuncturist applied all the interventions, and many of the studies did not specify whether the assessors or patients were blinded to the treatment. Additionally, the effect of the patient's belief in the treatment was not addressed in many of these studies.

11.6 What Do We Know Thus Far and What Is Next?

Although the ear has been an important structure in traditional Chinese medical acupuncture, many textbooks agree that auricular acupuncture is a recent development. The conceptual framework of auricular acupuncture does not relate to traditional Chinese medicine-based acupuncture. In addition, there are several issues related to auricular acupuncture, e.g., there are many auricular maps and little agreement exists between points, lacking definitive anatomic study on ear skin resistance and its somatotopic correspondences (Gori and Firenzuoli 2007; Wang et al. 2010). Dr. Nogier states in his book "Man in the Ear" that the auricle is not simply a piece of uneven cartilage. It has a close connection with the internal organs. However, there are no existing anatomical pathways that directly connect inner organs with the ear (Ernst 2007). The innervation of the central part of the auricle (triangular fossa and concha) comes from trigeminal, geniculate, and superior vagal ganglions, whereas the peripheral regions receive their innervation mainly from spinal nerves. The central parts of the neurons constituting the auricular branch of the vagal nerve are situated in the superior ganglion of the vagal nerve and nucleus tractus solitarii. Functionally, the duration and site of ear stimulation in healthy persons are associated with changed activity in the sympathetic and parasympathetic nervous system, depending on the site of stimulation and duration of observation (Haker et al. 2000). Researchers have found that stimulation of the sympathetic auricular acupuncture point significantly decreased the stimulus-evoked electrodermal responses compared with auricular acupuncture stimulation to a nonspecific point of the helix. In fact, several auricular acupuncture clinical studies also demonstrate that a better analgesic effect of auriculotherapy will result if auricular acupuncture stimulation is applied to the auricular point located in the central regions of the auricle rather than stimulations at the nonspecific points of the helix (Usichenko et al. 2005, 2007; Wang et al. 2009). Therefore, the autonomic systems may play an important role in the analgesic effect of auriculotherapy, and yet there is only paucity of data available in the current literature. Some research studies have demonstrated that the additional electrical stimulations may potentially affect the analgesic effect of auricular therapy. A recent development of portable auricular electrical stimulator (P-Stim) may provide a better analgesic effect than traditional manual needle stimulation and allow scientists to understand the basic mechanism of auriculotherapy. Lastly, a rigorous scientific design with special emphasis on blinding the operator, assessors, and patients as well as assessing the effect of patient's belief is needed in assessing the efficacy of auriculotherapy as treatment for various clinical entities.

11.7 Concluding Remarks

Compared to traditional body acupuncture, auricular acupuncture has limited history. The underlying mechanism of auriculotherapy remains unclear. In fact, there is paucity of scientific data to explore the potential mechanism and lack of validation of somatotopic mapping to address the discrepancy between Chinese and French ear maps. Nevertheless, auricular acupuncture has served as an easy and complementary treatment for several clinical conditions.

Acknowledgments The author would like to thank Mrs. Annette Forte for editing the manuscript.

References

Barker R, Kober A, Hoerauf K, Latzke D, Adel S. Kain ZN, Wang SM (2006) Out-of-hospital Auricular Acupressure in Elder Patients with Hip Fracture: A Randomized Double-blinded Trial. Acad Emer Med 13:19–23

Dale R (1976) The Micro-acupuncture System. Am J Acupuncture 4: 7–24

Eisenberg DM, Davis RB, Ettner SL, Appel S, Wilkey S, Van Rompay M, Kessler R (1998) Trends in Alternative Medicine Use in the United States, 1990–1997. Results of A Follow-up National Survey JAMA 280:1569–1575

Ernst E (2007) Auricular Acupuncture Can Med Assoc J 176: 1307

Fedoseeva O, Kalyuzhnyi L, Sudakov K (1990) New Peptide Mechanisms of Auriculo-acupuncture Electro-analgesia: Role of Angiotensin II. Acupuncture electro 15: 1–8

Frank BL, Soliman NE (2006) A History and Fortuitous Discovery. In Auricular Therapy: A Comprehensive Text Auricular Phases, frequencies, and blockage. First Edition, Author House ™, Bloomington IN 1–5
Gori L, Firenzuoli F (2007) Ear Acupuncture in European Traditional Medicine eCAM 4(S1) 13–16
Haker E, Egekvist H, Bjerring P (2000) Effect of Sensory Stimulation (Acupuncture) on Sympathetic and Parasympathetic Activities in Healthy Subjects. J Auton Nerv Syst 79: 52–59
Kalyuzhnyi L (1990) Analgesic Naloxone's Effect on Acupuncture Resistant, Acupuncture-tolerant, and Acupuncture-sensitive Rabbits. Acupuncture electro 15: 259
Karst M, Winterhalter M, Münte S, Francki B, Hondronikos A, Eckardt A, Hoy L, Buhck H, bernateck M, Fink M (2007) Auricular Acupuncture for Dental Anxiety: A Randomized Controlled Trial Anesth Analg 104: 295–300
Kober A, Scheck T, Schubert B, Strasser H, Gustorff B, Bertalanffy P, Wang SM, Kain, ZN, Hoerauf K (2003) Auricular Acupressure as A Treatment for Anxiety in Prehospital Transport Settings. Anesthesiology 98(6):1328–1332
Lee T (1994) Thalamic Neuron Theory: Theoretical Basis for The Role Played by The Central Nervous System (CNS) in The Causes and Cures of All Diseases. Med Hypotheses 3: 285–302
Lewis GB (1987) An Alternative Approach to Premedication: Comparing Diazepam with Auriculotherapy and A Relaxation Method. Am J Acupunct 15(3):205–214
Lewis SM, Clelland JA, Knowles CJ, Jackson JR, Dimick AR (1990) Effects of Auricular Acupuncture-like Transcutaneous Electric Nerve Stimulation on Pain Levels following Wound Care in Patients with Burns: A Pilot Study. J Burn Care Rehab 11: 322–329
Likar R, Jabarzadeh H, Kager I, Trampitsch E, Breschan C, Szeles J (2007) Elektrische Punktualstimulation (P-STIM) Mittles Ohrakupunktur. Der Schmerz 21: 154–159
Lu, HC (2005) The Healing Effects of Exercise in Traditional Chinese Medicine: An authoritative and Comprehensive Guide. Basic Health Publications. Laguna Beach, CA
Michalek-Sauberer A, Heinzl H, Sator-Katzenschlager SM, Monov G, Knolle E, Kress HG (2007) Perioperative Auricular Electroacupuncture Has No Effect on Pain and Analgesic Consumption after Third Molar Tooth Extraction. Anesth Analg 104: 542–7
Mora B, Iannuzzi M, Lang T, Steinlechner B, Barker R, Dobrovits M, Wimmer C, Kober A (2007) Auricular Acupressure as A Treatment for Anxiety Before Extracorporeal Shock Wave Lithotripsy in The Elderly. J Urol 178:160–164
Nogier P (1983) From Auriculotherapy to Auricular Medicine. Sainte-Ruffine, France: Maisonneuve
Ohman A (2000). Fear and Anxiety: Evolutionary, Cognitive, and Clinical Perspectives. In M. Lewis & J. M. Haviland-Jones (Eds.). Handbook of Emotions. New York: The Guilford Press 573–593
Oleson T, Kroening R, Bresler D (1980) An Experimental Evaluation of Auricular Diagnosis: Somatotopic Mapping of Musculoskeletal Pain at Ear Acupuncture Points. Pain 8: 217–229
Oleson T (1996) Theoretical Basis of Auriculotherapy. In Auriculotherapy Manual: Chinese and Western Systems of Ear Acupuncture. Second Edition. Health Care Alternatives, Los Angeles CA 9–19
Pilkington K, Kirkwood G, Rampes H, Cummings M, Richardson J (2007) Acupuncture for Anxiety and Anxiety Disorders-A Systematic Literature Review. Acupunct Med 25: 1–10
Sator-Katzenschlager S, Scharbert G, Kozek-Langenecker S, Szeles J, Finster G, Schiesser A, Heinze G, Kress H (2004) The Short- and Long-term Benefit in Chronic Low Back Pain through Adjuvant Electrical versus Manual Auricular Acupuncture. Anesth Analg 98:1359–1364
Sator-Katzenschlager SM, Wölfler MM, Kozek-Langenecker SA, Sator K, Stor P-G, Li B, Heinze G, Sator MO (2006) Auricular Electro-acupuncture as An Additional Peri-operative Analgesic Method during Oocyte Aspiration in IVF Treatment. Hum Reprod 21: 2114–2120
Soliman N, Frank BL (1999) Auricular Acupuncture and Auricular Medicine. Complement Ther Med 10: 547–554
Suen LKP, Wong TKS, Chung JWY, Yip VYB (2007) Auriculotherapy on Low Back Pain in The Elderly. Complement Ther Clin Prac 13: 63–69
Szeles J, Litscher G (2004) Objectivation of Cerebral Effects with A New Continuous Electrical Auricular Stimulation Technique for Pain Management. Neurol Res 26: 797–800

Usichenko TI, Dinse M, Hermsen M, Witstruck T, Pavlovic D, Lehmann C (2005) Auricular Acupuncture for Pain Relief after Total Hip Arthroplasty—A Randomized Controlled Study. Pain 114: 320–327

Usichenko TI, Kuchling S, Witstruck T, et al. (2007) Auricular Acupuncture for Pain Relief after Ambulatory Knee Surgery: A Randomized Trial. Can Med Assoc J 176: 179–83

Wang SM, Kain ZN (2001) Auricular Acupuncture: A Potential Treatment for Anxiety. Anesth Analg 92:548–553

Wang SM, Peloquin C, Kain ZN (2001) The Use of Auricular Acupuncture to Reduce Preoperative Anxiety. Anesth Analg 93:1178–1180

Wang SM, Maranets I, Weinberg ME, Caldwell-Andrews AA, Kain ZN (2004) Parental Auricular Acupuncture as An Adjunct for Parental Presence During Induction of Anesthesia. Anesthesiology 100:1399–1404

Wang SM, DeZinno P, Lin EC, Lin H, Yue JJ, Berman MR, Braveman F, Kain ZN (2009) Auricular Acupuncture as A Treatment for Pregnancy Women Suffering from Low Back Pain and Posterior Pelvic Pain: A Pilot Study. Am J Obstet Gynecol 201: 271e1–9

Wang SM, Maranets I, DeZinno P (2010) The Positive Predictive Value of Acupuncture Point Finder in Identifying the Active Auricular Reflex Point and Its Correlation with The Location of Pain, Abstract, Society of Acupuncture Research Annual Meeting. Chapel Hill, NC

Chapter 12
Abdominal Acupuncture for Nerve Root Cervical Spondylosis

Xiangrui Wang and Yongjun Zheng

Abstract Abdominal acupuncture is a new microneedle therapy system (MTS) for the treatment of nerve root cervical spondylosis based on Chinese ancient theories of Taiji, Bagua, and Yijing. Patients treated by abdominal acupuncture could regain the ability to work and live and enjoy a higher clinical cure rate. Its mechanism lies in the alleviation of vasospasm, the improvement of microcirculation, the changes of ischemia–hypoxia in tissues surrounding nerve roots, and the acceleration of absorption of inflammatory substances by downregulating the contents of serum interleukin-1 (IL-1), endothelin (ET), and thromboxane B2 (TXB2). Being a new microneedle technique and easy to operate, abdominal acupuncture shows a good clinical application prospect in the treatment of patients with nerve root cervical spondylosis.

Keywords Zhouyi • Holography • Nerve root cervical spondylosis • Thromboxane B2 (TXB2) • Endothelin (ET)

12.1 Introduction

Cervical spondylosis refers to the degenerative changes in cervical disc tissues, the secondary pathological changes involving the surrounding tissues (i.e., the nerve roots, spinal cords, vertebral arteries, and sympathetic nerves), and the correspondent clinical manifestations and symptoms. The incidence of cervical spondylosis in recent years has increased significantly than before. It has become a common and

X. Wang (✉) • Y. Zheng
Department of Anesthesiology and Pain Management, Renji Hospital,
Shanghai JiaoTong University School of Medicine, Shanghai, China
e-mail: xiangruiwang@gmail.com; zhengyongjun1971@126.com

Y. Xia et al. (eds.), *Current Research in Acupuncture*,
DOI 10.1007/978-1-4614-3357-6_12,

frequent disease influencing the adolescents and the elderly (Mazanec and Reddy 2007; Rao et al. 2008; Wang et al. 2011). The nerve root cervical spondylosis accounts for 50–60 % (Qin 2010; Xie et al. 2010) of all types of cervical spondylosis. Eighty percent of patients with cervical spondylosis suffer radicular pain (Binder 2007; Harrop et al. 2007; Jia et al. 2008). The nerve root cervical spondylosis is caused by the stimulation or oppression on the spinal nerve roots by the protrusion (intervertebral disc degeneration and spurs, etc.) at the intervertebral foramens, which leads to sensory and motor dysfunction of the upper limbs. It is one of the common causes for neck–shoulder–arm pain. There is a great variety of treatment methods for cervical spondylosis, which include two categories: the surgical therapy and the nonsurgical therapy. The primary problems for surgical therapy are uncertain surgical indications and improper surgical methods with a certain failure rate. The reoperation rate of cervical spondylosis reported in the literatures was generally 10–30 %. The nonsurgical therapy includes the oral administration of drugs and external application of acupuncture, iontophoresis, massage, manipulative reduction, physical therapy, and Qigong.

Acupuncture is an integral part of an ancient Chinese system of medicine that has been used for more than 2,500 years to treat diseases and relieve pain. According to radiation, the practice of acupuncture is based on a philosophy of balance and unity between the universe, living beings, and energy flow that penetrates everywhere and everything. Any imbalance, disruption, or energy-flow blockage within the body can cause disease or pain. The main concept and philosophy of acupuncture is to return the body to a harmonized, balanced state (Chernyak and Sessler 2005). There are numerous techniques and approaches to acupuncture. Warming needle at spinal nerves in front of transverse posterior tubercle of cervical vertebra, as a main therapy for treating cervical spondylosis of nerve root type, is a safe and effective method that is better than acupuncture at the acupoints of cervical Jiaii (EX-B2) (Qin 2010). Shu-needling therapy, as a main therapy, has a better therapeutic effect on cervical spondylosis of the nerve root (Wang 2008).

The traditional Chinese medicine is combined with Yijing acupuncture in abdominal acupuncture by using the theories of Zhouyi. The traditional meridian system, the Shenque (RN-8)–Bagua meridian system, and the body network's holographic system are integrated to build the basic structure of the abdominal acupuncture. It has the primary holographic resonance effects involving the Zang-fu organs and meridians, Taiji and Bagua as well as the body network. Based on the traditional meridian theory, the needling manipulation with the integration of modern human holographic theory is combined with the internal and external Bagua. It can stimulate the maximal potential energy within the human body by distributing the needles. At the same time, it can arouse the physiologically holographic resonance by a thorough communication and regulation among the internal, middle, and external Bagua to superimpose the stimulated energy information exponentially and improve the efficacy rapidly. The treatment of acupuncture on spondylosis is characterized by significant effects, simple operation, low cost, few side effects, etc. (Chernyak and Sessler 2005; Liao et al. 2011; Wang et al. 2011; Wu 2007).

12.2 Mechanism of Abdominal Acupuncture

Abdominal acupuncture is to stimulate the abdominal outpoints to treat systemic disease. The acupoints are mainly concentrated in the abdomen.

12.2.1 Anatomic Basis of Abdominal Acupuncture

According to anatomical knowledge, the human abdominal cavity has a thick covering of the anterior abdominal wall, skin, superficial fascia, muscle layer (and the deep fascia), transverse fascia, extraperitoneal fat, parietal peritoneum, and other components, including wealth of deep, superficial veins, lymphatic vessels, intercostal nerve, and lumbar nerve. The body's internal organs provide not only a rich material base to normal physical activity but also the regulatory body for the abdomen to provide for multilevel spatial structure. These nerves are connected with the spinal cord to reflect and convey a variety of information so that we can be different with the depth of differentiation, to influence and stimulate different peripheral systems, and to achieve the adjustment of local or overall effect.

There is a high number of superficial veins that contact each other to form a network on the anterior abdominal wall, especially in the umbilical area. Superficial anterior abdominal wall of the umbilical vein is bounded roughly, divided into upper and lower groups, but the umbilical area has a wide range of fit. Therefore, the anterior abdominal wall of the superficial veins forms the inferior vena cava, and its major tributaries block the establishment of collateral circulation of potential avenues. And umbilical vein, abdominal arteries, deep lymphatic vessels, intercostal nerve and lumbar nerve provide a rich material base for the diagnosis treatment of abdominal acupuncture.

12.2.2 The Physiological Basis of Abdominal Acupuncture

While of minimal functional importance, the umbilicus is a key aesthetic landmark of the anterior abdominal wall, and its absence or distortion is a frequent cause for concern and patient complaint. Umbilical fossa by newborn umbilical cord is stump to dry; the umbilical cord is connected with the abdominal wall skin cracks, occurring gradually, and from the abdominal wall, left after the formation of wound healing. Umbilical cord of the fetus is a channel that absorb oxygen and nutrients from maternal. The fetal abdominal wall connected with the umbilical cord by the umbilical wheel. Umbilical cord contains two umbilical arteries and one umbilical vein, during which the charge to the navel-like; it has the protection of umbilical blood vessel and promotes the blood flow. After birth, the umbilical cord as an organization will be abandoned in the neonatal period and dries off, and nutrient intake means a fundamental change, absorbing oxygen through breathing and absorption of intake of various nutrients through food digestion. Congenital means for the supply of oxygen is artificially cut off (the umbilical cord was ligated); the system's function is replaced by a new way, into the new system. Input distribution function of umbilical blood down to a second-class status, some of the blood vessels and blood vessels surrounding the establishment of new relationships, some organizations of atresia, some organization of connective tissue, was overlooked. However, the inherent Qi and blood still exists. Patients can be found some corresponding meridian points in the abdomen, many of the symptoms of the disease will be released in a very short period of time as long as these points are slightly stimulated. Found in the outer layer of umbilical cord, stem cells form into specific cells, such as skin, bone, or fat cells. Umbilical cord blood stem cells contain a large number of hematopoietic cells that can form blood cells; these umbilical cord stem cells that form the external layer of other potential cell types is remarkable and is the best. These are the navel and from another angle, the important relationship between human body and special significance.

12.2.3 *Nerve Reflex Mechanism*

Abdominal acupuncture treatment creates a very good condition; clinical observations suggest that acupuncture can lead to peripheral nerve, spinal cord, thalamus, limbic system, cerebral cortex, and other levels of bioelectrical parameter changes. Abdominal needle stimulate the abdomen to induce afferent impulses, and the impulse pass to the central nervous system at different levels, then the endogenous opioids, neurotransmitters and other substances can be secreted and regulate the energy metabolism, immune function, and other activities. The function of Zang-fu viscera to be adjusted to better help the body through acupuncture, regulates neurohumoral environment of stability, can increase the internal organs under stress and is relatively stable.

12.2.4 *The Regulation Mechanism of Abdominal Acupuncture*

As the body's structure is a three-dimensional shape, the abdominal anatomy, which is the body posture for belly regulation, provides for a multilevel spatial structure so that we can be different with the treatments to stimulate and influence the depth of the different peripheral systems, which is to adjust the local or global role. The human body shows the symmetry in the distribution of nerves and meridians on both sides of limbs, and the distribution of spinal nerves and meridians (except for the conception vessel) showed the same symmetry in the abdomen, and internal organs in the abdominal cavity though they are orderly but asymmetric. This will appear in the dual nature of the acupuncture points selected when there is a single take about the possibility.

And because the abdomen showed multiple hierarchies, each level of the nerve meridians has different effects, which has different shades of acupuncture. When the light is about the same name of puncture point in time, around the point, due to the impact of visceral nerves and surrounding tissue, is the difference in the two decidedly different clinical results. When abdomen is stabbed at a shallow point, the main impact to the deep abdominal wall, the superficial vein, the depth of lymphatic vessels and spinal nerves, the cutaneous branches of the abdominal section of the abdominal wall layers, and the number of other organizations primarily affects the peripheral system. When the abdomen is deeply punctured, the internal organs of the body cavity and peripheral nerve tissue are stimulated, the stimulation of information reaches the spinal cord through afferent nerves and uploads to the cerebral cortex then goes back to the corresponding internal organs which plays a therapeutic role.

12.3 Chinese Medicine Theory of Abdominal Acupuncture in the Treatment of Nerve Root Cervical Spondylosis

The Bagua (Chinese: *bāguà*) (Chen 2007) are eight diagrams used in Taoist cosmology to represent the fundamental principles of reality, seen as a range of eight interrelated concepts (Lore 2007). Each consists of three lines, each line either "broken" or "unbroken," representing yin or yang, respectively. Due to their tripartite structure, they are often referred to as "trigrams" in English. The trigrams are related to the Taiji philosophy and the Wu Xing philosophy, or "five elements." The relationships between the trigrams are represented in two arrangements: the Primordial, "Earlier Heaven" or "Fuxi" Bagua, and the Manifested, "Later Heaven," or "King Wen" Bagua. The trigrams have correspondences in astronomy, astrology, geography, geomancy, anatomy, the family, and elsewhere. Based on the Chinese ancient theories of Taiji, Bagua, and Yijing and after a multidimensional and multiangle synthesis, the new needling therapy–abdominal acupuncture is created. Under the guidance of theory of traditional Chinese medicine, it regards the regulation system of Taiji, Bagua, and meridians as the pivots, featured by a three-layer mode of Taiji and Bagua which coordinates with the 12 meridians and 8 extra meridians (see Fig. 12.1).

The regulation system of Taiji, Bagua, and meridian serves as the meridian theory of abdominal acupuncture, which has an internal connection with five Zang organs and six Fu organs as well as an external connection with the whole body. "Qi of yin and yang meridians communicates with each other; Qi of Zang-fu organs in the abdomen and chest corresponds mutually." Many diseases can be treated by puncturing the regulation system of Taiji, Bagua, and meridian as well as the correspondent regions. The internal Bagua and middle Bagua serve as the center and guidance, respectively, which are connected with the pathological regions. The important points including five transport points, source points, connecting points, alarm points, cleft points, crossing points, eight meeting points, and lower sea points are all combined in abdominal acupuncture. The Qi of internal Gagua, middle Gagua, and external Gagua is comprehensively connected and communicated by "four-step"

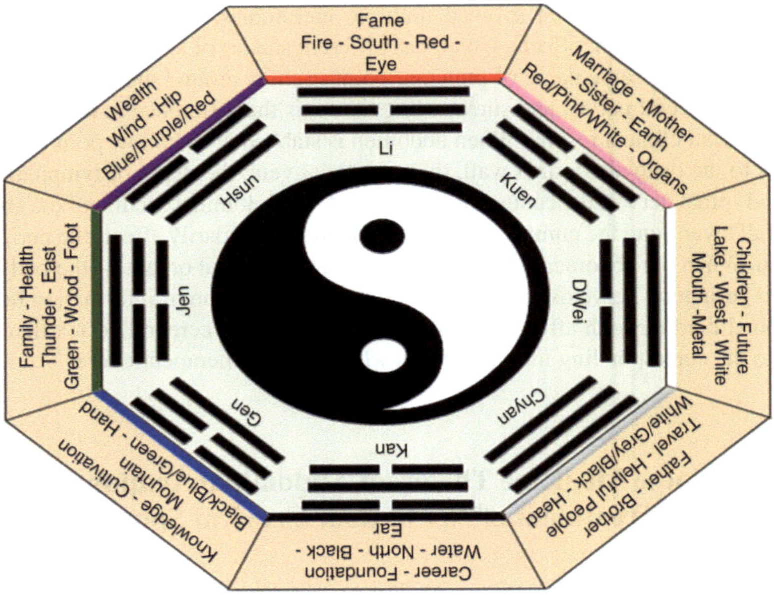

Fig. 12.1 The mode of Taiji and Bagua

needling manipulation. The ascending and descending of Qi movement in three energizers are regulated to promote the juncture of yin–yang Qi between Zang-fu organs in the abdomen and chest. Therefore, the interior and exterior are connected; the Ren and Du meridians are communicated thoroughly to ensure the harmony between the Qi and blood in the Zang-fu organs and meridians as well as the balance between yin and yang, rapidly curing the diseases.

Abdominal acupuncture is primarily indicated for pain and neurological disorders. It is particularly useful in the treatment of the sequelae of wind stroke; therefore, it is now used more often than scalp acupuncture in some clinics in China. It may also be helpful in treating cardiovascular and psychiatric disorders. Because the needling is superficial and there is only mild or no deqi obtained, abdominal acupuncture is indicated for any patients who are weak or who are particularly sensitive to acupuncture, including the very young and the elderly people.

12.3.1 Point Location

Abdominal acupuncture is applied at three levels of needle penetration: superficial, middle, and deep. Each level has its distinctive functions and therapeutic effects. The superficial, or "sky," level is situated subcutaneously between the skin and the fatty tissue. Microsystems use representative imagery whenever possible to facilitate organization and learning: in the case of Nogier's ear acupuncture, for example, an

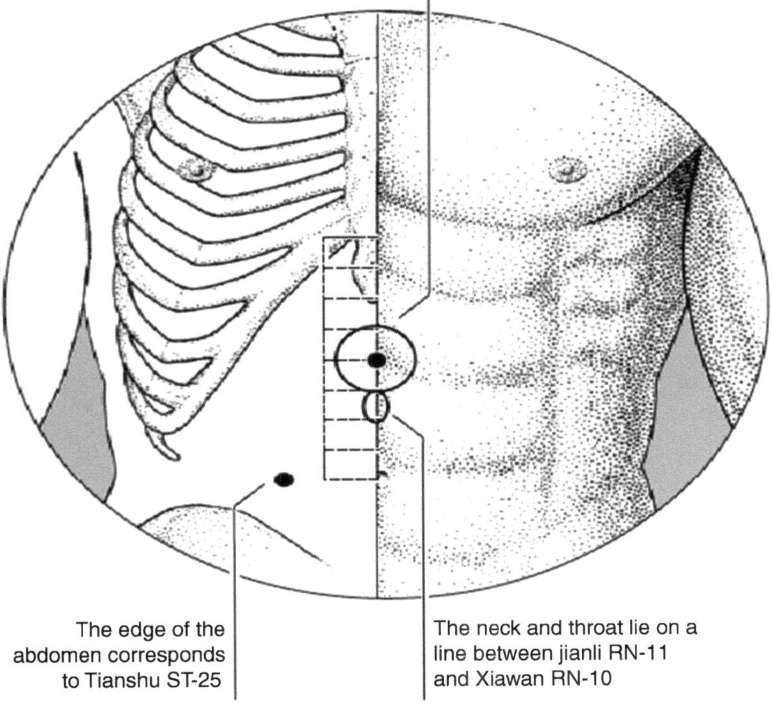

Fig. 12.2 The head is centered at Zhongwan RN-12, with the ears on the kidney channel on either side of Zhongwan RN-12. The neck and throat correspond to a line from Jianli RN-11 to Xiawan RN-10. The edge of the abdomen corresponds to Tianshu ST-25

inverted fetus is depicted. Abdominal acupuncture at the superficial level uses the image of a turtle or tortoise, one of the benevolent animal spirits and a prominent symbol in Chinese literature whose plastron is traditionally associated with the human abdomen. The tortoise is centered on the navel, and its head lies on the epigastrium.

It should be noted that abdominal acupuncture is two-dimensional, which means that the central line treats both the conception and governing vessels, for example, and the kidney channel also treats the bladder. Each area of the tortoise's body reflects a different part of the human body and corresponds to the following abdominal acupoints (see Figs. 12.2, 12.3, 12.4, and 12.5):

- The head is centered at Zhongwan RN-12, with the ears on the kidney channel on either side of Zhongwan RN-12.
- The neck and throat correspond to a line from Jianli RN-11 to Xiawan RN-10.
- The edge of the abdomen corresponds to Tianshu ST-25.

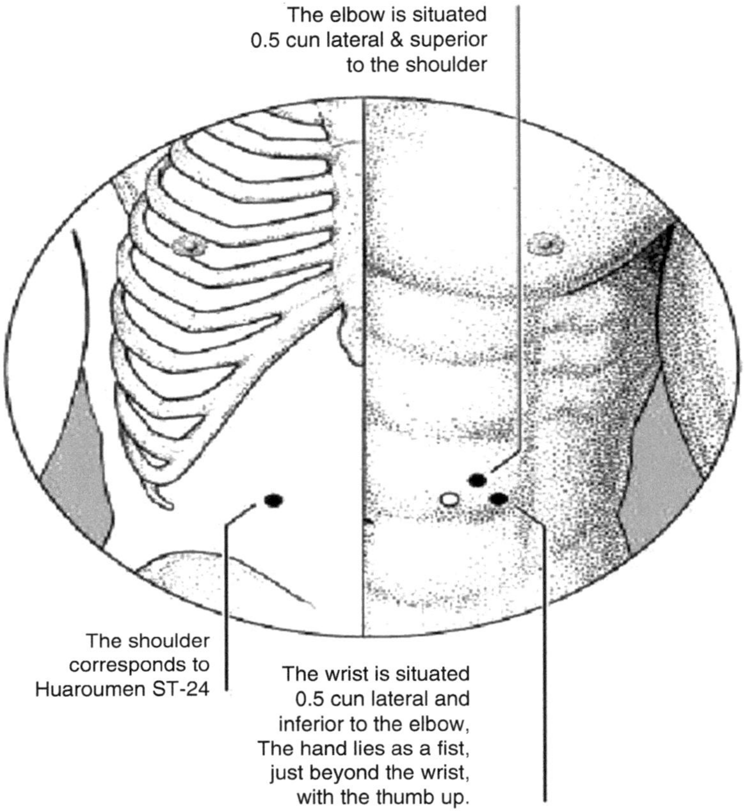

Fig. 12.3 The shoulder corresponds to Huaroumen ST-24. The elbow is situated 0.5 cun lateral and superior to the shoulder. The wrist is situated 0.5 cun lateral and inferior to the elbow. The upper limb forms an inverted "V"

- The shoulder corresponds to Huaroumen ST-24.
- The elbow is situated 0.5 cun lateral and superior to the shoulder.
- The wrist is situated 0.5 cun lateral and inferior to the elbow. The upper limb forms an inverted "V."
- The hand is represented as a fist just beyond the wrist, with the thumb up.
- The lower border of the first lumbar vertebra corresponds to Qihai RN-6, and Yaoyangguan DU-3 corresponds to Guanyuan RN-4.
- The coccyx corresponds to Zhongji RN-3.
- The hip corresponds to Wailing ST-26.
- The knee is situated 0.5 cun lateral and inferior to the hip. Note that a secondary knee point lies halfway between the primary one and Qihai RN-6. All three points lie on the same horizontal line.
- The ankle is situated 0.5 cun inferior and lateral to the knee, and the foot is situated just beyond, with the toes 45° inferior and medial to the foot.

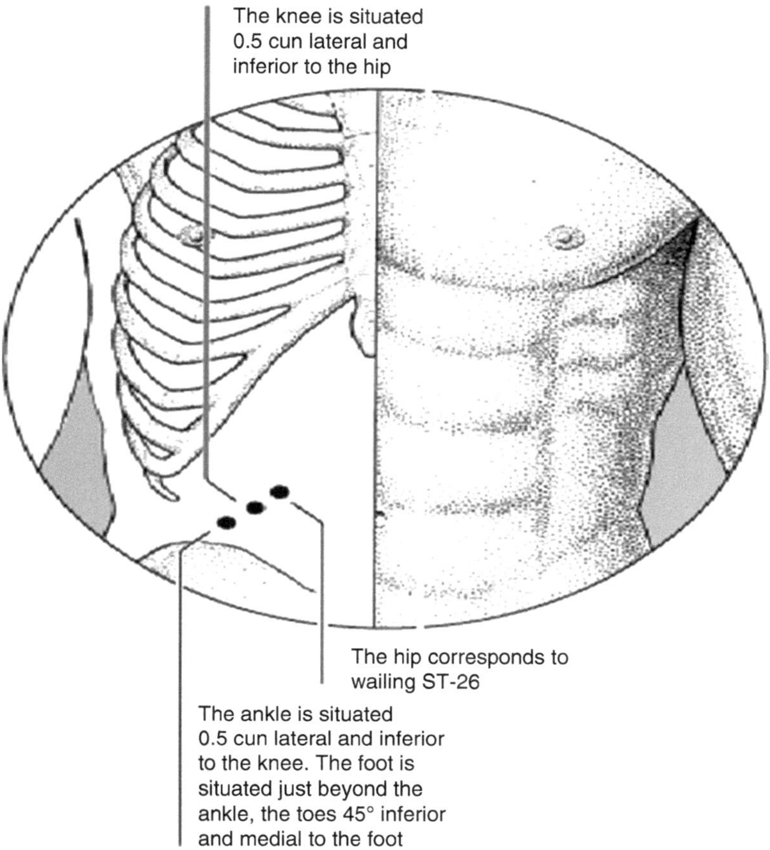

Fig. 12.4 The hip corresponds to Wailing ST-26. The knee is situated 0.5 cun lateral and inferior to the hip. Note that a secondary knee point lies halfway between the primary one and Qihai RN-6. All three points lie on the same horizontal line. The ankle is situated 0.5 cun inferior and lateral to the knee, and the foot is situated just beyond, with the toes 45° inferior and medial to the foot

12.4 Clinical Application of Acupuncture for Nerve Root Cervical Spondylosis

Cervical spondylosis is a result of cervical disc degeneration, secondary to the upper and lower vertebral bone hyperplasia, nerve root compression and spinal cord or vertebral artery caused by impact of a range of symptoms. The most common nerve root symptoms include the performance of neck and shoulder pain radiating to the arms, fingers or neck, limited mobility, severe weakness, and tinnitus. According to international statistics, acupuncture is about 90 % effective for the disease. A variety of acupuncture point stimulations, such as acupuncture, warm needle,

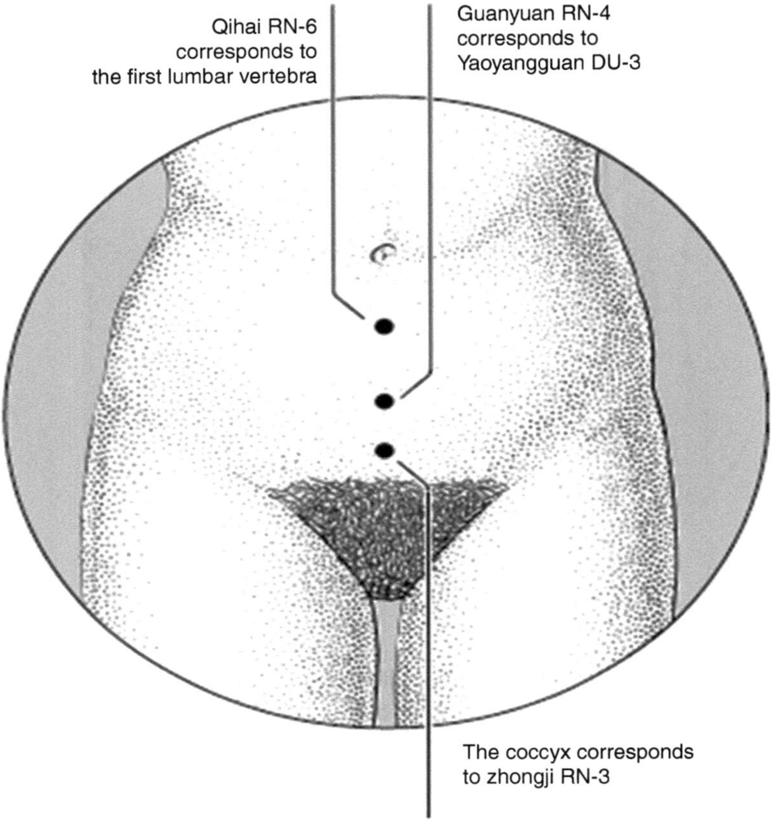

Fig. 12.5 The coccyx corresponds to Zhongji RN-3

medicine-separated moxibustion, bamboo pot method, laser acupuncture, point injection needle, and beryllium needle are used in the treatment of this disease. Abdominal acupuncture is also a very effective method for cervical spondylosis.

12.4.1 Treatment Method Used in Abdominal Acupuncture

Needle punched into the abdomen should avoid pores and blood vessels, and then the operation should be light and slow, only twist insert or light twist slowly in order to avoid torn viscera and visceral injury. Shi surgery has generally adopted three laws, namely, waiting gas, line gas, and gas practice reminders. The needle is kepted for 3–5 min in order to Deqi, that is a composite of unique sensations that is essential for clinical efficacy according to traditional Chinese medicine, and the needle is twisted to strengthen the Deqi feeling, then to the needle is remained for 30 min.

12.4.1.1 Selecting Points

Shenque (RN-8) serves as the center. Kan, Li, Zhen, and Kun are selected as the internal Bagua, and the orientation of them is correspondent to the 6 o'clock, 12 o'clock, 9 o'clock, and 1 o'clock, respectively. Kan, Li, Zhen, and Kun also serve as the middle Bagua. Both Houxi (SI-3) points are selected as the external Bagua.

12.4.1.2 Position

1. Position of internal Bagua.

   The internal Bagua is around the center of the Shenque (RN-8) with a diameter of about 3 cun. It is not limited and subject to the size of the umbilicus (which is the biggest difference compared with navel needling therapy) because the size of the umbilicus and its proportion on the abdomen vary among individuals; and the size of the umbilicus in the same person even changes with age. Therefore, the internal Bagua should be located in accordance with the proportion of the entire abdomen. As we all know, the distance from the middle point of linea Alba to the abdominal edge is 6 cun, the distance between the two abdominal edges is 12 cun. As the diameter of internal Bagua is about 3 cun, it accounts for about 1/4 of the abdominal transverse line.
2. Position of middle Bagua.

   After the position of internal Bagua is determined, it is very easy to localize the middle Bagua. In fact, the middle Bagua is the corresponding extension of internal Bagua. The size of the middle Bagua is similar to the Taiji figure on the abdomen. Firstly, the four due orientations are located as follows: The Zhongwan (RN-12) and Zhongji (RN-3) on the Ren meridian is the upper and lower border, respectively; Bilateral Dahengs (SP-15) are the left and right borders. Four corners of the middle Bagua are correspondent to the four due orientations. Because the localization of middle Bagua is totally same to the internal Bagua, it is only an outward radial extension of the exactly located internal Bagua.
3. Position of external Bagua.

   The external Bagua has a relatively extensive and complicated range including all the regions except the middle Bagua such as the head, the face, the trunk, and the four limbs.

12.4.1.3 Needling Instruments

Disposable needles used in the study had different specifications of 0.25×20, 0.25×30, and 0.25×40 mm. All patients were treated by needles with the same thickness. The needles with different lengths were selected depending on the thickness of adipose layer of patients' abdominal wall.

12.4.1.4 Manipulation

The operator carefully examined the muscular tension, tenderness, abdominal mass, and accumulation in each region by touching and pressing from up to down before the treatment. All points were conventionally sterilized with 75 % alcohol.

Acupuncture Manipulation. The operator inserted the needles into the abdomen and avoided the pores and vessels with gentle and slow manipulations. When the tip of the needle reached the expected depth, the method of twirling without lifting and thrusting the needles, or gentle twirling with slow lifting and thrusting, is generally adopted to give the greater omentum in the abdominal cavity enough time to react, avoiding pricking the internal organs. Generally three steps are followed, i.e., the manipulation of awaiting Qi, moving Qi, and hastening Qi. The step of inserting the needles and keeping them in place for 3–5 min is called awaiting Qi. The step of twirling the needles to produce local needling sensation is called moving Qi. And the step of manipulating the needles at an interval of 5 min to strengthen the needling sensation and make it spread all around or to the distance is called hastening Qi.

Needling Depth. For abdominal acupuncture, the needling depth is divided into three layers: sky, earth, and human. The shallow puncture to the sky layer with a depth of 0.3–0.5 cun is applicable for patients with a short disease course or attacked by superficial pathogenic factors. The middle puncture to the human layer with a depth of 0.6–1.0 cun is applicable for patients with a long disease course without affecting the interior. The deep puncture to the earth layer with a depth of 1.0–1.5 cun is applicable for patients with a long-term disease course involving the Zang-fu organs or with pathogen remaining in the interior.

Handling of Emergency. If the pores are pricked, the patients will feel pain, and then the operator should change the needle-insertion direction. If bleeding is caused by insertion, the operator should press the bleeding sites after withdrawing the needle.

The needles are withdrawn after retention for 30 min. The patients are treated once a day, and ten times are considered as a treatment course (see Fig. 12.6).

12.4.1.5 Control Group

Selecting Points: Cervical Jiaji (EX-B2), Fengchi (GB-20)

After the routine disinfection at the Jiaji (EX-B2), the operator inserted the needles perpendicularly 0.5–0.8 cun into patients in prone position. The needles were inserted 0.8–1.2 cun toward the nose tip with the tip of the needle slightly downward until the local swelling sensation is produced. The needles were remained for 30 min each time. Ten times is a treatment course. The patients are treated once a day for three successive courses of treatment with 2 days rest at the weekends.

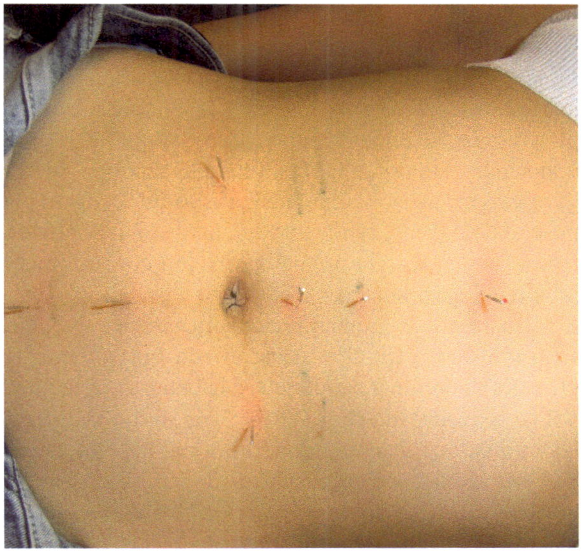

Fig. 12.6 Acupuncture manipulation

The patients in two groups stopped all the drugs and other therapies during the period of treatment.

12.4.2 Clinical Effect of Abdominal Acupuncture on Nerve Root Cervical Spondylosis

Abdominal acupuncture with invigorating the kidney and nourishing marrow method can improve the allomeric function of the brain cognition network of the central nervous system (Zhong et al. 2011). Twelve old healthy volunteers were treated by standardized abdominal acupuncture with invigorating the kidney and nourishing marrow method, and Zhongwan (CV-12), Xiawan (CV-10), Qihai (CV-6), Guanyuan (CV-4), Qixue (KI-13), Shangqu (KI-17), and Huaroumen (ST-24) were selected. They were examined using resting-state fMRI before and after treatment of abdominal acupuncture, and the brain functional connectivities were analyzed based on bilateral hippocampus chosen as seed points. (1) Before acupuncture, functional connectivities were found in the bilateral middle frontal gyrus, bilateral superior temporal gyrus, bilateral middle temporal gyrus, bilateral thalamus, bilateral subiculum hippocampi, bilateral insula, cupula and declivis of right cerebellum, left superior frontal gyrus, left head of caudate nucleus, left posterior cingulate gyrus, and left inferior parietal lobule (all $P<0.05$). (2) After acupuncture, functional connectivities were found in the bilateral superior frontal gyrus, bilateral superior temporal gyrus, bilateral middle temporal gyrus, bilateral thalamus, bilateral insula, bilateral subiculum hippocampus, bilateral inferior parietal lobule, bilateral postcentral gyrus, bilateral cingulate gyrus, right head of caudate nucleus, right middle occipital gyrus, and right posterior cingulate gyrus (all $P<0.05$). (3) After acupuncture, there were intensity functional connectivities greater than before acupuncture in the bilateral superior frontal gyrus, bilateral middle frontal gyrus, bilateral superior temporal gyrus, bilateral superior parietal lobule, right inferior frontal gyrus, right anterior cingulate gyrus, right fusiform gyrus, right precuneus, and right postcentral gyrus, while the intensity of functional connectivities of left middle temporal gyrus, right lingual gyrus, left cuneus/precuneus, and left inferior parietal lobule were attenuated.

Abdominal acupuncture can better reduce the pain of the patient caused by CS, with transient pain-alleviating effect, but whether or not the clinical therapeutic effect of abdominal acupuncture is better than the traditional acupuncture still cannot be proved (Guo et al. 2007). This study compares clinical therapeutic effects of abdominal acupuncture and traditional acupuncture on cervical spondylosis (CS). *Methods.* Sixty-two cases of neck or nerve root-type CS were randomly divided into an observation group ($n=32$) treated by abdominal acupuncture at Zhongwan (CV-12), Guanyuan (CV-4), and others and a control group ($n=30$) treated by traditional acupuncture at Fengchi (GB-20), cervical Jiaji (EX-B2), etc. Simplified McGill Pain Questionnaire (MPQ) and clinical therapeutic effects were served as the objective indexes. Their clinical therapeutic effects were compared after the first session of treatment, at the end of therapeutic course, and 3 months after the end of treatment. The two groups had a same effective rate of 100.0 %. All items of MPQ in these two groups after treatment and 3 months after the end of treatment significantly improved, and in the observation group, the differences in the PRI feeling score before and after the first treatment and the difference of the total PRI scores after the first treatment, at the end of therapeutic course, and 3 months after the end of treatment significantly improved as compared with the control group ($P<0.05$).

Abdominal acupuncture plus Long's bone-setting manipulation has significant efficacy of either short or long term on CSR, which is superior to the efficacy of either simple abdominal acupuncture or Long's bone-setting manipulation and indicates superimposed effect. Hence, it is one of the better approaches in CSR treatment (Fan et al. 2010). One hundred and eighty cases of CSR were randomly allocated into abdominal acupuncture plus bone-setting group (combined therapy group), bone-setting group, and abdominal acupuncture group, with 60 cases in each group. In combined therapy group, the abdominal acupuncture and Long's bone-setting were applied in combination. Abdominal acupuncture was applied to Zhongwan (CV-12), Guanyuan (CV-4), Shiguan (KI-18), Shangqu (KI-17), etc. Long's manipulation, such as bone-setting in head-upward posture and bone-setting in head-lateral posture, was adopted. In bone-setting group and abdominal acupuncture group, Long's bone-setting manipulation and abdominal acupuncture were adopted simply and respectively. The clinical therapeutic effects were compared after two courses of treatment (short-term) and 1 month after treatment (long-term) among groups. The short- and long-term curative and markedly effective rates in combined therapy group were 80.7 % (46/57) and 68.4 % (39/57), respectively, which were better than those of 63.64 % (35/55) and 30.9 % (17/55) in bone-setting group and 58.9 % (33/56) and 50.0 % (28/56) in abdominal acupuncture group, separately (all $P<0.05$). Moreover, the long-term curative and markedly effective rate in abdominal acupuncture group was superior to that in bone-setting group ($P<0.05$).

Alternative administration of scalp acupuncture, body acupuncture and abdominal acupuncture can effectively improve stroke patients' upper limb motor function, and its effect is obviously better than that of simple body acupuncture (Zhou et al. 2009). A total of 222 stroke patients were randomized into treatment group ($n=133$) and control group ($n=89$) according to the random number table method. For patients of treatment group, scalp point used was Motor Area (MS 6); body acupoints were Hegu (LI-4), Quchi (LI-11), Shousanli (LI-10), etc.; and abdominal acupoints Zhongwan (CV-12), Qihai (CV-6), etc. For patients in control group, only the same body acupoints were used. Acupuncture treatment was given once daily for 5 weeks. The therapeutic effects were evaluated by Fugl-Meyer motor function assessment (FMA) and Barthel index (BI) scores. In comparison with pretreatment, FMA score and BI score increased significantly 1 to 3 months in treatment group and 1 to 3 months in control group ($P<0.01$). Comparison between two groups showed that both FMA score and BI score of treatment group were markedly higher than those of control group 1 to 3 months after the treatment ($P<0.01$).

Abdominal acupuncture TENS has a better transient analgesic effect and can use less stimulation points to increase the analgesic effect (Wang et al. 2007). One hundred and twenty cases of pain on neck, shoulder, loin, and legs were randomly divided into four groups: abdominal acupuncture TENS group, acupoints TENS group, electroacupuncture (EA) group, and non-abdominal acupuncture TENS group, 30 cases in each group. All the cases were treated by the same stimulation parameters but different stimulation points. The VAS scores were recorded before and after treatment. The VAS scores were significantly different before and after treatment in abdominal acupuncture TENS group ($P<0.01$); the total effective rate of the transient analgesic effect was 96.7 % in the abdominal acupuncture TENS group, 93.3 % in the acupoints TENS group, and 96.7 % in the EA group with no significant difference among the three groups but with a very significant difference between the abdominal acupuncture TENS group and the non-abdominal acupuncture TENS group (10.0 %), $P<0.01$.

No significant differences were found between abdominal acupuncture and routine acupuncture [OR=3.29, 95 % CI (0.13, 82.99)], EA [OR=2.09, 95 % CI (0.36, 11.95)], and traction therapy [OR=6.06, 95 % CI (3.01, 12.18)] in the total effective rate, the pain rating index score [WMD=−2.24, 95 % CI (−5.29, 0.81)], and the present pain intensity score [WMD=−0.84, 95 % CI (−2.13, 0.44)]. *Conclusion.* At the present, there has been no sufficient evidence to ensure that in the treatment of cervical spondylosis, the abdominal acupuncture therapy is superior to routine acupuncture, EA, and traction therapy. Attention should be paid to the randomized controlled study of larger samples and qualified design (Wang et al. 2011).

The clinical curative rate, pain intensity (PI), sensation score in MPQ, and ability to work and live in the abdominal acupuncture group are all prior to the acupuncture

Table 12.1 Comparison of general conditions of patients

Group	Cases	Age (year)	Course of disease (year)
Treatment group	37	40.19 ± 11.56	2.58 ± 3.19
Control group	36	38.69 ± 11.46	2.29 ± 3.21

Table 12.2 Comparison of effect between the two treatment methods

Group	Cases	Cured	Better	Unrecovered	Curative rate (%)	Total effective rate (%)
Treatment group	37	16	18	3	43.24	91.89
Control group	36	11	19	6	30.56	83.33

Table 12.3 Comparison of PI between the two groups (mean ± standard deviation)

Group	Cases	Pretreatment	Posttreatment
Treatment group	37	3.444 ± 1.229	0.778 ± 1.198**, ***
Control group	36	3.482 ± 1.424	2.074 ± 1.517*

Notes: Self-comparison before and after the treatment: *$P<0.05$, **$P<0.01$; compared with the control group: ***$P<0.05$

There is no significant difference in the analysis of disease condition before the treatment between the two groups ($P<0.05$)

Table 12.4 Comparison of sensation score in SF-MPQ before and after the treatment between the two groups (mean ± standard deviation)

Group	Cases	Pretreatment	Posttreatment
Treatment group	37	3.611 ± 1.479	1.361 ± 1.175*, **
Control group	36	4.222 ± 1.672	2.852 ± 1.512*

Notes: Self-comparison before and after the treatment: *$P<0.05$; compared with the control group: **$P<0.05$

There is no significant difference in the analysis of disease condition before the treatment between the two groups ($P<0.05$)

group [Jiaji (EX- B2), and Fengchi (GB-20)]. There is no significant difference in disease course and age between the abdominal acupuncture group and the control group, so they are comparable ($P>0.05$) (Table 12.1).

The clinical and total curative rates in abdominal acupuncture group are 43.24 and 91.89 %, respectively, while that of control group is 30.56 and 83.33 %, respectively. There is a significant difference in the effect between the two groups ($P<0.05$) (Table 12.2).

The pain intensity (PI) of abdominal acupuncture group and control group decreased significantly after the treatment. There is a significant difference in PI between the two groups ($P<0.05$) (Table 12.3).

The SF-MPQ sensation score in the abdominal acupuncture group and control group decreased significantly after the treatment. There is a significant difference in the sensation score between the two groups ($P<0.05$) (Table 12.4).

Table 12.5 Comparison of emotion score in SF-MPQ before and after the treatment between the two groups (mean ± standard deviation)

Group	Cases	Pretreatment	Posttreatment
Treatment group	37	1.528 ± 0.506	0.611 ± 0.688*, **
Control group	36	1.704 ± 0.823	1.185 ± 0.879

Notes: Self-comparison before and after the treatment: *$P < 0.01$; compared with the control group: **$P < 0.05$

There is no significant difference in the analysis of disease condition before the treatment between the two groups ($P < 0.05$)

Table 12.6 Comparison of total score in SF-MPQ before and after the treatment between the two groups (mean ± standard deviation)

Group	Cases	Pretreatment	Posttreatment
Treatment group	37	5.056 ± 2.013	1.889 ± 1.670**, **
Control group	36	5.926 ± 2.269	4.037 ± 2.175*

Notes: * $P < 0.05$, Self-comparison before and after the treatment; **$P < 0.05$, compared with the control group; There is no significant difference in the analysis of disease condition before the treatment between the two groups ($P < 0.05$)

The SF-MPQ emotion score in abdominal acupuncture group decreased significantly and had a significant difference compared with the control group ($P < 0.05$) (Table 12.5).

The total score in SF-MPQ of abdominal acupuncture group and the control group decreased significantly. There is a significant difference in the total score between the two groups ($P < 0.05$) (Table 12.6).

12.5 Mechanism of Abdominal Acupuncture in the Treatment of Nerve Root Cervical Spondylosis

The clinical study showed that no significant positive sign was presented in some patients' X-ray films, but they had relatively grave clinical symptoms and signs of nerve root pain. The nerve root pain was relieved or cured in some patients after being treated by acupuncture or taking drugs, though no obvious improvement was seen in X-ray films. One of the main causes of nerve root cervical spondylosis was the inflammation and edema of cervical nerve roots as well as the damage of peripheral vascular nets and stasis in the small veins surrounding them. After being treated by abdominal acupuncture, the contents of vasoactive substances such as ET and TXB2 in patients with nerve root cervical spondylosis decreased. The abdominal acupuncture can relieve the vasospasm, improve the microcirculation, change the ischemia–hypoxia in tissues surrounding the nerve roots, and accelerate the absorption of inflammatory substances to relieve the clinical symptoms.

12.5.1 Influence of Abdominal Acupuncture on Serum IL-1 of Patients with Nerve Root Cervical Spondylosis

The activity of serum IL-1 in abdominal acupuncture group decreased significantly and had a significant difference compared to the control group ($P<0.05$) (Table 12.7).

12.5.2 Influence of Abdominal Acupuncture on Serum ET of Patients with Nerve Root Cervical Spondylosis

The activity of serum ET in abdominal acupuncture group decreased significantly and had a significant difference compared to the control group ($P<0.05$) (Table 12.8).

12.5.3 Influence of Abdominal Acupuncture on Serum TXB2 of Patients with Nerve Root Cervical Spondylosis

The activity of serum TXB2 in abdominal acupuncture group decreased significantly and had a significant difference compared to the control group ($P<0.05$) (Table 12.9).

Table 12.7 Changes in serum IL-1 content ($\bar{x} \pm s$, pq/ml)

Group	Cases	Pretreatment	Posttreatment
Abdominal acupuncture group	30	0.7468±0.1795	0.5567±0.1745*, **
Control group	16	0.74671±0.1124	0.6592±0.1557*

*Compared with pretreatment, $P<0.05$; **compared with the control group, $P<0.05$

Table 12.8 Changes in serum ET content ($\bar{x} \pm s$, pq/ml)

Group	Cases	Pretreatment	Posttreatment
Abdominal acupuncture group	30	91.4783±15.1566	68.4671±14.7624*, **
Control group	16	90.6821±16.1748	78.5649±15.1557*

*Compared with pretreatment, $P<0.05$; **compared with the control group, $P<0.05$

Table 12.9 Changes in serum TXB2 content ($\bar{x} \pm s$, pq/ml)

Group	Cases	Pretreatment	Posttreatment
Abdominal acupuncture group	30	60.4673±13.2815	41.2853±9.1703*, **
Control group	16	60.6913±20.9388	48.2038±13.4625*

*Compared with pretreatment, $P<0.05$; **compared with the control group, $P<0.05$

Table 12.10 Changes in serum SOD content ($\bar{x} \pm s$, U/ml)

Group	Cases	Pretreatment	Posttreatment
Abdominal acupuncture group	30	20.3517±3.2614	26.0971±7.0468*
Control group	16	19.2318±4.6719	23.5932±4.4625*

*Compared with pretreatment, $P<0.05$

Table 12.11 Changes in serum MDA contents ($\bar{x} \pm s$, nM/ml)

Group	Cases	Pretreatment	Posttreatment
Abdominal acupuncture group	30	4.3515±1.2618	3.0975±1.0461*
Control group	16	4.2358±1.6417	3.5935±1.4627*

*Compared with pretreatment, $P<0.05$

12.5.4 *Influence of Abdominal Acupuncture on Serum Superoxide Dismutase in Patients with Nerve Root Cervical Spondylosis*

The activity of serum SOD in abdominal acupuncture group increased significantly and had a significant difference compared to the control group ($P<0.05$) (Table 12.10).

12.5.5 *Influence of Abdominal Acupuncture on Serum Malondialdehyde in Patients with Nerve Root Cervical Spondylosis*

The activity of serum MDA in abdominal acupuncture group decreased significantly and had a significant difference compared to the control group ($P<0.05$) (Table 12.11).

12.6 Concluding Remarks

The optimized therapeutic program of the immediate effect of acupuncture for cervical spondylosis is electroacupuncture+C4–C7 Jiaji (EX-B2)+5 min. The importance of different factors for the immediate effect in acupuncture treatment of cervical spondylosis of vertebral artery type is acupoint selection>needling technique>time of needle retaining (Liao et al. 2011).

In the treatment of patients with nerve root cervical spondylosis, abdominal acupuncture can significantly improve the symptoms of pain and numbness while decreasing the general score. Its mechanism might be related with the alleviation of vasospasm, the improvement of microcirculation, the changes of ischemia–hypoxia in tissues surrounding the nerve roots, and the acceleration of absorption of inflammatory substances.

According to the theory of abdominal acupuncture, the umbilicus has a known circulatory system and a close relationship with Qi and blood flowing throughout the body. The abdomen also has an advanced regulating/controlling system for the whole body that was formed during the embryological period. This regulating/controlling system is a substantial foundation for abdominal acupuncture. In the past, old Qigong masters imagined a 3-cun Taiji (yin/yang) symbol centered on the umbilicus. Embraced in the center were two fishes, one being yang and one being yin. From this they produced the yin and yang, the ascending/descending, and entering/exiting of Qi and blood through the body.

12.6.1 *Zang-fu Organ Relationships*

According to TCM, humans are divided into three portions by the Sanjiao. Both the middle and lower Jiao are located on the abdomen and contain all the Zang-fu organs and inner generative organs, except the lungs and heart. But, the lungs and heart have a direct or indirect relationship with the other Zang-fu organs located in the middle and lower Jiao. This means that the abdomen has a direct or indirect relationship with all the Jingluo and Zang-fu organs.

12.6.2 *Abdominal Holograms*

Abdominal acupuncture is a microsystem of the whole body. The three different levels are similar but different to the heaven, human, and earth sequence of the standard meridian system. The superficial level is pictured as a hologram of a pronated tortoise. Although a tortoise, all its parts correspond to the human body. The neck begins to extend from bilateral Shangqu points (Kid-17) and extends up to Zhongwan (RN-12). The tail begins from bilateral Qipang Points (0.5 cun lateral to RN-6) and extends to Guanyuan (RN-4). The upper limbs (shoulder area) start from Huaroumen (ST-24), flex at the upper rheumatism point (0.5 cun lateral and superior to ST-24), and end at the upper lateral rheumatism points (1 cun lateral to ST-24). The lower limbs (hip) begin at bilateral wailing points (ST-26), extend across the lower rheumatism points (0.5 cun lateral and inferior to ST-26), and end at the lower lateral rheumatism points (1 cun lateral and inferior to ST-26). This top superficial layer relates to disorders of the head, torso, upper and lower limbs, disorders of the periphery, meridian and collateral disorders, and acute disorders.

The medial layer consists of the standard abdominal meridian system. It is the connection between the superficial and deep levels. The deep level is represented by the Bagua figure. The Bagua begins at the umbilicus and extends upward to RN-12, down to RN-4, and laterally to the Leg Taiyin meridian. The deep level corresponds to the Zang-fu organs and is chosen for the treatment of the internal viscera and long-term chronic disorders. Abdominal acupuncture is the new powerful microsystem acupuncture. In ancient times, the abdomen was used exclusively for diagnosis.

Today the abdomen is used to treat the entire body. Acupuncture performed on the abdomen was developed over the last 15 years and is becoming very popular. This new technique involves placing needles into acupuncture points on the abdomen related to the spinal region.

Ear, hand, and abdominal acupunctures work on the same theory that the abdomen and navel area has points related to the entire body. By stimulating special points on the meridians on the abdomen and using traditional meridian systems, immediate relief is provided to the patients.

The aim is to harmonize, adjust, and rebalance the body and mind. This increases energy and blood flow allowing the body to become harmonized. Since energy flows of the entire body are focused in the navel area, all parts of the body can be treated by this technique. It is a particularly powerful tool in treating chronic long-term diseases and problems that result in joint and muscle pain.

Since the needles are superficially placed, much shallower than conventional acupuncture, they are painless to the patients. Stiff necks; spinal, lumbar, thoracic, and cervical problems; shoulder pain; tendonitis; and rheumatoid arthritis have all been helped with abdominal acupuncture. It is frequently used to treat strokes, senile dementia, high blood pressure, and cerebellum problems in conjunction with scalp, ear, and traditional acupuncture.

Anxiety, depression, headaches, dizziness, arthritis, cataracts, glaucoma, and Parkinsonism and diabetic neuropathy are also helped with abdominal acupuncture.

It is particularly indicated in patients who are weak and sensitive to acupuncture, such as the very young and the very elderly. This procedure is contraindicated in pregnancy, abdominal infections, and enlargement of any abdominal organs, especially the spleen and liver.

Initially, the Chinese inventor, Bo, treated a patient with low back pain and sciatica by applying needles to RN-6 and RN-4, and the patient's pain disappeared in 5 min.

The system was formerly introduced in 1992 and was called the BMAA system. The first symposium in China on the BMAA system took place in 2007. This system is present in fetal and embryonic development. In the case of ear acupuncture by Nogier, an inverted fetus is depicted. In abdominal acupuncture, at the superficial level, the image is one of a turtle (which in Chinese literature is associated with the human abdomen). The turtle is centered at the navel, and the head lies high on the epigastrium.

The system is two dimensional, with its center line treating both the conception and governing vessels (e.g., the kidney and bladder channels can be treated simultaneously). Each area of the turtle's body reflects a different channel.

Abdominal Meridian points are located in the middle layer of the belly; these channels include abdominal acupuncture points as well as extraordinary points, with bands of points radiating from CV-8. It is called the miraculous tortoise because its shape looks like a turtle.

The superficial level, or sky level, is the most commonly used and is usually combined with conventional acupuncture. This level is subcutaneous between the skin and the fatty tissues. Musculoskeletal points are found very close to the surface of the abdominal wall just penetrating the skin. The effect of needling this area results in instantaneous flow of Qi and blood to the whole body. By contrast, the traditional 14-channel postnatal system is deeper in the fat tissue.

By deeper needling into the fat and muscle layers of the belly, the Zang-fu viscera (liver, spleen, and kidney) can be harmonized. Dr. Bo labels this a second system, the Zang-fu system.

Rather than needling distal points on the extremities to get visceral action, this method eliminates the possibility of blocked pathways and the need of protracted manipulation and additional needling to release these blocks.

This system combines needling the prenatal channel system (Heaven level), the postnatal system and the extraordinary vessels on the belly (Humanity level), and the Zang-fu organ system (Earth level). Needles are inserted from top to bottom, deep to shallow, inside to outside. The system has more than 20 pattern-based standard formulas for most diseases. By using very small needles just under the skin with light manipulations when inserted, the procedure is virtually painless.

For neck problems with cervical compression and numbness in the hands, CV-12 and CV-4 are used to generate Qi from the spleen and kidney (these govern bones and muscles). The muscles of the neck are relaxed, and the upper extremities are energized by treating K-17 and K-18 and ST-24, bilaterally. Ren points Ab1 and Ab2 are used on the affected side to relieve numbness of the hands.

12.6.3 *The "Turtle" Figure – A Meridian Theory in Combination with Bio-holographic Theory*

The turtle's head is centered at RN-12, with its ears on the kidney channel just lateral to RN-12. The neck and throat is on a line between RN-11 and RN-10. The edge of the abdomen is at ST-25, and the shoulders are at ST-24. The elbow is situated above the shoulder 0.5 cun laterally. The wrist is just inferior to the elbow and 0.5 cun lateral. The upper limb forms an inverted V. The hand is represented as a fist just beyond the wrist with the thumb up.

The lower portion of L1 is at RN-6, the coccyx is at RN-3, and the hip at ST-26.

The knee is 0.5 cun lateral and inferior to the hip. Another knee point is halfway between RN-6 and the primary one. All three points lie on the same horizontal line. The ankle is lateral to the knee and 0.5 cun inferior. The foot is just beyond this, with the toes 45° inferior and medial to the foot (see Figs. 12.7, 12.8, and 12.9).

The diamond treatment is the first needling protocol area, with RN-12, RN-6, and RN-4 combined with ST-25 bilaterally. This improves the circulation of the abdomen and enhances all subsequent needle points that will be treated.

The second step in treatment, (called the four gates) involves bilateral needling of ST-24 and ST-26. This promotes a circulation of Qi throughout the body as a conventional acupuncture involving LI-4 and LIV-3.

If the upper limb is affected, the first needling is to the shoulder point. If the lower limb is involved, the first point placed is at the hip area. This is then followed by direct acupuncture to the direct pain site to be needled.

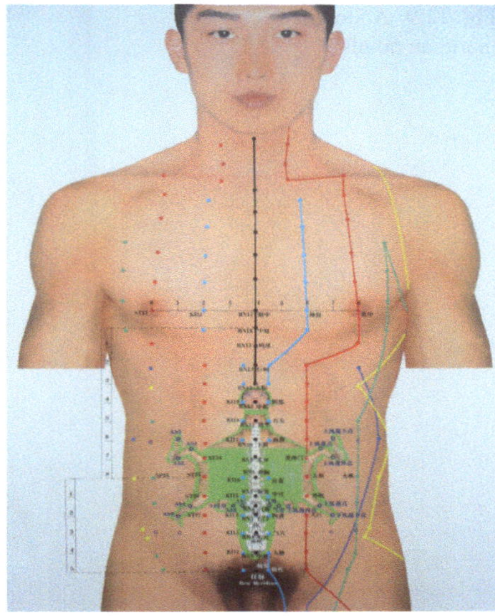

Fig. 12.7 Abdominal meridian points

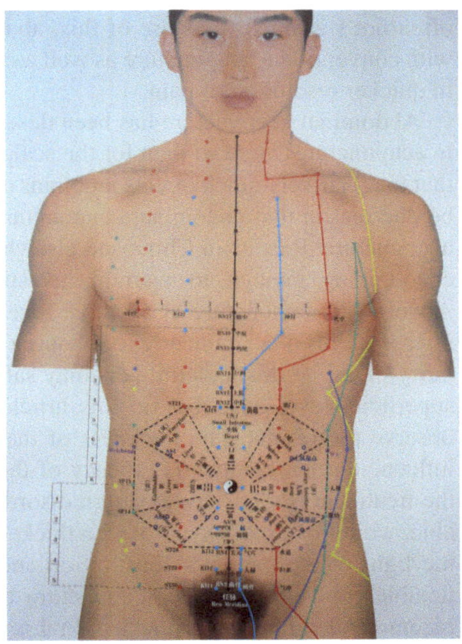

Fig. 12.8 Abdominal meridian points

Often conventional acupuncture treatment at ST-36, SP-6, and LI-11 is treated along with abdominal acupuncture. Other points used are GB-39, GB-30, ST-41, LI (4, 11, and 15), and SI-5. Abdominal acupuncture is a powerful microsystem similar to that of scalp and ear acupuncture. Despite quick dramatic effects, these effects are

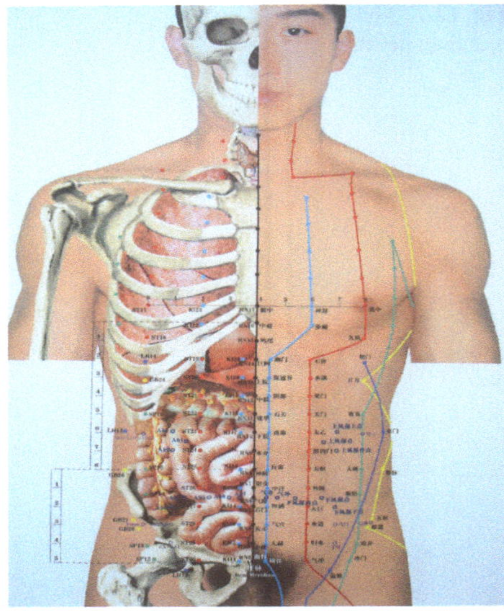

Fig. 12.9 Abdominal meridian points

often not sustained. Because of this, abdominal acupuncture is usually combined with conventional acupuncture as well as scalp, hand, and ear acupuncture resulting in quicker resolution of pain.

Abdominal acupuncture has been described as "the most powerful microsystem in acupuncture, more so than for the scalp or ear." It has also been noted, however, that the relatively quick effects it obtains are not necessarily sustained, and this may be one reason that abdominal acupuncture is usually combined with conventional acupuncture. Results in China and elsewhere (such as Southern Europe) have been encouraging enough to warrant investigation in the English-speaking world. Abdominal acupuncture at all three levels takes longer to learn than many other microsystems of acupuncture (one source recommends 3 years to attain basic familiarity). Despite this, practitioners may safely combine the treatment routines at the superficial level described in this article with conventional protocols, and then observe the results. The thickness of the abdominal needle is a negligible factor influencing the therapeutic efficacy of the reinforcing–reducing manipulations. In the treatment of deficiency-pattern disorders, a thinner acupuncture needle is suitable, while for excess syndrome, a thicker acupuncture needle is recommended. In the light of the patient's constitution and different state of clinical conditions, a flexible application of the acupuncture needle with different diameters is highly recommended in clinical practice. In this way, the clinical therapeutic effect would be improved much (Wang and Zhang 2010).

Wang assessed the effect and methodological quality of clinically randomized controlled studies on abdominal acupuncture therapy for cervical spondylosis and to make out its current situation, validity, and applicability. Using the PubMed, CNKI (China Academic Journals Full-text Database), VIP (VIP Chinese Science and Technology Periodicals Database), and Wanfang Digital Periodicals Electronic Database covering the period of 1989–2009, we did a literature search on the original articles of abdominal acupuncture treatment of cervical spondylosis and selected those accorded with the standards of randomized controlled studies. Animal studies, surveys, news articles, and those duplicated, being absent in diagnostic criteria and nonrandomized controlled trials, were excluded. The papers' quality was analyzed using the Jadad quality assessment scoring system, and the therapeutic effect evaluated using Review Manage 4.2.7 software. A total of eight papers written in Chinese containing 909 cervical spondylosis patients were included. These eight studies used the effective rate as the primary outcome; two of them used the MPQ scales at the same time. Meta-analysis showed that the abdominal acupuncture group was better than the control group in visual analogue scale score ($P<0.05$). No significant differences were found between abdominal acupuncture and routine acupuncture [OR=3.29, 95 % CI (0.13, 82.99)], EA [OR=2.09, 95 % CI (0.36, 11.95)], and traction therapy [OR=6.06, 95 % CI (3.01, 12.18)] in the total effective rate, the pain rating index score [WMD=−2.24, 95 % CI (−5.29, 0.81)], and the present pain intensity score [WMD=−0.84, 95 % CI (−2.13, 0.44)]. At present, there has been no sufficient evidence to ensure that in the treatment of cervical spondylosis, the abdominal acupuncture therapy is superior to routine acupuncture, EA, and traction therapy. Attention should be paid to the randomized controlled study of larger samples and qualified design (Wang et al. 2011).

Acknowledgments Chinese Ministry of Science and Technology (973) 2007CB512504: Basic Research of Specificity of Acupoints based on Clinical Acupuncture Anesthesia

References

Binder AI (2007) Cervical spondylosis and neck pain. BMJ 334: 527–531

Bo Songqi S, Justin Fontanini, (2006) Abdominal Acupuncture: Mechanisms and Formula. acupuncture today 6: 1–3

Chen X (2007) The Illustrated Canon of Chen Family Taijiquan. INBI Matrix Pty Ltd 12: 11

Chernyak GV, Sessler DI (2005) Perioperative acupuncture and related techniques. Anesthesiology 102: 1031–1049; quiz 1077–1038

Fan DH, Liu G, Wang TC, Huang F, Wang XY, Zeng HW, Zhou FX, Yao GX, Chen Xu XL, Li RD, (2010) Observation on short and long-term effects of cervical spondylotic radiculopathy treated with abdominal acupuncture plus Long's bone-setting manipulation. Zhongguo Zhen Jiu 30: 909–912

Guo YQ, Chen LY, Fu WB, Xu MZ, Ou XM (2007) Clinically randomized controlled study on abdominal acupuncture for treatment of cervical spondylosis. Zhongguo Zhen Jiu 27: 652–656

Harrop JS, Hanna A, Silva MT, Sharan A (2007) Neurological manifestations of cervical spondylosis: an overview of signs, symptoms, and pathophysiology. Neurosurgery 60: S14–20

Jia CS, Li XF, Liu EJ, Zheng LY (2008) Survey and prospects of mini-acupuncture system therapy for cervical spondylosis. Zhongguo Zhen Jiu 28: 313–316

Liao YY, Sun DM, Zhong CP, Lai XS (2011) Evaluation of the immediate effect of acupuncture on cervical spondylosis of vertebral artery type based on orthogonal design. Zhongguo Zhen Jiu 31: 499–502

Lore R (2007) Abdominal Acupuncture: A Practical Introduction. Journal of Chinese Medicine 83: 29–32

Mazanec D, Reddy A (2007) Medical management of cervical spondylosis. Neurosurgery 60: S43–50

Qin YG (2010) Clinical observation on therapeutic effect of warming needle at spinal nerves for treatment of cervical spondylosis of nerve root type. Zhongguo Zhen Jiu 30: 121–123

Rao RD, Currier BL, Albert TJ, Bono CM, Marawar SV, Poelstra KA, Eck JC (2008) Degenerative cervical spondylosis: clinical syndromes, pathogenesis, and management. Instr Course Lect 57: 447–469

Wang LP, Zhang B (2010) Comments on the thickness of abdominal acupuncture needles. Zhen Ci Yan Jiu 35: 236–240

Wang XL (2008) Observation on therapeutic effect of shu-needling therapy as main on cervical spondylosis of nerve root type. Zhongguo Zhen Jiu 28: 497–498

Wang YW, Fu WB, Ou AH, Fan L, Huang YF (2011) A systematic review of randomized controlled clinical trials of abdominal acupuncture treatment of cervical spondylosis. Zhen Ci Yan Jiu 36: 137–144

Wang ZL, Chen LF, Zhu WM (2007) Observation on the transient analgesic effect of abdominal acupuncture TENS on pain of neck, shoulder, loin and legs. Zhongguo Zhen Jiu 27: 657–659

Wu M (2007) Treatment of cervical spondylosis by puncturing both the local and distal points-a report of 76 cases. J Tradit Chin Med 27: 31–32

Xie YF, Ruan YD, Ning XJ, Ma CL, Chen Y (2010) Heat sensitive moxibustion for treatment of nerve root cervical spondylosis. Zhongguo Zhen Jiu 30: 379–382

Zhong ZP, Wu SS, Chen ZG, Bo L (2011) Study on response of resting-state functional magnetic resonance imaging induced by abdominal acupuncture with invigorating the kidney and nourishing marrow method. Zhongguo Zhen Jiu 31: 139–143

Zhou W, Liu H, Wang LP, Xu ZG, Feng YW, Lv H, Xie Y, Guo J, Liu M, Liu Y (2009) Effect of alternative administration of scalp-acupuncture, body-acupuncture and abdominal-acupuncture on the upper-limb dyscinesia in patients with stroke. Zhen Ci Yan Jiu 34: 128–131

Chapter 13
Electro-acupuncture Regulation of Central Monoamine Neurotransmitters in Ischaemia-Reperfusion

Mingshu Xu, Linbao Ge, and Dan Zhao

Abstract Cerebrovascular disease is one of the three primary diseases, with cerebral ischaemia accounting for more than 80 % of cerebrovascular events. Monoaminergic neurons are distributed throughout the central nervous system and play an important role in physiological functions. These neurons also affect the pathophysiological process of cerebrovascular disease. While the effects of monoamine neurotransmitters on cerebral ischaemia-reperfusion injury/recovery have been widely studied, to date, the exact role of these neurotransmitters is unclear. Based on previous reports, we speculate that monoamine neurotransmitters have deleterious effects in the early stage of ischaemia but that they promote recovery in the later stage of cerebral ischaemia-reperfusion. Due to the various stages of ischaemia and changes in the internal and external cellular environment, monoamine neurotransmitters function as a "double-edged sword". A large body of prior evidence has demonstrated that electro-acupuncture is safe and effective for the treatment of cerebral ischaemia. Our findings suggest that Fengchi electro-acupuncture can reduce the accumulation of acid base metabolin and can alleviate the injury caused by cerebral ischaemia-reperfusion. This method can regulate the levels of dopamine (DA) and norepinephrine

M.S. Xu (✉)
State Laboratory of Cerebral Neurobiology, Shanghai Research Institute of Acupuncture and Meridian, 650 Wanping South Road, Shanghai 200030, China
e-mail: mingshuxu@yahoo.com.cn

L. Ge
Shanghai Research Institute of Qigong, 650 Wanping South Road, Shanghai 200030, China

Shanghai Research Center for Acupuncture and Meridian, 199 Guoshoujing Road, Shanghai 201203, China
e-mail: gelinbao@vip.163.com

D. Zhao
Shanghai Research Institute of Qigong, 650 Wanping South Road, Shanghai 200030, China
e-mail: z_d_z_d_@hotmail.com

Y. Xia et al. (eds.), *Current Research in Acupuncture*,
DOI 10.1007/978-1-4614-3357-6_13,

(NE), as well as selectively strengthen the neuronal activity of noradrenergic and dopaminergic nerves in the rats with cerebral ischaemia-reperfusion, increasing the metabolic levels of NE and DA and promoting nerve cell recovery.

Keywords Electro-acupuncture • Brain • Ischaemia-reperfusion • Rats • Monoamine neurotransmitters

13.1 Introduction

Cerebrovascular injury is one of the main diseases in the world, especially in developed countries and developing countries with continuously increasing living standards, such as China. Studies found that among adults aged 20 and older, in 2004 an estimated 5,700,000 cases (2,400,000 males; 3,300,000 females) were reported in the United States (Rosamond et al. 2008). In 2008, Japan reported more than 400/100,000 annual incidence rate,while more than 120/100,000 were reported in China. Half of all premature deaths and disability in the world are attributed to nervous system diseases resulting from stroke (Hachinski 2009).

Cerebrovascular disease not only threatens the life of the patient but also seriously influences the quality of life following an ischaemic event. When brain injury occurs due to a lack of blood flow, the specific neurological functions controlled by the affected brain region can be damaged or even lost. The resulting neurological disabilities depend on the region affected and the severity of the infarction (Kandel et al. 2000).

Survivors of stroke are often beset by serious, long-term disabilities, including paralysis and disruption of higher cognitive functions, such as speech and memory. Among the survivors of ischaemic stroke, unilateral paralysis, an inability to walk without assistance, aphasia (trouble in speaking or understanding the speech of others), and the inability to perform daily tasks (e.g. grooming, eating, and bathing) are the most frequently observed disabilities in the 6 months following the stroke insult. Some of the patients may even have mental disorders, such as depressive symptoms. Individuals with such disabilities often require extensive long-term care by health care professionals and family.

In addition to the immediate consequences, cerebrovascular diseases also have a high rate of recurrence. Consequently, the governmental and personal cost of cerebrovascular diseases increases every year.

Due to the high and wide-ranging social impact of cerebrovascular disease, there is great interest in researching methods to increase the cure rate of cerebrovascular diseases, reduce the financial burden of both government agencies and affected families, and improve the quality of life of ischaemic patients.

Although the damage to neurons located in the ischaemic core is usually severe and typically results in cell death, the penumbra neurons can still be salvaged and functionally recovered by employing rapid interventional treatment.

These neurons can be rescued by re-establishing blood flow to the affected area as soon as possible. The "time window" for the interventional treatment is usually approximately 6 h. Although the current revascularisation and human tissue-type plasminogen activator (tPa) offer advantages for some patients, the "time window" for these clinical practices is severely limited. Beyond this window, these treatments may have only minimal efficacy and may even potentially cause further damage. Post-ischaemic reperfusion can harm cells and tissues, referred to as reperfusion injury. Therefore, there is great utility in the identification of methods to reduce potential tissue damage during cerebral ischaemia-reperfusion.

There is a wide gap between the current treatments and our expectations. Currently, no clinical modality has demonstrated promising efficacy in terms of stroke treatment. Therefore, new strategies should be developed to establish better preventative measures or treatment of this serious disease.

13.2 Clinical Application of Acupuncture on Stroke

Previous studies (Liu et al. 2009; Schaechter et al. 2007; Chau et al. 2009) have demonstrated that acupuncture has beneficial effects as an ischaemia treatment. After acupuncture therapy, the patients showed enhanced recovery and reduced infarction sizes, including beneficial effects on motor function, a reduction in spasticity, and improvement in post-stroke depression (Fan et al. 2001; Johansson et al. 1993; Naeser et al. 1994; Hu et al. 1993; Wong et al. 1999; Moon et al. 2003; Liu et al. 2008).

Currently, acupuncture is widely used for the treatment of stroke in China and other Asian countries, such as Japan and Korea, and is recommended by the World Health Organization (WHO) as one of the alternative and complementary strategies for stroke treatment. The National Institute of Health Consensus Development Conference on Acupuncture concluded that "acupuncture might serve as a useful adjunct or alternative treatment for stroke rehabilitation". In addition, some evidence has recently been published that shows that 3 weeks of transcutaneous electrical stimulation to lower leg acupuncture points administered five times per week within 10 days post-stroke significantly decreased ankle plantar flexor spasticity and increased dorsiflexor strength concomitantly with a decrease in antagonist co-contraction (Yan and Hui-Chan 2009).

However, there is evidence showing that acupuncture did not demonstrate specific efficacy over placebo (Hopwood et al. 2008), and some systematic reviews did not provide evidence to support the routine use of acupuncture for patients with subacute or chronic stroke (Wu et al. 2008). Thus, although scientific studies on acupuncture therapy have been performed for the last 50 years, a consensus on the utility of this approach has not yet been reached.

Randomised clinical trials have demonstrated that acupuncture may be effective in the treatment of post-stroke rehabilitation. However, poor study quality and the possibility of publication bias reduce the strength of this recommendation and argue for a large, transparent, well-conducted randomised clinical trial to support this claim and implement changes to clinical practice (Wu et al. 2010). Recently, studies investigating the factors affecting the efficacy of acupuncture have been performed. These factors include the intervention time point of acupuncture, the interval and frequency of acupuncture, the combination of acupuncture with other therapies, and the individual condition of patients (Guo et al. 2010).

13.2.1 Acupuncture for the Treatment of Paralysis Following a Stroke

In the 1970s, acupuncturists in China developed the method of scalp acupuncture, which uses filiform needles to topically puncture the scalp points. All the scalp points are representations of the underlying functional areas of the brain, such as the contralateral motor area, sensory areas, the foot motor sensory area, or speech areas. If part of the brain is damaged, for instance by a stroke, then the scalp is stimulated over the damaged area of the brain. Currently, scalp acupuncture is widely used to treat stroke patients, and some acupuncturists suggest that it could induce recovery better than ear acupuncture or body acupuncture therapy (Lu 1991). It has been shown that the scalp acupuncture can widen cerebrovasculature, improve microcirculation, reduce blood viscosity, and enhance central peripheral nerve conduction function.

Body acupuncture is the most commonly used acupuncture treatment for stroke. The traditional stimulation method is to insert fine needles at acupoints and manipulate them manually. Xuemin Shi (Shi 2005) adopted acupuncture treatment for 9005 stroke cases, primarily using Yin meridians and Du Channel (Governor Vessel) acupoints. Treatments were administered twice a day for a continuous 30–50 days. The total effective rate was 98.56 %. Clinical laboratory-related indexes have reported that acupuncture treatments for stroke can alter several haemodynamic parameters in stroke patients, including improving the cerebral blood flow condition, improving superoxide dismutase (SOD) activity, reducing lipoperoxide (LPO) content, and reducing brain injury, thereby promoting tissue repair. Some commonly used acupoints for stroke therapy in Chinese hospitals are listed in Table 13.1.

Another report using evidence-based medicine (EBM) methods found that acupuncture treatment can have curative effects on cerebral infarction. In particular, they reported a definite curative effect for acute cerebral infarction, which may reduce the long-term mortality of stroke patients.

Alternatively, these acupoints can be stimulated by electrodes conducting electrical currents through the needles (electro-acupuncture (EA)). Electrical stimulation has been used at a low frequency (2–4 Hz), where patients experience slight trembling but no pain. Following a regimen of 30 min treatments once a day, the total effective rate reached 94.45 % (Gao et al. 2004).

For both clinical and experimental studies, EA has become one of the primary treatments for stroke. In the application of EA, the wave frequency is one of the

Table 13.1 Commonly used acupoints in stroke treatment

Meridian name (abbreviation)	Yin/Yang	Acupoints
Taiyin lung meridian of hand (LU)	Yin	Shaoshang (LU-11), Yuji (LU-10), Taiyuan (LU-9), Lieque (LU-7), Chize (LU-5)
Shaoyin heart meridian of hand (HT)	Yin	Jiquan (HT-1), Shaohai (HT-3), Tongli (HT-5), Shenmen (HT-7), Shaochong (HT-9)
Jueyin pericardium meridian of hand (PC)	Yin	Quze (PC-3), Jianshi (PC-5), Neiguan (PC-6), Daling (PC-7), Laogong (PC-8), Zhongchong (PC-9)
Shaoyang triple energizer meridian of hand (TE)	Yang	Guanchong (TE-1), Yemen (TE-2), Zhongzhu (TE-3), Yangchi (TE-4), Waiguan (TE-5), Zhigou (TE-6), Huizong (TE-7), Jianliao (TE-14), Yifeng (TE-17), Sizhukong (TE-23)
Taiyang small intestine meridian of hand (SI)	Yang	Shaoze (SI-1), Houxi (SI-3), Wangu (SI-4), Jianzhen (SI-9)
Yangming large intestine meridian of hand (LI)	Yang	Shangyang (LI-1), Sanjian (LI-3), Hegu (LI-4), Yangxi (LI-5), Shousanli (LI-10), Quchi (LI-11), Binao (LI-14), Futu (LI-18), Yingxiang (LI-20)
Taiyin spleen meridian of foot (SP)	Yin	Yinbai (SP-1), Taibai (SP-3), Gongsun (SP-4), Sanyinjiao (SP-6), Yinlingquan (SP-9), Xuehai (SP-10)
Shaoyin kidney meridian of foot (KI)	Yin	Yongquan (KI-1), Taixi (KI-3), Zhaohai (KI-6)
Jueyin liver meridian of foot (LR)	Yin	Dadun (LR-1), Xingjian (LR-2), Taichong (LR-3), Xiguan (LR-7), Ququan (LR-8)
Shaoyang gallbladder meridian of foot (GB)	Yang	Tinghui (GB-2), Yangbai (GB-14), Fengchi (GB-20), Qubing (GB-7), Huantiao (GB-30), Fengshi (GB-31), Yanglingquan (GB-34), Yangjiao (GB-35), Yangfu (GB-38), Xuanzhong (GB-39), Qiuxu (GB-40)
Taiyang bladder meridian of foot (BL)	Yang	Cuanzhu (BL-2), Wuchu (BL-5), Chengguang (BL-6), Tongtian (BL-7), Tianzhu (BL-10), Dazhu (BL-11), Xinshu (BL-15), Geshu (BL-17), Ganshu (BL-18), Pishu (BL-20), Shenshu (BL-23), Dachangshu (BL-25), Ciliao (BL-32), Chengfu (BL-36), Weizhong (BL-40), Chengshan (BL-57), Feiyang (BL-58), Kunlun (BL-60), Shenmai (BL-62), Zhiyin (BL-67)
Yangming stomach meridian of foot (ST)	Yang	Sibai (ST-2), Juliao (ST-3), Daying (ST-5), Jiache (ST-6), Dicang (ST-4), Renying (ST-9), Shuitu (ST-10), Quepen (ST-12), Biguan (ST-31), Futu (ST-32), Tiaokou (ST-38), Fenglong (ST-40), Jiexi (ST-41), Neiting (ST-44), Zusanli (ST-36)
Du meridian (governor vessel, GV)	Yang	Yaoshu (GV-2), Yaoyangguan (GV-3), Mingmen (GV-4), Shenzhu (GV-12), Dazhui (GV-14), Yamen (GV-15), Fengfu (GV-16), Shangxing (GV-23), Renzhong (GV-26), Baihui (GV-20)-
Ren meridian (conception vessel, CV)	Yin	Zhongji (CV-3), Guanyuan (CV-4), Qihai (CV-6), Shenque (CV-8), Tanzhong (CV-17), Tiantu (CV-22), Lianquan (CV-23), Chengjiang (CV-24)-
Extra points (EX)		Yintang (EX-HN-3), Shixuan (EX-HN-11), Jinjin (EX-HN-12), Yuye (EX-HN-13), Jiaji (EX-B-2)-

Cited and modified from the work of Fan et al. (2001)

most important parameters. Recent studies have reported that a continuous wave effect affords the best results (Ernst et al. 2007).

13.2.2 *Acupuncture for Ischaemic Complications*

Urinary incontinence is a common consequence of stroke and has many causes. In early stroke rehabilitation, acupuncture shows promise in reducing the number of people with urinary incontinence. In the later phases of stroke recovery, the use of acupuncture may be helpful in reducing symptoms associated with urinary incontinence. Even after stroke rehabilitation, acupuncture targeted at specific causes of incontinence may be helpful (Liu and Wang 2006; Yang et al. 2003).

Patients who have difficulties swallowing (dysphagia) as a result of stroke are less likely to survive and be free of disability than stroke patients who are able to swallow normally. Acupuncture is commonly used to treat this complication in traditional Chinese medicine (Huang et al. 2010; Chau et al. 2010; Li et al. 2009).

The effect of acupuncture on depression, including depressive neurosis and depression following stroke, has been documented repeatedly in controlled studies (Wu 2010; Guo et al. 2009; Dong et al. 2007; He and Shen 2007). Acupuncture is comparable with amitriptyline in the treatment of depression but has fewer side effects. In addition, acupuncture has been found to be more effective in depressive patients with decreased excretion of 3-methyl-4-hydroxy-phenylglycol (the principal metabolite of the central neurotransmitter norepinephrine).

13.2.3 *Treatment of Various Stroke Stages*

In recent years, clinical specialists have begun to tailor treatments for the individual stroke stages. In 1986, the cooperation unit of urgent brain disease of the State Administration of Traditional Chinese Medicine made The Evaluation Standard of Diagnosis and Curative Effect of Stroke (Pilot), which divided the stroke process into three development stages: the acute period, the recovery period, and the sequela period. The aetiology and pathogenesis of each period is different; thus, different treatment methods have been adopted for each period.

In the acute phase, patients appear to suddenly faint, are often unconscious, suffer aphasia, and may be partially paralysed. In the first 3 days, hyperactivity of the liver yang is the primary syndrome traditional Chinese medicine (TCM) type. After 3 days, wind phlegm and blood stasis can cause internal obstruction, with the latter increasing gradually. Hyperactivity of the liver yang, wind phlegm and blood stasis, and phlegm-heat account for approximately 95 % of cases in the first 21 days. Therefore, approaches to pacifying the liver and subduing yang, eliminating phlegm, and extinguishing wind are the principal treatment components. Therefore, the following acupoints are used: Baihui (GV-20), Fengchi (GB-20), Quchi (LI-11), Hegu (LI-4), Zusanli (ST-36), and Taichong (LR-3).

During the recovery period (2 weeks to 6 months after stroke), the patient has reached a stable disease level. A large number of the reported acupuncture treatments for stroke patients have been concentrated in this period. This period is often characterised by meridian Qi obstruction. Acupuncture treatments can promote physical activity and regulate channels. The following acupoints are used: Jianyu (LI-15), Quchi (LI-11), Waiguan (SJ-5), Hegu (LI-4), Huantiao (GB-30), Yanglingquan (GB-34), Fenglong (ST-40), and Kunlun (BL-60). During this period, acupuncture treatments still focus primarily on Yang meridians so as to promote circulation, regulate meridians, and recover body function.

In the apoplectic sequela period (more than 6 months post-stroke), stroke patients are often Yang Qi deficient and weak, and sequela is slowly restored. Stimulation of the Yang meridians alone would not be beneficial at this point; thus, a Yin meridian treatment is often chosen. The following acupoints are used: Jianyu(LI-15), Chize (LU 5) to Xiaohai (SI 8), Neiguan (PC-6) to Waiguan (TE-5), Xuehai (SP 10) to Liangqiu (ST 34), Yinlingquan (SP 9) to Yanglingquan (GB-34), Sanyinjiao (SP-6) to Juegu (GB 39), and Taichong (LR-3) to Yongquan (KI-1).

13.3 The Relationship Between Monoamine Neurotransmitters and Pathological Injury During Cerebral Ischaemia-Reperfusion

Numerous studies have confirmed that acupuncture can be beneficial to patients during the convalescence of ischaemic apoplexy (Wu et al. 2010). Animal experiments (Shen et al. 2010) have also demonstrated that acupuncture can accelerate the restoration of function and ease the cerebral tissue lesion during cerebral ischaemia-reperfusion. Recent progress has been made toward identifying the mechanism of acupuncture effects on cerebral ischaemia-reperfusion injury. During ischaemia-reperfusion, there is an accumulation of hydroxyl radicals that alters cell membranes and surface proteins, thus leading to increased intracellular Ca^{2+} and encephaledema. This leads to neuronal excitotoxicity, resulting in neuronal injury and death. Monoamine neurotransmitters play an important role in this process.

The major categories of monoamine neurotransmitters are (1) norepinephrine (NE)/noradrenaline (NA), (2) epinephrine/adrenaline (A), (3) dopamine (DA), and (4) 5-hydroxytryptamine/serotonin (5-HT). NA, A, and DA all share a catechol structure; thus, they are referred to as the catecholamines (CA). CA and 5-HT are classical monoamine neurotransmitters in the central nervous system. Neurons that contain monoamine neurotransmitters make up the monoaminergic system. The cell bodies and fibres of these neurons are distributed throughout the central nervous system.

13.4 Distribution and Physiological Function of Monoamine Neurotransmitters

The cell bodies of the most prominent noradrenergic neurons are located in the pons and medulla oblongata and project to a variety of brain structures (such as locus coeruleus, central grey matter, stria terminalis, hypothalamus, amygdala, and cingulate gyrus). The physiological functions of norepinephrine include (1) the regulation of cardiovascular processes, (2) the regulation of body temperature, (3) the regulation of food intake, (4) sleep and arousal, (5) learning and memory, (6) analgesia, and (7) psychonosema.

The cell bodies of the dopaminergic neurons were originally identified in the midbrain. The primary dopaminergic tract is the mesolimbic tract, which runs from the ventral tegmental area to many parts of the limbic system. Later, dopaminergic tracts were found in diencephalon and telencephalon. The primary physiological functions of DA include the regulation of extrapyramidal movements, psychomotility, prolactin secretion, regulation of the cardiovascular system, regulation of the gastrointestinal system, emetic action, pain, and analgesia. Dopaminergic neurons play an important role in extrapyramidal movements. When the activity of dopaminergic neurons increases, these movements will also be increased. When dopaminergic activity decreases, the movements will likewise be reduced. The injury or loss of dopaminergic neurons results in deficient dopamine neurotransmission, leading to movement disorders, including Parkinson's disease, Huntington's disease, and tardive dyskinesia. The striatum is viewed as the control centre for extrapyramidal movements. It contains several kinds of neurotransmitters. In addition to DA, there are acetylcholine (Ach), gamma-aminobutyric acid (GABA), substance P (SP), cholecystokinin octapeptide (CCK), ENK (enkephalin), and somatotropin release-inhibiting factor (SRIF). These neurotransmitters (or modulators) regulate activities by means of various afferent and efferent connections. The main regulatory pathway involves the release of DA from dopaminergic nerve terminals projecting from the substantia nigra to the striatum. These terminals generate inhibitory effects that directly regulate the parallel feedback pathway. The majority of the inputs into the substantia nigra are GABAergic neurons [also SP, CCK, ENK, and DYN (dynorphin)]. DA can also inhibit the activities of Ach interneurons to decrease excitatory actions on GABAergic neurons. Nigral GABAergic synaptic interactions have been described in the substantia nigra pars compacta (SNc) and substantia nigra pars reticulate (SNr) and include the GABA–DA pathway and GABA–GABA–DA pathway. The activities of dopaminergic neurons can be inhibited through the GABA–DA pathway and can be activated through the GABA–GABA–DA pathway, inhibiting the local neurons of the SNr.

The cell bodies of most 5-HT neurons are located in medulla, pons, and the raphe nuclei and its adjacent areas. The neurons of the raphe nuclei are the principal source of 5-HT in the brain. The physiological function of 5-HT includes the regulation of the cardiovascular system, breathing movements, body temperature regulation, sleep, sexual behaviour, food intake, endocrine function, psychomotility, and pain and analgesia (Zou 1999).

13.5 Role of Monoamine Neurotransmitters in Cerebral Ischaemia

Some reports suggest that ischaemia induces the release of [3H]-DA via two mechanisms: Ca^{2+}-dependent exocytosis and transporter reversal (Kim et al. 1995). Ischaemia is induced by the deprivation of glucose and oxygen, resulting in the degradation of energy and an inability to sustain the ion gradient. Under these conditions, voltage-dependent calcium channels open, and the extracellular calcium ions move down the chemical gradient and into the cells. The ischaemia-evoked release of glutamate (Glu) activates receptors and opens calcium channels, thus leading to calcium entry. This calcium overload results in the release of DA (Lewen et al. 2000). At the same time, DAT reverses its transport of DA due to the degradation of energy, thus enhancing the release of DA (Buyukuysal and Mete 1999). The exact injury mechanism of DA released into the synaptic space during cerebral ischaemia-reperfusion injury is poorly understood. The current theories can be summed up as follows: local neurotoxicity (including the neurotoxicity of DA, oxidative stress, intensifying neurotoxicity, inducing cell apoptosis, and microglial activation) and cardiovascular toxicity (Fig. 13.1).

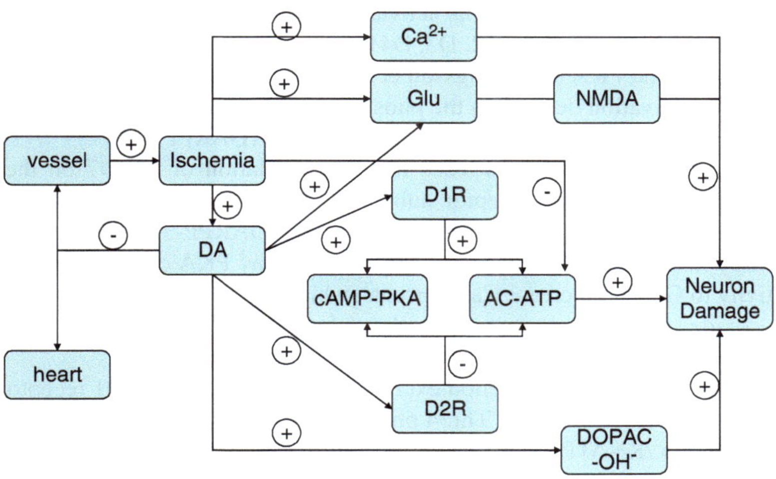

Fig. 13.1 Neuronal damage by monoamines during cerebral ischaemia and reperfusion. *DA* dopamine, *DOPAC* 3,4-Dihydroxyphenylacetic acid, *D1R* dopamine D1 receptor, *D2R* dopamine D2 receptor, *Glu* glutamate, *NMDA* *N*-methyl-D-aspartic acid receptor, *PKA* protein kinase A, *AC* adenylate cyclase, *ATP* adenosine-triphosphate. ⊕, active or promote the pathway; ⊖, decrease or inhibit the pathway

13.5.1 *Local Monoamine Neurotoxicity during Cerebral Ischaemia and Reperfusion*

In addition to Glu, other neurotransmitters released into the extracellular space during ischaemia can significantly influence the resulting brain injury. DA, which increases 500-fold in the extracellular space following global ischaemia, may contribute to striatal neuronal death. Furthermore, several lines of evidence suggest that CA release and metabolism might underlie the selective vulnerability of striatal neurons to ischaemic insult (Weinberger et al. 1983, 1985). For example, depletion of CA by α-methyl-paratyrosine exerts a strong protective effect on ischaemic damage to nerve terminals, and reduction of striatal DA content by lesioning the nigrostriatal tract protects intrinsic striatal neurons from global cerebral ischaemia injury (Globus et al. 1987). Moreover, the experimental reduction of DA release attenuates striatal injury in rodent global ischaemia models (Obrenovitch and Richards 1995).

Although the precise mechanism of neuronal injury by DA is unclear, by-products of its metabolism, such as hydrogen peroxide, superoxide ions, and hydrogen radicals, have been implicated in this deleterious process (Kahn et al. 1995).

13.5.1.1 Monoamine Neurotoxicity

DA is a major neurotransmitter in striatum and is known to regulate the function of several proteins, including those important in ischaemic injury, such as Glu receptors and Na^+/K^+-ATPase. DA acts on two classes of receptors: D1-like receptors (D1, D5) coupled to G_s proteins that activate adenylate cyclase and protein kinase A (PKA) and D2-like receptors (D2, D3, D4) coupled to $G_{i/o}$ proteins that inhibit adenylate cyclase and PKA (Svenningsson et al. 2004). Many of the effects of D1 and D2 receptor activation depend on the phosphorylation state of multiple sites on the DA- and cAMP-regulated phosphoprotein of 32 kDa (DARPP-32).

Stimulation of D1 receptors leads to phosphorylation of Ser897 on the NR1 (*N*-methyl-D-aspartic acid receptor subunit 1) subunit of NMDA (*N*-methyl-D-aspartic acid) receptors (Dudman et al. 2003) via a DARPP-32-dependent mechanism (Snyder et al. 1998). Phosphorylation of NR1 by PKA can increase receptor sensitivity to Glu and increase Ca^{2+} currents, which can then increase the activity of nitric oxide (NO) synthase and other downstream cascades involved in excitotoxicity(Flores-Hernandez et al. 2002).

Although D2 receptor activation exerts the opposite effects of D1 receptor activation on PKA and DARPP-32 Thr34 phosphorylation, D1 and D2 receptor activation reduces Na^+/K^+-ATPase activity in a synergistic, rather than antagonist, manner (Bertorello et al. 1990). Decreased Na^+/K^+-ATPase activity could have important consequences on ionic regulation in post-ischaemic neurons. Together, these findings open the possibility that DA receptors may influence ischaemic neuronal damage by modulating NMDA receptors, Na^+/K^+-ATPase, or other components involved in excitotoxicity.

Studies in rat brain hippocampal slices showed that DA activated the Ca^{2+}-dependent protease by activating D1-like receptors (Yurko-Mauro and Friedman 1996), thus further demonstrating that DA induces increased levels of intracellular calcium via D1-like receptors. Piomelli reported that in Chinese hamster ovary (CHO) cells transfected with D2 receptors, D2 agonists potently enhanced arachidonic acid release, suggesting that this release was initiated by increasing intracellular Ca^{2+}. In CHO cells expressing D1 receptors, D1 agonists show no such effect. When D1 and D2 receptors are co-expressed, however, activation of both subtypes results in a marked synergistic potentiation of arachidonic acid release (Piomelli et al. 1991). These results suggest that the facilitation of arachidonic acid release may be mediated by DA responses.

Whether D2 receptors exert opposing or similar actions to those of D1 receptors during ischaemia is uncertain. However, it is clear that D1/D2 receptor synergism may play a role in the cerebral ischaemia-reperfusion injury, and this possibility should be investigated further.

13.5.1.2 Intensifying Neurotoxicity

Many studies regarding the relationship between DA and excitatory amino acid (EAA) in the process of ischaemic brain damage have been reported. DA can inhibit the release of Glu by means of D2-like receptors on glutamatergic nerve terminals. However, both DA levels and Glu levels were reduced in ischaemic rats following lesions with 6-hydroxydopamine (6-OHDA) in the substantia nigra (Werling et al. 1993). This suggests that DA promotes the release of Glu in the ischaemic brain. Thus, the ability to enhance Glu receptor currents may contribute to the DA-induced potentiation of ischaemic injury. The release of DA activates D2-like receptors that inhibit the high-affinity Glu re-uptake system, thus leading to Glu accumulation. At the same time, Glu promotes the release of DA by means of the NMDA receptor. Furthermore, DA increases the levels of cAMP in use by D1 receptors. The increased levels of cAMP activate cAMP-dependent protein kinase (PKA) to phosphorylate DARPP-32 and inhibit protein phosphatase (PP-1), resulting in the continuous activation of Ca^{2+}/calmodulin-dependent protein kinase II (CaMPK II), which plays a role in phosphorylation and intensifies the excitotoxicity of glutamic acid (Strack et al. 1997).

13.5.1.3 Oxidative Stress

Ischaemia and reperfusion have been shown to increase the generation of reactive oxygen species (ROS). DA auto-oxidation may contribute to this increased ROS generation. Most scholars hypothesised that the free radicals and quinine present in the metabolic process of DA played an important role in neurotoxicity. Suzuki found that type A and type B monoamine oxidase (MAO) inhibitors decreased free radicals and 3,4-dihydroxyphenylacetic acid (DOPAC) levels in ischaemia and reperfusion

(Suzuki et al. 1995). Thus, MAOs participate in the generation of hydroxyl radicals during brain ischaemia-reperfusion. Furthermore, it is hypothesised that in Parkinson's disease (PD), neurons are exposed to oxidative reactions from DA metabolism during which hydrogen peroxide and toxic semiquinones are formed. Additionally, the brains of PD patients show decreased concentrations of defensive mechanisms, such as glutathione (Koziorowski and Jasztal 1999). Callaway found that middle cerebral artery (MCA) occlusion resulted in a marked increase of 2,3 DHBA and a secondary increase in all microdialysis analytes 180–300 min later. AM-36, a novel neuroprotective agent, significantly reduced the ischaemia-induced increases in 2,3 DHBA, DA, and infarct volume in the striatum. Significant improvements in a battery of behavioural tests were also found in AM-36-treated rats (Callaway et al. 2003).

13.5.1.4 Inducing Cell Apoptosis

There are two forms of ischaemia- and cerebral trauma-induced neuron death: necrosis and apoptosis. Both in vitro and in vivo experiments have demonstrated that DA induces apoptosis in a time- and concentration-dependent manner (Hattori et al. 1998). Over-expression of the anti-apoptotic gene bcl-2 reduces the decreased levels of DA-induced intracellular hydrosulphuryl to inhibit the DA-induced apoptosis. Luo found that DA-induced apoptosis by means of oxidative stress and the JNK (c-Jun N-terminal kinases) pathway. Recent studies have shown that the production of active oxygen following DA metabolism induced the expression of apoptotic genes and led to apoptosis via the MAPK (mitogen-activated protein kinase) pathway (Luo et al. 1998). Zhang found that NMDA receptor antagonists inhibited DA activity, thus indicating that excitatory amino acids promote DA-induced apoptosis (Zhang et al. 1998).

13.5.2 Cardiovascular Toxicity

Monoamines play a key role in the cardiovascular system. The effect of monoamines on the cardiovascular system is decreased following cerebral ischaemia. Focal cerebral ischaemia and reperfusion impaired vasoconstrictor responses in remote brain arteries of rats via a mechanism involving monoamines (Kovács et al. 2010). Acute CA release following cerebral ischaemia can promote the propagation of brain injury (Bhardwaj et al. 2003). DA has been associated with an increase in brain oedema. There is also evidence that DA has many adverse effects in critically ill patients due to its ability to suppress circulating concentrations of most anterior pituitary-dependent hormones. Therefore, attenuation of the ischaemia-induced CA release could be beneficial. Both cerebral and circulating (plasma levels) CA have been implicated in exacerbating ischaemic brain injury (Strittmatter et al. 2003).

Stroke can lead to cerebrogenic cardiac arrhythmias. Min found that the serum and heart levels of NA were significantly higher in mice with left permanent middle cerebral artery occlusion (pMCAO) and heart dysfunction. Liner regression analysis indicates a significant inverse correlation between the severity of left insular cortex damage and heart dysfunction. Their study shows that left focal cerebral ischaemia can produce cardiac dysfunction, which is associated with the extent of left insular cortex damage (Min et al. 2009). Monoamines, especially NA, play an important role in cerebrogenic cardiac diseases.

13.6 Protection Mechanism of Monoamine Neurotransmitters During the Recovery Period

Monoamine neurotransmitters play an important role in the normal development of the central nervous system and signal transduction in nerve cells. At the same time, these transmitters also have effects on cerebral ischaemia-reperfusion injury. Many scholars hypothesised that monoamine neurotransmitters can add to the injury caused by cerebral ischaemia-reperfusion. However, other scholars have suggested that monoamine neurotransmitters play a role in neuroprotection and promote the functional recovery of nerve cells. Monoamine neurotransmitters can directly reduce the injury caused by cerebral ischaemia-reperfusion, neurotoxicity, and promote neural repair and rehabilitation (Fig. 13.2).

13.6.1 Reduced Injury Following Cerebral Ischaemia-Reperfusion

Self-protection responses are initiated immediately during cerebral ischaemia-reperfusion. These responses consist of increasing blood pressure, promoting cerebral blood perfusion, and activating antioxidant enzymes. Some of these responses are related to monoamines, such as influencing cortical spreading depression (CSD), reducing neurotoxicity, promoting cerebral blood perfusion, and activating antioxidant enzymes.

13.6.1.1 Influencing Cortical Spreading Depression

CSD is a neuronal mechanism that expands the penumbra zone following focal brain ischaemia. Both adrenergic agonists and antagonists significantly influence the size of the penumbra zone. Richter found that the adrenergic agonist NA and the alpha(2)-agonist clonidine reversibly blocked the migration of CSD. The beta-blocker propranolol dose-dependently diminished migration velocity or even

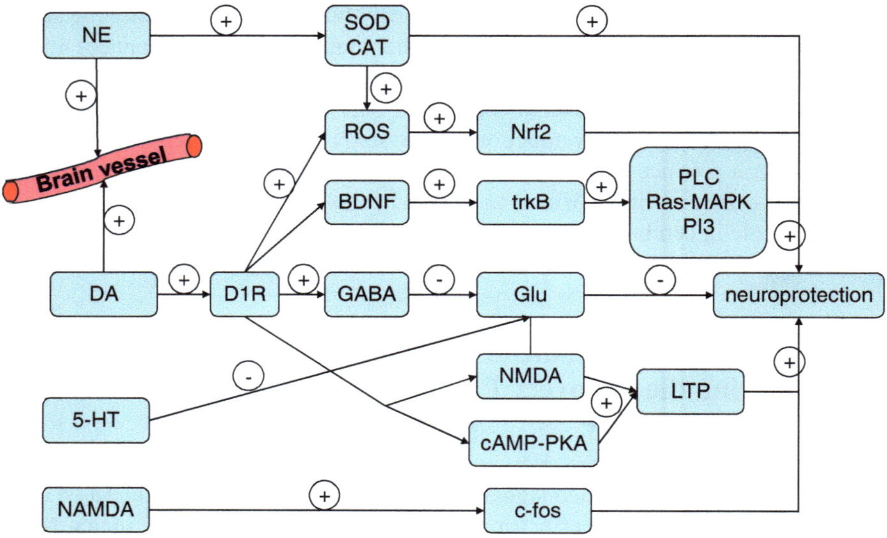

Fig. 13.2 Neuroprotection by monoamines during cerebral ischaemia and reperfusion. *DA* dopamine, *NE* norepinephrine, *5-HT* 5-hydroxytryptamine, *Glu* glutamate, *NMDA* *N*-methyl-D-aspartic acid receptor, *BDNF* brain-derived neurotrophic factor, *LTP* long-term potentiation, *D1R* dopamine D1 receptor, *SOD* superoxide dismutase, *CAT* catalase, *ROS* reactive oxygen species, *PKA* protein kinase A, *NAMDA* *N*-acetyl-3-*O*-methyldopamine, *GABA* gamma-aminobutyric acid, *PLC* phospholipase C, *MAPK* mitogen-activated protein kinase, *PI3* phosphatidylinositol 3-kinases. ⊕, active or promote the pathway; Θ, decrease or inhibit the pathway

blocked migration of CSD. None of the substances influenced regional cerebral blood flow or systemic arterial blood pressure at the concentrations used. The data suggest that the interference of these monoamines with CSD may contribute to their beneficial therapeutic effect (Richter et al. 2005).

13.6.1.2 Reducing Neurotoxicity

Recent studies have shown a high release of EAA and a corresponding increase in the levels of inhibitory amino acids in the early stage of ischaemia. D1 receptors can enhance GABA transfer from the basal forebrain into the cortex. In the process of cerebral ischaemia-reperfusion, the increased levels of GABA will strengthen the activity of neurons against excitotoxicity and inhibit the release of Glu to reduce neurotoxic injury (Crandall 2007). Hashimoto also found that D1-like receptors can mediate neuronal protection (Hashimoto et al. 1994).

Furthermore, in human cerebrocortical slices, Glu efflux caused by oxygen/glucose deprivation can also be controlled by 5-HT. The author found that the efflux of both [3H]D-aspartate and endogenous Glu evoked by ischaemia was reduced by at least 50 % in the presence of 5-HT. Moreover, the 5-HT$_{1A}$ receptor agonist

8-OH-DPAT mimicked the 5-HT effect. The result suggests that the activation of 5-HT_{1A} receptors might help to control Glu efflux and excitotoxic damage during ischaemia in the cerebral cortex and should be considered in a multi-pharmacological approach to neuroprotection in brain ischaemia (Marcoli et al. 2004).

13.6.1.3 Activating the Antioxidant Enzymes

Danielisová found that the activities of the antioxidant enzymes superoxide dismutase (SOD) and catalase (CAT) in the hippocampus, striatum, and cortex significantly increased 5 h after post-conditioning (5 min of ischaemia and intraperitoneal injections of NA). These results suggest that SOD and CAT could play a considerable neuroprotective role following post-conditioning (Danielisová et al. 2006).

The transcription factor Nrf2 controls the inducible expression of multiple antioxidant/detoxification genes. Previous reports have found that Nrf2–/– mice have increased sensitivity to in vivo mitochondrial stress and ischaemia. Although Nrf2 regulated these forms of neuronal toxicity, it was unclear which injury-triggered signal(s) led to Nrf2 activation in vivo.

Increased Nrf2 activity during ischaemia in vivo was specific to oxidative stress caused by catecholaminergic neurotransmitters, as A also induced Nrf2, but 5-HT had no significant effect. These in vitro results suggested that the Nrf2 activity in astrocytes and meningeal cells linked the neurotoxic actions of DA to neuroprotective pathways that may potentially modulate ischaemic injury and neurodegeneration (Shih et al. 2007)

13.6.1.4 Promoting Cerebral Blood Perfusion

Many studies have investigated the cerebrovascular effects of CA over a range of cerebral perfusion pressures above the range recommended by current guidelines. The relationship between patient outcome and the use of a specific substance to improve cerebral perfusion has not been investigated. DA, NA, and phenylephrine all seem to increase cerebral blood flow in various animal models and in patients. The data suggest that NA may be the most predictable. It is associated with an improved restoration of global and regional oxygenation when compared to DA. For all other CA and related substances, there are insufficient data on the cerebrovascular effects following head injury. This suggests that NA may be the CA that is the most suitable for maintaining or restoring adequate cerebral perfusion. The data, however, are insufficient to formulate a guideline. (Pfister et al. 2008) Other authors have concluded that induced arterial hypertension is feasible and safe in patients with acute stroke (Marzan et al. 2004).

Using a cortical impact injury model in rats, Kroppenstedt and colleagues (Kroppenstedt et al. 2000) showed an increase in regional cerebral perfusion measured by laser Doppler flowmetry after the mean arterial pressure (MAP) was increased with DA. Cerebral blood flow (CBF) increased significantly, and there was no evidence of DA-mediated vasoconstriction. In a later study by the same group using similar methods, the authors infused either NA or DA to increase MAP. Both substances significantly increased cerebral perfusion. However, despite similar MAP values, the effect on cerebral perfusion was more sustained under NA. (Kroppenstedt et al. 2002a). Based on electroencephalographic (EEG) activity and Glu levels, Kroppenstedt concluded that this sustained increase in MAP was due to a metabolically driven increase, which was found in NE but not in DA (Kroppenstedt et al. 2002b). In a further study, the same group compared the effects of early and late intravenous NA on cerebral perfusion. A rat model and a focal cortical contusion model were used. Early (4 h) and late (24 h) after-injury NA significantly improved perfusion (Kroppenstedt et al. 2003).

13.6.2 Promoting Neural Repair and Rehabilitation

In addition to reducing injury following cerebral ischaemia-reperfusion, monoamines have been shown to promote neural repair and rehabilitation during the recovery period by promoting early neural plasticity and nerve growth, increasing the levels of BDNF, inducing c-Fos expression, inducing neuroprotection by endocannabinoids, and promoting the recovery of neurological function.

13.6.2.1 Promoting Early Neural Plasticity and Nerve Growth

Synaptic plasticity is the very important ability for repairing the function of neurons damaged by nerve injury. Although the accepted hypothesis of synaptic plasticity suggests that the enduring changes of glutamatergic transmission are "physiological" synaptic correlates of learning and memory processes, many of the molecular pathways involved in the induction and maintenance of long-term potentiation (LTP) and long-term depression (LTD) are also observed during synaptic plastic changes induced by pathological stimuli. DA also seems to crucially influence these "pathological" forms of synaptic plasticity (Calabresi et al. 2007).

The D1 receptor is related to the neural plasticity of the central nervous system (Centonze et al. 2001). Activated D1 receptors can strengthen the excitability of striatal spiny neurons and glutamatergic signals (Surmeier et al. 2007). It is hypothesised that the activation of both NMDA and D1 receptors is the key to the initiation and regulation of the intracellular signal transduction and translation cascade reaction that induces gene expression and the changes that occur in early neural plasticity. The joint activation of the two receptors is required for early-LTP (E-LTP). And it plays a role in facilitation in late-LTP (L-LTP). All of these changes eventually lead to the remodelling of neural networks and behaviour changes (Abel and Lattal 2001; Koob and Le Moal 2001; O'Donnell 2003; O'Carroll and Morris 2004).

Experimental studies of brain injury in vitro have shown that neurons can respond to the acute energy deprivation associated with ischaemic insult by generating pathological forms of synaptic plasticity. In the striatum, transient oxygen and glucose deprivation (in vitro ischaemia) can induce an NMDA-dependent post-ischaemic LTP (i-LTP) at corticostriatal synapses (Calabresi et al. 2002, 2003). Pharmacological and genetic manipulations of DA-mediated signalling have shown that dopamine, through its interaction with D1 receptors, facilitates the expression of i-LTP on the AMPA-mediated component of excitatory postsynaptic potential (EPSP). Accordingly, blockade of the D1 receptor–cAMP–PKA pathway prevents the long-lasting increase in the EPSP amplitude caused by ischaemia (Saulle et al. 2002). Because a large release of DA occurs during ischaemic events in the striatum (Toner and Stamford 1996) and endogenous DA can amplify the neuronal damage induced by energy deprivation (Buisson et al. 1992), it is possible to hypothesise that DA could also facilitate both acute necrotic and delayed apoptotic neuronal death through its influence on this pathological form of synaptic plasticity.

13.6.2.2 Increasing the Levels of BDNF

Recent studies have also suggested that the trkB receptor of brain-derived neurotrophic factor (BDNF) regulates the expression of D1 receptors in a time- and concentration-dependent manner (Do et al. 2007). The expression of D1 receptors increases when the levels of BDNF increase. The cellular mechanisms involve the initiation of the expression of the neurotrophin family and their receptors caused by cell membrane depolarisation, the release of Glu, activation of Glu receptors, and Ca^{2+} entry in ischaemic animals. Several reports have shown that a rapid increase in the gene expression of BDNF is induced by cerebral ischaemia-reperfusion. BDNF is regarded as one of the members of the neurotrophin family and can stabilise the intracellular Ca^{2+} levels by means of inducing the expression of Calbindin (Cheng and Mattson 1994). When combined with trkB, BDNF can initiate intracellular signal transduction pathways, including phosphoesterase Cγ(PLC), the Ras-MAPK pathway, and the PI3 pathway, to produce related molecules to protect or promote the regeneration of neurons. The evidence also demonstrated that BDNF can reduce the gene expression of the NMDA receptor subunits NR2A and NR2C and down-regulate the function of NMDA receptors to inhibit the neurotoxicity of excitatory Glu (Rocha et al. 1999). There is also a close relationship between BDNF regulation of the D1 receptor and a series of early plasticity changes in the D1 receptor following cerebral ischaemia-reperfusion injury.

13.6.2.3 Inducing c-Fos Expression

In vivo and in vitro results have shown that a neuro-protectant, (*N*-acetyl-*O*-methyldopamine) NAMDA, in concert with another stimulus (e.g. ischaemia or PMA) up-regulates c-Fos expression. This suggests that the early rise of NAMDA-induced c-Fos expression in vulnerable CA1 neurons may account for the neuroprotective effect by means of up-regulating late gene expression for cell survival (Cho et al. 2001).

13.6.2.4 Neuroprotection Induced by Endocannabinoids

Given that endocannabinoids (eCBs) depress synaptic Glu transmission (Melis and Pistis 2007; Melis et al. 2004a, b) and limit further depolarisation, their release can be seen as a protective mechanism by which DA neurons reduce the strength of incoming excitation. These cells are exquisitely vulnerable to excitotoxicity and oxidative stress, and this vulnerability might be partially correlated with or even explained by the specific regulation of their excitability (Korotkova et al. 2004).

Study found evidence that eCBs released by DA neurons exert protective actions in a model of ischaemia/reperfusion (Melis et al. 2006). Perfusion of brain slices in oxygen and glucose deprivation for 7 min induced a progressive depolarisation of DA neuron membranes and an interruption in firing activity, which can be irreversible. This depolarisation would trigger the release of eCBs. In addition, the blockade of CB1 receptors led to a considerable worsening of the outcome of experimental ischaemia. Thus, eCBs might represent one of the neuroprotective mechanisms that reduce DA neuronal damage during episodes of energy deprivation. The current hypothesis posits that activation of CB1 receptors by eCBs might reduce Glu release in the ischaemic tissue and consequently decrease Ca^{2+} entry and excitotoxic damage.

Disturbances in the eCB system might be of considerable importance in the pathogenesis of neurological disorders involving the DA system. The enhancement of CB might be useful as a neuroprotective therapy to reduce cell death in the early stages of cerebral ischaemia. It is also possible that exogenous cannabinoids might mimic the eCBs by acting as neuroprotective agents in neurodegenerative diseases. This hypothesis is based on the observation that cannabinoids protect neurons from toxic insults, such as glutamatergic excitotoxicity (Shen and Thayer 1998), ischaemic stroke (Nagayama et al. 1999), hypoxia (Sinor et al. 2000), trauma (Panikashvili et al. 2001), and oxidative stress (Marsicano et al. 2003). Most of these protective effects appear to be mediated by the activation of the CB1 receptor (Parmentier-Batteur et al. 2002), although the contribution of other mechanisms (i.e. antioxidant and/or anti-inflammatory properties) cannot be ruled out (Grundy 2002; Grundy et al. 2001).

13.6.2.5 Promotion of the Recovery of Neurological Function

NA plays a definite supporting role in the recovery of neurological function (for relevant reports, see Gladstone and Black 2000). The monoamine agonists dextroamphetamine and amphetamine, which increase central nervous system NA levels, may facilitate the functional recovery of focal brain injury in animal models (Walker-Batson et al. 2001; Sonde et al. 2001; Long and Young 2003). Reports have confirmed that L-DOPA and carbidopa combined with physiotherapy enhance motor recovery in patients following a stroke, particularly in fine motor recovery (Scheidtmann et al. 2001).

Despite extensive research regarding the mechanism of monoamine neurotransmitters in cerebral ischaemia-reperfusion, the results are contradictory. However, this inconsistency by be caused by differences in the time period, animal model, and types of animals used. Based on the results mentioned above, we can see that monoamine neurotransmitters have damaging effects in the initial stages of ischaemia but play a role in cellular repair/restoration following post-ischaemia-reperfusion. Due to the varying time windows and the changes in the internal and external cellular environments, monoamine neurotransmitters play a "double-edged sword" role in the pathophysiological process.

13.7 Regulating the Effects of EA on the Monoamine Neurotransmitters Following Cerebral Ischaemia-Reperfusion

Our preliminary studies have confirmed that EA has a protective effect on experimental acute cerebral ischaemia-reperfusion injury (Xu et al. 2001a, b, c, 2002). These protective effects cannot be seen without the involvement of extracellular fluid. Extracellular fluid is the survival environment for cells, the source of essential energy for cell survival, and a carrier for material exchanges. A variety of pathological changes in cells are closely related with extracellular fluid, and they often influence each other. The changes in various materials in the extracellular fluid during cerebral ischaemia-reperfusion inevitably have some effects on cells. Given that previous studies primarily used the brain tissue homogenate method, the differences between the "cerebral ischaemia" period and the "reperfusion" period, as well as the differences between "intracellular fluid" and "interstitial fluid," cannot be finely studied. Our research, which applied microdialysis methods and focused on the dynamic changes of monoamine neurotransmitters in the "interstitial fluid" and the impact of "Fengchi" EA following reperfusion, avoided these limitations.

13.7.1 The Effects of EA on Monoamines During Cerebral Ischaemia-Reperfusion

Our results revealed that although the cerebral ischaemia and reperfusion did not significantly affect NA levels, EA treatment was able to specifically increase the levels of NA in a surgical rat model. Study also found that EA intervention immediately following reperfusion and 3 h after reperfusion could reduce DA and increase NA 24 h after reperfusion (Xu et al. 2001a, b, c, 2002). This was similar to our results. Other research (Gladstone and Black 2000) has shown that NA could contribute to neurological function recovery following cerebral ischaemia-reperfusion. EA increased the extracellular levels of NA, which may be one of the important mechanisms involved in reducing neuronal injury following ischaemia-reperfusion and promoting the recovery of neurological function.

Our results revealed that cerebral ischaemia and reperfusion did not significantly affect 5-HT levels, and EA treatment had no significant impact on 5-HT content.

Ischaemia and reperfusion lead to a transient increase in DA content in the striatum, and EA was able to reduce the extracellular DA content increase following reperfusion, thus reducing excitotoxicity.

The increase in DOPAC and HVA were found prior to DA. This might be because in the pathological state of cerebral ischaemia, the "cell interstitial fluid–blood" metabolic pathway was blocked, leading to an accumulation of DOPAC and HVA in the intercellular fluid. The essential energy for cell metabolism was then gradually exhausted, and DA re-uptake was blocked, resulting in an accumulation in the extracellular fluid.

Based on the above results, ischaemia and reperfusion could lead to disorders in DA oxidative metabolism both inside and outside of the neurons and an increase in DOPAC and HVA levels in the interstitial fluid. The accumulation of these acidic metabolites could further induce the generation of oxygen free radicals, thereby causing serious damage to neurons. The EA treatment using the "Fengchi point" effectively reduced the accumulation of DA acidic metabolites in reperfusion, which is one of the important mechanisms for EA treatment of cerebral ischaemia and reperfusion.

13.7.2 *The Effects of EA on the Turnover Ratio of Monoamine Neurotransmitters During Cerebral Ischaemia-Reperfusion*

Neurotransmitter turnover is dependent on the involvement of invertase, and the activity of this enzyme is closely related to the energy supply. The first and most immediate changes following cerebral ischaemia-reperfusion injury were ischaemia and hypoxia, energy depletion and the excessive oxidation accompanying reperfusion. Therefore, cerebral ischaemia-reperfusion will cause a significant change in neurotransmitter metabolism. The turnover ratio, which is closely related to neurotransmitter metabolism, could be used as a sensitive indicator to reflect neuronal activity (Guo et al. 2005; Macbeth et al. 2008).

NA/DA, DOPAC/DA, HVA/DA, and 5-HIAA/5-HT could be used to reflect the metabolism of NA, DA, and 5-HT both inside and outside neurons and could indirectly reflect NA and DA energy and 5-HT neuron activities.

After the EA treatment, the NE/DA, DOPAC/DA, and HVA/DA ratios were significantly raised 90 min after reperfusion. The ratio then remained near normal levels. This effect was not seen in the 5-HIAA/5-HT ratio.

Our results suggested that there is a similar effect on NA/DA, DOPAC/DA, and HVA/DA. An artery clip was used for the temporary occlusion of both the right common carotid and the internal carotid artery both in the complete modelling and control group, and a nylon suture was inserted into the internal carotid artery to block blood. Thus, there was a certain influence on the blood supply during the operation. However, the impact of the control group on blood supply was short and mild, while this effect was long and serious in the model group. And all these effects were reflected through the NA/DA, DOPAC/DA, and HVA/DA ratios, showing that the turnover ratio was sensitive to the blood supply.

Meanwhile, EA treatment was able to specifically enhance the activities of NA and DA neurons following cerebral ischaemia-reperfusion in rats, improved the metabolism of NA and DA, and had no significant impact on the sham-operated group. This suggested that the turnover ratio was sensitive to EA treatment.

Our results show that DA and NA were significantly increased shortly after EA treatment. This may be due to the fact that DA and NA in the synapse were rapidly depleted following EA treatment and cannot be added immediately. This does not mean that the EA treatment effect was short term. Our previous results have shown that acupuncture can reduce infarct size, increase brain electrical activity, and reduce neurological score. These results suggest that DA and NA may be central in the effects of acupuncture. Thus, acupuncture may play a role in damage resistance and promoting recovery via DA and NA.

13.7.3 The Cumulative Effects of EA Treatment Following Cerebral Ischaemia and Reperfusion

The early application of EA was able to promote blood supply to the brain following cerebral ischaemia-reperfusion and regulate the metabolic disorders of monoamine neurotransmitters. Based on this, we observed the cumulative effect of EA on stroke. We applied the Fengchi EA treatment for cerebral ischaemia once daily for 5 days in rats. The treatment was able to regulate changes of DA and its metabolites in the striatum and promote neurological function recovery. We speculated that the continuing EA treatment helped to establish collateral circulation in the perimeter zone of the ischaemic regions and reconstructed of neural pathways, thereby contributing to sensory and motor function recovery.

EA played a regulatory role to mobilise the factors that regulated the metabolism imbalance. The treatment not only adjusted the metabolism disorder in the early stage of stroke to avoid the injury caused by metabolite accumulation, it had a cumulative effect through sustained EA treatment and played a long-term positive regulatory role. In a clinical setting, we have not only used the EA treatment in the acute phase of stroke to alleviate the severity of sequelae but also carried out EA intervention in the ultra-early and recovery phase to improve the brain–blood circulation so as to avoid or alleviate sequelae.

13.8 Concluding Remarks

Along with the progress of existing treatment methods, early detection and timely treatment using tPA and other drugs for cerebral ischaemia have shown good curative effects. However, as a result of the limitations of existing medical treatments, the safety of using tPA and similar drugs cannot be assured in patients with more than 3 h ischaemia. It is important reduce the incidence of disease and alleviate the resulting dysfunction. This prompted us to adjust the intervention time earlier or later. The goals of intervention then became to prevent attacks and mitigate the consequences.

Some studies regarding early intervention have confirmed that acupuncture is able to produce ischaemic tolerance and reduce the symptoms of sequela (Han et al. 2009). Our studies were primarily focused on postponed intervention and the alleviation of dysfunction and promotion of functional recovery. We have carried out extensive studies on cerebral ischaemia animal models, infarct size volume, physical function, mental retardation, brain biochemistry in the prophase, and the effects of EA on these factors. The result showed that (1) the suppository line needed to insert at least 16.5 mm to generate a relatively stable model of cerebral ischaemia in rats (Xu et al. 2002); (2) EA has protective effects in the experimental model of acute cerebral ischaemia-reperfusion injury (Xu et al. 2001a) and could be used safely; (3) cerebral ischaemia-reperfusion can increase the brain MDA, NO, Glu, Ca^{2+}, Na^{+}, and K^{+} concentrations; (4) EA therapy had a combined effect of increased brain SOD activity and mAChR binding ratios; reduced NO, Ca^{2+}, and Na^{+} contents; and increased the contents of inhibitory amino acids, such as glycine (Gly) and taurine (Tau) (Xu et al. 2001b, c; Ge et al. 2009; Xu et al. 2009c) (Figs. 13.3 and 13.4).

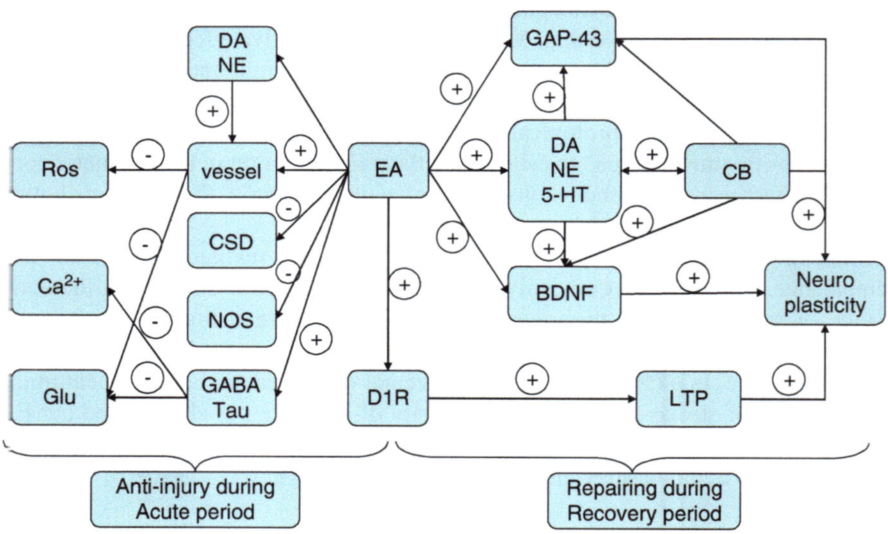

Fig. 13.3 Neuronal damage by monoamines during cerebral ischaemia and reperfusion. *EA* electroacupuncture, *DA dopamine*, *NE* norepinephrine; *5-HT* 5-hydroxytryptamine, *Glu* glutamate, *CSD* cortical spreading depression, *Glu* glutamate, *GABA* gamma-aminobutyric acid, *Tau* taurine, *ROS* reactive oxygen species, *NOS* nitrous oxide systems, *LTP* long-term potentiation, *BDNF* brain-derived neurotrophic factor, *GAP-43* growth associated protein 43, *CB* cannabinoid. ⊕, active or promote the pathway; Θ, decrease or inhibit the pathway

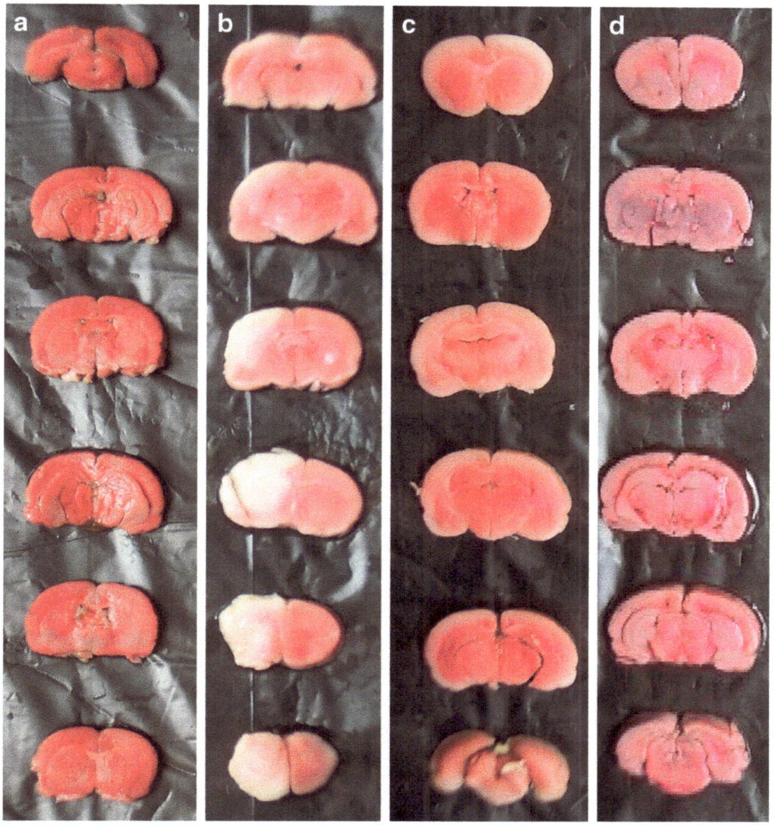

Fig. 13.4 Triphenyltetrazolium chloride (TTC) staining following MCAO and reperfusion. The infarction is represented by a pale white colour. Using EA before MCAO or after MCAO could decrease the ischaemic infarction area. EA can be used as a treatment and a preventative measure. *EA* electro-acupuncture, *MCAO* middle cerebral artery occlusion. (**a**) Control group, (**b**) MCAO group, (**c**) the group using EA before MCAO, (**d**) the group using EA after MCAO

Following our preliminary findings, we gradually focused on the monoamine neurons, which were closely related to neural plasticity (Xu et al. 2009a, b). We found that the changes of monoamine neurotransmitters across the pathological and physiological processes varied with the time window and changes in internal and external cellular environments. The EA treatment on the "Fengchi point" effectively reduced the accumulation of acidic metabolites and brain neuron injury following ischaemia-reperfusion. It also adjusted the disorder of DA and NA contents, specifically enhanced the activity of NA and DA neurons following cerebral ischaemia-reperfusion in rats, and improved the metabolism of NA and DA to promote the recovery of neurological function. The monoamine neurotransmitters had damaging effects during the initial ischaemic phase, but played a repair role during post-ischaemia-reperfusion.

Of course, monoamine transmitters exerted a variety of physiological functions and pathological effects by means of their corresponding receptors. To regulate the metabolism of monoamine neurotransmitters was one of the important mechanisms for EA to promote functional recovery in cerebral ischaemia; meanwhile, NE and DA receptors also played a positive role in ischaemia-reperfusion injury. After a preliminary study, we found that dopamine D2 receptors played a key role in the neural plasticity promoted by EA after cerebral ischaemia. The results provided a new path for the in-depth study of the EA-promoted neurological function recovery mechanism, a reference for improving the clinical efficacy by combined acupuncture with medicine, and for the use of acupuncture treatment in other neurodegenerative diseases.

At present, the incidence of cardiovascular and cerebrovascular diseases continues to rise. They have become one of the top health threats. Thus it is important for medical personnel to reduce cardiovascular and cerebrovascular disease morbidity, mortality, disability, and recurrence rates.

The results of preliminary studies suggest that acupuncture can not only fight against nerve injury in cerebral ischaemia and reperfusion and play a neuroprotective role but also promote cerebral ischaemia-reperfusion neuronal plasticity and neurological recovery.

Monoamine neurotransmitters have the properties of both neurotransmitter and vasoactive compounds. They play important roles in the acupuncture treatment of cerebral ischaemia-reperfusion injury and promote recovery of neurological function. Their receptors are widely distributed in both the circulatory and nervous systems. They play different roles under different physiological and pathological conditions. Monoamines exert biphasic effects. Preliminary finding adds a further layer of complexity to the balance between the neuroprotective/neurotoxic actions of monoamines. The balance between these two roles plays an important role in disease development.

Due to the wide distribution of these receptors, these compounds have a complex effect. They play a crucial role in the balance between different effects. If we want to reverse the balance in the direction of neuroprotective effects, we need in-depth study of these receptors. Due to receptor specificity, synthetic drugs may break the balance in a negative direction. If you carry out research on acupuncture combined with medicine, it is possible to reduce the dose of synthetic drugs, thus reducing the side effects, improving the clinical efficacy, and reducing the economic burden on families and society.

Acupuncture can mobilise the body's own ability to adjust the balance. Monoamine neurotransmitters and the sympathetic nervous system (SNS) both have a balance between different effects, and there are some links between SNS and MNs, such as NA and its receptors. Acupuncture may play a therapeutic role through the SNS–MN pathways. Research regarding the mechanism of acupuncture treatment will provide a basis for the use of acupuncture combined with medicine and acupoint optimisation.

In addition to monoamine neurotransmitters, acetylcholine, adrenomedullin, and cannabinoids have similar properties. Each of these substances has a balance between protecting and destroying effect, and these substances all have both neural and vascular activities. As a bridge between the central nervous system and blood vessels, they could be key points for research into the prevention and treatment of cerebrovascular disease using acupuncture.

For further improvement of the acupuncture-induced neurological rehabilitation and ischaemic tolerance, we suggest acupuncture treatment for the "prevention–treatment–recovery" three-stage intervention, for which "acupoints select–stimulation parameters–integrated therapy" should be optimised.

Acknowledgements This study was supported by the National Natural Science Foundation of China (NSFC) (Grant No. 81001547), Shanghai Science and Technology Commission (SSTC) (Grant No. 014319364), and Shanghai Education Commission (Grant No. 06CZ044). The authors thank Jia Xu, Changzhi Li, Cheng Fang, Xinhui Wang, Zhiqiang Wang, Yingru Chen, Chunyan Chen, Yin Shou, and Yingqian Zhao, who participated and contributed to this study.

References

Abel T, Lattal KM (2001) Molecular mechanisms of memory acquisition, consolidation and retrieval. Curr Opin Neurobiol 11: 180–187

Bertorello AM, Hopfield JF, Aperia A, Greengard P (1990) Inhibition by dopamine of (Na(+)+K+) ATPase activity in neostriatal neurons through D1 and D2 dopamine receptor synergism. Nature 347:386–388

Bhardwaj A, Alkayed NJ, Kirsch JR, Hurn PD (2003) Mechanisms of ischemic brain damage. Curr Cardiol Rep 5:160–167

Buisson A, Callebert J, Mathieu E, Plotkine M, Boulu RG (1992) Striatal protection induced by lesioning the substantia nigra of rats subjected to focal ischemia. J Neurochem 59: 1153–1157

Buyukuysal RL, Mete B (1999) Anoxia-induced dopamine release from rat striatal slices: involvement of reverse transport mechanism. J Neurochem 72: 1507–1513

Calabresi P, Centonze D, Pisani A, Cupini L, Bernardi G (2003) Synaptic plasticity in the ischaemic brain. Lancet Neurol 2: 622–629

Calabresi P, Picconi B, Tozzi A, Di Filippo M (2007) Dopamine-mediated regulation of corticostriatal synaptic plasticity. Trends Neurosci 30: 211–219

Calabresi P, Saulle E, Centonze D, Pisani A, Marfia GA, Bernardi G (2002) Post-ischaemic long-term synaptic potentiation in the striatum: a putative mechanism for cell typespecific vulnerability. Brain 125: 844–860

Callaway JK, Lawrence AJ, Jarrott B (2003) AM-36, a novel neuroprotective agent, profoundly reduces reactive oxygen species formation and dopamine release in the striatum of conscious rats after endothelin-1-induced middle cerebral artery occlusion. Neuropharmacology 44: 787–800

Centonze D, Picconi B, Gubellini P, Bernardi G, Calabresi P (2001) Dopaminergic control of synaptic plasticity in the dorsal striatum. Eur Neurosci 13: 1071–1077

Chau AC, Fai Cheung RT, Jiang X, Au-Yeung PK, Li LS (2010) An fMRI study showing effect of acupuncture in chronic stage stroke patients with aphasia. Journal of Acupuncture & Meridian Studies 3: 53–57

Chau CM, Cheung RTF, Jiang XY, Au-Yeung PKM, Li LSW (2009) Increased Brain Activation in Motor Cortex After Acupuncture Treatment for Motor Recovery in Chronic Stroke Patients. The Open Rehabilitation Journal 2: 89–94

Cheng B, Mattson MP (1994) NT23 and BDNF protect CNS nervous against metabolic excitotoxic insults. Brain Res 640: 56–67

Cho S, Park EM, Kim Y, Liu N, Gal J, Volpe BT, Joh TH (2001) Early c-Fos induction after cerebral ischemia: a possible neuroprotective role. J Cereb Blood Flow Metab 21:550–556

Crandall JE (2007) Dopamine receptor activation modulates GABA neuron migration from the basal forebrain to the cerebral cortex. Neurosci 27: 3813–3822

Danielisová V, Némethová M, Gottlieb M, Burda J (2006) The changes in endogenous antioxidant enzyme activity after postconditioning. Cell Mol Neurobiol 26:1181–1191

Do T, Kerr B, Kuzhikandathil EV. (2007) Brain-derived neurotrophic factor regulates the expression of D1 dopamine receptors. J Neurochem 100: 416–428

Dong JP, Sun WY, Wang S, Wu ZQ, Liu F (2007) Clinical observation on head point-through-point electroacupuncture for treatment of poststroke depression. Zhongguo Zhen Jiu 27:241–244 (In Chinese with English abstract)

Dudman JT, Eaton ME, Rajadhyaksha A, Macias W, Taher M, Barczak A, Kameyama K, Huganir R, Konradi C (2003) Dopamine D1 receptors mediate CREB phosphorylation via phosphorylation of the NMDA receptor at Ser897-NR1. J Neurochem 87:922–934

Ernst E, Pittler MH, Wider B, Boddy K (2007) Acupuncture: Its Evidence-Base is Changing. Am J Chin Med 35: 21–25

Fan GQ, Zhang DB, Luo W (2001) Treatment of Acupuncture on Stroke. Second Military Medical University Press, Shanghai (In Chinese)

Flores-Hernandez J, Cepeda C, Hernandez-Echeagaray E, Calvert CR, Jokel ES, Fienberg AA, Greengard P, Levine MS (2002) Dopamine enhancement of NMDA currents in dissociated medium-sized striatal neurons: role of D1 receptors and DARPP-32. J Neurophysiol 88:3010 - 3020

Gao BX, Sun Y, Cong AM, Yao J (2004) Observation on therapeutic effects of acupuncture in 162 cases of cerebrovascular diseases. Zhongguo Zhen Jiu 24(1) :17–19 (In Chinese with English abstract)

Ge LB, Fang C, Xu MS, Xu J, Li CZ (2009) Effects of electroacupuncture on the ability of learning and memory in rats with ischemia-reperfusion injury. Journal of Acupuncture and Tuina Science 7: 3–7 (In Chinese with English abstract)

Gladstone DJ, Black SE (2000) Enhancing recovery after stroke with noradrenergic pharmacotherapy: a new frontier? Can J Neurol Sci 27: 97–105

Globus MY, Ginsberg MD, Dietrich WD, Busto R, Scheinberg P (1987) Subsantia nigra lesion protects against ischemic damage in the striatum. Neurosci Lett 80: 251–256

Grundy RI (2002) The therapeutic potential of the cannabinoids in neuroprotection. Expert Opin Investig Drugs 11: 1365–1374

Grundy RI, Rabuffetti M, Beltramo M (2001) Cannabinoids and neuroprotection. Mol Neurobiol 24: 29–51

Guo J, Gao Y, Zhu LQ, Wang SR, Wang YY (2005) The effects of Xing Pi Kai Yu Prescription on central monoamine neurotransmitters, A rat model of depression. Journal of Beijing University of Chinese Medicine 28: 55–57 (In Chinese with English abstract)

Guo JC, Cheng JS, Xia Y (2010) Acupuncture Therapy for Stroke. In Xia Y, Cao XD, Wu GC, Cheng JS (eds): Acupuncture Therapy for Neurological Diseases: A Neurobiological View, 1st edn. Tsinghua University Press-Springer), Beijing, Berlin, Heidelberg

Guo RY, Su L, Liu LA, Wang CX (2009) Effects of Linggui Bafa on the therapeutic effect and quality of life in patients of post-stroke depression. Zhongguo Zhen Jiu 29: 785–790 (In Chinese with English abstract)

Hachinski V (2009) World Stroke Day 2009: What Can I Do? Stroke 40: 3165

Han XM, Ying C, Zhao Q, Liu SY (2009) The induction of acupuncture preconditioning on cerebral ischemia tolerance and erythropoietin. Chinese Journal of Gerontology 29: 1762–1764 (In Chinese with English abstract)

Hashimoto N, Matsumoto T, Mabe H, Hashitani T, Nishino H (1994) Dopamine has inhibitory and accelerating effects on ischemia-induced neuronal cell damage in the rat striatum. Brain Res Bull 33: 281–288

Hattori A, Luo Y, Umegaki H, Munoz J, Roth GS (1998) Intra-Striatal injection of dopamine results inDNA damage and apoptosis in rats, Neuroreport 9: 2569–2572

He J, Shen PF (2007) Clinical study on the therapeutic effect of acupuncture in the treatment of post-stroke depression. Zhen Ci Yan Jiu 32: 58–61 (In Chinese with English abstract)

Hopwood V, Lewith G, Prescott P, Campbell MJ (2008) Evaluating the efficacy of acupuncture in defined aspects of stroke recovery: A randomised, placebo controlled single blind study. J Neurol 255: 858–866

Hu HH, Chung C, Liu TJ, Chen RC, Chen CH, Chou P, Huang WS, Lin JC, Tsuei JJ (1993) A randomized controlled trial on the treatment for acute partial ischemic stroke with acupuncture. Neuroepidemiology 12: 106–113

Huang Z, Huang F, Yan HX, Min Y, Gao Y, Tan BD, Qu F (2010) Dysphagia after stroke treated with acupuncture or electric stimulation: a randomized controlled trial. Zhongguo Zhen Jiu 30: 969–973. (In Chinese with English abstract)

Johansson K, Kindgren I, Windener H, Wiklund I, Johansson BB (1993) Can sensory stimulation improve the functional outcome in stroke patients? Neurology 43: 2189–2192

Kahn RA, Weinberger J, Brannan T, Prikhojan A, Reich DL (1995) Nitric oxide modulates dopamine release during global temporary cerebral ischemia. Anesth Analg 80:1116–1121

Kandel ER, Schwartz JH, Jessell TM (2000) Principles of Neural Science. 4th edn. McGraw Hill, New York

Kim KW, Kim DC, Kim YH, Eun YA, Kim HI, Cho KP (1995) Ca2+-dependent and -independent mechanisms of ischemia-evoked release of [3 H]-dopamine from rat striatal slices, Cli Exp Pharmacol Physiol 22: 301–302

Koob GF, Le Moal M (2001) Drug addiction, dysregulation of reward, and allostasis. Neuropsychopharmacology 24: 97–129

Korotkova TM, Ponomarenko AA, Brown RE, Haas HL (2004) Functional diversity of ventral midbrain dopamine and GABAergic neurons. Mol Neurobiol 29: 243–259

Kovács A, Móricz K, Albert M, Benedek A, Hársing LG Jr, Szénási G (2010) Decreased vasoconstrictor responses in remote cerebral arteries after focal brain ischemia and reperfusion in the rat, in vitro. Eur J Pharmacol 644:154–159

Koziorowski D, Jasztal J (1999) Factors which can play important role in Pathogenesis of Parkinson disease. Neurol Neurochir Pol 33: 907–921

Kroppenstedt SN, Sakowitz OW, Thomale UW, Unterberg AW, Stover JF (2002) Norepinephrine is superior to dopamine in increasing cortical perfusion following controlled cortical impact injury in rats. Acta Neurochir Suppl 81: 225–227

Kroppenstedt SN, Sakowitz OW, Thomale UW, Unterberg AW, Stover JF (2002) Influence of norepinephrine and dopamine on cortical perfusion, EEG activity, extracellular glutamate, and brain edema in rats after controlled cortical impact injury. J Neurotrauma 19: 1421–1432

Kroppenstedt SN, Stover JF, Unterberg AW (2000) Effects of dopamine on posttraumatic cerebral blood flow, brain edema, and cerebrospinal fluid glutamate and hypoxanthine concentrations. Crit Care Med 28: 3792–3798

Kroppenstedt SN, Thomale UW, Griebenow M, Sakowitz OW, Schaser KD, Mayr PS, Unterberg AW, Stover JF (2003) Effects of early and late intravenous norepinephrine infusion on cerebral perfusion, microcirculation, brain tissue oxygenation, and edema formation in brain-injured rats. Crit Care Med 31: 2211–2221

Lewen A, Matz P, Chan PH (2000) Free radical pathways in CNS injury. J Neurotrauma 17: 871–890

Li M, Sun JH, Lu JW, Liu LY, Geng H (2009) Observation on therapeutic effect of five-needle- in-nape acupuncture for treatment of poststroke pseudobulbar paralysis dysphagia. Zhongguo Zhen Jiu 29: 873–875 (In Chinese with English abstract)

Liu HL, Wang LP. (2006) Randomized controlled study on ginger-salt-partitioned moxibustion at shenque (CV 8) on urination disorders poststroke. Zhongguo Zhen Jiu 26: 621–624 (In Chinese with English abstract)

Liu SY, Hsieh CL, Wei TS, Liu PT, Chang YJ, Li TC (2009) Acupuncture Stimulation Improves Balance Function in Stroke Patients: A Single-Blinded Controlled, Randomized Study. Am J Chin Med 37: 483–494

Liu W, Mukherjee M, Sun C, Liu H, McPeak LK (2008) Electroacupuncture may help motor recovery in chronic stroke survivors: a pilot study. J Rehabil Res Dev 45: 587–595

Long D, Young J (2003) Dexamphetamine treatment in stroke. QJM 96: 673–685
Lu SK (1991) Evaluation of the acupuncture treatment on stroke induced hemiplegia. Zhong Yi Za Zhi 10: 49 (in Chinese)
Luo Y, Umegaki H, Wang X, Abe R, Roth GS (1998) Dopamine induces apoptosis through an oxidation-involved SAPK/JNK activation pathway. J Biol Chem 273: 3756–3764
Macbeth AH, Gautreaux C, Luine VN (2008) Pregnant rats show enhanced spatial memory, decreased anxiety, and altered levels of monoaminergic neurotransmitters. Brain Res 1241: 136–147
Marcoli M, Cervetto C, Castagnetta M, Sbaffi P, Maura G (2004) 5-HT control of ischemia-evoked glutamate efflux from human cerebrocortical slices. Neurochem Int 45: 687–691
Marsicano G, Goodenough S, Monory K, Hermann H, Eder M, Cannich A, Azad SC, Cascio MG, Gutiérrez SO, van der Stelt M, López-Rodriguez ML, Casanova E, Schütz G, Zieglgänsberger W, Di Marzo V, Behl C, Lutz B (2003) CB1 cannabinoid receptors and on-demand defense against excitotoxicity. Science 302: 84–88
Marzan AS, Hungerbühler HJ, Studer A, Baumgartner RW, Georgiadis D (2004) Feasibility and safety of norepinephrine-induced arterial hypertension in acute ischemic stroke. Neurology 62: 1193–1195
Melis M, Perra S, Muntoni AL, Pillolla G, Lutz B, Marsicano G, Di Marzo V, Gessa GL, Pistis M (2004) Prefrontal cortex stimulation induces 2-arachidonoyl-glycerol-mediated suppression of excitation in dopamine neurons. J Neurosci 24: 10707–10715
Melis M, Pillolla G, Bisogno T, Minassi A, Petrosino S, Perra S, Muntoni AL, Lutz B, Gessa GL, Marsicano G, Di Marzo V, Pistis M (2006) Protective activation of the endocannabinoid system during ischemia in dopamine neurons. Neurobiol Dis 24: 15–27
Melis M, Pistis M, Perra S, Muntoni AL, Pillolla G, Gessa GL (2004) Endocannabinoids mediate presynaptic inhibition of glutamatergic transmission in rat ventral tegmental area dopamine neurons through activation of CB1 receptors. J Neurosci 24: 53–62
Melis M, Pistis P (2007) Endocannabinoid Signaling in Midbrain Dopamine Neurons: More than Physiology? Curr Neuropharmacol 5: 268–277
Min J, Farooq MU, Greenberg E, Aloka F, Bhatt A, Kassab M, Morgan JP, Majid A (2009) Cardiac dysfunction after left permanent cerebral focal ischemia: the brain and heart connection. Stroke 40: 2560–2263
Moon SK, Whang YK, Park SU, Ko CN, Kim YS, Bae HS, Cho KH (2003) Antispastic effect of electroacupuncture and moxibustion in stroke patients. Am J Chin Med 31: 467–674
Naeser MA, Alexander MP, Stiassny Eder D, Galler V, Bachman D (1994) Acupuncture in the treatment of paralysis in chronic and acute stroke patients: Improvement correlated with specific CT scan lesion sites. Acupunct Electrother Res 19: 227–249
Nagayama T, Sinor AD, Simon RP, Chen J, Graham SH, Jin K, Greenberg DA (1999) Cannabinoids and neuroprotection in global and focal cerebral ischemia and in neuronal cultures. J. Neurosci., 19, 2987–2995
O'Carroll CM, Morris RG (2004) Heterosynaptic co-activation of glutamatergic and dopaminergic afferents is required to induce persistent long-term potentiation. Neuropharmacology 47: 324–332
O'Donnell P (2003) Dopamine gating of forebrain neural ensembles. Eur J Neurosci 17: 429–435
Obrenovitch TP, Richards DA (1995) Extracellular neurotransmitter changes in cerebral ischaemia. Cerebrovasc Brain Metab Rev 7: 1–54
Panikashvili D, Simeonidou C, Ben-Shabat S, Hanus L, Breuer A, Mechoulam R, Shohami E (2001) An endogenous cannabinoid (2-AG) is neuroprotective after brain injury. Nature 413: 527–531
Parmentier-Batteur S, Jin K, Mao XO, Xie L, Greenberg DA (2002) Increased severity of stroke in CB1 cannabinoid receptor knock-out mice. J. Neurosci 22: 9771–9775
Pfister D, Strebel SP, Steiner LA (2008) Effects of catecholamines on cerebral blood vessels in patients with traumatic brain injury. Eur J Anaesthesiol 25: 98–103
Piomelli D, Pilon C, Giros B, Sokoloff P, Martres MP, Schwartz JC (1991) Dopamine activation of the arachidonic acid cascade as a basis for D1/D2 receptor synergism. Nature, 353: 164–167
Richter F, Mikulik O, Ebersberger A, Schaible HG (2005) Noradrenergic agonists and antagonists influence migration of cortical spreading depression in rat-a possible mechanism of migraine

prophylaxis and prevention of postischemic neuronal damage. J Cereb Blood Flow Metab 25: 1225–1235
Rocha M, Martins RA, Linden R (1999) Activation of NMDA receptors pro2tects against glutamate neurotoxicity in the retina: evidence for the involvement of neurotrophins. Brain Res 827: 79–92
Rosamond W, Flegal K, Friday G, Furie K, Greenlund K, Haase N, Hailpern SM, Ho M, Howard V, Kissela B, Kittner S, Lloyd-Jones D, McDermott M, Meigs J, Moy C, Nichol G, O'Donnell C, Roger V, Sorlie P, Steinberger J, Thom T, Wilson M, HongYL (2008) Heart Disease and Stroke Statistics-2008 Update A Report From the American Heart Association Statistics Committee and Stroke Statistics Subcommittee. Circulation 117:e25-e146
Saulle E, Centonze D, Martín AB, Moratalla R, Bernardi G, Calabresi P (2002) Endogenous dopamine amplifies ischemic longterm potentiation via D1 receptors. Stroke 33: 2978–2984
Schaechter JD, Connell BD, Stason WB, Kaptchuk TJ, Krebs DE, Macklin EA, Schnyer RN, Stein J, Scarborough DM, Parker SW, McGibbon CA, Wayne PM (2007) Correlated change in upper limb function and motor cortex activation after verum and sham acupuncture in patients with chronic stroke. J Altern Complement Med 13: 527–532
Scheidtmann K, Fries W, Muller F, Koenig E (2001) Effect of levodopa in combination with physiotherapy on functional motor recovery after stroke: a prospective, randomised, double-blind study. Lancet 358: 787–790
Shen M, Thayer SA (1998) Cannabinoid receptor agonists protect cultured rat hippocampal neurons from excitotoxicity. Mol Pharmacol 54: 459–462
Shen XY, Tang HT, Li J, Xiang T, Liu HF, and Yu WC (2010) Using microPET imaging in quantitative verification of acupuncture effect in ischemia stroke treatment. available from nature precedings. NeuroImage Netprint. http://hdl.handle.net/10101/npre.2010.4128.1. Accessed 4 Jan 2010
Shi XM (2005) Clinical studies on treatment of 9005 stroke cases with "Mind-refreshing & Orifice-opening" needling method. Guiding Journal of Traditional Chinese Medicine & Pharmacology 11: 3–5
Shih AY, Erb H, Murphy TH (2007) Dopamine activates Nrf2-regulated neuroprotective pathways in astrocytes and meningeal cells. J Neurochem 101: 109–19
Sinor AD, Irvin SM, Greenberg DA (2000) Endocannabinoids protect cerebral cortical neurons from in vitro ischemia in rats. Neurosci Lett 278: 157–160
Snyder GL, Fienberg AA, Huganir RL, Greengard P (1998) A dopamine/D1 receptor/protein kinase A/dopamineand cAMP-regulated phosphoprotein (Mr 32 kDa)/protein phosphatase-1 pathway regulates dephosphorylation of the NMDA receptor. J Neurosci 18: 10297–10303
Sonde L, Nordstrom M, Nilsson CG, Lokk J, Viitanen M (2001) A double-blind placebo-controlled study of the effects of amphetamine and physiotherapy after stroke. Cerebrovasc Dis 12: 253–257
Strack S, Barban MA, Wadzinski BE, Colbran RJ (1997) Differential inactivation of postsynaptic density-associated and soluble Ca2+/calmodulin-dependent protein kinase II by protein phosphatases 1 and 2A. J Neurochem 68: 2119–2128
Strittmatter M, Meyer S, Fischer C, Georg T, Schmitz B (2003) Location-dependent patterns in cardio-autonomic dysfunction in ischaemic stroke. Eur Neurol 50: 30–38
Surmeier DJ, Ding J, Day M, Wang Z, Shen W (2007) D1 and D2 dopamine-receptor modulation of striatal glutamatergic signaling in striatal medium spiny neurons. Trends Neurosci 30: 228–235
Suzuki T, Akaike N, Ueno K, Tanaka Y, Himori N (1995) MAO inhibitors,clorgyline and lazabemide, prevent hydroxyl radical generation caused by brain ischemia/reperfusion in mice. Pharmacology 50: 357–362
Svenningsson P, Nishi A, Fisone G, Girault JA, Nairn AC, Greengard P (2004) DARPP-32: an integrator of neurotransmission. Annu Rev Pharmacol Toxicol 44: 269–296
Toner CC, Stamford JA (1996) 'Real time' measurement of dopamine release in an in vitro model of neostriatal ischaemia. J Neurosci Methods 67: 133–140
Walker-Batson D, Curtis S, Natarajan R, Ford J, Dronkers N, Salmeron E, Lai J, Unwin DH (2001) A double-blind, placebo- controlled study of the use of amphetamine in the treatment of aphasia. Stroke 32: 2093–2098

Weinberger J, Cohen G, Nieves-Rosa J (1983) Nerve terminal damage in cerebral ischemia: greater susceptibility of catecholamine nerve terminals relative to serotonin nerve terminals. Stroke 14:986–989

Weinberger J, Nieves-Rosa J, Cohen G (1985) Nerve terminal damage in cerebral ischemia: protective effect of alpha-methyl-para-tyrosine. Stroke 16:864–870

Werling LL, Jacocks HM 3 rd, Rosenthal RE, Fiskum G (1993) Dopamine release from canine striatum following global cerebral ischemia/reperfusion. Brain Res 606: 99–105

Wong AMK, Su T, Tang F, Cheng P, Liaw M (1999) Clinical Trial of electrical acupuncture on hemiplegic stroke patients. Am J Phys Med Rehabil 78: 117–122

Wu HM, Tang JL, Lin XP, Lau JTF, Leung PC, Woo J, Li YP (2008) Acupuncture for Stroke Rehabilitation. Stroke 39: 517–518

Wu JP (2010) Clinical observation on acupuncture treatment of 150 cases of post-stroke depression according to syndrome differentiation. Zhen Ci Yan Jiu 35: 303–306 (In Chinese with English abstract)

Wu P, Mills E, Moher D, Seely D (2010) Acupuncture in Poststroke Rehabilitation: A Systematic Review and Meta-Analysis of Randomized Trials. Stroke 41: e171-e179

Xu B, Lu RY, Ni GX, Xu Y (2001) The impact of electro-acupuncture on monoamine neurotransmitters and their metabolites in rat's striatum, hypothalamus of cerebral ischemia-reperfusion. The Journal of Traditional Chinese Medicine Information 8:27–28 (In Chinese with English abstract)

Xu J, Ge LB, Chen HP (2001 c) A comparative study on the effects of near and far acupoints EA on Ca2+□Na+□K+contents in cerebral ischemia rat brain tissue. Shanghai Journal of Acupuncture 20: 36–37 (In Chinese with English abstract)

Xu J, Ge LB, Ju JH, Zheng JL, Cheng JS, Chen HP (2001 b) Comparison of Near and far points on the SOD, MDA in cerebral ischemia rat brain tissue. Liaoning Journal of Traditional Chinese Medicine 28: 237–238 (In Chinese with English abstract)

Xu J, Ge LB, Xu MS, Chen HP (2002) The study on the depth of MCAO model in rats using suppository-line. Shanghai Laboratory Animal Science 22: 209–212 (In Chinese with English abstract)

Xu J, Ge LB, Zheng JL, Ju JH, Chen HP (2001 a) A comparative study on the effects of near and far acupoints EA on Serum nitric oxide in transient cerebral ischemia in rat brain tissue. Acupuncture Research 26: 243–246 (In Chinese with English abstract)

Xu MS, Fang C, Xu J, Zhang GF, Ge LB (2009b) Dynamic observation of EA effects on striatum dopamine and its metabolites in rats with cerebral ischemia-reperfusion. Zhen Ci Yan Jiu 34: 230–235 (In Chinese with English abstract)

Xu MS, Ge LB, Wang XH, Xu J, Wang ZQ, Cui YJ (2009c) Dynamic changes of glycine and taurine in corpus striatum of rats with ischemia-reperfusion injury and effects of electroacupuncture. Journal of acupuncture and tuina science 7: 129–133 (In Chinese with English abstract)

Xu MS, Ge LB, Xu J, Fang C (2009a) The effects of EA on the turnover ratio of striatum monoamine neurotransmitters in rats with cerebral ischemia-reperfusion. Shanghai Journal of Acupuncture 28: 239–242 (In Chinese with English abstract)

Yan T, Hui-Chan CW (2009) Transcutaneous electrical stimulation on acupuncture points improves muscle function in subjects after acute stroke: A randomized controlled trial. Journal of Rehabilitation Medicine 41: 312–316 (In Chinese with English abstract)

Yang T, Liu Z, Liu Y (2003) Electroacupuncture at ciliao and huiyang for treating neuropathic incontinence of defecation and urination in 30 cases. J Tradit Chin Med 23: 53–54 (In Chinese with English abstract)

Yurko-Mauro KA, Friedman E (1996) Dopamine-stimulated changes in activated calpain I in rat hippocampal slices. J Neurosci Res 43: 476–481

Zhang J, Price JO, Graham DG, Montine TJ (1998) Secondary excitotoxicity contributes to dopamine induced apoptosis in dopaminergic neuronal cultures. Biochem Biophy Res Comm 248: 812–816

Zou G (1999) The basis of neuropharmacology. 2nd edn. Science Press, Beijing (In Chinese)

Chapter 14
Protective Effect of Acupuncture on Myocardium

Xianrui Wang and Yongjun Zheng

Abstract Perioperative myocardial ischemia–reperfusion, pulmonary interstitial edema, and postoperative central nervous system complications are the main risk factors that affect the postoperative rehabilitation of patients who have undergone cardiac surgery. Based on the principle of "where there is a meridian acupoint, there is the indication," we have selected two points of pericardium meridian of hand-Jue Yin [Neiguan (PC-6)] and lung meridian of hand-Tai Yin [Yunmen (LU-2), Lieque (lU-7)] to strengthen prevention and protection of perioperative myocardial ischemia, reduce perioperative risk factors effectively, and improve patient's postoperative rehabilitation through regulation of body's internal environment. The regulation mechanisms of circulatory function by acupuncture include (1) cardiac autoresuscitation rate, incidence of myocardial ischemia, vasoactive drug dosage, incidence of postoperative complications, ICU, and hospital stay days for cardiac surgery patients to whom the acupuncture anesthesia were significantly better than patients with endotracheal intubation anesthesia. (2) The studies show that the acupuncture anesthesia for cardiac surgery has reduced operative risk and improved surgical safety. Its main mechanisms include (a) regulation of β-EP, ACTH, cortisol, and plasma dopamine by electroacupuncture stimulation that improve impact of surgery on internal environment; (b) increase HSP70 expression by myocardial cells; (c) reduce mitochondrial membrane permeability; (d) inhibit cell adhesion process so as to reduce myocardial cell injury. (3) According to the principle of "where there is a meridian acupoint, there is the indication," we find that double points selection of lung meridian of hand-Tai Yin [Yunmen (LU-2), Lieque (lU-7)] has apparent effect on myocardial protection, suggesting local and remote synergy of acupoints along meridians. Acupuncture anesthesia is a combined technique, including acupuncture combined with hypothermic anesthesia, and acupuncture anesthesia combined drugs (salvia, nicardipine, L-arginine) which have a synergistic effect on circulation regulation.

X. Wang (✉) • Y. Zheng
Department of Anesthesiology and Pain Management, Renji Hospital, Shanghai JiaoTong University School of Medicine, Shanghai, China
e-mail: xiangruiwang@gmail.com; zhengyongjun1971@126.com

Y. Xia et al. (eds.), *Current Research in Acupuncture*,
DOI 10.1007/978-1-4614-3357-6_14, © Springer Science+Business Media New York 2013

Keywords Acupuncture anesthesia • Myocardial ischemia–reperfusion • Cardiopulmonary bypass • Cardiac protection • Immune regulation

14.1 Introduction

The patients with heart diseases, especially the elderly people with coronary heart disease, often suffer myocardial ischemia (the highest incidence rate is 78 %), myocardial infarction, and even severe cardiopulmonary insufficiency during perioperative period of cardiac surgery due to a variety of factors. Therefore, studies on perioperative myocardial ischemia–reperfusion injury and its mechanism are needed in order to enhance perioperative myocardial protection and prevention or mitigate the occurrence and development of myocardial ischemia, which shows important clinical significance in improving patient's postoperative rehabilitation and safety and saving medical costs. Over the years, a variety of methods for myocardial protection have been carried out both at home and abroad. Despite the apparent efficacy, side effects and medical cost have also increased significantly, which further aggravate the patient's economic burden.

In addition to analgesic effect of acupuncture, it can stimulate body's internal adjustment mechanism to enhance cellular immune function and resistance to ischemia. Acupuncture has clear efficacy in protection of ischemic myocardium. Heart surgery with acupuncture anesthesia is an innovative technology originally developed by our hospital. Clinical studies show that the acupuncture anesthesia patients gain quick recovery and suffer few complications after cardiac surgery. On this basis, we have observed the regulation of internal environment by acupuncture for cardiac surgery patients and analyzed the possible main factors. By monitoring hemodynamic parameters and measuring myocardial enzymes, cardiac metabolism, mitochondrial activity of myocardial cells, expression of heat shock protein HSP70 mRNA, myocardial apoptosis, and apoptosis-related gene expression, we have explored the effects of acupuncture, acupuncture combined with hypothermic anesthesia, and acupuncture anesthesia combined drugs on improving the patient's perioperative cardiac function and reducing the incidence of myocardial ischemia on the molecular, animal experiments and clinical level thoroughly for the first time, especially in patients who have undergone heart surgery under cardiopulmonary bypass (CPB) and noncardiac surgery. Using rat model of cardiopulmonary bypass established by us, we have demonstrated the protection mechanism of perioperative acupuncture on myocardium, including its antiapoptotic, antioxidation, antiadhesion, antistress, and immune regulatory role. We propose that acupuncture combined with hypothermic anesthesia has synergistic effect with ischemia perfusion, which has enhanced myocardial protection. Acupuncture-assisted salvia, nicardipine, and L-arginine can inhibit body stress response and have synergistic effect on the protection of myocardium injury caused by ischemia–reperfusion under cardiopulmonary bypass.

14.2 Clinical Application of Acupuncture for Myocardial Ischemia

Surgery causes an insult to the body, inducing responses such as pain, anxiety, proinflammatory reactions, and increase in resting HR, all of which are modulated by the autonomic nervous system (Yavuz et al. 2006). Vagal suppression which leads to increased HR has been described in patients after cardiac surgery. Since acupuncture exerts antisympathotonic, provagotonic, and anti-inflammatory effects as shown, it might be attractive and beneficial in addition to standard heart failure medication and myocardial ischemia (Fig. 14.1).

More than 500 patients underwent noncardiac surgery or cardiopulmonary bypass cardiac surgery, including atrial septal defect, ventricular septal defect repair, and mitral valvuloplasty. The patients were classified as Grade I–II according to ASA grading method. Preoperative blood biochemistry and coagulation function test were normal. No obvious infection, malignant lesion, and endocrine and immune system diseases were found. According to the experimental design, they were divided into three groups, including acupuncture anesthesia group, acupuncture combined with general anesthesia group, and general anesthesia group. There was no significant difference in general information such as age, height, weight, preoperative cardiac function, and cardiopulmonary bypass between groups (Table 14.1). All the patients were premedicated with 0.1 g phenobarbital, 50 mg pethidine, and 0.3 mg scopolamine by intramuscular injection 2 h before the induction of anesthesia. Anesthesia was induced with fentanyl 20 μg/kg, midazolam 0.15 mg/kg, and vecuronium bromide 0.25 mg/kg. Anesthesia was maintained with continuous infusions of fentanyl

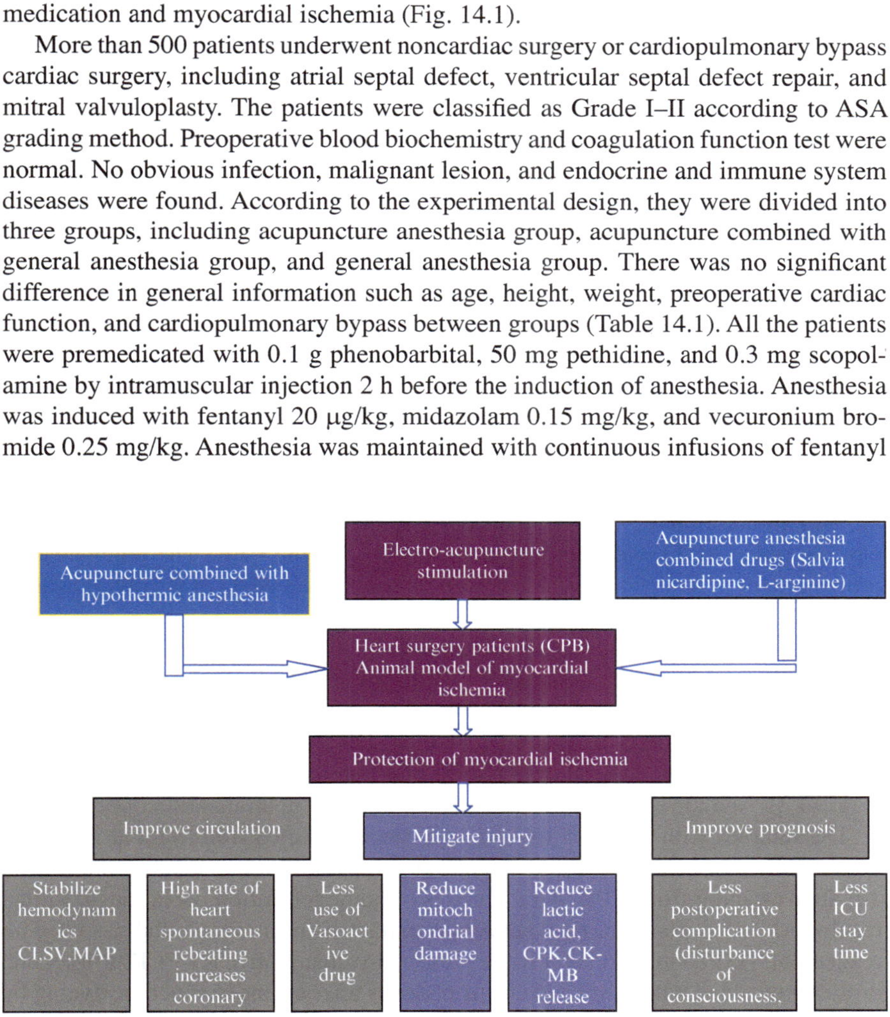

Fig. 14.1 Effect of acupuncture for myocardial ischemia

Table 14.1 General information of patients ($\overline{x} \pm s$)

	Acupuncture anesthesia group	Acupuncture combined with general anesthesia group	General anesthesia group
Age (years)	26.2±5.3	27.4±8.9	28.9±10.1
Height (cm)	163.4±4.1	163.2±9.4	163.7±6.2
Weight (kg)	49.2±4.7	50.7±12.2	51.2±6.4
Preoperative cardiac function	I–II	I–II	I–II

Table 14.2 Information about cardiopulmonary bypass ($\overline{x} \pm s$)

	Acupuncture anesthesia group	Acupuncture combined with general anesthesia group	General anesthesia group
CPB time (min)	44.5±5.7	49.4±10.1	48.4±11.4
Aortic cross-clamping time (min)	27.7±6.7	30.7±3.5	27.2±7.3
Heart arrest time (min)	36.7±2.9	35.4±4.3	44.7±7.2
Lowest temperature (°C)	35.2±0.8	32.7±1.9	33.1±1.5

and midazolam (0.15 and 0.6 μg/kg/min, respectively) until the end of operation. Isoflurane was administered as required during the operation.

During CPB, nonpulsatile pump flow and membrane oxygenation (Compactflo, Dideco-Shiley, Modena, Italy) were used in conjunction with tepid systemic hypothermia (nasopharyngeal temperature of 32–34°C). Pump flow was maintained at 2.4 L/min/m^2. For CPB, the patients received anticoagulation with heparin 300 IU/kg, and an additional 5,000 IU was added to the priming solution. The activated clotting time (ACT) was kept longer than 400 s during CPB. The bolus dose of protamine for neutralization of heparin effect was 1 mg/100 IU of the initial loading dose of heparin (Table 14.2).

The acupuncture points included Neiguan (PC-6), Lieque (LU-7), and Yunmen (LU-2). The needles (with a diameter of 0.3 mm and a length of 50 mm, made by Shanghai Medical Factory, China) were connected to a multichannel electroacupuncture device (G6805-2, Shanghai Medical Factory, China) which can deliver electric current (4 Hz and 10 mA) for 30 min before operation until heparinizing (Fig. 14.2). And the stimulation was resumed after the neutralization of protamine.

The samples with a size of 1×1×1 mm^3 were collected from right atria 10 min before CPB (after catheterization in superior vena cava), immediately, 30 min, and 1 h after weaning from CPB. After a series of homogenization, isolation, washing, resolution, and deposition, total RNA was extracted from myocardial tissues with reagent kit. HSP70 mRNA level was evaluated by quantitative RT-PCR. Primer and probe were added with 0.8 % gel electrophoresis after bromophenol blue dyeing by an ultraviolet spectrophotometer (Shimatsu UV-2100). Quantitative analysis of HSP70 mRNA was operated on a spot hybridization membrane after coloration by a density scanner (Shimatsu type CS-920, Japan).

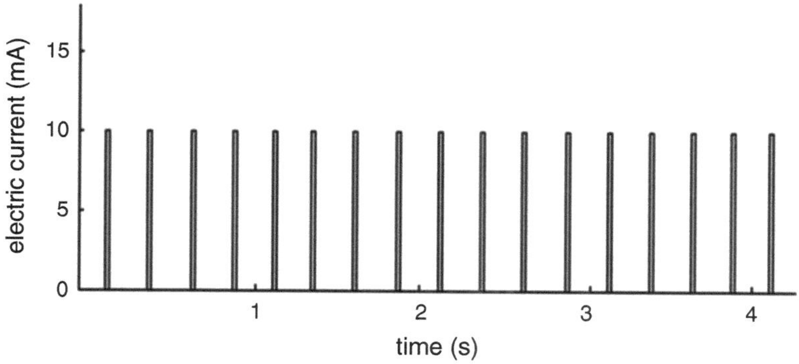

Fig. 14.2 Schematic diagram showing the electric current of electrical stimulator (G6805-2)

Blood samples were collected from right internal jugular vein catheter, and plasma CK-MB concentration was measured with Hitachi 7150 fully automatic analyzer. Time for cardiovascular support in the first 24 h after weaning from CPB, duration of mechanical ventilation, and ICU stay were recorded as main index for postoperative complications.

It indicates that the electroacupuncture could increase the gene expression of HSP70 in myocardial cells and plays a protective role in patients undergoing cardiac surgery. It is probably one of the protective mechanisms to ischemic cardiac muscles by improving the gene expression of HSP during surgery. In the field of cardiac surgery, it would be clinically meaningful to alleviate the ischemia–reperfusion injury of cardiac muscles and improve the prognosis of cardiac surgery by inducing the expression of HSP70 mRNA in myocardial cells through different methods. The leading factor in myocardial ischemia–reperfusion injury is the production of oxygen free radicals, which could injure the membrane, nucleolus, and ribose protein and inhibit enzyme system, thus blocking the cellular function. HSP could reduce the release of oxygen free radical and stabilize the cellular and lysosome membranes, thus avoiding the denaturalization of protein and reducing the ischemia–reperfusion injury.

14.2.1 Effects of Acupuncture Combined with General Anesthesia on Myocardial Enzyme Released by Ischemic Myocardium Under Cardiopulmonary Bypass (Wang et al. 1999a, b, 2000)

Several small studies have suggested that cardiac enzyme elevation in the 24 h following coronary artery bypass graft (CABG) surgery is associated with worse prognosis, but a definitive study is not available. The long-term prognostic impact of small increase of perioperative enzyme has not been reported either. Among patients

Table 14.3 CK-MB changes during operation in three groups ($\overline{x} \pm s$, U/L)

Group	Before bypass	After bypass	1 h after bypass
Acupuncture anesthesia group	11.30±2.95	27.81±5.35	38.41±20.76*
Acupuncture combined with general anesthesia group	12.55±3.42	31.82±18.27	46.78±24.21*
General anesthesia group	11.90±2.95	66.56±24.14	94.80±10.04*, **

*Compared with before bypass, $P<0.05$
**Comparison between groups, $P<0.05$

who have undergone CABG surgery, the elevation of CK-MB or troponin levels within the first 24 h is independently associated with increased intermediate- and long-term risk of mortality (Domanski et al. 2011).

CK-MB released after bypass and 1 h after bypass in the three groups were significantly higher than CK-MB before bypass ($P<0.01$). One hour after bypass, the increase of CK-MB in nonacupuncture group was significantly higher than that of acupuncture group (Table 14.3).

14.2.2 Effects of Acupuncture Combined with General Anesthesia on Hemodynamics Under Cardiopulmonary Bypass

Hemodynamics is an important part of cardiovascular physiology that deals with the forces, which the pump (the heart) has to develop to circulate blood through the cardiovascular system. Adequate blood circulation (blood flow) is a necessary condition for adequate supply of oxygen to all tissues, which, in return, is synonymous with cardiovascular health, survival of surgical patients, longevity, and quality of life. To an outside observer (a physician or a nurse), these hemodynamic forces demonstrate themselves as blood pressure and blood flow paired values at different nodes of the cardiovascular system. We will concentrate on systemic hemodynamics—the blood pressure and blood flow at the output of the left heart (Wang et al. 1999a, b).

After the induction, the CI, SV, and MAP in general anesthesia group showed a more significant drop than that of acupuncture anesthesia group and acupuncture combined with general anesthesia group ($P<0.05$). There was no significant difference of SVR and HR between the groups. After bypass, CI, SV, and MAP in three groups decreased significantly, but the decline in general anesthesia group was most evident. Decrease in acupuncture anesthesia group was less than that of other two groups ($P<0.05$). There was no significant difference of SVR between the groups. After the surgery, CI, SV, and MAP in general anesthesia group were significantly lower than that of other two groups ($P<0.01$). CI, SV, MAP, and HR changes in acupuncture anesthesia group were less than that of acupuncture combined with general anesthesia group (2.18±0.59 days, $P<0.05$). Hemodynamic parameters were listed in Table 14.4.

Table 14.4 Hemodynamic parameters during operation ($\overline{x} \pm s$)

		CO (L/min)	CI (L/min m^2)	SV (ml)	SVR (kPa s/l)	MAP (kPa)	HR (bpm)
General anesthesia group	Before operation	5.9±1.3	4.0±1.0	72.5±24.0	98.2±35.8	91±13	86±13
	After induction	5.0±1.2	3.4±0.9	63.5±25.4	101.8±44.1	81±17	82±16
	Skin incision	4.3±1.3*	3.2±1.0*	56.4±19.0	123.0±55.6	88±12	82±17
	Before bypass	4.3±1.0*	2.9±0.7*	45.1±15.1*	105.3±25.4	76±11*	96±16
	Stop bypass	4.1±0.6*	2.8±0.3*	41.1±6.1*	128.3±17.6*	89±10	105±19*
	After bypass	402±0.6*	2.8±0.4*	42.9±4.3*	117.2±20.4	79±9*	98±13*
	After operation	4.1±0.5*	2.8±0.4*	45.8±5.6*	124.3±28.8*	88±14	94±11
Acupuncture combined with general anesthesia group	Before operation	6.1±1.1	4.0±0.7	75.7±22.4	89.8±16.9	91±9	85±17
	After induction	5.7±1.5	3.8±1.0	81.8±26.4	84.3±22.6	79±9*	71±17
	Skin incision	5.3±1.2	3.5±0.8	79.3±23.2*	104.7±33.4	91±11	71±14
	Before bypass	5.0±1.1	3.3±0.6*	51.6±17.3*	101.9±23.6	84±12	100±13*
	Stop bypass	4.6±0.7*	3.1±0.6*	45.8±14.2*	111.5±24.5*	87±14	107±17*
	After bypass	4.8±0.6*	3.2±0.5*	48.9±13.1*	112.0±17.5*	91±8	101±16*
	After operation	5.0±0.7*	3.3±0.6*	50.8±7.0	104.0±13.9*	88±5	101±8*
Acupunctural anesthesia group	Before operation	4.5±0.5*	3.1±0.4	46.8±6.0	120.7±18.9	92±10	111±6
	After induction	4.5±0.4	3.1±0.3	51.6±5.9*	117.6±14.4	89±8	89±6*
	Skin incision	4.5±0.4	3.1±0.3	49.8±5.3	121.0±14.1	92±8	93±7*
	Before bypass	4.0±0.5*	2.8±0.4*	41.9±0.6	130.2±28.0	88±15	90±5
	Stop bypass	3.9±0.4*	2.6±0.3*	39.6±5.7*	127.2±18.8	86±12	99±6
	After bypass	4.1±0.5*	2.8±0.4*	49.5±7.8	128.6±16.7	88±10	81±5*
	After operation	4.3±0.5*	2.9±0.4	46.8±5.3	114.7±14.0	88±8	90±4*

*Comparison with before operation, $p<0.05$

Therapeutic strategies focusing on a decrease of sympathetic activity and an increase of parasympathetic activity may further reduce the morbidity and mortality of patients with open heart surgery.

In summary, acupuncture remains an intriguing yet understudied therapeutic modality in patients with open heart surgery. Animal data supports the hypothesis that acupuncture at specific sites could attenuate sympathetic nerve activation. Preliminary data in humans proves consistent with the hypothesis that acupuncture could activate cardiovascular centers in the central nervous system, which directly modulates central sympathetic neural outflow. Direct sympathetic nerve recordings in patients with heart surgery lend further support to the sympatholytic potential of acupuncture. Because it is known that the most effective pharmacologic therapies in heart surgery work through interactions with the autonomic nervous system, further studies of acupuncture efficacy and its mechanisms in heart failure are mandated.

14.2.3 Impact of Acupuncture Anesthesia on Vasoactive Drug in Cardiopulmonary Bypass, Rejump Approach, and Postoperative Complications

Circulatory failure is one of the most common organ dysfunctions in patients admitted to the cardiac surgery. These patients often require administration of intravenous vasoactive or inotropic medications to optimize or support their cardiovascular function. These potent agents have a very short half-life and are generally infused in high concentration with a low flow rate and a very narrow safety margin. For these reasons, high-precision infusion pumps are recommended to maintain a precise and constant flow rate in order to achieve both constant plasma concentrations and effects on the cardiovascular system. Achieving hemodynamic stability is a major therapeutic goal for treating patients undergoing cardiac surgery.

Nine patients in general anesthesia group, accounting for 56.3 % of the total patients, used intraoperative dopamine with per capita consumption 22.5 mg. Vasoactive drug was not used in acupuncture anesthesia group. Application rate of vasoactive drug in general anesthesia group was significantly higher than that of acupuncture anesthesia group and acupuncture combined with general anesthesia group ($p<0.01$).

Heart spontaneous rebeating rate in general anesthesia group was 3 %, 92 % for acupuncture combined with general anesthesia group, and 100 % for acupuncture anesthesia group. Heart spontaneous rebeating rate in acupuncture anesthesia group and acupuncture combined with general anesthesia group were significantly higher than in general anesthesia group ($p<0.01$).

Five patients in general anesthesia group, accounting for 31 % of the total patients, used vasoactive drug 1 day after operation due to instability of circulation. Two patients had disturbance of consciousness and cerebral embolism, accounting for 14.4 %. The average ICU days were 5.41 ± 0.82, which was significantly higher than that of acupuncture anesthesia group and acupuncture combined with general anesthesia group ($p<0.01$) (Wang et al. 2001a, b, c, d).

14.3 Mechanisms of Acupuncture Therapy for Myocardial Ischemia

Through animal experiments and clinical observation, in vitro and in vivo observation, from the monitoring of physiological function to pathological observation and from gene and protein level to tissue and organ level, we have revealed the protection mechanisms of acupuncture anesthesia on patients undergoing cardiopulmonary bypass, including antiapoptosis, antioxidation, antiadhesion, antistress, and immune regulation action.

14.3.1 Effect of Acupuncture on Hypothalamus–Pituitary–Adrenal Axis Response and Glucose Metabolism of Patients Undergoing Open Heart Surgery with Cardiopulmonary Bypass

The interactions among these organs constitute the hypothalamic–pituitary–adrenal (HPA) axis, a major part of the neuroendocrine system that controls reactions to stress and regulates many body processes, including digestion, the immune system, mood and emotions, sexuality, and energy storage and expenditure. Experimental studies have investigated many different types of stress and their effects on the HPA axis in many different circumstances. Glucocorticoids modulate the stress response at a molecular level by altering gene expression, transcription, and translation, among other pathways. The effect is the inhibition of the functions of inflammatory cells, predominantly mediated through inhibition of cytokines, such as IL-1, IL-6, and TNF-α. The central effectors of the stress response are the corticotropin-releasing hormone (CRH) and locus coeruleus-norepinephrine (LC-NE)/sympathetic systems. The CRH system activates the stress response and is subject to modulation by cytokines, hormones, and neurotransmitters. Glucocorticoids also modulate the growth, reproductive, and thyroid axes. Glucocorticoids (GCs) are essential for the maintenance of homeostasis and enable the organism to prepare for, respond to, and manage stress, either physical or emotional. Cortisol, the principal GC in humans, is synthesized in the adrenal cortex. It is released in the circulation in a pulsatile and circadian pattern. GC secretion is governed by hypothalamus and pituitary. The hypothalamus senses change in the external and internal environment that may disrupt the homeostatic balance of the organism (i.e., stressors) and respond by releasing CRH and arginine vasopressin (AVP) from parvocellular neurons projecting from the paraventricular nucleus to the median eminence. These neurohormones are released into the anterior pituitary where they act synergistically via specific receptors (CRH-R1 and V1B receptor, respectively) to trigger the release of the adrenocorticotropic hormone (ACTH) from the corticotrope cells into the systemic circulation. In turn, ACTH exerts its actions on the adrenal cortex via specific receptors, type 2 melanocortin receptors (MC2-R), to initiate the synthesis of cortisol, which is released immediately into the systemic circulation by diffusion. Hypothalamic CRH and AVP, pituitary ACTH, and adrenal GCs comprise the HPA axis. In this brief review, the HPA axis and the various factors that regulate its function are described Papadimitriou and Priftis (2009).

Acupuncture has a two-way role in regulation of stress and immune. It enhances body stress and anti-injury ability during operation. It also eliminates stress after operation in order to rehabilitate. Acupuncture-assisted anesthesia increases blood β-EP and blood sugar and reduces ACTH and cortisol levels during cardiopulmonary bypass surgery, but after bypass, β-EP, ACTH, cortisol, and blood glucose are significantly lower than that of nonacupuncture group (Yang et al. 2001a, b, c).

Cortisol in general anesthesia group before bypass increased significantly compared with that before anesthesia ($P<0.01$); after bypass, β-EP, ACTH, cortisol, and blood glucose increased further compared with that before anesthesia ($P<0.01$). Blood glucose in acupuncture anesthesia group before bypass was significantly higher than before anesthesia ($P<0.05$); after bypass, β-EP, ACTH, cortisol, and blood glucose were significantly higher compared with that before anesthesia ($P<0.01$). β-EP, ACTH, and cortisol in GAE group before bypass were significantly lower than before anesthesia ($P<0.05$); after bypass, no significant difference of β-EP and ACTH was found compared with that before anesthesia ($P>0.05$). Cortisol was significantly lower than before anesthesia ($P<0.05$), and blood glucose was significantly higher than before anesthesia ($P<0.01$).

Among the three groups, β-EP in acupuncture anesthesia group before bypass was significantly higher than in general anesthesia group ($P<0.01$) and GAE group ($p<0.01$). ACTH and cortisol in GAE group were significantly lower than in general anesthesia group ($P<0.05, P<0.01$) and acupuncture anesthesia group ($P<0.01$, $P<0.01$). Blood glucose in acupuncture anesthesia group was significantly higher than in general anesthesia group and GAE group ($P<0.01$). After bypass, β-EP, ACTH, cortisol, and blood glucose in GAE group were significantly lower than in general anesthesia group ($P<0.05$) and acupuncture anesthesia group ($P<0.05$). Intraoperative plasma β-EP, ACTH, serum cortisol, and blood sugar changes were listed in Table 14.5.

14.3.2 Effects of Acupuncture on HSP70 mRNA Content in Ischemic Myocardium in Patients Under Cardiopulmonary Bypass

The induction of heat shock proteins in the myocardium has been suggested as a possible intervention to allow for enhanced cardioprotection. Acupuncture can stimulate stress heat shock protein (HSP) expression and thereby reduce myocardial injury by regulating stress hormone, β-EP, ACTH, cortisol, and blood glucose (Wang et al. 2001a, b, c, d).

Table 14.5 Comparison of plasma β-EP, ACTH, serum cortisol, and blood sugar between three groups ($\overline{x} \pm s$)

Group	Cases		β-EP (ng/l)	ACTH (ng/l)	Cortisol (μg/l)	Blood sugar (mmd/l)
General anesthesia	10	Before anesthesia	35.03±7.20	16.21±3.52	96.8±12.7	4.49±0.53
		Before bypass	35.08±6.31	15.11±3.35	118.1±8.2*	4.38±0.58
		After Bypass	296.41±50.24*	124.59±21.76*	157.0±13.4*	9.09±2.01*
Acupuncture anesthesia	9	Before anesthesia	41.13±6.69	19.52±3.44	104.4±21.4	4.62±0.55
		Before bypass	43.43±7.48***	18.81±4.56	123.1±19.7	5.47±0.46***
		After Bypass	310.46±51.00*	117.61±12.66*	263.2±32.0**	9.87±2.63*
GAE	10	Before anesthesia	39.17±5.94	17.15±4.14	103.9±28.6	4.27±0.51
		Before bypass	33.33±5.42‡	11.55±1.29**,‡	58.0±16.1***,‡	4.04±0.38‡
		After Bypass	37.35±6.26***,‡	14.73±2.09***,‡	78.8±22.1***,‡	6.94±0.81*,**,†

GAE general anesthesia and electroacupuncture anesthesia group
Compared with that before anesthesia,*$p<0.05$; Compared with that of general anesthesia group under the same situation, **$\rho<0.05$, ***$\rho<0.01$; Compared with that of acupuncture anesthesia group under the same situation, †$\rho<0.05$, ‡$\rho<0.01$

Table 14.6 HSP70 mRNA changes during operation in two groups ($\overline{x} \pm s$)

Group	Before bypass	Stop bypass	30 min after bypass	1 h after bypass
Acupuncture anesthesia group	2.29±0.25	3.11±0.23	3.08±0.72	3.20±0.19*, **
Acupuncture combined with general anesthesia group	2.21±0.20	2.88±0.15	2.99±0.51	2.91±0.17
General anesthesia group	2.24±0.17	2.37±0.19	2.42±0.43	2.47±0.15

*Compared with that before bypass, $P<0.05$
**Comparison between group I and III, $P<0.05$

The discarded right atrial appendage during operation was used as specimen and placed in liquid nitrogen for preservation. Reagent RNA was used for extraction of total tissue RNA. After RT-PCR amplification by random primer, bromophenol blue staining and electrophoresis were performed. After transfer, 32P-labeled probe was used for hybridization, and autoradiography was done. A density scanner (Island Jin CS-920 type) was used for quantification of heat shock protein (HSP70) mRNA. HSP70 mRNA expression in acupuncture anesthesia group was higher than in control group, and the density values were listed in Table 14.6.

Acupuncture makes HSP70 mRNA expression increased, which is closely related to the inhibition of myocardial enzymes release. One hour after bypass, HSP70 mRNA in acupuncture anesthesia group increased by 39.7 % and CK-MB increased by 272.7 %. HSP70 mRNA in nonacupuncture group increased by 10.3 % and CK-MB increased by 696.6 %. The increase of HSP70 mRNA gene expression rate was related to CK-MB increase. As HSP70 mRNA increased, CK-MB increase declined markedly, showing that stress can activate heat shock protein expression, and upregulation of heat shock proteins has a clear link with the extent of myocardial injury.

14.3.3 *Effect of Acupuncture on Apoptosis of Ischemic Myocardium Cell of Rat Under Cardiopulmonary Bypass*

The positive correlation between AI obtained by TUNEL method and expression of Bak protein may suggest that apoptosis is activated mainly through mitochondrial activation pathway in ischemia–reperfusion injury. The results suggest that ischemia–reperfusion injury increases the AI in the right atrial tissue. If so, immunohistochemical expression of Bak protein could be used as a marker of myocardial ischemia-induced injury. Inflammatory mediators during CPB prolong the functional life span of neutrophils through modulation of apoptosis and potentiate the inflammatory response observed after coronary bypass operation.

Acupuncture can reduce apoptosis of ischemic myocardial cell, and HSP is one of the upstream regulatory pathways of apoptosis regulation. The number of TUNEL-positive cells in acupuncture anesthesia group was significantly less than in control

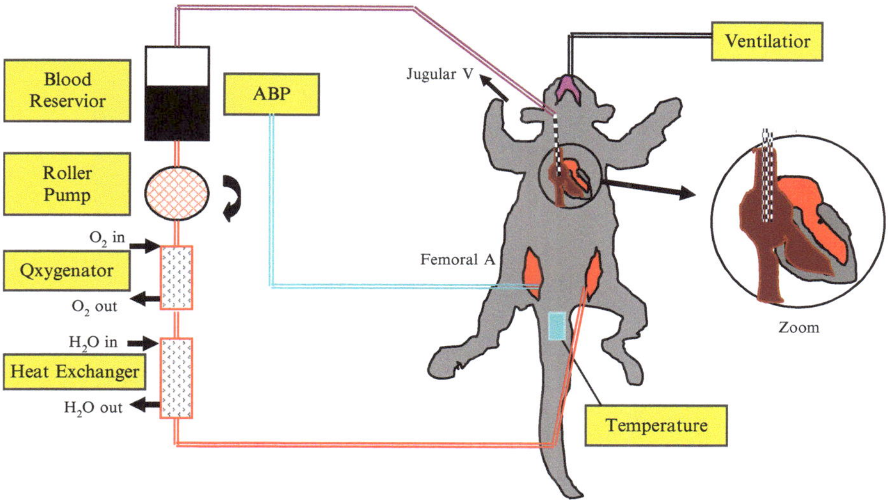

Fig. 14.3 Cardiopulmonary bypass of rat model

group. Apoptosis-related gene Bax mRNA and protein expression reduced significantly, but Bcl-2 mRNA and protein expression increased significantly. The above-mentioned changes had obvious relevance with increased expression of HSP70. After blocking the expression of HSP70, protective effect on apoptotic myocytes by electroacupuncture stimulation decreased significantly (see Figs. 14.3 and 14.4).

After myocardial ischemia–reperfusion injury, the number of myocardial apoptosis increased significantly compared with that of sham operation ($P<0.05$). The number of apoptotic cells in acupuncture anesthesia group reduced significantly compared with in myocardial ischemia group ($P<0.05$). After blocking HSP by KNK-437 (200 mg/kg), protective effect on apoptotic myocytes by electroacupuncture decreased significantly (Fig. 14.5).

Bcl-2 and *bax* mRNA expression in ischemic myocardial tissue were significantly higher than in sham operation group ($P<0.05$). Compared with the ischemia group, *bax* mRNA expression in acupuncture anesthesia group was significantly lower ($P<0.05$), and *bcl-2* mRNA expression was significantly higher ($P<0.05$). After the applications of Hsp inhibitor (KNK-437, 200 mg/kg), the inhibitory effect on *bax* mRNA and enhanced effect on *Bcl-2* mRNA by acupuncture decreased significantly (Fig. 14.6).

Bcl-2 and *bax* protein expression by cardiac tissue in ischemic group were significantly higher than in sham-operated group ($P<0.05$). Compared with ischemic group, *bax* protein expression in acupuncture anesthesia group was significantly lower ($P<0.05$), and *bcl-2* protein expression increased significantly ($P<0.05$). After the application of Hsp inhibitor (KNK-437, 200 mg/kg), the regulation of *bax* and *bcl-2* protein expression by acupuncture reduced significantly (Fig. 14.7).

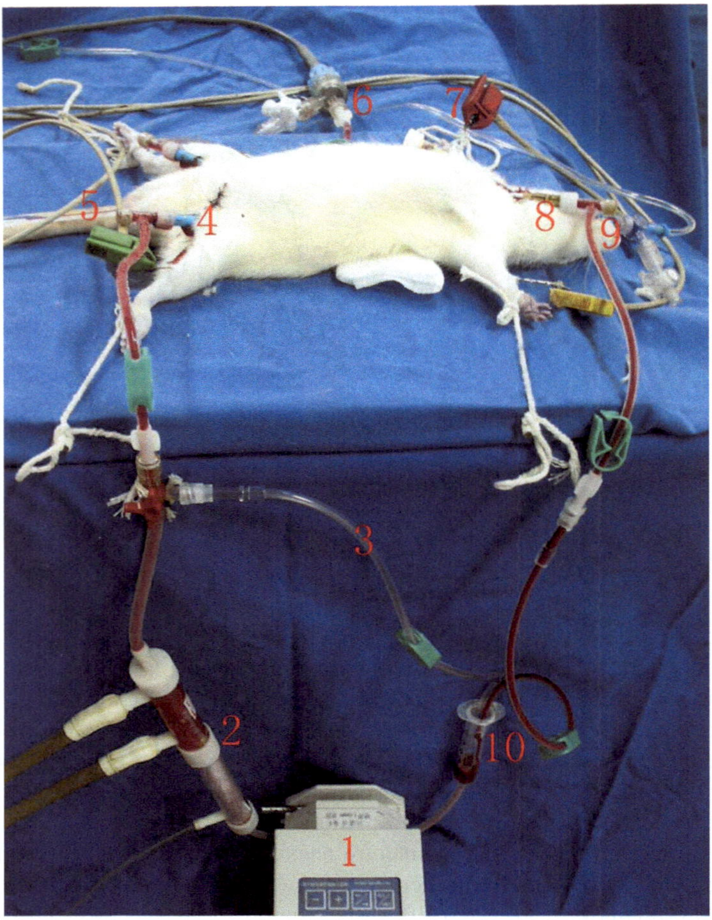

Fig. 14.4 Cardiopulmonary bypass of rat (1) blood pump, (2) oxygenator and heat exchanger, (3) short circuit bypass tube, (4) left femoral artery intubation, (5) measurement of rectal temperature, (6) artery blood pressure transducer, (7) electrode of ECG, (8) cervical vein intubations, (9) tracheal intubation, and (10) blood reservoir

14.3.4 *Effects of Electroacupuncture Stimulation on Leukocyte Cell Adhesion in Rat with Myocardial Ischemia During Cardiopulmonary Bypass*

During the inflammatory response, triggered by cardiopulmonary bypass, interaction between activated leukocytes, platelets, and endothelial cells is mediated through the expression of three main groups of adhesion molecules: the selectins, the integrins, and the immunoglobulin superfamily. The selectins, which mediate the initial rolling of the leukocyte on the endothelium, are divided in three

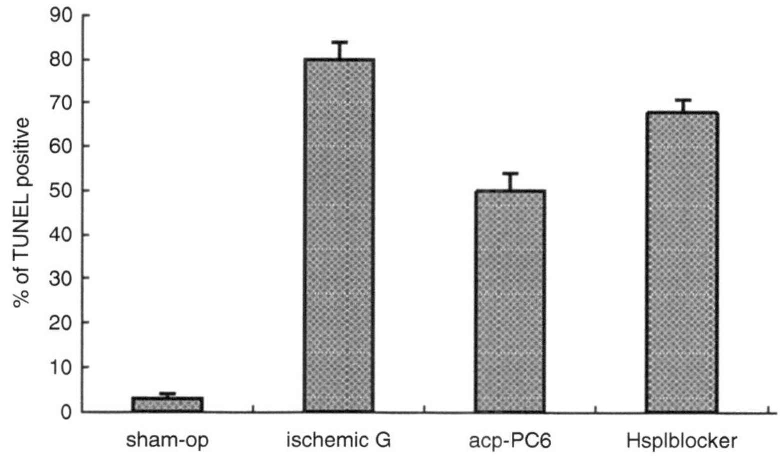

Fig. 14.5 Comparison of myocardial apoptosis between groups (TUNEL) *star* compared with sham operation group, $P<0.05$; *filled triangle* compared with myocardial ischemia group, $P<0.05$

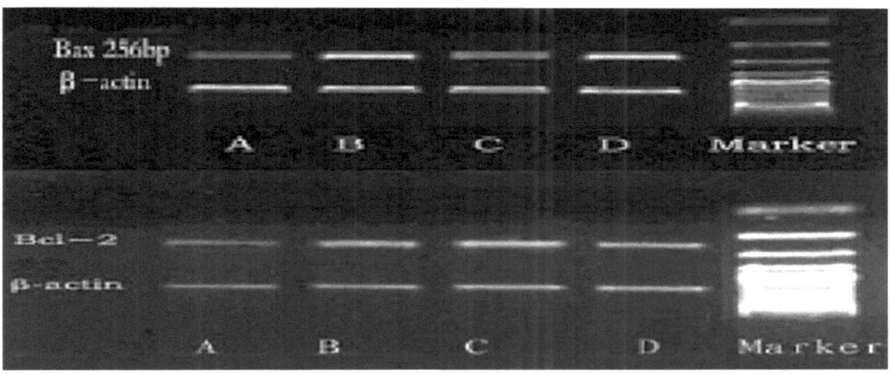

Fig. 14.6 Electropherogram of PCR products of Bcl-2 and bax in myocardial tissue. A = sham operation group, B = myocardial ischemia group, C = acupuncture group, and D = Hsp inhibitor group

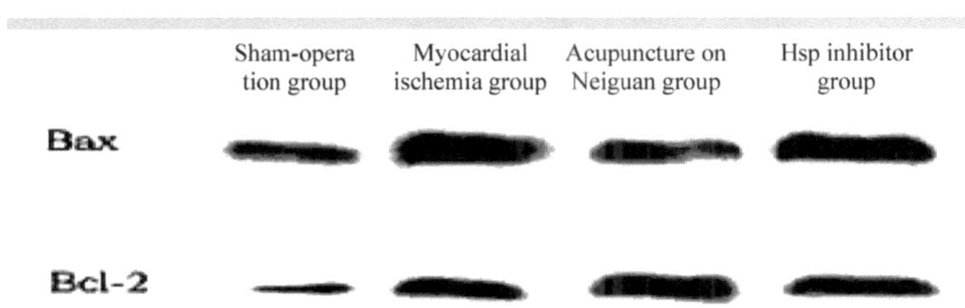

Fig. 14.7 Western blot detection of *bax* and *bcl-2* protein expression by myocardial tissues in each group

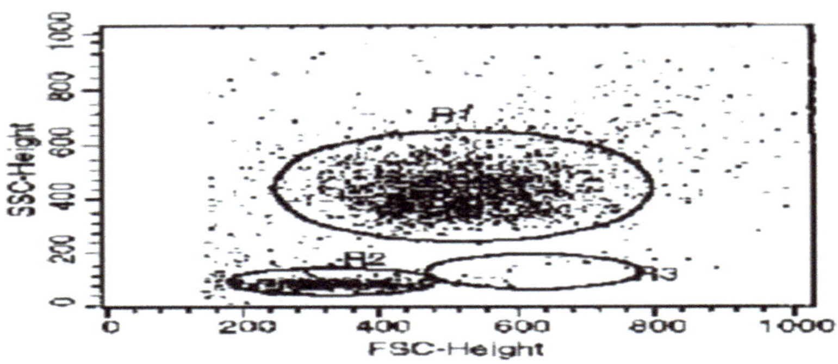

Fig. 14.8 Detection of adhesion molecule expression of neutrophil by flow cytometry. R1 for neutrophils, R2 for lymphocytes, and R3 for monocytes

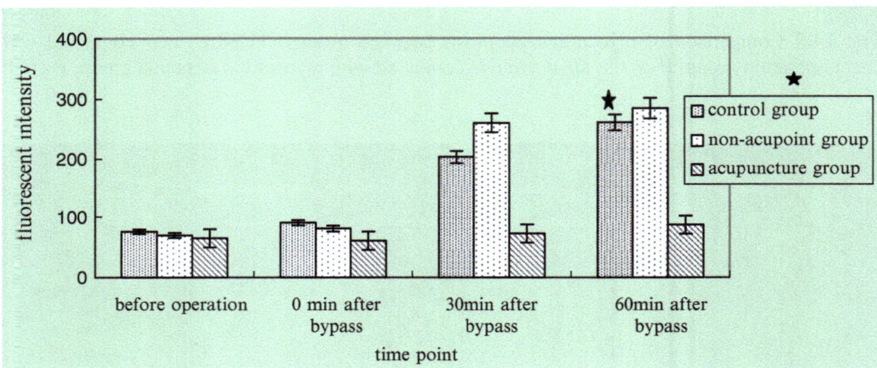

Fig. 14.9 Surface expression of CD11b by neutrophil cell in rat under cardiopulmonary bypass. Compared with the control group and nonacupoint group *star* $P<0.05$

subgroups: L-selectin is expressed on all three leukocyte types, P-selectin is expressed on platelets and endothelial cells, and E-selectin is only expressed on endothelial cells. Integrins can be found on most cell types, consisting of an alpha and a beta subunit and mediate firm adhesion of the leukocyte and migration into the tissue. The events of ischemia–reperfusion injury trigger a systemic inflammatory response and can lead to cellular injury and organ failure. Such effects are noted in the postoperative recovery, especially with the use of cardiopulmonary bypass. Nowadays, it is known that leukocytes play an important role in this process (Francischetti et al. 2010).

Acupuncture-caused anti-inflammatory cell adhesion is one of the important downstream pathways of myocardial protection regulation. In the first established rat model of cardiopulmonary bypass at room temperature, we found that at different time points during operation and after operation, PMN surface expression of adhesion molecule C11b in electroacupuncture group decreased significantly compared with in control group and nonacupuncture group ($P<0.05$). (1 h after bypass, CD11b fluorescence intensity in acupuncture anesthesia group was 83 ± 12 vs. nonacupuncture group 311 ± 22) (Figs. 14.8 and 14.9).

14.3.5 *Effects of Electroacupuncture Stimulation on Mast Cell in Myocardial Ischemia–Reperfusion Rats*

Eighteen male SD rats were randomly assigned to sham group, model (IR) group, and EA group ($n=6$/group). MI/R model was established by occlusion of the descending anterior branch of the coronary artery. Blood samples were taken from the femoral vein before MI (T0), EA for 30 min (T1), 30 min after MI (T2), 30 min after MI/R (T3), and 120 min after MI/R (T4) for assaying serum tumor necrosis factor (TNF)-alpha and histamine contents by using ELISA. Serum lactate dehydrogenase (LDH) and creatine kinase isoenzyme (CK-MB) levels were measured at T0, T3, and T4 by using an automatic biochemistry analyzer. The infarct size was detected by Evans blue and tetrazolium chloride (TTC) staining. Myocardial TNF-alpha and histamine contents were detected by ELISA. The percentage of mast cell degranulation was determined by toluidine blue staining. The result showed that following MI/R, serum LDH and CK-MB levels at phase T3 and T4, serum TNF-alpha and histamine contents at phase T2 and T3, and myocardial mast cell degranulation rate increased significantly, and myocardial TNF-alpha and histamine contents decreased in model group as comparison with pre-MI/R ($P<0.05$). Compared with IR model group, serum LDH and CK-MB levels at phase T3 and T4, myocardial TNF-alpha, and histamine contents all decreased significantly ($P<0.05$), but serum TNF-alpha infarct size was remarkably smaller in EA group than that in IR model group ($P<0.05$). "Neiguan" (PC-6)-EA preconditioning has a cardioprotective effect on the ischemia–reperfusion myocardium by promoting mast cell degranulation (Zhang et al. 2010).

14.3.6 *Acupuncture Pretreatment Protects Heart from Injury in Rats with Myocardial Ischemia and Reperfusion via Inhibition of the Beta(1)-Adrenoceptor Signaling Pathway*

Cardioprotective effect was produced by pretreatment with acupuncture at bilateral Neiguan acupoints (PC-6), and the effect of EA was diminished by propranolol, a nonspecific antagonist of beta-adrenoceptors (beta-ARs) which are the most powerful cardiac receptors, indicating an involvement of beta-ARs. Myocardial ischemia was achieved by ligating the left anterior descending coronary artery and reperfusion by releasing the ligation. Adult rats were divided into three groups, namely, a normal control (NC) group, a group subject to ischemia and reperfusion (IR) only, and a group given electroacupuncture (EA) before IR. For EA, bilateral Neiguan points (PC-6) of the rats were stimulated for 30 min once a day for 3 consecutive days. The ST segment of ECG, the ratio of infarct size over risk zone, and the contents of beta(1)-adrenoceptor [beta(1)-AR], Gs alpha protein, and cAMP in ischemic myocardium were compared among the three groups. IR increased the elevation of ECG ST segment, myocardial infarct size, contents of beta(1)-AR, Gs alpha protein, and cAMP. These effects were attenuated by EA pretreatment at bilateral Neiguan acupoints. The present results indicate that EA produces cardioprotective effect against IR which may be mediated via the beta(1)-AR–Gs-protein–cAMP pathway (Gao et al. 2007).

14.3.7 *The Protective Effect of Electroacupuncture and Ischemic Preconditioning on Ischemia–Reperfusion*

Studies on preconditioning have only recently demonstrated a reduction in infarct size by remote ischemia applied during transport of heart attack victims to the hospital before percutaneous coronary interventions (PCIs). The "conditioning" paradigm has been extended to include remote preconditioning and remote postconditioning. However, the biology of preconditioning is virtually unknown. Postconditioning has enjoyed enthusiastic attention from scientists that have done much to demonstrate that the model of triggers, mediators, and effectors used in preconditioning may also apply to postconditioning, with the addition and important contribution of physiological mechanisms resulting in cardioprotection, including gradual normalization of tissue pH, reduction in generation of reactive oxygen species, and avoidance of hypercontracture (Vinten-Johansen and Shi 2011).

Domestic pigs with 15–20 kg are supplied by the Agriculture Institute of Shanghai University. Pigs fasted for 12 h were injected thiopental (5 mg/kg) as abdominal anaesthesia. With intubation and mechanical ventilation, pigs were maintained with isoflurane and intermittent Norcuron. We controlled $PetCO_2$ and blood pH value within physiologic ranges through adjusting tidal volume and respiratory frequency on the basis of blood gas analysis. Eighteen pigs were allocated randomly into three groups: group I (control group, $n=6$), clamped the left coronary artery and loosened it 20 min later; group II (preconditioning with ischemia group, $n=6$), clamped the artery 5 min at first, loosened it for 10 min, and then clamped it twice for 20 min; group III (electroacupuncture and preconditioning with ischemia group, $n=6$), acupunctured the acupoints of inner pass, branching crevice and cloud gate of two sides, connected with G6805 acupuncture anesthetic apparatus, set pulse frequency at 3–4 Hz and the induction of acupuncture anesthesia for 20–30 min, with others same as the group II.

Right femoral artery was separated and inserted with a 22-gauge cannula for sampling. Mean artery pressure (MAP) was measured through a pressure transducer. A pulmonary artery catheter was inserted through left femoral venous to monitor cardiac output. Right internal carotid venous was used for transfusion. After median sternotomy, we dissociated coronary artery at the junction of left coronary artery and anterior interventricular branch. Ultrasound Doppler optode, connected with volumeter, was placed immediately under the coronary artery. When the artery was clamped by an unscathing clip, the volumeter showed no flow anymore, and local myocardium turned to dim red with ST elevated remarkably in ECG. Then the acute myocardial ischemia (AMI) pig model was finished.

Metabolic indexes: Samples from femoral artery were taken to measure the content of malondialdehyde (MDA) in plasma and to detect the activities of superoxide dismutase (SOD) by chemical colorimetry before clamped (as control values) and 20 and 60 min after loosened. Blood samples near right atrium were taken from right internal carotid venous catheter to detect CPK and CK-MB by an automatic analyzer. Coronary artery flow (CAF) was measured by ultrasound Doppler at 10 min

Table 14.7 Biochemical index changes of ischemic myocardium during reperfusion ($x \pm s$)

	Group	n	Control value	Reperfusion 20 min	Reperfusion 60 min
CPK (U/L)	I	6	653.5±136.7	1880.2±205.5**	1977.8±258.9**
	II	6	667.8±133.4	1017.5±341.3*	1032.5±238.8**
	III	6	685.6±119.7	991.6±164.3*, ***	988.9±287.7***
CK-MB (U/L)	I	6	568.5±198.7	1355.2±247.6**	1458.5±120.8**
	II	6	532.5±98.3	1057.2±238.9*	1124.3±104.5***
	III	6	495.8±101.2	799.6±111.4*, ***	898.9±111.9*, ***
MDA (nmol/g)	I	6	1.03±0.2	5.5±0.58	5.44±0.6
	II	6	1.28±0.7	2.18±0.1	3.15±0.8
	III	6	1.18±0.5	2.21±0.6	2.83±0.7***
SOD (μU/g)	I	6	60.5±12.1	51.2±10.2	38.1±10.5**
	II	6	55.4±8.9	65.4±9.6	67.2±15.6***
	III	6	58.2±9.4	66.5±10.4	69.8±17.8***

*vs. control $p<0.05$
**$p<0.01$
***vs. group I $p<0.05$

before ischemia and 20 and 60 min after reperfusion, respectively. Detection of heat shock protein of myocardial cells: Samplings from left auricle were reserved in liquefacient nitrogen at 10 min before and 20 min after ischemia and 20 and 60 min after reperfusion, respectively. After a series of homogenization, isolation, washing, resolution, and deposition, total RNA was extracted from myocardial tissues using Reagent Kit. The concentration of mRNA was measured through RT-PCR after adding random primer and probe with 0.8 % gel electrophoresis after bromophenol blue dyeing by an ultraviolet spectrophotometer. The molecular weight was marked by 123 mark. Quantitative analysis of HSP70 mRNA was operated on a spot hybridization membrane with color reaction by a density scanner (Table 14.7).

Severe decrease of CAF results in the storage of lactic acid and glycogen exhaustion. The contractive function should be damaged or lost totally because of the resolution or necrosis of myocardium if ischemia could not be corrected timely. There are many changes occurred in ischemic myocardium resulting from the activated endogenous protective mechanism, including the increase of Calcium concentration, membrane injury, creation of free radicals, decrease of ATP, oxygen exhaustion, etc. All of these lead to the metabolic or oxygen-inadequate stress. Some study indicated that injury from reperfusion period might be more severe than ischemic period. The mechanism of myocardial ischemia–reperfusion injury is complex. The eruption of oxygen free radicals is the most important factor. Oxygen free radicals act on extracellular matrix and make collagen and hyaluronic acid collapse, injure cell membrane directly through peroxidation of polyunsaturated fatty acid in phospholipid structure, and cause sarcoplasmic reticulum and mitochondrion rupture. The concentration of MDA is a symbol of severity of lipid peroxidation. The creation of oxygen free radicals mainly occurred after reperfusion. Our study Han Lin et al. (2002, 2003), indicated that MDA content increased and SOD

Table 14.8 Changes of HSP mRNA expression in ischemic myocardium during reperfusion ($x \pm s$)

		HSP70 mRNA(optical intensity)			
Group	n	Control	Ischemia for 20 min	20 min after reperfusion	60 min after reperfusion
I	6	2.32±0.45	2.11±0.33	2.08±0.52	2.69±0.19**
II	6	2.20±0.24	2.58±0.35	2.79±0.41	3.41±0.17*
III	6	2.25±0.27	2.49±0.29	3.01±0.33**	3.67±0.15*

*vs. control $p<0.05$
**Group I vs. group III $p<0.05$

decreased significantly in control group at the time of 20 and 60 min after reperfusion, which demonstrated that the balance system between creation and elimination of oxygen free radicals was damaged, which led to the obvious increase of MDA content directly. The result that the MDA content increase in electroacupuncture and preconditioning with ischemia group was less than that in control group, and SOD increased slightly, suggested that the preconditioning with ischemia enhanced the oxidative phosphorylation in myocardium and improved the oxygenic utilization and creation efficiency of ATP. All these availed the integration of cell membrane and reduction of mitochondrion injury and protected all kinds of enzyme system in it. It is concluded that the electroacupuncture enhances the improvement of oxygenic utilization due to preconditioning with ischemia through activating meridians and collaterals, promoting Qi to activate blood and coordinating deficiency and excess. These acupoints were elected through a lot of animals and clinic studies and had been applied in nearly 1,000 of heart operations successfully (Wang et al. 2003).

The protective effect of preconditioning with ischemia on ischemic heart involved the creation and releasing of many endogenous protective materials. When the system was stimulated by stress factors, the tissue created stress proton to protect them. The heat shock proton belongs to the first-degree stress proton. The phenomenon that acupuncture and precondition can induce the increase of endogenous protective material maybe is the substantial foundation of myocardial protective effect. Through reducing myocardial ischemia–reperfusion injury, this protective mechanism improves the mechanical function of ischemic heart, benefits the reservation of ATP, and protects the function of mitochondrion. Our study Xiangrui Wang et al. (2003a, b, 2005), Zhenhong Wang and Xiangrui Wang (2004a, b, 2005), Zhongping Lu et al. (2003) indicates that the expression of HSP70 mRNA in acupuncture and preconditioning with ischemic group is significantly higher than that of control group (Table 14.8). HSP70 protect ischemic myocardial cells by means of facilitating renaturation/hydrolysis of denaturated protein, maintaining the structure of protein, improving the replacement between new synthetic protein and aged 1, preserving the fourth structure of normal or partly denaturated protein, and helping big molecular weight protein or peptide to transport transmembrane in mitochondrion or endoplasmic reticulum. The protective effect functioned mainly at the ischemic period and early period after ischemia. The synthetic at later period cannot prevent the death of myocardial cells. The results that the recovery of CAF in

Table 14.9 CAF changes during reperfusion ($x \pm s$)

	Group	n	Control	Reperfusion 20 min	Reperfusion 60 min
CAF(ml/min)	I	6	159.0±80.8	46.6±26.0**	103.0±29.2*
	II	6	148.2±48.9	93.6±31.9**	129.8±40.7***
	III	6	146.4±46.5	110.5±31.3*, ***	134.1±24.5***

*vs. control $p<0.05$
**$p<0.01$
***vs. group I $p<0.05$

electroacupuncture and preconditioning with ischemia group was superior to that of only under precondition group demonstrated acupuncture can enforce the effect that preconditioning with ischemia protected ischemic myocardium through the creation and releasing of endogenous protective material (Table 14.9).

14.4 Concluding Remarks

Acupuncture is an alternative medicine that treats patients by insertion and manipulation of needles in the body. Its proponents variously claim that it relieves pain, treats infertility, treats disease, prevents disease, or promotes general health. Progresses of researches on the protective effect of acupuncture turn out with resisting acute myocardial ischemia–reperfusion injury in patients and animals from (1) resisting injury of oxygen-derived free radicals, (2) reducing intracellular Ca^{2+} overload, (3) relieving dysfunction of vascular endothelial cells of the myocardium, (4) regulating immune functional activity and reducing inflammatory injury, (5) reducing cardiocyte apoptosis, (6) upregulating the expression of heat shock protein (HSP70) in the myocardium, and (7) regulating activities of multiple cardiovascular protection factors and intracellular signaling pathways. The above-mentioned results are mainly obtained from experimental animal studies. For this reason, the authors hold that the coming researches on acupuncture therapy for cardiovascular disorders should pay more attention to its clinical application, for example, controlled blood pressure decline induced lower perfusion of the coronary artery, cardiac surgery (percutaneous transluminal coronary angioplasty, intracardiac operation under direct vision), etc. (Zhou and Shao 2010).

One of the unmet clinical needs in heart surgery is the prevention of myocardial stunning and necrosis that occurs as a result of ischemia–reperfusion. Myocardial stunning, a frequent consequence after heart surgery, is characterized by a requirement for postoperative inotropic support despite a technically satisfactory heart operation. While we now have a much better understanding of the molecular mechanisms underlying myocardial stunning and MI, we still have no effective way to prevent these complications, nor a consistently effective means to engage the well-studied endogenous mechanisms of cardioprotection. Electroacupuncture pretreatment may alleviate cardiac ischemia–reperfusion injury in adult patients undergoing heart valve replacements. This simple and convenient treatment has the potential to be used in the clinic for reducing myocardial injury in patients with heart valve replacement surgery (Yang et al. 2010). This study showed that the application of Acu-TENS to PC-6 augmented BP and was associated with a faster return to preoperative resting HR and RPP in patients after recent open heart surgery. Acu-TENS also reduced the postoperative symptoms of nausea and vomiting in this patient group. Acu-TENS is a noninvasive modality that does not require the presence of an acupuncturist and is easy to apply. This study suggests that Acu-TENS facilitates early recovery of hemodynamic variables and may reduce myocardial work, thereby providing a useful adjunctive therapy when optimizing postoperative cardiac rehabilitation (Maggie et al. 2011).

These reflexes consist of an afferent arc of homoeostatic autonomic reflexes activated by oxygen, glucose, and other metabolites, finally resulting in activation of the efferent motor neural arc that transmits the signal to modulate immune responses. Direct stimulation of the vagus nerve inhibits cytokine production by innate immune cells in different organs—for example, spleen, liver, gastrointestinal tract, and the heart (Tracey 2009). A feature of cardiac surgery is immune activation, with proinflammatory cytokines overexpressed both in the systemic circulation and locally in the failing myocardium. TNF-α has several properties that lead to metaboreceptor activation (Jankowska et al. 2006) and are particularly detrimental, such as negatively inotropic effects, the promotion of left ventricular remodeling, and the induction of dilated cardiomyopathy. Furthermore, TNF-α can cause skeletal muscle wasting and apoptosis and, therefore, may be important in the development of cardiac cachexia and exercise limitation.

Our study in animal experiments and clinical studies demonstrated that acupuncture anesthesia patients having undergone cardiac surgery had less postoperative cardiopulmonary complications and quicker recovery and significant reduced usage of perioperative anesthetics, vasoactive drugs, and antibiotics, which reduced medical costs effectively and shortened ICU and hospitalization time. Cai et al. (2010) observed the effect of electroacupuncture (EA) of "Neiguan" (PC-6) and "Taichong" (L-R3) on mechanical activities of the heart in AMI rabbits. Fifty chinchilla rabbits were randomized into normal control, model, EA-PC-6, EA-LR-3, and nonacupoint groups ($n = 8$/group). AMI model was duplicated by intravenous injection (femoral vein) of posterior pituitrin (3 U/kg). The maximal ascending velocity (+dp/dt max), maximal descending velocity (–dp/dt max), and left ventricular systolic pressure (LVSP) of the heart were recorded by using Biopac Bio-signal system. EA (1 mA, 2 Hz) was applied to PC-6, LR-3, and nonacupoint for 10 min. The result showed that compared with normal control group, + dp/dt max, – dp/dt max, and LVSP of the heart in model group reduced significantly ($P < 0.01$), while in comparison with model group, left ventricular + dp/dt max and – dp/dt max at 10 min after EA in EA-PC-6 and EA-LR-3 groups increased markedly ($P < 0.01$), and LVSP of EA-PC-6 group increased remarkably ($P < 0.05$). No significant differences were found between EA-nonacupoint group and model group in the above-mentioned three

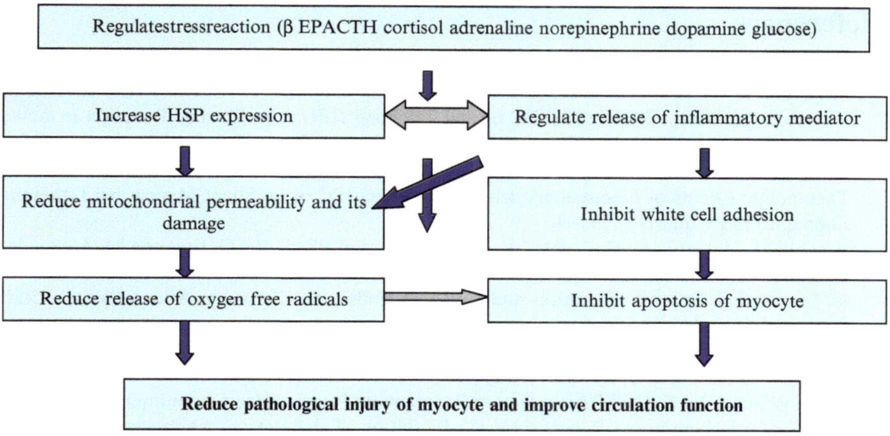

Fig. 14.10 The mechanism of acupuncture on myocardial protection

indexes (P>0.05). Both EA of PC-6 and LR-3 points can improve AMI-induced reduction of the left ventricular myocardial mechanical activities in the rabbit, and the therapeutic effect of EA-PC-6 is superior to that of EA-LR-3. Acupuncture at Neiguan (PC-6) has a good therapeutic effect on SMI. It can decrease the heart rate and blood pressure, reduce the afterload in left ventricle, and is superior to that of compound Danshen dripping pill (Diao et al. 2011).

The mechanism of acupuncture on myocardial protection may be related to the following factors: (1) regulate heat shock protein-related gene and protein expression and reduce cardiac myocytes apoptosis; (2) reduce the release of inflammatory cytokines, reduce adhesion of inflammatory cells in the ischemic myocardial region and the release of oxygen free radicals, and reduce mitochondrial damage; (3) antiapoptosis, antioxidation, antiadhesion, antistress, and immune regulation action (Fig. 14.10).

Acknowledgments This work was supported by the following grants: (1) Chinese Ministry of Science and Technology (973) 2007CB512504: Basic research of specificity of acupoints based on clinical acupuncture anesthesia; (2) National Natural Science Foundation of China (39670898): Protection of myocardium by electroacupuncture stimulation during open heart surgery; (3) Ministry of Health (Research Grant No.: 96-1-342): In vivo protection of myocardium by acupuncture during open heart surgery; (4) National Natural Science Foundation 2000-2002 (39970921): Protective effect of acupuncture-assisted hypothermia anesthesia on myocardial ischemia–reperfusion; (5) Ministry of Health (98-2-314): Myocardial protective effect of acupuncture-assisted hypothermia anesthesia on patients undergoing cardiac surgery; (6) Science and Technology Commission of Shanghai Municipality (05DZ19746): Clinical study of acupuncture; and (7) Science and Technology Commission of Shanghai Municipality (Grant No.: 07DZ19722-4): Effect of moderate activation of mast cells mediated ROS-HIF pathway in myocardial protection with acupuncture anesthesia and its mechanism.

References

Cai RL, Hu L, Wang KM, Wu ZJ, Chen YN, Long DH, He L, Liu J.(2010) Effects of Electroacupuncture of “Neiguan (PC-6) and Taichong (LR-3) on Cardiac Function in Acute Myocardial Ischemia Rabbits. Zhen Ci Yan Jiu. 30 (2):104–7.

Diao LH, Yang ZB, Zhou GX, Chen Y, Fan LY, Zhang YY, Liu H, Liu ST. (2011,)Observation on Therapeutic Effects of Electroacupuncture at Neiguan (PC 6) on Silent Myocardial Ischemia. Zhongguo Zhen Jiu 31(7):591–4.

Domanski MJ, Mahaffey K, Hasselblad V, Brener SJ, Smith PK, Hillis G, Engoren M, Alexander JH, Levy JH, Chaitman BR, Broderick S, Mack MJ, Pieper KS, Farkouh ME. (2011)Association of Myocardial Enzyme Elevation and Survival Following Coronary Artery Bypass Graft Surgery. JAMA 305(6):585–91.

Francischetti I, Moreno JB, Scholz M, Yoshida WB(2010) Leukocytes and the Inflammatory Response in Ischemia-reperfusion Injury. Rev Bras Cir Cardiovasc 25(4):575–84.

Gao J, Fu W, Jin Z, Yu X. (2007)Acupuncture Pretreatment Protects Heart from Injury in Rats with Myocardial Ischemia and Reperfusion via Inhibition of the Beta(1)-adrenoceptor Signaling Pathway. Life Sci 27;80(16):1484–9.

Han Lin, Xiangrui Wang, Zhenhong Wang.(2002) Protective Effects of Acupuncture Anesthesia Combined with Intravenous Injection of Salviae Miltiorrhizae Radix on Myocardial Ischemia-reperfusion Injury. Acupuncture Research 27 (3): 186–191.

Han Lin, Xiangrui Wang, Zhenhong Wang(2003). Protective Effects of Electro-Acupuncture Anesthesia Combined with Intravenous Injection of Salviae Miltiorrhizae Radix on Myocardial Ischemia-reperfusion Injury in Rats. Chinese Journal of Integrated Traditional and Western Medicine 23 (Supplement): 7–10.

Jankowska EA, Ponikowski P, Piepoli MF.(2006) Autonomic Imbalance and Immune Activation in Chronic Heart Failure—pathophysiological Links. Cardiovasc Res 70:434–45

Lifang Yang, Jian Yang, et al. (2010)Cardioprotective Effects of Electroacupuncture Pretreatment on Patients Undergoing Heart Valve Replacement Surgery: A Randomized Controlled Trial. The Annals of Thoracic Surgery 89(3):781–786.

Maggie C. S. Ng, Alice Y. M. Jones, and L. C. Cheng.(2011) The Role of Acu-TENS in Hemodynamic Recovery after Open-Heart Surgery. Evid Based Complement Alternat Med 21: 301–4.

Papadimitriou A, Priftis KN.(2009) Regulation of the Hypothalamic-pituitary-adrenal Axis. Neuroimmunomodulation 16(5):265–71.

Qingguo Yang, Yannan Hang, Dajin Sun.(2001) Effects of Electroacupuncture on Plasma Catecholamine and Angiotensin II in Open Heart Surgical Patients Undergoing Cardiopulmonary Bypass. World Journal of Acupuncture-Moxibustion 11 (1): 5–8.

Qingguo Yang, Yannan Hang, Dajin Sun, et al. (2001) Effect of Acupuncture Anesthesia Combined Drugs on Hypothalamus-pituitary-adrenal Axis Response and Glucose Metabolism of Patients Undergoing Open Heart Surgery with Cardiopulmonary Bypass. Chinese Journal of Integrative Medicine 21 (10): 729–731.

Qingguo Yang, Ximing Chen, Yannan Hang, et al. (2001)Changes of Plasma Catecholamines in Patients undergoing Posterior Cranial Fossa Surgeries Under Acupuncture Anesthesia Combined with Anesthetics. Chinese Acupuncture and Moxibustion 21 (12): 743–745.

Tracey KJ.(2009) Reflex Control of Immunity. Nat Rev Immunol 9:418–28.

Vinten-Johansen J, Shi W.(2011) Perconditioning and Postconditioning: Current Knowledge, Knowledge Gaps, Barriers to Adoption, and Future Directions. J Cardiovasc Pharmacol Ther 16(3–4):260–6.

Xiangrui Wang, Qinyan Yu, Dajin Sun.(2000) Regulatory Role of Electro-acupuncture Stimulation on Myocardial Ischemia in Patients Undergoing Cardiac Surgery. Chinese Acupuncture and Moxibustion, 20 (5): 261–263.

Xiangrui Wang, Yannan Hang, Dajin Sun.(1999) Observation of Functional Regulation of Acupuncture on Patients Undergoing Heart Surgery. Shanghai Journal of Acupuncture and Moxibustion 18 (5): 6–7.

Xiangrui Wang, Yannan Hang, Dajin Sun.(1999) Hemodynamic Changes in patients Undergoing Heart Surgery Using Acupuncture Anesthesia. Chinese Acupuncture and Moxibustion 19 (10): 628–630.

Xiangrui Wang, Zhongping Lu, Canran Xu. (2001) Protection by Acupuncture on Heart Surgery Patients. Acupuncture Research 26 (3): 173–174.

Xiangrui Wang, Tengfei Zhang, Shuliang Ma.(2001) Effects of Electroacupuncture on Myocardial Cellular HSP70 mRNA Gene Expression in Patients Undergoing Cardiac Surgery. Chinese Acupuncture and Moxibustion 20 (2): 99–101.

Xiangrui Wang, Qinyan Yu, Jun Yan, et al. (2003)Protective Effect of Electroacupuncture and Ischemic Preconditioning on the Circulatory Function in Pigs with Ischemia/Reperfusion Myocardial Injury CJIM, 9 (2):124–127.

Xiangrui.Wang, J. Xiao, D. Sun.(2003) Myocardial Protective Effects of Electroacupuncture and Hypothermia on Porcine Heart after Ischemia/Reperfusion. Acupuncture & Electro-Therapeutic RES 28 (3–4): 193–200.

Xiangrui.Wang, H Lin, Z Wang.(2003) Protective Effects of Electroacupuncture and Salviae Tiorrhizae on Myocardial Ischemia/Reperfusion in Rabbits. Acupuncture&Electro-Therapeutic RES 28 (3–4):175–82.

Xiangrui Wang, Zhenhong Wang, Han Lin, et al.(2005) Effect of Electroacupuncture on Rabbit Myocardial Cell after Ischemia and Reperfusion Chinese Remedies & Clinics 5 (5) :328–331.

Xiangrui Wang, Changzhi Chen, Jia Zhou.(2001) Myocardial Protective Effect of electroacupuncture and Ischemic Preconditioning on Functions of Ischemia-reperfusion Pig Heart. Chinese Acupuncture and Moxibustion 21 (12): 739–741.

Xiangrui Wang, Changzhi Chen, Qinyan Yu. (2001) Myocardial Protective Effect of Electroacupuncture and Hypothermia on Functions of Ischemia-reperfusion Pig Heart. Chinese Journal of Integrated Traditional and Western Medicine in Intensive and Critical Care 8 (2): 73–75.

Yavuz B, Duman U, Abali G, et al.(2006) Coronary Artery Bypass Grafting is Associated with a Significant Worsening of QT Dynamicity and Heart Rate Variability. Cardiology 106(1):51–55.

Zhang JL, Chen J, Wang XR, Li WW, Wang BL, Zhou J.(2010) Protective Effect of "Neiguan" (PC-6)-electroacupuncture Preconditioning on the Myocardium in Myocardial Ischemia/ Reperfusion Rats Zhen Ci Yan Jiu, 35(3):182–7.

Zhenhong Wang, Xiangrui Wang.(2005) Protection of Electroacupuncture and Compound Salviae in Coordination on Ischemia-reperfusion Myocardium. Chinese Journal of Clinical Rehabilitation 9(27):220–222.

Zhenhong Wang, Xiangrui Wang.(2004) Protective Effects of Acupuncture Anesthesia Combined with Intravenous Injection of Salviae Miltiorrhizae Radix on Myocardial Ischemia-reperfusion Injury in the Rabbit. Chinese Journal of Integrated Traditional and Western Medicine in Intensive and Critical Care 11 (3): 137–140.

Zhenhong Wang, Xiangrui Wang.(2004) Protective Effects of Acupuncture Anesthesia Combined with Nicardipine on Myocardial Ischemia-reperfusion on the Ischemia/Reperfusion. Chinese Journal of Critical Care Medicine 24 (7): 496–498.

Zhongping Lu, Xiangrui Wang, Dajin Sun.(2003) Involvement of IL-8 in the Protective Effect of Electroacupuncture on Myocardial Ischemia-reperfusion Injury. Acupuncture Research 28 (3): 210–213.

Zhou CL, Shao XM.(2010) Progress of researches on the protective effect of acupuncture in resisting myocardial ischemia-reperfusion injury. Zhen Ci Yan Jiu 35(2):156–60.

Chapter 15
Acupuncture's Role in Cardiovascular Homeostasis

Peng Li, Stephanie C. Tjen-A-Looi, and John C. Longhurst

Abstract This chapter summarizes data from more than 50 years of experimental and clinical studies describing the clinical effects and mechanisms of manual and electroacupuncture (EA) in cardiovascular disease, including hypertension and hypotension. These studies are based on concepts of anatomy, physiology, pharmacology, and neuroscience, utilizing many techniques, including electrophysiology, immunohistochemistry, and confocal microscopy, pharmacology, and molecular methods. We now understand the concept of meridians, acupoint specificity, stimulation parameters required to evoke optimal responses, factors influencing the action of EA, and many of its mechanisms of action. As a result of many experiments incorporating a number of animal models (dogs, rabbits, rats, cats), we now can describe the neural pathways and neurotransmitters involved in EA that inhibit sympathetic outflow to reduce hypertension and arrhythmias. We also have begun to evaluate mechanisms underlying the long-lasting inhibition of elevated blood pressure by EA. Recent preliminary studies have shown that the same acupoints and stimulation parameters that lower elevated BP also have the potential to reverse hypotension. Although EA is not effective in all subjects, we have preliminary clues about how to improve responsiveness so that nonresponders can be converted to responders. Finally, we provide suggestions for continued research in this potentially important area of treatment.

Keywords Arcuate nucleus • Ventrolateral periaqueductal gray • Rostral ventrolateral medulla • Nuclei raphé • Nucleus tractus solitaries • Splanchnic nerve • Renal nerve • Coherence • Antidromic stimulation • Homeostasis

P. Li (✉) • S.C. Tjen-A-Looi • J.C. Longhurst
Department of Medicine, Susan Samueli Center for Integrative Medicine,
School of Medicine, University of California-Irvine, Irvine, CA 92697, USA
e-mail: pengli@uci.edu; stjenalo@uci.edu; jcl@uci.edu

Y. Xia et al. (eds.), *Current Research in Acupuncture*,
DOI 10.1007/978-1-4614-3357-6_15, © Springer Science+Business Media New York 2013

15.1 Introduction

Acupuncture in China is a component of Traditional Chinese Medicine (TCM) that has been used in the treatment of a variety of diseases for at least 2,500 years. The Chinese ancients put forward the "meridian hypothesis" to explain the mechanism of acupuncture therapy. This theory was described in detail in the "Yellow Emperor's Classics of Internal Medicine" (Huang Di Nei Jing) (Unschuld 2003). As physiologists, we realize that the "meridian hypothesis" is merely the ancestors' understanding of the body's regulatory systems and their function, long before the discovery and realization of the integrative function of the central nervous and humoral systems. Since the long history of clinical practice has proven that acupuncture provides restorative effects, a physiological basis must be involved. The role of science is to study and clarify the mechanisms that underlie acupuncture's clinical actions. In the 1960s and 1970s most Chinese institutes were studying acupuncture's analgesic effects. One member of our team (Li) began to focus on the actions of acupuncture on normal and abnormal cardiovascular function. This work progressed smoothly. Working together with Chinese investigators and colleagues in other countries, more than 100 papers were published. This research was funded by a number of awards from the Chinese Ministry of Health and the Chinese National Education Commission.

In the USA the practice of acupuncture began to be accepted after President Nixon's trip to China. Today, because of the increased demand for acupuncture and other alternative therapies, it is important for scientists and physicians to have a clear understanding of the mechanism(s) by which acupuncture can exert its beneficial effects. Between 1992 and 1994 Dr. Longhurst began discussions with Dr. Li and they began their collaborations in 1995. In 1999 Dr. Li retired from Shanghai Medical University (SHMU), now affiliated with Fudan University, and traveled to the University of California, Irvine to continue work with Dr. Longhurst on a full-time basis. Over the last 12 years our group has published more than 30 papers and organized four international symposia related to the mechanisms of acupuncture regulation of cardiovascular function. As a result of this collaboration, we now understand more about the neural scientific basis of acupuncture, including acupoint specificity, optimal stimulation parameters, factors influencing the actions of acupuncture, the long-lasting action of EA in hypertension, central neural pathways that process input from somatic nerves activated during acupuncture, neurotransmitter systems, and autonomic cardiovascular modulation by EA, and have begun studies designed to see if it is possible to convert nonresponders into responders.

Some of the work cited was described originally in monographs published in 1992 (Li and Yao 1992) and 2001 (Cheung et al. 2001). This chapter includes work of our colleagues both in SHMU, China, and at the University of California, Irvine, USA.

The chapter describes our current understanding of how acupuncture and TCM work mechanistically. More research in the future will help to even better understand acupuncture's mechanisms of action.

15.2 What Are the Ancient Hypotheses of Acupuncture and Meridians?

For more than 2,500 years, acupuncture has been used by ancient Chinese doctors to treat many diseases. According to their knowledge gained from clinical experience, the "Meridian theory" was established to explain mechanisms of acupuncture therapy, which was described in detail in the ancient text "Yellow Emperor's Classics of Internal Medicine" (Huang Di Nei Jing). This book was published in the third century BC and described 12 bilateral and two midline channels that connect internal splanchnic organs to external regions on the extremities. TCM theory suggests that channels regulate vitality and death and can be used to treat disease, to modify deficiency and excess of energy (called *Qi*), and to balance yin and yang (two opposing forces contained within everyone's body, a concept similar as homeostasis). The meridian hypothesis was based on the clinical experience of ancient doctors and, as such, is anecdotal. The book "Huang Di Nei Jing" (Unschuld 2003) defines acupuncture points or acupoints as the site of energy (*Qi*) arising from splanchnic organs that is directed through channels as it is transferred to the body surface. They are the locations where acupuncture is applied. Each acupoint is located on a meridian, and each meridian is related to and named after a particular (Chinese) splanchnic organ. Therefore, acupoints, meridians, and splanchnic organs are connected to each other and are inseparable (chapter Lingshu needles in Huang Di Nei Jing). A great number of additional acupoints have been added over time. These are the extrameridian points. Clinical reports have suggested that acupuncture potentially can be used to treat many diseases and, as such, many scientists have attempted to identify meridians and the mechanisms that underlie acupuncture's actions.

15.2.1 Searching Meridians on the Body Surface

Many researchers have attempted to locate the 14 meridian channels and acupoints on the body surface. Early in the 1950s, skin resistance or impedance over acupoints was reported to be lower than surrounding areas. For example, Nakatani in Japan proposed that different areas of the body surface have abnormally high or low conductivity (Nakatani and Yamashita 1977). He speculated that the lower conductivity was closely related to the meridian lines. He called the meridians "Rydoraku channels," which in Japanese means "good conduction line." Based on his theory, a number of instruments that could measure skin resistance were designed and sold. Many scientists, including Chinese doctors bought these machines and tried to find the channels on body surface. When they were unable to find the meridians using the Rydoraku apparatus, they tried German devices or designed equipment manufactured by their own institutes and factories to measure the skin's electrical potential, resistance, and impedance. However, they were unable to repeat Nakatani's findings. Eventually they discovered that Nakatani had placed the indifferent electrode on the palm, the main source of the skin's galvanic reflex, related to the activity of sweat glands innervated by sympathetic cholinergic fibers, and that this measure is very sensitive to psychological changes. The potential or resistance change beneath an indifferent electrode on the palm is much larger than other skin areas (acupoints) (Li et al. 1962). Additionally, the moisture and pressure of electrodes applied to the acupoints and non-acupoints are not easily controlled. Therefore, reliable identification of acupoints using measures of skin resistance is not possible.

In 1963, Bong-Han Kim, a doctor in North Korea, claimed to have found meridians and acupoints in rabbits. He published a monograph containing beautiful color pictures to show the structure and physiological evidence of the acupoints and meridians. This was big news and many Chinese scientists went to his laboratory to learn techniques of identification of these structures. However, it was reported that the structures he identified did not satisfy the definition of meridians. Both the morphological and physiological evidence appear to have been technical errors. Furthermore, it has been suggested by others that some of Dr. Kim's results may have been manufactured (Cheung et al. 2001).

In 1985 Verneijoul in France injected a radioactive isotope subcutaneously into humans and claimed that vertical lines of migration were seen only from acupoints and that these migration patterns resembled meridian lines in their distribution. However, in this study, lymphatic and venous drainage were not excluded. These explanations for this isotope migration has raised serious doubts in the minds of many scientists about the legitimacy of claims about the ability of radioisotopes to identify meridians (Verneijoul et al. 1985).

More recent research groups have continued to use a number of different techniques to search for unique morphological evidence of meridians in the body. They not only measured electrical potential, resistance, and impedance of acupoints but also applied ultrasound, infrared imaging, radioactive isotopes, electron microscopy, computerized tomography (CT), and magnetic resonance imaging to scan the whole body and locate channels on or underneath the skin. Some investigators believe that they had found unique anatomical characteristics marking acupoints. For example, some have suggested that collagen fiber bands wind around acupuncture and may contribute to stimulation of meridians (Langevin et al. 2011), while others reported that channels on the skin contain nitric oxide (NO) or acetylcholine (Ahn et al. 2008). However, many of these reports have not been confirmed and, as such, their reliability is uncertain.

We therefore need to ask ourselves the question of how we can explain the concept of meridians rather than asking the question, what is the anatomical basis of meridians. For example, why do patients feel a dull aching, numbness or a propagated sensation during acupuncture, which TCM practitioners have called *deQi*? Why does needling sometimes cause a sensation to radiate along the meridian away from the needle? How does acupuncture applied at a remote location on the body act locally and distantly, and how does *deQi* relate to the action of acupuncture? To answer these questions we believe that it is not productive to continue to search for meridians along the body's surface. To gain perspective, it is important to understand that even ancient books like "Huang Di Nei Jing" do not mention meridians on the body surface. In fact, to put this issue to rest in the 1990s, the Society of Chinese Physiological

Sciences and Neuroscience Society advised members to stop trying to identify meridians on the body surface. To underscore this belief, the National Natural Science Foundation of China, equivalent to the National Institutes of Health in the USA, stopped supporting research focused on locating meridians. In fact, to truly understand the meaning of meridians, it is necessary to adopt other approaches incorporating modern physiology and neuroscience, as discussed below.

15.2.2 Understanding Meridians Through Anatomy, Physiology, and Neuroscience

In 1973 researchers at Shanghai First Medical College (Department of Anatomy 1973) reported a relationship between acupoints and the surrounding nerves. They showed that no unique structures existed beneath acupoints other than nerves and nerve endings. Yu et al. (1996) studied a three-dimensional structure of the Neiguan (P 6) acupoint and found that only normal tissue structures including nerve endings contributed to this acupoint, confirming that no special structures are related to these regions of stimulation.

Chiang et al. (1973) at Shanghai Institute of Physiology reported that vascular occlusion of the upper arm of normal adults did not prevent the analgesic effect of acupuncture needling of a point on the hand (Hegu, LI 4.) On the other hand, infiltration with the local anesthetic procaine in the deep tissues around the point of acupuncture entirely abolished the analgesic effect, suggesting that neural but not humoral factors were responsible for mediating the acupuncture response. Han et al. (Research group of acupuncture anesthesia, Beijing Medical College 1987) reported similar findings. Similarly, the analgesic response to acupuncture at LI 4 was present in hemiplegic patients when the needles were inserted in the healthy side but were absent on the paralyzed side of the body. These results indicate involvement of peripheral sensory nerves and afferent pathways in the spinal cord in the analgesic response to acupuncture. This conclusion has been supported by numerous publications using multiple approaches including denervation, local injection of transmitter antagonists, and focal lesions in the central nervous system (Han and Terenius 1982; Wang et al. 2008; Zhao 2008).

Kline et al. (1978) reported that low-frequency, low-intensity (2 mA, 2 ms, 4 Hz) electrical stimulation of the Zusanli (S 36) acupoint elicited significant decreases in arterial BP and heart rate when the tip of the acupuncture needle was placed close to branches of the deep peroneal nerve (DPN). Cutting the sciatic nerve abolished these responses, whereas direct stimulation of the intact peroneal nerve to mimic acupuncture elicited a decrease in BP and heart rate in paralyzed animals. These results suggest that the decrease of BP and heart rate induced by acupuncture is related to direct sensory nerve stimulation and is not due to the activation of motor fibers resulting in muscle contraction and subsequent excitation of muscle and/or joint receptors, which might induce an inhibitory reflex. Their results also showed that there is no depressor response if the acupuncture needle is inserted into a control point not situated in close proximity to a somatic nerve.

Dung (1984) listed a number of anatomical structures that can be found in the vicinity of acupoints. These include large peripheral nerves, nerves emerging from a deep to a more superficial location, cutaneous nerves emerging from deep fascia, nerves emerging from bone foramina, motor points, neuromuscular attachments, blood vessels in the vicinity of neuromuscular attachments, bifurcation points of peripheral nerves, ligaments rich in nerve endings, and suture lines of the skull. This list confirms that, other than neural components, no particular structure dominates acupuncture points. Rather, the major structures that are present at acupoints are nerves, including large nerve bundles and nerve endings.

Heine (1988) also suggested that 80 % of acupoints correlate with perforations in the superficial fascia. Through these holes, cutaneous nerve vessel bundles penetrate the skin. More details about structures underneath acupoints were described in two anatomical atlases of acupoints (The Cooperative Group of Shandong Medical College and Shandong College of Traditional Chinese Medicine 1990). A summary of their observations follows:

1. Acupoints on the face and forehead region are located along terminals or cutaneous branches of the trigeminal and facial nerves. They are located either at nerve trunks where they leave the foramen or at the end of its terminal branch. Sometimes an acupoint is situated at the site of anastomosis of two different nerves or at the point at which a nerve branches bilaterally. On the posterior aspect of the head and neck, most acupoints are found along occipital nerves. Acupoints in the front of the neck lie along terminal branches of the cervical plexus.
2. On the external ear, many acupoints can be used to treat different kinds of diseases, and in this small area, both the medial and lateral auricle receive innervation from the vagus and glossopharyngeal nerves and branches of the facial and occipital minor nerve. The medial region of the ear also receives branches from auriculotemporal, auricular magnus, and occipital major nerves. Acupuncture applied to the auricle has the potential to stimulate any of these nerves.
3. On the trunk, many spinal nerves have a number of cutaneous branches that reach the skin of the thorax and abdomen. Each branch corresponds to acupoints along a meridian.
4. On the forearm and hand, acupoints commonly are related to the ulnar, median, and radial nerves and often are used to treat disease located in the area that these nerves innervate, as well as areas on the head, face, and chest. Acupoints on the lower limbs often are related to the sciatic, peroneal, tibial, and femoral nerves. These acupoints are usually used to treat the disease in the lower limbs, abdomen, or pelvis.

Histological examination by Croley (1996) showed twice as many dermal papillae in the region of acupuncture points as in regions of non-acupuncture points. There was a concentration of papillae containing capillary loops with sympathetic nerve endings within the loci of the acupuncture points. Han (1996) showed that acupoints are endowed with more nerve fibers including pressure and stretch receptors and that there are more myelinated group II fibers in the area of acupuncture points when compared to non-acupoints. Lu (1983) found that in the region of the Zusanli

(S 36) acupoint, there are 1,550 myelinated fibers and 577 nonmyelinated fibers with a ratio of 2.7:1. In the non-acupoint area, he observed 640 myelinated fibers and 860 nonmyelinated fibers, at a ratio of 0.7:1. He also showed that the ratio of Aα, Aβ to Aδ at Zusanli was 3.1:1, while at non-acupoints the ratio was 1.6:1. Although these studies were not performed with electron microscopy (the best method for identification of unmyelinated fibers), they do suggest that myelinated fibers in the region of S 36 acupoint are more numerous than in the non-acupoints. Their physiological significance is discussed in the next section.

Some authors (Wong and Cheng 1987) have suggested that the acupoints are motor points, where somatic nerves enter muscle, and that they approximate but are not identical to the end-plate zone of motor nerve endings. For example, some acupoints are located just proximal to the Achilles tendon or immediately proximal to the tendon insertion on a bony prominence. Other acupoints are located at the arterial arch of the hand or foot, e.g., Hegu (LI 4) on the back of the hand and Taichong (Liv 3) on the back of the foot. Needling these acupoints is an effective way to influence the sympathetic activity (see below).

15.3 What Type of Nerve Fibers Is Activated by Acupuncture?

Toda and Ichioka (1978) reported that activation of Group II somatic afferents was sufficient for acupuncture analgesia in the rat. The following year, Pomeranz and Paley (1979), using multiunit recordings of sensory nerve activity at the Hegu acupoint (LI 4, located in the first dorsal interosseous muscle) of mice, suggested that stimulation of Group II afferents produced acupuncture analgesia. However, multiunit recordings do not accurately assess unmyelinated afferent activity. Single-unit recordings with conduction velocities are the best method to assess the role of thin fiber afferents in acupuncture-related responses. Wang et al. (1985) recorded activity of median nerve (MN) fibers in humans using single-unit recordings during acupuncture. They observed that during *deQi*, numbness was related to the activation of Group II muscle afferent fibers, while heaviness, distension, and aching were in response to activation of Group III fibers and soreness was related to the activation of Group IV unmyelinated fibers. Acupuncturists frequently observe the sensations of *deQi* during stimulation. These studies suggest that both myelinated and unmyelinated somatic afferents have the potential to contribute to acupuncture responses.

In the clinic and recommended in acupuncture books, different acupoints typically are selected to treat different diseases. Treatment of cardiovascular diseases such as hypertension, coronary heart disease, angina pectoris, and heart failure is best accomplished by stimulating Neiguan (P 6), Shemen (H 7), Jueyingshu (UB 14), Xinshu (UB 15), Ximen (P 4), Zusanli (S 36), Sanyingjiao (Sp 6), Quchi (LI 11), and Taichong (Liv 3) as well as certain auricular acupoints (Shi and Zhang 1998). On the other hand, treatment of hypotension and shock may be best accomplished by stimulating Renzhong (Du 26), Chengjiang (Ren 24), and Shixuan (tips of the fingers and toes) (Cheung et al. 2001; Shi and Zhang 1998). This later group of acupoints is innervated mostly by superficial unmyelinated fibers in somatic nerves (Li 2001). Interestingly, Li et al. (1998) from our group using single-unit afferent recording methods to examine the response to acupuncture in a feline model demonstrated that EA at P 5-6 inhibits the reflex-induced increases in BP in association with activation of both Group III and IV afferent fibers in a ratio of 7:3. More recently, Tjen-A-Looi et al. (2005) also from our group have shown that Group IV fiber stimulation during EA is essential to the modulation of reflex increases in BP. Her studies in rodents used capsaicin administered neonatally to eliminate unmyelinated fibers in rats capsaicin. Therefore, although the earlier literature emphasized the essential role of myelinated fibers, recent studies have demonstrated that unmyelinated fibers also are important in acupuncture-cardiovascular responses.

15.4 Acupoint Specificity

Some scientists believe that because acupuncture releases opioids systemically and has the potential to exert widespread actions, acupoint specificity does not exist. For example, Cho et al. (1998)initially claimed that acupoint specificity occurs with respect to stimulation of the visual cortex. However, this paper was later withdrawn (Cho et al. 2006) because he could not confirm that acupuncture point specificity existed with regard to its analgesic effect. To clarify this concept, we evaluated the effectiveness of activating different acupuncture points on reflex-induced increases in BP, caused by visceral organ nerve stimulation in cats (Tjen-A-Looi et al. 2004a, b). This visceral reflex leads to stimulation of the sympathetic nervous system through activation of cardiovascular premotor sympathetic neurons in the rostral ventral lateral medulla (rVLM), an important brain stem region that regulates sympathetic outflow. We observed that EA at P 5–6 (Jianshi-Neiguan, pericardial meridian, overlying the MN) and LI 10–11 (Shousanli-Quchi, large intestine meridian, overlying the deep radial nerve) is most effective in its influence on cardiovascular function. EA at LI 4–L 7 (Hegu-Lique, large intestine and lung meridians, overlying branches of median and superficial radial nerves) and S 36–37 (Zusanli-Shangjuxu, stomach meridian overlying the DPN) is less effective, while EA at LI 6–7 (Pianli-Wenliu) and K 1-B 67 (Yongquan-Zhiyin) does not influence elevated BP at all. Furthermore, direct stimulation of the deep or superficial nerves underneath these acupoints produces similar results (Li et al. 1998; Tjen-A-Looi et al. 2004a, b). Similar observations have been made in a rat model employing gastric distension (Zhou et al. 2005a, b). Neuronal recordings in the arcuate nucleus also have shown that more afferent inputs originate from P 5-6, LI 4-11, H 5-7 (Tongli-Shenmen, overlying ulnar nerve), and cranial nerves (Sibai, S2 branch of trigeminal nerve and Tinghui G 2, branch of facial nerve) than from LI 6-7 and G 37-39 (Guangming-Xuanzhong) (Li et al. 2006a, b).

In a study of humans, we observed that bicycle-exercise-induced increases in BP were inhibited by EA at P 5-6 or LI 4 for 30 min before exercise, while EA at G 37-39 did not alter the exercise pressor response (Li et al. 2004). Furthermore, in hypertensive patients we have observed that P 5-6 and S 36-37 more effectively decrease BP than acupuncture at LI 6 7 and G 37-39 (Li and Longhurst 2007). Therefore, there is strong evidence to show clear acupuncture point specificity with respect to its action on cardiovascular function.

15.5 What Is the Best Method for Selecting Acupoints?

15.5.1 Local Acupoints: Ah Shi *Points*

To treat pain in any part of our body, needles are usually placed close to or at the site of pain. These acupoints are called *Ah Shi* points, which in Chinese means "these acupoints are where the patient has pain." This method of treatment is effective probably because these acupoints are near the site of pain, and during stimulation, their afferent input often converges on neurons in the same or nearby segment of the spinal cord or higher centers of the brain. Afferent stimulation by needling thus has the potential to inhibit afferent input from the area of pain at several different locations. Additionally, the local action of needle stimulation should be considered. In this regard, Goldman et al. (2010) recently reported analgesic responses to needle trauma induces local release of adenosine, which interacts with A_1 receptors (when ATP is released in response to mechanical stimulation, it rapidly degraded to adenosine and can suppress pain through adenosine A_1 receptors) located on the nearby afferent nerves, thus interfering with the transmission of nociceptive signals. Nevertheless, *Ah Shi* points are not the only points that should be selected for stimulation. In fact, if only *Ah Shi* points are stimulated, the response may be suboptimal.

15.5.2 Acupoint Stimulation in the Same or Nearby Spinal Segment

Angina pectoris can be treated by needling Neiguan (P 6) and Xinshu (UB 15) (Ballegaard et al. 1990; Kurono et al. 2002; Xu and Wang 2005). Genitourinary diseases are better treated by stimulating the Sanyinjiao (Sp 6) acupoint (Shi et al. 2011; Stener-Victorin et al. 2000). These acupoints are on extremities. They are not near the organ manifesting pain, but rather are located in the area of referred pain. The ability to stimulate acupoints with the same dermatomal segment to control pain that likely relates to the gate control theory of stimulating afferent fibers to suppress the neuronal transmission of pain at the spinal cord level may, in part, explain the mechanism analgesia by this form of acupuncture (Melzack and Wall 1984). According to Chinese meridian theory, Neiguan (P 6) is on the channel that connects to the pericardium, which surrounds the heart. P 6 sends projections to nearby spinal segments (C 6–T 1) receiving input from sympathetic preganglionic

Table 15.1 Innervations of acupoint

Acupoints	Nerves beneath the acupoint	Spinal segment
Neiguan (P-6)	Median nerve	C6-T1
Ximen (P-4)	Median nerve	C6-T1
Shenmen (H-7)	Ulnar nerve	C8-T1
Quchi (LI-11)	Radial nerve	C5-T1
Zusanli (S-36)	Deep peroneal nerve	L4-S3
Sanyinjiao (Sp-6)	Tibial nerve	L4-S3
Jueyingshu (B-14)	4th and 5th thoracic nerves	T2-4
Xinshu (B-15)	5th and 6th thoracic nerves	T3-5
Renzhong (Du-26)	2nd branch of the trigeminal nerve and buccal branch of the facial nerve	
Chenjiang (Ren-24)	3 rd branch of the trigeminal nerve	
Shixuan (Ex-30)	Cutaneous nerves	

neurons innervating the myocardium (T 2-5) (Liu et al. 1994). Our studies (Zhou et al. 2005a, b, 2006a, b) show that stimulation of P6 acupoints activates afferent discharge of neurons that project to spinal or supraspinal levels where inhibitory neurotransmitter systems modulate the central transmission of pain. This action lowers sympathetic output to decrease cardiac oxygen demand, which reduces myocardial ischemia and so relieves angina. Similarly, Sanyinjiao (Sp 6), innervated by the tibial nerve, projects to L 4-S 3 and is near the spinal center of the micturition reflex. The innervations of several principal acupoints used to treat cardiovascular disease are listed in Table 15.1.

15.5.3 Responses to Stimulation of Distant Acupoints

Sometimes, acupoints at a site distant from the affected region are stimulated to achieve optimal results. Traditional Chinese doctors usually make their differential diagnosis according to the patient's history as well as through pulse and tongue examination. Using this pattern diagnosis they then choose acupoints according to classical meridian theory, which often leads to stimulation of acupuncture points located far from the diseased organ. The sensory innervation of these acupoints does not project to the same or a nearby spinal segment. The effectiveness of these acupoints is likely related to convergent afferent input activated during somatic nerve stimulation on neurons in regions of the central nervous system that regulate the manifested symptoms, e.g., pain or high BP (Li and Longhurst 2010). It is known, for example, that neurons in the reticular formation of the brain stem such as rVLM receive afferent input from many different regions of the body (Lin et al. 1991; Gao and Li 1994; Peng et al. 1998). Furthermore, there are extensive interconnections among sensory, motor, and other integrative nuclei with nuclei in the brain stem. Our study suggestes that the neurons in the rVLM, arcuate nucleus in hypothalamus, ventrolateral periaqueductal gray (vlPAG), and nucleus raphe pallidus (NRP) receive inputs both from different brain areas and a variety of somatic nerves (Tjen-A-Looi et al. 2003, 2006). In this regard, we have found that cardiovascular neurons in the rVLM, arcuate nucleus, and vlPAG receive projections from the greater splanchnic nerve, P 5-6, and S 36-37 acupoints (MN and DPN) (Tjen-A-Looi et al. 2004a, b; Li et al. 2006a, b). With such widespread convergent input to so many CNS regions that regulate cardiovascular function, it is not surprising that stimulation of the MN and DPN can profoundly influence sympathoexcitatory cardiovascular responses induced during stimulation of the gallbladder with bradykinin (BK) or with gastric distension (Li et al. 2006a, b).

15.6 Responses to Manual and Electroacupuncture

15.6.1 Stimulation Parameters in EA-Cardiovascular Responses

Previous studies using compound action potential recording techniques suggested that finely myelinated Group III fibers are the predominate pathway stimulated during low-current, low-frequency EA (Cheung et al. 2001; Li and Yao 1992). In our feline model, we have demonstrated that both myelinated and unmyelinated nerve fibers are activated during EA by stimulating the MN (P 5–6 acupoints) with very small currents (0.1–3.0 mA) and using motor threshold as a guide to achieve similar stimulus intensities between animals. Of the 62 fibers recorded, 37 % were C fibers (≤2.5 m/s) and 63 % Aδ fibers (>2.5 m/s) (Li et al. 1998). Thus, both myelinated and unmyelinated fibers may participate in the acupuncture-cardiovascular response. Recently, we observed that neonatal capsaicin-treated rats depleted of substance P from unmyelinated afferents in the dorsal root ganglia are insensitive to the inhibitory action of EA during gastric distension. Thus, inhibition by EA at P 5–6 of cardiovascular excitatory reflex increases in BP requires input from unmyelinated Group IV fibers (Tjen-A-Looi et al. 2005). In our rat model of reflex-induced hypertension, we have shown that sham acupuncture involving needle insertion without manipulation at P 5–6 or LI 6–7 acupoints does not attenuate gastric-distension-induced reflex increases in BP, thus demonstrating that this procedure can serve as a control for EA. However, EA at P 5–6, H 6–7 (overlying the ulnar nerve), or S 36–37 with low current (2 mA) and low frequency (2 Hz) for 30 min inhibits the reflex increase in BP for 30–40 min. Similar decreases in elevated blood pressure are observed when manual acupuncture (MA) was applied (Zhou et al. 2005a, b), indicating that when EA and MA are matched for frequency, they induce very similar responses. Increasing the EA stimulation frequency to 40 or 100 Hz shows less and often no inhibition of elevated BP. Thus, we observed a reciprocal relationship between the frequency of stimulation and the afferent response. Recording premotor sympathetic neuron activity in the rVLM also has shown that 100 Hz stimulation induces little input to rVLM neurons. Thus, it appears that low-frequency, low-current EA in a point-specific manner optimally influences reflex-induced hypertension (Zhou et al. 2005a, b).

15.6.2 Stimulation Duration

Usually acupuncture is applied for 30 min (20–40 min) to achieve optimal therapeutic responses (Han 1987; Li et al. 2004, 2009; Li and Longhurst 2007). In animal experiments EA inhibition of sympathoexcitatory reflexes starts after 10–20 min, gradually increases, and is maintained for about an hour after termination of EA. Some acupuncturists and researchers apply acupuncture or EA for very brief periods, often <1 min, and claim to observe decreases in BP and HR (Uchida et al. 2010). It is likely that hemodynamic responses to these forms of stimulation are due to different mechanisms than those that underlie more prolonged stimulation. In this regard, short-term somatic nerve stimulation is known to induce somatosensory responses that can influence BP and HR (Sato and Schimdt 1987). The major difference is that with brief stimulation hemodynamic responses return to normal within a few minutes (Li and Li 1992). Conversely, 20–40 min of EA leads to activation of long-loop pathways in the brain and induces the release of a number of neurotransmitter systems, including β-endorphin, enkephalins, γ-aminobutyric acid (GABA), serotonin or 5-hydroxytryptamine (5-HT), acetylcholine, nociceptin, and endocannabinoids, in different regions of the hypothalamus, midbrain, and medulla that contribute to the long-lasting BP modulatory response (see text below). Thus, a duration of approximately 30 min of acupuncture stimulation seems to be optimal in achieving prolonged cardiovascular modulation following a single application.

15.7 Acupuncture's Normalizing Influence on BP

In traditional Chinese medical practice it is emphasized that acupuncture does not influence normal subjects but lowers high BP in hypertensive patients and raises BP in hypotensive conditions (Hu Nan Med College 1973). Similarly our experimental studies have shown that stimulation of S 36 acupoints with low-frequency, low-intensity EA reduces BP in hypertensive animals and raises BP in hypotensive animals (Lin and Li 1981; Xiao et al. 1983). The mechanism is complicated and is discussed in the following section.

15.7.1 EA in Hypertension

15.7.1.1 Clinical Studies

Hypertension affects approximately one billion individuals worldwide (Hajjar and Kotchen 2003). Although a number of treatment strategies have been developed for this disease, treatment is not perfect and often is associated with adverse side effects. Acupuncture is a potential alternative to pharmacological therapy (World Health Organization 2003). Drug therapy indiscriminately blocks many receptors that can lead to a multiplicity of side effects. Acupuncture, on the other hand, is associated with a very low (0.13 %) incidence of side effects (Mayer 2000; MacPherson et al. 2001) and is relatively inexpensive. As such, there has been increasing interest worldwide in exploring alternative medical therapies like acupuncture for treating chronic conditions, including hypertension (Eisenberg et al. 1993, 1998).

One study, the SHARP trial, concluded that acupuncture does not reverse hypertension (Macklin et al. 2006). Conversely, a number of other clinical reports have suggested that acupuncture can reduce BP, although frequently the reduction is quite modest (Zhang 1956; Tam and Yiu 1975; Alshelevich et al. 1985; Akhmedov et al. 1993; Dan 1998; Flachskampf et al. 2007). However, there are weaknesses in prior acupuncture research. For example, power tests to determine proper sample size, randomization with adequate concealment in the application of acupuncture in control and test subjects, and appropriate control groups and/or control acupoints often are missing. In many studies, only intermittent cuff blood pressures have been taken and the medication status has not been addressed. Furthermore, the period of follow-up after acupuncture treatment generally has been short. Thus, to evaluate more definitively the effectiveness of acupuncture in the management of hypertension, rigorously designed and conducted randomized clinical trials are still needed to overcome methodological limitations of previous investigations.

In 2004 we reported that EA typically does not change BP significantly in healthy subjects at rest (Li et al. 2004). In contrast, ~70 % of healthy humans during exercise demonstrate inhibition of stress-induced increases in systolic and mean blood pressures (SBP, MBP) as well as the double product (SBP × heart rate, reflecting myocardial oxygen demand) (Longhurst et al. 1980) following 30 min EA at P 5-6 acupoints (Li et al. 2004). These observations are consistent with those of Cao et al. (1983) in Shanghai and Han and Terenius (1982) in Beijing who reported that the EA analgesic response occurs in 70 % of patients. Our work in Irvine likewise has demonstrated that EA inhibition of exercise stress-induced increases of BP also is effective in ~70 % of subjects. In contrast to EA at cardiovascularly active acupoints, EA at control acupoints G 37-39 does not alter exercise-related increases in blood pressure or the double product (cite Dr. Li's exercise study as a reference). This clinical study confirmed the efficacy of P 5-6 and LI 4 acupoints in lowering elevated BP in humans and reinforced the role of G 37-39 as suitable control acupoints for evaluating nonspecific placebo responses. The strong effect at P5-6 on exercise-induced elevations in BP and the absence of an influence of acupuncture at G 37-39 reinforces the concept of point-specific actions of acupuncture on cardiovascular function that result from stimulation of some but not other acupoints (Fig. 15.1).

Recently, in a preliminary study, we evaluated the influence of EA in patients with mild to moderate hypertension (BP 140–180/90–110 mmHg) (Li and Longhurst 2007). In this investigation, a group of 22 hypertensive patients without medication were assessed with 24 h ambulatory BP monitoring and low-frequency, low-intensity EA at P 5-6 and ST 36-37 once weekly for 30 min for 8 weeks of therapy. Over the 8 week period, treatment in 16 patients (73 % of the total population studied) reduced peak and 24 h average SBP by 18 and 11 mmHg, respectively; the 24 h peak and average diastolic BP decreased 9 and 6 mmHg, respectively, while the heart rate was

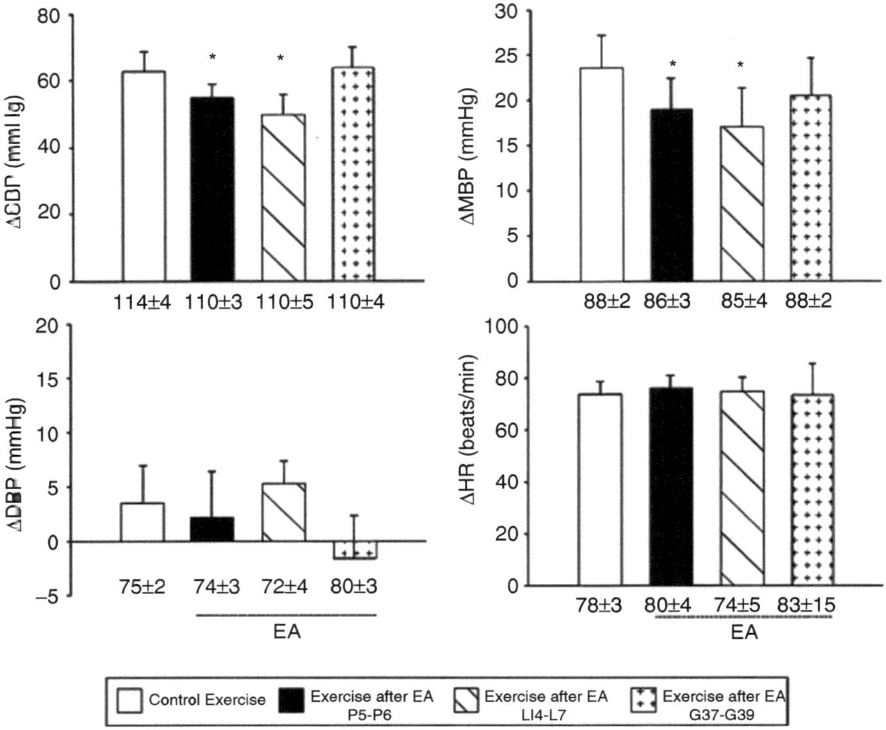

Fig. 15.1 Action of EA on hemodynamic responses to bicycle exercise during a ramp protocol. Numbers at the bottom of each figure indicate resting blood pressure (mmHg). $*P<0.05$ compared to control (pre EA) exercise. Control exercise, $n=12$; exercise after EA at P 5-6, $n=12$; LI 4–L 7, $n=9$; G 37-39, $n=5$. Modified from Li et al. (2004)

not significantly changed by EA. Following termination of EA treatment both peak and average SBP remained low for 4 additional weeks, but after 8 weeks, the BP of most patients had returned to near their pretreatment level. Six other non-responsive patients (27 %) receiving EA at the P5-6 and S 36-37 showed no significant decrease of BP over the 2 month period. On the other hand 14 additional hypertensive patients receiving EA for 30 min at control acupoints LI 6-7 and G 37-39 once weekly for 8 weeks showed no significant change in blood pressure. These data suggest that EA at select acupoints known to have a strong cardiovascular action, performed once weekly for 4–8 weeks, has the potential in ~70 % of patients to significantly reduce BP. EA at a control acupoints inconsistently lowered elevated BP, again demonstrating acupoint specificity in human subjects. These data also raise a series of important questions that need to be addressed in future studies. First, why is diastolic BP in hypertensive patients less influenced by EA? Second, why do only 70 % patients manifest a beneficial effect (responders), whereas 30 % are insensitive to the EA (nonresponders)? Third, why is the beneficial effect of EA on BP slow in onset, and why does it persists for prolonged periods of time. We will discuss these questions in light of our experimental studies in the following paragraphs.

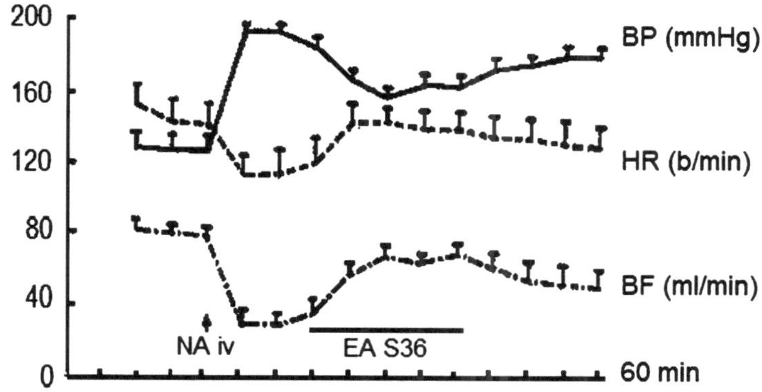

Fig. 15.2 Effect of EA at S 36-37 (2 mA, 2 Hz, 0.5 ms duration) on BP, heart rate (HR), and blood flow in the mesenteric artery (BF) during norepinephrine (NA) intravenous infusion in conscious dogs (means ± SE, $n = 10$). Modified from Lin and Li (1981) (in Chinese)

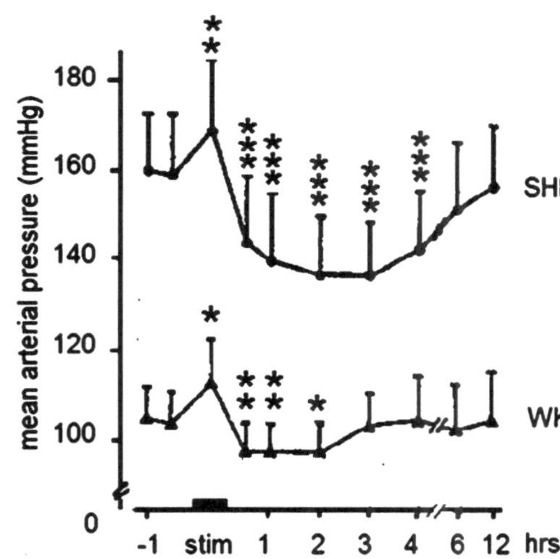

Fig. 15.3 Long-lasting depressor response to acupuncture-like stimulation of sciatic nerve in spontaneously hypertensive and Wister–Koyoto control rats (SHR and WKY). Values are expressed as means ± SD. $*P<0.05$, $**P<0.01$, $***P<0.001$. Modified from Yao et al. (1982)

15.7.1.2 Mechanistic Studies

Studies on the mechanism of the inhibitory effect of EA on cardiovascular function began in the 1980s. Li and Yao (Li 1984; Li and Yao 1992) evaluating several experimental hypertensive animal models, including conscious dogs (with norepinephrine infusion), spontaneously hypertensive rats, and stress-induced hypertension in Sprague Dawley (SD) rats (foot shock twice daily for 2 weeks to induce hypertension for 2 weeks), have demonstrated that EA and acupuncture-like stimulation reduce BP through activation of opioid receptors in the brain (Li and Longhurst 2010; Li and Yao 1992; Xie et al. 1997) (Figs. 15.2 and 15.3). In these early studies specific nuclei in the brain were not identified precisely nor were the actions of different opioid receptors examined.

Additionally, Chiu et al. (1997) reported that manual acupuncture at S 36 and G 34 for 30 min in 50 hypertensive patients transiently decreased systolic and diastolic BP during and for 30 min after acupuncture. Heart rate decreased slightly. Plasma renin likewise decreased slightly after acupuncture. However, they did not measure the duration of the acupuncture inhibitory response nor the long-term effect of acupuncture. Importantly, they did not use control acupoints for the EA, which substantially limits the value of the study.

In the late 1990s Dr. Longhurst and Dr. Li began their collaboration resulting in a series of experimental studies on mechanisms involved in low-frequency EA regulation of cardiovascular function.

15.7.1.3 Rostral Ventrolateral Medulla in the Cardiovascular Response to Acupuncture

The rVLM is an important source of premotor sympathetic neurons that regulate sympathetic outflow from the thoracic spinal cord (Dampney et al. 1985; Lovick 1985). In 1985–1988 Huangfu and Li performed several morphological (horseradish peroxidase) and electrophysiological experiments showing that the rVLM is essential in EA inhibition of defense-reaction-induced elevations in BP during stimulation of hypothalamic or midbrain defense areas. They also showed that iontophoresis of morphine inhibited the evoked excitation of rVLM premotor sympathetic neurons and that EA inhibition of these neurons could be blocked by iontophoresis of naloxone (Huangfu and Li 1985, 1986, 1987). However, they did not determine which opioid receptor(s) in rVLM played a role in the EA inhibition.

Over the last 10 years, our laboratory has demonstrated that EA induces c-Fos expression in the rVLM and vlPAG, two nuclei that produce enkephalin (Guo et al. 2004). Our in vivo experiments have shown that EA at select acupoints, such as Neiguan–Jianshi along the pericardial meridian (P 5–6) inhibits sympathoexcitatory-related increases in BP (Li et al. 1998, 2002). Furthermore, low-frequency, low-intensity EA exerts a beneficial effect by reversing demand-induced myocardial ischemia through its reduction of myocardial oxygen demand. EA-induced inhibition of elevated BP as well as premotor sympathetic neural discharge in the rVLM is long-lasting, since it is maintained for at least an hour after termination of EA in most studies (Tjen-A-Looi et al. 2003). We have found that administration of naloxone to nonspecifically block opioid receptors, or gabazine to block $GABA_A$ receptors in the rVLM abolishes the prolonged inhibition of reflex-induced neuronal activity by EA (Tjen-A-Looi et al. 2003, 2007a, b). We furthermore have shown that EA inhibition of elevated BP and rVLM activity is related to activation of μ and δ but not κ opioid receptors in the rVLM (Li et al. 2001).

We also have examined the central neural mechanisms that underlie point-specific cardiovascular responses. As noted above, we observed that the rVLM is an important brain stem region that processes somatic input during acupuncture stimulation. Electrophysiological studies of neural activity in the rVLM have shown that, compared to cardiovascular inactive points (LI 6–7, G 37–39), P 5–6 and LI 4-11 acupoints along the large intestine meridian (all located over deep somatic neural pathways) provide more convergent afferent input to cardiovascular premotor sympathetic neurons in the rVLM (Tjen-A-Looi et al. 2004a, b). Similarly, as noted below, stimulation of cardiovascular active points (as compared to less active or inactive points) provides more input to nuclei that constitute a long-loop pathway (see below), which appears to be critical to the sympathoinhibition occurring during electroacupuncture (Li et al. 2009).

15.7.1.4 Hypothalamic and Midbrain Nuclei in the Acupuncture Response

Previous studies by Li and his colleagues (Cheung et al. 2001) suggested that EA as well as direct stimulation of the DPN inhibited elevated BP induced by stimulation of the hypothalamic defense region and that this modulation could be prevented by destruction of the arcuate nucleus either electrolytically or by decerebration caudal to this region. These experiments do not unequivocally prove that the arcuate plays a role in the EA response since these techniques may have destroyed cell bodies and fibers in passage. However, because the cardiovascular response elicited by the defense reaction also could be inhibited by stimulation of the arcuate with the excitatory amino acid DL-homocysteic acid (DLH), it seemed that at least part of the response to arcuate destruction could be mediated by EA-induced deactivation of cells in this hypothalamic region (Huangfu and Li 1987, 1988). These data in aggregate thus suggest that the arcuate, which contains a sizeable population of β-endorphin neurons (Guo and Longhurst 2007), contributes to the inhibitory action of EA (Huangfu and Li 1987) with respect to its influence in the cardiovascular defense reaction.

Recently we have further evaluated the role of the hypothalamic arcuate nucleus and its interaction with the midbrain vlPAG and rVLM in the EA-cardiovascular sympathoexcitatory responses. Guo and Longhurst demonstrated that EA activates c-Fos expression in the arcuate nucleus, which contains opioids and frequently colocalizing with VGlut3, a vesicular glutamate transporter (Guo and Longhurst 2007). This morphological information is useful since the arcuate is activated by EA (through a glutamatergic mechanism) and because we have shown that the modulatory action by EA of reflex elevations in BP is related to inhibition of cardiovascular neurons in the rVLM through a mechanism that involves opioids. This inhibition of sympathoexcitation is long-lasting and involves a long-loop neural supraspinal pathway, including in part the arcuate nucleus, an important site of opioid neurotransmitter synthesis (Li et al. 2009, 2010a, b, c).

To evaluate the role of the arcuate and its interaction with the vlPAG in the EA BP response, we have used reflex hemodynamic studies to elicit short-term elevations in BP in the whole animal. It is necessary to increase BP because, as noted earlier, EA does not have an action on BP under resting normotensive conditions. Furthermore, whole animal reflex studies provide a way to evaluate the role of various brain regions (nuclei) with respect to the action of EA on cardiovascular function, including using pharmacological manipulation to identify specific neurotransmitter systems operative in EA. We found that, similar to the response to EA, microinjection of DLH into the arcuate nucleus causes prolonged inhibition (30 min) of reflex increases in BP induced by visceral afferent stimulation. Conversely, microinjection of kianic acid (KA) into the arcuate to induce depolarization blockade of cell bodies (and not axons in passage) blocked the inhibitory influence of EA at P5-6 on the reflex sympathoexcitatory response. These results suggest that excitatory projections from the arcuate to the vlPAG are essential to EA inhibition of the reflex increase in BP induced by visceral afferent stimulation.

Detailed electrophysiological studies have shown that spontaneous activity of neurons in the arcuate and vlPAG is low (4–5 imp/s) and that they respond to both visceral and somatic acupoint stimulation. As noted earlier, we also have observed a gradation of responses to stimulation of different acupoints, providing a cellular basis for point-specific responses. In this regard, evoked responses in both regions were greatest during stimulation of the P 5–6, LI 4–11, H 5–6, and S 2–G 2 acupoints (named for the pericardial, large intestine, heart, small intestine, and gallbladder meridians, located over the median, deep radial, ulnar, and cranial nerves, respectively) compared to responses resulting from stimulation of LI 6–7 (large intestine, overlying the superficial radial nerve) and G 37–39 (gallbladder meridian, over the superficial peroneal nerve) (Li et al. 2006a, b, 2009). Thus, we have found a similar differential in cellular evoked response of central nuclei responsible for the EA-cardiovascular response as we have observed during our hemodynamic acupoint specificity studies (see Sect. 15.3 above) during stimulation of acupoints along several meridians that overlie different mixed somatic nerves.

Microinjection of DLH, into the arcuate nucleus augments the responses of vlPAG neurons, while microinjection of KA to transiently inactivate neurons in this region of the ventral hypothalamus causes a decrease in the vlPAG responses to visceral (SN) stimulation (Li et al. 2006a, b). Additionally, 30 min of EA at P 5–6 increases SN-evoked discharge of vlPAG neurons, a response that can be blocked by microinjection of KA into the arcuate nucleus. In aggregate, these results suggest that excitatory projections from the arcuate nucleus to the vlPAG appear to be essential to the inhibitory influence of EA on reflex increases in BP induced by SN and gallbladder visceral afferent stimulation.

15.7.1.5 Direct and Indirect vlPAG–rVLM Projections

We have shown that bilateral stimulation with EA at P 5-6 acupoints reduces increases in BP for 60 min or more. Unilateral depolarization blockade of neuronal activity in the vlPAG with KA restores the BP reflex responses for 30 min. DLH in the vlPAG mimics the EA response, significantly reducing the reflex BP responses.

Since blockade of the caudal vlPAG reverses EA-associated modulation of evoked rVLM activity, we believe that the long-loop-mediated cardiovascular inhibitory influence of EA principally involves the caudal vlPAG. Additional studies from our laboratory show that the vlPAG provides inhibitory input to sympathetic premotor cells in the rVLM (Tjen-A-Looi et al. 2006).

We also have investigated the nature of the inhibitory connection between the vlPAG and the rVLM, with respect to acupuncture's modulation of elevated BP. Direct projections from the vlPAG to the medulla have been documented in tract-tracing studies (Loewy 1990). However, anatomical studies from our laboratory show that EA activates neurons in the medullary raphé and particularly the Nucleus Raphe pallidus (NRP) (Guo et al. 2008) has the potential to form an indirect pathway from the vlPAG to the rVLM. In this regard, depolarization blockade of the NRP with KA reverses increased neuronal activity in the rVLM during activation of the vlPAG as well as during EA modulation of visceral excitatory reflexes (Li et al. 2010a, b, c). These studies show that the NRP is a critical part of the vlPAG–rVLM portion of the long-loop pathway activated during EA.

Because our anatomical studies showed that many neurons in the NRP contain serotonin or enkephalin (or both), we have examined the possibility that serotonin plays a role in NRP–rVLM regulation of sympathetic outflow during EA. In this respect, serotonergic projections operating through a 5-HT_{1A} receptor mechanism between the raphé and the rVLM have been shown to modulate cardiovascular activity and EA regulation of sympathetic outflow from the rVLM (Dean 2005; Dean and Woyach 2004; Moazzami et al. 2007; Tjen-A-Looi et al. 2007a, b; Guo et al. 2008). Thus, an indirect connection from the vlPAG to the rVLM that involves a serotonergic projection from the NRP to the rVLM plays an important role in long-loop modulation of cardiovascular sympathetic outflow during reflex visceral stimulation (Tjen-A-Looi et al. 2006). These studies do not eliminate the possibility that direct projections between the vlPAG and the rVLM also serve a functional role in EA-cardiovascular modulation. However, indirect projections from the vlPAG to the rVLM complete the long-loop pathway and provide an important circuit for the inhibitory influence of EA on rVLM premotor neurons and ultimately sympathoexcitatory cardiovascular responses (Li et al. 2006a, b).

To complete our studies of interactions between nuclei concerned with the long-loop pathway, we have activated rVLM neurons by stimulating the SN to investigate whether the arcuate directly or indirectly inhibits rVLM activity (Li et al. 2006a, b, 2009; Tjen-A-Looi et al. 2003, 2004a, b, 2006, 2007a, b). Microinjection of KA into the arcuate reduces EA inhibition of SN-induced reflex increases in the rVLM neuronal activity. Furthermore, microinjection of DLH into the arcuate, like EA, inhibits reflex increases of rVLM neuronal discharge (Li et al. 2009). As such, during EA the arcuate clearly regulates not only PAG but also rVLM activity.

15.7.1.6 Direct and Indirect Arcuate–rVLM Projections

As noted previously, neurons in the vlPAG receive convergent input from a number of somatic nerves during stimulation of various acupoints with EA, as well as from the arcuate nucleus. Bilateral microinjections of KA into the rostral vlPAG partially reverse rVLM neural responses and cardiovascular inhibition during DLH stimulation of the arcuate. Conversely, depolarization blockade of the caudal vlPAG completely reverses arcuate evoked rVLM responses (Li et al. 2009). In parallel studies, we have observed that arcuate neurons can be antidromically activated from the rVLM and that arcuate perikarya are labeled with a retrograde tracer microinjected into the rVLM (Li et al. 2009). Many neurons from the arcuate that project to the rVLM are activated during EA stimulation (c-Fos positive), and they frequently contain opioid peptides, particularly β-endorphin (Guo et al. 2004). As such, the vlPAG, particularly the caudal vlPAG, appears to be required for inhibition of rVLM neuronal activation by the arcuate and subsequent EA-related cardiovascular activation. However, direct projections from the arcuate nucleus to the rVLM likely serve as an important source of β-endorphin (Li et al. 2009). This latter observation is consistent with our earlier anatomical study showing that cells in the rVLM contain enkephalin but not β-endorphin (Guo et al. 2004). Hence, EA-cardiovascular responses that result from the action of β-endorphin on μ-opioid receptors located on rVLM sympathoexcitatory premotor neurons (Li et al. 2001) depend on this hypothalamic–medullary projection.

15.7.1.7 Neurotransmitters Responsible for EA Modulation of BP

Glutamate, Opioids, and Nociceptin in the rVLM

In the 1980s and 1990s preliminary studies in several models of hypertension suggested that EA inhibition of pressor responses might be related to the action of a number of modulatory neurotransmitters in the rVLM (Huangfu and Li 1988; Li and Yao 1992; Zhang et al. 1992). More recently Zhou and colleagues (2006a, b) in our laboratory using microdialysis to study neurotransmitter release in the rVLM showed that acupuncture modulates visceral-reflex-induced release of glutamate through an opioid mechanism in during stimulation of visceral afferents.

We have found that μ- and δ- but not κ-opioid receptors in the rVLM underlie EA inhibition of reflex autonomic sympathoexcitatory responses in cats (Li et al. 2001). These data suggest that endorphins, enkephalins, and perhaps endomorphin but not dynorphin are responsible for EA modulation of cardiovascular responses. We also have shown that nociceptin, a modulatory peptide that generally is thought to mediate pain (Rossi et al. 1998), is involved in the early but not the late influence of EA on cardiovascular excitatory reflexes (Crisostomo et al. 2005; Tjen-A-Looi et al. 2007a, b). As noted above, the sources for these opioid peptides have been revealed by immunocytochemistry combined with confocal microscopy, which have shown the presence of enkephalinergic neurons in the rVLM and endorphinergic neurons in the arcuate nucleus that project directly to the rVLM (Guo et al. 2004, 2008). Identification of the early-immediate gene, c-Fos, demonstrates that neurons containing both neurotransmitter systems in rVLM and arcuate neurons are activated by EA.

Glutamate, Acetylcholine, Endocannabinoids, and GABA in the Arcuate and vlPAG

Electrophysiological studies (Li et al. 2008, 2009, 2010a, b, c) have documented reciprocal excitatory glutamatergic (NMDA and non-NMDA) projections between the arcuate nucleus and the vlPAG, which participate in the long-term inhibition of cardiovascular function. This network includes a cholinergic component in the arcuate nucleus but not in the vlPAG. Furthermore, EA, through presynaptic endocannabinoid CB1 receptor stimulation, reduces vlPAG release of GABA but not glutamate during EA (Fu and Longhurst 2009). Reduced GABA release disinhibits vlPAG neurons which, indirectly through the NRP inhibit rVLM cardiovascular presympathetic neurons and related sympathoexcitatory reflex responses (Tjen-A-Looi et al. 2009; Li et al. 2010a, b, c). Therefore, it is clear that a variety of neurotransmitter systems underlie the cardiovascular action of EA. The list includes both excitatory and inhibitory neurotransmitters with their importance varying from one nucleus to another.

15.7.2 Mechanism of the Long-Lasting EA Effect

EA inhibition of rVLM cardiovascular sympathetic premotor neuron activity lasts for more than 1 h after cessation of EA (Huangfu and Li 1986; Tjen-A-Looi et al. 2003). EA inhibition of cardiovascular excitatory reflex responses in anesthetized rabbits, rats, and cats lasts from 1 to 6 h (Guo et al. 1981; Lovick et al. 1995; Li et al. 1998, 2002). In conscious animals (dogs and spontaneous hypertensive rats) Modulation of blood pressure by EA lasts for 1–12 h (Lin and Li 1981; Yao et al. 1982; Lovick et al. 1995). As noted earlier, a preliminary clinical study in our laboratory has shown that, following repetitive EA once weekly for 8 weeks, BP in hypertensive patients is lowered after 4–6 weeks of acupuncture for a period lasting 4 weeks after discontinuation of treatment (Li and Longhurst 2007). Thus, inhibition of cardiovascular function by acupuncture cannot be explained simply by a short-term neural occlusive response originating from somatic–visceral convergent interaction on common rVLM interneurons (Tjen-A-Looi et al. 1997). As discussed below, we have identified several mechanisms that likely are involved in the prolonged action of EA on cardiovascular function.

15.7.2.1 Neurotransmitter Systems

The inhibitory action of EA on rVLM sympathetic premotor neuronal responses to excitatory visceral input lasts for 30–40 min after the cessation of a single 30 min application of EA as a result of modulation by opioids and GABA but not by nociceptin (Tjen-A-Looi et al. 2007a, b).

15.7.2.2 Arcuate–vlPAG Neuronal Circuit

Recently we found that prolonged inhibition of rVLM neurons by EA requires an intact arcuate nucleus since microinjection of the glutamatergic antagonist, kynurenic acid, into this hypothalamic region blocks EA inhibition of reflex-induced

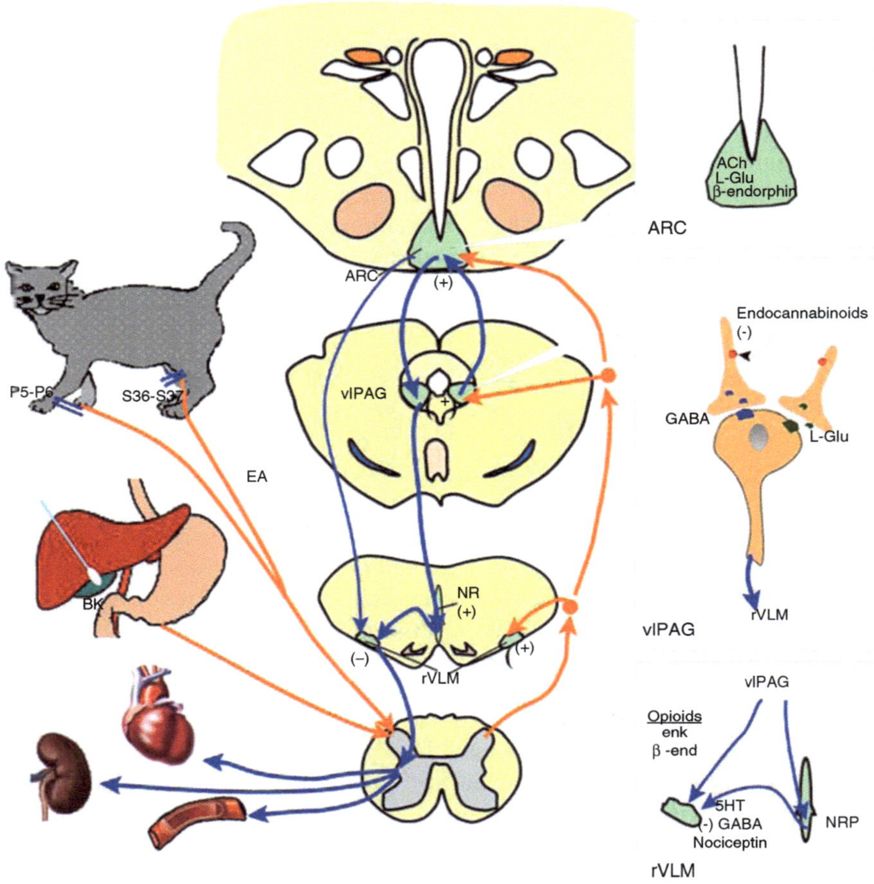

Fig. 15.4 Diagram of long-loop pathway responsible for EA inhibition of rVLM neurons and related neurotransmitters. Modified from Li and Longhurst (2010)

hypertension, including the prolonged action of EA in this hypertensive reflex response (Li et al. 2009, 2010a, b, c). Furthermore, as noted earlier, we have observed reciprocal excitatory projections between the arcuate nucleus and the vlPAG (Guo and Longhurst 2010) that form a reinforcing circuit contributing to prolongation of EA-cardiovascular modulation for as long as 30–60 min (Li and Longhurst 2010) (Fig. 15.4).

15.7.2.3 Transcriptional Regulation During EA

Still other mechanisms may be evoked by EA to induce prolonged inhibition of BP that can last for several weeks in patients with mild to moderate hypertension. In this regard, although limited in scope, studies have suggested that acupuncture may increase mRNA expression of opioid precursors in the brain for 24–72 h (Guo et al. 1996; He et al. 1995). Recent data from our laboratory using real-time PCR show that preproenkephalin in the rVLM is increased 90 min after completion of a single 30 min application of EA at P 5-6 acupoints of rats (Li et al. 2010a, b, c). Moreover, in conscious rats repeated EA may increase preproenkephalin in the rVLM for more than 48 h (unpublished data). The roles of opioid mRNA expression and other neurotransmitter precursors, as well as neurotransmitters released in the hypothalamus, midbrain, and other regions of the medulla, particularly in studies involving repetitive EA over time, are worthy of further investigation.

The long-lasting inhibition of EA in hypertensive patients may last for more than a month. This may be related to humoral factors. Chiu et al. (1997) have reported that renin is decreased during for 30 min after acupuncture. The role of the renin–angiotensin–aldosterone system in the long-term EA regulation of cardiovascular function is yet another area that deserves more investigation.

15.7.3 EA Effect on Hypotension

In Chinese traditional medical practice, acupuncture also is used to treat symptoms caused by low blood pressure or circulatory shock. For example, in 1973 the No. 2 Hospital of Hu Nan Medical College suggested that acupuncture elevates low BP and had therapeutic effect in shock (Hu Nan Med College 1973). Thus, acupuncture may be useful in elevating low BP in hypotensive patients in addition to its better studied action to lower high BP in hypertensive patients. In the 1980s in China, a series of experimental studies of acupuncture's action on hypotension were conducted, as described below.

15.7.3.1 Action of EA in Experimental Hypotension

When acute experimental hypotension is produced by intravenous infusion of nitroprusside (NP), 20 min of low-frequency, low-current EA applied to Zusanli (S 36) or Neiguan (P 6) elevates BP within 5–10 min, in association with an increase in cardiac output. However, the response is short-lived, reversing quickly after the cessation of EA (Xiao et al. 1983) (Fig. 15.5). EA also elevates BP in other hypotensive models, like endotoxic shock in dogs and hemorrhage in rats (Cheung et al. 2001; Sun et al. 1983; Xiao et al. 1985). EA at S 36 is capable of increasing BP and prolonging survival times in these models.

Similarly, Zhou and coworkers in 2003 (Syuu et al. 2003) studying hemorrhagic hypotension reported that EA at Neiguan (P 6) elevated BP by increasing left ventricle end-diastolic and stroke volume without changing heart rate. They concluded that EA improved left ventricle performance by increasing venous filling to restore BP following hemorrhage. Further investigation is needed to determine the actions of acupuncture on venous return, cardiac preload as well as neurohumoral mechanisms that serve to increase circulating blood volume in hypotensive conditions.

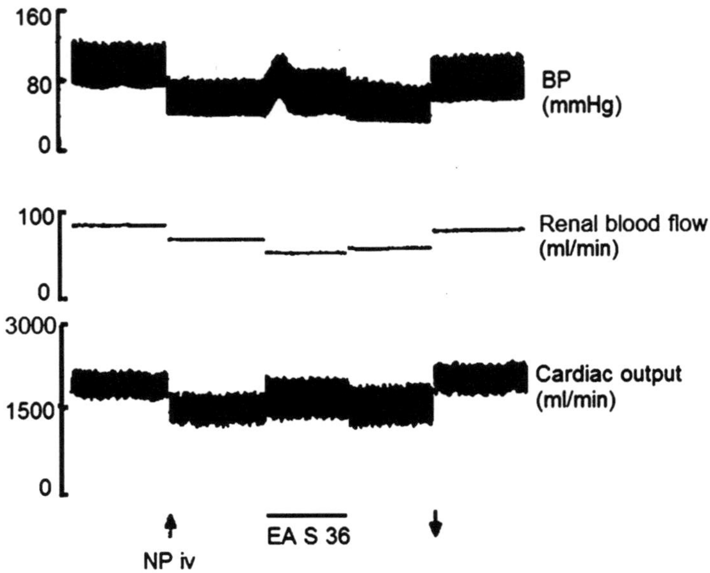

Fig. 15.5 Influence of EA on the BP, renal blood flow, and cardiac output during hypotension induced by nitroprusside i.v. Modified from Xiao et al. (1983) (in Chinese)

15.7.3.2 Cholinergic and Opioid Mechanisms in the Action of EA in Hypotension

Previous studies in China suggest that intravenous naloxone does not influence the action of EA in NP-induced hypotension (Xiao et al. 1983; Xiao and Li 1985). Conversely, intravenous or microinjection of atropine and scopolamine into the midbrain PAG, reticular formation, and ventromedial hypothalamus appear to reverse the BP raising effect of EA during NP-induced hypotension (Xiao and Li 1985). As such, the midbrain PAG and reticular formation and ventromedial hypothalamus may be sites of cholinergic action in the EA-cardiovascular response during NP infusion. Other neurotransmitters that might take part in the BP raising action of EA deserve further investigation.

15.8 Concluding Remarks

Research from our laboratories over the last 50 years has provided a framework to describe how EA modulates BP. We now understand that cardiovascular function during EA stimulation of somatic Group III and IV fibers is related to activation of a long-loop pathway that includes a number of nuclei in the medulla, midbrain, and hypothalamus as well as the dorsal horn and possibly the IML of the spinal cord.

Many neurotransmitter systems are involved, including opioids (β-endorphin, met and leu-enkephalin), nociceptin, GABA, serotonin, glutamate, acetylcholine, and endocannabinoids. However, a number of fertile areas remain for future investigation. These include:

1. The long-lasting inhibition of high BP by acupuncture. Although we currently understand that this response is related to the action of certain neurotransmitters such as opioids and GABA, reciprocal circuitry in the hypothalamus and midbrain, as well as transcriptional regulation of opioid peptides, we currently do not know why the inhibition can persist for more than 1 month following repeated stimulation. Perhaps a humoral system such as renin–angiotensin–aldosterone plays a role.
2. How does EA modulate both increases and decreases in BP, i.e., why does it normalize BP? Although acupuncture quiets (i.e., down modulates) activity of cardiovascular regions of the brain, including both sympathetic and parasympathetic areas that can elevate and reduce BP, additional research is necessary to define its mechanisms of action in different conditions that alter BP.
3. It is unclear why EA is effective in 70 % of patients (responders) but ineffective in 30 % of patients (nonresponders). The inaction of acupuncture may be related to antagonism of the opioid system by CCK in the brain. Future study of this mechanism is required and may allow practitioners to covert nonresponders to responders.

Acknowledgements This work was supported by the National Natural Science Foundation of China, Science fund of the Chinese Ministry of Health, and National Educational Commission of China, as well as the IBRO Foundation for international collaboration. Over that last decade research has been supported by several continuing grants from the National Heart, Lung and Blood Institute, including HL-36527, HL-51428, HL-63313 and HL-72125.

References

Ahn AC, Colbert A Anderson BJ, Martinsen OG, Hammerschlag R. Electrical properties of acupuncture points and meridians: a systematic review. Bioelectricmagnetics 29: 245–256. 2008.

Akhmedov TI, Vasil'ev IuM, Masliaeva LV. The hemodynamics and neurohumoral correlaties of the changes in status of hypertension patients under the influence of acupuncture. Ter Arkh 65: 22–24, 1993. (in Russian)

Alshelevich IuV, Merson MA, Afanas'eva GA. Serum aldosterone level in patients with hypertension during treatment by acupuncture. Ter Arkh 57: 42–45, 1985. (in Russian)

Ballegaard S, Pedersen F, Pietersen A, Nissen VH, Olsen NV. Effects of acupuncture in moderate stable angina pectoris. J Intern Med 227: 25–30, 1990.

Cao XD, Xu SF, Lu WX. Inhibition of sympathetic nervous system by acupuncture. Acupunct Electrother Res 8: 25–35, 1983.

Cheung L, Li P and Wong C. The Mechanism of Acupuncture Therapy and Clinical Case Studies. Taylor and Francis, 2001.

Chiang CY, Chang CT, Chu H L, Yan LF Peripheral afferent pathway for acupuncture analgesia. Sci Sin 16(210), 217. 1973.

Chiu YJ, Chi A and Reid IA. Cardiovascular and endocrine effects of acupuncture in hypertensive patients. Clin Exp Hypertens 19: 1047–1063, 1997.
Cho ZH, Chung SC, Jones JP, Park JB, Park HJ, Lee L, Wong EK, Min BI. Correction for Cho et al., New findings of the correlation between acupoints and corresponding brain cortices using functional MRI. Proc Natl Acad Sci U S A 103: 10527, 2006.
Cho ZH, Chung SC, Jones JP, Park JP, Park HJ, Lee HJ, Wong EK, Min BI. New findings of the correlation between acupoints and corresponding brain cortices using functional MRI. Proc Natl Acad Sci USA 95: 2670–2673, 1998.
Crisostomo M, Li P, Tjen-A-Looi SC, Longhurst JC. Nociceptin in rVLM mediates electroacupuncture inhibition of cardiovascular reflex excitatory response in rats. J Appl Physiol 98: 2056–2063, 2005.
Croley T. Acupuncture points-histological properties. In: Acupuncture the past and the present, edited by Hung KC. New York: Vontage Press, 1996, p. 47.
Dampney RAL, GR and Tan E. Vasopressor neurons in the rostral ventrolateral medulla of the rabbit. J auton Nerv Syst 14: 239–254, 1985.
Dan N. Clinical observation on the effect of acupuncture on hypertension by ambulatory blood pressure monitor. Chin J Combination of Chinese and Western Medicine 18: 26–27, 1998.
Dean C. Sympathoinhibition from ventrolateral periaqueductal gray mediated by the caudal midline medulla. Am J Physiol Regul Integr Comp Physiol 289: R1477-R1481, 2005.
Dean C and Woyach VL. Serotonergic neurons of the caudal raphe nuclei activated in response to hemorrhage in the rat. Brain Res 1025: 159–168, 2004.
Department of Anatomy, Shanghai First Medical College. The relationship between channels, acupoints and their surrounding nerves. Shanghai Science and Technology Publisher, 1973, p. 251–264.
Dung, HC. Anatomical Features Contributing to the Formation of Acupuncture Points. Am J Acupuncture 12, 139–143. 1984.
Eisenberg DM, Davis R, Ettner S, Appel S, Wilkey S, Van Rompay M, Kessler RC. Trends in alternative medicine use in the United States, 1990–1997: results of a follow-up national survey. JAMA 280: 1569–1575, 1998.
Eisenberg DM, Kessler RC, Foster C, Norlock FE, Calkins DR, Delbanco TL. Unconventional medicine in the United States: prevalence, costs and patterns of use. N Engl J Med 328: 246–252, 1993.
Flachskampf FA, Gallasch J, Gefeller O, Gan J, Mao J, Pfahlberg AB, Wortmann A, Klinghammer L, Pflederer W and Daniel WG. Randomized trial of acupuncture to lower blood pressure. Circulation 115: 3121–3129, 2007.
Fu L-W and Longhurst C. Electroacupuncture modulates vlPAG release of GABA through presynaptic cannabinoid CB_1 receptor. J Appl Physiol 106: 1800–1809, 2009.
Gao KM and Li P. Responses of neurones in the nucleus paragigantocellularis lateralis to somatic nerve stimulation. Neurosci Lett 178: 15–18, 1994.
Goldman N, Chen M, Fujita T., Xu Q, Peng W, Liu W, Jensen T, Pei Y, Wang FH, X, Chen J, Schnermann J, Takano T, Bekar L, Tieu K, Nedergaard M. Adenosine A1 receptors mediate local anti-nociceptive effects of acupuncture. Nature Neuroscience Advanced on line publication: 1–7, 2010.
Guo, HF, Tian J H, Wang X M, Fang Y, Hou Y P, Han J S Brain substrates activated by electroacupuncture of different frequencies (I): comparative study on the expression of oncogene c-fos and genes coding for three opioid peptides. Mol. Brain Res. 43, 157–166. 1996.
Guo XQ, Jai RJ, Cao QY, Guo ZD and Li P. Inhibitory effect of somatic nerve afferent impulses on the extrasystole induced by hypothalamic stimulation. Acta Physiol Sin 33: 343–350, 1981. (in Chinese with English abstract
Guo ZL, Longhurst JC Activation of reciprocal pathways between arcuate nucleus and ventrolateral periaqueductal gray during electroacupuncture: involvement of VGLUT3. Brain Res 1360: 77–88, 2010.
Guo ZL, Moazzami A, Longhurst J. Electroacupuncture induces c-Fos expression in the rostral ventrolateral medulla and periaqueductal gray in cats: relation to opioid containing neurons. Brain Res 1030: 103–115, 2004.

Guo ZL, Moazzami A, Tjen-A-Looi S, Longhurst J. Responses of opioid and serotonin containing medullary raphe neurons to electroacupuncture. Brain Res 1229: 125–136, 2008.
Guo Z and Longhurst J. Expression of c-Fos in arcuate nucleus induced by electroacupuncture: Relations to neurons containing opioids and glutamate. Brain Res 1166: 65–76, 2007.
Hajjar I and Kotchen TA. Trends in prevalence, awareness, treatment, and control of hypertension in the United States. 1988–2000. JAMA 290: 199, 2003.
Han CC. Acupuncture points-histological properties. Vontage Press, New York, 1996, p. 47.
Han JS and Terenius L. Neurochemical basis of acupuncture analgesia. Annu Rev Pharmacol Toxicol 22: 193–2 20, 1982.
Han JS. The neurochemical basis of pain relief by acupuncture. Beijing: China Medical and Pharmaceutical Technology, 1987.
He LF, Yu YH and Gao M. Temporal alterations of proopiomelanocortin mRNA level in rat hypothalamic arcuate nucleus following electroacupuncture. World J Acu-Mox 5: 36–41, 1995.
Heine H. Akupunkturtherapie-perforationen der oberflachlichen korperfaszie durch kutane GafaB-Nerverbundel. Therapentikon 4, 230–244. 1988.
Hu Nan Med College. Analysis of 160 cases with shock treated by new acupuncture. New Med J 2: 10, 1973. (in Chinese)
Huangfu DH and Li P. The inhibitory effect of the deep peroneal nerve inputs on defence reaction elicited by brainstem stimulation. Chin J Physiol Sci 1 (2): 176–184, 1985.
Huangfu D and Li P. Effect of deep peroneal nerve inputs on ventral medullary defence-reaction-related neurons. Chin J Physiol Sci 2: 123–131, 1986
Huangfu DH, Huang Q and Li P. Afferent connections of the ventrolateral medulla in the rabbit-studied with HRP techniques. Chin J Physiol Sci 3: 86–95, 1987.
Huangfu DH and Li P. The role of nucleus arcuatus in the inhibitory effect of deep peroneal nerve inputs on defence reaction. Chin J Physiol Sci 3: 37–46, 1987.
Huangfu DH and Li P. The inhibitory effect of ARC-PAG-NRO system on the ventrolateral medullary neurons in the rabbit. Chin J Physiol Sci 4: 115–125, 1988.
Kline, RL, Yeung KY, and Calaresu FR. Role of somatic nerves in cardiovascular response to stimulation of an acupuncture point in anesthetized rabbits. Exp Neurol 61, 561–570. 1978.
Kurono Y; Egawa m; Yano T; Shimoo K. The effect of acupuncture on the coronary arteries as evaluated by coronary angiography: a preliminary report. Am J Chin Med 30: 387–96, 2002.
Langevin HM, Wayne PM, MacPherson H, Schnyer R, Milley RM, Napadow V, Lao L, Park J, Harris RE, Cohen M, Sherman KJ, Haramati A, Hammerschlag R. Paradoxes in acupuncture research: strategies for moving forward. Evid Based Complement Alternat Med 2011: 180805. Epub 2010 Oct 11: 2011.
Li J and Li P. Effect on blood pressure by stimulating nerves under acupoint. Shanghai J Acup Moxib 11: 13–16, 1992.
Li M, Tjen-A-Looi SC and Longhurst JC. Electroacupuncture enhances preproenkephalin mRNA expression in rostral ventrolateral medulla of rats. Neurosci Lett 477: 61–65, 2010.
Li P, Zheng XZ, Jin WQ, Zhao XL. Studies on skin potential and skin galvanic reflex in human subjects. Acta Physiol Sin 25, 171–181. 1962 (in Chinese with English abstract).
Li P. Mechanism of the modulatory effect of acupuncture on acute experimental hypertension, hypotension and extrasystoles. Acupunct Res 9: 241–249, 1984.
Li P and Yao T. Mechanism of the modulatory effect of acupuncture on abnormal cardiovascular functions. Shanghai, China: Shanghai Medical University Press, 1992.
Li P, Pitsillides K, Rendig S, Pan HL, Longhurst J. Reversal of reflex-induced myocardial ischemia by median nerve stimulation: a feline model of electroacupuncture. Circulation 97: 1186–1194, 1998.
Li P. The Anatomical and Physiological Rule of Acupoints. In: Mechanism of Acupuncture Therapy and Clinical Case Studies, CRC Press, 2001, p. 34–43.
Li P, Tjen-A-Looi S and Longhurst J. Rostral ventrolateral medullary opioid receptor subtypes in the inhibitory effect of electroacupuncture on reflex autonomic response in cats. Autonomic Neuroscience 89: 38–47, 2001.
Li P, Rowshan K, Crisostomo M, Tjen-A-Looi S and Longhurst J. Effect of electroacupuncture on pressor reflex during gastric distention. Am J Physiol 283: R1335–R1345, 2002.

Li P, Ayannusi O, Reed C and Longhurst J. Inhibitory effect of electroacupuncture (EA) on the pressor response induced by exercise stress. Clinical Autonomic Research 14: 182–188, 2004.
Li P, Tjen-A-Looi S and Longhurst J. Excitatory projections from arcuate nucleus to ventrolateral periaqueductal gray in electroacupuncture inhibition of cardiovascular reflexes. Am J Physiol 209: H2535–H2542, 2006.
Li P, Tjen-A-Looi S, Longhurst J. Role of Arcuate Nucleus and Ventrolateral Periaqueductal Gray in Electroacupuncture Inhibition of Sympathoexcitatory Cardiovascular Reflex Response. FASEB J 20: A734, 2006.
Li, P and Longhurst, JC. Long-lasting inhibitory effect of EA on blood pressure in patients with mild to moderate hypertension. Society for Neuroscience 35. 2007. Ref Type: Abstract
Li P, Tjen-A-Looi S, Guo Z and Longhurst JC. Reciprocal neuronal projections between arcuate and ventrolateral periaqueductal gray participate in long-lasting EA inhibition of reflex blood pressure elevation. FASEB J 22: 2008.
Li P, Tjen-A-Looi SC, Guo Z, Fu L, Longhurst JC. Long loop pathways in cardiovascular electroacupuncture response. J Appl Physiol 106: 620–630, 2009.
Li P and Longhurst JC. Neural mechanism of electroacupuncture's hypotensive effects. Autonomic Neuroscience: Basic & Clinical 157: 24–30, 2010.
Li P, Tjen-A-Looi SC, Guo Z-L and Longhurst J.C. An arcuate-ventrolateral periaqueductal gray reciprocal circuit participates in electroacupuncture cardiovascular inhibition. Autonomic Neuroscience: Basic & Clinical 158: 13–23, 2010.
Li P, Tjen-A-Looi SC, Longhurst J.C. Nucleus raphé pallidus participates in midbrain-medullary cardiovascular sympathoinhibition during electroacupuncture. Am J Physiol Regul Integr Comp Physiol 299: R1369–R1376, 2010.
Lin RJ, Gong QL and Li P. Convergent inputs to neurones in the nucleus paragigantocellularis lateralis in the cat. Neuroreport 2: 281–284, 1991.
Lin SX and Li P. Mechanism of inhibitory effect of electro-acupuncture on noradrenaline hypertension. Acta Physiol Sin 33: 335–342, 1981. (in Chinese with English abstract).
Liu J, Han Z, Chen S, and Cao Q. Influence of electroacupuncture of neiguan (PC 6) on ami-induced changes in electrical activity of dorsal horn neurons Zhen Ci Yan Jiu. 19(1), 37–41. 1994. (in Chinese).Ref Type: Abstract
Loewy AD. Central autonomic pathways. In: Central regulation of autonomic functions, edited by Loewy AD and Spyer KM. New York: Oxford University Press, 1990, p. 88–103.
Longhurst JC, Kelly AR, Gonyea WJ et al. Cardiovascular responses to static exercise in distance runners and weight lifters. J Appl Physiol 49: 676–683, 1980.
Lovick TA. Descending projections from the ventrolateral medulla and cardiovascular controls. Pflulgers Arch 404: 197–202, 1985.
Lovick TA, Li P and Schenberg LC. Modulation of the cardiovascular defence response by low frequency stimulation of a deep somatic nerve in rats. J Auton Nerv Syst 50: 347–354, 1995.
Lu GW. Characteristics of afferent fiber innervation on acupuncture points zusanli. Am J Physiol 245: R606–R612, 1983.
Macklin EA, Wayne PM, Kalish, LA, Valaskatgis P, Thompson J, Pian-Smith, MC, Zhang, Q, Stevens, S, Goertz, C, Prineas RJ, Buczynski, B, Zusman, RM. Stop Hypertension with the Acupuncture Research Program (SHARP): results of a randomized, controlled clinical trial. Hypertension 48(5), 838–845. 2006.
MacPherson H, Thomas K, Walters S, Fitter M. A prospective survey of adverse events and treatment reactions following 34,000 consultations with professional acupuncturists. Acupunct Med 19: 93–102, 2001.
Mayer DJ. Acupuncture: An evidence-based review of the clinical literature. Annu Rev Med 51: 49–63, 2000.
Melzack R and Wall PD. Acupuncture and transcutaneous electrical nerve stimulation. Postgrad Med J 60: 893–896, 1984.
Moazzami A, Tjen-A-Looi SC, Guo ZL, Longhurst JC. Serotonergic projections from nucleus raphe pallidus to rostral ventrolateral medulla participate in acupuncture modulation of cardiovascular excitatory reflexes. FASEB J 582.23, 2007.

Nakatani Y and Yamashita H. Ryodoraku acupuncture. Ryodoraku Research Institute, Osaka, Japan. 1977.
Unschuld P U. Huang Di Nei Jing Su Wen : nature, knowledge, imagery in an ancient Chinese medical text, with an appendix, the doctrine of the five periods and six qi in the Huang Di nei jing su wen. University of California Press, 2003.
Peng YJ, Gong QL, Li P. Convergence of midbrain, visceral and somatic inputs onto neurons in the nucleus paragigantocellularis lateralis in rats. Acta Physiol Sin 50: 575–580, 1998.
Pomeranz B and Paley D. Electroacupuncture hyperalgesia is mediated by afferent nerve impulses: an electrophysiological study in mice. Exp Neurol 66: 398–402, 1979.
Research group of acupuncture anesthesia Peking Medical College. Effect of acupuncture on pain threshold of human skin. In: The neurochemical basis of pain relief by acupuncture, edited by Han JS. China Medical & Pharmaceutical Technical Publisher, 1987, p. 21–28.
Rossi G, Mathis J and Pasternak G. Analgesic activity of orphanin FQ2, murine prepro-orphanin FQ 141–157 in mice. Neuroreport 1165–1168, 1998.
Sato A and Schimdt RF. The modulation of visceral functions by somatic afferent activity. Jpn J Physiol 337: 1–17, 1987.
Shi GX, Liu CZ, Zhu J, Guan LP, Wang DJ, Wu MM. Effects of acupuncture at Sanyinjiao (SP6) on prostaglandin levels in primary dysmenorrhea patients. Clin J Pain. 27(3), 258–61. 2011. Ref Type: Abstract
Shi XM and Zhang MC. A Chinese-English dictionary of acupuncture and moxibustion. Huaxia Publisher, Beijing., 1998.
Stener-Victorin E, Waldenstrom U, Tagnfors U, Lundeberg T, Lindstedt G and Janson PO. Effects of electro-acupuncture on anovulation in women with polycystic ovary syndrome. Acta Obstet Gynecol Scand 79: 180–188, 2000.
Sun X, Yu J, Yao T. Pressor effect produced by stimulation of somatic nerve on hemorrhagic hypotension in conscious rats. Acta Physiol Sin 35: 264–270, 1983.
Syuu Y, Matsubara H, Hosogi S and Suga H. Pressor effect of electroacupuncture on hemorrhagic hypotension. Am J Physiol Heart Circ Physiol 285: R1446–R1452, 2003.
Tam KC and Yiu HH. The effect of acupuncture on essential hypertension. Am J Chin Med 3: 369–375, 1975.
The Cooperative Group of Shandong Medical College and Shandong College of Traditional Chinese Medicine. Anatomical Atlas of Chinese Acupuncture Points. Jinan, China: Shandong Science and Technology Press, 1990.
Tjen-A-Looi S, Bonham A and Longhurst J. Interactions between sympathetic and vagal cardiac afferents in nucleus tractus solitarii. Am J Physiol 272: H2843–H2851, 1997.
Tjen-A-Looi SC, Li P and Longhurst JC. Prolonged inhibition of rostral ventral lateral medullary premotor sympathetic neuron by electroacupuncture in cats. Autonomic Neuroscience 106: 119–131, 2003.
Tjen-A-Looi SC, Li P and Longhurst JC. Medullary substrate and differential cardiovascular response during stimulation of specific acupoints. Am J Physiol 287: R852–R862, 2004.
Tjen-A-Looi, SC, Li, P, and Longhurst, JC. Midbrain vlPAG inhibits rVLM cardiovascular sympathoexcitatory activity during low frequency electroacupuncture. Society for Neuroscience 34. 2004.Ref Type: Abstract
Tjen-A-Looi S, Fu LW, Zhou W, Longhurst JC. Role of unmyelinated fibers in electroacupuncture cardiovascular responses. Autonomic Neuroscience 118: 43–50, 2005.
Tjen-A-Looi SC, Li P and Longhurst JC. Midbrain vIPAG inhibits rVLM cardiovascular sympathoexcitatory responses during acupuncture. Am J Physiol 209: H2543–H2553, 2006.
Tjen-A-Looi S C, Li P, Longhurst J C Endocannabinoid/GABA system in vlPAG during electroacupuncture-related cardiovascular responses. Neuroscience 32, 417.12. 2007.Ref Type: Abstract
Tjen-A-Looi SC, Li P and Longhurst J.C. Role of medullary GABA, opioids, and nociceptin in prolonged inhibition of cardiovascular sympathoexcitatory reflexes during electroacupuncture in cats. Am J Physiol 293: H3627–H3635, 2007.

Tjen-A-Looi S, Li P, Longhurst C. Processing cardiovascular information in the vlPAG during electroacupuncture in rats: Roles of endocannabinoids and GABA. J Appl Physiol 1793–1799, 2009.
Toda K & Ichioka M. Electroacupuncture relation between forelimb afferent impulses and suppression of jaw opening reflex in the rat. Exp Neurol 61: 464–470. 1978.
Uchida S, Kagitani F, Watanabe N, Hotta H. Sympatho-inhibitory response of the heart as a result of short-term acupuncture-like stimulation of the rat hindlimb is not augmented when sympathetic tone is high as a result of hypercapnia. J Physiol Sci 60: 221–225, 2010.
Verneijoul P, Parras J and et al. Etude des meridiens d'acupuncture par les traccurs radioactifs. Bull Acad Natl Med (Paris) 169: 1071–1075, 1985.
Wang SM, Kain ZN, and White P. Acupuncture analgesia: I. The scientific basis. Anesth Analg 106, 602–610. 2008.
Wang KM, Yao SM, Xian YL and Hou ZL. A study on the receptive field of acupoints and the relationship between characteristics of needling sensation and groups of afferent fibres. Scientia Sinica Series B, 28: 963–971, 1985.
Wong J and Cheng R. The science of acupuncture therapy. Hong Kong: Kola Mayland Co., 1987.
World Health Organization. Acupuncture: Review and Analysis of Reports on Controlled Clinical Trials. Geneva : World Health Organization, 2002: WHO Press, 2003.
Xiao YF and Li P. Influence of microinjection of scopolamine into various brain areas on pressor effect of electroacupuncture in nitroprusside induced hypotensive dogs. Acta Physiol Sin 37: 31–36, 1985. (in Chinese with English abstract).
Xiao YF, Lin S, Deng Z, Li P. Mechanism of pressor effect of electroacupuncture on nitroprusside hypotension in dogs. Acta Physio Sin 35: 257–263, 1983. (in Chinese with English abstract).
Xiao YF, Sun XY, Yao T and Li P. Cardiovascular effects of deep peroneal nerve stimulation in hypotensive or shock dogs. Chin J Appl Physiol 1: 269–273, 1985.
Xie GZ, Zhu DN and Li P. The depressor effect on stress induced hypertensive rat by electroacupuncture applied on deep peroneal nerve underneath Zusanli (St 36). Shanghai J Acup &Moxib 1 G: 32–33, 1997. (in Chinese).
Xu FH and Wang JM. Clinical observation on acupuncture combined with medication for intractable angina pectoris. Zhongguo Zhen Jiu. 25, 89–91. 2005. (in Chinese)Ref Type: Abstract
Yao T, Andersson S and Thoren P. Long-lasting cardiovascular depression induced by acupuncture-like stimulation of the sciatic nerve in unanaesthetized spontaneously hypertensive rats. Brain Res 240: 77–85, 1982.
Yu AS, Zhao YX, Li XL. Morphological research in Neiguan (Pe 6)'s three dimensional structure. Shanghai J Acup Moxib 15, 30–31. 1996. (in Chinese)
Zhang CL. Clinical investigation of acupuncture therapy. Clin J Med 42: 514–517, 1956.
Zhang Z, Lin RJ, Fan W, Gong QL and Li P. Involvement of 5-HT and GABA in the inhibition induced by NRO stimulation of rVLM -defense-reaction-related neurons. Chinese Journal Physiological Sciences 8: 208–215, 1992.
Zhao ZQ. Neural mechanism underlying acupuncture analgesia. Prog Neurobiology 85, 355–375. 2008.
Zhou W, Fu LW, Tjen-A-Looi SC, Li P and Longhurst JC. Afferent mechanisms underlying stimulation modality-related modulation of acupuncture-related cardiovascular responses. J Appl Physiol 98: 872–880, 2005.
Zhou W, Tjen-A-Looi S and Longhurst JC. Brain stem mechanisms underlying acupuncture modality-related modulation of cardiovascular responses in rats. J Appl Physiol 99: 851–860, 2005.
Zhou W, Hsiao I, Lin V and Longhurst J. Modulation of cardiovascular excitatory responses in rats by transcutaneous magnetic stimulation: role of the spinal cord. J Appl Physiol 100: 926–932, 2006.
Zhou W, Fu LW, Tjen-A-Looi SC, Guo ZL and Longhurst JC. Role of glutamate in a visceral sympathoexcitatory reflex in rostral ventrolateral medulla of cats. American Journal of Physiology 291: H1309–H1318, 2006.

Chapter 16
Acupuncture in Polycystic Ovary Syndrome: Potential and Challenge

Yi Feng, Elisabet Stener-Victorin, and Boying Chen

Abstract Polycystic ovary syndrome (PCOS) is the most common endocrine and metabolic disorder affecting 5–10 % of women of reproductive age. Conventional pharmacological therapy and surgery are effective but with obvious side effects and limited long-term treatment. Therefore, acupuncture therapy, as a complement or alternative to conventional therapies, may be an alternative for women with PCOS by reducing their symptoms. Experimental, mechanistic trials clearly indicate positive effects of acupuncture on reproductive and metabolic function. However, lack of randomized controlled trials hinders the translation and application of acupuncture in the treatment of PCOS. In this chapter, we summarize the PCOS diagnostic criteria and characters, PCOS-like animal models as well as the possible physiological basis and mechanism of the effect of acupuncture in PCOS. Clinical data from studies in women with PCOS and experimental observations in PCOS-like rat models suggest that acupuncture exerts long-lasting beneficial effects on metabolic, endocrine, and neuroendocrine systems.

Y. Feng (✉) • B. Chen
Department of Integrative Medicine and Neurobiology, State Key Lab of Medical Neurobiology, Shanghai Medical College, Fudan University, Shanghai 200032, China

Institute of Acupuncture Research (WHO Collaborating Center for Traditional Medicine), Institutes of Brain Science, Fudan University, Shanghai 200032, China
e-mail: fengyi17@fudan.edu.cn; bychen@shmu.edu.cn

E. Stener-Victorin
Department of Physiology/Endocrinology, Institute of Neuroscience and Physiology, The Sahlgrenska Academy at University of Gothenburg, Gothenburg, Sweden

Department of Obstetrics and Gynecology, The First Affiliated Hospital, Heilongjiang University of Chinese Medicine, Harbin 150040, China
e-mail: elisabet.stener-victorin@neuro.gu.se

Y. Xia et al. (eds.), *Current Research in Acupuncture*,
DOI 10.1007/978-1-4614-3357-6_16,

Keywords Acupuncture • Polycystic ovary syndrome • Metabolic syndrome • Androgen receptor • Gonadotropin-releasing hormone • Corticotropin-releasing hormone • Hypothalamic–pituitary–ovarian axis

16.1 Introduction

Polycystic ovary syndrome (PCOS), an endocrine and metabolic disorder, is a significant women's health issue due to its high incidence (Ehrmann 2005). It is estimated to affect 5–10 % of women of reproductive age worldwide (Chapman et al. 2009; Norman et al. 2007). The disorder presents a range of clinical signs and symptoms that include hyperandrogenism, polycystic ovarian morphology, chronic anovulation, infertility, hyperinsulinemia, and insulin resistance with a high prevalence of obesity (Adams et al. 2004; Chapman et al. 2009; Norman et al. 2007; Taylor 1998). Given that neither etiological factors nor mechanisms of this disorder progression are completely clear, understanding of the pathophysiology of human PCOS is challenging.

Many women with PCOS require prolonged treatment. Pharmacological approaches are effective but have adverse effects. First-line therapy in PCOS is often hormonal medication, which reduces hirsutism and acne but adversely affects glucose tolerance, coagulability, and fertility (Lanham et al. 2006). Therefore, treatment strategies such as acupuncture or physical exercise need to be evaluated in PCOS.

Acupuncture, a treatment that dates back 3,000–5,000 years, is an integral part of traditional Chinese medicine and has become more established in Western medicine as a complement or alternative to conventional therapies. Repeated acupuncture treatment in women with PCOS and women with undefined ovulatory dysfunction has been shown to exert long-lasting beneficial effects on endocrine parameters and ovulation with no negative side effects (Chen and Yu 1991; Stener-Victorin et al. 2000a).

16.2 Clinical Application of Acupuncture for PCOS

16.2.1 Human PCOS: Diagnostic Criteria

In 1935, Stein and Leventhal published the first paper about seven cases of women with amenorrhea, hirsutism, obesity, and a characteristic polycystic appearance of ovaries (Stein and Leventhal, 1935). The syndrome is known as PCOS today.

The first diagnostic criteria for PCOS were set up by the National Institutes of Health (NIH) in 1990 (Zawadski and Dunaif 1992). The major criteria include (1) hyperandrogenism and/or hyperandrogenemia, (2) menstrual dysfunction, and

Table 16.1 The progress of diagnostic criteria for PCOS

Conference	Diagnostic criteria for PCOS	References
NIH 1990	(1) Hyperandrogenism and/or hyperandrogenemia (2) Menstrual dysfunction (3) Exclusion of other known disorders	(Zawadski and Dunaif 1992)
Rotterdam 2003	(1) Oligo-ovulation or anovulation (usually manifested as oligomenorrhea or amenorrhea) (2) Elevated levels of circulating androgens (hyperandrogenemia) or clinical manifestations of androgen excess (hyperandrogenism) (3) Polycystic ovaries as defined by ultrasonography	(Revised 2003 consensus … (2004))
AE–PCOS 2006	(1) Hyperandrogenism: hirsutism and/or hyperandrogenemia (2) Ovarian dysfunction: oligo-anovulation and/or polycystic ovaries (3) Exclusion of the other androgen excess or related disorders	(Azziz et al. 2009)

(3) exclusion of other known disorders. With the development of ultrasonography and the ability to examine the hormone concentrations, the diagnostic criteria were revised in a conference in 2003 in Rotterdam, Netherlands. PCOS should be diagnosed according to at least two of the following criteria: (1) oligo-ovulation or anovulation (usually manifested as oligomenorrhea or amenorrhea), (2) elevated levels of circulating androgens (hyperandrogenemia) or clinical manifestations of androgen excess (hyperandrogenism), and (3) polycystic ovaries as defined by ultrasonography (2004). This broader definition increased significantly the affected population. Therefore, Androgen Excess and PCOS (AE–PCOS) Society suggested new criteria in 2006 as follows: (1) hyperandrogenism: hirsutism and/or hyperandrogenemia, (2) ovarian dysfunction: oligo- or anovulation and/or polycystic ovaries, and (3) exclusion of the other androgen excess or related disorders (Azziz et al. 2009). The clinical presentation requires an evaluation to rule out other endocrine disorders (e.g., virilizing tumors, hyperprolactinemia, and Cushing's syndrome) (Ehrmann 2005) (Table 16.1).

Polycystic ovaries are not necessary to be present to make a diagnosis of PCOS (Adams et al. 2004), and conversely, their presence alone does not establish the diagnosis (Adams et al. 2004; Ehrmann 2005). Because of diverse and heterogeneous nature of PCOS, the phenotypic expression of the disorder is variable (Taylor et al. 1997). Though it has already occurred at prepuberty or adolescence (Norman et al. 2007), the syndrome is often found and diagnosed at adulthood. The main symptom is menstrual irregularity, oligomenorrhea (fewer than 9 menses per year) or amenorrhea, and infertility. The clinical presentation has been assigned to increased ovarian androgen production and high levels of luteinizing hormone (LH) or a high LH to follicle stimulating hormone (FSH) ratio (Norman et al. 2007). Chronic anovulation may lead to dysfunctional uterine bleeding and decrease fertility. In addition, hypersecretion of androgens into circulation stimulates the pilosebaceous unit and produces cutaneous manifestations including hirsutism, acne, and male pattern hair loss (androgenic alopecia) (Zacche et al. 2009). At the same time, metabolic dysfunction is highly prevalent with PCOS patients. Over 50 % of women with PCOS are overweight or obese, and most of them have the abdominal phenotype, which is partly responsible for insulin resistance and associated hyperinsulinemia (Nardo et al. 2008). Furthermore, the prevalence of impaired glucose tolerance, type 2 diabetes, and dyslipidemia was increased (Abbott et al. 2009). In addition, the risk of cardiovascular disease (CVD) is also potentially increased (Moran and Teede 2009).

16.2.2 Conventional Therapies for Patients with PCOS

The best therapy for metabolic disturbances in women with PCOS is lifestyle adjustment. The strategies aim to reduce weight—including different diets, e.g., low glycemic index, and exercise—and to stop smoking in order to improve the metabolic and reproductive profile.

The first-line pharmacological therapy for ovulation induction in PCOS is clomiphene citrate (CC). CC plays an anti-estrogenic effect on the hypothalamic level, on the endometrium, and on the development of hostile cervical mucus. It is effective in approximately 50–70 % of cases, resulting in pregnancy in half of these cases (Legro 2007; Legro et al. 2007; Moll et al. 2006; Neveu et al. 2007). Gonadotropin therapy is often used as second-line therapy in anovulatory women with PCOS who either are resistant to ovulation induction or ovulate but do not conceive. Further, insulin-sensitizing drugs including metformin, rosiglitazone, pioglitazone, and D-chiro-inositol alone or in combination with CC are also applied in women with PCOS. Flutamide, an antiandrogen treatment, is used to decrease symptoms of hyperandrogenemia. However, both the effects and risks of long-term use are questionable.

16.2.3 Acupuncture Therapy in Clinic for PCOS

We conducted one study including 11 anovulatory women (nine with PCOS) that received low-frequency EA (3 days/cycle, 13 cycles) to induce ovulation (Chen and Yu 1991). Ovulation was induced in 6 of 13 menstrual cycles. EA had no effect in controls. In the anovulatory women, high plasma β-endorphin levels and low hand skin temperature, indicating increased sympathetic nervous activity, were improved by EA, likely reflecting inhibition of the sympathetic nervous system (Chen and Yu 1991).

In a German study, auricular acupuncture, once a week during 3 months, given to 45 infertile women with hormonal disturbances was compared to hormone treatment in 45 women in a nonrandomized trial (Gerhard and Postneek 1992). In the 45 women, auricular acupuncture yielded pregnancy rates equivalent to those induced by hormonal treatment but with fewer side effects and miscarriages.

In another Chinese study (Xiaoming et al. 1993), 12 of 24 women with undefined ovulatory dysfunction treated with manual acupuncture (average, 30 treatments) had marked improvements in menstruation and biphasic basal body temperature (BBT) for more than two cycles or became pregnant. Regulatory effects on LH, FSH, and estradiol (E_2) were shown, indicating an influence on the hypothalamic–pituitary–gonadal axis (HPGA).

In 2000, we conducted another study on women with well-defined and diagnosed PCOS and anovulation to elucidate the effect of repeated EA treatments on endocrine and neuroendocrine parameters as well as on anovulation (Stener-Victorin et al. 2000b). Twenty-four women (age 24–40 years) with PCOS and oligo-/amenorrhea were included in this nonrandomized, longitudinal, prospective study. The study period was defined as the period extending from 3 months before the first EA treatment to 3 months after the last EA treatment (10–14 treatments), in total 8–9 months. The menstrual and ovulation patterns were confirmed by recording of vaginal bleedings and by daily registrations of the BBT. Blood samples were collected within a week before the first EA, within a week after the last EA, and 3 months after EA. Nine women (38 %) experienced a good effect. They displayed a mean of 0.66 ovulations/woman and month in the period during and after the EA period compared to a mean of 0.15 before the EA period ($p<0.01$). Before EA, women with a good effect had a significantly lower body mass index (BMI) ($p<0.001$), waist-to-hip circumference ratio (WHR) ($p<0.01$), serum testosterone concentration ($p=0.0098$), serum testosterone/sex hormone-binding globulin (SHBG) ratio ($p<0.05$), and serum basal insulin concentration ($p<0.01$) and a significantly higher concentration of serum SHBG ($p<0.05$) than did those women with no effect. This study showed that repeated EA exerts long-lasting effects on both endocrine parameters as well as anovulation. These results are in accordance with previous results, but it is obvious that randomized, comparative studies are needed to verify the results and to exclude nonspecific effects. However, these studies do not enlighten possible underlying mechanisms of EA.

Recently, for the first time it was demonstrated in a randomized controlled trial that low-frequency EA and physical exercise lower high sympathetic nerve activity in women with PCOS. Thus, treatment with low-frequency EA or physical exercise with the aim to reduce sympathetic nerve activity may be of importance for women with PCOS (Stener-Victorin et al. 2009).

Further, fourteen low-frequency EA treatments during 16 weeks and 16 weeks of physical exercise compared with no intervention, improved menstrual bleeding pattern and decreased high levels of sex steroid precursors, estrogens, androgens, and glucuronidate androgen metabolites in women with PCOS (Jedel et al. 2011). Acupuncture was superior to physical exercise when compared directly after the treatment. In the most recent RCT, twelve treatments of true acupuncture was compared with sham acupuncture (so called placebo needles) during eight weeks and they found similar ovulation frequency and improvement in LH/FSH ratio in both groups of women with PCOS (Pastore et al. 2011). Thus, they were unable to demonstrate differences between true and sham acupuncture and they did not include a non-intervention group.

16.3 Animal Model for PCOS Study

Since the mechanism and course of this disorder are poorly understood, the combined information from relevant experimental models and human disease phenotype may help to define mechanisms for the initiation and development of PCOS-related pathological events. The establishment of animal models compatible with human PCOS is challenging, and the dissection of this complicated syndrome is a difficult task.

Common PCOS animal models include estrogen-induced, androgen-induced, or constant-illumination-induced PCOS in female rats (Table 16.2) and in rhesus monkey and sheep.

16.3.1 Estrogen-Induced PCOS Rat

A single or repeated injection of 2 mg *estradiol valerate* (*EV*) to adult female rat develops polycystic ovary morphology within 8–9 weeks. The ovaries become multifollicles and contain no corpora lutea (Brawer et al. 1986). The granulosa cells degenerate, whereas theca cells hypertrophy. The concentrations of dehydroepiandrosterone (DHEA), testosterone (T), and androstenedione (A) as well as 17β-estradiol (E_2) and leptin do not change significantly. However, the plasma LH increases significantly, while the FSH is not as dramatic as LH (Mahajan 1988). Progesterone increases significantly, which is different from human PCOS. Another estrogen-induced model is young female rats (175–200 g) with regular estrous cycle receive an E_2 injection. The ovaries were large and smooth surfaced. Small cystic follicles were distributed largely in the peripheral cortical region of the ovary (McCarthy and Brawer 1990). The E_2-treated rat exhibits a complex plasma LH pattern. The FSH pattern shows large scale conformity with the LH pattern.

16.3.2 Androgen-Induced PCOS Rat

Androgen-induced PCOS rats include *dehydroepiandrosterone* (*DHEA*)-treated, *letrozole*-treated, and *dihydrotestosterone* (*DHT*)-treated rats. Though the ovary presents polycystic morphology, DHEA model (60 mg/kg BW, 27-day-old female

Table 16.2 General rat animal models with PCOS-like phenotypes

	Human	Rat							
		EV-treated	E2-treated	DHEA-treated	Letrozole-treated	DHT-treated	RU486-treated	JCR:LA-cp strain	Constant illumination
Ovary									
Ovarian size/weight	↑	↓	↑	↑	↑	↓	↑	↓	–
Follicular cysts	+	+	+	+	↑	+	+	+	+
Follicular atresia	+	+	n.d.	+	+	+	+	+	+
Theca layer	↑	–	↑		↑	↑	↑	–	–
Granulosa layer	↓	↑	↓		↓	↓	–	↓	–
No. of follicles	↑	↑	↑	↑	↑	↑	↑	↑	↑
Ovulation disturbances	+	+	+	+	+	+	↑	+	+
Infertility	+	+	+	+	+	+	+	+	+
Hormones									
GnRH	↑	n.d.	n.d.	↑	n.d.	↑	n.d.	n.d.	↓
LH	↑	↑	complex	↓	n.d.	n.d.	↑	n.d.	↑
FSH	↑/–	–	complex	↓	↑	n.d.	↓	n.d.	↓
LH/FSH	↑	n.d.	n.d.	↑	n.d.	n.d.	↑	n.d.	↑
T	↑	↓/–	n.d.	↑	↑	–	↑	↑(total)	–
DHT	↑	–	n.d.	↑	–	–	n.d.	n.d.	–
P4	↓	↑	n.d.	↑	↓	↓	↓	n.d.	↓
E2	↑/–	↑/–	–	n.d.	↓	–	↑	–	↑
A	↑/–	–	n.d.	n.d.	n.d.	n.d.	n.d.	n.d.	–
Leptin	↑	–	n.d.	n.d.	–	↑	n.d.	↑	n.d.

(continued)

Table 16.2 (continued)

	Human	Rat							
		EV-treated	E2-treated	DHEA-treated	Letrozole-treated	DHT-treated	RU486-treated	JCR:LA-cp strain	Constant illumination
BMI/body weight	↑	–	n.d.	n.d.	↑	↑	n.d.	↑	n.d.
Obesity	+	–	n.d.	n.d.	–	+	n.d.	+	n.d.
Insulin resistance	+	–	n.d.	n.d.	–	+	n.d.	+	n.d.
References	(Ehrmann 2005; Manneras et al. 2007)	(Mahajan 1988; Manneras et al. 2007)	(McCarthy and Brawer 1990)	(Mahesh et al. 1987; Roy et al. 1962)	(Kafali et al. 2004; Manneras et al. 2007)	(Manneras et al. 2007)	(Ruiz et al. 1996; Sanchez-Criado et al. 1993)	(Shi et al. 2009; Wu-Peng et al. 1997)	(Mahesh et al. 1987)

+ Presence, – no changes, ↑ increased, ↓ decreased, n.d. not determined

rats for 3 days) differs from the human PCOS because FSH is slightly higher and LH is suppressed (Mahesh et al. 1987). This model displays high circulating levels not only of DHEA but also of androstenedione and T as in women with PCOS (Mahajan 1988). Six-week-old female rats were administered *letrozole* at concentrations of 0.1 or 0.5 or 1 mg/kg p.o. daily or by implantation of a subcutaneous slow-releasing pellet [30]. Histology of ovaries exhibited very striking similarity to human PCOS. In both situations, it is possible to see the abundance of subcapsular cysts lined with a thin layer of granulosa cells and hyperplasia of theca interna cells. The thickened fibrous capsule of ovary observed in PCOS was also present in letrozole-treated rat (Manneras et al. 2007). These rats display higher androgen concentrations which is in line with human PCOS but decreased estrogen concentrations together with increased FSH concentrations, not typical with human PCOS, which is a drawback of this model (Kafali et al. 2004). The letrozole seems to be a suitable model to study the ovarian part of the syndrome. Twenty one-day-old immature female rats were implanted s.c. with 90-day continuous-release pellets containing 7.5 mg DHT (daily dose, 83 μg) (Manneras et al. 2007). DHT rats develop irregular cycles. Their ovaries had large, atretic antral follicles, follicular cysts with a thickened theca interna cell layer, a diminished granulosa cell compartment, and low plasma testosterone and progesterone concentrations. In addition, DHT rats also display metabolic features similar to human PCOS, including increased body weight, increased body fat, and enlarged mesenteric adipocytes, as well as increased leptin levels and insulin resistance (Manneras et al. 2007). Therefore, the DHT model is suitable for studies of both ovarian and metabolic features of the syndrome.

16.3.3 Constant Illumination PCOS Rat

The rat estrous cycle is related with the circadian system (Schwartz 1969). By continuous exposure to light, the ovarian follicles develop to the preovulatory stage and are able to synthesize and secrete estrogen. By the end of 2 months of exposure to *Constant Light*, atretic follicles are found, and most of the tertiary follicles are decidedly distended and cystic (Mahajan 1988).

Because the ovulation is stopped and no corpora lutea form in the ovary, the level of progesterone is extremely low. In contrast, the serum level of E_2 is higher than that on diestrous. Whereas theca cells of cystic follicles and atretic follicles are not hypertrophic and do not secrete excessive amounts of androgens, the amount of T is normal (Mahajan 1988). Chronic exposure to constant light has an inhibitory effect on the metabolic system of the hypothalamic–pituitary–ovarian axis. This inhibition could result in a permanent suppression of the release of gonadotropin-releasing hormone (GnRH) and eventually lead to a decline in LH output (Mahajan 1988). Although *constant illumination* rats show some changes in hypothalamus, pituitary, and ovaries, these lesions are easily reversed to normal if deprived of the constant light. Moreover, constant illumination rats have low LH and androgen levels, which are different from the human PCOS (Mahajan 1988).

16.3.4 Genetic Modifications for Molecular and Pathological Studies in Human PCOS

While the abnormal gene expression profiles in the ovary and theca cells for human PCOS have been identified (Jansen et al. 2004; Wood et al. 2003), transgenic and knockout (KO) mouse models with ovarian pathological phenotypes mimicking those found in human PCOS are of great interest.

Animal studies have demonstrated that both estrogens and androgens contribute to folliculogenesis, ovarian remodeling, and development of several diseases (Kaipia and Hsueh 1997). There is in vivo and in vitro evidence to support the concept that theca cells in growing follicles are a significant site of androgen biosynthesis (Ehrmann 2005); however, estrogens have been shown to inhibit androgen production in estrogen receptor (ER) α-expressing theca cells (Leung and Armstrong 1979; Leung et al. 1978; Marins et al. 1981). Understanding the role of the ERα in the regulation of theca cell function has been aided by the ablation of the global and theca cell-specific *ERα* gene in mice (Couse et al. 1999; Dupont et al. 2000; Lee et al. 2009; Lubahn et al. 1993). The initial analyses of ovarian phenotypes of these KO mice indicate that formation of hemorrhagic cystic follicles is likely to arise from the ablation of ovarian ERα action in theca cells. However, the notion of hyperplasia of theca cells in human PCOS (Legro et al. 1998) is consistent with histological findings in overexpression of hCG/LH (Matzuk et al. 2003; Risma et al. 1995; Risma et al. 1997), or plasminogen activator inhibitor-1 (PAI-1) KO mice (Devin et al. 2007) and ERαKO mice (Couse et al. 1999; Dupont et al. 2000; Lee et al. 2009; Lubahn et al. 1993; Schomberg et al. 1999), but it is not supported by the observation when female mice are lacking P450arom (ArKO), although ovaries of ArKO mice show hemorrhagic cystic follicles (Britt et al. 2000; Ito et al. 1993). Because the human PCOS is often associated with a metabolic disturbance (Azziz et al. 2009; Norman et al. 2007), caution should be taken when selecting appropriate transgenic or KO models. In addition to impaired ovarian function and fertility, the onset of insulin resistance and diabetes is observed in adult female ERαKO mice (Heine et al. 2000) with similar phenotypes as seen in some PCOS patients (Azziz et al. 2009; Ehrmann 2005). Thus, activation of ERα may serve as a critical link between reproduction and metabolic paradigms. Although defects in theca cell-specific ERαKO resulting in various metabolic changes are needed to be determined, ERαKO mouse model can be useful for progressive studies of ovarian dysfunction and metabolism changes or can be expanded to interrogate complex PCOS signature.

Manipulations of endogenous estrogen levels by treatment with EV or pregnant mare's serum gonadotropin (PMSG) have previously been shown to increase the production of ovarian nerve growth factor (NGF) in rat theca cells in association with the induction of follicular cysts formation (Dissen et al. 2000; Lara et al. 2000). Interestingly, transgenic NGF mice were found to be susceptible to challenge with gonadotropins. After treatment with gonadotropins, they presented a detectable ovarian phenotype, including the formation of massive follicular cysts (Dissen et al. 2009), suggesting that endogenous steroid hormones participate in NGF-mediated ovarian dysfunction.

Another example is the glycoprotein PAI-1, which is highly related with human PCOS. Several studies have shown that increases in PAI-1 levels and activities are detected in women with PCOS(Atiomo et al. 1998; Orio et al. 2004; Sampson et al. 1996; Tarkun et al. 2004) and PAI-1 4G/5G polymorphism is significantly associated with the risk of developing PCOS(Bagos 2009). Both pathological and histological ovarian phenotypes were observed in PAI-1-overexpressing mice (Devin et al. 2007). They showed the similar characters of PCOS, and had increased testosterone levels compared to human PCOS. Thus, transgenic PAI-1 mouse model could be used for the future experimental investigation to understand many gaps of metabolic interactions in human PCOS.

16.4 Mechanism Research of Acupuncture Treatment on PCOS

16.4.1 Effect of Acupuncture: A Real or a Placebo Effect?

The control situation in acupuncture studies has been widely discussed, and the following biases challenge the concept of acupuncture study designs. When a needle penetrates the skin, it should be considered as a form of sensory stimulation that activates afferent nerve fibers. Recently, light touch of the skin has been shown to stimulate mechanoreceptors coupled to slow-conducting unmyelinated (C) afferents (Olausson et al. 2002) that modulate activity in the central nervous system (Olausson et al. 2002).

In many acupuncture studies, two different acupuncture techniques have been compared (e.g., superficial skin acupuncture needling and deep muscle acupuncture needling). We believe that it is the intensities of acupuncture that are compared; the acupuncture procedure itself is not controlled because the control treatment is not a physiologically inert treatment. It is likely that the control procedures in many acupuncture studies, which are meant to be inert, actually activate C tactile afferents and consequently elicit physiological responses (Lund and Lundeberg 2006). This could explain why superficial, sham acupuncture is as effective as deep muscle acupuncture (Linde et al. 2005); superficial, sham acupuncture and placebo needling acupuncture cannot be considered as placebo procedures because they are not inert.

16.4.2 Physiological Basis for Acupuncture

Physiological responses to acupuncture depend on a number of parameters employed during its application. The placement and stimulation of the acupuncture needles activate muscle afferents to the spinal cord and the central nervous system. In EA, low-frequency (1–15 Hz) electrical stimulation excites ergoreceptors in the muscles, which are physiologically activated during muscle contractions. Low-frequency EA causes release of a large number of neuropeptides, which seem to be essential for inducing functional changes in different organ systems (Stener-Victorin et al. 2008). Figure 16.1 illustrates a hypothetical model of the effects and mechanisms of acupuncture.

We have recently demonstrated that EA at low (2–10 Hz) but not EA at high (30–100 Hz) frequency with needle placed in abdomen or in combination of abdomen and hind limb significantly increases ovarian blood flow responses and that the response was mediated via ovarian sympathetic nerves, via the central nervous system (Stener-Victorin et al. 2003, 2006). These studies demonstrate that low-frequency electrical stimulation, which activates somatic afferents (Aδ and C fibers), modulates sympathetic output to the ovaries. Furthermore, these studies demonstrate that the strongest response was when needles were placed in abdominal muscles in combination with hind limb muscles.

Evidence exists that the physiological responses produced by acupuncture vary depending on site, intensity, and duration of stimulation (Stener-Victorin et al. 2002). Acupuncture intensity can vary enormously depending on the form of stimulation used (e.g., superficial skin needling, sham non-acupuncture point needling, or placebo needling where a blunt tip of a needle touches skin without penetrating) (Streitberger and Kleinhenz 1998; White et al. 2003). Other acupuncture confounders that may influence outcome are duration of stimulation, number of needles inserted, and needle diameter, as well as psychological factors and the environment.

16.4.3 *Acupuncture Treatment Was Not Stressful*

We have measured plasma corticosterone concentrations before and after 4 weeks of treatment with low-frequency EA in DHT-induced PCOS rats to evaluate potential stress induced by treatment of conscious rats. Four weeks of treatment did not affect corticosterone concentrations (PCOS, 229 ± 35 ng/ml vs. PCOS EA, 160 ± 34 ng/ml). Furthermore, there was no difference on the weight of adrenal glands (PCOS, 45.0 ± 2.1 vs. PCOS EA, 48.6 ± 1.6 mg). This shows that handling and EA treatments were not stressful for the rats (Manneras et al. 2008).

16.4.4 *Comparison of Acute Effects of Low-Frequency and High-Frequency EA*

We have showed that treatment with electroacupuncture (EA) of low frequency (2 Hz) induced regular ovulation in more than one-third of the women with PCOS anovulation (Stener-Victorin et al. 2000b). In animal model, evidence also suggested that low-frequency EA has an inhibitory effect on the activity of the ovarian

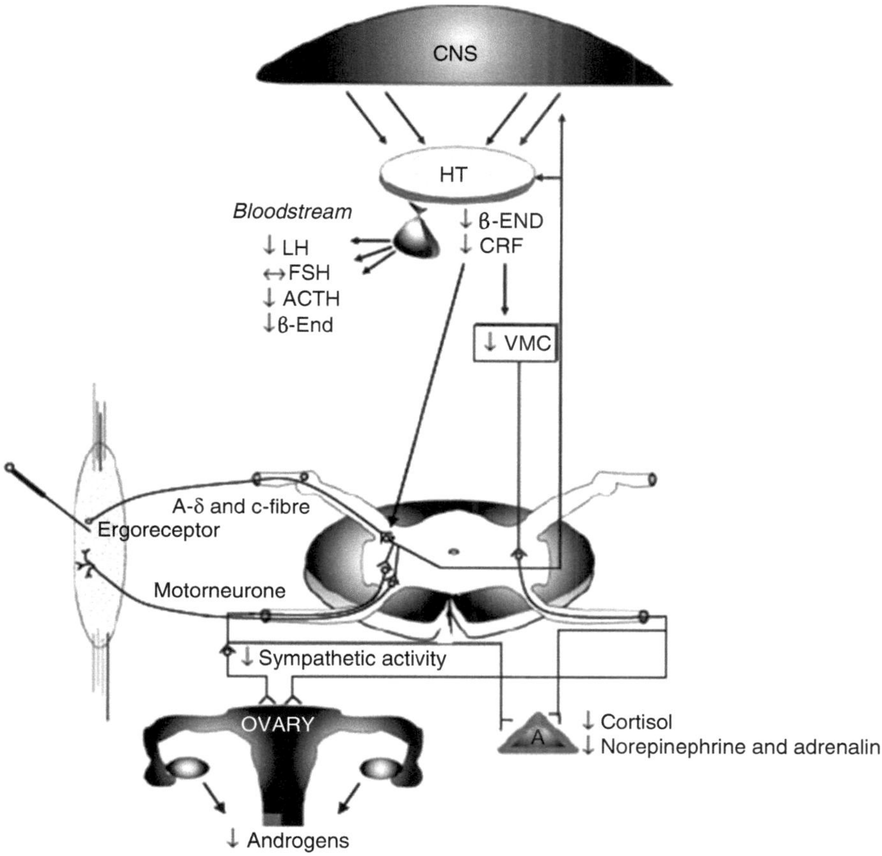

Fig. 16.1 A hypothetical model of the effects of low-frequency electroacupuncture (EA) in polycystic ovary syndrome (PCOS) (Stener-Victorin et al. 2008). Needle insertion in to the skin and muscle excites ergoreceptors and causes afferent activity in Aδ and C fibers. Needles placed and stimulated in the same somatic innervation area as the ovary decrease sympathetic nerve activity, which leads to decreased secretion and the release of ovarian androgens. In parallel, the activity of higher control systems is modulated by the release of opioids, in particular β-endorphin (β-END) that induces functional changes in different organ systems. In women with PCOS, we suggest an increased sympathetic nerve activity and an increased β-END production/release. Low-frequency EA decreases the central β-END, providing decreased sympathetic tone and decreased LH and release of β-END into the bloodstream. Low-frequency EA may further decrease hypothalamic–pituitary–adrenal (HPA) axis activity by inhibiting release of corticotropin-releasing factor (CRF), causing decreased adrenocorticotropic hormone (ACTH) release from the pituitary gland and decreased cortisol/dehydroepiandrostenedione (sulfate) [DHEA (S)] release from the adrenal cortex and a decrease in sympathetic adrenal (SA) axis activity. The adrenal medulla will then decrease noradrenaline and adrenaline secretion. *A* Adrenal gland, *CNS* central nervous system, *FSH* follicle stimulating hormone, *HT* hypothalamus, *LH* luteinizing hormone, *P* pituitary, *VMC* vasomotor center

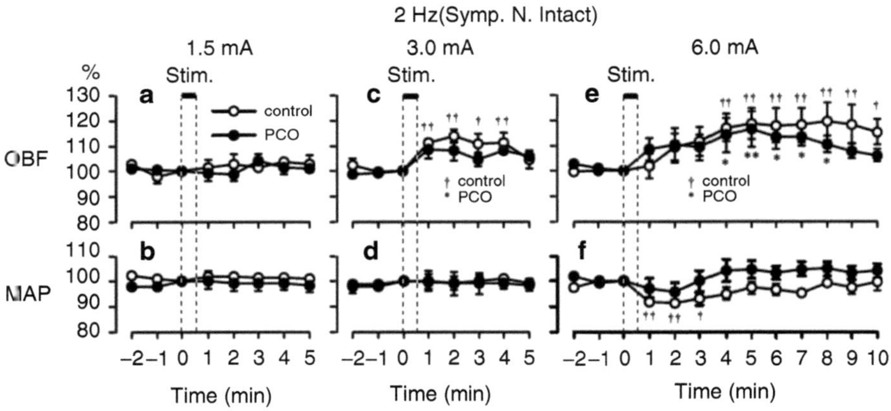

Fig. 16.2 Low-frequency (2 Hz) EA, OBF and MAP. Summarized OBF (**a**, **c**, **e**) and MAP (**b**, **d**, **f**) responses in both the control rats and the PCOS rats when stimulating with low-frequency (2 Hz) EA (Stener-Victorin et al. 2003). Changes in OBF and MAP were calculated every minute (1 s record), and response magnitude is expressed as percentage of the prestimulus value at the time point 0 min. The vertical lines and thick horizontal bar indicate the time of stimulation. The onset of stimulation is indicated as zero. The data is expressed as mean ± SEM. †$p<0.05$, ††$p<0.01$ (control rats), *$p<0.05$, **$p<0.01$ (PCOS rats), compared with the prestimulus control value

sympathetic nerves in rats with PCOS (Stener-Victorin et al. 2006). However, the exact impact of different stimulation frequencies and intensities on sympathetic activity in rats with PCOS needs to be well understood. Therefore, the involvement of the two ovarian sympathetic nerves, superior ovarian nerve (SON) and plexus ovarian nerve (OPN), in ovarian blood flow (OBF) responses was elucidated by severance of SON and OPN in both control (oil) and PCOS rats (EV-induced). How systemic circulatory changes affect OBF was evaluated by continuous recording of the blood pressure. OBF was measured on the surface of the ovary using laser Doppler flowmetry. Acupuncture needles were inserted bilaterally into the abdominal and hind limb muscles and connected to an electrical stimulator. Two frequencies—2 Hz (low) and 80 Hz (high)—with three different intensities—1.5, 3, and 6 mA—were applied for 35 s. The results suggested that low-frequency EA at intensities of 3 and 6 mA elicited significant increases in OBF compared to baseline. In the PCOS group the increases in OBF were significant only when stimulating with low-frequency EA at 6 mA. After severance of the ovarian sympathetic nerves, the increased response of OBF that had been induced by low-frequency EA in both the control and PCOS groups was abolished, indicating that the OBF response is mediated via the ovarian sympathetic nerves (Fig. 16.2). High-frequency EA at 6 mA significantly decreased OBF and mean arterial blood pressure (MAP) in the control group compared to baseline. In the PCOS group, the same stimulation produced similar decreases in MAP but not in OBF (Fig. 16.3) (Stener-Victorin et al. 2004).

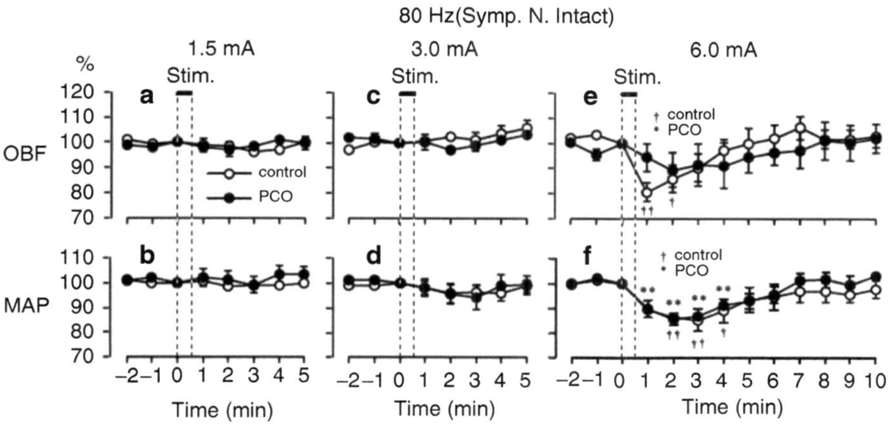

Fig. 16.3 High-frequency (80 Hz) EA, OBF and MAP. Summarized OBF (**a**, **c**, **e**) and MAP (**b**, **d**, **f**) responses in both the control rats and the PCOS rats when stimulating with high-frequency (80 Hz) EA (Stener-Victorin et al. 2003). See Fig. 16.2 for other details

16.4.5 Repeated Low-Frequency EA Treatment in Rats with DHT-Induced PCOS

Low-frequency EA was given to conscious rats daily from Monday to Friday for 4–5 weeks (20–25 treatments in total). The treatment duration was 15 min in week 1, 20 min in weeks 2 and 3, and 25 min thereafter. Acupuncture needles were inserted in the rectus abdominis (stomach) (Herbison et al. 1995) and in the triceps surae muscles (spleen) (Feng et al. 2009) bilaterally in somatic segments that correspond to the innervation of the ovaries (i.e., from spinal levels T10 to L2 and at the sacral level). The needles (HEGU Svenska, Landsbro, Sweden) were inserted to a depth of 0.5–0.8 cm and attached to an electric stimulator (CEFAR ACU II; Cefar-Compex Scandinavia, Malmo, Sweden). The points were electrically stimulated with a low frequency of 2 Hz with 0.1-s, 80-Hz burst pulses. The intensity was adjusted to produce local muscle contractions and varied from 0.8 to 1.4 mA.

Before handling or needle insertion, all rats were lightly anesthetized with isoflurane (2 % in a 1:1 mixture of oxygen and air; Isoba vet; Schering-Plough, Stockholm, Sweden) for 2–3 min. One investigator inserted all needles. During EA treatment, the rats were placed in a fabric harness and suspended above the desk. To avoid potential acute effects of EA, no treatment was performed 24 h before examinations and blood sampling. Rats in the control and PCOS groups were anesthetized, suspended in a harness, and handled in the same way as rats in the PCOS EA group but without needle insertion or electrical stimulation. All rats were conscious during handling and treatment.

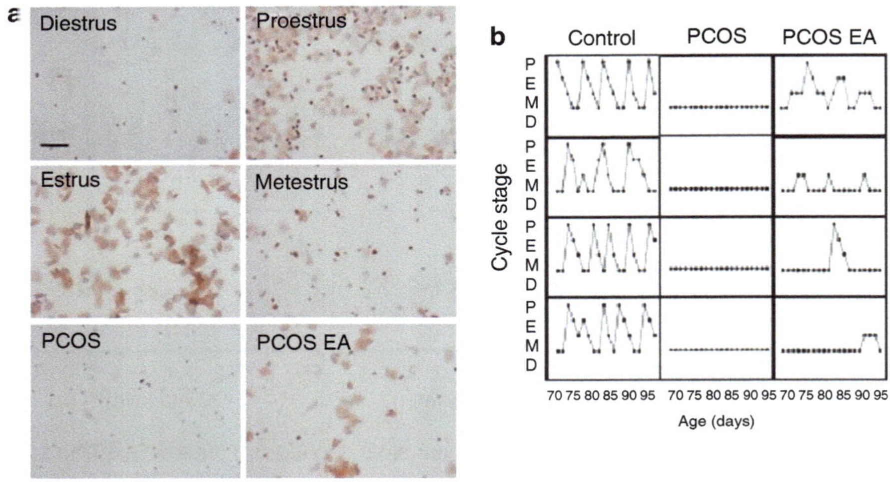

Fig. 16.4 Vaginal smears and estrous cycle patterns of control, PCOS, and PCOS EA rats (Feng et al. 2009). (**a**) Representative vaginal smears from a normal cycling control rat at different stages in the estrous cycle diestrus (*top left*), proestrus (*top right*), estrus (*middle left*), and metestrus (*middle right*). Representative hematoxylin–eosin-stained vaginal smear from a PCOS rat exhibiting predominantly leukocytes, the main cell type during diestrus stage (*bottom left*). Representative vaginal smear from a PCOS EA rat exhibiting epithelial keratinocytes, the main cell type during estrus stage (*bottom right*). Scale bar, 100 mm (*top left*). (**b**) Estrous cycle patterns at 70–95 days of age (i.e., 49–84 days after pellet implantation) in four representative rats from each group. *P* proestrus, *E* estrus, *M* metestrus, *D* diestrus

16.4.6 Low-Frequency EA Improves Estrous Cyclicity of DHT-Induced PCOS Rats

Normal adult female rats have 4–5-day estrous cycles, comprising diestrus, proestrus, estrus, and metestrus (Fig. 16.4a). Rats with DHT-induced PCOS had no dynamic change in estrous cycle, were constantly in diestrus, and exhibited predominantly leukocytes. During the first week of treatment, some rats in the PCOS EA group started to exhibit epithelial keratinocytes, the main cell type during estrus, indicating estrous cycle changes. After 4–5 weeks of low-frequency EA treatment, 11 of 12 rats (91.7 %) in the PCOS EA group exhibited epithelial keratinocytes (Fig. 16.4a, b).

16.4.7 Hypothalamic AR Protein Expression: Effect of EA

Exogenous androgen exposure undoubtedly results in androgenic actions at all levels of the female reproductive axis (Foecking et al. 2008). Clinical and animal studies suggest that PCOS has a developmental origin, in which androgen excess during fetal or prepubertal life reprograms multiple tissues to manifest the syndrome in adolescence and adulthood (Steckler et al. 2009; Walters et al. 2008). Interestingly,

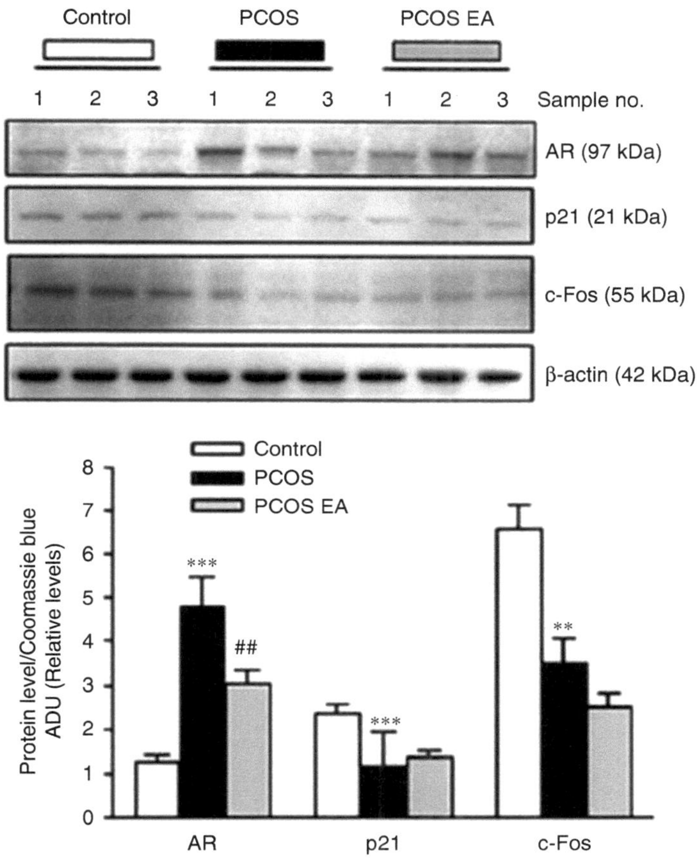

Fig. 16.5 Western blot analysis of AR, p21, and c-Fos protein expression in the hypothalamus (Feng et al. 2009). Regulation of AR, p21, and c-Fos protein expression in the hypothalamus in the control ($n=7$), PCOS ($n=6$), and PCOS EA ($n=6$) groups. *Top*: total protein (50 mg) was isolated and used for western blot analysis. The blot is representative of each run with independent samples. *Bottom*: densitometric analysis of the levels of AR, p21, and c-Fos protein expression. Equal sample loading was confirmed by Coomassie blue staining. Relative levels of AR, p21, and c-Fos proteins were expressed as a ratio of densitometric value to whole proteins in Coomassie blue-stained gels. Values are mean ± SEM of two independent experiments ($n=3$ pools/group)

female rats that received DHT continuously, starting before puberty, had arrested cycles. Previously, we showed that these rats develop ovarian characteristics of PCOS in adulthood (Manneras et al. 2007). Recently, we found hypothalamic AR protein expression was higher in PCOS rats than in controls ($p<0.001$) (Fig. 16.5), and so was AR immunoreactivity (AR-ir) in the medial preoptic area (MPO) ($p<0.05$) (Fig. 16.6). Low-frequency EA decreased hypothalamic AR protein expression ($p<0.05$, Fig. 16.5) and AR-ir in the MPO and the ventromedial hypothalamus (VMH) compared with the PCOS group ($p<0.01$, $p<0.05$, Figs. 16.5 and 16.6).

Functional androgen response elements are present in the promoter sequences of p21 (Lu et al. 1999), and the regulation of p21 expression is AR dependent (Hu

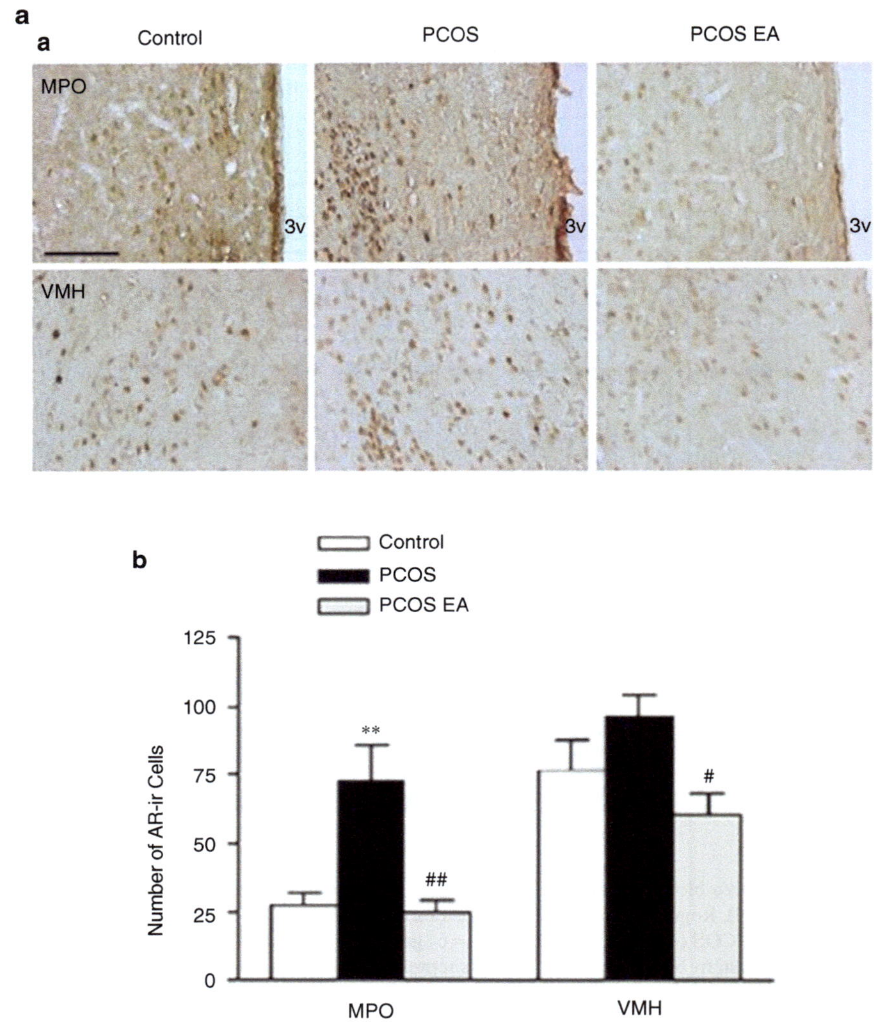

Fig. 16.6 AR-ir cells in the medial preoptic area (MPO) and ventromedial hypothalamus (VMH) (Feng et al. 2009). (**a**) Light micrographs of AR-ir cells, detected with polyclonal antibody as described in Materials and Methods. Scale bar, 100 mm. (**b**) Quantification of AR-ir cells in the different regions in the control ($n=5$), PCOS ($n=6$), and PCOS EA ($n=6$) groups. Values are mean ± SEM. $^{**}p<0.01$ vs. control; $^{\#\#}p<0.01$ vs. PCOS; $^{\#}p<0.05$ vs. PCOS

et al. 2004). To determine whether the AR was functionally active, we analyzed hypothalamic samples for AR activation in response to expression of p21, which regulates cell-cycle progression and correlates inversely with the levels of active AR. In both PCOS groups, p21 levels were lower than in controls by western blot analysis (Fig. 16.5). Western blot analysis revealed similar decreases in the expression of c-FOS, an early marker of neuronal activation (Sheng and Greenberg 1990), in the PCOS groups (Fig. 16.5). However, low-frequency EA did not elicit additional effects on p21 or c-FOS expression in PCOS rat hypothalamus.

This effect is most likely due to AR activation. We also showed that the circulating estradiol concentration is unchanged in DHT-treated rats (Manneras et al. 2007). Thus, there is no cross talk in MPO between AR and ER in DHT-treated rats.

16.4.8 GnRH-Immunoreactive Hypothalamic Cells: Effect of EA

In the brain, GnRH is the primary upstream regulator of reproductive function, including estrous cyclicity in females. Intrinsic and extrinsic stimulation, such as circulating estrogen and androgen, physical exercise, or stress, can activate GnRH signaling to regulate the hypothalamic–pituitary–gonadal axis (Naftolin et al. 2007). The effect of excessive androgens on the activity of the GnRH pulse generator has been explained in part (Blank et al. 2006; Foecking et al. 2008). However, it is unclear whether androgens exert physiological or pathophysiological actions in females by activating AR expression in GnRH-expressing neurons (Foecking et al. 2008). But rodent brain has few GnRH neurons, about 600–2,000 cells (Silverman et al. 1994; Wu et al. 1997), and the highest density of the cells is in the MS, diagonal band of Broca, and MPO (Witkin 1999). Under most conditions, it is hard to detect the GnRH protein because of the limited sensitivity of general western blotting methods. Therefore, immunohistochemistry and PCR are the most commonly used methods to study GnRH synthesis and expression.

GnRH-immunoreactive (GnRH-ir) cells are abundant in the MPO, rostral medial septum (MS), and nucleus of the horizontal limb of the diagonal band (HDB) of the hypothalamus (Silverman et al. 1994). PCOS rats had more GnRH-ir cells in the MPO and HDB than control rats ($p<0.05$) (Fig. 16.7a, b). After 4–5 weeks of EA treatment, PCOS EA rats had fewer GnRH-ir cells in those areas ($p<0.05$) (Fig. 16.7a, b) than the PCOS group. However, there was no obvious difference in the MS between the control, PCOS, and PCOS EA groups (Fig.16.7a, b). Even though we detected GnRH-ir cells in the hypothalamus, we failed to detect a GnRH band by western blot analysis because of low GnRH expression. The low expression was confirmed by immunoprecipitation and western blot (Fig. 16.7c).

16.4.9 Hypothalamic GnRH-R Expression: Effect of EA

The GnRH-receptor (GnRH-R) is mainly distributed in the pituitary gland and hypothalamus in the central nervous system. Continuous DHT exposure decreased both the number of GnRH-R-ir cells in the pituitary, MPO, and HDB ($p<0.01$) and hypothalamic GnRH-R protein expression ($p<0.05$ vs. controls) (Fig. 16.8a). There was no difference in the MS between groups (Fig. 16.8a). Low-frequency EA treatments did not affect the number of GnRH-R-ir cells or the level of GnRH-R expression (Fig. 16.8b, c).

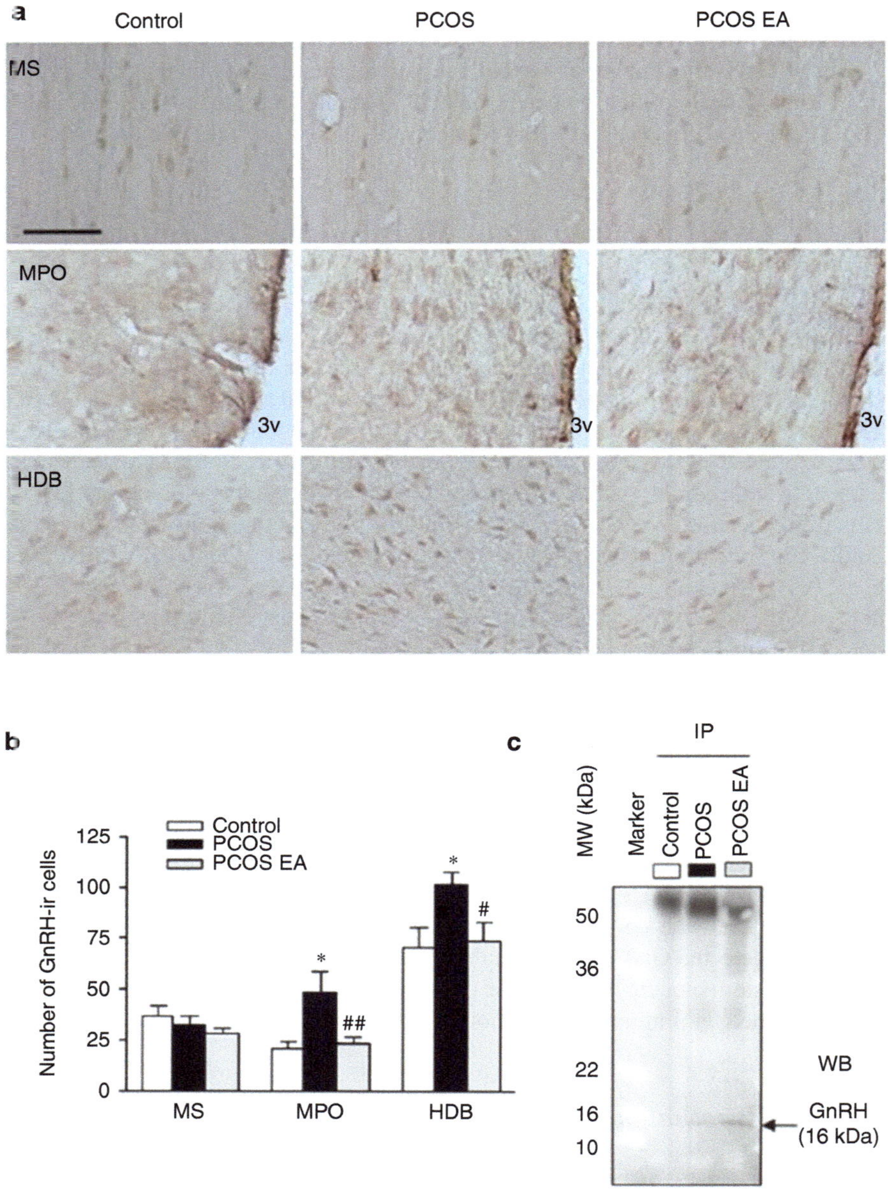

Fig. 16.7 GnRH-ir cells in the rostral MS, MPO, and HDB of the hypothalamus (Feng et al. 2009). (**a**) Light micrographs of GnRH-ir cells, detected with polyclonal antibody as described in Materials and Methods. Scale bar, 100 mm. (**b**) Quantification of GnRH-ir cells in the different regions in the control ($n=5$), PCOS ($n=6$), and PCOS EA ($n=6$) groups. Values are mean ± SEM. *$p<0.05$ vs. control; $^{\#}p<0.05$ vs. PCOS. (**c**) Immunoprecipitation and western blot (WB) of GnRH protein in the hypothalamus

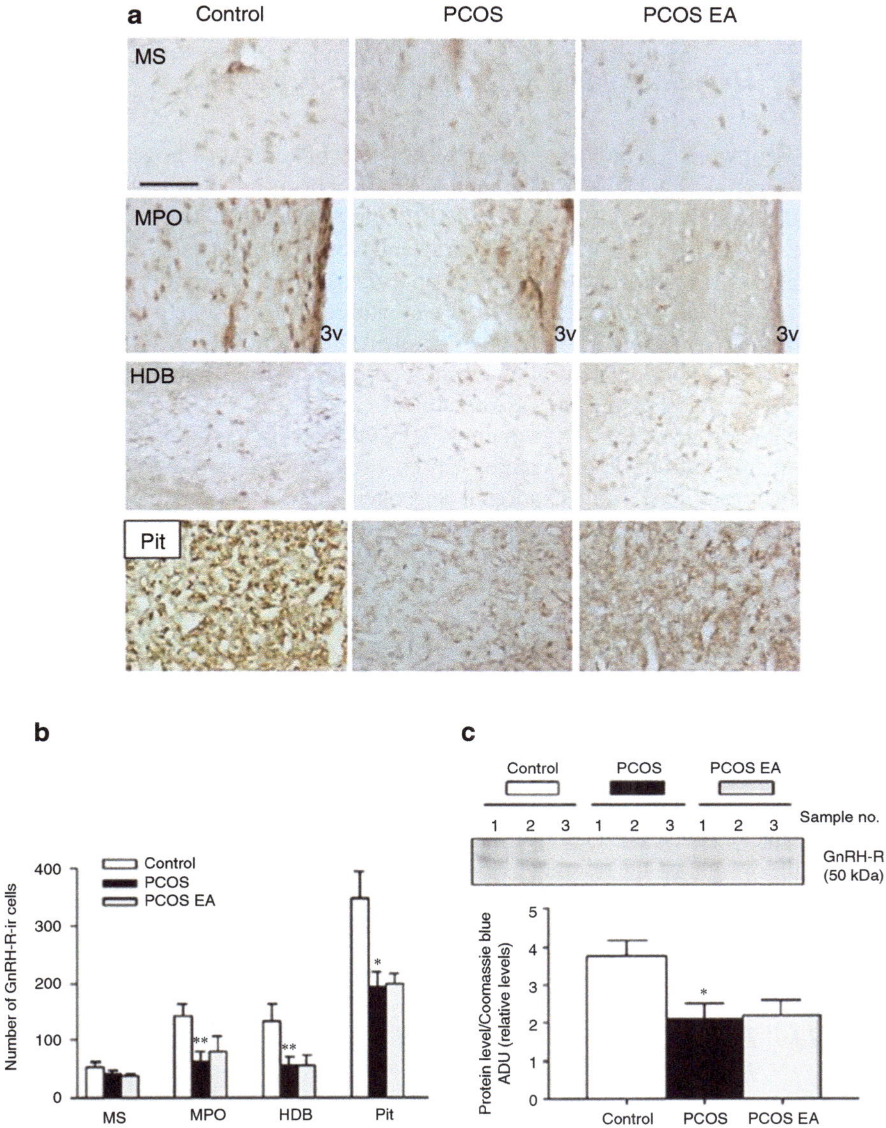

Fig. 16.8 GnRH-R-ir cells in the rostral MS, and nucleus of the HDB of the hypothalamus and pituitary (Pit) (Feng et al. 2009). (**a**) Light micrographs of the GnRH-R-ir cells, detected with polyclonal antibodies as described in Materials and Methods. Scale bar, 100 μm. (**b**) Quantification of GnRH-R cells in the control ($n=5$), PCOS ($n=6$), and PCOS EA ($n=6$) groups. Values are mean ± SEM. $^{*}p<0.05$ vs. control; $^{**}p<0.05$ vs. PCOS. (**c**) Western blot of GnRH-R protein in the hypothalamus. Total protein (50 μg) was isolated and used for western blot analysis. The blot is representative of two essentially similar experiments, each run with independent samples. Densitometric analysis of GnRH-R protein expression in two independent experiments. Equal sample loading was confirmed by Coomassie blue staining. Relative levels of GnRH-R proteins were expressed as a ratio of densitometric value to whole proteins in Coomassie blue-stained gels. Data are expressed as ADU; values are the mean ± SEM of two independent experiments ($n=3$ pools/group). $^{*}p<0.05$ vs. control

16.4.10 Hypothalamic CRH Expression: Effect of EA

Over 50 % of women with PCOS have excess levels of adrenal androgens (Lanzone et al. 1995), but it is not known whether this excess reflects hypothalamic–pituitary–adrenal axis dysfunction due to exaggerated pituitary secretion of adrenocorticotropic hormone (ACTH) in response to hypothalamic corticotropin-releasing hormone (CRH), excess responsiveness of adrenal androgens to ACTH stimulation, or both (Azziz et al. 1998; Lanzone et al. 1995). Two main regions in the hypothalamus, the paraventricular nucleus (PVN) and MPO, showed CRH immunoreactivity (CRH-ir). However, CRH-ir and CRH protein expression did not differ between PCOS and control rats or after 4–5 weeks of EA treatment.

Adrenal androgen excess in women with PCOS may not be related to an altered pituitary response to CRH or to increased sensitivity to ACTH (Azziz et al. 1998). Rather, increased secretion of adrenal androgen may be due to increased zona reticularis mass or to P450c17 alpha activity. This possibility needs to be assessed. Consistent with these observations in humans, hypothalamic CRH protein expression was not affected in rats with DHT-induced PCOS. Therefore, the decreased circulating corticosterone concentrations in these rats in our previous study (Manneras et al. 2008) is most likely due to a local effect in the adrenal gland.

Low-frequency EA did not affect hypothalamic CRH concentrations, which is in line with our finding that EA does not affect corticosterone concentrations (Manneras et al. 2008). In the EV-induced PCOS model, however, the CRH concentration was increased in the median eminence, indicating increased activity in the hypothalamic–pituitary–adrenal axis, but was restored to normal by repeated low-frequency EA (Stener-Victorin et al. 2001). Furthermore, in the present study, there was no difference in the weight of the adrenal glands between the PCOS and PCOS EA groups (PCOS, 45.0 ± 2.1 mg; PCOS EA, 48.6 ± 1.6 mg). Notably, these results support the findings that handling and treatment are not stressful for the rats.

16.4.11 Expression of AR in GnRH-ir or CRH-ir Neurons in the Hypothalamic MPO

The effect of excessive androgens on the activity of the GnRH pulse generator has been analyzed (Blank et al. 2006). However, it is unclear whether androgens exert physiological or pathophysiological actions in females by activating AR expression in GnRH-expressing neurons (Foecking et al. 2008). Given the failure of initial efforts to detect ARs or estrogen receptors in GnRH-expressing neurons (Herbison et al. 1995; Herbison and Theodosis 1992), many investigators concluded that steroid hormones affect GnRH indirectly (Kallo et al. 2001; Skynner et al. 1999). Later, coexpression of GnRH and estrogen receptor β was demonstrated, indicating a direct action of sex steroids in regulating GnRH (Azziz et al. 1998). Nevertheless, evidence that the AR has a role in regulating the function of GnRH-expressing neurons in vivo has not been presented. Our findings demonstrate that DHT exposure increases both GnRH and AR expression in the MPO. The distribution

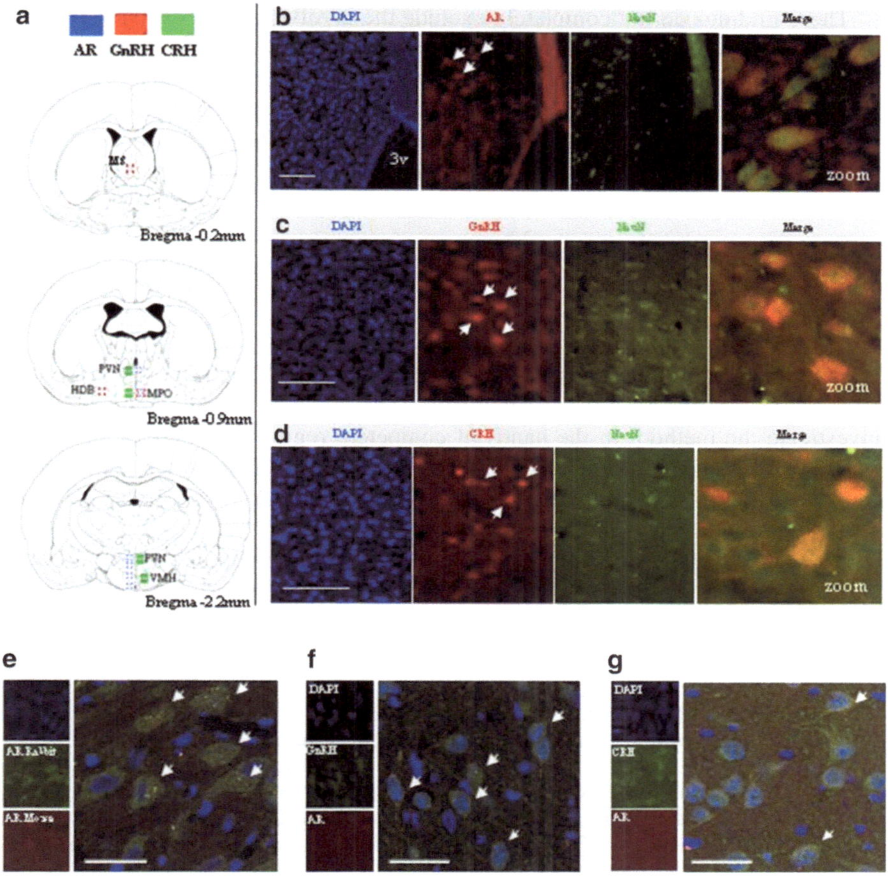

Fig. 16.9 Co-localization of AR, GnRH, and CRH in MPO neurons, determined by dual-fluorescence immunohistochemistry and confocal laser scanning microscopy (Feng et al. 2009). (**a**) Main distribution of AR, GnRH, and CRH in control female rat brain (3v = third ventricle). Adapted from reference (Paxinos and Watson 2005). (**b**–**d**) Co-localization of AR, GnRH, CRH, and NeuN immunoreactivity in MPO neurons. (**e**) Co-localization of rabbit polyclonal AR antibody and mouse monoclonal AR antibody in hypothalamic MPO. (**f**, **g**) Co-localization of mouse monoclonal AR antibody with GnRH or CRH immunoreactivity in hypothalamic MPO neurons. In panels (**b**–**g**), *arrows* indicate co-localization. Scale bars, 100 μm

of AR and GnRH overlapped, indicating a potential mechanism for the regulation of GnRH-expressing neurons by androgens. This possibility was further confirmed by the co-localization of AR and GnRH in the MPO, including both cytoplasmic and nuclear expression.

Dual-fluorescence immunohistochemistry and confocal analysis showed co-localization of AR, GnRH, and CRH with the neuronal marker NeuN in controls. In addition, AR and GnRH expression in the MPO co-localized in the cytoplasm and nucleus, but few cytoplasmic ARs co-localized with CRH-expressing neurons (Fig. 16.9a–g).

These findings do not completely exclude the involvement of estrogen receptor (ER) β activation in the MPO since DHT can be metabolized into 5α-androstane-3β and 17β-diol (3β-diol), which can act via the ERβ receptor (Handa et al. 2008). In the hypothalamus, the VMH expresses mainly ERα (Shughrue et al. 1996), whereas the PVN contains only ERβ (Patisaul et al. 1999). In DHT-treated rats in the present study, AR expression was mainly increased in the MPO and to a lesser extent in the VMH but was unaltered in the PVN.

16.5 Concluding Comments

Despite the lack of a great number of evidence, we should not ignore the fact that many women with PCOS use acupuncture. This alone is a compelling reason to investigate the method. In the hands of competent registered health practitioners, acupuncture is safe (Vincent 2001). Clinical and experimental evidence shows that acupuncture can be a suitable alternative or complement to pharmacological induction of ovulation in women with PCOS and may also relieve other symptoms without adverse side effects. Clearly, acupuncture modulates endogenous regulatory systems, including the sympathetic nervous system, the endocrine system, and the neuroendocrine system. However, randomized controlled trials are needed to further evaluate the effect of acupuncture in women with PCOS.

Acknowledgments We thank the Swedish Institute and China Scholarship Council for their support and the Center for Mouse Physiology and Bio-Imaging, University of Gothenburg.

This program was supported by grants from the Swedish Medical Research Council (Project No. 2008-72VP-15445-01A); Novo Nordisk Foundation; Wilhelm and Martina Lundgren's Science Fund; Hjalmar Svensson Foundation; Tore Nilson Foundation; Åke Wiberg Foundation; Adlerbert Research Foundation; Ekhaga Foundation; Swedish federal government under the letters of understanding agreement of Medical Education (ALFFGBG-10984) and Regional Research and Development agreement (VGFOUREG-5171, 11296, and 7861); Special grant of China Postdoctoral Science Foundation (No. 200801170); and National Natural Science Foundation of China (No. 81001544).

References

(2004). Revised 2003 consensus on diagnostic criteria and long-term health risks related to polycystic ovary syndrome (PCOS). Hum Reprod *19*, 41–47.

Abbott, D.H., Tarantal, A.F., and Dumesic, D.A. (2009). Fetal, infant, adolescent and adult phenotypes of polycystic ovary syndrome in prenatally androgenized female rhesus monkeys. Am J Primatol *71*, 776–784.

Adams, J.M., Taylor, A.E., Crowley, W.F., Jr., and Hall, J.E. (2004). Polycystic ovarian morphology with regular ovulatory cycles: insights into the pathophysiology of polycystic ovarian syndrome. J Clin Endocrinol Metab *89*, 4343–4350.

Atiomo, W.U., Bates, S.A., Condon, J.E., Shaw, S., West, J.H., and Prentice, A.G. (1998). The plasminogen activator system in women with polycystic ovary syndrome. Fertil Steril *69*, 236–241.

Azziz, R., Black, V., Hines, G.A., Fox, L.M., and Boots, L.R. (1998). Adrenal androgen excess in the polycystic ovary syndrome: sensitivity and responsivity of the hypothalamic-pituitary-adrenal axis. J Clin Endocrinol Metab *83*, 2317–2323.

Azziz, R., Carmina, E., Dewailly, D., Diamanti-Kandarakis, E., Escobar-Morreale, H.F., Futterweit, W., Janssen, O.E., Legro, R.S., Norman, R.J., Taylor, A.E., *et al.* (2009). The Androgen Excess and PCOS Society criteria for the polycystic ovary syndrome: the complete task force report. Fertil Steril *91*, 456–488.

Bagos, P.G. (2009). Plasminogen activator inhibitor-1 4 G/5G and 5,10-methylene-tetrahydrofolate reductase C677T polymorphisms in polycystic ovary syndrome. Mol Hum Reprod *15*, 19–26.

Blank, S.K., McCartney, C.R., and Marshall, J.C. (2006). The origins and sequelae of abnormal neuroendocrine function in polycystic ovary syndrome. Hum Reprod Update *12*, 351–361.

Brawer, J.R., Munoz, M., and Farookhi, R. (1986). Development of the polycystic ovarian condition (PCO) in the estradiol valerate-treated rat. Biol Reprod *35*, 647–655.

Britt, K.L., Drummond, A.E., Cox, V.A., Dyson, M., Wreford, N.G., Jones, M.E., Simpson, E.R., and Findlay, J.K. (2000). An age-related ovarian phenotype in mice with targeted disruption of the Cyp 19 (aromatase) gene. Endocrinology *141*, 2614–2623.

Chapman, J.C., Min, S.H., Freeh, S.M., and Michael, S.D. (2009). The estrogen-injected female mouse: new insight into the etiology of PCOS. Reprod Biol Endocrinol *7*, 47.

Chen, B.Y., and Yu, J. (1991). Relationship between blood radioimmunoreactive beta-endorphin and hand skin temperature during the electro-acupuncture induction of ovulation. Acupunct Electrother Res *16*, 1–5.

Couse, J.F., Bunch, D.O., Lindzey, J., Schomberg, D.W., and Korach, K.S. (1999). Prevention of the polycystic ovarian phenotype and characterization of ovulatory capacity in the estrogen receptor-alpha knockout mouse. Endocrinology *140*, 5855–5865.

Devin, J.K., Johnson, J.E., Eren, M., Gleaves, L.A., Bradham, W.S., Bloodworth, J.R., Jr., and Vaughan, D.E. (2007). Transgenic overexpression of plasminogen activator inhibitor-1 promotes the development of polycystic ovarian changes in female mice. J Mol Endocrinol *39*, 9–16.

Dissen, G.A., Garcia-Rudaz, C., Paredes, A., Mayer, C., Mayerhofer, A., and Ojeda, S.R. (2009). Excessive ovarian production of nerve growth factor facilitates development of cystic ovarian morphology in mice and is a feature of polycystic ovarian syndrome in humans. Endocrinology *150*, 2906–2914.

Dissen, G.A., Lara, H.E., Leyton, V., Paredes, A., Hill, D.F., Costa, M.E., Martinez-Serrano, A., and Ojeda, S.R. (2000). Intraovarian excess of nerve growth factor increases androgen secretion and disrupts estrous cyclicity in the rat. Endocrinology *141*, 1073–1082.

Dupont, S., Krust, A., Gansmuller, A., Dierich, A., Chambon, P., and Mark, M. (2000). Effect of single and compound knockouts of estrogen receptors alpha (ERalpha) and beta (ERbeta) on mouse reproductive phenotypes. Development (Cambridge, England) *127*, 4277–4291.

Ehrmann, D.A. (2005). Polycystic ovary syndrome. The New England journal of medicine *352*, 1223–1236.

Feng, Y., Johansson, J., Shao, R., Manneras, L., Fernandez-Rodriguez, J., Billig, H., and Stener-Victorin, E. (2009). Hypothalamic neuroendocrine functions in rats with dihydrotestosterone-induced polycystic ovary syndrome: effects of low-frequency electro-acupuncture. PLoS One *4*, e6638.

Foecking, E.M., McDevitt, M.A., Acosta-Martinez, M., Horton, T.H., and Levine, J.E. (2008). Neuroendocrine consequences of androgen excess in female rodents. Horm Behav *53*, 673–692.

Gerhard, I., and Postneek, F. (1992). Auricular acupuncture in the treatment of female infertility. Gynecol Endocrinol *6*, 171–181.

Handa, R.J., Pak, T.R., Kudwa, A.E., Lund, T.D., and Hinds, L. (2008). An alternate pathway for androgen regulation of brain function: activation of estrogen receptor beta by the metabolite of dihydrotestosterone, 5alpha-androstane-3beta,17beta-diol. Horm Behav *53*, 741–752.

Heine, P.A., Taylor, J.A., Iwamoto, G.A., Lubahn, D.B., and Cooke, P.S. (2000). Increased adipose tissue in male and female estrogen receptor-alpha knockout mice. Proc Natl Acad Sci U S A *97*, 12729–12734.

Herbison, A.E., Horvath, T.L., Naftolin, F., and Leranth, C. (1995). Distribution of estrogen receptor-immunoreactive cells in monkey hypothalamus: relationship to neurones containing luteinizing hormone-releasing hormone and tyrosine hydroxylase. Neuroendocrinology *61*, 1–10.

Herbison, A.E., and Theodosis, D.T. (1992). Localization of oestrogen receptors in preoptic neurons containing neurotensin but not tyrosine hydroxylase, cholecystokinin or luteinizing hormone-releasing hormone in the male and female rat. Neuroscience *50*, 283–298.

Hu, Y.C., Wang, P.H., Yeh, S., Wang, R.S., Xie, C., Xu, Q., Zhou, X., Chao, H.T., Tsai, M.Y., and Chang, C. (2004). Subfertility and defective folliculogenesis in female mice lacking androgen receptor. Proc Natl Acad Sci U S A *101*, 11209–11214.

Ito, Y., Fisher, C.R., Conte, F.A., Grumbach, M.M., and Simpson, E.R. (1993). Molecular basis of aromatase deficiency in an adult female with sexual infantilism and polycystic ovaries. Proc Natl Acad Sci U S A *90*, 11673–11677.

Jansen, E., Laven, J.S., Dommerholt, H.B., Polman, J., van Rijt, C., van den Hurk, C., Westland, J., Mosselman, S., and Fauser, B.C. (2004). Abnormal gene expression profiles in human ovaries from polycystic ovary syndrome patients. Molecular endocrinology (Baltimore, Md *18*, 3050–3063.

Jedel E, Labrie F, Oden A, et al. (2011). Impact of electro-acupuncture and physical exercise on hyperandrogenism and oligo/amenorrhea in women with polycystic ovary syndrome: a randomized controlled trial. Am J Physiol Endocrinol Metab *300*, E37–45.

Kafali, H., Iriadam, M., Ozardali, I., and Demir, N. (2004). Letrozole-induced polycystic ovaries in the rat: a new model for cystic ovarian disease. Arch Med Res *35*, 103–108.

Kaipia, A., and Hsueh, A.J. (1997). Regulation of ovarian follicle atresia. Annual review of physiology *59*, 349–363.

Kallo, I., Butler, J.A., Barkovics-Kallo, M., Goubillon, M.L., and Coen, C.W. (2001). Oestrogen receptor beta-immunoreactivity in gonadotropin releasing hormone-expressing neurones: regulation by oestrogen. J Neuroendocrinol *13*, 741–748.

Lanham, M.S., Lebovic, D.I., and Domino, S.E. (2006). Contemporary medical therapy for polycystic ovary syndrome. Int J Gynaecol Obstet *95*, 236–241.

Lanzone, A., Petraglia, F., Fulghesu, A.M., Ciampelli, M., Caruso, A., and Mancuso, S. (1995). Corticotropin-releasing hormone induces an exaggerated response of adrenocorticotropic hormone and cortisol in polycystic ovary syndrome [see comments]. Fertil Steril *63*, 1195–1199.

Lara, H.E., Dissen, G.A., Leyton, V., Paredes, A., Fuenzalida, H., Fiedler, J.L., and Ojeda, S.R. (2000). An increased intraovarian synthesis of nerve growth factor and its low affinity receptor is a principal component of steroid-induced polycystic ovary in the rat. Endocrinology *141*, 1059–1072.

Lee, S., Kang, D.W., Hudgins-Spivey, S., Krust, A., Lee, E.Y., Koo, Y., Cheon, Y., Gye, M.C., Chambon, P., and Ko, C. (2009). Theca-specific estrogen receptor-alpha knockout mice lose fertility prematurely. Endocrinology *150*, 3855–3862.

Legro, R.S. (2007). Pregnancy considerations in women with polycystic ovary syndrome. Clin Obstet Gynecol *50*, 295–304.

Legro, R.S., Barnhart, H.X., Schlaff, W.D., Carr, B.R., Diamond, M.P., Carson, S.A., Steinkampf, M.P., Coutifaris, C., McGovern, P.G., Cataldo, N.A., *et al.* (2007). Clomiphene, metformin, or both for infertility in the polycystic ovary syndrome. The New England journal of medicine *356*, 551–566.

Legro, R.S., Spielman, R., Urbanek, M., Driscoll, D., Strauss, J.F., 3rd, and Dunaif, A. (1998). Phenotype and genotype in polycystic ovary syndrome. Recent progress in hormone research *53*, 217–256.

Leung, P.C., and Armstrong, D.T. (1979). Estrogen treatment of immature rats inhibits ovarian androgen production in vitro. Endocrinology *104*, 1411–1417.

Leung, P.C., Goldenberg, S.R., and Armstrong, D.T. (1978). Changes with age in aromatization of testosterone in immature female rats. Biol Reprod *19*, 1036–1039.

Linde, K., Streng, A., Jurgens, S., Hoppe, A., Brinkhaus, B., Witt, C., Wagenpfeil, S., Pfaffenrath, V., Hammes, M.G., Weidenhammer, W., *et al.* (2005). Acupuncture for patients with migraine: a randomized controlled trial. Jama *293*, 2118–2125.

Lu, S., Liu, M., Epner, D.E., Tsai, S.Y., and Tsai, M.J. (1999). Androgen regulation of the cyclin-dependent kinase inhibitor p21 gene through an androgen response element in the proximal promoter. Mol Endocrinol *13*, 376–384.

Lubahn, D.B., Moyer, J.S., Golding, T.S., Couse, J.F., Korach, K.S., and Smithies, O. (1993). Alteration of reproductive function but not prenatal sexual development after insertional disruption of the mouse estrogen receptor gene. Proc Natl Acad Sci U S A *90*, 11162–11166.

Lund, I., and Lundeberg, T. (2006). Are minimal, superficial or sham acupuncture procedures acceptable as inert placebo controls? Acupunct Med *24*, 13–15.

Mahajan, D.K. (1988). Polycystic ovarian disease: animal models. Endocrinology and metabolism clinics of North America *17*, 705–732.

Mahesh, V.B., Mills, T.M., Bagnell, C.A., and Conway, B.A. (1987). Animal models for study of polycystic ovaries and ovarian atresia. Advances in experimental medicine and biology *219*, 237–257.

Manneras, L., Cajander, S., Holmang, A., Seleskovic, Z., Lystig, T., Lonn, M., and Stener-Victorin, E. (2007). A New Rat Model Exhibiting Both Ovarian and Metabolic Characteristics of Polycystic Ovary Syndrome. Endocrinology *148*, 3781–3791.

Manneras, L., Jonsdottir, I.H., Holmang, A., Lonn, M., and Stener-Victorin, E. (2008). Low-frequency electro-acupuncture and physical exercise improve metabolic disturbances and modulate gene expression in adipose tissue in rats with dihydrotestosterone-induced polycystic ovary syndrome. Endocrinology *149*, 3559–3568.

Marins, N., da Silva, C.T., da Motta, V.P., Scianni, C.C., Arcaldi, N., and de Matos, J.V. (1981). [Hemodynamic study in subjects with Chagas' disease without apparent heart disease]. Arquivos brasileiros de cardiologia *37*, 463–466.

Matzuk, M.M., DeMayo, F.J., Hadsell, L.A., and Kumar, T.R. (2003). Overexpression of human chorionic gonadotropin causes multiple reproductive defects in transgenic mice. Biol Reprod *69*, 338–346.

McCarthy, G.F., and Brawer, J.R. (1990). Induction of Stein-Leventhal-like polycystic ovaries (PCO) in the rat: a new model for cystic ovarian disease. The Anatomical record *228*, 137–144.

Moll, E., Bossuyt, P.M., Korevaar, J.C., Lambalk, C.B., and van der Veen, F. (2006). Effect of clomifene citrate plus metformin and clomifene citrate plus placebo on induction of ovulation in women with newly diagnosed polycystic ovary syndrome: randomised double blind clinical trial. BMJ (Clinical research ed *332*, 1485.

Moran, L., and Teede, H. (2009). Metabolic features of the reproductive phenotypes of polycystic ovary syndrome. Hum Reprod Update *15*, 477–488.

Naftolin, F., Garcia-Segura, L.M., Horvath, T.L., Zsarnovszky, A., Demir, N., Fadiel, A., Leranth, C., Vondracek-Klepper, S., Lewis, C., Chang, A., *et al.* (2007). Estrogen-induced hypothalamic synaptic plasticity and pituitary sensitization in the control of the estrogen-induced gonadotrophin surge. Reprod Sci *14*, 101–116.

Nardo, L.G., Patchava, S., and Laing, I. (2008). Polycystic ovary syndrome: pathophysiology, molecular aspects and clinical implications. Panminerva medica *50*, 267–278.

Neveu, N., Granger, L., St-Michel, P., and Lavoie, H.B. (2007). Comparison of clomiphene citrate, metformin, or the combination of both for first-line ovulation induction and achievement of pregnancy in 154 women with polycystic ovary syndrome. Fertil Steril *87*, 113–120.

Norman, R.J., Dewailly, D., Legro, R.S., and Hickey, T.E. (2007). Polycystic ovary syndrome. Lancet *370*, 685–697.

Olausson, H., Lamarre, Y., Backlund, H., Morin, C., Wallin, B.G., Starck, G., Ekholm, S., Strigo, I., Worsley, K., Vallbo, A.B., *et al.* (2002). Unmyelinated tactile afferents signal touch and project to insular cortex. Nature neuroscience *5*, 900–904.

Orio, F., Jr., Palomba, S., Cascella, T., Tauchmanova, L., Nardo, L.G., Di Biase, S., Labella, D., Russo, T., Savastano, S., Tolino, A., *et al.* (2004). Is plasminogen activator inhibitor-1 a cardiovascular risk factor in young women with polycystic ovary syndrome? Reprod Biomed Online *9*, 505–510.

Pastore LM, Williams CD, Jenkins J, Patrie JT. (2011). True and Sham Acupuncture Produced Similar Frequency of Ovulation and Improved LH to FSH Ratios in Women with Polycystic Ovary Syndrome. J Clin Endocrinol Metab *96*, 3143–3150.

Patisaul, H.B., Whitten, P.L., and Young, L.J. (1999). Regulation of estrogen receptor beta mRNA in the brain: opposite effects of 17beta-estradiol and the phytoestrogen, coumestrol. Brain Res Mol Brain Res *67*, 165–171.

Paxinos, G., and Watson, C. (2005). The rat brain in stereotaxic coordinates, 5th edn (Elsevier academic press).

Risma, K.A., Clay, C.M., Nett, T.M., Wagner, T., Yun, J., and Nilson, J.H. (1995). Targeted overexpression of luteinizing hormone in transgenic mice leads to infertility, polycystic ovaries, and ovarian tumors. Proc Natl Acad Sci U S A *92*, 1322–1326.

Risma, K.A., Hirshfield, A.N., and Nilson, J.H. (1997). Elevated luteinizing hormone in prepubertal transgenic mice causes hyperandrogenemia, precocious puberty, and substantial ovarian pathology. Endocrinology *138*, 3540–3547.

Roy, S., Mahesh, V.B., and Greenblatt, R.B. (1962). Effect of dehydroepiandrosterone and delta4-androstenedione on the reproductive organs of female rats: production of cystic changes in the ovary. Nature *196*, 42–43.

Ruiz, A., Aguilar, R., Tebar, A.M., Gaytan, F., and Sanchez-Criado, J.E. (1996). RU486-treated rats show endocrine and morphological responses to therapies analogous to responses of women with polycystic ovary syndrome treated with similar therapies. Biol Reprod *55*, 1284–1291.

Sampson, M., Kong, C., Patel, A., Unwin, R., and Jacobs, H.S. (1996). Ambulatory blood pressure profiles and plasminogen activator inhibitor (PAI-1) activity in lean women with and without the polycystic ovary syndrome. Clin Endocrinol (Oxf) *45*, 623–629.

Sanchez-Criado, J.E., Sanchez, A., Ruiz, A., and Gaytan, F. (1993). Endocrine and morphological features of cystic ovarian condition in antiprogesterone RU486-treated rats. Acta endocrinologica *129*, 237–245.

Schomberg, D.W., Couse, J.F., Mukherjee, A., Lubahn, D.B., Sar, M., Mayo, K.E., and Korach, K.S. (1999). Targeted disruption of the estrogen receptor-alpha gene in female mice: characterization of ovarian responses and phenotype in the adult. Endocrinology *140*, 2733–2744.

Schwartz, N.B. (1969). A model for the regulation of ovulation in the rat. Recent progress in hormone research *25*, 1–55.

Sheng, M., and Greenberg, M.E. (1990). The regulation and function of c-fos and other immediate early genes in the nervous system. Neuron *4*, 477–485.

Shi, D., Dyck, M.K., Uwiera, R.R., Russell, J.C., Proctor, S.D., and Vine, D.F. (2009). A unique rodent model of cardiometabolic risk associated with the metabolic syndrome and polycystic ovary syndrome. Endocrinology *150*, 4425–4436.

Shughrue, P.J., Komm, B., and Merchenthaler, I. (1996). The distribution of estrogen receptor-beta mRNA in the rat hypothalamus. Steroids *61*, 678–681.

Silverman A.J., L.I., Witkin J.W. (1994). The gonadotropin-releasing hormone (GnRH) neuronal systems: Immunocytochemistry and in situ hybridization (New York, Raven Press).

Skynner, M.J., Sim, J.A., and Herbison, A.E. (1999). Detection of estrogen receptor alpha and beta messenger ribonucleic acids in adult gonadotropin-releasing hormone neurons. Endocrinology *140*, 5195–5201.

Steckler, T.L., Herkimer, C., Dumesic, D.A., and Padmanabhan, V. (2009). Developmental programming: excess weight gain amplifies the effects of prenatal testosterone excess on reproductive cyclicity--implication for polycystic ovary syndrome. Endocrinology *150*, 1456–1465.

Stein, I.F., and Leventhal, M.L. (1935). Amenorrhea associated with bilateral polycystic ovaries. Am J Obstet Gynecol *29*, 181–191.

Stener-Victorin, E., Fujisawa, S., and Kurosawa, M. (2006). Ovarian blood flow responses to electroacupuncture stimulation depend on estrous cycle and on site and frequency of stimulation in anesthetized rats. J Appl Physiol *101*, 84–91.

Stener-Victorin, E., Jedel, E., Janson, P.O., and Sverrisdottir, Y.B. (2009). Low-frequency electroacupuncture and physical exercise decrease high muscle sympathetic nerve activity in polycystic ovary syndrome. Am J Physiol Regul Integr Comp Physiol *297*, R387–395.

Stener-Victorin, E., Jedel, E., and Manneras, L. (2008). Acupuncture in polycystic ovary syndrome: current experimental and clinical evidence. J Neuroendocrinol *20*, 290–298.

Stener-Victorin, E., Kobayashi, R., and Kurosawa, M. (2003). Ovarian blood flow responses to electro-acupuncture stimulation at different frequencies and intensities in anaesthetized rats. Auton Neurosci *108*, 50–56.
Stener-Victorin, E., Kobayashi, R., Watanabe, O., Lundeberg, T., and Kurosawa, M. (2004). Effect of electro-acupuncture stimulation of different frequencies and intensities on ovarian blood flow in anaesthetized rats with steroid-induced polycystic ovaries. Reprod Biol Endocrinol *2*, 16.
Stener-Victorin, E., Lundeberg, T., Waldenstrom, U., Bileviciute-Ljungar, I., and Janson, P.O. (2001). Effects of electro-acupuncture on corticotropin-releasing factor in rats with experimentally-induced polycystic ovaries. Neuropeptides *35*, 227–231.
Stener-Victorin, E., Lundeberg, T., Waldenstrom, U., Manni, L., Aloe, L., Gunnarsson, S., and Janson, P.O. (2000a). Effects of electro-acupuncture on nerve growth factor and ovarian morphology in rats with experimentally induced polycystic ovaries. Biol Reprod *63*, 1497–1503.
Stener-Victorin, E., Waldenstrom, U., Tagnfors, U., Lundeberg, T., Lindstedt, G., and Janson, P.O. (2000b). Effects of electro-acupuncture on anovulation in women with polycystic ovary syndrome. Acta Obstet Gynecol Scand *79*, 180–188.
Stener-Victorin, E., Wikland, M., Waldenstrom, U., and Lundeberg, T. (2002). Alternative treatments in reproductive medicine: much ado about nothing. Acupuncture-a method of treatment in reproductive medicine: lack of evidence of an effect does not equal evidence of the lack of an effect. Hum Reprod *17*, 1942–1946.
Streitberger, K., and Kleinhenz, J. (1998). Introducing a placebo needle into acupuncture research. Lancet *352*, 364–365.
Tarkun, I., Canturk, Z., Arslan, B.C., Turemen, E., and Tarkun, P. (2004). The plasminogen activator system in young and lean women with polycystic ovary syndrome. Endocr J *51*, 467–472.
Taylor, A.E. (1998). Polycystic ovary syndrome. Endocrinology and metabolism clinics of North America *27*, 877–902, ix.
Taylor, A.E., McCourt, B., Martin, K.A., Anderson, E.J., Adams, J.M., Schoenfeld, D., and Hall, J.E. (1997). Determinants of abnormal gonadotropin secretion in clinically defined women with polycystic ovary syndrome. J Clin Endocrinol Metab *82*, 2248–2256.
Vincent, C. (2001). The safety of acupuncture. BMJ (Clinical research ed *323*, 467–468.
Walters, K.A., Allan, C.M., and Handelsman, D.J. (2008). Androgen actions and the ovary. Biol Reprod *78*, 380–389.
White, P., Lewith, G., Hopwood, V., and Prescott, P. (2003). The placebo needle, is it a valid and convincing placebo for use in acupuncture trials? A randomised, single-blind, cross-over pilot trial. Pain *106*, 401–409.
Witkin, J.W. (1999). Synchronized neuronal networks: the GnRH system. Microsc Res Tech *44*, 11–18.
Wood, J.R., Nelson, V.L., Ho, C., Jansen, E., Wang, C.Y., Urbanek, M., McAllister, J.M., Mosselman, S., and Strauss, J.F., 3 rd (2003). The molecular phenotype of polycystic ovary syndrome (PCOS) theca cells and new candidate PCOS genes defined by microarray analysis. J Biol Chem *278*, 26380–26390.
Wu-Peng, X.S., Chua, S.C., Jr., Okada, N., Liu, S.M., Nicolson, M., and Leibel, R.L. (1997). Phenotype of the obese Koletsky (f) rat due to Tyr763Stop mutation in the extracellular domain of the leptin receptor (Lepr): evidence for deficient plasma-to-CSF transport of leptin in both the Zucker and Koletsky obese rat. Diabetes *46*, 513–518.
Wu, T.J., Gibson, M.J., Rogers, M.C., and Silverman, A.J. (1997). New observations on the development of the gonadotropin-releasing hormone system in the mouse. J Neurobiol *33*, 983–998.
Xiaoming, M.O., Ding, L.I., Yunxing, P.U., Guifang, X.I., Xiuzhen, L.E., and Zhimin, F.U. (1993). Clinical studies on the mechanism for acupuncture stimulation of ovulation. Journal of Traditional Chinese Medicine *13*, 115–119.
Zacche, M.M., Caputo, L., Filippis, S., Zacche, G., Dindelli, M., and Ferrari, A. (2009). Efficacy of myo-inositol in the treatment of cutaneous disorders in young women with polycystic ovary syndrome. Gynecol Endocrinol, 1–6.
Zawadski, J.K., and Dunaif, A. (1992). Diagnostic criteria for polycystic ovary syndrome: towards a rational approach (Oxford, Blackwell Scientific Publications).

Chapter 17
Auricular Acupuncture for Pain and Inflammation

Shi Ping Zhang, Wai Yeung Chung, and Yi Jie Zhan

Abstract Auricular acupuncture (AA) is a special acupuncture treatment system that can be used on its own or in conjunction with other acupuncture treatment systems. Review of the literature showed that AA had been used extensively in treatment of various pain and inflammatory conditions. However, its mechanism of action remained poorly understood, and it was not clear whether the location of stimulation at the auricle was important for its efficacy. By using rat models of experimentally induced inflammation, we explored the issue of acupuncture point location specificity and tested the roles of opioid and muscarinic receptors in mediating the anti-inflammatory effects of AA. We also compared the efficacy of AA with that of electroacupuncture at meridian points. We found that there was no strict functional topography for AA, although certain regions appeared to produce more consistent anti-inflammatory effects. Muscarinic receptors played an important role in both the analgesic and antiedematous actions of AA, whereas opioid receptors were only involved in the analgesic action. We also found that the efficacy of both AA and electroacupuncture at meridian points were comparable. Our findings suggest that the action of AA may be mediated by a unique mechanism and it should be used in conjunction with meridian acupuncture to enhance the efficacy of acupuncture treatment.

Keywords Ear acupuncture • Vagus nerve • Acupuncture stimulation • Anti-inflammation • Mechanism of action • Cranial afferent nerves

---

S.P. Zhang (✉) • Y.J. Zhan
School of Chinese Medicine, Hong Kong Baptist University, Kowloon Tong, Kowloon, Hong Kong
e-mail: spzhang@hkbu.edu.hk; greencabin@hotmail.com

W.Y. Chung
School of Chinese Medicine, The Chinese University of Hong Kong, Shatin, New Territory, Hong Kong
e-mail: chungwym@gmail.com

Y. Xia et al. (eds.), *Current Research in Acupuncture*,
DOI 10.1007/978-1-4614-3357-6_17,

17.1 Introduction

Acupuncture involves various treatment modalities, and auricular acupuncture (AA) is one of the commonly used ones. AA is a special acupuncture treatment system that can be used on its own or in conjunction with other acupuncture treatment systems. It involves applying various types of physical stimulation on the auricle to treat certain diseases, especially diseases involving the nervous and endocrine system. Particularly, it has been used in treatment of various pain conditions. Compared with meridian acupuncture, which has been better studied, little is known about the mechanism of AA action. In this chapter we reviewed our recent studies on the mechanism of AA in treatment of pain and inflammation and discussed the efficacy of AA with reference to the more commonly used meridian acupuncture.

17.2 Historical Use of Auricular Acupuncture for Pain and Inflammation

In one of the earliest medical texts discovered from the Han (202 BC–220 AD) tomb in Ma Huang Dui, Chang Sha, China, meridians connecting the ear and other body regions were mentioned in several places. According to the description in the text, a great number of inflammatory and painful conditions, including pain in front of the ear, eye pain, painful cheek, swollen throat, head and neck pain, swollen ribs, leg pain, and knee pain, could be treated by applying moxibustion to the meridian, presumably including the meridian segments on the auricular surface. More definite description of the use of AA can be found in the Chinese medical classic Neijing, which is also written around the Han Dynasty. In this classic, a specific point called Er Zhong was mentioned, and dark veins that appeared on the auricular surface were described as sites for acupuncture. Documentation of specific auricular points for pain- and inflammatory-related conditions first appeared in medical texts of the Tang Dynasty (618–907) (Zhang 1995). In the following centuries, stimulation of the auricle as a treatment for various conditions involving pain and inflammation occasionally appeared in ancient Chinese medical texts, but it was not until 1956 that systematic use of AA for pain conditions was described by the French doctor P. Nogier. Nogier reignited the interest of AA by proposing a theory that there was a functional somatotopic and viscerotopic representation on the auricle, forming a homunculus of inverted fetus. In his schema, stimulation of representation sites of different body regions and organs on the ear could treat problems arising from the corresponding region (Oleson 2003). Since then, AA has been used for a great number of pain and inflammation conditions, as well as other diseases.

Many diseases have been treated with AA in China (Chen 1993; Li 1993). Of these, conditions involving pain and inflammation were herpes zoster, cholelithiasis, gastritis, ulcerative colitis, allergic rhinitis, chronic rhinitis, pharyngitis, sty, conjunctivitis, toothache, skin lesion, warts, acne, pruritus, urticaria, dysmenorrhea, etc.

Also, researchers outside China reported the use of AA for pain conditions such as chronic spinal pain (Sator-Katzenschlager et al. 2003, 2004), cancer pain (Alimi et al. 2003), phantom limb pain (Katz and Melzack 1991), and wound care in patients with burns (Lewis et al. 1990), as well as for postoperative pain (Asher et al. 2010). In view of the comprehensive use of AA for pain and inflammation, it is important to understand the physiological basis of these treatments.

17.3 Is There a Functional Topography in the Auricle?

As with other modalities of acupuncture treatment, the physiological basis of AA is far from clear. The peripheral and central nervous systems are believed to play a crucial role in integrating the effects of AA. Anatomical studies concerning the connections of the ear have revealed some interesting and unique features. That is, several nerves with distinct origins innervate the ear to form an overlapping sensory network on the auricular surface. In humans, the afferent nerves come from the great auricular nerve and the lesser occipital nerve of the cervical plexus, the auriculotemporal branch of the mandibular nerve, which originated from the trigeminal nerve, and the auricular branches of the vagus, facial, and glossopharyngeal nerves. The auriculotemporal nerve innervates the upper region of the auricle, the cervical plexus innervates the outer lower region and the ear lobe, and the vagal nerve innervates the cavities (see Fig. 17.1), and it has been theorized that they connect to the skeletal, neuronal, and visceral components of the body, respectively (Oleson 2003).

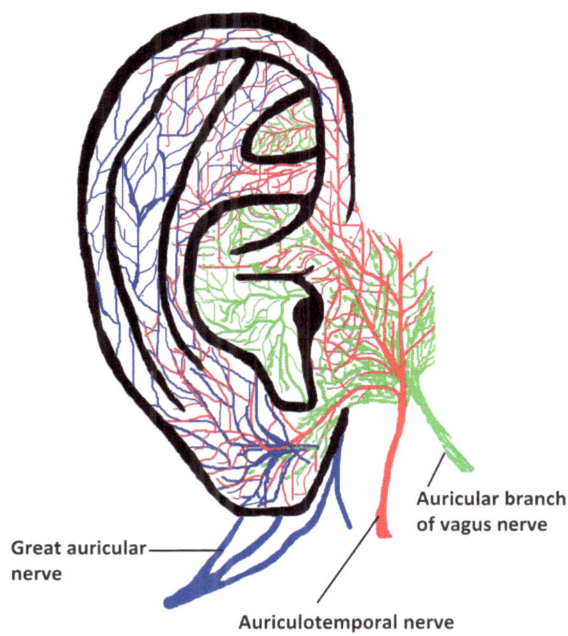

Fig. 17.1 Schematic drawing of the auricular surface showing the pattern of preferential nerve innervation in different areas (adopted from Oleson 2003)

Study of the rat ear with tracer injection also found differential patterns of innervation (Zhang et al. 2002). That is, injection into the *conchae* resulted in more labeled neurons in the superior vagal ganglia than injection into the *helix*, and injection into the *helix* labeled more neurons in the dorsal root ganglia of cervical plexus than injection into the *conchae*. Similar findings have been reported in the dog (Chien et al. 1996, 1998). Taken together, both human and animal studies suggest that there exists a topographical organization in auricular innervation.

It has also been suggested that there is a linkage between auricular points and specific body parts. By means of palpation for tender point, Oleson et al. (1980) showed that the origin of pain could be identified in about 75 % of patients suffering from different pain complaints. However, Andersson et al. (2007) reported no correlation between the locations of musculoskeletal pain and tender ear zones, although very tender points are found in patients with chronic musculoskeletal pain. On the other hand, reduction of electrical impedance at specific loci has also been reported in patients with hepatitis, tuberculosis, myocardial infection, and nephritis (Chen 1981) and in patients after orthopedic surgery (Usichenko et al. 2003).

There is inconsistent evidence regarding the topographical organization in terms of output function. It has been reported that stimulation at the oral point was 20 % more effective than stimulation at the buttock point in treatment of Bell's palsy (Chen 1981) and that stimulation at the appropriate points for experimental wrist pain is more effective than inappropriate points (Oliveri et al. 1986; Noling et al. 1988). However, Johnson et al. (1991) reported no difference between wrist point and face point in analgesic effect produced by TENS. It has also been suggested that selecting points with low impedance is essential for pain relief (Alimi et al. 2003). Thus, more concrete evidence is required for establishing the topography of auricular point functions.

17.4 Possible Mechanisms in Auricular Acupuncture Anti-inflammation

There have been very few studies concerning the mechanism of AA in treatment of pain and inflammation. On the other hand, previous studies of meridian acupuncture have identified a number of possible mechanisms, including opioids acting both locally and systemically (Sekido et al. 2004; Huang et al. 2004; Zhang et al. 2004a; although see Zhang et al. 2004b), activation of the hypothalamus–pituitary–adrenal axis (Li et al. 2008; Zhang et al. 2005) and the sympathetic–adrenal system (Kim et al. 2007), and release of adenosine locally (Goldman et al. 2010). The cholinergic anti-inflammatory pathway, which was discovered at the turn of this century, had also been proposed to play a role in acupuncture in general (Tracey 2007). Interestingly, AA has a tendency to activate the vagal efferent nerve (Gao et al. 2008; La Marca et al. 2010), which is a major source of cholinergic output. For this reason, we have examined the effects of AA on both acute and chronic models of inflammation in the rat, with specific reference to the role of cholinergic receptors (Chung et al. 2011), and the findings are reviewed below.

17.5 The Effects of Auricular Acupuncture on Acute Inflammation

17.5.1 Methods

Because electroacupuncture can provide controlled stimulation to auricular points, we used this method of stimulation in all our acute experiments. Adult male Sprague-Dawley rats were anesthetized with chloral hydrate (Fluka, 400 mg/kg, i.p.), for 60–80 min. After the initial anesthesia, animals were either allowed to wake up for behavioral testing or kept anesthetized by receiving additional anesthetics (20 mg of chloral hydrate) when showing blinking reflex or movements to handling. Under anesthesia, local inflammation was induced by subcutaneous injection of 50 μl of 2 % carrageenan lambda (CA, Sigma) in 0.9 % saline at the center of the left paw on the planter surface. Paw volume was measured at the level of the calcaneus with a plethysmometer, and nociception thresholds were evaluated by measuring responses to thermal and pressure stimuli before anesthesia and at least 4 h after the initial anesthetic injection, the time at which the rats had completely recovered from anesthesia and displayed normal walking and exploratory behaviors, allowing reliable algesic tests to be carried out.

17.5.1.1 Finding a Suitable Intensity for Electroacupuncture Stimulation

We believed that it was important to apply stimulation with clinical relevance in our experiments, so we first determined the range of electrical intensity acceptable to humans in six healthy human subjects (three male and three female). In 21 tests, acupuncture needles (Carbo, 0.2 × 13 mm) were inserted about 2 mm deep bilaterally at one pair of the auricular points indicated in Fig. 17.2. They were connected to an electroacupuncture machine (Cafar Acus II, Sweden), which was set to deliver 4 Hz rectangular pulses (0.45 ms in duration). The median (and interquartile) ranges of thresholds for sensory detection, pain sensation, and maximum tolerable sensation were 0.3 (0.3–0.4) mA, 0.8 (0.7–1.1) mA, and 1.3 (1.1–1.5) mA, respectively. We therefore tested the effects of AA in the rat between 0 and 1.5 mA. Five groups of animals ($n = 10–12$ per group) were given AA (0, 0.5, 0.7, 1.0, or 1.5 mA) at a point that had been shown good anti-inflammatory effect. It was found that the 0.7 and 1.0 mA groups had the strongest inhibitory effect on paw edema. Therefore, for all later experiments, the stimulation intensity was set to 0.7–1.0 mA.

17.5.2 Is There Location Specificity for AA in Anti-inflammation?

Both the theories of Chinese medicine and topographical innervation of the ear as reviewed above suggest that there may be location specificity for the effect of AA. Furthermore, it had also been suggested that auricular acupoints typically correspond

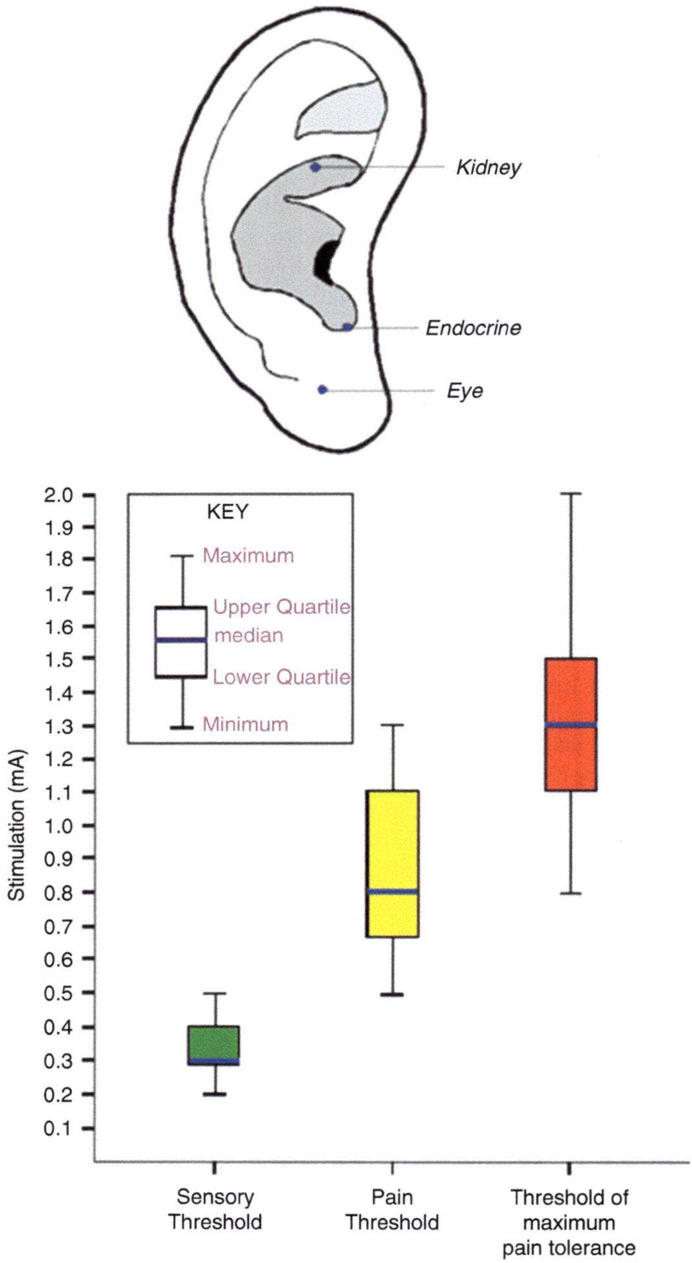

Fig. 17.2 *Box* plots showing the various sensation thresholds for auricular acupuncture with electrical stimulation in human subjects. Drawing of the ear indicates the location of points tested. The median (and interquartile) for sensory threshold, pain threshold, and maximum tolerable sensation was 0.3 (0.3–0.4) mA, 0.8 (0.7–1.1) mA, and 1.3 (1.1–1.5) mA, respectively ($n=20$)

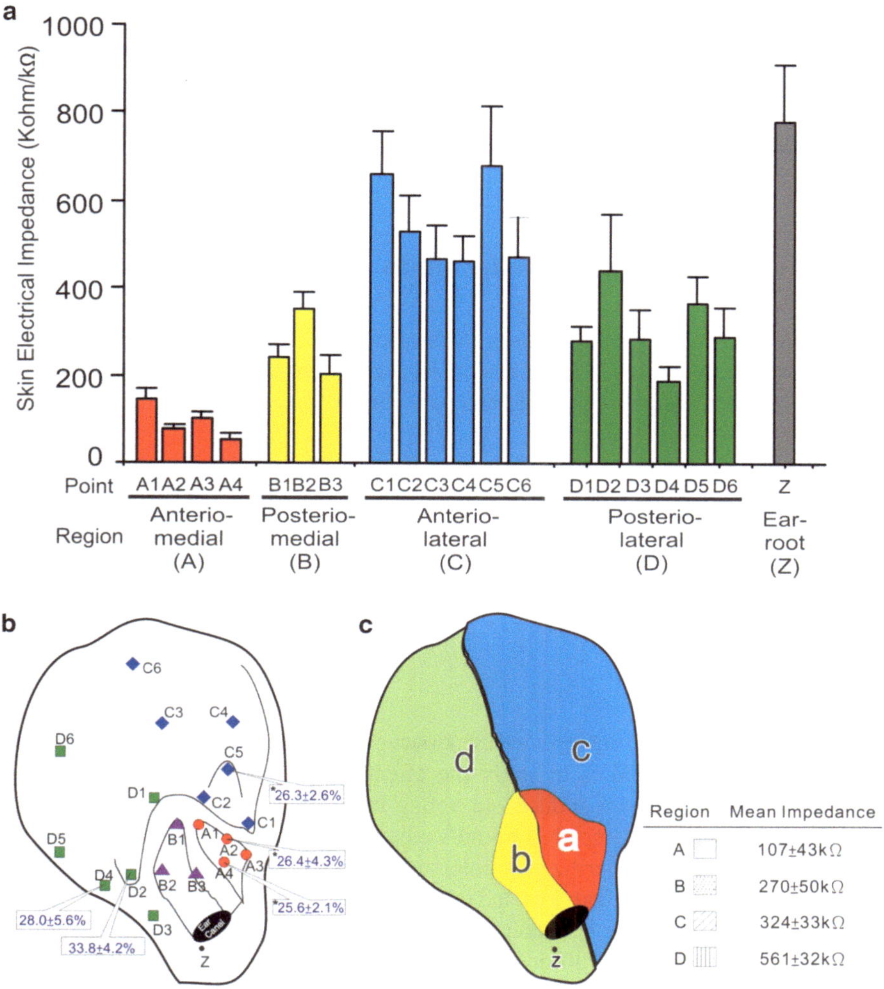

Fig. 17.3 (**a**) *Bar chart* showing the average of 36 measurements of auricular skin electrical impedance taken in 6 rats. (**b**) Schematic drawing showing the location of 20 points selected for measurements. The paw volumes after CA injection and electroacupuncture at sites A2, A4, C5, D2, and D5 are shown in the respective boxes, and they are indicative of the anti-inflammatory efficacy of the given site. (**c**) Schematic drawing showing the impedance of four anatomical region of the auricle

to regions of low electrical impedance; we thus identified low impedance points in six rats based on the method described previously (Kawakita et al. 1991). We found that low impedance points were mainly found in area A (Fig. 17.3), which corresponded to the cymba conchae in humans.

We then compared the antiedematous effect of AA at different auricular points in fully anesthetized animals. It was found that points in area A produced the best antiedematous effect (Fig. 17.3). However, some points outside area A also produced good effect (e.g., Fig. 17.3b: D4 in area D and C5 in area C). This finding is in agreement with our previous study on AA modulation of cardiovascular function (Gao et al. 2008). Taken together, our studies indicate that strict functional topography does not exist for AA.

17.5.3 *Analgesic and Antiedematous Effects of AA*

In conscious animals we tested both thermal and mechanical analgesic effects of AA, as well as the antiedematous effect. As seen in Fig. 17.4, AA was effective in reducing mechanical pain threshold and paw edema, without affecting thermal pain threshold.

17.6 Effects of Auricular-Embedded Needle on Chronic Inflammation

17.6.1 *Methods*

Due to the difficulty of using repeated electroacupuncture at the ear in chronic animals, we used embedded needles to study the effect of AA on chronic inflammation. Chronic inflammation was induced in young male Wistar rats with subcutaneous injection of 250 μg of complete Freund's adjuvant (CFA) into the left tibial–tarsal joint. For AA treatment, sterilized stainless needles (3.0 × 0.15 mm) were embedded into the inferior concha of both ears and left in place for the duration of the experiment. For comparison, electroacupuncture was applied to a separate group of animals once daily at acupoints ST36 and SP6 contralateral to the side of CFA injection using the electroacupuncture parameters described above in conscious condition. The contralateral side was chosen because pilot experiments found that electroacupuncture at the same acupoints ipsilateral to the side of CFA injection was ineffective in this model.

17.6.2 *Auricular-Embedded Needles Inhibited Swelling and Joint Destruction in Chronic Inflammation*

AA treatments reduced paw swelling in the CFA-injected paw as well as in the contralateral paw (Fig. 17.5). The differences appeared at day 11. X-ray results taken at day 20 also showed less joint destruction in the treatment group but had not reached

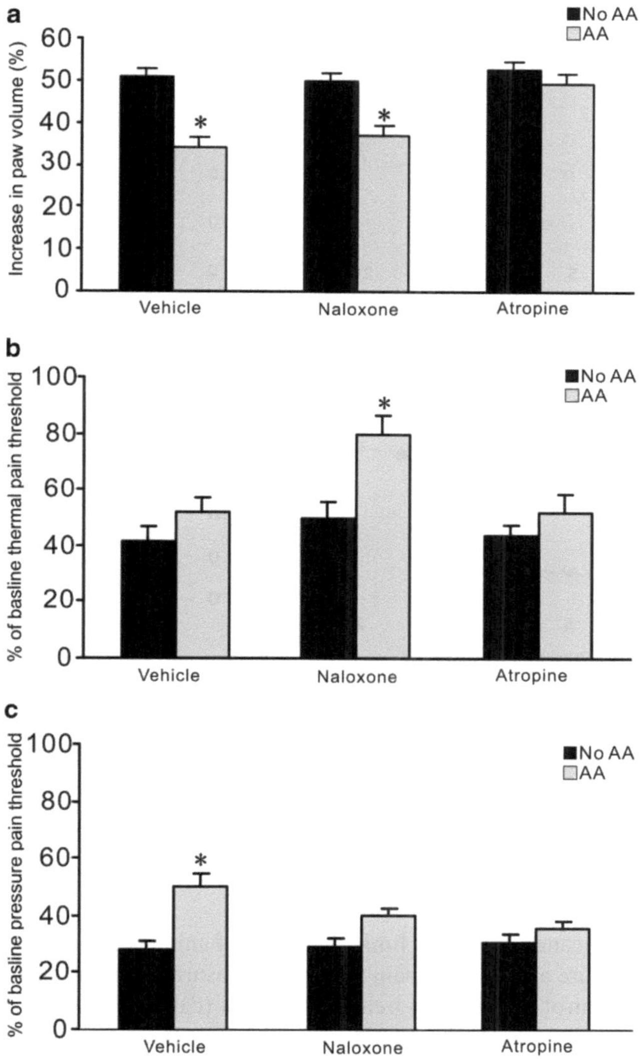

Fig. 17.4 *Histograms* showing the effects of auricular acupuncture (AA) with electrical stimulation on (**a**) paw volume, (**b**) thermal pain threshold and (**c**) pressure pain threshold in conscious animals with CA-induced inflammation. Paw volume is expressed as percentage increase from the baseline, and thermal or pain threshold is shown as percentage change from the baseline. The effects of opioid and muscarinic receptors blockades were examined using a 2×3 full-factorial design ($n=12$ per group) with methyl atropine (2 mg/kg, ip) and naloxone (5 mg/kg followed by 2.5 mg/kg, ip), respectively. *$P<0.05$, comparison between AA and no AA with the same ip injection (vehicle, naloxone, or atropine)

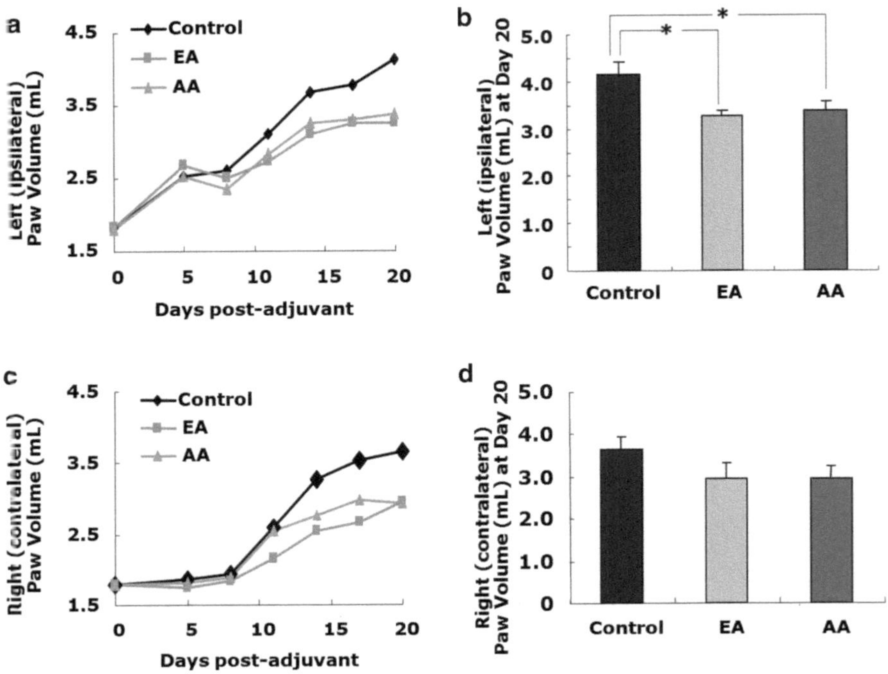

Fig. 17.5 *Line graphs* and *histograms* showing changes in paw volume of animals with chronic arthritic inflammation induced by injection of complete Freund's adjuvant (CFA) into the left tibiatarsal joint. AA, animals received embedding needles at the inferior concha bilaterally ($n=11$). EA, animals received electroacupuncture at acupoints ST36 and SP6 at the right side ($n=5$). (**a**), left (ipsilateral paw volume; (**c**), right (contralateral) paw volume. At day 20 (**b&d**) ipsilateral paw volumes of the EA and AA groups were significantly different from that of the control (*$P<0.05$), but there was no difference between EA and AA groups

statistical significance due to the limited number of animals tested ($n=5$, Fig. 17.6). However, pressure and thermal pain threshold measurements showed no difference between the control and the two treatment groups (data not shown). The extent of inhibition in paw swelling and joint destruction was similar between AA and body electroacupuncture.

17.7 Peripheral Muscarinic Receptors Had an Important Role in Mediating Auricular Acupuncture Anti-inflammation

To examine the involvements of opioid and peripheral muscarinic receptors in the action of AA, the acute carrageenan-induced inflammation model described in 17.5.1 was used ($n=12$ in each group). As seen in Fig. 17.4, blockade of peripheral muscarinic receptors with atropine methyl bromide (AT) antagonized the effects of AA on edema reduction and on pressure pain threshold elevation. In contrast,

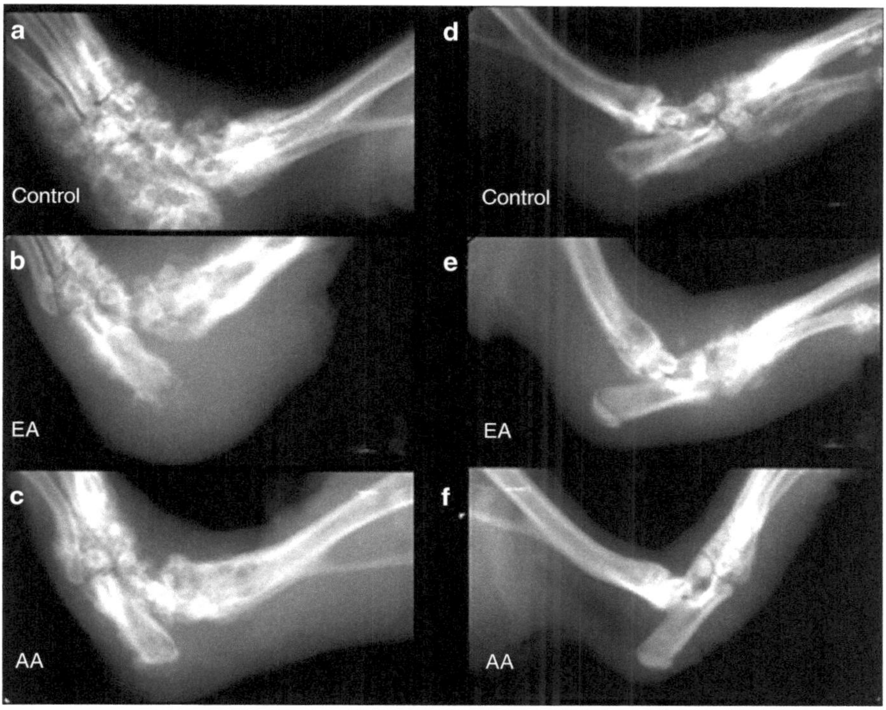

Fig. 17.6 Representative X-ray samples of the rat ankle joints. All animals received complete Freund's adjuvant (CFA) injection at the left side (**a**– **c**). (**d**–**f**) indicate changes at the contralateral side. AA, animals received embedding needles at the inferior concha bilaterally. EA, animals received electroacupuncture at acupoints ST36 and SP6 at the right side

administration of opioid receptor antagonist naloxone hydrochloride had no antagonistic effect on edema reduction but antagonized the analgesic effect of AA on mechanical hyperalgesia. Furthermore, the antiedematous effect of AA was blocked by local injection of a small dose of AT (12.5 μg, data not shown). These results indicated that peripheral muscarinic receptors were involved in mediating the anti-inflammatory effects of AA.

17.8 Auricular or Meridian Acupuncture in Treatments of Pain and Inflammation

The foregoing results have demonstrated that AA has a moderate efficacy in reducing both acute and chronic inflammation. Considering the results of our previous studies of meridian electroacupuncture on acute inflammation (Zhang et al. 2004) together, it appears that AA and meridian electroacupuncture have similar efficacies

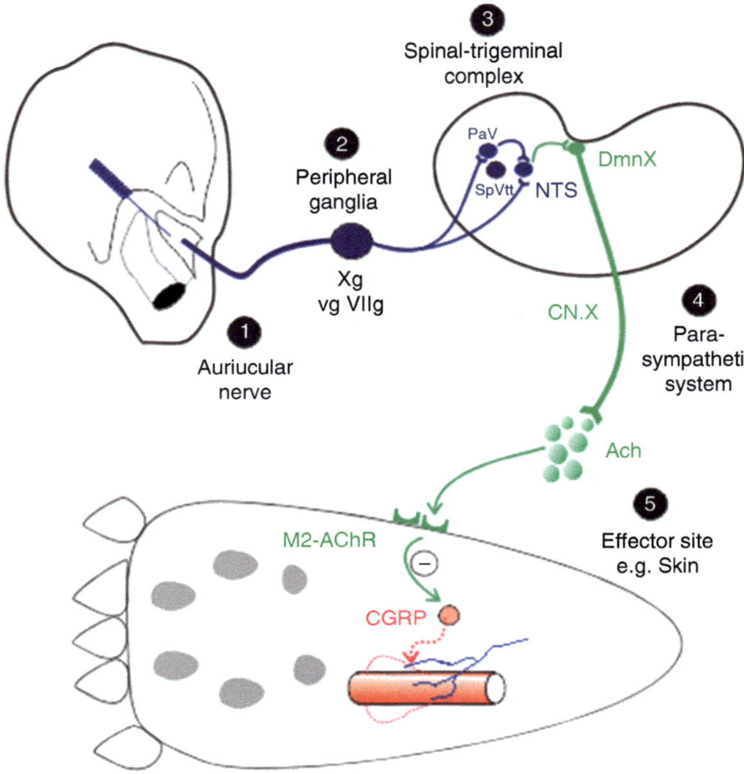

Fig. 17.7 Schematic diagram illustrating the hypothetical pathway for the cholinergic mechanism of auricular acupuncture (AA) in anti-inflammation. Upon AA stimulation, signals from the auricular skin are transmitted along the auricular nerves (1) via peripheral ganglia (2) to the spinal trigeminal complex and the solitary tract nucleus (NTS) (3). Through synaptic transmission the parasympathetic system, especially the dorsal motor nucleus of vagus (DmnX) (4), is activated, which leads to acetylcholine (ACh) being released from nerve endings. ACh then acts on M2 receptors of C fibers, resulting in the suppression of calcitonin gene-related peptide (CGRP) release and reducing vasodilation and nociceptor sensitization at nerve endings

for both acute and chronic inflammation. The mechanisms underlying AA and meridian acupuncture treatments, however, can be different in some aspects.

Regarding the anti-inflammatory actions observed in AA, it is unlikely that local adenosine release plays a significant role as there is no local stimulation. Activation of the hypothalamus–pituitary–adrenal axis cannot be ruled out as a previous study has shown activation of hypothalamic neurons during AA (Shiraishi et al. 1995). Opioid receptors appear to play a role in AA-induced analgesia but are not important for its antiedematous effect.

We have identified a novel mechanism for AA treatment of pain and inflammation that is mediated via peripheral muscarinic receptors. Cholinergic agonists acting on M2 receptors at C fiber terminals can inhibit calcitonin gene-related peptide (CGRP) release and lead to nociceptor desensitization (Bernardini et al. 2002). As the anti-inflammatory effect of AA is best evoked from the concha region, which has preferential innervation from vagal sensory afferents and can induce vagal efferent activation upon stimulation, we hypothesize that AA activates vagal efferent nerves to produce the anti-inflammatory effects (see Fig. 17.7). On the other hand, meridian acupuncture on the limbs may be less effective in activating vagal efferents. Future studies with direct measurement of acetylcholine at the site of inflammation would be useful for testing that hypothesis. Nonetheless, it is possible that, like meridian acupuncture, AA involves several physiological mechanisms in regulation of inflammation.

17.9 Concluding Remarks

In summary, we have shown that AA can relieve pain and inflammation via a muscarinic mechanism, which may be a unique feature of AA. Nevertheless, the overall magnitude of the anti-inflammatory effects of AA is not very large, and it is comparable to meridian acupuncture. In clinical practice, the concurrent use of AA and meridian acupuncture may enhance the efficacy of acupuncture treatment as the two modalities may have complementary therapeutic actions. Future studies should investigate the efficacy of combining auricular and meridian acupuncture for pain and inflammation.

Acknowledgements The authors would like to thank Nickie Chan for technical assistance. This study was supported by a Faculty Research Grant (FRG 30-08-318) of Hong Kong Baptist University.

References

Alimi D, Rubino C, Pichard-Leandri E, Fermand-Brule S, Dubreuil-Lemaire ML, Hill C. (2003) Analgesic effect of auricular acupuncture for cancer pain: a randomized, blinded, controlled trial. J Clin Oncol 21: 4120–4126.

Andersson E, Persson AL, Carlsson CP. (2007) Are auricular maps reliable for chronic musculoskeletal pain disorders? A double-blind evaluation. Acupunct Med. 25(3): 72–79.

Asher GN, Jonas DE, Coeytaux RR, Reilly AC, Loh YL, Motsinger-Reif AA, Winham SJ. (2010) Auriculotherapy for pain management: a systematic review and meta-analysis of randomized controlled trials. J Altern Complement Med. 16(10): 1097–1108.

Bernardini N, Roza C, Sauer SK, Gomeza J, Wess J, Reeh PW. (2002) Muscarinic M2 receptors on peripheral nerve endings: a molecular target of antinociception. J Neurosci. 22(12): RC229:1–5.

Chen G. (1981) Guanyu Erxue teyixing de Yanjiu. Zhenci Yanjiu Jinzan (Progress in Acupuncture Research). Beijing: People's Health Press. 72–79. (In Chinese)

Chen H. (1993) Recent Studies on Auriculoacupuncture and Its Mechanism. J Tradit Chin Med. 13(2): 129–143.

Chien CH, Shieh JY, Liao MH, Ling EA, and Wen CY. (1998) Neuronal Connections Between the Auricular Skin and the Sympathetic Pre- and Postganglionic Neurons of the Dog As Studied by Using Pseudorabies Virus. Neurosci Res. 30(2): 169–175.

Chien CH, Shieh JY, Ling EA, Tan CK, and Wen CY. (1996) The Composition and Central Projections of the Internal Auricular Nerves of the Dog. J Anat. 189 (Pt 2): 349–362.
Chung WY, Zhang HQ, Zhang SP. (2011) Peripheral muscarinic receptors mediate the anti-inflammatory effects of auricular acupuncture. Chin Med. 6(1): 3.
Gao XY, Zhang SP, Zhu B, Zhang HQ. (2008) Investigation of specificity of auricular acupuncture points in regulation of autonomic function in anesthetized rats. Auton Neurosci 138(1-2): 50–56.
Goldman N, Chen M, Fujita T, Xu Q, Peng W, Liu W, Jensen TK, Pei Y, Wang F, Han X, Chen JF, Schnermann J, Takano T, Bekar L, Tieu K, Nedergaard M. (2010) Adenosine A1 receptors mediate local anti-nociceptive effects of acupuncture. Nat Neurosci. 13(7): 883–888.
Huang C, Hu ZP, Long H, Shi YS, Han JS, Wan Y. (2004) Attenuation of mechanical but not thermal hyperalgesia by electroacupuncture with the involvement of opioids in rat model of chronic inflammatory pain. Brain Res Bull. 63(2): 99–103.
Johnson MI, Hajela VK, Ashton CH, Thompson JW. (1991) The effects of auricular transcutaneous electrical nerve stimulation (TENS) on experimental pain threshold and autonomic function in healthy subjects. Pain. 46(3):337–42.
Katz J, Melzack R. (1991) Auricular transcutaneous electrical nerve stimulation (TENS) reduces phantom limb pain. J Pain Symptom Manage 6: 73–83.
Kawakita K, Kawamura H, Keino H, Hongo T, Kitakohji H. (1991) Development of the low impedance points in the auricular skin of experimental peritonitis rats. Am J Chin Med, 19:199–205.
Kim HW, Kang SY, Yoon SY, Roh DH, Kwon YB, Han HJ, Lee HJ, Beitz AJ, Lee JH. (2007) Low-frequency electroacupuncture suppresses zymosan-induced peripheral inflammation via activation of sympathetic post-ganglionic neurons. Brain Res. 1148: 69–75.
La Marca R, Nedeljkovic M, Yuan L, Maercker A, Ehlert U. (2010) Effects of auricular electrical stimulation on vagal activity in healthy men: Evidence from a three-armed randomized trial. Clinical Science. 118: 537–546.
Lewis SM, Clelland JA, Knowles CJ, Jackson JR, Dimick AR. (1990) Effects of auricular acupuncture-like transcutaneous electric nerve stimulation on pain levels following wound care in patients with burns: a pilot study. J Burn Care Rehabil 11: 322–329.
Li A, Lao L, Wang Y, Xin J, Ren K, Berman BM, Tan M, Zhang R. (2008) Electroacupuncture activates corticotrophin-releasing hormone-containing neurons in the paraventricular nucleus of the hypothalamus to alleviate edema in a rat model of inflammation. BMC Complement Altern Med. 12: 8–20.
Li H (1993) Erya Kangyan [Anti-Inflammation With Auricular Pressure]. Zhen Jiu Xue Bao. 23(1): 51–55.
Noling LB, Clelland JA, Jackson JR, and Knowles CJ. (1988) Effect of Transcutaneous Electrical Nerve Stimulation at Auricular Points on Experimental Cutaneous Pain Threshold. Phys Ther. 68(3): 328–332.
Oleson TD, Kroening RJ, Bresler DE. (1980) An experimental evaluation of auricular diagnosis: the somatotopic mapping or musculoskeletal pain at ear acupuncture points. Pain. 8(2):217–229.
Oleson T. (2003) Auriculotherapy manual : Chinese and Western systems of ear acupuncture. Edinburgh: Churchill Livingstone.
Oliveri AC, Clelland JA, Jackson J, and Knowles C. (1986) Effects of Auricular Transcutaneous Electrical Nerve Stimulation on Experimental Pain Threshold. Phys Ther. 66(1): 12–16.
Sator-Katzenschlager SM, Scharbert G, Kozek-Langenecker SA, Szeles JC, Finster G, Schiesser AW, Heinze G, Kress HG. (2004) The short- and long-term benefit in chronic low back pain through adjuvant electrical versus manual auricular acupuncture. Anesth Analg 98:1359–1364.
Sator-Katzenschlager SM, Szeles JC, Scharbert G, Michalek-Sauberer A, Kober A, Heinze G, Kozek-Langenecker SA. (2003) Electrical stimulation of auricular acupuncture points is more effective than conventional manual auricular acupuncture in chronic cervical pain: a pilot study. Anesth Analg 97:1469–1473.
Sekido R, Ishimaru K, Sakita M. (2004) Corticotropin-releasing factor and interleukin-1beta are involved in the electroacupuncture-induced analgesic effect on inflammatory pain elicited by carrageenan. Am J Chin Med. 32(2):269–279.

Shiraishi T, Onoe M, Kojima T, Sameshima Y, Kageyama T. (1995) Effects of auricular stimulation on feeding-related hypothalamic neuronal activity in normal and obese rats. Brain Res Bull. 36(2):141–148.

Tracey KJ. (2007) Physiology and immunology of the cholinergic antiinflammatory pathway. J Clin Invest 117: 289–296.

Usichenko TI, Lysenyuk VP, Groth MH, Pavlovic D. (2003) Detection of ear acupuncture points by measuring the electrical skin resistance in patients before, during and after orthopedic surgery performed under general anesthesia. Acupunct Electrother Res. 28(3-4): 167–173.

Zhang N (1995) Dun huang yi shu zhong de er yue yu er kong jiu fa. Chinese Journal of Medical History 25(3): 156. (In Chinese)

Zhang RX, Lao L, Wang L, Liu B, Wang X, Ren K, Berman BM. (2004a) Involvement of opioid receptors in electroacupuncture-produced anti-hyperalgesia in rats with peripheral inflammation. Brain Res. 1020(1-2): 12–17.

Zhang SP, Zhang JS, Yung KK, Zhang HQ. (2004b) Non-opioid-dependent anti-inflammatory effects of low frequency electroacupuncture. Brain Res Bull. 15;62(4): 327–334.

Zhang RX, Lao L, Wang X, Fan A, Wang L, Ren K, Berman BM. (2005) Electroacupuncture attenuates inflammation in a rat model. J Altern Complement Med. 11(1): 135–142.

Zhang SX, Zhou KF, Jiang WF, and Hua JL. (2002) Projection of ear points on the brain and spinal ganglia. Shanghai J Acu-Mox. 21(3): 35–36. (In Chinese)

Chapter 18
Glial–Neuronal Interactions in Electroacupuncture Analgesia

Hong Cao, Zhi-Qi Zhao, Gen-Cheng Wu, and Yu-Qiu Zhang

Abstract Electroacupuncture (EA) can produce analgesic effects on a diversity of acute or chronic pain. Although neuronal mechanisms to account for these phenomena have been sought for many decades, their bases remain unresolved. Within the past 4 years, a novel nonneuronal candidate has been uncovered that opposes EA analgesia. This novel candidate is spinal cord glia, which importantly contributes to the initiation and maintenance of exaggerated pain. New data suggest that disruption of glial function can enhance EA-induced analgesic effect, which might provide new insight for EA analgesia.

Keywords Electroacupuncture analgesia • Spinal cord glia • Inflammatory cytokines • Pain

18.1 Introduction

In traditional Chinese medicine, acupuncture is considered as one of the most effective medical treatments to treat a variety of diseases and symptoms, especially disorders accompanied with severe pain (Cheng and Cheng 2008). There are two different strategies of performing acupuncture therapy, that is, manual acupuncture

---

H. Cao • Z.-Q. Zhao • Y.-Q. Zhang (✉)
Institute of Neurobiology, Institutes of Brain Research, Fudan University, Shanghai 200032, China

State Key Laboratory of Medical Neurobiology, Shanghai Medical College of Fudan University, Shanghai 200032, China
e-mail: hongcao@fudan.edu.cn; zqzhao@fudan.edu.cn; yuqiuzhang@fudan.edu.cn

G.-C. Wu
Department of Integrative Medicine and Neurobiology, Shanghai Medical College of Fudan University, Fudan University, Shanghai, China
e-mail: gcwu@shmu.edu.cn

Y. Xia et al. (eds.), *Current Research in Acupuncture*,
DOI 10.1007/978-1-4614-3357-6_18,

(MA) and electroacupuncture (EA). EA, a modified form of traditional MA, has been investigated extensively for several decades in studies in human subjects and experimental animals with acute and chronic pain (Dai et al. 2001; Huang et al. 2002; Kim et al. 2005, 2007).

Inflammation and injury are the common causes of chronic pain which manifests multiple classic characters including spontaneous pain, increased responsiveness to painful stimuli (hyperalgesia), and pain perceived in response to normally non-noxious stimuli (allodynia). The dominant theme in pathological pain research has been investigating alterations in neurons and neuronal functioning in the peripheral and central nervous systems. So traditionally, pathological pain is viewed as being modulated solely by neurons. However, mounting evidence has proved that spinal cord glia (both astrocyte and microglia) play very important roles in the pain processing, which has advanced our understanding for pain. In addition to providing physical support and housekeeping for the neurons, glia also actively contributes to the modulation of neuronal activity and synaptic transmission (Araque et al. 2011). Studies from our laboratory and other researchers have demonstrated that spinal cord glia are key players in the initiation and maintenance of pathological pain (Cao and Zhang 2008; Hains and Waxman 2006; Hashizume et al. 2000; Ledeboer et al. 2005b; Milligan et al. 2003; Raghavendra et al. 2003; Sun et al. 2006, 2007b).

Acupuncture analgesia is termed as the acupuncture-induced analgesic effect which could alleviate diverse types of pain, especially chronic pain. Lots of efforts have been made to explain analgesic effect of acupuncture. One of the primary mechanisms is to stimulate the central nervous system (CNS) to release neurotransmitters or neuromodulators including opioid peptides, serotonin and noradrenaline, which either change the experience of pain or promote release of other chemicals, such as neurohormones that influence neuroimmune function (Dickenson and Sullivan 1986; Malmberg and yaksh 1995; Ulett et al. 1998; Cheunsuang et al. 2002). The second possibility is that acupuncture-induced changes in spinal chemistry influence spinal immune-like glial activation and resultant release of algesic substances under conditions of pain enhancement (Sun et al. 2006, 2007b, 2008). Work described in this chapter indicates that (1) glia are activated by electroacupuncture and (2) spinal glial dysfunction potentiates electroacupuncture analgesic effects. These data predict that the clinical efficacy of electroacupuncture, and probably manual acupuncture, could be enhanced by targeting these newly recognized, pain-facilitatory effects of glial activation.

18.2 Historical Overview

Acupuncture is now accepted as a complementary analgesic treatment in clinical practice. The efficiency on pain relief was impressive, especially in pathological situations. On the other hand, in addition to providing physical support and housekeeping for neurons, glia are accepted to contribute to the modulation of neuronal excitability, synaptic transmission, and chronic pain in the last decade.

18.2.1 Electroacupuncture Analgesia

Mounting evidence from clinical observations and laboratory research points to the overwhelming recognition that the stimulating current at various parameters applied to acupoints through acupuncture needles can produce analgesic effects on a diversity of acute or chronic pain (Han et al. 1983; Kim et al. 2000; Lao et al. 2004; Takakura et al. 1995). Despite the involvement of psychological factors in clinical treatment of patients and stress in animal behavioral experiments, a large amount of evidence has demonstrated that acupuncture analgesia has its physiological, anatomical, and neurochemical bases.

De-Qi is an intricate sensation of soreness, numbness, heaviness, and distension during acupuncture or electroacupuncture manipulation (Hui et al. 2005; Haker and Lundeberg 1990). A considerable amount of evidence from clinical research suggests that such a De-Qi feeling mainly originates from acupuncture-induced impulses from the deep tissue beneath the acupoints and is very essential to acupuncture analgesia (Cao 2002; Zhou et al. 1979). Studies from electrophysiology, animal behavior, and human experiments demonstrate that electrical current via acupuncture needles at intensities sufficient to excite Aβ-type afferents is capable of producing analgesia. However, excitation of some Aδ-type afferent induces a much more potent analgesia. Whether excitation of C fiber is involved in EA analgesia is controversial. Liu and colleagues found that collision with the orthodromic EA stimulation decreased the amplitude of the antidromic C wave of the compound action potential in the peroneus nerve, suggesting excitation of C fibers in EA (Liu et al. 1990). Also, if there is predestruction of unmyelinated fibers by neonatal capsaicin treatment in rats (Nagy 1982), research found that EA analgesia was significantly reduced compared to the control group, supporting the involvement of C-type afferents in EA analgesia. On the contrary, using the same treatment, Pan et al. found that Fos expression after EA to the pads of the hindpaw was unaffected after degeneration of C afferent fibers by neonatal capsaicin treatment. These results suggest that expression of Fos was induced by EA in dorsal horn neurons via capsaicin-insensitive afferents, presumably Aδ rather than C afferents. The conflicts may come from the variety of stimulation parameters, such as the intensity, frequency, duration, and interval between stimuli.

Following peripheral activation of different afferents induced by EA, the information of analgesia conveys to the central. Clinical observations and experimental studies suggest that the neuronal pathways of acupuncture analgesia are integrated with pain pathways. Nociceptive information is transmitted from the spinal cord to the brain via several different pathways. There are three ascending funiculi involved, and they are the ventrolateral funiculus (VLF), dorsolateral funiculus (DLF), and dorsal column (DC). Among them, there are two leading ascending pathways. A projection from the superficial dorsal horn in the spinal cord to the parabrachial nucleus connecting to brain areas mainly involved in processing pain emotion is called the spinoparabrachial tract. Another pathway originates in the superficial and deep dorsal horn and projects to the thalamus connecting to the cortical areas involved in the sensory and discrimination and emotion of pain, called the spinothalamic tract. Lines of studies suggest that impulses from acupoints ascend mainly through the VLF (Cao 2002). Consequently, many studies show that a complex network of multiple brain structures is involved in the modulation of EA analgesia, including the rostral ventromedial medulla (RVM) mainly nucleus raphe magnus (NRM), periaqueductal gray (PAG), locus coeruleus (LC), arcuate nucleus (Arc), preoptic area (Po), centromedian nucleus (CM), nucleus submedius (Sm), anterior pretectal nucleus (AptN), habenular nucleus (Hab), nucleus accumbens (Ac), caudate nucleus (Cd), septal area (Sp), amygdale, anterior cingulate cortex (ACC), and hypothalamic paraventricular nucleus (PVH) (Zhao 2008). Importantly, most of nuclei are constitutive parts of the endogenous descending inhibitory system in the central nervous system. Except for the Hab and LC, lesions of these nuclei attenuate EA analgesia, while excitation potentiates it.

Various transmitters and modulators are implicated in EA analgesia. Endogenous opioid peptides are the most important among them. Three main groups of opioid peptides, β-endorphin, enkephalins, and dynorphins, and their μ-, δ-, and κ-receptors are widely distributed in parts of the peripheral and central nervous systems related to nociception and pain (Basbaum and Jessell 2001; Fields et al. 2005). Early in the 1970s, researchers found that naloxone, a specific opioid receptor antagonist, partially attenuated the analgesic effect of acupuncture on electrical stimulation-induced tooth pulp pain in humans (Mayer et al. 1997) and noxious heat stimulation-induced pain in mice (Pomeranz 1976), strongly suggesting the involvement of endogenous opioid peptides in EA analgesia. Consequently, many investigations proved the definite mechanism of morphine-like substrate in CNS underlying EA analgesia. Clinical studies showed that EA increased the content of β-endorphin-like immunoreactive substances in the lumbar cerebrospinal fluid (CSF) with chronic pain (Sjollund et al. 1997) and in the ventricular CSF of patients with brain tumor (Chen and Pan 1984). Also, increased endorphins occur in the PAG after acupuncture in rabbit (Zhang et al. 1981). Notably, there was a linear correlation between the percentage upregulation of β-endorphin and the efficiency of EA analgesia.

A remarkable finding among relationships between EA analgesia and opioid peptides is that the release of different opioid peptides is dependent on the frequency of EA stimulation. Studies from Han's group showed that EA at low frequency (2 Hz) facilitates the release of β-endorphin, enkephalin, and endomorphin within the network of the CNS, while EA at high frequency (100 Hz) stimulates the release of dynorphin (Fei et al. 1987; Han 2003; Huang et al. 2000). A combination of the two frequencies (2 Hz/100 Hz) can produce a simultaneous release of all four opioid peptides, resulting in a maximal analgesic effect (Ghoname et al. 1999; Han 2003). Using different subtypes of specific opioid receptor antagonists, their further studies showed that 2- and 100-Hz EA-induced analgesic effects were differentially reduced by blockade of μ-, δ-, and κ-receptors, strongly confirming that low- and high-frequency EA are mediated by μ-, δ-, and κ-receptors in the rat spinal cord, respectively, under the physiological pain conditions (Fig. 18.1) [reproduced from (Han 2003) with permission from Elsevier]. However, κ-receptors seem not to be involved in EA analgesia in pathological pain states (both inflammatory pain and neuropathic pain) (Zhang et al. 2004; Kim et al. 2004; Sun et al. 2004).

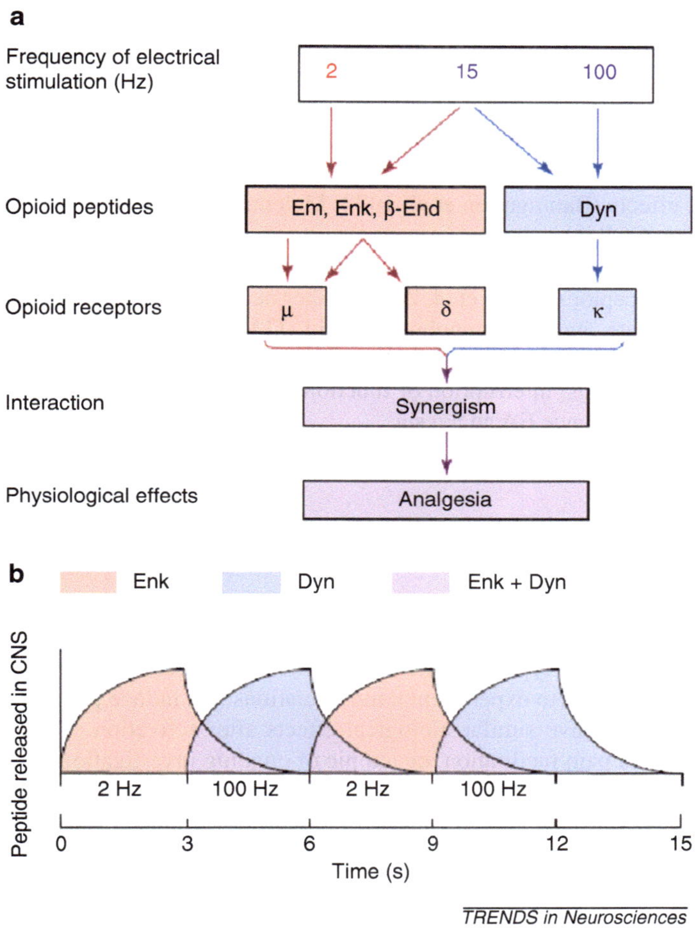

Fig. 18.1 Possible mechanisms for electroacupuncture analgesia. (**a**) Opioid peptides and opioid receptors involved in analgesia elicited by electroacupuncture of different frequencies: 2 Hz (*red*), 100 Hz (*blue*), 15 Hz (*purple*). Abbreviations: *Dyn* dynorphin A, *β-End* β-endorphin, *Em* endomorphin, *Enk* enkephalins. Synergism: Simultaneous activation of all three types of opioid receptor elicits a synergistic analgesic effect. (**b**) Model for the synergistic analgesic effect produced by alternating low- and high-frequency stimulations. Stimulation at 2 Hz facilitates the release of enkephalin (*red*); that at 100 Hz stimulates the release of dynorphin (*blue*). The overlapping areas (*purple*) indicate the synergistic interaction between the two peptides [reproduced from (Han 2003) with permission from Elsevier]

As mentioned above, many supraspinal regions are involved in analgesic effect induced by EA. There are also opioid peptides and related opioid receptors in these regions including Po, Sp, Sm, Cd, Ac, Arcu, amygdala, Hab, and PAG. When there is blockade of activity of opioid receptor by microinjection of opioid receptor antagonist into these given brain areas, the analgesic effect of EA is decreased (He 1987).

Beyond opioid peptides, other neuromodulators or neurotransmitters also play a very important role in EA analgesia. 5-Hydroxytryptamine (5-HT) and its receptors are widely distributed in the CNS and exert many physiological functions. EA increases the concentration of 5-HT and its metabolic products in the NRM and the spinal cord (Han et al. 1979a, b; Ye et al. 1979; Zhu et al. 1997). Selective depletion of 5-HT with 5, 6-dihydroxytryptamine (5, 6-DHT) significantly reduced EA analgesic effects (Baumgarten et al. 1972; Mclennan et al. 1997). Cholecystokinin octapeptide (CCK-8) is the most potent neuropeptide involved in anti-opioid effects. Behavioral studies showed that CCK-8 can antagonize EA analgesia via the spinal cord CCK8 receptor (Huang et al. 2007). Other neurotransmitters like noradrenalin (NA), glutamate, gamma-aminobutyric acid (GABA), or other neuromodulators like substance P (SP) and somatostatin (SOM) are all complicated in EA analgesia (Zhao 2008). That is, interruption of functions of these substances could positively or negatively influence EA analgesia.

18.2.2 Glial Activation and Pain

Glia, we mentioned here, is referring to both astrocyte and microglia. Although we are fully aware that they are very different, to date most studies of pathological pain showed that both cell types are activated by the same kinds of stimuli and both are activated in response to experimental manipulations that induce pain enhancement. In addition, they have similar biological effects after activation. Their respective contributions to pain facilitation are a topic of ongoing investigation.

The first indication that glia modulate pain arose from studies of neuropathic pain. In 1991, Garrison and colleagues reported that chronic constriction injury of sciatic nerve (CCI), an animal model of neuropathic pain, induced spinal astrocyte activation (Garrison et al. 1991). Using immunohistochemistry, they examined the expression of glial fibrillary acidic protein (GFAP), an astrocyte-specific activation marker. The results showed that (1) sciatic nerve constriction, which creates enhanced pain, also activated astrocytes, and (2) a drug that pharmacologically blocked enhanced pain also blocked astrocytic activation. From then on, almost each animal model of pathological pain studied to date has revealed a positive correlation between exaggerated pain responses and glial activation in the spinal cord. These procedures include spinal nerve injury (Gao et al. 2009; Raghavendra et al. 2003; Sweitzer et al. 2001), peripheral inflammation (Bao et al. 2001; Chacur et al. 2004; Gao et al. 2010 Pain; Sun et al. 2006; Sweitzer et al. 1999; Watkins et al. 1997), peripheral tissue injury (Fu et al. 2006; Obata et al. 2006), peripheral nerve inflammation (Ledeboer et al. 2005b; Milligan et al. 2003), spinal cord injury (Hains and Waxman 2006; Popovich et al. 1997), and bone cancer (Zhang et al. 2005). Mounting evidence from the mentioned models accrued that spinal cord glia importantly contribute to the creation and maintenance of exaggerated pain states. First, an early microglial and delayed astrocytic activation in the spinal cord was observed in response to trauma to, or inflammation of, peripheral nerves or tissue resulting in chronic pain. This is reflected by increased expression of activation markers on both mRNA and protein levels, including GFAP and S100B for astrocytes, and OX-42, ITGAM, TLR4, CD14, and MHC-II for microglia (Colburn et al. 1999; Hashizume et al. 2000; Tanga et al. 2004). Second, blockade of glia by fluorocitrate treatment, which at appropriate dosage (1 nmol), selectively inhibits the tricarboxylic acid cycle in astrocytes and microglia by blocking aconitase, a metabolic enzyme not used by neurons (Bergjohnsen et al. 1993; Hassel et al. 1992; Paulsen et al. 1987), inhibited persistent pain, whereas minocycline that inhibits microglia activation without direct effects on astrocytes or neurons by blockade of p38 mitogen-activated protein kinase, caspase-1, and caspase-3 in microglia (Tikka and Koistinaho 2001; Yrjanhelkki et al. 1988; Zhu et al. 2002) blocked the development of pain hypersensitivity behaviors but had no effect on the existing mechanical allodynia and hyperalgesia (Meller et al. 1994; Raghavendra et al. 2003). Third, intrathecal (i.t.) administration of HIV-1 envelope glycoprotein gp120 or spinal implantation of activated microglia produces a robust activation of spinal glia and exaggerated pain responses (Holguin et al. 2004; Milligan et al. 2000, 2001). Thus, glial activation is both necessary and sufficient for pain enhancement. And early microglial activation seems to be important in initiation and development of behavioral hypersensitivity, whereas astrocytic activation, which succeeds microglial activation, may play a role in maintaining exaggerated pain in animal models of both neuropathic and inflammatory pain.

It is important to note a finding from studies of spinal glia. That is, glial involvement in nociception is selective, as blockade of glial function does not affect acute pain (Meller et al. 1994; Milligan et al. 2000, 2003; Raghavendra et al. 2003; Sun et al. 2006). Several parallels will become evident between these data and those we will describe for EA-induced glial alterations.

What causes glia to become activated is another important issue. Tissue/nerve inflammation/damage results in immune activation and release of various neuroactive substances to peripheral terminals of sensory neurons, or along the length of the degenerating fibers. Spinal glia express an array of receptors for a host of substances released by primary sensory afferents, for example, excitatory amino acids (EAAs), substance P (SP), calcitonin gene-related peptide (CGRP), and ATP (Aicher et al. 1997; Araque et al. 1999; Marriott 2004; Tsuda et al. 2003). Peripheral neuropathy-induced degeneration of central terminals of the sensory neurons can directly recruit and activate spinal resident microglia (Watkins and Maier 1999). In addition, they may be activated in response to substances released by excited postsynaptic neurons in the spinal dorsal horn, such as PGs, NO, and fractalkine (Watkins et al. 2001, 2005).

Fractalkine is a newly found novel neuron-to-glia signal. It is a member of the chemokine family, which binds only one receptor (CX3CR1) and its receptor binds only fractalkine (Hughes et al. 2002; Jung et al. 2000). Spinal fractalkine is constitutively expressed by primary afferent terminals and spinal neurons where it is tethered to the extracellular membrane by a mucin stalk. Intriguingly, CX3CR1 is

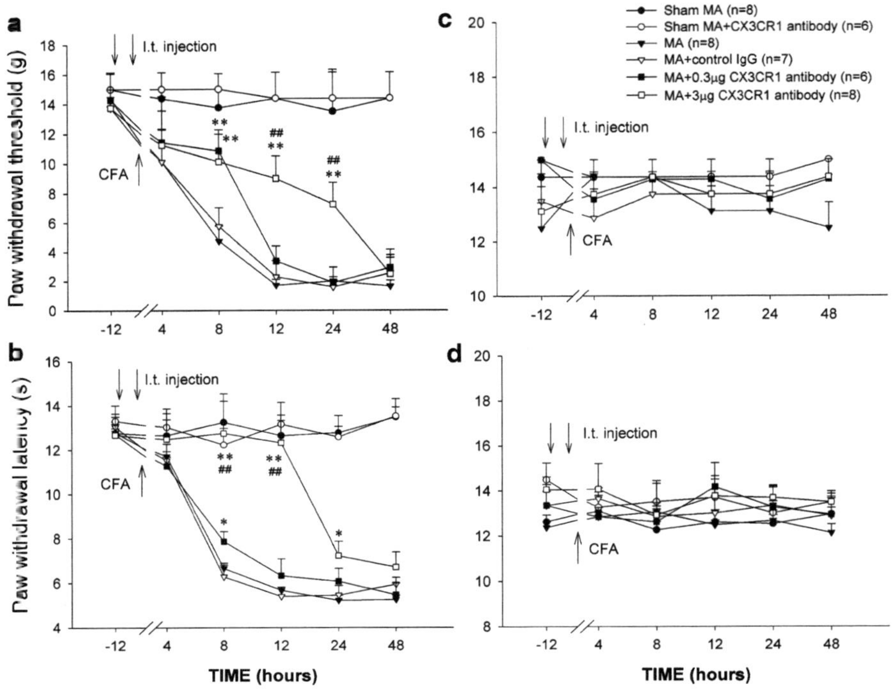

Fig. 18.2 Intrathecal pre-administration of anti-CX3CR1 neutralizing antibody (3 μg) blocked the onset of mechanical allodynia (**a**) and thermal hyperalgesia (**b**) in the ipsilateral hindpaw of monoarthritis (MA) rats, but did not altered the PWTs and PWLs in the contralateral hindpaw (**c**, **d**) of MA rats and both hindpaws of sham MA rats. Anti-CX3CR1 neutralizing antibody or control IgG was injected intrathecally 12 h and 1 h prior to MA. $*p<0.05$, $**p<0.01$ vs. control IgG; $\#\#p<0.01$ vs. 0.3 μg CX3CR1 antibody [reproduced from (Sun et al. 2007a) with permission from Elsevier]

expressed predominantly by spinal microglia (Verge et al. 2004). Excitation of sensory afferents and intrinsic spinal cord neurons after inflammation/injury can release fractalkine, then activating nearby microglia (Chapman et al. 2000). Studies from our group showed that CX3CR1 expression level was significantly increased in the spinal cord following CFA-induced rat monoarthritis (Sun et al. 2007a). Intrathecal injection of fractalkine produced thermal hyperalgesia in the bilateral hindpaws of naive rats. Intrathecal injection of neutralizing antibody against CX3CR1 blocked both induction (Fig. 18.2) and maintenance (Fig. 18.3) [reproduced from (Sun et al. 2007a) with permission from Elsevier] of monoarthritis-induced pain hypersensitivity behaviors. Interestingly, blockade of CX3CR1 suppressed MA-evoked activation of spinal glia, especially microglia (Figs. 18.4 and 18.5) [reproduced from (Sun et al. 2007a) with permission from Elsevier], which was consistent with findings from CCI and SIN models (Milligan et al. 2004, 2005; Verge et al. 2004). In addition, Milligan et al. (2005) also reported that blockade of microglial activation with minocycline prevented intrathecal fractalkine-induced mechanical allodynia and thermal hyperalgesia. Taken together, neuronally released fractalkine might

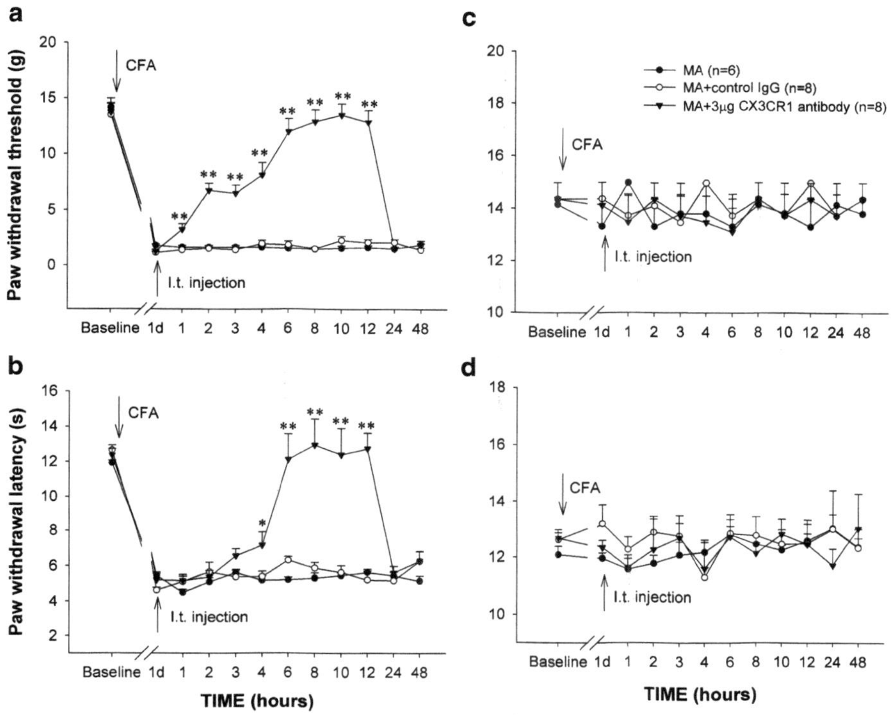

Fig. 18.3 Intrathecal injection of anti-CX3CR1 neutralizing antibody (3 μg) inhibited established mechanical allodynia (**a**) and thermal hyperalgesia (**b**) in the ipsilateral hindpaw of monoarthritis (MA) rats, whereas the PWTs and PWLs in the contralateral hindpaw were not altered (**c** and **d**). Anti-CX3VR1 antibody or control IgG was given 1 day after MA. $^{*}p<0.05$, $^{**}p<0.01$ vs. control IgG [reproduced from (Sun et al. 2007a) with permission from Elsevier]

exert its pain-facilitatory effects via the activation of CX3CR1 receptor on microglia, leading to microglia activation and consequent pain enhancement.

Taken together, there are multiple routes by which neuronal activation could lead to glial activation.

The question subsequently is how activated glia induce pain facilitation. Following nerve injury or inflammation, spinal microglia express or upregulate several ligand-gated ion channels (e.g., P2X4 and P2X7) and G-protein-coupled receptors (e.g., CX3CR1, TLR4, CCR2, and P2Y receptors). The activation of these receptors evokes a raised intracellular Ca^{2+} level that leads to activation of intracellular Ca^{2+}-sensitive signaling cascades, such as p38 mitogen-activated protein kinase (p38 MAPK) (Tsuda et al. 2005; Zhuang et al. 2007), and nuclear transcription factor-κB (NF-κB) (Deleo et al. 2004). As a consequence of intracellular signaling pathway activation, the transcriptional activation of proinflammatory cytokines, chemokines, enzymes for COX-2 and iNOS, adhesion molecules, and immune receptors ensues (Kumar et al. 2003; Ledeboer et al. 2005a; Rivest 2003). All these molecules are involved in central sensitization and persistent pain states.

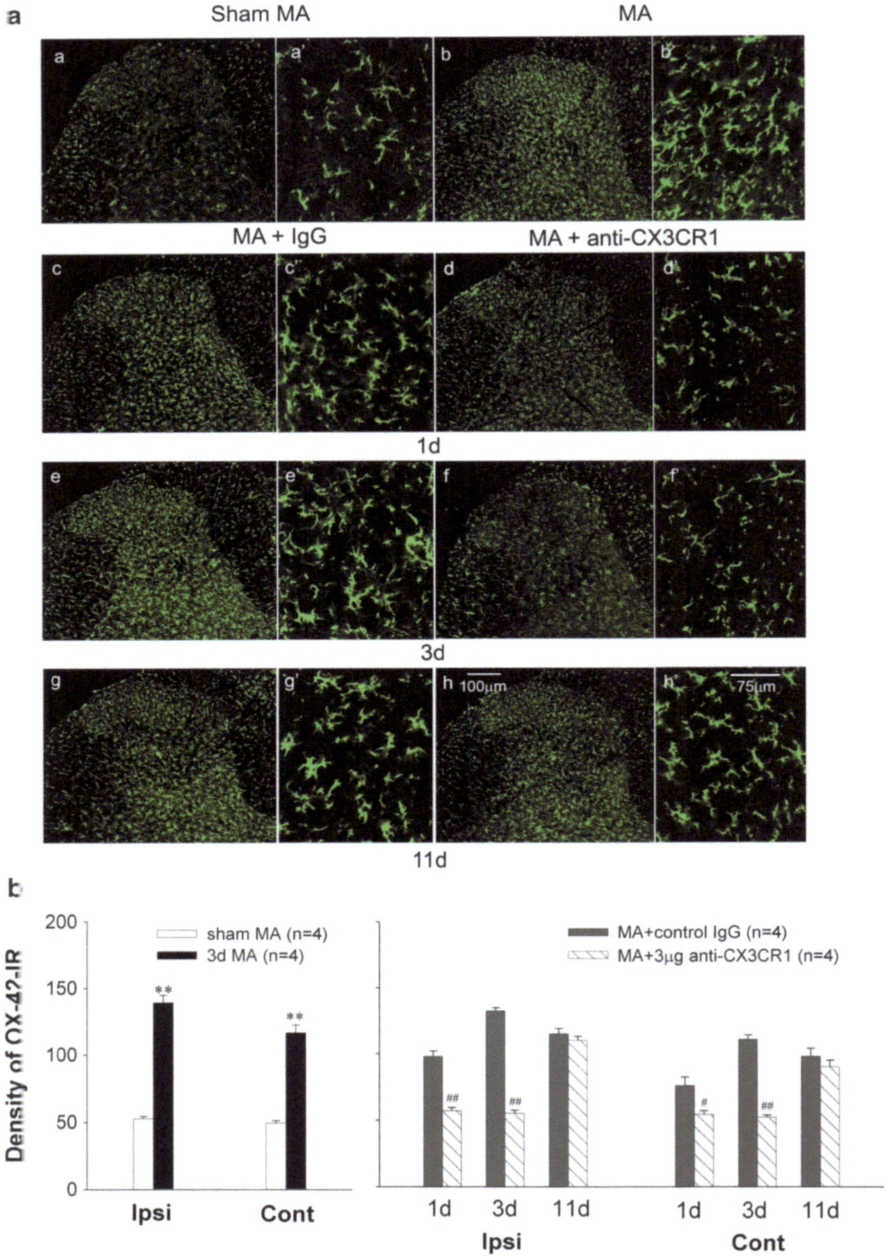

Fig. 18.4 (**a**) Photomicrographs showing OX-42 (microglial marker) immunoreactivity on the ipsilateral spinal cord. (a) Sham MA; (b) MA; (c, e, g) control IgG (3 μg)+MA on day 1 (c), day 3 (e), and day 11 (g) after MA; (d, f, h) CX3CR1 neutralizing antibody (3 μg)+MA on day 1 (d), day 3 (f), and day 11 (h) after MA. (a–h) ×10; (a′–h′) ×40. (**b**) Quantification of OX-42 immunoreactivity on the both sides of the spinal cord. Anti-CX3CR1 antibody or control IgG was given 12 and 1 h prior to MA. $^{**}p<0.01$ vs. sham MA; $^{\#\#}p<0.05$, $^{\#\#}p<0.01$ vs. control IgG [reproduced from (Sun et al. 2007a) with permission from Elsevier]

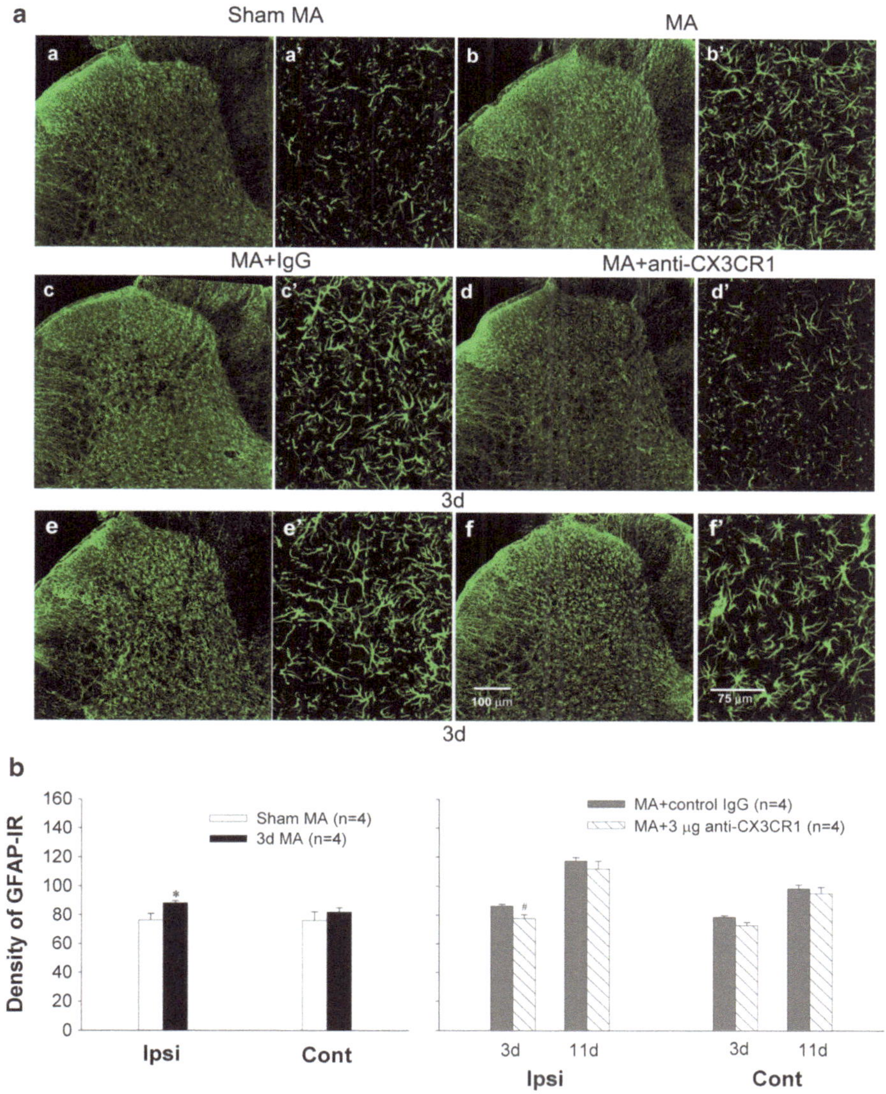

Fig. 18.5 (**a**) Photomicrographs showing GFAP (astrocytic marker) immunoreactivity on the ipsilateral spinal cord. (a) Sham MA; (b) MA; (c, e) control IgG (3 μg) + MA on day 3 (c) and day 11 (e) after MA; (d) CX3CR1 neutralizing antibody (3 μg) + MA on day 3 (c) and day 11 (e) after MA. (a–f) ×10; (a′–f′) ×40. (**b**) Quantification of GFAP immunoreactivity on the both sides of the spinal cord. Anti-CX3CR1 antibody or control IgG was given 12 and 1 h prior to MA. $^{*}p<0.05$ vs. sham MA; $^{\#}p<0.05$ vs. control IgG [reproduced from (Sun et al. 2007a) with permission from Elsevier]

Of the various substances released by activated glia, the proinflammatory cytokines have been most intensively studied. Pretreatment with minocycline, a microglial inhibitor, blocks gp120-induced IL-1β and TNF-α mRNA in the spinal cord, suggesting that activated microglia might be a major source of these cytokines, or alternatively, that activated microglia release substances that, in turn, cause the release of these cytokines from some other cell types such as astrocytes (Ledeboer et al. 2005a). There are expressions of receptors for these proinflammatory cytokines in both spinal glia and neurons. Activation of these receptors induces alteration of neuronal excitability, cellular proliferation, remodeling of the extracellular matrix, and further release of proinflammatory cytokines, in turn accelerate the process of exaggerated pain states (Han et al. 2001; Obraja et al. 2002; Saha and Pahan 2003).

18.3 Glia as a Counter-Regulator of Electroacupuncture

Since electroacupuncture produces analgesic effect and glial activation amplified pain, could the glia be a regulator of electroacupuncture? In the following sections, the relationship between glial activation and electroacupuncture analgesia, also its possible underlying mechanism, will be further demonstrated.

18.3.1 Electroacupuncture and Glial Activation

The first report that glia would be involved in acupuncture process is provided by Yang and their colleagues (1999). Their study implied that, in striatum, astrocytes might play an important role in the protection of neuron via the expression of bFGF, whereas in cortex, neurons may exert autoprotection through secreting bFGF themselves. One possible protective effect of electroacupuncture lies in regulating the endogenous expression of bFGF. Actually, the first indications that EA modulates glia function arose not from the studies of analgesia but rather its neuronal protective effect.

Spinal cord glia plays a key role for creating exaggerated pain following tissue inflammation or injury. Electroacupuncture can effectively control the exaggerated pain both in humans with inflammatory disease and animals with experimental pain. However, little is known about the relationship between spinal glia activation and EA analgesia. The involvement of spinal cord glia in pathological pain gives us great hint that glia might be a key modulator during electroacupuncture analgesia. Our group concentrated to this question in rat inflammatory pain model for the first time. Complete Freund's adjuvant (CFA) has been used to induce rat arthritic immunopathological disease that manifests many of the pathological features of human rheumatoid arthritis (Colpaert 1987; Schaible et al. 2002). A series of our studies show the following: (1) Unilateral intra-articular injection of CFA produces

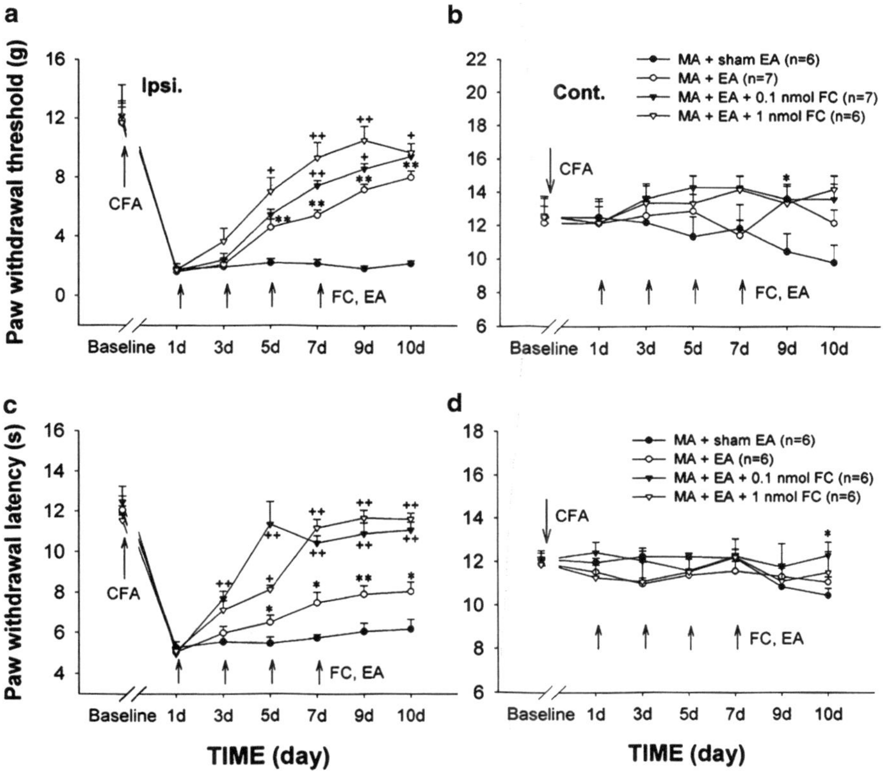

Fig. 18.6 Effects of repeated sham electroacupuncture (EA), EA, or EA plus intrathecal (i.t.) fluorocitrate (FC) (0.1, or 1 nmol) on paw withdrawal threshold (PWT) (**a**, **b**) and paw withdrawal latency (PWL) (**c** and **d**) of the hindpaws ipsilateral (**a**, **c**) and contralateral (**b**, **d**) to monoarthritis (MA). Sham EA, EA, and EA plus i.t. fluorocitrate were given directly after behavioral tests on days 1, 3, 5, and 7 post-MA. Repeated EA attenuated MA-induced mechanical allodynia (**a**) and thermal hyperalgesia (**b**) in the hindpaw ipsilateral to MA. Repeated co-applications of fluorocitrate (0.1, or 1 nmol) and EA produced an enhanced inhibition, as compared with repeated EA alone. $^{*}p<0.05$, $^{**}p<0.01$ vs. sham EA; $^{+}p<0.05$, $^{++}p<0.01$ vs. EA alone. *CFA* complete Freund's adjuvant, *Cont* contralateral, *Ipsi* ipsilateral [reproduced from (Sun et al. 2008) with permission from Elsevier]

enhanced pain within hours and lasts for more than 2 weeks in injected paw, and upregulation of immunostaining of GFAP, a marker for astrocyte, and OX-42, a marker for microglia, in the spinal cord, implicating a robust astrocytic and microglia activation. (2) EA stimulation of ipsilateral "Huantiao" (GB30) and "Yanglingquan" (GB34) acupoints significantly suppressed CFA-induced behavioral hypersensitivity (Fig. 18.6) [reproduced from (Sun et al. 2008) with permission from Elsevier] and spinal glial activation (Figs. 18.7 and 18.8). (3) Combination of EA with minocycline, a microglia inhibitor, or fluorocitrate, an inhibitor for both astrocyte and microglia, significantly enhanced the inhibitory effects of EA

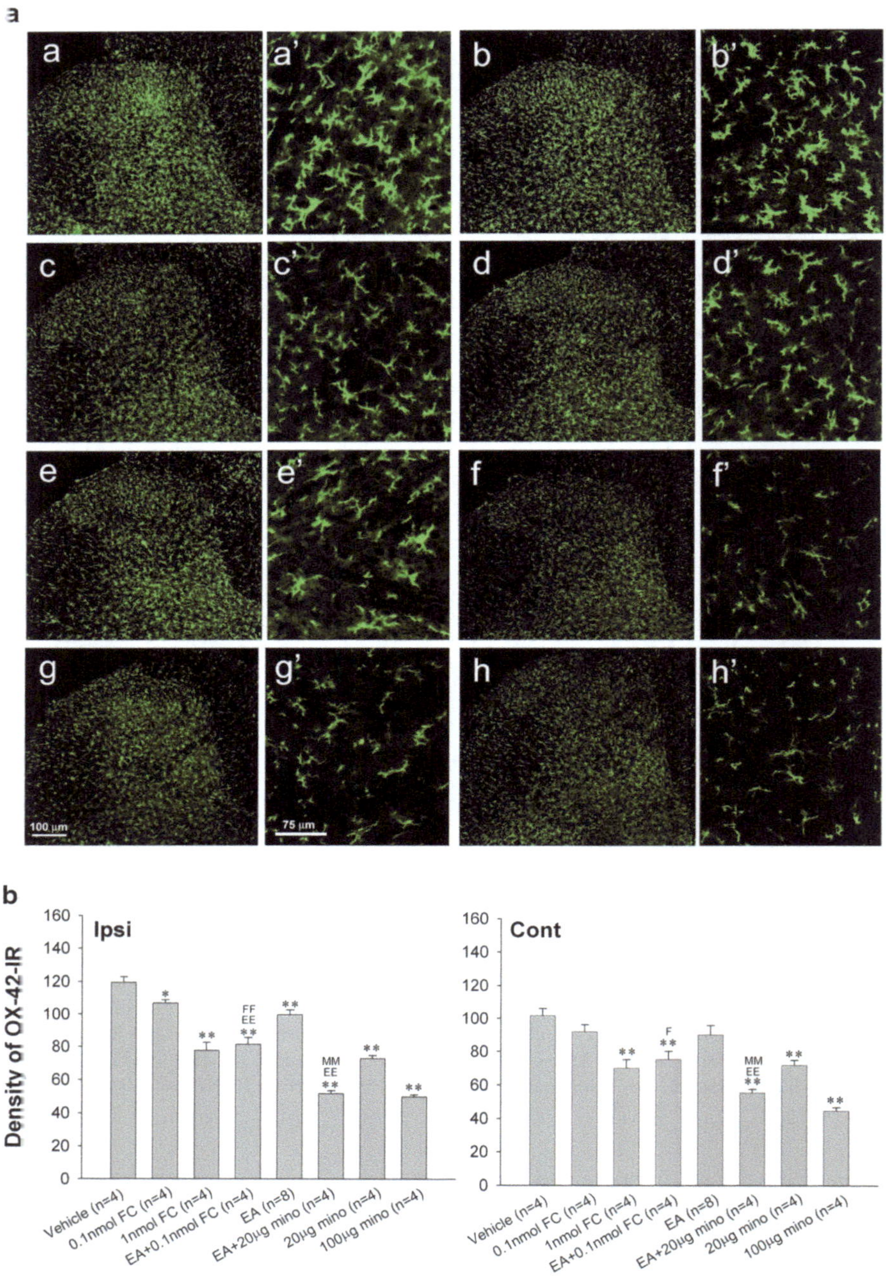

Fig. 18.7 Effects of repeated intrathecal (i.t.) injections of fluorocitrate (FC) or minocycline, electroacupuncture (EA), co-application of EA, and 0.1 nmol FC or 20 μg minocycline on increase in OX-42-immunoreactivity (OX-42-IR) expression on the spinal dorsal horn induced by monoarthritis (MA). (**a**) Photomicrographs showing OX-42-IR on the ipsilateral spinal dorsal horn by 10 days post-MA. (a, a′) vehicle; (b, b′) 0.1 nmol FC; (c, c′) 1 nmol FC; (d, d′) EA; (e, e′) 20 μg minocycline; (f, f′) 100 μg minocycline; (g, g′) 0.1 nmol FC plus EA; (h, h′) 20 μg minocycline plus EA. (a–h) ×10; (a′–h′) ×40. (**b**) Quantification of OX-42-IR showing that repeated i.t. FC, minocycline, and EA suppressed OX-42-IR expression on the spinal dorsal horn. Repeated co-applications of

on allodynia and hyperalgesia (Fig. 18.6). (4) Combination of EA with a subeffective dosage of fluorocitrate markedly suppressed the MA-induced spinal cord astrocytic and microglial activation (indicated by the immunostaining of GFAP and OX-42) (Figs. 18.7 and 18.8). Our results give the direct evidence that EA might modulate glial function and thereby produce analgesic effects (Sun et al. 2006, 2007b, 2008).

The involvement of glial activation in neuropathic pain is much more widely and extensively studied. However, so far, there is no report about the effect of EA on glial activation induced by nerve injury or inflammation. Our recent study showed a preliminary finding between EA analgesia and glial function in rat neuropathic pain model. Tetanic sciatic stimulation (TSS) can induce long-term potentiation (LTP) of C-fiber-evoked field potential or WDR neurons in the spinal dorsal hone and is considered as a novel model for neuropathic pain. Following TSS, the rat produced a long-lasting low threshold mechanical allodynia bilaterally and showed upregulated expression of GFAP and GLT-1 astrocytic transporter for glutamate. EA to "Huantiao" (GB30) and "Yanglingquan" (GB34) acupoints (2-/100 Hz alternation, 1-2-3 mA, 30 min) on day 7 after TSS, mechanical allodynia was remarkably blocked compared with sham EA group. Intrathecal administration of low dose (0.1 μg) of propentofylline, a glial-modulating drug, without EA had no significant effects on mechanical allodynia. However, combination of EA with a subeffective dose of propentofylline (0.1 μg) enhanced EA-induced anti-allodynic effects. Furthermore, the co-application of 1 μg propentofylline and EA produced a much more potent inhibition of allodynia than 1 μg propentofylline or EA treatment alone. Thus, PPF and EA synergistically blocked the spinal cord glial activation and provided the robust analgesic effect (Liang et al. 2010).

18.3.2 Possible Mechanisms of Glial Regulation of Electroacupuncture Analgesia

Mechanisms underlying glial regulation of EA analgesia are likely to be complex. As mentioned above, the release of proinflammatory cytokines and other algesic

Fig. 18.7 (continued) EA and low dose of FC or minocycline exhibited more intense suppression on MA-induced OX-42-IR upregulation on both sides of the spinal cord, as compared with EA, 0.1 nmol FC, or 20 μg minocycline alone. Repeated i.t. FC and FC plus EA were applied directly after behavioral tests on days 1, 3, 5, and 7 post-MA, whereas repeated i.t. minocycline and minocycline plus EA were given once daily for 5 days with the first administration 30 min before MA on day 0 (days 0–4). Repeated EA were applied directly after behavioral tests either on days 0–4 or on days 1, 3, 5, and 7 post-MA. Since there was no detectable difference in suppression on OX-42-IR expression between the two protocols of EA treatment, data from the two EA groups were merged in the figure (column 5). Rats were perfused for immunohistochemistry analysis on day 10 after behavioral tests. *$p<0.05$, **$p<0.01$ vs. vehicle control; EE $p<0.01$ vs. EA alone; F $p<0.05$, FF $p<0.01$ vs. 0.1 nmol FC alone; MM $p<0.01$ vs. 20 μg minocycline alone. *Cont* contralateral, *Ipsi* ipsilateral, *Mino* minocycline

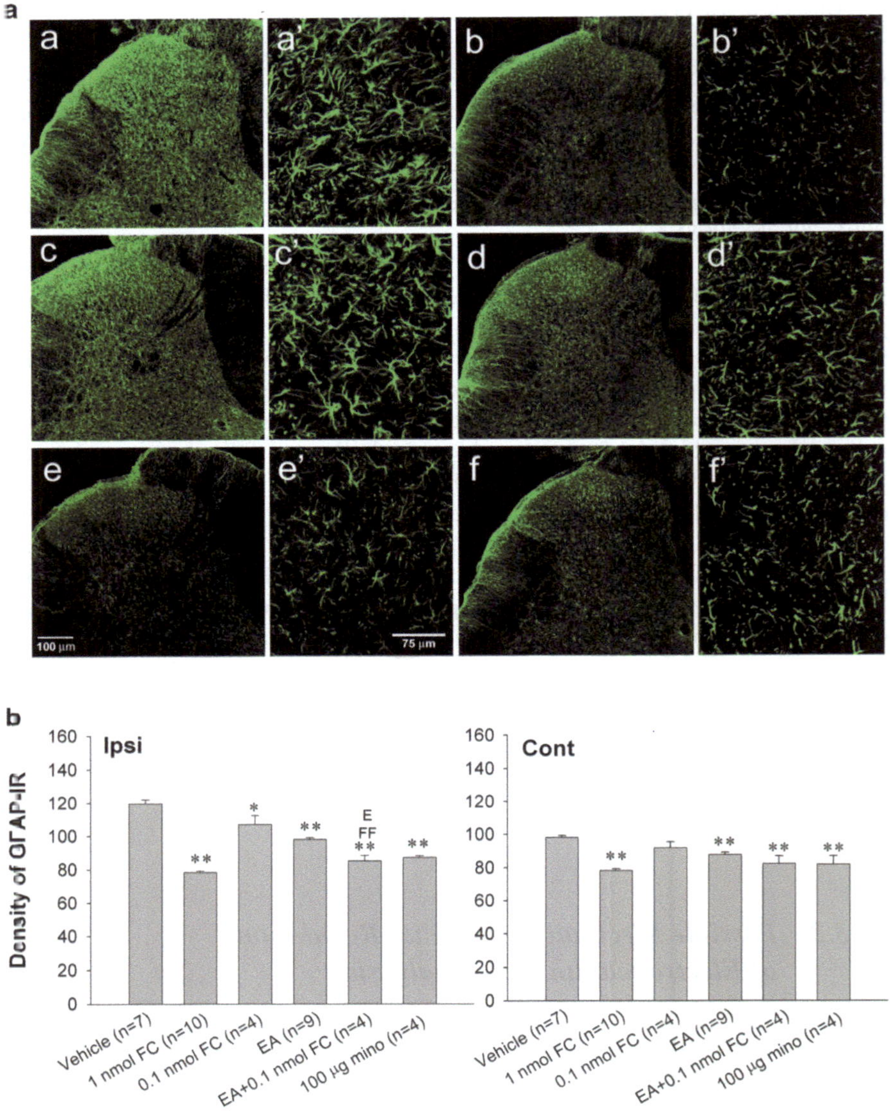

Fig. 18.8 Effects of repeated intrathecal (i.t.) injections of fluorocitrate (FC) or minocycline, electroacupuncture (EA) or i.t. FC combined with EA on increase in glial fibrillary acidic protein (GFAP)-immunoreactivity (GFAP-IR) expression on the spinal dorsal horn induced by monoarthritis (MA). (**a**) Photomicrographs showing GFAP-IR on the ipsilateral spinal dorsal horn by 10 days post-MA. (a, a′) Vehicle; (Ab, b′) 1 nmol FC; (c, c′) 0.1 nmol FC; (d, d′) EA; (e, e′) 0.1 nmol FC plus EA; (f, f′) 100 μg minocycline. (a–f) ×10; (a′–f′) ×40. (**b**) Quantification of GFAP-IR showing that repeated i.t. FC, minocycline, and EA significantly suppressed GFAP-IR expression on both sides of the spinal dorsal horn. Co-application of low dose of FC and EA exhibited more intense suppression on MA-induced GFAP-IR upregulation on the ipsilateral spinal cord, as compared with either EA or 0.1 nmol FC alone. Repeated i.t. injections and EA were given directly after behavioral tests on days 1, 3, 5, and 7 post-MA, and rats were perfused for immunohistochemistry analysis on day 10 after behavioral tests. *$p<0.05$, **$p<0.01$ vs. vehicle control; E $p<0.05$ vs. EA alone; FF $p<0.01$ vs. 0.1 nmol FC alone. *Cont* contralateral, *Ipsi* ipsilateral, *Mino* minocycline

substances from activated glia in the spinal cord contributes to the creation and development of exaggerated pain. Whether EA influences the level of these proinflammatory cytokines under normal or chronic pain conditions is unknown. Using semiquantitative RT-PCR analysis, we evaluate level of IL-1, IL-6, and TNF-α mRNA in the spinal cord following repeated EA on CFA-induced monoarthritis rat. The results showed that repeated EA once daily for 5 successive days with the initial application 30 min before MA markedly reduced MA-induced upregulation of IL-1, IL-6, and TNF-α mRNA in the spinal cord, implicating involvement of proinflammatory cytokines in EA analgesia (Sun et al. 2007b). In the meanwhile, Zhang and colleagues investigated the effect of EA and expression of IL-1 on rat cancer pain models. Consistently, their results indicated that EA with 10 Hz/2 mA/0.4 ms for 30 min daily at Huantiao (GB30) between days 14 and 18 after cancer cell inoculation could significantly inhibit the upregulation of IL-1 and its mRNA induced by cancer cell inoculation (Zhang et al. 2007). These findings strongly support the hypothesis that EA might be a counter-regulator of glia activation by suppressing the production of proinflammatory cytokines, thereby producing analgesic effects.

Another important mechanism underlying inhibitory effect of EA on spinal glial activation might be activating glial opioid receptors. Glia express multiple opioid receptors or subtypes, including μ-, κ-, and δ-opioid receptors, and respond to opioid stimulation by changing their morphology and proliferation rat (Ruzicka et al. 1996; Ruzicka and Akil 1997; Tryoen et al. 2000). Studies on the mechanisms of EA revealed that endogenous opioid peptides in the CNS play a pivotal role in mediating analgesic effect of EA. Low-frequency EA stimulates the release of β-endorphin, enkephalin, and endomorphin, which in turn activates the μ- and δ-opioid receptors, while high-frequency EA stimulates dynorphin which activates the κ-opioid receptor. EA-induced release of endogenous opioid peptides might activate the spinal cord glia via different opioid receptors, thereby reducing the release of glial-dependent inflammatory mediator including proinflammatory cytokines and some enzymes (COX-2, inducible nitric oxide syntheses) via modulating intracellular signaling pathway such as p38/MAPK cascades (watkins and Maier 2003).

When referring to the involvement of endogenous opioid peptides in glial regulation of EA analgesia, it cannot be neglected that all three classes of opioid receptors are also expressed on primary nociceptive afferent terminals and on the dendrites of nociceptive neurons in the spinal dorsal horn. Presynaptic activation of opioid receptors on the primary afferent terminals results in a direct reduction of calcium conductance or indirect in potassium conductance and in turn decreases release of transmitters such as excitatory acids and SP (Dickenson and Sullivan 1986; Malmberg and Yaksh 1995; Ulett et al. 1998; Cheunsuang et al. 2002). Extracellular levels of excitatory amino acids, nitric oxide, and prostaglandins which, as already noted, are all neuroexcitatory substances implicated in both glial activation and EA analgesia under pathological pain states. It has been reported that spinal glia express an array of receptors including NMDA, AMPA/KA, NK-1, ATP, and CGRP receptors (Tsuda et al. 2003; Araque et al. 1999; Svensson et al. 2003; Marriott 2004; Reddington et al. 1995). All the neurotransmitters or neuromodulators can activate spinal microglia and astrocytes, leading to pain facilitation. Thus, one of the major mechanisms of EA suppressing spinal glial activation might be to diminish the release of some "pain" neurotransmitters or neuromodulators (such as excitatory amino acids and neuropeptides) from primary afferent terminals and spinal nociceptive neurons and disrupt neuron-to-glia-to-neuron excitatory circuit.

It is also important to note that electroacupuncture does not influence glial function in normal conditions. There are no significant morphological changes on GFAP and OX-42 immunostaining following EA stimulation on "Huantiao" and "Yanglingquan" acupoints in naïve animals (unpublished data).

Lastly, it should be noted that there might be distinct differences between the glial involvement in electroacupuncture analgesia and pathological pain. This warrants further investigation.

18.4 Concluding Remarks

Acupuncture, a traditional therapeutic modality, has been used in China for thousands of years to treat various disorders, especially chronic pain. Electrical or manual stimulation of an acupuncture point (acupoint) is the most popular form of acupuncture therapy. Lots of efforts have been made to explain analgesic effect of acupuncture. One of the primary mechanisms is to stimulate the CNS to release neurotransmitters or neuromodulators, for example, opioid peptides. The other candidate might be spinal cord glia. Spinal cord glia are sensitive to many perturbations. Activation of glia can occur in response to peripheral or central inflammation/injury, or in response to electroacupuncture stimulation (Fig. 18.9). Whether glia are activated in their role as immune-like cells in response to neuron-to-glia signals or in response to EA, the underlying mechanism is strikingly similar. In all cases, proinflammatory cytokines appear to be crucial. Glial activation is essential for exaggerated pain, and EA can deactivate glial function and downregulate the pain-response neuroactive substance, thereby producing analgesic effects.

Besides pathological pain, EA is also involved in many kinds of other functional disorders. For example, EA stimulation has neuroprotective effects and alleviates behavioral deficits in rats with Parkinson's disease (PD). Importantly, Kang et al. proved that the neuroprotective effects of EA are related with disruption of glial activation. Their studies demonstrated that acupuncture on Yanglingquan (GB34) and Tainchong (LR3) inhibited the decrease of tyrosine hydroxylase (TH) immunoreactivity, attenuated the increase of macrophage antigen complex-1 (MAC-1), a marker of microglial activation at 1 and 3 days, and downregulated the expressions of cyclooxygenase-2 (COX-2) and inducible nitric oxide syntheses (iNOS) on days 1, 3, and 7 in the striatum (ST) and the substantial nigra (SN) (Kang et al. 2007). Also, Chen and colleagues found that repeated 2 Hz electric stimulations can improve the spatial memory ability in rats (Chen et al. 2005). These fascinating findings leave us with the question as to how electroacupuncture regulates higher cortical functions by changing the function of glia.

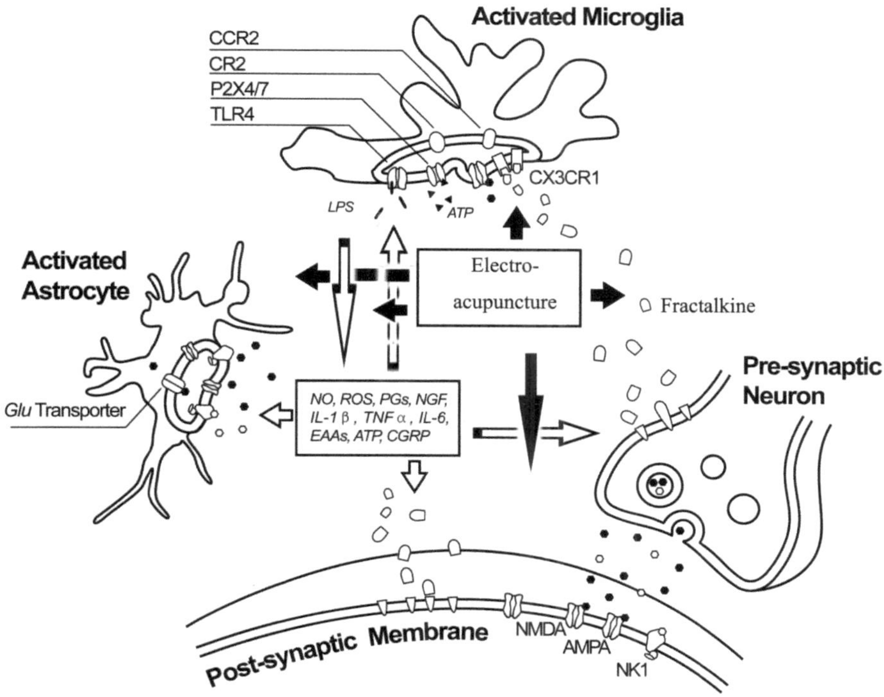

Fig. 18.9 Schematic illustration of probable mechanisms underlying glial counter-regulation in EA analgesia during pathological pain. In pathological pain states, spinal cord glia are activated by some "pain" neurotransmitters or neuromodulators and neuron-to-glia signals, such as EAA, SP, ATP, and fractalkine released from primary afferent terminals. Once activated, they release several proinflammatory cytokines [interleukin-1β (IL-1β), tumor necrosis factor-α (TNF-α), and interleukin-6 (IL-6)], inflammatory mediators [nitric oxide (NO), reactive oxygen species (ROS), prostaglandins (PGs), and nerve growth factor (NGF)]. This leads, in turn, to cause exaggerated release of neurotransmitters from presynaptic neuron, sensitize the postsynaptic membrane, and also enhance spinal cord glia themselves propagating of neuromediators. Such positive feedback loops sustain the perseverant release of pain mediators, facilitating the development of neuronal hypersensitivity, which leads to exaggerated pain. Following electroacupuncture stimulation with 2/100 Hz alternation frequencies, different kinds of neuropeptides including β-endorphin, enkephalin, endormorphine, and dynorphin are released. They might directly inhibit spinal cord glial activation via their respective receptors expressed on glia and reduce the products secreted by activated glia. On the other hand, they might activate the opioid receptors expressed on presynaptic primary afferent terminals and postsynaptic nociceptive neurons and suppress the release of pain-related neurotransmitters/neuromodulators. These, in turn, prevent the development of hyperalgesia and produced analgesic effects

Finally, electroacupuncture could produce tolerance after its prolonged application, which is also called EA tolerance. Previous studies proved that CCK-8 plays an important role in the development of tolerance to morphine analgesia and electroacupuncture (EA) analgesia in the rat, implying that there may be similar mechanism underlying tolerance to morphine and EA analgesia. A series of evidence has showed that glial activation actively contributes to morphine analgesic tolerance. Whether EA tolerance is also regulated by glial function remains unknown.

Acknowledgements This work is supported by National Natural Science Fund of China (NSFC, 30870835, 30821002, and 30900444), National Basic Research Program of China (2007CB512303, 2007CB512502, and 2006CB500807), Postdoctoral Science Fund of China (20080440578, 200902207), and Postdoctoral Natural Science Foundation of Shanghai (09R21411200). We are grateful to Elsevier publisher for permissions to use Figs. 18.2, 18.3, 18.4, 18.5, and 18.6.

References

Aicher SA, Sharma S, Cheng PY, Pickel VM (1997) The N-methyl-D-aspartate (NMDA) receptor is postsynaptic to substance P-containing axon terminals in the rat superficial dorsal horn. Brain Research 772 (1-2): 71–81.

Araque A, Carmignoto G, Haydon PG (2011) Dynamic signaling between astrocytes and neurons. Annual Review of Physiology 63:795–813.

Araque A, Sanzgiri RP, Parpura V, Haydon PG (1999) Astrocyte-induced modulation of synaptic transmission. Canadian Journal of Physiology and Pharmacology 77 (9): 699–706.

Bao L, Zhu Y, Elhassan AM, Wu QY, Xiao BG, Zhu J, Lindgren JU (2001) Adjuvant-induced arthritis: IL-1 beta, IL-6 and TNF-alpha are up-regulated in the spinal cord. Neuroreport 12 (18): 3905–3908.

Basbaum AI, Jessell TM (2001) The perception of pain. In: Kandel ER, Schwartz, JH, Jssell, TM (Eds), Principles of Neural Science McGraw-Hall 472–491.

Baumgarten HG, Bjorklund B, Lachenmayer L, Nobin A, Stenevi U (1972) Long lasting selective depletion of brain 5-HT by 5,6-dihydroxytryptamine. Acta Physiology Sinica 84(Suppli. 373): 1–15.

Bergjohnsen J, Paulsen RE, Fonnum F, Langmoen IA (1993) Changes in evoked potentials and amino-acid content during fluorocitrate action studied in rat hippocampal cortex. Experimental Brain Research 96:241–246.

Cao H and Zhang YQ (2008) Spinal glial activation contributes to pathological pain states. Neuroscience and Biobehavioral Reviews 32 (5) : 972–983.

Cao XD. Scientific bases of acupuncture analgesia. Acupuncture & Electro-Therapeutics Research 2002;27(1):1–14.

Chacur M, Gutierrez JM, Milligan ED, Wieseler-Frank J, Britto LRG, Maier SF, Watkins LR, Cury Y (2004) Snake venom components enhance pain upon subcutaneous injection: an initial examination of spinal cord mediators. Pain 111 (1-2) : 65–76.

Chapman GA, Moores K, Harrison D, Campbell CA, Stewart BR, Strijbos P (2000) Fractalkine cleavage from neuronal membranes represents an acute event in the inflammatory response to excitotoxic brain damage. Journal of Neuroscience 20 (15): U2–U6.

Chen JH, Liang J, Wang GB, Han JS, Cui CL (2005) Repeated 2 Hz peripheral electrical stimulations suppress morphine-induced CPP and improve spatial memory ability in rats. Experimental Neurology 194 (2) : 550–556.

Chen XH and Han JS (1992) Analgesia induced by electroacupuncture of different frequencies is mediated by different type of opioid receptors: another cross-tolerance study. Behavior Brain Research 47 : 143–149.

Cheng XR and Cheng K (2008) Survey of studies on the mechanism of acupuncture and moxibustion treating diseases abroad. Zhongguo Zhen Jiu 28 (6) : 463–467.

Cheunsuang O, Maxwell D, Morris R (2002) Spinal lamina I neurones that express neurokinin 1 receptors: II. Electrophysiological characteristics, responses to primary afferent stimulation and effects of a selective mu-opioid receptor agonist. Neuroscience 111 (2) : 423–434.

Colburn RW, Rickman AJ, DeLeo JA (1999) The effect of site and type of nerve injury on spinal glial activation and neuropathic pain behavior. Experimental Neurology 157 (2): 289–304.

Colpaert FC (1987) Evidence that adjuvant arthritis in the rat is associated with chronic pain. Pain 28: 201–222.

Dai Y, Kondo E, Fukuoka T, Tokunaga A, Miki K, Noguchi K (2001) The effect of electroacupuncture on pain behaviors and noxious stimulus-evoked fos expression in a rat model of neuropathic pain. Journal of Pain 2 (3): 151–159.

Deleo JA, Tanga FY, Tawfik VL (2004) Neuroimmune activation and neuroinflammation in chronic pain and opioid tolerance/hyperalgesia. Neuroscientist 10: 40–52.

Dickenson AH and Sullivan AF (1986) Electrophysiological studies on the effects of intrathecal morphine on nociceptive neurons in the rat dorsal horn. Pain 24: 211–222.

Fei H, Xie GX, Han JS (1987) Low and high frequency electroacupuncture stimulation release [Met5] enkephalin and dynorphin A in rat spinal cord. Science Bulletin China 32: 1496–1501.

Fields HL, Basbaum AI, Heinricher MM (2005) Central nervous system mechanisms of pain modulation. In: Wall, Melzack, (Eds), Text book of Pain fifth ed Elsevier.

Fu D, Guo QL, Ai YH, Cai HW, Yan JQ, Dai RP (2006) Glial activation and segmental upregulation of interleukin-1 beta (IL-1 beta) in the rat spinal cord after surgical incision. Neurochemical Research 31 (3): 333–340.

Gao YJ, Xu ZZ, Liu YC, Wen YR, Decosterd I, Ji RR (2010) The c-Jun N-terminal kinase 1 (JNK1) in spinal astrocytes is required for the maintenance of bilateral mechanical allodynia under a persistent inflammatory pain condition. Pain 148 (2): 309–319.

Gao YJ, Zhang L, Samad OA, Suter MR, Yasuhiko K, Xu ZZ, Park JY, Lind AL, Ma Q, Ji RR (2009) JNK-Induced MCP-1 Production in Spinal Cord Astrocytes Contributes to Central Sensitization and Neuropathic Pain. Journal of Neuroscience 29: (13): 4096–4108.

Garrison CJ, Dougherty PM, Kajander KC, Carlton SM (1991) Staining of Glial Fibrillary Acidic Protein (Gfap) in Lumbar Spinal-Cord Increases Following a Sciatic-Nerve Constriction Injury. Brain Research 565: (1): 1–7.

Ghoname EA, Craig WF, White PF (1999) Use of percutaneous electrical nerve stimulation (PENS) for treating ECT-induced headaches. Headache 39: (7): 502–505.

Hains BC and Waxman SG (2006) Activated microglia contribute to the maintenance of chronic pain after spinal cord injury. Journal of Neuroscience 26: (16): 4308–4317.

Haker E and Lundeberg T (1990) Acupuncture treatment in epicondylalgia: a comparative study of two acupuncture techniques. Clinical Journal of Pain 6: 221–226.

Han JS (2003) Acupuncture: neuropeptide release produced by electrical stimulation of different frequencies. Trends in Neurosciences 26: (1): 17–22.

Han JS, Chou PH, Lu CH, Yang TH, Lu LH, Jen MF (1979) The role of central 5-HT in acupuncture analgesia. Science Sinica 22: 91–104.

Han JS, Ren MF, Tang J, Fan SG, Xu JM, Guan XM (1979) The role of catecholamine in acupuncture analgesia. Chinese Medicine 92: 793–800.

Han JS, Zhou Z, Xuan Y (1983) Acupuncture has an analgesic effect in rabbits. Pain 15: 83–91.

Han YP, Tuan TL, Wu HY, Hughes M, Garner WL (2001) TNF-alpha stimulates activation of pro-MMP2 in human skin through NF-kappa B mediated induction of MT1-MMP. Journal of Cell Science 114: 131–139.

Hashizume H, DeLeo JA, Colburn RW, Weinstein JN (2000) Spinal glial activation and cytokine expression after lumbar root injury in the rat. Spine 25: (10): 1206–1217.

Hassel B, Paulsen RE, Johnsen A, Fonnum F (1992) Selective-inhibition of glial cell metabolism in vivo by fluorocitrate. Brain Research 576:120–124.

He LF (1987) Involvement of endogenous opioid peptides in acupuncture analgesia. Pain 31: 99–121.

Holguin A, O'Connor KA, Biedenkapp J, Campisi J, Wieseler-Frank J, Milligan ED, Hansen MK, Spataro L, Maksimova E, Bravmann C, Martin D, Fleshner M, Maier SF, Watkins LR (2004) HIV-1 gp120 stimulates proinflammatory cytokine-mediated pain facilitation via activation of nitric oxide synthase-1 (nNOS). Pain 110: (3): 517–530.

Huang C, Hu ZP, Jiang SZ, Li HT, Han JS, Wan Y (2007) CCKB receptor antagonist L365,260 potentiates the efficacy to and reverses chronic tolerance to electroacupuncture-induced analgesia in mice. Brain Research Bulletin 71: (5): 447–451.

Huang C, Wang Y, Chang JK, Han JS (2000) Endomorphin and mu-opioid receptors in mouse brain mediate the analgesic effect induced by 2 Hz but not 100 Hz electroacupuncture stimulation. Neuroscience Letters 294: (3): 159–162.

Huang C, Wang Y, Han JS, Wan Y (2002) Characteristics of electroacupuncture-induced analgesia in mice: variation with strain, frequency, intensity and opioid involvement. Brain Research 945: (1): 20–25.

Hughes PM, Botham MS, Frentzel S, Mir A, Perry VH (2002) Expression of fractalkine (CX3CL1) and its receptor, CX3CR1, during acute and chronic inflammation in the rodent CNS. Glia 37: 314–327.

Hui KK, Liu J, Marina O, Napadow V, Haselgrove C, Kwong KK, Kennedy DN, Makris N (2005) The integrated response of the human cerebro-cerebellar and limbic systems to acupuncture stimulation at ST 36 as evidenced by fMRI. Neuroimage 27: 479–496.

Jung S, Aliberti J, Graemmel P, Sunshine MJ, Kreutzberg GW, Sher A, Littman DR (2000) Analysis of fractalkine receptor CX(3)CR1 function by targeted deletion and green fluorescent protein reporter gene insertion. Molecular and Cellular Biology 20:4106–4114.

Kang JM, Park HJ, Choi YG, Choe IH, Park JH, Kim YS, Lim S (2007) Acupuncture inhibits microglial activation and inflammatory events in the MPTP-induced mouse model. Brain Research 1131: (1): 211–219.

Kim JH, Min BI, Na HS, Park DS (2004) Relieving effects of electroacupuncture on mechanical allodynia in neuropathic pain model of inferior caudal trunk injury in rat: mediation by spinal opioid receptors. Brain Research 998: (2): 230–236.

Kim JH, Min BI, Schmidt D, Lee HJ, Park DS (2000) The difference between electroacupuncture only and electroacupuncture with manipulation on analgesia in rats. Neuroscience Letters 279: (3): 149–152.

Kim SK, Moon HJ, Park JH, Lee G, Shin MK, Hong MC, Bae H, Jin YH, Min BI (2007) The maintenance of individual differences in the sensitivity of acute and neuropathic pain behaviors to electroacupuncture in rats. Brain Research Bulletin 74: (5): 357–360.

Kim SK, Park JH, Bae SJ, Kim JH, Hwang BG, Min BI, Park DS, Na HS (2005) Effects of electroacupuncture on cold allodynia in a rat model of neuropathic pain: Mediation by spinal adrenergic and serotonergic receptors. Experimental Neurology 195: (2): 430–436.

Kumar S, Boehm J, Lee JC (2003) p38 map kinases: key signalling molecules as therapeutic targets for inflammatory diseases. Nature Reviews Drug Discovery 2: 717–726.

Lao LX, Zhang RX, Zhang G, Wang XY, Berman BM, Ren K (2004) A parametric study of electroacupuncture on persistent hyperalgesia and Fos protein expression in rats. Brain Research 1020: (1-2): 18–29.

Ledeboer A, Gamanos M, Lai W, Martin D, Maier SF, Watkins LR, Quan N (2005) Involvement of spinal cord nuclear factor kappa B activation in rat models of proinflammatory cytokine-mediated pain facilitation. European Journal of Neuroscience 22: (8): 1977–1986.

Ledeboer A, Sloane EM, Milligan ED, Frank MG, Mahony JH, Maier SF, Watkins LR (2005) Minocycline attenuates mechanical allodynia and proinflammatory cytokine expression in rat models of pain facilitation. Pain 115: (1-2): 71–83.

Liang LL, Yang JL, Lu N, Gu XY, Zhang YQ, Zhao ZQ (2010) Synergetic Analgesia of Propentofylline and Electroacupuncture by Interrupting Spinal Glial Function in Rats. Neurochem Res 35: 1780–1786

Liu X, Jiang MC, Huang PB, Zou T (1990) Role of afferent C fibers in electroacupuncture of "Zusanli" point in activation nucleus raphe magnus. Acta Physiology Sinica 42:523–533. In Chinese with English abstract

Malmberg AB and Yaksh TL (1995) The effect of morphine on formalin-evoked behavior and spinal release of excitatory amino acids and prostaglandin E2 using microdialysis in conscious rats. British Journal of Pharmacology 114:1069–1075.

Marriott I (2004) The role of tachykinins in central nervous system inflammatory responses. Frontiers in Bioscience 9:2153–2165.

Mayer DJ, Price DD, Raffi A (1997) Antagonism of acupuncture analgesia in man by narcotic antagonist naloxone Brain Research 121: 368–372.

Mclennan H, Gilfillan K, Heap Y (1997) Some pharmacological observation on the analgesia induced by acupuncture in rabbits. Pain 3: 229–238.

Meller ST, Dykstra C, Grzybycki D, Murphy S, Gebhart GF (1994) The possible role of glia in nociceptive processing and hyperalgesia in the spinal-cord of the rat. Neuropharmacology 33: 1471–1478.

Milligan E, Zapata V, Schoeniger D, Chacur M, Green P, Poole S, Martin D, Maier SF, Watkins LR (2005) An initial investigation of spinal mechanisms underlying pain enhancement induced by fractalkine, a neuronally released chemokine. European Journal of Neuroscience 22: (11): 2775–2782.

Milligan ED, Mehmert KK, Hinde JL, Harvey LO, Martin D, Tracey KJ, Maier SF, Watkins LR (2000) Thermal hyperalgesia and mechanical allodynia produced by intrathecal administration of the human immunodeficiency virus-1 (HIV-1) envelope glycoprotein, gp120. Brain Research 861: (1): 105–116.

Milligan ED, O'Connor KA, Armstrong CB, Hansen MK, Martin D, Tracey KJ, Maier SF, Watkins LR (2001) Systemic administration of CNI-1493, a p38 mitogen-activated protein kinase inhibitor, blocks intrathecal human immunodeficiency virus-1 gp120-induced enhanced pain states in rats. Journal of Pain 2: (6): 326–333.

Milligan ED, Twining C, Chacur M, Biedenkapp J, O'Connor K, Poole S, Tracey K, Martin D, Maier SF, Watkins LR (2003) Spinal glia and proinflammatory cytokines mediate mirror-image neuropathic pain in rats. Journal of Neuroscience 23: (3): 1026–1040.

Milligan ED, Zapata V, Chacur M, Schoeniger D, Biedenkapp J, O'Connor KA, Verge GM, Chapman G, Green P, Foster AC, Naeve GS, Maier SF, Watkins LR (2004) Evidence that exogenous and endogenous fractalkine can induce spinal nociceptive facilitation in rats. European Journal of Neuroscience 20: (9): 2294–2302.

Nagy LI (1982) Capsaicin action on nervous system. TINS 5: 362–365.

Obata H, Eisenach JC, Hussain H, Bynum T, Vincler M (2006) Spinal glial activation contributes to postoperative mechanical hypersensitivity in the rat. Journal of Pain 7: (11): 816–822.

Obraja O, Rathee PK, Lips KS, Dister C, Kress M (2002) IL-1β potentiates heatactivated currents in rat sensory neurons: involvement of IL-1R1, tyrosine kinase, and protein kinase C. The FASEB Journal 16: 1497–1503.

Paulsen RE, Contestabile A, Villani L, Fonnum F (1987) An in vivo model for studying function of brain-tissue temporarily devoid of glial-cell metabolism - the use of fluorocitrate. Journal of Neurochemistry 48: 1377–1385.

Pomeranz BC (1976) Naloxone blockade of acupuncture analgesia: endorphin implicated. Life Science 19:1757–1762.

Popovich PG, Wei P, Stokes BT (1997) Cellular inflammatory response after spinal cord injury in Sprague-Dawley and Lewis rats. Journal of Comparative Neurology 377: (3): 443–464.

Raghavendra V, Tanga F, DeLeo JA (2003) Inhibition of microglial activation attenuates the development but not existing hypersensitivity in a rat model of neuropathy. Journal of Pharmacology and Experimental Therapeutics 306: (2): 624–630.

Reddington M, Priller J, Treichel J, Haas C, Kreutzberg GW (1995) Astrocytes and microglia as potential targets for calcitonin gene related peptide in the central nervous system. Canadian Journal of Physiology and Pharmacology 73: (7): 1047–1049.

Rivest S (2003) Molecular insights on the cerebral innate immune system. Brain Behavior and Immunity 17:13–19.

Ruzicka BB and Akil H (1997) The interleukin-1beta-mediated regulation of proenkephalin and opioid receptor messenger RNA in primary astrocyte-enriched cultures. Neuroscience 79: 517–524.

Ruzicka BB, Thompson RC, Watson SJ, Akil H (1996) Interleukin-1 beta-mediated regulation of mu-opioid receptor mRNA in primary astrocyte-enriched cultures. Journal of Neurochemistry 66: (1): 425–428.

Saha RN and Pahan K (2003) Tumor necrosis factor-alpha at the crossroads of neuronal life and death during HIV-associated dementia. Journal of Neurochemistry 86: (5): 1057–1071.

Schaible HG, Ebersberger A, Von Banchet GS. (2002) Mechanisms of pain in arthritis. Ann. N. Y. Acad. Sci. 966, 343–354.

Sjollund B, Trenius L, Eriksson M (1997) Increased cerebro-spinal fluid leveled of endorphin after electroacupuncture. Acta Physiology Scand 100: 382–384.

Sun RQ, Wang HC, Wan Y, Jing Z, Luo F, Han JS, Wang Y (2004) Suppression of neuropathic pain by peripheral electrical stimulation in rats: mu-opioid receptor and NMDA receptor implicated. Experimental Neurology 187: (1): 23–29.

Sun S, Cao H, Han M, Li TT, Pan HL, Zhao ZQ, Zhang YQ (2007a) New evidence for the involvement of spinal fractalkine receptor in pain facilitation and spinal glial activation in rat model of monoarthritis. Pain 110: (1-2): 64–75.

Sun S, Cao H, Han M, Li TT, Zhao ZQ, Zhang YQ (2008) Evidence for suppression of electroacupuncture on spinal glial activation and behavioral hypersensitivity in a rat model of monoarthritis. Brain Research Bulletin 75: (1): 83–93.

Sun S, Chen WL, Wang PF, Zhao ZQ, Zhang YQ (2006) Disruption of glial function enhances electroacupuncture analgesia in arthritic rats. Experimental Neurology 198: (2): 294–302.

Sun S, Mao-Ying QL, Cao H, Li TT, Han M, Wang YQ, Zhao ZQ, Zhang YQ (2007b) Is functional state of spinal microglia involved in the anti-allogynic and anti-hyperalgesic effects of electroacupuncture in rat model of monoarthritis? Neurobiology of Disease 26: (3): 558–568

Svensson CI, Hua XY, Protter AA, Powell HC, Yaksh TL (2003) Spinal p38 MAP kinase is necessary for NMDA-induced spinal PGE(2) release and thermal hyperalgesia. Neuroreport 14: (8): 1153–1157.

Sweitzer SM, Colburn RW, Rutkowski M, DeLeo JA (1999) Acute peripheral inflammation induces moderate glial activation and spinal IL-1 beta expression that correlates with pain behavior in the rat. Brain Research 829: (1-2): 209–221.

Sweitzer SM, Schubert P, DeLeo JA (2001) Propentofylline, a glial modulating agent, exhibits antiallodynic properties in a rat model of neuropathic pain. Journal of Pharmacology and Experimental Therapeutics 297: (3): 1210–1217.

Takakura N, Ogawa H, Iijima S, Kanamaru A, Sibuya M, Homma I (1995) Effect of acupuncture at the Hoku point on vibration-induced finger flexion reflex in man: comparison between press needle technique, electroacupuncture, and in-situ technique. The American Journal of Chinese Medicine 23: 313–318.

Tanga FY, Raghavendra V, DeLeo JA (2004) Quantitative real-time RT-PCR assessment of spinal microglial and astrocytic activation markers in a rat model of neuropathic pain. Neurochemistry International 45: (2-3): 397–407.

Tikka TM and Koistinaho JE (2001) Minocycline provides neuroprotection against N-methyl-D-aspartate neurotoxicity by inhibiting microglia. Journal of Immunology 166: (12): 7527-7533.

Tryoen P, Gavenraux C, Labourdette G (2000) Down-regulation of mu-opioid receptor expression in rat oligodendrocytes during their development in vitro. Journal of Neurosience Research 60: 10–20.

Tsuda M, Inoue K, Salter MW (2005) Neuropathic pain and spinal microglia: a big problem from molecules in 'small' glia. Trends in Neurosciences 28: (2): 101–107.

Tsuda M, Shigemoto-Mogami Y, Koizumi S, Mizokoshi A, Kohsaka S, Salter MW, Inoue K (2003) P2X(4) receptors induced in spinal microglia gate tactile allodynia after nerve injury. Nature 424: (6950): 778–783.

Ulett GA, Han SP, Han JS (1998) Electroacupuncture: Mechanisms and clinical application. Biological Psychiatry 44: (2): 129–138.

Verge GM, Milligan ED, Maier SF, Watkins LR, Naevel GS, Foster AC (2004) Fractalkine (CX3CL1) and fractalkine receptor (CX3CR1) distribution in spinal cord and dorsal root ganglia under basal and neuropathic pain conditions. European Journal of Neuroscience 20: (5): 1150–1160.

Watkins LR, Hutchinson MR, Johnston IN, Maier SF (2005) Glia: novel counter-regulators of opioid analgesia. Trends in Neurosciences 28: (12): 661–669.

Watkins LR and Maier SF (1999) Implications of immune-to-brain communication for sickness and pain. Proceedings of the National Academy of Sciences of the United States of America 96: (14): 7710–7713.

Watkins LR and Maier SF (2003) Glia: A novel drug discovery target for clinical pain. Nature Reviews Drug Discovery 2: (12): 973–985.

Watkins LR, Martin D, Ulrich P, Tracey KJ, Maier SF (1997) Evidence for the involvement of spinal cord glia in subcutaneous formalin induced hyperalgesia in the rat. Pain 71: (3): 225–235.

Watkins LR, Milligan ED, Maier SF (2001) Glial activation: a driving force for pathological pain. Trends in Neurosciences 24: (8): 450–455.

Yang WO, Huang YL, Da CD, Cheng JS (1999) Electroacupuncture reduces rat's neuronal ischemic injury and enhances the expression of basic fibroblast growth factor. Acupuncture Electrotherapy Research 24: (1): 1–10.

Ye WL, Feng XC, Chao DF, Chang JW, Chang C (1979) Effect of electric stimulation of "Du-channel" point on 5-HT and 5-HIAA contents in the caudated nucleus in the rabbit Kexue Tongbao 24: 235–236. In Chinese with English abstract

Yrjanhelkki J, Keinanen R, Pellikka M, Hokfelt T, Koistinaho J (1988) Tetracyclines inhibit microglial activation and are neuroprotective in global brain ischemia. Proceedings of the National Academy of Sciences of the United States of America 95:15769–15774.

Zhang AZ, Huang DK, Zeng DY, Zhang LM, Wang DL, Zhu GG (1981) The changes of endorphins in perfusate of certain nuclei in rabbits during acupuncture analgesia. Aca Physiology Sinica 33:8–18. In Chinese with English abstract

Zhang RX, Lao LX, Wang XY, Ren K, Berman BB (2004) Electroacupuncture combined with indomethacin enhances antihyperalgesia in inflammatory rats. Pharmacology Biochemistry and Behavior 78: (4): 793–797.

Zhang RX, Li A, Liu B, Wang L, Ren K, Qiao JT, Berman BM, Lao L (2007) Electroacupuncture attenuates bone cancer pain and inhibits spinal interleukin-1βexpression in a rat model. Anesthesia & Analgesia 105: (5): 1482–1488.

Zhang RX, Liu B, Wang LB, Ren K, Qiao HT, Berman BM, Lao LX (2005) Spinal glial activation in a new rat model of bone cancer pain produced by prostate cancer cell inoculation of the tibia. Pain 118: (1-2): 125–136.

Zhao ZQ (2008) Neural mechanism underlying acupuncture analgesia. Progress in Neurobiology 85: (4): 355–375.

Zhou PH, P.D. Q, Huang DK, Gu HY, Wang HY (1979) A study of the relationships between the points of the changes and peripheral nerves. In: National Symposia of Acupuncture-Moxibustion & Acupuncture Anesthesis, Beijing pp302.

Zhu CB, Li XY, Zhu YH, Wu GC, Xu SF (1997) Alteration of monoamine contents in microdialysate following droperidol enhanced electroacupuncture. Acta Physiology Sinica 49:382–388. In Chinese with English abstract

Zhu S, Stavrovskaya IG, Drozda M, Kim BYS, Ona V, Li MW, Sarang S, Liu AS, Hartley DM, Du CW, Gullans S, Ferrante RJ, Przedborski S, Kristal BS, Friedlander RM (2002) Minocycline inhibits cytochrome c release and delays progression of amyotrophic lateral sclerosis in mice. Nature 417:74–78.

Zhuang ZY, Kawasaki Y, Tan PH, Wen YR, Huang J, Ji RR (2007) Role of the CX3CR1/p38MAPK pathway in spinal microglia for the development of neuropathic pain following nerve injury-induced cleavage of fractalkine. Brain, Behavior, and Immunity 21: 642–651.

Chapter 19
Meridian–Viscera Correlationship

Yiping Zhou, Keming Wang, Ling Hu, Meiqi Zhou, Zijian Wu, Ronglin Cai, Zibing Liu, and Lu He

Abstract Meridians and collaterals are considered to be objective entities in the human body. Meridian theory is a foundational component of traditional Chinese medicine (TCM). In traditional acupuncture, doctors are guided by the theory of meridian–viscera correlativity (MVC). In fact, MVC theory is multifaceted and explains the physiology of organ systems, the pathophysiology of disease processes,

Y.P. Zhou (✉)
Research Center for Meridian-Viscera Correlationship, Anhui University of Traditional Chinese Medicine,103 Meishan Road, Hefei, China

Anhui Institute of Acupuncture and Meridian, Anhui University of Traditional Chinese Medicine, 103 Meishan Road, Hefei, China
e-mail: zhouyp@foxmail.com

K.M. Wang
Research Center for Meridian-Viscera Correlationship, Anhui University of Traditional Chinese Medicine, 103 Meishan Road, Hefei, China
e-mail: kmwanghe@126.com

L. Hu
Anhui Institute of Acupuncture and Meridian, Anhui University of Traditional Chinese Medicine, 103 Meishan Road, Hefei, China

School of Acupuncture and Osteology, Anhui University of Traditional Chinese Medicine, 103 Meishan Road, Hefei, China
e-mail: hulingtcm@126.com

ZJ. Wu • RL. Cai • ZB. Liu • L. He
Research Center for Meridian-Viscera Correlationship, Anhui University of Traditional Chinese Medicine, 103 Meishan Road, Hefei, China

Anhui Institute of Acupuncture and Meridians, Anhui University of Traditional Chinese Medicine, 103 Meishan Road, Hefei, China

School of Acupuncture and Osteology, Anhui University of Traditional Chinese Medicine, 103 Meishan Road, Hefei, China
e-mail: vagabondowl@163.com; ronglincai@163.com; zibingliu@163.com; helu828@163.com

Y. Xia et al. (eds.), *Current Research in Acupuncture*,
DOI 10.1007/978-1-4614-3357-6_19, © Springer Science+Business Media New York 2013

and the potential management of illness. With respect to physiology, the function of the meridians and their corresponding organs are related closely. Secondly, pathological changes in an organ system can be reflected in the associated meridian, and disease processes can be deduced by the analysis of affected meridians. Based on studies of meridian theory, we published the view that correlation between meridians and viscera is the essential concept that should guide future research. It is our belief that ultimately meridian theory will be the key to understanding TCM. In this chapter, we discuss the basic concepts of meridian theory, how these concepts apply to clinical situations, current understanding of MVC, future directions for MVC research, and the perceived barriers to furthering our understanding.

Keywords Meridian–nerve correlation • Sympathetic reflex • Limbic system • Autonomic nervous system • Body–viscera connection • Neural regulation

19.1 Introduction

Meridians and collaterals are considered to be objective entities in the human body (Shen 2003). Meridian theory is a foundational component of traditional Chinese medicine (TCM); it highlights the nature of human physiology, disease pathology, and the process by which the body adjusts. There have been many classic monographs that have explained meridian theory, including "11 feet-arm meridian-moxibustion," "Yin-Yang 11 meridian-moxibustion," and "Lingshu." Although the number, branches, starting and ending positions, and associated viscera of each meridian are different, this was considered acceptable, as these changes reflected the development of meridian theory.

The complex relationship between the meridians and viscera is explained in the theory of acupuncture. In traditional acupuncture, doctors are guided by the theory of meridian–viscera correlativity (MVC). In the Suwen, the relationship between the two is discussed in the following way: "the pathway of the viscera is the meridian, which runs in the blood and qi, if the qi and blood are not in harmony, diseases occur." Additionally, in the Linshu, another classic TCM monograph, the 12 meridians are described as the link between the inner viscera and the joints and structures of the outer body. All of these discussions show the characteristic and function of 12 meridians. So some scholars thought it was the earliest published theory in the world that illustrates the body–organ correlation. In addition, in the Linshu, the relationship between the 12 meridians and the organs, viscera, and 5 sense organs is described in detail. Finally,

M.Q. Zhou
Research Center for Meridian-Viscera Correlationship, Anhui University of Traditional Chinese Medicine, 103 Meishan Road, Hefei, China

School of Chinese Medicine, Anhui University of Traditional Chinese Medicine, 103 Meishan Road, Hefei, China
e-mail: meiqizhou@163.com

there are records of theory-guided acupuncture treatment in clinical practice. For example, in the Suwen, there are records of heart disease being characterized as being reflected in the path of the heart meridian. More recent theories include Meridian diagnosing, which has promoted the development of acupuncture clinics.

In summary, MVC theory is multifaceted and explains the physiology of organ systems, the pathophysiology of disease processes, and the potential management of illness. With respect to physiology, the function of meridians and their corresponding organs are related closely. Second, the pathological changes in an organ system can be reflected in the meridian, and disease processes can be deduced by the analysis of affected meridians. For example, in the article "Nine needle and twelve resources" in Linshu, visceral disease was thought to be reflected into the 12 resources. Third, the physical and chemical stimuli within the meridian can regulate the corresponding function of organs; this is the key mechanism underlying acupuncture treatment. The mechanism by which acupuncture regulates visceral functioning is closely related to the three kinds of biological relationships in the body, which are as follows: segmental, intersegmental, and the spinal cord sympathetic reflex. Thus, the theory of MVC is important given its two major implications. The theory integrated the body into one organism, as it related the organs and viscera to the trunk of the body. In addition, it provided a means by which to relate basic theory and acupuncture, thus providing a means by which acupuncture practitioners could diagnose disease, select points, and manipulate needles.

19.2 Scientific Fundamental and Progress of MVC

Since from 1950s, research on the meridian has been considered a particularly active area within biological research. The success of acupuncture anesthesia promoted a global increase in the use of acupuncture as well as the emergence of basic and clinical research in acupuncture. With respect to the basic science research, discussions concerning the essence of the meridian have been the primary focus, and in a symposium on meridian research in 1977 (Hefei, China), scholars highlighted the importance of research on propagated sensation along channels (PSC). Since that time, researchers have worked a great deal to understand the PSC.

19.2.1 Morphology Support on MVC

Morphological studies of the meridians indicate that the distribution of the meridian closely matches that of the nervous system; as in or near meridian, there is generally a neural stem or larger nerve branch present (Shen 2003).

The distribution of meridian sections is often the same as that of the neural stem and their main branches, especially in the limbs below the elbow and knee. Collateral points often lie in the nerve branches, and the exterior–interior meridians are often dominated by the same spinal cord segments.

There is a close relationship between the actions of the meridian and its corresponding spinal nerve and autonomic nervous functions. Acupuncture on three-hand yin meridian could treat diseases of heart and lung because anatomically, this area corresponds to the cervical and brachial plexus of the spinal cord, which has a close relationship with the portion of the sympathetic nervous system that predominately impacts the lung and heart. Similarly, acupuncture on three-hand yang meridian could treat diseases of the head, as there is a relationship between these meridians and organs in the head via the sympathetic ganglion of the spinal nerve, the carotid artery, and the nerves coursing through the neck. With respect to foot meridians, their functions depend on the close relationship between the meridian and its corresponding autonomic nervous system segments.

Back-shu points are special positions that the qi of organs transmits. They lie in the foot taiyang meridian, which is in the back of the body. Anatomically, many studies had confirmed that most of the sympathetic trunk and the cross point of contact of the surface projection of the ridge are closely related, thus providing further evidence for the meridian–viscera correlativity theory.

In addition, animal experiments have shown close relationship between the MVC and the nervous system. For example, 18 healthy, pigmented rabbits received point injection at the Neiguan (PC-6), Quze (PC-3), Tianquan (PC-2), Shenmen (HT-7), and Shaohai (HT-3) points along the center line of the palmar forelimb, elbow, and upper arm and at the "nonpoint point" at the midpoint of the upper arm as well, resulting in a total of 36 points that were injected with 25% HRP 5 μl/point, in order to assess the bilateral C_4–C_2 spinal ganglion of the upper, middle, and lower portions of the neck, and one-to-five sections of sympathetic neurons were labeled and counted with Edwards OD (o-United anise amine) Green reaction under microscopy. Other ten Wistar rats were injected with 20% HRP 5 μl each time along the selected bilateral brachial plexus medial cord. Microscopic examination of the bilateral C_2–C_4 spinal ganglion occurred under microscopic examination using the TMB (tetramethylbenzidine) to developing, and the neurons were then counted. The results showed sensory neurons to occur in a segmental distribution. That is to say, near the point area in heart and pericardium meridians, mostly segmental sensory neurons are distributed (C_5–T_2 spinal ganglia) and the main density occurring primarily at C_8. However, the heart meridian is relatively concentrated in the C_7–T_2 distribution, and the pericardium meridian is relatively concentrated in C_6–T_3. The heart sensory area dominated in the C_5–T_1 segment, and the heart pain and skin area involved the C_5–T_2 segment in approximately the same ratio, although there was a slight difference between the two meridians. Also, the above results take a certain morphological evidence for the pathway of MVC. The second finding concerns the classification of sensory neurons in size. The ratio of the small and large sensory neurons at corresponding point area at the heart and pericardium meridians indicates that a large number of neurons were less than that in small areas (in the Heart meridian, 21.4% was large, 35.3% was medium, and 43.3% was small, while in the Pericardium meridian, the corresponding ratio was 1.47%, 12.19%, and 86.44%). This indicates that at the meridian area, sensory afferent mainly depends on the nerves with smaller diameters, and because of smaller-diameter nerves and also disposable blood vessels, the wall of blood vessels and nerves in the body acupuncture afferent unit plays an important role, and this also is one related way of MVC that cannot be ignored. The third finding concerns sympathetic postganglionic neurons. The distribution of labeled postganglionic neurons of corresponding point area in heart and pericardium meridian in the rabbit was mainly in subcervical and the thoracic sympathetic section and especially in the subcervical sympathetic nervous system. This is overlap to the dominance of sympathetic postganglionic cardiac neurons (upper cervical, cervical, subcervical, and the major part of 1–5 thoracic nerves) that may be another morphological basis of MVC.

19.2.2 Research on Meridian Phenomena

Meridian phenomenon is any kind of physiological or pathological phenomenon in the body that occurs along the classic meridian lines for some reason, including PSC, dermatosis, or sensory disturbances along channels and any phenomena detected by physical, chemical, or histological techniques. Reports on the existence of the meridian have recently focused on the PSC, the occurrence of dermatosis along meridians, and the emergence of disease along meridians.

19.2.3 Propagated Sensation Along Channels

PSC is an important and common meridian phenomenon and refers to the "de qi" sensation caused by acupuncture, the electrical pulse, and other ways to stimulate acupuncture points, including acids, hemp, inflation, and other special senses, as well as the conduction phenomenon alone along the channel from the beginning of stimulation. PSC is very close to the effects of acupuncture.

As early as 1948, a famous Japanese acupuncturist mentioned that PSC could be related to acupuncture points in his book Introduction to Acupuncture Medicine, which followed research reports concerning PSC and aroused wide attention in China. From 1972 to 1978, in more than 20 provinces, municipalities, and autonomous regions, as well as 30 U in China, experts worked together using uniform methods and a sample size of 63,228 people; they found that PSC is a normal phenomenon in life. They also found that it was a phenomenon associated with skin disease and other phenomena of the meridian. The results showed that when transmission occurred, although the feeling was subjective, the causes were objective. Additionally, the visible phenomenon of the meridian proved it. Such visible phenomenon could be seen not only in the 12 meridians but also in the 8 extra channels and in the kidney, large intestine, lung, and pericardial meridian. A few people could cause the dermatovascular nervous reactions along channels to disappear when stimulating their points. From the pathological changes, the vascular changes in the upper dermis, perivascular inflammatory infiltration, and epidermal changes may be followed by the common features of skin diseases. Later on, as assessed by the traditional objective response of the body and its sense of mass, the effect of changes in organ function further confirmed

PSC using traditional objectivity. The PSC has specific characteristics such as strongly along the channels, two-way transmission, slower conduction velocity, blockage, effectivity, symmetry, acupoint phenomenon, pan-meridian phenomenon, tendency toward the disease, sense of the phenomenon of mass return, phantom limb phenomenon in patients with congenital or acquired short limbs, acupuncture-induced phantom limb, and magical sensations. (Li 1999; Lin et al. 1995a, b, c).

All cases introduced above could be diagnosed with PSC, as such phenomenon on meridian could be elicited on a few of people who were sensitive to meridians when their points are stimulated consistently over time. The opposite was recessive transmission, which is also named LPSC, with percussion by stimulating the meridian points, and other aids induced the PSC using the traditional phenomenon, which is common in the population. Under the meridian heat, disease, and other conditions, LPSC can be transformed into PSC; similarly, PSC can also be transformed to LPSC. The findings of PSC and LPSC greatly strengthen the theory of the meridian. Subsequent to this data, various lines of research have been conducted, resulting in considerable progress.

19.2.4 Physical Phenomena of Meridian

The physical character of the meridian mainly includes low resistance, high conductivity, the phenomenon of temperature distribution, high-conductivity acoustics, high luminosity, and biological and other physical phenomena.

The earliest finding of the low-resistance, high-conductivity characteristic of the meridian was discovered by Yoshio Nakatani in Japan. He used 12-V DC to measure the surface of nephritis patients and found areas of high conductivity; after he linked these points, he found a line in the kidney similar to the Chinese meridian in the kidney, which he termed the "Ryodoraku." In the 1950s, work at Tianjin Medical University indicated that if you passed a small, direct electric current through the body, in the corpse all points detected showed a relationship between resistance and distance, and such phenomenon was symmetrical. In the living body, there was variance between the same points in different areas of the body. Additionally, in the Shanghai Public Health Services hospital, experts observed 100 cases of liver disease and found that changes in the meridian were due to imbalances and the value of the measured meridian was changed by the impact of various physical activities, which were also related to the measured pressure, compression time, and moisture of the skin. Studies by Zhu Zongxiang and Zhang Renji showed that the meridian itself is a low-resistance line, and even in anesthesia, blood loss still exists in the skin.

Jiang et al. (1980), who used thermal imaging on the subjects tested, found that stimulating acupoints in 5–10-min intervals resulted in a screen showing a bright band. The subjects report that they experienced a transmission line that is consistent with the classical record, and this image is not occasional but can be repeated many times. Wang et al. (2002) used infrared instruments and measured infrared radiant tracks that were similar to classic meridian lines. Acupuncture can increase the skin temperature along channel, and animal experiments have further confirmed that such increased temperature is related to local substance metabolism. They proposed the idea that "there is a meridian line that is multi-structured and is regulated in relation to human body material, energy and information transmission channels" after deep study of its organizational structure, function, and metabolism. It has been reported (Zhu 1989) that with a stethoscope on the large intestine point [such as Binao (LI-14)], one can hear the reflex hammer achieves this through the other points [such as Hegu (LI-4)]. The percussive sound increases the volume and makes the tone more clear. This changed when the subjects complained of numbness or a sense of simultaneous transmission. This could be understood as the phenomenon of high-conductivity sound. Another experiment showed that in the area of Shangyang (LI-1) and Laogong (PC-2), ultraweak visible light was constantly being emitted, and the wavelength was detected at approximately 380–420 nm, indicating that the meridian has high photoluminescence (Yan et al. 1979).

19.2.5 Research on Physical and Chemical Characteristics of Acupoints and Related Substances

In recent years, research on the meridian and physical properties of the material basis has focused on the study of the character of resistance, magnetic properties, acoustic properties, infrared radiation, energy metabolism, self-organizing structure, and chemical ions carried through the distribution of angles in the meridian.

19.2.5.1 Low-Impedance Characteristics

Scholars in many countries have considered the low-resistance characteristics of acupoints extensively. In 1951, the French scholar Niboyet applied resistance measurements with an ohmmeter finding low resistance at the point of the phenomenon. In 1970, he repeated his work with a Bridge DC Resistance Tester on the limb and skin. The results showed that under 5–10 V voltage conditions, the general resistance of acupuncture points is lower than nonacupuncture points, with 50% of control points showing lower resistance and 90% in line with the Meridian showing lower resistance. Reichmanis et al. (1979) took a bridge balance circuit electrode in the rolling resistance detection mode, which was arranged in a small T-type combination method and proved that the majority of acupuncture points had resistance lower than that of skin resistance. Zhang Renji and Yang Weisheng used four electrodes in 28 healthy adults and placed them approximately 2 mm deep in the skin impedance distribution. This resulted in six low-resistance meridians. The meridian line and the low-resistance line are very similar to the traditional meridian. Hu Xianglong et al. used temperature electrodes point by point and then electrode forward scanning; samples were obtained continuously, after which the data were processed. This allowed for the visual processing of the data. The results showed that low-resistance points went primarily through the classic meridian. The overwhelming majority of subjects could detect the distribution through low-resistance points; however, the degree of difficulty varied. Low-resistance points were generally located in the meridian or within the 0.5 cm of it on both sides, not much deviation from more than 0.5 cm in the test meridian; meridians with sides of adjacent area between the control points have very few low resistance. Each meridian is generally tested at the level for 15–25 months. Most subjects testing the scope of the basic accessibility are accessible or meridians of the whole. Low-impedance characteristics of the skin and skin sympathetic nerve activity may be related. However, in recent years, there has been little data concerning the function of its generating mechanism and the significance of further research. There is data that does contradict the above data. Noordengraaf (1973) by series of studies used skin resistance method of data points, depending on the shape and the electrode surface area, pressure, electrode angle, and the physical properties of the skin, such as temperature.

19.2.5.2 Magnetic Properties of Meridians

Research on the magnetic properties of meridians has made much progress, and thus, acupuncture points could result in changes in the frequency and intensity of magnetic signals. Since 1992, the French scholar Lagrange made an oscillator that could emit single-phase square-wave signals according to experiments on the magnetic properties of the meridian. The principle was interrupted when the current flowed through the solenoid by a series of magnetic fields generated when the solenoid was closer to the body, as assessed by the observation of whether the department vibrated. If there was vibration, the magnetic vibration of the body and the solenoid magnetic resonance effect followed. Thus, the body can be used to detect the existence and distribution of the magnetic field.

Based on research, the distance between the track of magnetic vibration and the body surface varies, thus there may be three kinds of magnetic lines. The first category is far from the surface of the skin, with an external vibration line as Baihui (GV-20) to be the center, which is related to the two poles of the head and foot and the contact point with skin. The second is an internal vibration line that is related to the meridian and is close to the skin surface. The third category is closer to the meridian and is nearly in contact with the skin. Magnetic vibration that is near to skin and track of line was related to 12 channels, but still, there was some magnetic vibration existence that is left to the channel, which is related to pathological functions of the body.

19.2.5.3 Acoustic Characteristics of Meridian

Sun Shengping et al. (1988)) input certain intensity acoustic signals in the Shangyang (LI-1) or Hegu (LI-4) of the large intestine meridian and found that such signals could transmit to the Yinxiang (LI-20) along channel, although the signal would eventually decay. The bilateral sides of the control point, however, did not have such a reaction, which clearly demonstrates the low-collar spread using acoustic signal features.

With this sound input method, the authors also studied the stomach and bladder meridians applying voice information transmission, which tended to have the ability to transmit sound. As for the transmission of such audio signals through the matrix, it remains unclear whether this occurs.

19.2.5.4 Infrared Radiation Characteristics of Meridian Line

In recent years, the infrared radiation character of the meridian line and infrared radiant track along meridians (IRRTM) and its mechanism have been researched extensively. Currently, the use of infrared radiant imaging techniques to show meridian root makes it possible to characterize the infrared characteristics of the meridian and has proven that IRRTM is an important biological phenomenon in normal people and is closely relevant to function regulation. Stimulating acupoints could induce or change the IRRTM state and travel, and diseases induced in organs along the meridians appear to be related to the high-temperature zone as well. Meridian line-related acupoint organization is significantly better than nonpoint positioning by the control site with respect to thermal conductivity. The formation of IRRTM is relevant to the microcirculation of skin and the heat transfer channel under the meridian, with the latter being particularly important. The channel is deep and shallow and has a higher PO_2 and tissue temperature, stronger microcirculation, and active energy metabolism. IRRTM is only an expression of its surface (Liu et al. 2009).

Further observation has shown that the peak point of the infrared radiation occurs in the vicinity of 7.5 μm in the human body but is also observed to occur in two peaks, 2–2.5 μm and 15 μm. Presumably, this is relevant to the organization of energy metabolism. In the human body, acupuncture points deep in the connective tissue in the arrangement of collagen fibers are consistent with the direction of meridians to the superficial fascia and skin connection. Along the fiber direction of collagen fibers on the 9–20 μm infrared peak, there is high permeability. Infrared information could easily reflect the collagen fibers oriented along the meridian direction of the transfer surface. When heated in the acupuncture points, infrared radiation is also easier to access along the collagen fiber transmission lines, thus resulting in the formation of the infrared radiant track.

19.2.5.5 Characteristics of Meridians Based on Organizational Structure

Biological liquid crystal is a kind of dynamic ordered structure, which is comprised of corresponding biological molecules that self-organized. After acupuncture on Jiexi (ST-41) of rabbit, then to observed liquid crystal particles at five test points that along stomach meridian. As a result, particles could be found at Zusanli (ST-36) and Shangjuxu (ST-37), while very few particles could be found in the control point. Meanwhile, in the stomach liquid, liquid crystal increased accordingly. There was also liquid crystal in the points of Zhongting (CV-16), Zhongwan (CV-12), and Xiawan (CV-10) and a one nonpoint on the meridian, along the conceptive meridian, after acupuncture and moxibustion on Danzhong (CV-17) in rabbit (Chen et al. 2006; Lu 1996). In addition, after acupuncture on Zusanli (ST-36), Waiguan (TE-5), Neiguan (PC-6), and Kunlun (BL-60) points, liquid crystal-like material was found in the corresponding points being Quchi (LI-11), Waiguan (TE-5), Quze (PC-3), and Weizhong (BL-40). These preliminary results show that when applied during acupuncture, liquid crystal material releasing is also a property of the meridian (Jiang et al. 1997).

19.2.5.6 Chemical Ion Distribution in the Meridian

Research on the biochemical basis of the meridian is also important. Research on the chemical substances along the meridian line (i.e., ions, oxygen, neurotransmitters, intercellular communication, etc.) shows that the activity of molecules is closely related to the meridians. Of the molecules, the ions that regulate acid–base balance also maintain the normal activities of life within the body (Liu et al. 2007; Wang et al. 1996; Wang et al. 1998). Modern studies have shown that calcium, potassium, and sodium ions are critical for nerve and muscle excitability, and "de qi" by acupuncture meridian is bound to these materials involved (Jin et al. 2001). In recent years, many scholars took ion as a breakthrough on studying correlativity of peripheral ion activities along meridian; meridians and corresponding organs of the ion-related and other aspects were systematically studied, and the hypothesis have been put forward and more from the ion to verify the organs of the specific acupoint meridian scientific theories (Yang et al. 2007). Concerning calcium, calcium levels in the rabbit meridian are higher than in the non-Meridian Branch; viscera lesions of the corresponding meridian line of peripheral Ca^{2+} concentrations are decreased with a certain degree of specificity. Acupuncture can belong to other points along the meridian where calcium is also seen to be increased; when combined with EDTA treatment, acupuncture is less effective, which shows that calcium interacts with acupuncture's effect to a certain extent (Guo et al. 1996, 1998, 2002, 2004).

19.2.5.7 Energy Metabolism of Meridian

Measurement of tissue oxygen consumption is one of the basic methods of studying energy metabolism. Using tissue oxygen partial-pressure sensor needles inserted into the deep tissue under the meridian, the Hu Xianglong laboratory was able to detect physiological states and changes in tissue oxygen partial pressure; as a result, the deep tissue's oxygen tension was significantly higher on the left side of the control line with features along meridians (Yuan et al. 2005). On the acupuncture meridian lines, tissue oxygen tension was lower when compared to that of the control area. Declines in tissue oxygen tension may reflect the meridian line of cells or the increase in oxygen consumption under the action of acupuncture.

Similarly, Zhang Weibo et al. (1995) used highly sensitive CO_2 detector to examine exhaled quantity of CO_2 of skin in the forearm segment of six levels of pericardium meridian, large intestine meridian, and stomach meridians of 30 healthy people and pericardium meridians of 22 healthy people individually and found that (1) the pericardium meridian by the forearm segment of six levels in four levels of online RCO_2 was significantly higher than outside the line ($P<0.05$, later is the same), RCO_2 from the distal (room to point) to the proximal end (Quze, PC-3) gradually increased, and a significant linear regression effect was noted. (2) The large intestine and stomach meridians by the three levels were significantly higher than out of meridians of RCO_2, while other levels were not significantly different. RCO_2 showed U-distribution from the wrist and ankle. (3) In all pericardium meridians, only 19 horizontal wrist elbow segments of the six levels of RCO_2 were significantly higher after the elbow over nonspecific. (4) RCO_2 is significantly a higher part of the majority (75%) of the meridian. All measurement points after acupuncture were significantly increased with respect to RCO_2, while the amplitude by the line through things and in the proximal reach was significantly different. Through continuous measurement, we found RCO_2 to have significant fluctuations. Five minutes after inserting a needle, the RCO_2 was significantly restored, and there was almost no recovery warp. Thus, acupuncture can be described as a process of adjusting blood carbon dioxide and the corresponding changes in energy metabolism, which then changes the volatility of the activities. This change is transmitted via a wave through the blood and, as a result, is more likely to impact cardiac conduction. Meridian line of energy metabolism activity has the potential to increase heat production, thus rendering body temperature changes.

19.2.5.8 Study on Drug Response of Point and MVC

Studies have shown that injecting drugs into acupuncture points and nonpoints was higher than injecting into bilateral control points or muscle. Drug response of points also showed the characteristics of an alone channel. In certain areas, points seemed to distinguish and have increased affinity for some drugs. For example, some points could reduce the toxic response to some harmful drugs, such as aconitine, increasing the efficacy and benefit of some drugs. Recent literature has shown that there is no relationship between the response to injected drugs at certain points and the concentration of the drugs in the blood or the integral nature of the nervous system. However, if certain points are preconditioned with collagen IV, the response to drugs injected in these points could be blocked. It has been demonstrated that extracellular matrix such as collagen protein may be an important part of the response to drugs injected at various points. Zhou et al. (2001) studied material of drug signals of the initial effect after injection of drugs, efficacy occurs via the transmission of information by collagenase IV in Zusanli (ST-36) 6 mm above the same area (Stomach Meridian of Foot-Yangming); after 30 min of pretreatment, gentamicin injection Zusanli (ST-36) observed after administration of gentamicin on the impact of gastric electrical and with the same amount of collagenase IV in the top of 6-mm Zusanli (ST-36) by side (side of foot Yangming stomach open 4 mm) pretreatment group and saline Zusanli (ST-36) top of the pretreated group compared with the results found that collagenase IV Zusanli (ST-36) top of the same pretreatment completely blocked by the Zusanli (ST-36) the initial injection of gentamicin inhibition of gastric electrical role, while the foot Yangming stomach next to the pretreatment group and the saline pretreated group no such blocking effect. Prompt injection of gentamicin on Zusanli (ST-36) can inhibit gastric electrical activity.

Cheng et al. (2002) based their study on the theory of the specific regulation of acupuncture and Chinese herbs. They then studied the effect of acupuncture at the Neiguan (PC-6) and Taichong (LR-3) points to determine the effect of Chinese herbs in animals. Acupuncture in healthy adult Wistar rats at the "Neiguan (PC-6)," "Taichong (LR-3)," and other points was conducted, and the rats were fed TMP (TMP) with high-performance liquid chromatography (HPLC). The TMP in vivo absorption and step-targeting rules were observed, and the TMP found at the same time in the corresponding acupuncture points in the corresponding organs of rats was significantly greater than in the simple perfusion group. This was particularly significant in the heart and liver content of TMP in the rats, suggesting that acupuncture results in organ-specific adjustments and promotes the distribution of drugs to particular targets.

19.2.5.9 Molecular Biological Study on MVC

The role of acupuncture was closely related to the neuroendocrine-immune system (NEI networks) and refers to cellular messengers, neurotransmitters, quenching, neuropeptides, cytokines, endocrine hormones, and other factors; however, the real process on how acupuncture signal transfers intracellularly and intercellularly is complicated and remains unclear. Gene chip techniques could be helpful in understanding the expression of genes in the viscera and may reveal principles concerning acupuncture.

Zhou YP team (Zhou et al. 1997, 2004, 2007; Zhang 2002; Wu 2006) studied the impact of acupuncture on the heart and lung meridian on cardiovascular function in an acute myocardial ischemia model. They used gene chip technology to analyze gene expression in the heart and hypothalamus after myocardial ischemia and studied the effect of acupuncture of the heart meridian and small intestine meridian. Comparing the results of acupuncture of the lung meridian and determining the variations in expressed genes, the authors attempt to confirm the relative relationship between the heart meridian and the small intestine meridian with the goal of indirectly illustrating the regulatory mechanisms of acupuncture.

Yan Xingke et al. applied SDS-PAGE technology to find, under the pathological state, the protein expression of the lung and large intestine meridian that was changed. In constipation, there is loss of 27.24 kDa and 22.79 kDa molecular weight compared with normal group of protein map of large intestine and increase of a 17.74 kDa band of lung meridian. In asthma, the 31 kDa bands of protein map of lung were desalinated and lack of 23.05 band of protein map of large intestine. In the heart and small intestine meridian, there was no evidence of an abnormal protein pattern. The tips of the lung and the large intestine were associated with a specific protein expression pattern when there was abnormal tone of these organs. After acupuncture constipation model, there was increased 18.01 kDa protein band of lung protein map, while in acupuncture asthma model, there were increased 17.65 kDa–20.56 kDa protein bands of lung protein map and no changes in large intestine map. There is a suggestion that acupuncture on the pulmonary and intestinal diseases and related disorders may have some regulatory role. This is consistent with the clinical manifestations of lung and bowel disease and the practice of acupuncture treatment of pulmonary intestinal disease.

Yao Wen et al. applied surface-enhanced laser desorption ionization time-of-flight mass spectrometry (SELDI-TOF-MS) and WCX_2 (weak cation exchange chip) technology to analyze protein fingerprints of three experimental cases of normal and electroacupuncture stomach meridian of rat serum, in order to obtain acupuncture response protein and further to elucidate the mechanism of acupuncture effect of body fluids and provide a theoretical basis. The results showed that compared with normal rat serum, stomach EA serum protein mass to charge ratio of 2,000–50,000 in 25 protein peaks showed significant differences, of which 19 marker proteins in the stomach EA serum showed high expression of six marker proteins in the stomach EA serum low level of expression; quality of analysis that they may be related to general acupuncture analgesia, stress response, and damage repair-related factors is closely related to the release. That EA stomach with normal serum levels of low molecular weight serum proteins was obvious, such proteins might promote gastric mucosal injury and repair the electroacupuncture effect, of which it is closely related to, but the specific identification of differentially purified proteins still needs to be explored in further study.

Epidermal growth factor receptors (EGFRs) are closely related with gastric mucosal injury and repair of gastric mucosal protection, and damage repair has a very important role. EGFR tyrosine kinases are a transmembrane receptor family, and EGFR activation can activate IP_3/Ca^{2+} and DG/PKC double signal systems and Fos/Jun-related Ras/Raf/MAPK-dependent pathway, causing cell proliferation, differentiation, adhesion, and migration (Liu et al. 2003).

19.2.5.10 Other Studies on the Meridian

In recent years, China has three research groups that were from the "low flow resistance through the channel by," "follow the flow line through the hemodynamic," and "along the perivascular space" camps. The perspective of the three different studies is that they considered that "the meridian line presence a material transfer fluid channel." If the transmission of body fluids contains certain biologically active substances, can be caused by traditional, and so on, the meridian phenomenon may constitute a material part of the basis meridian. Chinese scholars have observed that injection, at a human meridian point, with the radionuclide $^{99\,m}$TC can be followed at a migration rate of about 17 cm/s in the limbs of the tracer trajectory consistent with the classical meridian line rate of about 78%. The track of this tracer is not directly related with the lymphatic system but with a more complex relationship between it and the blood circulation (Li et al. 2005). PET technology was recently applied on the spatial orientation of radionuclide tracer trajectory, showing initial success. On the β-ray observations of tracer, 86% of subjects appeared in 3–5 mm wide tracing image, and the depth about 2–3 mm under the skin or subcutaneous tissue. Judging from the time of imaging, the tracer trajectory has nothing to do with blood vessels and lymphatic vessels, and their exact nature remains to be explored (Liu et al. 1999).

Yi et al. (2004) showed that in a rabbit model of acupuncture ulcer disease in "three yang meridians of foot," RIA, and PC-PCR method, the epidermal growth factor in gastric contents, the gastric mucosal damage index, the somatostatin content, and the SSR mRNA expression were significantly decreased. It was also proved by electroacupuncture points to improve gastric function, and acupuncture increases gastric antral tissue CCKA-R-mRNA expression. Lu (2008) referenced other research material proposed a composition model meridian, and he thought that meridian is a body fluid environment of the life the main material composition is a solution in the vicinity of the critical concentration of hyaluronic acid crystals (HA) of liquid crystal, which is characterized by body fluids: When life is over, it tends to depolymerize, allowing the LCD to phase transition to the gel state. At this time, the solvent water (tissue fluid) will partially collapse and flow out.

Point tenderness is a disease state of the reaction and, to a certain extent, reflects the dysfunction of the body; the meridian is the organs of the reaction zone, while the point is the point of disease response. Acupuncture diagnosis is based on meridian theory, hyperalgesia, and another positive point and is based on exploring the meridians and organ lesion in new diagnostic techniques. Such a treasure house of Chinese medicine diagnosis is one of the valuable heritages. Superficial tenderness and organ and meridian point lesions exist between specific internal relations; their contact information is disease → organs → meridian point tenderness (i.e., the response surface organ diseases often have followed through with features). Points through the diagnosis are Point → organs → Meridian (Meridian, organs may have other links along the way).

A large number of clinical practices show that the body of illness often appears in the relevant tender points such as pathological response, and this reaction will rise and fall with the change of the disease. The performance point of pathological reactions occurs mainly as the following: First is the point palpation. Gently press your fingers on the point where the patient has felt acids, hemp, swelling, pain, or hypersensitivity, in which tenderness is most common, as well as points of the surface with cool or hot feeling, points of organizational boards that are hard or soft or with depression or eminence, and acupoints on skin with vacuoles, nodules, or article cords. Second is inspection by the acupoints. These are points of skin color change or petechia, rash, desquamation, and muscle and other uplift or depression.

19.2.6 Modern Scientific Basic Research on Meridian–Viscera Correlation

Meridian–viscera correlation (MVC) reflects the internal organs and external appearance of the close ties between viscera and meridians, which is the key in meridian and "viscera state" theory of Chinese medicine and, also, the basis of holism and from an overall perspective to understand the world (Dong 1992). "Viscera" refers to the body organs; "state" indicates the surface of various physiological and pathological phenomena. A sentence in Linshu that "12 meridians, which belong to corresponding viscera and organs in the body and relate limbs and joints in the surface of the body" emphasized the role of the meridian: In anatomy and physiology aspect, meridian was considered to be a channel that relates five sense organs, seven holes of the surface body, and viscera in the body; when disease of viscera and organs occurred, it can be distributed by the color of surface features, temperature, pain threshold, sensitivity, pulse, and other aspects of the manner shown, embodied in the meridians and acupoints corresponding above. From therapy aspect, through many kinds of stimulus on meridians and points which showed abnormal changes such as acupuncture appliance, manipulations, moxibustions, to regulate viscera and organs in double way, and achieve the purpose of treat disease. So this is a powerful self-regulating method and has several advantages that are simple, effective, clear, economical, and so on (Zhu 1998).

19.2.6.1 Contact Forms of Body Organs

The surfaces between internal organs and the body surface are an organic link and are related to nerve segmental distribution. In Ishikawa, Japan (1962), a research team that studied the autonomic body surface made outstanding contributions to our understanding of the visceral connection. They identified the visceral relationship with the spinal segment and the movement and secretory responses, which were called reflexes, while the feelings associated with associated abnormal reactions are called references. The body surfaces between internal organs and the external body as well as between body surface reflections and their associated forms are numerous and are as follows: body–visceral association (somato–visceral reference), viscero–physical association (viscera–somatic reference), body–visceral reflex (somato–visceral reflex), visceral–sports reflex (viscero–motor reflex), the body–glands reflex (somato–secretor reflex), internal organ–body nutrition reflex (viscera–somatic trophic reflex), internal organs–body autonomic reflex (viscera–somatic autonomic reflex), internal organs–visceral reflex (viscero–visceral reflex), skin–muscle reflex (cutane–muscular reflex), and the skin–skin association (cutane–cutaneous reference).

19.2.6.2 Neurological Mechanisms of MVC

Although nervous system is not the same as the meridian, some of the features of meridians are closely related to the nervous system. The nervous system is undoubtedly the introduction of acupuncture information and integration and is the site most implicated in the relationship between the meridians and the internal organs. The twentieth century has been termed the "brain decade," as a result of the numerous advances that have been made in the neurosciences and the other brain sciences.

Starting from the brain sciences, in-depth study related to the meridians and organs will provide a link between the right brain and a feasible direction for future research. Meridians are related to the functional activities of organs, and the meridian table is based on the consistency of observations concerning this interaction. Modern acupuncture research also shows that the effects that acupuncture produces depend on the structure and function of the nervous system integrity, as well as the health of the endocrine and immune systems, which are closely related to the function of the nervous system.

Segmental innervations of nervous system, in the chain of the nervous system with lower animals, have shown the section on morphology. In humans, the distribution of spinal nerves or the brain also shows the characteristics of segmental dominance. The segmental distribution of animals is referred to as the body segment (somite). In general, the body section should also include joints, muscle sections, skin sections, and the corresponding nerve segment (neurotome), which also dominates the trunk, limbs, blood vessels, sweat glands, and arrector pili muscle sympathetic nerve segmental distribution and corresponds to the same spinal cord segment-dominated organs.

Each body segment has a nerve, a "segment" that is a primitive neural tube of the row unit. It is in contact with body parts of the section and includes the section corresponding to the body organs arranged. A spinal segment of the visceral nerve (afferent and efferent) dominates at the level of the viscera; it simultaneously innervates the somatic nerves and skin, muscles, etc. This shows that the body segment during embryonic development may be very close in proximity during development of the body surface and the viscera. The location of the viscera within the body changes, as does its relative position against the body wall. However, the same nerve segments remain.

Internal organs are dominated by the autonomic nervous system, which also has a segmental distribution of features, but because of changes in individual organs during great position changes, organ size increases; especially organs of the body, the Segmental distribution becomes less clear. Disposable skin, vascular smooth muscle, skin glands, and internal organs of the sympathetic nerve and preganglionic fibers are fixed in the spinal cord from the sympathetic nerve cell bodies. However, through sympathetic postganglionic nerves, there is a fixed relationship with the spinal nerve by which it is accompanied. Therefore, there is a relatively clear relationship between the distribution segment (see Tables 19.1 and 19.2). Parasympathetic nervous dominated the visceral efferent fibers such as vasodilation and sweat gland inhibitory fibers. The parasympathetic fiber also has its segmental distribution, location, and morphology in the skin of the same segment.

Visceral organs are rich in receptors and sense changes in the internal organs due to physical activity and changes in the environment. Through the central visceral afferent, these receptors help the organs maintain homeostasis. In general, it is believed that nerve conduction in the vagus and pelvic nerve yields a unique feeling as well as visceral feelings, while sympathetic nerve conduction via the spinal nerve is responsible for the sensory fibers of the gastrointestinal tract and other major thoracic, abdominal, and pelvic organ pain (Zhu et al. 2009).

Table 19.1 The segmental domains of the sympathetic efferent nerve

Organs and parts of organs	Segment of the spine	Parts of the spinal cord	Segment of the spine
Head and neck	T_1~T_5	Spleen	T_6~T_{10}
Upper limbs	T_2~T_5, T_3~T_6	Pancreatic	T_6~T_{10}
Lower limbs	T_{10}~L_2	Kidney	T_{10}~L_4
Heart	T_1~T_5	Ureter	T_{11}~L_2
Lung, bronchus	T_2~T_4	Adrenal	T_8~L_1
Esophageal	T_5~T_6	Testes or ovaries	T_{10}~T_{11}
Stomach	T_6~T_{10}	Epididymis, vas deferens and seminal vesicle	T_{11}~T_{12}
Small intestine	T_9~T_{10}	Bladder	T_{11}~L_2
Cecum–splenic flexure	T_{11}~L_4	Prostate and prostatic urethra	T_{11}~L_1
Splenic flexure–rectal	L_1~L_2	Uterus	T_{12}~L_1
Liver, gallbladder	T_7~T_9	Oviduct	T_{10}~L_1

Table 19.2 The site distribution of skin nerve segments

Parts of body	Nerve segment	The regional distribution of the various segments
Head and neck	Trigeminal nerve	Trigeminal nerve in the forehead and face
	C_2~C_4	C_1-free distribution of dermatomes
		C_2 occipital, auricular half, submental
		C_3 neck, nuchal
		C_4 lower neck, shoulder
Upper arm	C_5~T_2	C_5 lateral (deltoid muscle area)
		C_6~C_7 forearm, radial side of hand (palm, back side)
		C_8, T_1 forearm, ulnar hand (palm, back side)
		Medial arm and armpit T_2 (T_3 is also distributed)
Trunk	T_2~L_2	T_5 male nipple
	Root distribution	T_7 xiphoid level
		T_{10} umbilical
		L_1 groin area
		L_2 iliac crest of a narrow waist extended leg
Lower limb	L_2~S_2	L_2~L_3 front thigh, lateral, medial
		L_4~L_5 anterior leg, dorsal medial
		S_1~S_2 foot, dorsal lateral, posterior thigh and calf
Perineal	S_3~S_5	S_3 skin around the anus
		S_3~S_5 followed by distribution of a concentric circle

The segmental nervous system is primarily responsible for sensation at the body surface and is connected to the viscera via visceral referred pain. Unlike the clear, precise position induced by skin damage or a pain stimulus, visceral pain is often dull, and its positioning is generally not clear. In 1888, Ross divided visceral pain into two categories. One is authentic visceral pain from internal organs and sensory organs, and the second is body pain, although the feeling comes from the visceral reaction of the body, such as a positioning error in the body pain, which Head (1893) reported as being involved in visceral pain (referred pain). Involved in visceral pain is an impulse from the internal organs by the autonomic nervous system's transmission of pain. The site of referred pain and disease often appears in the organs of the body surface, some distance from the nerve segment. It is generally believed that lesions involving visceral pain lie mainly in the trunk area of the dorsal and ventral side; however, when stimulated by excessive visceral pain, the afferent impulses in the spinal segments can be spread up and down; thus, visceral pain could spread from the extremity nerve segment junction to the limbs. In current opinion, the occurrence of visceral pain and referred pain is more closely related to the neural segmental distribution. The concentration of various chemicals may be responsible for disease occurrence within internal organs, thus passing a strong impulse that generates excitement via spinal cord lesions and reduces the stimulation threshold, making the same skin section of the normal incoming sensory impulses result in pain. In the body, the sympathetic nerve reflex theory says that the visceral stimulation in disease pulses through sympathetic afferent fibers to reach the spinal cord segments after their roots and produces excitement through the sympathetic efferent fibers, which is the same segment of the skin caused by local vasoconstriction and ischemic or nutritional disturbance, resulting in allergies at areas involved in pain.

Modern research suggests that the impact of physical stimulation on the regulation of visceral function activity by the autonomic nervous system, especially the sympathetic efferent system, has its core in the body of sensory stimulation by the sympathetic reflex sympathetic nerve activity, which plays a regulatory role and is the core of the soma–visceral connection (Mao et al. 1994). We could thought that there were segmental, intrasegmental or intraspinal cord and above segmental or spinal cord regulative mechanism of somato-viscera correlation according to the characteristic of early, later and delayed reflex of body stimuli in pre- and pro-ganglionic fibers. The meridian can be imagined as a pathway that conveys information from the nervous system to the ganglia and spinal dorsal horn, which is the first level segment in the integration of visceral information. Its purpose is to regulate and control the transmission of visceral information to the central nervous system. On the other hand, in the body, the intrasegmental regulation of viscera organs were through synapse relations between collaterals and autonomic nervous in spinal cord, and the mainly way was through neurotransmitter, cellular electronic activities and so on. Furthermore, of great importance, the region that is stimulated sends this information to the spinal dorsal horn, and the information is then sent to higher areas of processing, such as the limbic system and hypothalamus, by the ventrolateral central beam. In addition, the dorsal lateral beam acts as a regulator of sensory input during visceral injury and in acupuncture by integrating various sensory inputs. This not only reflects the effect of acupuncture via the meridian mechanism but also reflects the role that holistic acupuncture plays in adjusting the homeostatic mechanisms of the body.

Not only do sympathetic and parasympathetic efferent fibers contain a vegetative system, they also deliver visceral sensory information via incoming fibers. The presence of visceral afferent fibers was first recognized in 1894 by Langely and Anderson. They estimated that the ratio of outgoing and incoming fiber was approximately 10:1. The cell bodies of these afferent fibers are located in the spinal ganglia. Later, histological and physiological methods were used to prove the existence of sympathetic and parasympathetic nerve fibers in the visceral pain signal transmission fibers.

Much of this early work confirmed the existence of dual sensory organs. Accompanied by a number of visceral afferent fibers in the sympathetic nerve (such as visceral nerves), the same organ also has afferent fibers that accompany the parasympathetic nerve (such as the vagus nerve and pelvic nerve) with the line afferent fibers. The sympathetic and parasympathetic function of visceral afferent fibers for each piece of visceral sensory information is generally believed that visceral afferent activity may be double visceral stimuli (such as nonawareness and awareness of sexual reflexes in the sense of experience) that are induced. However, it is recognized that the visceral sensory experience, especially pain, which is passed by the sympathetic nerve fibers in transmission and is accompanied by parasympathetic nerves in activating afferent fibers does not lead to the visceral sensory and visceral functions, rather it only regulates reflexive activities. In the abdominal organs of the parasympathetic system, the vagus nerve and pelvic nerve afferent fibers may also pass information on concerning nonpain sensory stimuli.

Sympathetic afferent fibers pass pain information from the abdominal area, which has been confirmed by many experiments. In 1909, Sherrington reported for the first time that gallbladder injury caused by a noxious stimulus, as assessed by blood pressure changes, still existed after vagal nerve apocope, confirming the transmission of pain via nociceptive afferent fibers to the sympathetic nerve. Schrager and Ivy in 1928 confirmed that the use of an embedded balloon filling in the gallbladder or bile duct could induce behavioral responses in conscious dogs during chronic experiments. Abnormal motor responses of animals to pain signals and respiratory depression due to bile duct dilatation via the right splanchnic nerves were cut off entirely, and bile duct dilation caused increased salivation, vomiting, and other symptoms, which did not occur when the visceral nerve was removed. However, after vagotomy, there was inhibition of these reactions alone. This experiment in 1978 was repeated by Stulrajter. He observed that in cats and dogs, the placement of a balloon into the gallbladder resulted in pain, which was caused by nerve integrity only in the internal organs of animals.

Clinical observation also supports the fact that abdominal visceral pain is transmitted via sympathetic afferent fibers. In 1939, Ljriche reported that stimulating local anesthesia in patients at the visceral nerve could cause severe pain. In 1943, White also reported that blocking or cutting off the visceral nerve could alleviate this pain, but that dry cutting or blocking of parasympathetic nerves was not effective.

These behavioral, physiological, and clinical research data show that abdominal visceral pain information is transmitted via sympathetic afferent fibers. However, there are also data that report that lower abdominal pain (mainly pelvic pain) may also occur via visceral parasympathetic transmission.

19.2.6.3 Sympathetic Nervous System and Correlation of Meridian and Vessels and Entrails

We believe that the human neural connections throughout the body surface and internal organs are connected through the autonomic nervous system. Therefore, the autonomic nervous system (ANS) is the body surface. Thus, a very important visceral component of acupuncture is to adjust the effects on the viscera.

The sympathetic nervous system is an autonomic nervous system, with neurons from four different areas: (1) preganglionic neurons, (2) the former motor neuron (premotor neuron) which regulates the activities of preganglionic neurons independently, (3) afferent neurons which conduct signals from peripheral receptors, and (4) the more advanced middle of the central neurons which then connect the incoming. Sympathetic premotor neurons in the ventral medullary external front, before the ventral medullary central raphe nucleus, pons, and hippocampus within the paraventricular hypothalamic nucleus, which is located outside the ventral medullary sympathetic premotor neurons are important in the maintenance of blood pressure. Sympathetic premotor neurons in the outgoing path go down to the first thoracic and to the second or third lumbar spinal lateral horn of gray matter and change transactions being preganglionic neurons in the anterior horn of the spinal cord. Sympathetic preganglionic neurons then send nerve fibers to ganglion formations in three ways, which are as follows: paired paravertebral sympathetic chain, all unpaired, and remote plexus ganglions near target organs. Preganglionic sympathetic fibers leave the spinal cord in the spinal anterior horn with the spinal neural stem into the paravertebral sympathetic ganglia; 22 pairs of sympathetic ganglia are arranged in pairs on both sides of the spine, and the ganglion sympathetic chain links these ganglia to create a formation of ganglia. Sympathetic preganglionic fibers in the section are once again replaced with postganglionic neurons and sympathetic postganglionic fibers and are issued the corresponding direct effect of spinal nerve organs.

The sympathetic nervous system is a living system that relies on regulation of vascular smooth muscle (Zhu et al. 2008). These smooth muscles adjust the caliber of small arteries, which can then rapidly adjust the blood perfusion of various organizations to meet the metabolic needs of the individual organ. To continue calibrating the caliber of small arteries, the sympathetic nervous system of animals in the resting state, despite a certain degree of activity and tension to maintain the force of small arteries, has a control mechanism that changes the caliber of small arteries in response to the metabolic needs of the tissue. Because of this, sympathetic nervous system tissues and organs can regulate both blood perfusion and, in the whole animal model, any change in the sympathetic nervous system activity. Thus, it is difficult to avoid changes in blood flow caused by the impact of manipulation of this system.

Numerous studies have shown that under psychological and emotional pressure, the sympathetic nervous system activity increases, parasympathetic activity decreases, and the heart, immune, and hormone systems become increasingly active. As sympathetic nerve activity increases, there is a decrease in the ventricular fibrillation threshold, and the risk of ventricular fibrillation increases. The myocardial and coronary vascular distribution of sympathetic nerve terminals is dense in myocardial ischemia and sympathetic afferent stimulation; sympathetic nerve function, angina, myocardial infarction, severe arrhythmia, and coronary artery disease are all closely related. It is unclear whether hypertension, coronary heart disease, arrhythmias, or heart failure can be regarded as a pathological state of excessive sympathetic nervous system activation. The sympathetic nervous system is important in maintaining homeostatic regulation of the following activities: sympathetic nerve activity-induced gene expression, cold, pain, movement, and activation of specific disease states, including high sympathetic activity in cardiovascular disease. In a series of hypertensive patients with risk factors for coronary artery disease, high sympathetic activity was an important risk factor. The dorsal spines on both sides of the back are important acupoints for the treatment of both visceral and local disease. Segment arrangement is also used to describe the distribution of sympathetic skin sections and a special segment called the sympathetic spinal nerve dermatomes; skin ganglion is the most common feature. Head's allergy belt showed that the corresponding allergy area of certain viscera, was similar to the position of sympathetic chain that dominated. Stimulating the Xinshu (BL-15) acupoint regulates heart rhythm, likely due to stimulation of the autonomic nerve fibers. The autonomic nervous that dominate liver organs was from thoracic 6 to 10 nervous, while the point of Ganshu (BL-18) was exactly lied on the 9 thoracic nervous. In the abdominal splanchnic nerves, the internal organs innervated by small nerves, the celiac ganglion, the superior mesenteric nerve ganglion, and the pelvic nerve plexus, have wide contacts, participating in the regulation of abdominal and pelvic organs. Thus, when acupoints are stimulated by the relevant organs, it results in readjustment. Thus, the bladder meridian is the second core, and the "internal organs are" the root cause. Based on modern anatomy and physiology, they can treat diseases related to internal organs by stimulating the autonomic nervous system.

The correlation of the meridian, its vessels and entrails and acupuncture meridian theory and the organs is related to clinical studies. A large number of studies have shown that the sympathetic autonomic nervous system in particular organs within the meridian is linked to the important pathways. In the 1970s and throughout the twentieth century, needle-puncture analgesia and its mechanism of inducing clinical anesthesia have been studied. The integrity of the sympathetic nervous system and the independence of its balance are the key. In the acupuncture meridians and at the acupuncture points, the form and function of organs are maintained, as is the sympathetic nerve integrity. Thus, the structural basis of meridian theory is the body sympathetic reflex. Sympathetic nerve fibers are widely distributed in blood vessels, lymphatics and skin sweat glands, sebaceous glands, arrector pili muscle, and other tissues; when the needle is inserted into a point or living acupoint, it is able to stimulate the sympathetic nervous system and the sympathetic innervation of skin sections. Traced to its source, the bladder input flows through the spinal nerves to the spinal nerve fibers of the sympathetic chain, and the whole body sympathetic nerve fibers respond to the acupuncture, thus all of the sympathetic nerves end up playing an important role in acupuncture. Bladder full sun is the 12 meridians of the core theory of status that cannot be ignored, with Chinese characteristics, as it is the theory by which human organs and autonomic function are regulated and should be carefully studied. Some scholars who have studied the origin of the thoracic sympathetic trunk and the greater splanchnic nerve have found that the stomach stimulates acupoints via central areas, such as hippocampal neurons. They also found that stimulation of the splanchnic nerve sensory fibers and pain fibers from the impulse can be spread to the hippocampus and other limbic systems.

19.2.6.4 The Limbic–Hypothalamus–Autonomic Nervous System Regulation and Correlation of Meridian and Vessels and Intestines

The hypothalamus and the limbic forebrain as well as the brain stem reticular formation and function form close ties in the regulation of the function of internal organs. These impulses reach the hypothalamus from the limbic forebrain, thalamus, and brain stem reticular formation; the outgoing impulse can also reach these parts but may also go through the pituitary portal system and the hypothalamic–pituitary and pituitary beam adjustment posterior lobe activity. The pituitary portal is the median eminence (gray nodules on the inside front), and there are many nerve endings containing secretory granules in the median eminence to terminate its discharge through the portal system to the pituitary gland, thus regulating the activities of the pituitary.

We believe that the edge of the brain, the hypothalamic autonomic nervous system, and the brain stem reticular formation in the acupuncture visceral process play an important role in the integration of the effects of acupuncture. As an antagonism, Wangshu Ju et al. have found that the ventrolateral medulla and the paraventricular nucleus are protected electrically by Neiguan (PC-6), one of the central pathways of myocardial ischemia. The solitary nucleus participates at relevant points on the protective effect of myocardial ischemia in the process. Cai Dingjun et al. suggest that acupuncture can increase 5-HT_7 receptor activity, as well as γ-aminobutyric acid, c-Fos, and nitric oxide synthase expression at the hypothalamic suprachiasmatic nucleus, while also increasing the expression of nitric oxide synthase at the level of the ventral dorsomedial nucleus of the hypothalamus. There are studies demonstrating that the existence of the suprachiasmatic nucleus is a two-way adjustable gastric electrical acupuncture premise. Liu Zhicheng et al. found that acupuncture in the hypothalamic arcuate nucleus of type 2 diabetic rats functioned in a benign manner to correct insulin resistance and the abnormal metabolism of the central nervous–endocrine system. Furthermore, it was found that the arcuate nucleus participates in the inhibitory effect of acupuncture on the stomach. Acupuncture is also found to impact the paraventricular nucleus of obese rats, and the positive role of the amygdala may be its central role in achieving weight loss effects. Wangshu Ju and Li Jiangshan et al. found that NTS is the center of acupuncture's effects and that the integration of visceral (stomach) noxious stimulation occurs at this site, providing evidence that NTS neurons are closely associated with acupuncture. The acupuncture treatment of diabetes, depression, and cerebral ischemia is closely related to acupuncture's impact on the hippocampus. Chen Zebin also found that acupuncture pretreatment of cerebral ischemia of the central amygdala impacted CRF-like protein expression. Ji Laixi et al. found that acupuncture at the "Neiguan (PC-6)," "Zhongwan (CV-11)," and "Zusanli (ST-36)" acupoints resulted in a significant effect on the medullary dorsal reticular nucleus neurons; they considered that the effects of acupuncture might be mediated by the integration of information coming from these various channels.

Recently, fMRI studies have found that different meridians of acupuncture points can specifically activate the cerebral cortex-specific functional area. The meridian (point)–organ–brain-specific link exists between the systems, representing the means by which acupuncture results in functional regulation of the viscera. The results show that manipulation of points to the right of relevant acupuncture points is associated with the bilateral hemispheres of the temporal lobe; frontal lobe changes; activation of the right superior temporal gyrus, middle temporal gyrus, left middle temporal gyrus, bilateral frontal back, inferior frontal gyrus, cingulate gyrus, and the anterior lobe of the cerebellum; and no significant activation of other brain regions. In addition, the ventrolateral medulla opioid receptor antagonist naloxone appears to decrease the acupuncture effect of the Neiguan (PC-6) acupoint to improve the myocardial ischemia, suggesting that the ventrolateral medulla is relevant for acupuncture-related effects. Acupuncture in the Shenmen (HT-7) acupoint in patients with senile dementia resulted in changes in the functional activities of the contralateral cerebral cortex, as well as the specific performance of the left middle frontal gyrus, superior temporal gyrus, transverse gyrus, and the central front and rear back, where the hypothalamus and shell nuclei are located. In addition, the right middle frontal gyrus and superior temporal gyrus were activated. Additionally, we found that electroacupuncture on Neiguan (PC-6) stimulates frontal lobe activation, while electroacupuncture on Shenmen (HT-7) activates the temporal lobe. The left Waiguan (TE-5) point may induce activity of the cingulate gyrus, dorsal thalamus and the temporal lobe, cerebellum, and occipital lobe and other activation. In Hegu (LI-4) acupuncture, among persons who reported having a needle feeling, there was decreased NMR signal activity in the amygdala, hippocampus, parahippocampus, hypothalamus, ventrolateral areas, anterior cingulate cortex, caudate nucleus, putamen, temporal pole, insula, and other regions; there were also decreased somatosensory NMR signals. Acupuncture activated the bilateral central gyrus, the cingulate gyrus, the left insula, the supplementary motor area, the bilateral temporal lobe, the bilateral occipital lobe, the bilateral frontal lobe, the bilateral cerebellar hemispheres, the bilateral parietal lobe, the right medial frontal gyrus, the left caudate head, and the left putamen. Acupuncture points in the example also caused activation of the following brain areas to varying degrees: the right side of the central gyrus, the cingulate gyrus, the bilateral supplementary motor area, the bilateral temporal area, the bilateral occipital area, the bilateral frontal area, the bilateral cerebellar hemispheres, the bilateral top leaves, and the bilateral putamen. The EA Quchi (LI-11) acupoint on the right activated the ipsilateral primary motor areas, supplementary motor area, bilateral primary somatosensory area, and the secondary somatosensory area of bilateral superior parietal lobule, inferior parietal lobule, contralateral cingulate gyrus, both sides of the precuneus, bilateral middle temporal gyrus, superior temporal gyrus, substantia nigra at the bottom, the midbrain, hippocampus, hypothalamus, and caudate nucleus (Yan et al. 2005).

Acupuncture on Zusanli (ST-36) can activate the frontal lobe, temporal lobe, thalamus, and paraventricular nucleus. Central Zusanli (ST-36) acupuncture can also activate the posterior and right inferior frontal gyrus and inhibits the bilateral parahippocampal gyrus, hippocampus, cingulate gyrus, precuneus, and cerebellum. It can also lead to increased glucose metabolism and increased cerebral blood flow to the hypothalamus, paraventricular nucleus, and the ipsilateral and bilateral temporal lobe (Zhang et al. 2007). Acupuncture points in the Taichong (LR-3) area can activate the limbic system and subcortical gray matter structures, the cerebellum, and other brain regions. fMRI studies of brain activity after Taichong (LR-3) acupuncture have been found to correlate with the differential activation of the cerebellum, prefrontal cortex, superior parietal lobule, inferior parietal lobe, occipital lobe, parahippocampal gyrus, insula, thalamus, putamen, and contralateral temporal pole and the posterior cingulate. Acupuncture at the Yanglingquan (GB-34) point did not result in increased cortical signal area; however, there was reduced signal area in the bilateral parahippocampal gyrus, hippocampus, cingulate gyrus, precuneus, and cerebellum. Acupuncture at the Sanyinjiao (SP-6) to varying degrees altered the activation of the central gyrus, right inferior frontal gyrus, left inferior frontal gyrus, left insula, left inferior parietal lobe, left cerebellum top, left middle temporal gyrus, and left middle frontal gyrus and inhibited the bilateral parahippocampal gyrus, hippocampus, cingulate gyrus, precuneus, and cerebellum. Acupuncture at the Shenmai (GB-62) point caused activation of the limbic system (thalamus, cingulate gyrus, parahippocampal gyrus) and frontal and temporal lobe. Baihui (GV-20) acupuncture activated the left and right side of the superior temporal gyrus, right precentral gyrus, postcentral gyrus, supramarginal gyrus, middle temporal gyrus, dorsal thalamus, and caudate nucleus.

Therefore, the effects of acupuncture are a compilation of the interaction of multiple brain areas, thus, the effect of acupuncture may not lie in a single brain area but may be linked by a number of functional brain areas formed by a complex liquid. The relationship between the brain and acupuncture is still an area of literature that is still worth exploring more deeply.

19.2.6.5 Research on the Relative Specificity of Correlation Between Meridians and Viscera

Much research has been carried out with respect to the relative specificity of the correlation between meridians and viscera and has focused on the heart meridian and the heart, the pericardial meridian and the heart, the stomach meridian and the stomach, the bladder meridian and the bladder, and so on. Skin electrical conductance of meridian measurements and acoustic technology have been used to detect the lung, large intestine, pericardium, triple burner, spleen, stomach, heart, and intestine through the liver and gallbladder meridian acupoints of rabbits (Cao et al. 2003; Wang et al. 2003; Wu et al. 2003; Xu et al. 2003; Zhen et al. 2003). The points with large, high amplitude sound waves were identified as meridian points. It has been found that the skin conductivity along the meridian is higher than that of skin adjacent to control points, and the sound wave amplitude of points along the meridian is higher than that of control points near the meridian.

There are relative specificity correlations between the heart meridian and the heart, which occur primarily through the body-sympathetic reflex and the closer sympathetic relationship (Rong et al. 2005). Acupuncture heart meridian can increase the discharge frequency of the left cardiac sympathetic nerve of rabbits, thus raising the threshold for symptoms reflecting cardiac sympathetic nerve activity, which is just superior to the lung meridian. Acupuncture points on the heart meridian with aconitine-induced arrhythmia in rabbits are effective, while acupuncture points on the lung meridian effect are not. After most of the cardiac sympathetic nerve is cut off, the effect of acupuncture disappears. Acupuncture at the heart meridian may regulate sympathetic nerve activity, which could then regulate the functioning of other organ. The effect of acupuncture disappears if the sympathetic nerve is cut off. On the other hand, the organs of patients with a pathological heart response were seen in the pericardium meridian and accounted for 38.10%, heart meridian 28.57%, and the two together account for 66.67%. The organs responsible for pulmonary pathological response in patients can be accounted for by the lung meridian in 64.28%, indicating the close relationship between organ pathology and pathology of the meridian. After acupuncture of the heart meridian, there is evidence that angina is relieved, left ventricular functioning is improved, levels of 5-HT are NE significantly reduced, the TXB_2 content and T/P ratio are significantly reduced, ET is significantly reduced, and CGRP and NO are significantly increased. Acupuncture of the lung and heart meridian can raise NO levels and decrease ET levels, although plasma CGRP levels after acupuncture of the heart meridian are significantly decreased, while acupuncture of the lung meridians has no significant effect on this parameter. The difference between the two groups was significant ($P<0.001$). The effect of acupuncture on the heart meridian and the lung is relatively specific, and the prompt treatment of coronary heart disease by acupuncture of the heart meridian is regulated by sympathetic nerve-related monoamine neurotransmitters and peptides that are related to thromboxane levels (Fang et al. 2000; Zhang et al. 2002). Zhou YP has observed in rats that the "Heart meridian" of labeled cells in the distribution of the peak occurs mainly at T_3, while the "lung meridian" distribution of the peak is at C_{6-7}, the "Heart meridian" and the heart of double-labeled cells occur in the peak T_1–$_4$, and the "Heart meridian" and "lung meridian" in the distribution of double-labeled cells occur at peak C_7. It has been shown that different meridians are different to varying degrees. The two institutions have found some new results using electrophysiology, radioimmunoassays, and immunohistochemistry. They found that thoracic spinal ganglion cells of SP and small CGRP, glutamyl CGRP in the spinal cord, and the SP of rat and rabbit spinal cord opioid receptors, acid NMDA receptor, and non-NMDA receptors participate in EA pericardium "Ximen (PC-4)" and adjust the ischemic heart of the functional activities of the process. In addition, with gene chip technology in acupuncture of the heart and lung, as well as the small intestine meridian for a number of conditions, including acute myocardial ischemia in cardiac tissue, researchers have found that genes are differentially expressed in the hypothalamus, and the study found that in the myocardium, genes were differentially expressed when compared to the group who received acupuncture of the small intestine meridian. Both groups showed up- and downregulation of genes, although the nature of the expression was different for the two groups. The group who received acupuncture of the heart meridian had significantly more anaerobic metabolism, which was related to gene upregulation (in particular, lipid catabolism-related genes went up the most) and downregulation of aerobic metabolism genes, which facilitated the improvement in myocardial cell's hypoxic tolerance, significantly boosted immune-related gene expression, and significantly promoted apoptosis and stress response genes associated with a reduction in protein expression, ion channel, and transport protein-related genes down-expression; thus, gap junction membrane channel protein expression was significantly reduced, which may be related to antimyocardial ischemia. The group that received acupuncture to the small intestine had no such gene expression (Wu et al. 2006).

The differential expression of genes in the acupuncture lung meridian group was significantly less than that of the heart meridian group, which showed 20 downregulated genes and 14 upregulated genes. Some genes were similar to that seen in the acupuncture heart meridian group, although the expression signal was different, indicating that the mechanism of acupuncture of the lung meridian and heart meridian was different. The ion channels and protein transport genes of acupuncture of the heart meridian were superior to those of the acupuncture lung meridian. The group receiving acupuncture to the heart meridian had an improved cytoskeleton, cell differentiation and combination, transcription and DNA transcription by gene expression, and subclass apoptosis and protein expression of genes in response to stress, while the group receiving acupuncture to the lung meridian did not have this result. There were significant differences between the two groups, and the difference may have been closely related to the function of myocardial cells. The effect of acupuncture of the heart meridian and lung meridian has relative specificity.

In the hypothalamus of patients receiving acupuncture at the heart meridian, there were 224 differentially expressed genes, while in the small intestine acupuncture group, there were only 34. The difference was expressed not only in terms of number but in terms of differences in gene function. The number of differentially expressed genes related to the autonomic nerve activity of the acupuncture heart meridian group was significantly higher than in the acupuncture lung meridian group, indicating the differential expression of myocardial ischemia in the hypothalamus among patients with heart meridian acupuncture versus lung meridian acupuncture.

There is also relative specificity between the heart meridian and the pericardium meridian. The research of Liu Jun-ling and other scientists showed that, using electrophysiological methods, electrical stimulation of the left ventricular cat pericardium "Tianquan (PC-2)" point district via EMG and afferent nerve discharges in C_8–T_1 is spread mainly through C_8. EA Neiguan (PC-6) of cat and the afferent nerve of the left ventricular nerve may converge to the thoracic spinal cord dorsal horn of the same neurons. Using tracer tracking technology, we found afferents of the Daling (PC-7), Quze (PC-3), and Tianquan (PC-2) point areas of the pericardium meridian in rats, and the Yangchi (TE-4), Tianjin (TE-10), and Naohui (TE-13) points of triple energizer meridian marked the dorsal root ganglion cells in C_5–$_8$ with some differences in the distribution at the same segment of the dorsal root ganglia, from the pericardium by the mark more than triple burner. The central synapses of labeled cells from the pericardium meridian are found in the C_3–$_8$ section, mainly C_3–$_5$. In the C_6–$_7$ section of ethanol injection, thoracic segments of the T_1–$_5$ horn, and central tube around the side, there were many marks. This data demonstrates that the afferent from the pericardium meridian through the spinal cord can go to these areas via sympathetic preganglionic neurons and their connection with the heart.

Using hybridization and immunohistochemical techniques in rats, it has been shown that the anterior descending coronary artery can cause myocardial ischemia via C_8-spinal-cord injection of ethanol production at the I, II, III, and horn-side c-fos neurons. These results provide evidence for the relatedness of the meridian and the viscera via the spinal cord, adding additional data to the body of work in this area.

The stomach meridian has relative specificity for the stomach. In a study administering acupuncture at 12 sites, 10 of which were stomach sites along the stomach meridian [i.e., Sibai (ST-2), Touwei (ST-8), Tianshu (ST-25), Biguan (ST-31), Futu (ST-32), Liangqiu (ST-34), Zusanli (ST-36), Shangjuxu (ST-37), Chongyang (ST-42), Neiting (ST-44)]; 1 at 1 cm below Sibai (ST-2), 1 at 2 cm below the Zusanli (ST-36) points, and about the same point in paragraph 1–2 cm-open next control point were randomly divided into 6 groups of 15 people, after the acupuncture was twisting gas (per hole 3 min), no mention of interpolation; accumulation area element method using computer programming (per 2′ sampling time) was analyzed. The results showed that the antral area increased at the eight point after acupuncture that is, Sibai (ST-2)>Zusanli (ST-36)>Neiting (ST-44)>Liangqiu (ST-34)>Chongyang (ST-42)>Biguan (ST-31). The area surrounding the hole near the open area of the control point had little effect on the gastric antrum. In patients with gastrointestinal disease, gastric pyloric sphincters with low contraction amplitude, and slow contraction frequency, the Yangming meridian could promote gastric motility and increase contraction strength and frequency, thus promoting gastric emptying.

Yan Jie (Yan et al. 1983) observed that acupuncture at the Sibai (ST-2), Liangmen (ST-21), and Zusanli (ST-36) points could significantly reduce the experimental gastric mucosa damage index as well as improve MTL content and SP in the gastric antrum. It could also decrease the concentrations of GAS and increase concentrations of PGE_2, EGF, SS, and serum NO. However, manipulation of the control point did not result in changes to these parameters (Li et al. 2004). Acupuncture on Sibai (ST-2) and Zusanli (ST-36) of the stomach meridian can improve the EGG frequency, and frequency power significantly increased the normal rhythm band. At the same time, the ratio of frequency to energy was also significantly increased. Liruiwu reports that after Zusanli (ST-36) acupuncture, they observed decreases in the electrical activity of the stomach as well as decreases in the gastric fundal activity of nitric oxide synthase (NOS). Yan et al. (2007a, b, c) also reported that acupuncture in the stomach meridian in healthy patients and those with functional dyspepsia can reduce the relative spectral power of the stomach, decrease the total power of gastric impedance, increase GAS, and reduce elevated plasma SS. However, acupuncture control points adjacent to "Zusanli (ST-36)" did not have the same effect.

19.3 Barriers to Research on the Meridian–Viscera Correlationship

The current research focus in meridian theory has been on the phenomenon of sense conduction along the meridian and the mechanism by which it does this, the biophysical and morphological basis for the meridian, the use of radionuclide studies of the meridians, the use of low flow resistance channels, the ions of the importance and volatility of the meridian, calcium stores in the meridian, connective tissue conduction, the infrared detection, and study of the organs and meridians systems theory and the related laws. There has been much debate among meridian scholars concerning function versus structure and whole versus local. Some researchers have advocated using neurophysiology to understand the meridian, others have advocated the doctrine of the physiological fluid cycle, and others have promoted the biological field theory of biological physics, while others have advocated the doctrine of the overall structure of the connective tissue gap. However, scholars have, as of yet, been unable to form a comprehensive theory of the meridian, thus enabling the public to accept a uniform interpretation of the meridian.

We believe that these views reflect the activity of academic thoughts both at home and abroad and also reflect the policies of "competition and flourishing." However, it is obvious that the absence of thought and study on the meridian theory of holism has made it difficult to obtain satisfactory progress. In summary, "in meridian research, there has been progress but no breakthrough."

In China, the study of the meridian began in the 1950s. Under the influence of studies by Nagahama (Japan) and FengHan Kim (South Korea) on the material basis of 12 meridians and the impact of Western medicine, the requirements of long-term dominance of the ruling and the launching of the "meridians followed by the material basis" and other phenomena of the real exploration of the meridian. Just as mentioned above, we have attempted to explain the meridian using the singular perspectives of sound, light, magnetic fields, electricity, physics, electrochemistry, collagen fibers, the musculoskeletal system, and genetics, while others have interpreted the meridian as a "part of the pending development" of modern physiology and have used the perspective of physics to explore long-term clinical practice using humanistic philosophies, the concept of nature and man, and the yin and yang balance of the holism. However, it has been difficult to obtain adequate data to support the theory, thus making it impossible to make significant progress and breakthroughs.

Second, limited by domestic research and policy, there has been a lack of long-term planning in meridian research. Much of the work on the "meridians substance and material basis" has followed that of South Korea, Japan, and France; thus, there is no specific Chinese vantage point in the meridian research literature. In the 1970s, scholars in Anhui proposed that the study of the meridian should begin with the functioning of the meridian and that the "meridian–viscera correlationship" should be the main focus of research. In recent studies, there have been only three teams that have worked on "the meridian–viscera correlationship," including the group in Anhui (heart meridian–heart correlation), Beijing (pericardium meridian–heart correlation), and Hunan (stomach meridian–stomach correlationship); due to the lack of funds and long-term policy support, the progress in getting ample data has been slow.

Third, the meridian theory is holistic, and the meridian is systematic. However, domestic research institutions at all levels demonstrate an apparent lack of research on the holism that drives meridian theory and often lacks an organizing philosophy. Thus, we have not been able to organize a real "multidisciplinary, multiangle" meridian system research team or to obtain national financial support for our work, which has created a large barrier for the production and dissemination of meridian research.

19.4 Future Prospects and the Great Significance of Meridian Theory

Traditional Chinese medicine and ancient Chinese acupuncture theory are medical theories that observe basic human physiological and pathological mechanisms and use these observations to understand and treat human diseases. They also represent the integration of ancient Chinese medical philosophy and practice, the origin of which is centuries old. In recent years, modern biotechnology has pushed modern biomedical informatics into the genomic and postgenomic era. The use of computer technology in biology has greatly expanded the depth and breadth of biomedical research. A series of mathematical models and analysis methods have allowed for the modern biomedical analysis of both local effects and systemic integration, thus improving the practical application and reproducibility of clinical research. These new methods from the bioinformatics could help to support the theories underlying acupuncture and may also introduce new data analysis based on mathematical methods, thus expanding the current body of clinical research and improving the efficiency of research and data integration.

In recent years, many studies have shown that acupuncture acupoints may result in corresponding brain response and activity. Different acupuncture points could affect different areas of brain function and may reflect the presence of a specific correspondence between acupoints and the brain. Research has found that the Taichong (LR-3) or Hegu (LI-4) acupoints may activate the temporal gyrus and the cerebellum. In addition, acupuncture on Hegu (LI 4) may activate the central post back, posterior cingulate gyrus, the hippocampus, and Brodmann's areas (BA) 7, 19, and 41, while acupuncture on Taichong (LR-3) points may activate the temporal cortex (Yan B. et al. 2005). The change in brain function when acupuncture is conducted on the Sanyinjiao (SP-6), Zusanli (ST-36), and Yanglingquan (GB-34) points has been observed in the same areas that are activated in Zusanli (ST-36) and Sanyinjiao (SP-6) acupuncture; these include the posterior central gyrus and the right inferior frontal gyrus. The different activation areas included the left inferior frontal gyrus, left insula, left inferior parietal lobule, left middle temporal gyrus, and left middle frontal gyrus, while the acupuncture on Yanglingquan (GB-34) point did not increase the cortical signal area (Fang et al. 2006). A study of the activation of brain regions by the acupuncture of the Five Shu points of the liver meridian and lung meridian resulted in a specific findings concerning the brain activity distribution associated with these acupoints. The acupuncture of the Five Shu points of the liver meridian tended to activate the following areas: Ipsilateral BA 34, BA 47, red nucleus, contralateral BA 19, BA 30, BA 39, superior parietal lobule, cerebellar slopes on both sides of BA 3, and the cerebellar roof, while acupuncture of the Five Shu points of the lung meridian activated the following brain areas: the same side of BA 2, BA 18, BA 35, contralateral BA 19, and the substantia nigra (Xu et al. 2007). A particular cortical functional areas can be activated when a stimulus is applied to a specific single point, e.g., the use of laser stimulation on the left Zhiyin (BL 67) can activate the ipsilateral medial occipital visual cortex and the central wedge leaf (Siedentopf et al. 2002). The bilateral frontal cortex and left superior temporal gyrus were activated on electrical stimulation of the language-related Sanjangluo (TE-8) and Yamen (GV-15) (Li et al. 2003). Using manganese-enhanced fMRI technology, animal studies have also observed that acupuncture on Zusanli (ST-36) caused hippocampal excitation in rabbits, while the acupuncture on Yanglingquan (GB-34) point activates the hypothalamus, insula, and motor cortex (Chiu et al. 2001). The above studies show the relationship between acupoints and various brain areas, thus the effect of acupuncture may depend on (Jin et al. 2003) the excitement caused in different brain regions and the extent of excitation seen in various brain areas.

fMRI imaging of the acupuncture on Zhiyin (BL-67) point of the bladder meridian showed activation of the occipital cortex, and Zutonggu (BL-66), Sugu (BL-65), and Kunlun (BL-60) points were similar with respect to this. However, the nonacupuncture point located 2–5 cm adjacent to Zhiyin (BL-67) did not result in specific activation of the occipital cortex (Cho et al. 1998). When the needles were simply inserted (i.e., they were not twisted) into the Taichong (LR-3) acupoint, the Qiuxu (GB-40) acupoint, and a nonpoint, fMRI imaging showed that the Qiuxu (GB-40) and Taichong (LR-3) acupoints resulted in increased somatosensory cortex activation; however, there was no activation from needle stimulation of nonpoints (Fang et al. 2004a, b). In Fang's study (Fang et al. 2004a, b) using acupuncture with twist (flat-up-flat diarrhea) stimulation at the Taichong (LR-3) acupoint, the Qiuxu (GB-40) acupoint, and false points on healthy adults, the author found that only real acupuncture needlepoint twisting resulted in a strong activation effect, while study of twisting and nontwisting at nonpoints did not result in significant activation of these areas. Fang's continuing study (Fang et al. 2005, 2008) found that acupuncture at Taichong (LR-3), Xingjian (LR-2), and Neiting (ST-44) acupoints and at nonacupoints using fMRI showed that the activation of nonacupoints was smallest and weakest and the most prominent inhibition occurred in the nonacupoint distribution. The differences in Taichong (LR-3) and nonmeridian points in the same plane resulted in the largest differences.

Stimulation using different methods and different stimulation parameters acting on acupoints results in different changes in brain function. For example, the fMRI response of the human brain to acupuncture points at different frequencies of electrical stimulation found that 2- or 100 Hz electric nerve stimulation acting on the left Zusanli (ST-36) and Sanyinjiao (SP-6) can activate the primary and secondary somatosensory area. However, low-frequency electric nerve stimulation selectively activated the ventral hypothalamus, thalamus, and entorhinal bottom leaves, and high-frequency stimulation caused the dorsal thalamus and cingulate activation. This result suggests that the same points of electrical stimulation at different frequencies cause different reactions in the brain and different neural pathways of activation (Jin Z. et al. 2001). Cho et al. (1998) found that fMRI brain imaging of manual acupuncture at Zhiyin (BL 67) and the results of laser stimulation at Zhiyin (BL-67), which were mentioned earlier, are different. Both studies found that BA18 and 19, areas of the visual cortex, could be activated. However, the laser needle stimulation failed to find primary visual cortex activation or activation of the ipsilateral brain regions, while the hand acupuncture stimulation activated bilateral brain regions. A study using stimulation of hand acupuncture and electroacupuncture on the left side of the Hegu (LI-4) observed that fMRI imaging of electroacupuncture stimulation activated the precentral gyrus, postcentral gyrus, and insula, while hand acupuncture reduced the activity in the posterior cingulate gyrus, temporal gyrus, and insula (Kong et al. 2002). Using 2–100 Hz electroacupuncture and hand acupuncture on Zusanli (ST-36), fMRI demonstrated that the signal enhancement of low-frequency electroacupuncture stimulation had a broader range than that of hand stimulation. The pontine raphe area signals with 2 Hz electroacupuncture stimulation changed significantly (Napadow et al. 2005). These differences may have been caused by differing stimulations with different patterns of neural excitability. Yoo et al. (2004) observed healthy people receiving acupuncture on Neiguan (PC-6), acupuncture adjacent to Neiguan (PC-6), and stimulation of the surface of the skin; fMRI imaging data showed that Neiguan (PC-6) activation of the left frontal gyrus, anterior cingulate back, and dorsal thalamic nucleus was different between the acupuncture group and the other two groups. Acupuncture stimulation and nontoxic general tactile stimulation were administered on the Hegu (LI-4) as was fake acupuncture on the adjacent of Hegu (LI-4). The fMRI imaging observed that the real points played a role in regulating neural responses regardless of the common deep tissue stimulation or sensory stimulation that occurred (Yoo et al. 2007). A study using deep and shallow acupuncture of the right Hegu (LI-4) of 17 right-handed healthy subjects found that brain activation and inhibition of deep acupuncture and shallow acupuncture did not result in significantly different fMRI activation (Mac Pherson et al. 2008).

Fu's (Fu et al. 2005) study of Neiguan (PC-6) electroacupuncture among elderly healthy people and patients with Alzheimer disease (AD group) found that the bilateral frontal and temporal lobes were activated to varying degrees in normal patients, while the AD group showed differential levels of frontal lobe, temporal lobe, cingulate gyrus, cerebellum, and other brain area activation; however, there was no significant activation of other brain areas. The study confirmed that acupuncture impacts brain function, thus providing a theoretical basis for the bidirectional regulation of acupuncture.

Based on the numerous fMRI data concerning the effect of acupuncture, we can see that the results can be affected by many factors and are directly related to the different acupuncture conditions, including needles, acupuncture operating practices, stimulus, etc. The status of the body, the physical condition, and the environment, as well as disturbances in vision, hearing, and the psyche, are also important. To be effective, the data must be rigorous and the research design must be good, i.e., it must control for confounding factors, have uniform experimental conditions, and must be reproducible.

At the same time, fMRI can be combined with ERP or PET–CT technology to bolster the weaknesses of ERP or PET–CT, thus allowing for the study of brain metabolism and brain cell function as well as the observation of a range of activities. In addition, we can also utilize a wide range of techniques, including molecular biology, metabolomics, computer technology, biological engineering, and technologies from other disciplines of knowledge, as a comprehensive means by which to study the meridian effect using clinical acupoints along meridians, acupoint syndromes, assessment of the time from needle insertion to "de qi" response, and needle retention time range issues, to further standardize the clinical acupoints acupuncture program, thus laying the foundation for the development of standardized treatment protocols.

Modern biological methods, such as molecular biology, have resulted in several advances in our understanding of the microbiology of the system. However, there is also a need to understand an organism as a whole, thus researchers must also take on a systems perspective. This macroresearch on life processes is an important "return to trend." The development of meridian research coincides with this regression to systems-based, integrative study. This is the great advantage of Chinese medicine, as it is integrative in nature, and the meridian theory as a subset of theories within TCM only increases its utility to be integrative.

Modern meridian researchers should answer the following three questions under modern biology: First, is the meridian an objective phenomenon of life? Second, what is Substance of the meridian in the modern scientific sense? Third, how do we achieve modernization of the meridian theory with respect to clinical diagnosis and treatment? The essence of meridian theory is that the upper and lower body are connected to each other through a dynamic system of interconnections and that the meridians are actually an ancient but intuitive guess of how these interconnections are mapped onto the body. Thus, the current research on meridians explores the ancient meridian theory in the direction of the body's up and down links within and outside the scientific value of the law and its relationship with the modern life sciences.

Meridian research is promising, as it will give further insight into the functioning of the meridian system. The meridian is a complex system whereby the reaction of the system is constantly adjusted to reflect the state of the blood. On the basis of the world's first theory concerning the relationship between the body and its organs and the meridians and the viscera and after 30 years of scientific exploration, we have proposed that the relationship between meridians and organs is the foundation of the meridian theory (Zhou 1999; Zhou et al. 2008). We propose that five key directions be made in research in acupuncture in the future. They are as follows: a longitudinal study of one meridian with several viscera, a horizontal study of one viscus with several meridians, the study of heart and the small intestine as a paired relationship, the study of the lung and the large intestine as a paired relationship, the study of the back-shu point of the bladder meridian and its corresponding organs, and a study on the corelationship between meridians and viscera with the brain.

The following interconnections are a part of the complex network of connections that make up the meridian system: one meridian with several viscera, one viscus with several meridian, visceral and fu-organ, fu-organ and fu-organ, viscera and viscera, inner and outer, outer and inner, upper and lower, and lower and upper. To resolve these complex links, researchers are beginning to use modern technologies and methods from systems biology and are able to study the overall functioning of neurons, thus, these techniques enable research to clarify the basic meridian's complex network of contacts among relevant organs. The further study of the mechanism underlying the theory of the correlationship between meridians and viscera, e.g., "heart and small intestine," "lung and large intestine," and "the bladder meridian is the core of twelve meridians," will play an important role in the innovation and further development of the treatment of internal organ diseases in Chinese medicine.

In a sense, the function of a meridian may be a comprehensive reflection of nervous–endocrine–immune (NEI) networks in some respects. Therefore, the study of acupuncture from a neuroscience perspective may provide additional insights and an exciting direction for research in the near future. In the twenty-first century, neuroscience will likely continue to be an important perspective by which to explore acupuncture. The structure and function of the human brain, which has the most complex organization of all of the organ systems, are both complex and precise with respect to its ability to process and integrate information.

Since entering the twenty-first century, there have been several advances in the techniques used in molecular biology. Acupuncture research has become an increasingly popular area of research internationally. Because the mechanism of acupuncture at various meridian acupoints related to neurological, endocrine, immune, and other systems involves intracellular messengers, neurotransmitters, neuromodulators, neuropeptides, cytokines, hormones, and other endocrine factors, acupuncture at various meridian points may have different efficacy. Thus, research in acupuncture and meridian is still difficult to develop in depth. Over the past 10 years, there have been many new imaging technologies, such as positron emission tomography technique (PET), functional magnetic resonance imaging technique (fMRI), nuclear magnetic resonance spectroscopy technique (MRS), and single photon emission computerized scanning technique (SPECT) techniques, that have made the study of acupuncture considerably easier. They provide a noninvasive means by which to analyze chemical changes within the nervous system and neural activity and offer insights into the overall function of the brain.

The application of EEG, event-related potentials (ERPs), and other technologies related to the central meridian organs and the dynamic functions of the brain is being used more often to understand the effect of acupuncture. ERP is a special kind of brain evoked potential that is due to multiple potentials from a variety of brain stimuli; they are a record of the total brain surface potential and are recorded from the skull. They reflect the cognitive processes in the brain, as evidenced by the neural electrophysiological changes that occur in the brain, also known as cognitive potential. ERP traceability analysis techniques, which have the characteristics of high time resolution, can map the dynamic process of awareness. Through the establishment of an ERP workstation, researchers can study how the brain extracts information concerning the integration of the dynamic changes that occur during acupuncture.

Cerebral microdialysis is the most common method of microdialysis (Zhang et al. 2004). It is currently available for the cortex, hippocampus, and certain areas of the brain stem, periaqueductal gray matter, and even some nuclei. Using in vivo microdialysis probes in animals, researchers were able to study brain function in different links. Microdialysis techniques in the brain and spinal cord are becoming an important and popular research tool for understanding the neurochemical basis of acupuncture. The combination of microdialysis techniques and electrical stimulation may also be used to study the functional connections between brain regions. Recently, improvements in both the microdialysis probe and the silver chloride electrode have resulted in the ability to simultaneously measure extracellular currents and neurotransmitter release in same brain regions. Currently, this technology has been used to study the neurochemistry of general anesthesia; more specifically, it has been used to determine the levels of particular neurotransmitter and neuropeptides. It is also starting to be used in research on the brain and spinal cord mechanisms of pain. In addition, there are also reports of the clinical use of the technology to monitor cerebral metabolism.

There has been some data exploring the use of quantum dot nanometer technology in research on the structure of the neural connections of meridian organs and the brain. The composition, distribution, and interaction of cells have been investigated using a microperspective in vivo, in situ, and in real time to explore cellular and biological molecules, with the aim of obtaining early diagnoses for serious diseases and clarifying the pathogenesis of important disease processes. In the study of gene and protein expression, biological imaging technologies using fluorescent probes provide an important means by which to access and use biochemical information. However, the optical stability of organic fluorescent probes is poor, and optical degradation can occur quite easily under the excitatory action of light. Thus, it would be a poor marker for observing the dynamics of molecular markers; it is also likely to cause overlapping of wide spectrum emission spectroscopy with the simultaneous detection of several fluorescent-labeled biomolecules being difficult. Quantum dots have been used as biological fluorescent probes due to their excellent optical properties, and their good applicability in biological imaging has increased its use in the biomedical field (Zhao et al. 2008; Yang H. et al. 2007). The use of advanced micropositioning technology can be quite useful in studying the course of acupuncture meridians and can also be used to observe the gene of interest (or protein) that underlies the changes caused by acupuncture, thus providing additional experimental support for acupuncture.

The aim of acupuncture and Chinese or Western medicine is to achieve a new dynamic equilibrium. Chinese medicine emphasizes "harmony" and the tenet that "Yin ping yang mi" is the basis of health. This concept is reflected in the microenvironment within the steady state in the last. Acupuncture treatment aims to use the acupuncture points (points, lines) of the meridians as a benign stimulation to identify abnormal signals from the body using two-way adjustments to adjust the negative feedback of information to the limbic–hypothalamic central nervous system, thus integrating all of the body's messages to a corresponding target organ. In so doing, acupuncture stabilizes the environment and coordinates the yin and yang. The mechanism underlying acupuncture can be explained by its stress effect. It yields the largest body response effect for the smallest amount of energy/effort expended, thus regulating the body by using the entire response system, including the nervous, endocrine, circulatory, respiratory, and digestive system; it is also an efficacious treatment for pain (Shen et al. 2008).

Numerous studies have indicated that the autonomic nervous system, particularly the sympathetic nervous system, plays an important role in the relationship between meridians and organs (Ma et al. 2000). In the clinical and basic science acupuncture research conducted in the 1970s, the integrity and balance of sympathetic autonomous nervous system function were thought to be the key to acupuncture-induced anesthesia. By studying the acupuncture meridians and acupuncture points and their relationship to the form and function of organs, the integrity of the sympathetic nervous and body sympathetic reflex is thought to be the primary unit of function and structure. Sympathetic nerve fibers are widely distributed in blood vessels, lymphatics and skin sweat glands, sebaceous glands, erector pili muscle, and other tissues. When the needle is inserted into a point or "A shi" points, it can stimulate the sympathetic system and the sympathetic innervations in skin sections. When traced to its source, the spinal nerve and spinal nerve fibers that are related to the bladder meridian are composed of nerves of the sympathetic chain, which play an important role in the acupuncture response. The theory that the bladder meridian has at its core 12 meridians cannot be ignored; the theory that human organ regulation may be based on the meridian regulation of autonomic function should be carefully studied.

19.5 Concluding Remarks

In summary, the limbic system and autonomic nervous system of hypothalamus play leading roles for the treatment of diseases, for the steady state and stress in the body, and in neuroendocrine-immune network. Therefore, we thought of that to study the complex channels of meridians and organs from the marginal brain system to the autonomic nervous system of the hypothalamus (Fig. 19.1).

On the way of exploring brain–mind relations, we are faced with revolutionary opportunities and challenges. Research method of life science is experiencing change from genomics to proteomics and to the nerve connections. With the postgenome era coming, life sciences had been greatly changed and expressed the dialectical relationship between the diversity and consistency of life. The development of brain and cognitive sciences could promote our understanding of nature and human being, and innovation, psychological and spiritual health, information science and artificial intelligence, culture, social science, education, and linguistics also will be surely affected.

As to acupuncture, one task in the twenty-first century is to improve the level throughout the world and achieve acupuncture internationalization and modernization. Generally, four tasks should be concerned first: to promote acupuncture to be one part of world medicine, so that all humans in the world could accept acupuncture treatment; to improve meridian diagnosis and treatment techniques of acupuncture; to improve clinical effect, in order to solve more difficult cases; and to enhance theory study as meridian theory to be the key study. In the twenty-first century, we should give more attention on overcoming the bias that theory from clinical practice gives emphasis on the guide of theoretical studies to clinical practice and give more attentions on the acupuncture stimulation parameters and signal regulations, meridian relations and meridian–viscera correlativity, and objective and material study of meridian as well. Also, we advocate application of different methods in studying in order to improve credibility. With the development of research on acupuncture theory, qi and blood theory should be given more attention, and more experimental study should go deep into relations between meridian and nerve regulation

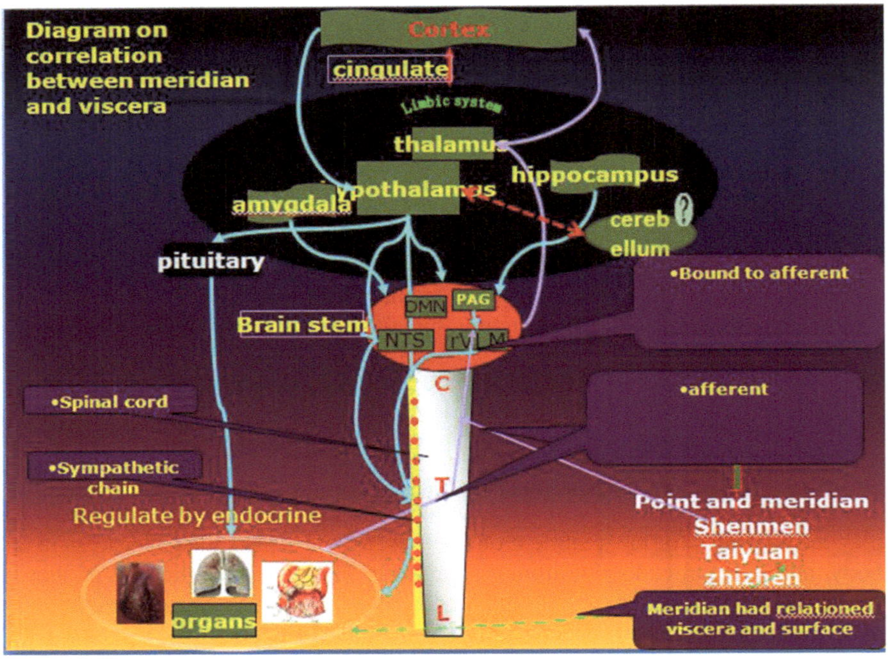

Fig. 19.1 Meridian–viscera correlativity schematic diagram

and immune regulation, in order to construct new frame of acupuncture theory, which will indicate the birth of modern acupuncture and laying the foundation for the modernization of the basic theory of traditional Chinese medicine and the promotion of modern biological concepts.

The theory that the "correlation between meridians and viscera is the core of the meridian" was put forward to change the way in which long-term study of finding the meridians passes through the material basis for a sense of direction became the basis for acupuncture research direction. Research on the relative specificity between the heart meridian and the lung meridian has provided new research approaches for the study of correlation between meridians and viscera.

Given the specific challenges in the current design of scientific research in this area, future work should also address the following questions:

1 Is the relationship between meridian acupoints and viscera "one to one" or "one to several?"
2 Is the relationship of links between meridian acupoints and viscera a point-to-point contact, a whole meridian contact, or a particular region of the contact sheet?
3 Is the relationship between the viscera and organs "one to one," "one to several," or "exterior and interior?"
4. Is the relationship between the meridians exterior and interior, or do they occur simultaneously or via a streamer contact?

Acknowledgements This work was partially supported by NSFC (30801484, 81072873), National Basic Research Program of China (973 Program) (2005CB523306, 2005CB0523308), and Anhui Natural Scientific Foundation (090413112).

References

Cao R, Wu CJ, Zhang KX, Wang X (2003) Detective and Effect Verification of Commonly Used Acupoints in Lung and Large Intestine Meridian of Rabbit. Liaoning Journal of Traditional Chinese Medicine30: 305–306. (In Chinese)

Chen YG, Cheng ZD, Zhang LD (2006) Special functional protein system of identical resonance structure compose meridian. Liaoning Journal of Traditional Chinese Medicine 33:276–277. (In Chinese with English abstract)

Cheng ZD, Chen YG, Zhang LT (2002) Experimental Study on Effect of Acupuncture on Tropism of Effective Compositions of the Chinese Drug in the Rat. Chinese Acupuncture & Moxibustion 22:51–53. (In Chinese with English abstract)

Chiu JH, Cheng HC, Tai CH, Hsieh JC, Yeh TC, Cheng H, Lin JG, Ho LT (2001) Electroacupuncture-induced neural activation detected by use of manganese-enhanced functional magnetic resonance imaging in rabbits. American journal of veterinary research, 62:178

Cho ZH, Chung SC, Jones JP, Park JB, Park HJ, Lee HJ, Wong EK, Min BI (1998) New findings of the correlation between acupoints and corresponding brain cortices using functional MRI. Proc Natl Acad Sci USA, 95:2670

Dong Z (1992) Discussion on Body Surface and Viscera Correlativity. Beijing: China Science and Technology Publishing House (In Chinese)

Fang J, Jin Z, Wang Y, Li K, Kong J, Nixon EE, Zeng Y, Ren Y, Tong H, Wang Y, Wang P, Hui KK (2008) The salient characteristics of the central effects of acupuncture needling: Limbic-paralimbic-neocortical network modulation. Hum Brain Mapp, 6:1

Fang JL, Jin Z, Wang Y, Li K, Zeng YW, Tong HB, Reng YS, Gao B, Wang YH, Chen MK (2005) Comparison of central effects of acupuncturing Taichong and nearby two acupoints by functional MRI. Chinese Journal of Medical Imaging Technology 21:1332 (In Chinese with English abstract)

Fang JL, Krings T, Weidemann J, Meister IG, Thron A (2004) Functional MRI in healthy subjects during acupuncture: different effects of needle rotation in real and false acupoints. Neuroradiology. 46:359

Fang JL, Timo K, Juergen W, Ingo M, Armin T (2004) Different central effects of needle rotation in false and real acupoints on functional MRI. Chinese Journal of Radiology 12:1281. (In Chinese with English abstract)

Fang SH, Zhang SZ, Liu H, Shou YQ, Zhang JW, Cui HW, Hu XY (2006) Study on Brain Response to Acupuncture by Functional Magnetic Resonance Imaging — Observation on 14 Healthy Subjects. Chinese Journal of Integrated Traditional and Western Medicine 26:965. (In Chinese with English abstract)

Fang ZB, Wang KM, Wang YL, Zhou YP (2000) Effect of Electroacupuncture of the Heart Meridian Acupoints on Ischemic Cardiac Function in the Rabbit. World Journal of Acupuncture-Moxibustion 12:35–38

Fu P, Jia JP, Zhu J, Huang JP (2005) Effects of acupuncture at Neiguan (PC-6) on human brain functional imaging in different functional states. Chinese Acupuncture & Moxibustion 25:784. (In Chinese with English abstract)

Guo Y, Chen SB, Zhang CX, Wang XY, Miao WF, Shi LP, Zhang GZ (2002) Specific Distribution of Ca^{2+} Concentration in Acupoints of Healthy Human Bodies. Shanghai Journal of Acuuncture and Moxibustion 21:37–38.(In Chinese with English abstract)

Guo Y, Hu LM, Zhang YJ, Wang XY, Zhang CX (2004) Effect of Acupuncture of Zusanli, Shangjuxu on Ca^{2+} Concentration in Other Acupoints of Stomach Meridian. Tianjin Journal of Traditional Chinese Medicine 21:476–478. (In Chinese with English abstract)

Guo Y, Liu YM, Zhang YJ, Wang XY, Xu TP(1996) Investigation on the relationship between acupuncture effect and calcium handling. China Journal of Basic Medicine in Traditional Chinese Medicine 2:52–54.(In Chinese)

Guo Y, Zhang YJ, Wang XY, Xu TP, Miao WF, Zhang CX(1998) The Effects of Complexing the Ca^{2+} at Zusanli upon Inhibiting the Motion of Stomach of the Normal Rabbits by Acupuncture at Zusanli. Acupuncture Research 23:146–148. (In Chinese with English abstract)

Jiang L, Chen ZX, Song GM, Liu YL (1980) Initial observation on infrared image of propagated sensation along channel. Journal of Traditional Chinese Medicine,2:46. (In Chinese)

Jiang ZS, Chen EQ (1997) The theory discussion on meridian essence. Journal of Clinic al Acupuncture and Moxibustion 13:15–17. (In Chinese)

Jin SY, Sun SX, Sang XS (2001) Approach to Acupoint-Viscera Correlativity from Bile Flow and Na^{+}, K^{+}, Cl^{-} Contents. Chinese Acupuncture & Moxibustion 21:490–492.(In Chinese with English abstract)

Jin XL, Yin L, Liu ML, Zhang X, Wu YG, Yao B, Zhu K (2003) fMRI pilot study on rat brain function while puncturing the acupoint ST36(Housanli). Chinese Journal of Rehabilitation Medicine 18:715. (In Chinese with English abstract)

Jin Z, Zhang WT, Luo F, Zhang KL, Zhang W, Zeng YW, Han JS (2001) Frequency-specific responses of human brain to peripheral transcutaneous electric nerve stimulation: a functional magnetic resonance imaging study. Acta Physiological Sinica 53:275. (In Chinese with English abstract)

Kong J, Ma L, Gollub RL, Wei J, Yang X, Li D, Weng X, Jia F, Wang C, Li F, Li R, Zhuang D (2002) A pilot study of functional magnetic resonance imaging of the brain during manual and electroacupuncture stimulation of acupuncture point (LI-4 Hegu) in normal subjects reveals differential brain activation between methods. J Altern Complement Med, 8:411

Li DZ, Fu ST, Li XZ (2005) The discussion for meridian-the third of meridian theory and clinical application research. Chinese Acupuncture & Moxibustion 25:53–59. (In Chinese with English abstract)

Li G, Liu HL, Cheung RT, Hung YC, Wong KK, Shen GG, Ma QY, Yang ES (2003) An fMRI study comparing brain activation between word generation and electrical stimulation of language-implicated acupoints. Hum Brain Mapp, 18:233

Li JS, Yan J, Chang XR (2004) A Study on the Specificity of Acupoint via the Observation on the Effects of Electroacupuncture to the Points of Foot-Yangming Meridian on the Gastric Motion Function of Rabbits. Journal of Hunan University of Traditional Chinese Medicine 24:43–5. (In Chinese with English abstract)

Li Y (1999) Information theory of meridian-a brief talk on meridian essence which radiating sensations of the meridians. Journal of Clinical Acupuncture and Moxibustion 15:1–2. (In Chinese)

Li ZP, Liu SW (2009) Clinical Central Nervous Anatomy(The Second Edition). Beijing: Sciences Publishing House. (In Chinese)

Lin WZ, Guo HY (1995a) The hypothesis of medulla and brain stem neural network for meridian transmission. Journal of Shanghai acupuncture and moxibustion 6:277. (In Chinese)

Lin XZ (1995b) The form mechanism of human meridian transmission. Yunnan Journal of TCM 3:32. (In Chinese)

Lin XZ, Xiao Y (1995c) The form mechanism of human meridian transmission. Yunnan Journal of TCM 18:32–38. (In Chinese)

Liu F, Huang GY (2007) Advances of Studies on the Biophysical and Biochemical Properties of Meridians. Acupuncture Research 30:281–284(In Chinese with English abstract)

Liu HG, Zhu L, Jiang DZ (1999) Analysis of modern science and technology theory for ancient acupuncture and moxibustion theory. Journal of Xi'an Jiaotong University (social sciences) 19:42–48,65. (In Chinese)

Liu JP, Bo T, Bai J L, Hou XM, Li ZG, Han Y, Zong QH, Chen ZQ, Niu BZ (2003) The expression and significance of EGF,TGF-α and EGFR in the natural concrescence course of rats gastric ulcer. Chinese Pharmacological Bulletin19:1185–1188. (In Chinese with English abstract)

Liu XF, Ruan YE, Chen H (2009) Discussion on the Relations between Lung and Large Intestine under ATA Infrared Thermal Automatic Detect System. China Journal of Traditional Chinese Medicine and Pharmacy, supplement 50–52. (In Chinese)

Lu LS (1996) Meridian essence approach and analysis. Chinese Acupuncture & Moxibustion (4):20–22. (In Chinese)

Lu LS (2008) A hypothesis about component model of substances of channels and collaterals. Chinese Acupuncture & Moxibustion 28:764–766. (In Chinese with English abstract)

Ma GY, Pan CC, Wang KM, Wang YL, Zhou YP(2000) Research on Connection Pathway between Channels and Viscera and Its Relation with Neuropeptide-like Substances. Acupuncture Research 25:117–120. (In Chinese with English abstract)

Mac Pherson H, Green G, Nevado A, Lythgoe MF, Lewith G, Devlin R, Haselfoot R, Asghar AU (2008) Brain imaging of acupuncture: comparing superficial with deep needling. Neurosci Lett, 434:144

Mao YJ, Gao XY (1994) Autonomic nervous reflex relay-meridian essence. Chinese medicine of inner-Mongolia 1:37–38. (In Chinese)

Napadow V, Makris N, Liu J (2005) Effects of electroacupuncture versus manual acupuncture on the human brain as measured by fMRI. Hum Brain Mapp 24:193

Noordengraaf A (1973) Electro-acupuncture. IEEE Trans Biomed Eng BME, 20:364–366

Reichmanis M, Marino AA, Becker RO (1979) Laplace plane analysis of impedance on the H meridian. Am J Chin Med. 7:188–193

Rong PJ, Zhu B (2005) Electrophysiological study on the relationship between heart meridian and heart. Acupuncture Research 30:238–242. (In Chinese with English abstract)

Shen J, Yan J (2008) Progress of study on relative effect between meridians and internal organs. Journal of Traditional Chinese Medicine University of Hunan28:76–78. (In Chinese with English abstract)

Shen XY (2003) chief editor Science of meridian and acupuncture points, china press of Traditional Chinese Medicine. (In Chinese)

Siedentopf CM, Golaszewski SM, Mottaghy FM, Ruff CC, Felber S, Schlager A (2002) Functional magnetic resonance imaging detects activation of the visual association cortex during laser acupuncture of the foot in humans. Neurosci Lett. 327:53

Pingsheng S, Yuzhou Z, Yulan L, Qinglian Y, Hong L (1988) The study on conduction of acoustic information along meridians. Acupuncture Research 2:139–142 (In Chinese with English Abstract)

Wang PQ, Hu XL, Xu JS, Wu BH, Xu XY (2002) Infrared Image-Displaying of the Fourteen Meridian Courses. Acupuncture research, 27:260. (In Chinese with English abstract)

Wang W, Zhang YJ, Guo Y, Miao WF, Wang XY, Xu TP (1996) The Detection and Studies on the Change of H^+ Concentration in the Regular Points of the Rabbit Suffering from Arrhythmia Induced by Aconitine. Acupuncture Research 21:59–63. (In Chinese with English abstract)

Wang X, Zhang KX, Zheng LY, Zhen XC (2003) Detective and Effect Verification of Commonly Used Acupoints in Pericardium and Triple Energizer Meridian of Rabbit Liaoning Journal of Traditional Chinese Medicine30: 390–391. (In Chinese)

Wang XY, Guo Y (1998). Study on an Ionic Mechanism of Peripheral Acupoint on Acupuncture Analgesia. Acupuncture Research 23:263–265. (In Chinese with English abstract)

Wang XY, Guo Y, Zhang YJ, Xu YP, Zhang CX (1998).Study on the effects of K+ concentration in rabbits acupoints with acupuncture. China Journal of Basic Medicine in Traditional Chinese Medicine 4:48–50. (In Chinese with English abstract)

Wu CJ, Cao R, Zheng LY, Wang X, Zhang KX, Cao FY, Zhen XC, Xu ZX (2003) Detective and Effect Verification of Commonly Used Acupoints in Heart and Small Intestine Meridian of Rabbit. Liaoning Journal of Traditional Chinese Medicine30: 488–489. (In Chinese)

Wu ZJ, Wang KM, Wang YL, Hu L, Zhou YP (2006) Effect of Electroacupuncture on Signal Transduction of G-protein in Rat Ischemic Myocardial Cells. Acupuncture Research 31: 264–267. (In Chinese with English abstract)

Xu CX, Zheng LY, Zhen XC, Wu CJ (2003) Detective and Effect Verification of Commonly Used Acupoints in Liver and Gallbladder Meridian of Rabbit. Liaoning Journal of Traditional Chinese Medicine 30: 142–143. (In Chinese)

Xu FM, Xie P, Lv FJ, Mou J, Li YM, Zhao JN, Chen WJ, Gong QY, Zhao LB, Liu QJ, Shen L, Zhai H, Yang DY (2007) Study on Corresponding areas the Liver and Lung Channels in brain with fMRI. Chinese Acupuncture & Moxibustion, 27:749. (In Chinese with English abstract)

Yan B, Li K, Xu J, Wang W, Li K, Liu H, Shan B, Tang X (2005) Acupoint-specific fMRI patterns in human brain. Neurosci Lett, 2005, 383:236

Yan J, Diao LH, Yi SX, Chang XR, Lin YP, Yang XF (2007) Myocardial energy metabolism and morphological changes in myocardial ischemia/reperfusion injury rats by electroacupuncture on Neiguan Journal of Clinical Rehabilitative Tissue Engineering Research 11:9443–47. (In Chinese with English abstract)

Yan J, Yang XF, Yi SX, Chang XR, Lin YP, Wang C, Diao LH (2007) Effect of Electroacupuncture of "Neiguan"(PC 6) on Na^+-K^+-ATPase Activity and Its Gene Expression in Cardiocyte Membrane in Rats with Myocardial Ischemia-reperfusion Injury. Acupuncture Research 32:296–300. (In Chinese with English abstract)

Yan J, Yang ZB, Chang XR, Yi SX, Lin YP, Zhong Y (2007) Expressions of epidermal growth factor receptor signaling substances in gastric mucosal cells influenced by serum derived from rats treated with electroacupuncture at stomach meridian acupoints. Journal of Chinese Integrative Medicine,5: 338–342. (In Chinese with English abstract)

Yan J, Yi SX, Wang LH, Li WL, Deng EM, Wang JJ, Yao XX (1983) Myoelectricity observation on meridian transmission phenomenon. Journal of Hunan College of Traditional Chinese Medicine 3:56–62. (In Chinese)

Yan LP, Sun ZR, Xie B, Ma X (2005) fMRI Research on Brain of Electro-acupuncture Quchi (LI-11). Journal of Clinical Acupuncture and Moxibustion 21:61. (In Chinese)

Yan ZQ, Zhang XL (1979) Initial study on glow of human body surface. Progress in biochemistry and biophysics 2:48–52. (In Chinses)

Yang H, Gao HJ, Liang QS, Luo XD, Luo H (2007) Application of quantum dots technology in biological imaging. Journal of Clinical Rehabilitative Tissue Engineering Research 11: 6248–51. (In Chinese with English abstract)

Yang XF, Wang C, Yi SX, Chang XR, Diao LH, Lin YP, Yan J (2007) Effect of electro-acupuncture at Neiguan on Ca2+ concentration in cardiomyocytes of rats with ischemia/reperfusion injury. Journal of Clinical Rehabilitative Tissue Engineering Research 11:6759–61. (In Chinese with English abstract)

Yang YQ (2007) Strategy, Conception and Methodology of Acupuncture Research. Shanghai Journal of Traditional Chinese Medicine, 41:3–5. (In Chinese with English abstract)

Yi SX, Yang RD, Yan J, Chang XR, Lin YP (2004) Comparison of effects of acupuncture at acupoints of the three -yang meridians of foot on expression of somatostatin receptor gene in rabbits with gastric ulcer. Chinese Acupuncture & Moxibustion 24:785–788. (In Chinese with English abstract)

Yoo SS, Kerr CE, Park M, Im DM, Blinder RA, Park H, Kaptchuk TJ (2007) Neural activities in human somatosensory cortical areas evoked by acupuncture stimulation. Complement Ther Med, 15:247

Yoo SS, Teh EK, Blinder RA (2004) Modulation of cerebellar activities by acupuncture stimulation: evidence from fMRI study. Neuroimage, 22:932

Yuan L, Jiao PF, Tang L, Huang WH (2005) Anatomical Basic of Traditional Chinese Medical Channels. Chinese Basic Science 3:44–47. (In Chinese)

Zhang CY, Du GY, Wang W, Ye ZG (2004) Application of microdialysis in the study of neurotransmitters in cerebral ischemia animals. Chinese Pharmacological Bulletin 20:1209–11. (In Chinese with English abstract)

Zhang FB, Zhou YP, Wang YL, Wang KM (2002) Effect of Electroacupuncture on Blood Free Radicals, ET and CGRP Contents in Rats with Acute Myocardial Ischemia. Acupuncture Research 27:192–196. (In Chinese with English abstract)

Zhang WB, Jing XH, Xu RM, Zhu ZX, Li CH, Li H (1995) Character of the Body Skin Exhaled Carbon Dioxide on Large Intestine and Stomach Meridians. Chinese Journal of Integated Traditional and Western Medicine 15:625–627 (In Chinese)

Zhang R, Zou YQ, Huang SQ, Chen ZG, Liang BL, Li Y, Zhong JL (2007) MRI cerebral function imaging following acupuncture at Hegu, Zusanli, Neiguan and Sanyinjiao points. Journal of Clinical Rehabilitative Tissue Engineering Research 11:4271–4274. (In Chinese with English abstract)
Zhao W, Zhang ZL, Pang DW (2008) Application of quantum dots as fluorescent labels in medicine and biology. Chemical Sensor 28:8–13. (In Chinese with English abstract)
Zheng LY, Zhen XC, Xu ZX, Wu CJ, Cao R, Wang X, Zhang KX, Cao FY (2003) Detective and Effect Verification of Commonly Used Acupoints in Spleen and Stomach Meridian of Rabbit. Liaoning Journal of Traditional Chinese Medicine30 : 60–62. (In Chinese)
Zhou AL, Shao ZY, Luo L, He M (2001) Effect of Pretreatment of Collagenase IV on Gastric Electricity Induced by Gentamicin Injecting to Zusanli Point. China Journal of Basic Medicine In Traditional Chinese Medicine 7:54–57. (In Chinese with English abstract)
Zhou MQ, Zhou YP, Wang KM, Wang YL, Chen YN (2004) Effects of Electroacupuncture Stimulation of Three Segments of the Heart Meridian on Cardiac Function in Rabbits with Acute Myocardial Ischemia Acupuncture Research 29:179–182. (In Chinese with English abstract)
Zhou MQ, Zhou YP, Xu GS, Gao XG (1997) Protective Effect of Acupuncture at Different Channels on injury or Heart Induced by Semi-Ligation of Small Intestine in Rabbits. Chinese Acupuncture & Moxibustion 22:410–412. (In Chinese with English abstract)
Zhou YP (1999) Correlation between Channels-collaterals and Viscera-the Focus in the Study of the Theory of Channels and Collaterals. Acupuncture Research 24:238–241. (In Chinese with English abstract)
Zhou YP, Wang KM, Hu L, Wu ZJ, Zhou MQ, Wang YL, Yang HT, Zhang FB, Sun L (2007) Research on the Differentially Expressed Genes Related to the Relatively Specific Connection Between the Heart Meridian and the Heart. Acupuncture Research 32:3–8. (In Chinese with English abstract)
Zhou YP, Zhou MQ, Wang KM, Hu L, Wu ZJ, Wang YL, Chen YN (2008) Research on the Correlation between Channels-Viscera and Brain Is the Breakthrough of Theory Combination between Traditional Chinese Medicine and Western Medicine. Journal of Anhui Traditional Chinese Medical College 27:1–8. (In Chinese)
Zhu B. The Scientific Basis of Acupuncture (1998) Qingdao: Qingdao Publishing House. (In Chinese)
Zhu CG, Li YQ, Gu XS (2009) Neuroanatomy (The Second edition).Beijing: People's Medicine Publishing House. (In Chinese)
Zhu DN, Wu BW, Fan XL (2008) Physiology. Beijing: People's Medicine Publishing House.
Zhu ZX (1989) Biophysics for acupuncture and meridian. 1st edition. Beijing: the people's medical publishing house 450. (In Chinese)

Chapter 20
Theory and Methodology of Evidence-Based Medicine in Acupuncture Research

Fanrong Liang, Tingting Ma, Wenjing Huang, Xi Wu, Ying Li, Yulan Ren, Hui Zheng, Li Fang, Jie Yang, Mailan Liu, and Lei Lan

Abstract As an ancient empirical medicine, acupuncture has profound theory basis and extensive clinical application. With the popularity of evidence-based medicine (EBM), rigorous challenges have been brought to acupuncture: (1) How to translate the tremendous source of evidence into an optimal treatment schema in clinical practice? (2) How to integrate the theory and methodology of the empirical medicine with that of EBM into a theory and methodology which will benefit the development of acupuncture? The aim of this chapter is to encourage reflection on the combination of the theory and methodology of EBM with the features of acupuncture so as to guide acupuncture clinical practice, health policy, and scientific research.

Keywords Evidence-based medicine • Science of acupuncture and moxibustion • Methodology • Clinical trial design • Clinical decision making • Clinical practice guideline

---

F. Liang (✉) • W. Huang • X. Wu • Y. Li • Y. Ren • H. Zheng • J. Yang • M. Liu • L. Lan
College of Acupuncture and Moxibustion, Chengdu University of TCM,
37 Shierqiao Road, Chengdu, Sichuan 610075, China
e-mail: lfr@cdutcm.edu.cn; sunnyhwj@yahoo.com; wuxi403@hot.comail.com;
jialee@mail.sc.cninfo.net; renxg2468@163.com; zhhui126@yahoo.com.cn;
jenny_yang_jie@126.com; liumailan2006@yahoo.com.cn; lanleicdutcm281@139.com

T. Ma
Department of Treatment for Undiseased and Sub-health,
Teaching Hospital of Chengdu University of TCM, 39 Shierqiao Road,
Chengdu, Sichuan 610075, China
e-mail: matingtingcn@yahoo.com.cn

L. Fang
Department of Acupuncture and Moxibustion, The Third Affiliated Hospital
of Zhejiang University of Traditional Chinese Medicine,
219 Moganshan Road, Hangzhou, Zhejiang 310005, China
e-mail: lilyfang126@126.com

Y. Xia et al. (eds.), *Current Research in Acupuncture*,
DOI 10.1007/978-1-4614-3357-6_20,

20.1 Introduction

As one of the traditional empirical medicines, acupuncture is widely applied in disease prevention and treatment as a special natural therapy and has made great contribution to human health care over thousands of years. With the popularity of traditional medicine and natural therapy in western countries, acupuncture is accepted and applied in over 100 countries as an important complementary or alternative therapy. Nowadays, the great challenges to the development of acupuncture are improvement of effectiveness, evaluation of ancient evidence, development of therapeutic guidelines, and translation of evidence into clinical practice. In some western countries, such as Germany, a series of acupuncture studies have been done. Most of the studies focused on comparing verum acupuncture with nonacupoint puncture in order to observe the specific effect of acupuncture. Greater efficacy was only observed in the treatment of osteoarthritis of the knee, instead of migraine, tension-type headache, and chronic low back pain, which used to be considered as the predominance disease of acupuncture. The question is why is there variance in the results, and how does one make objective evaluation of acupuncture?

Evidence-based medicine (EBM) aims to apply the best available evidence gained from the scientific method to clinical decision making and practice (Sackett et al. 1996). To guarantee every practice made on the basis of best current clinical evidence, knowledge and technique of clinical practitioners, and patient's requirements, EBM uses techniques from science, engineering, and statistics, such as the systematic review of medical literature, meta-analysis, risk–benefit analysis, and randomized controlled trials (RCTs). With the integration of EBM into western and Chinese medicines, it is an inevitable trend to bring the EBM theory and methodology into acupuncture research to standardize acupuncture clinical research and evaluate acupuncture therapeutic efficacy and effectiveness.

Reflection on the combination of the theory and methodology of EBM with the features of acupuncture is important. First, it can promote the development of acupuncture by moving it from empirical medicine into EBM. Second, it will help the practitioners of acupuncture make better clinical decision so as to optimize therapeutic effectiveness. Third, it can also improve the methodology of acupuncture clinical research through implementation of well-designed RCTs and safety evaluations.

20.2 Development of Acupuncture Research Methodology Based on Theory of Evidence-Based Medicine

Evidence is the core of EBM. There are different types of clinical evidence and ranking the quality of evidence is foundational. The strongest evidence for therapeutic interventions is provided by systematic review of randomized, triple-blind, placebo-controlled trials with allocation concealment and complete follow-up involving a homogeneous patient population and medical condition (Timmermans and Mauck 2005). In contrast, case reports and expert opinion have little value as proof and are classified as low-level evidence. Whether the evidence rating of EBM is totally suitable for acupuncture is still a matter of discussion. The reason for this is that, as an interventional treatment, placebo effect can't be completely excluded in acupuncture research, and ancient clinical experience is of great importance in the development of acupuncture practice.

20.2.1 Ancient Evidence of Acupuncture

The ancient evidence of acupuncture refers to ancient monographs on acupuncture and moxibustion, comprehensive medical literatures and records of acupuncture and moxibustion in nonmedical works. Skills of data gathering and analysis have been used to collect ancient evidence on Traditional Chinese Medicine from the Pre-Qin period to the end of the Qing dynasty. Detailed methods will be introduced in the book *Data Mining and Clinical Application of Acupuncture and Moxibustion* (Liang and Ren 2009).

20.2.2 Quality and Strength of Acupuncture Evidence

As no gold standard for acupuncture evidence exists, methodologists and acupuncture guideline developers are giving thought and effort to the criteria and approaches needed to make an optimal method for rating the strength of acupuncture evidence.

20.2.2.1 Rating the Strength of Acupuncture Evidence

The standard of evidence quality and strength of EBM is established by the working group of GRADE (Grading of Recommendations Assessment, Development and Evaluation), which is used for reference by acupuncture methodologists and guideline developers (Atkins et al. 2004). Although the strength of acupuncture case reports in ancient books is low according to EBM, to a certain extent this long-standing human experience, which is one kind of sources of clinical evidence, can substitute and supplement the present evidence. Moreover, China is a developing country with relatively few health resources; it is difficult to repeat acupuncture research for experience-proved disease in high quality. Therefore, considering the history and current status of acupuncture, it is recommended to regard the classical literatures and experiences from famous doctors as relatively high-level evidence (Liang et al. 2006) (Table 20.1).

Table 20.1 Recommendation of evidence level of acupuncture research

Recommendation level		Evidence level
High	Ia	Classical literature records or long-term existence of treatment schemas confirmed by meta-analysis based on high-quality RCTs
	Ib	(a) Classical literature records
		(b) Long-term existence of treatment schemas through the years
		(c) Meta-analysis based on high-quality RCTs
	Ic	Ancient literature records or experience from famous TCM doctors confirmed by RCTs
Moderate		(a) Ancient literature records
		(b) Experience from famous TCM doctors
		(c) RCTs
		(d) Nonrandomized clinical trials
Low		Case series studies
Very low		Case report

Classic literature refers to *SuWen, LingShu, NanJing, JiaYijing*, etc.
Long-term existence of treatment schemas refers to acupuncture prescription repeatedly used in all generations

20.2.2.2 Criteria of Evidence Quality in Acupuncture Research

The assessment of the acupuncture research quality includes the study design, conduct, measurement and confounding biases. The Cochrane Collaboration has well established the quality assessment criteria for the randomized concurrent controlled trials, nonrandomized concurrent controlled trials, and case series studies, which is the reference for acupuncture research.

20.2.2.3 Systematic Review of Acupuncture and Moxibustion

Systematic review (SR), as an assessment tool of trials, can offer valuable support to evidence-based decision making and clinical practice (Sackett et al. 2000). In the following, an example of acupuncture for herpes zoster is taken (Liang and Wu 2009) to explain the method and problems in current acupuncture RCTs.

To answer the question as to whether acupuncture is helpful in increasing the cure rate of herpes zoster and more effective in relieving pain than western medicine, a literature review is required based on available research. The following published and unpublished literatures were carefully searched:

- RCTs.
- Participants with herpes zoster suffer, regardless of age and gender.
- Intervention was acupuncture therapy, controlled by western medicine.
- Outcome measurements were cure rate, the period of pain relief.

All the data was researched in CNKI, VMIS, and Wanfang databases by computer and hand searched in nine sorts of journals from January 1996 to November 2006. After quality assessment according to the Jadad score, RevMan 5.0 software was used to perform statistical analysis.

In the end, 43 RCTs in which acupuncture was used to treat herpes zoster were found, among which only one was published in English. Thirty-three studies were excluded by two reviewers (13 studies adopted different acupuncture manipulation as control; the sample size of seven trials was less than 30, 11 trials could not identify the effect of acupuncture due to miscellaneous interventions, one got false randomization, and one had no description of baseline comparison). Only ten studies meeting the aforementioned criteria were included. The total number of participants in the trails was 816. All of them compared acupuncture therapy with western medicine. Among them, five used acupuncture and cupping, three acupuncture and moxibustion, one electroacupuncture, and one acupuncture with cupping and moxibustion as the acupuncture therapy intervention, while mainly adopting acyclovir as the control western medicine. All of them used cure rate and effectiveness as the outcome measurement. Six of them reported the period of pain relief.

By combing the results, the cure rate of all the included trials containing a total of 816 participants indicated that acupuncture therapy is superior to western medicine (OR = 4.27, 95 %CI [2.90–6.29], $\overline{X} = 7.33$, $P<0.01$), using fixed-effect model with homogeneity ($P>0.05$). The same result was also showed in the period of pain relief of four included trials containing a total of 358 participants: Acupuncture therapy is superior to western medicine (OR = 7.64, 95 %CI [−8.12 to −7.15], $\overline{X} = 30.85$, $P<0.01$) using fixed-effect model with homogeneity ($P>0.05$).
The limitations of the SR were as follows:

- The number of included trials was not enough to provide definite evidence.
- Due to the difference of overall outcome in the trials, the comparison of cure rate was the only factor studied; the outcome of response rate and markedly improved rate were ignored.
- The sample was not enough when adopting period of pain relief as a measurement of outcome.

Although the SR-identified acupuncture can achieve a relatively higher cure rate and shorten the pain duration compared with western medicine in the treatment of herpes zoster, all RCTs were in low quality. For example, only one described the sequence generation, and none described allocation concealment, design of blindness, follow-up, and dropout or loss of follow-up. Consequently more high-quality trials are needed to verify the superiority of acupuncture therapy. The similar problem occurred in other SRs of acupuncture. Improvement in trial quality which is crucial to EBM is needed in acupuncture research in order to provide reliable evidence.

20.3 Evidence-Based Medicine in Acupuncture Research

20.3.1 Standardization of Acupuncture Clinical Trials

Design, performance, and report are three key phases which determine the quality of clinical trials. For study design, a randomized, controlled, blinded (if possible), sample-size calculated study with a preplanned statistical analysis and trial monitoring is the gold standard for a therapeutic clinical study (Sackett et al. 2000). It should be applied to all trials, including acupuncture study. In the phase of performance, trials registration, standard operation procedure (SOP) development, researchers training, and quality control are essential to maintain the quality (Li et al. 2009). Additionally, the report of acupuncture clinical trials should conform to CONSORT (Consolidated Standards of Reporting Trials) statement and STRICTA (STandards for Reporting Interventions in Controlled Trials of Acupuncture) recommendations (http://www.consort-statement.org/; http://www.stricta.info/checklist.htm).

20.3.1.1 Design of Acupuncture Clinical Trials

Sample Size

To provide the trial with reliable result, an adequate sample should be planned before the study performance. Generally, sample size is determined according to the data provided by pilot trial or relevant references, with test level α of 0.05 or 0.01 and power of 0.80 or 0.90. After deciding on a one-side or two-side test, sample size calculation formula should be used to determine the sample size.

An example of sample size is the acupuncture trial for migraine in China funded by 973 Program. Linde et al. (2005) claimed an improvement for the cure of headache days as being 2.4 days was considered a sham. A multiple sample size calculation formula was used (Wang 2001) and set α as 0.05 and $1-\beta$ as 0.90. Finally, 103 subjects should be recruited in each group, and a total of 480 subjects were equally recruited in four groups considering 15 % of the dropout rate (Li et al. 2008).

Randomization

The random-number table kept in serial-numbered, opaque, sealed envelopes is traditionally applied for randomization. Central randomization is a kind of permuted-block randomization procedure, which is automatically under the control of central computer system, in order to avoid or reduce the possibility of research bias. For the advantage in randomization and complete allocation concealment, central randomization system is widely used in multicenter RCTs of the ongoing 973 Program and key projects in the National Science and Technology Pillar Program during the eleventh 5-year plan period, managed by Chengdu Good Clinical Practice (GCP) Center (Fig. 20.1). In the study of acupuncture for treating acute attacks of migraine, an RCT (Li et al. 2009) with central randomization performed by the National Clinical Trial Center of Chinese Medicine (Chengdu, China) was used. Patients meeting the inclusion criteria were randomly assigned in a 1:1:1 ratio to the verum acupuncture treatment group or to one of the two control groups.

Email Response of Central Randomization system↵

Subjects: MXHO successfully registered ↵

Project Number: 234 ↵

Project Name: Perpheral facial paralysis ↵

Abbreviation of Name: MXHO ↵

Sex: Female ↵

Date of Birth: 1989-6-4 ↵

Random Number: 702 ↵

Group: D ↵

Registration: Telephone ↵

Date: 2008-11-22 ↵

Central Randomization System in National Clinical Trial of Traditional Chinese Medicine ↵

Fig. 20.1 Randomization information feedback e-mail

Control

Several methods are used to set up controls: acupuncture vs. medicine (TCM medicine or western medicine), acupuncture vs. electroacupuncture or other kinds of acupuncture, acupoints vs. nonacupoints, verum acupuncture vs. sham acupuncture, acupuncture vs. minimal acupuncture, acupuncture vs. waiting list, etc. The former two methods are commonly used in Chinese research, which aim at comparing acupuncture therapy with medicine or finding the best acupuncture treatment schema. The ongoing project of acupuncture for Bell's palsy funded by the National Science and Technology Pillar Program is a good example, which compares different types of acupuncture with stage differentiation (Li et al. 2004). Many foreign acupuncture clinical researches use nonacupoints, sham acupuncture, minimal acupuncture, and waiting list as common controls for acupuncture, which focuses on the efficacy of verum acupuncture. For example, Alastair Forbes used nonacupoints as control in their RCT of acupuncture for irritable bowel syndrome (Forbes et al. 2005). Meanwhile, Linde used nonacupoints and waitlist as controls in their RCT of acupuncture for migraine (Linde et al. 2005), whereas Nabeta used pain-sensitive points as control for the treatment of chronic neck pain and frozen shoulder (Nabeta and Kawakita 2002), and Kazunori Itoh used acupuncturing tender point as control for elders' lumbar pain (Itoh et al. 2006). Both trials of Nabeta and Kazunori Itoh had used placebo acupuncture with blunt needle tips. Furthermore, Kleinhenz used collapsible placebo acupuncture as control for shoulder pain in their RCT (Kleinhenz et al. 1999).

Focusing on comparing the efficacy of verum acupoints vs. nonacupoints for treating acute migraine pain, Li conducted a RCT (Li et al. 2009a, b, c) with two control groups to find the better proper location of nonacupoints. In the treatment group, acupoints along shaoyang meridian were chosen (SJ-5, GB-34, GB-40, SJ-20, GB-20). The two control groups were given on nonacupoint puncture treatment.

The nonacupoints used in control group 1 were as follows:	The nonacupoints used in control group 2 were (Fig. 20.2) as follows:
(1) Halfway between the triple energizer and small intestine meridians lateral to Waiguan (SJ-5) horizontally	(1) Medial arm on the anterior border of the insertion of the deltoid muscle at the junction of the deltoid and biceps muscles
(2) Halfway between the line from Qiuxu (GB-40) to Jiexi (ST-41)	(2) Halfway between the tip of the elbow and the axillae
(3) Halfway between the gallbladder and bladder meridians lateral to Yanglingquan (GB-34) horizontally	(3) Halfway between the epicondylus medialis of the humerus and ulnar side of the wrist
(4) Halfway between the line from Jiaosun (SJ20) to Shuaigu (GB-8)	(4) The edge of the tibia 1–2 cm lateral to the Zusanli (ST-36) point horizontally
(5) Halfway between the line from Fengchi (GB-20) to Anmian (extra point)	(5) The inside of the mid-thigh region 2 cm lateral to half the distance from the anterior superior iliac spine to the lateral superior corner of the patella on the rectus femoris bilaterally

Blinding

Blinding is set to avoid bias, which consists of single-blind, double-blind, and triple-blind. However, blinding in acupuncture trials is difficult to perform because of the particularity of acupuncture intervention. It is impossible to blind acupuncture providers, but it is realizable to blind outcome assessors.

20.3.1.2 Acupuncture Trials Performance

Registration of Clinical Trials

Registration of acupuncture clinical trials contributes to international communication and decrease or elimination of publication bias. Since 2005, the member magazines of International Committee of Medical Journal Editors (ICMJE) have only published the registered trials. In January 2007, 52 member magazines of Chinese Clinical Trial Registration and Publication Collaboration (ChiCTRPC) announced to publish the registered trials in priority and will gradually give up publication of nonregistered trials. The trial of acupuncture for Bell's palsy has registered the protocol in ClinicalTrials.gov in January 2008 (ClinicalTrials.gov NCT00608660). Another two trials of acupuncture, respectively for migraine and functional dyspepsia funded by 973 Program, which is the first Acupuncture 973 Program in China, have also registered for unique ID (ClinicalTrials.gov NCT00599586, ClinicalTrials.gov NCT00599677).

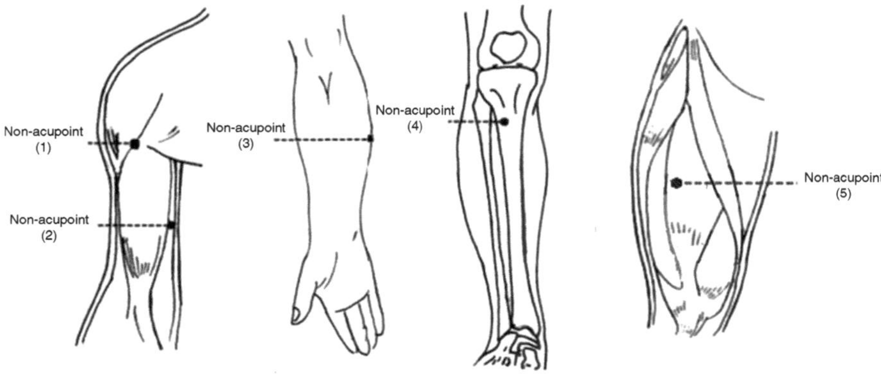

Fig. 20.2 Nonacupoints in control group 2

Developing Standard Operation Procedure

Developing SOP ensures unified and standardized rules for acupuncture manipulations. According to STRICTA, the SOP should include standards for acupuncture manipulations, standards for locations of acupoints, name of acupuncture needles, manufacturers, specification of the needles, etc.

Research Training

CONSORT statement suggests reporting the experience of the acupuncture providers. It indicates that training for researchers is very important because it will determine whether the trial can be carried out in high quality. The training should include acupuncture manipulation training according to the research guideline and SOP, training for trial protocol, training for case record and assessment, etc.

Quality Monitoring

Quality monitoring over the progress aims to ensure that the conduct, record, and report are in accordance with the protocol and SOP, so the results are true and reliable. In December 2007, the Chinese government held the conference of controlling and ensuring the quality of clinical trials for traditional Chinese medicine, which finally developed standardizations for quality control (Ministry of Technology State Administration of TCM of PRC 2007). The ongoing 973 Program in China has set up two levels of quality control system according to the rules made by the government. The first level is quality inspection, and the second level is quality monitoring. Two independent persons were appointed as quality inspector or quality monitor. Issues to be monitored every month or every 3 months. Relevant records should be written according to the quality control examination (Zhao et al. 2009).

20.3.1.3 Standardization of Reports for Acupuncture Trials

Recently, most of systematic reviews of acupuncture showed that the following issues hindered the evaluation and application of acupuncture: nonstandardized report, deficient details for methodology, and inaccuracy in reporting intervention procedure (including control group). So reporting acupuncture trials in standardized form is important to ensure the quality of clinical trials.

CONSORT statement is an international guideline for reporting clinical trials. This statement includes recommendations for reporting titles, abstracts, background, methods, results, and discussion. Because of the particularity of acupuncture trials, STRICTA (MacPherson et al. 2010) is recommended for acupuncture report.

20.3.2 Acupuncture Clinical Decision-Making Process

It is thought that decision making in clinical practice are decided by personal experience, knowledge, etc. Actually, sound decision making depends upon fully informed assessments and astute, analytic judgments. The process of clinical decision making includes identification of cues or of relevant clinical features, data gathering, assimilation, analysis, evaluation and choice, and finally producing an operational decision.

Assessment is the first stage of the decision-making process which consists of data gathering, assimilation, and analysis. To make accurate clinical judgments and reliable operational decisions, assessment is fundamental and crucial. As stated before, the reliable and accurate sources of acupuncture are not only from high-quality RCTs but also from case reports or knowledge in ancient classical literatures. In addition, acupuncturists make the final and effective clinical decisions depending on sound judgments which take account of the patient's clinical needs, priorities, and values. In the following, the acupuncture treatment for acute cerebral infarction is taken as an example to illustrate the procedures of making clinical decision (Liang and Wu 2009).

20.3.2.1 Data Gathering

The evidence from ancient records was searched manually, including all the documents about acute apoplexy treatment with acupuncture from the Qin and Han dynasties to the late Qing dynasty in the *Acupuncture and Moxibustion Section, Medical Volume, China Canon*. The text words for searching were "stroke," "apoplexy," and so on.

The evidence from modern researches was searched by computer and manual work. The following online databases were included: Cochrane Library (2008, Issue 1), PubMed (1966 to March 2008), Cochrane DSR, ACP Journal Club, DARE, CCTR (1991–2008), CBM disc (1978 to March 2008), VIP (1989 to March 2008),

CNKI (1979 to March 2008), and Wanfang Data (1983 to March 2008). The following journals (1990 to March, 2008) were manually searched: Chinese Acupuncture and Moxibustion, Acupuncture Research, Shanghai Journal of Acupuncture and Moxibustion, Journal of Clinical Acupuncture and Moxibustion, Journal of New Chinese Medicine, Foreign Medical Sciences (Section of Chinese Medicine), and Journal of Traditional Chinese Medicine. The searching targets were limited to SR and RCT on acute cerebral infarction treatment with acupuncture, with language limitation of Chinese and English.

Retrieval strategy and inclusion criteria were as follows:

- Referring to search strategy of systematic review and RCTs on Cochrane Collaboration and combining the result with keywords as "acupuncture" or "cerebrovascular accident" or other free words
- SR or RCT on acute cerebral infarction treatment with acupuncture; no limitation to age, sex, and race; disease course within a month

Finally, there were 33 ancient articles, 1 SR, and 35 RCTs totally.

20.3.2.2 Analysis on Modern Evidence

The SR was from Cochrane in 2005, which compared the acupuncture treatment with nonacupuncture treatment for acute stroke. In the article, it concluded that acupuncture appeared to be safe but didn't show clear evidence of benefit. There were statistical differences between acupuncture therapy and other therapies on death and disability rates, demand for conventional treatment, and improvement in neurological deficit scores. However, the sample size was too small to determine the curative effect of acupuncture for acute ischemic stroke.

In the reference of Higgins 2008 on Cochrane Web site, the quality of the 21 RCTs was assessed. All of them were included because two trails were assessed as A levels, four as B levels, and 15 as C levels. The analysis results showed that acupuncture therapy had a relatively high safety for acute cerebral infarction; acupuncture with conventional treatment had some advantages in bringing down death rate and improving neurological deficit for acute cerebral infarction. In addition, acupuncture intervention in an early stage (within 72 h) tended to exert better efficacy. However, due to poor quality of the available researches and small sample size, the conclusions above still require larger and methodologically sound randomized controlled trials to support.

20.3.2.3 Evaluation on Characteristics of Point-Selection and Needling Methods of Ancient and Modern Evidences

Thirty-three articles were involved in the ancient evidence and 21 RCTs were included in the modern evidence, both of which put an emphasis on the combination of acupuncture therapy and medicinal application to improve efficacy. Points selected in the ancient and the modern were similar, but not to needling methods. In ancient times, moxibustion was more favorable than acupuncture to treat stroke, and Shenque (RN-8) (with the function of warming yang to rescue from counterflow) and the use of 12 jing-well points (with the function of opening the orifices and inducing resuscitation) were stressed. While in modern times, therapies such as acupuncture and electroacupuncture are more common to treat acute cerebral infarction, especially scalp acupuncture.

Upon the synthesized analysis of the ancient and modern evidences, characteristics of acupuncture therapy for acute cerebral infarction treatment were as follows:

- Points on yang meridians such as large intestine meridian, governor vessel, gallbladder meridian, stomach meridian, and triple energizer meridian were often selected to treat this disease in either ancient or modern times.
- Points for emergency were emphasized. For instance, some modern researches reveal that it is effective to apply the manipulation of "inducing resuscitation and opening the orifices" on Shuigou (DU-26), Neiguan (PC-6), and Sanyinjiao (SP-6) as the main points to treat acute cerebral infarction.
- Usually, points on head and limbs, such as Baihui (DU-20), Fengchi (GB-20), and scalp points and lines, are combined to treat the disease.
- Moxibustion, filiform needle, and three-edged needle were preferable to be used in ancient researches. For modern researches, the filiform needle, electroacupuncture, and three-edged needle are more commonly used.

Generally, the current acupuncture researches on acute ischemic stroke are generally poor in quality. As there are deficits in methodology and heterogeneity of the interventions, current evidence is not sufficient to support the superiority of point-selection or needling methods. According to the present analysis, the treatment of acupuncture combined with conventional therapy is safe, convenient, and more effective than pure medication treatment. Therefore, it is better to use acupuncture combined with conventional therapy to treat acute cerebral infarction.

20.3.2.4 Judgment

Acupuncture combined with conventional therapy to treat acute cerebral infarction is suggested.

Protocol 1: Filiform Needle

- Acupoints:
  - Baihui (DU-20), Shuigou (DU-26), anterior oblique line of vertex-temporal on affected side, posterior oblique line of vertex-temporal on affected side, middle line of vertex on affected side, and lateral line 1 of vertex on affected side
  - Upper limbs: Quchi (LI-11), Waiguan (SJ-5), Neiguan (PC-6), Hegu (LI-4), Jiquan (HT-1), Jianyu (LI-l5), and Shousanli (LI-10)
  - Lower limbs: Zusanli (ST-36), Huantiao (GB-30), Yanglingquan (GB-34), Jiexi (ST-41), Sanyinjiao (SP-6), Kunlun (BL-60), Taixi (KI-3), and Taichong (LR-3)
- Manipulations: Strong manipulation should be applied to points on head. The strong manipulation applied to Shuigou (DU-26) should be to the extent of tearing or eyes moistened. Rapid twirling method (200 circles per minute) is applied to scalp needle. Routine needling or the needling method of "inducing resuscitation and opening the orifices" created by Prof. Shi Xue-min (including Shuigou (DU-26), Neiguan (PC-6), Sanyinjiao (SP-6) as the main points) is applied to points on limbs of both sides.
- Treatment course: once a day and five sessions a week.

Protocol 2: Electroacupuncture

- Acupoints:
  - Baihui (DU-20) and Shuigou (DU-26)
  - Upper limbs: Jianyu (LI-l5), Quch (LI-11), Hegu (LI-4), Waiguan (SJ-5), and Neiguan (PC-6)
  - Lower limbs: Zusanli (ST-36), Huantiao (GB-30), Sanyinjiao (SP-6), Yanglingquan (GB-34), or Fenglong (ST-40)
- Manipulations: After arrival of *qi* by needling, put the electrodes on points of the affected side, Quchi (LI-11)–Hegu (LI-4) and Zusanli (ST-36) or Yanglingquan (GB-34)–Fenglong (ST-40). Choose dilatational wave and tolerable intensity, about 3–7.5 mA with a frequency of 5/45 Hz.
- Treatment period: once a day and five sessions a week.

Protocol 3: Bleeding Therapy with Three-Edged Needle

According to ancient documents, obstruction syndrome of acute apoplexy can be treated by using three-edged needle on 12 jing-well points or Shixuan (Ex-UE-11). For example, *The Great Compendium of Acupuncture and Moxibustion, Further Added Therapies, Needling Methods for Emergency of Stroke At an Early Stage* records that "Cited from book *Indication of The Heaven And Earth*, for all the patients of stroke at an early stage, with symptoms of falling, sudden dizziness and heaviness, phlegm accumulation, loss of

(continued)

(continued)

consciousness, lock-jaw and inability to swallow herbal decoctions, it is emergent to let bleeding on the 12 jing-well points with the three-edged needle, thus removing blood stasis." There is also available evidence in modern research to support the effect of this method.

- Acupoints: 12 jing-well points on both sides, or Shixuan (Ex-UE-11), Erjian (EX-HN-6), or posterior ear veins.
- Manipulations: Select 4–6 points each time. After the needles and point locations are sterilized, bloodletting is administrated to points until 2–3 drops of blood are let out.
- Treatment period: once a day and seven sessions in a row.

Protocol 4: Moxibustion

In ancient documents, direct moxibustion or indirect moxibustion was applied to prevent or treat acute stroke, but it is not so commonly used in modern times.

- Acupoints:
  - Tinghui (GB-2), Jiache (ST-6), Dicang (ST-4), Baihui (DU-20), Jianyu (LI-15), Fengshi (GB-31), Zusanli (ST-36), Xuanzhong (GB-39), hairline, DaiZhui (DU-14), and Fengchi (GB-20)
  - Transverse creases under the first toe of both feet
  - Shenque (RN-8), Guanyuan (RN-4), and Qihai (RN-6)
- Manipulations: Direct moxibustion from the upside downwards is applied to Tinghui (GB-2), Jiache (ST-6), Dicang (ST-4), Baihui (DU-20), and so on. As for the transverse creases under the first toe of both feet, apply direct moxibustion or moxa cones, whose number is the same to the patient age. Moxibustion on salt or ginger is used for Shenque (RN-8) without amount limitation of moxa cones. Apply moxibustion on ginger or suspended moxibustion on Guanyuan (RN-4) and Qihai (RN-6).

20.3.3 *Evidence-Based Clinical Practice Guideline of Acupuncture and Moxibustion*

A Clinical Practice Guideline is a recommendation developed by experts for individual health care, based on the best available evidence. A guideline is made to help health care professionals in practice but does not replace their knowledge and skills (http://www.nice.org.uk/). There are three ways to formulate a clinical practice guideline: informal consensus development, formal consensus development, and evidence-based guideline development (EBD). At present, to develop a guideline based on evidence development is becoming a main trend.

EBD refers to the connection between recommendations and evidence quality, and it is a process of making recommendation based on evidence evaluation. The development of evidence-based guideline has strict procedural requirements, which include raising clinical questions, retrieval of clinical evidence, evaluation of the strengths and quality of evidence, and making final recommendation based on evidence evaluation. A high-quality clinical practice guideline can provide health care providers with reliable recommendation on treatment help patients informed and improve patient–doctor relationship. Additionally, it can also be used for formulating assessment criteria of clinical practice, educating and training for health care workers, optimizing the health care process, and improving efficiency.

The development of evidence-based guideline follows the rules from SIGN (Scottish Intercollegiate Guidelines Network) and NICE (the National Institute for Health and Clinical Excellence), including confirming clinical questions, target users, and target environment; founding of the guideline development group and task allocation; data extraction, evaluation of the strengths and quality of evidence, and grading of recommendations; assessment of social economics and resource conflict, and making group decisions and reaching consensus; reviewing and consulting of manuscript; and updating and correcting.

The standardization and normalization of acupuncture clinical practice is a worldwide requirement and also one of the actions to put the WHO "traditional medicine strategy" into practice. The development and application of acupuncture clinical practice guideline will regularize acupuncture activity in a better way, improve therapeutic effect, optimize the allocation of acupuncture resource, and minimize unnecessary waste of medical measurement. And in the last, it can promote acupuncture treatment quality and cost-effectiveness ratio to build a better acupuncture health care environment in which more people can enjoy benefit. In the following, an example of acupuncture and moxibustion for Bell's palsy supported by WHO Western Pacific Region, State Administration of TCM, and the World Federation of Acupuncture–Moxibustion Societies in China will introduce steps towards acupuncture clinical practice guidelines.

20.3.3.1 Founding the Guideline Development Group

The guideline development group consists of the head of the group (a person in charge of the guideline), specialist committee (specialists of clinical acupuncture, neurologists, epidemiologists and specialists of EBM and medical informatics, statisticians), guide-writing committee (associate head of the group, literature retrieval and guide-writing faculty), patient representatives, and medical care faculty representatives.

20.3.3.2 Confirming Clinical Questions

Guide-writing committee proposed the original clinical questions, following PICO (Patient, Intervention, Comparison, and Outcome) model and medical economics, and confirmed the final clinical questions after screening by questionnaire and discussion.

20.3.3.3 Retrieval of Clinical Evidence

The search methods and strategy should be established by medical mathematicians. For ancient acupuncture evidence, bibliography specialists of Traditional Chinese Medicine made the inclusion criteria of checklist and edition of acupuncture ancient books. Except for modern literature, collections of expertise were also searched. Finally, pertinent literatures were manually searched to ensure inclusion of all potential evidence. Language was limited to Chinese, English, Japanese, and Korean. Evidence retrieval was managed by specific people according to defined inclusion and exclusion criteria.

20.3.3.4 Evaluation of the Strengths and Quality of Evidence

The evidence evaluating criterion was made by expert consultation and discussion among groups, formal consensus development on the basis of CONSORT Scale and Jadad Scale. The evaluating group evaluated and graded the literature according to the evaluating criterion. One paper was evaluated by two independent reviewers. The reviewers contacted the authors for the missing information by phone contact. Discrepancies in quality ratings were resolved by the specialist committee.

20.3.3.5 Grading of Recommendations

The interventions recommended here have been developed through long-term clinical practice, and clinical evidence of acupuncture is composed of confirmatory research, observational study, classical literature, and experience of experts. Following the methodology of evidence-based medicine, the grading system of evidence and recommendation was launched finally after discussions among experts and consultation by questionnaires based on the characteristics of acupuncture (Table 20.2).

20.3.3.6 Making Group Decisions and Reaching Consensus

The group grade and classify all included literature according to different phrases of a disease, different therapies, and types of studies to develop interventions for different patients. The final recommendations for the management of Bell's palsy

Table 20.2 Main recommendations

Principle and stage	Recommendations of acupuncture and moxibustion for Bell's palsy (grades of recommendation)
Therapeutic principles	(a) Select acupuncture–moxibustion, western medicine, or combined use of acupuncture–moxibustion and western medicine for patients with mild facial paralysis within 3 months (A) (b) Select acupuncture–moxibustion or combined use of acupuncture–moxibustion and western medicine for patients with severe facial paralysis within 3 months (A) (c) Select acupuncture–moxibustion therapy after 3 months (A) Receive acupuncture–moxibustion therapy as early as possible (A) (a) Syndrome differentiation in each stage and symptomatic treatment are important for acupuncture–moxibustion therapy (GPP) (b) The syndromes in acute or subacute stage are wind–cold or wind–heat assailing the collaterals (GPP) (c) The syndromes in convalescence and residual stage are blood stasis–phlegm obstructing collaterals, *qi* deficiency with blood stasis, and yin deficiency engendering wind (GPP) Acupoint selection is characterized as local points, points along meridians, and according to pattern (A) Combine at least two of the needling or moxibustion methods (filiform needling methods, moxibustion, and electroacupuncture methods) in clinical practice (A)
Acute stage	Receive acupuncture–moxibustion therapy as early as possible (A) Select local points and Hegu (LI-4) bilaterally, differentiate syndrome as pattern of external invasion of wind–cold or wind–heat, and apply filiform needling method mainly (A)
Subacute stage	Select local points and Hegu (LI-4) bilaterally, differentiate syndrome as pattern of external invasion of wind–cold or wind–heat, and apply filiform needling method or electroacupuncture method (A)
Convalescent stage	Select local points and Hegu (LI-4) bilaterally, differentiate syndrome as pattern of phlegm and blood stasis obstructing collaterals, *qi* deficiency with blood stasis, or yin deficiency engendering wind, and apply penetrating method with filiform needles or electroacupuncture method (A)
Sequelae stage	Select local points, Hegu (LI-4) and Zusanli (ST-36) bilaterally, differentiate syndrome as pattern of phlegm and blood stasis obstructing collaterals, *qi* deficiency with blood stasis, or yin deficiency engendering wind, and apply penetrating method with filiform needles plus electroacupuncture method (B)

were established by members of guide-writing committee through discussing each recommendation by approaches such as questionnaire, telephone, and conference.

20.3.3.7 Reviewing and Consultation Process

To consummate the revisal of guideline and finalize the manuscript, after the first draft was completed, the working group obtained the opinions and suggestions from the experts on the recommendation of main treatment protocols of the guideline by three round of soliciting comments from experts in China. The final version of guideline was made based on feedback opinions from physicians, patients, and medical care faculty after guideline draft delivery in grade A hospitals.

20.3.3.8 Dissemination and Propagation

The guideline will be disseminated internationally by the WHO Western Pacific Region. In addition, the related academic and administrative organizations take responsibility for the normal propagating through doctors.

20.3.3.9 Implementation

This guideline aims to provide doctors with the at-present best evidence and help them make correct decisions on the acupuncture treatment for Bell's palsy. But in clinics, based on the guideline, doctors are welcomed to make the reasonable treatment protocols by considering the state of the illness, the willingness of the patient, and the situation of the country, the level, characteristic, resources of local medical treatment, as well as their own experiences and knowledge. In addition, doctors and patients are invited to give advice, suggestions, or feedback information to the guideline development group during the implementation of guidelines in order to assist guideline development group review and update the guideline periodically.

20.3.3.10 Updating

In normal occasions, the working group will revise the content and evidence of the guideline at least every 3 years. In particular, the guideline needs updating when changes in any of the following occur:

- The available interventions come into being.
- Evidence on the benefits/harms of existing interventions comes into being.
- Important outcomes come into being.
- Evidence that verifies the current practice is optimal comes into being.
- Values place on outcomes.
- Resources for Bell's treatment available for health care come into being.

20.4 Concluding Remarks

The prevailing trend of EBM gives opportunity for the development of acupuncture because the principle of humanistic care, individual-oriented treatment, and social economics which EBM demands is in accordance with the features and advantages of acupuncture. Moreover, the advanced medical concepts and scientific means of

EBM provide a higher platform for the improvement of acupuncture therapeutic effectiveness, the development of acupuncture research, education, and health care decision making. The reflection on the combination of the theory and methodology of EBM with the feature of acupuncture will encourage the thoughts adjustment in acupuncture clinics, expand acupuncture treatment spectrum, and improve its therapeutic efficacy and research quality as well as education.

Acknowledgments We very appreciate the invitation from Prof. Xia that we could introduce our research team's ideas and results on combination of evidence-based medicine with acupuncture. Most of the researches were funded by Key Project of Chinese National Programs for Fundamental Research and Development (973 Program). We want to thank all the participants, especially the Hunan University of Traditional Chinese Medicine and Hubei University of Chinese Medicine, for their full support. Special appreciation goes to all clinical centers for technical, administrative, and material support and Chengdu GCP for statistical analysis. At last, we would like to make a grateful acknowledgment to all participants and authors for their contribution in developing evidence-based acupuncture and moxibustion.

References

Atkins D, Best D, Briss PA, Eccles M, Falck-Ytter Y, Flottorp S, Guyatt GH, Harbour RT, Haugh MC, Henry D, Hill S, Jaeschke R, Leng G, Liberati A, Magrini N, Mason J, Middleton P, Mrukowicz J, O'Connell D, Oxman AD, Phillips B, Schünemann HJ, Edejer TT, Varonen H, Vist GE, Williams JW Jr, Zaza S; GRADE Working Group (2004) Grading quality of evidence and strength of recommendations. BMJ 328(7454):1490–1494.

Forbes A, Jackson S, Walter C, Quraishi S, Jacyna M, Pitcher M (2005) Acupuncture for irritable bowel syndrome: a blinded placebo-controlled trial. World J Gastroenterology 11(26): 4040–4044.

Itoh K, Katsumi Y, Hirota S, Kitakoji H (2006) Effects of trigger point acupuncture on chronic low back pain in elderly patients-a sham-controlled randomized trial. Acupuncture in medicine 24(1):5–12.

Kleinhenz J, Streitberger K, Windeler J, Güssbacher A, Mavridis G, Martin E (1999) Randomized clinical trial comparing the effects of acupuncture and a newly designed placebo needle in rotator cuff tendinitis. Pain 83(2):235–241.

Liang FR, Wu Xi (2009) Evidence-Based Acupuncture [M]. Beijing: People's Publishing House. 20–25. (In Chinese)

Liang FR, Ren YL (2009) Data mining and clinical decision making of acupuncture [M]. Chengdu: Sichuan publishing Group Bashu publishing House.15–25. (In Chinese)

Liang FR, Wu X, Li Y (2006) Current State and Prosperity of Evidence-Based Acupuncture. Tianjin Journal of Traditional Chinese Medicine 23(6): 441–444. (In Chinese)

Li Y, Liang FR, Yang XG, Tian XP, Yan J, Sun G, Chang X, Tang Y, Ma TT, Zhou L, Lan L, Yao W, Zou R (2009) Acupuncture for Treating Acute Attacks of Migraine: A Randomized Controlled Trial. Headache 49:805–816.

Li Y, Liang FR, Yu SG, Li CD, Hu LX, Zhou D, Yuan XL, Li Y, Xia XH (2004) Efficacy of acupuncture and moxibustion in treating Bell's palsy: a multicenter randomized controlled trial in China. Chinese Medical Journal 117 (10): 1502–1506. (In Chinese with English abstract)

Li Y, Liang F, Yu S, Liu X, Tang Y, Yang X, Tian X, Yan J, Sun G, Chang X, Zheng H, Zhang H, Ma T (2008) Randomized controlled trial to treat migraine with acupuncture: design and protocol. Trials 9:57–64.

Li Y, Liang FR, Zhao L (2009) The Design, Implementation, Supervision and Reporting of Multi-centered Acupunctural Clinical Trial. Chinese Journal of Integrated Traditional and western medicine 29(7):652–655. (In Chinese with English abstract)

Linde K, Streng A, Jürgens S, Hoppe A, Brinkhaus B, Witt C, Wagenpfeil S, Pfaffenrath V, Hammes MG, Weidenhammer W, Willich SN, Melchart D (2005) Acupuncture for patients with migraine: a randomized controlled trial. JAMA 293(17):2118–2125.

MacPherson H, Altman DG, Hammerschlag R, Youping L, Taixiang W, White A, Moher D (2010) Revised Standards for Reporting Interventions in Clinical Trials of Acupuncture (STRICTA): Extending the CONSORT Statement. Acupuncture in Medicine 28:83–93.

Ministry of technology State Administration of TCM of PRC (2007) Key Projects in the National Science and Technology Pillar Program during the Eleventh Five-Year Plan Period, Specification of Quality control and maintenance of TCM research [M]. 12:15. (In Chinese)

Nabeta T, Kawakita K (2002) Relief of chronic neck and shoulder pain by manual acupuncture to tender points-a sham-controlled randomized trial. Complementary Medicine 10(4):217–222.

Sackett DL, Rosenberg WM, Gray JA, Haynes RB, Richardson WS (1996) Evidence based medicine: what it is and what it isn't. BMJ 312(7023):71–72.

Sackett DL, Straus SE, Richardson WS, Rosenberg W, Haynes RB (2000) Evidence based medicine: how to practice and teach EBM [M]. Second Edition. Churchill Livingstone: Edinburgh, 1–2.

Timmermans S, Mauck A (2005) The promises and pitfalls of evidence-based medicine. Health Aff (Millwood) 24(1):18–28.

Wang JL (2001) Clinical Epidemiology –design and evaluation [M]. Shanghai: Shanghai technology and science publishing house, 47, 142. (In Chinese)

Zhao L, Liang FR, Li Y, Zhang FW, Zheng H, Wu X (2009) Improved quality monitoring of multicenter acupuncture clinical trials in China. Trials 27(10):123.

Chapter 21
Chinese Medicine as a Model of System Biology: Diagnosis as the Foundation of Acupoint Selection

Henry Johannes Greten

Abstract Acupuncture did not fall from heaven. Instead, it was developed by human beings based on (1) clinical findings and (2) a background of a pre-existing regulatory theory. In recent years, our ideas of the ancient corpus medicus of Chinese Medicine have become more complete. One of the main findings is that yang and yin, like one and zero, can form binary numbers which can be used to describe regulatory processes on an almost mathematical basis. This basis was already present in the *I Ging*, a book at least 700 years older than the *Yellow Emperor's Classic of Internal Disease*. It contains mathematical systems based on circular functions (sinus functions) that may be used to form a tool to roughly describe the ups and downs often seen in regulatory processes.

From a nowadays standpoint it seems that, based on such theories, the clinical findings and signs and the palpatory features of the skin had been categorized by the ancient inventors of Chinese Medicine in order to practically select and search for useful acupuncture points for the individual case.

This chapter gives you insight in the basic structure of such systems based on the teaching of so-called organ patterns (orbs/functional circles) of Chinese Medicine. Clinical signs and symptoms are grouped systematically to select adequate acupuncture points.

From a western point of view this is comparable to a theory on how to find skin reflexes to treat clinical patterns of vegetative dysregulation. This contemporary understanding of acupuncture and its background in Chinese Medicine offers rational access to TCM, has proven its efficacy by double- and triple-blinded studies and is frequently referred to as the Heidelberg Model of Chinese Medicine.

H.J. Greten (✉)
Heidelberg School of Chinese Medicine, Karlsruher Str. 12, Heidelberg 69126, Germany

Institute of Biomedical Sciences Abel Salazar (ICBAS), Porto University, Lg. Prof. Abel Salazar, 2, Porto 4099-003, Portugal
e-mail: heidelbergschool@aol.com

Y. Xia et al. (eds.), *Current Research in Acupuncture*,
DOI 10.1007/978-1-4614-3357-6_21,

Keywords Heidelberg model • Chinese medicine • CM • TCM • Ear • Nose and throat medicine • ENT • System biology • Acupuncture

21.1 Introduction

In the first section the method of acupuncture is defined as reflex therapy and presented in its therapeutic context. It is emphasized that acupuncture as a reflexological method derives from the *context of Chinese Medicine* and includes the *stimulation of skin points with various other stimuli*. The data available so far as to the proof of efficacy are explained, and the role of functional diagnostics of acupuncture is underlined.

In the second section a short overview is given of the inner logics of functional diagnostics of Chinese Medicine as a vegetative functional diagnosis (reflexological status) as a basis for acupoint selection. Chinese Medicine is represented as a cohesive scientific system, the principles of which are based on a system of binary numbers yin and yang. From this model of circular functions, further technical terms are derived such as the phases ("elements"), organ patterns (orbs), and others. The full diagnosis consists of four modules: constitution (vegetative functional type and personality type of the patient), agent (pathogenic factor, the trigger of disease), "organ pattern," and the so-called guiding criteria.

In the third section some practical aspects of acupuncture in selected clinical conditions are given. Key symptoms of common pain syndromes, which are significant for the selection of points and the stimulation technique in practice, are translated into Chinese medical language.

21.2 Acupuncture as a West–Eastern Reflex Therapy

Three aspects should be emphasized here:

1. *Tissue Activation.* The extimal "full" (repleted) points are predominantly associated with *increased tissue activation* (in Chinese Medicine: *more "qi," "fullness" or repletio*) of a target tissue of the body, where the functionally activated target tissue sensorially appears as extended or "full." This sensation can be explained as a purely sensory phenomenon in which the functional increase in microcirculation may contribute to the sensation of expansion. For pain therapy, this primarily means that pain increases on pressure. According to the classical rules, such conditions require deactivating interventions at the sensitive skin areas, such as the needling technique of "dispelling" (*dispulsion*).

   On the contrary, the "emptied" points of the skin are associated with *decreased tissue activation* (in Chinese Medicine: *less qi, "emptiness" or depletio*) of the target tissue. For pain therapy, this means that pain, one of the key symptoms which go along with this functional deactivation, becomes better on pressure, for example, when a patient intuitively presses his aching stomach with his hands. Such conditions demand stimulating interventions on these rather indolent skin areas such as the needling technique of *suppletion* which is also sometimes called tonification.

This example shows that specific key symptoms are associated with the functional properties hypothesized by Chinese Medicine. These key symptoms then lead to the specific selection of acupoints and stimulation techniques for an individual clinical condition.

2. *Spatial Pattern of Sensations*. The functional finding of palpation and, by that, the functional state of the skin therefore correlate—comparable to Head's concept—with a vegetative functional state of the tissue within the interior of the body. Accordingly, the related sensations of "over-fullness" or "emptiness" principally correlate with the picture of vegetative activation. Chinese Medicine holds that not all regions of the body (body islands) have the same degree of activation (content of qi) at the same time but that there are different regions that are regularly activated in specific sequences. This is a functional picture, quite analogous to the vegetative activation patterns of western physiology in which, according to certain functional necessities, certain tissues may be activated according to the functional challenges of life. For instance, the region of the heart will be stimulated by the sympathetic nervous system during physical exercise, and the region of the heart (the body island) may be felt more intensely than before. According to this principle, even less dramatic challenges like keeping up with the regular homoeostasis of vegetative functions may be seen as a complex pattern of a spatial activation of parts of the body resulting in a spatial shift of activation and therefore minute sensations of "fullness" and "emptiness." As most vegetative functions work on a subconscious level, this may not regularly be "felt" unless irregular patterns of activation take place. Consecutive spatial changes in the skin lead to the thesis of a "flow of qi" or activation patterns.

   Chinese diagnosis may be understood as an analysis of physical sensations in different parts of the body, and as these sensations reflect vegetative functional changes, this leads—from a western point of view—to a functional vegetative diagnosis through a systematic categorization of physical sensations. Functional activation or inactivation of the tissues within the body is part of complex vegetative patterns accompanied by various functional signs, as explained below.

   Chinese Medicine has a regulatory theory on vegetative activation which is quite compatible with western physiology. As a general idea, Chinese Medicine describes physical sensations of the patient, whereas western physiology describes the single vegetative mechanisms underneath these sensations. Spatial patterns of activation may lead to the sensation or assumption of "flows" of qi which may be interpreted as spatial shifts of activation patterns and the vegetative capacity to function of defined tissues.

3. *Qi Flow in the Conduits*. As the vegetative activity in the skin areas reflects the internal spatial shifts of vegetative activation, such a sensation of flow may be

felt on the skin, too. The connections between such skin areas are named conduits or "meridians" as they look similar to the meridians on geographical maps. If the functional spatial transitions of the vegetative steering of tissues within the body are disturbed, this is, according to the classical teaching, mirrored within the consecutively activated points of the skin and may be felt or palpated. This is sometimes referred to as a functional blockade or an "energetic" block and is regarded, according to Chinese Medicine, as the main reason of pain. If, by therapeutic intervention, these functional transitions are restored, this may lead to a temporary enhancement of the vegetative functional sensation on the skin which is known as the *deqi sensation*. According to a western understanding, this may most likely be a transmission of the vegetative functional sensation by C-fibres which may be propagated along the conduits (meridians).

Activation patterns of parts of the body with consecutive key symptoms, according to Chinese Medicine, are due to a functional power called qi. Regarding the key symptoms, these signs, which in Chinese medical understanding are hypothetically due to the disturbed functional power qi, are almost entirely due to vegetative activation or deactivation patterns according to western medical understanding. Therefore, in a first approach and—reduced to the medical context—we can define the term of qi in an auxiliary manner:

Neurovegetative Auxiliary Definition of Qi. Qi is the vegetative capacity to function of an organ or tissue which may sensorily present as a sensation of tearing, pressure, or flow.

As the normal vegetative activation patterns take place automatically, "nothing special" happens, and the individuals in most of the cases do not have any extraordinary sensations. However, "non-normal," so to speak,"pathological" vegetative patterns of activation inside the body may be individually noticed as some sort of hard-to-describe movement, like a "whiff" or "breath," like a "cloud"; in fact, all these meanings can be detected in the classical Chinese emblem of qi.

The etymological explanation for the word qi 氣, written in its traditional form, is "steam 气 arising from rice 米 as it cooks."

As qi may be perceived by the senses as a sensation of flow, tearing or pressure, the traditional Chinese concept of disease as a disturbance of the flow of qi in parts of the skin which goes along with functional symptoms can approximately be translated into western physiology. All these analogues lead to the thesis that those symptoms which are called disturbance of qi in Chinese Medicine may be vegetatively induced symptoms, for example, in the depth of the tissue, which go along with sensory changes of skin areas.

It is well known that within the human being all these vegetative patterns are regulated as part of the homoeostasis of the body. One vegetatively induced functional state passes over into the next, and all these countless vegetative processes together form a functional vegetative continuum which on the whole leads to a great deal to the homoeostasis of the body (cf. periodicity of homoeostasis). This is why homoeostasis in western medicine is not understood as a steady state but as a continuous and balanced flow of functions in which the individual function is constantly superseded by the next. This makes it difficult to detect, for example, by measurement, what the vegetative nervous system actually does in a specific moment.

Chinese Medicine understands disease as a block of qi flow, and one might literally translate this roughly as a disturbance of functional vegetative transitions with the consequence that the specific function becomes sensorily detectable, becomes "unnormal" to the individual and therefore appears to be symptomatic. "Something is wrong within the body," but this is sometimes hard to express verbally for the patient. The individual symptomatic function emerges from the continuum of the innumerable processes within the body. This is interpreted as an obstacle, a blockade of the transitions. According to Chinese Medicine, this is called a block of qi flow, and this can be translated into western functional categories as a block of vegetative functional transitions. According to the fact that vegetative functions are mirrored by the skin, this leads to functional changes in skin areas which are called conduits or meridians.

Chinese Medicine has developed a couple of theories which describe, categorize, and systemize the functional processes which we call the vegetative functional continuum of homoeostasis in western medicine. They explain quite rationally what actually happens in the body and are in no way esoteric or mystical in its origin. However, all too often this complex theory needs an exact wording and usage of its terminology, and all too often it is presented to us in a drastically shortened or imprecise form as it is introduced to us in short curricula for social, economic, and historical reasons. This theory is decisive for the accurate use and definite success of therapy, as it can explain to us the indications, effects and combinations of acupuncture and the pharmacological knowledge it.

21.2.1 Peripheral and Central Effects of Acupuncture

The effects of acupuncture can nowadays be well explained by neurophysiological mechanisms underlying this kind of therapy. One can divide them into peripheral and central effects.

- *Peripheral effects* of acupuncture are those effects which directly affect the symptomatic target tissues involved. They include improvement of microcirculation, relaxation of muscles, and activation and deactivation of tissue functions such as peristalsis or effects on nasal and bronchial secretions and other functions of mucous membranes, etc.
- *Central nervous effects* are those effects of acupuncture which have been shown within the brain or the spinal cord and which are believed to change the functional perception of our body and by that of the symptoms. These functions include the release of endorphins, the stimulation of complex pain suppression mechanisms, for example, as part of the gate-control theory of pain and other mechanisms,

and the influence on neurohumoral mechanisms of regulation. Some important effects take place on the level of bone marrow where C-fibres mediate sensations on the laminae 3 and 4, which lead to further mechanisms of pain reduction in the periaqueductal grey and are especially important for the perception of pain.

21.2.2 Unspecific and Specific Effects of Acupuncture

- *Unspecific effects* of needling are effects which may basically be induced by penetration of a needle by any kind on any part of the skin. They consist of a bunch of physiological and also suggestive effects, for example, the momentary elevation of the sympathetic tone, release of adrenalin and changes in vigilance by stimulation of the reticular arousal system (formatio reticularis). Some of these mechanisms may also lead to a phenomenon that is sometimes called peripheral counter-stimulation which is understood as some sort of masking a sensation by eliciting a second one which, within our perception, covers the first one and hides sensations like pain and others.
- *Specific effects*, which can only be elicited by the stimulation of acupuncture points, go far beyond these mechanisms. Since the 1970s, these are known from animal experiments and clinical applications.[1] These specific effects are the reason why we need criteria for the selection of acupoints to determine the individual optimum combination of points. The specific effects may be seen as a local enhancement of peripheral microcirculation, control of heart rate (up- and down-regulation), of blood pressure, effects on peristalsis and the vagal tone (tilt table), of peripheral pain, and numerous other vegetative functions which may affect the local origins of pain and also induce functional changes in specific areas of the brain (Cho et al. 2001).

21.2.3 Effects According to the Technique of Stimulation

The effect of acupuncture as a reflexology phenomenon is also determined by the kind of skin stimulus. As physiology knows the term of an adequate stimulus for sensory organs and structures, the diversity of the effects which may be elicited by acupoints, according to Chinese Medicine and western physiology, may tremendously be increased by this. An acupuncture point may even unfold opposite effects according to the kind of stimulation and may or may not have an effect depending on the stimulation frequency (Ying et al. 2007).

[1] A comprehensive list of studies on specific effects of acupuncture points can be downloaded from http://www.dgtcm.de.

Private photo

Fig. 21.1 *Moxibustion*. The term moxibustion stands for the burning of moxa, a weed called mugwort in English. It is distributed in cigarette-like rolls: The needle is pinned through this moxa cone which mediates the healing warmth into the depth of the reflex points. By this, long-term histamine release via mast cells is activated, so the reflex point is active for a longer time

The most important kinds of stimulation are:

- Vegetatively *activating techniques*, which are called "suppletive" or tonifying methods, replenish the functional capacities ("energies").[2]
- *Deactivating techniques*, which are called "dispulsive" methods, distribute the functional capacity of an over-reactive target tissue and thereby return it to normal.
- *Warming techniques*, for example, moxibustion (Fig. 21.1), increase microcirculation within the target tissue and thereby treat the most frequent origin of pain in daily practice, the so-called algor/"cold" reflex (see below).
- *Cupping, bloody cupping and bloody needling*. They all have an effect on the content of fluid within a tissue and seemingly affect the venous outflow of microcirculation.
- *Manual techniques* of Chinese manual therapy (*tui na an mo*). There are more than 50 different methods to influence target tissues, zones of the connective tissue and acupoints. (Downloads of video clips, e.g. on http://www.dgtcm.de).
- *Physical stimulation techniques* like the application of electric current in various frequencies or the application of magnetic balls (the so-called Gecko Ball) (Zhou 2004).
- *Pestle acupunture* is an interesting application of stimulatory metal pestles of various kinds.
- *Scratching or scraping* which is frequently combined with different oils, partly with etheric components.

[2]The word "energy" was only introduced into the language of Chinese Medicine in 1932 by Soulié de Morant as a tentative translation; it does not occur in the classical literature. A more appropriate translation is, for example, "functional power" or "capacity to function".

– *Taping*
– *Application of ointments and injection of various substances* like V-poison, ant poison, or *homoeopathic substances (sino-homoeopathy).*

21.2.4 Clinical Findings of Palpation

It seems that the major theoretical concepts of acupuncture, especially the concept referring to the findings of the surface ("extima"), have been developed by the clinical findings of *palpation*. Chinese Medicine holds that so-called external pathogenic agents may also affect the skin and via the skin may influence internal findings of the body, for example.

– Palpable changes such as *swelling* of the skin have been found to be connected with functional changes of internal functional tissue (the yin) postulated to be induced by humor/"humidity." These changes are similar to the western concept of *oedema* reducing tissue functions.
– Palpable *coolness* of the tissue is associated with functional alterations postulated to be originated by algor/"cold." This vegetative reaction is defined by internal functional changes similar to the consequences of a *regional lack of microcirculation*. The key symptoms are sensations of
  • Coolness
  • Stiff muscles and connective tissue
  • Loss of functional capacity ("energy") of the tissue
– Palpable changes in skin areas which give the impression of
  • Warmth
  • Sweat
  • Reddishness
  • Underlying spasticityare found together with clinical changes of the functional tissue as if induced by a draught of air. This reaction is called ventus/"wind." These reactions are somewhat comparable to the effect of driving a cabriolet without appropriate protection:
– Red eyes and runny nose
– Swollen aching tonsils
– Spastic muscular overfunction

These changes may be explained by vegetative reactions such as the well-known *substance-P mast-cell reflex* induced via the skin and the mucous membranes over the *pterygopalatine ganglion* and its connections to the *Waldeyer's lymphoid ring* (the lymphatic tissue surrounding the "entrance" of the body: the mouth, nose, and larynx) such as the tonsils.

Furthermore, the pattern of muscular spasms may be interpreted as being originated by certain patterns of motion and an inborn pattern of motion which can still be observed in swimming human babies. Eliciting these patterns of *swimming motions* may induce symptoms such as:

- Sudden torticollis
- Shoulder pain
- Sciatic pain

We refer to these patterns of vegetative reactions in more detail below.

The idea that functional changes of the skin are associated with vegetative clinical changes in various other functional tissues is illustrated by these findings. They contribute to the understanding of CM as a medical concept focusing on the vegetative functional state of the body which in brief can be described as "full" and "empty" or induced by *environmental factors*.

These vegetative changes are again connected with:

- Both regional and systemic changes of microcirculation.
- Functions of the capillary bed.
- The distribution of body fluids.
- The interaction of plasmatic and cellular blood functions.
- The endothelium.
- The surrounding connective tissue and spaces.
- Parenchymal target cells.
- The *immune system* as well is influenced by the vegetative system and vice versa. This leads to the specific interpretation of *defence mechanisms* described by the *model of six stages*.

A contemporary and scientific understanding of Chinese Medicine therefore focuses on the *three major basic regulatory mechanisms* of the body. These three systems are:

- The vegetative system
- The microcirculatory system
- The immune system

For which three distinct models of key symptoms have been developed in Traditional Chinese Medicine. They are all interconnected with the functional tissue (the yin) through which their functions (the yang) arise.

21.2.5 Selection Criteria

There have always been differing opinions about the criteria of acupoint selection and the selection of stimulation techniques, both in China and the west. In general, in China, various schools argue about the exact exegesis and interpretation of classical texts or about traditional family schools, whereas in the west, an ambivalent understanding of acupuncture developed.

A core aspect of this argumentation always was the significance and weight of the concepts of Chinese Medicine for the selection of acupoints. Of course, for such a discussion the quality and mediation of the concepts of TCM play an important role which appeared—for historical and linguistic reasons—to be sometimes reduced to an ethnomedical export product (Müller 2000). As majority of the western medical doctors were not convinced of the seemingly fluffy categories of yin, yang, Wood, Fire, etc., in many western countries a type of acupuncture developed which we might call the western topographical or schematic acupuncture.

This school of western topographic acupuncture tries to "clean up" acupuncture from the remainders of "mystical" teachings and was partly supported by Chinese specialists who appeared in TCM schools within the course of the Cultural Revolution.

A second school, however, tried to take over functional concepts from China, mostly with a reduced background of the technical meaning of scientific Chinese terminology. Language problems and the fact that this terminology seemingly contradicts Chinese Medicine contributed to the image of TCM appearing as a second incompatible kind of medicine which could only exist in parallel with western medicine but could hardly share the functional language and concepts (Porkert 1997).

The first fraction of the acupuncture experts were more numerous and influential, and they performed major acupuncture studies which were financed by the social security system of Germany during the last 10 years. Studies involved several thousands of patients but unfortunately showed that the western topographic criteria of acupoint selections are not superior to randomly chosen skin points. Sometimes it is extended to the thesis that the effect of acupuncture may even not be due to specific effects of specific skin areas but to unspecific suggestive or placebo effects. Of course, such an interpretation is not compatible with the specific effects shown since the 1970s in animal and other clinical arrangements.

Therefore, we must take a closer look to some aspects of these studies below.

21.2.6 *Significance of Acupuncture Within the Multimodal Treatment Concept of TCM*

Acupuncture is part of Chinese medical methodology and is generally combined with other therapeutic methods. Insofar, one can say that in the early classical times the still modern concept of a multimodal therapy of pain had already been developed in which acupuncture is combined with manual therapy (tuina) and pharmacotherapy, for example, by herbs, minerals, and other pharmacological agents. Another component of therapy may be the treatment by specific exercises like the methods of qigong or taiji which may be considered as concentrative breathing and movement exercises which are the major predecessor of the autogenic training of western medicine. These techniques of self-perceptions can, from a western standpoint, be understood as traditional vegetative biofeedback treatments as they induce vegetative functional changes and an enhancement of self-perception at the same time. An overview of the therapeutic contexts in Chinese Medicine is given in Table 21.1.

Table 21.1 Overview of methods combined in TCM to form a multimodal concept of therapy

Method	Execution	Traditional ratio of overall TCM	Western understanding
Acupuncture	Stimulation of the skin by needles and burning of moxa	50 %	Reflexes suppress the local development of pain, transmission of pain signal, and pain perception by the brain; regulation of numerous vegetative mechanisms
Chinese pharmacotherapy	Application of herbs, minerals, and other components as concoctions, pills, or ointments	80 %	Application of bioactive substances with pharmacological activity in systematic combinations; proven effects such as anti-inflammatory action, induction of apoptosis, suppression of macrophage functions, and effects on the nervous system
Chinese manual therapy (tuina)	Treatment by the hands	Not quantifiable	Systematic effect on reflex points, connective tissue zones, peripheral receptors of muscles, muscle tension, microcirculation of soft tissues, and mobilization of joints
Chinese dietetics	Treatment by food	Not quantifiable	Vegetative and functional effects of food (Chinese functional food) as a complement to the western description of the effect as induced by carbohydrates, lipids, proteins, vitamins, etc.
Taiji qigong	By exercises (traditional vegetative feedback therapy)	Not quantifiable	Concentrative breathing and moving exercises, vegetative feedback therapy (Greten 2004), training effect on stress, depression, over-stimulation, and in anxiety syndromes; lowers cortisol, enhances immunoglobulins (Kreutz et al. 2004), enhances microcirculation within the capillaries, brain metabolism, and QoL, used in drug withdrawal and detox (Li et al. 2002), and applied in cancer therapy (Chen and Yeung 2002)
PTTCM	Psychotherapy of Chinese medicine	Not quantifiable	A TCM-specific way to describe somatopsychic interactions, patterns, and symptoms of the locomotor system, and vegetative patterns are detected and understood as a subconscious evaluation system; based on the model of the four-layered ontology (functional anatomy of the "Es" of psychoanalysis)

In western pain therapy as well, acupuncture as a method of reflexology is mainly not used as a stand-alone method. Pharmacotherapy, manual therapy, physiotherapy, relaxation methods, psychosomatic treatment, lifestyle changes, and other concepts are parts of the mosaic which forms the multimodal therapy of pain in the west. Some countries have decided to pay for acupuncture by the social security system for certain indications which further augments the position in the multimodal concept of pain therapy.

21.2.7 *Controversial Data on Efficacy*

21.2.7.1 Aspects of the Methodology of Research

The data for the proof of efficacy of acupuncture are controversial. In animal experiments and clinical applications the specificity of the effects of acupuncture has already been proven to a high extent. The basic phenomenon of acupuncture effects depending on the kind of stimulation, such as the frequency of electro-acupuncture (Ying et al. 2007), has also been shown extensively. Naturally the proof of efficacy in man reveals some methodological difficulties. For instance, in acupuncture research—like in all other natural sciences—the quality of scientific studies largely depends on the controls and double-blinding. Some of the problems which make it difficult to evaluate acupuncture on a scientific basis are:

- There is no real placebo which means a placebo without real physiological effects (Streitberger and Vickers 2004).
- Some of the so-called sham acupunctures as used in some of the main studies have putative effects on the symptomatology by segmental mechanisms, the zones of Head, and other mechanisms (Hummelsberger and Ernst 2005).
- Blinding of punctures is a specific problem of acupuncture research, as especially in studies with a lot of participants even single-blinding ended up in premature unblinding (Wettig 2006).

So in certain studies with longer intervention intervals, the participants could inform themselves about controls and verum over the internet and exchange their information in specific chat rooms via the net. By this there was no blinding and no controls in the respective studies that could be taken seriously.

In the GerAc (Molsberger et al. 2002; Endres et al. 2007) and ART (Brinkhaus et al. 2003; Melchart et al. 2003; Hummelsberger et al. 2003) studies there was no significant difference between needling just any unspecific points of the skin (sham acupuncture) and real acupoints. Nevertheless, both kinds of needling, for example, in the treatment of gonarthrosis, were superior to non-steroidal analgesics. This superiority of acupuncture over non-steroidal painkillers even led to the reimbursement by the social security system in Germany. Surprisingly, sham acupuncture is not reimbursed, although it is comparable in the effects and easier and cheaper to establish in practice.

21.2.7.2 Studies into Schematic Acupuncture

Furthermore, a meta-analysis of 33 acupuncture studies (Linde 2003) revealed that verum acupuncture and needling of unspecific skin points (non-acupoints) or sham acupuncture had only minor differences in the success of treatment. Sometimes this is interpreted in the sense that the location of the puncture has no influence on the effects of acupuncture. This interpretation, however, contradicts the numerous animal experiments and clinical data in which specific effects were already objectified.

The results of these biggest acupuncture studies ever made are more understandable if we look at the selection criteria of the points. In fact, the allocation of acupoints for the treatment depended largely on western diagnoses without prior definition of the key symptoms and the guiding findings of Chinese Medicine.

According to tradition, the *selection of points* and the selection of the kind of stimulation are made on the basis of *key symptoms* arising from the patient's history and examination. These key symptoms are then ordered according to an objectifiable hierarchy—at least in the scientific form of Chinese Medicine. To current understanding,[3] a vegetative functional status (Greten 2006) is derived from this order of key symptoms, i.e. the (pathological) reflexological state in which the patient presents. This is the basis for selecting the point and the stimulation pattern for healing reflexes (acupuncture) in the individual case.

21.2.7.3 Studies into Acupuncture with Functional Diagnosis

Doubts as to the methodology of the large and, in part, politically motivated acupuncture studies contributed to the development of the first *double-blind design in acupuncture research* which allowed a better comparison of the effects of different acupuncture methods (Karner et al. 2005).

In fact, the first double-blind study into acupuncture in gonarthrosis already showed that the benefit of treating acupoints that had been selected and stimulated on an individualized basis according to previously defined criteria was approximately twice as high in terms of movability of the joint than that of acupoints selected according to those of the GerAc study (Molsberger et al. 2002). The increase in efficacy of acupuncture by the previously determined key symptoms for point selection and stimulation technique, that is the use of the diagnostic rules of TCM, could already be established for a number of further parameters and diagnoses:

- *Pain and walking distance* in peripheral arterial occlusive disease (PAOC) (Buhlmann et al. 2006)
- *Pain after tonsillectomy* (Sertel et al. 2009a, b)
- *Airway resistance (rhinomanometry) in rhinitis* (Sertel et al. 2009a, b)
- *Pain after sternotomy* (Kick et al. 2006; Sack et al. 2007)
- *Heart failure* (Greten et al. 2008)
- *Polyneuropathy* (Schröder et al. 2007, 2008)

---

[3] State Administration of Traditional Chinese Medicine, Beijing, download under http://www.hscm.asia

The fact that the therapeutic response depends on the quality of the functional diagnostics is an argument in favour of promoting the classical criteria of selection and stimulation.

21.3 The Functional Diagnostics of TCM

21.3.1 Historical Aspects

Chinese Medicine, which had been pushed into the background, has been recompilated again since the 1950s in order to provide basic medical care for the population in China which at that time was a developing country. The result was a brief version of Chinese Medicine which then was called Traditional Chinese Medicine (Fig. 21.2).

Traditional Chinese Medicine is the recompilation of medical applications which developed in the 1950s–1970s and was aimed at the rapid distribution of pragmatic concepts. The result was a reduction of its theoretical and diagnostic foundations.

21.3.2 Diagnosis of Four Components

A more recent recompilation generated in cooperation with Chinese scientists is the so-called Heidelberg Model which combines the essential classical schools of Chinese Medicine and also includes ancient sources such as the *I Ging* (Fig. 21.3).

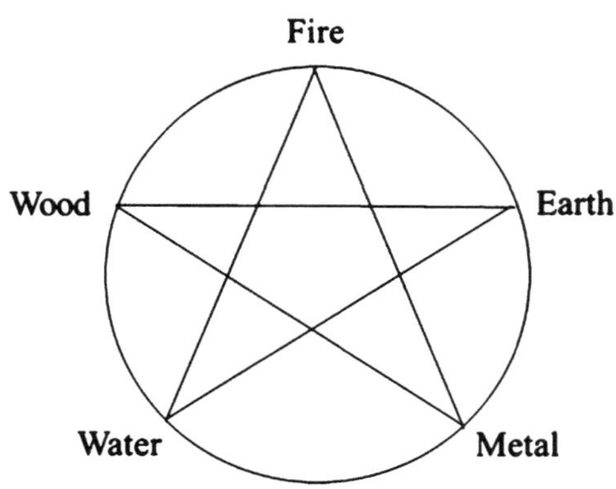

Fig. 21.2 *Reduced "philosophy" of TCM*. The so-called sheng or co-cycle shown above stems from the third century before Christ. Feature constellations of clinical signs were arranged according to "elements" and balanced against each other. This representation has so many restrictions in its origin that it simply cannot be seen as the core of the theory of TCM. It was in the 1950s–1970s that the theory of TCM began to be based on this representation. Acupuncture diagnostics was thus reduced to providing a tool of basic medical care for masses instead of providing a complex regulative theory (Modified from Greten HJ, Kursbuch Traditionelle Chinesische Medizin, 2nd ed., Thieme 2007)

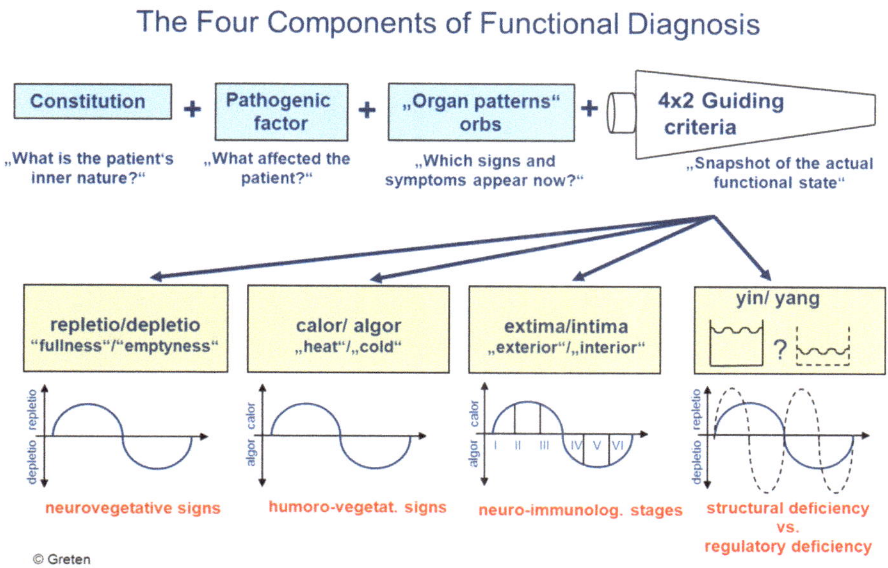

Fig. 21.3 *Components of diagnosis* (Modified from Greten HJ, Kursbuch Traditionelle Chinesische Medizin, 2nd ed., Thieme 2007)

A complete Chinese diagnosis therefore inevitably consists of the components constitution, agent (pathogenic factor), orb, and guiding criterion.

1. *The constitution* of a patient in western language describes the vegetative reaction type of the patient including at the same time a behavioural and emotional typology (the "inner nature" of the patient). In pain therapy this is above all decisive for conflict processing patterns which play a central role in pain processing. According to Chinese Medicine, the constitution also significantly determines expressivity (including pain experience), body tension, and the guidance of joints by connective tissue (e.g. in hypermobile spine), and also bondage intensity up to bondage addiction, co-addiction, and a tendency to particular forms of compulsion (so-called pulmonary and renal compulsions).
2. *The agent*/pathogenic factor causes the dysregulated, symptomatic state of the patient. In pain therapy the agent primarily determines the modalities of pain, such as worse with cold, change of weather, and draught. Therefore, the respective agent is defined by key symptoms (Table 21.2). In most cases the agent also determines the *stimulation technique of needling*. The *palpation finding* of the skin, subcutaneous tissue, and muscles is for instance the primary criterion for selecting the *stimulation technique in pain management* by local points. It is accompanied by typical forms of general condition and typical tongue and pulse patterns (Xu et al. 2008). Specific reflectory mechanisms can already be allocated to these findings whose exact mechanisms, however, are still under research.

Table 21.2 The three most important agents and their key symptoms

Agents/pathogenic factor	Ventus/"wind"	Algor/"cold"	Humor/"humidity"
Symptoms	"As if you had been exposed to a draught of air"	"As if you had been exposed to environmental cold"	"As if you had been exposed to environmental humidity"
Reflex pattern	Reaction of the draught defence type	Reaction of the cold defence type	Reaction of the humidity defence type
Postulated mechanisms	Substance-P mast-cell reflex Old reflexes of motor control as known from fish and other species	Local microcirculation disorder	Pre-oedema or oedema
Reaction pattern in the direction of	Yang (repletio, Wood)	Yin (depletio, metal, and water)	Yin
Skin	Warm and sweaty	Cold and dry	Bloated, doughy, sticky
Tongue	"Raspberry dots"	Hyaline coating (bloodless, gel-like)	Sticky coating, tooth impressions
Character of pain	Sudden onset, shooting	Gradual onset, tearing	dull
Muscles	Spasm, myogelosis	Muscles not stretchable, rigid	Swollen, feeling of heaviness
General condition	Driven	Exhausted	Dyspnoea, feeling of heaviness
Stimulation technique	Dispulsion	Suppletion/warming	Generally suppletion, in Wood–Earth imbalance also dispulsion
Exemplary points	F 20 *stagnum venti* Ic 4 *valles coniunctae* Ic 15 *fossa subacromiale* It 6 *senectus felix* It 10 *inductorium lacerti* "All points in the face" Points of the guillotination line and others	Tk 5 *clusa externa* L 9 *fons tumuli yin* S 32 *lepus subreptus*	All points of the lienal conduit (except L10) S 40 *abundantia* S 36 *vicus tertius pedis* L 3 *candidum maius* and others
Exemplary drugs	r. ledebouriellae	ram. cassiae	rh. alismatis, poria alba

In the presence of the key symptoms it is important to select both the right point and the right needling technique in order to achieve the response desired.

External agents are vegetative reaction types of the defence type against cold, draught, humidity, etc. (Table 21.3); *internal agents* are emotions. *Neutral agents* are, for example, overwork, physical effects, and poisoning.
3. *The orb* ("organ pattern" or "functional circle" or circle for short) is the current disturbance pattern which also includes the current main symptom as can be seen from the example of the hepatic orb/"liver" circle in Fig. 21.5. Like this, the symptom is regarded as part of a group of predefined diagnostic signs. These groups of signs ("organ patterns" or "functional circles" or circles for short, Latin: orbs) are an expression of the body regulation (see below) and are named after organ regions (body islands) which can become symptomatic within the overall pattern. Examples are the hepatic orb, cardiac orb, pulmonal orb, and renal orb. They are manifestations of a vegetative functional tendency which is called phase (see below).

Like this, the symptoms are allocated in a first approach to these three categories, and the exact individual symptom pattern is defined. Then the symptoms are interpreted in the context of the regulatory overall status of the patient. For *determination of the overall status* four fundamental physiological models are used on which the so-called guiding criteria (Fig. 21.3 and Table 21.3) are based. According to current understanding, like this, the status of the neurovegetative system (Fig. 21.3 below left, outside); the humoro-vegetative system, i.e. the fluid phase with microcirculation pattern (below left); the neuroimmunological system (below right) and the functional tissue (below right, outside) is determined by using the key symptoms. This allows to evaluate from what region of control the current symptom most likely results.

All components of diagnosis are exactly determined as part of a reflex pattern or vegetative functional state, and adequate reflex points are selected which are then needled according to their individual significance for the main complaint.

In the *functional diagnostics of Chinese Medicine* the selection of acupoints and the stimulation technique are invariably derived from the following components:

1. *Constitution* (vegetative reaction type of the patient)
2. *Agent* (pathological factor, see also Table 21.2)
3. *Orb*/functional cycle (current symptomatic orb or localization of the symptom)
4. *Guiding criterion* (overall evaluation of the regulatory status)

21.4 TCM as System Biology

The inner logic of the functional Chinese diagnosis is shown in the following by the example of the teaching of the so-called first guiding criterion. This first guiding criterion repletion ("fullness")/depletion ("emptiness") can be interpreted as an *interpretation structure of signs and symptoms* which, according to western understanding, are of neurovegetative origin. In Chinese Medicine they are allocated to the functional power qi and categorized as groups of diagnostically relevant signs, functional circles, or rather *orbs* (Chin.: *zhao*).

Table 21.3 Guiding criteria and their significance in determining the overall regulatory status

Chinese guiding criterion	Level of regulation	Physiological process	Chinese perception
Repletio ("fullness")	1. Neurovegetative regulation	Increased activation by the nervous system	"To much capacity to function ('qi') in body and circuits"
Depletio ("emptiness")		Decreased activation by the nervous system	"Insufficient capacity to function ('qi') in body and circuits"
Calor/heat	2. Humoro-vegetative regulation ("interplay of plasma, endothelium and parenchyma")	Increased microcirculation with pro-inflammatory effects and raised sympathetic tone	Excess activation of xue/blood (blood and its effects)
Algor/cold		Decreased microcirculation with lack of inflammatory effects and sympathetic tone	Insufficient activation of xue (blood and its effects)
Extima/outside	3. Neuroimmunological regulation	Early stages of disease	Defence mechanisms active on the surface of the body
Intima/inside		Late stages of disease	Defence mechanisms active in the depth of the body
Yang (problem of the degree of unfolding function)	4. Cellular mechanisms	Control of functional tissue defect from outside	Problem on the levels named above
Yin (problem of the functional tissue)		Defect in the cellular control of the functional tissue	Problem resides in the cellular functional tissue, in the degree of hydration or in tissue supply

Levels of regulation (in western language) and their analogues in Chinese medical theory in a first approach (easy entry for western readers)

It is one of four physiological models of TCM which taken together can be understood as a system-biological model of body regulation. Like this, the principle of diagnostics does not at all rest on mythology but rather on an almost mathematical model of regulation.

We want to work out this fundamentally different approach of Chinese medical theory in a brief overview. For further details please see the references cited. After a brief description some regional and symptomatically simplified treatment concepts are given at the end of this chapter which nevertheless have proven to have some value in important pain scenarios in daily practice.

21.4.1 Periodicity of Homoeostasis

Homoeostasis (literally: staying equal) is a fundamental property of biological systems. This staying equal is not to be understood as a stationary state but rather as a continuously changing steady state equilibrium. If this equilibrium is disturbed, symptoms develop.

Definition. Homoeostasis means "the maintenance of relatively stable internal physiological conditions... under fluctuating environmental conditions" (Roche Lexikon Medizin 2010).

In order to illustrate the processes of regulation in homoeostasis in a simple model, homoeostasis can be explained by the example of an incubator (Fig. 21.4).

As shown in Fig. 21.5, regulative processes are subject to periodic changes which, according to the understanding of western measuring and control technology, are a fundamental property of regulated systems. One can describe this up and down in a simplified manner as a sinus wave and characterize the quadrants of the sinus wave, the phases I–IV, as the phase of providing potential, the functional phase, the deactivation phase, and the regeneration phase.

In Chinese Medicine these deviation tendencies are described with a technical language of its own. The major defining terms of TCM, yin (below target value), yang (above target value), and the phases (phases I–IV), are also called directional norm conventions (Porkert 1974); this means an understanding of directed ("vectorial") description.

Thus, the regulation-induced fluctuations of homoeostasis in this model lead to periodicity of the actual state with a certain deviation direction from the target value. Human beings are also subject to the periodicity of homoeostasis, and the symptoms are described as an expression of the deviation from the (individual) target value.

The order and categorization of symptoms in TCM follows an exactly classified hierarchical principle. Symptoms such as pain develop as a deviation from the target value upward (yang) or downward (yin). Then, the symptoms of people can be further defined, namely, as functional deviation tendencies of the whole system from the target value according to the phases I–IV which are called evolutive phases or just phases.

To simplify, the symptoms can be understood as part of an overall pattern of diagnostic signs which indicate the functional deviation pattern from the target

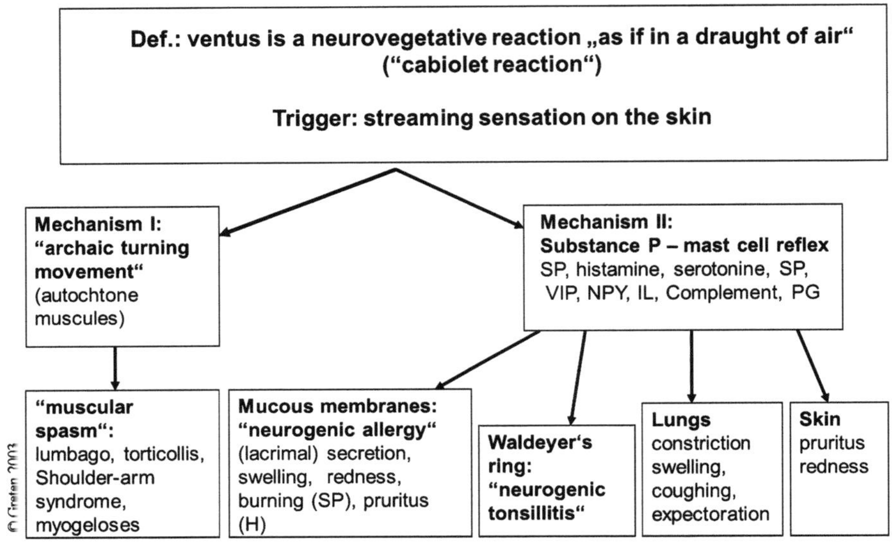

Fig. 21.4 *Ventus/“wind” as an example of an external agent.* Chinese Medicine defines this vegetative reaction of the draught defence type by key symptoms such as those that may occur after driving in a cabriolet. They include sudden spasm (e.g. torticollis, headache), reddening of the skin and mucosa, tonsillitis, and cough. Western medicine attributes these symptoms to a combination of reflectory archaic swimming motions and the substance-P mast-cell reflex. In comparison to other points, the needling of ventus/wind points can reduce nasal resistance in rhinomanometry (Sertel et al. 2008) (Modified from Greten HJ, Kursbuch Traditionelle Chinesische Medizin, 2nd ed., Thieme 2007)

value. In order to better understand these patterns one should know that the phases have developed from binary numbers by which circular processes and sinus curves can be described.

21.4.2 Yin, Yang, and the Phases as Binary Numbers

Apart from general and philosophical meanings, yang and yin have also found a technical application as a numbering system. This had already been completely represented in the *I Ging*, the oldest book of mankind on which Chinese Medicine is based.

The German philosopher and mathematician Leibniz (1646–1716) decoded this binary number system of the *I Ging*. He also derived the arithmetic rules for calculating with binary numbers (Fig. 21.7) and developed from it both the mathematical foundations of information technology and the decisive main ideas for his most important work, the Theodicee (Fig. 21.6).

Basically these numbers can be composed by yang and yin (I/0) to form so-called bigrammes (numbers I–IV), trigrammes (1–8 see Fig. 21.12), etc. These numbers

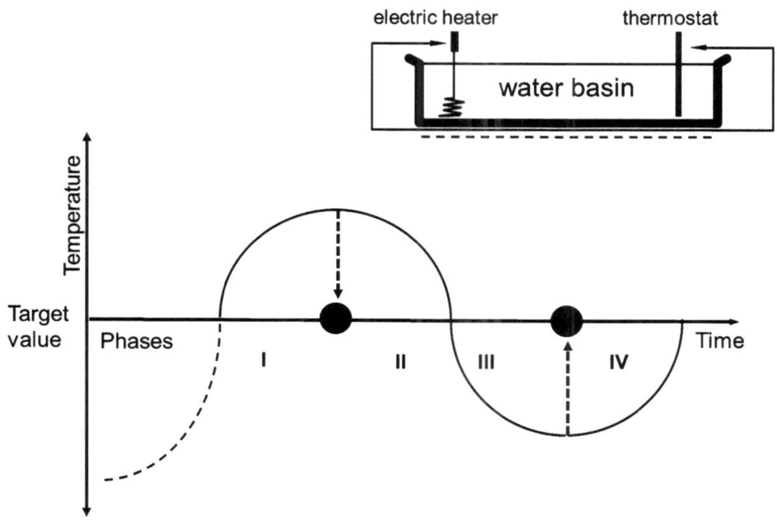

Fig. 21.5 *Fundamental aspects of regulation by the example of an incubator.* The temperature of a water basin or an incubator in a lab is regulated via a thermostat and an electric heater. A target value of for example, 37 °C is adjusted, and periodic changes of the actual value are observed. *Phase 0 (dotted line)*: the temperature rises from room temperature to reach the target value. *Phase I*: reaching the target value means turning off the heater by the thermostat. Afterheat is present in the heater like an electric stove when you turn off the current. Consequently the temperature rises above the target value. Thus, potential above target value, so-called overgrade potential, is accumulated. *Phase II*: temperature descends back to target value as the capacity of the heater is used up after the peak value. Heat within the basin is then given away to the environment. Overgrade potential of phase I is then transformed to heating the surroundings. This phase therefore is featured by transforming potential into function, here heating up the environment. *Phase III*: temperature and the physical potential of the system reach a minimum. Turning the system into an undergrade content of energy (potential) of this phase is featured by the turning point of energy content and a rhythmical capacity, as well as lack of energy. Keywords are the following: turning point, rhythm, and functional relaxation. *Phase IV*: energy content recovers as the heater has warmed up over the actual temperature. The key feature of this phase is regeneration of energy content (Modified from Greten HJ, Kursbuch Traditionelle Chinesische Medizin, 2nd ed., Thieme 2007)

by tradition are represented as solid line (yang) or broken line (yin) (Fig. 21.7 above left). Circular processes have been described in this way since classical times.

21.4.3 Phases are Part of a Circular Process

In mathematics the "regulatory sinus wave" can also be transformed into a circular function (Fig. 21.7). In China this teaching of cycles had been developed in great detail and initially described with binary numbers. It is interesting to note that these numbers were largely replaced from the third century before Christ by the emblems Wood, Fire, Metal, and Water (the so-called phases) (Fig. 21.7 above right).

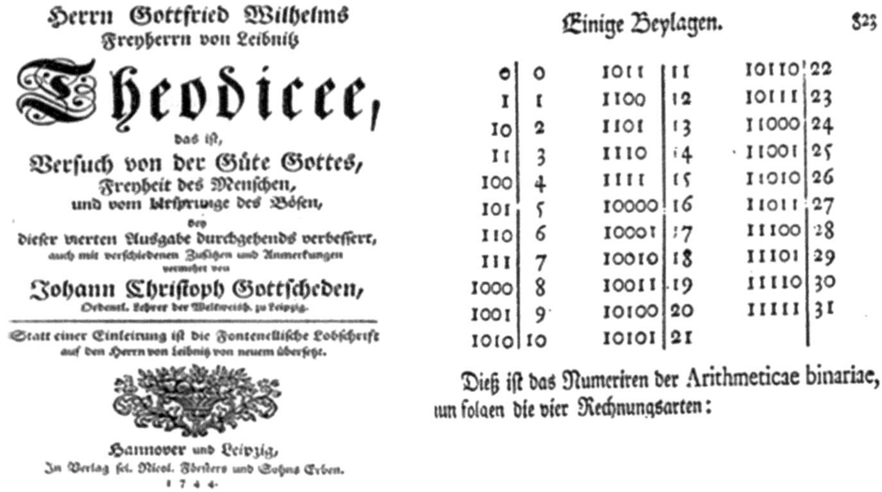

Einige Beylagen. 523

0	0	1011	11	10110	22
1	1	1100	12	10111	23
10	2	1101	13	11000	24
11	3	1110	14	11001	25
100	4	1111	15	11010	26
101	5	10000	16	11011	27
110	6	10001	17	11100	28
111	7	10010	18	11101	29
1000	8	10011	19	11110	30
1001	9	10100	20	11111	31
1010	10	10101	21		

Dieß ist das Numeriren der Arithmeticae binariae, nun folgen die vier Rechnungsarten:

Fig. 21.6 *Left:* cover of the Theodicee, the main work of Leibniz. *Right:* extract from Leibniz' work showing the character of 0/I. He is quoting the arithmetic rules of binary numbers ("arithmeticae binariae") Leibniz, Theodicee, 1744

Phases were therefore initially parts of a circular process. The Earth (axis of the system, the target value) by tradition is placed in the centre of the cycle which in Fig. 21.7 below corresponds to the *x*-axis.

21.4.4 Phases are Regulatory Terms: "Harmony" of Yang and Yin

As we have seen, the circular function with the quadrants I–IV (Wood, Fire, Earth, Metal, and Water) also describes the functional state of regulated systems. This view of cyclicity of natural processes therefore implicitly also means that they are subjected to regulation. Thus yin, yang, and the phases are regulatory ("cybernetic") terms.

When looked at in this perspective, systems are in fact well regulated if the areas above (yang) and below (yin) the *x*-axis in Fig. 21.7 are equal (yang and yin are in "harmony" what is meant technically rather than romantically).

The directions of movement

- Wood (in the yang but still below = yang in yin)
- Fire (in the yang but still above = yang in yang)
- Metal (in the yin but still above = yin in yang)
- Water (in the yin but still below = yin in yin)are also balanced and "harmonious."

The writing as a cycle or sinus wave can also be transferred to a common symbol. The target value of the function is then symbolized as a dot on the *x*-axis.

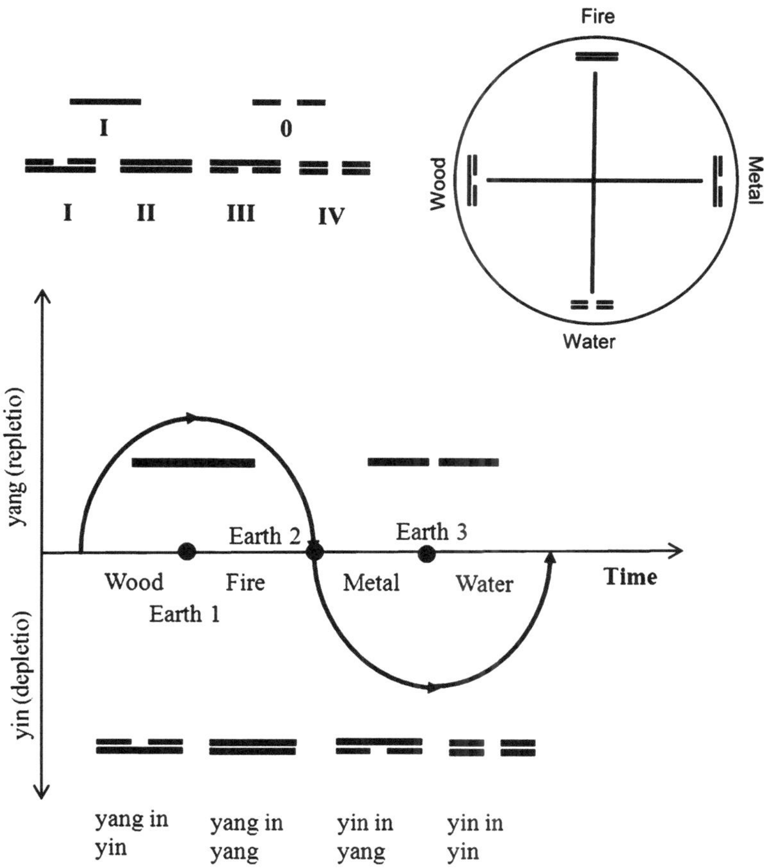

Fig. 21.7 *Above left:* I (*solid line*) and 0 (*broken line*) can be combined to form binary numbers. *Above right:* the quadrants of a circle can be described by them. In the third century before Christ, they were replaced by the emblems Wood, Fire, Metal, and, Water (so-called phases). Phases were therefore initially parts of circular process. The Earth (axis of the system, target value) is by tradition placed into the centre of the circle. *Below*: transformation into a sinus curve which constitutes a simple model to represent periodic changes of regulated systems. The circular function with the quadrants I–IV (Wood, Fire, Metal, and Water) now describes the functional state of regulated systems. Yin, yang, and the phases are therefore regulative (cybernetic) terms. In the schematic representation the Earth (target value) is marked as a dot on the x-axis according to the axis of the cycle. Between Wood and Fire, the Earth is seen as down-regulation of excess values and between Metal and Water, as up-regulation (Modified from Greten HJ, Kursbuch Traditionelle Chinesische Medizin, 2nd ed., Thieme 2007)

This emblem is referred to as taiji sign, foci sign, or yin–yang sign (Fig. 21.8). Leibniz uses the term monad (sign of the unity of the opposites I and II). This mathematical emblem symbolizes the unity of yang and yin, target value, and actual value. It emblematizes the mathematical and regulatory meaning of yang and yin when describing cybernetic systems.

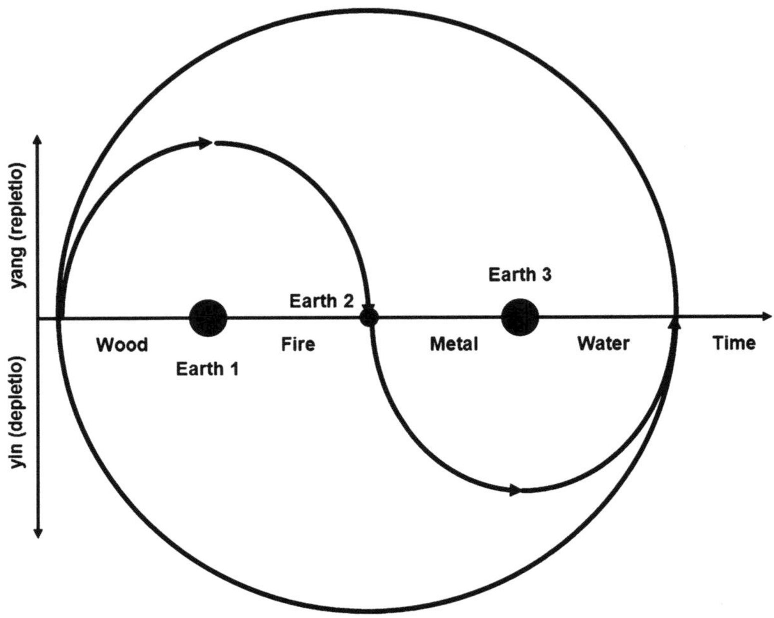

Fig. 21.8 Both forms of writing a circle, the circle and the sinus wave, are united in one symbol. The *two dots* show the target value of a regulated system (Modified from Greten HJ, Kursbuch Traditionelle Chinesische Medizin, 2nd ed., Thieme 2007)

21.4.5 Phases Are Vegetative Functional Tendencies

Transferring this curve to the regulation of the vegetative overall activity of man results in the following picture (Fig. 21.9):

In the above picture the parameters of western medicine such as transmitters and vegetative mechanisms are shown. They form a complex interplay, a concert of transmitters and mechanisms which determine the degree of regulatory activity.

In the phases of Wood and Fire (quadrants I and II of the curve) they predominantly represent activating mechanisms, and in the phases Metal and Water, deactivating and regenerative mechanisms which are known in western medicine. Meanwhile, further 132 mechanisms of this kind have been allocated to the regulatory curve from a western point of view.

21.4.6 Phases Manifest in Clinical Signs (Orbs)

The drawings in the upper part of Fig. 21.9 represent the four clinical pictures in which the effect of these transmitter-induced mechanisms can also be detected clinically. The patterns correspond to the functional patterns of Wood, Fire, Metal, and

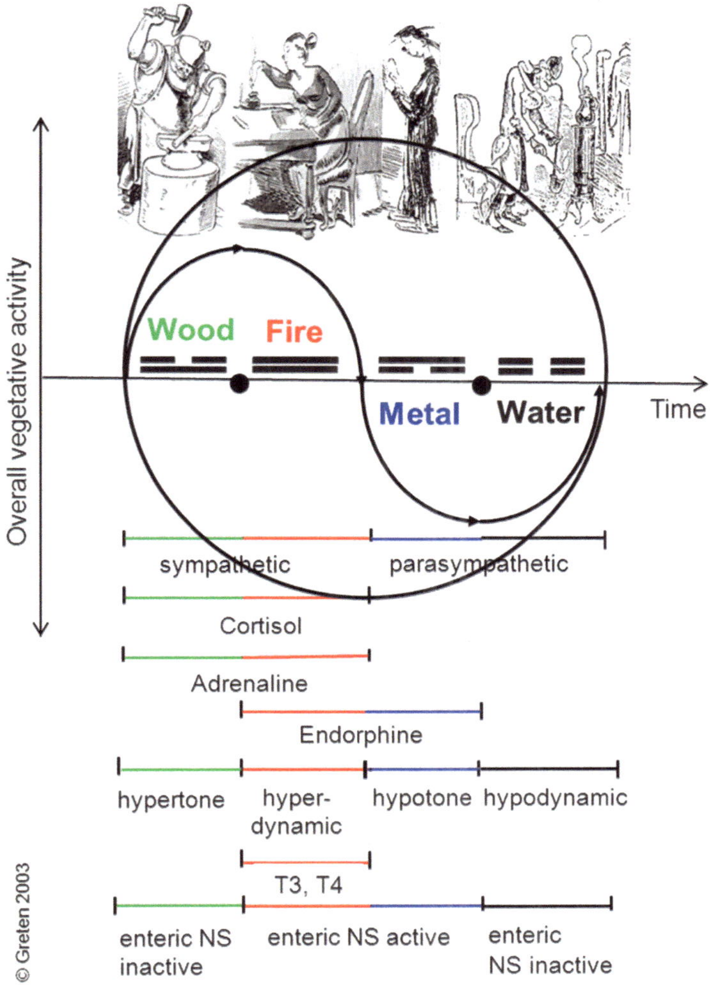

Fig. 21.9 The upper part of the picture shows human functional states which can be differentiated by their key symptoms (diagnostically relevant signs). Such groups of key symptoms are called orbs in Chinese Medicine. The lower part of the picture shows a selection of vegetative functional mechanisms which in western medicine are seen as the causes of these signs. The upper part shows the symptoms in the language of Chinese Medicine; the lower part shows the same symptoms in the language of western medicine. For orientation, the yin–yang code (*yin solid line, yang broken line*) has also been included again in the form of bigrammes. So, TCM really is based on yang and yin (Modified from Greten HJ, Kursbuch Traditionelle Chinesische Medizin, 2nd ed., Thieme 2007)

Water which represent *vegetative functional tendencies*. These vegetative functional tendencies are expressed in groups of diagnostic signs which are called circles of function or rather orbs ("organ patterns" or circles of diagnostically relevant signs).

Symptoms such as pain are interpreted in this way as an expression of a "derailed" vegetative functional direction.

Note: Chinese Medicine uses phases to describe vegetative functional tendencies which, in western medicine, are described by a concert of vegetative transmitters, control circuits, and mechanisms. They manifest in specific diagnostically relevant clinical signs which are grouped together (orbs).

They are named after organ regions ("body islands") which, as part of these patterns, may become symptomatic. These body regions are part of the interior of the body which react together with the skin regions mentioned at the beginning. Therefore, they are not "organs" in Chinese Medicine but—in western medical language—neuroaffective activation patterns of man. This term refers to the fact that such an "organ pattern" also has an emotional dimension (see Fig. 21.10 "hepatic orb").

The teaching of the "organs" of TCM (*zhang fu*) actually is a teaching of vegetative functions and their clinical signs rather than a teaching of organs in western understanding. In these vegetative functional activation patterns, a *body island* ("sensation region") also shows in symptoms (e.g. by the sensation of being stretched or emptied). Groups of signs are named after these "organ regions." This clearly shows that the "liver," for example, has another meaning in Chinese Medicine than in western medicine. In western medicine the liver is an organ with clearly defined functions and measurable margins, whereas the orb "liver" in Chinese Medicine is a vegetative functional state which also *can* but not necessarily has to manifest itself in unpleasant abnormal sensations in the liver region.

For pain therapy this means that with the hepatic orb, for instance, the pain may be localized:

- On the hepatic conduit or
- Approximately in the region of the liver.

Moreover, the mechanisms connected with this pattern may become functionally symptomatic. The diagnostically relevant signs of the orbs, for the hepatic orb, for example, clenching one's fist, activation of the eyes, and straightening up of the body initially are not necessarily disease signs as well. However, if their activation is "overgrade," they may become symptomatic as in our example (Fig. 21.10) and present for instance as:

- Writing cramp
- Sensation of pressure of the eyeball
- Muscle cramp of the erecting muscles (physiological extensors)

Reflex points (acupuncture points) can then treat the whole activation system and by this the clinical signs as well as the symptoms of the respective body island (organ region). *Reflex points with an effect on the mentioned clinical signs or the body island of an orb are called conduits* ("meridians").

According to classical understanding orbs are therefore defined as:

- Groups of diagnostically relevant *signs*.
- These *signs* indicate the functional state of a "*body island*" (organ region).
- The state of the body island correlates with the functional state of the *conduit*.

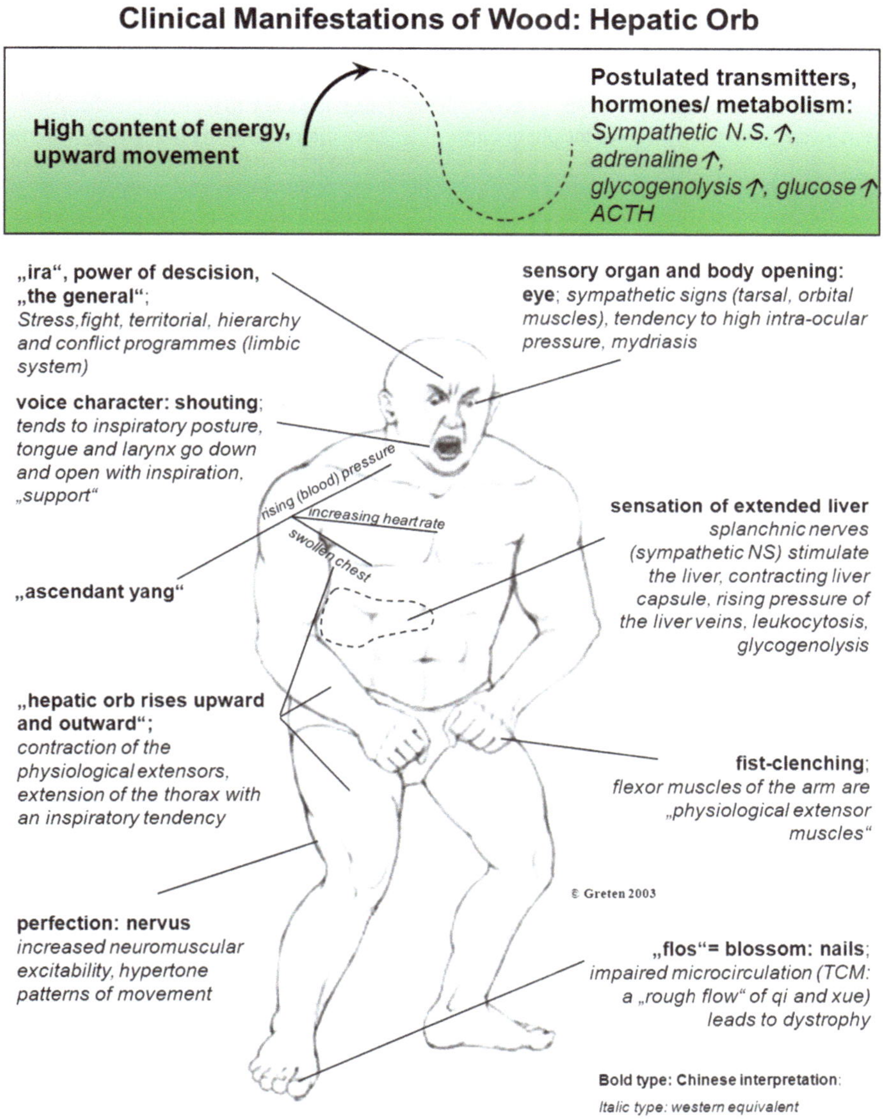

Fig. 21.10 *Hepatic orb ("liver pattern").* The example of the hepatic orb shows how a vegetative functional tendency (here the phase of Wood) manifests in signs that can be recognized from the outside. In the *green part* above, the position of the pattern in the *regulatory curve* is seen together with the most important western transmitter constellations. Therefore the neurological mechanisms which underlie such an activation pattern can be explained today. If such signs are seen, this indicates the state of the body island (here: the region of the liver). Furthermore, reflexes have been observed which act on these signs and have been grouped as acupuncture points on conduits (Modified from Greten HJ, Kursbuch Traditionelle Chinesische Medizin, 2nd ed., Thieme 2007)

Table 21.4 Lienal and stomachal orb

Lienal and stomachal orb	
Lienal orb/"spleen circle"	Stomachal orb/"stomach circle"
Yin, tendency to depletio	Yang, tendency to repletio
Storage orb generating the qi of the phase Earth	Transition orb storing the "energy" (qi) of the phase Earth, releasing it according to requirements and transporting it
Transforms, metabolism; brings "the clear" upward. If transformation is disturbed, "postprandial syndrome"	Intermediate storage, brings "the turbid" downward and directs "energies" into the orbs
Excretes continuously "waste substances" (in western terms), i.e. humor and pituita (TCM)	Needs to be moisturized by fluids
Analogous to metabolism	Analogous to digestive tract
Brings upward	Brings downward
"Holds the flesh"	–

Fig. 21.11 *Lienal orb/"spleen circle" and stomachal orb/"stomach circle"* (Modified from Greten HJ, Kursbuch Traditionelle Chinesische Medizin, 2nd ed., Thieme 2007) (From Wilhelm Busch)

21.4.7 Orbs of the Phase Earth

The down-regulation and the up-regulation can also be interpreted as vegetative functional tendencies (phases). These "inner movements" can manifest in clinical signs which are called stomachal orb/"stomach circle" and lienal orb/"spleen circle." The most important characteristics are shown in the Table 21.4 above Fig. 21.11.

So far, six orbs can be explained by the phases, i.e. one orb each for the four phases Wood, Fire, Metal, Water, and two orbs for the phase Earth.

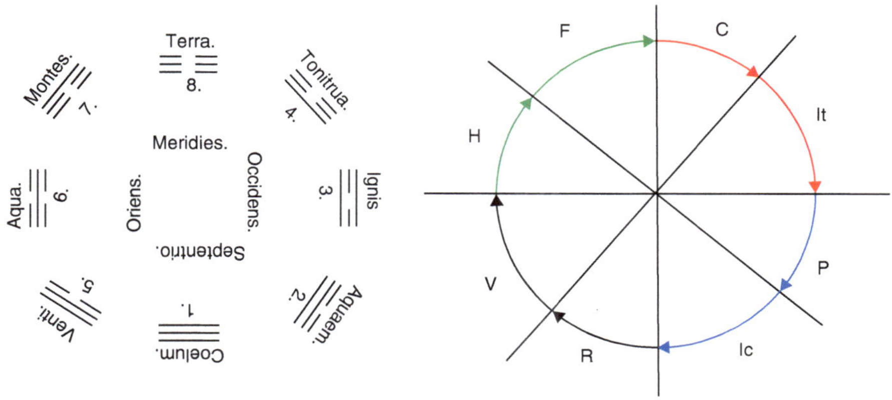

Fig. 21.12 *Trigrammes in the representation of Leibniz*. The trigrammes are represented here as a division of the cycle movement into eight sections. They bear symbolic names which are not discussed here in more detail. Apart from the above-mentioned four orbs/circles, four additional intermediate steps are obtained at the transition into the quadrants of the phases which are called transition orbs. Later, they will be extended by two further orbs/circles. Apart from the teaching of the eight basic orbs/circles, the theory of the "eight directions" (*ba gua*) has survived in some variants of CM. In feng shui, the teaching of spatial perception, this representation is still of central importance *Left*: Leibniz, Theodicee, 1744; *right* © Greten

21.4.8 System of Ten Orbs/Functional Circles: Transition Orbs

So far we have divided the regulatory cycle function into four quadrants (bigrammes of yang and yin), arriving at four main orbs/circles. Of course, this cycle function can also be divided into eight trigrammes. This corresponds to dividing the regulatory cycle function into eight sections (Fig. 21.12) resulting in additional intermediate steps of the functional transition of homoeostasis.

In this way we obtain four additional orbs/circles which represent the transition between the phases. This is why they are called transition orbs/circles.

From these eight regulatory states "functional themes" of the orbs/circles arise which are to "control," i.e. regulate specific functions (Fig. 21.13). This can be shown both in circular form (Fig. 21.12 right) or on the level of the sinus curve (Fig. 21.13). In addition there are two further Earth orbs resulting in a primary system of ten orbs ("organ patterns" or functional circles).

From this, *hierarchical pathways of diagnostic decisions* can be derived, which may considerably improve diagnostics of vegetative patterns in patients.

If a patient is, for example, in the yang state ("over-fullness with qi" i.e. "fullness," *repletio*), the next question is whether he rather expresses Wood or Fire symptoms. If he expresses, for example, more Wood symptoms than Fire symptoms, the next decision to be taken is between the hepatic orb/"liver" circle and the felleal orb/"gallbladder" circle.

In this way the diagnosis can exactly be made by directional steps (directionalities) within a *decision arbogram of yang and yin*. This ultimately corresponds to a

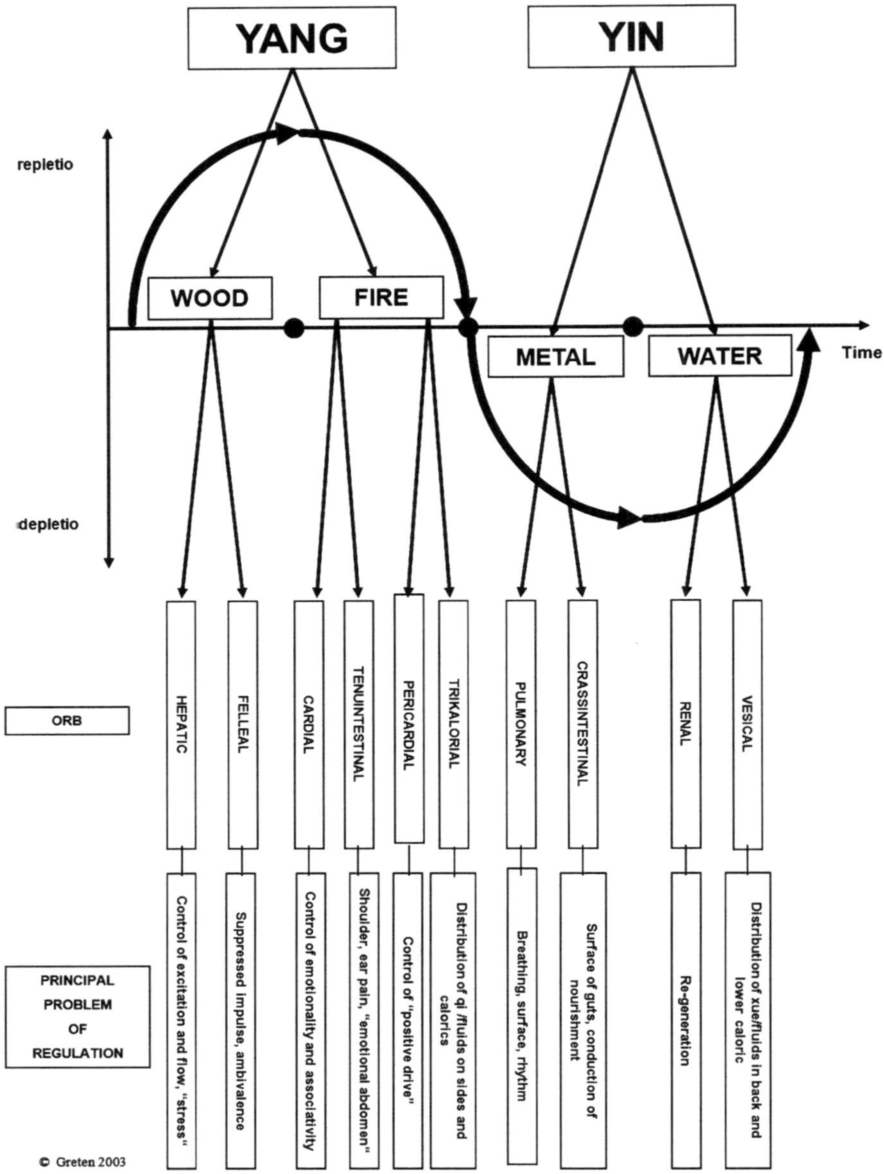

Fig. 21.13 *Order of the orbs ("organs") as a decision arbogram for diagnostics.* For technical reasons, the pericardiac and the tricaloric orb/"triple burner" circles were included in the phase Fire (pericardiac circle and triple burner circle). Yang and yin (in man "fullness"/repletion and "emptiness"/depletion) are differentiated by the phases into ascending and descending movements in the yang and yin. These phases are then further differentiated into orbs/functional circles (Modified from Greten HJ, Kursbuch Traditionelle Chinesische Medizin, 2nd ed., Thieme 2007)

differentiation of the vegetative sinus curve of the phases and, similar to a computer language, opens up the possibility to find an exact diagnosis by "simple questions" (Chin.: Su Wen). This, however, requires an exact knowledge of the system which for lack of space is only presented in a brief survey here.

21.4.9 The Present System of 12 Orbs/Functional Circles

Originally in the phase Fire, symptoms were described as well which are the result of excessive activation of xue/"blood." More and more detailed functions had consequently been allocated to xue/"blood" so that the teaching of the functions of xue/"blood" is expressed in a guiding criterion, i.e. a system interpretation structure of this own, the second guiding criterion.

In the further course of history, the *pericardiac orb*/"pericardiac" circle and the *tricaloric orb*/"triple burner" circle were added resulting in the present order of 12 orbs/functional circles.

21.4.10 Further Guiding Criteria: Calor/"Heat" and Algor/"Cold"

The regulatory approach can be seen in the example of the phases and the orbs/functional circles. Such regulatory curves do not only exist for the overall vegetative regulation but also for the up- and down-regulation of microcirculation. This determines, for instance, the degree of inflammability of a disease and is referred to as the criteria calor/"heat" with increased microcirculation and increased central nervous excitation on the one hand and algor/"cold" with markedly diminished microcirculation and reduced vigilance and unspecific defence reactions on the other.

Definition: The excess dynamics (over-activity, "yang") of xue/"blood" in Chinese Medicine is called calor/"heat"; the diminished dynamics (under-activity, "yin") is called algor/"cold." The symptoms indicating these conditions are summarized in Table 21.3.

Xue, "moved structivity" or "blood," is a term which describes all fluids moving in the body and their effects, i.e. their functions. In a narrower sense, the term xue comprises in western understanding the effects of microcirculation (xue/"blood") on tissue function and interstitium.

The second guiding criterion can therefore be interpreted as the interpretation structure of signs which, according to western understanding, are generated by *effects of microcirculation*, the biochemical effects in the interplay between blood cells, plasma, endothelium, and functional tissue.

The term calor/"heat" describes the excess functional activation of these factors, whereas the term algor/"cold" describes symptoms resulting from diminished

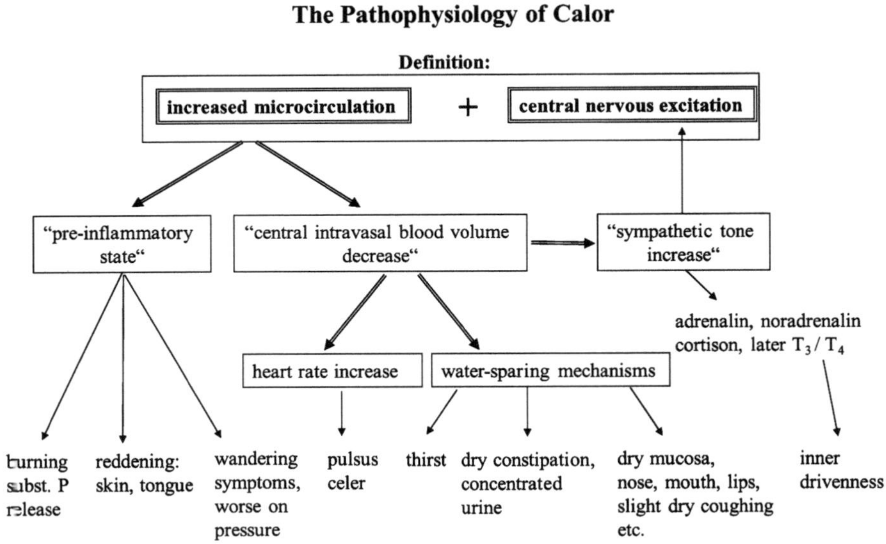

Fig. 21.14 *The signs of calor/"heat"* can today be explained by physiological mechanisms. An increase in microcirculation in the capillaries of the tissues leads to a pre-inflammatory or inflammatory state and to a volume shift of fluids in the direction of peripheral tissues. This results in a lack of volume in the large vessels and consequently in an increase in heart rate, water-sparing mechanisms, sparse urine, and dry mucosa, as well as sympathetic stimulation (Modified from Greten HJ, Kursbuch Traditionelle Chinesische Medizin, 2nd ed., Thieme 2007)

Table 21.5 Symptoms of calor/"heat" and algor/"cold"

Calor/"heat"	Algor/"cold"
Reddish tongue	Pale tongue
Reddish skin	Pale skin
Pulsus celer (rapid pulse)	Pulsus tardus (slow pulse)
Inner drivenness	Poor drive
Small quantity of dark urine, may smell	Large quantity of light, clear urine, almost odourless
Dry constipation	
Tendency to body odour	Body smell unremarkable
Possibly thirst	No thirst
Dry mucosa of mouth, nose, airways	Moist mucosa of mouth, nose, airways
Pain worse on pressure, wandering location	Pain better on pressure, localized
Pain may be burning	Pain better with warmth
Frequently used points in calor/"heat" include	Frequently used points in algor/"cold" include
H3/Lr3, S44/St44 (dispulsive)	Tk5/Tb5, S36/St36, L3/Sp3, R3/Ki3 (all suppletive or moxa)
R7/Ki7 (suppletive)	

activation. Therefore this can be called an *interpretation structure of signs originating from the humoro-vegetative system* (Fig. 21.14).

The clinical signs and symptoms indicating the presence of calor/"heat" and algor/"cold" can be explained today by physiological mechanisms (Table 21.5, Figs. 21.15 and 21.16).

Fig. 21.15 *Calor/"heat."* Sepsis causes, for example, symptoms like feeling hot, red face, rapid pulses, intensely red tongue, thirst, a tendency to constipation, and possibly delirious talking to oneself. The blanket is felt as too warm and thrown away because the patient needs cooling. He appears restless and agitated (From Wilhelm Busch)

Fig. 21.16 *Guiding criterion algor/"cold."* General dislike of cold, ineloquent patient and patient withdraws socially (From Wilhelm Busch)

21.4.11 *Guiding Criterion Extima/Outside or Intima/Inside*

A similar sinus curve exists for those functions which in western medicine are understood as immunological control involving neuronal mechanisms. In brief, this comes down to "Chinese neuroimmunology and rheumatology" which is of particular importance for rheumatic diseases and especially in pain management of joints.

According to this teaching, six functional defence mechanisms are overcome by the agent algor/"cold" which are characterized by *six typical stages*.

If the agent algor/"cold" persists, these stages may become chronic, and the "six syndromes" develop to disease patterns of their own. They are of particular importance for the treatment of infections and immunological residual states following virus attacks, for example, for the treatment of chronic inflammations and collagenoses. Many chronic pain diseases of the locomotor system can reasonably be described and treated by this model. The conduits affected are the same as in acute affections so that the dysfunctional patterns seen correspond to the orbs ("organ patterns" or "functional circles" or circles for short) affected. Examples of western diseases and their relation to the six stages or syndromes are given in Fig. 21.17 together with exemplary treatment options. This algor laedens theory (ALT) (Shang Han Lun, the Theory on Cold Damage) is of utmost importance for symptoms such as fibromyalgia, arthrosis pattern, and posttraumatic disorders and can only briefly be mentioned here.

21.4.12 *Guiding Criterion Yang and Yin*

The fourth partial teaching of Chinese Medicine is the theory of yin, the functional tissue or substrate to be regulated, and yang, the over- or undergrade function arising from it. It is a law of cybernetics that in case of diminished substrate a labile actual/target curve results. This is comparable to the regulation of temperature in a laboratory incubator and the great fluctuations seen when the volume of water, i.e. the substrate to be regulated, is going to be used up. This complex theory leads to the mathematical speculation about the influence of constitutional factors on regulation.

21.5 Concluding Remarks

Acupuncture is the most well-known treatment method of Chinese Medicine. From a contemporary perspective, TCM is a traditional model of system biology based on a vegetative functional and regulatory theory. Its basis may not be only philosophical speculation but a unique model of natural sciences based on regulatory theories. Four subsets of technical terms called guiding criteria are the physiological backbone of TCM. The understanding of the key symptoms of these guiding criteria is the basis of acupoints selection.

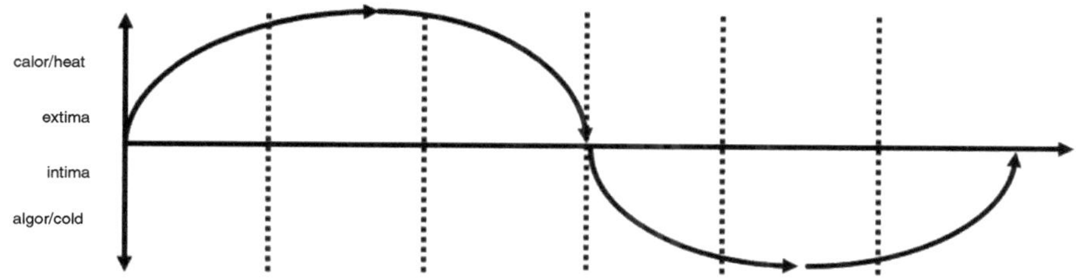

	Stage I	Stage II	Stage III	Stage IV	Stage V	Stage VI
Name	Yang major/ large yang (tai yang)	Splendor yang/ splendid yang (yang ming)	Yang minor/ small yang (shao yang)	Yin major/ large yin (tai yin)	Yin flectens/ yielding yin (jue yin)	Yin minor/ small yin (shao yin)
Pictograms						
Examples of disease in the acute state	Sciatic and ear pain stage	Acute sinusitis (thick head stage)	Hot-cold-nausea stage	Bronchitis with phlegm	„Flat-down" phase (pneumonia)	convalescence stage
Conduits and orb/functional circles involved	V-It / Bl-Si	S-Ic / St-Li	F-Tk / Gb-Tb	P-L / Lu-Sp	H-Pc /Lr-Pc	R-C / Ki-Ht
Coupled conduits and orbs (circles)	R-C / Ki-Ht 2nd circuit	L-P / Lu-sp 1st circuit	H-Pc / Lr-Pc 3rd circuit	Ic-S / Lu-sp 1st circuit	F-Tk / Gb-Tb 3rd circuit	V-It / Bl-Si 2nd circuit
Functional power involved ("energy")	qi defensivum (wei qi) / defensive "energy"	Conduit qi	Conduit xue / "blood"	Body island qi	Body island xue / "blood"	Yin / functional tissue
Exemplary decoction	Dec. ramulorum cassiae Dec. ephedrae	Dec. duorum veterum Dec. ancillae yue	Dec. parvum bupleuri	Dec. ephedrae Dec. fritillarieae	Dec. octo gemmarum with r. bupleuri and r. astragali	Pil. rehmanniae sex saporum and derivatives
Back pain	Lumbalgia (worse when lying on the back or fastening shoe laces)	Lumbalgia (worse when lying on the belly, when sitting, driving tendency to protrusion or prolapse)	Lumbalgia. lumboischialgia (can easily lie on one side, while lying on the other is impossible)	---	----	Back pain worse on prolonged standing, often ice-cold)
Knee pain	Baker cyst, retropatellar chondropathy with the key symtoms of going downstairs worse than upstairs	Most common form of knee pain, arthrosis pain or chondropathy with the key symptom of going upstairs worse than downstairs	Pain worse when turning the foot inward, lateral meniscus	Pain at the origin of the medial ligament, medial meniscus		Pain at night and worse at rest
Frequent joint pain	Achilles tendon, calcaneal spur	Knee, thumb	Hip, coxarthrosis, ankel joint, finger middle joints	Rhizarthrosis	Wrist	Finger basic joints, Achilles tendon
Typical point combination	V40/Bl40 and I66/Si6 "Tsingtao Hoola"	S34/St34 and S25/St25 with Ic4/Li4	Tk5/Tb5 with local points at the joint	L10/Sp10 and L6/Sp6 with P10/Lu 10	H3/Li3 with Pc5 or Pc7	C3/Ht3 or C6/Ht6 with R7/Ki7 or R3/Ki3

Fig. 21.17 The teaching of the six stages (ALT) of the third guiding criterion and western diseases according to ALT stages (see below). The six stages are shown by examples of airway infection and pain. In the outer stages (*I–III*), in most cases, symptoms are seen which occur in the course of the conduits, for example, cold sensations on the extremities and stiff limbs (*I*), pain in the head (*II*) and tearing pressure pain in the epigastric region (*III*). This always bears the risk that the symptoms go over to the inner body regions (body islands) which depend on the conduits. This is why a preventive activation of functions in most cases starts at the same time in the interior as well (mostly an activation of microcirculation, so-called reactive calor/"heat"). As a rule, the stimulation technique to be selected in most cases is "warm externally and dispel internally" (Modified from Greten HJ, Kursbuch Traditionelle Chinesische Medizin, 2nd ed., Thieme 2007)

Acknowledgments Thanks to Petra Fröschen for typing and revising the manuscript.

References

Brinkhaus B, Becker-Witt C, Jena S, Linde K, Streng A, Wagenpfeil S, Irnich D, Hummelsberger J, Melchart D, Willich SN (2003) Acupuncture Randomised Trials (ART) in patients with chronic low back pain and osteoarthritis of the knee design and protocols. Forsch Komplementärmed Klass Naturheilkd 10:185–191

Buhlmann et al (2006). AMSTCAR, Shanghai 2006

Chen K, Yeung R (2002) Exploratory Studies of Qigong Therapy for Cancer in China. Integrative Cancer Therapies 1(4):345–70

Cho ZH, Wong EK, Fallon JH (2001) Neuroscience Basics. Qpuncture

Endres HG, Victor N, Haake M, Witte S, Streitberger K, Zenz M (2007) Akupunktur bei chronischen Knie- und Rückenschmerzen. Dtsch Ärzteblatt 104:A123–A130

Greten HJ (2004) Arbeit mit Qi - Nonsens oder Therapie? Vortrag auf den Deutschen Qi Gong Tagen

Greten HJ (2006) Kursbuch Traditionelle Chinesische Medizin 2. Auflage, Stuttgart, S. 138

Greten J, Kick A, Remppis A et al. (2008) Acupuncture Effects on Heart Failure - How to create objective Study Designs. J. Acupunct. Tuina. Sci 6: 307–308

Hummelsberger J, Melchart D, Willich SN (2003) Acupuncture Randomized Trials (ART) in Patients with Chronic Low Back Pain and Osteoarthritis of the Knee - Design and Protocols. Forsch Komplementärmed Klass Naturheilkd 10:185–191 (DOI: 10.1159/000073474).

Hummelsberger J, Ernst E (2005) Wie geht es weiter mit der Akupunktur? MMW Fortschr. Med. Nr. 25 (147. Jg.) pp.12–14

Karner et al (2005) Heidelberg Blinding Assay. AMSTCAR 2005

Kick et al (2006). AMSTCAR, Shanghai 2006

Kreutz G, Bongard S, Rohrmann S, Hodapp V, Grebel D (2004) Effects of Choir Singing or Listening on Secretory Immunoglobulin A, Cortisol, and Emotional State. Journal of Behavioral Medicine (27), No. 6

Li M, Chen K, Mo Z (2002) Use of Qigong Therapy in the Detoxification of Heroin Addicts. Alternative Therapies in Health and Medicine Jan/Feb 2002; 8 (1): 50–59

Linde, K (2003) Systematische Übersichtsarbeiten und Meta-Analysen—Anwendungsbeispiele und empirisch-methodische Untersuchungen, Habilitationsschrift Humboldt Universität Berlin

Melchart D, Linde K, Streng A, Reitmayr S, Hoppe A, Brinkhaus B, Becker-Witt C, Wagenpfeil S, Pfaffenrath V, Hammes M, Willich SN, Weidenhammer W (2003) Acupuncture Randomized Trials (ART) in Patients with Migraine or Tension-Type Headache - Design and Protocols. Forsch Komplementärmed Klass Naturheilkd 10:179–184 (DOI: 10.1159/000073473).

Molsberger A, Diener HC, Krämer J, Michaelis J, Schäfer H, Trampisch HJ, Victor N, Zenz M (2002) GERAC-Akupunktur-Studien: Modellvorhaben zur Beurteilung der Wirksamkeit. Dtsch Arztebl 99(26): A-1819/B-1539/C-1435

Müller RS (2000) Heil aus dem Osten. Der Spiegel 18, 01.05.2000, S. 156

Porkert M (1974) Theoretical Foundations of Chinese Medicine: Systems of Correspondence. MIT Press (July 15, 1974)

Porkert M (1997) Chinese Medicine Debased: Essays on Methodology. Phainon Editions

Roche Lexikon Medizin, Urban & Fischer Berlin (2010) Glossary, http://www.roche.com

Sack F, Lichtenberg A, Kick A, Remppis A et al. (2007) Controlled Integration of TCM in Heart Surgery. Scientific Approaches to Chinese Medicine, Heidelberg School Editions

Schröder S, Liepert J, Remppis A, Greten HJ (2007) Acupuncture Treatment Improves Nerve Conduction in Peripheral Neuropathy. European Journal of Neurology 2007, 14: 276–281

Schröder S, Remppis A, Greten T, Brazkiewicz F et al. (2008) Quantification of Acupuncture Effects on Peripheral Neuropathy of Unknown and Diabetic Cause by Nerve Conduction Studies. J. Acupunct. Tuina Sci 6: 312–314

Sertel S, Gaugel D, Choudhry A, Giannakaki K, Baumann I, Remppis A, Goncalves M, Wendt I, Greten HJ, Plinkert P (2008) Nasal Airway Resistance Measured by Rhinomanometry: a Novel Tool to Measure Acupuncture Effects J. Acupunct. Tuina Sci. 6: 9

Sertel S, Herrmann S, Greten HJ, Haxen V, El-Bitar S, Simon CH, Baumann I, Plinkert PK (2009a) Additional Use of Acupuncture to NSAID Effectively Reduces Post-Tonsillectomy Pain. Eur Arch Otorhinolaryngol 266 (6)

Sertel S, Bergmann Z, Ratzlaff K, Baumann I, Greten HJ, Plinkert PK (2009b) Acupuncture for Nasal Congestion: A Prospective, Randomized, Double-Blind, Placebo-Controlled Clinical Study. Am J Rhinol Allergy 23(6):23–8

Streitberger K, Vickers A (2004) Placebo in Acupuncture Trials. Pain 109 1–2: 195

Wettig, D (2006) Die GERAC-Migräne-Akupunktur-Studie: Wertlos durch vorzeitige Entblindung. Z. ärztl. Fortbild. Qual. Gesundh.wes. (100); 156–157

Xu L, Meng MQH, Shi C, Wang K, Li N (2008) Quantitative Analyses of Pulse Images in Traditional Chinese Medicine. Medical Acupuncture 20(3): 175–189

Ying X, Fei Z, Jingchung G, Jieshi C, Gencheng, W (2007) Optimal Conditions for Electro-Acupuncture to Protect the Brain from Experimental Stroke. Scientific Approaches to Chinese Medicine. Heidelberg School Editions

Zhou, J (2004) Chinesische Medizin, Kompendium. OZV

Chapter 22
Modernization of Acupuncture Using High-Tech Methods: Teleacupuncture Bridges Science and Practice

Gerhard Litscher

Abstract Acupuncture has been practiced successfully for thousands of years. The Western demand for other treatment methods to complement classical Western medicine has also been increasing over the last years. Since 1997 the Research Unit of Biomedical Engineering in Anesthesia and Intensive Care Medicine and the TCM Research Center Graz (http://tcm-graz.at) of Graz Medical University has been dealing with the demystification of acupuncture and examining, using noninvasive methods, how different stimulation modalities (manual needle acupuncture, laser needle acupuncture, electroacupuncture) affect central and peripheral functions. Laser acupuncture is also very important for evidence-based high-tech acupuncture research. One only needs to mention the treatment of children or of patients with needle phobia. The content of this book chapter includes the latest biomedical investigations of measuring the effects of acupuncture and contains some case studies to demonstrate the application of the concept in practice. In general, this chapter is a summary of some of the scientific results obtained at the Medical University of Graz and explains innovative methods like teleacupuncture which bridges acupuncture science and practice.

Keywords Traditional Chinese medicine • Complementary medicine • Integrative medicine • Modernization of acupuncture • High-tech acupuncture • Laser acupuncture • Electroacupuncture • Teleacupuncture

---

G. Litscher (✉)
TCM Research Center Graz and Research Unit of Biomedical Engineering in Anesthesia and Intensive Care Medicine, Medical University of Graz, Auenbruggerplatz 29, A-8036 Graz, Austria
e-mail: gerhard.litscher@medunigraz.at

Y. Xia et al. (eds.), *Current Research in Acupuncture*,
DOI 10.1007/978-1-4614-3357-6_22, © Springer Science+Business Media New York 2013

22.1 Introduction

Traditional Chinese Medicine (TCM), especially acupuncture, is one of the oldest medical procedures in the world. In China and neighboring countries and regions, it has served as the basis of medical knowledge for thousands of years (Wang et al. 1999). The World Health Organization stated recently that "acupuncture has been proven effective in relieving postoperative pain, nausea during pregnancy, nausea and vomiting resulting from chemotherapy, and dental pain with extremely low side effects. It can also alleviate anxiety, panic disorders, and insomnia." Basic and translational research in traditional and innovative acupuncture has been performed at the TCM Research Center in Graz/Austria (http://tcm-graz.at) at the Medical University since 1997. This book chapter focuses on a review and the latest aspects that underline the further enhancement and development of modern acupuncture. High-tech acupuncture comprises many different forms of stimulation and recording techniques (Litscher 2009a). In Graz, a broad spectrum of future-oriented bioengineering methods is used. One can find an exemplary listing of important noninvasive procedures in Fig. 22.1.

The goal of this book chapter is to describe innovative acupuncture stimulation methods, e.g., the first results obtained with the blue laser needle stimulation.

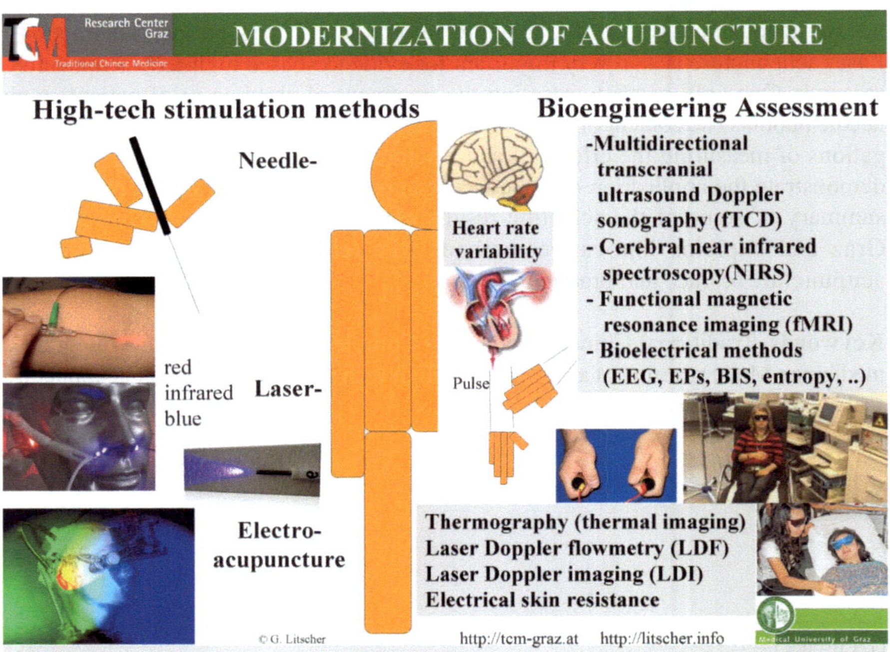

Fig. 22.1 High-tech biomedical methods for computer-controlled quantification of peripheral and cerebral effects of acupuncture. All these noninvasive procedures are used at the Medical University of Graz/Austria (modified from Litscher 2009a)

Furthermore, different recording methods for objectivation and quantification of specific effects of acupuncture are also presented as a review. Heart rate variability (HRV) completes the spectrum of evidence-based methods for acupuncture research.

22.2 High-Tech Acupuncture Stimulation Methods

In the Eastern and Western world, metal needles are the most important tools for stimulating different acupoints. Within the last years new optical and also electrical stimulation methods were scientifically investigated by our research group. These procedures are described within the next three sections.

22.2.1 Red and Infrared Laser Acupuncture

The ability to heal illnesses and relieve pain with light is a fascinating vision. One of the most recent technologic developments in the field of acupuncture has been the introduction of multichannel laser acupuncture (Litscher and Schikora 2005a). The basic idea of the so-called needles of light or laser needles originated from an analysis of laser acupuncture. Similar to needle acupuncture in TCM, simultaneous stimulation of different acupoints using laser beams and optical fibers was possible for the first time in 2001. These laser needles were developed at the University of Paderborn/Germany (Dr. Detlef Schikora), and the first clinical investigations were performed in Lauenförde/Germany (Dr. Michael Weber). The first scientific experiments and publications on this field of research started in 2002 at our Medical University in Graz/Austria (Litscher 2009b). A part of a laser acupuncture system based on red and infrared laser light is shown in Fig. 22.2.

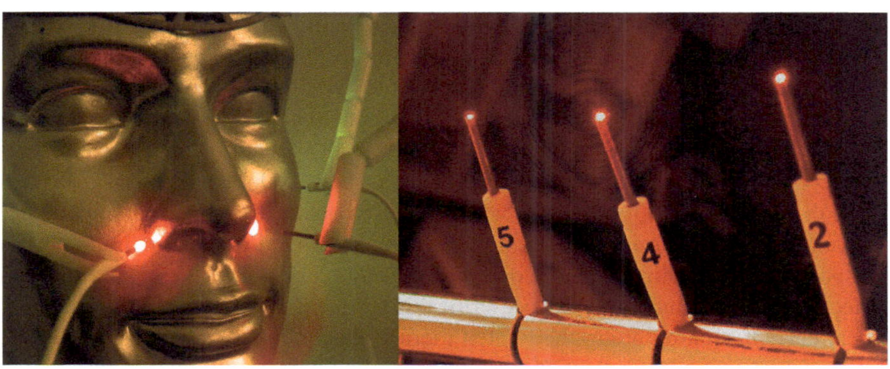

Fig. 22.2 Multichannel laser acupuncture using bichromatic laser needles with *red* (685 nm) and infrared (785 nm) light

Multichannel laser needle acupuncture allows the simultaneous stimulation of individual point combinations (Litscher 2008). Variation and combination of acupuncture points on the body are possible according to TCM or at the ear and hand using Korean or Chinese hand acupuncture. The bichromatic laser needle method is based on systems with 8–12 separate semiconductor laser diodes and emission wavelengths of 685 nm and 785 nm. The system consists of flexible optical light fibers, which conduct the laser light with almost no loss to the laser needle. Thus, a high optical density at the distal end of the laser needle is achievable. The intensity of the laser needles is optimized in such a way so that the volunteer or patient does not feel the activation of the needle immediately (30–40 mW per needle; diameter 500 μm; duration 10 min; power density ~ 20 J/cm^2 per acupuncture point). More details regarding this method are described in previous studies and books (Litscher and Schikora 2002a, 2005a; Litscher et al. 2004a).

22.2.2 *Violet Laser Acupuncture*

Blue or Violet lasers are utilized in just a few special areas in medicine (Mittermayr et al. 2007; Enwemeka et al. 2009) because it is a new and still expensive invention. In acupuncture research in particular, violet laser was applied only in two scientific investigations until now, which were published by our research group in autumn 2009 (Litscher 2009a; Litscher et al. 2009c). Violet laser needle acupuncture is a new optical method for stimulating different acupoints continuously and simultaneously. Violet laser light with a wavelength of 405 nm, an output power of 110 mW, and a diameter of 500 μm was used for our preliminary investigations. These violet laser needles are special light conductors which are placed vertically at the skin and trigger painless but perceptible stimulation at the acupoint (Fig. 22.3).

Violet laser acupuncture was made possible only due to a new laser invention (Nakamura et al. 2000). His group developed small, convenient blue and violet lasers which had not been available before. The acupuncture laser equipment used in our first studies operates, as already mentioned before, at a wavelength of 405 nm. It is worth noticing that this wavelength is not in fact blue but appears to the eye as violet, a color for which the human eye has a very limited sensitivity.

Although the blue or violet laser does not have the same penetration depth in human skin as the red and infrared laser [violet: approximately 2 mm vs. red/infrared: 2–3 cm (Litscher and Schikora 2005a; Litscher 2009a; Anderson and Parrish 1981)], the evoked deQi sensation, which is a prerequisite for effective acupuncture stimulation, may warrant the scientific investigation of violet laser stimulation in acupuncture. According to TCM one must first obtain deQi for acupuncture to be effective. Patients and volunteers describe this feeling as a heaviness or like an electrical current running along the treated meridians, which are "channels" in TCM. DeQi sensation is also achieved with violet laser light of sufficient intensity (for technical parameters see above). It is important to understand that laser light may be concentrated to achieve high intensities yet still be given at relatively low average

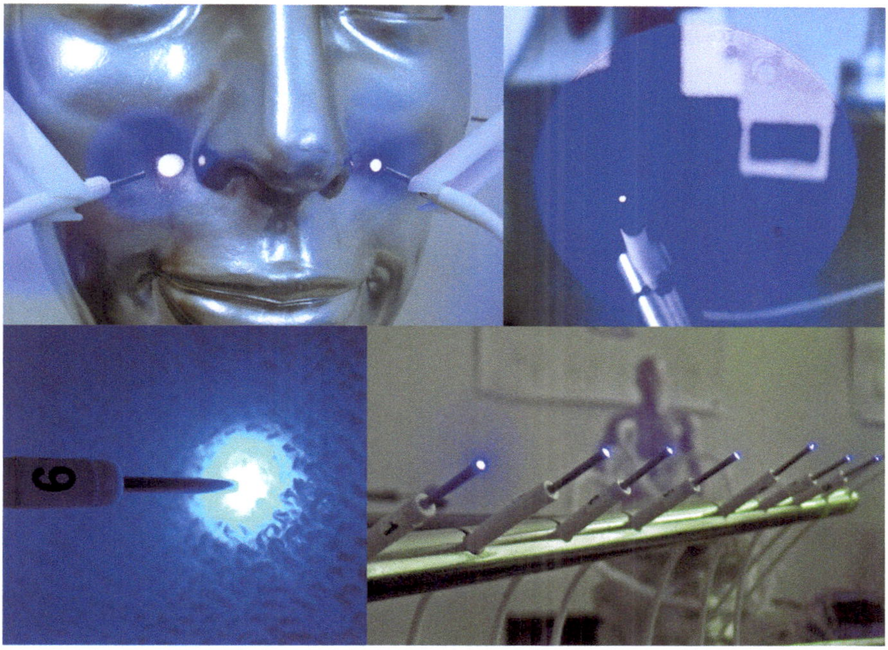

Fig. 22.3 Innovative acupuncture with violet laser needles

output powers. This is accomplished by reducing the area of the laser beam. If red (685 nm) or infrared (785 nm) lasers are used, the patients normally do not notice when the laser is started. So in the beginning of the treatment they also do not feel any deQi sensation. Several minutes later (5–10 min) many patients report a pleasant warm and sometimes vibrating feeling in some treated areas (Rindge 2009). On the contrary, using the violet laser acupuncture volunteers described that the violet laser induced feelings similar to an "ant-like bite" followed by a feeling similar to a slight electrical current which is similar to deQi evoked by a metal needle. Up to now, Chinese people were investigated in order to get results from volunteers who already had extensive experience with needle acupuncture. In follow-up studies we will compare the data with European volunteers with and without any experience concerning acupuncture.

In our first study we found that violet laser stimulation increases temperature (mean ~1.5 °C) and microcirculation (mean ~20 %) at the acupoint Hegu (LI-4), significantly and immediately (1 min) after stimulation onset (Litscher 2009a). The main interesting finding of our second publication concerning violet laser acupuncture was that heart rate decreases significantly within an interval of 5 min after violet laser stimulation onset at the acupoint Neiguan (PE-6) (Litscher et al. 2009c). Heart rate in this study did not change significantly during and after stimulation with deactivated laser (Fig. 22.4).

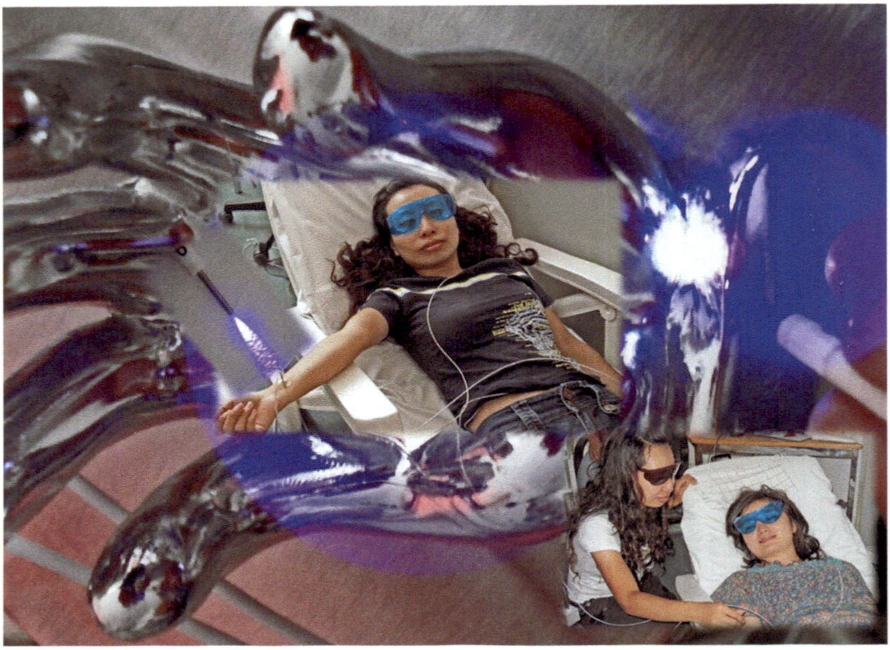

Fig. 22.4 Violet laser acupuncture at the TCM Research Center in Graz/Austria (modified from Litscher 2009a)

22.2.3 Electroacupuncture

A new method for pain treatment using ear acupuncture is the electrical point stimulation system developed by the Medical University of Vienna (P-STIM®; Biegler, Austria). This kind of electroacupuncture has also been investigated at our TCM Research Center (Széles and Litscher 2004). Ultrathin permanent needles are applied at the ear. A generator, located behind the acupunctured ear, produces electrical stimulation impulses, which are transferred via the needles to the acupuncture areas (comp. Fig. 22.1). In the past, traditional acupuncture and electroacupuncture were characterized by the fact that these methods were confined to the clinic and/or the physician's practice. P-Stim allows continuous, intermitting stimulation up to several days combined with absolute mobility of the patient (Sator-Katzenschlager et al. 2003).

After selection of the ear acupoints, a position tape previously prepared with the P-Stim application pointer is applied. This procedure is repeated until all acupuncture points are marked. Then, the needles can be taken up by the application pointer and applied. The actual device is adhered behind the acupunctured ear with the integrated adhesive electrode. Then, the wires are connected to the permanent needles by snapping over the plastic rings and are fastened with adhesive tape. Finally, P-Stim is activated by removing the adhesive foils from the batteries. P-Stim administers continuous electrical stimulation impulses (current intensity: 2 mA, frequency: 1 Hz, impulse duration: 20 ms, maximum possible stimulation duration: 4×24 h).

Manual needle acupuncture, laser needle stimulation, and electroacupuncture are based on totally different physical principles, and the stimulus intensities are not directly comparable. However, all three methods can induce similar effects in the brain and in the periphery. This has been shown in studies comparing metal needle acupuncture and laser needle stimulation (Litscher 2006a) as well as in investigations comparing optical and electrical stimuli (Litscher et al. 2007a).

22.3 Bioengineering Assessment of Acupuncture Using High-Tech Methods

Over the last years, a great variety of methods of bioengineering assessment were used. In this chapter a division into registration methods for the periphery and recording procedures for the central nervous system has been performed. HRV measurements and analyses complete the methodological spectrum (comp. Fig. 22.1). The different methods are presented in respect to single- and double-blind, controlled, randomized crossover studies.

22.3.1 Effects in the Periphery

22.3.1.1 Thermography

Quantitative thermal imaging is becoming an important method in acupuncture research. Infrared thermography enables the measurement of patients' or healthy volunteers' skin surface temperature profiles without influences caused by direct contact of probes to the skin. Therefore thermography is also a useful method for evaluating peripheral effects of acupuncture.

Using infrared thermography the effects of changes in peripheral temperature during the initial phase of manual needle and laser acupuncture were evaluated under standardized conditions. According to TCM, the combination of the acupoints Neiguan (PE-6) and Quchi (LI-11) leads to a general increase in energy and is applied when circulatory problems in the upper extremities are present (Litscher 2006b; Litscher and Wang 1999a) (Fig. 22.5).

For the investigations shown in Fig. 22.5, the Agema ThermaCAM™ P640 infrared camera system (Flir Systems Inc., Portland, USA) was used. The camera operates at a wavelength range from 7.5 to 13 μm. The temperature measurement range lies between −40°C and +500°C and can be increased to 2,000°C using filters.

In addition, thermography is also a useful method for measuring the temperature during moxibustion (Litscher et al. 2009e). An example is shown in Fig. 22.6.

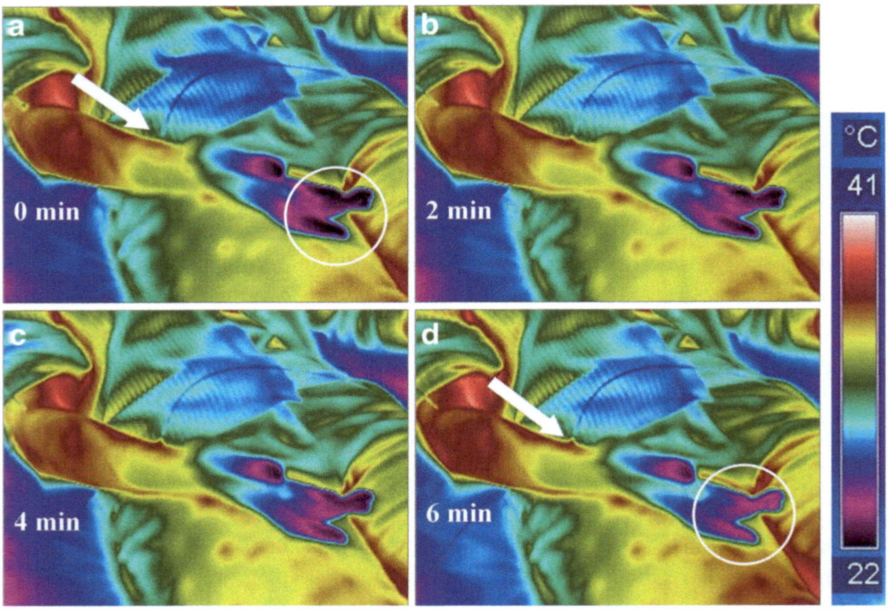

Fig. 22.5 Typical example of the thermographic evaluation of peripheral effects during laser needle acupuncture in a young female. Note that the difference between (**a**) and (**d**) on the index finger is about +2.1 °C (modified from Litscher 2009d)

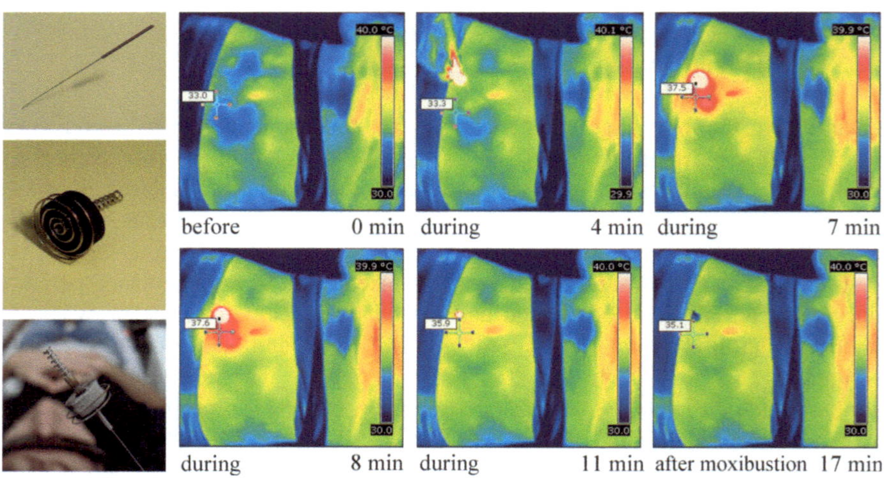

Fig. 22.6 Thermal images during moxibustion with a new needle-moxa system (*right panels*; modified from Litscher et al. 2009e)

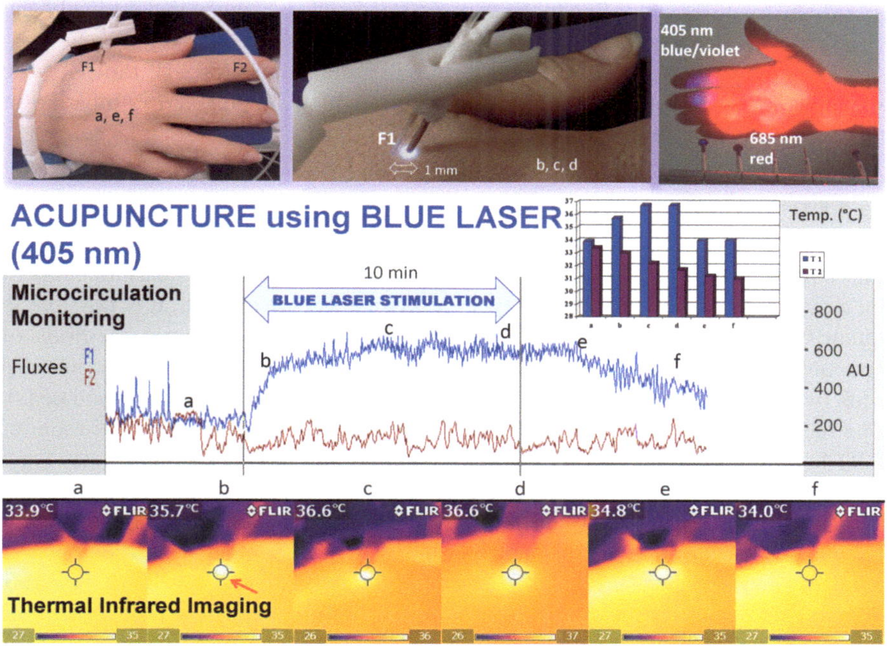

Fig. 22.7 The first "blue or violet laser acupuncture" using a wavelength of 405 nm. Note the significant increase in microcirculation (Flux1, F1; *blue curve*) 1 mm beside the laser needle at the acupoint Hegu (LI-4). Temperature at this location (T1, same as F1) was measured using thermal infrared imaging (Flir i5). Note also the increase of temperature from 33.9 °C to 36.6 °C in the region of interest. Flux2 (F2) and temperature (T2, location same as F2) were recorded at the index finger (modified from Litscher 2009a)

For these investigations an infrared-thermographic camera ThermaCAM B200 9 Hz (Flir Systems Inc., Portland, USA) was used for thermographic documentation of the study. This camera is maintenance-free, does not require cooling, and allows continuous measurements. Its spectral range lies between 7.5 and 13 μm, temperature range of the instrument is –20 °C to +120 °C, picture repetition frequency is 9 Hz, and geometric resolution 2.18 mrad, and it has a thermal sensitivity of 80 mK at +30 °C. The registered temperatures are shown on a color-coded display so that color pictures of temperature distribution are generated.

For the investigations in Fig. 22.7, a new FLIR i5 infrared camera (Flir Systems Inc., Portland, USA) was used which operated at a wavelength range from 7.5 to 13 μm. Temperature measurements were possible in the range between 0 °C and +250 °C. The data were transferred to a notebook using ThermaCAM Researchers Pro 2.8 software (Flir Systems Inc., Portland, USA).

There are many possible uses of thermal imaging in the field of TCM in general and acupuncture research in particular, but there are still methodological limitations of this modern measuring procedure. The validity of the method for proving meridian structures according to the view of TCM must be considered critically and analyzed scientifically (Litscher 2005b; Litscher and Ammer 2007b).

Thermographic methods such as infrared cameras are effective tools for the visualization of effects in acupuncture research which support demystification of this ancient medical treatment method.

22.3.1.2 Laser Doppler Flowmetry

Laser Doppler flowmetry (LDF) provides noninvasive real-time measurements of local tissue blood flow. Using this type of bioengineering method, it is possible to quantify peripheral changes in microcirculation during different methods of acupuncture stimulation (Litscher 2006c). An important parameter in this context is Flux (= product of mean blood flow velocity and concentration of red blood cells). The measurements described in Fig. 22.7 were performed with the LDF-monitor DRT4 (Moore Instruments Ltd., Millwey, Axminster, UK). Edge frequencies were 20 Hz and 22.5 kHz. The temperature unit (5–50 °C) had a resolution of 0.2 °C and an accuracy of 0.5 °C. A special probe (DP3s S/N 00166, diameter 1.3 mm) was designed for recordings at a distance of 1 mm to a blue laser needle, and a commercially available probe (DPIT, diameter 8 mm, length 7 mm) was used for the index finger. This point was chosen as a further recording location for control purposes following previous investigations (Litscher 2009d).

LDF is a technique used for investigating changes in microcirculation. It allows objectifying circulation within the microcapillary range without strongly influencing tissue structures. This principle is based on the Doppler shifting of light when light hits moving particles (erythrocytes). LDF is particularly suitable in the field of pharmacology for comparison of measures which influence circulation, for controlling transplants and flaps in plastic surgery, and for objectifying and classifying the stages of disease in angiological and dermatological research and industrial medicine.

22.3.1.3 Laser Doppler Imaging

Using the noninvasive laser Doppler imaging (LDI) method, circulation in surface tissue structures can be determined and illustrated. Like LDF, this method is also based on the laser Doppler principle, whereby data can be obtained directly from the reflected light and is displayed as a color-coded picture showing the distribution of tissue perfusion. LDI is a further advancement of punctual LDF. Here, laser light with low output is directed to the surface of the skin. A servomotor directs the laser light to a maximum of 4,096 different positions from one measurement point to the next (Litscher et al. 2002b).

The "PIM II" laser Doppler perfusion measurement system from Lisca AB (Linköping, Sweden) includes a scanning head, an opto isolator unit, an AD converter, and a computer. The scanning head comprises a laser unit, an optical system for gradual scanning, and an optode detector unit. The optode isolator unit includes circuits for signal evaluation, stepper motor drive, and circuits to guarantee galvanic isolation between the scanning head and the computer.

A solid laser (670 nm) scans the tissue surface step by step. The reflected light is checked for Doppler effects at 4,096 positions of the tissue surface. Monochromatic laser light penetrates each tissue area to a depth of several hundred micrometers. When a moving particle such as a red blood cell is hit, one part of the photons experiences a frequency shift according to Doppler's principle; the remaining photons in stationary tissues do not underlie these changes. A small part of the reflected light comprises changed and unchanged light and thus influences the photodetector in the scanning head. The combination of light on the photodetector surface results in a change in photocurrent determined by the Doppler-shifted particles of light. Major Doppler frequency shifts caused by the high velocity of moving blood cells usually increase the frequency of photocurrent, whereas the peak of fluctuation reflects the number of moving blood cells. If this fluctuation element of photocurrent is guided through a filter and the immediate energy is calculated, a signal for tissue circulation can be derived (i.e., mean velocity multiplied by the concentration of blood cells within tissue distribution).

Any relative movement between laser light and tissue must be avoided during Doppler signal registration in order to prevent artifacts. The system guarantees this with its patented step-by-step scanning method. The laser light remains fixed in relationship to tissue, while the reflected light is analyzed. For registration of light reflection, laser light remains at each single measurement point for about 50 ms (analogous to the lower bandwidth of 20 Hz).

Figure 22.8 shows four pictures (upper series) of changes in perfusion in a 37-year-old female volunteer, during stimulation with a new moxibustion method (comp. Fig. 22.6; right panels), recorded at the China Academy of Chinese Medical Sciences (Litscher et al. 2009e). A significant increase in perfusion (circular structures in the middle parts of Fig. 22.8, upper line) could be observed during the active (smoldering) phase of moxibustion stimulation. The lower series shows the original photos which correlate chronologically with the pictures in the upper series. Notice that a large amount of the pressed mugwort coal cylinder in photo no. 3 (Fig. 22.8) has been burned off (remaining part see photo 3, Fig. 22.8, lower line, center).

LDI studies on acupuncture show that needle acupuncture and also laser needle acupuncture can lead to quantifiable changes in microcirculation parameters in addition to the already known changes in temperature of the surface of the skin.

22.3.1.4 Electrical Skin Resistance

A scientific literature research (http://www.pubmed.gov) shows about 500 articles (December 2009) referring to electrical skin resistance measurements. Most of them are case reports. Only 18 studies meet the criteria for further evaluation with regard to evidence-based medicine (Ahn et al. 2008; Kramer et al. 2008). These studies can be divided into acupuncture studies ($n=9$) and so-called meridian studies ($n=9$). Five out of nine acupuncture point studies showed positive association between acupuncture point and lower electrical resistance and impedance. Four studies on this topic showed contrary results. The present limitations of the method are based on the one hand on the small tip of a pen-shaped device which is pressed manually

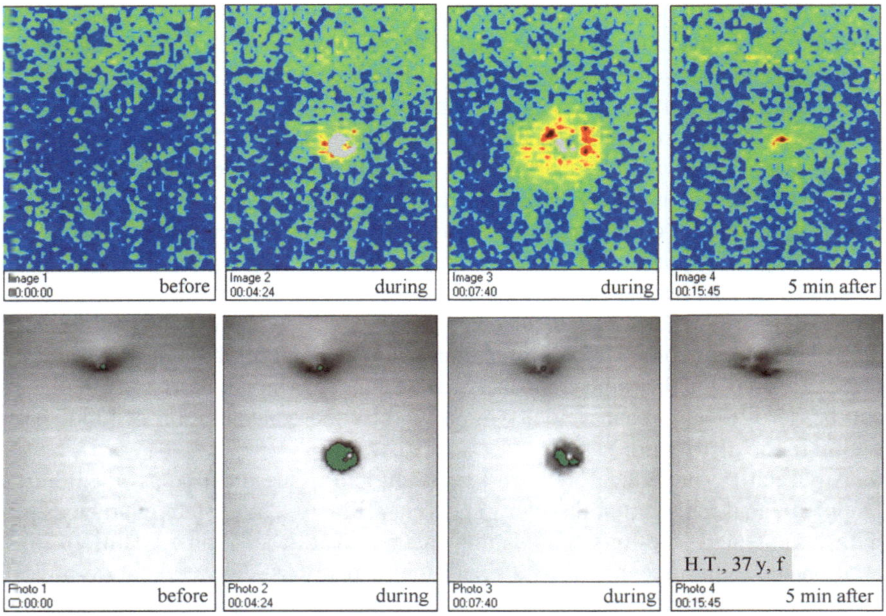

Fig. 22.8 Course and changes in perfusion in a 37-year-old female volunteer during moxibustion at acupoint Qihai. This test was performed at the China Academy of Chinese Medical Sciences in Beijing (modified from Litscher et al. 2009e)

against the skin by the examiner, and on the other hand on the repeated pressing of the electrode tip into the skin or scratching over the skin surface. Therefore an electrical skin resistance mapping consisting of many electrodes is desirable and has already been performed recently (Kramer et al. 2008).

In our TCM Research Center, some pilot measurements concerning electrical skin resistance were carried out using a new prototype from Mentaltech (St. Martin i.I., Austria). Figure 22.9 shows the electrodes (upper panel) and the protocol of a recording in a healthy volunteer (lower panel).

Based on existing literature and pilot measurements, the evidence does not conclusively support the claim that acupuncture points or meridians are electrically distinguishable. Future research is absolutely necessary (Ahn et al. 2008).

22.3.2 *Effects in Central Areas*

22.3.2.1 Multidirectional Transcranial Ultrasound Doppler Sonography (fTCD)

Using the Doppler principle, cerebral blood flow velocity can be registered. Usually the transmission frequencies lie between 1 and 20 MHz in vascular diagnostics or during monitoring. Low frequencies have a high penetration depth but low sonic

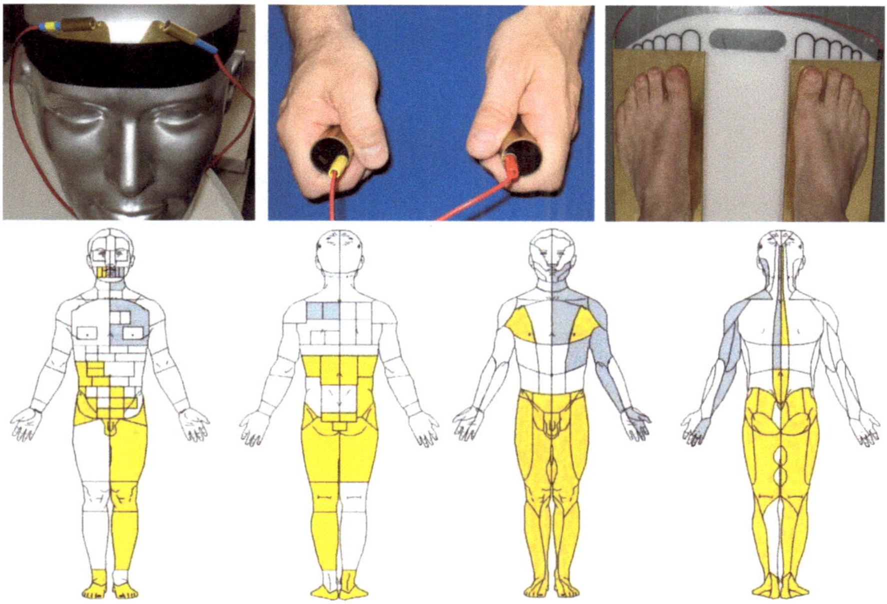

Fig. 22.9 Electrical skin resistance for acupuncture research. Measurement electrodes at the head, hands, and feet and graphical presentation of an example of a measurement with normal electrical skin resistance

reflection of blood cells, whereas high transmission frequencies have a lower penetration depth but better sonic reflection. Ultrasound waves are mechanic, matter-bound density waves with frequencies of >20 KHz. These are produced by electric alternating voltage being applied to piezoelectric crystals ("transducer"). The waves propagate in biologic tissue (with the exception of bone) at a nearly constant speed (~1,550 m/s). The waves are totally or partially reflected and weakened by scattering and absorption at biologic–acoustic border regions. Ultrasound waves of low intensity (<10 mW/cm^2; diagnostics) are considered to be harmless (Marshall 1996).

Spectrum analysis is generally required for intracranial TCD vascular monitoring due to the weak reflection of Doppler signals. A pulse curve above the zero line shows the course of blood flow to the probe in basal cerebral arteries, whereas the opposite is usually true for extracranial vessels. This curve was selected for reasons of definition, since the important intracranial vessel, the middle cerebral artery (MCA), shows the direction to the temporally applied ultrasound probe.

The MCA with its different branches can usually be investigated at a depth of 3–4 cm, the media main trunk usually lies at a depth of 5.5–6.5 cm. Blood flow velocity in the internal carotid artery (distal part at a depth of 60–65 mm) as well as in the MCA can be determined by transtemporal entry. With increasing depth of measurement volumina, a part belonging to the anterior cerebral artery (ACA), can be determined also. If the ultrasound probe is turned slightly in the dorsal and caudal direction, the origin of the posterior cerebral artery (PCA) can be reached at a depth of 65–75 mm.

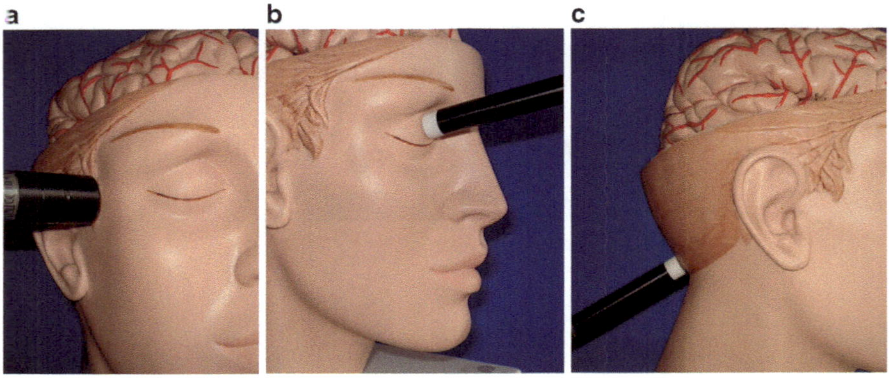

Fig. 22.10 Acoustic windows for cerebral ultrasound measurements in acupuncture research (**a**) transtemporal, (**b**) transorbital, and (**c**) transnuchal

One restriction of intracranial TCD results could be ultrasound transmission through the skull. According to literature, particularly older women could only be examined sufficiently in 50 % of the cases. However, optimal transtemporal registration was not possible in 5–10 % of men and younger women either (Widder 1999).

In addition to the three transtemporal acoustic windows (front, middle, and rear), transorbital and transnuchal entry for TCD-monitoring is also possible (see Fig. 22.10).

Registration of blood flow profiles in the ophthalmic artery (OA) was performed with a probe, applied lateral to the cornea on the closed bulb of the eye (comp. Fig. 22.10b). Reduction in transmission energy should be achieved as far as possible in order to avoid direct exposure to sonic waves of the lens. A signal from the OA is registered at a depth of 40–50 mm by a probe turned slightly in the central direction.

The foramen magnum offers a further path of entry for ultrasound measurements (comp. Fig. 22.10c). Dependent upon anatomical variations, the basilar artery (BA) lies at a depth between 70 and 110 mm. The vertebral artery and the BA show blood flow directed away from the probe.

For the ultrasound measurements in acupuncture research, we have developed new helmet constructions in the late 1990s (Fig. 22.11).

Back in the year 1997, our interdisciplinary research group was able to scientifically prove that acupuncture needles can increase blood flow velocity in the brain (Litscher et al. 1998a). The computer- and robotic-controlled biosensors and probes integrated in a special measurement helmet, coupled with light, ultrasound, and highly sensitive bioelectrical monitoring methods, yield reproducible results indicating that the blood flow velocity in the MCA is higher and the oxygen supply in the brain is increased by acupuncture in healthy volunteers (Litscher et al. 1998b).

In 1999, a further step for objectifying the old Chinese treatment method was achieved. In tests with healthy volunteers, proof that acupuncture does not only lead to *general* changes in blood flow velocity in the brain but to *specific* changes (i.e., different regional localization) could be documented. Thus, stimulation of distant acupoints on the hand or outer side of the foot, which according to TCM are often

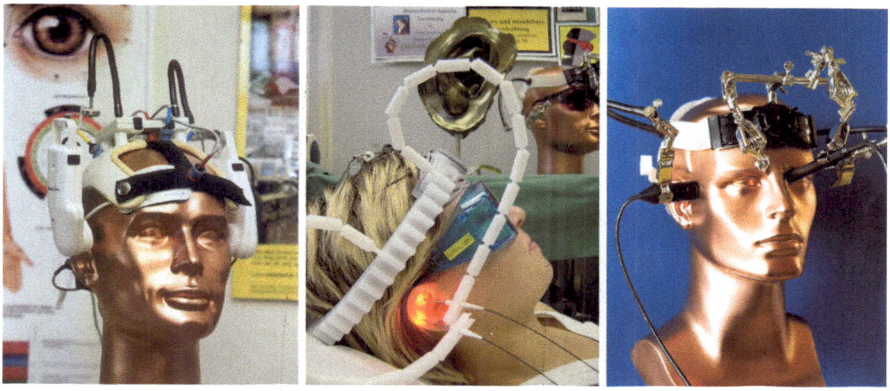

Fig. 22.11 Multifunctional helmet constructions for acupuncture research. (*Left*) Helmet containing computer-controlled ultrasound measuring devices for recording cerebral blood flow velocity in eight different depths, active electrodes for electroencephalographic recordings and sensors for near-infrared spectroscopy. (*Center*) Construction for the application of laser needles. (*Right*) Multidirectional ultrasound probe holder construction for simultaneous measurement of blood flow velocity in different arteries in the brain (modified from Litscher 2009a)

connected to the optical system, lead to an increase in blood flow velocity in the cerebral artery responsible for vision. At the same time, the blood flow velocity in other cerebral arteries remains nearly unchanged. These effects can only be registered up to now when light stimulation is performed. Comparative investigation of points at the inner edge of the foot did not show changes in blood flow velocity in this particular cerebral artery (Litscher et al. 1999b).

Crossover studies were performed to master the placebo problem during acupuncture. Each person was treated with a special optic acupuncture scheme A, which according to TCM should improve vision and with another scheme B to improve perfusion in the MCA. When using the optic scheme, a significant increase in mean blood flow velocity in the supratrochlear artery and ophthalmic artery occurred, whereas the flow velocity in the MCA remained nearly unchanged. In reverse, scheme B led to a significant increase in flow velocity in the MCA with nearly unchanged flow profile patterns in the optic arteries. Several crossover studies were performed and confirmed the initial results of selective changes in cerebral perfusion after acupuncture (Litscher 2006d).

22.3.2.2 Cerebral Near-Infrared Spectroscopy

The discovery that the functional activation of the human cerebral cortex can be explored by near-infrared spectroscopy (NIRS) has added a new dimension also to acupuncture research. NIRS is a noninvasive approach that allows the evaluation of cerebral oxygenation through the intact skull. First in vivo results in humans were reported by Jöbsis (1977). Figure 22.12 shows the NIRS principle schematically.

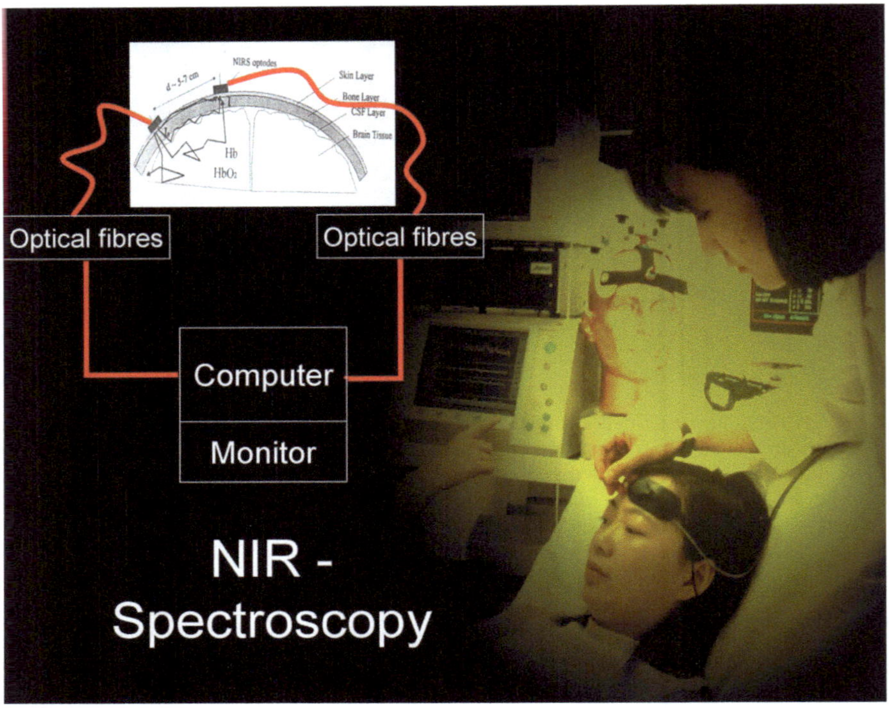

Fig. 22.12 Schematic diagram showing the principles for regional cerebral near-infrared spectroscopy (modified from Litscher 2006e)

Different instrumental components are used in our NIRS acupuncture research. One system is the NIRO 300 (Hamamatsu Photonics, Hamamatsu, Japan). In addition to standard parameters including changes in oxyhemoglobin (ΔO_2Hb), deoxyhemoglobin (ΔHHb), complete hemoglobin ($\Delta cHb = \Delta O_2Hb + \Delta HHb$), and oxidized cytochrome aa3 ($\Delta CtOx$) in μmol, tissue oxygenation or TOI (Tissue Oxygenation Index) is determined in %. The absorption spectrometer measures the difference in optical densities at sequential measuring times during four selective wavelengths in the near-infrared range (4 laser diodes: approximately 760, 820, 850, 910 nm).

Only absolute changes in chromophore can be determined (micromolar), however not the level (absolute concentration) at which these changes (positive or negative direction) occur. The measurement value is zero as long as no changes in chromophore concentration take place (Litscher and Schwarz 1997).

Measurement of TOI is done with the SRS (Spatial Resolved Spectroscopy) method, in which the absorption coefficient of tissue is determined from the regionally dependent weakening of light. A detector especially developed for the SRS method registers the emitted near-infrared light from the light source (glass fiber) into tissue at three different intervals and allows calculation of TOI with the SRS algorithm.

Multichannel measurements were made possible by separating the central unit and measuring unit. Using a silicone holder, the sensor and emitter can be easily and reproducibly fixed on the head. Data (ΔO_2Hb, ΔHHb, ΔcHb, ΔCtOx, TOI) are shown on a color LC display and color printer.

The INVOS 5100 Oximeter (Somanetics Corp., Troy, USA) was also used in various acupuncture studies. Here, regional cerebral oxygen saturation (rSO_2) was calculated in percent from the proportion of oxygenated and deoxygenated hemoglobin. Normal values lie between 60 and 80 % (Litscher and Schwarz 1997).

Important results of our NIRS studies can be summarized as follows (Litscher et al. 2002c). Needling and laser stimulation of a placebo point did not lead to marked changes in cerebral NIRS parameters during or 5 min after acupuncture. The combination of Korean hand acupuncture (E2) and Chinese hand acupuncture (Yan Dian) as well as TCM body acupuncture [Zhanzhu (UB-2) and Yuyao (EX-3)] and combined body, ear, and hand acupuncture leads to a marked increase in O_2Hb and a simultaneous decrease in HHb. This effect was still present 5 min after removing the needles or deactivating laser needle stimulation. One case of minimal contrary behavior in O_2Hb and HHb is present when needling or performing laser stimulation at both ear points (eye and liver). Standard monitoring parameters such as blood pressure did not show any significant changes during all types of acupuncture or combined acupuncture methods.

NIRS studies in combination with laser needle stimulation are of great importance, not only for the field of spectroscopy and laser medicine but also for acupuncture research in general. Since the volunteer or patient does not feel the red or infrared laser stimulation, double-blind randomized controlled crossover studies are possible in acupuncture research and indicate that cerebral effects of optical stimulation concerning the NIRS parameters are nearly equivalent to those of metal needle stimulation.

22.3.2.3 Functional Magnetic Resonance Imaging

Functional magnetic resonance experiments are based on the indirect representation of neuronal activity and the resulting metabolic and circulatory changes, particularly the relative changes in concentration of oxygenated and deoxygenated hemoglobin. This method has been successfully used to investigate changes in cerebral activity during acupuncture (Litscher 2006f).

Since the late 1990s, functional magnetic resonance imaging (fMRI) has been used to investigate the underlying mechanism of the Chinese medical treatment of acupuncture. The research group of Prof. Cho from Korea was the first to investigate effects of acupuncture using fMRI (Cho et al. 1998; Litscher and Cho 2000). However, five of the eight authors retracted the paper from Cho et al. 1998, on June 21, 2006. In this retraction they stated that there is no point specificity, at least for pain and analgetic effects.

In the meantime, there are nearly 200 publications (http://www.pubmed.gov; December 2009) on the topic "fMRI and acupuncture." The point specificity is still discussed controversially; for each positive result, there is at least one other paper

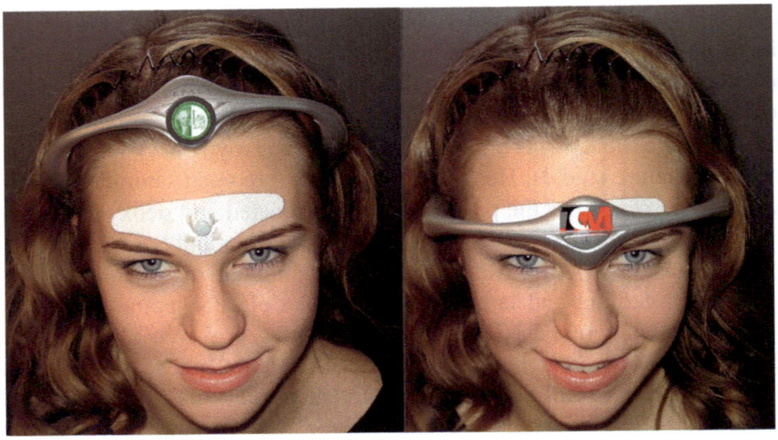

Fig. 22.13 First pilot measurement using electrical stimulation at the Yintang acupoint area at the TCM Research Center Graz (Medical University of Graz/Austria). Courtesy of the healthy volunteer (modified from Litscher 2009g)

with a negative result. A review of fMRI papers was published in Critical Reviews in Biomedical Engineering about 3 years ago (Litscher 2006f). The recently published review by Beissner and Henke 2009 should be mentioned especially as it deals with the topic in a critical way, showing the limitations of fMRI investigations in acupuncture research.

22.3.2.4 Bioelectrical Methods

Neurophysiological monitoring, especially electrophysiological monitoring, is gaining more and more attention in acupuncture research (Litscher 2007c, 2009f). The bispectral index (BIS) and entropy are important numerical descriptors of the electroencephalogram (EEG) and are mainly used for assessing depth of anesthesia.

In previous studies (Litscher 2004b, 2009g), the group of the authors of this chapter found that acupressure at Yintang in healthy awake volunteers results in statistically significant and clinically relevant reductions in BIS while needle acupuncture, laser needle acupuncture, and acupressure at a control point resulted in statistically significant but clinically unimportant reductions. Electrical stimulation (Cefaly, STX-Med, Liege, Belgium) around the acupoint Yintang on the forehead (constant current: max. 30 mA; duration of rectangular electrical impulses: 100–500 μs; max. frequency: 150 Hz; stimulus duration: 20 min) can induce a state of deep relaxation and sedation as shown in our preliminary studies (Kim et al. 2006; Litscher 2009g) (Fig. 22.13). The results highlight the electroencephalographic similarities of non-pharmacologically induced sedation and anesthesia.

As mentioned above, EEG entropy is another parameter which is used for quantification of depth of anesthesia and sedation. For the first time, this method of entropy in EEGs was applied in an acupuncture study, and evidence regarding an acupuncture scheme consisting of "sedative acupoints" compared to an acupuncture scheme that is considered to activate energy was validated (Litscher 2006a). The study was conducted in a group of healthy volunteers ($n = 9$) to determine whether manual needle acupuncture and laser acupuncture stimulations applied to two groups of acupuncture points (sedative vs. enhancement of energy) have effects on either state entropy (SE) or response entropy (RE) or both.

In the same volunteer, two different acupuncture schemes were tested on different days. The first acupuncture scheme (A) contained so-called sedative points and included the following acupuncture points: Shenting (DU-24), Yintang (EX-1), Sedative point 1 (EX-8, Anmian I), Sedative point 2 (EX-9, Anmian II), and Shenmen (HE-7). Points Shenting and Yintang and their cerebral effects are described comprehensively in the scientific literature (Litscher 2005c). Sedative point 1 is located at the median point of the connecting line between SJ-17 (Yi Feng) and Yi Ming (EX-7). Sedative point 2 lies at the median point of the connection line between GB-20 (Feng Chi) and Yi Ming (EX-7). The main indication for the latter two points is sleeplessness.

Acupuncture scheme two (B) consists of points, which according to TCM should support the "general availability of Qi-energy": Neiguan (PE-6), Qihai (REN-6), Zusanli (ST-36), and Sanyinjiao (SP-6). This scheme has been investigated by our study group a number of times, and it also leads to an increase in blood flow velocity of the MCA (Litscher 2006d).

For the first time, we have proved that two different acupuncture schemes can influence different parameters in the brain. The so-called sedation point scheme (A) showed a significant decrease in two entropy parameters (state entropy and response entropy) which can be interpreted as a sedation effect in the EEG (Fig. 22.14). In the same volunteer, a different acupuncture scheme (B), which according to TCM leads to a general increase in Qi-energy, did not yield this effect. Moreover, several preliminary tests showed that the latter scheme leads to a significant increase in blood flow velocity in the MCA (Litscher 2006d). In addition, this study (Litscher 2006a) proves that laser needle acupuncture is needle equivalent regarding the values of entropy described (Fig. 22.15) and has an important ranking as a painless method of acupuncture.

Neurophysiologic monitoring provides functional evaluation of the nervous system. Although numerous questions concerning acupuncture remain still unsolved, the present findings could be a further step to discover the complex mechanism underlying the effects of acupuncture.

22.3.3 *Effects in the Autonomic Nervous System*

Mental status and hypnotic agents are known to influence the autonomic nervous system (ANS). In the last years, several studies showed that also acupuncture can influence the ANS (Litscher 2007d).

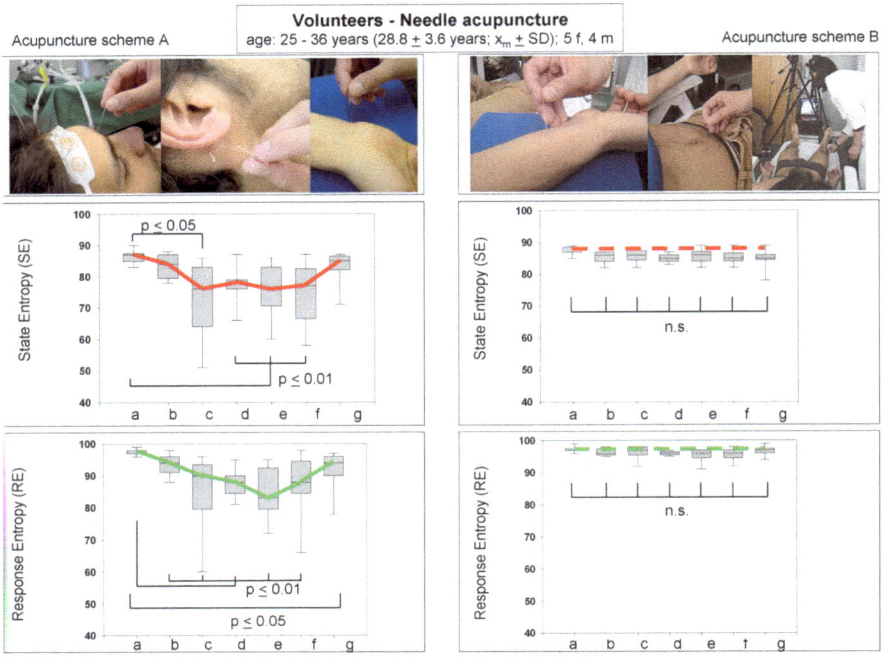

Fig. 22.14 Entropy—needle acupuncture. Box-plot illustration of alteration in state (SE) and response (RE) entropy values in nine healthy volunteers 2 min before (a), during [(b) immediately after applying the needles; (c) 2 min after b; (d) 5 min after b; (e) 10 min after b; and (f) 15 min after applying needles] and 2 min after (g) needle acupuncture using two different acupuncture schemes (**A**, **B**). Note the significant decrease in entropy when using scheme **A** (sedation points). The horizontal line in the box indicates the position of the median. The end of the bars defines the 25th and 75th percentile, and the error bars mark the 10th and 90th percentile (modified from Litscher 2006a)

HRV is an index value of the neurocontrol of the heart and is used more and more in recent scientific research and practice. Today innovative research including the latest recording technology and also artificial intelligence techniques are used for data acquisition and data analysis of HRV in acupuncture research.

Already in the third century the Chinese medical doctor and scientist Wang-Shu Ho realized that variable heartbeats are a sign of good health. He stated: "If the pattern of the heart beat becomes regular as the tapping of woodpecker or the dripping of rain from the roof, the patient will be dead in 4 days."

HRV is measured as the percentage change in sequential chamber complexes, so-called RR intervals, in the electrocardiogram (ECG). The registration of HRV is performed using three electrodes on the chest. It is important to simultaneously record respiration and if possible continuous blood pressure. The RR intervals in the ECG are controlled by the blood pressure control system, influenced by the hypothalamus and in particular controlled by the vagal cardiovascular center in the lower brainstem. HRV can be quantified over time using registration of percentage changes

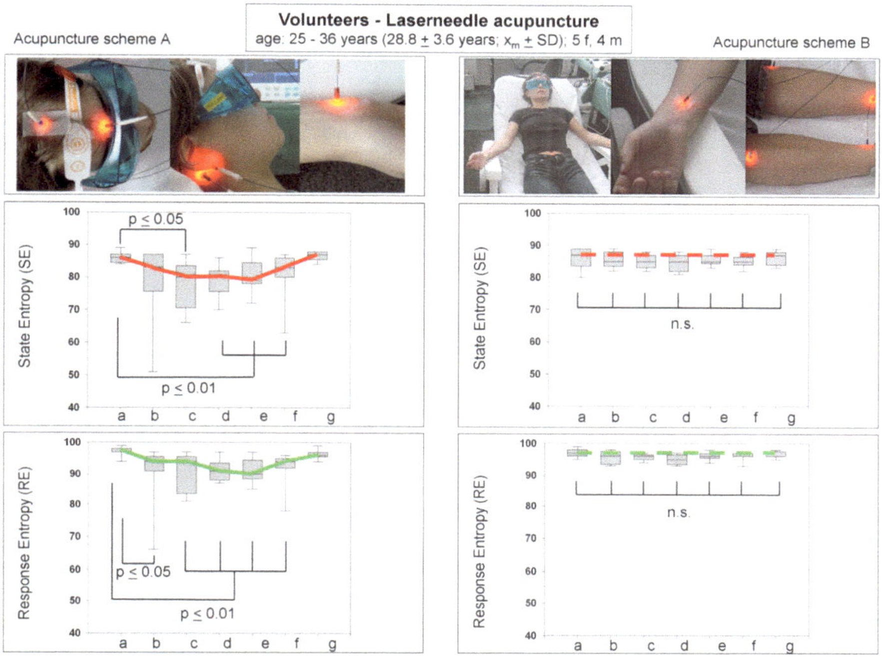

Fig. 22.15 Entropy—laser needle acupuncture. Box-plot illustration of changes in state (SE) and response (RE) entropy values in nine healthy volunteers (a) before, (b–f) during and (g) after laser needle acupuncture using two different acupuncture schemes (**A**, **B**). For further explanations see Fig. 22.14 (modified from Litscher 2006a)

in RR intervals in the time domain as well as the changes in the frequency range by analysis of electrocardiographic power spectra. Parameters are recommended by the task force of the European Society of Cardiology and the North American Society of Pacing and Electrophysiology and are standard since 1996 (Task force of the European Society of Cardiology and the North American Society of Pacing and Electrophysiology: heart rate variability 1996). Calculation of ECG power spectra is also possible. This kind of analysis is thought to provide an understanding of the effects of sympathetic and parasympathetic systems on HRV. Early work pointed out a few bands in the spectrum of HRV that could be interpreted as markers of physiological relevance. Associated mechanisms are thermoregulation which can be found in the very-low-frequency band (so-called VLF band), blood pressure (LF band), and respiratory effects (HF band). In addition, total power of the heart rate spectrum is calculated for total HRV, and as an important parameter, the ratio of the low-frequency and high-frequency band has become accepted.

Beside power spectral analysis a totally new method of visualization of HRV is the so-called Fire of Life analysis. Within this diagram, frequencies are not shown at the *X*-axis but on the *Y*-axis. Therefore the VLF band can be recognized at the bottom.

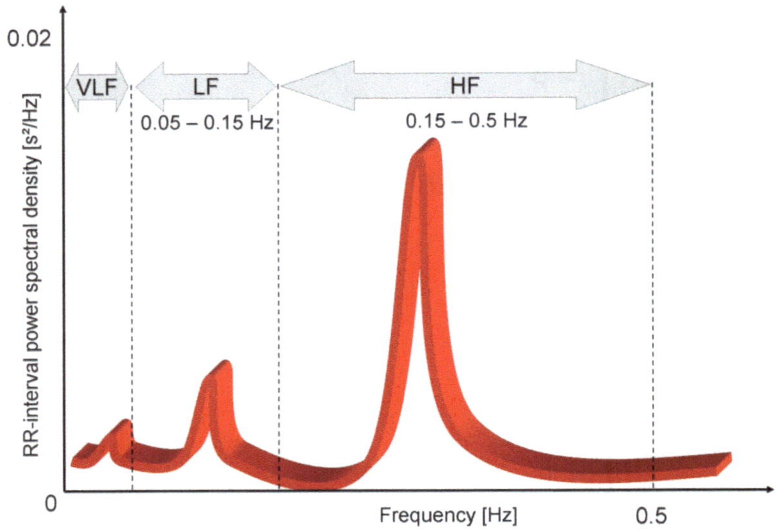

Fig. 22.16 Schematic presentation of RR-interval power spectral density [very-low-frequency (VLF) band, low-frequency (LF) band, and high-frequency (HF) band]

A low-frequency component at around 0.1 Hz is represented by the LF band between 0.05 and 0.15 Hz. The power output in the low-frequency band is partly dependent on the sympathetic tone because of baroreceptor activity. Blood pressure waves of third order prove the connection of this so-called 10-s rhythm.

A high-frequency component which is represented by the HF band, generally between 0.15 and 0.5 Hz, is associated with the breathing frequency and considered an indicator of vagal activity (Figs. 22.16 and 22.17).

HRV related to respiration defined as high-frequency spectral component is a marker of vagal modulation. HRV and arterial pressure variability defined as the low-frequency spectral component is a marker of sympathetic modulation. The ratio of these two bands, as mentioned before, is an important numerical parameter.

It has to be stated that there are different methods for calculating HRV spectra. Each method has its own advantages and disadvantages. The comparability of different studies is complicated and each research group has to collect their own normal values.

Telecommunication technology includes the assisted transmission of signals and biological data over a distance. In this context telemedicine like telesurgery and also teleanesthesiology has become more interesting. However, *teleacupuncture* has not yet been performed. In Fig. 22.18 the first results from a present teleacupuncture pilot study together with the China Academy of Chinese Medical Sciences in Beijing are presented.

For the first time we used a new HRV analysis technique, the "Fire of Life" analysis. The example shows the first teleacupuncture measurement between Asia and Europe. A 24-h ECG was recorded using a system partly developed in Austria.

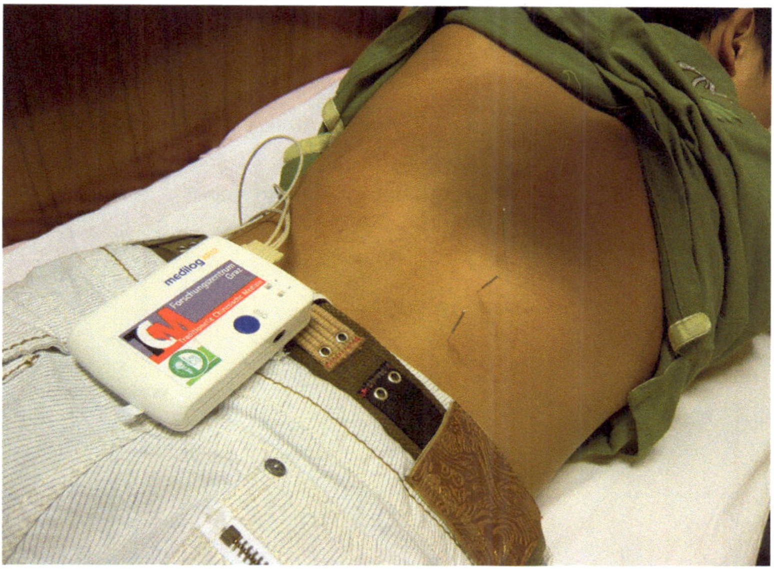

Fig. 22.17 HRV recording in China with equipment from the TCM Research Center Graz (modified from Litscher 2009h).

The raw data were transferred using the internet to the TCM Research Center Graz in Austria from the patients' computer in Beijing to the control computer in Graz over a distance of 7,650 km. Data analysis of different parameters of heart rate and HRV was performed immediately for control of possible therapeutic effects of acupuncture. The acupuncturists in China were informed about the findings immediately, and the success of the therapy could be demonstrated objectively. This could probably be useful under special circumstances, for example, for cooperations between experts from different continents as demonstrated in our Sino-Austrian collaboration (Litscher 2009a, 2009h, 2010).

Of course, follow-up measurements are possible and can vividly demonstrate the success of different therapies like acupuncture using HRV. In this Chinese patient an obvious sleep–wake cycle appears during 10 acupuncture treatments within a time period of about 9 weeks. The initially reduced "Fire of Life" starts to burn more brightly already after the fourth treatment (Fig. 22.19).

The scope of HRV is not yet completely clear, but it is known that there are intraindividual and interindividual variances and that heart rate variation depends on age. It becomes less random with the aging process and the appearance of age-related diseases. Apart from age, circadian variations (sleep–wake cycle), physical condition, and mental and physical exertion are important influencing factors. HRV can also be affected by diverse conditions such as age-related diseases like diabetic neuropathy, renal failure, essential hypertension, cardiac disorders, coronary artery disease, or intracranial lesions. In all cases, different medications have to be taken into account.

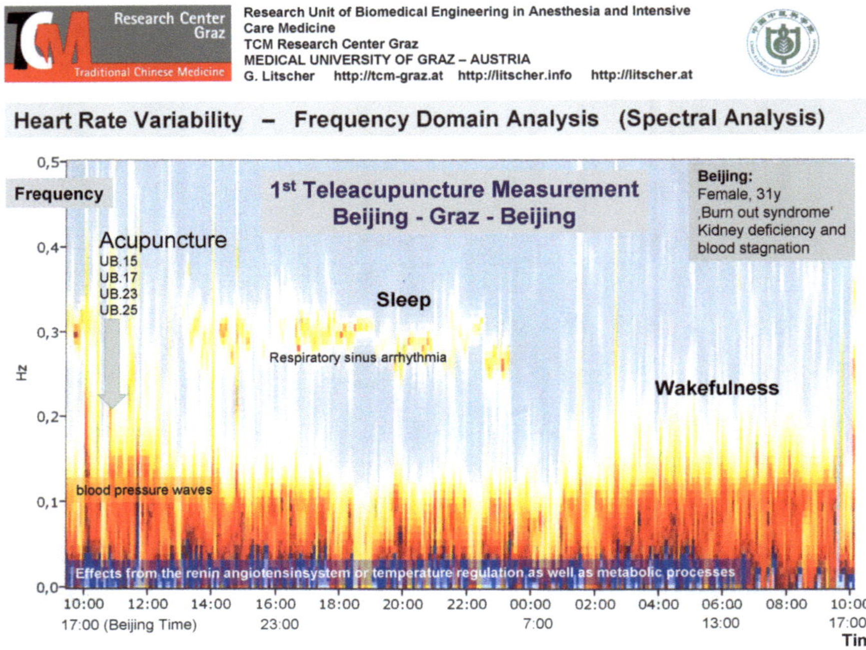

Fig. 22.18 First teleacupuncture measurement between Europe (Graz) and Asia (Beijing). The graphic shows sympathetic and vagal activity, and this can be used to indicate the patient's health and quality of sleep. The success of the manual needle acupuncture therapy can be quantified using modern technology. Acupuncture points: Xinshu (UB-15), Geshu (UB-17), Shenshu (UB-23), and Dachangshu (UB-25). Up to now (January 2010), 30 teleacupuncture experiments have been successfully performed (modified from Litscher 2009a)

HRV can be used as a reliable indicator of the state of health. However, it could be demonstrated that in special syndromes like stress one can counteract this process using different preventive methods like acupuncture. This has been shown in recent investigations concerning patients with burnout syndrome as performed in a common teleacupuncture study between China and Austria (Litscher 2010).

22.4 Concluding Remarks

TCM, especially acupuncture, has made many important contributions to the medicine of the world. A large number of empirical data are available, but the technical objectivation of effects of acupuncture has been intensified within the last years. Using needle, laser needle, or electrical stimulation and modern biomedical techniques, changes in peripheral and cerebral activities can be quantified.

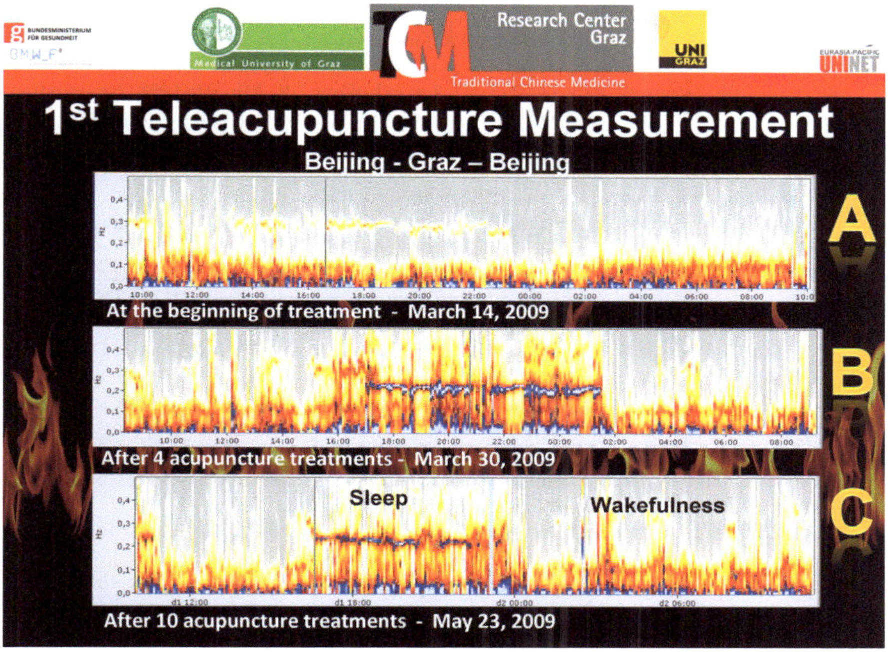

Fig. 22.19 Follow-up investigations during a total of ten acupuncture sessions in China. Note the appearance of an obvious sleep–wake cycle already after the fourth acupuncture treatment (modified from Litscher 2009h)

Acknowledgments This book chapter was presented by the author in part as a lecture at the Shanghai Research Center for Acupuncture and Meridians in China on October 30, 2009.

The author thanks Ms. Lu Wang, MD LA, for performing acupuncture and Ms. Ingrid Gaischek, MSc, for her valuable support in data registration and analysis, both from Research Unit of Biomedical Engineering in Anesthesia and Intensive Care Medicine (http://litscher.info) and TCM Research Center Graz (http://tcm-graz.at), Medical University of Graz.

The investigations were supported partially by the Zukunftsfonds of the Styrian Government (project 4,071) and are part of the project "Bioengineering and clinical assessment of high-tech acupuncture—a Sino-Austrian research pilot study," supported by the Austrian Federal Ministries of Science and Research and of Health and the Eurasia Pacific Uninet.

The following scientists delivered valuable contributions: Weibo Zhang (Beijing, China), Seung-Ho Yi (Seoul, Korea), Lu Wang (Graz, Austria), Tao Huang (Beijing, China), Ingrid Gaischek (Graz, Austria), Zheng Xie (Beijing, China, and Graz, Austria), and Jan Valentini (Beijing, China, and Graz, Austria). The author wishes to thank everyone sincerely.

References

Ahn AC, Colbert AP, Anderson BJ, Martinsen ØG, Hammerschlag R, Cina S, Wayne PM, Langevin HM (**2008**) Electrical properties of acupuncture points and meridians: A systematic review. Bioelectromagnetics 29: 245–256.

Anderson RR, Parrish JA (**1981**) The optics of human skin. J Invest Dermatol 77(1): 13–19.

Beissner F, Henke C (**2009**) Methodological problems in fMRI studies on acupuncture: A critical review with special emphasis on visual and auditory cortex activations. Evid Based Complement Alternat Med, epub ahead of print, Sep 30, doi 10.1093/ecam/nep154

Cho ZH, Chung SC, Jones JP, Park JB, Park HJ, Lee HJ, Wong EK, Min BI (**1998**) New findings of the correlation between acupoints and corresponding brain cortices using functional MRI. Proc Natl Acad Sci USA 95: 2670–2673.

Enwemeka CS, Williams D, Enwemeka SK, Hollosi S, Yens D (**2009**) Blue 470-nm light kills methicillin-resistant staphylococcus aureus (MRSA) in vitro. Photomed Laser Surg 27(2): 221–226.

Jöbsis FF (**1977**) Noninvasive, infrared monitoring of cerebral and myocardial oxygen sufficiency and circulatory parameters. Science 198(4323): 1264–1267.

Kim MS, Soh KS, Nam TC; Seo KM, Litscher G (**2006**) Evaluation of sedation on electroencephalographic spectral edge frequency 95 in dogs sedated by acupuncture at GV20 or Yintang and sedative combination. Acupunct Electrother Res 31(3-4): 201–212.

Kramer S, Zaps D, Wiegele B, Irnich D (**2008**) Changes in electrical skin resistance at Gallbladder 34 (GB34). J Acupunct Meridian Stud 1(2): 91–96.

Litscher G (**2010**) Translational research in acupuncture – Teleacupuncture bridges science and practice. Health 2(1): 17–20.

Litscher G (**2009a**) Modernization of traditional acupuncture using multimodal computer-based high-tech methods – Recent results of blue laser and teleacupuncture from the Medical University of Graz. J Acupunt Meridian Stud 2(3): 202–209.

Litscher G (**2009b**) Laser and high-tech acupuncture. In: Simunovic Z (ed.) Lasers in medicine. Science and praxis. Zrinski, Rijeka, 627–643.

Litscher G, Xie Z, Wang L, Gaischek I (**2009c**) Blue 405 nm laser light mediates heart rate – investigations at the acupoint Neiguan (Pe.6) in Chinese adults. North Am J Med Sci 1: 226–231.

Litscher G (**2009d**) Ten Years Evidence-based High-Tech Acupuncture - A Short Review of Peripherally Measured Effects. Evid Based Complement Alternat Med 6(2): 153–158.

Litscher G, Zhang W, Yi SH, Wang L, Huang T, Gaischek I, Tian YY, Wang GJ (**2009e**) The future of acupuncture moxibustion – a transcontinental three-center pilot study using high-tech methods. Medical Acupuncture 21(2): 115–121.

Litscher G (**2009f**) Ten years evidence-based high-tech acupuncture - A short review of centrally measured effects. Evid Based Complement Alternat Med 6(3):305–314.

Litscher G (**2009g**) Bioengineering assessment of acupuncture. In: Dössel O, Schlegel WC (eds) WC 2009, IFMBE Proceedings 25/XII, 134–137.

Litscher G (**2009h**) Transkontinentale Akupunkturforschung mittels Teleakupunktur. promed komplementär 3: 8–11 (German).

Litscher G (**2008**) High-tech laser acupuncture is Chinese medicine. Medical Acupuncture 20(4): 245–254.

Litscher G, Wang L, Gaischek I (**2007a**) Electroencephalographic responses to laserneedle and punctual stimulation quantified by bispectral (BIS) monitoring: A pilot study to evaluate methods and instrumentation. Internet J Laserneedle Med 1(1). URL: http://www.ispub.com/journal/the_internet_journal_of_laserneedle_medicine/volume_1_number_1_36/article/electroencephalographic_responses_to_laserneedle_and_punctual_stimulation_quantified_by_bispectral_bis_monitoring_a_pilot_study_to_evaluate_methods_and_instrumentation.html

Litscher G, Ammer K (**2007b**) Visualization of equipment dependent measurement errors, but not of meridian-like channels in complementary medicine – a thermographic human cadaver study. Thermology International 17(1): 32–35.

Litscher G (**2007c**) Bioengineering assessment of acupuncture, part 6: monitoring - neurophysiology. Crit Rev Biomed Eng 35(1-2): 1–36.

Litscher G (**2007d**) Bioengineering assessment of acupuncture, part 7: heart rate variability. Crit Rev Biomed Eng 35(3-4): 183–195.

Litscher G (**2006a**) Electroencephalogram-entropy and acupuncture. Anesth Analg 102: 1745–1751.

Litscher G (**2006b**) Bioengineering assessment of acupuncture, part 1: thermography. Crit Rev Biomed Eng 34(1): 1–22.

Litscher G (**2006c**) Bioengineering assessment of acupuncture, part 2: monitoring of microcirculation. Crit Rev Biomed Eng 34(4): 273–293.
Litscher G (**2006d**) Bioengineering assessment of acupuncture, part 3: ultrasound. Crit Rev Biomed Eng 34(4): 295–325.
Litscher G (**2006e**) Bioengineering assessment of acupuncture, part 5: cerebral near-infrared spectroscopy. Crit Rev Biomed Eng 34(6): 439–457.
Litscher G (**2006f**) Bioengineering assessment of acupuncture, part 4: functional magnetic resonance imaging. Crit Rev Biomed Eng 34(4): 327–345.
Litscher G, Schikora D (eds) (**2005a**) Laserneedle-Acupuncture. Science and practice. Pabst Science Publishers, Lengerich.
Litscher G (**2005b**) Infrared thermography fails to visualize stimulation-induced meridian-like structures. Biomed Eng Online 4: 38. URL: http://www.biomedical-engineering-online.com/content/4/1/38
Litscher G (**2005c**) Shenting and Yintang: quantification of cerebral effects of acupressure, manual acupuncture, and laserneedle acupuncture using high-tech neuromonitoring methods. Medical Acupuncture 16(3): 24–29.
Litscher G, Wang L, Schikora D, Rachbauer D, Schwarz G, Schöpfer A, Ropele S, Huber E (**2004a**) Biological effects of painless laserneedle acupuncture. Medical Acupuncture 16(1): 24–29.
Litscher G (**2004b**) Effects of acupressure, manual acupuncture and Laserneedle acupuncture on EEG bispectral index and spectral edge frequency in healthy volunteers. Eur J Anaesthesiol 21(1): 13–19.
Litscher G, Schikora D (**2002a**) Cerebral effects of noninvasive laserneedles measured by transorbital and transtemporal Doppler sonography. Lasers Med Sci 17: 289–295.
Litscher G, Wang L, Huber E, Nilsson G (**2002b**) Changed skin blood perfusion in the fingertip following acupuncture needle introduction as evaluated by laser Doppler perfusion imaging. Lasers Med Sci 17(1): 19–25.
Litscher G, Schikora D (**2002c**) Near-infrared spectroscopy for objectifying cerebral effects of needle and laserneedle acupuncture. Spectroscopy 16: 335–342.
Litscher G, Cho ZH (eds) (**2000**) Computer-Controlled Acupuncture. Pabst Science Publishers, Lengerich.
Litscher G, Wang L (**1999a**) Thermographic visualization of changes in peripheral perfusion during acupuncture. Biomed Tech 44(5): 129–134.
Litscher G, Wang L, Yang NH, Schwarz G (**1999b**) Computer-controlled acupuncture. Quantification and separation of specific effects. Neurol Res 21: 530–534.
Litscher G, Schwarz G, Sandner-Kiesling A, Hadolt I, Eger E (**1998a**) Effects of acupuncture on the oxygenation of cerebral tissue. Neurol Res 20(S1): S28–S32.
Litscher G, Schwarz G, Sandner-Kiesling A, Hadolt I (**1998b**) Robotic transcranial Doppler sonography probes and acupuncture. Int J Neurosci 95: 1–15.
Litscher G, Schwarz G (eds) (**1997**) Transcranial cerebral oximetry. Pabst Science Publishers, Lengerich.
Marshall M (**1996**) Praktische Doppler-Sonographie. Springer, Berlin (German).
Mittermayr R, Osipov A, Piskernik C, Haindl S, Dungel P, Weber C, Vladimirov YA, Redl H, Kozlov AV (**2007**) Blue laser light increases perfusion of a skin flap via release of nitric oxide from hemoglobin. Mol Med 13(1-2): 22–29.
Nakamura S, Pearton S, Fasol G (**2000**) The blue laser diode. The complete story. 2nd ed. Springer, Berlin Heidelberg New York.
Rindge D (**2009**) Lasers and deQi. Acupuncture Today 10(5). URL: http://www.acupuncturetoday.com/mpacms/at/article.php?id=31944 (Aug 16th, 2011)
Sator-Katzenschlager SM, Széles JC, Scharbert G, Michalek-Sauberer A, Kober A, Heinze G, Kozek-Langenecker SA (**2003**) Electrical stimulation of auricular acupuncture points is more effective than conventional manual auricular acupuncture in chronic cervical pain: A pilot study. Anesth Analg 97: 1469–1473.
Széles JC, Litscher G (**2004**) Objectivation of cerebral effects with a new continuous electrical auricular stimulation technique for pain management. Neurol Res 26: 797–800.

Task force of the European Society of Cardiology and the North American Society of Pacing and Electrophysiology: heart rate variability (**1996**) Standards of measurement, physiological interpretation and clinical use. Eur Heart J 17: 354–381.

Wang Z, Chen P, Xie P (**1999**) History and development of traditional Chinese medicine. Science Press, Beijing.

Widder B (**1999**) Doppler- und Duplexsonographie der hirnversorgenden Arterien. Springer, Berlin (German).

Chapter 23
Future Research in Acupuncture: Better Design and Analysis for Novel and Valid Findings

Tetsuya Asakawa and Ying Xia

Abstract Acupuncture has gradually found its way into the mainstream medical academia as a main therapeutic method of Traditional Chinese Medicine. Both clinicians and scientists have an increasing interest in acupuncture research. However, an important issue that holds back the progress of acupuncture is the weaknesses in the experimental design and statistical analysis, especially in the field of clinical research. This reduces the reliability of the studies that evaluate the effectiveness, safety, and mechanism of acupuncture; the aim of this chapter is to provide a methodological "guide" which shall prove beneficial for the conduct of acupuncture research. First, we have summarized the major methodological problems presented by the previous acupuncture studies and then have discussed the appropriate methods for the experimental design based on the principles of clinical epidemiology. We have also detailed some important statistical problems associated with the acupuncture study. At the end of this chapter, we advocate establishment of a new interdisciplinary, namely Clinical Acupunctural Epidemiology that will focus on the methodology of acupuncture studies and aid the future of acupuncture research.

T. Asakawa (✉)
Department of Neurosurgery, Hamamatsu University School of Medicine,
1-20-1, Handayama, Higashi-ku, Hamamatsu, Shizuoka 431-3192, Japan
e-mail: asakawat1971@gmail.com

Y. Xia (✉)
Department of Neurosurgery, The University of Texas Medical School at Houston,
Houston, TX 77030, USA

The University of Texas Graduate School of Biomedical Sciences at Houston,
Houston, TX 77030, USA

Yale University School of Medicine, New Haven, CT 06520, USA
e-mail: Ying.Xia@uth.tmc.edu

Y. Xia et al. (eds.), *Current Research in Acupuncture*,
DOI 10.1007/978-1-4614-3357-6_23,

Keywords Experimental design • Clinical epidemiology • Clinical trial • Evidence-based medicine • Randomization • Double-blind control • Placebo • Bias • Confounding • Jadad index • Consort • Clinical acupunctural epidemiology

23.1 Introduction

As we began writing other chapters of this book, we came across numerous studies conducted to assess the effectiveness and/or safety of acupuncture therapy. However, it was amazing that we could not draw any firm conclusions from these trials owing to the poor quality of their experimental designs. Not only do these methodological flaws reduce the reliability of these studies, worse still, they bring the effectiveness of acupuncture therapy under question as well. Improving the methodology of the experimental design has always been a tricky issue in acupuncture research. We keenly feel the necessity of introducing a methodological approach from the standpoint of clinical epidemiology.

The studies to assess the efficacy and safety of acupuncture substantially fall under the category of clinical trials. Thus, it is imperative that they follow the principles of evidence-based medicine. In other words, to judge whether acupuncture is "effective" or "ineffective," we must have strong evidence from rigorously designed trials based on the principles of clinical epidemiology (PCE). From this viewpoint, there is no difference in the principles between acupuncture and the mainstream therapies. However, there are some specific issues we want to stress on in acupuncture research. In this chapter, we will systematically review the methodological flaws found in previous acupunctural studies, and then we will introduce the appropriate approaches to overcome these flaws and redesign the experimental methodology based on the PCE. In addition, we will discuss some common statistical problems faced in acupunctural studies and present possible solutions for these problematic issues. In the end, we would like to present a new discipline—clinical acupunctural epidemiology, for conducting the acupunctural research in a more scientific and orderly fashion. We do not intend to provide a statistical or epidemiological tool here. Instead, we hope to present some scientific concepts for the acupuncture researchers to better explore the mystery of acupuncture with novel and solid findings in future research. Although we mainly focus on the issues of clinical studies and their solutions, the principles presented here can be applied to all acupuncture studies, including basic science research.

23.2 Improvement of Experimental Design

Despite the plethora of acupuncture studies available, it is difficult to derive firm conclusions on the effectiveness of acupuncture therapy for some diseases. The most vexing issue with acupuncture studies, as with other alternative therapies, is the weakness in the methodology of the study. In this section, we will review the flaws in the acupuncture study and introduce the possible approaches to improve the experimental design based on the PCE.

23.2.1 Common Problems Seen in Previous Studies

The aim of classic epidemiology is to explore the causes of a disease by investigating the health event and its distribution pattern in a population. With the progress in epidemiology, numerous new scientific methods and theories have developed. Application of these principles and methods to a clinical study gave way to a new division called the clinical epidemiology. John Paul first defined the concept of clinical epidemiology as a "new basic science for preventive medicine and public health began with the study of individual patients" (Paul 1938; Sackett 2002), which emphasizes that the exploration of relevant aspects of human ecology and public health began with the study of individual patients. Sackett (1969) described that the chief aim of clinical epidemiology is to provide appropriate epidemiological and statistical methods for clinical trials. The most widely accepted clinical epidemiological principle is the Design, Measurement, and Evaluation (D.M.E.) system initiated by McMaster University (Thabane et al. 2008). It plays a central role in five evolutions of health care: in evidence generation, its rapid critical appraisal, its efficient storage and retrieval, evidence-based medicine, and evidence synthesis (Sackett 2002). Nowadays, the PCE have been accepted as a general standard in the clinical trials. Thus, a "rigorous" experimental design will strictly follow the PCE and will be accepted by the mainstream medical academia. On the contrary, the studies whose experimental design does not meet the marked standards of PCE are labeled as "flawed," and the data/findings gathered from these studies are generally not accepted owing to the weakness in their experimental design. Recently, several standards have been established/recommended on evaluation of the quality of a clinical trial, such as *Consolidated Standards of Reporting Trials* (*CONSORT*) statements (Bian and Shang 2011) and *Standards for Reporting Interventions in Clinical Trials of Acupuncture* (*STRICTA*) recommendation (Hammerschlag et al. 2011). Based on the strict standards laid by these recommendations, a significant amount of published acupunctural papers may not be considered "rigorous study" for many reasons, as is the case with other alternative therapeutics.

The traditional "experience-based" approach followed in the TCM studies sometimes conflicts with the PCE. In ancient China, the efficacy and safety of acupuncture was evaluated based on certain subjective criteria/estimates observed by a TCM grandmaster, or subjective experience of a patient rather than an objective standard. Such subjective solutions tend to influence the TCM researches, which make the results derived from these studies disagreeable with the mainstream medical academia. Replacing these "experience-based" traditional ideas with experimental designs following the principles of evidence-based medicine has been a big challenge. In this chapter, we will make a systematic review on the methodology of acupuncture studies. In order to capture a wider picture and gather more information on the methodological weaknesses, we performed a systematic literature search to look for many articles on acupuncture research published in English and Chinese languages.

23.2.1.1 The Undefined Sample Size

Sample size is a crucial aspect of formulating an experimental design. Too small sample size will cause statistical bias and degrade the reliability of the study, whereas too large sample size will result in wastage of manpower and resources. The determination of appropriate sample size is, therefore, the first critical point in all the acupunctural researches, which should be followed by appropriate statistical testing (Sect. 23.3.1). A satisfactory study, especially a clinical trial, should describe the method used for estimating the sample size. Unfortunately, there are not many papers at hand that report the rationale for sample size estimation. The sample size in these studies was set arbitrarily and did not utilize statistical methods and is by far the most significant flaw in the acupunctural studies (Lam et al. 2008; Lee et al. 2008; Wu et al. 2000; Yang et al. 2009; Zhao et al. 2009).

23.2.1.2 Lack of "Blinding"

The second well-known flaw is the lack of blinding. Although there is triple blinding described in literature, the two commonly adopted methods of blinding in most clinical trials are single and double blinding (Sect. 23.2.4.2). As with most procedural therapies, in acupuncture too it is sometimes difficult to blind the therapist. Single blinding with objective indices for evaluation is therefore regarded as an acceptable solution. However, in the acupuncture practice technical difficulties exist in order to perform blinding. This could explain the lack of blinding in most of the acupunctural studies. Lee et al. (2008) reviewed 103 studies, and Lam et al. (2008) reviewed 784 papers on acupuncture for Parkinson's disease (PD); they found only one study (Cristian et al. 2005) that described the use of blinding (Lam et al. 2008). None of the Chinese reviews reported even a single paper that described the use of strict blinding, other than a few studies that mentioned limited and unverified blinding methods. For instance, Yang and Ma (2005) reported 0 of 206 studies on Myocardial Ischemia; Zhao et al. (2009) reported 6 of 38 studies on Injures of Ankle Joint (unverified); and Wu et al. (2000) reported 15 of 263 studies published in the Journal of Chinese Acupuncture & Moxibustion (unverified).

23.2.1.3 Insufficient Randomization or Pseudo-Randomization

Almost all of the studies published recently claimed to have used "randomization" in selection of the patients, but only a handful described a robust procedure for randomization, whereas the rest used insufficient randomization, even pseudo-randomization in some cases. The most popular method of "randomization" reported in these studies is selection of the patients based on their sequence of visit to the hospital. Such a rough "randomization" protocol that has a high potential for selection bias cannot be accepted as a rigorous approach.

Lam et al. (2008) reported only 5 of the 784 studies on acupuncture for PD used a strict randomization method; Cristian et al. (2005) used coin toss sampling;

Fu et al. (2004) used lot drawing; Wang et al. (2003) and Yang et al. (2006) used random number table sampling (Lam et al. 2008). But none of them reported the concealment of allocation, which is also important for randomization (Sect. 23.2.4.1). It is reported that the efficacy might be exaggerated by 30–40% without incorporating an appropriate method for concealment of allocation (Altman et al. 2001; Schulz 2000). In Lee's review, only 6 of 103 papers used the correct method of randomization (Lee et al. 2008), and 1 of them reported the concealment of allocation (Chen et al. 2007). Wu et al. (2000) reported that only 2 of the 263 studies adequately described the method of randomization, but none of them reported the concealment of allocation.

After reviewing the above-mentioned studies, we observed that although most of acupunctural researchers had a fair idea of randomization, however, seldom did they use proper methods of randomization or describe concealment of allocation.

23.2.1.4 Inappropriate Setting of Control Group

This is a big problem associated with the design of acupuncture study. We have discussed the flaws of design using the protocol like C+A vs. C (C=classical therapy, A=acupuncture) in the PD chapter (please see Sect. 7.2.6.2). Hence use of placebo acupuncture is a probable solution. However, there are certain technical and ethical difficulties with setting up of placebo acupuncture, and the use of placebo acupuncture is controversial and against the ethics/principles of some researchers (Alecrim-Andrade et al. 2008; Diener et al. 2006). We will further discuss placebo acupuncture in the Sect. 23.2.3.2.

Another problem is setting up of a control group without balancing the confounding factors, which can cause a biased conclusion. While some studies failed to compare the baseline characteristics of the patients before the start of therapy (Yang and Ma 2005), others set up their control groups much casually and without fully accounting for the confounding factors which can exaggerate or reduce the acupuncture efficacy. We will further discuss this problem in a later section.

23.2.1.5 Other Methodological Flaws

Besides the major problems in the experimental design described above, there are some other minor methodological flaws which can also be found in the acupuncture studies.

Lack of Objective Evaluation Standard

Acupuncture is considerably influenced by the traditional ideology of TCM that practices experience-based medicine. Such subjective thinking can result in lack of an objective standard for evaluation in these acupuncture studies, which is one of the salient causes that renders barring of acupuncture from mainstream medicine.

There are at least 3 causes for a "lack of objective evaluation standard" in "old" acupuncture studies. *Firstly, there is a lack of an objective standard for acupoint selection*. The acupoint selection follows the theory of differential symptoms (Bian, Zheng, Shi, Zhi) of TCM that is very subjective and experience-based, which therefore lacks the capacity to be standardized. *Secondly, there is no standardized method of performing manual acupuncture*. We all know that acupuncture is a quite complicated procedure that involves insertion or penetration in the body. The needle may be twisting/twirled and moved up and down at different speeds and depths. Sometimes, it is heated or charged with a small electric current until the "De qi" sensation occurs. All the operations, including the strength and angle of the needle insertion, are based on the experience of the therapist and the subjective feeling of the patient. Therefore, there is huge individual difference in order to achieve the particular feeling of "De qi" (a special feeling including tingling, numbness, heaviness, and others) that occur after an acupuncture needle has hit the appropriate place in the body. *Thirdly, the evaluation criteria are also subjective*. In most of the previous studies, some nonstandardized evaluation indices, namely "effective rate" and/or "totally effective rate," were employed for evaluation of efficacy of acupuncture. However, they did not properly define "effective" or "ineffective" treatments using an objective standard, instead they used subjective judgments to assess the efficacy.

Over time, most of the acupunctural researchers have come to understand these problems. More and more objective standards are now being established. One example is replacement of traditional manual acupuncture by electroacupuncture (EA). In comparison with traditional acupuncture, EA has relatively objective indices and is easily controlled by different therapists. Another example is inculcating certain objective indices of modern medicine to evaluate the efficacy of acupuncture. Indices of molecular biology, biochemistry, immunology, electrophysiology, and functional imaging are gradually being employed to assess the efficacy of acupuncture. It is likely to bring major improvements in standardizing and objectifying major indices in acupuncture research.

Weakness in Statistical Design

In early acupuncture studies (before 1990s), we can easily find primary statistical errors like incorrect use of rate and ratio, use of nonparametric statistics for normally distributed data, etc. Although with an improved understanding and application of statistical knowledge, such errors are seldom encountered nowadays, but still some of the available studies fail to describe the data analysis methods, thus making it difficult for us to evaluate the statistical appropriateness of these studies.

A minor problem is that some acupunctural report listed only raw data in tables without any figure with statistical results. It is our view that such figures will provide more useful information to the readers.

Neglect in Recording Adverse Events and Withdrawals

In Lam's study, only 2 of 784 papers reported the adverse events, and two papers mentioned withdrawals (Lam et al. 2008); Lee's study also reported only five papers that mentioned withdrawals out of the 103 papers reviewed (Lee et al. 2008).

In addition to efficacy, safety is another important aspect of evaluating a new therapy. Without adequately reporting the adverse events and withdrawals, the results from a clinical trial cannot be accepted.

We will now discuss some solutions to overcome these flaws and improve the quality of an acupuncture study in the following sections.

23.2.2 *The Principle of Randomization*

Randomization is a core principle in statistical theory and PCE. It was first described by Charles S. Peirce in "Illustrations of the Logic of Science" (1877–1878) and "A Theory of Probable Inference" (1883) (Dehue 1997). The principle of randomization is especially important in experimental design and in survey sampling. The aim of randomization is to avoid selection bias (Sect. 23.2.5.1) and to reduce the difference between the control and treatment groups at baseline, which means to make the two groups identical as far as possible. Every research subject has the same probability to enter the treatment group or control group. This principle holds true on ethical grounds as well as it provides an equal to each patient to try the new therapy.

23.2.2.1 Common Methods of Randomization in Acupuncture Study

With the development of the statistics, new and more complex methods of randomization have developed as well. Here, we will introduce few practical methods that can be used in designing an acupuncture study.

Simple Randomization

Randomization by a single sequence of random assignment is described as simple randomization (Altman and Bland 1999). The most commonly employed procedures for simple randomization are coin toss (Cristian et al. 2005) and lot drawing (Fu et al. 2004).

For example, in a trial to compare the efficacy of acupuncture in two groups, we will predefine what each side of the coin represents, say heads = acupuncture group and tails = nonacupuncture group, then allocate each participant to their respective group based on the result of their coin toss. This method is easier to perform. However, if the sample size is small ($n < 100$), it can cause an unequal distribution of participants in the two groups (Kang et al. 2008) and can potentially affect the results later.

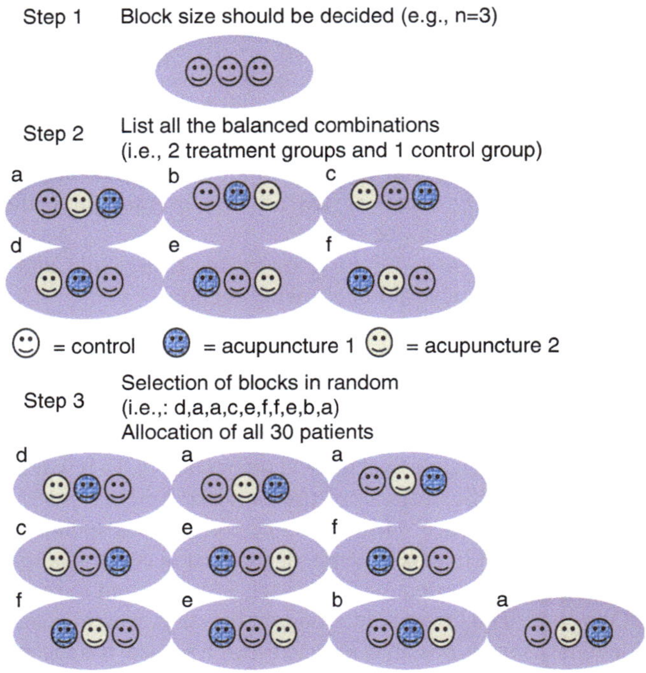

Fig. 23.1 Block randomization method for small sample size

Block Randomization

If the sample size is small ($n < 100$) and there is no covariate that needs to be controlled (such as age, sex, smoking, etc.), then in order to avoid an unequal distribution of participants, we can use the method of block randomization.

Block randomization involves setting up of blocks of size that is usually a multiple of the number of groups (i.e., block size of four or six with two treatment groups) (Altman and Bland 1999) and listing all possible combinations of group allocation within that block. Blocks are selected at random for allocation of cases to the treatment groups. To illustrate this method further, we use an example of allocation of 30 patients to two treatment groups (acupuncture 1 and 2) and one control group using a block size of three (Fig. 23.1). This method can achieve balance in a study with small sample size without other covariates.

Stratified Randomization

In most cases, however, a study is affected by several covariates. In order to avoid confounding bias (Sect. 23.2.5.2), we will have to stratify the participants by these covariates. The stratified randomization method is designed for controlling the influence of covariates.

		Stroke Types			Total
		CH	CI	SAH	
Age	Young < 50y	8	6	4	18
	Senile ≥ 50y	7	★9	6	22
Total		15	15	10	40

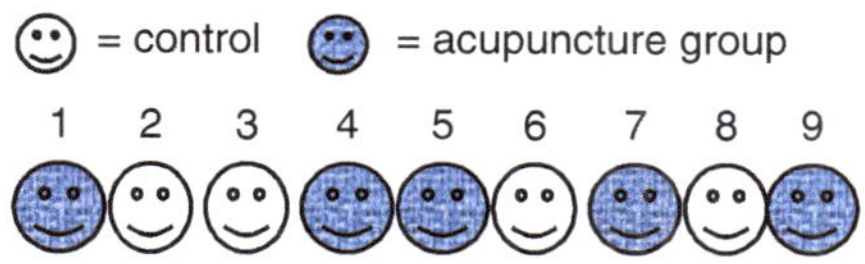

The allocation of 9 patients is determined by a coin toss method

Fig. 23.2 A sample to introduce the method of stratified randomization. Items of each block are assigned by a simple randomization procedure

For example, we want to investigate the efficacy of acupuncture in 40 poststroke patients, and we know that the type of stroke and the age of the patients can profoundly influence the efficacy. Therefore, the stratified randomization method is applied to control the covariates—stroke type [three levels: cerebral hemorrhage (CH), cerebral infarction (CI), and subarachnoid hemorrhage (SAH)] and age (two levels: young <50 years and senile >50 years). We can use a simple randomization procedure to assign the patients within each block to one of the treatment groups. For example, we choose senile CH patients ($n=9$) as shown in Fig. 23.2.

There are two limitations to this method: (1) if there are too many covariates, the procedure becomes very complicated and susceptible to errors, and (2) this method can only be used if all patients were identified simultaneously at the point of entry into the cohort. However, in an actual clinical trial, the patients are recruited in real-time as and when the present. This means that we have to select a covariate adaptive method for randomization.

Covariate Adaptive Randomization

Covariate adaptive randomization is designed for the clinical trial when there is a sequential enrollment of the patients. Once the patient has consented to join the trial, he/she is allocated randomly based on the covariates and previous assignment of the patients. Taves (1974) was the first to report the covariate adaptive method. Covariate adaptive randomization uses the method popularly known as "minimization" of samples that adjusts the imbalances in sample size based on several covariates.

Taves used the above-described patient group allocation method and then makes further decisions on individual basis when a new participant steps into the trial. For example, when studying the efficacy of acupuncture after stroke, we set up a control group and an acupuncture group. The covariates include stroke type (CI, CH, and

Control group		Stroke Types			Total
		CH	CI	SAH	
Age	Young < 50y	1	0	0	1
	Senile ≥ 50y	0	1	1	2
Total		1	1	1	3

Acupuncture group		Stroke Types			Total
		CH	CI	SAH	
Age	Young < 50y	1	0	1	2
	Senile ≥ 50y	1	2	0	3
Total		2	2	1	5

Fig. 23.3 A sample to introduce the covariate adaptive randomization. The allocation information of the first 8 patients. In this example, the characteristics of the 9th patient are senile and CI. The summation of samples of senile and CI in control group is 3 (one for CI category plus two for senile category) while five in the treatment group (two for CI category plus three for senile category). Thus the summation in the control group is less than that in the treatment group ($3<5$), and in order to minimize the imbalance, the ninth patient should be assigned to the control group

SAN) and age (young and senile). Suppose the first 8 patients have been randomly assigned by a simple randomization method, as shown in Fig. 23.3. When the 9th patient, who is senile and CI, is enrolled into the trial, according to Taves' method and considering the patient's characteristics, we will assign this patient to the group with small samples to minimize the imbalance (Fig. 23.3). This method has been further modified and developed by the recent researchers (Frane 1998; Kang et al. 2008; Pocock and Simon 1975).

Through these illustrations we have introduced the most common randomization methods that can be used for an acupuncture study. But it is essential to select the most appropriate method for an acupuncture study. Kang et al. (2008) introduced an algorithm for the appropriate selection of a randomization procedure (Fig. 23.4.). Due to the space limitations we have included only the most commonly used methods. Please refer to specific reference books if more complicated randomization methods are needed.

23.2.3 *Critical Issues in the Control*

With an increasing trend of evidence-based medicine, "to set up control group in a clinical trial" has become a general practice. We collected 2,354 Chinese papers on acupunctural clinical trials published between 1964 and 2011. Only 12 of 323 published before 1980 had set up a control group, while 654 of 789 published between

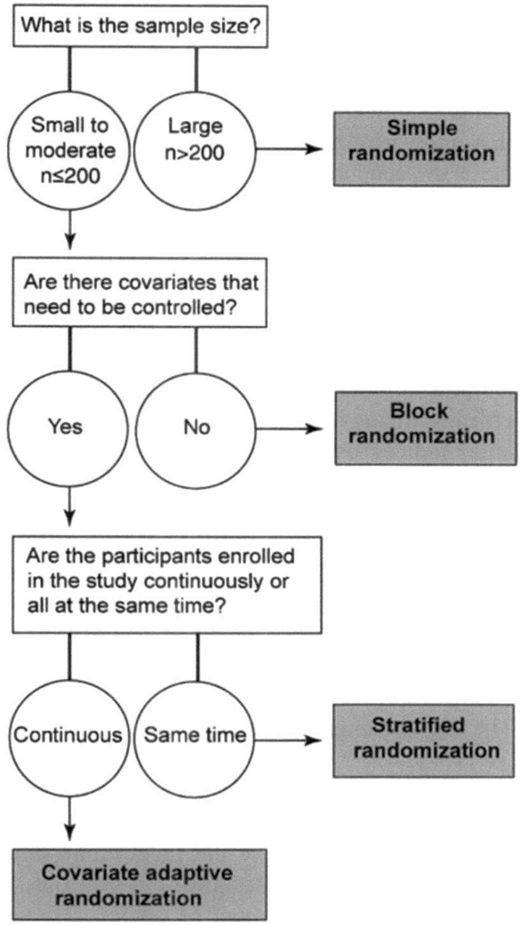

Fig. 23.4 The selection of the randomization methods by the feature of the trial (Kang et al. 2008)

1980 and 2000 described setting up a control group, and almost 100% used controlled design in the trials after 2000. The basic knowledge of "setting up control group" is well known nowadays. We therefore do not need to spend much time in introducing it here. In this section, we will focus on discussing the potential problems using the most popular protocol, namely C+A vs.C, since such flaws likely reduce the value of the acupunctural study.

Another important issue is that in order to avoid confounding (Sect. 23.2.5.2), more covariates should be taken into account when setting up the controls. Sometimes control groups have to be set up by classification according to the characteristics of the samples. Matching is also adopted to achieve a higher quality (Nartey et al. 2007).

23.2.3.1 The Most Popular Control-Setting Protocol: A+C vs. C

Acupuncture is usually adopted as an adjuvant treatment with the established mainstream therapy. Ethically, it is impossible to design the clinical trial as A vs. E (E=empty treatment). As previously discussed in the PD chapter (Sect. 7.2.6.2), of all the 2,354 acupunctural trials at hand, 1,855 papers (78.8%) adopted the controlled design of A+C vs. C (A=acupuncture, C=classic therapeutics). However, such a protocol can still suffer from bias even if evaluating it by the strict standards of PCE.

Observation Bias May Exaggerate the Efficacy

Despite introduction of new objective indices for the evaluation of acupuncture efficacy, owing to its long association with TCM and traditional ideas, some evaluation indices used in acupuncture studies are still subjective. In addition, the "improvement of symptoms" being quite a subjective feeling reported by the patients is difficult to analyze. The words "I feel better after acupuncture" will encourage both patient and acupuncturist greatly, which can result in an overly optimistic assessment of the efficacy. Practically it is hard to achieve "double blinding" in acupunctural studies (Sect. 23.2.4.2). Therefore, it is inevitable to encounter an observation bias with A+C vs. C design.

The Phenomenon of Autosuggestion

Acupuncture induces a positive/stimulatory effect in patients and is believed to possess mysterious healing powers in most countries of East Asia. Such experience will invariably induce positive autosuggestion. Although this thought is controversial as well, many rigorously designed clinical trials have confirmed the effectiveness of autosuggestion in treatment of several diseases (Ikezuki et al. 2002; Stetter and Kupper 2002). In the other words, even if there is no direct efficacy of autosuggestion, it will have some indirect benefit to the patient such as improve the self-confidence, make the patient hopeful, and accept the treatment. Obviously, these "good" effects result from the nonspecific autosuggestion, instead of a direct effect of acupuncture.

In the absence of a well-defined objective standard, even the A+C vs. C design is not a perfect solution. The ideal answer to this problem is a rigorous double-blinded protocol, but unfortunately, it is not always possible to achieve it for all disease processes.

23.2.3.2 A+C vs. C+P: A Better Design?

Since the trial design of "A+C vs. C" can easily lead to bias and it is difficult to conduct a double-blinded acupunctural clinical trial, researchers have come up with a modified version of this design and introduced A+C vs. C+P, where P=placebo acupuncture. We believe it is the most acceptable protocol at present until we invent a more satisfactory way to employ double blinding. The basic structure of this design is that treatment group will be given both standard treatment and acupuncture, whereas the control group will undergo standard treatment and placebo acupuncture. Placebo acupuncture provides a similar acupunctural experience as the real acupuncture is performed on the control patients. Since the patients remain blinded at all times, such a design is single-blinded by definition. This protocol can reduce confounding of results due to observational bias and nullify the effect of autosuggestion. It appears to be better than A+C vs. C; however, the only technical problem is engineering placebo acupuncture.

Adequate placebo acupuncture should have two features: (1) per definition, "placebo" acupuncture should have no effectiveness. (2) Apart from effectiveness, all other features of placebo and real acupuncture should be comparable, including the needle shape, acupuncturist, timing, and course of therapy. Patients should be blinded that means that a patient should not know if he/she is being treated with an effective acupuncture or placebo acupuncture.

Several types of placebo acupuncture are reported in various acupuncture studies:

Sham Acupuncture

Such methods are described by many literatures (Biernacki and Peake 1998; Fink et al. 2002; Fireman et al. 2001; Shapira et al. 2002). Using the same needle and same depth for needle insertion (1–2 cm) as real acupuncture, however, different points are used that are located at a distance from the known effective acupoints.

Advantages of this approach. The patients experience the same subjective feeling as with the real acupuncture; it is the most common method used as placebo acupuncture nowadays (Zhang and Tang 2003).

Drawbacks of this approach. Lewith reported such method of inserting the needle into the nonacupoint location can have some therapeutical effect, which reduces its reliability as a placebo (Vincent and Lewith 1995). In addition, the depth of needle insertion is over 1 cm, which can potentially injure a blood vessel or nerve if the operator is not familiar with the anatomy of the area.

Nonspecific Acupoint Acupuncture

Using the same needle and depth of insertion (1–2 cm) as in the real acupuncture, placebo acupuncture is performed at the points which are irrelevant to the treatment (Park et al. 2003).

Advantages. This method will provide the same experience to the patients as the real acupuncture (such as the feeling of "De qi").

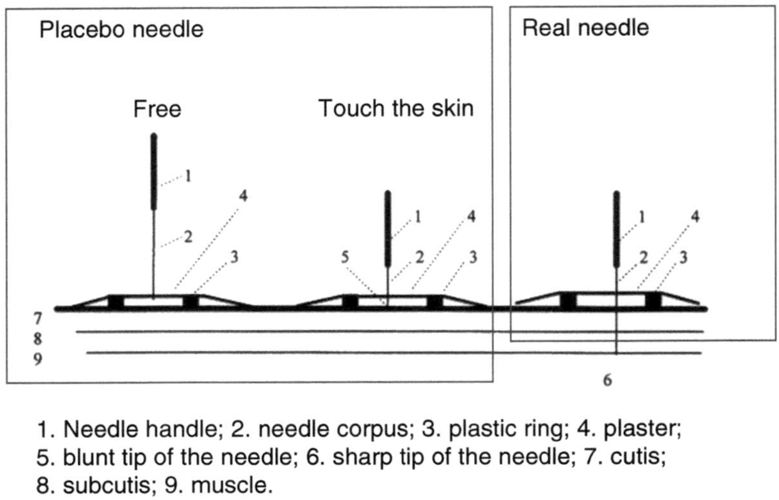

Fig. 23.5 The placebo needles made for acupuncture study (Kleinhenz et al. 1999)

Drawbacks. This method cannot be used for those who have knowledge of TCM. Sometimes, it is quite tricky to select the placebo acupoints for some patients.

Placebo Needle

This is a specially designed needle with a blunt tip. When the needle touches the skin, a pricking sensation is felt, but the body of the needle is not fixed inside the copper handle. The handle of the needle moves over the needle; the needle is shortened so that the patients "see" the needle moving inside their body (Fig. 23.5) (Kleinhenz et al. 1999; Streitberger and Kleinhenz 1998).

Advantages. Because the patients "see" the needle "penetrating" into the skin, the acupunctural experience is similar to real acupuncture. Placebo efficiency is perfect.

Drawbacks. The needle is not available in the market and thus needs to be custom made and can be expensive. In addition, the operator needs a special training for using this needle.

Minimal Acupuncture

Such method is described by several studies (Carlsson et al. 2000; Goddard et al. 2002; Tavola et al. 1992; Zaslawski et al. 1997). This involves the use of the needle that is similar to the real acupuncture at the acupoints, but the depth of needle insertion is much less than that of real one. The depth of needle is 1–4 mm, yet some reports advocate that 1–2 mm is better (Zaslawski et al. 1997).

Advantages. It is safe without therapeutic effect.

Drawbacks. The subjective feeling is different with real acupuncture; the placebo efficiency is weaker than sham acupuncture.

Other Types of Placebo Acupuncture

Moore and Berk (1976) suggested another placebo approach by inserting the needle in the scarfskin. It is safe and without any potential therapeutic effect. But the pattern of the needle insertion is quite different from real acupuncture, thus the placebo efficiency is low and can be only used at the locations that are not in sight of the patients. Mock electroacupuncture is also mentioned in some studies (Zhang and Tang 2003), but it is quite different from the classical acupuncture and is beyond the scope of discussion here.

The method of using placebo acupuncture in the control group is also called "placebo-based control." Although use of placebo acupuncture for acupunctural clinical trial (Alecrim-Andrade et al. 2008; Diener et al. 2006) is often disputed, we believe it is the best solution currently available to set up a controlled trial.

23.2.4 Allocation Concealment and Blinding

Allocation concealment and blinding are two completely different concepts that are commonly confused with each other and hence improperly used. In most of the reported acupunctural trials, both allocation concealment and blinding are neglected almost in entirety. Thus, in this section, we will discuss allocation concealment and blinding together.

23.2.4.1 Allocation Concealment

Allocation concealment is a concept closely related to randomization and is employed to reduce the artificial errors of randomization. Allocation concealment is defined as "the procedure to protect the randomization methods so that the treatment to be allocated remains unknown both to the patients and the investigators before the patient is enrolled in the study" (Forder et al. 2005). The aim of allocation concealment is to reduce the potential selection bias by reducing potential selection bias by limiting the influence or interference of the patients and/or investigators with the process of selection or treatment allocation. It has been widely reported that trials using inadequate or unclear allocation concealment can affect the results by yielding approximately 40 % larger estimates of the effect when compared to those that used adequate concealment (Juni et al. 2001; Moher et al. 1998; Schulz et al. 1995; Schulz and Grimes 2002b). Wood et al. (2008) investigated 146 meta-analyses, including 1,346 trials, and examined a wide range of interventions and outcomes. He found that in trials which lack adequate methods for allocation concealment tend to produce overoptimistic estimates of the effect of interventions and are biased toward beneficial effects. It is therefore recommended that allocation concealment methods should be elaborated in detail in the protocol of a trial. However, we *did not find even a single paper* that mentions allocation concealment in the large number of acupuncture studies that we reviewed. Likewise, this problem exists in the mainstream academia. Pildal et al. (2005) investigated 102 trials and found only 15 of these trials had described adequate allocation concealment. The use of allocation concealment in an experimental design is largely being promoted and is being considered as an index to judge the quality of a trial (Bian and Shang 2011; Hammerschlag et al. 2011).

Several methods of confirming allocation concealment are in practice today, including sequentially numbered opaque-sealed envelopes (SNOSE), sequentially numbered containers, pharmacy-controlled randomization, and central randomization (Schulz and Grimes 2002a). All in all, it is not practically possible for researchers to independently employ allocation concealment. Therefore, it is recommended that a detailed allocation concealment protocol should be established and supervised by independent epidemiologists and be followed upon strictly in different research centers. For this reason, large and multicentric acupuncture clinical trials are recommended because they can accrue larger sample size as well as manage strict execution of a uniform allocation concealment protocol.

23.2.4.2 Blinding

Blinding is defined as "an approach that prevents the investigators, participants, outcome assessors, and other study-related personnel from knowing what intervention is undertaken" (Wood et al. 2008). This is different from allocation concealment as it is to be carried out during the whole length of the study. Blinding plays a crucial role in experimental studies since unblinded trials tend to be biased toward beneficial results (Wood et al. 2008). Blinding is traditionally classified as "single-blind (only the participants are blinded), double-blind (both participants and investigators are blinded), and triple-blind (participants, investigators, and outcome assessors are all blinded)." The 2010 CONSORT statement appeals for gradual abolition of these items since several researchers have reported these terms to be confusing (Devereaux et al. 2001; Haahr and Hrobjartsson 2006). We still use the terms of "single-blind" and "double-blind" in this chapter since we have defined their meanings here. Blinding can be sometimes difficult to perform in acupuncture studies because acupuncture requires active interaction and coordination between acupuncturist and patients.

Most published acupuncture studies that describe use of blinding are single-blinded. Development of the concept of placebo acupuncture (see Sect. 23.2.3.2) makes the single-blind design easier to install. If the assessment indices are objective enough and are not affected by subjective factors, for example, the improvement in functional status of D_2 receptor in PD patients or hippocampal AchE activity levels in AD cases, such a design with "single-blind+objective indices (SBOI)" is widely accepted. Besides placebo acupuncture, several approaches such as using barrier (such as gauze), selection of the acupoints beyond the patient's sight are also used to perform patient blinding (Cristian et al. 2005; Hammerschlag 1998; McManus et al. 2007).

In pragmatic acupuncture studies, sometimes it is difficult to use SBOI design, and both patients and the acupuncturists are unblinded. Such studies can still be considered acceptable if one can effectively keep the outcome assessors blinded to obtain objective data for evaluation (Zwarenstein et al. 2008).

23.2.5 Bias and Confounding

Bias and confounding are considered as the two goliaths in the field of epidemiology. All of the approaches like randomization, control, and blinding in an experimental design are directed toward the singular aim of preventing bias. Both bias and confounding should always be taken into account while structuring an experimental design of acupuncture studies. In this section, we will discuss the introduction of bias and confounding in an acupuncture study.

23.2.5.1 Bias

Bias is defined as "any systematic error in an experimental design which results in an incorrect estimation of the association between exposure and risk of a disease." In an acupunctural study, any factor that can exaggerate or reduce the efficacy and/or safety estimation of acupuncture can be attributed to bias. Bias is a systematic error rather than a human error and cannot be eliminated by improving subjective/objective attitude toward work. Bias can be introduced at any stage of research (Sackett 1979) (Table 23.1). Once a bias is introduced in a study it cannot be "corrected" by statistical treatments. Thus the only way to "avoid" it is through appropriately structured experimental design. An important criterion to evaluate the quality of a study is to determine if the study design can effectively "avoid" biases.

The most commonly encountered biases in an experimental study are selection bias, measurement bias, and analysis bias. Confounding bias is a controversial item, and we will discuss it in the confounding section.

Selection Bias

A selection bias occurs during inappropriate enrollment of participants into scientific study. It is the most common bias seen with experimental design. There are several types of selection bias, and the most important one is sampling bias. If the samples

Table 23.1 Types of bias at different stages of a research study (Sackett 1979)

Stage of the study	Items of biases	Examples
Literature review	Foreign language exclusion bias	
	Literature search bias	
	One-sided reference bias	
	Rhetoric bias	
Study design	Selection bias	
	Sampling frame bias	Centripetal bias, hospital access bias, etc.
	Nonrandom sampling bias	Autopsy series bias, diagnostic work-up bias
	Noncoverage bias	Early comer bias, response bias
	Noncomparability bias	Ecological (aggregation) bias, length bias
Study execution	Bogus control bias	
	Contamination bias	
	Compliance bias	
Data collection	Instrument bias	Diagnostic vogue bias, diagnostic vogue bias
	Data source bias	Hospital discharge bias, competing death bias
	Observer bias	Exposure suspicion bias, expectation bias
	Subject bias	Subject bias, faking good/bad bias
	Recall bias	Reporting bias
	Data handling bias	
Data analysis	Confounding bias	Multiple exposure bias, latency bias
	Analysis strategy bias	Distribution assumption bias, Overmatching bias
	Post hoc analysis bias	
Interpretation of results	Correlation bias	
	Assumption bias	
	Cognitive dissonance bias	
	Generalization bias	
	Magnitude bias	
	Significance bias	
	Under-exhaustion bias	
Publication	Positive result bias	
	All's well-literature bias	
	Hot topic bias	

from a certain population are not selected at random, sampling bias can occur. For example, if we want to evaluate the efficacy of acupuncture in PD patients, selection of patients from a setting of a small clinic as opposed to an advanced medical center is likely to yield better results. This is because most of the patients in a small clinic are mild, early-stage PD patients, whereas most of those in a medical center are severe and advanced-stage patients. We know acupuncture will achieve a much better efficacy in early-stage patients (see PD chapter). This reported/observed "better efficacy" of the patients selected from a small clinic roots from the error in sampling. This is a typical illustration for introduction of selection bias. In other words, the investigators will always have the tendency to select more serious patients for the control group, while the milder cases for treatment group. In this regard, adequate randomization with allocation concealment is the best way to avoid sampling bias.

Measurement Bias

Measurement bias is usually introduced during the stage of data collection. It may arise due to technical reasons (e.g., inaccuracy of the measurement instrument) or subjective reasons, e.g., bias due to the expectations (for positive results) of patients or researchers called observational bias. Observational bias is another important factor that can exaggerate the efficacy and can be introduced in the early stages of acupuncture studies. For example, when observing the efficacy of acupuncture in PD patients, the potential effect of autosuggestion and expectations for a good outcome can affect the assessment of the observer. In addition, the subjective understanding of PDRS varies among different observers/researchers. Using the same PDRS to assess the symptoms of same PD patients can yield different results among different observers/researchers. Such a bias is also attributed to observational bias. There two probable solutions available to avoid observational bias: one is adopting an objective index for evaluation and the other is use of blinding (please see Sect. 23.2.4.2).

There are other types of biases (Table 23.1) that can potentially influence the results of acupuncture study. Acupuncture researchers should keep the term of "bias" in mind and always adopt appropriate measures to avoid bias while conducting an experimental trial.

23.2.5.2 Confounding

Confounding is another major concept of epidemiology that researchers should be wary of. A confounder or confounding variable is an extraneous variable that correlates with both the exposure (risk factors or intervention methods) and the disease and can distort the observed relationship between the disease and exposure factors that are under study. This systemic error is called confounding. The use of the term "confounding bias" is still controversial, and most of the researchers tend to accept keeping confounding and bias as separate concepts and use the term "confounding" instead of "confounding bias." Confounding factors can be positive or negative, known or unknown. Figure 23.6 shows several situations that may or may not result in confounding.

If we neglect the confounding factors in a study, sometimes we end up with absurd conclusions. For example, if we investigate the relationship between the acupuncture and tooth loss in PD patients, we may get the results that those with mild tooth loss achieve better acupunctural efficacy, whereas those who have severe tooth loss will have less improvement in symptoms, which concludes that tooth loss has a negative effect on acupuncture efficacy in PD that needless to mention is an absurd conclusion. It is more like a jeer at the basic medical knowledge. We know that the efficacy of acupuncture in PD is related to the age of patients (see PD chapter); we also know that tooth loss will increase with aging. Thus, aging, instead of tooth loss, is the real factor that influences the efficacy of acupuncture. In this case, tooth loss is an exposure factor, while age is a nonexposure extraneous variable.

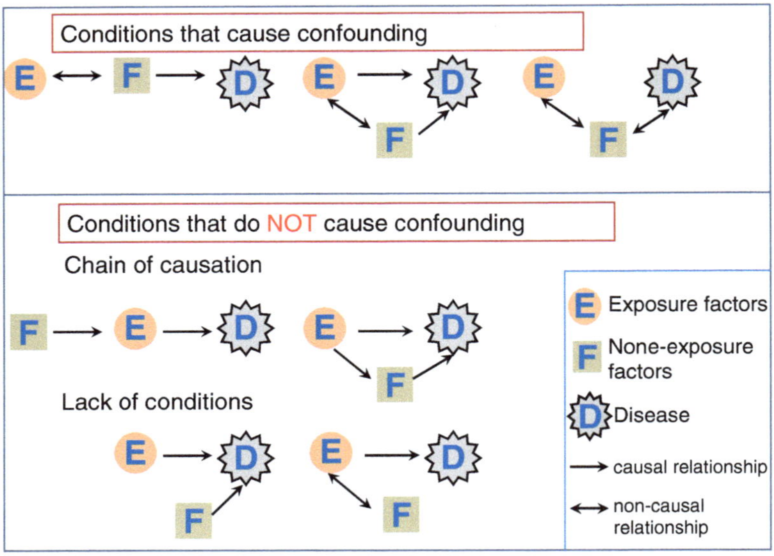

Fig. 23.6 The situations that may/may not cause confounding

Since "age" distorts the actual relationship between tooth loss and acupuncture efficacy, here "age" acts as a confounding factor.

Confounding can only be controlled when the confounding factors are known and measurable. The effective solutions to avoid confounding are however limited. *Case–control studies* are often used for the trials with previously known confounding factors. Let's go back to that example. If we think age is a possible confounding variable in evaluation of the relationship between tooth loss and acupuncture efficacy, a 55-year-old patient must be matched with a healthy 55-year-old control. Case–control studies are appropriate for those trials where it is easy to find controls with all known potential confounding factors comparable to the cases/patients. However, when there are rare or too many confounding factors, it becomes difficult to find well-matched controls for the patients. *Cohort studies* are also adopted to separate the participants into several groups according to the potential confounding factors. For our example, we may set up several age-stratified groups or set up an age-specific cohort (e.g., 45–50). In this case only 45–50 years old PD patients would be involved in a cohort study to assess the relationship of tooth loss and acupuncture efficacy. If the study has been executed without accounting for the potential confounding factors, stratification analysis can be employed as a solution. In the above example, the association between tooth loss and acupuncture efficacy should be analyzed per each age-stratified group. Some statistical tools such as Mantel–Haenszel method can be effectively utilized in such a situation.

Bias and confounding are very important factors that can affect the quality of a study. We hope this section will help acupuncture researchers to better understand ways to avoid bias and confounding in experimental studies. Methods employed to counter bias and confounding should be described in detail in the protocol of an acupuncture trial. This will improve the study quality and make the data dependable.

23.2.6 A Convincing and Efficient Trial Design

Acupuncture is a form of alternative medicine therapy and needs to be validated by a clinical trial, as with other mainstream therapies. Only the results obtained from a well-designed study can provide powerful evidence to prove or disprove the worthiness of acupuncture.

The rudiment of a clinical trial was firstly mentioned at *The Book of Daniel* that describes a planned experiment to judge if "partook" of "the King's Meat" in two groups the meaning is not clear at all (Meinert and Tonascia 1996). The book of "The Canon of Medicine" in 1024 AD first delineated the rules for conducting an experiment to prove the efficacy of a new drug or therapy (Brater and Daly 2000). The methodology of a modern clinical trial was first established by Frederick Akbar Mahomed (1849–1884), who organized the first collaborative clinical trial (O'Rourke 1992). Based on the research methods, there are two types of clinical trials, namely observational study and interventional study. The more popular classification by NIH is on the basis of the purpose of the study and classifies as prevention trials, screening trials, diagnostic trials, treatment trials, quality of life trials, and compassionate use trials.

From the evidence-based medicine standpoint, observational studies such as cohort and case–control studies provide less powerful evidence as opposed to the randomized controlled trials (RCT). RCT is therefore considered to be the strongest tool for evaluating the efficacy of acupuncture since it can provide a robust evidence on the efficacy and safety of acupuncture, albeit there are several known drawbacks, such as the cost, limited external validity, time, and conflict of interest. (Black 1996; Sanson-Fisher et al. 2007).

An ideal RCT is one that is adequately randomized with allocation concealment, triple- or double-blind, and placebo-controlled. The researcher should take the potential biases and confounding factors under serious consideration before setting up the treatment and control groups. The sample size and statistical tools must be selected according to the specific aims of the study. A single-blind study design using placebo acupuncture is the recommended protocol if the evaluation indices follow objective standards. The process of allocation concealment and blinding, along with the withdrawal and adverse events, should be described in the study protocol.

In the future, large, multicenter and transnational RCTs are recommended for the following reasons: (1) larger RCTs can accumulate larger sample size, which will increase the power of statistical analysis. (2) A multicenter design can reduce the occurrence of biases that are prone in a smaller unit. (3) Such large-scale projects easily accrue grant support from government. Epidemiologists and statisticians can be employed to design the study/trial, supervise the trial execution, and analyze the data. This will improve the quality of the study. We hope that such large, multinational trials will help unveil the enigma of the efficacy and safety of acupuncture in various diseases.

23.3 Reliable Statistical Analysis

Statistics is a big problem faced by many acupunctural researchers. Without appropriate statistical methods we cannot draw proper conclusions even if the original data are collected from a rigorous RCT. In this section we will discuss the most common statistical problems confronted in acupuncture studies. We hope this section will help the acupunctural researchers to better understand few important statistical principles associated with the experimental design, data analysis and reporting. Owing to the limited scope of this book, we cannot discuss the detailed procedures of statistical tools here. The details of all statistical procedures mentioned in this section are available in statistical textbooks.

23.3.1 Sample Size

When a new acupuncture research is started, the first problem to be encountered is "what is a reasonable sample size?" This is indeed a complicated problem. The error in determining the sample size has been a major problem in acupuncture study (Sect. 23.2.1.1). Theoretically, larger sample size usually achieves better precision when estimating unclear parameters; however, a very large sample size is ineffective and will increase the cost of the study. In practice, we should choose a "reasonable" sample size that can balance sufficient statistical power and efficiency of the study.

There are several methods used for calculating the sample size.

23.3.1.1 Formulas

We will try to illustrate this method with an example.

To investigate the efficacy of acupuncture in PD, we set up two groups, a control group undergoing classic treatment (L-dopa) as well as placebo acupuncture and a treatment group undergoing classic treatment and acupuncture (A+C vs. C+P).

Step 1. *Specify the outcome measure of the study*. For example, we use *Unified Parkinson's Disease Rating Scale* (UPDRS) to determine the outcomes of the patients. We set the standard for assessing symptom improvement as a 3-point reduction (Δ) on UPDRS, which means if the scores on UPDRS decrease by three points, we regard the treatment as "effective." A review of the previous studies or

Table 23.2 The function table of formulas to estimate the sample size

		β			
		0.05	0.1	0.2	0.3
α	0.1	10.8	8.6	6.2	2.7
	0.05	13.0	10.5	7.9*	3.8
	0.02	15.8	13.0	10.0	5.4
	0.01	17.8#	14.9	11.7	6.6

*f(α, β) = 7.9.
#f(α, β) = 17.8.

estimate of preliminary studies shows the observed standard variation (SD) in the PD patients is six points (this parameter is usually determined in a large population).

Step 2. *Determine the degree of certainty (DC) for the study*. In statistical terms, DC is also known as type II error (β, we will discuss it in the Sect. 23.3.2). Here we have set it as ≤20 %. Thus, the power of this study is 1–20 % = 80 % (Sect. 23.3.2).

Step 3. *Set up a significance level*. This parameter denotes the probability of causing a type I error (α, see Sect. 23.3.2). We set the significance level as 0.05 here.

Thus, for this study we have the following parameters: α=0.05, β=0.20, power = (1–0.20) = 0.80, clinically significant difference (Δ) = 3 and SD = 6.

Using the following formula to estimate the sample size:

$$n = 2(\mathrm{SD})^2 \times \mathrm{f}(\alpha, \beta) / (\Delta)^2 \text{ (Altman 1980)}, \tag{23.1}$$

f(α, β) can be obtained from Table 23.2.

In this case, f(α, β) = 7.9.

Thus, in this study, $n = 2 \times (6)^2 \times 7.9/3^2 = 63.2$, we should select 64 patients in each group.

If the data are enumeration data (see Sect. 23.3.3.1), which can be recorded as rate, we should select another formulate:

$$n = [p_1 \times (100 - p_1) + p_2 \times (100 - p_2)] \times \mathrm{f}(\alpha, \beta) / (p_1 - p_2)^2 \text{ (Altman 1980)}. \tag{23.2}$$

We still use the example, namely (A+C vs. C+P), if we define an "effective" acupuncture will have to improve at lease 88% of the patients, then p_1 = the expected effective rate for acupuncture; and we set the improvement rate of standard L-dopa treatment (from previous/preliminary studies) as 80%, then p_2 = the effective rate of a classic treatment. Then, for a 95 % power and at a significance level of 0.01, we have

$p_1 = 88$ (please pay attention to the unit, do not use "%"),
$p_2 = 80$, $\alpha = 0.01$, $\beta = 0.05$.
From Table 23.2 we have, f(α, β) = 17.8.
Thus, $n = [88 \times (100 - 88) + 80 \times (100 - 80)] \times 17.8 / (88 - 80)^2 = 738.7$.

We should select 739 patients in each group to attain study results with 95 % power and p-value <0.01.

Table 23.3 An experienced table for estimating sample size (Kenny 1987)

	Cohen's d		
Power	0.2	0.5	0.8
0.25	84	14	6
0.50	193	32	13
0.60	246	40	16
0.70	310	50	20
0.80	393	64	26
0.90	526	85	34
0.95	651	105	42
0.99	920	148	58

23.3.1.2 Using Experienced Table

For analyzing the data using a two-sample *t* test, when the treatment group and the control group are equal in size, we can use an experienced table reported by Kenny (1987).

(a) Step 1: Calculate the value of Cohen's d.

Cohen's d is defined as the difference between two methods divided by a standard deviation for the data.

$$\text{Cohen's d} = (\bar{X}_1 - \bar{X}_2) / \text{SD (Kenny 1987).} \qquad (23.3)$$

(b) Step 2: Use the experienced table.

If we set the significance level $(\alpha) = 0.05$, we can get the minimum sample size from Table 23.3.

In addition to the two methods above that are commonly used in estimation of sample size for clinical trials, there are many other complicated methods used to estimate the minimum sample size. These methods are designed for more complex situations that are beyond the level of discussion here.

23.3.2 *Type I Error and Type II Error*

Type I error and Type II error are two important statistical concepts closely associated with sample size estimation (see Sect. 23.3.1) and selection of appropriate statistical tools. An acupuncture researcher must consider the Type I error and Type II error while designing the experimental trial. We will further elaborate on them in this section.

The statistical analysis is an essential procedure to validate the statistical hypothesis. In the statistical analysis, we generally have two hypotheses:

Hypothesis H_0. Null hypothesis, which states that all the data are derived from the same population or the tested treatment, is ineffective; in other words, "there is no difference between the two groups."

Hypothesis H_1. Alternative hypothesis, which states that two groups of data are not derived from the same population or the tested treatment, is effective; in other words, "there is a difference between the two groups." Thus there are two kinds of conclusions that can be drawn: accept H_0 and reject H_1, which means that acupuncture is an ineffective therapy; or accept H_1 and reject H_0, which means that acupuncture is an effective therapy. Statistical decision to "accept" or "reject" a hypothesis is based on the user-defined confidence interval.

23.3.2.1 Type I Error

This type of error occurs if we reject a H_0 hypothesis (accept a false H_1 hypothesis) when the H_0 is actually true. Illustrating this concept in our acupuncture study example, a type I error will occur if we draw a conclusion that the acupuncture therapy is effective in the event when it is actually ineffective. Implication of this wrong conclusion is overexaggeration of the actual efficacy, or in other words the result is "false positive."

In acupuncture studies, results/conclusions are commonly reported as "acupuncture *significantly* improved the PD symptoms ($p<0.05$)." The "p-value" here means if we repeat the test for 100 times, the likelihood or the probability to erroneously exaggerate the efficacy of acupuncture is 5 times out of 100. In other words, there is a 5 % chance to draw a wrong conclusion that the acupuncture is effective (the truth in fact is that it is ineffective). Thus the probability of making a type I error or probability that the results are attributed to chance is less than 5 %.

Thus, the statistical conclusion is a probability-based conclusion. Here, the "p-value" is also called "significance level" and usually equals to the rate of Type I error which is recorded as "α." Increasing the significance level will increase the confidence level of the study ($p<0.01$ is more reliable than $p<0.05$ since $p<0.01$ means only 1 % chance of Type I error).

23.3.2.2 Type II Error

A type II error occurs if we reject an H_1 hypothesis in error (accept a false H_0 hypothesis) when the H_1 is actually true. In terms of an acupuncture study, if we draw a conclusion that the acupuncture is ineffective when in reality it is effective, we have made a type II error. Such an incorrect conclusion tends to reduce the efficacy of acupuncture and increases the "false negative."

The rate of occurrence of Type II error in a study is denoted as "β." The power of a study is calculated as "$1-\beta$."

From acupuncture study standpoint, the relationship of Type I error and Type II error is shown as Table 23.4.

Mathematically speaking, on increasing the significance level in a study with small sample size will enhance the reliability of the study; however, the chance of occurrence of Type II error will also increase proportionately, which will reduce the power of the study ("$1-\beta$" is reduced). In an experimental design, the only way to

Table 23.4 Relationship between Type I error and Type II errors in acupuncture study

	H_0 is true (acupuncture is ineffective)	H_0 is false (acupuncture is effective)
Reject H_0	Type I error False positive	Correct outcome True positive
Fail to reject H_0 (accept H_1)	Correct outcome True negative	Type II error False negative

reduce both Type I error and Type II error is to increase sample size of the study. However, in real-time it is sometimes difficult to increase the sample size due to time and monetary constraints, thus we have to achieve a realistic balance of Type I error and Type II error. For example, set the significance level as 0.05 instead of 0.01 or 0.001.

23.3.3 *Selection of Statistical Methods*

To select appropriate statistical methods for data analysis is another tough call for acupunctural researchers. We will briefly summarize the statistical methods used in published acupunctural literature. We will not elaborate on the procedures of these statistical methods since they can be found in all statistical textbooks.

Typically, the selection of statistical methods is based on the type of the data to be analyzed. According to the characterization and distribution of the data, two types of methods—parametric and nonparametric methods—are generally employed. Apart from these methods, there is a so-called resampling method that is seldom used in an acupuncture study, thus we will omit it in this section.

23.3.3.1 Types of Data

It is essential to understand and distinguish between the various types of data. Data can be classified as enumeration data, measurement data, and ranked data.

Enumeration Data

Enumeration, attribute, or qualitative data refer to the number of samples belonging to a certain category. For example, the data in Table 23.5 are typical enumeration data that show the number of patients selected for an acupuncture study from different countries. Enumeration data can be reported/recorded as rate. Enumeration data are usually classified as a type of Discrete Data.

Table 23.5 An example of enumeration data in an acupuncture study

	Treatment		
Countries	L-dopa + acupuncture	L-dopa + placebo acupuncture	Total
Korea	13 (26 %)	9 (17.65 %)	22
China	25 (50 %)	25 (49.02 %)	50
Japan	12 (24 %)	17 (33.33 %)	29
Total	50	51	101

The number of cases from different countries involved in an acupuncture study is listed here. They can also be recorded as rate (such as the data in the brackets)

Measurement Data

Quantitative data are a specific form of data that always represent a measurable value, for example, the points on UPDRS in PD patients, the scores on MMSE in AD patients, the concentration of homovanillic acid (HVA) in PD patients, etc. All these kinds of data are measured by specific methods and have specific units. Measurement data can be divided into two types, namely *Discrete Data* and *Continuous Data*. Discrete data represent countable events, such as tremors in PD patients or movements of MPTP monkey per day. Discrete Data can be zero or positive whole number and cannot be a negative number or a decimal figure. Continuous Data represents special indices such as body weight of patients and theoretically exists as all possible values between any two numbers.

Ranked Data

Between typical enumeration and measurement data, there exists another form of data called ranked data which represent certain ranked indices, such as the efficacy of acupuncture (good, common, or ineffective, see Table 23.6) or (−, ±, +, ++) of urine glucose (UGLU) in a patient.

23.3.3.2 The Distribution of Original Data

The data of acupunctural studies sometimes accord with certain mathematical distribution model. We usually select the statistical methods according to the distribution of the data. Here we will introduce the binomial distribution and normal distribution, which are associated with acupuncture research.

Distribution of Discrete Data: Binomial and Poisson Distribution

We will use an example of acupuncture treatment for PD to elaborate the concept of this type of distribution. Suppose there are only two possible outcomes, i.e., effective and ineffective acupuncture treatments, the effective rate for one patient is π,

Table 23.6 Ranked data in an acupuncture study

Efficacy	Treatment		
	L-dopa+ acupuncture	L-dopa+placebo acupuncture	Total
Good efficacy (60–90 % reduction in UPDRS)	13	9	22
Common efficacy (30–59 % reduction in UPDRS)	25	25	50
No efficacy (<30 % reduction in UPDRS)	12	17	29
Total	50	51	101

The numbers of PD patients treated by acupuncture are recorded based on the different levels of efficacy. Please compare the difference between this table and Table 23.5

and the ineffective rate is therefore $1-\pi$. For one patient, if the outcome is effective, then π; if ineffective, then $1-\pi$. For two patients ("a" and "b"), if effective in both of them, then π^2; if "a" effective while "b" failed, then $\pi\times(1-\pi)$; if "b" effective while "a" failed, then $(1-\pi)\times\pi$; if both failed, then $(1-\pi)^2$.

Therefore, the total probability

$$(\mathrm{P}) = \pi^2 + \pi\times(1-\pi) + (1-\pi)\times\pi + (1-\pi)^2 = [\pi + (1-\pi)]^2$$

...

If we have n patients, the total probability

$$(\mathrm{P}) = [\pi + (1-\pi)]^n = \sum_{x=0}^{n} \frac{n!}{x!(n-x)!}\pi^X(1-\pi)^{n-x}$$

$$= \frac{n!}{0!(n-0)!}\pi^0(1-\pi)^n + \frac{n!}{1!(n-1)!}\pi^1(1-\pi)^{n-1}$$

$$+ \frac{n!}{2!(n-2)!}\pi^2(1-\pi)^{n-2} + \cdots + \frac{n!}{n!(n-n)!}\pi^n(1-\pi)^0. \quad (23.4)$$

The probability of achieving effective outcome of No. x individuals in the population (total number $=n$) is

$$\mathrm{P}(x) = \frac{n!}{x!(n-x)!}\pi^x(1-\pi)^{n-x}. \quad (23.5)$$

There are two requirements that require to be met in order to apply binomial distribution. First is that each event should be independent. In our example, the efficacy of acupuncture in one patient does not affect its efficacy in another patient. Second requirement is that the outcomes should be mutually exclusive. In our example, there are only two possible outcomes: effective and ineffective.

To illustrate the use of binomial distribution in acupuncture study:

Consider that the effective rate of administration of L-dopa for certain groups of AD patients is 60 %, the ineffective rate is hence 1–60 % = 40 % (L-dopa is a standard therapy for PD, please see the PD section).

Question 1. Given the above information, if we perform acupuncture + L-dopa on five patients and all the patients are effective, then, is the combined treatment of acupuncture + L-dopa better than L-dopa alone?

So, if we assume that the efficacy of acupuncture + L-dopa is NOT better than L-dopa alone, there are only two outcomes, namely acupuncture is effective or ineffective. Such data distribution accords with binomial distribution. According to formula (23.4), the P that all five patients have effective outcomes is given by

$$P(5)=\frac{5!}{5!(5-5)!}\pi^{5}(1-\pi)^{5-5}=(0.6)^{5}=0.0778>0.05.$$

Thus, at a significance level of 0.05, no significant difference exists between acupuncture + L-dopa group and L-dopa alone group. That is to say although five patients have effective outcomes, we cannot draw a solid conclusion that acupuncture + L-dopa is more effective than only use L-dopa.

Question 2. How would it differ if ten patients underwent combined treatment and nine had effective outcome while only one patient failed treatment?

Basically, in this case we have to calculate the probability of effective treatment in nine patients + probability of effective treatment in ten patients, that is,

$$P(10)+P(9)=(0.6)^{10}+10\times(0.6)^{9}\times(1-0.6)^{(10-9)}=0.00605+0.04031=4.6\%$$

i.e., $p<0.05$.

Thus, there is significant difference, if $\alpha=0.05$, between the two treatment groups. That is to say, if nine patients have an effective outcome, we can conclude that acupuncture + L-dopa is more effective than L-dopa alone.

Poisson distribution is a special type of binomial distribution. Since it is seldom used in the acupuncture study, we will not discuss it in this section.

Distribution of Continuous Data: Normal Distribution

The most important distribution for continuous data in acupuncture study is the normal distribution. Continuous random variables, having any value on the real-number-line, are said to be normally distributed in nature as they fluctuate and cluster around a single mean value. The probability function of such a distribution is called probability density function, namely $f_x(x)$ (23.6). The graph of the probability density function looks like a bell and is known as the Gaussian function or bell curve (Fig. 23.7):

$$f_{x}(x)=\frac{1}{\partial\sqrt{2\pi}}e^{-\frac{(x-\mu)^{2}}{2\sigma^{2}}}\quad -\infty<x<\infty. \tag{23.6}$$

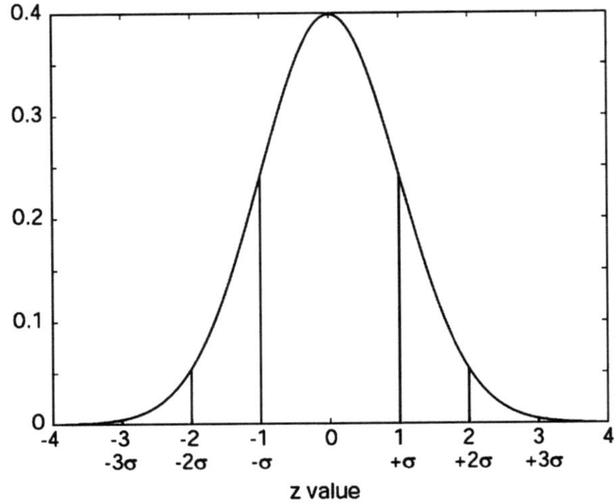

Fig. 23.7 Graphical representation of a normal distribution (Delorme 2006). The normal probability density function shows symmetry about a vertical line through μ; σ is a variability parameter. The vertical bars show ± σ, ±2σ and ±3σ (Delorme 2006)

Continuous data follows the laws of normal distribution where μ is the mean (location of the peak) and σ^2 is the variance (the width of the distribution). The distribution with $\mu=0$ and $\sigma^2=1$ is called the normal standard distribution.

Since most statistical methods are based on the assumption that data is normal, it is essential to understand the concepts of normal distribution. The statistical thought concerning the normal distribution is that any linear mixture distribution of two or more independent random variables is "more normal" than the distribution of the random variables themselves (Delorme 2006). Analysis of variance, *t* test, and regression are based on this assumption. We will discuss these in the next section.

Sometimes the data can also be mathematically approximated to normal distribution by using mathematical tools, for example, we always use logarithms of the original data if they are distributed geometrically. (The most common sample is the serum bactericidal titer in the clinical biochemical examination.) Such approximation methods are detailed in almost all statistical textbooks. There are other distribution models that describe probability distribution of continuous variables. Since they are seldom used in the acupuncture studies, therefore we will not discuss them here.

23.3.3.3 Selecting an Appropriate Statistical Method

Selection of proper statistical methods in an acupuncture study is crucial step. Incorrect statistical analysis will lead to biased conclusions even if the acupunctural trial has been designed and executed rigorously with randomization, blinding, and placebo-based control all applied in perfect order. Based on our discussion in the previous sections, this section will focus on the steps to select a proper statistical method (Fig. 23.8).

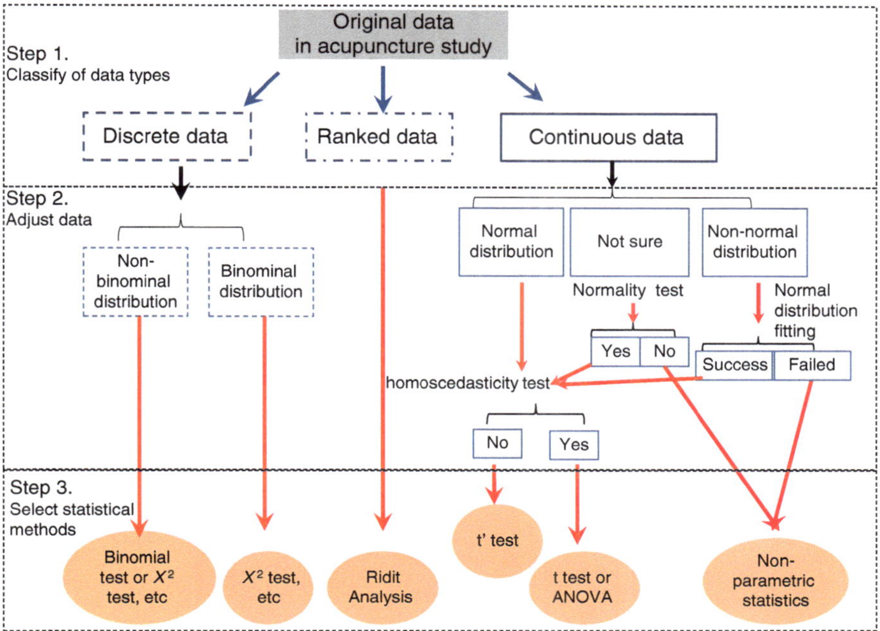

Fig. 23.8 Procedures of selection statistical methods

There are two important issues that need to be dealt with during the selection of the statistical methods.

First is the selection of the one-tailed test or two-tailed test. In a one-tailed test, the null hypothesis (H_0) will be rejected only in one of the rejection areas, namely "sufficiently large" or "sufficiently small," while in a two-tailed test, H_0 will be rejected when the data is both sufficiently small and large. In an acupuncture study, only if we know for certain that acupuncture will improve (or worsen) the symptoms, we can select the one-tailed test. But in majority of the cases, we cannot ensure the effect of acupuncture, thus a two-tailed test is recommended for acupuncture study.

The second problem is concerning the nonparametric statistics. We know that nonparametric statistics can be used for any type of data without considering the data distribution, thus it can be used for a wider range than that of parametric statistics. Certainly data that is normally distributed can also be analyzed by nonparametric statistics; however, using nonparametric methods for normally distributed data will result in loss of statistical information and will reduce the statistical efficiency. Thus, use of nonparametric methods is not recommended for normally distributed data.

Here we have listed some common statistical methods in Table 23.7; the examples and procedural details of these methods can be consulted from a statistical textbook.

Table 23.7 Commonly used statistical methods based on the type of study and data

	Types of data			
			Continuous data	
Aim of study	Binomial or discrete data	Ranked data	Normal distribution	Nonnormal distribution
Describe one data sample	Proportions	Ranked proportions	Mean, SD	Median
Compare one data sample to a hypothetical distribution	χ^2 test or binomial test	Ridit analysis or Wilcoxon test	One-sample *t* test	Sign test or Wilcoxon test
Compare two paired samples	Sign test	Ridit analysis or Wilcoxon test	Paired *t* test	Sign test or Wilcoxon test
Compare two unpaired samples	χ^2 Fisher's exact test	Ridit analysis or Mann–Whitney test	Unpaired *t* test	Mann–Whitney test
Compare three or more unmatched samples	χ^2 test	Ridit analysis or Kruskal–Wallis test	One-way ANOVA	Kruskal–Wallis test
Compare three or more matched samples	Cochrane Q test	Friedman test	Repeated-measure ANOVA	Friedman test
Quantify association between two paired samples	Contingency coefficients	Spearman correlation	Pearson correlation	Spearman correlation

Partly cited and modified from Delorme (2006). Note that most of the acupuncture studies are subject to statistical comparison between two unpaired samples (*gray areas*)

23.4 Evaluation of Acupuncture Study

We can evaluate an acupuncture study based on the principles of scientific designing of an acupuncture study by using randomization, placebo-controlled, and blinding; proper selection of statistical tests is also an important issue to analyze the quality of an acupuncture study. There are several widely accepted evaluation systems involved in the acupuncture study, namely *Jadad Scale* (Jadad et al. 1996a, b; White and Ernst 1999), CONSORT statements (Bian and Shang 2011), and STRICTA recommendations which was set up specifically for evaluation of acupunctural trials following the recommendations of CONSORT (Hammerschlag et al. 2011). Hammerschlag reported an *Oregon CONSORT STRICTA Instrument* (*OCSI*) which is basically modified STRICTA recommendations for acupuncture trials based on the revised CONSORT guidelines (Hammerschlag et al. 2011). Albeit the organizers of these systems are different, their evaluation criteria are grossly similar. They have made checklists that can be conveniently used by the researchers as well as journals reviewers. Irrespective of the standard that is followed, a strict adherence to the guidelines can establish a well-designed acupunctural trial. In addition, this makes it easier for a researcher to evaluate the power of acupunctural papers based on these standards.

Here we use an example to introduce how to evaluate an acupuncture study. We will illustrate how to evaluate an acupuncture study based on Jadad scale (Jadad et al. 1996b), which is the simplest of the all standards. We will use acupunctural RCT by Cristian et al. (2005) for illustration. According to Jadad scoring, the scores are awarded as follows: a randomized study—one point; appropriate methods for randomization—one point; inappropriate randomization—deduct one point; single-blind (to patients)—one point; investigator blinding—one point; description of withdrawals and dropouts—one point. The maximum possible score is five (Jadad et al. 1996b). Cristian et al. (2005) performed EA in 14 mild or severe PD patients. A simple randomization method—coin toss—was used for treatment group allocation. The acupoints used in the treatment group were Taixi (KI-3), Yingu (KL-10), Kunlun (BL-60), Taichong (LR-3), Jiexi (ST-41), Zusanli (ST-36), Yanglingquan (GB-34), Scalp stimulation area (MS-6), Hegu (LI-4), and Baihui (GV-20). The points of needle insertion in the control group were randomly selected areas on arms, legs, and scalp, which did not coincide with any known acupoints. The EA was delivered at a sensory threshold level of 4 Hz for 20 min. Single-use disposable needles were used for both groups. The same acupuncturist performed the whole protocol, thus the participants in both groups had same experience. (The patient did not know if he underwent a treatment EA or a placebo EA.) The same research team that was initially blinded to patient allocation performed outcome measurements. UPDRS, Parkinson's disease questionnaire (PDQ), and geriatric depression scale (GDS) were used to assess the symptoms of PD before and after EA. This trial concluded that there is no significant difference in the improvement of PD symptoms between the EA group and control group (Cristian et al. 2005).

Based on Jadad scoring criteria, this is a high-quality study as it scored five points: used randomization—one point; had described use of simple randomization method (coin toss)—one point; performed double-blind, therefore for blinding the patients—one point and blinding the outcome assessors—one point; and described the withdrawal events—one point.

By the CONSORT standards, however, this published study would be even better if they described (1) how the sample size ($n = 14$) was calculated and (2) the procedure for allocation concealment.

This example shows a simple way to evaluate the quality of an acupuncture study, which may provide an idea for find valuable references with high quality of work. We recommend using the CONSORT checklist when performing the experimental design. We have added the checklist of CONSORT 2011 in the appendix for readers' information.

23.5 A New Discipline: Clinical Acupunctural Epidemiology?

Finding solutions to these baffling problems and devising ways to improve the credibility of acupuncture research are indeed some of the big challenges faced by acupuncture researchers. We have highlighted the most common methodological problems encountered in acupuncture study in terms of the study design and briefly discussed the potential solutions. To better evaluate the efficacy of acupuncture and discover the underlying mechanisms in future research, it is of utmost importance to conduct acupuncture in a scientific way with maximal reduction of methodological weakness. In this regard, we advocate to establish a new discipline—*Clinical Acupunctural Epidemiology* (*CAE*), which aims at improving the research methodology in acupunctural trials and provides special methodological support for acupuncture research.

CAE is formulated by merging together the epidemiological and statistical theories for acupuncture. It is an offshoot of clinical epidemiology. The aim of CAE is to develop systematic theories and methods for acupuncture studies and provide a robust system of evaluating acupuncture studies. The field of CAE covers the entire process of designing an acupuncture study encompassing the review of previous studies (to screen high-quality papers for references), guidance for experimental study design (how to set randomization, blind, and control in acupunctural study), data collection, statistical analysis, and interpreting and reporting the results. CAE lays down the experimental methodology for practical acupuncture. The objectives of CAE includes (1) determining the efficacy and safety of acupuncture (clinical trials), (2) evaluating new observational/objective indices for the acupunctural investigation, (3) providing a methodological guideline for acupuncture research, and (4) assessing the quality of a given acupuncture study by the viewpoints of PCE and Health Economics. The researchers of CAE can be both clinical acupuncturists and basic acupuncture scientists with basic understanding of epidemiology and statistics or the epidemiologists and statisticians who are interested in acupuncture research.

Currently, evaluation systems for acupuncture studies such as STRICTA have been established with a recently revised version being put together by a group of 7 acupuncture researchers known as the OCSI group (Hammerschlag et al. 2011). In generally, this falls in the field of CAE. To establish CAE, there are many things that need to be done, including training the specialists, editing the textbook, setting up the congress, and formally establishing a journal for the field of CAE. We hope that CAE will greatly enhance the acupuncture research in near future.

23.6 Concluding Remarks

Through intensive worldwide research in past five decades, our understanding of the efficacy and mechanisms of acupuncture has largely increased. Substantial evidence has suggested that acupuncture can significantly affect various functions and treat certain diseases, especially ones based on neural mechanisms, through needle stimulation that initiates a serial of events at cellular and molecular levels in the nervous system (Xia et al. 2010). However, the current knowledge is far from complete understanding of acupuncture therapy. There exists a plethora of controversies in the present literature. In a way it reflects upon the complexity of acupuncture-induced effects on human body. On the other hand, the problematic research designs and execution may largely contribute to the previous controversies. Therefore, it is now important to clear our thoughts and strengthen scientific design with modern approaches in the field of acupuncture research.

In this chapter, we have targeted the major problems presented by the previous acupuncture studies in terms of research design and performance and have discussed the appropriate methods for scientific design of acupuncture studies based on the PCE. We have also addressed several important issues of statistical problems associated with the acupuncture study. Although we mainly focus on the clinical acupuncture research, the principles presented in this chapter are generally applicable to all acupuncture studies, including basic research.

It is worthy to stress upon the importance of critical thinking in acupuncture research. In the previously published literature, including some recent reports, some investigators measured one or two biomedical indices (e.g., change in the level of a substance in blood) during acupuncture and then claimed without any other evidence that the acupunctural mechanisms were mediated by regulation of these substances. On the other hand, some researchers blindly cover acupuncture with a mysterious aura and try to explain it with ancient concepts and terms. All these pose obstacles and confusions in the field of acupuncture research.

Also, it is important to learn from previous lessons in acupuncture research. In some oriental countries, for example, huge efforts and resources were invested for dissecting out "unique" structures of meridians and acupoints only because the ancient TCM literature mentioned their existence. All the related work ended up as a failure since other independent researchers could not repeat/reproduce any of their "findings." Fortunately, most current investigators are well-trained clinicians and scientists with solid biomedical background and current knowledge and conduct acupuncture in a scientific way.

We would like to end this chapter at the following note: (1) acupuncture is effective in certain diseases but is not universal for all; (2) acupuncture is amazing, but not mysterious; and (3) acupuncture mechanism is difficult to learn but can be explored by using modern scientific approaches. It is our belief that with a better design and analysis of the acupunctural study, we will make novel and solid findings that would find practical implications in future acupuncture research.

Acknowledgements TA was supported by Japan Society for the Promotion of Science (Grant-in-Aid for Young Scientists, Type B, #20791025 and Grant-in-Aid for Scientific Research General, #24592157). YX was supported by the National Institutes of Health (AT-004422 and HD-034852) and the Vivian L. Smith Neurologic Foundation.

References

Alecrim-Andrade, J., Maciel-Junior, J.A., Carne, X., Severino Vasconcelos, G.M., Correa-Filho, H.R., 2008. Acupuncture in migraine prevention: a randomized sham controlled study with 6-months posttreatment follow-up. Clin J Pain. 24, 98–105.

Altman, D.G., 1980. Statistics and ethics in medical research: III How large a sample? Br Med J. 281, 1336–8.

Altman, D.G., Bland, J.M., 1999. How to randomise. BMJ. 319, 703–4.

Altman, D.G., Schulz, K.F., Moher, D., Egger, M., Davidoff, F., Elbourne, D., Gotzsche, P.C., Lang, T., 2001. The revised CONSORT statement for reporting randomized trials: explanation and elaboration. Ann Intern Med. 134, 663–94.

Bian, Z.X., Shang, H.C., 2011. CONSORT 2010 statement: updated guidelines for reporting parallel group randomized trials. Ann Intern Med. 154, 290–1; author reply 291–2.

Biernacki, W., Peake, M.D., 1998. Acupuncture in treatment of stable asthma. Respir Med. 92, 1143–5.

Black, N., 1996. Why we need observational studies to evaluate the effectiveness of health care. BMJ. 312, 1215–8.

Brater, D.C., Daly, W.J., 2000. Clinical pharmacology in the Middle Ages: principles that presage the 21st century. Clin Pharmacol Ther. 67, 447–50.

Carlsson, C.P., Axemo, P., Bodin, A., Carstensen, H., Ehrenroth, B., Madegard-Lind, I., Navander, C., 2000. Manual acupuncture reduces hyperemesis gravidarum: a placebo-controlled, randomized, single-blind, crossover study. J Pain Symptom Manage. 20, 273–9.

Chen, X.H., Li, Y., Kui, Y., 2007. [Clinical observation on abdominal acupuncture plus Madopa for treatment of Parkinson's disease]. Zhongguo Zhen Jiu. 27, 562–4 (in Chinese with English abstract).

Cristian, A., Katz, M., Cutrone, E., Walker, R.H., 2005. Evaluation of acupuncture in the treatment of Parkinson's disease: a double-blind pilot study. Mov Disord. 20, 1185–8.

Dehue, T., 1997. Deception, efficiency, and random groups. Psychology and the gradual origination of the random group design. Isis. 88, 653–73.

Delorme, A., 2006. Statistical methods. Encyclopedia of Medical Device and Instrumentation, Vol. 6, pp 240–64, Wiley interscience.

Devereaux, P.J., Manns, B.J., Ghali, W.A., Quan, H., Lacchetti, C., Montori, V.M., Bhandari, M., Guyatt, G.H., 2001. Physician interpretations and textbook definitions of blinding terminology in randomized controlled trials. JAMA. 285, 2000–3.

Diener, H.C., Kronfeld, K., Boewing, G., Lungenhausen, M., Maier, C., Molsberger, A., Tegenthoff, M., Trampisch, H.J., Zenz, M., Meinert, R., 2006. Efficacy of acupuncture for the prophylaxis of migraine: a multicentre randomised controlled clinical trial. Lancet Neurol. 5, 310–6.
Fink, M., Wolkenstein, E., Karst, M., Gehrke, A., 2002. Acupuncture in chronic epicondylitis: a randomized controlled trial. Rheumatology (Oxford). 41, 205–9.
Fireman, Z., Segal, A., Kopelman, Y., Sternberg, A., Carasso, R., 2001. Acupuncture treatment for irritable bowel syndrome. A double-blind controlled study. Digestion. 64, 100–3.
Forder, P.M., Gebski, V.J., Keech, A.C., 2005. Allocation concealment and blinding: when ignorance is bliss. Med J Aust. 182, 87–9.
Frane, J.W., 1998. A method of biased coin randomization, its implementation, and its validation. Drug. 32, 423–32.
Fu, X.L., Li, R., Lin, W.D., 2004. Electroacupuncture at head and du pulse acupoints for treatment of Parkinson disease. Chin J Chin Rehabil. 8, 4524–5 (in Chinese).
Goddard, G., Karibe, H., McNeill, C., Villafuerte, E., 2002. Acupuncture and sham acupuncture reduce muscle pain in myofascial pain patients. J Orofac Pain. 16, 71–6.
Haahr, M.T., Hrobjartsson, A., 2006. Who is blinded in randomized clinical trials? A study of 200 trials and a survey of authors. Clin Trials. 3, 360–5.
Hammerschlag, R., 1998. Methodological and ethical issues in clinical trials of acupuncture. Journal of Alternative and Complementary Medicine. 4, 159–171.
Hammerschlag, R., Milley, R., Colbert, A., Weih, J., Yohalem-Ilsley, B., Mist, S., Aickin, M., 2011. Randomized Controlled Trials of Acupuncture (1997–2007): An Assessment of Reporting Quality with a CONSORT- and STRICTA-Based Instrument. Evid Based Complement Alternat Med. 2011.
Ikezuki, M., Miyauchi, Y., Yamaguchi, H., Koshikawa, F., 2002. [Development of Autogenic Training Clinical Effectiveness Scale (ATCES)]. Shinrigaku Kenkyu. 72, 475–81 (in Japanese).
Jadad, A.R., Carroll, D., Moore, A., McQuay, H., 1996a. Developing a database of published reports of randomised clinical trials in pain research. Pain. 66, 239–46.
Jadad, A.R., Moore, R.A., Carroll, D., Jenkinson, C., Reynolds, D.J., Gavaghan, D.J., McQuay, H.J., 1996b. Assessing the quality of reports of randomized clinical trials: is blinding necessary? Control Clin Trials. 17, 1–12.
Juni, P., Altman, D.G., Egger, M., 2001. Systematic reviews in health care: Assessing the quality of controlled clinical trials. BMJ. 323, 42–6.
Kang, M., Ragan, B.G., Park, J.H., 2008. Issues in outcomes research: an overview of randomization techniques for clinical trials. J Athl Train. 43, 215–21.
Kenny, D.A., 1987. Statistics for the social and behavioral sciences. Vol., Little, Brown, Boston.
Kleinhenz, J., Streitberger, K., Windeler, J., Gussbacher, A., Mavridis, G., Martin, E., 1999. Randomised clinical trial comparing the effects of acupuncture and a newly designed placebo needle in rotator cuff tendinitis. Pain. 83, 235–41.
Lam, Y.C., Kum, W.F., Durairajan, S.S., Lu, J.H., Man, S.C., Xu, M., Zhang, X.F., Huang, X.Z., Li, M., 2008. Efficacy and safety of acupuncture for idiopathic Parkinson's disease: a systematic review. J Altern Complement Med. 14, 663–71.
Lee, M.S., Shin, B.C., Kong, J.C., Ernst, E., 2008. Effectiveness of acupuncture for Parkinson's disease: a systematic review. Mov Disord. 23, 1505–15.
McManus, C.A., Kaptchuk, T.J., Schnyer, R.N., Goldman, R., Kerr, C.E., Nguyen, L.T., Stason, W.B., 2007. Experiences of acupuncturists in a placebo-controlled, randomized clinical trial. J Altern Complement Med. 13, 533–8.
Meinert, C.L., Tonascia, S., 1996. Clinical trials: design, conduct, and analysis. Vol., Oxford University Press.
Moher, D., Pham, B., Jones, A., Cook, D.J., Jadad, A.R., Moher, M., Tugwell, P., Klassen, T.P., 1998. Does quality of reports of randomised trials affect estimates of intervention efficacy reported in meta-analyses? Lancet. 352, 609–13.
Moore, M.E., Berk, S.N., 1976. Acupuncture for chronic shoulder pain. An experimental study with attention to the role of placebo and hypnotic susceptibility. Ann Intern Med. 84, 381–4.

Nartey, L., Huwiler-Muntener, K., Shang, A., Liewald, K., Juni, P., Egger, M., 2007. Matched-pair study showed higher quality of placebo-controlled trials in Western phytotherapy than conventional medicine. J Clin Epidemiol. 60, 787–94.

O'Rourke, M.F., 1992. Frederick Akbar Mahomed. Hypertension. 19, 212–7.

Park, H.J., Lim, S., Joo, W.S., Yin, C.S., Lee, H.S., Lee, H.J., Seo, J.C., Leem, K., Son, Y.S., Kim, Y.J., Kim, C.J., Kim, Y.S., Chung, J.H., 2003. Acupuncture prevents 6-hydroxydopamine-induced neuronal death in the nigrostriatal dopaminergic system in the rat Parkinson's disease model. Exp Neurol. 180, 93–8.

Paul, J.R., 1938. President's Address Clinical Epidemiology. J Clin Invest. 17, 539–41.

Pildal, J., Chan, A.W., Hrobjartsson, A., Forfang, E., Altman, D.G., Gotzsche, P.C., 2005. Comparison of descriptions of allocation concealment in trial protocols and the published reports: cohort study. BMJ. 330, 1049.

Pocock, S.J., Simon, R., 1975. Sequential treatment assignment with balancing for prognostic factors in the controlled clinical trial. Biometrics. 31, 103–15.

Sackett, D.L., 1969. Clinical epidemiology. Am J Epidemiol. 89, 125–8.

Sackett, D.L., 1979. Bias in analytic research. J Chronic Dis. 32, 51–63.

Sackett, D.L., 2002. Clinical epidemiology. what, who, and whither. J Clin Epidemiol. 55, 1161–6.

Sanson-Fisher, R.W., Bonevski, B., Green, L.W., D'Este, C., 2007. Limitations of the randomized controlled trial in evaluating population-based health interventions. Am J Prev Med. 33, 155–61.

Schulz, K.F., Chalmers, I., Hayes, R.J., Altman, D.G., 1995. Empirical evidence of bias. Dimensions of methodological quality associated with estimates of treatment effects in controlled trials. JAMA. 273, 408–12.

Schulz, K.F., 2000. Assessing allocation concealment and blinding in randomized controlled trials: why bother? ACP J Club. 132, A11–2.

Schulz, K.F., Grimes, D.A., 2002a. Generation of allocation sequences in randomised trials: chance, not choice. Lancet. 359, 515–9.

Schulz, K.F., Grimes, D.A., 2002b. Allocation concealment in randomised trials: defending against deciphering. Lancet. 359, 614–8.

Shapira, M.Y., Berkman, N., Ben-David, G., Avital, A., Bardach, E., Breuer, R., 2002. Short-term acupuncture therapy is of no benefit in patients with moderate persistent asthma. Chest. 121, 1396–400.

Stetter, F., Kupper, S., 2002. Autogenic training: a meta-analysis of clinical outcome studies. Appl Psychophysiol Biofeedback. 27, 45–98.

Streitberger, K., Kleinhenz, J., 1998. Introducing a placebo needle into acupuncture research. Lancet. 352, 364–5.

Taves, D.R., 1974. Minimization: a new method of assigning patients to treatment and control groups. Clin Pharmacol Ther. 15, 443–53.

Tavola, T., Gala, C., Conte, G., Invernizzi, G., 1992. Traditional Chinese acupuncture in tension-type headache: a controlled study. Pain. 48, 325–9.

Thabane, L., Walter, S.D., Hanna, S., Goldsmith, C., Charles, H., Pullenayegum, E., 2008. Developing a Biostatistical Collaboration Course in a Health Science Research Methodology Program. J Stat Educ. 16, 1–16.

Vincent, C., Lewith, G., 1995. Placebo controls for acupuncture studies. J R Soc Med. 88, 199–202.

Wang, S., Zhou, Z.K., Hu, B.C., 2003. Clinical study on head point-through-point electroacupuncture for treatment of Parkinson's disease. Chin Acupunct Moxibustion. 23, 129–32 (in Chinese).

Wang, S., Cai, Y.Y., Shang, Y.J., Jin-rong, L., 2006. [Effects of head point-through-point electroacupuncture on SOD and LPO in the patient of Parkinson's disease]. Zhongguo Zhen Jiu. 26, 240–2.

White, A.R., Ernst, E., 1999. A systematic review of randomized controlled trials of acupuncture for neck pain. Rheumatology (Oxford). 38, 143–7.

Wood, L., Egger, M., Gluud, L.L., Schulz, K.F., Juni, P., Altman, D.G., Gluud, C., Martin, R.M., Wood, A.J., Sterne, J.A., 2008. Empirical evidence of bias in treatment effect estimates in controlled trials with different interventions and outcomes: meta-epidemiological study. BMJ. 336, 601–5.

Wu, B., He, J., Li, N., Liu, Y., Xiong, S.Q., 2000. Quality Evaluation of Randomized Controlled Trails Papers published by Chinese Acupuncture & Moxibustion. Chinese Acupuncture & Moxibustion. 20, 504–6 (in Chinese).

Xia, Y., Cao, X.D., Wu, G.C., Cheng, J.S., 2010. Acupuncture Therapy for Neurological Diseases: A Neurobiological View (Research Monograph). Vol., Springer-Tsinghua Press, Heidelberg, London, New York, Beijing,.

Yang, D.H., Shi, Y., Jia, Y.M., 2006. Influence of acupuncture plus drug in the amelioration of symptoms and blood antioxidant system of patients with Parkinson's disease. Chin J Chin Rehabil. 10, 14–6 (in Chinese).

Yang, L.H., Li, B., Xiong, J., Du, Y.H., 2009. Quality Evaluation of Randomized Controlled Trails Papers on Acupuncture and Moxibustion for Vertebrobasilar Insufficiency. Chinese Journal of Information on Traditional Chinese Medicine. 16, 104–6 (in Chinese with English abstract).

Yang, X., Ma, T.M., 2005. Quality Evaluation of studies on acupuncture in treating myocardial ischemia. Journal of Liaoning college of TCM. 7, 31–2 (in Chinese).

Zaslawski, C., Rogers, C., Garvey, M., Ryan, D., Yang, C.X., Zhang, S.P., 1997. Strategies to maintain the credibility of sham acupuncture used as a control treatment in clinical trials. J Altern Complement Med. 3, 257–66.

Zhang, H.W., Tang, J.L., 2003. [Designing and applying of the placebo-needle in clinical trial of acupuncture]. Zhongguo Zhong Xi Yi Jie He Za Zhi. 23, 247–50.

Zhao, X., Du, Y.H., Xiong, J., Li, B., 2009. Quality Evaluation of Randomized Controlled Trails Papers on Acupuncture and Moxibustion for Injures of ankle joint. Journal of Clinical Acupuncture and Moxibustion. 25, 1–2 (in Chinese).

Zwarenstein, M., Treweek, S., Gagnier, J.J., Altman, D.G., Tunis, S., Haynes, B., Oxman, A.D., Moher, D., 2008. Improving the reporting of pragmatic trials: an extension of the CONSORT statement. BMJ. 337, a2390.

About the Editors

Dr. Ying Xia is the Professor and Vice-Chairman (Research) of the Department of Neurosurgery at the University of Texas Medical School at Houston. He obtained his medical training at Suzhou Medical College, China. After working as a clinical physician for almost 4 years, he pursued his graduate study and obtained a Master's degree in Medicine followed by a Doctoral degree in Medical Science (Ph.D.) in Neurobiology and Integrative Medicine at Shanghai Medical University (now Shanghai Medical College of Fudan University), China. Later, he joined as a faculty when he was then appointed as a lecturer in the department of Integrative Medicine and Neurobiology. In 1988, Dr. Xia joined Yale University School of Medicine to obtain his postdoctoral training and following which he initiated his own laboratory at Yale in 1994. During his 22-year academic career at Yale University School of Medicine, he held faculty position since 1993 and was promoted to Associate Professor in 2002. In May 2010, he joined the University of Texas Medical School at Houston. He is the Guest Professor of the Laboratory of Molecular Neurology at Shanghai Research Center for Acupuncture and Meridians.

His early research chiefly focused on cardiovascular regulation, interactive modulation of endogenous opioids and other neurotransmitters, and the mechanism of acupuncture. Currently, Dr. Xia's laboratory utilizes molecular, transgenic, and

DOI 10.1007/978-1-4614-3357-6, © Springer Science+Business Media New York 2013

electrophysiological approaches to investigate the hypoxic dysfunction of cortical neurons, neuroprotection against ischemic injury, and inhibition of epileptic hyperexcitability, besides in addition to studying the effects of acupuncture on neurological diseases (e.g., stroke, epilepsy, and hypoxic encephalopathy). More recently, Xia's Laboratory has initiated new explorations on the mechanisms of deep brain stimulation for treatment of Parkinson's disease and the relationship between hypoxic/ischemic injury and Alzheimer's disease. His research has been consecutively supported by NIH, America Heart Association, March of Dimes Foundation, Cerebral Palsy Foundation and the Vivian L. Smith Neurologic Foundation since 1994.

Dr. Xia is the Guest Professor of the Laboratory of Molecular Neurology at Shanghai Research Center for Acupuncture and Meridians. He has been invited to deliver speeches by many academic institutions (e.g., NIH and Harvard) and international conferences (e.g., INRC and iSAMS). He has been serving as a manuscript-reviewer for over 50 international journals and as a grant-reviewer for NIH, US National Science Foundation, America Heart Association, other American foundations (e.g., US Civilian Research and Development Foundation), European foundations (e.g., Scotland Medical Research of UK), Hong Kong University Grants Committees, and Natural Science Foundation of China.

Professor Guanghong Ding is the Director of Biomechanics Institute of Fudan University and the Director of Shanghai Research Center for Acupuncture and Meridians. He is a member of the Council of Chinese Society of Acupuncture and Meridians, Executive Member of Shanghai Society Biomedical Engineering, and the member of the Council of Chinese Society of Mechanics. He is also the Associate Chairman of Shanghai Society of Mechanics, Associate Chairman of Shanghai Society of Acupuncture and Moxibustion, Associate Chairman of Chinese-German Research Foundation for Traditional Chinese Medicine, Chairman of Sponsors of the Chinese-German Research Foundation for Traditional Chinese Medicine. He is the editor of the several esteemed journals including the Journal of Fudan University (Natural Science), Shanghai Journal of Biomedical Engineering, Journal of Medical

Biomechanics, journal for Acupuncture Research and of Mathematical Methods in the Applied Sciences.

He has been investigating the underlying scientific mechanisms of acupuncture since 1990s. His work has clearly demonstrated that acupuncture application induces mast cells degranulation in the area of acupoints. He has also researched on the physical mechanism of acupoints and meridians, such as infrared radiation, calcium accumulation, interstitial fluid flow, and connective tissue at acupoints.

He has published more than 120 papers in peer-reviewed journals and has been an invited speaker for more than 30 times at various International Congresses in USA, Canada, France, Germany, Italy, Hong Kong, and China.

Dr. Gen-Cheng Wu is the Professor of Neurobiology; Chairman of the Department of Integrative Medicine and Neurobiology; Director of the Institute of Acupuncture Research; Director, WHO Collaborating Center for Traditional Medicine, at the Shanghai Medical College of Fudan University, Shanghai, China. He obtained his M.D. from Shanghai First Medical College in 1969 and later his M.S. from the Department of Neurobiology at the Shanghai First Medical College in 1982. He worked as a Research Fellow in the First Department of Physiology, Showa University, Tokyo, Japan (1986–1987) and a visiting Researcher at the National Institute of Environment Health Science (NIEHS), NIH, USA (1992–1994).

He is a member of Disciplinary Appraisal Panels, Academic Degrees Committee of the State Council, member of the Specialist Panels of the National Postdoctoral Administration, member of Standing Committee, Chinese Association of Acupuncture and Moxibustion, and the member of the editorial board of the America Journal of Chinese Medicine, the International Journal in USA. His research interests mainly involve pain modulation and investigating the underlying mechanisms of acupuncture.

He has published 154 research articles, including 80 international SCI papers. He has been honored with 3 awards from the Chinese Central Government and 15 awards from several ministries of the Chinese Government.

Index

Y. Xia et al. (eds.), *Current Research in Acupuncture*,
DOI 10.1007/978-1-4614-3357-6, © Springer Science+Business Media New York 2013

T

V